FLOW
MEASUREMENT
ENGINEERING
HANDBOOK

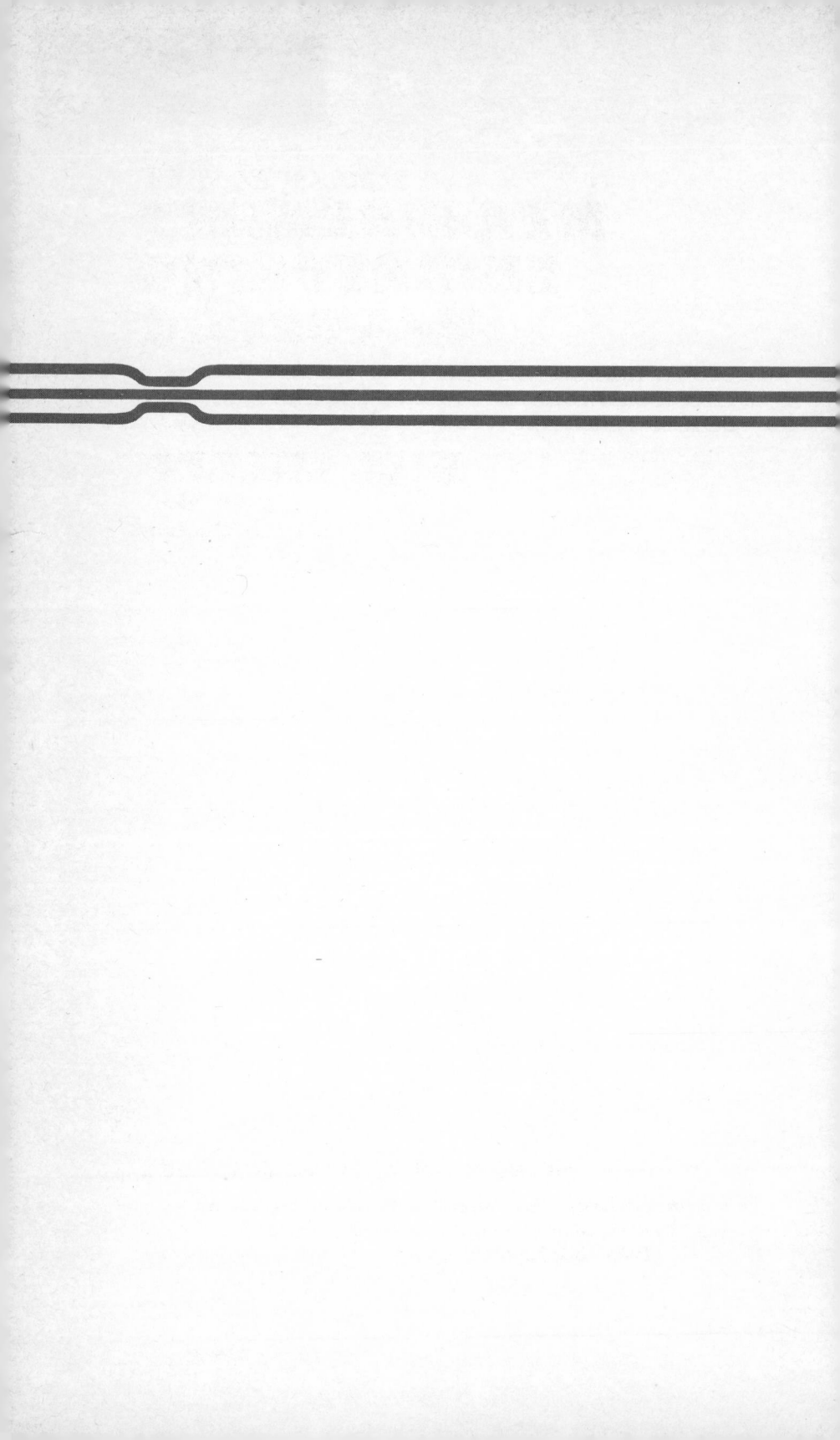

FLOW MEASUREMENT ENGINEERING HANDBOOK

R. W. MILLER
(Richard W.)

Flow Consultant
The Foxboro Company

McGRAW-HILL BOOK COMPANY

New York   St. Louis   San Francisco   Auckland   Bogotá   Hamburg
Johannesburg   London   Madrid   Mexico   Montreal   New Delhi   Panama
Paris   São Paulo   Singapore   Sydney   Tokyo   Toronto

To my wonderful wife,
BARBARA

Library of Congress Cataloging in Publication Data

Miller, R. W. (Richard W.)

Flow measurement engineering handbook.

Includes index.
1. Flowmeters—Handbooks, manuals, etc.
I. Title.
TC177.M56   681'.2   82-6590
ISBN 0-07-042045-9   AACR2

    567890 KGP/KGP   89876

ISBN 0-07-042045-9

The editors for this book were Diane D. Heiberg, Edward M.
Millman, and Margaret Lamb, the designer was Elliot Epstein,
and the production supervisor was Thomas G. Kowalczyk. It
was set in Century Schoolbook by University Graphics, Inc.

Printed and bound by The Kingsport Press.

CONTENTS

PREFACE

This handbook covers in detail the flowmeters most commonly used in process measurement and control. The discussions are limited to closed-conduit flowmeters; nonindustrial specialty meters, laboratory-type meters, and open-channel flow measurement are not treated. For the meters that are covered, both theoretical and practical information is presented, in an attempt to provide a text suitable for students and for entry-level and practicing flow measurement engineers.

The primary aim of the handbook is to provide a single reference source that includes all the material needed to perform flow calculations in both U.S. customary and SI units. A major part of the preparation of the handbook was to unravel the many equation forms existing in the literature, so that the ubiquitous hand calculator (or computer) could be more effectively used in performing these calculations. All equations were derived from theory, engineering equations were tabulated for easy reference, and the resulting flow-rate conversion constants are given to seven places. Although measurement accuracy can never be expected to be that fine, this will allow agreement between U.S. customary and SI conversions and computer calculations. In developing text examples, intermediate calculations are sometimes truncated to save space. The reader may calculate slightly different results in the fourth or fifth decimal place depending on the degree of precision of the calculator or computer used. In all cases, however, the final result is within .01 percent, which is well within any flowmeter's accuracy.

Every attempt has been made to carefully check equations, constants, tables, examples, and graphs for accuracy. But errors can easily occur in a text of this magnitude. It is hoped that such errors will be brought to my attention, and that my apology will be accepted for any inconvenience they may cause.

This book is a continuation of The Foxboro Company policy to provide process measurement and control engineers with the latest information on flowmetering. This policy dates from the first (1930) edition of *Principles and Practice of Flowmeter Engineering* by L. K. Spink, published by The Foxboro Company. This work, published in nine editions, is considered the flow measurement bible by many. The *Flow Measurement Engineering Handbook,* undertaken with the support of Foxboro's management, is a completely new text that is intended to replace the work of L. K. Spink.

Normally, three ingredients are essential to the publication of a technical text: moral support, technical assistance, and manuscript preparation. In the case of this text, a fourth and significant ingredient was provided—the Foxboro management. It was this ingredient that brought the others together to make the book possible. I am deeply indebted to Charles A. McKay, John Bernard, and Peter McCrea for their sustained management support.

The technical support of Loy Upp and Ray Teyssandier of Daniel Industries, who spent endless hours on the telephone, was invaluable. Winston Lee of Rockwell, Bill Reese of Fluidic Techniques, Elmer Mannherz and John Yard of Fischer & Porter, Joe Baumoel of Controlotron, E. A. Spencer of NEL(G.B.), Jean Stolz of Electricité de France, Ron Brunkalla of Badger Meter, Dan Sullivan of Fern Engineering, Larry Lynnworth of Panametrics, Desi Halmi of BIF, Norm Alston of Dieterich Standards, Jim Adam of Dresser Industries, Roger Dowdell of URI, Ed Blanchard of UGC, and Ray Owens and Dennis Zientara of Taylor are but a few of the flow measurement engineers who provided important data and engineering information. A telephone call was all that was required to immediately obtain, without restriction, use of the resources of their respective companies.

I am particularly indebted to Ken Petros, Mark Freeman, George Koslow, and Donna Connors of The Foxboro Company. Ken and Mark proofread the manuscript in detail and offered numerous helpful suggestions; George provided the invaluable computer programming necessary for the development of graphs and tables; and Donna painstakingly transcribed my original chapters into legible form. The list of other Foxboro employees who contributed to this book is difficult to compile, because so many people helped in so many ways. Such a list would have to include Greg Shinskey, Bruce Hainsworth, Jim Vignos, John Zifcak, John Thorp, John Boynton, Bill Kaphaem, Sheila Lynch, Jane Ouimet, Don Bridges, Paul Lefebvre, Bill Cunningham, Wade Mattar, Dale Ihnat, Pat Rohner, Pete Hansen, Mead Bradner, Walter Wiktorowicz, Phil Scott, Harry Blomberg, Ron Stott, Ed Marcoux, Jean Hainsworth, Fred Ezekiel and many more.

Of special note are the contributions of the late Bill Howe and Lew Emerson. Bill was always an excellent critic and technical advisor. Lew edited and provided technical support during the initial compilation of the manuscript. To Evelyn Gulzinski I owe a special expression of appreciation. Her dedication, skill, and creativity in typing tables, typing the final text, and organizing it all, and her tenacity in getting things done, made possible the completion of this work.

LIST OF SYMBOLS

Symbol and meaning		U.S. units	SI units†
a	Constant in Hall-Yarborough equation of state		
a	Constant in specific-heat equation	Btu/(lb$_m$ · mol · °R)	J*/(kg · mol · K)
a	Constant in gas viscosity equation		
a_S	Acceleration along a stream tube	ft/s^2	m/s^2
A	Area	ft^2	m^2
A_a	Annular area between float and tapered wall of a variable-area flowmeter	ft^2	m^2
A_p	Deadweight-tester piston area	in^2	mm^2
A_t	Throat area of a critical nozzle	ft^2	m^2
A_{fl}	Effective area of float in a variable-area flowmeter	ft^2	m^2
A_P	Pipe area	ft^2	m^2
Acc	Accuracy, combined precision and bias errors	%	%
(Acc)$_{ref}$	Reference-condition accuracy	%	%
A_{plate}	Plate area in viscosity derivation equation	ft^2	m^2
$\mathbf{A}$	Constant in Redlich-Kwong equation of state		
$\mathbf{A}$	Constant in Ostwald power-law equation		
$\mathbf{A}_L$	Constant in liquid viscosity equation		
b	Constant in Hall-Yarborough equation of state		
b	Constant in equation for specific heat at constant pressure		

†Except for dimensionless or defined SI unit symbols, as in T_K, symbols that apply to SI units are shown in the text with a superscript asterisk, as in F_a^*.

Symbol and meaning		U.S. units	SI units†
b	Constant in general form of discharge-coefficient equation		
b_c	Slope constant in liquid-bulk-modulus equation		
b_P	Frequency coefficient for pulsating flow		
B	Directional bias error	%	%
B_{var}	Directional bias error, where subscript var (variable) is denoted as q, G_F, G_b, Z, etc.		
$\pm B$	Bias-error range	%	%
$\pm B_{var}$	Bias-error range, where subscript var (variable) is denoted as q, G_b, G_F, Z, etc.		
B_f	Magnetic flux density	G	T
$\mathbf{B}$	Constant in Redlich-Kwong equation of state		
$\mathbf{B}_L$	Constant in liquid-viscosity equation		
c	Constant in liquid-density equation	$(°F)^{-1}$	
c	Constant in Hall-Yarborough equation of state		
C_p	Specific heat at constant pressure	Btu/(lb$_m$·mol·°R)	J*/(kg·mol·K)
$C_{p,i}$	Specific heat at constant pressure for ideal gas	Btu/(lb$_m$·mol·°R)	J*(kg·mol·K)
$C_{p,mix}$	Specific heat at constant pressure of a gas mixture	Btu/(lb$_m$·mol·°R)	J*(kg·mol·K)
$C_{p,p}$	Specific heat at constant pressure of a perfect gas	Btu/(lb$_m$·mol·°R)	J*/(kg·mol·K)
C_p/C_v	Ratio of specific heats of a real gas		
$(C_p/C_v)_i$	Ratio of specific heats of an ideal gas		
$(C_p/C_v)_p$	Ratio of specific heats of a perfect gas		
C_v	Specific heat of a gas at constant volume	Btu/(lb$_m$·mol·°R)	J*/(kg·mol·K)
C	Discharge coefficient, true flow rate divided by theoretical flow rate		
C_∞	Discharge coefficient at infinite Reynolds number		
C_{DH}	Discharge coefficient for a drain (vent) hole through a primary element		
C_N	Discharge coefficient at normal flowing conditions		
C_{max}	Discharge coefficient at $R_D = 10^4$ for an eccentric orifice		
C_{min}	Discharge coefficient at $R_D = 10^6$ for an eccentric orifice		

†Except for dimensionless or defined SI unit symbols, as in T_K, symbols that apply to SI units are shown in the text with a superscript asterisk, as in F_a^*.

Symbol and meaning		U.S. units	SI units†
C_{mp}	Mean molecular heat of a pure gas	$(°F)^{-1}$	$(°C)^{-1}$
$C_{mp,\text{mix}}$	Mean molecular heat at constant pressure of a gas mixture	$(°F)^{-1}$	$(°C)^{-1}$
C_{WA}	Temperature correction factor in Hall-Yarborough equation of state	°R	K
d	Constant in Hall-Yarborough equation of state		
d	Bore of a differential producer [measured at 68°F (20°C) when used in an equation with thermal-expansion factor F_a]	in	mm
d_h	Pressure-tap-hole diameter	in	mm
d_w	Diameter of a thermal well or other protrusion into a pipe	in	mm
d_T	Target (disk) diameter	in	mm
d_{DH}	Diameter of a drain or vent hole	in	mm
D	Internal pipe diameter [measured at 68°F (20°C) when used in an equation with thermal-expansion factor F_a]	in	mm
D_c	Diameter of circle containing the segment of a segmental orifice	in	mm
D_F	Internal pipe diameter	ft	
D_M	Internal pipe diameter		m
e	Enthalpy	Btu/lb$_m$	J*/kg
e_s	Magnetic-flowmeter signal voltage	V	V
E	Velocity-of-approach factor, $1/\sqrt{1-\beta^4}$		
f	Friction factor in Darcy-Weisbach equation		
f	Constant in Hall-Yarborough equation of state		
f_f	Fanning friction factor		
$(f_f)_c$	Ryan-Johnson critical friction factor for power law fluids		
f_{Hz}	Frequency	Hz	Hz
$(f_{Hz})_{CW}$	Constant wave frequency	Hz	Hz
$(f_{Hz})_{DN}$	Downstream frequency of sonic pulses	Hz	Hz
$(f_{Hz})_{UP}$	Upstream frequency of sonic pulses	Hz	Hz
Δf_{Hz}	Frequency difference	Hz	Hz
F	Function used in Newton's solution		
F'	Derivative of function in Newton's solution of a zero root equation		
F_a	Thermal-expansion-factor correction for differential producers	in/(in·°F)	mm/(mm·°C)
F_k	Correction for real gas in an isentropic expansion		

†Except for dimensionless or defined SI unit symbols, as in T_K, symbols that apply to SI units are shown in the text with a superscript asterisk, as in F_a^*.

Symbol and meaning	U.S. units	SI units†
F_g Specific-gravity factor in gas-factor equation, $\sqrt{1/G}$		
F_p Correction for liquid compressibility, ρ_f/ρ_F		
F_{pv} Supercompressibility factor, $\sqrt{1/Z_f}$		
F_B Bias-error correction factor, $(1 + B/100)^{-1}$		
$F_{B,\mathrm{var}}$ Bias-error correction factor, where subscript var (variable) is denoted as q, G_b, G_F, Z, etc.		
F_K Flow-coefficient Reynolds-number correction, K/K_{ref}		
F_L Local gravity correction, g_l/g_c	$\mathrm{lb_f/lb_m}$	N/kg
F_M Manometer correction factor		
F_P Flow-rate correction for pulsating flow		
F_T Target (disk) force, target flowmeter	$\mathrm{lb_f}$	N
F_Y Gas-expansion-factor correction, Y/Y_N		
F_{DH} Drain-hole (vent-hole) correction factor		
F_{EL} Elevation correction for gas-filled leads		
F_{FB} Buoyancy correction factor for force determination		
F_{MB} Buoyancy correction factor for mass determination		
F_{MC} Meter coefficient factor, including all unmeasured variables		
F_{PB} Base-pressure correction factor in gas-factor equation, $14.69595/p_b$ or $101.325/p_b^*$		
F_{RD} Reynolds-number correction factor, C/C_N or K/K_N		
F_{RF} Recovery factor for dynamic pressure		
F_{TB} Base-temperature correction factor in gas-factor equation, $T_b/518.67$ or $T_{Kb}/288.15$		
F_{TF} Flowing-temperature correction factor in gas factor equation, $\sqrt{518.67/T_f}$ or $\sqrt{288.15/T_K}$		
F_{TP} Factor correcting static pressure to total pressure		
F_{VA} Factor correcting variable-area-flowmeter reading from design conditions to flowing conditions		
$F_{WV,\mathrm{dry}}$ Factor converting wet-gas volume to dry-gas volume		

†Except for dimensionless or defined SI unit symbols, as in T_K, symbols that apply to SI units are shown in the text with a superscript asterisk, as in F_a^*.

Symbol and meaning		U.S. units	SI units†
$F_{WVM,dry}$	Factor converting wet-gas mass to dry-gas mass		
F_{X}	Wet-steam quality correction factor		
F_{ARC}	Pen-arc correction for a radial planimeter		
$\mathbf{F}_{N,l}$	Net force exerted by a liquid	lb_f	N
$\mathbf{F}_{N,m}$	Net force exerted by a mass	lb_f	N
$\mathbf{F}_{plate}$	Plate force in viscosity derivation equation	lb_f	N
$F_{\gamma p}$	Correction for pressure to specific heat at constant pressure		
$F_{\gamma R}$	Real-gas correction factor to ratio of specific heats		
$F_{\mu p}$	Viscosity pressure-correction factor for an oil or gas		
g	Constant in Hall-Yarborough equation of state		
g_c	Dimensional conversion constant, $32.17405 \ lb_m \cdot ft/(lb_f \cdot s^2)$ or $1 \ kg \cdot m/(N \cdot s^2)$	$lb_m \cdot ft/(lb_f \cdot s^2)$	$kg \cdot m/(N \cdot s^2)$
g_l	Local gravitational constant	ft/s^2	m/s^2
g_0	Standard acceleration due to gravity, $32.17405 \ ft/s^2$ or $9.806650 \ m/s^2$	ft/s^2	m/s^2
G	Gas (vapor) specific gravity, $M_{w,gas}/M_{w,air}$		
G_b	Liquid base specific gravity, $\rho_b/(\rho_w)_{60,g_0}$		
G'_b	Liquid base specific gravity at a hydrometer temperature other than $60°F$ ($15.6°C$)		
G_f	Flowing specific gravity of a liquid, $\rho_f/(\rho_w)_{60,g_0}$		
G_{fl}	Specific gravity of float in a variable-area flowmeter, $\rho_{fl}/62.3663$ or $\rho^*_{fl}/999.012$		
G_{wv}	Specific gravity of water vapor, 0.6220		
G_{dry}	Specific gravity of dry gas in a gas–water vapor mixture		
G_{mix}	Specific gravity of a gas mixture, $M_{w,mix}/M_{w,air}$		
G_{wet}	Specific gravity of a gas mixed with water vapor		
G_F	Flowing liquid specific gravity uncorrected for pressure, $\rho_F/(\rho_w)_{60,g_0}$		
G_R	Real specific gravity of a gas, ρ_{gas}/ρ_{air}		
G_{SM}	Shear modulus of a solid	lb_f/in^2	N/m^2

†Except for dimensionless or defined SI unit symbols, as in T_K, symbols that apply to SI units are shown in the text with a superscript asterisk, as in F^*_a.

Symbol and meaning		U.S. units	SI units†
h_l	Permanent pressure loss in inches of water	in	
h_s	Step height between mating pipe	in	mm
h_w	Differential pressure in inches of water at 68°F, 14.696 psia, and $g_0 = 32.17405$ ft/s^2		
$(h_w)_g$	Differential produced by gas phase in two-phase (or two-component) flow	in	
$(h_w)_{ss}$	Steady-state differential pressure in pulsating flow	in	
$(h_w)_{ind}$	Indicated differential pressure, uncorrected for fluid head in lead lines	in	
$(h_w)_{max}$	Maximum differential-pressure amplitude in pulsating flow	in	
$(h_w)_{min}$	Minimum differential-pressure amplitude in pulsating flow	in	
$(h_w)_N$	Differential pressure at normal operating flow rate	in	
$(h_w)_{URV}$	Upper-range value of differential pressure corresponding to upper-range flow rate	in	
h_F	Vortex-element barrier width	ft	m
H	Manometer reading	ft	
H_s	Segmental-orifice segment height	in	mm
H_L	Pressure loss in feet of flowing fluid	ft	
H_{EL}	Elevation above a datum	ft	m
H_{LL}	Lead-line pressure-tap elevation above a dry- or wet-type differential-pressure measuring device	ft	m
HP	Horsepower	500 ft·lb$_f$/s	
i	ith data point in a series of points		
I_i	Indicated value of a measurement		
I_t	True value of a measured variable		
$(I_\%)_i$	Percentage difference between indicated value and average of indicated values	%	%
$\bar{I}$	Average of indicated values		
I_P	Pulsation index		
I_{PT}	Threshold pulsation index		
J	Joules: energy, work, quantity of heat	ft·lb$_f$/Btu	N·m
k	Isentropic exponent for a real gas		
k_i	Ideal-gas isentropic exponent. $(C_p/c_v)_i$		
k_p	Perfect-gas isentropic exponent $(C_p/c_v)_p$		

†Except for dimensionless or defined SI unit symbols, as in T_K, symbols that apply to SI units are shown in the text with a superscript asterisk, as in F_a^*.

Symbol and meaning		U.S. units	SI units†
$k_1, k_2, k_3, \ldots$	Constants		
k_{FS}	Combined constant for a falling-sphere viscometer		
K	Flow coefficient, $C/\sqrt{1-\beta^4} = EC$		
K_∞	Flow coefficient at infinite Reynolds number		
K_{ref}	Flow coefficient at reference Reynolds number		
K_N	Flow coefficient at normal operating Reynolds number		
K_{BM}	Liquid bulk modulus	lb_f/in^2	N/m^2
$\overline{K}_{BM}$	Liquid average bulk modulus	lb_f/in^2	N/m^2
K_{BM0}	Liquid bulk modulus at zero pressure, zero intercept	lb_f/in^2	N/m^2
K_{FC}	Permanent pressure-loss coefficient for flow conditioners in inches of water	$(in \cdot ft \cdot s^2)/lb_m$	
$K_{FC,H}$	Permanent pressure-loss coefficient for flow conditioners, in feet of flowing fluid	$(lb_f \cdot s)/(ft^2 \cdot lb_m)$	
$K_{FC,\Delta P}$	Permanent pressure-loss coefficient for flow conditioners		
K_{LT}	Permanent pressure-loss coefficient for a target flowmeter	$1/in^2$	$1/m^2$
K_{UF}	Flow coefficient for ultrasonic flowmeter		
$K_{F,v}$	K factor for pulse-type flowmeter; subscript v may be gal, ft^3, m^3, l, etc.	pulses/v	pulses/v
$\overline{K}_{F,v}$	Mean K factor for pulse-type flowmeter; subscript v may be gal, ft^3, m^3, l, etc.	pulses/v	pulses/v
$(K_{F,v})_{calib}$	Laboratory determined K factor for pulse-type meter; subscript v may be gal, ft^3, m^3, l, etc.	pulses/v	pulses/v
$(K_{F,v})_{flow}$	K factor at flowing conditions	pulses/v	pulses/v
$(K_{F,v})_{max}$	Maximum value of K factor over designated linear range	pulses/v	pulses/v
$(K_{F,v})_{min}$	Minimum value of K factor over designated linear range	pulses/v	pulses/v
$K_{MF,v}$	Meter factor for pulse-type meter; subscript v may be gal, ft^3, m^3, l, etc.	v/pulse	v/pulse
$(K_{MF,v})_{max}$	Maximum value of meter factor over designated linear range	v/pulse	v/pulse
$(K_{MF,v})_{min}$	Minimum value of meter factor over designated linear range	v/pulse	v/pulse

†Except for dimensionless or defined SI unit symbols, as in T_K, symbols that apply to SI units are shown in the text with a superscript asterisk, as in F_a^*.

Symbol and meaning		U.S. units	SI units†
$\overline{K}_{MF,v}$	Mean meter factor for pulse-type flowmeters; subscript v may be gal, ft^3, m^3, l, etc.	v/pulse	v/pulse
K_0, K_1	Constants in API 2054 liquid-petroleum equation		
K_{VA}	Flow coefficient for a variable-area flowmeter		
$\overline{L}$	Linearity percentage about mean K factor	%	%
L	Development length for velocity profile	ft	m
L_p	Path length for a sonic pulse in an ultrasonic flowmeter	ft	m
L_s	Length of straight pipe following a step between two pipes	ft	m
L_P	Length of straight pipe	ft	m
LC	Lu-diagram ordinate value for liquid-compressibility determination		
m	Mass	lb_m	kg
m	Exponent in specific-heat equation		
m	Exponent in Ostwald power-law fluid equation		
m_l	Mass of liquid	lb_m	kg
M	M factor in Reynolds-number correction factor F_{RD} with discharge coefficient C		
M_K	M factor in Reynolds-number correction factor F_{RD} with flow coefficient K		
M_w	Molecular weight	$lb_m/(lb_m \cdot mol)$	$kg/(kg \cdot mol)$
$M_{w,air}$	Molecular weight of air, 28.96247	$lb_m/(lb_m \cdot mol)$	$kg/(kg \cdot mol)$
$M_{w,gas}$	Molecular weight of a gas	$lb_m/(lb_m \cdot mol)$	$kg/(kg \cdot mol)$
$M_{w,mix}$	Molecular weight of a gas mixture	$lb_m/(lb_m \cdot mol)$	$kg/(kg \cdot mol)$
MV	Measured variable; pressure, temperature, flow rate, etc.		
$(MV)_{LRV}$	Lower-range value of measured variable		
$(MV)_{URV}$	Upper-range value of measured variable		
n	Number of moles		
n	Exponent in gas viscosity equation		
n	Exponent in specific heat equation		
n	Exponent in power-law velocity profile equation		
n	Number of data points		
n	Coefficient in Pai velocity profile equation		

†Except for dimensionless or defined SI unit symbols, as in T_K, symbols that apply to SI units are shown in the text with a superscript asterisk, as in F_a^*.

Symbol and meaning		U.S. units	SI units†
N	Exponent in Goldhammer density equation		
N_t	Time correction for hours planimeter, chart hours ÷ planimetered hours		
N_{vG}	N factor for flowing volume with specific-gravity determination, liquids		
$N_{v\rho}$	N factor for flowing volume with density determination, liquids and gases (vapors)		
N_{vhp}	N factor for flowing volume in gas-factor equation		
N_{vpT}	N factor for flowing volume in pvT equation		
N_{MG}	N factor for mass flow with a specific-gravity determination, liquids		
$N_{M\rho}$	N factor for mass flow with a density determination, liquids and gases (vapors)		
N_{Mhp}	N factor for mass flow, gas-factor equation		
N_{MpT}	N factor for mass flow, pvT equation		
N_{VG}	N factor for base volume with specific-gravity determination; liquids at 60°F (15.6°C) and 14.696 psia (101.325 kPa)		
$N_{V\rho}$	N factor for base volume with density determination, liquids and gases (vapors)		
N_{Vhp}	N factor for standard (ISO 5024) gas base volume, gas factor equation; $p_b =$ 14.69595 psia ($p_b^* = 101.325$ kPa), $T_b = 518.67°$R ($T_{Kb} = 288.15$ K)		
$(N_{Vhp})_b$	N factor for nonstandard base volume at selected base pressure and temperature, gas-factor equation		
N_{VpT}	N factor for standard (ISO 5024) gas base volume, pvT equation; $p_b =$ 14.69595 psia ($p_b^* = 101.325$ kPa), $T_b = 518.67°$R ($T_{Kb} = 288.15$ K)		
$(N_{VpT})_b$	N factor for nonstandard base volume at selected base pressure and temperature, pvT equation		
p_b	Base absolute pressure for gas volume	lb_f/in^2	kPa
p_c	Critical absolute pressure of a substance	lb_f/in^2	kPa

†Except for dimensionless or defined SI unit symbols, as in T_K, symbols that apply to SI units are shown in the text with a superscript asterisk, as in F_a^*.

Symbol and meaning		U.S. units	SI units†
p_{ca}	Pseudocritical absolute pressure, Hall-Yarborough equation of state	lb_f/in^2	kPa
p_d	Pressure of dry gas in a wet gas mixture	lb_f/in^2	kPa
p_f	Absolute pressure at flowing conditions	lb_f/in^2	kPa
p_{f1}	Upstream-tap absolute pressure at flowing conditions	lb_f/in^2	kPa
p_{f2}	Downstream-tap absolute pressure at flowing conditions	lb_f/in^2	kPa
$(p_f)_{des}$	Absolute pressure at design flowing conditions	lb_f/in^2	kPa
p_{pc}	Gas (vapor) mixture pseudocritical absolute pressure	lb_f/in^2	kPa
p_{pr}	Gas (vapor) mixture pseudocritical reduced pressure ratio, p_f/p_{pc}		
p_r	Reduced absolute pressure of a gas (vapor), p_f/p_c		
p_t	Critical-nozzle throat absolute pressure	lb_f/in^2	kPa
p_v	Vapor pressure	lb_f/in^2	kPa
p_{wv}	Absolute pressure of water vapor in gas–water vapor mixture	lb_f/in^2	kPa
p_{sat}	Absolute pressure of water vapor in gas–saturated water vapor mixture	lb_f/in^2	kPa
p_B	Barometric pressure	lb_f/in^2	kPa
p_G	Gauge pressure, $p_f - p_B$	lb_f/in^2	kPa
$(\Delta p^*)_g$	Differential pressure of gas phase in two-phase or two-component flow		kPa
$(\Delta p^*)_l$	Permanent pressure loss		kPa
$(\Delta p^*)_N$	Differential pressure at normal operating conditions		kPa
$(\Delta p^*)_{URV}$	Upper-range value of differential pressure corresponding to upper-range flow rate		kPa
P_f	Absolute pressure at flowing conditions	lb_f/ft^2	Pa
P_{f1}	Upstream-tap absolute pressure at flowing conditions	lb_f/ft^2	Pa
P_{f2}	Downstream-tap absolute pressure at flowing conditions	lb_f/ft^2	Pa
P_{f1}'	Upstream-tap absolute pressure at lead-line elevation H_{LL}	lb_f/ft^2	Pa
P_{f2}'	Downstream-tap absolute pressure at lead-line elevation H_{LL}	lb_f/ft^2	Pa

†Except for dimensionless or defined SI unit symbols, as in T_K, symbols that apply to SI units are shown in the text with a superscript asterisk, as in F_a^*.

Symbol and meaning		U.S. units	SI units†
P_t	Critical-nozzle throat absolute pressure	lb_f/ft^2	Pa
$(P_{rdg})_{MV}$	Measured-variable planimeter reading; pressure, temperature, flow rate, etc.		
$(P_{rdg})_0$	Circular-chart inner-radius planimeter reading		
P_B	Barometric pressure	lb_f/ft^2	Pa
P_D	Dynamic pressure, $\rho_f V_p^2/2g_c$ or $\rho_f^* V_p^{*2}/2$	lb_f/ft^2	Pa
P_G	Gauge pressure, $P_f - P_B$	lb_f/ft^2	Pa
P_T	Total (stagnation) pressure, $P_f + P_D$	lb_f/ft^2	Pa
ΔP	Differential pressure, $P_{f1} - P_{f2}$	lb_f/ft^2	Pa
$(\Delta P)_{ind}$	Indicated differential pressure, $P'_{f1} - P'_{f2}$	lb_f/ft^2	Pa
$(q)_{URV}$	Flow-rate upper-range value in mass or volume units		
q_v	Volumetric flow rate at flowing conditions; subscript v may be cfs, cfh, gpm, lpm, etc.		
$q_{acfs}, q_{acfm}, q_{acfh}, q_{acfd}$	Gas (vapor) volumetric flow rate at flowing conditions	$ft^3/s, ft^3/min, ft^3/h, ft^3/24h$	
$q_{acms}^*, q_{acmm}^*, q_{acmh}^*, q_{acmd}^*$	Gas (vapor) volumetric flow rate at flowing conditions		$m^3/s, m^3/min, m^3/h, m^3/24h$
$q_{alps}^*, q_{alpm}^*, q_{alph}^*, q_{alpd}^*$	Gas (vapor) volumetric flow rate at flowing conditions		$L/s, L/min, L/h, L/24h$
$q_{bps}, q_{bpm}, q_{bph}, q_{bpd}$	Liquid volumetric flow rate at flowing conditions	$bbl/s, bbl/min, bbl/h, bbl/24h$	
$q_{cfs}, q_{cfm}, q_{cfh}, q_{cfd}$	Liquid volumetric flow rate at flowing conditions	$ft^3/s, ft^3/min, ft^3/h, ft^3/24h$	
$q_{cms}^*, q_{cmm}^*, q_{cmh}^*, q_{cmd}^*$	Liquid volumetric flow rate at flowing conditions		$m^3/s, m^3/min, m^3/h, m^3/24h$
$q_{lps}^*, q_{lpm}^*, q_{lph}^*, q_{lpd}^*$	Liquid volumetric flow rate at flowing conditions		$L/s, L/min, L/h, L/24h$
$(q_v)_{av}$	Average of maximum and minimum volumetric flow rates in pulsating flow		
$(q_v)_{max}$	Maximum volumetric flow rate in pulsating flow		
$(q_v)_{min}$	Minimum volumetric flow rate in pulsating flow		
q_M	Mass flow rate; subscript M may be, PPH, KPD, KPS, etc.		
$q_{KPS}^*, q_{KPM}^*, q_{KPH}^*, q_{KPD}^*$	Liquid, gas (vapor) mass flow rate		$kg/s, kg/min, kg/h, kg/24h$

†Except for dimensionless or defined SI unit symbols, as in T_K, symbols that apply to SI units are shown in the text with a superscript asterisk, as in F_a^*.

Symbol and meaning		U.S. units	SI units†
q_{PPS}, q_{PPM}, q_{PPH}, q_{PPD}	Liquid, gas (vapor) mass flow rate	lb_m/s, lb_m/min, lb_m/h, $lb_m/24h$	
$(q_M)_{TC}$	Mass flow rate of two-component or two-phase liquid-gas (vapor) mixture; subscript M may be PPS, PPH, KPS, etc.		
q_V	Volumetric flow rate calculated at standard (gas) or base (liquid) temperature and pressure		
$(q_V)_b$	Gas (vapor) volumetric flow rate calculated at selected base pressure and temperature		
q_{BPS}, q_{BPM}, q_{BPH}, q_{BPD}	Liquid volumetric flow rate at $T_F = 60°F$ and $p_b = 14.696$ psia	bbl/s, bbl/min, bbl/h, bbl/24h	
q_{CFS}, q_{CFM}, q_{CFH}, q_{CFD}	Liquid volumetric flow rate at $T_F = 60°F$ and $p_b = 14.696$ psia	ft^3/s, ft^3/min, ft^3/h, $ft^3/24h$	
q_{GPS}, q_{GPM}, q_{GPH}, q_{GPD}	Liquid volumetric flow rate at $T_F = 60°F$ and $p_b = 14.696$ psia	gal/s, gal/min, gal/h, gal 24h	
q^*_{LPS}, q^*_{LPM}, q^*_{LPH} q^*_{LPD}	Liquid volumetric flow rate at $T_{°C} = 15.6°C$ and $p^*_b = 101.3$ kPa		L/s, L/min, L/h, L/24h
q_{SCFS}, q_{SCFM}, q_{SCFH}, q_{SCFD}	Standard gas (vapor) volumetric flow rate at ISO-5024 base: $T_b = 518.67°R$ and $p_b = 14.69595$ psia	ft^3/s, ft^3/min, ft^3/h, $ft^3/24h$	
$(q_{SCFS})_b$, $(q_{SCFM})_b$, $(q_{SCFH})_b$, $(q_{SCFD})_b$	Standard gas (vapor) volumetric flow rate at selected base temperature and pressure	ft^3/s, ft^3/min, ft^3/h, $ft^3/24h$	
q^*_{SCMS}, q^*_{SCMM}, q^*_{SCMH}, q^*_{SCMD}	Standard gas (vapor) volumetric flow rate at ISO-5024 base: $T_{Kb} = 288.15$ K and $p^*_b = 101.325$ kPa		m^3/s, m^3/min, m^3/h, $m^3/24h$
$(q^*_{SCMS})_b$, $(q^*_{SCMM})_b$, $(q^*_{SCMH})_b$, $(q^*_{SCMD})_b$	Standard gas (vapor) volumetric flow rate at selected base temperature and pressure		m^3/s, m^3/min, m^3/h, $m^3/24h$
q^*_{SLPS}, q^*_{SLPM}, q^*_{SLPH}, q^*_{SLPD}	Standard gas (vapor) volumetric flow rate at ISO-5024 base: $T_{Kb} = 288.15$ K and $p^*_b = 101.325$ kPa		L/s, L/min, L/h, L/24h
$(q^*_{SLPS})_b$, $(q^*_{SLPM})_b$, $(q^*_{SLPH})_b$, $(q^*_{SLPD})_b$	Standard gas (vapor) volumetric flow rate at selected base temperature and pressure		L/s, L/min, L/h, L/24h
Q	Total mass or volume units		

†Except for dimensionless or defined SI unit symbols, as in T_K, symbols that apply to SI units are shown in the text with a superscript asterisk, as in F^*_a.

Symbol and meaning		U.S. units	SI units†
Q_v	Total flow in volume units at flowing conditions; subscript v may be gal, ft³, m³, etc.		
Q_{acf}	Gas (vapor) total volume at flowing conditions	ft³	
Q^*_{acm}	Gas (vapor) total volume at flowing conditions		m³
Q_{bbl}	Liquid total volume at flowing conditions	bbl	
Q_{cf}	Liquid total volume at flowing conditions	ft³	
Q^*_{cm}	Liquid total volume at flowing conditions		m³
Q_{gal}	Total volume at flowing conditions	gal	
Q^*_l	Total volume at flowing conditions		L
Q_M	Total flow in mass units; subscript M may be lb_m, kg, g, etc.		
Q^*_{kg}	Total mass		kg
Q_{lbm}	Total mass	lb_m	
Q_V	Total volume at standard (gas) or base (liquid) temperature and pressure	ft³	m³
$(Q_V)_b$	Gas (vapor) total volume at selected pressure and temperature	ft³	m³
Q_{BBL}	Liquid total volume at $T_F = 60°F$ and $p_b = 14.696$ psia	bbl	
Q_{GAL}	Liquid total volume at $T_F = 60°F$ and $p_b = 14.696$ psia		gal
Q^*_L	Liquid total volume at $T_{°C} = 15.6°C$ and $p^*_b = 101.3$ kPa		L
Q_{SCF}	Gas (vapor) total volume at ISO-5024 base: $T_b = 518.67°R$ and $p_b = 14.69595$ psia	ft³	
$(Q_{SCF})_b$	Gas (vapor) total volume at selected base temperature and pressure	ft³	
Q^*_{SL}	Gas (vapor) total volume at ISO-5024 base: $T_{Kb} = 288.15$ K and $p^*_b = 101.325$ kPa		L
$(Q^*_{SL})_b$	Gas (vapor) total volume at selected base temperature and pressure		L
Q_{SCM}	Gas (vapor) total volume at ISO-5024 base: $T_{Kb} = 288.15$ K and $p^*_b = 101.325$ kPa		m³
$(Q^*_{SCM})_b$	Gas (vapor) total volume at selected base temperature and pressure		m³
$(Q_{SCF})_{wet}$	Total volume of wet gas at standard conditions	ft³	

†Except for dimensionless or defined SI unit symbols, as in T_K, symbols that apply to SI units are shown in the text with a superscript asterisk, as in F^*_a.

Symbol and meaning		U.S. Units	SI units†
$(Q^*_{\text{SCM}})_{\text{wet}}$	Total volume of wet gas at standard conditions		m^3
$(Q_{\text{lbm}})_{TC}$	Total mass of two-component gas mixture	lb_m	
$(Q_{\text{lbm}})_g$	Total mass of dry gas in a two-component mixture	lb_m	
r	Radius to a point	in	mm
r_b	Elbow radius at centerline	in	mm
r_p	Pipe radius	in	mm
R_d	Bore Reynolds number		
R_{temp}	Resistance of a metal at measured temperature	Ω	Ω
R_{ref}	Reference resistance value at 0°C	Ω	Ω
R_0	Universal gas constant, 10.73151 psia· $\text{ft}^3/(\text{lb}_m \cdot \text{mol} \cdot °\text{R})$ or 8.31441 J*/(g· mol·K)		
R_D	Pipe Reynolds number		
$(R_{\text{GA}})_c$	Govier-Aziz critical Reynolds number for a Bingham fluid		
R_{MR}	Metzner-Reed Reynolds number for a power-law fluid		
$(R_{MR})_c$	Metzner-Reed critical Reynolds number for a power-law fluid		
RH	Relative humidity, p_{wv}/p_{sat}		
s	Coefficient in Pai profile equation (5.16)		
S	Strouhal number, $f_{\text{Hz}} \cdot h_F/V_{\text{free}}$		
S_s	Fluid shear stress	lb_f/ft^2	N/m^2
$(S_s)_w$	Fluid wall shear stress	lb_f/ft^2	N/m^2
$(S_s)_y$	Bingham-fluid yield shear stress	lb_f/ft^2	N/m^2
$\dot{S}$	Fluid shear rate	s^{-1}	s^{-1}
$\dot{S}_w$	Fluid shear rate at wall	s^{-1}	s^{-1}
dS	Differential displacement in direction of flow	ft	m
S_M	Sizing factor for differential producer, a constant		
SH	Specific humidity, $\rho_{wv}/\rho_{\text{dry}}$		
t	Time	s	s
t_{tF}	Time for test fluid to flow through a capillary viscometer	s	s
t_{ref}	Reference time for a capillary viscometer	s	s
t_{DN}	Time for a sonic pulse to travel downstream	s	s
t_{FS}	Time for sphere to fall in a falling-sphere viscometer.	s	s

†Except for dimensionless or defined SI unit symbols, as in T_K, symbols that apply to SI units are shown in the text with a superscript asterisk, as in F^*_a.

Symbol and meaning		U.S. units	SI units†
t_{RA}	Admiralty Redwood seconds, viscosity determination	s	s
t_{RS}	Standard Redwood seconds, viscosity determination	s	s
t_{ST}	Student's t statistic		
t_{UP}	Time for a sonic pulse to travel upstream	s	s
t_{SSF}	Saybolt Furol seconds, viscosity determination	s	s
t_{SSU}	Saybolt universal seconds, viscosity determination	s	s
T_b	Base absolute temperature for a gas volume	°R	
T_c	Critical temperature of a substance	°R	K
T_f	Flowing absolute temperature, $T_F + 459.67$	°R	
T_{f1}	Flowing absolute temperature measured at upstream tap	°R	
T_{f2}	Flowing absolute temperature measured at downstream tap	°R	
T_{fi}	Indicated flowing absolute temperature	°R	
T_r	Reduced temperature of a gas (vapor), T_f/T_c		
$(T_f)_{des}$	Flowing absolute temperature at design conditions	°R	
T_{pc}	Pseudocritical temperature of a mixture of gases	°R	K
T_{pr}	Pseudocritical reduced temperature of a mixture of gases, T_f/T_{pc}		
T_{stag}	Stagnation temperature	°R	K
T	Reciprocal of reduced temperature, $1/T_{pr}$		
T	Period of oscillation in pulsating flow	s	s
T_B	Boiling point, absolute temperature	°R	K
$T_{°C}$	Temperature in degrees Celsius		°C
T_F	Flowing temperature in degrees Fahrenheit	°F	
ΔT_F	Difference in temperature, $T_F - 60$	°F	
T_K	Flowing absolute temperature, Kelvin scale		K
T_{K1}	Absolute temperature measured at upstream tap		K
T_{K2}	Absolute temperature measured at downstream tap		K

†Except for dimensionless or defined SI unit symbols, as in T_K, symbols that apply to SI units are shown in the text with a superscript asterisk, as in F_a^*.

Symbol and meaning		U.S. units	SI units†
T_{Kb}	Base absolute temperature for a gas volume		K
$T_{°R}$	Absolute temperature in degrees Rankine	°R	
T_{WB}	Wet-bulb temperature	°F	°C
u	Internal energy	Btu/lb$_m$	J*/kg
v	Specific volume, $1/\rho$	ft³/lb$_m$	m³/kg
v_{wv}	Specific volume of water vapor	ft³/lb$_m$	m³/kg
$V_{f/B}$	Fluid velocity with respect to blade for turbine flowmeters	ft/s	m/s
V_p	Point velocity along pipe radius	ft/s	m/s
V_{free}	Free-stream velocity, no confining walls	ft/s	m/s
V_{max}	Maximum (centerline) velocity	ft/s	m/s
$\overline{V}_f$	Average pipe velocity	ft/s	m/s
$(\overline{V}_f)_{av}$	Average of minimum and maximum velocities in pulsating flow	ft/s	m/s
$(\overline{V}_f)_C$	Critical velocity for transition to turbulent flow for Bingham fluid	ft/s	m/s
$(\overline{V}_f)_{max}$	Maximum average pipe velocity in pulsating flow	ft/s	m/s
$(\overline{V}_f)_{min}$	Minimum average pipe velocity in pulsating flow	ft/s	m/s
$(\overline{V}_f)_{VC}$	Average velocity at minimum flow area (vena contracta) of an orifice	ft/s	m/s
$\overline{V}_t$	Average throat velocity for a contoured-inlet primary element	ft/s	m/s
$\overline{V}_{son}$	Sonic velocity at throat of a critical nozzle	ft/s	m/s
ΔV_{plate}	Velocity difference between two parallel plates	ft/s	m/s
V_B	Turbine-flowmeter blade velocity	ft/s	m/s
V_P	Plug velocity in flow of a Bingham fluid	ft/s	m/s
V_{SO}	Velocity of sound in a liquid	ft/s	m/s
V	Volume	ft³	m³
Δ**V**	Change in volume with pressure	ft³	m³
V$_0$	Liquid volume at zero pressure	ft³	m³
V$_l$	Liquid volume	ft³	m³
V$_m$	Volume of a standard mass	ft³	m³
V$_{fl}$	Volume of float in a variable-area flowmeter	ft³	m³
V$_{dry}$	Volume of dry gas in a wet (water) gas mixture	ft³	m³
V$_{wet}$	Volume of wet (water) gas in a wet gas mixture	ft³	m³

†Except for dimensionless or defined SI unit symbols, as in T_K, symbols that apply to SI units are shown in the text with a superscript asterisk, as in F_a^*.

Symbol and meaning	U.S. units	SI units†
W Energy	W	W
W Weight force	lb_f	N
W_i Weighting function along the sonic path of an ultrasonic flowmeter		
W_s Annular-slot width for corner-tapping	in	mm
$\mathbf{W}$ Work	Btu/lb_m	J^*/kg
x Mole fraction in gas (vapor) phase		
x_m Mass fraction, mass of component ÷ mass of total mixture		
x_p Differential-pressure amplitude ratio in pulsating flow		
$\mathbf{x}_1$ Pressure ratio based on upstream tap pressure, $\Delta p/p_{f1}$		
$\mathbf{x}_2$ Pressure ratio based on downstream pressure, $\Delta p/p_{f2}$		
X Sensitivity coefficient of a measured variable		
X_{var} Sensitivity coefficient, where subscript var (variable) is denoted as G_b, G_F, Z, etc.		
$\mathbf{X}$ Mixture quality, mass of gas phase ÷ mass of total mixture		
y Elevation above sea level	ft	
y Mole fraction in liquid phase		
$\bar{y}$ Distance from pipe wall to point of average velocity	in	mm
Y Gas expansion factor		
Y_1 Gas expansion factor based on upstream pressure		
Y_2 Gas expansion factor based on downstream pressure		
Y_N Gas expansion factor at normal flowing conditions, usually design conditions		
Y_{CR} Critical flow function		
ΔY Spacing between parallel plates	ft	
Y_{Arn} Critical flow function derived by Arnberg	$\dfrac{lb_m}{lb_f \cdot s}\sqrt{°R}$	
Z Gas (vapor) compressibility factor		
Z_b Gas (vapor) compressibility factor at base temperature and pressure		
Z_c Gas (vapor) compressibility factor at critical point		
Z_f Gas (vapor) compressibility factor at flowing conditions		
Z_{pc} Gas (vapor) pseudocritical compressibility factor for a mixture		

†Except for dimensionless or defined SI unit symbols, as in T_K, symbols that apply to SI units are shown in the text with a superscript asterisk, as in F_a^*.

Symbol and meaning	U.S. units	SI units†
Z_{wv} Water-vapor compressibility factor in a wet gas		
Z_{air} Compressibility factor for air		
Z_{dry} Compressibility factor of dry components in a wet gas		
Z_{wet} Compressibility factor of a wet gas		
Z_L Liquid compressibility factor		
$Z_{R/K}$ Redlich-Kwong equation-of-state compressibility factor		
$(Z_{R/K})_f$ Redlich-Kwong equation-of-state compressibility factor calculated at flowing conditions		
$(Z_{R/K})_{des}$ Redlich-Kwong equation-of-state compressibility factor calculated at design conditions		
Z^0 Edmister-Pitzer simple fluid compressibility factor		
Z^1 Edmister-Pitzer compressibility-factor correction for deviation from simple fluid		
α Waveform coefficient for pulsating flow		
α_b Thermal-expansion coefficient in API-2540 liquid-petroleum density equation		
α_b' Derivative of thermal-expansion coefficient in API-2540 liquid-petroleum density equation		
α_H Hydrometer cubical coefficient of expansion	$(°F)^{-1}$	$(°C)^{-1}$
α_{HO} Thermal-expansion coefficient for meter housing	in/(in · °F)	mm/(mm · °C)
α_P Thermal-expansion coefficient for pipe material	in/(in · °F)	mm/(mm · °C)
α_{PE} Thermal-expansion coefficient for primary-element material	in/(in · °F)	mm/(mm · °C)
β Beta ratio, d/D		
β_c Segment beta ratio for a segmental orifice, $\beta/0.98$		
β_T Target or annular-orifice beta ratio, d_T/D		
γ_f Specific weight of a fluid, liquid, or gas (vapor), $(g_l/g_c)\rho_f$	lb_f/ft^3	N/m^3
ϵ Average depth of pipe-wall roughness	in/in	mm/mm
η Efficiency of motor and pump		

†Except for dimensionless or defined SI unit symbols, as in T_K, symbols that apply to SI units are shown in the text with a superscript asterisk, as in F_a^*.

Symbol and meaning		U.S. units	SI units†
θ	Sonic-path angle for an ultrasonic flowmeter	degrees	degrees
μ_{app}	Apparent viscosity, $S_s/\dot{S}$, absolute viscosity units	$lb_f \cdot s/ft^2$	cP‡
$(\mu)_a$	Absolute viscosity at atmospheric pressure,	$lb_m/(ft \cdot s)$	cP‡
$(\mu)_p$	Absolute viscosity corrected for pressure	$lb_m/(ft \cdot s)$	cP‡
μ_{cP}	Absolute viscosity in centipoises		cP‡
$\mu_{cP,mix}$	Absolute viscosity of a mixture in centipoises		cP‡
$\mu_{cP,ref}$	Reference viscosity for a capillary viscometer		cP‡
μ_{mix}	Absolute viscosity of a mixture	$lb_m/(ft \cdot s)$	cP‡
$(\mu_f)_e$	Absolute English-system viscosity, force units	$lb_f \cdot s/ft^2$	
$(\mu_m)_e$	Absolute English-system viscosity, mass units	$lb_m/(ft \cdot s)$	
μ_P	Absolute viscosity in poises		P‡
$\mu_{Pa \cdot s}$	Absolute viscosity in pascal seconds		Pa·s
ν_e	Kinematic viscosity in English units	ft^2/s	
ν_{cSt}	Kinematic viscosity in centistokes		cSt‡
ν_{St}	Kinematic viscosity in stokes		St‡
ρ_b	Density at base conditions: liquids, 60°F (15.6°C) and 14.7 psia (101.3 kPa); gases, 59°F (15°C) and 14.69595 psia (101.325 kPa); or at other selected base values	lb_m/ft^3	kg/m^3
ρ_{air}	Air density at time of calibration	lb_m/ft^3	kg/m^3
$\rho_{air,c}$	Air density for calibrating a weigh tank	lb_m/ft^3	kg/m^3
ρ_f	Density at flowing conditions	lb_m/ft^3	kg/m^3
ρ_{f1}	Upstream density at flowing conditions	lb_m/ft^3	kg/m^3
ρ_{f2}	Downstream density at flowing conditions	lb_m/ft^3	kg/m^3
$(\rho_f)_{des}$	Density at design conditions	lb_m/ft^3	kg/m^3
ρ_{fl}	Density of float in a variable-area flowmeter	lb_m/ft^3	kg/m^3
ρ_{ft}	Density at throat of a critical flow nozzle	lb_m/ft^3	kg/m^3
ρ_{g1}	Upstream density of gas in a two-component or two-phase flow	lb_m/ft^3	kg/m^3

†Except for dimensionless or defined SI unit symbols, as in T_K, symbols that apply to SI units are shown in the text with a superscript asterisk, as in F_a^*.

‡The poise (P) and the stokes (St) are cgs metric units, not SI metric; 1 P = 0.1 Pa·s; 1 St = 0.0001 m^2/s.

Symbol and meaning		U.S. units	SI units†
ρ_l	Density of liquid in a two-component or two-phase flow	lb_m/ft^3	kg/m^3
ρ_m	Manometer-fluid density	lb_m/ft^3	kg/m^3
ρ_{mass}	Density of a standard mass	lb_m/ft^3	kg/m^3
ρ_v	Density of water vapor at saturation	lb_m/ft^3	kg/m^3
ρ_r	Hall-Yarborough equation-of-state reduced density	lb_m/ft^3	kg/m^3
ρ_{r0}	Initial estimate for Hall-Yarborough reduced density	lb_m/ft^3	kg/m^3
ρ^*_{ref}	Fluid reference density in a falling-sphere viscometer		kg/m^3
ρ_s	Manometer seal-fluid density	lb_m/ft^3	kg/m^3
ρ^*_{sph}	Sphere density in a falling-sphere viscometer		kg/m^3
ρ_{wet}	Density of a wet gas	lb_m/ft^3	kg/m^3
ρ_F	Density of fluid at flowing conditions, uncorrected for pressure	lb_m/ft^3	kg/m^3
ρ_{Hg}	Density of mercury	lb_m/ft^3	kg/m^3
ρ_{TP}	effective density of a two-phase flow of same substance in liquid and gas phases	lb_m/ft^3	kg/m^3
$(\rho_w)_{T,g0}$	Density of water at standard gravity (32.174) and any temperature	lb_m/ft^3	kg/m^3
$(\rho_w)_{68,g0}$	Density of water at 68°F, standard gravity, and atmospheric pressure: 62.31572 lb_m/ft^3 (998.2019 kg/m^3)	lb_m/ft^3	kg/m^3
$(\rho_w)_{60,g0}$	Density of water at 60°F, standard gravity, and atmospheric pressure: 62.36630 lb_m/ft^3 (999.0121 kg/m^3)	lb_m/ft^3	kg/m^3
σ	Standard deviation	%	%
σ_P	Precision, $t_{ST}\,\sigma$	%	%
σ_c	Cavitation number		
$(\sigma_c)_i$	Incipient cavitation number		
ϕ	Degrees latitude		
$(\phi_f)_e$	Coefficient of rigidity for a Bingham fluid	$lb_f\cdot s/ft^2$	
ψ	Angle of swirl	degrees	degrees
ω	Acentric factor, Edmister-Pitzer diagram		
ω_{pc}	Pseudoacentric factor for a gas mixture		

†Except for dimensionless or defined SI unit symbols, as in T_K, symbols that apply to SI units are shown in the text with a superscript asterisk, as in F^*_a.

1

INTRODUCTION

The purpose of this handbook is to provide flow measurement engineers with a single reference for engineering equations, physical-property data, accuracy estimation, and installation requirements for the most commonly used industrial flowmeters, in both United States customary (U.S.) and SI units. Equations are developed for sizing and flow-rate calculations for differential producers and linear-output flowmeters. This information is presented in tables and in graphical form for ready reference. Orifices, nozzles, venturis, Annubars, elbow flowmeters, integral orifices, target flowmeters, critical-flow nozzles and orifices, and positive-displacement, turbine, vortex, magnetic, variable-area, and ultrasonic flowmeters are covered. Separate chapters are devoted to accuracy, measurement units, flowmeter selection, fluid properties, and influence quantities such as cavitation and pulsating flow.

OVERVIEW

The handbook is organized into three distinct sections: the first deals with factors common to all flowmeters, and the second with specific flowmeters; the third section is a physical properties appendix applicable to all flowmeters. The reader is first introduced to fluid properties, measurement units, accuracy calculations, and common influence quantities. The principle of operation for each flowmeter is then briefly explained, and the reasons for selecting a particular flowmeter are given. Differential producers are next introduced and explained. Engineering equations are developed, and examples presented. Design information and graphical information for sizing primary elements and fixed-geometry devices conclude the section on differential producers. Critical flowmeters are then discussed, and equations are developed for both gases (vapors) and choked liquid flow. Linear flowmeters are introduced and grouped according to the equations used to calculate flow rate. Their principle of operation is explained, and practical considerations are discussed. Flow equations are tabulated, and examples are presented along with the necessary graphical material.

The recently adopted ISO/ASME† orifice equation is used throughout. Methods and graphs pertaining to this equation and to other ISO/ASME recommendations are developed into working engineering equations for both U.S. and SI units. Additionally, installation and design requirements given in ISO and ANSI‡ standards are presented. Equations are given for calculating pressure loss and energy cost.

The β_0 method, which simplifies the sizing calculation for differential producers (orifices, nozzles, Lo-Loss tubes, venturis, etc.), is introduced. This method is then used with either an iteration or with Newton's solution for a zero-root equation to exactly *size* these primary devices. Newton's method is also used to develop a single flow-rate equation that eliminates present iterative calculations.

In the chapter on critical flowmeters, equations are developed, design information is presented, and detailed calculations are given. There is also a section on choked liquid flow.

Reference installation conditions and primary constants are defined. These primary constants are then used to develop the flow-rate-equation constants to seven places. This eliminates the calculation errors associated with previous equation constants that were often rounded to three significant figures. Nonnewtonian fluids, pulsating flow, and cavitation are also discussed.

Information is presented on Annubars, magnetic flowmeters, ultrasonic flowmeters, and Universal Venturi Tubes. The accuracy chapter simplifies the calculations required by present standards. These are tabulated, with examples given in subsequent chapters.

Organization

Throughout, the text material is organized for ready reference to equations and physical data in both U.S. and SI units. The individual chapters are organized as follows.

- In Chap. 2, the fluid properties used in the calculation of flow rate for all flowmeters are discussed, the equations are developed, and the use of the appendix data is discussed and exemplified.

- In Chap. 3, the measurement of pressure and temperature is explained, the relationship between U.S. and SI flow-rate units is discussed, and the subscripting system to be followed in the handbook is described.

- In Chap. 4, flow measurement accuracy is defined, and single-point accuracy and accuracy over a range are discussed in detail; sensitivity coefficients are derived and, through an example, used to calculate bias and accuracy.

---

†International Standards Organization/American Society of Mechanical Engineers.

‡American National Standards Institute.

■ In Chap. 5, the conditions required for reference measurement accuracy are presented; the influence of nonnewtonian fluids, pulsating flow, and cavitation on flowmeter performance is described.

■ In Chap. 6, the principles of operation, reasons for selection, and a selection guide are presented for flowmeters covered in the handbook, and pressure-cost and energy-cost equations are given.

■ Chapter 7 is a brief introduction to the many flowmeters classified as differential producers.

■ In Chap. 8, installation requirements for differential producers are given and recommended practice for installing lead lines is detailed.

■ In Chap. 9, the volumetric and mass flow-rate equations are developed; primary constants, correction factors, Reynolds-number equations, and procedures for bore sizing and flow-rate calculations are given, along with examples for liquid, gas (vapor), and quality (wet) steam measurements.

■ In Chap. 10, design requirements for the differential producer are presented, along with examples for devices that require graphical solutions and graphs based on the equations developed in Chap. 9.

■ In Chap. 11, fixed-geometry flowmeters, for which the differential or target force must be determined, are described and their equations are tabulated; these devices include arithmetic-progression orifices, target flowmeters, integral orifices, Annubars, and elbow flowmeters.

■ In Chap. 12, measured and unmeasured variables are defined; measurement equipment is then described for analog computers, digital flow computers, and for planimetering charts to calculate total flow.

■ In Chap. 13, engineering equations for critical nozzle venturis and thick-orifice flowmeters are developed and tabulated, and the critical-flow function for many common gases is given in tables; liquid choked-flow equations for restrictive venturis and orifices are also presented, along with examples.

■ In Chap. 14, the principles of operation, flow equations, and examples are given for turbine, vortex, and positive-displacement pulse-output type meters. Magnetic, ultrasonic, and variable-area flowmeters are discussed, and measurement practices and equations are given for these devices.

Open-Channel Measurement

Open-channel measurement is almost a separate field from closed-conduit flow measurement. The terminology, equations, and theory, although similarly based on hydraulics, are completely different. Many open-channel flowmeters (weirs, flumes) must be fabricated in situ; and they often require on-site hydraulic engineering to assure suitable results. This part of the field of flow measurement is rapidly changing, with newer technology being used to obtain more accurate results. For these reasons, open-channel measurement is not discussed in this book. However, texts have been written on the more traditional meters, and numerous technical publications have appeared in the literature over the last 10 years. The reader is referred to the following texts and standards for the proper use of open-channel flow measurement devices:

Bureau of Reclamation: *Water Measurement Manual,* U.S. Department of the Interior, Bureau of Reclamation, catalog no. I 27.19/2:W29/2/974, U.S. Government Printing Office, Washington, D.C., 1975.

National Bureau of Standards: *A Guide to Methods and Standards for the Measurement of Water Flow,* NBS Special Publication 421, U.S. Department of Commerce/NBS, code XNBSAV, 1975.

Ackers, P., W. R. White, J. A. Perkins, and A. J. M. Harrison: *Weirs and Flumes,* Wiley, New York, 1978.

ISO Standard 1438, *Liquid Flow Measurement in Open Channels Using Thin Plate Weirs and Venturi Flumes,* ISO 1438-1975(E), Geneva, 1975.

ISO Standard 1438/1, *Water Flow Measurement in Open Channels Using Weirs and Venturi Flumes,* pt. 1, *Thin-Plate Weirs,* ISO 1438/1-1980(E), Geneva, 1980.

FLOWMETER DEFINITION

Defining a flowmeter is almost as difficult as classifying flowmeters. The only standard definition available is that given in ISA 51.1 (1976). In it, the following definitions are given:

Flowmeter—a device that measures the rate of flow or quantity of a moving fluid in an open or closed conduit. It usually consists of both a primary and a secondary device.

> *Note:* It is acceptable in practice to further identify the flowmeter by its applied theory, such as differential pressure, velocity, area, force, etc., or by its applied technology, such as orifice, turbine, vortex, ultrasonic, etc. Examples include turbine flowmeter, magnetic flowmeter, etc.

Flowmeter primary device—the device mounted internally or externally to the fluid conduit which produces a *signal* with a defined relationship to the fluid flow in accordance with known physical laws relating the interaction of the fluid to the presence of the primary device.

> *Note:* The primary device may consist of one or more *elements* necessary to produce the primary device signal.

Flowmeter secondary device—the device that responds to the signal from the primary device and converts it to a display or to an output signal that can be translated relative to flow rate or quantity.

> *Note:* The secondary device may consist of one or more *elements* as needed to translate the primary device signal into standardized or nonstandardized display or transmitted units (recorder, indicator, totalizer, etc.).

By these definitions, the primary device of an orifice flowmeter includes upstream and downstream piping, flow conditioner, orifice plate, and pressure taps. The differential-pressure transmitter, manifold valves, and connecting tubing are elements of the secondary device. The combination of these two devices is the orifice flowmeter.

FLOWMETER CLASSIFICATION

Flowmeters have been classified in numerous ways; in the most common classification system, meters are separated into quantity (total-flow) and rate meters, and then further subdivided by operating principle. This essentially divides meters into those that measure discrete volume (positive-displacement meters) and those that directly or indirectly utilize the movement of the fluid to actuate a secondary element. Although this system reasonably describes the two meter classes, in practice rate meters are often used as quantity meters (vortex, turbine, orifice, etc.), and quantity meters are sometimes used as rate meters.

For the purpose of calculating flow rate or total flow, the classification of meters as either square-root (differential-producer) or linear meters is more convenient since, in general, all meters for which the flow is not a function of the square root of the differential pressure are essentially linear-scale meters. This grouping gives the same engineering equations for all meters in each group and avoids much confusion in flow calculations. Although linear meters have completely different operating principles from square-root meters, scaling to base volume and to mass flow are the same for both types.

REFERENCE

ISA 51.1, Instrument Society of America, Research Triangle Park, NC, 1976.

2

FLUID
PROPERTIES

Data on physical properties is often required for calculations of base flow rates and pipe Reynolds numbers, and to predict the properties of a gas (vapor) after an expansion. The physical properties of liquids and gases change with pressure and temperature, and whether corrections need to be considered depends on the design objective. In many cases, properties are assumed constant at design conditions, and corrections are not applied. While there is no substitute for experimental data, estimates of the properties of a mixture may often have to be used in calculations. This requires theory, common sense, and experience.

Accuracy in predicting the properties of pure substances is considerably better for liquids and gases than for mixtures. In many applications, particularly for high inert mole fractions in natural gas, large errors can occur, and the estimated value should first be properly verified by test.

This chapter is a discussion of the most commonly used fluid properties and the estimation of these properties at various pressures and temperatures, for both pure substances and mixtures. For illustrative purposes shaded areas on graphs in this chapter are expanded and are not scaled.

THE pvT RELATIONSHIP

The pvT Behavior of a Pure Substance

Fluid density can be measured with a liquid or gas densitometer, but it is more common to use temperature and pressure measurements to calculate density. The reciprocal of the specific volume is the fluid's mass density, and it can be determined from pressure and temperature measurements using the pvT relationship. The interrelationships of pressure, temperature, and specific volume are also important because of the law of corresponding states. From these relationships, the fluid state can be defined, or the density of an unknown mixture can be calculated.

Depending on temperature and pressure, a substance may be either a solid, a solid-liquid mixture, a liquid, a liquid-vapor mixture, a vapor, or a gas. The words vapor and gas are often used interchangeably because they are thermo-dynamically identical. Historically, the term *vapor* has been used to designate a substance, such as water, that exists as a solid or liquid at room temperature and atmospheric pressure, and the term *gas* to designate a substance that exists in the gaseous state under the same conditions (air, oxygen, etc.). At and above the saturated-vapor line, all substances are thermodynamically gases and contain no liquid, as the term vapor implies. The term vapor has also been used for gases in the region between the saturated-vapor line and the critical isotherm, where gases are more compressible.

Figure 2.1*a* shows the pressure-temperature diagram for water. The well-known characteristics of water will be used to explain the relationships among pressure, temperature, and specific volume.

Three regions, corresponding to fluid states, are shown separated by heavy lines in Fig. 2.1*a*. Ice at atmospheric pressure (14.7 psia) and 28°F is indicated by point *A*. When heat is added, water remains in the solid phase (ice) until the temperature reaches 32°F (point *B*). The temperature remains at 32°F (point *B* on the heavy line) until sufficient heat has been added to convert all the ice into liquid (water) at 32°F; liquid and solid coexist as a mixture to this point, with the percentages of solid and liquid determined by the exchange of heat to or from the external environment. If the amount of ice decreases, the amount of liquid must increase to maintain the initial mass.

When the temperature rises above 32°F, the solid phase completely disap-pears, leaving only liquid. Additional heat brings the temperature to 212°F, shown at point *C* on the second heavy line. At this point, some of the liquid is transformed into a vapor (gas). As long as the temperature remains at point *C*, two phases, liquid and gas, coexist. When the liquid disappears, the fluid is in the saturated-vapor (gas) phase.

Increasing the temperature to the critical temperature of 705.5°F (point *D*) brings the saturated vapor (gas) through the vapor region. In this region no liq-uid is present, and, although it is called a vapor, the fluid acts like a *real gas*. (If the pressure is increased to 3198 psia, the ice-to-liquid temperature at point *B* is not significantly changed.) At 705.5°F (374°C), it is impossible to discern where the liquid phase ends and the gas phase begins. At this point, the specific volumes (or densities) of the two phases are identical.

The point of equal specific volumes is referred to as the *critical point*. The temperature and pressure at the critical point are the critical temperature T_c and and critical pressure p_c; these are the values used to correlate fluids on a single diagram. The point indicated as the *triple point* is where solid, liquid, and gas are all present in varying amounts.

Figure 2.1*b* shows the pressure–specific volume diagram, with fluid-density values given below the specific-volume axis. The points *A*, *B*, *C*, and *D* denote regions of changing density. This figure also shows the *thermodynamically* defined vapor region between the saturated-vapor line and the critical-temper-ature isotherm.

In a single phase (solid, liquid, or gas), density can be defined through pres-sure and temperature measurements. For a single phase, any two of pressure,

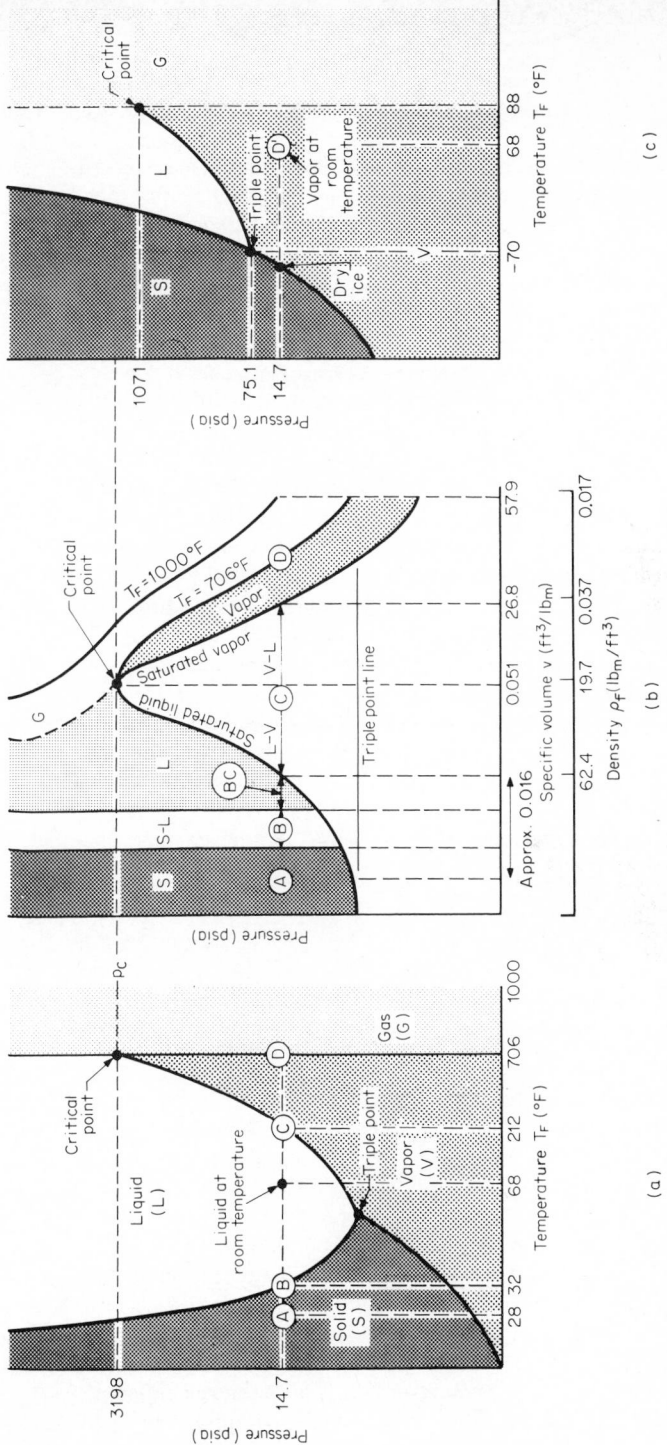

Figure 2.1 Pressure-temperature and pressure-volume diagrams. (a) Water. (b) Water. (c) Carbon dioxide.

temperature, and specific volume are independent, but any two will define the third. In the two-phase region, temperature and pressure are interdependent; pressure cannot be changed without changing temperature, and the state of the fluid cannot be defined by these measurements alone.

Two-phase mixtures contain differing amounts of each phase, and each has a substantially different density from the other. In two-phase pipe flows, liquid and vapor may be moving at different velocities, or they may occupy differing volumes of the pipe at different locations. This is a flow region to be avoided when possible, because of problems in calculating density and in separating the effects of the two phases on flowmeter performance.

The phase diagram for carbon dioxide is shown in Fig. 2.1c. This diagram is similar in appearance to the water diagram, but the critical pressure and temperature are substantially lower [1070 psia (7380 kPa) and 88°F (31.1°C)]. At room temperature and atmospheric pressure, the fluid is in the vapor phase (point D'). For this reason, carbon dioxide is usually referred to as a gas, although by thermodynamic definition it should be called a vapor because it is below the critical isotherm at 68°F (20°C).

Thermodynamically, there is no difference between point D on the water diagram and point D' on the carbon dioxide diagram. Note that dry ice (solid carbon dioxide) at atmospheric pressure does not pass through the liquid region but *sublimates* from a solid directly into a gas.

All fluids (except helium, which has no defined triple point) exhibit phase diagrams and pressure-volume-temperature diagrams similar to those in Fig. 2.1.

Law of Corresponding States

First premised by Van der Waals, the *law of corresponding states* allows experimental data on some fluids to be used to approximate the properties of all fluids. At their critical points, the two fluids shown in Fig. 2.1a and c would be in corresponding states, although they would have different temperatures, specific volumes, and pressures. With critical temperature and pressure as correlating parameters, properties are reduced to corresponding states by defining the reduced temperature, pressure, and volume as

$$p_r = \frac{p_f}{p_c} \tag{2.1}$$

$$T_r = \frac{T_f}{T_c} \tag{2.2}$$

$$v_r = \frac{v_f}{v_c} \tag{2.3}$$

where p_c, T_c, and v_c are the respective values of pressure (in pounds per square inch absolute), temperature (in degrees Rankine), and specific volume (in cubic feet per pound-mass or per pound-mass-mole) at the critical point.

In Fig. 2.2, 10 fluids are shown on a plot of the compressibility factor Z versus

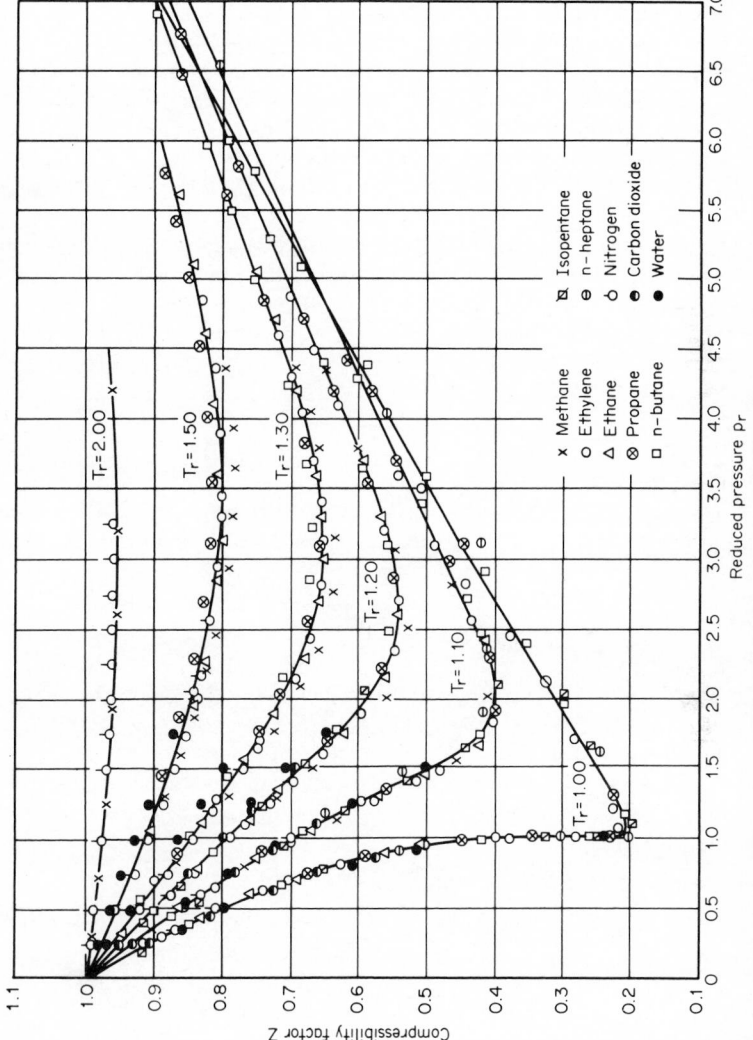

Figure 2.2 Generalized compressibility diagram showing data for various gases *(from Obert, 1948).*

the reduced pressure and temperature for each. (The compressibility factor, which is a correction factor applied to the ideal-gas equation to adjust for real-gas effects, will be discussed in more detail later.) The previous examples, carbon dioxide and water, are included on this curve.

The region between the critical pressure ratios of 0 and 1 has been enlarged in Fig. 2.3, to show the water and carbon dioxide points of Fig. 2.1. At room temperature and atmospheric pressure, the reduced pressure and temperature for water places point B in the shaded liquid region. Increasing the temperature brings the fluid to the saturated-liquid line. In the two-phase region denoted by line C, it is impossible to determine the proportions of the mixture from any diagram, since the reduced pressure and temperature will locate the two phases everywhere along line C. The fluid could be a saturated liquid, a saturated vapor (gas), or any mixture of the two.

When the reduced temperature is above the point where the reduced-temperature isotherm intercepts the saturated-vapor line, the state of the fluid is

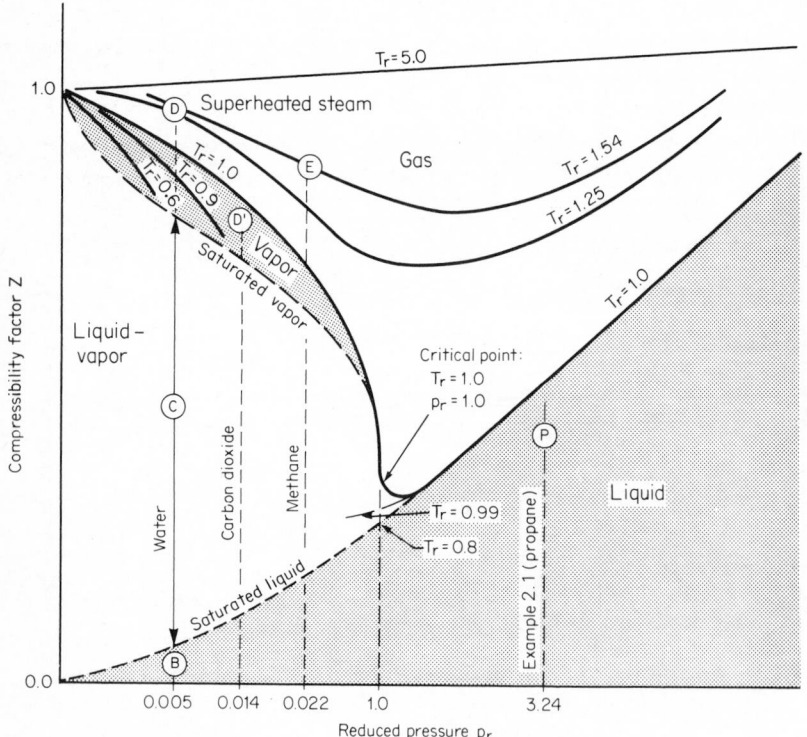

Figure 2.3 Corresponding states for water, carbon dioxide, and methane at room temperature and pressure.

again defined by its temperature and pressure. Increasing the temperature further to 1000°F (538°C) brings the fluid to point D, in the superheated region.

Carbon dioxide has a higher reduced pressure at atmospheric pressure, and the reduced temperature places it in the vapor region. Methane (point E) is shown as a third example; at room temperature and atmospheric pressure, it is a gas.

EXAMPLE 2.1

Propane is at 2000 psia (13,790 kPa) and 50°F (10°C). Determine whether it is a liquid or a gas. If it is a liquid, at what temperature, with the same pressure, will it become a gas?

From Table D.1, the critical properties are

$$p_c = 615.8 \text{ psia}, \qquad T_c = 665.6°\text{R}$$

The reduced pressure and temperature are calculated with Eqs. (2.1) and (2.2) as

$$p_r = \frac{2000}{615.8} = 3.24$$

$$T_r = \frac{459.67 + 50}{665.6} = 0.766$$

These are shown as point P in the liquid region in Fig. 2.3. In this region, for the change from a liquid to a gas, the critical temperature must be reached; from Eq. (2.2),

$$T_r = \frac{459.67 + T_F}{665.6} = 1.0$$

and the required temperature is

$$T_F = 205.9°\text{F } (96.6°\text{C})$$

The pvT Gas-Density Equation

The pressure, temperature, and volume relationship for a real gas can be expressed as

$$p_f V = nZ_f R_0 T_f \tag{2.4}$$

In the equation, n is the number of moles, and a mole is defined as a mass of gas equal to its molecular weight. A pound-mole of methane thus has a mass of 16.043 lb$_m$ (the molecular weight), and a pound-mole of air has a mass of 28.963 lb$_m$. Note that a mole of one gas has a different mass than a mole of another gas (16.043 versus 28.963 lb$_m$, for example). The mole is, then, a mass unit that defines the same number of molecules; 1 mol of methane contains the same number of molecules as 1 mol of nitrogen. For this reason, it is the most convenient mass unit to use, particularly when the densities of gas mixtures are to be considered. In addition, use of the mole eliminates the need for determining individual gas constants, and the *universal* gas constant R_0 can be used in the English engineering system:

$$R_0 = 10.73151 \frac{\text{psia} \cdot \text{ft}^3}{\text{lb}_m \cdot \text{mol} \cdot {}^\circ\text{R}} \qquad (2.5)$$

The relationship among the number of moles, the pounds-mass of gas, and the molecular weight of the gas is

$$n = \frac{m}{M_w} \qquad (2.6)$$

The factor Z in Eq. (2.4) is the compressibility factor, which corrects the equation for real-gas effects. When the compressibility factor is 1.0, the gas is called either *ideal* or *perfect*. If the specific heat at constant pressure is assumed independent of temperature and pressure, the gas is called perfect. When the specific heat is assumed to be only temperature-dependent, the gas is referred to as ideal.

By defining the ideal specific gravity as the ratio of the molecular weight of the gas to that of air, Eq. (2.4) may be rearranged into a density equation. The *specific gravity* of a gas (sometimes referred to as the *ideal specific gravity* or *relative density*) is defined as

$$G = \frac{M_{w,\text{gas}}}{M_{w,\text{air}}} \qquad (2.7)$$

Substituting Eqs. (2.6) and (2.7) into Eq. (2.4) gives

$$p_f \mathbf{V}_f = \frac{m}{G M_{w,\text{air}}} Z_f R_0 T_f \qquad (2.8)$$

which, when rearranged to a density equation, becomes

$$\rho_f = \frac{m}{\mathbf{V}_f} = \frac{G M_{w,\text{air}} p_f}{Z_f R_0 T_f} \qquad (2.9)$$

Substituting the universal gas constant R_0, from Eq. (2.5) and the molecular weight of air into Eq. (2.9) (Jones, 1978) gives the density equation

$$\rho_f = \frac{28.96247}{10.73151} \frac{G p_f}{Z_f T_f} = 2.698825 \frac{G p_f}{Z_f T_f} \qquad (2.10)$$

in U.S. units and

$$\rho_f^* = \frac{28.96247}{8.31441} \frac{G p_f^*}{Z_f T_K} = 3.483407 \frac{G p_f^*}{Z_f T_K} \qquad (2.11)$$

in SI units.

Equations (2.10) and (2.11) are used to derive all flow equations that calculate flow rate from temperature T_f, pressure p_f, specific gravity G, and compressibility Z.

GAS SPECIFIC GRAVITY

Real and Ideal Specific Gravity

Historically, two almost identical specific-gravity values have been used in Eq. (2.10). The early definition of specific gravity was the ratio of gas density to air density when both are at the same temperature and pressure; this is now called the *real* specific gravity in the natural gas industry. The *ideal* specific gravity now in use is defined as the ratio of the molecular weights.

Specific-gravity measurement devices determine the density ratio at pressures and temperatures close to atmospheric conditions. It had been assumed that the measurement would be correct for any pressure and temperature. However, where improved accuracy is required, the slight effects of compressibility Z on both gas and air must be considered.

The real specific-gravity equation is

$$G_R = \left(\frac{\rho_{\text{gas}}}{\rho_{\text{air}}} \right)_{T,p} \tag{2.12}$$

where T and p are close to ambient temperature and pressure. Substituting the density equation (2.10) or (2.11) into Eq. (2.12) gives the relationship between ideal and real specific gravity as

$$G = \left[\left(\frac{ZT}{p} \right)_{\text{gas}} \left(\frac{p}{ZT} \right)_{\text{air}} \right]_{T,p} G_R \tag{2.13}$$

If a real specific-gravity measurement is made, the specific gravity G must be calculated with Eq. (2.13) before substitution into Eq. (2.10) or (2.11).

In most practical applications, the temperature and pressure of the sampled gas and the air are the same within the specific-gravity measuring device, and Eq. (2.13) reduces to

$$G = \left(\frac{Z_{\text{gas}}}{Z_{\text{air}}} \right)_{T,p} G_R \tag{2.14}$$

In this equation, the compressibilities of the gas and the air are calculated at the same pressure and temperature.

Specific-Gravity and Density Measuring Devices

Figure 2.4 shows several real specific-gravity measuring devices and a vibrating-type densitometer. In the buoyancy gas balance, illustrated in Fig. 2.4*a*, a displacer is first balanced in air, and the pressure and temperature are recorded. Air is then purged from the chamber using the unknown gas, and a new balance is obtained by adjusting the gas pressure. The ratio of the pressures is used to infer the specific gravity of the unknown gas. Obviously, this laboratory device is not well suited for on-line measurements.

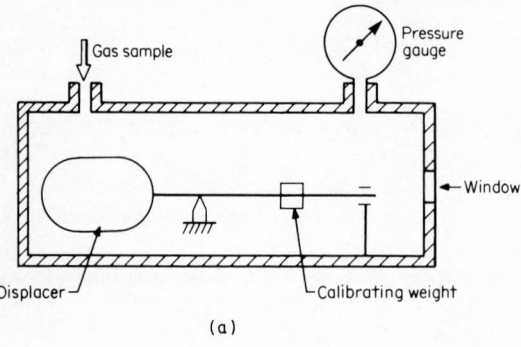

(a)

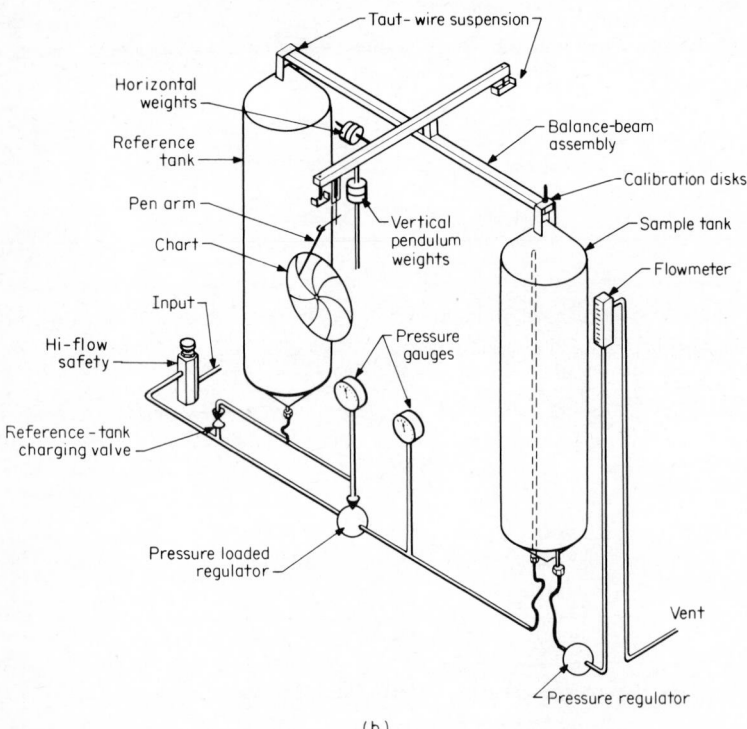

(b)

Figure 2.4 Density and specific-gravity measuring devices. (*a*) Buoyancy gas balance. (*b*) **Recording gravitometer** (*courtesy UGC Industries, Inc.*). (*c*) **Momentum gravitometer: Ranarex Gas Analyzer** (*courtesy Permutit Co., Inc.*). (*d*) **Densitometer** (*courtesy ITT Barton*).

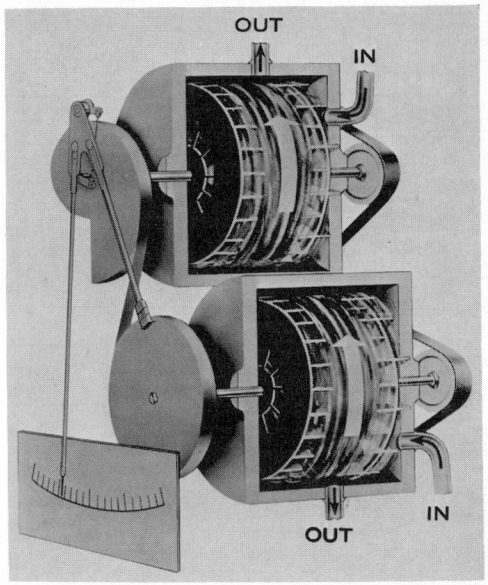

(c)

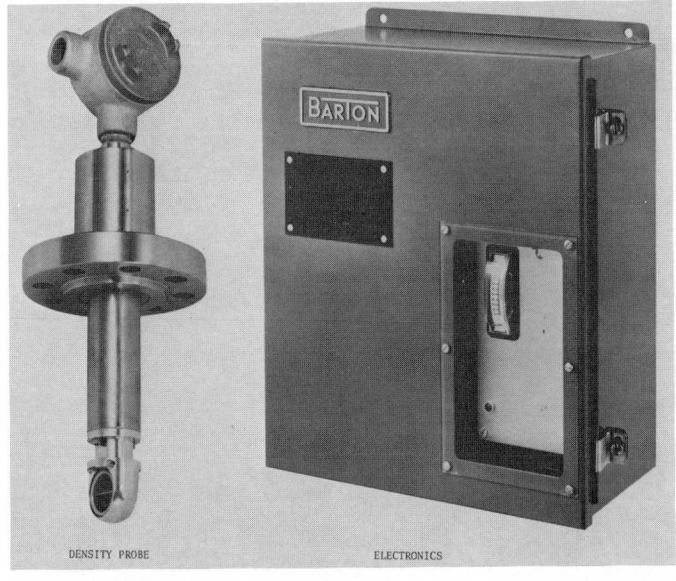

DENSITY PROBE   ELECTRONICS

(d)

Figure 2.4 (*Continued*).

For continuous recording, the gravitometer shown in Fig. 2.4b is used. A steady flow of the unknown gas is passed through a fixed-volume float that is balanced by a float of equal volume filled with a reference gas, with the two floats regulated to the same pressure. The difference between the densities of the reference gas and the sample gas causes an imbalance that is recorded on a calibrated specific-gravity chart.

The two rotating drums shown in Fig. 2.4c provide an on-line specific-gravity measuring device. Air is drawn in by the upper drum, and the lower drum brings in the gas sample. The two drums rotate in opposite directions. Adjacent to each impeller is a torque wheel that picks up the momentum of the spinning gas and transmits it to a lever. Since momentum is a direct function of density, for constant rotor speed the torque difference is a measure of specific gravity. With suitable linearization, the torque difference is recorded as real specific gravity.

EXAMPLE 2.2

Calculate the density of hydrogen at 588 psia (4054 kPa) and $-10°F$ ($-23°C$). Also determine the reading of a device being used to measure the real specific gravity of hydrogen when the pressure and temperature of the hydrogen and air are 14.7 psia (101.3 kPa) and 80°F (26.7°C).

From Table G.6, at $p_f = 588$ psia and $T_f = 459.67 - 10 = 449.67$, the compressibility factor is $Z_f = 1.0264$. From Table D.1, $(M_w)_{H2} = 2.0158$. Then, from Eq. (2.7),

$$G = \frac{2.0158}{28.963} = 0.06960$$

The density is calculated from Eq. (2.10) as

$$\rho_f = 2.6988 \frac{(0.06960)(588)}{(1.0264)(449.67)} = 0.2393 \text{ lb}_m/\text{ft}^3$$

The real specific gravity reading is calculated by rearranging Eq. (2.14) as

$$G_R = \frac{Z_{air}}{Z_{gas}} G = \frac{0.9999}{1.0006} 0.06960 = 0.06955$$

where Z_{air} and Z_{gas} are the values at 14.7 psia and 80°F.

The densitometer, shown in Fig. 2.4d, employs a thin rectangular plate in a cylindrical support structure as the density-sensing element. The mechanical components are assembled into a probe that is readily inserted through the pipe wall such that the resonating plate is exposed to the process gas. In operation, the plate oscillates in simple harmonic motion, which accelerates the surrounding gas. By a closed feedback system utilizing a detector, an amplifier, and a driver coil, the plate is driven to maximum amplitude. The mass of surrounding fluid that vibrates with the plate varies with density, and the plate resonant frequency decreases with increasing gas density. The feedback-system resonant frequency then indicates the fluid density. Although the relationship between frequency and density is nonlinear, electronic linearization provides a standardized output. The device can be calibrated for use with both liquids and gases.

GAS COMPRESSIBILITY

Compressibility-Factor Diagrams

Deviation from the ideal gas law is taken into account with a multiplying factor called the compressibility factor, which can be found in tables, calculated from several available equations of state, or determined through generalized diagrams. For desk calculations, the generalized Nelson-Obert reduced two-parameter diagram is widely used. For improved accuracy, the three-parameter Edmister-Pitzer diagrams are used.

Nelson-Obert Diagram Using experimental data for 26 single-component gases, Nelson and Obert (1954) developed the reduced two-parameter temperature and pressure diagram illustrated in Fig. 2.5a. For these gases, the smoothed curves fitted to within 1 percent. When used for other gases, except those near the critical point or strongly polar fluids, the accuracy of the diagram ranges approximately from 1 to 2 percent for compressibility factors greater than 0.6, and from 4 to 6 percent for factors in the range of 0.3 to 0.6. This diagram is not recommended for helium, hydrogen, or neon.

Edmister-Pitzer Diagrams The Nelson-Obert diagram is based on a single two-parameter correlation. To improve the prediction, Pitzer (1955) introduced the *acentric factor* ω as a third parameter, to correct for the nonspherical nature of the molecular force field. The acentric factor is zero for the defined spherical simple fluids argon, krypton, and xenon. The Pitzer correlation yields good results for slightly polar and nonpolar gases, with an accuracy of approximately ± 1 percent when the compressibility factor ranges from 0.6 to 1.0, and ± 3 percent for factors in the range 0.2 to 0.8.

Edmister (1974) graphically presented Pitzer's results on two separate reduced-parameter diagrams. The first is the *simple-fluid* diagram illustrated in Fig. 2.5b; this is similar to the Nelson-Obert diagram, except that the reduced pressure axis has a logarithm scale. The second diagram is the *generalized compressibility correction for deviation from a simple fluid.* Assuming fluids with the same acentric factor have the same pvT relationship, and further assuming that the deviation function can be used in a linear manner to adjust for the acentric nature of the fluid, the compressibility can be calculated as

$$Z = Z^0 + \omega Z^1 \tag{2.15}$$

The acentric factors for many fluids are given in Table D.1. The two Edmister-Pitzer diagrams are presented in Figs. G.7 and G.8.

Edmister (1974) proposed an equation for estimating the acentric factor from critical pressure, critical temperature, and boiling temperature as

$$\omega = \frac{3}{7} \frac{\log (p_c/14.7)}{T_c/T_B - 1} - 1 \tag{2.16}$$

This relationship correlates well with measured values and is especially useful in determining the compressibility factor of a gas mixture based on molar values.

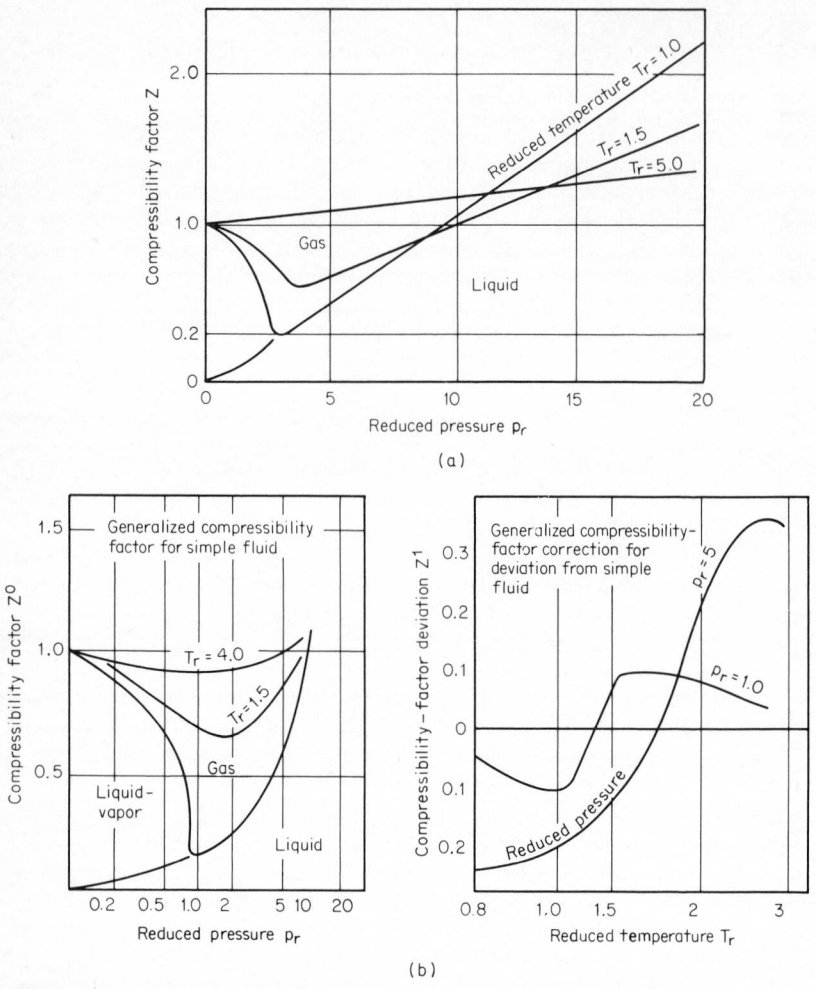

Figure 2.5 Generalized compressibility diagrams (not to scale). (*a*) Nelson-Obert (see Figs. G.1 through G.6). (*b*) Edmister (see Figs. G.7 and G.8).

EXAMPLE 2.3

Calculate the density of methane at $-99.7°F$ ($-73.2°C$) and 1470 psia (10,135 kPa) using (*a*) the Nelson-Obert diagram and (*b*) the Edmister-Pitzer diagrams.

From Table D.1, for methane,

$$p_c = 667.2 \text{ psia} \qquad (M_w)_{\text{CH4}} = 16.043 \text{ lb}_m/\text{lb}_m \cdot \text{mol}$$

$$T_c = 343.1°R \qquad \omega = \quad 0.008$$

The reduced parameters are then, from Eqs. (2.1) and (2.2),

$$p_r = \frac{1470}{667.2} = 2.20 \qquad T_r = \frac{459.67 - 99.7}{343.1} = 1.05$$

a. Nelson-Obert diagram

From Fig. G.4, $Z = 0.365$. The specific gravity is, from Eq. (2.7),

$$G = \frac{16.043}{28.963} = 0.5539$$

The density is then, from Eq. (2.10),

$$\rho_f = (2.6988) \left[\frac{(0.5539)(1470)}{(0.365)(459.67 - 99.7)} \right] = 16.72 \text{ lb}_m/\text{ft}^3$$

b. Edmister-Pitzer diagrams

From Figs. G.7 and G.8, the simple-fluid compressibility and the correction for deviation from a simple fluid are

$$Z^0 = 0.372 \qquad Z^1 = -0.08$$

Then, from Eq. (2.15),

$$Z = 0.372 + (0.008)(-0.08) = 0.371$$

The density is calculated with Eq. (2.11):

$$\rho_f = 2.6988 \frac{(0.5539)(1470)}{(0.371)(459.67 - 99.7)} = 16.45 \text{ lb}_m/\text{ft}^3$$

Calculating Compressibility

Numerous equations of state have been developed to relate pressure, temperature, and specific volume. Each has particular advantages and a range over which it can be applied. The more accurate equations require a change of constants for each gas and are most often used for thermodynamic studies and specific flow situations. In cases where mixture-combination rules are not available, a verification test is required if accuracy is important.

For many gas applications, the compressibility factor is seldom below 0.85, and it is most often between 0.95 and 1.0. Within this range, iterative solution of the equation of Redlich and Kwong (1949) is suggested for general gas applications. For natural gas the AGA† NX-19 (1963) equations, the Pacific Energy Association equations (fitted to NX-19), or the Hall-Yarborough equation (fitted to the Standing-Katz diagrams) are commonly used.

Redlich-Kwong Equation Redlich and Kwong (1949) developed a two-constant state equation that can be rearranged to calculate the compressibility factor from the reduced pressure and temperature as

$$Z = 1 + A - \frac{AB}{A + Z} + \frac{A^2B}{AZ + Z^2} \tag{2.17}$$

where

$$A = \frac{0.0867 p_r}{T_r} \qquad B = \frac{4.934}{T_r^{1.5}} \tag{2.18}$$

†American Gas Association.

Equation (2.17) is nonlinear, and an iterative solution is required. Newton's method for solving for the zero root of an equation is the simplest to use. By rearranging Eq. (2.17) into the function form given in Eq. (A.2) in App. A, the second estimate for the compressibility factor may be written as

$$Z_1 = Z_0 - \frac{F_0}{F_0'} \tag{2.19}$$

where Z_0 is the initial estimate, F_0 is

$$F_0 = 1 + \mathbf{A} - Z_0 - \frac{\mathbf{AB}}{\mathbf{A} + Z_0} + \frac{\mathbf{A^2B}}{\mathbf{A}Z_0 + Z_0^2} \tag{2.20}$$

and the derivative of the function F_0 is

$$F_0' = \frac{\mathbf{AB}}{(\mathbf{A} + Z_0)^2} - \frac{\mathbf{A^2B}(2Z_0 + \mathbf{A})}{(Z_0^2 + \mathbf{A}Z_0)^2} - 1 \tag{2.21}$$

With an initial estimate of $Z_0 = 1$, Eq. (2.19) will provide the compressibility factor to within 0.2 percent when this factor is in the range 0.9 to 1.0. A third estimate may be obtained from the second estimate by replacing the subscript 0 with 1 in the equations and repeating the calculation. This gives a calculation accuracy of 0.001 percent over the same range of compressibility factors.

Figure 2.6 shows the percentage difference between compressibility factors computed with the Redlich-Kwong equation and the actual factors for argon, methane, and oxygen. For reduced pressures below 0.6, the bias error is less than 1 percent. For lower reduced pressures, the error is 0.5 percent or less.

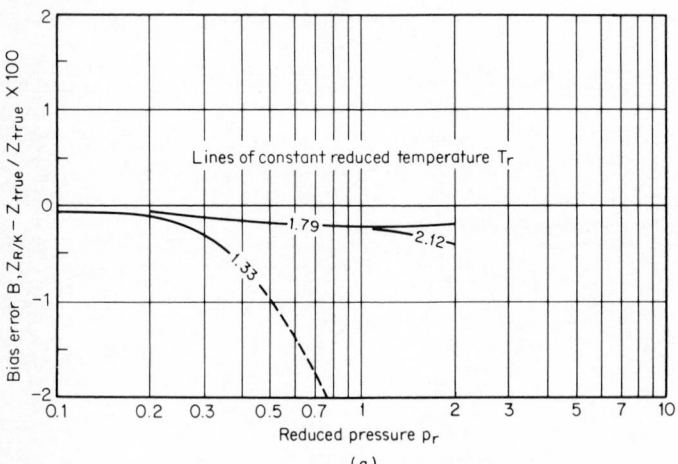

Figure 2.6   Bias error for Redlich-Kwong equations. (a) Argon. (b) Methane. (c) Oxygen.

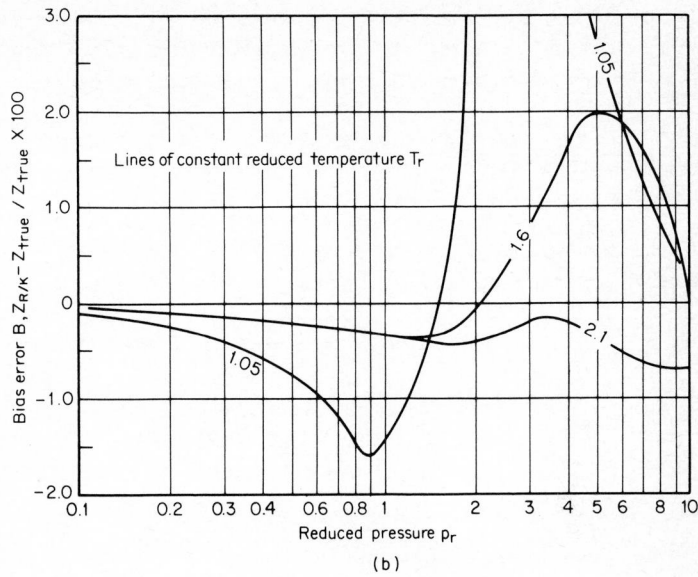

(b)

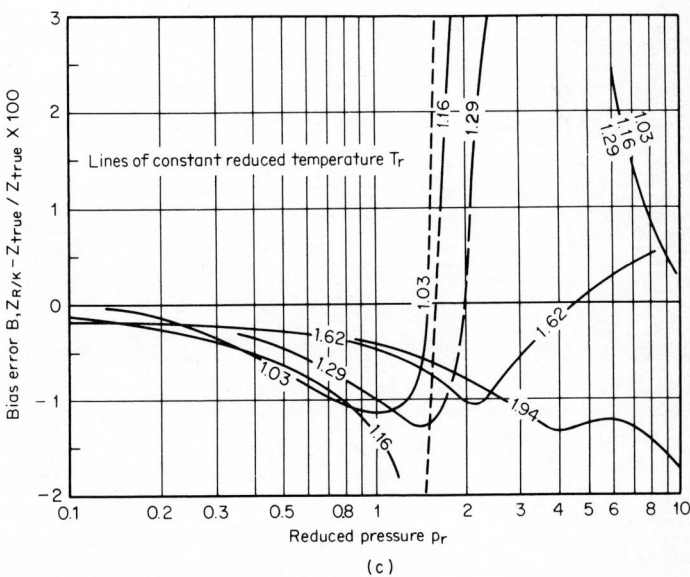

(c)

Figure 2.6 (*Continued*).

Interpolation Using the Redlich-Kwong Equation When the compressibility factor is known from tables or by measurement at a given design reduced pressure and temperature, the Redlich-Kwong equation can be used to predict operating values. For example, many on-site computers use a preset, dialed-in compressibility factor based on average daily pressure, temperature, and specific gravity. If the pressure or temperature varies from these values, a bias error will be introduced in the calculated flow rate.

If it is assumed that the compressibility factor calculated from the Redlich-Kwong equation is biased from the actual compressibility factor, then a ratio can be used to continuously adjust the dialed-in factor as follows:

$$Z_f = \frac{(Z_{R/K})_f}{(Z_{R/K})_{des}} Z_{des} \tag{2.22}$$

In Eq. (2.22) the tabular or measured compressibility factor Z_{des} is multiplied by the ratio of the compressibility factor calculated with the Redlich-Kwong equation at measured flowing temperature and pressure, to that calculated at the design conditions $(Z_{R/K})_{des}$.

In many gas (vapor) flow applications, density is read from tabular values at a design pressure and temperature. Under flowing conditions, the pressure and temperature are not always equal to design values, and the flow rate must be corrected. If the gas (vapor) is superheated and approximates an ideal gas ($Z = 1.0$), the correction is made with the ratio of pressure to temperature in accordance with Eq. (2.11). For real gas (vapor) and for accurate work, the compressibility of the gas should also be calculated at flowing conditions.

For computer applications, the Redlich-Kwong equation can be used to continually correct for the density at flowing conditions

$$\rho_f = \frac{(Z_{R/K})_{des}}{(Z_{R/K})_f} \frac{(T_f)_{des}}{T_f} \frac{p_f}{(p_f)_{des}} (\rho_f)_{des} \tag{2.23}$$

where the subscript des means design conditions, and the subscript f means flowing conditions. The Redlich-Kwong compressibility factor $(Z_{R/K})_{des}$ is calculated from design conditions for use in Eq. (2.23).

EXAMPLE 2.4

The output of a vortex flowmeter is scaled to read steam flow in pounds-mass per hour at 400°F (204°C) and 120 psia (827 kPa), based on a density of 0.2452 lb_m/ft^3 (3.928 kg/m^3). Using the Redlich-Kwong equation with a single iteration ($Z_0 = 1$), calculate the bias error if the steam temperature and pressure change to 410°F (210°C) and 140 psia (965 kPa).

A vortex flowmeter measures volumetric flow at flowing conditions q_{acfh}. The calculated mass flow is then

$$(q_{PPH})_{des} = 0.2452(q_{acfh})_{vor}$$

at design conditions, and

$$(q_{PPH})_{true} = \rho_f(q_{acfh})_{vor}$$

at flowing conditions, where the subscript vor stands for "vortex." The bias error is calculated with Eq. (4.5):

$$B = \frac{(q_{PPH})_{des} - (q_{PPH})_{true}}{(q_{PPH})_{true}} 100 = \frac{0.2452 - \rho_f}{\rho_f} 100$$

At design conditions of 400°F and 120 psia the critical properties, from Table D.1, are

$$p_c = 3197.9 \text{ psia} \qquad T_c = 1165.1°R$$

and the reduced parameters are, from Eqs. (2.1) and (2.2),

$$p_r = \frac{120}{3197.9} = 0.0375 \qquad T_r = \frac{459.67 + 400}{1165.1} = 0.738$$

The Redlich-Kwong coefficients are, from Eq. (2.18),

$$A = \frac{(0.0867)(0.0375)}{(0.738)} = 0.004405$$

$$B = \frac{4.934}{(0.738)^{1.5}} = 7.782$$

With $Z_0 = 1$ and a single iteration, the function F_0 [Eq. (2.20)] becomes

$$F_0 = 1.0 + 0.004405 - 1.0 - \frac{(0.004405)(7.782)}{0.004405 + 1.0} + \frac{(0.004405)^2(7.782)}{(0.004405)(1.0) + (1.0)^2}$$

$$= -0.02958$$

and the derivative of the function is, from Eq. (2.21),

$$F_0' = \frac{(0.004405)(7.782)}{(0.004405 + 1.0)^2} - \frac{(0.004405^2)(7.782)[(2)(1.0) + 0.004405]}{[(1)^2 + (0.004405)(1.0)]^2} - 1$$

$$= -0.9663$$

From Eq. (2.19) the first iteration gives, with $Z_0 = 1$, the design value

$$(Z_{R/K})_{des} = 1 - \frac{-0.02958}{-0.9663} = 0.9694$$

At flowing conditions, the reduced pressure and temperature are, from Eqs. (2.1) and (2.2),

$$p_r = \frac{140}{3197.9} = 0.0438 \qquad T_r = \frac{459.67 + 410}{1165.1} = 0.746$$

The Redlich-Kwong coefficients for flowing conditions are

$$A = 0.005090 \qquad B = 7.658$$

so that $F_0 = -0.03350$ and $F_0' = -0.9618$, and the flowing compressibility factor becomes, from Eq. (2.19),

$$Z_f = 1.0 - \frac{-0.03350}{-0.9618} = 0.9652$$

From Eq. (2.23), the density at flowing conditions via the Redlich-Kwong equation is then

$$\rho_f = 0.2452 \, \frac{0.9694}{0.9652} \, \frac{459.67 + 400}{459.67 + 410} \, \frac{140}{120} = 0.2840$$

The bias error is calculated as

$$B = \frac{0.2452 - 0.2840}{0.2840} \, 100 = -13.7 \text{ percent}$$

Hall-Yarborough Equation   For general natural gas measurement, where the Standing and Katz (1942) compressibility diagram is used, the closed-form solution proposed by Hall and Yarborough (1973, 1974) is useful. The Hall-Yarborough has an accuracy of ± 0.3 percent for reduced temperatures above 1.05 and for very high reduced pressures. To derive the equation, a hard-sphere equation of state was modified by a generalized real-gas deviation function that fitted the Standing-Katz Z-factor diagram. The hard-sphere equation approximates real-gas characteristics at high pressures and temperatures and allows for a generalized closed-form equation, without specific constants for each gas.

To calculate the compressibility factor, two equations are required. The first is a zero-root equation that is iterated to find a reduced density value; from this, the compressibility factor is calculated.

First the mixture's pseudocritical temperature is calculated from known mole fractions as

$$T_{pc} = \Sigma x_i T_{ci} \tag{2.24}$$

This is corrected for the mole fraction of carbon dioxide and hydrogen sulfide content by computing

$$T_{ca} = T_{pc} - C_{WA} \tag{2.25}$$

where $C_{WA} = 120[(x_{CO2} + x_{H2S})^{0.9} - (x_{CO2} + x_{H2S})^{1.6}] + 15(x_{H2S}^{1/2} - x_{H2S}^{4})$ $\quad$ (2.26)

The mixture's pseudocritical pressure is then determined as

$$p_{pc} = \Sigma x_i p_{ci} \tag{2.27}$$

which is adjusted for the mole fraction of carbon dioxide and hydrogen sulfide by computing

$$p_{ca} = \frac{p_{pc} T_{ca}}{T_{pc} + C_{WA} + x_{H2S}(1 - x_{H2S}) C_{WA}} \tag{2.28}$$

The pseudocritical reduced temperature and pressure are then obtained using

$$T_{pr} = \frac{T_f}{T_{ca}} \qquad p_{pr} = \frac{p_f}{p_{ca}} \tag{2.29}$$

The reduced density ρ_r is then calculated by Newton's method (see Sec. A.1) as

$$(\rho_r)_n = (\rho_r)_{n-1} - \frac{F_{n-1}}{F'_{n-1}} \tag{2.30}$$

where
$$F = -b + \frac{\rho_r + \rho_r^2 + \rho_r^3 - \rho_r^4}{(1 - \rho_r)^3} - c\rho_r^2 + d\rho_r^f \tag{2.31}$$

and
$$F' = \frac{1 + 4\rho_r + 4\rho_r^2 - 4\rho_r^3 + \rho_r^4}{(1 - \rho_r)^4} - 2c\rho_r + fd\rho_r^g \tag{2.32}$$

The constants in Eqs. (2.31) and (2.32) are derived from pseudocritical properties as

$$b = 0.06125 p_{pr} T \exp a \qquad\qquad f = 2.18 + 2.82T$$
$$c = 14.76T - 9.76T^2 + 4.58T^3 \qquad g = 1.18 + 2.82T$$
$$d = 90.7T - 242.2T^2 + 42.4T^3$$

where
$$T = \frac{1}{T_{pr}} \quad \text{and} \quad a = -1.2(1 - T)^2$$

As the first estimate for the reduced density, the approximation

$$(\rho_r)_0 = b \tag{2.33}$$

is used. Three iterations are usually required to determine the reduced density to a closure accuracy of 0.1 percent. From this value the compressibility factor is calculated as

$$Z_f = \frac{b}{\rho_r} \tag{2.34}$$

EXAMPLE 2.5

Calculate, using the Hall-Yarborough equation, the Standing-Katz compressibility factor for the mixture detailed below, at a flowing temperature of 100°F (37.8°C) and a pressure of 1000 psia (6895 kPa).

Component	Mole fraction x_i	Critical temperature T_{ci},† °R	$x_i T_{ci}$	Critical pressure p_{ci},† psia	$x_i p_{ci}$
Methane	0.6	343.1	205.9	667.2	400.3
Carbon dioxide	0.2	547.6	109.5	1069.9	214.0
Hydrogen sulfide	0.2	671.8	134.4	1269.2	253.8
			449.8		868.1

†From Table D.1.

From Eqs. (2.24) and (2.27), respectively,

$$T_{pc} = \Sigma x_i T_{ci} = 449.8 \qquad p_{pc} = \Sigma x_i p_{ci} = 868.1$$

From Eq. (2.26),

$$C_{WA} = 120[(0.2 + 0.2)^{0.9} - (0.2 + 0.2)^{1.6}] + 15(0.2^{1/2} - 0.2^4) = 31.6°F$$

The adjusted pseudocritical temperature is then, from Eq. (2.25),

$$T_{ca} = 449.8 - 31.6 = 418.2°R$$

The adjusted pseudocritical pressure is, from Eq. (2.28),

$$p_{ca} = \frac{(868.1)(418.2)}{449.8 + 31.6 + (0.2)(1 - 0.2)(31.6)} = 746.3 \text{ psia}$$

The pseudocritical reduced temperature and pressure are, from Eq. (2.29),

$$T_{pr} = \frac{459.67 + 100}{418.2} = 1.338$$

$$p_{pr} = \frac{1000}{746.3} = 1.340$$

The Hall-Yarborough constants are, with $T = 1/T_{pr}$,

$$a = -0.07658 \qquad d = -49.80$$
$$b = 0.05682 \qquad f = 4.288$$
$$c = 7.490 \qquad g = 3.287$$

The initial estimate for the reduced density is, from Eq. (2.33),

$$(\rho_r)_0 = 0.0952 \text{ lb}_m/\text{ft}^3$$

and the iterative solution is tabulated as follows:

Iteration n	Reduced density $(\rho_r)_{n-1}$	Function F_{n-1} [Eq. (2.31)]	Function F'_{n-1} [Eq. (2.32)]	Reduced density $(\rho_r)_n$ [Eq. (2.30)]
1	0.0952	0.0150	0.5890	0.0699
2	0.0699	-0.000759	0.6498	0.07105
3	0.0705	-0.000002	0.6462	0.07105

According to Eq. (2.34), the compressibility factor then is

$$Z_f = \frac{b}{(\rho_r)_n} = \frac{0.05682}{0.07105} = 0.7997$$

Pacific Energy Association Equations Orifice flowmeters are used almost exclusively for the measurement of high-pressure natural gas. In developing the flow equation for calculating standard cubic feet per hour, a correction factor called the *supercompressibility factor* was introduced. For convenience it is derived as

$$F_{pv} = \sqrt{\frac{1}{Z_f}} \qquad (2.35)$$

In the natural gas industry the term supercompressibility is almost always used, even with reference to vortex, turbine, or other linear-output flowmeters. For these flowmeters, Eq. (2.35) is rewritten as

$$Z_f = \frac{1}{F_{pv}^2} \qquad (2.36)$$

For natural gas containing small amounts of inerts such as carbon dioxide and nitrogen, the supercompressibility factor may be found in the AGA (1963) tables. There are six volumes that cover pressures from 0 to 3000 psig (0 to 21,000 kPa). A seventh volume contains the corrections for nitrogen and carbon dioxide content. In the *Manual for the Determination of Supercompressibility Factors for Natural Gas* (AGA, 1963), the range is extended by computer calculations to a pressure of 5000 psig (34,475 kPa), temperatures of -40 to $240°F$ (-40 to $116°C$), specific gravities of 0.554 to 1.000, and 15 percent carbon dioxide and nitrogen contents.

These tables and the manual computations are too extensive to publish in this handbook. For noncustody transfer flow calculations, the empirical equations and tabular corrections for inert constituent tables developed by the Pacific Energy Association (PEA, 1977) are used for computer computation. The supercompressibility factor is calculated from the generalized equation

$$F_{pv} = \sqrt{1 + \frac{k_1 p_G (10^{5+k_2 G})}{T_f^{3.825}}} \qquad (2.37)$$

Values of k_1 and k_2 are specified for selected specific-gravity ranges, and these are given in Table 2.1.

In Tables G.11 through G.16, values from Eq. (2.37) at $60°F$ ($15.6°C$) are listed. These must be adjusted for flowing temperature by first using Table G.17 and then algebraically adding the correction for inerts given in Table G.18.

EXAMPLE 2.6

The specific gravity of a natural gas containing 5 percent by volume of carbon dioxide is measured as 0.73. The flowing pressure is 400 psig (2758 kPa), and the flowing temperature is $40°F$ ($4.44°C$). Determine the compressibility factor Z_f (a) by using the Pacific Energy Association equation and inert correction table, and (b) from the tables given in App. G.

a. By calculation and inert table

From Table 2.1, $k_1 = 4.66$ and $k_2 = 1.600$. The supercompressibility factor is, from Eq. (2.37),

$$F_{pv} = \sqrt{1 + \frac{(4.66)(400)(10^{5+(1.6)(0.73)})}{(459.67 + 40)^{3.825}}}$$
$$= 1.063$$

From Table G.19, the correction for 5 percent CO_2 is $+0.006$, which is an additive correction; therefore,

$$F_{pv} = 1.063 + 0.006 = 1.069$$

The compressibility factor is then given by Eq. (2.36):

$$Z_f = \frac{1}{(1.069)^2} = 0.875$$

b. Tabular solution

From Table G.13, $F_{pv} = 1.055$ at 400 psig and 60°F. Then, from Table G.17, the corrected value at 40°F is $F_{pv} = 1.064$. This value is corrected for 5 percent CO_2 by the addition of 0.004, so that

$$F_{pv} = 1.064 + 0.006 = 1.070$$

The compressibility factor is, again from Eq. (2.36),

$$Z_f = \frac{1}{(1.070)^2} = 0.873$$

Measuring Compressibility

Several laboratory-type instruments have been designed to measure compressibility. These are shown schematically in Fig. 2.7. In the National Bureau of Stan-

Table 2.1 Constants for Natural Gas Supercompressibility Equation; $p_f \leq 600$ psia

Range of specific gravity G	Constants	
	k_1	k_2
$0.600 \leq G$	2.48	2.020
$0.601 \leq G \leq 0.650$	3.32	1.810
$0.651 \leq G \leq 0.750$	4.66	1.600
$0.751 \leq G \leq 0.900$	7.91	1.260
$0.901 \leq G \leq 1.100$	11.63	1.070
$1.101 \leq G \leq 1.500$	17.48	0.900

SOURCE: Pacific Energy Association (1977).

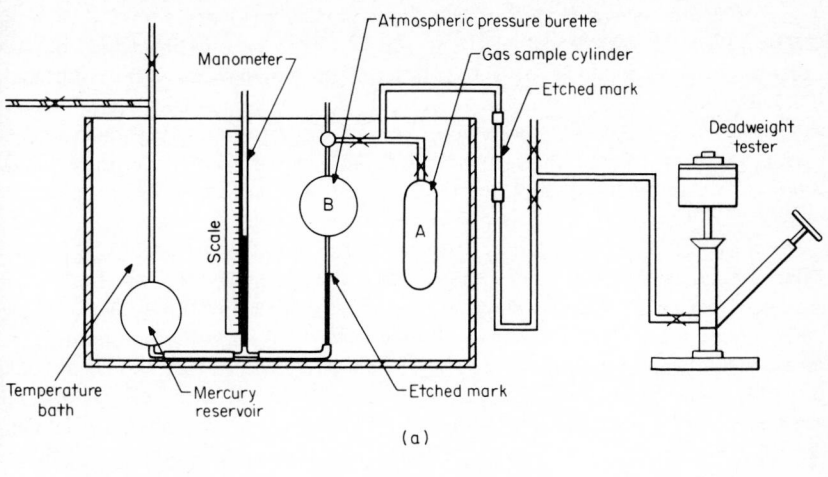

(a)

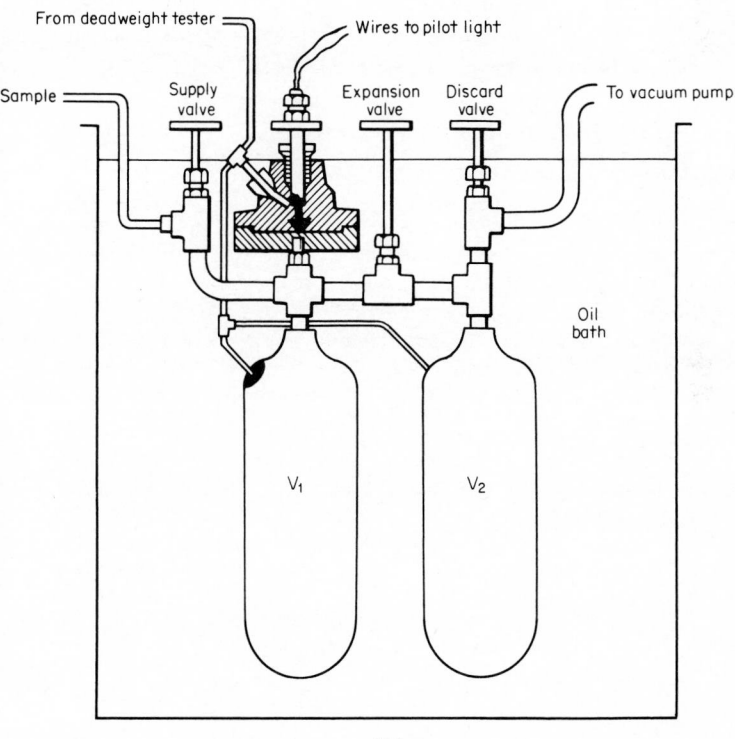

(b)

Figure 2.7 Compressibility-factor measuring devices. (a) National Bureau of Standards apparatus. (b) U.S. Bureau of Mines apparatus.

dards apparatus (Bean, 1929) in Fig. 2.7a, a known gas volume A under pressure is repeatedly, and very carefully, discharged to a lower pressure in a fixed-volume burrette B. This displaces mercury, under slight pressure, to an etch-marked position below burette B, which fixes the volume. The burrette gauge pressure is then read on the adjacent manometer scale. The initial and final deadweight readings establish the initial and final gas sample conditions. By successive fixed temperature and volume reduction to measured lower pressure, the initial compressibility of the high-pressure gas sample is calculated.

In the U.S. Bureau of Mines–Burnett compressibility apparatus (Burnett, 1936), chamber V_1 is initially charged with the unknown gas, which then is repeatedly discharged into a chamber V_2 that is evacuated after each fill. If an ideal gas is expanded to an increased volume $V_1 + V_2$, and the temperature is returned to the original value, the ratio of final to initial pressure should be constant for successive expansions. This ratio is not constant for a real gas during the initial higher-pressure expansions, but the real gas does approach ideal-gas behavior at lower pressures. In a plotting of the ratio of initial pressure to reduced pressure versus initial pressure, a sloping line is obtained, rather than the horizontal line obtained for an ideal gas. By applying the real-gas equation, the compressibility factor at various pressures can be derived.

GAS MIXTURES

Few tables or charts exist that can be used to determine density, or other fluid properties, for gas mixtures. The psychrometric charts for air and water and the supercompressibility tables for natural gas mixtures are two examples of tables that are available.

The density behavior of a flowing multicomponent gas is similar to that of a single-component gas, and the temperature and pressure are measured in the conventional manner. However, specific gravity and compressibility must be calculated on the basis of certain empirically determined combination rules. These are, in turn, based on certain mixture laws, and experimental evidence is often used to select one method over another. When no specific equations are given, the following may be used.

Specific Gravity

The number of moles in a mixture of gases is defined by

$$n = \Sigma n_i = n_1 + n_2 + n_3 + \cdots \tag{2.38}$$

where n_1, n_2, n_3, ... are the number of moles of each gas. The *mole fraction* of each component is defined as

$$x_1 = \frac{n_1}{n} \qquad x_2 = \frac{n_2}{n} \qquad x_3 = \frac{n_3}{n} \qquad \cdots \tag{2.39}$$

where the mole-fraction summation is further defined by

$$\Sigma x_i = 1 \tag{2.40}$$

The average molecular weight of the mixture is calculated by dividing the total mass, in pounds-mass, by the number of moles contained in the mixture; from Eq. (2.6),

$$M_{w,\text{mix}} = \frac{m}{n} \tag{2.41}$$

The mass of the total mixture is then the sum of the individual masses, which can be determined from the molecular weight of each gas with

$$m = \Sigma n_i M_{wi} = n_1 M_{w1} + n_2 M_{w2} + n_3 M_{w3} + \cdots \tag{2.42}$$

Substituting Eq. (2.42) into Eq. (2.41) yields

$$M_{w,\text{mix}} = \frac{\Sigma n_i M_{wi}}{n} \tag{2.43}$$

which can be reduced to a mole-fraction equation by using Eqs. (2.39):

$$M_{w,\text{mix}} = \Sigma x_i M_{wi} = x_1 M_{w1} + x_2 M_{w2} + x_3 M_{w3} + \cdots \tag{2.44}$$

The mole fraction of each component is obtained by either gravimetric or volumetric analysis, and the ideal specific gravity is calculated by dividing the mixture's averaged molecular weight by the molecular weight of dry air. From Eq. (2.44), this is

$$G_{\text{mix}} = \frac{\Sigma x_i M_{wi}}{M_{w,\text{air}}} = \frac{\Sigma x_i M_{wi}}{28.96247} \tag{2.45}$$

As an example, Table 2.2 presents the mole fractions and molecular weights of the 11 constituents of *standard* dry air (Jones, 1978). The average molecular weight is shown calculated in accordance with Eq. (2.44).

Amagat's Volume Law

Amagat's volume law is derived under the supposition that the volume of a gas mixture can be divided into separable unique volumes, one for each gas, each volume existing at the mixture's pressure and temperature. The total volume is the sum of the individual volumes and is expressed as

$$V = \Sigma V_i = V_1 + V_2 + V_3 + \cdots \tag{2.46}$$

For ideal gases, Eq. (2.46) may be rearranged to allow solution for the volume of each constituent, at the pressure and temperature of the mixture, as

$$V = \frac{nR_0T_f}{p_f} \qquad V_1 = \frac{n_1R_0T_f}{p_f} \qquad V_2 = \frac{n_2R_0T_f}{p_f} \qquad V_3 = \frac{n_3R_0T_f}{p_f} \qquad \cdots \tag{2.47}$$

Table 2.2 Composition of Dry Air and
Computation of Average Molecular Weight

Component	Mole fraction x_i	Molecular weight M_{wi}	$x_i M_{wi}$
Nitrogen	0.78102	28.0134	21.87903
Oxygen	0.20946	31.9988	6.70247
Carbon dioxide	0.00033	44.0098	0.01452
Argon	0.00916	39.948	0.36592
Neon	0.00001818	20.179	0.00037
Helium	0.00000524	4.0026	0.00002
Krypton	0.00000114	83.80	0.00010
Xenon	0.000000087	131.30	0.00001
Hydrogen	0.000005	2.0158	0.00001
Methane	0.0000015	16.0426	0.00002
Nitrous oxide	0.0000003	44.0128	0.00001
Average molecular weight $(M_w)_{air} = \Sigma x_i M_{wi}$			28.96247

SOURCE: Jones (1978).

Since the ratio $R_0 T_f / p_f$ appears in each expression, the volumes of the components may be written as

$$\mathbf{V}_1 = \frac{n_1}{n} \mathbf{V} \qquad \mathbf{V}_2 = \frac{n_2}{n} \mathbf{V} \qquad \mathbf{V}_3 = \frac{n_3}{n} \mathbf{V} \qquad \cdots \qquad (2.48)$$

The ratio of each volume to the total volume defines the mole fraction, so that

$$x_1 = \frac{n_1}{n} = \frac{\mathbf{V}_1}{\mathbf{V}} \qquad x_2 = \frac{n_2}{n} = \frac{\mathbf{V}_2}{\mathbf{V}} \qquad x_3 = \frac{n_3}{n} = \frac{\mathbf{V}_3}{\mathbf{V}} \qquad \cdots \qquad (2.49)$$

For these equations, the ratios of volumes are experimentally determined by volumetric analysis, in which the gas is passed through various reagents that sequentially remove known constituents. By measuring the initial volume and the volume after each reduction, the ratios of volumes, and hence the mole fractions, are determined. Since this reduction takes place at atmospheric pressure, it is reasonable to assume negligible compressibility.

Dalton's Law of Partial Pressure

Rather than assume a unique volume for each constituent, Dalton assumed each gas occupies the total volume at the mixture's temperature. Under these conditions, the following relationships may be written for an ideal gas:

$$\frac{p_f V}{R_0 T_f} = \sum \frac{p_{fi} V}{R_0 T_f} = \frac{p_{f1} V}{R_0 T_f} + \frac{p_{f2} V}{R_0 T_f} + \frac{p_{f3} V}{R_0 T_f} + \cdots \qquad (2.50)$$

This equation reduces to

$$p_f = \Sigma p_{fi} = p_{f1} + p_{f2} + p_{f3} + \cdots \qquad (2.51)$$

Dalton's law of partial pressure states that *the pressure of a mixture of gases is equal to the sum of the partial pressures which each component of the gas would exert if it existed singularly in the mixture's volume at the mixture's temperature.* If Dalton's law is applied to each component, a useful relationship between the partial pressures and mole fractions is derived:

$$n = \frac{p_f V}{R_0 T_f} \qquad n_1 = \frac{p_{f1} V}{R_0 T_f} \qquad n_2 = \frac{p_{f2} V}{R_0 T_f} \qquad n_3 = \frac{p_{f3} V}{R_0 T_f} \qquad \cdots \qquad (2.52)$$

These equations reduce further to

$$p_{f1} = x_1 p_f \qquad p_{f2} = x_2 p_f \qquad p_{f3} = x_3 p_f \qquad \cdots \qquad (2.53)$$

Relationship between Mass (Gravimetric) and Volumetric Analysis

Two analyses are made to determine constituent percentages in a gas mixture. The volumetric analysis reports the results as a volume ratio, and the mass (gravimetric) analysis gives results based on mass determination, with the mass ratio defined by

$$x_{m1} = \frac{m_1}{m} \qquad x_{m2} = \frac{m_2}{m} \qquad x_{m3} = \frac{m_3}{m} \qquad \cdots \qquad (2.54)$$

where $\qquad\qquad m = \Sigma m_i = m_1 + m_2 + m_3 + \cdots \qquad (2.55)$

EXAMPLE 2.7

A mixture of carbon dioxide and methane has a volumetric analysis of 5 and 95 percent, respectively. The line pressure is 500 psia (3450 kPa), and the temperature is 40°F (4.4°C). Assuming an ideal gas mixture, determine (a) the gravimetric analysis, (b) the mixture density, and (c) the pressure exerted by each component.

The molecular weights of carbon dioxide and methane are, from Table D.1, $(M_w)_{CH4} = 16.043$ and $(M_w)_{CO2} = 44.010$.

a. Gravimetric analysis

According to Eq. (2.49), the mole fraction is written

$$x_i = \frac{\text{moles (component)}}{\text{moles (mixture)}} = \frac{n_i}{n}$$

so that $\qquad\qquad x_{CO2} = \frac{n_{CO2}}{n} = 0.05 \qquad x_{CH4} = \frac{n_{CH4}}{n} = 0.95$

The masses of the components are *assumed,* from Eq. (2.6), to be

$$m_{CO2} = (0.05)(44.01) = 2.201 \text{ lb}_m$$

$$m_{CH4} = (0.95)(16.043) = 15.241 \text{ lb}_m$$

The total mass is then, by Eq. (2.55),

$$m = \Sigma m_i = 2.201 + 15.241 = 17.442 \text{ lb}_m$$

Finally, from Eq. (2.54), the gravimetric analysis is

$$(x_m)_{CO2} = \frac{2.201}{17.442} = 0.126 = 12.6 \text{ percent}$$

$$(x_m)_{CH4} = \frac{15.241}{17.442} = 0.874 = 87.4 \text{ percent}$$

b. Density of mixture

The specific gravity of the mixture is, according to Eq. (2.45),

$$G_{\text{mix}} = \frac{\Sigma x_i M_{wi}}{M_{w,\text{air}}} = \frac{\Sigma x_i M_{wi}}{28.96247}$$

Substituting the mole fractions x from part a and the molecular weights gives

$$G_{\text{mix}} = \frac{(0.95)(16.043) + (0.05)(44.01)}{28.96247} = 0.6022$$

The density of the mixture is then, from Eq. (2.10),

$$\rho_f = 2.698825 \frac{(0.6022)(500)}{(1.0)(459.67 + 40)} = 1.626 \text{ lb}_m/\text{ft}^3$$

where $Z_f = 1.0$ for an ideal gas.

c. Pressure exerted by each component

Dalton's law of partial pressure [Eq. (2.53)] gives

$$p_{CO2} = (0.05)(500) = 25 \text{ psia}$$

$$p_{CH4} = (0.95)(500) = 475 \text{ psia}$$

General Mixture Rules   When no specific mixture rules are given, the accepted rules for general applications are Kay's (1936) rule for the pseudocritical temperature, the modified Prausnitz and Gunn (1958) combination rule for the pseudocritical pressure, and the Joffe (1971) approximation for the acentric factor. With these pseudocritical properties, the Nelson-Obert, Edmister-Pitzer, or Redlich-Kwong state equation is used in the conventional manner to obtain the compressibility factor.

Kay's rule for the pseudocritical temperature is a mole-fraction average method that calculates values to within 2 percent of other proposed rules. This rule is

$$T_{pc} = \Sigma x_i T_{ci} = x_1 T_{c1} + x_2 T_{c2} + x_3 T_{c3} + \cdots \tag{2.56}$$

The pseudocritical pressure is calculated with the Prausnitz-Gunn combination rule as

$$p_{pc} = \frac{(\Sigma x_i Z_{ci}) R_0 T_{pc}}{\Sigma x_i V_{ci}} \quad (2.57)$$

where Z_{ci} and V_{ci} are, respectively, the compressibility factor and volume of each component at the critical temperature and pressure.

In using the Edmister diagram, the acentric factor is approximated with a mole-fraction average given by Joffe (1971):

$$\omega_{pc} = \Sigma x_i \omega_i = x_1 \omega_1 + x_2 \omega_2 + x_3 \omega_3 + \cdots \quad (2.58)$$

The pseudocritical reduced temperature and pressure are then calculated as

$$T_{pr} = \frac{T_f}{T_{pc}} \qquad p_{pr} = \frac{p_f}{p_{pc}} \quad (2.59)$$

EXAMPLE 2.8

Applying the mixture rules, use the Edmister diagrams to calculate the compressibility factor of the mixture of Example 2.5.

The pertinent data from Table D.1 are as follows:

Component	Mole fraction x_i	Critical temperature T_{ci}	Critical pressure P_{ci}	Critical volume V_{ci}	Critical compressibility factor Z_{ci}	Acentric factor ω
Methane	0.6	343.1	667.2	1.589	0.288	0.008
Carbon dioxide	0.2	547.6	1069.9	1.505	0.274	0.225
Hydrogen sulfide	0.2	671.8	1269.2	1.613	0.284	0.100

The pseudocritical properties of the mixture are, from Eq. (2.56),

$$T_{pc} = (0.6)(343.1) + (0.2)(547.6) + (0.2)(671.8) = 449.8°R$$

and, from Eq. (2.57),

$$p_{pc} = \frac{[(0.6)(0.288) + (0.2)(0.274) + (0.2)(0.284)](10.7315)(449.8)}{(0.6)(1.589) + (0.2)(1.505) + (0.2)(1.613)} = 870.5 \text{ psia}$$

The acentric factor is found with Eq. (2.58):

$$\omega_{pc} = (0.6)(0.008) + (0.2)(0.225) + (0.2)(0.100) = 0.0698$$

The reduced pseudocritical pressure and temperature are, from Eq. (2.59),

$$T_{pr} = \frac{559.67}{449.8} = 1.244 \qquad p_{pr} = \frac{1000}{870.5} = 1.149$$

From Edmister diagrams in Figs. G.7 and G.8 give

$$Z^0 = 0.792 \qquad Z^1 = 0.095$$

The compressibility factor of the mixture is then, according to Eq. (2.15),

$$Z_f = 0.792 + (0.0698)(0.095) = 0.799$$

SPECIAL CASE OF A WET GAS

Water vapor in a gas stream is one component of a mixture's specific gravity, and it is measured as part of the total flow. In many cases, however, the volume or mass of dry gas must be calculated—for example, to establish heating value.

Psychrometric Principles

A gas mixed with a vapor behaves differently from a mixture of gases. Since a vapor is, by definition, a gas that lies between the saturated-vapor line and the critical isotherm (see Fig. 2.1b), changes in pressure and temperature result in condensation or vaporization. When the vapor is water, the mixture may range between dry and saturated gas. A water vapor-gas mixture is referred to as a *psychrometric* mixture. Properties used to define gas-vapor mixtures are specific humidity, relative humidity, and dew point.

Specific Humidity The *specific humidity* or *humidity ratio* is defined as the ratio of the mass of water vapor to the mass of dry gas. Under the assumption that the water vapor is an ideal gas at the low pressure existing in mixtures, Eq. (2.10) can be used to derive the following specific-humidity relationships for gas and vapor occupying the same volume:

$$\text{SH} = \frac{(\text{lb}_\text{m})_{wv}}{(\text{lb}_\text{m})_{\text{dry gas}}} = \frac{\rho_{wv}}{\rho_{\text{dry gas}}} \approx \frac{0.6220}{G_{\text{dry}}} \frac{p_{wv}}{p_f - p_{wv}} \tag{2.60}$$

Relative Humidity The *relative humidity* is defined as the ratio of the water vapor pressure p_{wv} in the mixture to the water vapor pressure if the mixture were at saturation at the same temperature:

$$\text{RH} = \left[\frac{p_{wv}}{p_{\text{sat}}} \right]_T \tag{2.61}$$

Dew-Point Temperature The water vapor in a gas may be superheated steam, or it may be at saturation. If the mixture is cooled at constant pressure, and if the vapor was superheated, each component will be cooled at constant partial pressure because the composition (mass) remains constant. At some reduced temperature the water vapor will reach saturation, and any further temperature decrease will result in condensation. The temperature at which liquid first appears is the *dew-point temperature*. Because the cooling takes place at constant pressure, the dew-point temperature is used to enter the steam tables to find the *partial pressure* of the water vapor in an unsaturated mixture; this partial pressure is the pressure corresponding to the saturation temperature, that is, the dew-point temperature.

Saturated and Unsaturated Gas

The quantity of water vapor in a gas can vary from zero to saturation. Shown in Fig. 2.8 are the pressure-volume (density) curves for carbon dioxide and water at saturation (Fig. 2.8a) and in an unsaturated state (Fig. 2.8b). These curves are useful in discussing saturation and nonsaturation.

Saturated Gas   Shown in Fig. 2.8a is the pressure–specific volume (density) diagram for carbon dioxide saturated with water vapor. The mixture pressure p_f is 100 psia (690 kPa) at a temperature of 100°F (37.8°C). The carbon dioxide pressure is, by Dalton's law, the mixture pressure less the pressure of the saturated water vapor. This vapor pressure is found in the steam tables and for 100°F (37.8°C) is 0.95 psia (6.6 kPa). The density shown for the dry carbon dioxide is calculated with Eq. (2.10) at a pressure of 99.05 psia. The ratio of the water vapor density to the density of the dry carbon dioxide is 0.0029/0.729 = 0.00398, which is the specific humidity. Since the carbon dioxide is saturated, the relative humidity, which is the ratio of the water vapor pressure in the mixture to that at saturation, is 1. Because the mixture is saturated, any further decrease in temperature results in condensation; therefore, the wet-bulb temperature is the temperature of the mixture [100°F (37.8°C)].

Unsaturated Gas   In Fig. 2.8b the mixture is shown unsaturated at the same 100°F (37.8°C) temperature. The arrow indicates cooling at constant pressure to the 75°F (23.9°C) dew-point temperature, at which liquid first appears. The pressure 0.43 psia (2.96 kPa) corresponds to the saturation pressure at 75°F (23.9°C) as found in the steam tables. Since the mixture is cooled at constant pressure, this is the *partial pressure* of the water vapor existing in the mixture at 100°F (37.8°C). Since the water vapor pressure is lower in the unsaturated case, the density of the dry carbon dioxide is higher; hence, the mass of water vapor per pound of dry gas, the specific humidity, is less than in the saturated case. The specific humidity is then 0.00129/0.732 = 0.00175, and the relative humidity is 0.43/0.95 = 0.45.

Equation Development

Under the assumption that water vapor is an ideal gas at the low saturation pressures existing in mixtures, the mole-fraction equation (2.53) can be substituted into Eq. (2.45) to yield the wet-gas specific gravity as

$$G_{\text{wet}} = \frac{p_f - p_{wv}}{p_f} G_{\text{dry}} + \frac{p_{wv}}{p_f} G_{wv} \tag{2.62}$$

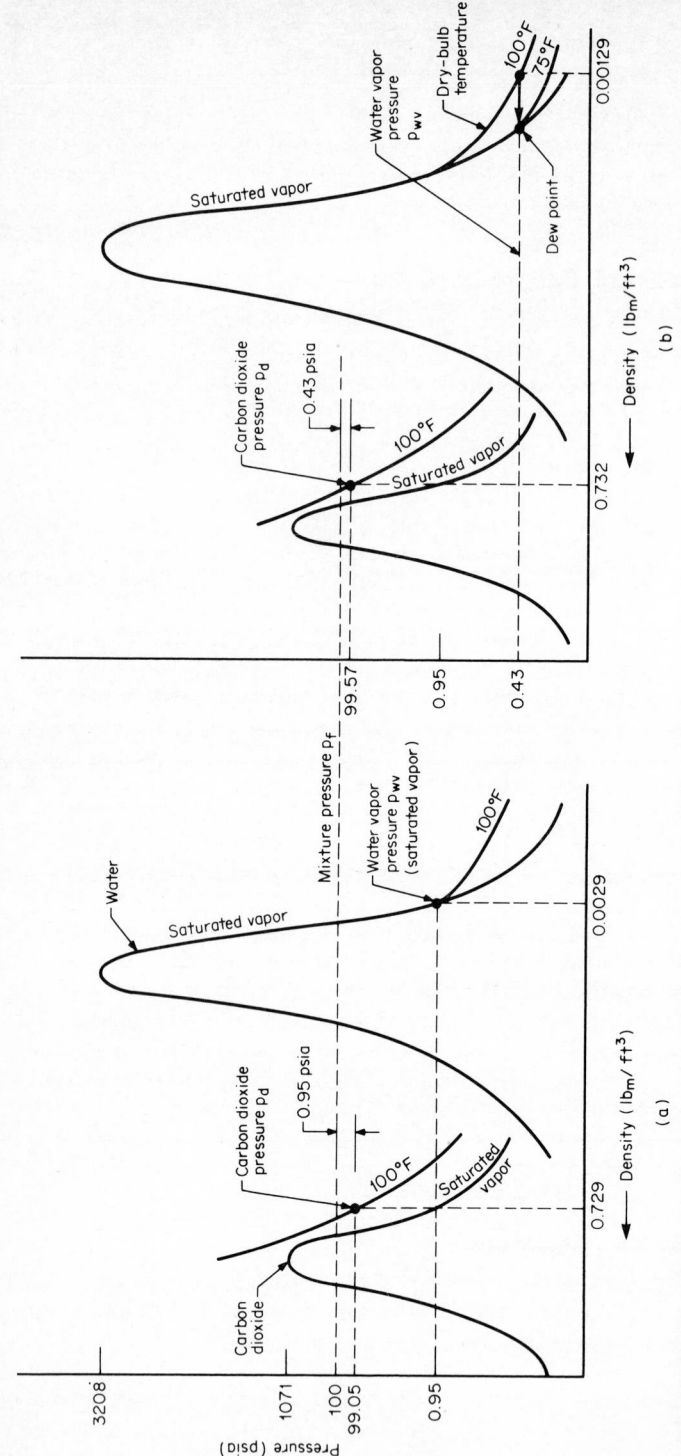

Figure 2.8 Pressure-density curves for carbon dioxide and water (not to scale). (a) Saturated mixture. (b) Unsaturated mixture.

The pressure difference in the first term is the partial pressure of the dry gas as defined by Dalton's law. The ratios of pressures in both terms are the respective mole fractions: dry gas and water vapor.

Since the specific gravity of water vapor is 0.6220, Eq. (2.62) may be rewritten as

$$G_{wet} = \left[1 + \frac{p_{wv}}{p_f} \left(\frac{0.6220}{G_{dry}} - 1 \right) \right] G_{dry} \tag{2.63}$$

The relationship between the water vapor pressure p_{wv}, dry-gas specific gravity G_{dry}, and wet-gas specific gravity G_{wet} is shown in Fig. 2.9. As the specific gravity of the dry gas approaches that of water vapor, the wet-gas specific gravity is well approximated by the dry-gas specific gravity. However, with gases such as hydrogen, where the specific gravity is significantly different, it is important that the relationship given by Eq. (2.63) be used.

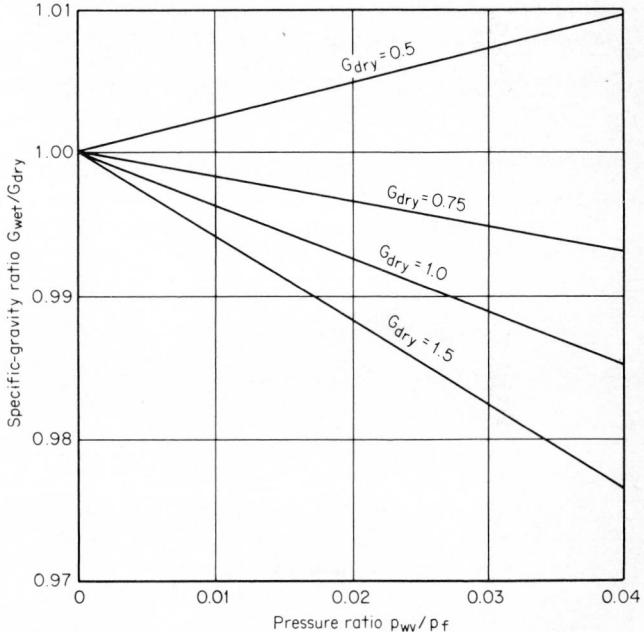

Figure 2.9 Ratio of dry-gas to wet-gas specific gravity.

The compressibility factor of the wet gas can be estimated with a simple mole-fraction average as

$$Z_{wet} = \left(1 - \frac{p_{wv}}{p_f} \right) Z_{dry} + \frac{p_{wv}}{p_f} Z_{wv} \tag{2.64}$$

In this equation, Z_{wv} is usually assumed to be 1.0 because the water vapor pressure is low.

To adjust a wet volume to a dry-volume basis, Amagat's volume law is first written as

$$V_{wet} = V_{dry} + V_{wv} \tag{2.65}$$

When rearranged, this gives, as the relationship between the dry volume and wet volume,

$$F_{WV,dry} = \frac{V_{dry}}{V_{wet}} = 1 - \frac{V_{wv}}{V_{wet}} \tag{2.66}$$

which, after substitution of the relationships in Eqs. (2.49) and (2.53), reduces to

$$F_{WV,dry} = 1 - \frac{p_{wv}}{p_f} \tag{2.67}$$

To convert a volume of wet gas from one moisture condition to another, the partial pressure of the water vapor must be considered at the two differing temperatures. The pertinent relationship is

$$\frac{V_2}{V_1} = \frac{(1 - p_{wv}/p_f)_1}{(1 - p_{wv}/p_f)_2} \frac{p_{f1}}{p_{f2}} \frac{Z_{f2}}{Z_{f1}} \frac{T_{f2}}{T_{f1}} \tag{2.68}$$

where the subscript 1 refers to the measured or initial volume, and 2 to the converted volume.

For adjusting a wet-gas mass flow to a dry mass flow, a mass balance yields

$$(\text{lb}_m)_{wet} = (\text{lb}_m)_{dry} + (\text{lb}_m)_{wv} \tag{2.69}$$

This gives the relationship

$$F_{WVM,dry} = \frac{(\text{lb}_m)_{dry}}{(\text{lb}_m)_{wet}} = 1 - \frac{(\text{lb}_m)_{wv}}{(\text{lb}_m)_{wet}} \tag{2.70}$$

Rearranging Eq. (2.8) to solve for the mass of water vapor and for the mass of the wet mixture, and substituting the mole-fraction relationships given by Eq. (2.53), reduces Eq. (2.70) to

$$F_{WVM,dry} = 1 - \frac{0.6220}{G_{wet}} \frac{p_{wv}}{p_f} \frac{Z_{wet}}{Z_{wv}} \tag{2.71}$$

The following may be used to determine the density of a wet gas: the expression is

$$\rho_{wet} = \frac{2.698825 p_f G_{wet}}{Z_{wet} T_f} \qquad \rho_{wet}^* = \frac{3.483407 \, p_f^* G_{wet}}{Z_{wet} T_K} \tag{2.72}$$

Moisture Determination

There are many humidity- and moisture-measuring devices. Generally, these devices either measure moisture content directly or infer moisture content from relative humidity, specific humidity, dew-point temperature, or hygrometer measurements. They may be summarized as follows:

1. Inferential-measurement devices

    a. Dew-point meters

        ■ Vapor equilibrium meter

        ■ Mechanized dew-point meter

        ■ Fog chamber

        ■ Electrolytic hygrometer

        ■ Manually operated dew cup

    b. Hygrometers

        ■ Hair hygrometer

        ■ Salt-conductivity hygrometer

    c. Psychrometers

        ■ Sling psychrometer

        ■ Assman psychrometer

        ■ Wet- and dry-bulb psychrometer

2. Moisture-measurement methods

    a. Electric conductivity

    b. Electric capacitance

    c. Radio-frequency absorption

    d. Microwave absorption

    e. Infrared absorption

    f. Equilibrium methods

Three commonly used devices are shown in Fig. 2.10.

Wet- and Dry-Bulb Psychrometer   If two thermometers are inserted in the flow line (Fig. 2.10*a*), and one is covered with a continuously wetted wick, the temperature T_{WB} of the wet thermometer will be lower than that of the uncovered thermometer. If the line velocity is greater than 15 ft/s (4.6 m/s), the wetted thermometer will read the adiabatic saturation temperature.

Within the range of 32 to 100°F (0 to 38°C), the ratio of water vapor pressure to mixture pressure can be determined as

$$\frac{p_{wv}}{p_f} = \frac{p_{sat}}{p_f} - 0.000048 C_{mp}(T_F - T_{WB}) \qquad (2.73)$$

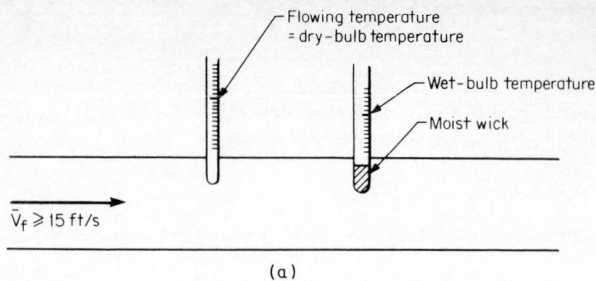

(a)

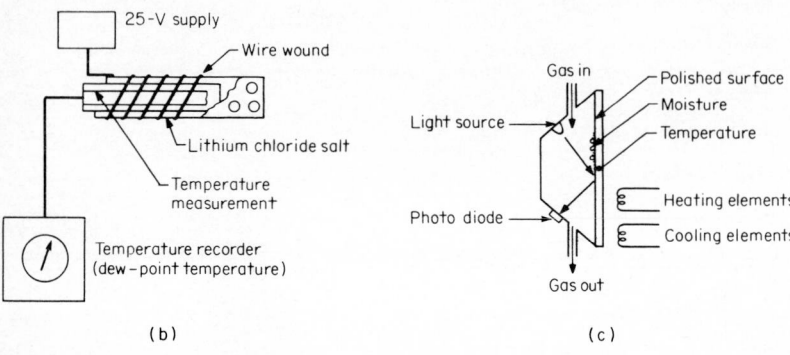

(b)

(c)

(d)

Figure 2.10 Commonly used humidity-measuring devices. (*a*) Wet- and dry-bulb measurements. (*b*) **Dewcel** *(courtesy The Foxboro Co.).* (*c*) **Dew-point measuring device.** (*d*) **Dewscope** *(courtesy UGC Industries, Inc.).*

Values of the mean molecular heat C_{mp} are given in Table 2.3. The saturated vapor pressure p_{sat} is obtained from the steam tables using the line (wet-bulb) temperature T_{WB}.

If the mole fractions are known for a mixture of gases, the mean molecular heat can be calculated on a mole-fraction basis as

$$C_{mp,\text{mix}} = \Sigma x_i C_{mpi} = x_1 C_{mp1} + x_2 C_{mp2} + \cdots \tag{2.74}$$

Table 2.3 **Mean Molecular Heats of Some Gases**

Gas	Mean molecular heat C_{mp}
Acetylene	10.7
Air	7.7
Carbon dioxide	10.0
Carbon monoxide	7.7
Ethane	13.8
Ethylene	11.3
Hydrogen	7.7
Methane	9.4
Nitrogen	7.7

Dewcel The dew-point temperature, as well as the specific humidity and partial pressure of the water vapor, can be determined with the Dewcel shown in Fig. 2.10b. A thin metal tube is covered with a woven glass cloth that is saturated with lithium chloride. A voltage is applied to a double winding of silver wire over the glass cloth. The electrical conductivity between the wires is directly proportional to the moisture that the salt takes on from the surrounding gas. When the moisture is low, there is a small current flow, and the temperature rise within the cell due to Joule heating is low. When the moisture is high, there is a high current flow and the temperature is high. A thermometer placed inside the tube will measure the cell temperature and, therefore, the dew point or absolute humidity. The speed of response of the salt, in changing from moist (as a result of moisture in the stream) to dry is quite high.

Moisture-Condensation (Dew-Point) Instrument If a small flow of moist gas is passed over a mirror whose temperature is controlled by refrigeration or heating (Fig. 2.10c), the moisture will just condense when the dew-point temperature is reached. This results in mirror fogging, which may be detected as a change in the output of a photocell and used in a feedback loop to control the mirror temperature. Figure 2.10d shows a commercially available dew-point instrument that operates in this way. Care should be taken in the case of natural gas, since the mirror may fog when hydrocarbons condense out before the water vapor.

EXAMPLE 2.9

The flow rate of a carbon dioxide and water vapor mixture is 20 lb_m/h (9.07 kg/h). The flowing temperature is 100°F (37.8°C), the flowing pressure is 100 psia (689 kPa), and the dew-point temperature is 75°F (23.9°C). What is the dry carbon dioxide flow rate?

Properties of dry gas

From Table D.1, $(M_w)_{CO2}$ = 44.010. The specific gravity of the dry gas is then, according to Eq. (2.7),

$$G_{dry} = \frac{44.010}{28.963} = 1.5195$$

The compressibility factor of the dry gas (carbon dioxide) is, from Table G.4, Z_{dry} = 0.9699.

Properties of water vapor

From the steam tables, the pressure of saturated steam (saturated vapor) at 75°F is p_{wv} = 0.42964 psia. Now, rearranging Eq. (2.10) for compressibility-factor determination and substitution yield

$$Z_{wv} = 2.69883 \frac{G_{wv} p_{wv}}{\rho_{wv} T_f} = 2.69883 \frac{(0.6220)(0.42964)(704.3)}{459.67 + 100} = 0.9076$$

where, from the steam tables,

$$\rho_{wv} = \frac{1}{v_{wv}} = \frac{1}{704.3} \ lb_m/ft^3$$

Properties of wet gas

The compressibility factor of the wet gas is, from Eq. (2.64),

$$Z_{wet} = \left(1 - \frac{0.42964}{100} \right) 0.9699 + \frac{0.42964}{100} 0.9076 = 0.9696$$

The specific gravity of the wet gas is then, from Eq. (2.63),

$$G_{wet} = \left[1 + \frac{0.42964}{100} \left(\frac{0.6220}{1.5195} - 1 \right) \right] 1.5195 = 1.5156$$

The moisture correction factor for mass flow rate is given by Eq. (2.71):

$$F_{WVM,dry} = 1 - \frac{0.6220}{1.5156} \frac{0.42964}{100} \frac{0.9696}{0.9076} = 0.9982$$

The mass flow of dry gas is, then, by Eq. (2.70),

$$q_{PPH,CO2} = F_{WVM,dry} q_{M,wet} = (0.9981)(20) = 19.97 \ lb_m/h$$

EXAMPLE 2.10

Manufactured gas has the dry-component volumetric analysis tabulated below. The flowing temperature is 100°F (37.8°C) at a mixture pressure of 25 psia (172 kPa). A wet-bulb reading of 72°F (22°C) is measured at a flow rate of 10,000 standard ft^3/h (283 SCMH). Assuming an ideal gas mixture, what is the dry specific gravity and the dry-gas flow rate?
Required data from Tables D.1 and 2.3 are as follows:

Component	Mole fraction x_i	Molecular weight M_{wi}	Molecular heat C_{mpi}
Ethylene	0.004	28.054	11.3
Ethane	0.006	30.070	13.8
Methane	0.1139	16.043	9.4
Carbon monoxide	0.2024	28.011	7.7
Hydrogen	0.3790	2.016	7.7
Nitrogen	0.0083	28.013	7.7
Carbon dioxide	0.2863	44.010	10.0

The mean molecular heat of the mixture is, by Eq. (2.74),

$$C_{mp,mix} = \Sigma x_i C_{mpi} = 8.60$$

The specific gravity of the mixture, from Eq. (2.45), is

$$G_{mix} = \frac{\Sigma x_i M_{wi}}{28.96247} = 0.738$$

From the steam tables, the saturation pressure at 72°F is $p_{sat} = 0.39$ psia. Then, from Eg. (2.73),

$$\frac{p_{wv}}{p_f} = \frac{0.39}{25} - 0.000048(8.60)(100 - 72) = 0.0040$$

The moisture correction factor for volume flow rate is then, by Eq. (2.67),

$$F_{WV,dry} = 1 - \frac{p_{wv}}{p_f} = 1 - 0.004 = 0.996$$

The dry-volume flow rate is then, from Eq. (2.66),

$$q_{SCFH,dry} = F_{WV,dry} q_{SCFH,wet}$$
$$q_{SCFH,dry} = (0.996)(10,000) = 9960 \text{ standard ft}^3/\text{h}$$

LIQUID DENSITY AND SPECIFIC GRAVITY

Liquid density decreases with increasing temperature and, to a much smaller degree, increases with pressure. Depending on how the density is determined, corrections may be needed if the value at flowing conditions is to be obtained.

The density of a liquid is relatively easy and inexpensive to measure, and for most fluids at least one experimental data point is available. The density of a mixture of liquids is usually calculated on a molar volume-average basis, preferably with an experimentally determined point to ensure accuracy.

Water Density (Reference Equation)

The literature contains a number of slightly different values for the density of water at both 60 and 68°F. Since the density of water at these temperatures is a basic reference (for specific gravity at 60°F, and for the inch-of-water differential pressure unit at 68°F), it is important to select one source as the reference standard. In this handbook, the PTB (1968) equation will be used. This equation expresses mass density, in kilograms per cubic meter, as a function of temperature:

$$\rho_w^* = 9.998395639 \times 10^2 + 6.798299989 \times 10^{-2} T_{°C}$$
$$- 9.106025564 \times 10^{-3} T_{°C}^2 + 1.005272999 \times 10^{-4} T_{°C}^3 \qquad (2.75)$$
$$- 1.126713526 \times 10^{-6} T_{°C}^4 + 6.591795606 \times 10^{-9} T_{°C}^5$$

From this equation, the reference density of water at 60°F (15.6°C) and 14.696 psia (101.325 kPa) is

$$(\rho_w)_{60,14.696} = 62.36630 \ \text{lb}_m/\text{ft}^3 \ (999.0121 \ \text{kg/m}^3) \qquad (2.76)$$

and at 68°F (20°C) and 14.696 psia (101.325 kPa) it is

$$(\rho_w)_{68,14.696} = 62.31572 \ \text{lb}_m/\text{ft}^3 \ (998.2019 \ \text{kg/m}^3) \qquad (2.77)$$

These values will be used in developing all subsequent equations for which the density of water is a reference.

Liquid Specific Gravity

The liquid specific gravity (relative density) of a fluid is the ratio of the density of the fluid, at its temperature, to the density of water at a specified reference temperature. In scientific work, the reference is double-distilled water at 4°C (39.2°F); for general engineering work, the reference temperature is 60°F (15.6°C), with no specification as to the condition of the water (although distilled water is implied). ISO 5024 (1976) is standardized on 15°C (59°F) for petroleum liquids. However, API 2540 (1980) adopts 60°F (15.6°C) as the reference for density corrections [see Eq. (2.97)]. Since all published data, curves, and tables are based on 60°F (15.6°C), that temperature will be used in this handbook.

In flow measurement, two liquid specific gravities are used: the specific gravity at base conditions G_b and the specific gravity at flowing conditions G_F. Base conditions are defined as 60°F (15.6°C) and 14.696 psia (101.325 kPa), or standard atmospheric pressure. The base specific gravity is the ratio of fluid density to water density when both are at base conditions; it is written as

$$G_b = \left(\frac{\rho_b}{\rho_w} \right)_{60,14.696} = \left(\frac{\rho_b}{62.36630} \right)_{60,14.696} = \left(\frac{\rho_b^*}{999.0121} \right)_{15.6,101.33} \qquad (2.78)$$

The flowing specific gravity is defined as the ratio of fluid density at line temperature to the density of water at base conditions and is defined by

$$G_F = \frac{\rho_F}{(\rho_w)_{60,14.696}} = \frac{\rho_F}{62.36630} = \frac{\rho_F^*}{999.0121} \qquad (2.79)$$

Measurement of Specific Gravity and Density

Hydrometers   Hydrometers (Fig. 2.11) are the most convenient and widely used devices for determining the specific gravity of relatively transparent liquids. One tenth of one percent accuracy is usually achievable with a calibrated device.

The hydrometer displaces a volume of the unknown liquid until an equilibrium is reached between upward buoyant force and downward hydrometer weight force. By suitable graduation, calibration with known fluids, and the addition of lead-pellet weights, the scale reading is calibrated to read either specific gravity or some related unit (such as API degrees or Baumé degrees).

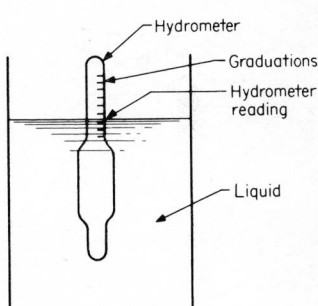

Figure 2.11   Liquid hydrometer.

API Degrees   The scale of the American Petroleum Institute (API) hydrometer is calibrated in degrees API (°API). The specific gravity at base conditions [60°F (15.6°C)] is calculated from the API hydrometer reading as

$$G_b = \frac{141.5}{131.5 + °\text{API}} \qquad (2.80)$$

API-degree hydrometers read correctly only when the liquid temperature is 60°F (15.6°C), since the glass volume changes with changes in temperature. A correction for temperature is included in the tabular values given in Table 5A of ASTM Standard 2540.

Baumé Degrees   In 1768 the French chemist Baumé proposed two hydrometer scales—one for liquids heavier than water, and the other for liquids lighter than water. These scales are widely used to measure the specific gravity of syrups, acids, and other light and heavy liquids. With 60°F (15.6°C) as the reference water density, the Baumé scales provide base readings via

$$G_b = \frac{140}{°\text{Bé} + 130} \tag{2.81}$$

for lighter-than-water liquids, and

$$G_b = \frac{145}{145 - °\text{Bé}} \tag{2.82}$$

for heavier-than-water liquids.

Westphal Balance   Shown in Fig 2.12 is the Westphal balance, a specific-gravity-measuring instrument. Unlike the hydrometer, which floats in the liquid, the Westphal balance plummet is fully submerged. The cord tension is the difference between the downward plummet weight force and the upward buoyant force. The moment of this resultant force is balanced by placing calibrated rider weights in notches on the balance arm. Calibration with distilled water at 60°F (15.6°C) allows the instrument to be calibrated to read the specific gravity directly.

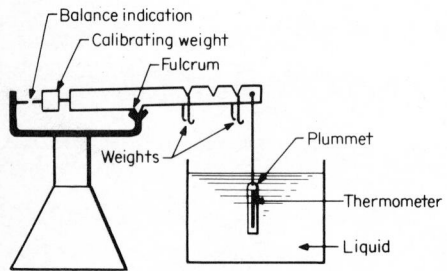

Figure 2.12   Westphal balance.

Glass Volumetric Correction   Hydrometers are designed and calibrated to read correctly at a specified base temperature. Usually this temperature is 60°F (15.56°C), but other base temperatures may be indicated.

To calculate the specific gravity at the test fluid temperature, the cubical expansion coefficient given in Table 2.4 can be used. The specific gravity G_b' based on the hydrometer reading is first calculated from the proper conversion equation [Eqs. (2.80) to (2.82)]; then G_b' is used in the following equation to calculate the specific gravity:

$$G_F = [1 - \alpha_H(T_F - 60)]G_b' \tag{2.83}$$

Table 2.4 Cubical Coefficients for
Hydrometer Glass (Direct-Reading
Hydrometers)

Hydrometer	Cubical coefficient α_H	
	$(°F)^{-1}$	$(°C)^{-1}$
Normal Pyrex	0.0000056	0.000010
Jena III, API†	0.0000128	0.000023
New hydrometers	0.0000139	0.000025
Sinkers	0.0000150	0.000027
Sinker no. 234	0.00000594	0.0000107

†For API-degree hydrometer correction, $G_F = [1 - 0.00001278(T_F - 60) - 0.0000000062(T_F - 60)^2]G_b'$.

For the API-degree hydrometer, a more exact expression is

$$G_F = [1 - 0.00001278(T_F - 60) - 0.0000000062(T_F - 60)^2] \tag{2.84}$$
$$\times \frac{141.5}{131.5 + °API}$$

where $°API$ is the hydrometer reading at temperature T_F.

Pycnometer The most accurate way to determine specific gravity and density is by using a pycnometer bottle (Fig. 2.13). The bottle and stopper are weighed twice—first empty, and then filled with distilled water—to determine the bottle volume. The bottle is then filled with the unknown liquid and weighed. The mass density is the recorded mass divided by the bottle volume, and the specific gravity is the ratio of the fluid mass to the distilled-water value. Care must be exercised to ensure that the temperature is recorded and used to adjust to the reference-density temperature. Measurement accuracy to the fourth decimal place (approximately 0.01 percent) is possible with this instrument. However, in measuring light hydrocarbons, care must be taken to obtain a representative sample and to clean the bottle properly between tests.

Liquid Densitometer Shown in Fig 2.14 is an on-line liquid densitometer that provides continuous measurement of liquid density. The stainless steel sensor consists of a single, straight, unobstructed polished bore tube. Operating on the mass-resonance principle, the sensor is maintained in oscillation by self-contained electronics requiring ac power input. Density is measured as a function of sensor-liquid

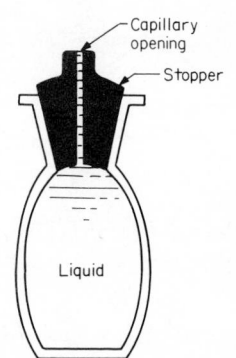

Figure 2.13 Pycnometer
bottle.

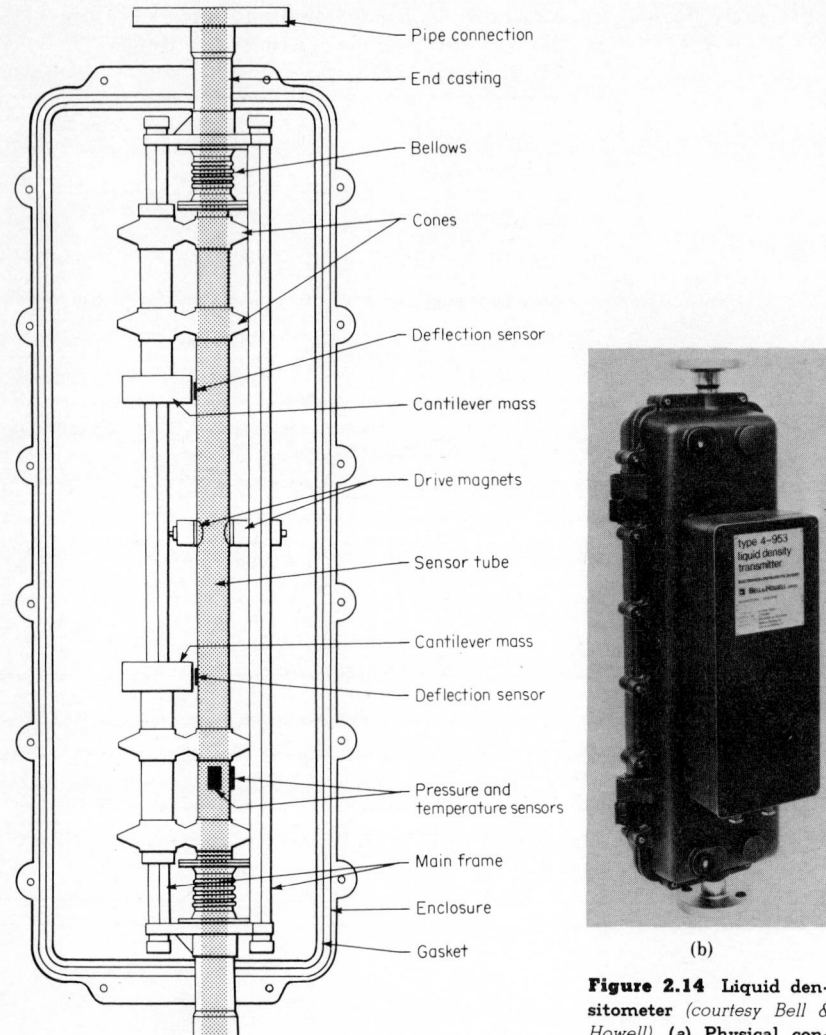

(a)

(b)

Figure 2.14 Liquid densitometer *(courtesy Bell & Howell).* **(a)** Physical construction. **(b)** Transmitter.

natural frequency. The transmitter provides either a current or voltage analog that is directly proportional to density.

Calculating Liquid Density and Specific Gravity

When measurements are not available, it is sometimes necessary to estimate density. Several generalized equations are useful both for this purpose and for estimating the effect of a temperature change on the flow calculations. These equations usually provide estimates to within only 1 or 2 percent and should be used only in the absence of data.

Goldhammer Equation   The simplest extrapolation equation, based on the fluid's critical temperature T_c, is the early Goldhammer (1910) ratio equation, which is

$$\frac{(\rho_F - \rho_v)_{TF}}{(\rho_b - \rho_v)_{60°F}} = \left(\frac{T_c - T_f}{T_c - 519.67} \right)^{1/N} \tag{2.85}$$

where ρ_v is the density of the vapor at T_F and at 60°F (15.6°C). The value of N was found by Fishtine (1963) to vary slightly with compound type. Several values are given in Table 2.5.

Table 2.5   Exponent Values for Goldhammer Equation

Compound type	Exponent N
Water and alcohol	4
Hydrocarbons and ethers	3.45
Other organic compounds	3.23
Inorganic compounds	3.03

Under the assumption that the vapor density ρ_v is negligible at both base conditions and the flowing temperature, Eq. (2.85) reduces to

$$\rho_F = \rho_b \left(\frac{T_c - T_f}{T_c - 519.67} \right)^{1/N} \tag{2.86}$$

Substituting Eqs. (2.78) and (2.79) into Eq. (2.86) gives the relationship between flowing and base specific gravity as

$$G_F = G_b \left(\frac{T_c - T_f}{T_c - 519.67} \right)^{1/N} \tag{2.87}$$

Goyal-Doraiswamy Equation   Goyal and Doraiswamy (1966) presented an equation for calculating density based on critical properties and molecular weight. The equation can be rearranged in terms of specific gravity as

$$G_F = \frac{p_c M_w}{T_c} \left(\frac{0.008}{Z_c^{0.773}} - 0.01102 \frac{T_f}{T_c} \right) \tag{2.88}$$

The compressibility factor Z_c is the value at the critical point, which is fluid-dependent; it may range from 0.22 for methyl alcohol to 0.312 for air. Values of p_c, T_c, M_w, and Z_c are given in Table D.1.

When no data is available on the critical compressibility factor, the relationship developed by Edmister (1974) can be used:

$$Z_c = 0.371 - 0.0343 \frac{\log{(p_c/14.7)}}{T_c/T_B - 1} \tag{2.89}$$

where T_B is the boiling-point temperature in degrees Rankine.

Numerous equations have been suggested for the computation of density using critical properties, based on atomic and structural contributions of the substances. These have an uncertainty of 1 to 5 percent and are used only when laboratory data is unavailable and flow measurement accuracy is not required. The works by Perry and Chilton (1973), Reid et al. (1977), and Smith and Van Ness (1978) present these equations in detail.

Specific Gravity of a Mixture

When two or more liquids are mixed, the specific gravity of the mixture may be estimated by substituting pseudocritical properties into Eq. (2.88). With known mole fractions y_i, the pseudocritical pressure, temperature, and compressibility factor are calculated using Kay's (1936) rule:

$$p_{pc} = \Sigma y_i p_{ci} = y_1 p_{c1} + y_2 p_{c2} + \cdots \tag{2.90}$$

for pseudocritical pressure,

$$T_{pc} = \Sigma y_i T_{ci} = y_1 T_{c1} + y_2 T_{c2} + \cdots \tag{2.91}$$

for pseudocritical temperature, and

$$Z_{pc} = \Sigma y_i Z_{ci} = y_1 Z_{c1} + y_2 Z_{c2} + \cdots \tag{2.92}$$

for the pseudocritical compressibility factor at the critical point. The average molecular weight can be estimated as

$$M_{w,\text{mix}} = \Sigma y_i M_{wi} = y_1 M_{w1} + y_2 M_{w2} + \cdots \tag{2.93}$$

EXAMPLE 2.11

Using Goyal's generalized equation, estimate the specific gravity of a 22 percent ethylene and 78 percent isobutane mixture at (a) 60°F (15.6°C) and (b) 105°F (40.6°C).

From Table D.1, the pertinent properties are as follows:

Property	Ethylene	Isobutane
p_c	730.4 psia	529.1 psia
T_c	508.3°R	734.6°R
Z_c	0.276	0.283
M_w	28.054	58.124

The pseudocritical properties of the mixture are then, from Eqs. (2.90) to (2.93),

$$p_{pc} = (0.22)(730.4) + (0.78)(529.1) = 573.4 \text{ psia}$$

$$T_{pc} = (0.22)(508.3) + (0.78)(734.6) = 684.8°R$$

$$Z_{pc} = (0.22)(0.276) + (0.78)(0.283) = 0.282$$

$$M_{w,\text{mix}} = (0.22)(28.054) + (0.78)(58.124) = 51.509 \text{ lb}_m/(\text{lb}_m \cdot \text{mole})$$

a. *Specific gravity at 60°F (15.6°C)*

From Eq. (2.88),

$$G_b = \frac{(573.4)(51.509)}{684.8} \left[\frac{0.008}{(0.282)^{0.773}} - 0.01102 \frac{459.67 + 60}{684.8} \right]$$

$$= 0.557$$

b. *Specific gravity at 105°F (40.6°C)*

From Eq. (2.88),

$$G_F = 43.13 \left(0.02128 - 0.01102 \frac{459.67 + 105}{684.8} \right)$$

$$= 0.526$$

Thermal-Expansion Equation For moderate temperatures [0 to 100°F (−18 to 38°C)], the density at atmospheric pressure can be approximated by

$$\rho_F = [1 - c(T_F - 60)]^2 \rho_b \approx [1 - 2c(T_F - 60)]\rho_b \qquad (2.94)$$

and the specific gravity at flowing conditions by

$$G_F = [1 - c(T_F - 60)]^2 G_b \approx [1 - 2c(T_F - 60)]G_b \qquad (2.95)$$

The constant c is given in Table 2.6 for various fluids.

API 2540 Equation The American Petroleum Institute, in a joint program with the National Bureau of Standards, developed a density equation based on 463 samples of five different oil products. The results of this work are incorporated into Chapter 11.1, "Volume Correction Factors," of API Standard 2540 (1980).

The density equation is based on the thermal-expansion coefficient of the product at 60°F (15.6°C) base temperature, which is calculated from the base density as

$$\alpha_b = \frac{K_0}{\rho_b^{*2}} + \frac{K_1}{\rho_b^*} \qquad (2.96)$$

where the base density ρ_b^* is in kilograms per cubic meter. The empirically derived constants K_0 and K_1 for the five product groups are given in Table 2.7. The density of the product at flowing temperature is then calculated as

$$\rho_F^* = \rho_b^* \exp\left[-\alpha_b \, \Delta T_F \left(1 + 0.8\alpha_b \, \Delta T_F \right) \right] \qquad (2.97)$$

Table 2.6 Thermal-Expansion Constant c for Some Fluids

Liquid	c	Liquid	c
Acetic acid	0.000298	Ether	0.00046
Acetone	0.000413	Glycerine	0.0001403
Amyl alcohol	0.000251	HCl, 33.2%	0.0001264
Ethyl alcohol	0.0003111	Mercury	0.0000505
Methyl alcohol	0.0003331	Olive oil	0.0002003
Benzene	0.000344	Potassium chloride, 24.3%	0.0000981
Bromine	0.0003144	Phenol	0.000303
Calcium chloride, 5.8%	0.0000694	Sodium chloride, 20.6%	0.000115
Calcium chloride, 40.9%	0.0001272	Sodium sulfate, 24%	0.000114
Carbon disulfide	0.000383	Sulfuric acid, 10.9%	0.0001075
Carbon tetrachloride	0.0003433	Sulfuric acid, 100%	0.000155
Chloroform	0.0003536	Turpentine	0.0002703

Table 2.7 Constants K_0 and K_1 for Five Product Groups

Product group	K_0	K_1
Crude oils	341.0957	0.0
Jet fuels, kerosenes, solvents	330.3010	0.0
Gasolines and naphthenes	192.4571	0.2438
Lubricating oils	144.0427	0.1895
Diesel oil, heating oils, fuel oils	103.8720	0.2701

NOTE: Pentanes and hydrocarbons lower in the hydrocarbon chain are *not* covered by this data.

where $\Delta T_F = T_F - 60$. The specific gravity at flowing or measured temperature is then

$$G_F = G_b \exp\left[-\alpha_b \Delta T_F (1 + 0.8\alpha_b \Delta T_F)\right] \tag{2.98}$$

In many installations, the specific gravity or density is measured at flowing temperature, and the base specific gravity is required for the calculation of base volumes. Since Eq. (2.98) cannot be solved directly for base density, an iterative solution is required. Newton's method can be applied by rearranging Eq. (2.97) into a zero-root equation (see Sec. A.1) to give the function F as

$$F = \ln \rho_b^* - \ln \rho_F^* - \alpha_b \Delta T_F (1 + 0.8\alpha_b \Delta T_F) \tag{2.99}$$

The derivative of this function is

$$F' = \frac{1}{\rho_b^*} + \alpha_b' \, \Delta T_F + 1.6\alpha_b\alpha_b' \, \Delta T_F^2 \tag{2.100}$$

where
$$\alpha_b' = \frac{2K_0}{\rho_b^{*3}} + \frac{K_1}{\rho_b^{*2}} \tag{2.101}$$

The iteration for the base density is then

$$(\rho_b^*)_n = (\rho_b^*)_{n-1} - \frac{F_{n-1}}{F'_{n-1}} \tag{2.102}$$

where F and F' are the solutions of Eqs. (2.99) and (2.100), for an initial estimate of base density use

$$(\rho_b^*)_0 = \rho_F^* \, [1 + \Delta T_F \exp (0.0106 \times {}^\circ\text{API} - 8.05)] \tag{2.103}$$

EXAMPLE 2.12

For gasoline at a temperature of 80.5°F (26.9°C), the API hydrometer reading is 63.5°API. Using the API 2540 equations, determine (a) its specific gravity G_F at 80.5°F, (b) its specific gravity G_b at 60°F, and (c) its API-degree reading at 60°F (or $\text{API}_{60/60}$).

a. *Flowing specific gravity*

The flowing specific gravity, uncorrected for pressure, is obtained by correcting the hydrometer for volumetric expansion. From Eq. (2.84),

$$G_F = \left[1 - (0.00001278)(80.5 - 60) - (0.0000000062)(80.5 - 60)^2 \right] \frac{141.5}{131.5 + 63.5}$$

$$= 0.72545$$

b. *Specific gravity at 60°F*

Now, from Eq. (2.79), the density is

$$\rho_F^* = (0.72545)(999.012) = 724.73 \text{ kg/m}^3$$

The initial estimate of the base density is, by Eq. (2.103),

$$(\rho_b^*)_0 = 724.73 \, \{1 + (80.5 - 60) \exp [(0.0106)(63.5) - 8.05] \}$$

$$= 734.02 \text{ kg/m}^3$$

The next step is to calculate α_b and α_b'. From Table 2.7, $K_0 = 192.4571$ and $K_1 = 0.2438$. Then, by Eq. (2.96),

$$\alpha_b = \frac{192.4571}{(734.02)^2} + \frac{0.2438}{734.02} = 0.0006893$$

and by Eq. (2.101),

$$\alpha_b' = \frac{(2)(192.4571)}{(734.02)^3} + \frac{0.2438}{(734.02)^2} = 0.0000014258$$

Now the function F and its derivative F' are calculated and used to compute a second estimate, as follows: Since $\Delta T_F = 80.5 - 60 = 20.5°F$, Eq. (2.99) gives

$$F_0 = \ln 734.02 - \ln 724.73 - (0.0006893)(20.5)[(1 + (0.8)(0.0006893)(20.5)]$$

$$= -0.001553$$

and Eq. (2.100) gives

$$F'_0 = \frac{1}{734.02} + (0.0000014258)(20.5) + (1.6)(0.0006893)(0.0000014258)(20.5)^2$$

$$= 0.001392$$

The second estimate of the base density is then, from Eq. (2.102),

$$(\rho_b)_1 = 734.02 - \frac{-0.001553}{0.001392} = 735.14$$

A second iteration provides an estimate that is different from this value by only 0.0001 percent. The specific gravity at 60°F is then, by Eq. (2.78),

$$G_b = \frac{735.14}{999.012} = 0.7359$$

c. API degrees at 60°F

Rearranging Eq. (2.80) and substituting yield

$$\mathrm{API}_{60/60} = \frac{141.5}{G_b} - 131.5 = \frac{141.5}{0.7359} - 131.5 = 60.78$$

LIQUID COMPRESSIBILITY

Liquids are not normally considered compressible, and the values for density ρ_F or specific gravity G_F found in tables or measured at atmospheric pressure are seldom increased. There are cases, however, when compressibility must be considered, and a correction made. For example, a liquid becomes more compressible with increases in pressure and temperature; as the critical-temperature isotherm is approached, the liquid acts much like a gas and becomes quite compressible. This can occur for pentanes and lighter hydrocarbons.

The pvT Relationship

Using water as an example, Fig. 2.15 illustrates the relationship among temperature, pressure, and density in the liquid region. Point A (Fig. 2.15a) lies on the saturated-liquid line, at a reduced temperature of 0.7 and a reduced pressure of 0.045. Increasing the pressure at constant temperature to a reduced pressure of 1.0 [3200 psia (22,000 kPa)] increases the density by approximately 2 percent (Fig. 2.15b). When the reduced pressure is further increased to 4.8 (point B), there is a 6 percent increase in density.

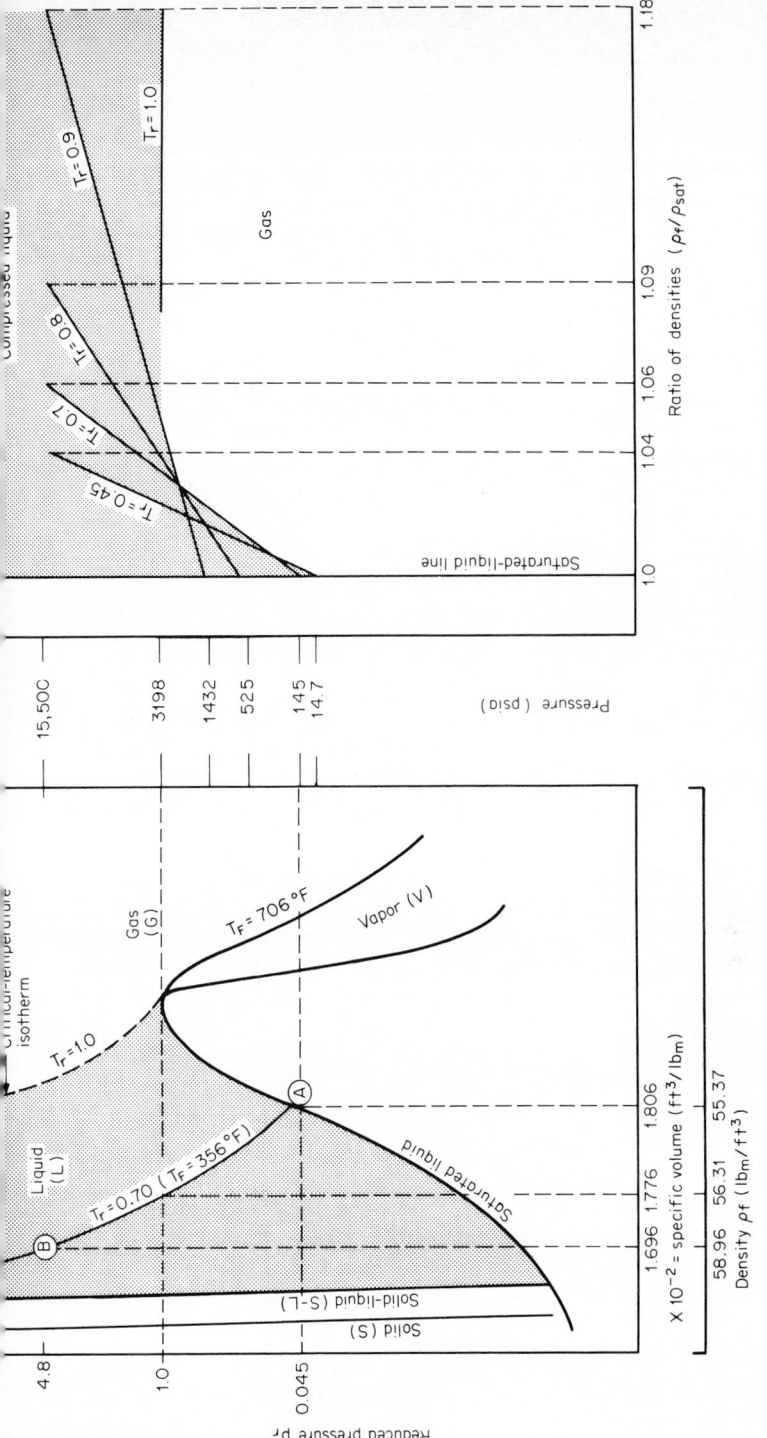

Figure 2.15 Effect of pressure and temperature on the density of water. (a) Pressure versus density. (b) Ratio of densities.

Compressibility is approximately linear over wide pressure ranges, and it can, with reasonable accuracy, be represented by the straight lines shown in Fig. 2.15*b*. With this figure, the relationship among pressure, temperature, and density is easily visualized. As the critical isotherm is approached ($T_r = 1.0$), the ratio of the density to the density at saturation increases substantially. For example, at a reduced temperature of 0.9 ($T_F = 590°F$), there is an 18 percent density increase for the 10:1 pressure increase from 1400 to 15,500 psia (9700 to 107,000 kPa).

Correction Methods

Four methods are suggested for estimating compressibility corrections:

1. Bulk-modulus method

2. Generalized reduced temperature and pressure curves of Lu

3. Generalized liquid compressibility factor

4. Average liquid hydrocarbon compressibility factor

Bulk-Modulus Method   The ratio of the pressure stress to the volumetric strain defines a fluid's bulk modulus of elasticity. The equation relating the pressure change to the volumetric change is then

$$K_{BM} = - \frac{\mathbf{V}_0 \Delta P}{\Delta \mathbf{V}} \tag{2.104}$$

where K_{BM} is the bulk modulus, and $\mathbf{V}_0$ the initial, uncompressed volume. Unfortunately, the value of K_{BM} is a function of both temperature and pressure, as can be seen from Fig. 2.16*a*. At constant temperature, the change in the bulk modulus with pressures up to 15,000 lb/in² is well approximated by the tangent bulk-modulus equation (Hayward, 1967):

$$\overline{K}_{BM} = K_{BM0} + b_c p_f \tag{2.105}$$

where K_{BM0} is the zero-pressure intercept. Values for K_{BM0} and b_c are given in Table E.1 for various substances at 68°F (20°C).

By defining the liquid compressibility correction factor as the ratio of the density at line pressure to the density at atmospheric pressure, both at the same temperature, the following equation may be derived from Eq. (2.105):

$$F_p = \frac{\rho_f}{\rho_F} = \frac{G_f}{G_F} = \frac{\overline{K}_{BM}}{\overline{K}_{BM} - p_f} \tag{2.106}$$

where F_p is the *liquid compressibility correction factor*. Over the pressure ranges normally encountered in pipelines [up to 6000 lb/in² (41,400 kPa)], the effect of

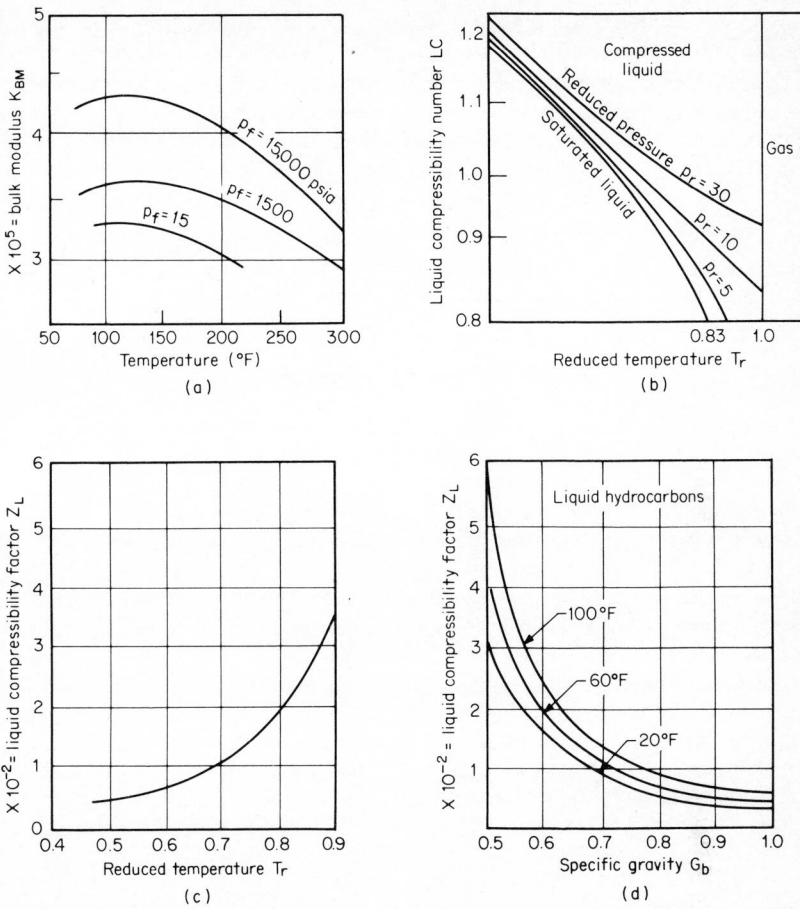

Figure 2.16 Liquid compressibility curves. (a) Bulk modulus for water (see Table E.1). (b) Lu's generalized diagram (see Fig. E.23). (c) Generalized liquid compressibility factor (see Fig. E.24). (d) Average liquid-hydrocarbon compressibility factor (see Fig. E.22).

pressure on the bulk modulus is negligible, and Eq. (2.106) can be reduced to a simpler equation that includes a liquid compressibility factor Z_L:

$$F_p = 1 + \frac{p_f}{1000} Z_L \qquad (2.107)$$

The liquid compressibility factor Z_L, derived by equating Eq. (2.106) to Eq. (2.107), is then

$$Z_L = \frac{1000}{K_{BM0} + (b_c - 1)p_f} \qquad (2.108)$$

For moderate pressures [up to 2500 lb/in² (17,200 kPa)], the compressibility factor Z_L at the intercept is suggested for general use; this is defined by

$$Z_L = \frac{1000}{K_{BM0}}$$ (2.109)

Values of Z_L are given in Table E.1 for some fluids.

Lu Diagram For estimating liquid density changes with pressure and temperature, when the critical pressure and temperature are known, the generalized curves of Lu (1959) can be used. These curves are presented in Fig. E.23 and illustrated in Fig. 2.16*b*.

If a density is known at a given temperature and pressure, the density at a second pressure and temperature can be calculated as

$$\rho_{f2} = \frac{LC_2}{LC_1} \rho_{f1}$$ (2.110)

where LC_2 and LC_1 are read from the appropriate curve intersections for the reduced temperature and pressure coordinates.

Generalized Liquid Compressibility Factor The Lu generalized curves can be used to construct a generalized liquid compressibility factor curve. Under the assumptions that (1) the bulk modulus changes slightly with temperature up to the saturated-liquid line (Fig. 2.16*b*), and (2) the compressibility factor Z_L is independent of pressures up to 10,000 lb/in² (69,000 kPa), the liquid compressibility factor curve illustrated in Fig. 2.16*c* and presented in Fig. E.24 can be drawn for reduced temperatures up to 0.9. Equation (2.107) can then be rewritten in terms of the reduced pressure as

$$F_p = 1 + Z_L p_r$$ (2.111)

where Z_L is determined from Fig. E.24 or calculated with the series expansion formula

$$Z_L = 0.269 T_r - 0.5163 T_r^2 + 0.3521 T_r^3 - 0.0461$$ (2.112)

Liquid-Hydrocarbon Compressibility Curves Compressibility test work on various liquid hydrocarbons has shown that generalized curves can be constructed (API 1101, 1960), based on base specific gravity G_b and temperature T_F. Illustrated in Fig. 2.16*d* and presented in detail in Fig. E.22 is the Z_L curve for liquid hydrocarbons. The value of Z_L is an average, and test data on the fluid should be obtained if accuracy is important.

EXAMPLE 2.13

Determine the density of water at 1500 and 15,000 psia (10,342 and 103,421 kPa) at 300°F (149°C) using (*a*) the bulk-modulus method, (*b*) the Lu diagram, and (*c*) the liquid compressibility factor derived from the Lu diagram.

From the steam tables, the following data is available:

- Saturated liquid: $\rho_f = 57.31$ lb$_m$/ft^3 (300°F, 67 psia)
- Compressed liquid: $\rho_f = 57.64$ lb$_m$/ft^3 (300°F, 1500 psia)
- Compressed liquid: $\rho_f = 60.24$ lb$_m$/ft^3 (300°F, 15,000 psia)

a. Bulk-modulus method

From Table E.1, $Z_L = 0.00262$ at 68°F and 14.7 psia. At 1500 psia, the liquid compressibility factor is, from Eq. (2.107),

$$F_p = 1 + 0.00262 \frac{1500 - 67}{1000} = 1.0038$$

The flowing density at 1500 psia is then, by Eq. (2.106),

$$\rho_f = F_p \rho_F = (1.0038)(57.31) = 57.53 \text{ lb}_m/\text{ft}^3$$

which differs from the steam-table value by -0.2 percent.
   At 15,000 psia the liquid compressibility factor is, from Eq. (2.107),

$$F_p = 1 + 0.00262 \frac{15,000 - 67}{1000} = 1.0391$$

The flowing density is, by Eq. (2.106),

$$\rho_f = F_p \rho_F = (1.0391)(57.31) = 59.55 \text{ lb}_m/\text{ft}^3$$

which differs from the steam-table value by -1.1 percent.

b. Lu generalized diagram

At 1500 psia the reduced pressure is, from Eq. (2.1),

$$p_r = \frac{1500}{3198} = 0.469$$

and the reduced temperature is, from Eq. (2.2),

$$T_r = \frac{459.67 + 300}{1165.1} = 0.652$$

In the Lu diagram (Fig. E.23), from the saturated-liquid line the ordinate is read as LC_1 = 0.960, and for the reduced properties $p_r = 0.469$ and $T_r = 0.652$, $LC_2 = 0.964$. The density at 1500 psia is now calculated from Eq. (2.110) as

$$\rho_{f2} = \rho_{f1} \frac{LC_2}{LC_1} = 57.31 \frac{0.964}{0.960} = 57.55 \text{ lb}_m/\text{ft}^3$$

which differs from the steam-table value by -0.2 percent.
   At 15,000 psia the reduced pressure is

$$p_r = \frac{15,000}{3198} = 4.690$$

and the reduced temperature is

$$T_r = \frac{459.67 + 300}{1165.1} = 0.652$$

At these values, the Lu diagram gives

$$LC_1 = 0.960 \qquad LC_2 = 0.998$$

and the density at 15,000 psia is, by Eq. (2.110),

$$\rho_{f2} = \rho_{f1} \frac{LC_2}{LC_1} = 57.31 \frac{0.998}{0.960} = 59.58 \text{ lb}_\text{m}/\text{ft}^3$$

which differs from the steam-table value by -1.1 percent.

c. Liquid compressibility factor

At 1500 psia, by Eqs. (2.112), (2.111), and (2.106),

$$Z_L = (0.269)(0.652) - (0.5163)(0.652)^2 + (0.3521)(0.652)^3 - 0.0461 = 0.0074$$

$$F_p = 1 + (0.0074)(0.469) = 1.0035$$

$$\rho_f = F_p \rho_F = (1.0035)(57.31) = 57.51 \text{ lb}_\text{m}/\text{ft}^3$$

which shows a bias error of -0.2 percent.

At 15,000 psia,

$$F_p = 1 + (0.0074)(4.690) = 1.035$$

$$\rho_f = (1.035)(57.31) = 59.30 \text{ lb}_\text{m}/\text{ft}^3$$

for a bias error of -1.3 percent.

EXAMPLE 2.14

Use the Goldhammer equation and the generalized liquid compressibility factor to estimate the density of water at 500 psia (3450 kPa) and 100°F (38°C). *Note:* The density of water at 500 psia (3450 kPa) and 100°F (38°C) is 62.07 lb$_\text{m}$/ft^3.

According to Eq. (2.86), the density at 100°F, with $T_c = 1165.1$ from Table D.1 and $N = 4$ from Table 2.6, is

$$\rho_F = 62.366 \left[\frac{1165.1 - (459.67 + 100)}{1165.1 - 519.67} \right]^{1/4} = 61.38 \text{ lb}_\text{m}/\text{ft}^3$$

The reduced temperature at 100°F is, from Eq. (2.2),

$$T_r = \frac{459.67 + 100}{1165.1} = 0.480$$

The liquid compressibility factor is then, by Eq. (2.112),

$$Z_L = (0.269)(0.480) - (0.5163)(0.480)^2 + (0.3521)(0.480)^3 - 0.0461 = 0.003$$

The reduced pressure at 500 psia, with $p_c = 3198$, is from Eq. (2.1),

$$p_r = \frac{500}{3198} = 0.156$$

The liquid compressibility factor is then, from Eq. (2.111),

$$F_p = 1 + Z_{LP_r} = 1 + (0.003)(0.156) = 1.0005$$

and Eq. (2.106) gives the density at flowing conditions as

$$\rho_f = F_p \rho_F = (1.0005)(61.38) = 61.41 \text{ lb}_m/\text{ft}^3$$

which shows a bias error of -1.1 percent.

EXAMPLE 2.15

Estimate the flowing density of the hydrocarbon of Example 2.12 if the line pressure is 1000 psia (6895 kPa).
    From Example 2.12,

$$\rho_F^* = 724.73 \text{ kg/m}^3$$

From Fig. E.22, the average liquid hydrocarbon compressibility is $Z_L = 0.0088$. The liquid compressibility factor is then, by Eq. (2.107),

$$F_p = 1 + Z_L \frac{p_f}{1000} = 1 + 0.0088 \frac{1000}{1000} = 1.0088$$

The flowing density then is, from Eq. (2.106),

$$\rho_f^* = F_p \rho_F^* = (1.0088)(724.73) = 731.12 \text{ kg/m}^3$$

VISCOSITY

Viscosity is the measure of a fluid's internal, or intermolecular, resistance to shear stress. For pipe flows this property causes a velocity profile which can affect flowmeter performance.

Absolute Viscosity

Figure 2.17a shows two parallel plates of equal area, separated by a small distance, with fluid between them. A constant force applied to the top plate causes both plate and fluid to move at constant velocity. The fluid in contact with the bottom fixed plate has no velocity. The relationship among applied force, plate area, and linear displacement is given by

$$F_{\text{plate}} = (\mu_f)_e A_{\text{plate}} \frac{\Delta V_{\text{plate}}}{\Delta Y} \tag{2.113}$$

The constant $(\mu_f)_e$ is the fluid's *absolute viscosity*, which depends primarily on interactions among fluid molecules. For liquids such as molasses, its value can be quite high, and for gases very low.
    The applied force divided by the plate area is the shear stress S_s. For newtonian fluids the relationship between shear stress and deformation rate is linear; the ratio of the two, the absolute viscosity, is constant. There are, however, fluids

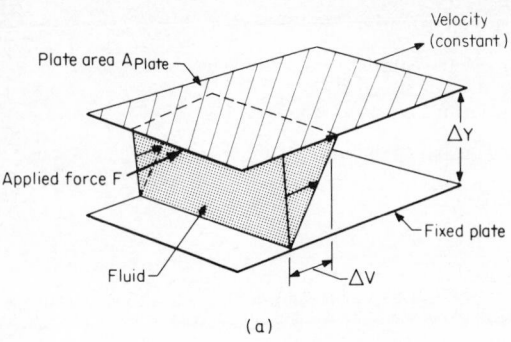

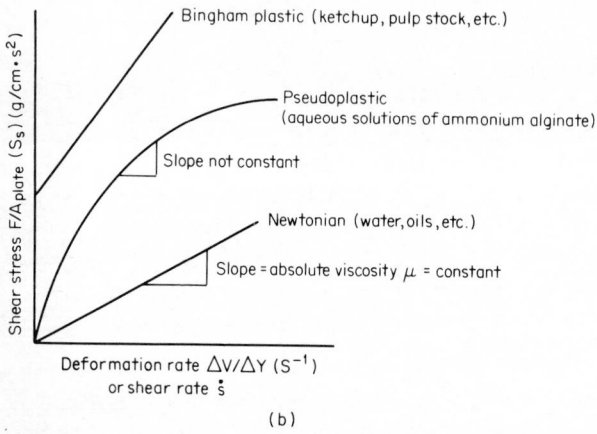

Figure 2.17 Fluid viscosity. (*a*) Newtonian fluid between parallel plates. (*b*) Newtonian and nonnewtonian fluids: shear stress versus shear strain rate (time-independent).

that are nonnewtonian, and their viscosities are not constant, as illustrated in Fig. 2.17*b*. These fluids are covered in more detail in Chap. 5.

Absolute-viscosity units are derived from Eq. (2.113). In the English engineering system, two sets of units are often found in tables. One is based on the pound-mass, and the other on the pound-force. For clarity, these will be denoted by the subscripts m and f. The absolute viscosity in force units is defined as

$$(\mu_f)_e = \frac{\mathbf{F}_{\text{plate}}}{A_{\text{plate}}} \frac{\Delta Y}{\Delta V_{\text{plate}}} = \frac{\text{lb}_f \cdot \text{s}}{\text{ft}^2} \tag{2.114}$$

and in mass units as

$$(\mu_m)_e = \frac{\text{lb}_m}{\text{ft} \cdot \text{s}} \tag{2.115}$$

The dimensional gravitational constant g_c converts force units to mass units:

$$(\mu_m)_e = g_c(\mu_f)_e = \frac{\text{lb}_m}{\text{lb}_f}\frac{\text{ft}}{\text{s}^2}\frac{\text{lb}_f \cdot \text{s}}{\text{ft}^2} = \frac{\text{lb}_m}{\text{ft} \cdot \text{s}} \tag{2.116}$$

In the SI system the absolute viscosity unit is the Pascal second (Pa·s), which is defined dimensionally by

$$\mu_{\text{Pa·s}} = \frac{F^*_{\text{plate}}}{A^*_{\text{plate}}} \frac{\Delta Y^*}{\Delta V^*_{\text{plate}}} = \frac{\text{N} \cdot \text{m}}{\text{m}^2(\text{m/s})} = \frac{\text{N} \cdot \text{s}}{\text{m}^2} = \text{Pa·s} \tag{2.117}$$

The more commonly used viscosity unit is the poise (or centipoise), which has the units

$$\mu_P = \frac{g}{\text{cm} \cdot \text{s}} \tag{2.118}$$

Equations (2.117) and (2.118) indicate that no gravity correction is required in the SI system between force and mass absolute-viscosity units. While some texts continue to use both English viscosity units, the poise (or centipoise) is now widely used and eventually should replace the two English units.

Kinematic Viscosity

Mass is eliminated from the viscosity unit if the absolute viscosity is divided by the fluid density. The result is referred to as the *kinematic viscosity*. In the English engineering system, the absolute viscosity $(\mu_m)_e$ should be divided by the density in pounds-mass per cubic foot. The viscosity in poises divided by the density in grams per cubic centimeter gives the kinematic viscosity in *stokes*. It should be noted that the density in grams per cubic centimeter is *approximately* the fluid's specific gravity, and, for this reason, some conversion equations appear with specific gravity as the divisor.

The kinematic-viscosity unit in the English engineering system is

$$\nu_e = \frac{(\mu_m)_e}{\rho_f} = \frac{32.174(\mu_f)_e}{\rho_f} = \frac{\text{ft}^2}{\text{s}} \tag{2.119}$$

The SI kinematic-viscosity unit (stokes) is

$$\nu_{\text{St}} = \frac{1000\mu_P}{\rho_f^*} = \frac{1.00099\mu_P}{F_P G_F} \approx \frac{\mu_P}{G_F} = \frac{\text{cm}^2}{\text{s}} \tag{2.120}$$

EXAMPLE 2.16

Steam tables give the absolute viscosity of steam at 800°F (427°C) and 1500 psia (10,340 kPa) in the English force system as 5.37×10^{-7} lb$_f$·s/ft^2. Convert this viscosity unit to centipoises and centistokes using the basic conversion equations.

a. Viscosity in centipoises

The viscosity in mass units is first calculated with Eq. (2.116) as

$$(\mu_m)_e = g_c(\mu_f)_e = (32.174)(5.37 \times 10^{-7}) = 1.72 \times 10^{-5} \text{ lb}_m/(\text{ft} \cdot \text{s})$$

From Table C.1, the conversion to the SI system is

$$\text{lb}_m \times 453.6 = g \qquad \text{and} \qquad \text{ft} \times 30.48 = \text{cm}$$

The viscosity in poises is then calculated by Eq. (2.118) as

$$\mu_P = 1.72 \times 10^{-5} \frac{453.6}{30.48} = 2.57 \times 10^{-4} \text{ g/(cm} \cdot \text{s)}$$

and in centipoises it is

$$\mu_{cP} = 100\mu_P = 2.57 \times 10^{-2} = 0.0257 \text{ cP}$$

b. Viscosity in centistokes

The density from Table G.1 is $\rho_f = 2.299 \text{ lb}_m/\text{ft}^3$. From Table C.1, the conversion to SI units gives $\rho_f^* = (2.299)(16.02) = 36.9 \text{ kg/m}^3$. The kinematic viscosity in stokes is then, from Eq. (2.120),

$$\nu_{St} = \frac{1000\mu_P}{\rho_f^*} = \frac{(1000)(2.57 \times 10^{-4})}{36.9} = 0.00696 \text{ St}$$

and the viscosity in centistokes is

$$\nu_{St} = 100\nu_{St} = 0.696 \text{ cSt}$$

Liquid Viscosity

The estimation of liquid viscosity, except for simple cases, is based on empirical equations. Viscosity data at two points is necessary to predict the effects of both pressure and temperature. An increase in temperature reduces the viscosity while an increase in pressure increases the viscosity, but to a much lesser degree. Petroleum products and liquids with complex molecular structures, being relatively more compressible than other liquids, are more sensitive to pressure.

Pressure Correction for Liquid Hydrocarbons In most flow applications, the effect of pressure on viscosity is not significant. The small change in viscosity affects only the Reynolds number, which, for most normal flow rates, has a slight effect on derived flow coefficients. For water, the viscosity at 5000 lb/in² (34,500 kPa) is 1.14 times the viscosity measured at atmospheric pressure, while for amyl alcohol it is 1.35 times higher.

For liquid hydrocarbons with $\mu_{cP} > 1.2$, the viscosity at high pressure can be estimated with Kouzel's (1965) relationship:

$$F_{\mu p} = \frac{(\mu_{cP})_p}{(\mu_{cP})_a} = 10^{(p_f/1000)[0.0239 + 0.01638(\mu_{cP})_a^{0.278}]} \qquad (2.121)$$

where the absolute viscosity is in centipoises, and the subscripts p and a refer to the higher-pressure and atmospheric-pressure values, respectively. Figure 2.18a illustrates the pressure correction; a more detailed curve is presented in Fig. F.27.

Temperature Correction from Two Known Values The absolute viscosities of liquids decrease with temperature, with the larger changes occurring for the more complex-structured fluids. Viscosity values are normally read from tables, graphs, or nomographs that relate viscosity to temperature, based on prior measurements for the more common fluids. When necessary, viscosity measurements of newtonian fluids are easily made.

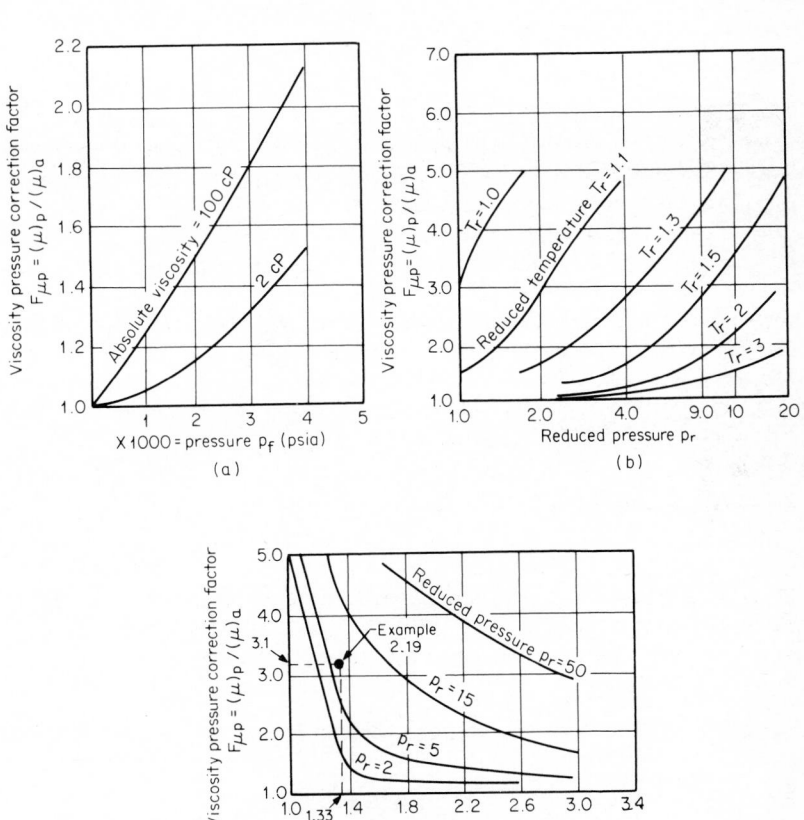

Figure 2.18 Effects of pressure on liquid-hydrocarbon viscosity and on gas viscosity. (*a*) Pressure correction for liquid hydrocarbons (see Fig. F. 27). (*b*) Generalized curves for pressure corrections for gas viscosity (see Fig. H.17). (*c*) Cross-plot of part *b* for interpolation (see Fig. H.18).

When two viscosity values are known from either measurements or tables, Andrade's (1930) equation is commonly used to predict intermediate viscosities or for data extrapolation. Andrade's equation is

$$\mu = \mathbf{A}_L \exp \frac{\mathbf{B}_L}{T_{°\mathrm{R}}} \qquad (2.122)$$

The constants $\mathbf{A}_L$ and $\mathbf{B}_L$ are calculated from two known viscosity values. With the subscripts 1 and 2 referring to the known values, the constants $\mathbf{A}_L$ and $\mathbf{B}_L$ can be calculated as

$$\mathbf{B}_L = \frac{T_{°\mathrm{R}1} T_{°\mathrm{R}2} \ln (\mu_1/\mu_2)}{T_{°\mathrm{R}2} - T_{°\mathrm{R}1}} \qquad (2.123)$$

and

$$\mathbf{A}_L = \frac{\mu_1}{\exp (\mathbf{B}_L/T_{°\mathrm{R}1})} \qquad (2.124)$$

In these equations, the absolute viscosity can be in either the English or SI system of units, and the absolute temperature in degrees Rankine or kelvins. Over reasonable temperature limits, the kinematic viscosity ν_{cSt} can be used, provided the change in specific gravity with temperature is small.

Work by Doolittle (1951) and others has shown that the constant $\mathbf{A}_L$ is a function of the free molar volume of the fluid and the compressed molar volume. The constant $\mathbf{B}_L$ has been related to the diffusional motion activation-energy level and the universal gas constant. These relationships are sometimes used to predict the combined effect of temperature and pressure. However, in most practical flow measurements, the additional complexity and required data do not warrant adjustment for the combined effect. Rather, a separate adjustment for pressure and then a temperature correction are suggested.

Temperature Correction from a Single Value When only a single viscosity value is available, the best approximation of the viscosity at other temperatures is the curve illustrated in Fig. 2.19 and given in Fig. F.26. To use the curve, one locates the known viscosity value, in centipoises, on the vertical axis and then moves along the temperature scale to the *difference* between the desired temperature and the temperature for which the viscosity is known.

Mixtures If no viscosity data is available for a mixture, Reid et al. (1977) recommend that the viscosity be calculated on the basis of mole fractions, as

$$\mu_{\mathrm{cP,mix}} = \exp \left[\Sigma y_i \ln (\mu_{\mathrm{cP}})_i \right] \qquad (2.125)$$

where the $(\mu_{\mathrm{cP}})_i$ are the viscosity values of the *pure* components at the mixture temperature. The bias errors to be expected with Eq. (2.125) may vary widely for chemically similar substances, but they are usually below 15 percent (Reid et al., 1977).

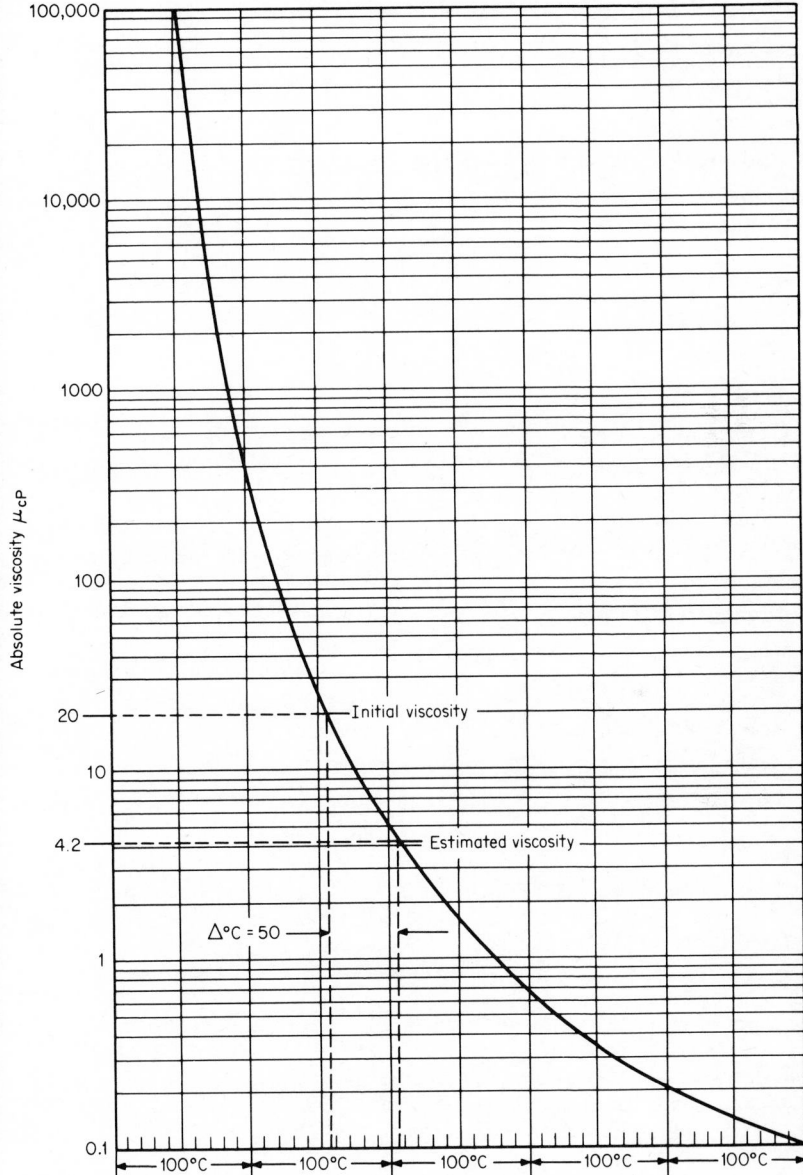

Figure 2.19 Curve for estimating viscosity from a single measured value. Values shown are for Example 2.17 *(From Gambill, 1959).*

EXAMPLE 2.17

The viscosity of an oil is measured as 20 cP at 50°F (10°C), and 2 cP at 198°F (92°C). Estimate the viscosity at 140°F (60°C), using (a) Andrade's equation and (b) the generalized curve given in Fig. 2.19, and (c) estimate the viscosity at 500 psia (3448 kPa) and 140°F (60°C).

a. *Andrade's equation*

Andrade's equation (2.122), written for centipoises, is

$$\mu_{cP} = A_L \exp \frac{B_L}{T_{°R}}$$

The coefficients B_L and A_L are calculated from Eqs. (2.123) and (2.124) with $T_{°R1} = 509.67$ and $T_{°R2} = 657.67$:

$$B_L = \frac{(509.67)(657.67) \ln (20/2)}{657.67 - 509.67} = 5215$$

$$A_L = \frac{20}{\exp (5215/509.67)} = 0.00072$$

The viscosity at 140°F (599.67°R) is then, by Eq. (2.122),

$$\mu_{cP} = 0.00072 \exp \frac{5215}{T_{°R}} = 0.00072 \exp \frac{5215}{599.67} = 4.31 \text{ cP}$$

b. *Viscosity from generalized curve*

The change in temperature is $\Delta°C = 60 - 10 = 50$. For this change in temperature, the viscosity is read on Fig. 2.19 as $\mu_{cP} = 4.2$.

c. *Viscosity at 500 psia and 140°F*

The correction for pressure is given by Eq. (2.121):

$$F_{\mu p} = 10^{(P_f/1000)[0.0239 + 0.01638(\mu_{cP})_a^{0.278}]}$$
$$= 10^{(500/1000)[0.0239 + (0.01638)(4.31)^{0.278}]}$$
$$= 1.057$$

The viscosity is then, by Eq. (2.121),

$$(\mu_{cP})_p = F_{\mu p}(\mu_{cP})_a = (1.057)(4.31) = 4.56 \text{ cP}$$

EXAMPLE 2.18

Estimate the viscosity of a liquid mixture of ethyl benzoate and benzyl benzoate at 77°F (25°C). The mole fraction of benzyl benzoate is 0.606. The following data is given in Reid et al. (1977), Example 9-18, for the pure components at 77°F:

- Ethyl benzoate: $(\mu_{cP})_1 = 2.01$ cP
- Benzyl benzoate: $(\mu_{cP})_2 = 8.48$
- Measured viscosity: 4.95 cP

The mixture's viscosity is calculated with Eq. (2.125):

$$\mu_{cP,mix} = \exp\left[(1 - 0.606)\ln 2.01 + 0.606 \ln 8.48\right] = 4.81 \text{ cP}$$

Viscosities of Gases

With a temperature increase, the viscosity of a gas (vapor) increases; this behavior is opposite that of liquids. At very high pressures, however, the viscosity mechanism inverts, and gases behave like liquids. For most gases, classical kinetic theory and experimental results show little viscosity change for pressures up to 1500 psia (10,340 kPa). Viscosity changes are not of particular concern in most flow measurements, for which pressures do not reach this value. Additionally, gas flows are usually at very high Reynolds numbers, where even large viscosity changes are usually negligible in the flow computation.

Exponential Equation for Two Known Viscosities   Over normally encountered temperature ranges, an exponential equation can be used to interpolate or extrapolate to unknown viscosities. The simplest such equation is

$$\mu_{cP} = a T_K^n \tag{2.126}$$

where the values of n and a are derived from two known viscosities as

$$n = \frac{\ln\left[(\mu_{cP})_2/(\mu_{cP})_1\right]}{\ln\left(T_{K2}/T_{K1}\right)} \tag{2.127}$$

and

$$a = \frac{(\mu_{cP})_1}{T_{K1}^n} \tag{2.128}$$

Temperature Correction from a Single Value   When the viscosity is known at a single temperature, the viscosity at another temperature can be estimated by rewriting Arnold's (1933) correlations as

$$(\mu_{cP})_2 = \left(\frac{T_{\circ R2}}{T_{\circ R1}}\right)^{1.5} \frac{T_{\circ R1} + 1.47 T_B}{T_{\circ R2} + 1.47 T_B} (\mu_{cP})_1 \tag{2.129}$$

where the subscript 1 refers to the known value, and subscript 2 to the unknown value. In this equation, the temperature can be in either degrees Rankine or kelvins.

Equation (2.129) may also be written in terms of the critical temperature by substituting the approximate equation relating the boiling temperature T_B to the critical temperature T_c. That substitution gives

$$(\mu_{cP})_2 = \left(\frac{T_{\circ R2}}{T_{\circ R1}}\right)^{1.5} \frac{T_{\circ R1} + 0.9 T_c}{T_{\circ R2} + 0.9 T_c} (\mu_{cP})_1 \tag{2.130}$$

Mixtures   The square-root rule used in the chemical industry provides the simplest and most useful equation for finding the viscosity of a mixture of gases. For 20 percent hydrogen-rich mixtures other equations are sometimes used, but for general flow measurements the square-root rule is adequate. This mixture rule is

$$\mu_{\text{mix}} = \frac{\Sigma x_i \mu_i M_{wi}^{1/2}}{\Sigma x_i M_{wi}^{1/2}} \tag{2.131}$$

EXAMPLE 2.19

Ammonia has viscosities of 0.0108 cP at 100°F (37.8°C), and 0.0285 cP at 1000°F (538°C). Estimate its viscosity at 300°F (149°C) with (a) the exponential equation and (b) Arnold's equation.

a. Exponential equation

The exponential equation (2.126) is

$$\mu_{\text{cP}} = a T_K^n$$

With $T_{K1} = 273.15 + 37.8 = 311$ K and $T_{K2} = 273.15 + 538 = 811$ K, the exponent n and coefficient a are, from Eqs. (2.127) and (2.128),

$$n = \frac{\ln (0.0285/0.0108)}{\ln (811/311)} = 1.012$$

$$a = \frac{0.0108}{(311)^{1.012}} = 0.0000324$$

The viscosity at 300°F (149°C) or $T_K = 273.15 + 149 = 422$ K is then

$$\mu_{\text{cP}} = (0.0000324)(422)^{1.012} = 0.0147 \text{ cP}$$

b. Arnold's equation

From Table D.1, the boiling temperature for ammonia is $T_B = 431.5°$R. Also, for use in Arnold's equation (2.129),

$$T_{\text{°R1}} = 459.67 + 100 = 560°\text{R}$$

$$T_{\text{°R2}} = 459.67 + 300 = 760°\text{R}$$

Substitution then gives

$$(\mu_{\text{cP}})_{300°\text{F}} = (0.0108) \left(\frac{760}{560} \right)^{1.5} \frac{560 + (1.47)(431.5)}{760 + (1.47)(431.5)} = 0.0146 \text{ cP}$$

EXAMPLE 2.20

The mole fractions of a methane-ethane mixture are 0.7 and 0.3, respectively. Estimate the viscosity of the mixture at 80°F (26.7°C).

From the nomograph in Fig. H.1, at 80°F the viscosity of methane is $\mu_{\text{cP}} = 0.0108$, and that of ethane is $\mu_{\text{cP}} = 0.0093$. From Table D.1, for methane, $(M_w)_{\text{CH4}} = 16.04$, and for ethane, $(M_w)_{\text{C2H6}} = 30.07$. The mixture's viscosity is then, from Eq. (2.131),

$$(\mu_{\text{cP}})_{\text{mix}} = \frac{\Sigma x_i \mu_i M_{wi}^{1/2}}{\Sigma x_i M_{wi}^{1/2}} = \frac{(0.7)(0.0108)(16.04)^{1/2} + (0.3)(0.0093)(30.07)^{1/2}}{(0.7)(16.04)^{1/2} + (0.3)(30.07)^{1/2}}$$

$$= 0.0102 \text{ cP}$$

Pressure Correction   The effect of high pressure on a gas is to increase its viscosity. Below 1500 psia (10,340 kPa) this effect is negligible, but in high-pressure applications the viscosity may be increased 2 to 5 times, depending on the temperature; the largest increases take place at temperatures approaching the critical isotherm. Figure 2.18b and c illustrate the viscosity-pressure correction factor $F_{\mu p}$, based on critical temperature and pressure ratios (Perry and Chilton, 1973); detailed curves are presented in Figs. H.17 and H.18.

To use these curves, the reduced pressure and temperature are first calculated with Eqs. (2.1) and (2.2), and the viscosity-pressure correction factor is read. The viscosity at the high pressure is then obtained as

$$(\mu)_p = F_{\mu p}(\mu)_a \tag{2.132}$$

where the subscripts p and a refer to the pressure-corrected viscosity and the atmospheric-pressure viscosity, respectively.

EXAMPLE 2.21

Estimate the viscosity of the methane-ethane mixture of Example 2.20 at 4000 psia (27,580 kPa).
Properties from Table D.1 and Example 2.20 are as follows:

Property	Methane	Ethane
x_i	0.7	0.3
p_c	667.2 psia	708.4 psia
T_c	343.1°R	549.7°R
V_c	1.589 ft³/lb$_m$·mol)	2.374 ft³/(lb$_m$·mol)
Z_c	0.288	0.285

The pseudocritical properties of the mixture are then, from Eqs. (2.56) and (2.57),

$$T_{pc} = \Sigma x_i T_{ci} = (0.7)(343.1) + (0.3)(549.7) = 405.1°R$$

$$p_{pc} = \frac{(\Sigma x_i Z_{ci}) R_0 T_{pc}}{\Sigma x_i V_{ci}} = \frac{[(0.7)(0.288) + (0.3)(0.285)](10.7315)(405.1)]}{(0.7)(1.589) + (0.3)(2.374)} = 684.1 \text{ psia}$$

where $R_0 = 10.73151$ psia·ft³/lb$_m$·mole·°R. The reduced properties are then, from Eq. (2.59),

$$T_{pr} = \frac{459.67 + 80}{405.1} = 1.33 \qquad p_{pr} = \frac{4000}{684.1} = 5.85$$

From the viscosity-pressure correction curve, the factor $F_{\mu p}$ is read as $F_{\mu p} = 3.1$. The viscosity at 4000 psia is then, by Eq. (2.132),

$$(\mu_{cP})_{4000} = F_{\mu p}(\mu_{cP})_a = (3.1)(0.0102) = 0.0316 \text{ cP}$$

Viscosity-Measuring Devices

There are two common viscometer types and numerous special types for high-pressure tests, in-line measurement (Fig. 2.20), or rheology studies. The most common and inexpensive is the efflux type, in which a known volume of fluid is discharged through an orifice or capillary, and the measured time to discharge the volume is substituted into an empirical equation to calculate the viscosity. In the second type, the measured torque on a constant-speed rotating cylinder (Fig. 2.21) or cone, immersed in the fluid, is taken as a direct indication of the viscosity of the fluid.

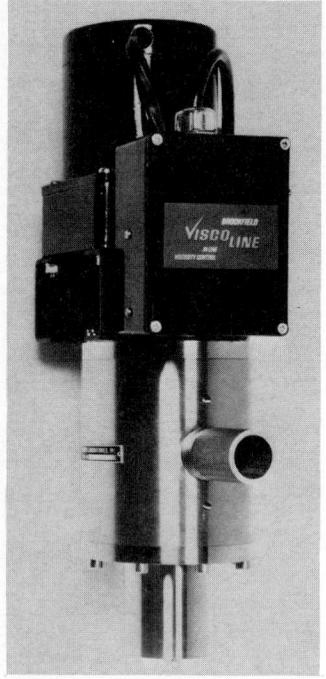

Figure 2.20 In-line viscometer (*courtesy Brookfield Engineering Laboratories*).

Of the special types, those that make use of a falling or rolling ball or the measured time for a bubble to rise through the liquid are the most common for field testing. Vibrating reeds, mixers with torque dynamometers, and plunger types are also used, either as continuous-measurement devices, for special measurements of inks and asphalt, or in the measurement of the softening characteristics of nonnewtonian fluids.

Orifice-Efflux Viscometers The most commonly used viscosity-measuring instrument in the United States is the Saybolt viscometer (see Fig. 2.22a). The sample fluid is held in a universal oil tube which is immersed in a controlled constant-temperature oil or water bath. The fluid is initially prevented from flowing by a small cork. When the temperature becomes constant, the cork is

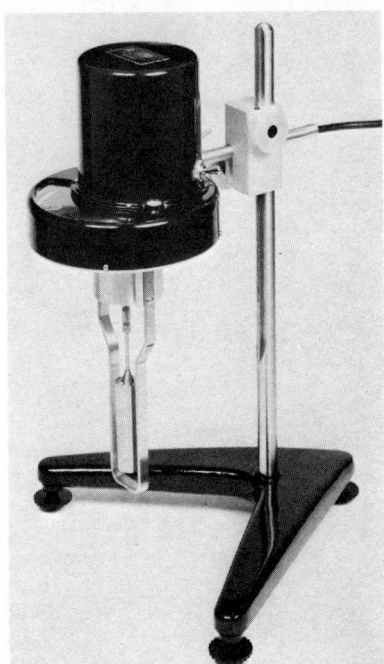

Figure 2.21 Rotational viscometer for rheology studies *(courtesy Brookfield Engineering Laboratories).*

removed, and a 60-cm³ flask is filled by discharging the sample through a 0.0695 in (1.77 mm) orifice located at the bottom of the oil tube. The time required to fill the 60-cm³ flask is the viscosity in *Saybolt universal seconds.* The device should be used only when laminar flow occurs in the tube, and for discharge times above 30 s. For very viscous oils a larger orifice is used, and the time unit is referred to as in the *Saybolt Furol second.*

The following empirical equations relate the measured Saybolt universal times to the centistoke viscosity unit; exact relationships are given in ASTM D445-71 (1971):

$$\nu_{cSt} = 0.226t_{SSU} - \frac{195}{t_{SSU}} \qquad \text{for } 32 \text{ s} < t_{SSU} < 100 \text{ s} \qquad (2.133)$$

$$\nu_{cSt} = 0.220t_{SSU} - \frac{135}{t_{SSU}} \qquad \text{for } t_{SSU} > 100 \text{ s} \qquad (2.134)$$

For highly viscous fluids, Saybolt Furol times are substituted into the following equations to calculate centistoke viscosities

$$\nu_{cSt} = 2.24t_{SSF} - \frac{184}{t_{SSF}} \qquad \text{for } 25 \text{ s} \leq t_{SSF} \leq 40 \text{ s} \qquad (2.135)$$

$$\nu_{cSt} = 2.16t_{SSF} - \frac{60}{t_{SSF}} \qquad \text{for } t_{SSF} > 40 \text{ s} \qquad (2.136)$$

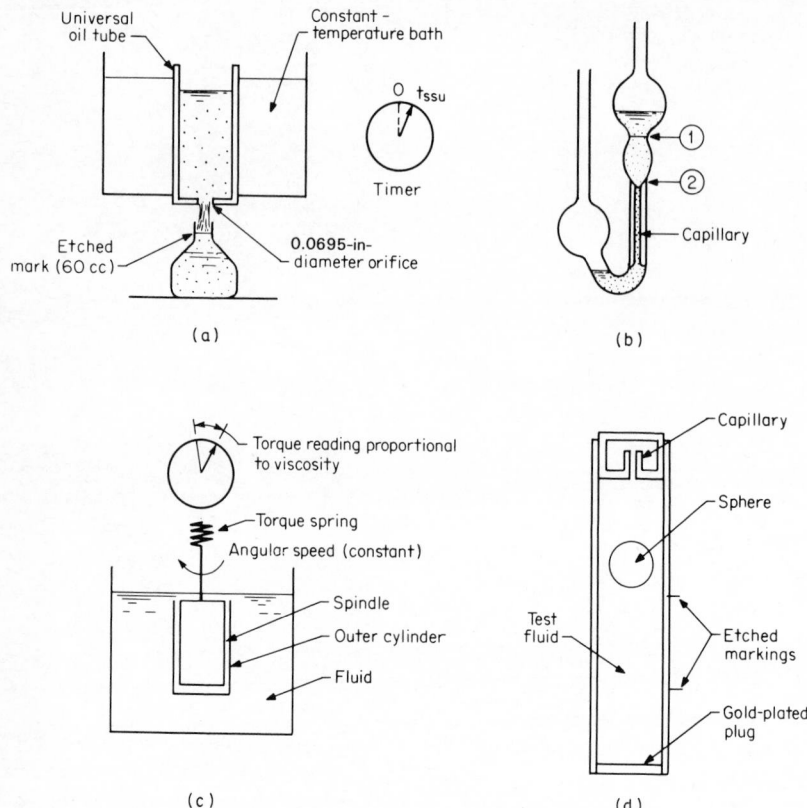

Figure 2.22   Viscometers. (*a*) Saybolt second. (*b*) Ostwald capillary. (*c*) Rotational. (*d*) Hoeppler (liquid and gas).

Other orifice-efflux-type viscometers include the Zahn, Cox, Scott, and Ford-cup. These were designed either for simpler operation or for the viscosity measurement of special fluids such as paint, varnish, and tar. The Redwood viscosity scale is commonly used in Great Britain. The Redwood viscosity equations are similar in form to the Saybolt equations and are as follows:

Redwood No. 1 (Standard)

$$\nu_{cSt} = 0.26 t_{RS} - \frac{179}{t_{RS}} \qquad \text{for } 34 \text{ s} \le t_{RS} \le 100 \text{ s} \qquad (2.137)$$

$$\nu_{cSt} = 0.247 - \frac{50}{t_{RS}} \qquad \text{for } t_{RS} > 100 \text{ s} \qquad (2.138)$$

Redwood No. 2 (Admiralty)

$$\nu_{\text{cSt}} = 2.46t_{\text{RA}} - \frac{100}{t_{\text{RA}}} \qquad \text{for } 32 \text{ s} \leq t_{\text{RA}} \leq 90 \text{ s} \qquad (2.139)$$

$$\nu_{\text{cSt}} = 2.45t_{\text{RA}} \qquad \text{for } t_{\text{RA}} > 90 \text{ s} \qquad (2.140)$$

Capillary-Efflux Viscometers Because of their short orifice length, orifice-efflux viscometers are not particularly well suited for rheological studies. End effects and the transition from laminar to turbulent flow within the tube or orifice are not easily analyzed or interpreted. For these reasons, a device with a well-defined capillary passage is best for high accuracy. The most familiar such device is the Ostwald viscometer tube, shown in Fig. 2.22b. As with the orifice type, the time required for a known volume of fluid to move from etched mark 1 to etched mark 2 is measured. This time is then compared to that for a fluid with known viscosity under the same conditions. For steady, incompressible flow, no wall slippage, a tube long enough to negate end effects, and negligible thermal gradients and pressure changes, the unknown viscosity is calculated as

$$\mu_{\text{cP}} = \frac{\mu_{\text{cP,ref}} \, t_{tF} \rho_F^*}{t_{\text{ref}} \rho_{r\,\text{ef}}^*} \qquad (2.141)$$

where the subscript F refers to the test fluid, and *ref* to the reference fluid.

The use of capillary-type viscometers requires a knowledge of the rheological properties of fluids and the many specialized empirical and theoretical equations used to accurately determine the behavior of both newtonian and nonnewtonian fluids. Many specialized capillary viscometers are available; for further information, the reader is referred to the text by Van Wazer et al. (1963).

Rotating Viscometers Rotating one cylinder inside another (Fig. 2.22c), with the fixed clearance space filled with the test fluid, creates a shear force proportional to the fluid viscosity. The torque, measured at constant rotational speed, is a direct indication of the viscosity; by suitable scaling and calibration with known fluids, the absolute viscosity may be read. Depending on the viscosity range, many combinations of cylinders, rotational speeds, and torque-measuring devices (often strain gauges) are available.

Falling-Sphere Viscometer A ball falling through a fluid through the action of the difference between ball density and fluid density obeys Stokes' law relating downward velocity to viscosity. The Hoeppler viscometer (Fig. 2.22d) can measure, to within ± 0.1 to ± 0.5 percent, the viscosities of gases or very viscous fluids (over a range of 0.01 to 1,000,000 cP) with the same tube, but with different calibrated spheres. The general equation for calculating the viscosity is

$$\nu_{\text{cSt}} = k_{FS}(\rho_{\text{sph}}^* - \rho_F^*)t_{FS} \qquad (2.142)$$

where k_{FS} is a combination constant determined with a known fluid, ρ_{sph}^* is the density of the sphere, ρ_F^* is the density of the test fluid, and t_{FS} is the measured time (in seconds) for the sphere to pass between the enclosing cylinder's two etched marks.

THE ISENTROPIC EXPONENT

Gases expand as they flow through the reduced area of a differential producer. The expansion to a lower pressure is assumed to follow a polytropic path, for which the relationship between pressure and volume is defined by

$$pV^n = \text{constant} \tag{2.143}$$

Because of the short expansion length and usually small pressure differences, an idealized one-dimensional isentropic expansion is assumed; the expansion is reversible, and there is no heat loss (adiabatic expansion). Under these assumptions, the relationship between the two pressures and volumes is

$$\frac{p_1}{p_2} = \left(\frac{V_2}{V_1} \right)^k = \left(\frac{\rho_{f1}}{\rho_{f2}} \right)^k \tag{2.144}$$

where the exponent k is the isentropic exponent. In the literature, three different exponents have been used in this equation; which one is selected depends on whether the gas is assumed to be perfect, ideal, or real.

Within the normal operating ranges of differential pressure, flowing pressure, and temperature, the flow equations are not particularly sensitive to the value of the isentropic exponent. The assumption of a perfect or ideal gas is therefore reasonable.

Perfect Gas

A *perfect* gas has a compressibility factor of 1, and its specific-heat ratio is assumed to be independent of both temperature and pressure. The relationship between the specific-heat ratio and the universal gas constant may be expressed as

$$(C_p - C_v)_p = 1.98586 \text{ Btu/(lb}_m \cdot \text{mol} \cdot {}^\circ\text{R}) \tag{2.145}$$

Rearranging Eq. (2.145) and rounding the universal gas constant yield the following expression for the ratio of specific heats for a perfect gas:

$$k_p = \left(\frac{C_p}{C_v} \right)_p = \frac{(C_p)_p}{(C_p)_p - 1.986} \tag{2.146}$$

where $(C_p)_p$ refers to the specific heat of the gas at 14.696 psia (101.325 kPa) and 59°F (15°C).

It should be noted that many tables, nomographs, and other sources give the specific heat at constant pressure in terms of pounds-mass or kilograms, and not on a mole basis. These values should be multiplied by the molecular weight of the gas to satisfy the units of Eq. (2.146):

$$C_p \text{ [in Btu/(lb}_m \cdot \text{mol} \cdot °R)] } = M_w C_p \text{ [in Btu/(lb}_m \cdot °R)]} \qquad (2.147)$$

For gas mixtures, the specific heat at constant pressure can be *approximated* by summing mole fractions as follows:

$$C_{p,\text{mix}} \approx \Sigma x_i C_{pi} \qquad (2.148)$$

Ideal Gas

An *ideal* gas has a compressibility factor of 1, and its specific heat at constant pressure C_p is a function of temperature but not pressure. For an ideal gas, the isentropic exponent is

$$k_i = \left(\frac{C_p}{C_v} \right)_i = \frac{(C_p)_i}{(C_p)_i - 1.986} \qquad (2.149)$$

where $(C_p)_i$ is a function of the absolute temperature. Data on many common gases has been fitted to equations of the general form

$$(C_p)_i = a + bT^m + cT^n \qquad (2.150)$$

These equations are given in Table I.3. Tables I.1 and I.2 list the specific heats $(C_p)_i$ for some gases.

EXAMPLE 2.22

Assuming air is an ideal gas mixture ($Z = 1.0$) consisting of 80 percent nitrogen and 20 percent oxygen, use the temperature-specific heat nomograph in App. I to calculate the isentropic exponent k_i for this ideal gas at (a) 68°F (20°C) and (b) 1000°F (538°C).
From Fig. I.17, at 68°F,

$$(C_p)_{N2} = 0.25 \text{ Btu/(lb}_m \cdot °F) \qquad (C_p)_{O2} = 0.22 \text{ Btu/(lb}_m \cdot °F)$$

and at 1000°F,

$$(C_p)_{N2} = 0.27 \text{ Btu/(lb}_m \cdot °F) \qquad (C_p)_{O2} = 0.26 \text{ Btu/(lb}_m \cdot °F)$$

From Table D.1,

$$(M_w)_{N2} = 28.0134 \text{ lb}_m/(\text{lb}_m \cdot \text{mol}) \qquad (M_w)_{O2} = 31.998 \text{ lb}_m/(\text{lb}_m \cdot \text{mol})$$

and the mixture's molecular weight is, from Eq. (2.44),

$$M_{w,\text{mix}} = \Sigma x_i M_{wi} = (0.8)(28.013) + (0.2)(31.998) = 28.810 \text{ lb}_m/(\text{lb}_m \cdot \text{mol})$$

a. *Ideal-gas isentropic exponent at 68°F*

The mixture's specific heat at 68°F is approximated with Eq. (2.148):

$$C_{p,\text{mix}} = \Sigma x_i C_{pi} = (0.8)(0.25) + (0.2)(0.22) = 0.244 \text{ Btu/(lb}_m \cdot °F)$$

The ideal-gas isentropic exponent for the mixture at 68°F is then, from Eq. (2.149) with M_w included for consistent units,

$$k_i = \frac{(0.244)(28.810)}{(0.244)(28.810) - 1.986} = 1.394$$

b. Ideal-gas isentropic exponent at 1000°F

The mixture's specific heat at 1000°F is, again from Eq. (2.148),

$$C_{p,\text{mix}} = (0.8)(0.27) + (0.2)(0.26) = 0.268 \text{ Btu/(lb}_\text{m} \cdot \text{°F)}$$

Then the isentropic exponent at 1000°F is, by Eq. (2.149), again with M_w included,

$$k_i = \frac{(0.268)(28.810)}{(0.268)(28.810) - 1.986} = 1.346$$

Real Gas

In the expansion of a *real* gas the effects of compressibility must be considered for correct prediction of the density change from upstream pressure and temperature measurements. Several isentropic flow models have been proposed (Sullivan, 1979) that predict the isentropic exponent from equations of state. Figures I.5 through I.16 present the isentropic exponents for many gases, based on these models.

A numerical solution that may also be used is based on the isentropic-exponent equation given in *Fluid Meters* (ASME, 1971); this equation is

$$k = \left\{ \frac{1}{1 - [(\partial Z/Z)/(\partial p/p)]_T} \right\} \frac{C_p}{C_v} = F_k \frac{C_p}{C_v} \tag{2.151}$$

A three-step calculation procedure is necessary to solve Eq. (2.151) numerically. The generalized diagrams for these steps are illustrated in Fig. 2.23; these diagrams are not to scale; Figs. I.1 through I.3 should be used for actual calculations. The appendix diagrams are for a simple fluid with acentric factor equal to 1.0. For other fluids, the diagrams presented in the text by Edmister (1974) are suggested for better accuracy.

Step 1: Pressure Correction   Shown in Fig. 2.23a is the Edmister (1974) generalized diagram that is used to correct the specific heat for pressure. This is an additive correction that approximates the pressure effect to within 5 to 10 percent for most fluids. The corrected (for pressure) specific heat is

$$C_p = (C_p)_i + F_{\gamma p} \tag{2.152}$$

where C_p is the specific heat corrected for pressure, and $(C_p)_i$ is the specific heat of an ideal gas at the temperature of the subject gas.

For the graphical determination of the correction for pressure, the reduced temperature T_r and pressure p_r are first calculated; then the value of $F_{\gamma p}$ is read from the curve.

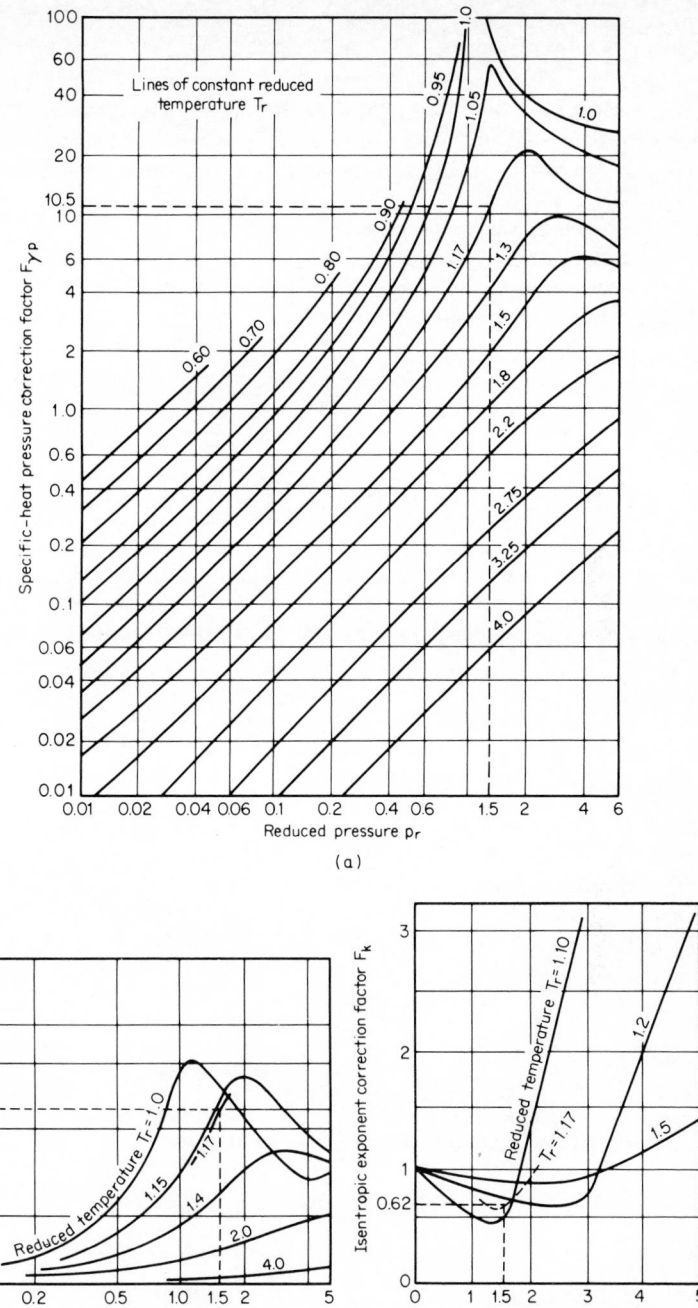

Figure 2.23 Generalized diagrams used to calculate the isentropic exponent (not to scale); dashed lines are for Example 2.24. (*a*) Pressure correction (see Fig. I.1). (*b*) Real-gas correction (see Fig. I.2). (*c*) Isentropic-exponent correction (see Fig. I.3).

2-77

Step 2: Specific-Heat Real-Gas Correction $F_{\gamma R}$ Edmister's (1974) generalized-properties diagram, shown in Fig. 2.23b, corrects for the effect of compressibility. It can be viewed as a modification to the universal gas constant shown in Eq. (2.146). Note from the curves that as the reduced-pressure ratio p_r decreases, the real-gas correction factor approaches 1.0.

The reduced pressure and temperature are calculated, and the value of the real-gas correction factor is read from the appropriate curve. The specific-heat ratio, corrected for pressure and temperature, is then calculated as

$$\frac{C_p}{C_v} = \frac{C_p}{C_p - 1.986F_{\gamma R}} \tag{2.153}$$

Step 3: Isentropic-Exponent Calculation Illustrated in Fig. 2.23c is a generalized-properties curve developed from the Hall-Yarborough equation of state by successively incrementing the pressure and calculating the compressibility change for an isothermal expansion in accordance with the term in braces in Eq. (2.151). The exact curve, given in Fig. I.3, can be used to find F_k for the calculation of the isentropic exponent by substitution into Eq. (2.151):

$$k = F_k \frac{C_p}{C_v} \tag{2.154}$$

EXAMPLE 2.23

Assuming carbon dioxide is an ideal gas, use the equation given in Table I.3 to calculate the ideal isentropic exponent at 1000°F (538°C).

The specific heat at constant pressure is

$$(C_p)_i = 16.2 - \frac{6.53 \times 10^3}{T_{\circ R}} + \frac{1.41 \times 10^6}{T_{\circ R}^2}$$

At 1000°F, $T_{\circ R} = 1459.7°R$. Then substitution gives

$$(C_p)_i = 16.2 - \frac{6.53 \times 10^3}{1459.7} + \frac{1.41 \times 10^6}{(1459.7)^2} = 12.39 \text{ Btu/(lb}_m \cdot \text{mol} \cdot °R)$$

The ideal isentropic exponent is then, by Eq. (2.149),

$$k_i = \frac{12.39}{12.39 - 1.986} = 1.191$$

EXAMPLE 2.24

The expansion factor for gas flow through an orifice is discussed below. For an orifice for which $\beta = 0.75$, calculate the bias error in measuring the flow rate of methane at 1000 psia (6895 kPa) and $-60°F$ ($-51.1°C$) if the gas is assumed to be (a) perfect and (b) ideal. The differential pressure h_w is 200 in (50 kPa).

The expansion factor for an orifice is

$$Y_1 = 1 - (0.41 + 0.35\beta^4)\,\frac{\mathbf{x}}{k}$$

where

$$\mathbf{x} = \frac{h_w}{27.73 p_{f1}} = \frac{200}{(27.73)(1000)} = 0.0072$$

Then,

$$Y_1 = 1 - [0.41 + (0.35)(0.75)^4]\,\frac{0.0072}{k} = 1 - \frac{0.00376}{k}$$

The flow-rate equaton for an orifice is

$$q_{\text{PPH}} = F_{MC}\,Y_1\,\sqrt{h_w \rho_{f1}}$$

at a measured differential pressure and density, the expansion-factor sensitivity coefficient is 1.0; a $\pm$ 1 percent expansion-factor change represents a ± 1 percent flow-rate bias error.

For methane, from Table D.1,

$$T_c = 343.1°R \qquad p_c = 667.1 \text{ psia}$$

From Eq. (2.59), the reduced properties are

$$T_r = \frac{459.67 - 60}{343.1} = 1.17 \qquad p_r = \frac{1000}{667.1} = 1.50$$

From Table I.1, at 59°F,

$$(C_p)_p = 8.45 \text{ Btu/(lb}_m \cdot \text{mol} \cdot °R)$$

and at $-60°F$,

$$(C_p)_i = 8.08 \text{ Btu/(lb}_m \cdot \text{mol} \cdot °R)$$

Perfect gas

The isentropic exponent of the perfect gas is, by Eq. (2.146),

$$k_p = \frac{(C_p)_p}{(C_p)_p - 1.986} = \frac{8.45}{8.45 - 1.986} = 1.307$$

Its expansion factor is, then,

$$Y_1 = 1 - \frac{0.00376}{k_p} = 1 - \frac{0.00376}{1.307} = 0.9971$$

Ideal gas

For the ideal gas, the isentropic exponent is calculated from the specific heat at the flowing temperature of $-60°F$, using Eq. (2.149):

$$k_i = \frac{(C_p)_i}{(C_p)_i - 1.986} = \frac{8.08}{8.08 - 1.986} = 1.326$$

The expansion factor for the ideal gas is then

$$Y_1 = 1 - \frac{0.00376}{k_i} = 1 - \frac{0.00376}{1.326} = 0.9972$$

Real gas

The real-gas isentropic exponent is calculated using Eq. (2.151),

$$k = F_k \frac{C_p}{C_v}$$

where, from Eq. (2.153),

$$\frac{C_p}{C_v} = \frac{C_p}{C_p - 1.986 F_{\gamma R}}$$

The specific heat corrected for pressure and temperature is, by Eq. (2.152),

$$C_p = (C_p)_i + F_{\gamma p}$$

From Fig. I.1, for $T_r = 1.17$ and $p_r = 1.50$, $F_{\gamma p} = 10.5$ Btu/(lb$_m$·mol·°R). Then, at $-60°$F with $(C_p)_i = 8.08$ Btu/(lb$_m$·mol·°R),

$$C_p = 8.08 + 10.5 = 18.58 \text{ Btu/(lb}_m\text{·mol·°R)}$$

From Fig. I.2, the real-gas correction for $T_r = 1.17$ and $p_r = 1.50$ is $F_{\gamma R} = 5.7$. The ratio of specific heats is then

$$\frac{C_p}{C_v} = \frac{18.58}{18.58 - (1.986)(5.7)} = 2.559$$

From Fig. I.3, the correction for a real gas is $F_k = 0.62$. The isentropic exponent is then, from Eq. (2.154),

$$k = F_k \frac{C_p}{C_v} = (0.62)(2.559) = 1.587$$

The expansion factor for the real gas is

$$Y_1 = Y_{\text{real}} = 1 - \frac{0.00376}{k} = 1 - \frac{0.00376}{1.587} = 0.9976$$

Bias error

The bias error is calculated as

$$B_\% = \frac{Y_{\text{ind}} - Y_{\text{true}}}{Y_{\text{true}}} 100$$

where it is assumed that $Y_{\text{true}} = Y_{\text{real}}$.

a. Perfect gas

$$B = \frac{0.9971 - 0.9976}{0.9976} 100 = -0.05 \text{ percent}$$

b. Ideal gas

$$B = \frac{0.9972 - 0.9976}{0.9976} 100 = -0.04 \text{ percent}$$

REFERENCES

AGA: *Gas Measurement Manual,* American Gas Association, New York, 1963.

AGA: *Manual for the Determination of Supercompressibility Factors for Natural Gas,* American Gas Association, New York, 1963.

Andrade, E. N. da C.: "The Viscosity of Liquids," *Nature,* vol. 125, pp. 309, 582, 1930.

API Standard 1101, *Measurement of Petroleum Liquid Hydrocarbons by Positive Displacement Meter,* p. 52, American Petroleum Institute, Washington, D.C., 1960.

API Standard 2540, *Manual of Petroleum Measurement Standards,* vol. 11, chap. 11.1, tables 5A, 5B, 6A, 6B, American Petroleum Institute, Washington, D.C., 1980.

Arnolds, J. H.: *J. Chem. Phys.,* vol. 1, p. 170, 1933.

ASME: *Fluid Meters,* 6th ed., ASME, New York, 1971.

ASTM Standard 445-71, ASTM, Washington, D.C., 1971.

Bean, H. S.: "An Apparatus and Methods for Determining the Compressibility of a Gas and the Correction for 'Supercompressibility,'" *J. Res. Nat. Bur. Stand.,* vol. 4, pp. 645–661, 1929.

Burnett, E. S.: "Compressibility Determination without Volume Measurements," *J. Appl. Mech.,* vol. 3, pp. A136–A140, 1936.

Doolittle, A. K.: "Studies in Newtonian Flow," *J. Appl. Phys.,* vol. 22, p. 1471, 1951.

Edmister, W. C.: *Applied Hydrocarbon Thermodynamics,* vol. 1, Gulf, Houston, Texas, 1974.

——: *Applied Hydrocarbon Thermodynamics,* vol. 2, Gulf, Houston, Texas, 1974.

Eirich, F. R.: *Rheology—Theory and Application,* vol. 1, pp. 258, Academic, New York, 1956.

Fishtine, S. H.: "Estimates of Saturated Fluid Densities and Critical Constants," *Ind. Eng. Chem. Fundam.,* vol. 2, no. 2, pp. 149–155, 1963.

Gambill, W. R.: "How P and T Change Gas Viscosity," *Chem. Eng.,* p. 157, 1958.

Goldhammer, D. A.: "Studien über die Theorie der übereinstimmenden Zustände," *Phys. Chem.,* vol. 71, p. 577, 1910.

Goyal, P., and L. K. Doraiswamy: "Estimating Liquid Densities," *Hydrocarbon Process. Pet. Refiner,* vol. 45, p. 200, 1966.

Hall, K. P., and L. Yarborough: "A New Equation of State for Z-Factor Calculation," *Oil Gas J.,* vol. 77, no. 25, pp. 82–92, 1973.

Hayward, A. T. J.: *Compressibility Equations for Liquids—A Comparative Study,* National Engineering Laboratory, Ministry of Technology, East Kilbride, Scotland, 1967.

ISO Standard 5024, *Petroleum Liquids and Gases—Measurement—Standard Reference Conditions,* International Standards Organization, Geneva, 1976.

Joffe, J.: "Combining Rules for the Third Parameter in the Pseudocritical Method for Mixtures," *Ind. Eng. Chem. Fundam.,* vol. 10, p. 532, 1971.

Jones, F. E.: "The Air Density Equation and the Transfer of the Mass Unit," *J. Res. Nat. Bur. Stand.,* vol. 83, no. 5, pp. 419–428, 1978.

Kay, W. B.: "Density of Hydrocarbon Gases and Vapors," *Ind. Eng. Chem.,* vol. 28, no. 9, pp. 1014–1019, 1936.

Kouzel, B.: "How Pressure Affects Liquid Viscosity," *Hydrocarbon Process. Pet. Refiner,* vol. 44, no. 3, p. 120, 1965.

Lu, B. C.: "Estimate Specific Liquid Volumes," *Chem. Eng.*, vol. 66, pp. 137–138, 1959.

Nelson, L. C., and E. F. Obert: "Generalized pvT Properties of Gases," *Trans. ASME*, vol. 76, p. 1057, 1954.

Obert, E. F.: *Thermodynamics*, McGraw-Hill, New York, 1948.

PEA: *Tentative Standard for the Application of Gas Measurement Procedures to Computer Usage*, Pacific Energy Association Bulletin T.S.-622-77, Los Angeles, 1977.

Perry, J. H., and C. H. Chilton: *Chemical Engineers' Handbook*, 5th ed., McGraw-Hill, New York, 1973.

Pitzer, K. S., B. Z. Lippmann, and R. F. Curl: "The Volumetric and Thermodynamic Properties of Fluids," *J. Am. Chem. Soc.*, vol. 77, pp. 3427–3433, 1955.

Prausnitz, J. M., and R. D. Gunn: "Volumetric Properties of Nonpolar Gaseous Mixtures," *AIChE J.*, vol. 4, pp. 430, 494, 1958.

PTB: "Die Dichte des Wasser im internationalen Einheitensystem und der internationalen praktischen Temperaturskala von 1968," *PTB Mitt.*, vol. 81, no. 6, pp. 412–414, 1971.

Redlich, O., and J. N. S. Kwong: "On the Thermodynamics of Solutions. V An Equation of State. Fugacities of Gaseous Solutions," *Chem. Rev.*, vol. 44, p. 233, 1949.

Reid, R. C., J. K. M. Prausnitz, and T. K. Sherwood: *The Properties of Gases and Liquids*, 3d ed., McGraw-Hill, New York, 1977.

Smith and Van Ness: *Introduction to Chemical Thermodynamics*, McGraw-Hill, New York, 1978.

Standing, M. B., and D. L. Katz: "Density of Natural Gas," *AIME Pet. Dev. Tech.*, vol. 146, pp. 140–149, 1942.

Sullivan, D. A.: "Historical Review of Real-Fluid Isentropic Flow Models," Paper 79-WA-FM-1, ASME Winter Annual Meeting, New York, 1979.

Van Wazer, J. R., J. W. Lyons, K. Y. Kim, and R. E. Colwell: *Viscosity and Flow Measurement*, Interscience/Wiley, New York, 1963.

Yarborough, L., and K. P. Hall: "How to Solve Equation of State for Z-Factors," *Oil Gas J.*, vol. 78, pp. 86–88, 1974.

3

MEASUREMENT

The purpose of this chapter is to present the basic measurement units used in flow measurement, and to discuss typical temperature- and pressure-measuring devices. This information will be used in subsequent chapters in the development of the engineering flow equations.

MASS, FORCE, WEIGHT

The English Engineering System of Units

Table 3.1 summarizes the five fundamental systems of units that have been constructed from Newton's second law of motion to relate force F, mass m, length L, and time t. While any system can be developed from three fundamental quantities, the four quantities of the English engineering system—the foot (ft), pound-force (lb_f), pound-mass (lb_m), and second (s)—will be used here to develop the U.S. customary unit equations.

To relate the pound-force to the pound-mass, a proportionality equation can be written between the engineering and the absolute units. Using the definition that 1 lb_f will accelerate 1 lb_m at 32.17405 ft/s², a dimensional conversion constant can be derived as

$$\mathbf{F} = \frac{1}{g_c}\, ma \tag{3.1}$$

$$lb_f = \frac{1}{lb_m \cdot ft/(lb_f \cdot s^2)}\, lb_m\, \frac{ft}{s^2}$$

The constant g_c has the same value as standard gravity g_0, defined at sea level and 45° latitude, but it has the dimensions of $lb_m \cdot ft/(lb_f \cdot s^2)$. It is, therefore, a

Table 3.1   **Dimensional Systems**

System	Mass (m)	Force (F)	Length (L)	Time (t)
English Engineering (F, m, L, t)	pound-mass (lb_m)	pound-force (lb_f)	foot (ft)	second (s)
Absolute (m, L, t)	pound-mass (lb_m)	poundal ($lb_m \cdot$ ft/s^2)	foot (ft)	second (s)
Technical (F, L, t)	slug ($lb_f \cdot s^2$/ft)	pound-force (lb_f)	foot (ft)	second (s)
Metric Absolute (m, L, t)	gram (g)	dyne ($g \cdot$ cm/s^2)	centimeter (cm)	second (s)
International SI (m, L, t)	kilogram (kg)	Newton ($kg \cdot$ m/s^2)	meter (m)	second (s)

dimensional conversion factor to relate pounds-force and pounds-mass. Substituting local gravity (g_l) for acceleration a in Eq. (3.1) gives the relationship between mass and weight force as

$$W = \frac{g_l}{g_c} m$$

$$lb_f = \frac{g_l}{g_c} lb_m \tag{3.2}$$

The ratio of local gravity g_l to the dimensional constant g_c can be approximated to within 0.005 percent with an expression given by Benedict (1977):

$$F_L = \frac{g_l}{g_c} = 1 - 2.637 \times 10^{-3} \cos 2\phi - 9.6 \times 10^{-8} y - 5 \times 10^{-5} \quad lb_f/lb_m \tag{3.3}$$

where ϕ is the latitude in degrees, and y the altitude in feet above sea level.

Some confusion results from the fact that the word *weight* has long been used to mean both mass (lb_m) and force (lb_f). When weighed on a beam scale, an object is calibrated against a known mass. Since both are influenced by the same local gravity, the scale reading is corrected to the mass value and, hence, mass is measured. However, a spring scale calibrated with a 1-lb mass at sea level and 45° latitude (standard gravity) will indicate *one pound* for this mass only at this location. At other gravities ($g_l \neq g_0$) the 1-lb mass will not "weigh" 1 lb_f on this scale.

The SI System of Units

In the International System of units (SI), the force unit is the newton (N). It is the force that, when applied to a 1-kg mass, will cause an acceleration of 1 m/s². The relationship between force, mass, and acceleration is

$$\mathbf{F}^* = \frac{m^* a^*}{g_c^*}$$

(3.4)

$$1 \text{ N} = \frac{1 \text{ kg} \times 1 \text{ m/s}^2}{g_c^*}$$

or

$$g_c^* = 1 \frac{\text{kg} \cdot \text{m}}{\text{N} \cdot \text{s}^2}$$

The *weight* force in newtons of a 1-kg mass is then

$$W^* = \frac{g_l^*}{g_c^*} m^* = 9.80665 F_L m^*$$

(3.5)

The kilogram mass is, by international agreement, the mass of a certain bar of platinum-iridium located in Sevres, France. The pound-mass in the English engineering system is exactly 0.45359237 kg.

Buoyancy Correction for Mass Determination

Standard masses (kilogram, pound-mass) are used to calibrate scales and as force references. However, the volume of the mass displaces surrounding air, creating an upward buoyant force that reduces the applied force. Mass standards are usually corrected for air buoyancy to values in a vacuum; when a standard mass is placed on a scale at any location, the scale reading is corrected to read the value of the applied mass. The correction thus includes the effects of both buoyancy and local gravity. However, the correction will apply to an object other than the standard mass only if the object displaces the same volume of air as the standard mass. If the standard mass displaces more air than the object, the corrected scale reading must be increased; if less, it must be decreased.

A buoyancy force is not exerted on a fluid in a pipeline. Therefore, if a liquid is collected in a weigh tank, and the reading is compared to a flowmeter's mass-flow indication, a 0.1 to 0.3 percent bias-error correction should be made to account for the difference between flowing liquid density and the density of the standard masses used to calibrate the scale.

Figure 3.1a shows a 10-lb standard mass placed on a scale at any geographic location. The upward buoyant force reduces the weight force W, and the scale reading is adjusted to read 10 lb_m. The net force transmitted is then

$$\mathbf{F}_{N,m} = F_L m - F_L \mathbf{V}_m \, \rho_{\text{air},c}$$

(3.6)

The first term in Eq. (3.6) is the force of the standard mass, its weight, and the second is the upward buoyant force created by displacing the volume of air occu-

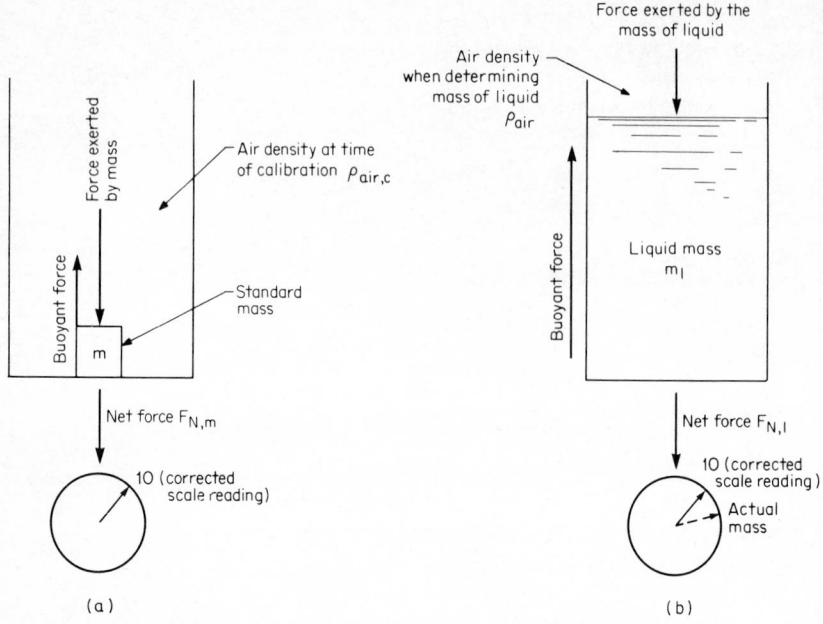

Figure 3.1   Effects of buoyancy in the determination of mass.

pied by the standard mass m. Equation (3.6) can be rewritten by substituting the density of the standard mass m as

$$\mathbf{F}_{N,m} = F_L m - F_L \frac{\rho_{\text{air},c}}{\rho_{\text{mass}}} m \tag{3.7}$$

where the subscript air,c refers to the density of air at the time of calibration.
Equation (3.6) can also be written as

$$\mathbf{F}_{N,m} = \left(1 - \frac{\rho_{\text{air},c}}{\rho_{\text{mass}}} \right) F_L m \tag{3.8}$$

If the weigh tank is filled with liquid (Fig. 3.1b), and the corrected scale reading is 10, the upward buoyant force is increased by the liquid volume, and the net force exerted by the liquid is

$$\mathbf{F}_{N,l} = \left(1 - \frac{\rho_{\text{air}}}{\rho_l} \right) F_L m_l \tag{3.9}$$

where the subscript air refers to the air density at the time the liquid mass is determined. Since the corrected scale readings are the same, Eq. (3.8) can be equated to Eq. (3.9), and the buoyancy correction becomes

$$F_{MB} = \frac{m_l}{m} = \left(\frac{\rho_{\text{mass}} - \rho_{\text{air},c}}{\rho_l - \rho_{\text{air}}} \right) \frac{\rho_l}{\rho_{\text{mass}}} \tag{3.10}$$

The mass of the collected liquid is then related to the standard mass (in pounds-mass or kilograms) by

$$m_l = F_{MB}m \tag{3.11}$$

Buoyancy Correction for Force Determination

A buoyancy correction is also required when standard masses are used as a force reference. Figure 3.2 shows standard masses placed on a platform to produce a force on the piston of a deadweight tester. The force exerted on the surface of the oil is given by Eq. (3.8), and the buoyancy correction is

$$F_{FB} = 1 - \frac{\rho_{air}}{\rho_{mass}} \tag{3.12}$$

The net force transmitted is then

$$\mathbf{F}_{N,m} = F_L F_{FB}m \tag{3.13}$$

In the SI system, the net force (in newtons) is

$$\mathbf{F}_{N,m}^* = 9.80665 F_L F_{FB}m^* \tag{3.14}$$

where m^* is the mass in kilograms.

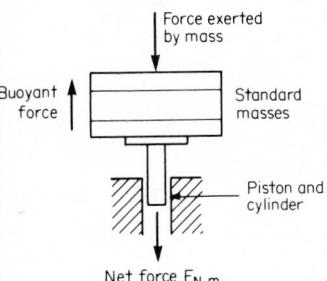

Force exerted by mass

Buoyant force

Standard masses

Piston and cylinder

Net force $F_{N,m}$

Figure 3.2 Effect of buoyancy when standard masses are used as a force reference.

EXAMPLE 3.1

A gallon of alcohol is weighed on a scale that was calibrated with steel masses when the air density was 0.07500 lb_m/ft^3. At the time of the weighing, the air density is 0.07490 lb_m/ft^3, and the scale reading is 6.572 lb_m. Determine the density of the alcohol if the density of the steel masses is 489 lb_m/ft^3.

The volume of the alcohol is

$$\mathbf{V}_l = (231 \text{ in}^3/\text{gal})(\text{ft}^3/1728 \text{ in}^3) = 0.13368 \text{ ft}^3$$

The initial estimate for the density of the alcohol is

$$(\rho_l)_0 = \frac{m_l}{\mathbf{V}_l} = \frac{6.572}{0.13368} = 49.162 \text{ lb}_m/\text{ft}^3$$

The initial estimate for the buoyancy correction factor is, from Eq. (3.10),

$$F_{MB} = \frac{489 - 0.07500}{49.162 - 0.07490} \frac{49.162}{489} = 1.00137$$

The corrected mass of the alcohol is, from Eq. (3.11),

$$m_l = (1.00137)(6.572) = 6.581$$

A second iteration gives $(\rho_l)_1 = 49.230$, and, from Eq. (3.10),

$$F_{MB} = \frac{489 - 0.07500}{49.230 - 0.07490} \frac{49.230}{489} = 1.00137$$

as previously. The mass of alcohol is then $m_l = 6.581$ lb$_m$, and its density is

$$\rho_l = \frac{6.581}{0.13368} = 49.230 \text{ lb}_m/\text{ft}^3$$

EXAMPLE 3.2

In Fairbanks, Alaska, at a latitude of 65° north and an altitude of 5000 ft (1524 m) above sea level, a 10-lb$_m$ (4.54-kg) mass is placed on a deadweight tester. What is the applied force if the air density is 0.07458 lb$_m$/ft^3? (Assume steel masses, $\rho_{mass} = 489$ lb$_m$/ft^3.)

The correction for local gravity is, by Eq. (3.3),

$$F_L = 1 - 2.637 \times 10^{-3} \cos 130 - (9.6 \times 10^{-8})(5000) - 5 \times 10^{-5} = 1.00117 \text{ lb}_f/\text{lb}_m$$

and the correction for buoyancy is, by Eq. (3.12),

$$F_{FB} = 1 - \frac{0.07458}{489} = 0.99985$$

Net force in English engineering system. From Eq. (3.13),

$$\mathbf{F}_{N,m} = (1.00117)(0.99985)(10) = 10.010 \text{ lb}_f$$

Net force in SI system. The mass in the SI system is

$$m^* = (0.4535924)(10) = 4.535924 \text{ kg}$$

and, from Eq. (3.14),

$$\mathbf{F}_{N,m}^* = (9.80665)(1.00117)(0.99985)(4.535924) = 44.527 \text{ N}$$

PRESSURE

Definitions

Pressure is defined as intensity of force and is evaluated as the force exerted on a unit area. In the English engineering system of units, the pressure unit is the pound-force per square inch (lb$_f$/in^2). In the SI system, the unit is the newton per square meter (N/m²) or pascal (Pa). From these units are derived such convenient units of measurement as the inch of water, bar, and standard atmosphere. Nine pressure terms are used to define pressure levels and pressure differences. These are discussed below and illustrated in Fig. 3.3.

Absolute Zero Pressure If all molecules were removed from within a chamber, a perfect vacuum would exist, and no pressure forces would be exerted on the chamber walls. This idealized state defines the condition of zero pressure and is referred to as *absolute zero*.

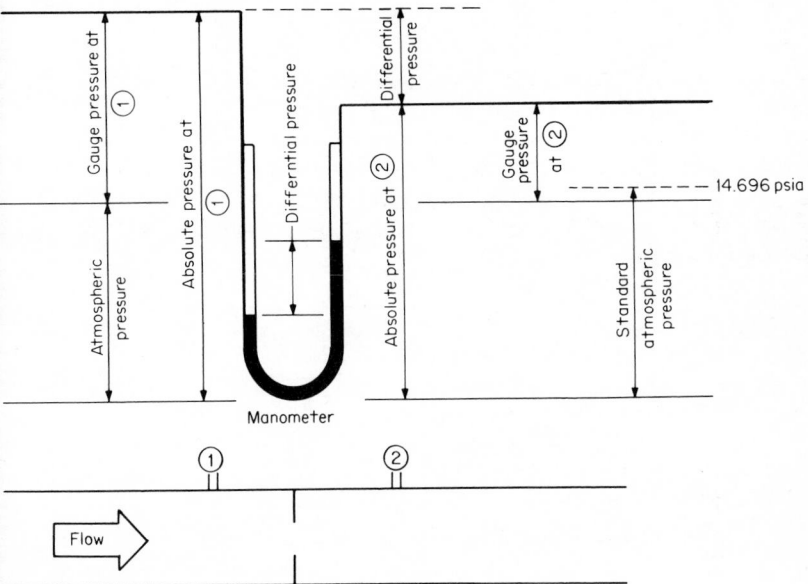

Figure 3.3 Pressure levels and terminology.

Absolute Pressure The absolute pressure is the pressure above absolute zero. The static absolute pressure defines the molecular activity of a gas. It is the pressure used in the calculation of gas density.

Atmospheric Pressure The pressure exerted by the atmosphere above absolute zero is defined as atmospheric pressure. Although this pressure varies with location, it is convenient to define a standard atmosphere of pressure at sea level as 14.696 psia (101.325 kPa), and to use this value as a reference in computing gas volumes.

Actual atmospheric pressure is measured with a barometer and varies with altitude. At 5000 ft (1524 m), it is approximately 12.3 psia (84.8 kPa); at 10,000 feet (3048 m), it is 10 psia (69 kPa).

Gauge Pressure Pressure gauges measure the difference between the pressure inside a pressure element and the surrounding atmospheric pressure. To obtain absolute pressure, atmospheric pressure must be added to the gauge reading.

Vacuums A vacuum-gauge reading is a reading below atmospheric pressure, usually expressed in inches of mercury (vacuum).

Differential Pressure A differential pressure is the difference between two pressures. It is measured by either separating the two pressures with a diaphragm and measuring the force or motion of the diaphragm, or by observing the height of a column of liquid in a manometer.

Static Pressure The actual pressure exerted by a fluid either at rest or in motion is its static pressure. Either a piezometer ring or a small radial hole in a pipe wall will allow the static pressure to be measured.

In obtaining the static pressure of a flowing fluid, it is important that the hole be drilled perpendicular to the pipe, with no burrs and no rounded corners. Rayle (1954) has shown (Fig. 3.4) that departure from recommended hole size, inclination, or edge condition results in bias errors of -0.5 to $+1.1$ percent of the static pressure.

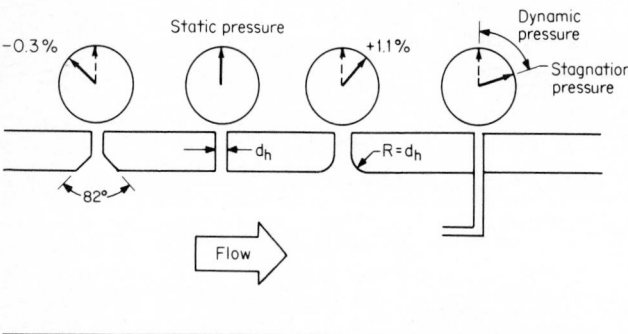

Figure 3.4 Bias error caused by edge condition of wall pressure-tap hole *(from Rayle, 1959).*

Dynamic Pressure If a tube is bent perpendicular to a flow (Fig. 3.4), the static pressure is increased by the directed kinetic energy of the stream. At zero velocity the pressure reading is the same as the static pressure, but as the velocity increases the difference is observed to increase by the square of the velocity. This difference in pressure levels is due to *dynamic pressure.*

Total Pressure The sum of the static and dynamic pressures is the stagnation, or total, pressure. The stagnation pressure may be read with a pressure gauge connected to a pitot tube as shown in Fig. 3.4.

Pressure Relations Because of the many pressure units involved, it is important that a consistent set of units be used in conjunction with the following pressure relationships.

For flow-line pressure in pounds-force per square foot,

$$P_f = P_G + P_B \tag{3.15}$$

the total pressure (pounds-force per square foot) is the sum of the dynamic pressure and the static pressure; that is

$$P_T = P_f + P_D \tag{3.16}$$

where the dynamic pressure P_D may be expressed in terms of the fluid density and velocity as

$$P_D = \frac{\rho_f V_P^2}{2g_c} \tag{3.17}$$

The differential pressure between two pressure levels is

$$\Delta P = P_{f1} - P_{f2} \tag{3.18}$$

Units

Numerous units and pressure scales have been developed to express pressure. Some of the more common ones are as follows:

1. Atmospheric pressure

- Standard atmospheres (atm)
- Atmospheric pressure (atm)
- Millimeters of mercury at 0°C (mm Hg)

2. Absolute pressure

- Pounds per square inch absolute (psia)
- Bars (bar)
- Pascals (Pa, kPa)

3. Differential pressure

- Feet of flowing fluid
- Inches of flowing fluid
- Inches of water (in H_2O) at flowing temperature
- Inches of water (in H_2O) at 39.2°F, at 60°F, at 68°F
- Inches of mercury (in Hg) at 32°F, at 60°F, at 68°F

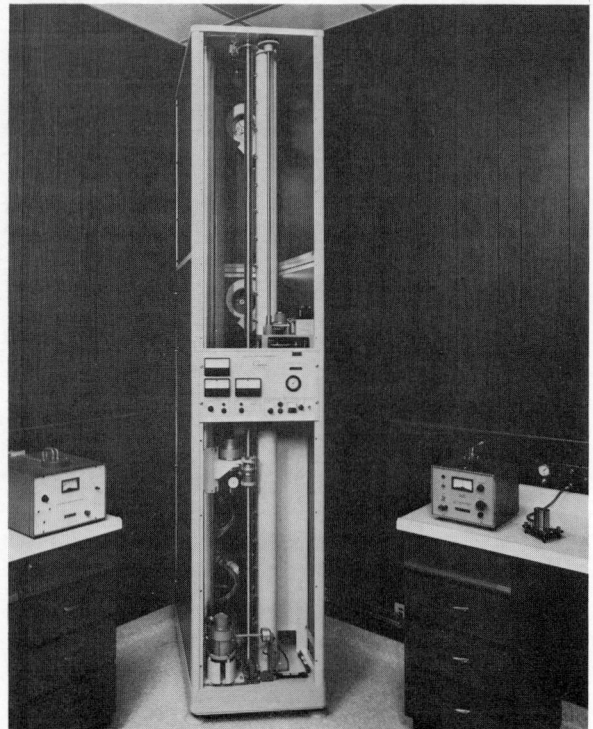

Figure 3.5 **High-accuracy manometer** *(courtesy Schwien Engineering, Inc.).*

- Bars (bar)
- Pascals (Pa, kPa)
- Pounds per square inch, differential (psid)
4. General terms
   - Pounds per square inch gauge (psig)
   - Inches of mercury (in Hg) at 32°F, at 60°F, at 69°F
   - Deciboyles
   - Torr (torr)

The SI conversion factors between these units and the pascal are given in Table C.1. For conversion between two non-SI pressure units, the known units are first converted to pascals; then, by division, the desired conversion is obtained. For example, the conversion of inches of mercury at 32°F to kilograms-force per square centimeter would be expressed as

$$(\text{in Hg})_{32°\text{F}}(3.38638\text{E} + 03) = \text{Pa} = (\text{kg}_\text{f}/\text{cm}^2)(9.806650\text{E} + 04)$$

$$\text{or} \quad \text{kg}_\text{f}/\text{cm}^2 = \left(\frac{3.38638\text{E} + 03}{9.806650\text{E} + 04}\right)(\text{in Hg})_{32°\text{F}} = (0.03453147)(\text{in Hg})_{32°\text{F}}$$

Standards

The deadweight tester and simple manometer are the two basic pressure standards. The U-tube or cistern-type manometer is used from approximately 3 lb$_\text{f}$/in^2 (21 kPa) to 100 lb$_\text{f}$/in^2 (690 kPa) with an accuracy of ±0.1 percent. Manometers can, however, be designed to be highly accurate. Shown in Fig. 3.5 is a high accuracy, ±0.0003-in-Hg, ±0.003 percent of reading, mercury manometer. The deadweight tester defines pressures from 0.01 to 10,000 psig (0.07 to 69,000 kPa) within an accuracy range of 0.01 to 0.15 percent, depending on design. A commercially available deadweight tester is shown in Fig. 3.6.

Figure 3.6 **Precision deadweight tester** *(courtesy Chandler Engineering Co.).*

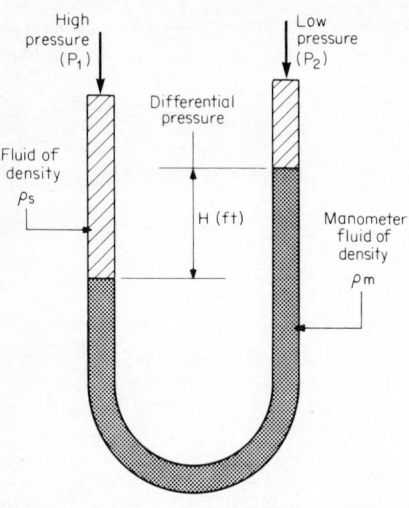

$$P_1 - P_2 = F_L(\rho_m - \rho_s)H$$

Figure 3.7 Basic U-tube-manometer principle.

Manometer Because of its inherent simplicity, the simple glass U-tube manometer (Fig. 3.7) has been the basic pressure standard for many years. Manometers measure pressure by balancing pressure forces directly against a liquid column. The liquid selected depends on the pressure difference and may range from a light-density fluid such as kerosene to a very dense liquid such as mercury.

For the manometer shown in Fig. 3.7, the indicated differential pressure is

$$\Delta P = \frac{g_l}{g_c}(\rho_m - \rho_s)H = F_L(\rho_m - \rho_s)H \qquad \text{lb}_\text{f}/\text{ft}^2 \qquad (3.19)$$

where H is in feet, ρ_m is the manometer-fluid density, and ρ_s is the fluid density in the high-pressure side. In the SI system, the differential pressure is given by

$$\Delta P^* = 9.806650 F_L(\rho_m^* - \rho_s^*)H^* \qquad \text{Pa} \qquad (3.20)$$

The *inch-of-water* differential-pressure unit is used in the differential-producer flow equation. It is derived as the equivalent pressure read on a water-filled manometer at a specified temperature and at standard gravity (g_0) or, from Eq. (3.19),

$$\Delta P = \frac{g_l}{g_c}(\rho_w)_{T,g_0}H = (\rho_w)_{T,g_0}\frac{h_w}{12} \qquad \text{lb}_\text{f}/\text{ft}^2 \qquad (3.21)$$

Substituting the density of water at standard gravity, given by Eq. (2.76), yields the relationship between an inch of water (pressure) at 68°F and the differential pressure as

$$\Delta P = \frac{62.31572}{12} h_w = 5.192977 h_w \qquad \text{lb}_\text{f}/\text{ft}^2 \qquad (3.22)$$

Deadweight Tester   Shown in Fig. 3.8 are the basic elements of a deadweight gauge used to produce a reference calibration pressure. An accurately honed piston of known area is inserted into a cylinder, and standard masses are then placed on the platform. When the oil pump supplies sufficient pressure to raise the masses, the force exerted by the oil pressure over the piston area is balanced by the weight force. The pressure is then defined by

$$p_G = \frac{\mathbf{F}_{N,m}}{\text{piston area}} = \frac{\mathbf{F}_{N,m}}{A_p} \tag{3.23}$$

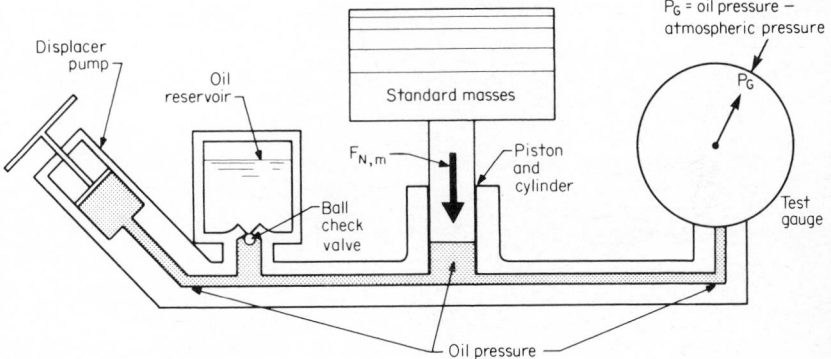

Figure 3.8   Oil deadweight tester.

Deadweight testers using air instead of oil (Fig. 3.9) are also available, as well as compact portable pneumatic calibrators for field use (Fig. 3.10).

Industrial Devices

The secondary devices used to measure differential pressure are commonly divided into two types—wet and dry. In the dry type (Fig. 3.11), the process fluid is isolated by a diaphragm; in the wet type, mercury usually is the separator. Mercury meters were the mainstay of the process and natural gas industries for many years. However, they have been completely replaced by the dry type in the process industry, and dry-type bellows meter (Fig. 3.12), which require no external power source, are rapidly replacing mercury meters in natural gas applications.

Dry Type   Dry-type devices are conveniently separated into motion and direct-measuring types. In the motion type (Fig. 3.12), the pressure difference across a diaphragm causes a bellows to move against a restraining spring. The motion, which can be recorded directly, is proportional to the differential pressure. In the direct-measuring type (Fig. 3.13a), small deflections of a diaphragm are either measured or restrained by a feedback force. The deflections may be detected via induction, capacitance, a strain gauge, or a taut resonating wire. In the force-

Figure 3.9 Pneumatic deadweight tester *(courtesy Ametek).*

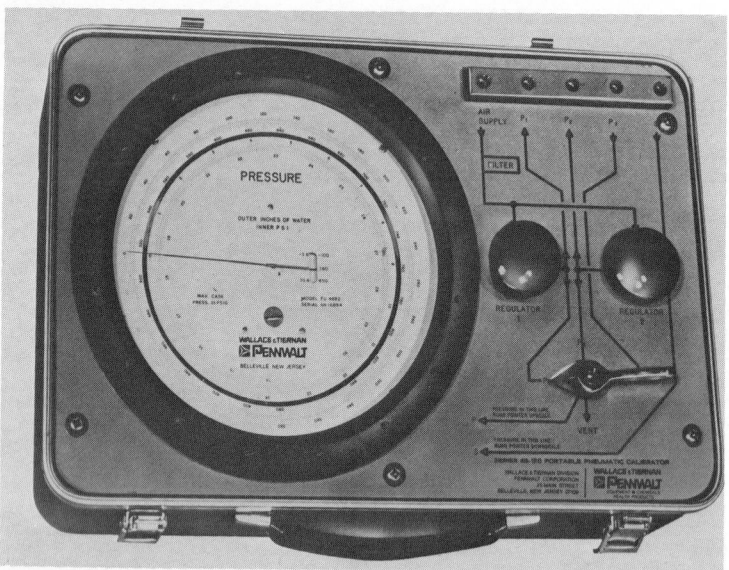

Figure 3.10 Portable pneumatic calibrator *(courtesy Wallace & Tiernan-Pennwalt).*

3-14

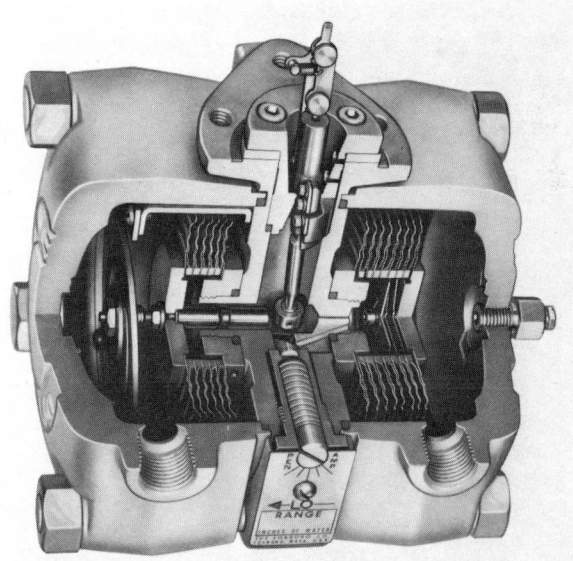

3-15

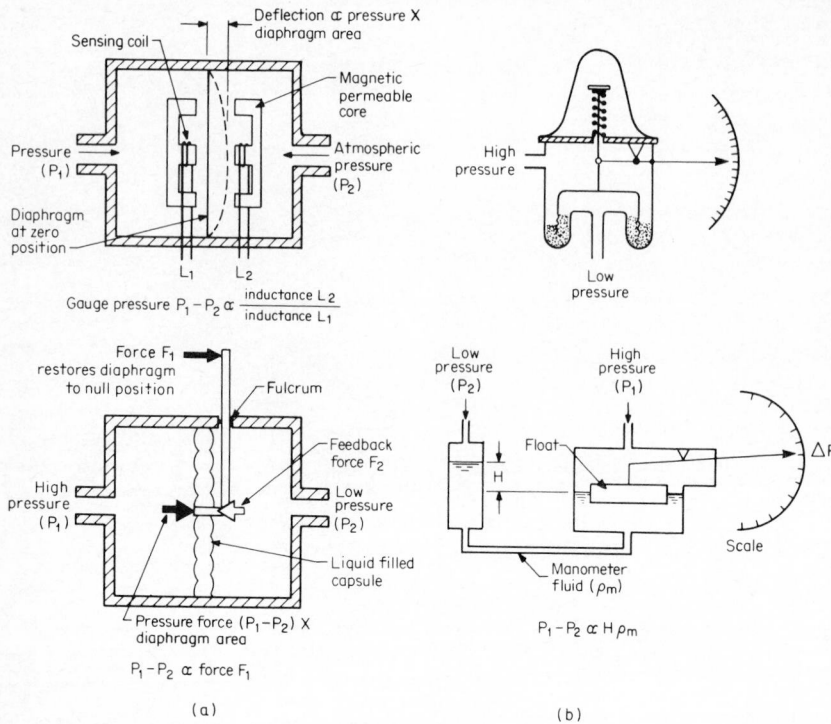

Figure 3.13 Dry and wet pressure or differential-pressure measuring devices. (*a*) Dry types. (*b*) Wet types.

feedback device, the differential pressure is proportional to the feedback force. If the low-pressure side is evacuated, then the absolute pressure is measured (Fig. 3.14). Dry-type *transmitters* have an accuracy in the range of ± 0.2 to ± 1 percent of the upper range value (URV).

Wet Type   Shown in Fig. 3.13*b* are the two basic wet-type meters—the inverted bell and the float type. In the inverted bell, the force developed by the differential pressure acting on the bell is opposed by a spring force. The motion of the bell is a direct measure of the differential pressure; and, with a sizeable bell, a considerable force can be developed and, hence, low differential pressures can be measured. Typical inverted-bell designs operate in the range of 0 to 10 in H_2O (0 to 2.5 kPa).

In the many float types, a cylindrical steel float chamber forms one side of a manometer and contains a steel disk floating on mercury. A cylindrical steel reference chamber forms the second side of the manometer. The area of this chamber is selected to give the desired mercury-level change for the operating differential pressure, which usually is from 20 to 200 in H_2O (5 to 50 kPa).

Correction Factors

In practice, wet- and dry-type differential-pressure measuring devices are sometimes located a substantial distance above or below the pressure taps. The indicated pressure is then other than that at the taps, and an *elevation correction* is required. Bean (1957) gives an (extreme) illustration of the possible effects of elevation for a manometer located 50 ft (15.2 m) below an orifice. In this example, a +0.6 percent bias error is calculated.

For the wet type, the displacement of the manometer fluid by the lower-density process or seal fluid requires that a *manometer correction factor* be applied to the indicated reading. These two corrections are considered here.

Elevation Correction   The pressures at location H_{LL} below the pressure sources in Fig. 3.15 are

$$P_{f1'} = P_{f1} + \rho_{f1}F_L H_{LL} \qquad P_{f2'} = P_{f2} + \rho_{f2}F_L H_{LL} \qquad (3.24)$$

The relationship between the *true* and *indicated* differentials is then

$$\Delta P = (\Delta P)_{\text{ind}} - (\rho_{f1} - \rho_{f2})F_L H_{LL} \qquad (3.25)$$

For gas (vapor) and liquid flows with liquid-filled leads at the same temperature, no elevation correction is required ($\rho_{f1} = \rho_{f2}$). However, with gas-filled leads, the downstream lead is always at a lower density ($\rho_{f1} > \rho_{f2}$). Substituting

Figure   3.14 Absolute-pressure measuring device *(courtesy The Foxboro Co.).*

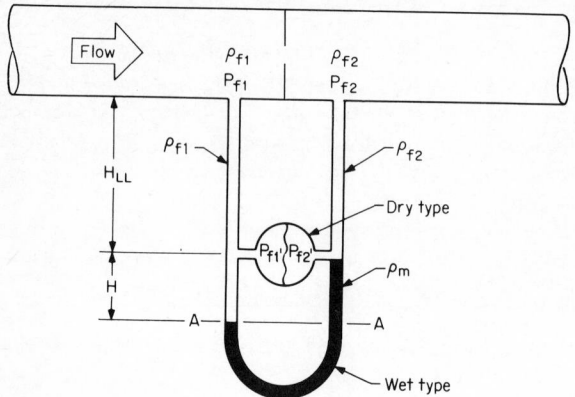

Figure 3.15 Schematic of wet- or dry-type device installation.

the gas density equation (2.10) into Eq. (3.25) and expressing the pressure differences in inches of water h_w give the relationship between indicated and true differential pressure as

$$h_w = (h_w)_{ind} - 0.01874 \frac{F_L G H_{LL}}{Z_{f1} T_{f1}} h_w \tag{3.26}$$

An iterative solution is then necessary to calculate the true differential from the indicated reading. A negligible error is, introduced, however, if it is assumed that the indicated and true differentials are equal; then

$$h_w = \left(1 - 0.01874 \frac{F_L G H_{LL}}{Z_{f1} T_{f1}} \right) (h_w)_{ind} \tag{3.27}$$

The elevation correction factor is then defined by

$$F_{EL} = 1 - 0.01874 \frac{F_L G H_{LL}}{Z_{f1} T_{f1}} \tag{3.28}$$

In this expression H_{LL} is positive for locations below the differential producer, and negative for locations above. In the SI system, the elevation correction factor is similarly derived as

$$F_{EL}^* = 1 - 0.03416 \frac{F_L G H_{LL}^*}{Z_{f1} T_{K1}} \tag{3.29}$$

Manometer Factor   The manometer-factor correction is obtained by equating pressures at location AA as

$$P_{f1'} + F_L \rho_{f1} H = P_{f2'} + F_L \rho_m H \tag{3.30}$$

Rearranging yields

$$P_{f1'} - P_{f2'} = F_L \left(1 - \frac{\rho_{f1}}{\rho_m} \right) H \rho_m \qquad (3.31)$$

Defining the pressure difference $P_{f1'} - P_{f2'}$ via Eq. (3.24) then yields

$$P_{f1} - P_{f2} = F_L F_{EL} \, \rho_m \left(1 - \frac{\rho_{f1}}{\rho_m} \right) H \qquad (3.32)$$

Since wet-type devices are scaled to read in inches of water, the relationship between the true and indicated differential pressures becomes

$$h_w = F_{EL} F_M (h_w)_{\text{ind}} \qquad (3.33)$$

for U.S. units, and

$$\Delta p^* = F_{EL}^* F_M^* (\Delta p^*)_{\text{ind}} \qquad (3.34)$$

for SI units, where F_M is the *manometer factor* defined by

$$F_M = \left(1 - \frac{\rho_{f1}}{\rho_m} \right) F_L \qquad (3.35)$$

for U.S. units, and

$$F_M^* = \left(1 - \frac{\rho_{f1}^*}{\rho_m^*} \right) F_L \qquad (3.36)$$

for SI units.

In the development of these equations, the fluid density over the manometer is assumed to be that of the upstream process fluid. If a seal liquid is used, then its density ρ_s should be substitued into Eq. (3.32) or (3.33). For gases (vapors), density is calculated with Eq. (2.10) for U.S. units and with Eq. (2.11) for SI units.

If the manometer fluid is water, Eq. (2.75) can be used to calculate density. If the fluid is mercury, the density (ANSI/API 2530, 1978) is given by

$$\rho_{Hg} = [1 - 0.000101(T_F - 60)]846.324 \qquad (3.37)$$

EXAMPLE 3.3

A deadweight tester with weights *trimmed* at a location with a local gravity of 32.164 ft/ s^2 is used to calibrate a pressure gauge at a location 2000 ft (609.6 m) above sea level and at 10° latitude. What is the actual pressure if 100-psig weights are used? (Assume the air-buoyancy correction is the same for both locations.)

At the new location, the local gravity correction is, by Eq. (3.3),

$$(F_L)_2 = 1 - 2.637 \times 10^{-3} \cos(2)(10) - (9.6 \times 10^{-8})(2000) - 5 \times 10^{-5}$$
$$= 0.99728 \text{ lb}_f/\text{lb}_m$$

At the original location, the local gravity correction is

$$(F_L)_1 = \frac{32.164}{32.174} = 0.99969 \text{ lb}_f/\text{lb}_m$$

Since weights of the same mass are used, the pressure at the new location can be obtained by combining Eqs. (3.13) and (3.23) as

$$\frac{(p_G)_1 A_p}{(F_L)_1 F_{FB} m_s} = \frac{(p_G)_2 A_p}{(F_L)_2 F_{FB} m_s}$$

Since the masses, buoyancy, and piston areas are the same,

$$(p_G)_2 = 100 \frac{0.99728}{0.99969} = 99.76 \text{ psig}$$

EXAMPLE 3.4

A dry-type differential-pressure transducer is located 50 ft (15.2 m) above an orifice flow-meter measuring the natural gas of Example 2.6. If the indicated differential pressure is 100 in H_2O, what is the true differential pressure for lead lines at the same temperature as the process? (Use $g_l = 32.162 \text{ ft/s}^2$.)

From Example 2.6, $G = 0.73$, $Z_f = 0.878$, and $T_f = 499.7°R$ (40°F). The location correction factor for density within the leads is, from Eq. (3.3),

$$F_L = \frac{32.162}{32.174} = 0.9996 \text{ lb}_f/\text{lb}_m$$

From Eq. (3.27) the true differential pressure is

$$h_w = \left[1 - 0.01874 \frac{(0.9996)(0.73)(-50)}{(0.878)(499.7)} \right] (100) = 100.16 \text{ in } H_2O$$

EXAMPLE 3.5

A wet-type mercury-filled meter based on the manometer principle is manifolded at the same position as the dry-type transducer of Example 3.4. What is its reading?

The density of the gas over the mercury is calculated with Eq. (2.10) as

$$\rho_{f1} = 2.69883 \frac{(0.73)(400 + 14.7)}{(0.878)(499.7)} = 1.862 \text{ lb}_m/\text{ft}^3$$

The density of the mercury at 40°F is, by Eq. (3.37),

$$\rho_{Hg} = 846.324[1 - (0.000101)(40 - 60)] = 848.034 \text{ lb}_m/\text{ft}^3$$

Equation (3.33) can be written as

$$(h_w)_{ind} = \frac{h_w}{F_{EL} F_M}$$

where, from Eq. (3.35),

$$F_M = \left(1 - \frac{1.862}{848.03} \right) (0.9996) = 0.9974$$

Substitution then gives

$$(h_w)_{ind} = \frac{100.16}{(1.0016)(0.9974)} = 100.26 \text{ in } H_2O$$

TEMPERATURE

Scales

In 1968, the International Committee on Weights and Measures adopted several changes in the empirical temperature scale. This work is reported in "The International Practical Temperature Scale of 1968." In it, the kelvin (K) was adopted as the basic thermodynamic temperature unit.

Although this new scale better approximates the thermodynamic scale, it has yet to replace the empirical equations used to calculate absolute temperatures. These are

$$T_{°R} = T_F + 459.67 \qquad (3.38)$$

for degrees Rankine, and

$$T_K = T_{°C} + 273.15 \qquad (3.39)$$

for kelvins, with the relationship between the Fahrenheit and Celsius scales being defined by

$$T_F = \tfrac{9}{5}T_{°C} + 32 \qquad (3.40)$$

and

$$T_{°C} = \tfrac{5}{9}(T_F - 32) \qquad (3.41)$$

The equation relating Rankine temperatures to the kelvin temperature scale is

$$T_{°R} = \tfrac{9}{5}T_K \qquad (3.42)$$

Measurement

Many temperature-measuring devices are based on the thermal expansion of a solid, liquid, or gas, on a thermoelectric measurement in which a thermally induced electromotive force (emf) is used to infer temperature, or on the measurement of a resistance change in either a precision resistor or a thermistor. Of these, the most commonly used are the mercury-in-glass thermometer, the gas (or vapor) expansion thermometer, the thermocouple, and the resistance thermometer.

Mercury-in-Glass Thermometer   The mercury-in-glass thermometer (Fig. 3.16*a*) is widely used in both laboratory and industry because of its basic simplicity. The typical thermometer consists of a bulb-reservoir and a capillary. As heat is transferred into the bulb, the mercury rises due to thermal expansion and displaces a sealed inert gas. The temperature is then read on a scale calibrated in the units of interest.

It is important that the thermometer be used under the conditions that prevailed during calibration. Specifically, thermometers are calibrated either partially or completely immersed in a temperature bath. If they are used under other conditions, then a stem correction is required. Benedict (1977) gives several

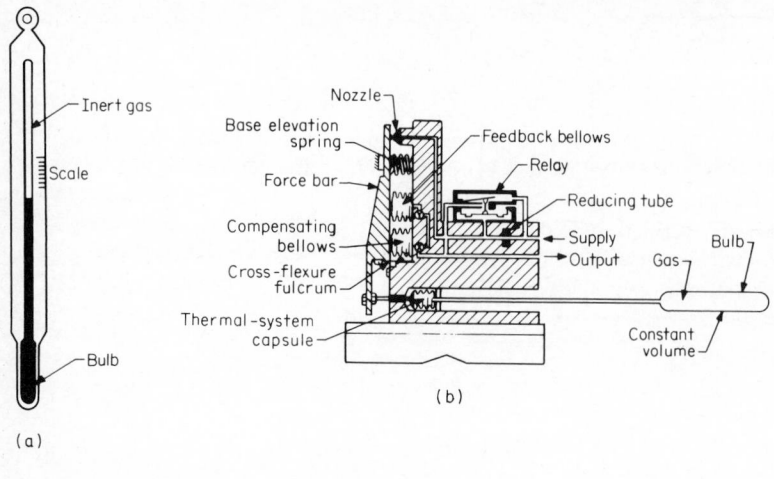

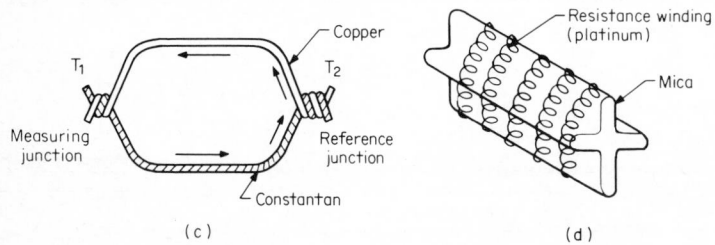

Figure 3.16 Thermometers. (*a*) Mercury in glass. (*b*) Gas-expansion thermometer *(courtesy The Foxboro Co.)*. (*c*) Simple thermocouple. (*d*) Platinum resistance thermometer.

examples concerning stem correction. In one, a total-immersion thermometer immersed only to the 200°F (93°C) mark is in error by 4 percent when used to measure an 814°F (434°C) temperature.

Gas (or Vapor) Expansion Thermometer   This thermometer operates on the principle that a gas expands at constant volume as its temperature is increased. In the industrial device shown in Fig. 3.16*b*, a change in temperature at the sensor causes a change in gas pressure. This pressure change is transferred by the thermal-system capsule to a change in force on the force bar. The design of the force-balance mechanism allows the bar to pivot imperceptibly about a pair of cross-flexure fulcrums when the gas pressure changes. This small bar motion causes a change in the clearance between the nozzle and the top of the force bar, resulting in a change in the relay output pressure. In addition to changing the output signal of the transmitter, this change in pressure is applied to the force bar by the feedback bellows. Thus the moments about the cross-flexure fulcrums are kept in equilibrium, and the output signal of the transmitter is proportional to the measured temperature.

Nitrogen is almost always used as the fill gas since it can be used over a wide temperature range and because it is available as a high-quality gas. The scale is nearly linear, and the temperature limits range from -350 to $1400°F$ (-220 to $760°C$).

Since a fluid's vapor pressure depends only on the free-surface temperature of a liquid, a measured vapor pressure can also be used to indicate temperature. The *vapor-actuated* thermometer is similar to the gas thermometer, with the bulb casing being larger to contain all the liquid when the temperature is below that of the capillary and thermal-system capsule.

Thermocouples   The thermocouple provides a reliable and accurate temperature indication for many industrial applications. In its simplest form it consists of a pair of dissimilar conductors joined together at both ends, as shown in Fig. 3.16c. If the temperature is to be measured at T_1, a reference temperature is maintained at T_2, and the emf produced is related to the temperature by a polynomial fitting equation. Depending on the combination of conductors and the desired temperature range, the order of the polynomial may extend from second to ninth.

The temperature-emf characteristics of a thermocouple depend upon the materials used in the element of the thermocouple and the temperature to which they are subjected. The selection of materials for thermocouple elements is subject to at least the following requirements:

1. The materials should be able to withstand the extremes of the temperature to be measured, without significant deterioration, for a suitable period of time.

2. A relatively large emf should be developed per degree change in temperature so that the instrument can be used to determine small temperature changes.

3. The emf should increase with increasing temperature continuously over the range of use.

4. The materials should maintain their original temperature-emf characteristics for long periods. Consistency of calibration is largely dependent upon freedom from contamination and mechanical strains. These introduce nonhomogeneities in the thermocouple elements.

5. The materials should be homogeneous and capable of easy standardization. They should be commercially available to permit replacement without the necessity of recalibrating the temperature-measuring instrument.

6. The materials should be capable of being formed into an adequately strong thermocouple assembly to meet installation and application requirements.

Table 3.2 **Designations and Limits of Error of Thermocouples and Thermocouple Wires**

ISA designation	Material (positive vs. negative element)	Temperature	Limit of error, %	
			°F	°C
Type T	Copper vs. copper-nickel	−328 to 32°F (−200 to 0°C)	±2.7	±1.5
	(copper-constantan)	32 to 700°F (0 to 350°C)	±1.35	±0.75
Type J	Iron vs. copper-nickel	32 to 1400°F (0 to 1400°C)	±1.35	±0.75
	(iron-constantan)			
Type E	Nickel-chromium vs. copper-nickel	−328 to 32°F (−200 to 0°C)	±1.8	±1.0
	(chromel-constantan)	32 to 2300°F (0 to 1250°C)	±0.9	±0.5
Type K	Nickel-chromium vs. nickel-aluminum	−328 to 32°F (−200 to 0°C)	±3.6	±2.0
	(chromel-alumel)	32 to 2300°F (0 to 1250°C)	±1.35	±0.75
Type R	Platinum 13% rhodium vs. platinum	32 to 2700°F (0 to 1450°C)	±0.34	±0.25
	(Pt 13 Rh–Pt)			
Type S	Platinum 10% rhodium vs. platinum (Pt 10 Rh–Pt)	1600 to 3100°F (870 to 1700°C	±0.34	±0.25
Type B	Platinum 30% rhodium vs. platinum 6% rhodium (Pt 30 Rh–Pt 6 Rh)	1600 to 3100°F (870 to 1700°C)	±0.9	±0.5

The ISA designations, temperature ranges, and limits of error for several of the more common combinations of thermocouple conductors are presented in Table 3.2. In Fig. 3.17 are shown several thermocouple types.

Resistance Thermometers   The industrial resistance thermometer is widely used because of its accuracy and basic simplicity; temperature changes of 0.03°F

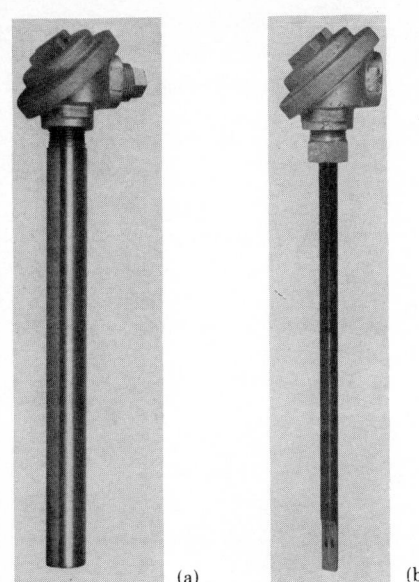
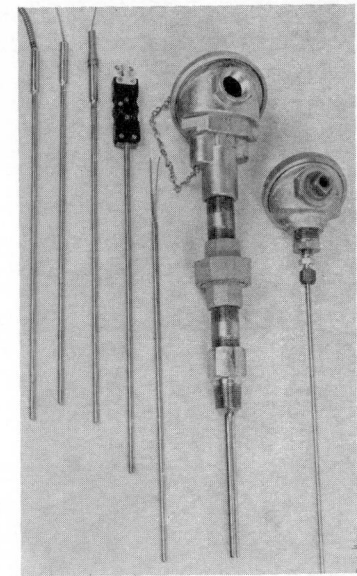

(a)   (b)   (c)

Figure 3.17 Thermocouple types *(courtesy The Foxboro Co.).* (*a*) Wire type. (*b*) PYOD type. (*c*) Minox type.

(0.02°C) are easily detected in industrial processes. The measurement range of this instrument is about the same as that of the copper-constantan thermocouple, iron-constantan thermocouple, and expansion thermometer; these limits are approximately −300 to 1200°F (−184 to 689°C).

The resistance thermometer requires no reference junction, since it operates on a measured change in the resistance of a metal or semiconductor (thermistor) with temperature. Usually the metal is platinum, copper, or nickel, and the semiconductor of a metallic oxide.

For a pure metal, the relationship between temperature and resistance is expressed as a series expansion in temperature with constant coefficients:

$$R_{\text{temp}} = (1 + aT_{°C} + bT_{°C}^2 + cT_{°C}^3 + \cdots)R_{\text{ref}} \tag{3.43}$$

where R_{ref} is the resistance, in ohms, at a reference temperature, usually 0°C (32°F), and R_{temp} is the the measured resistance. For many applications, an average coefficient of resistance is assumed, and all terms with powers higher than the first are assumed zero. For platinum, a second-order equation accurately predicts behavior at temperatures up to 1200°F (649°C). Copper sensors are less expensive than platinum sensors and quite linear. However, these are limited to temperatures up to 250°F (121°C). Nickel is quite nonlinear and has an upper temperature limit of 600°F (316°C). Because of its inherent stability and linearity, the basic platinum resistance temperature detector (RTD), shown in Figs. 3.16*d* and 3.18, continues to dominate the field in resistance thermometers.

The thermistor is classified as a semiconductor because its electric conductivity falls between that of an insulator and that of a conductor. The name thermistor comes from *thermally sensitive resistor*—the resistance of a thermistor varies as a function of temperature.

The thermistor is a solid semiconductor with a high temperature coefficient of resistivity. The variation of resistance with temperature can be approximated with a simple exponential equation of the form

$$R_{\text{temp}} = a \exp \frac{b}{T_f} \qquad (3.44)$$

where R_{temp} is the resistance at the measured absolute temperature T_f. The constants a and b are typically 0.06 and 8000, respectively.

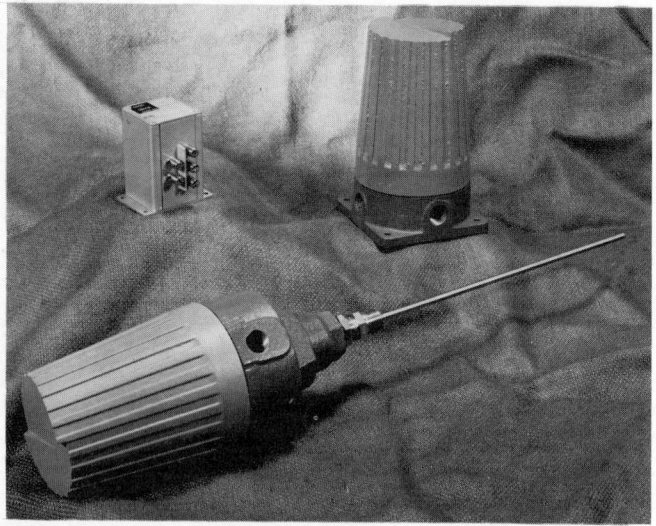

Figure 3.18 Industrial RTD temperature transmitter *(courtesy The Foxboro Co.).*

Thermometer Wells

In numerous applications, it is neither desirable nor practical to expose a temperature sensor directly to a process material. Thermometer wells are, therefore, used to protect against damage from corrosion, erosion, abrasion, and high-pressure processes. A well is also useful in protecting a sensor from physical damage during handling and normal operation. When the ambient temperature is significantly higher than the flowing temperature or when the well is exposed to direct sunlight, heat conduction along the well can raise the indicated temperature by 2 to 3°F. In these cases, care must be taken to properly insulate the thermometer well.

Wells are provided in many configurations; some of these are depicted in Fig.

3.19. The open-end type can either be plain or have a lagging extension. A lagging length of 3 in (75 mm) is standard, but other lengths are available. For mounting purposes, there is a choice of external threads or process flanges. The external threads can be obtained in either NPT or SI metric sizes, and the process flanges are made to either ANSI or ISO specifications. The closed end of a well can have either a straight or tapered tip, or a straight tip with a reinforced neck. A sanitary well is also available for use in food industries. This special-purpose well has a tapered tip but no external threads or flange. It is usually welded in a sanitary fitting, pipeline, or storage tank.

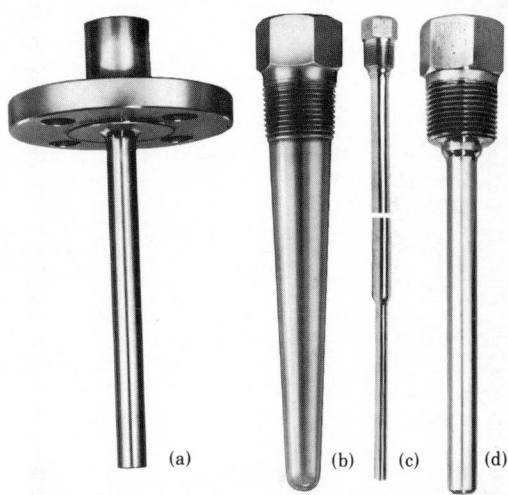

Figure 3.19 Thermometer wells *(courtesy The Foxboro Co.).* (*a*) Plain flanged solid. (*b*) Lagged threaded tapered. (*c*) Plain threaded welded. (*d*) Plain threaded solid.

Temperature of a Moving Stream

The true (static) temperature of a moving fluid is recorded only if the thermometer is located in a large reservoir or is moving at the same velocity as the stream (see Fig. 3.20*a* and *b*). However, practical measurements of temperature are made with a thermometer that is at rest with respect to the fluid, and the mass flow impact on the measuring device gives a reading higher than the static temperature. The maximum difference would be observed for an isentropic compression of the fluid to the stagnation temperature (Fig. 3.20*c*). In practice, this maximum is seldom achieved because of thermometer losses, fluid-frictional effects, thermometer-well design, and other factors. The indicated temperature is then somewhere between the static and theoretical stagnation temperatures (Fig. 3.20*d*).

For liquids, the difference between the indicated and static temperatures is quite small and can be ignored. Benedict (1977) gives an example concerning 200°F (93°C) water flowing at 20 ft/s (6.1 m/s); the difference between static and indicated temperatures was calculated to be only 0.002°F (0.001°C).

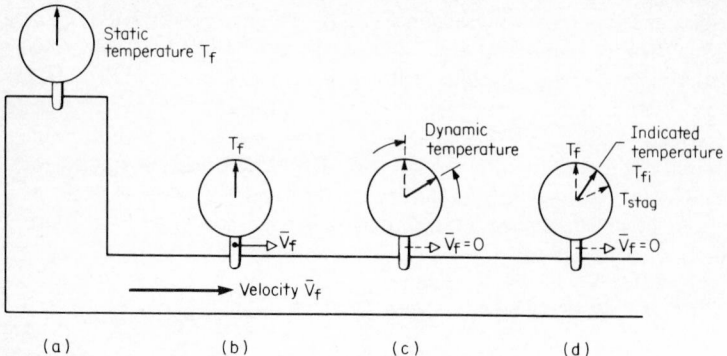

Figure 3.20 Temperatures of a moving stream.

For gases, however, this effect is not always insignificant, and the indicated temperature is corrected to true temperature by the introduction of a recovery factor as expressed in

$$T_f = T_{fi} - \frac{M_w \overline{V}_f^2}{2 g_c C_p J} F_{RF} \tag{3.45}$$

The recovery factor F_{RF} ranges from 0.5 to 0.9 for the ordinary straight or tapered wells used extensively in flow measurement.

EXAMPLE 3.6

Air is moving at 200 ft/s (61 m/s) in a 2-in (50-mm) line. The indicated temperature is 50°F (10°C). What is the static temperature if the thermometer well has a recovery factor of 0.9?

From Fig. I.17, $C_p/M_w = 0.25$ Btu/(lb$_m$ · °F). The static temperature is then, by Eq. (3.45)

$$T_f = 459.67 + 50 - \frac{(200)^2}{(2)(32.174)(0.25)(778)} \, 0.9 = 506.8°\text{R}$$

where $J = 778$ ft · lb$_f$/Btu.

FLOW MEASUREMENT UNITS

Flow rates are calculated in either mass or volumetric units. The mass flow-rate unit is time-dependent only—that is, in pounds-mass per hour, per second, or per day. However, the volumetric unit may be calculated in gallons, cubic feet, or cubic meters per unit time at flowing temperature or at a specified base temperature. To distinguish between mass-flow, base-volume, and actual volumetric units, an uppercase subscript will be used for mass-flow and base-volume units, and lowercase for flowing volumetric flow-rate units. Table 3.3 lists the symbols and subscripting that will be used in this book.

Table 3.3 Subscripting for Flow Rate and Total Flow

Symbols	Meaning	Units	
		U.S.	SI
	Flow rate		
q_{acfs}, q_{acfm}, q_{acfh}, q_{acfd}	Gas (vapor) volumetric flow rate at *flowing* conditions	ft³/s, ft³/min, ft³/h, ft³/24h	
q_{acms}, q_{acmm}, q^*_{acmh}, q^*_{acmd}	Gas (vapor) volumetric flow rate at *flowing* conditions		m³/s, m³/min, m³/h, m³/24h
q_{alps}, q^*_{alpm}, q_{alph}, q^*_{alpd}	Gas (vapor) volumetric flow rate at *flowing* conditions		L/s, L/min, L/h, L/24h
q_{bps}, q_{bpm}, q_{bph}, q_{bpd}	Liquid volumetric flow rate at *flowing* conditions	bbl/s, bbl/min, bbl/h, bbl/24h	
q_{cfs}, q_{cfm}, q_{cfh}, q_{cfd}	Liquid volumetric flow rate at *flowing* conditions	ft³/s, ft³/min, ft³/h, ft³/24h	
q_{cms}, q^*_{cmm}, q^*_{cmh}, q^*_{cmd}	Liquid volumetric flow rate at *flowing* conditions		m³/s, m³/min, m³/h, m³/24h
q_{lps}, q^*_{lpm}, q_{lph}, q^*_{lpd}	Liquid volumetric flow rate at *flowing* conditions		L/s, L/min, L/h, L/24h
q^*_{KPS}, q^*_{KPM}, q^*_{KPH}, q^*_{KPD}	Liquid and gas (vapor) mass flow rate		kg/s, kg/min, kg/h, kg/24h
q_{PPS}, q_{PPM}, q_{PPH}, q_{PPD}	Liquid and gas (vapor) mass flow rate	lb$_m$/s, lb$_m$/min, lb$_m$/h, lb$_m$/24h	
q_{BPS}, q_{BPM}, q_{BPH}, q_{BPD}	*Liquid* volumetric flow rate at $T_F = 60°F$, $p_b = 14.696$ psia	bbl/s, bbl/min, bbl/h, bbl/24h	
q_{CFS}, q_{CFM}, q_{CFH}, q_{CFD}	*Liquid* volumetric flow rate at $T_F = 60°F$, $p_b = 14.696$ psia	ft³/s, ft³/min, ft³/h, ft³/24h	
q_{GPS}, q_{GPM}, q_{GPH}, q_{GPD}	*Liquid* volumetric flow rate at $T_F = 60°F$, $p_b = 14.696$ psia	gal/s, gal/min, gal/h, gal/24h	
q^*_{LPS}, q^*_{LPM}, q^*_{LPH}, q^*_{LPD}	*Liquid* volumetric flow rate at $T_{°C} = 15.56°C$, $p_b = 101.3$ kPa		L/s, L/min, L/h, L/24h

Table 3.3 Subscripting for Flow Rate and Total Flow (Continued)

Symbols	Meaning	Units	
		U.S.	SI
	Flow rate		
q_{SCFS}, q_{SCFM}, q_{SCFH}, q_{SCFD}	Standard gas (vapor) volumetric flow rate at *ISO 5024 base*, $T_b = 518.67°R$, $p_b = 14.69595$	ft³/s, ft³/min, ft³/h, ft³/24h	
$(q_{SCFS})_b$, $(q_{SCFM})_b$, $(q_{SCFH})_b$, $(q_{SCFD})_b$	Standard gas (vapor) volumetric flow rate at *selected* base temperature and pressure	ft³/s, ft³/min, ft³/h, ft³/24h	
q^*_{SCMS}, q^*_{SCMM}, q^*_{SCMH}, q^*_{SCMD}	Standard gas (vapor) volumetric flow rate at *ISO 5024* base, $T_{KB} = 288.15$ K, $p^*_b = 101.325$ kPa		m³/s, m³/min, m³/h, m³/24h
$(q^*_{SCMS})_b$, $(q^*_{SCMM})_b$, $(q^*_{SCMH})_b$, $(q^*_{SCMD})_b$	Standard gas (vapor) volumetric flow rate at *selected* base temperature and pressure		m³/s, m³/min, m³/h, m³/24h
q^*_{SLPS}, q^*_{SLPM}, q^*_{SLPH}, q^*_{SLPD}	Standard gas (vapor) volumetric flow rate at *ISO 5024* base, $T_{Kb} = 288.15$ K, $p^*_b = 101.325$ kPa		L/s, L/min, L/h, L/24h
$(q^*_{SLPS})_b$, $(q^*_{SLPM})_b$, $(q^*_{SLPH})_b$, $(q^*_{SLPD})_b$	Standard gas (vapor) volumetric flow rate at *selected* base temperature and pressure		L/s, L/min, L/h, L/24h
	Total flow		
Q_{acf}	Gas (vapor) total volume	ft³	
Q^*_{acm}	Gas (vapor) total volume		m³

Symbol	Description	Units
Q_{bbl}	Liquid total volume at flowing conditions	bbl
Q_{cf}	Liquid total volume at flowing conditions	ft^3
Q^*_{cm}	Liquid total volume at flowing conditions	m^3
Q_{gal}	Liquid total volume at flowing conditions	gal
Q^*_l	Liquid total volume at flowing conditions	L
Q^*_{kg}	Total mass—liquid, gas (vapor)	kg
Q_{lbm}	Total mass—liquid, gas (vapor)	lb$_m$
Q_{BBL}	Liquid total volume at $T_F = 60°F$, $p_b = 14.696$ psia	bbl
Q^*_L	Liquid total volume at $T_{°C} = 15.56°C$, $p^*_b = 101.3$ kPa	L
Q_{SCF}	Gas (vapor) total volume at *ISO 5024 base*, ft^3 $T_b = 518.67°R$, $p_b = 14.69595$ psia	ft^3
$(Q_{SCF})_b$	Gas (vapor) total volume at *selected* base temperature and pressure	ft^3
Q^*_{SL}	Gas (vapor) total volume at *ISO 5024 base*, $T_{Kb} = 288.15$ K, $p^*_b = 101.325$ kPa	L
$(Q^*_{SL})_b$	Gas (vapor) total volume at *selected* base temperature and pressure	L
Q^*_{SCM}	Gas (vapor) total volume at *ISO 5024 base*, $T_{Kb} = 288.15$ K, $p^*_b = 101.3245$ kPa	m^3
$(Q^*_{SCM})_b$	Gas (vapor) total volume at *selected* base temperature and pressure	m^3

Liquid flow rates are almost always calculated in volume units (gallons per minute), and gas flows in standard cubic feet per hour. Vapor flows, such as of steam and ammonia, are usually calculated in mass units of pounds-mass per hour.

Fundamental Volumetric Flow-Rate Unit

In the English engineering system, the fundamental volume flow-rate unit is the cubic foot per second. Under the assumption that the velocity profile is one-dimensional and is represented by an average value (see Fig. 3.21), the flow-rate may be calculated as

$$q_{\text{cfs}} = A_p \overline{V}_f = \frac{\pi}{4}\left(\frac{D}{12}\right)^2 \overline{V}_f = 5.454154 \times 10^{-3} D^2 \overline{V}_f \tag{3.46}$$

where D is in inches, and $\overline{V}_f$ is in feet per second. In the SI system the corresponding equation is

$$q^*_{\text{cms}} = A^*_p \overline{V}^*_f = \frac{\pi}{4}\left(\frac{D^*}{1000}\right)^2 \overline{V}^*_f = 7.853982 \times 10^{-7} D^{*2} \overline{V}^*_f \tag{3.47}$$

where D^* is in millimeters, and $\overline{V}^*_f$ in meters per seconds.

Relationship between Mass and Volumetric Units

Flowmeters may be grouped in four broad classes, depending on their principle of operation:

1. Velocity-measuring (vortex, turbine, ultrasonic, magnetic)

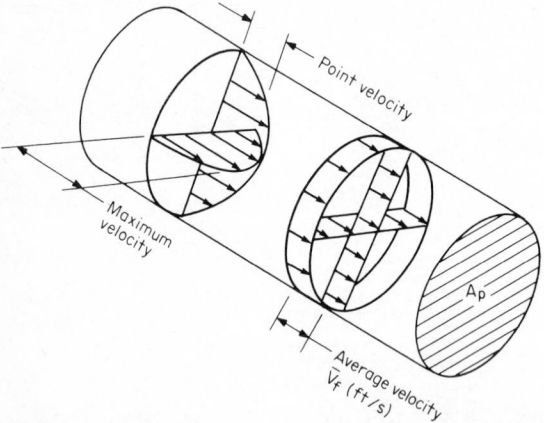

Figure 3.21 Average velocity used to derive volumetric flow-rate unit.

2. Discrete volume-measuring (positive displacement)

3. Velocity- and density-dependent (orifice, nozzle, venturi)

4. True mass flowmeters

Depending on the desired flow-rate unit, mass-flow continuity may be used to derive an equation relating a mass flow to a volumetric flow or base to flowing volume units. The relationship between the volumetric and mass flow-rate units for a density measurement is

$$q_{PPS} = \rho_f q_{\text{acfs}} \tag{3.48}$$

for the U.S. system, and

$$q_{KPS}^* = \rho_f^* q_{\text{a cms}}^* \tag{3.49}$$

for the SI system.

If density is calculated using the pvT equation, substitution of Eq. (2.10) into Eq. (3.48) gives the relationship for U.S. units as

$$q_{PPS} = 2.698825 \frac{GP_f}{Z_f T_f} q_{\text{acfs}} \tag{3.50}$$

A similar substitution of Eq. (2.11) into Eq. (3.49) yields the SI-unit relationship

$$q_{KPS}^* = 3.483407 \frac{GP_f^*}{Z_f T_k} q_{\text{a cms}}^* \tag{3.51}$$

Flowing Gallons per Minute

The gallon is a volumetric unit defined as exactly 0.13368056 ft³ (0.003785412 m³). The flow rate in gallons per minute at flowing conditions is then defined in terms of the flowing volumetric flow rate in cubic feet per second as

$$q_{\text{gpm}} = \frac{60}{0.13368056} q_{\text{cfs}} = 448.8312 q_{\text{cfs}} \tag{3.52}$$

and in terms of the average pipeline velocity as

$$q_{\text{gpm}} = 2.447994 D^2 \overline{V}_f \tag{3.53}$$

Base Gallons per Minute

In the United States, liquid flow rates are often calculated in either gallons or barrels per unit time, referred to a base temperature of 60°F (15.6°C) rather than at the flowing temperature. ISO Standard 5024 (1976) on petroleum liquids adopts 59°F (15°C) as the reference temperature, and it is expected that this will eventually replace 60°F (15.6°C).

A flowmeter can be scaled to either base or actual volumetric flow-rate units. The relationship between actual and base gallons is illustrated in Fig. 3.22. In Fig. 3.22a, the flowmeter is scaled to actual gallons per minute, and the tank level

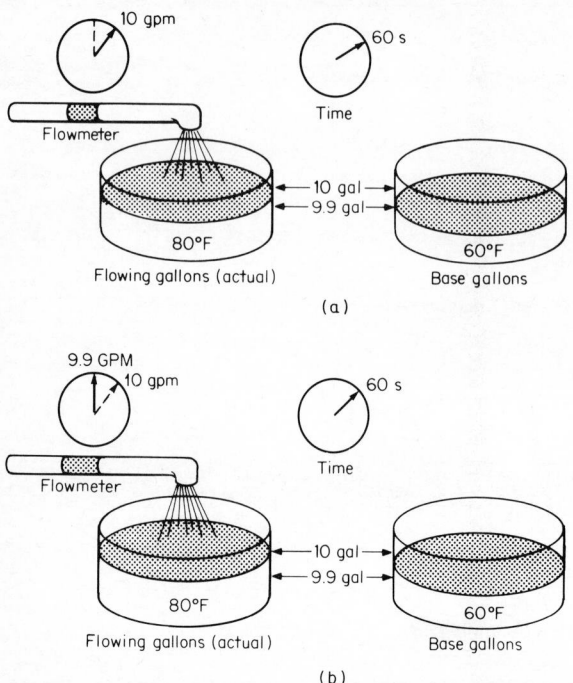

Figure 3.22 Flowmeters scaled to indicate flowing or base volumetric units. (a) Flowmeter scaled to flowing gallons per minute q_{gpm}. (b) Flowmeter scaled to base gallons per minute q_{GPM}.

will indicate the same volume as the integrated flowmeter's output. This volume is, however, lower when the liquid temperature is reduced to a base of 60°F (15.6°C) (the values shown are exaggerated for emphasis). In Fig. 3.22b the flowmeter is scaled to indicate base volume. In this case, for a temperature higher than the base temperature, the tank level will always be higher than the integrated meter output. In both cases the base volume remains the same, provided the liquid has the same density at 60°F (15.6°C).

Equating the mass flow at flowing temperature to the mass flow rate at base temperature gives the following relationship between flowing and base volumes:

$$q_{PPS} = \rho_f q_{cfs} = \rho_b q_{CFS} \tag{3.54}$$

or, the base volume may be defined via the ratio of flowing to base density as

$$q_{CFS} = \frac{\rho_f}{\rho_b}\, q_{cfs} \qquad q_{CMS} = \frac{\rho_f^*}{\rho_b^*}\, q_{cms}^* \tag{3.55}$$

Equation (3.55) is the basic equation that relates the measured flow rate to any base volume, in both U.S. and SI units and for both liquids and gases (vapors).

For base gallons per minute, the equation can be written in terms of density or specific gravity as

$$q_{GPM} = \frac{\rho_f}{\rho_b} q_{gpm} = \frac{G_f}{G_b} q_{gpm} = \frac{F_p G_F}{G_b} q_{gpm} \tag{3.56}$$

Provided consistent volume and time units are used, any volume units, such as barrels, cubic meters, or liters, can be used in Eq. (3.56) to convert between volume at flowing conditions and a selected base volume.

The Standard Cubic Foot (Meter)

In the United States, the standard cubic foot is the most commonly used gas volume unit. It is standardized in ISO 5024 (1976) for petroleum gas (natural gas) as the volume the flowing gas would occupy at a pressure of 101.3250 kPa (14.69595 psia) and a temperature of 15°C (59°F). However, other pressure and temperature bases are used for natural gas, oxygen, and nitrogen volumes. The base selected depends on the industry, the country, long-term usage, and often contractual requirements.

If density is measured, the relationship between actual and standard volumetric flow at a selected base is defined by Eq. (3.55) as

$$(q_{SCFS})_b = \frac{\rho_f}{\rho_b} q_{acfs} = \frac{\rho_f}{\rho_b} q_{cfs} \tag{3.57}$$

In this equation, the lowercase subscript acfs has been added because gas flows are commonly referred to as *actual* cubic feet rather than as cubic feet, although technically there is no difference.

In the SI system, the flow in standard cubic meters per second at a selected base is calculated as

$$(q^*_{SCMS})_b = \frac{\rho^*_f}{\rho^*_b} q^*_{acms} \tag{3.58}$$

If the density is calculated from the pvT equation, the substitution of Eq. (2.10) into Eq. (3.57) gives the selected-base-volume flow rate as

$$(q_{SCFS})_b = \frac{Z_b}{Z_f} \frac{T_b}{T_f} \frac{p_f}{p_b} q_{acfs} \tag{3.59}$$

for U.S. units, and

$$(q^*_{SCMS})_b = \frac{Z_b}{Z_f} \frac{T_{Kb}}{T_K} \frac{p^*_f}{p^*_b} q^*_{acms} \tag{3.60}$$

for SI units.

At the ISO Standard 5024 base of 101.325 kPa (14.69595 psia) and 15°C (59°F), Eqs. (3.59) and (3.60) become

$$q_{SCFS} = 35.29340 \frac{Z_b}{Z_f} \frac{p_f}{T_f} q_{acfs} \tag{3.61}$$

for U.S. units, and

$$q^*_{\text{SCMS}} = 2.843819 \frac{Z_b}{Z_f} \frac{P^*_f}{T_K} q^*_{\text{acms}}$$ (3.62)

for SI units.

Primary Constants and Standard Conditions for Gas Flows

To make exact comparisons among mass-flow, flowing-volume, and standard-volume flow rates, it is necessary to specify the *standard* conditions selected for pressure, temperature, air density, and the primary constants used in the gas density equation. Table 3.4 lists those that will be used for gases in this handbook in the development of the engineering equations.

Table 3.4 ISO Standard Conditions and Primary Constants for Developing Gas Flow Equations

STANDARD CONDITIONS†	
Pressure	14.69595 psia (101.325 kPa)
Temperature	59°F (518.67°R, 15°C, 288.15 K)

PRIMARY CONSTANTS	
Universal gas constant‡	10.73151 psia · ft³/(lb$_m$ · mol · °R) [8.31441 J*/(g · mol · K)]
Acceleration of gravity g_0	32.17405 ft/s² (9.806650 m/s²)

AIR AT STANDARD CONDITIONS	
Compressibility Z§	0.9995824
Molecular weight‡	28.96247 lb$_m$/(lb$_m$ · mol)
Density (dry)	0.0765002 lb$_m$/ft³ (1.225416 kg/m³)

†ISO 5024 (1976).

‡Jones (1978).

§Interpolated to seventh place using Redlich-Kwong equation of state.

EXAMPLE 3.7

The gasoline of Example 2.15 is flowing in a 2-in (50-mm) schedule-80 pipe. The measured flow rate is 100 gal/min (6.31 L/s) at base conditions. Determine (*a*) the flow rate in actual gallons per minute, (*b*) the flow rate in actual cubic feet per second, (*c*) the mass flow rate in pounds per day, and (*d*) the average pipeline velocity in feet per second. (See Examples 9.1 and 9.3 for the sizing of an orifice for these conditions.)

a. Actual gallons per minute

From Examples 2.12 and 2.15,

$$G_F = 0.7255 \qquad G_b = 0.7359 \qquad F_p = 1.0088$$

Substitution into Eq. (3.56), rewritten to solve for flowing gallons per minute, gives

$$q_{gpm} = \frac{G_b}{F_p G_F} q_{GPM} = \frac{0.7359}{(1.0088)(0.7255)}(100) = 100.55 \text{ gal/min}$$

b. Actual cubic feet per second

With Table C.1, the conversion equation can be written as

$$q_{cfs}(2.831685 \text{ E} - 02) = q_{cms} = q_{gpm}(6.309020 \text{ E} - 05)$$

$$q_{cfs} = \frac{6.309020 \times 10^{-5}}{2.831685 \times 10^{-2}} q_{gpm} = (2.228009 \times 10^{-3})(100.55) = 0.22403 \text{ ft}^3/\text{s}$$

or Eq. (3.52) may be rearranged to yield

$$q_{cfs} = \frac{q_{gpm}}{448.8312} = \frac{100.55}{448.8312} = 0.22403 \text{ ft}^3/\text{s}$$

c. Mass flow rate in pounds-mass per day

From Eq. (3.48), $q_{PPS} = \rho_f q_{cfs}$. The flowing density ρ_f is, by Eqs. (2.106) and (2.79),

$$\rho_f = F_p \rho_F = 62.3663 F_p G_F = (1.0088)(0.7255)(62.3663)$$
$$= 45.6449 \text{ lb}_m/\text{ft}^3$$

then,
$$q_{PPD} = (24)(3600)\rho_f q_{cfs} = (24)(3600)(45.4562)(0.22403)$$
$$= 883{,}512 \text{ lb}_m/\text{day}$$

d. Average pipeline velocity

From Table B.1, for 2-in Schedule-80 pipe, $D = 1.939$; then, by Eq. (3.46) rearranged to solve for $\overline{V}_f$,

$$\overline{V}_f = \frac{q_{cfs}}{(\pi/4)(D/12)^2} = \frac{0.22403}{(\pi/4)(1.939/12)^2} = 10.93 \text{ ft/s}$$

EXAMPLE 3.8

Natural gas is flowing at 340 standard m³/h (ISO Standard 5024 base) in a 52.50-mm (2.067-in) pipe. For the design information given below, determine (*a*) the actual flow rate in actual cubic meters per hour, (*b*) the mass flow rate in kilograms per hour, (*c*) the average pipeline velocity in meters per second, and (*d*) the standard cubic feet per hour. The design information is

$$p_f^* = 184 \text{ kPa} \qquad T_K = 300 \text{ K} \qquad G = 0.652 \qquad Z_f = 0.9995 \qquad D^* = 52.50 \text{ mm} \qquad Z_b = 1.0$$

a. Actual cubic meters per hour

Equation (3.58) can be rewritten for an hourly basis as

$$q^*_{\text{acmh}} = \frac{\rho^*_b}{\rho^*_f} q^*_{\text{SCMH}}$$

where ρ^*_f and ρ^*_b are calculated at flowing and base conditions using Eq. (2.11) as

$$\rho^*_f = 3.483407 \frac{Gp^*_f}{Z_f T_K} = 3.483407 \frac{(0.652)(184)}{(0.9995)(300)} = 1.39369 \text{ kg/m}^3$$

and

$$\rho^*_b = 3.483407 \frac{Gp^*_b}{Z_b T_{Kb}} = 3.483407 \frac{(0.652)(101.325)}{(1.0)(288.15)} = 0.79864 \text{ kg/m}^3$$

Substitution then yields

$$q^*_{\text{acmh}} = \frac{0.79864}{1.39369} \, 340 = 194.83 \text{ m}^3/\text{h}$$

b. Mass flow rate

Equation (3.49) in hourly flow-rate units is

$$q^*_{\text{KPH}} = \rho^*_f q^*_{\text{acmh}}$$

$$= (1.39369)(194.83) = 271.54 \text{ kg/h}$$

c. Average pipeline velocity

Rearranging Eq. (3.47) and substituting yields

$$\overline{V}^*_f = \frac{q^*_{\text{cms}}}{(\pi/4)(D^*/1000)^2}$$

$$= \frac{194.83/3600}{(\pi/4)(52.50/1000)^2} = 25.00 \text{ m/s}$$

d. Standard cubic feet per hour

Table C.1 shows the conversion $(\text{ft}^3/\text{s})(2.831685 \text{ E} - 02) = \text{m}^3/\text{s}$. Then

$$q_{\text{SCFH}} = \frac{q^*_{\text{SCMH}}}{2.831685 \times 10^{-2}} = \frac{340}{2.831685 \times 10^{-2}}$$

$$= 12{,}007 \text{ standard ft}^3/\text{h}$$

REFERENCES

ANSI/API 2530, *Orifice Metering of Natural Gas,* American Gas Association, New York, 1978.

Bean, H. S.: "Correction Factors," *Gas,* vol. 27, pp. 48–50, August 1957.

Benedict, R. P.: *Fundamentals of Temperature, Pressure and Flow Measurement,* Wiley, New York, 1977.

"The International Practical Temperature Scale of 1968," a committee report, *Metrologia,* vol. 5, p. 2, April 1969.

ISO Standard 5024, *Petroleum Liquids and Gases—Measurement—Standard Reference Conditions,* ISO, Geneva, 1976.

Jones, F. E.: "The Air Density Equation and the Transfer of the Mass Unit," *J. Res. Nat. Bur. Stand.,* vol. 83, no. 5, pp. 419–428, 1978.

Rayle, R. E.: "Influence of Orifice Geometry on Static Pressure Measurements," ASME Paper 59-A-234, 1959.

4

ACCURACY

Errors are inherent in all measurements. Even under ideal conditions, repeated measurements of a reference standard, such as a standard mass placed on a scale, will give slightly different instrument readings. These errors may be random or biased, stem from the reference standard, or change with time. Environmental factors, such as temperature, pressure, and relative humidity, can also affect measurements.

Accuracy, overall uncertainty, systematic error, bias error, repeatability, hysteresis, and reproducibility are but a few of the terms used by standardizing organizations and trade associations to define the many sources of error. Although called by different names, many of these have the same meaning. Because of the widespread use and understanding of the terms *accuracy, bias error,* and *precision,* they will be used exclusively in this book. Accuracy, which is the combination of bias and precision errors, will be defined in accordance with the author's interpretations of the available flow measurement uncertainty standards and current draft documents, including ANSI (1982), ISO/TC30 (1981), and ISO 5168 (1979).

The measurement of flow rate requires not only a primary device but usually additional instruments to measure temperature, pressure, density, differential pressure, etc. To determine flow-rate measurement accuracy, the accuracy of the primary device must be combined with the individual accuracies of these additional measuring devices and then properly weighed in the flow-rate calculation.

The following brief introduction to the subject presents a simple procedure for calculating overall flow-rate accuracy. The combining of errors by statistical methods is now widely accepted. For engineering purposes, it is difficult to apply the rigorous and complex statistical approaches presented in the standards for the purpose of evaluating second- and third-order effects. In most cases a suitable accuracy statement can be derived through the use of simplifying assumptions, with the more complete determination reserved only for contractual use.

FUNDAMENTAL CONSIDERATIONS

The International Organization for Legal Metrology (OIML, 1971) defines the accuracy of a measuring instrument as "the quality which characterizes the ability of a measuring instrument to give indications approximating" the true value of the quantity measured. A footnote to this definition reads: *"Accuracy is an overall quality of a measuring instrument from the point of view of errors. Accuracy is greater when the indications are closer to the true value."*

By definition, an accuracy of 99 percent is an inaccuracy of 1 percent, and, as accuracy improves, inaccuracy is reduced. For convenience, however, inaccuracy has almost always been referred to as "accuracy."

Accuracy statements for industrial and laboratory instruments are usually given as a percentage of either the upper range value (URV; formerly called full-scale value) or the true value, rather than in terms of the units of measurement. Since percentages are easier to use, particularly in the calculation of flowmeter accuracy, they will be utilized here to derive the accuracy equation.

At a single measurement point, there are three sources of error. The average of many readings might be offset from the true value (bias error); the readings may be randomly scattered about the offset (precision error); and one reading might fall well outside the majority of the readings (illegitimate, or outlier error). It is the combination of the first two that establishes the *accuracy* or quality of the instrument.

Two methods are commonly used to specify accuracy over a range of values. In the first, a calibration curve is plotted over the intended range of use, and each reading is then corrected by reference to a correction curve. If desired, a mathematical correction-factor equation is derived by regression analysis, and subsequent corrections are calculated by computer. In the second method, an accuracy value based on typical calibration curves for production units is given. While the first method has the advantage of eliminating bias, the considerable expense of individually calibrating each instrument is seldom warranted.

Flowmeters and auxiliary instruments are usually assigned two accuracies. The first is called the *reference-condition accuracy* and is obtained under laboratory conditions. A range of temperatures, pressures, supply voltages, flow conditions, etc., is specified over which the user can expect the device to be within a specified reference-accuracy envelope. The second refers to the effect of influence quantities on the reference accuracy. It applies when the device is used outside the referenced range for temperature, relative humidity, static pressure, flow-pulsation level, etc., and is usually specified as a *limit of error*, such as ± 1 percent per 100°F. For example, a pressure-measuring device would be within its ± 1 percent reference-accuracy envelope if used in an air-conditioned room, but a bias error, and possible increased precision error, might be observed if it were installed in the harsh temperature environment of a steel mill.

MEASUREMENT ACCURACY

Accuracy Components

As an illustration of the two separate errors that define accuracy, consider three temperature-measuring devices that have been placed in boiling water and have provided the readings given in Table 4.1. The three sets of temperature data are also shown plotted on targets in Fig. 4.1, where the bull's-eye is 212°F, the *true* value.

Data for the first device (Fig. 4.1a) are tightly clustered about an average value, but offset from the center. The difference between the average and true values is the instrument's bias error, and the scatter about the average is the precision error. Note that one point is significantly different from the rest. This is an outlier, because it is not part of the normal population. Statistical methods are available for determining whether to reject it or include it in calculations.

Table 4.1 Calibration Data for Three Temperature-Measuring Devices

| Measurement | Reading | | |
	Device 1	Device 2	Device 3
1	210.1	210	212.0
2	210.0	213	212.2
3	209.8	212	212.1
4	210.2	214	212.0
5	209.9	210	211.8
6	200.0†		

†Possible illegitimate error or an outlier.

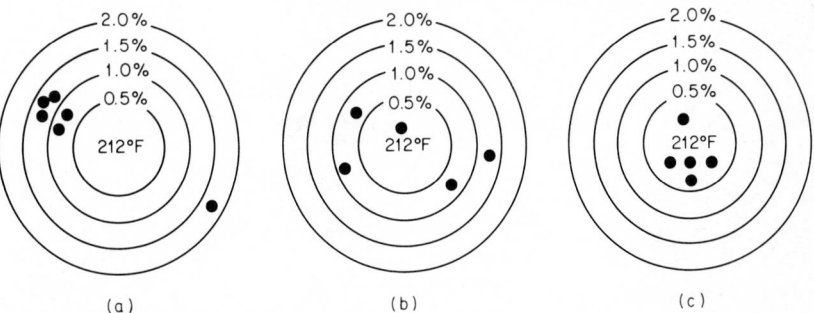

Figure 4.1 Targets showing data scatter with respect to true value. (a) Device 1. (b) Device 2. (c) Device 3.

A bias error is directional and must be added or subtracted from the instrument's reading to obtain the true value. A precision error is random about the bias. Precision can be improved only by selecting another measuring device, but bias error, if known, can be eliminated by correction. This particular device is considered precise but biased; it is not considered accurate without correction.

The second device (Fig. 4.1b) has a wide scatter about the bull's-eye. While the average is centered, indicating no significant bias, the inability to read values close to each other makes the device imprecise, or of poor precision; the chances of reading 212°F are quite poor. This device would not be considered accurate because of the large precision error.

The third device (Fig. 4.1c) is both precise and unbiased. The average of the five readings is 212°F, and there is little scatter. Because this device is both precise and unbiased, it is considered accurate.

This example suggests three distinct cases regarding accuracy:

1. Bias error is not negligible, but precision is good (device 1).

2. Bias error is negligible, but precision is poor (device 2).

3. Bias error is small and precision is good (device 3); this is an accurate device.

Depending on how sensitive the flow-rate calculation is to the particular temperature that is being measured, device 1 could be used for control; devices 1 and 2 could be used for measurement and control if the temperature sensitivity of the flow calculation were negligible; and device 3 could be used for both measurement and control. In all cases, device 3 is to be preferred, unless economic considerations outweigh measurement needs.

Precision and bias errors must first be treated separately, and then combined to obtain the final accuracy value. Precision errors follow the laws of chance, in that they are purely random in nature. The chance (probability) of reading either 211°F or 213°F when a particular thermometer is placed in boiling water may be only 5 percent, but the chance of reading 212°F may be 20 percent. The chances of reading intermediate values increase as the readings approach the average; these chances follow the familiar bell-shaped curve of gaussian (normal) distribution, as illustrated in Fig. 4.2 for device 1. Using statistically derived

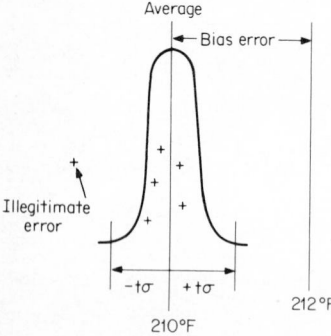

Figure 4.2 Data for device 1 shown on a bell curve.

equations, a range within which 95 percent of the readings would be expected to fall can be calculated. The 95 percent confidence range (or level) is used in the ISO 5168 (1979) and ANSI(MFFCC) (1981) flow-measurement standards.

Bias errors do not follow the laws of chance, and no confidence level can be derived for them. There are two types of bias errors, variable and constant. Typical sources are as follows:

Sources of Variable Bias Errors

1. Drift in the voltage of a standard cell used periodically to standardize a potentiometer

2. Progressive wear on the linkage of a recorder (a time-dependent wear function)

3. Progressive wear of the orifice edge in a dirty stream

4. An uncorrected zero shift in a differential-pressure transmitter (or other measuring device)

Sources of Constant Bias Errors

1. An unknown bias error in a reference standard that is used to calibrate secondary devices

2. Deadweight-tester weights not corrected for local gravity

3. Use of the assumption of an ideal gas

4. A scale used to determine mass but not corrected for liquid buoyancy

5. A flowmeter installed too close to an elbow or other flow disturbance

6. Incorrect measurement of an orifice or pipe bore

A source of error that can never be explained, calculated, or estimated (or overlooked) is the outlier (sometimes called human error, illegitimate error, or blunder) that occurs when an operator records 221°F instead of 212°F, or when the water was not at the boiling point as readings were recorded, or when a known source of bias error was not considered—such as not correcting the boiling point of water for atmospheric pressure. Illegitimate errors should always be reported, since they might be indicative of a meaningful phenomenon. But, for engineering purposes, outliers must be discarded after a reasonable effort to locate their source. Several statistical methods for rejecting outliers are covered in ISO 5168 (1979) and ANSI(MFFCC) (1981).

Calculating Precision

Precision is the ability of a measuring device to give (repeat) the same readings for the same input. Because variations in readings are random, statistical theory can be used to calculate a precision value at the 95 percent confidence level from the variance of the readings.

To calculate the precision, an approximate standard deviation is first calculated for the readings (data points). This is multiplied by a correction factor derived from Student's t distribution for the 95 percent confidence level.

This factor depends on the number of data points and, for an infinite number of points, approaches 1.96. Student's t values are plotted in Fig. 4.3 versus sample size, and the values are given in Table 4.2.

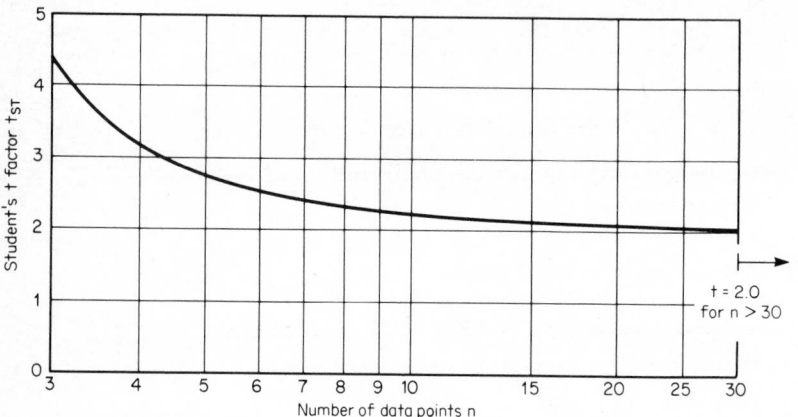

Figure 4.3 Student's t for the 95 percent confidence level.

The standard deviation, in percent, is

$$\sigma = \left[\frac{\Sigma(I_{\%})_i^2}{n-1} \right]^{1/2} \tag{4.1}$$

where n is the number of data points, and $(I_{\%})_i$ is the percentage deviation of each data point I_i from the average $\bar{I}$ of the indicated readings:

$$(I_{\%})_i = \frac{I_i - \bar{I}}{\bar{I}} 100 \tag{4.2}$$

with
$$\bar{I} = \frac{\Sigma I_i}{n} = \frac{I_1 + I_2 + I_3 + \cdots}{n} \tag{4.3}$$

The precision at the 95 percent confidence level is then

$$\sigma_P = t_{ST}\sigma \tag{4.4}$$

where t_{ST} is found in Table 4.2 or Fig. 4.3.

Table 4.2 Two-Tailed Student's t Values for 95 Percent Confidence Level

Number of data points n	Degrees of freedom ($n - 1$)	t_{ST}	Number of data points n	Degrees of freedom ($n - 1$)	t_{ST}
2	1	12.706	18	17	2.110
3	2	4.303	19	18	2.101
4	3	3.182	20	19	2.093
5	4	2.776	21	20	2.086
6	5	2.571	22	21	2.080
7	6	2.447	23	22	2.074
8	7	2.365	24	23	2.069
9	8	2.306	25	24	2.064
10	9	2.262	26	25	2.060
11	10	2.228	27	26	2.056
12	11	2.201	28	27	2.052
13	12	2.179	29	28	2.048
14	13	2.160	30	29	2.045
15	14	2.145	31	30	2.042
16	15	2.131	32	31	2.0
17	16	2.120			

Calculating Bias Error

The difference between the most likely value (the average) and the *true* value, as established by a reference standard, is the bias error (see Fig. 4.2). While the true value is never exactly known (because of inherent reference-standard precision and bias errors), engineering practice is to assume that the reference standard has negligible error.

The difference between the average of the instrument's readings and the reference-standard reading gives the *directional bias error,* defined by

$$B = \frac{\bar{I} - I_t}{I_t} \, 100 \tag{4.5}$$

where the subscript t refers to the true, or reference-standard, value. With a positive bias, the instrument is said to be reading high, and with a negative bias reading low.

It is common to use a correction factor to adjust readings for known bias errors. This factor is derived in terms of the bias error as

$$F_B = \left(1 + \frac{B}{100} \right)^{-1} \tag{4.6}$$

and the correction for directional bias is made as follows:

$$I_t = F_B \bar{I} \tag{4.7}$$

An added complication enters the bias-error calculations because of the confidence associated with the average value. Confidence in the average of 1000 readings is higher than that for an average of only five readings. It has been shown that precision can be used to calculate a 95 percent confidence level for the average. This establishes a range centered on the average and determined as

$$\pm B = \frac{\sigma_P}{\sqrt{n}} \tag{4.8}$$

It can be seen from Eq. (4.8) that as the number of readings increases, the bias-error range $\pm B$ shrinks toward zero. For many devices, precision error is small ($\sigma_P < 0.10$ percent), and the bias-error range can be neglected in the accuracy calculation.

Accuracy at a Single Point

Accuracy is *closeness* to the true value, and it includes the effects of both precision and bias error. These are combined by considering the precision [Eq. (4.4)] and the bias-error range $\pm B$ [Eq. (4.8)] as random, adding their squares, and taking the square root of their sum. The directional bias error is then algebraically added, and the accuracy is calculated as

$$\text{Acc} = B \pm \sqrt{\sigma_P^2 + \frac{\sigma_P^2}{n}} = B \pm \sqrt{\left(1 + \frac{1}{n} \right)\sigma_P^2} \tag{4.9}$$

When the directional bias error is known, each reading should be corrected with Eq. (4.6). The accuracy calculation then reduces to

$$\text{Acc} = \pm \sqrt{\left(1 + \frac{1}{n} \right)\sigma_P^2} \tag{4.10}$$

Many devices have good precision (<0.1 percent), and for a reasonable number of data points ($n > 5$), Eq. (4.10) further reduces to

$$\text{Acc} = \pm \sigma_P \tag{4.11}$$

Equation (4.11), in which the accuracy is set equal to the precision of the device at the given test value, also includes the following assumptions:

1. The accuracy of the reference standard is five times better than that of the instrument whose accuracy is being determined.

2. All readings have been corrected for known directional bias errors.

EXAMPLE 4.1

Determine the precision, directional bias error, correction factor, bias-error range, and accuracy for device 1 in Table 4.1.

Precision

With the outlier omitted, the calculation may be tabulated as follows:

i	I_i	$(I_\%)_i$, % [Eq. (4.2)]	$(I_\%)_i^2$
1	210.1	+0.040	0.0023
2	210.0	0.0	0.0
3	209.8	−0.095	0.0090
4	210.2	+0.095	0.0090
5	209.9	−0.048	0.0023
	1050.0		0.0226

In constructing the table, Eq. (4.3) was used to compute

$$\bar{I} = \frac{\Sigma I_i}{n} = \frac{1050}{5} = 210.0$$

The standard deviation is, by Eq. (4.1),

$$\sigma = \left(\frac{\Sigma (I_\%)_i^2}{n-1}\right)^{1/2} = \left(\frac{0.0226}{5-1}\right)^{1/2} = \pm 0.0753 \text{ percent}$$

From Table 4.2 for $n = 5$, $t_{ST} = 2.776$. The precision is then, by Eq. (4.4),

$$\sigma_P = t_{ST}\sigma = \pm(2.776)(0.0753) = \pm 0.21 \text{ percent}$$

Directional bias error

The directional bias error is calculated with Eq. (4.5):

$$B = \frac{\bar{I} - I_t}{I_t} 100 = \frac{210 - 212}{212} 100 = -0.94 \text{ percent}$$

Correction factor

The correction factor can now be calculated with Eq. (4.6):

$$F_B = \left(1 + \frac{-0.94}{100} \right)^{-1} = 1.0095$$

Bias-error range

The bias-error range is, from Eq. (4.8),

$$\pm B = \frac{\pm 0.21}{\sqrt{5}} = \pm 0.094 \text{ percent}$$

Accuracy

The accuracy is, by Eq. (4.9),

$$Acc = -0.94 \pm \sqrt{\left(1 + \frac{1}{5} \right)(0.21)^2} = -0.94 \pm 0.23$$
$$= -0.7 \text{ to } -1.2 \text{ percent}$$

If each reading had been multiplied by the correction factor F_B, the accuracy would be calculated with Eq. (4.10) as

$$Acc = \sqrt{\left(1 + \frac{1}{5} \right)(0.21)^2} = \pm 0.23 \text{ percent}$$

ACCURACY OVER A RANGE

If a temperature-measuring device is calibrated with different inputs, directional bias errors and precision errors similar to those shown in Fig. 4.4 are obtained. A *calibration curve* may be drawn through the directional-bias-error points, and each reading corrected to the true value I_t. Then the calculation of accuracy reduces to the calculation of precision as in Eq. (4.11).

To establish a calibration curve for each instrument is usually economically unjustified. Instead, most instruments are type-tested to establish a reference *accuracy envelope* that incorporates precision, directional bias, and bias-error range over a specified range of the measured variable. The limits of the envelope are expressed as a percentage of the upper range value (URV) or as a percentage of the reading. Accuracy envelopes are specified for reference conditions, and they apply within stated limits on ambient temperature, input power, relative humidity, etc.

Influence Quantities

If an installation's ambient temperature, relative humidity, etc., are outside the stated limits of the reference conditions for an accuracy envelope, additional bias

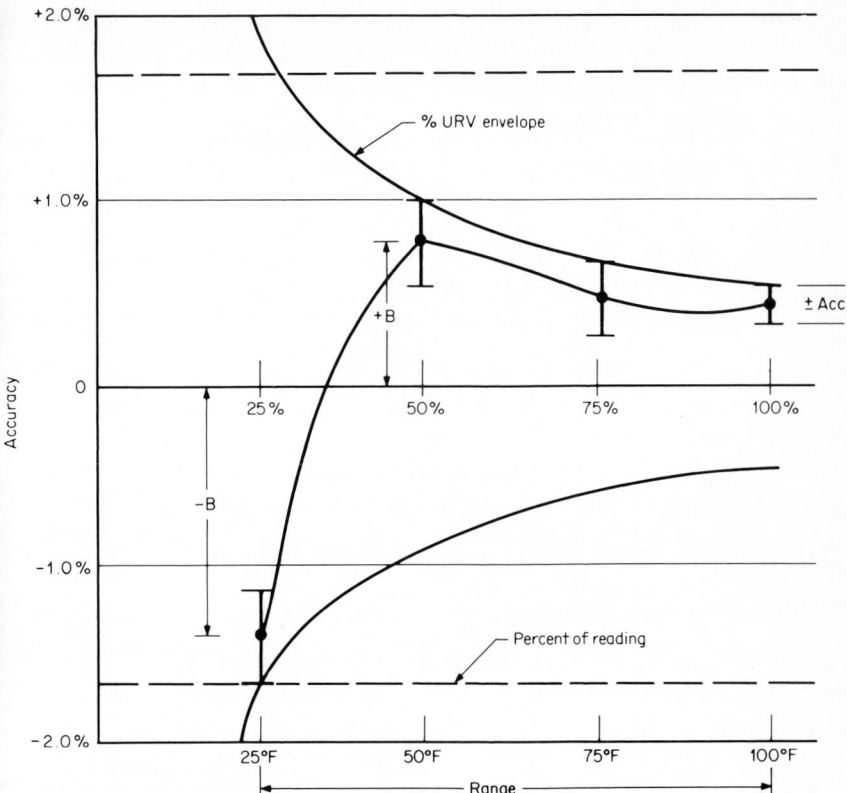

Figure 4.4 Accuracy envelopes over a range.

or precision errors may result. These sources of errors are referred to as *influence quantities,* and they give rise to the most difficult errors to assess, because of their possible interrelationships. For example, an additional bias of +0.5 percent may be indicated on the basis of a separate test for temperature, and a −0.5 percent bias based on a similar relative-humidity test. Combined, these would indicate no additional error. However, if both effects are tested for simultaneously, the error might be −0.25 percent. In many cases, the effect of a particular influence quantity is to directionally bias the reading and not to significantly affect the precision. In some cases, these influences change the instrument's zero and have negligible effect on span. By zeroing under flowing conditions, these errors are then minimized.

The ANSI(MFFCC) (1982) standard and ISO/TC30 (1982) draft of *Flow Measurement Uncertainty* suggest that, "When it is known that a bias results from a particular cause, special calibrations may be performed allowing the cause to perturbate through its complete range to determine the range of bias." They further suggest, "In practice most measurements will have many sources of bias limits from calibration (reference conditions), data acquisition and data reduction. As long as none of them are extremely large (10 to 1) relative to the others, the quadrature sum (root-sum-square) is a very good approximation of the combination of such errors."

For instruments, a limiting specification is usually given over a range, such as "±1.25 percent of the URV per 100°F temperature change." Within this range, the ±1.25 percent may be assumed to be linearly related to the difference between reference and operating temperatures. And, if it is assumed that each influence quantity is independent of other influence quantities, the root-sum-square method can be used to combine the errors. The operating-condition accuracy is then calculated as

$$\text{Acc} = \pm (\text{Acc})_{\text{ref}} \pm \sqrt{B_T^2 + B_p^2 + B_{RH}^2 + \cdots} \qquad (4.12)$$

where the subscript ref refers to reference-condition accuracy, and the subscripts of T, p, and RH refer to the influences of temperature, pressure, and relative humidity. Other influences should be included as they may arise in a particular measurement situation.

EXAMPLE 4.2

A differential-pressure-measuring device has a reference-condition accuracy specification of ±0.2 percent of the URV. For the reference and operating conditions given below, estimate the accuracy at an ambient temperature of 100°F (37.8°C) and a line static pressure of 1000 psia (6895 kPa).

	Reference condition	Operating condition
Temperature	75 ±3°F	±1.25% for 100°F (total error)†
Pressure	14.7 psia	±0.5% to 3000 psia

†Includes a zero error which can be corrected for by "zeroing" at temperature (100°F).

The influence quantities are temperature and pressure. Their bias errors are computed with Eq. (4.5), assuming linear relationships. They are

$$B_T = \pm \frac{100 - 75}{100} 1.25 = \pm 0.31 \text{ percent URV}$$

for temperature, and

$$B_p = \pm \frac{1000 - 14.7}{3000} 0.5 = \pm 0.16 \text{ percent URV}$$

for pressure. The accuracy at operating conditions is then estimated with Eq. (4.12):

$$\text{Acc} = \pm 0.2 \pm \sqrt{(0.31)^2 + (0.16)^2} = \pm 0.55 \text{ percent URV}$$

SENSITIVITY COEFFICIENTS

General

Flow measurement often requires that a number of measurements be combined. In the case of an orifice flowmeter measuring natural gas, for example, line pressure, differential pressure, temperature, and gravitometer measurements are combined in an equation to calculate the flow rate. The form of such an equation depends on the flowmeter's principle of operation; the mathematical relationships among the measured variables establishes the *sensitivity* of the final calculation to each of these variables.

Each measurement has a specific *sensitivity coefficient* that can be derived from a mathematical relationship or estimated from curves, charts, or tables. It is the proper combination of these coefficients that allows individual accuracies to be included in an overall flow-rate accuracy value. Sensitivity coefficients are also useful in determining what effect an individual instrument's directional bias error has on the calculated flow rate and in deciding whether a more precise device is necessary.

Mathematical Considerations

When an equation is used to calculate a quantity based on measured values of two or more variables, a mathematical entity called the *total differential* can be used to determine the individual effect of each variable on the final result. If the pertinent variables are independent, then the general functional relationship for the differential producer's flow-rate equation is

$$q = f(p,T,h_w,D,d, \ldots) \tag{4.13}$$

The total differential is the sum of the partial differentials of the independent variables:

$$dq = \frac{\partial q}{\partial p}\, dp + \frac{\partial q}{\partial T}\, dT + \frac{\partial q}{\partial h_w}\, dh_w + \frac{\partial q}{\partial D}\, dD + \cdots \tag{4.14}$$

Dividing Eq. (4.14) by the flow rate q and rearranging yield

$$\frac{dq}{q} = X_p \frac{dp}{p} + X_T \frac{dT}{T} + X_{hw} \frac{dh_w}{h_w} + X_D \frac{dD}{D} + \cdots \tag{4.15}$$

where X_p, X_T, X_{hw}, X_D, ... are the sensitivity coefficients for the variables, equal to the partial derivatives in Eq. (4.14). The terms they multiply in Eq. (4.15) represent relative changes in the variables. The percentage change in flow rate may thus be calculated as the sum of the products formed when each sensitivity coefficient is multiplied by the percentage change in its variable.

The variables in Eq. (4.13) are not, in general, independent. For example, the bore of a flow nozzle is a function of temperature and the thermal-expansion factor. Including these additional functional relationships significantly complicates the calculation but has little effect on calculational accuracy. An example in ISO 5168 (1979) shows that the combined uncertainty (accuracy) differs by less than 0.01 percent between the simplified and more detailed calculations.

Determining Sensitivity Coefficients

As an illustration of the procedure for determining sensitivity coefficients, consider only the temperature, dimensional, and differential-pressure measurements used in calculating an airflow rate using a flow nozzle. Bernoulli's equation and the conservation of mass flow can be combined to give

$$q_{\text{PPH}} = \frac{F_{MC}d^2}{\sqrt{1 - (d/D)^4}} \sqrt{h_w \rho_f} \tag{4.16}$$

where F_{MC} is assumed to be constant with a known accuracy, d is the nozzle bore, and D is the pipe diameter. The density ρ_f can be calculated from the temperature T_f, pressure p_f, compressibility Z_f, and specific gravity G with Eq. (2.10):

$$\rho_f = 2.698825 \frac{Gp_f}{Z_f T_f} \tag{2.10}$$

Substituting Eq. (2.10) into Eq. (4.16) yields

$$q_{\text{PPH}} = \frac{F_{MC}d^2}{\sqrt{1 - (d/D)^4}} \sqrt{\frac{h_w}{T_f}} \tag{4.17}$$

where F_{MC} now assumes a new value that includes the specific gravity G, compressibility factor Z_f, constant 2.698825, and a constant pressure p_f value.

In Eq. (4.17), the flow equation has been reduced to a form that includes only the effects of bore d, pipe size D, temperature T_f, and differential pressure h_w on the calculated mass flow. These will be referred to in this handbook as the *measured variables*.

First consider the sensitivity of the flow rate to temperature, which will be measured in degrees Fahrenheit. If the diameters d and D and the differential pressure h_w are temporarily assumed to be constant and the total differential of Eq. (4.17) is taken with respect to temperature, the partial derivative of the mass-flow equation with respect to temperature is obtained as

$$dq_{\text{PPH}} = \frac{F_{MC}d^2\sqrt{h_w}}{\sqrt{1 - (d/D)^4}} \left(-\frac{1}{2}\right)(459.67 + T_F)^{-3/2} \, dT_F \tag{4.18}$$

Dividing both sides of Eq. (4.18) by the mass flow [Eq. (4.17)] and multiplying by T_F/T_F yield an equation of the form of Eq. (4.15), from which the temperature sensitivity coefficient is found to be

$$X_{TF} = -\frac{1}{2} \frac{|T_F|}{459.67 + T_F} \tag{4.19}$$

The absolute-value symbol in the numerator of this equation indicates that the minus sign is to be removed from negative temperatures such as $-60°F$.

The sensitivity coefficient X_{TF} is multiplied by the temperature directional bias error to calculate the temperature bias error in mass flow, and by the temperature accuracy ($\pm$) to calculate the accuracy-range contribution of the temperature device. That is, for directional bias

$$(q_{PPH})_\% = X_{TF}B_{TF} \tag{4.20}$$

and for accuracy

$$\pm(q_{PPH})_\% = X_{TF}(\text{Acc})_{TF} \tag{4.21}$$

In the same way, sensitivity coefficients may be obtained as

$$X_D = -\frac{2\beta^4}{1 - \beta^4} \tag{4.22}$$

for pipe diameter D, where $\beta = d/D$;

$$X_d = \frac{2}{1 - \beta^4} \tag{4.23}$$

for flowmeter bore d, again with $\beta = d/D$;

$$X_{hw} = \tfrac{1}{2} \tag{4.24}$$

for differential pressure h_w; and finally

$$X_{FMC} = 1 \tag{4.25}$$

if F_{MC} is not constant as assumed.

Combining Sensitivity Coefficients

Sensitivity coefficients can be used for two purposes. First, if a measurement is made and a directional bias error is later found, the directional flow-rate bias can be calculated from the sensitivity coefficients. Second, the flow-rate measurement accuracy can be determined by properly combining the individual measuring-device accuracies with the sensitivity coefficients.

Directional Bias Errors Directional bias errors are combined by summation:

$$B_q = X_{TF}B_{TF} + X_D B_D + X_d B_d + X_{hw}B_{hw} + \cdots \tag{4.26}$$

where B_q is the percent directional bias error in flow rate for directional bias errors in temperature, and B_{TF}, B_D, B_d, ... are defined similarly. No "units" subscript is shown on the flow-rate bias error because all terms in Eq. (4.26) are dimensionless (percentages); the flow rate can be measured in any units.

Accuracy If the directional bias of each instrument is corrected for, or each instrument is operating within its accuracy envelope, the accuracies of the various measuring instruments may be combined by the root-sum-square method to estimate the flow-rate measurement accuracy:

$$(\text{Acc})_q = \pm \{[X_{TF}(\text{Acc})_{TF}]^2 + [X_D(\text{Acc})_D]^2$$
$$+ [X_d(\text{Acc})_d]^2 + [X_{h_w}(\text{Acc})_{h_w}]^2 + \cdots\}^{1/2} \quad (4.27)$$

For engineering purposes, this is an extremely useful equation. It provides the combined effect of all influence quantities, including reference-condition accuracies and the sensitivities of the measured variables.

EXAMPLE 4.3

Calculate the accuracy of a determination of mass flow using a flow nozzle, given the following accuracy statements:

Measurement	Indicated	Accuracy, %
Temperature T_F	60°F	± 0.5
Differential pressure h_w	100	± 0.2
Pipe diameter D	7.981 in	± 0.2
Nozzle bore d	5.5867 in	± 0.05
Flow constant F_{MC}	55	± 0.75

The calculations may be tabulated as follows:

Source	Sensitivity coefficient X	$[X(\text{Acc})]^2$
Eq. (4.19)	$X_{TF} = -\frac{1}{2}\dfrac{\|60\|}{460 + 60} = -0.058$	$[(-0.058)\,(\pm 0.5)]^2 = 0.0008$
Eq. (4.24)	$X_{h_w} = +\frac{1}{2}$	$[(+0.5)\,(\pm 0.2)]^2 = 0.0100$
Eq. (4.22)	$X_D = -\dfrac{(2)\,(0.7)^4}{1 - (0.7)^4} = -0.632$	$[(-0.632)\,(\pm 0.2)]^2 = 0.0160$
Eq. (4.23)	$X_d = +\dfrac{2}{1 - (0.7)^4} = +2.632$	$[(+2.632)\,(\pm 0.05)]^2 = 0.0173$
Eq. (4.25)	$X_{FMC} = +1.0$	$[(1.0)\,(\pm 0.75)]^2 = \underline{0.5625}$
		0.6066

The accuracy is then, from Eq. (4.27),

$$(\text{Acc})_q = \pm\{\Sigma[X(\text{Acc})]^2\}^{1/2} = \pm(0.6066)^{1/2} = \pm 0.78 \text{ percent}$$

EXAMPLE 4.4

For Example 4.3, the true values tabulated below were determined. Compute the correction factor to be applied to the flow rate that was calculated from the indicated values.

Variable	Indicated	True	Bias error, %
Temperature T_F	60°F	65°F	−7.69
Differential pressure h_w	100	99	+1.01
Pipe diameter D	7.981 in	7.9012 in	+1.01
Nozzle bore d	5.5867 in	5.6006 in	−0.25

The calculations may be tabulated as follows:

Variable	Sensitivity coefficient X†	Bias error B‡	XB
Temperature	−0.058	−7.69	+0.444
Differential pressure	+0.5	+1.01	+0.505
Pipe diameter	−0.632	+1.01	−0.638
Nozzle bore	+2.632	−0.25	−0.658
			−0.347

†From Example 4.3.
‡From Eq. (4.5).

The bias error in the flow rate is then, from Eq. (4.26),

$$B_q = \Sigma\, XB = -0.35 \text{ percent}$$

and the correction factor is, from Eq. (4.6),

$$F_{B,q} = \left(1 + \frac{-0.35}{100} \right)^{-1} = 1.0035$$

REFERENCES

ANSI(MFFCC): *Fluid Flow Measurement Uncertainty* (draft), ASME MFFCC S/C 1, 1982.

ISO/TC30/SC9, *Fluid Flow Measurement Uncertainty* (draft), NEL, East Kilbride, Scotland, 1981.

ISO Standard 3534, *Statistics—Vocabulary and Symbols,* ISO 3534-1977(E/F), Geneva, 1977.

ISO Standard 5168, *Measurement of Fluid Flow—Estimation of Uncertainty of a Flow Rate Measurement,* ISO 5168, Geneva, 1979.

OIML: *Vocabulary of Legal Metrology* (official version in French; unofficial English translation, PD 6461, BSI), International Organization of Legal Metrology, Paris, 1971.

INFLUENCE QUANTITIES

Accuracy statements for flowmeters are based on the steady flow of a homogeneous, single-phase newtonian fluid with an approach velocity profile that does not alter the coefficient obtained in long, straight runs of pipe. Departures from these reference conditions are called flowmeter *influence quantities*. Velocity-profile deviations, nonhomogeneous flow, pulsating flow, and cavitation are the four major influence quantities affecting all flowmeters. The errors associated with a particular influence quantity depend on the sensitivity of a particular flowmeter to that quantity and whether or not a calculation correction can be made. For newtonian fluids, velocity profiles can usually be brought into acceptable limits by the installation of sufficient straight pipe or, for shorter lengths, with flow conditioners. However, other influence quantities may require the installation of pulsating dampers or the use of a less sensitive flowmeter to achieve the desired degree of accuracy. The major influence quantities and their effects are discussed in detail in the following sections.

VELOCITY PROFILE

Velocity profile is probably the most important (and least understood) influence quantity. The effects of swirl, nonnewtonian fluids, and nonaxisymmetric profiles on a meter's performance are not only difficult to analyze, but they cannot easily be duplicated in a laboratory.

Newtonian Fluids

The rheological behavior of a fluid determines whether it is classified as newtonian or nonnewtonian. A newtonian fluid is defined as a fluid which, when acted upon by an applied shearing stress, has a velocity gradient that is solely proportional to the applied stress. The constant of proportionality is the absolute viscosity defined in Chap. 2. All gases, most liquids, and fine mixtures of spherical particles in liquids and gases are newtonian fluids.

The velocity profile established by a newtonian fluid is the basic reference condition for all flowmeters, and from this profile all corrections are made. Special laboratory tests are required to establish the effects of nonnewtonian fluids on flowmeters, and little published data is available because of the many types of nonnewtonian fluids.

Flow Characteristics   If there were no viscosity, the velocity of a flowing fluid would be uniform across a pipe section. The presence of even the small absolute viscosity of a gas induces a shearing action between adjacent fluid particles that reduces the velocity to zero at the pipe wall and thus forms a nonuniform velocity profile. Osborne Reynolds observed that, in the absence of swirl, a newtonian fluid had two distinctly different velocity profiles.

In Reynolds' experiment, a dye stream was injected into a flow of water in a glass pipe. At low velocities the dye stream was observed to remain parallel to the pipe axis along the length of pipe. However, when the velocity was increased, the dye stream initially oscillated, but eventually mixed completely with the water, at some distance from the dye injection point. In the first profile (Fig. 5.1*a*), the fluid moves in layers, or laminae, with one layer sliding over another. In the second (Fig. 5.1*b*), dye mixes with the fluid, and turbulent agitation occurs. These profiles were observed to be separated by a transition regime where both laminar and turbulent conditions may exist at differing pipe radii.

At lower velocities, or for more viscous fluids, viscous forces restrain fluid particles into parallel-layer motion. At higher velocities, or for less viscous fluids, inertia forces overcome viscous forces and particles move in a turbulent, almost random manner. The laminar profile is easily analyzed, but turbulent profiles with their complex and random motions are not well understood.

In Reynolds' experiments, the ratio of inertia to viscous forces was observed to be dimensionless and related to viscosity, average pipeline velocity, and geometrically similar boundary conditions. For a homogeneous newtonian fluid, this dimensionless ratio is

$$R_D = \frac{\overline{V}_f \, D_F \, \rho_f}{(\mu_m)_e} = \frac{\overline{V}_f^* \, D_M \, \rho_f^*}{\mu_{\text{Pa·s}}} \tag{5.1}$$

Reynolds Number   From an engineering viewpoint, the many variables that affect velocity profile cannot be evaluated for all possible flowmeters and for all pipeline conditions. For this reason, steady flow and a fully developed flow profile, as defined by a newtonian, homogeneous fluid, are initially assumed. Coefficient variation can then be predicted with the dimensionless Reynolds number. This number has been found to be an acceptable correlating parameter that combines the effects of viscosity, density, and pipeline velocity. A flow coefficient that is obtained for water at a specified Reynolds number will be the same for oil or gas at the same Reynolds number.

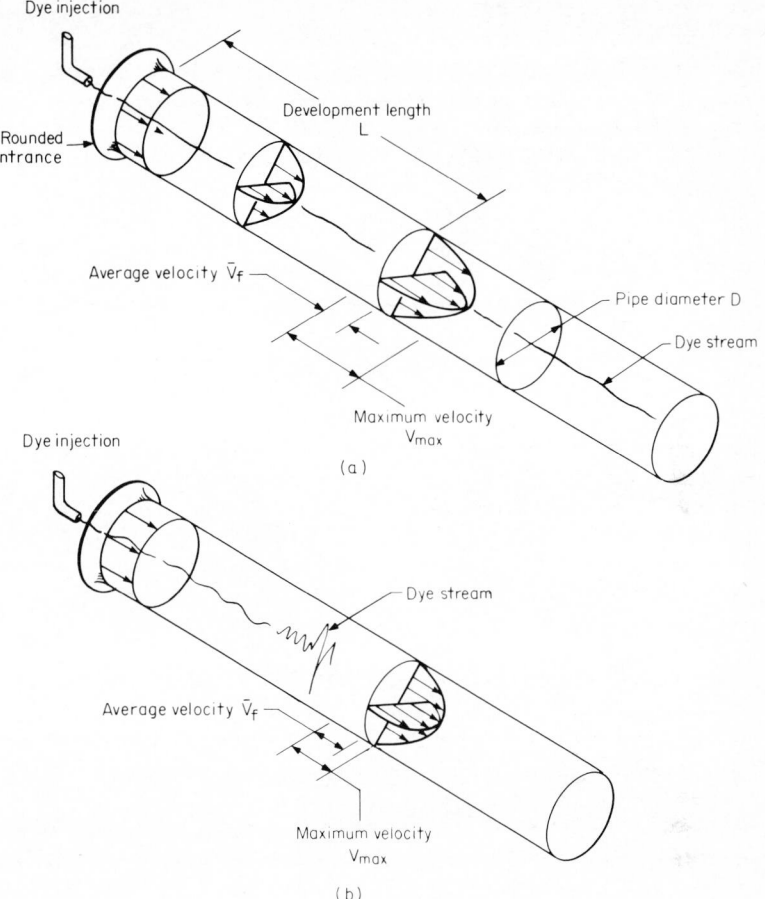

Figure 5.1 Laminar and turbulent profiles. (a) Laminar flow, $R_D \le 2000$. (b) Turbulent flow, $R_D \ge 4000$.

Table C.4 presents equations for calculating pipe Reynolds numbers for various U.S. flow-rate and viscosity units, and Table C.5 lists similar equations for SI units. In some flowmeter correlations, the *bore* Reynolds number is used; this is calculated by replacing D in Eq. (5.1) with the bore d of the primary element.

The laminar and turbulent profiles shown in Fig. 5.1 are symmetric about diametral planes and are conveniently shown two-dimensionally as in Fig. 5.2. Experimental evidence indicates that they have the following characteristics:

Laminar Profile The fully developed laminar flow profile is parabolic for pipe Reynolds numbers below 2000 and unaffected by wall roughness. The average

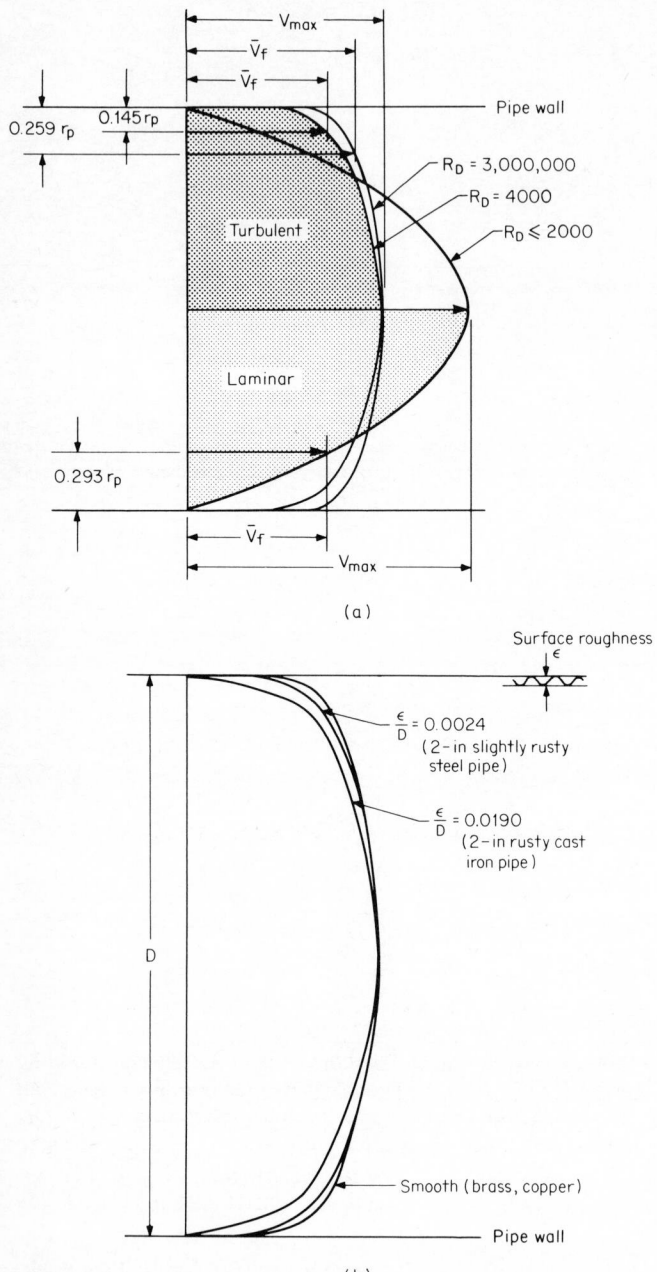

Figure 5.2 Velocity profiles in laminar and turbulent flow. (*a*) Smooth pipe. (*b*) Rough pipe.

5-4

pipeline velocity is one-half the centerline velocity, and the relationship between point and maximum velocities is given by

$$V_p = V_{max}\left[1 - \left(\frac{r}{r_p}\right)^2\right]$$ (5.2)

The average velocity is obtained by integration of Eq. (5.2):

$$\overline{V}_f = \tfrac{1}{2}V_{max}$$ (5.3)

Substituting Eq. (5.3) into Eq. (5.2) yields the distance of the average velocity from the pipe wall as

$$\overline{y} = 0.293r_p$$ (5.4)

The laminar profile develops from an almost blunt profile to a fully developed parabolic shape in a development length (Govier and Aziz, 1977) of

$$L = 0.028DR_D$$ (5.5)

For flow measurements to be taken in the laminar regime ($R_D < 2000$), sufficient upstream pipe must be available to provide the length defined by Eq. (5.5). Although laminar flow is unaffected by pipe roughness, pipe-entrance conditions (sharp edge versus rounded inlet) or other conditions may cause a need for additional length.

Turbulent Profile   At Reynolds numbers between approximately 2000 and 4000, a transition regime is crossed, and the parabolic geometry is altered. As shown in Fig. 5.2a, at Reynolds numbers of 4000 or larger the profile becomes blunter, and with increasing Reynolds number the profile becomes increasingly flatter until all particles, except at the walls, are moving at the same velocity. In practice, this profile is approached most nearly by gases at high velocities. This flat profile is referred to as the *infinite Reynolds-number profile,* the *rectilinear profile,* or sometimes the *plug-flow profile.*

The profile is complex and unstable in the narrow transition regime. Depending on whether the velocity is decreasing or increasing, it may be parabolic, blunt, or a combination of both. The Reynolds number for the onset of turbulence has been observed to range from 4000 to 7000, depending on whether the flow rate was slowly increased or decreased. Transition flow is a flow regime over which accurate flow measurement is quite difficult.

The turbulent profile is not of fixed geometry, but rather changes with wall roughness and Reynolds number. The simplest equation for calculating the velocity at a point in *smooth* pipes is the empirical power-law equation,

$$V_p = V_{max}\left(1 - \frac{r}{r_p}\right)^{1/n}$$ (5.6)

where the exponent n can be calculated with reasonable accuracy as

$$n = 1.66 \log R_D$$ (5.7)

Power-law profiles [Eq. (5.6)] adequately describe turbulent flow, but do not allow exact calculation of the centerline or wall velocity.

For smooth pipe, the point of average velocity $\overline{V}_f$ for Reynolds numbers larger than 10,000 is well approximated by

$$\overline{y} = \left[\frac{2n^2}{(n+1)(2n+1)} \right]^n r_p \tag{5.8}$$

As shown in Fig. 5.2b, a rough wall holds back additional fluid via increased shear at the wall. The *relative roughness* is defined as the ratio of average depth of pipe roughness ϵ to pipe diameter D. It is this ratio and the pipe Reynolds number that determine whether the turbulent profile is more or less blunted. Figure 5.2b shows that as the relative roughness increases, the profile becomes less blunt.

The ratio of average velocity to centerline velocity is graphed in Fig. 5.3. In smooth pipe, this ratio increases with increasing Reynolds number; the profile becomes increasingly blunt. For rough pipe, the ratio approaches a constant, which implies that the profile remains fixed above some minimum Reynolds number. Flows with Reynolds numbers above this minimum are said to be in the *complete-turbulence regime.*

Many equations have been proposed for calculating turbulent velocity profiles, using relative roughness and Reynolds number as correlating parameters. Many of these were developed from Nikuradse's (1932, 1933) smooth and sand-roughened pressure-loss data. Logan and Townes (1962), who supplemented this data, found no significant differences, except for very rough pipes.

The friction factor f, defined by the Darcy-Weisbach (1854) equation

$$f = \frac{8g_c(S_s)_w}{\rho_f \overline{V}_f^2} = -\frac{dP}{dL_P} \frac{2g_c D_F}{\rho_f \overline{V}_f^2} = \frac{H_L}{L/D\overline{V}_f^2/2g_c} \tag{5.9}$$

is the usual correlating parameter between Reynolds number and point velocity. Friction factors are usually presented in graphical form, based on the work of Moody (1944) (see Fig. B.3). The Colebrook (1939) partially rough wall friction-factor equation is widely used for calculating friction factors. For parameters defined in accordance with Eq. (5.9), this equation is

$$\frac{1}{\sqrt{f}} = -2\log\left(\frac{\epsilon/D}{3.7} + \frac{2.51}{\sqrt{f}R_D}\right) \tag{5.10}$$

Equation (5.10) is nonlinear; it can be solved with Newton's method (see Sec. A.1) by defining the function

$$F = -f^{-1/2} - 2\log\left(\frac{\epsilon/D}{3.7} + \frac{2.51}{\sqrt{f}R_D}\right) \tag{5.11}$$

The derivative of the function is

$$F' = \frac{1}{2}f^{-3/2} + \frac{4.034}{(\epsilon/D)R_D f^{3/2} + 9.287f} \tag{5.12}$$

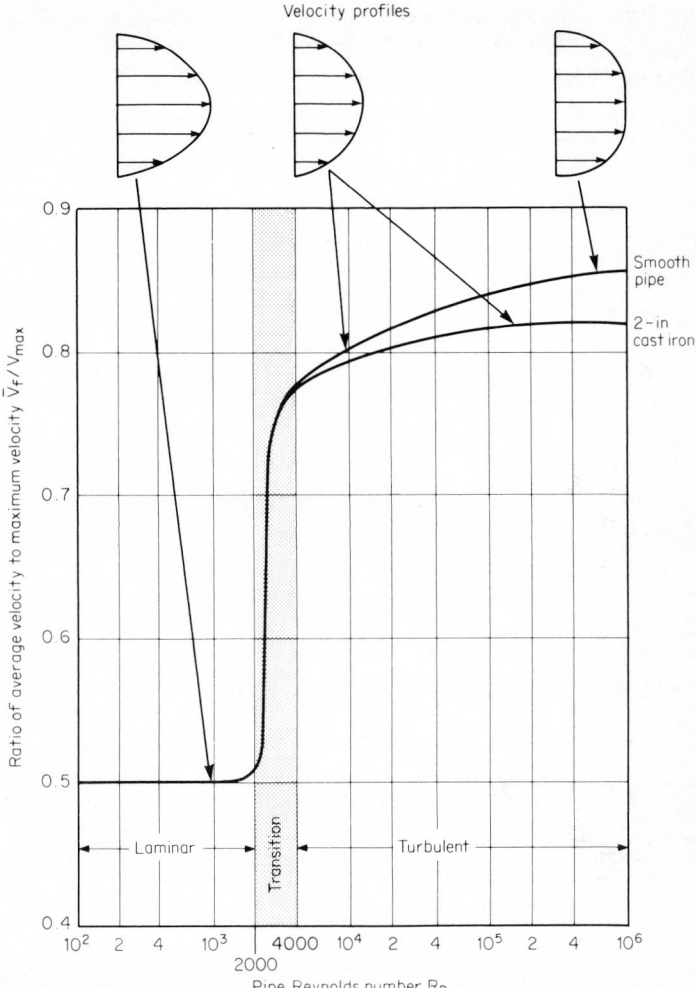

Figure 5.3 Ratio of average to maximum (centerline) velocity for smooth and rough pipe.

and the friction factor is calculated by iteration as

$$f_n = f_{n-1} - \frac{F_{n-1}}{F'_{n-1}} \qquad (5.13)$$

A single iteration will produce a result within ± 0.01 percent of the limit if the initial estimate is calculated as

$$f_0 = 0.25 \left[\log \left(\frac{0.2703\epsilon}{D} + \frac{5.74}{R_D^{0.9}} \right) \right]^{-2} \qquad (5.14)$$

Teyssandier (1975) found that the Pai (1953) profile equation best correlated empirical orifice-coefficient data. Pai's equation can be written for average pipeline velocity as

$$\overline{V}_f = \frac{V_{\max}}{1 + 1.44\sqrt{f}} \tag{5.15}$$

and for point velocity as

$$V_p = \left[1 + \frac{s - n}{n - 1} \left(\frac{r}{r_p} \right)^2 + \frac{1 - s}{n - 1} \left(\frac{r}{r_p} \right)^{2n} \right] V_{\max} \tag{5.16}$$

where

$$s = \frac{fR_D}{32 + 46.08\sqrt{f}} \tag{5.17}$$

and

$$n = \frac{2 - fR_D/32}{1.44\sqrt{f} - 1} \tag{5.18}$$

Another widely used equation relating friction factor to average velocity is given by Pao (1961) as

$$\frac{V_p}{\overline{V}_f} = 1 + \left(1.43 + 2.15 \log \frac{y}{r_p} \right) \sqrt{f} \tag{5.19}$$

which provides the velocity ratio as a function of relative roughness after substitution of Eq. (5.14):

$$\frac{V_p}{\overline{V}_f} = 1 + \frac{0.715 + 1.075 \log (y/r_p)}{\log (0.2703\epsilon/D + 5.74/R_D^{0.9})} \tag{5.20}$$

The location $\overline{y}$ of the average velocity is obtained by setting $V_p/\overline{V}_f = 1.0$ in Eq. (5.19) to obtain

$$\frac{\overline{y}}{r_p} = 10^{-1.43/2.15} = 0.216 \tag{5.21}$$

In Fig. 5.4a the ratio of average to centerline velocity, based on the Pai profile, is plotted against pipe Reynolds number for various relative-roughness values. In Fig. 5.4b the effect of relative roughness on the location of the average velocity is shown. The location of the average pipe velocity was obtained by integration of the Pai profile and iteration to locate the average velocity. This figure shows that at a given Reynolds number the average velocity is located further into the pipe with increased roughness. Also, for a given relative roughness, it is located further into the pipe with increased Reynolds number.

In Fig. 5.5 the point of average velocity is shown for power-law and Pao profiles for pipes normally used in flow measurement. This curve explains why a velocity traverse is usually necessary to properly locate a single-point velocity-measuring device (pitot tube). The fixed location for pitot tube recommended by ISO/DIS 7145 (1981) for ± 3 percent accuracy is shown on this curve.

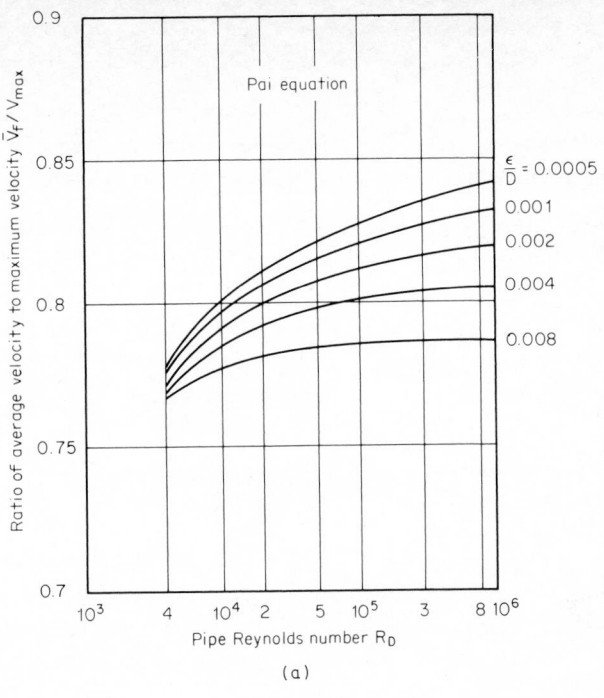

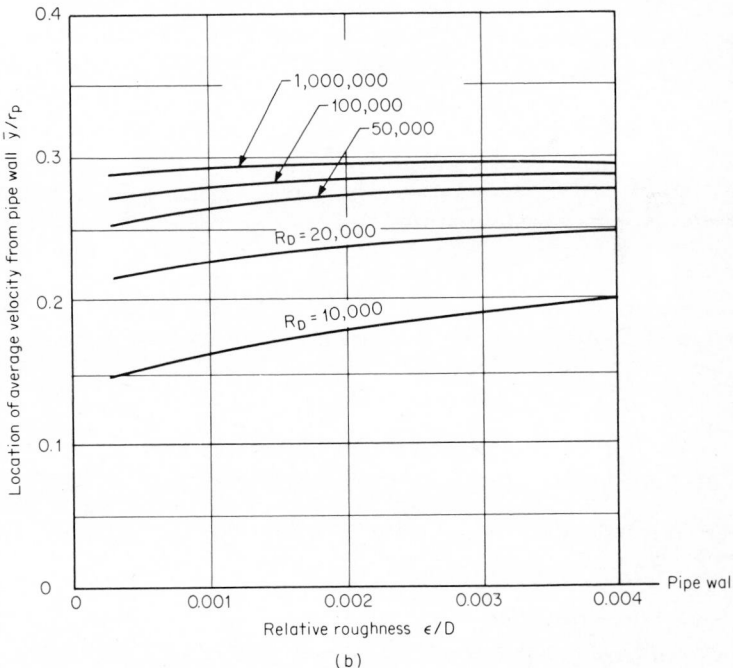

Figure 5.4 Effect of pipe roughness on the location and ratio of average velocity to centerline velocity. (*a*) Velocity ratio. (*b*) Location of average velocity.

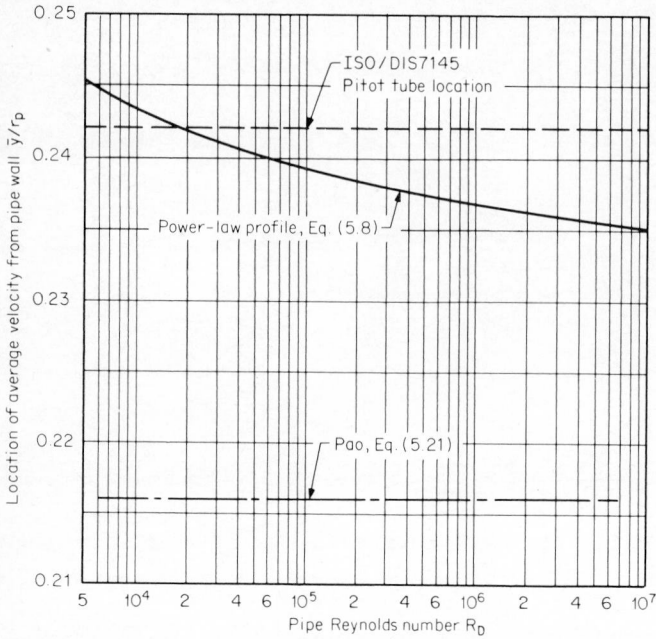

Figure 5.5 Location of average velocity from pipe wall.

Nonnewtonian Fluids

Reynolds-number corrections are developed for newtonian fluids, for which the ratio of shear stress to shear rate is constant. This ratio is easily obtained with a single viscometer measurement. For nonnewtonian fluids, the ratio changes with stress; it is, therefore, called the *apparent viscosity*. Nonnewtonian fluids are subdivided into two major classifications—*time-independent* and *time-dependent*. Under constant stress, a time-independent fluid's apparent viscosity will not change with time, while the apparent viscosity of a time-dependent fluid will change with time.

Among the more common time-independent nonnewtonian fluids are the following:

1. Plug-flow fluids

    a. Bingham (or Bingham-like) fluids

        ■ Aluminum soap grease (at 250°F)

        ■ Some asphalts

        ■ Some bitumens

        ■ Calcium soap grease (at 250°F)

        ■ Cement rock (aqueous suspensions, 92 percent under 74 μm)

- Chocolate
- Some emulsions
- Hydrocarbon grease, thickened
- Catsup
- Lithium stearate grease, 14G8
- Luting (dispersion of synthetic wax in polyisobutane)
- Mayonnaise
- Drilling mud
- Mustard
- Ointments
- Paint
- Some long-chain polymers (carboxymethylene)
- Paper pulp
- Toothpaste
- Water suspension of fly ash, clay (Milicz), coal char, finely divided minerals, kaolin, metallic oxides, paint, thorium oxide, titanium oxide, quartz, sewage sludge

b. Yield pseudoplastics

- Carboxypolymethylene (Carbopol), water dispersions
- Many Bingham-like suspensions at intermediate concentrations

2. Power-law fluids

a. Pseudoplastics ($m < 1$)

- Ammonium alginate (aqueous solution)
- Emulsions, e.g., incendiary bomb thickened with aluminum soaps and polymethyl methacrylate
- Ethyl cellulose, 11 percent solution in cyclohexanone
- 12 percent gel of aluminum soaps in gasoline
- Blood plasma
- Molten polyethylene [266°F (108°C)]
- Water suspensions† of clay (Attasol), ferrosilicon, galena, latex, magnetite, titania

†Some of these may be classified as dilatant or newtonian in certain concentrations and at high shear rates.

b. Dilatant fluids‡ ($m > 1$)

- Cornstarch in ethylene glycol
- Ethylene glycol–glycerine with small amount of water
- Starch in water
- Titania in 42-cP sucrose
- Titanium dioxide, 27 to 47 percent in water

The time-dependent nonnewtonian fluids include:

1. Thixotropic

- Water suspensions of bentonitic clay
- Some crude oils at low temperatures
- Pseudoplastic emulsions of soaps
- High polymers in gasoline

2. Rheopectic

- Printer's ink
- Saturated polyester ($M_w = 2000$)

3. Viscoelastic

- Carbopol solutions
- Flour dough
- Some jellïes
- Nylon
- Polymer solutions: carboxymethylcellulose (CMC), polyisobutylene in decalin, polyox, polyacrylamide, polyethylene oxide, sodium carboxymethyl cellulose

Nonnewtonian velocity profiles are not the same as newtonian profiles. The required zero wall velocity causes varying shear rates in a nonnewtonian fluid, and nonnewtonian fluids distribute the velocity differently among fluid layers. Although nonnewtonian profiles are not easily predicted, they are symmetrical, and the number of pipe lengths required to establish profiles are usually one-third to one-half those for a newtonian fluid.

In many applications, nonnewtonian fluids are in the laminar flow regime, and it is convenient to categorize their profiles in terms of departure from the newtonian laminar profile. In the turbulent regime, the distinction between profiles

‡Observed only in certain concentration ranges of suspensions of irregularly shaped solids.

is not so great, and *the apparent viscosity may be used in Eq. (5.1) for the Reynolds number.*

Time-independent pseudoplastics have blunter laminar profiles than newtonian fluids, while dilatant fluids produce pronounced, almost conical, peaking profiles. The Bingham fluids produce almost rectilinear profiles. Time-dependent thixotropic and rheopectic fluids exhibit time-variable pseudoplastic or yield-pseudoplastic behavior. This results in profile changes with both flow rate and time. Viscoelastic fluids are the most complex, having partial elastic recovery and time-dependent viscous properties.

Flow measurement data for nonnewtonian fluids in the laminar regime is almost nonexistent. In many applications, except for venturi measurements of slurry or sewage or where Reynolds-number corrections are not required, the magnetic flowmeter is used because its output is essentially an average of the profile.

Time-Independent Fluids Newtonian, pseudoplastic, dilatant, and Bingham-plastic-like fluids are time-independent fluids whose apparent viscosity does not change with the duration of an applied stress. A graph of shear stress versus shear rate for these fluids is shown in Fig. 5.6. In the figure, all the fluids have the same *apparent viscosity* as water (0.0095 P) at an applied shear stress of 200 g/(cm·s²).

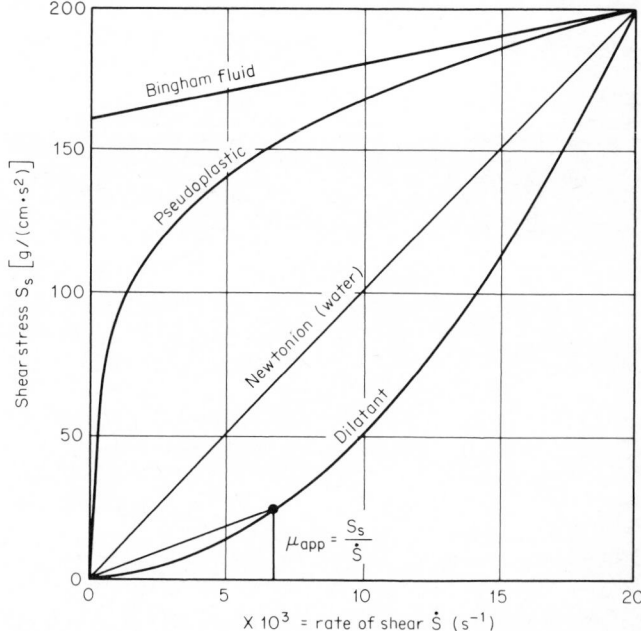

Figure 5.6 Shear stress versus shear rate for time-independent fluids.

Apparent viscosity is the ratio of shear stress to shear rate. Figure 5.6 shows that, except for newtonian fluids, the apparent viscosity changes with the shear stress. For pseudoplastics the apparent viscosity decreases with increasing shear stress, while for dilatant fluids it increases. For Bingham plastics an initial stress must be applied to initiate motion. After this yield stress is reached, a Bingham fluid exhibits a linear relation between shear stress and shear rate, similar to newtonian fluids.

If the shear stress is plotted against shear rate logarithmically, the straight-line relationships shown in Fig. 5.7 are obtained for dilatant and pseudoplastic fluids. That is, the shear stress is related to the shear rate raised to a power. The Ostwald equation is widely used in engineering calculations to describe the shear behavior of these *power-law* fluids. This equation is

$$S_s = AS^m \tag{5.22}$$

where the constant A and exponent m are called the *consistency index* (or *power-law coefficient*) and *flow-behavior index* (or *power-law exponent*). Both constants are determined by multiple viscosity measurements. Dilatant fluids have exponents greater than 1; newtonian fluids have an exponent of 1; and pseudoplastics have a power-law exponent smaller than 1. Table 5.1 presents typical constants for some power-law fluids.

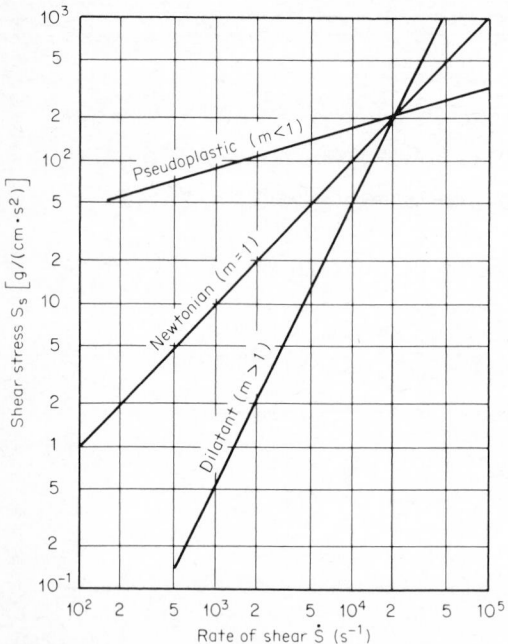

Figure 5.7 Logarithm plot of shear stress versus shear rate.

Table 5.1 Typical Constants for Power-Law Fluids

Fluid	Power-law constant A'	Power-law exponent $m' = m$
Water suspensions		
14.3% clay	0.0441	0.35
21.2% clay	0.0106	0.335
25% clay	0.0345	0.185
31.9% clay	0.0611	0.251
36.8% clay	0.1845	0.176
40.4% clay	0.4346	0.132
23% lime	0.1786	0.178
4% pulp	0.4611	0.575
Carboxymethylcellulose (3% water solution)	0.2146	0.566
10% napalm in kerosene	0.0995	0.520

SOURCES: Metzner and Reed (1955) and Govier and Aziz (1977).

By definition, the apparent viscosity is

$$\mu_{\text{app}} = \frac{S_s}{S} \tag{5.23}$$

Substituting Eq. (5.22) into Eq. (5.23) yields the apparent-viscosity equation for power-law fluids as

$$\mu_{\text{app}} = AS^{m-1} \tag{5.24}$$

For newtonian fluids, where $m = 1$, this equation reduces to the constant A—the absolute viscosity. Shown in Fig. 5.8 are apparent viscosities for the two power-law fluids involved in Fig. 5.7.

Laminar Flow of Power-Law Fluids Like newtonian fluids, nonnewtonian fluids exhibit a transition from laminar to turbulent flow profile with increasing Reynolds number. For newtonian fluids the transition begins at a Reynolds number of 2000; for power-law fluids the Metzner-Reed (1955) Reynolds number is used to predict the onset of turbulence.

Figure 5.9 shows a section of flow pipe of known diameter with two pressure taps at a fixed spacing. A force summation can be used to find the wall shear stress $(S_s)_w$ in terms of the differential pressure, pipe diameter, and pipe length:

$$\pi L_P D_F (S_s)_w = \frac{\pi}{4} D_F^2 \, \Delta P \tag{5.25}$$

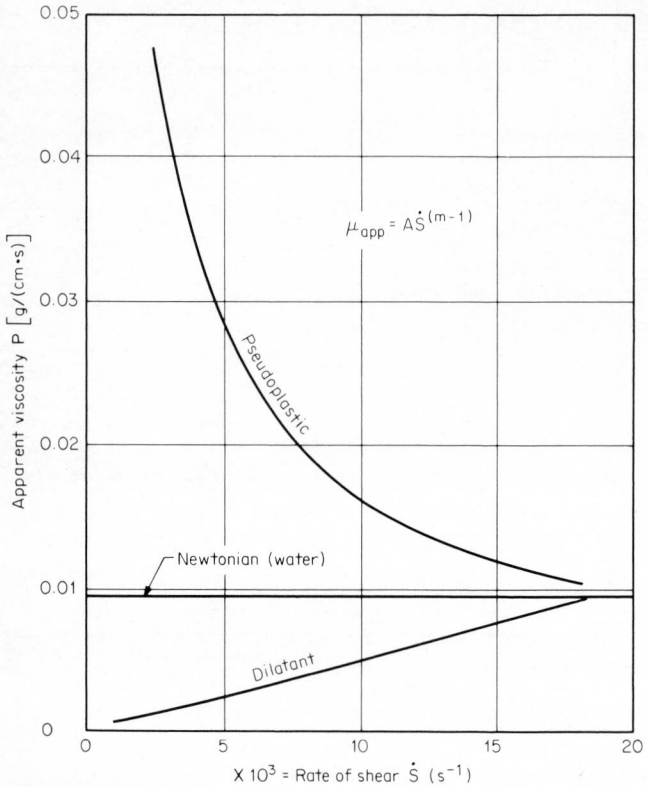

Figure 5.8 Apparent viscosity of power-law fluids.

where the right-hand side is the pressure force required to overcome the wall fluid shear force (shear stress times wall area). Rearrangement of Eq. (5.25) yields

$$(S_s)_w = \frac{D_F \, \Delta P}{4L_P} \tag{5.26}$$

The wall shear rate $\dot{S}_w$ can be determined from a log-log plot of the measured differential pressure versus velocity as

$$\dot{S}_w = \frac{1 + 3m'}{4m'} \frac{8\overline{V}_f}{D_F} \tag{5.27}$$

where m' is the slope of the plotted line. Combining Eqs. (5.23), (5.26), and (5.27) gives

$$\frac{D_F \, \Delta P}{4L_P} = A' \left(\frac{8\overline{V}_f}{D_F} \right)^{m'} = (S_s)_w \tag{5.28}$$

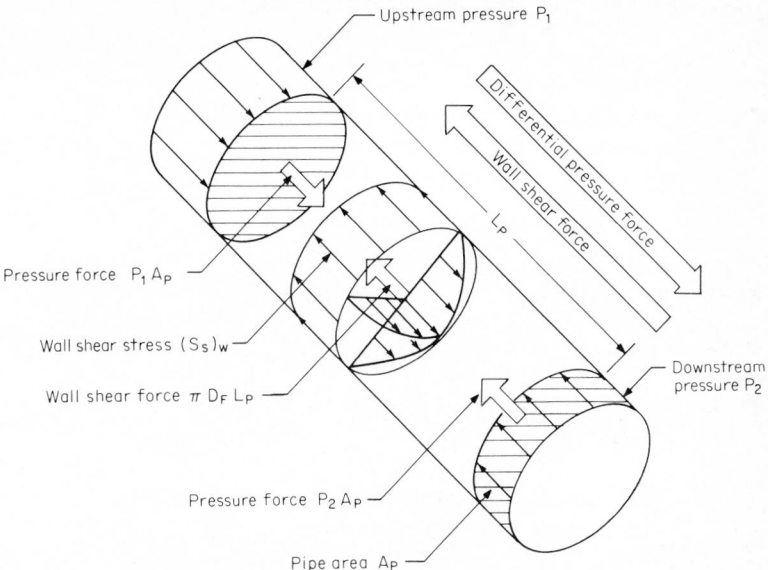

Figure 5.9 Wall shear and pressure force balance.

The term $8\overline{V}_f/D_F$ is called the *Poiseuille flow index,* and the constants A' and m' are obtained from pressure-loss measurements. In laminar flow the Fanning friction factor f_f is related to Reynolds number and independent of wall roughness. If the Reynolds number for nonnewtonian fluids is defined to be the Metzner-Reed Reynolds number, then the following relationship is developed:

$$f_f = \frac{16}{R_{MR}} = \frac{g_c D_F \,\Delta P}{2\rho_f \overline{V}_f^2 L_P} \tag{5.29}$$

where f_f is by definition one-quarter of the Darcy-Weisbach friction factor generally used for pressure-loss calculations. Substituting Eq. (5.28) into Eq. (5.29) yields the Metzner-Reed Reynolds number as

$$R_{MR} = \frac{D_F^{m'} \overline{V}_f^{2-m'} \rho_f}{g_c A' 8^{m'-1}} \tag{5.30}$$

Equations (5.28) through (5.30) are written for the fundamental English engineering units of feet, pounds-force, pounds-mass, and seconds. If the pipe size D is in inches, then the Metzner-Reed Reynolds number is

$$R_{MR} = \frac{8D^{m'} \overline{V}_f^{2-m'} \rho_f}{g_c A' 96^{m'}} \tag{5.31}$$

For power-law and newtonian fluids, m' equals m, and the relationship between the constants given in Eqs. (5.22) and (5.30) is

$$A' = \frac{A}{g_c} \left(\frac{1 + 3m}{4m} \right)^m \tag{5.32}$$

For the special case of newtonian fluids, $m = m' = 1$, and the relationship becomes

$$A' = (\mu_f)_e = \frac{(\mu_m)_e}{g_c} = \frac{A}{g_c} \tag{5.33}$$

That is, the absolute viscosity equals the constant A. The Metzner-Reed Reynolds-number equation then reduces to the familiar newtonian pipe Reynolds-number equation (5.1).

Figure 5.10 is a plot of Eq. (5.28) for a 118-cp newtonian fluid $[(\mu_f)_e =$

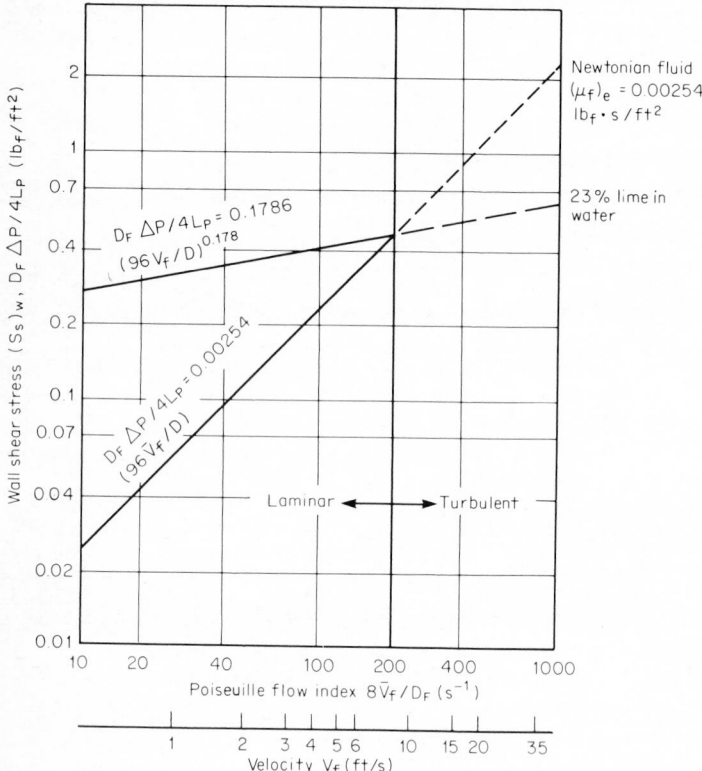

Figure 5.10 Poiseuille pressure and flow index for 4-in pipe flow of a newtonian and a power-law fluid.

0.00254] and a 23 percent lime in water power-law fluid. The values of A' and m' were selected from Table 5.1.

The Metzner-Reed Reynolds number was derived from the laminar friction-factor equation and, when it is combined with the Ryan-Johnson (1965) critical-friction-factor equation, an equation can be developed that will predict the onset of turbulence. The Ryan-Johnson critical friction factor is

$$(f_f)_c = \frac{1}{404} \frac{(1 + 3m')^2}{m'} \left(\frac{1}{2 + m'} \right)^{(2 + m')/(1 + m')} \tag{5.34}$$

For a newtonian fluid ($m' = 1.0$), equating Eq. (5.34) to Eq. (5.29) results in a pipe Reynolds number of 2099 for the onset of turbulence for a newtonian fluid. Use of the value 384 [rather than the 404 in Eq. (5.34)] to obtain the accepted Reynolds-number transition value of 2000 gives the critical Metzner-Reed Reynolds number as

$$(R_{MR})_c = \frac{6158m'(2 + m')^{(2+m')/(1+m')}}{(1 + 3m')^2} \tag{5.35}$$

The *critical velocity* for the onset of turbulence is then

$$(\overline{V}_f)_c = \left[770 \frac{g_c A' m'}{\rho_f} \left(\frac{8}{D_F} \right)^{m'} \frac{(2 + m')^{(2+m')/(1+m')}}{(1 + 3m')^2} \right]^{1/(2 - m')} \tag{5.36}$$

The critical Metzner-Reed Reynolds number is plotted against the power-law exponent in Fig. 5.11.

Many flow measurements are made in small-diameter pipes for flows of high-apparent-viscosity fluids. The Metzner-Reed Reynolds number for such flows is, therefore, below the values shown in Fig. 5.11. For such fluids the relationship between point and average pipeline velocity is derived from the power-law-model equation as

$$V_p = \frac{3m + 1}{m + 1} \left[1 - \left(\frac{r}{r_p} \right)^{(m+1)/m} \right] \overline{V}_f \tag{5.37}$$

Again, for a newtonian fluid ($m = 1$), this equation reduces to the parabolic velocity-profile Eqs. (5.2) and (5.3).

The ratio of point to average velocity is shown in Fig. 5.12. For pseudoplastics, the profile becomes increasingly blunt with decreasing power-law exponent, approaching a rectilinear profile. For dilatant fluids, the profile becomes conical in shape with increasing exponent. The location of the average velocity from the pipe wall is shown in Fig. 5.13.

Laminar Flow of Bingham Fluids   Bingham (or Bingham-like) fluids require some minimum stress to initiate motion between fluid layers. Until this *yield stress* is reached, adjacent layers retain a zero velocity gradient (shear rate).

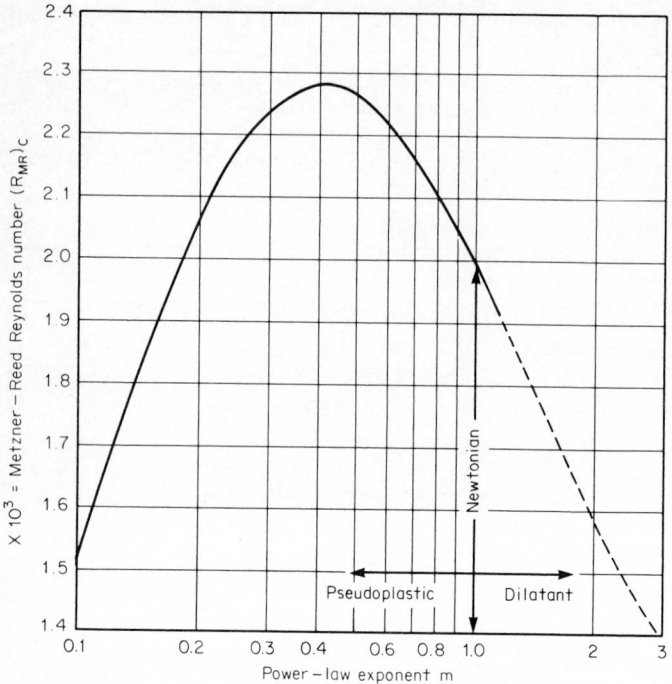

Figure 5.11 Critical Metzer-Reed Reynolds number for pseudoplastic and dilatant fluids.

Figure 5.14 shows shear stress plotted against shear rate for a typical Bingham fluid. The slope, as well as the yield shear stress, is seen to be a function of the volume fraction of solids. After the yield stress S_s is reached, the shear rate increases linearly. (For some fluids, called *yield-pseudoplastics*, this portion of the curve follows the power-law rather than a linear relationship.) The slope of the linear portion of the curve is defined as the *coefficient of rigidity* $(\phi_f)_e$ and, because it is linear, is sometimes referred to as the *plug-flow viscosity*.

Figure 5.15 illustrates the velocity profile and gives the coordinate system for the flow of a Bingham fluid. To initiate motion, a minimum pressure force is required to overcome the yield shear stress. Wall shear is transmitted only through a small fluid layer until equilibrium is reached between the pressure and the combined wall and fluid yield shear forces.

The velocity profile resembles a cylindrical plug of uniform-velocity fluid surrounded by a fluid annulus. An increase in the upstream pressure increases the plug velocity, which in turn increases the wall shear. This decreases the plug diameter until a laminar profile is approached.

The plug velocity $\overline{V}_P$, plug diameter D_P, velocity from the outside of the plug to the wall V_p, and average pipeline velocity $\overline{V}_f$ are derived from the equation for shear stress in cylindrical coordinates and the rheological shear stress–shear rate

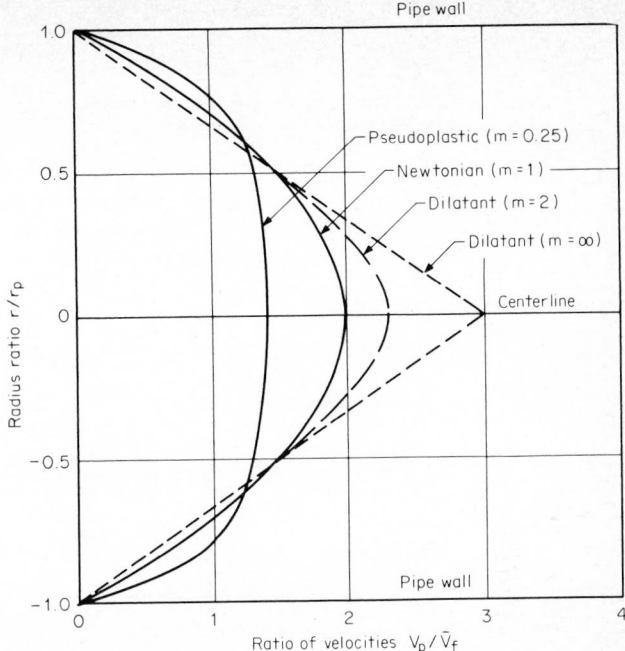

Figure 5.12 Velocity profile in laminar flow.

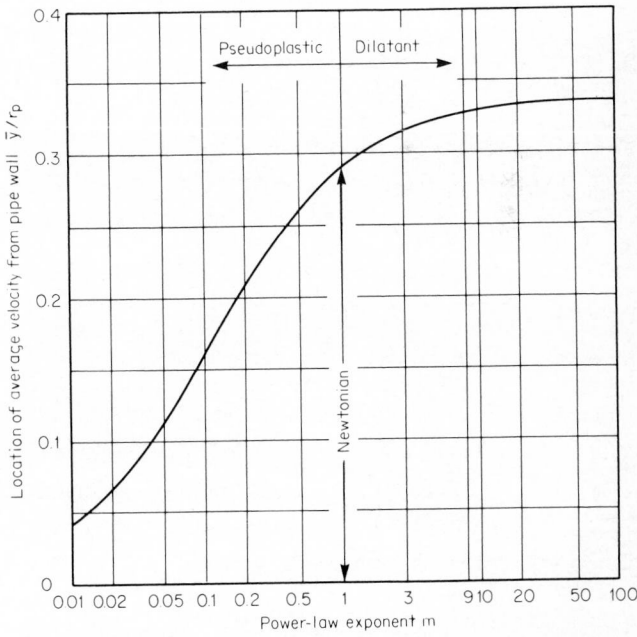

Figure 5.13 Location of average velocity from pipe wall in laminar flow.

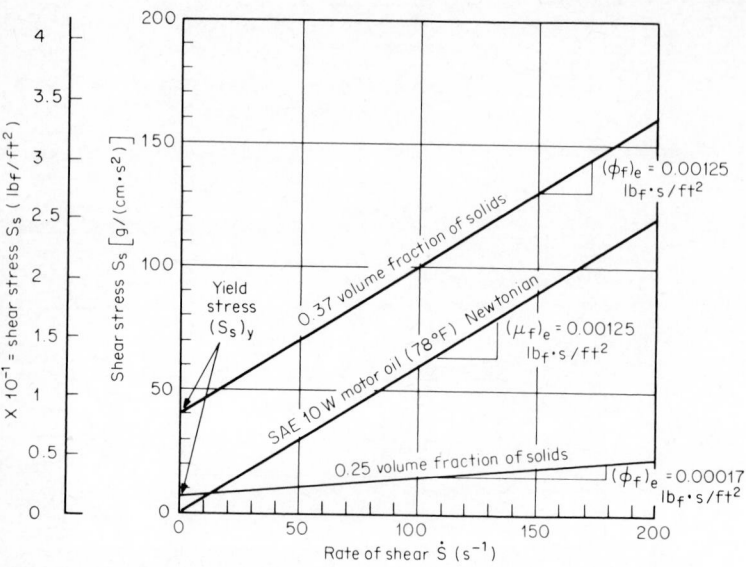

Figure 5.14 Shear stress versus shear rate for water suspensions of finely divided galena (data from Govier and Aziz, 1977).

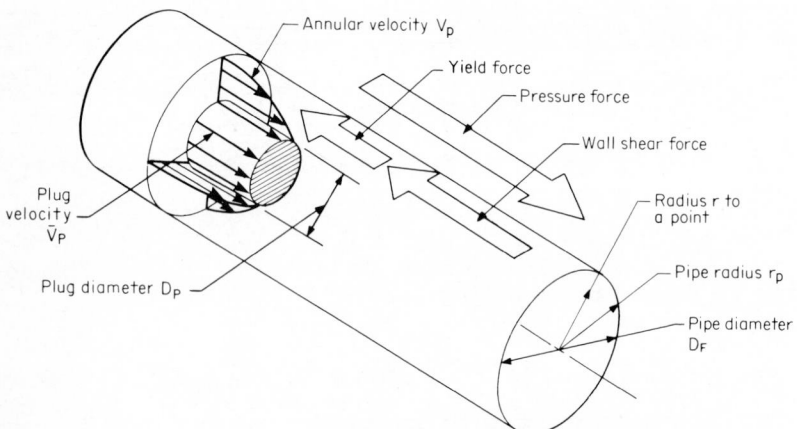

Figure 5.15 Pipe flow of Bingham or Bingham-like fluids.

equations for a Bingham fluid. For cylindrical coordinates and steady flow, the shear stress is defined by

$$S_s = -\frac{r}{2}\frac{dP}{dL_P} \qquad (5.38)$$

The shear stress is also the sum of the yield shear stress $(S_x)_y$ and shear rate:

$$S_s = (S_s)_y + (\phi_f)_e S \qquad (5.39)$$

When the shear stress S_s is less than the yield shear stress $(S_s)_y$, the shear rate $\dot{S}$ must be zero; no velocity gradients exist across the pipe section. Since the shear rate is the change in velocity divided by the incremental radius in cylindrical coordinates, the diameter of the plug can be derived from Eqs. (5.38) and (5.39) by setting shear rate equal to zero:

$$D_P = \frac{4(S_s)_y}{\Delta P / L_P} \tag{5.40}$$

where $\Delta P / L_P$ is the measured pressure loss over a given length of pipe. The plug velocity then is

$$\overline{V}_P = \frac{(S_s)_y^2}{(\phi_f)_e \, \Delta P / L_P} \left(\frac{D_F}{D_P} - 1 \right)^2 \tag{5.41}$$

and, in the region outside the constant-velocity plug, the point velocity is

$$V_p = \frac{\Delta P / L_P}{4(\phi_f)_e} (r_p^2 - r^2) - \frac{(S_s)_y}{(\phi_f)_e} (r_p - r) \tag{5.42}$$

The average pipeline velocity is the sum of the plug flow and integrated annular flow divided by the pipe area and is expressed as

$$\overline{V}_f = \frac{D_F}{8} \frac{1}{(\phi_f)_e} \frac{D_F \, \Delta P}{L_P} \left[1 - \frac{4}{3} (S_s)_y \left(\frac{4L_P}{D_F \, \Delta P} \right) + \frac{(S_s)_y^4}{3} \left(\frac{4L_P}{D_F \, \Delta P} \right)^4 \right] \tag{5.43}$$

This turns out to be the Buckingham equation when it is rearranged in the form

$$q_{\text{cfs}} = \frac{\pi D_F^3}{32(\phi_f)_e} \frac{D_F \, \Delta P}{4L_P} \left[1 - \frac{4}{3} (S_s)_y \left(\frac{4L_P}{D_F \, \Delta P} \right) + \frac{(S_s)_y^4}{3} \left(\frac{4L_P}{D_F \, \Delta P} \right)^4 \right] \tag{5.44}$$

Figs. 5.16 and 5.17 show wall shear stress versus the Poiseuille flow index for the Bingham fluids shown in Fig. 5.14.

The Reynolds number at the transition from laminar to turbulent flow can be estimated with the Reynolds number proposed by Govier and Aziz:

$$(R_{GA})_c = \frac{\rho_f (\overline{V}_f)_c D_F}{g_c(\phi_f)_e [1 + (S_s)_y D_F / 6(\phi_f)_e (\overline{V}_f)_c]} \approx 2100 \tag{5.45}$$

The bracketed term modifies the newtonian Reynolds number with the yield shear stress and coefficient of rigidity $(\mu_f)_e$. When the yield shear stress is zero, the coefficient of rigidity equals the absolute viscosity, and Eq. (5.45) reduces to the newtonian Reynolds-number equation.

For line sizes greater than 1 in, the bracketed term of Eq. (5.45) is greater than unity, and the critical velocity is well approximated by

$$(\overline{V}_f)_c = 106 \sqrt{\frac{(S_s)_y}{\rho_f}} \tag{5.46}$$

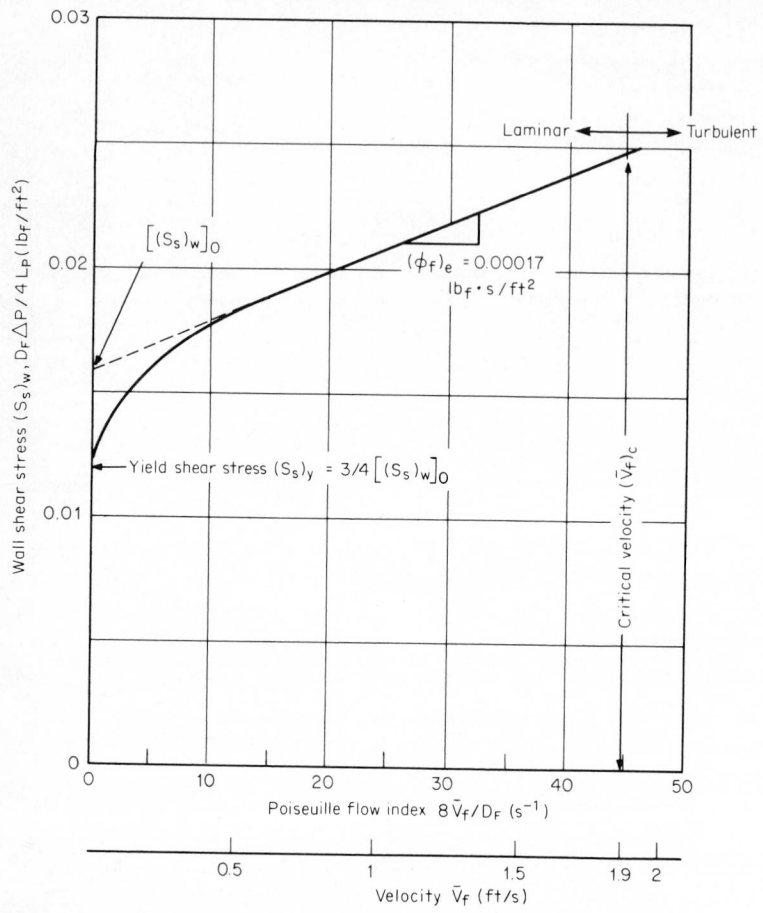

Figure 5.16   Flow with 0.25 percent volume fraction of solids in a 4-in pipe.

In this equation, the yield stress has the units of pounds-force per square foot. For most Bingham fluids, the flow remains laminar because the critical velocity exceeds 5 to 10 ft/s for yield stresses above approximately 0.1 lb_f/ft².

Table 5.2 gives some typical values of coefficient of rigidity and shear stress for Bingham-like fluids. Table 5.3 gives the critical velocities for some line sizes based on Eq. (5.46). Some of the smaller line sizes shown are not normally used for Bingham-like fluids, but values are given to illustrate the fact that Bingham fluids require higher velocities than newtonian fluids to reach transition.

Prediction of the flow profile for a Bingham fluid is obviously more complex than for other time-independent fluids. The plug diameter D_P is a function of wall shear stress and coefficient of rigidity $(\phi_f)_e$. To illustrate the possible combinations, curves for two Bingham fluids with identical coefficients of rigidity but

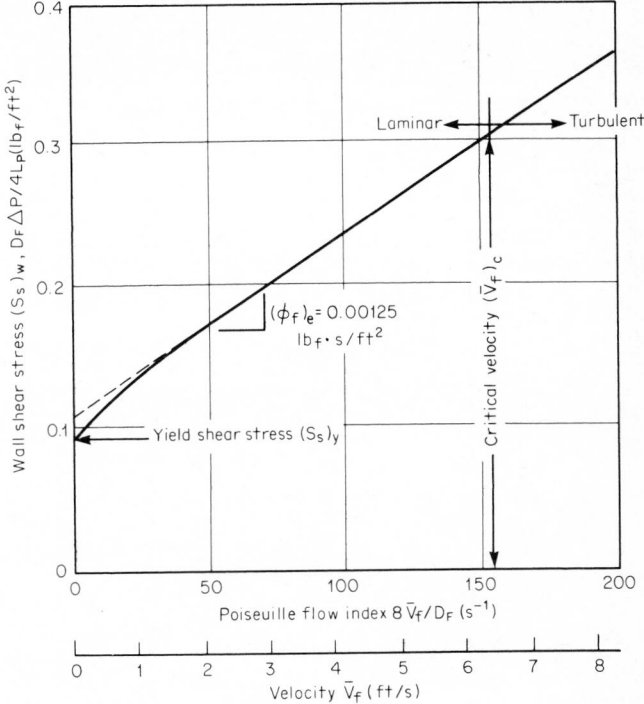

Figure 5.17 Flow with 0.37 percent volume fraction of solids in a 4-in pipe.

different yield shear stresses are shown in Fig. 5.18. In the region from point A to point C in Fig. 5.18, increasing pressure increases the velocity, and the profiles shown in Fig. 5.19a are successively produced for low-yield-stress fluids. For higher-yield-stress fluids (points A', B', C'), a larger plug diameter is formed for the same pipe velocity (Fig. 5.18b). For a fluid with zero yield stress (a newtonian fluid), the profile is parabolic.

The ratio of plug diameter to pipe diameter can be expressed in terms of the Poiseuille index as

$$\frac{8\overline{V}_f}{D_F} = \frac{(S_s)_w}{(\phi_f)_e}\left[1 - \frac{4}{3}\frac{D_P}{D_F} + \frac{1}{3}\left(\frac{D_P}{D_F}\right)^4\right] \tag{5.47}$$

Ratios of plug to pipe diameter for various values of the ratio of yield shear stress to the coefficient of rigidity are shown versus the Poiseuille index in Fig. 5.20. For the low-yield-stress fluids the plug diameter rapidly decreases until, at 0.1 ft/s, the profile is essentially newtonian. For the higher-yield-stress Bingham fluids, the plug diameter remains at one-half the pipe diameter.

Table 5.2 Yield Stress and Coefficient of Rigidity for Some Bingham (or Bingham-like) Fluids†

Substance	Metric units			English units			
	Specific gravity G_F	Yield stress $(S_s)_y$ g·cm/s²	Coefficient of rigidity ϕcP, g·cm/s	Yield stress $(S_s)_y$ lbf/ft²	lbf/in²	Coefficient of rigidity η_R $(\phi 0)_e$, lbf·s/ft²	$(\phi m)_e$ lbm/(ft·s)
Luting (dispersion of synthetic wax in polyisobutene)		15,700–19,600	1,000,000	32.8–40.9	0.227–0.284	20.8	670.0
Calcium soap grease (at 120°C)		7,100	6,500	14.8	0.103	0.136	4.37
Aluminum soap grease (at 120°C)		5,100	5,000	10.7	0.074	0.104	3.36
Lithium stearate grease, 14G8		6,500	100	13.6	0.094	0.0021	0.067
Flocculated aqueous china-clay, suspensions (80% under 1 μm):							
No. 6	1.149	78	4.0	0.16	0.0011	0.000084	0.0027
No. 4	1.207	250	6.7	0.522	0.0036	0.00014	0.0045
No. 1	1.280	590	13.1	1.32	0.0086	0.00027	0.0088
Aqueous suspensions of Milicz clay ($d_{av} = 70$ μm):							
13.9 weight % solids		23	8.7	0.048	0.00033	0.00018	0.0058
16.8 weight % solids		53	13.6	0.111	0.00077	0.000284	0.0091
19.6 weight % solids		130	25.0	0.215	0.00189	0.000522	0.0168

Aqueous clay suspensions:

No. V	1.36	66.5	19.4	0.139	0.00096	0.000406	0.0130
No. III	1.44	200	32.8	0.418	0.0029	0.000685	0.0220
No. I	1.49	345	44.7	0.721	0.0050	0.000834	0.0300

Aqueous suspensions of cement rock (92% under 74 μm), 54.3 weight % solids

	1.52	38	6.86	0.0794	0.0055	0.000143	0.0046

Sewage sludges, 10 weight % solids:

Chicago plant		6.7	52	0.0140	0.000007	0.00109	0.0349
Stuttgart plant		48	115	0.100	0.0007	0.00240	0.0773

Sludge No. 5 (Imhoff tank), 14 weight % solids

	1.06	31	24.5	0.065	0.00045	0.000512	0.0165

Aqueous suspensions of coal char:

Weight %								
Total solids	325-mesh solids							
45.5	28.6	1.23	26.2	31	0.055	0.00038	0.00065	0.0208
50.4	47.5	1.22	20	28	0.0418	0.00029	0.00058	0.0188

†Data from Govier and Aziz (1977).

Table 5.3 Critical Velocity by Line Size for Bingham (or Bingham-like) Fluids

Substance	Critical velocity $(\overline{V}_{fl})_c$, ft/s; for line size D of:					
	0.5 in	1 in	1.5 in	2 in	3 in	4 in
Luting (dispersion of synthetic wax in polyisobutene)			$(\overline{V}_{fl})_c > 40$ ft/s			
Calcium soap grease (at 120°C)						
Aluminum soap grease (at 120°C)						
Lithium stearate grease, 14G8						
Flocculated aqueous china-clay, suspensions (80% under 1 μm):						
No. 6	6.1	5.5	5.3	5.3	5.2	5.1
No. 4	10.5	9.6	9.4	9.2	9.0	9.0
No. 1	16.7	15.1	14.6	14.4	14.1	14.0
Aqueous suspensions of Milicz clay (d_{av} = 70μm):						
13.9 weight % solids	6.1	4.3	3.8	3.6	3.3	3.2
16.8 weight % solids	9.5	6.7	5.7	5.4	5.1	5.0
19.6 weight % solids	10.5	10.5	8.9	8.2	7.5	7.1

Aqueous clay suspensions:

No. V	9.6	6.6	5.8	5.4	5.0	4.8
No. III	15.7	11.0	9.6	9.0	8.3	8.1
No. I	20.5	14.3	12.5	11.6	10.8	10.4

Aqueous suspensions of cement rock (92% under 74 μm), 54.3 weight % solids

	4.5	3.7	3.5	3.4	3.3	3.2

Sewage sludges, 10 weight % solids:

Chicago plant	28.3	14.3	9.7	7.4	5.2	4.1
Stuttgart plant	>40	31.8	21.7	16.7	11.9	9.7

Sludge No. 5 (Imhoff tank), 14 weight % solids

	13.4	7.7	6.0	5.3	4.5	4.2

Aqueous suspensions of coal char

Weight %							
Total solids	325-mesh solids						
45.4	28.6	14.2	7.9	5.9	5.0	4.2	3.8
50.4	47.5	12.9	7.1	5.3	4.5	3.7	3.4

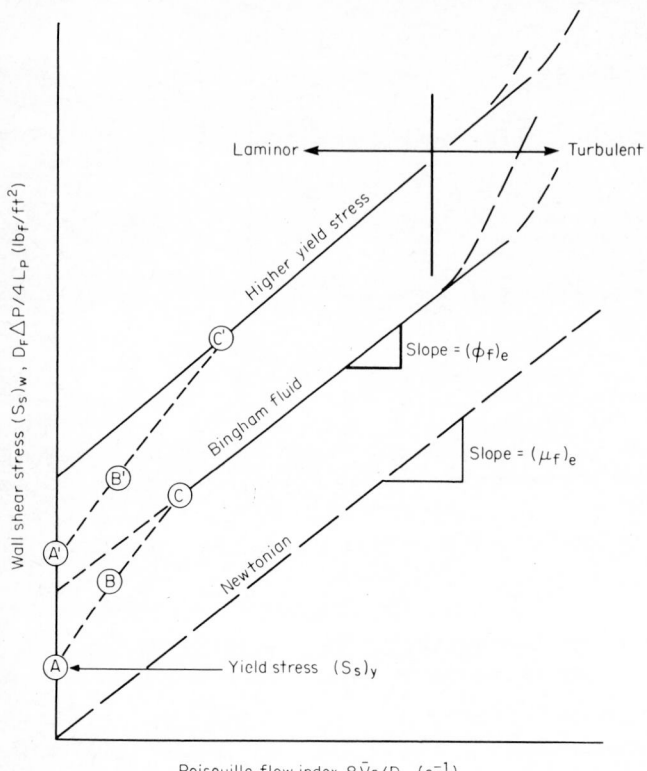

Figure 5.18 Typical pressure index versus Poiseuille index curve for Bingham fluids.

Time-Dependent Fluids The profiles established by time-independent fluids are, at least to some degree, predictable. There are, however, two broad classes of nonnewtonian fluids that change apparent viscosity with time. These time-dependent fluids are referred to as *thixotropic* if the apparent viscosity decreases with time, and *rheopectic* if the apparent viscosity increases with time. Thixotropic fluids become more pseudoplastic, and rheopectic fluids more dilatant, with time. In both cases shear stress–shear rate curves, although time-dependent, generally follow the power-law relationship of pseudoplastics; the exponent is less than 1. This results in profiles that approach, in time, the newtonian parabolic profile for rheopectic fluids, and blunter profiles for thixotropic fluids. Most thixotropic and rheopectic fluids have high apparent viscosities and, therefore, are in the laminar regime characterized in the earlier discussion of pseudoplastics.

Thixotropic Fluids Thixotropic fluids have a structure that breaks down under constant shear. Low-temperature crude oils, drilling mud with the consistency of

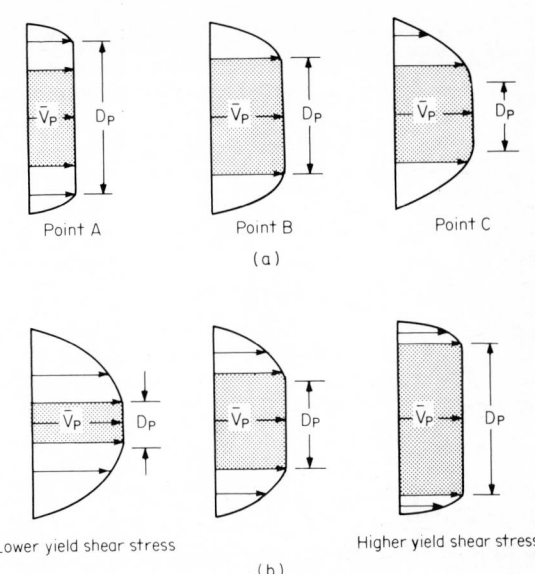

Point A Point B Point C

(a)

Lower yield shear stress Higher yield shear stress

(b)

Figure 5.19 Velocity profiles for Bingham fluids. (*a*) Development of laminar profile. (*b*) Effect of yield stress on plug diameter at point *B*.

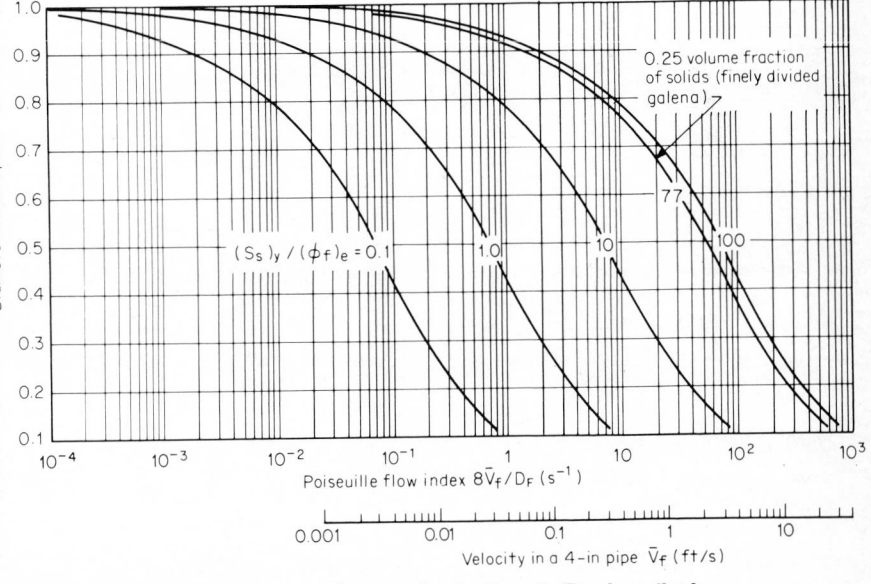

Figure 5.20 Ratio of plug to pipe diameter for the flow of a Bingham fluid.

bentonitic clay in water, paints, ink, and mayonnaise are examples of thixotropic fluids.

In Fig. 5.21 is a generalized shear stress–shear rate curve for thixotropic fluids. As the shear rate is increased over a short period of time, the upper curve *AB* is followed. An immediate decrease in shear causes the fluid to follow the lower portion of the hysteresis curve, from *B* to *A*. This results in the flow-profile changes shown in the lower portion of Fig. 5.21.

If the shear rate is held constant at point *B*, the structure continually breaks down and undergoes an apparent viscosity decrease. Decreasing the shear rate from that at point *C* to zero can set up a yield-stress value at point *D*. The resulting profiles can differ considerably at all flow rates, depending on whether the flow rate is increasing (*A* to *B*), remains fixed (*B* to *C*), or decreases (*C* to *D*). Most thixotropic fluids recover their original viscosity (point *D* to *A*) if allowed to stand for a sufficient length of time; some fluids require hours, and others restore within seconds. Figure 5.22 shows the decrease in apparent viscosity with constant shear rate of a thixotropic fluid.

Rheopectic Fluids   Time-dependent rheopectic fluids exhibit behavior opposite to that of thixotropic fluids in that, with constant stirring, the shear stress increases (Fig. 5.23). For this reason they are sometimes referred to as negative

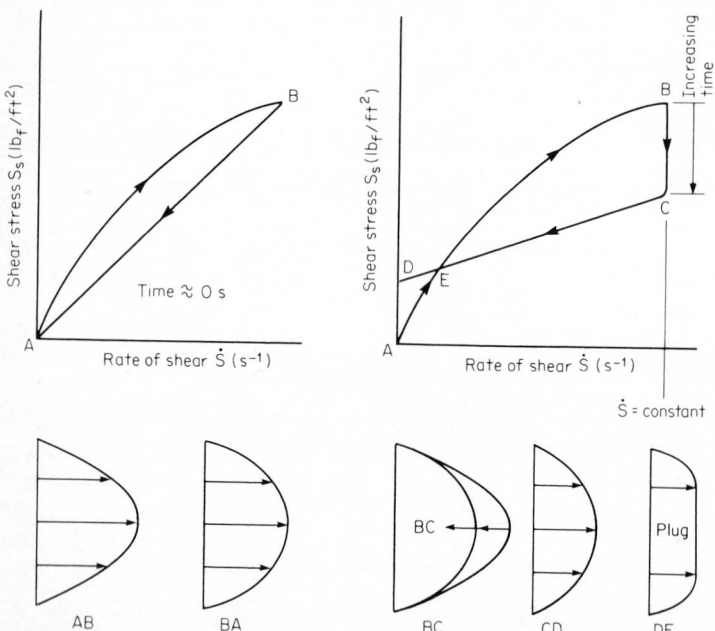

Figure 5.21   Characteristic shear stress–shear rate curves and possible velocity profiles for thixotropic fluids.

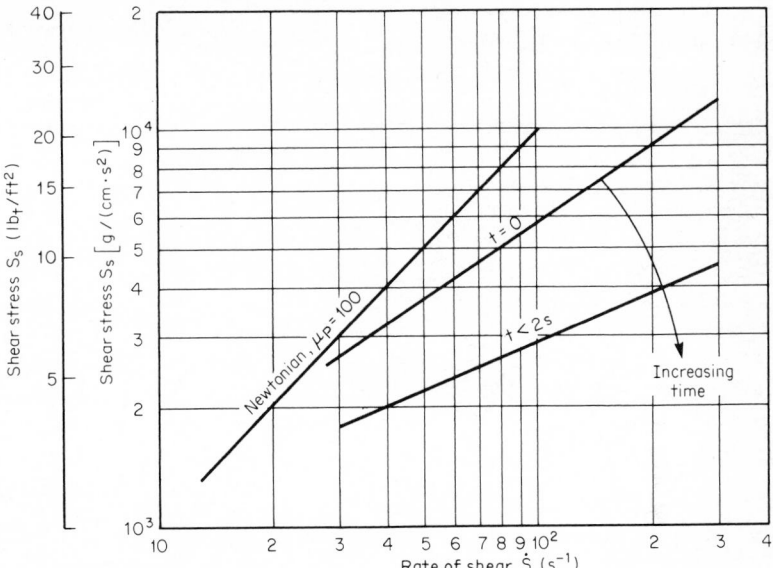

Figure 5.22 Thixotropy in an emulsion of soaps and higher polymers in gasoline (*data from Govier and Aziz, 1977*).

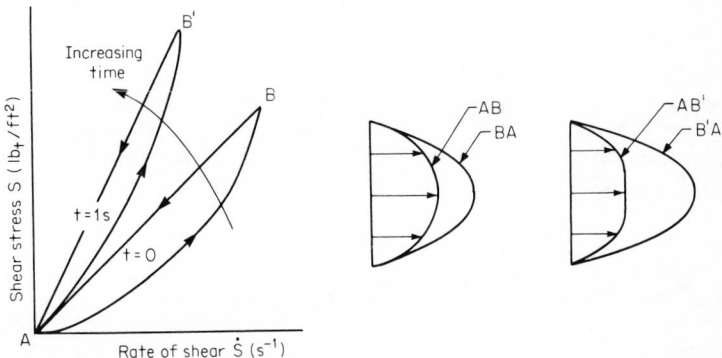

Figure 5.23 Characteristic shear stress–shear rate curves and possible velocity profiles for a rheopectic fluid.

thixotropic fluids. As the slope of the shear stress–shear rate curve increases with time, these fluids become more dilatant.

In rheopectic fluids a small shearing motion accelerates the formation of a more rigid higher-apparent-viscosity fluid structure. If the shear rate is high the viscosity may not increase, but, in general, the apparent viscosity increases with time to some limiting value. If the shear strain rate goes to zero, the initial viscosity and fluid structure are restored. In Fig. 5.24 is a shear stress–shear rate

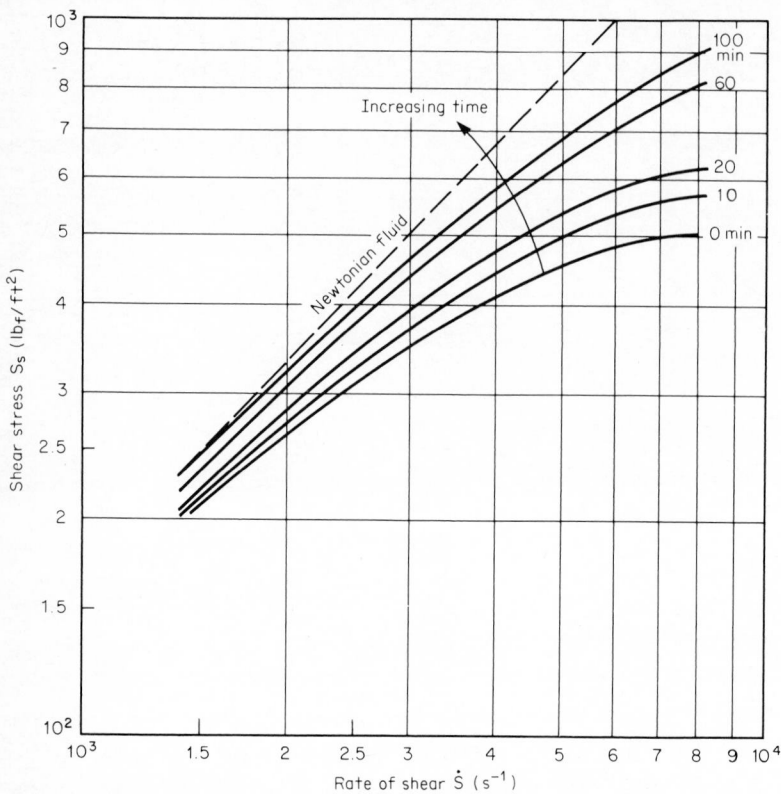

Figure 5.24 Time-dependent rheopectic fluid, a 2000-molecular weight polyester *(data from Govier and Aziz, 1977).*

curve for a 2000-molecular-weight polyester that exhibits rheopectic behavior in some ranges. In Fig. 5.25 are the expected profile alterations for rheopectic fluids. At constant flow the profile becomes more newtonian.

Viscoelastic Fluids Polymeric liquids, jellies, flour dough, and many fluids normally considered newtonian will, under certain conditions, not immediately return to a state of zero shear rate upon the removal of stress. This retention of shear stress accounts for the Weissenberg effect, in which flour dough climbs up the shaft of the stirrer due to stress buildup in the fluid, and the fact that an extruded polymer has a diameter that is larger than the die bore.

In most fluids the energy of deformation is dissipated, but viscoelastic fluids retain energy, like an elastic solid. If a shear stress is applied to a solid (Fig. 5.25*a*, a defined deformation angle is observed. But for liquids, even a minute shear stress causes a continually increasing shear angle (strain) because viscous liquids cannot support shear. Unlike elastic solids, for which shearing resistance

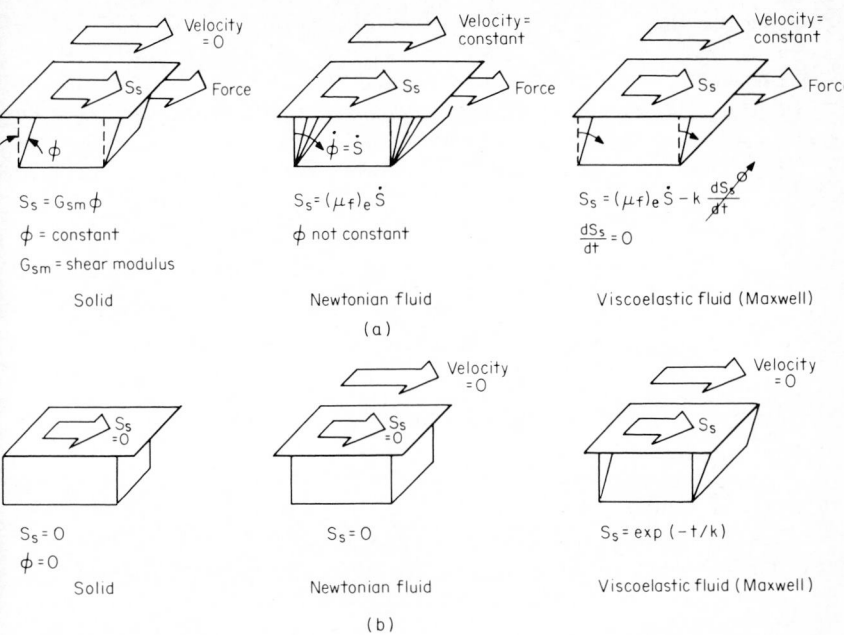

Figure 5.25 Effect of a force on a solid, newtonian fluid and on a viscoelastic fluid. (*a*) Application of force. (*b*) Removal of force.

depends on deformation angle, the shearing resistance of a viscous fluid is proportional to the rate of deformation. For solids the angle is uniquely defined by Hooke's law relating shear stress to strain angle. But for nonviscoelastic fluids there can be no elastic behavior.

Figure 5.25*a* and *b* presents a simplified illustration of the difference between a newtonian and a viscoelastic fluid under flowing conditions. For newtonian fluids, the only normal stresses are pressure-related, and wall shear-stress forces cannot induce elastic strain. But with viscoelastic fluids at the same velocity, a strain similar to that for a solid is present. This produces an additional normal stress within the fluid. At constant velocity, shear stress is constant, and both fluids exhibit newtonian flow behavior. When the flow rate is suddenly brought to zero, or in oscillatory flow, the change in shear stress is immediately accommodated by the newtonian fluid, and the normal pressure stresses are equalized. But, for the viscoelastic fluid, the strain energy resulting from elastic behavior requires time to dissipate. This time-dependent relaxation of both shear and normal stresses results in the liquid having both *viscous* and *elastic* properties.

The simplest mathematical model for this behavior is the Maxwell fluid equation, in which shear stress is separated into two terms:

$$S_s = (\mu_f)_e \dot{S} - k_1 \frac{dS_s}{dt} \qquad (5.48)$$

The first right-hand term is the linear newtonian viscosity relationship, and the second accounts for the relaxation of shear and normal stresses; k_1 is a constant that depends on viscosity and a rigidity modulus.

At steady flow the rate of change of shear stress dS_s/dt is zero (wall shear stress is constant), and the flow profile can be expected to be newtonian. If the flow rate is brought to zero (Fig. 5.25b,) the shear rate S returns to zero but the shear stress decays exponentially: Separating the variables in Eq. (5.48) gives

$$\frac{dS_s}{S_s} = \frac{-1}{k_1} \, dt \tag{5.49}$$

Upon integration, the shear stress is found to be

$$S_s = e^{-t/k_1} \tag{5.50}$$

The constant $1/k_1$ is called the *relaxation time;* it is the time constant for the exponential decay of normal stresses at constant strain.

Many modern fluids exhibit viscoelastic properties and have been the subject of extensive rheological investigation. Numerous equations have been proposed, of varying complexity. The reader is referred to the texts by Van Wazer et al. (1963) and Wilkinson (1960).

Profile Distortion

Pipe fittings, reducers, expanders, strainers, and elbows—necessary for normal plant piping—all affect profile. The many types and combinations of fittings result in changes that are difficult to predict. Axial velocity vectors may be altered by one or a combination of the following:

1. A pure swirl that causes rotation about the pipe axis.

2. Secondary flows, such as two or more counterrotational vortices in a plane perpendicular to the pipe axis, or a bound, recirculating, secondary vortex flow caused by separation (a sudden enlargement in pipe size). These are usually in a plane parallel to the pipe axis.

3. An asymmetrical profile that peaks near the wall.

4. A symmetrical profile that has a high core velocity resulting from a sudden or too rapid reduction in pipe area.

In combination, these usually result in radial, tangential, and axial velocity vectors that are not symmetrical. While some flowmeters may be insensitive to radial velocity components (magnetic flowmeters), others are highly susceptible (single-path ultrasonic flowmeters). Also, insensitivity to tangential components (swirl) varies among flowmeters. For example, Fig. 5.26 shows the discharge-coefficient bias error (Kinghorn, 1977) for a venturi and an orifice when pure swirl is introduced into the line.

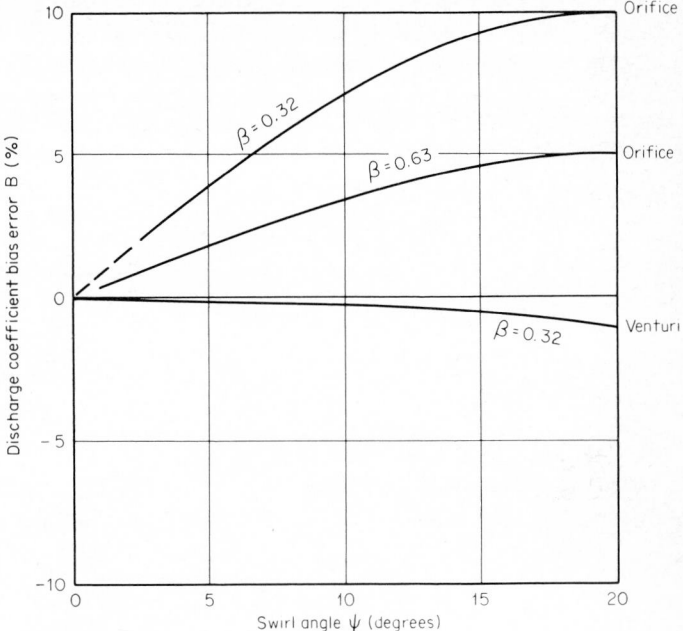

Figure 5.26 Bias error caused by swirl.

The straight lengths of pipe required to eliminate swirl are different for gases and for liquids; the higher viscosities of liquids dissipate both swirl and profile distortion sooner, so liquids require shorter pipe lengths. Figures 5.27 and 5.28 show the pipe lengths required to dissipate an entry swirl angle at various Reynolds numbers. These curves were derived from the equations presented by Kreith and Sonju (1965).

In general, upstream fittings can be grouped into two broad categories: those that distort profile but produce little swirl, and those that both distort and cause bulk swirl. A distorted profile can usually be brought into an acceptable form by adding sufficient straight lengths of pipe or by using flow conditioners in combination with reduced straight lengths.

A single elbow (Fig. 5.29) distorts the profile in the elbow plane, with the asymmetric profile progressively becoming double-peaked as the plane is rotated into the horizontal. Secondary flows, consisting of double vortices, are also produced, and these give rise to a radial velocity component. A single elbow, or any fitting that produces an asymmetric profile, is considered to be in the first category—producing little swirl. A single bend, two elbows in the same plane, or a partially opened valve will distort the profile but impart little rotation.

If elbows are arranged such that the flow must change direction (Fig. 5.30), swirl is produced in addition to the asymmetric profile. This causes the profile

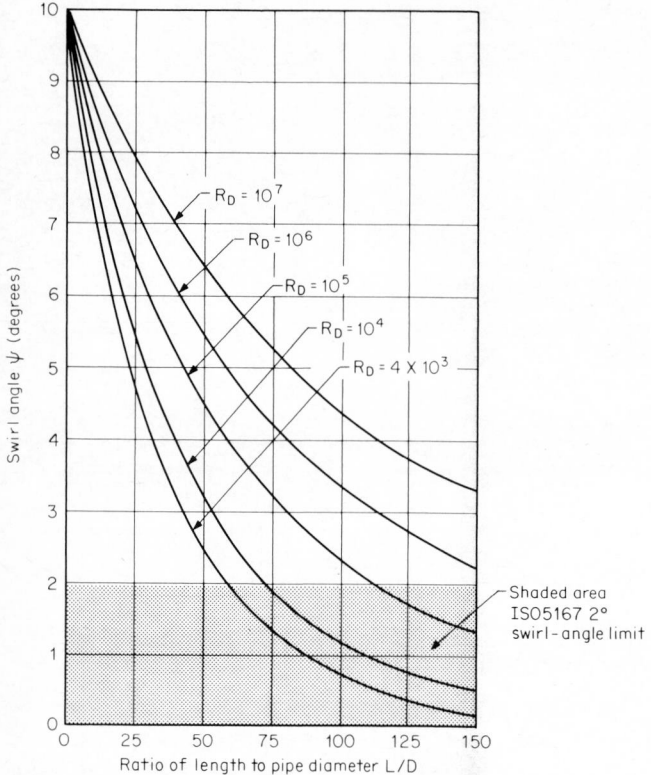

Figure 5.27   Pipe lengths to reduce 10° initial swirl angle.

to corkscrew down the pipe, as shown in Fig. 5.31. Swirl is caused by low-energy fluid moving from the inside of the first elbow to the inside of the second. When intersecting elbow planes form an angle between 30 and 90°, most of the low-energy fluid follows the shortest path, causing fluid rotation rather than producing secondary flows. The strongest swirl occurs when elbow planes are at 60° to each other. However, the addition of 2 to 5 (and preferably 10 to 30) diameters of straight pipe between elbows substantially reduces interaction and considerably reduces swirl.

Swirl and similar distorted profiles are also produced when a high-velocity stream is blended at right angles into a lower-velocity stream.

The analysis of such profiles is difficult. The swirl angle, profile distortion, and resulting three-component velocity vectors are related to the bend radius, entering profile, entering swirl, secondary flows, spacing between elbows, and angle between elbow planes. Figure 5.32 shows some typical fittings and piping conditions, grouped into the two flow-disturbance categories. In general, fittings that cause secondary flow in a plane parallel to the pipe axis produce higher fluid

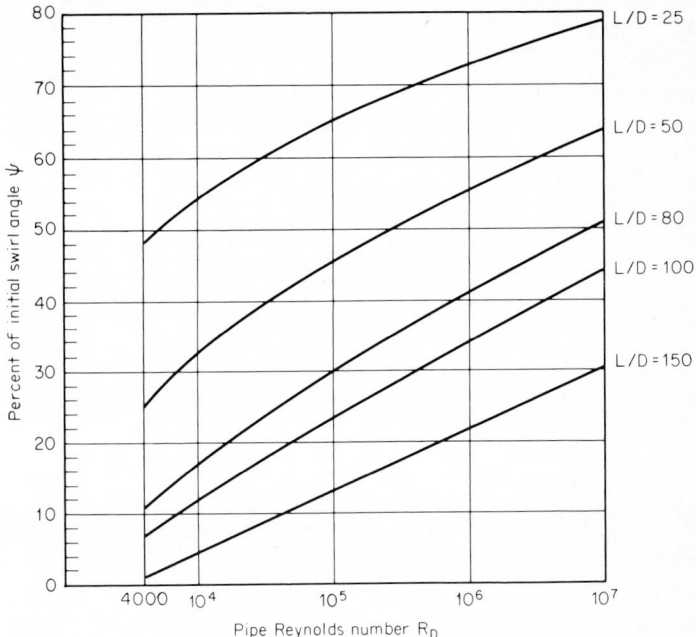

Figure 5.28 Pipe lengths to reduce initial swirl angle, percent.

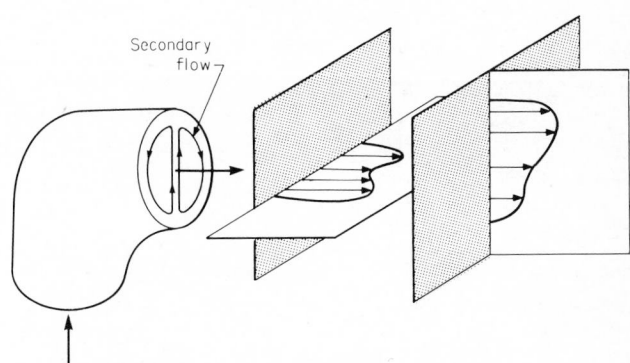

Figure 5.29 Velocity distribution following a single elbow.

turbulence, which assists in recovering the profile in shorter lengths. Contoured fittings, such as a long sweeping bend, more strongly establish both secondary flows and profile distortion.

Flow Conditioners In many installations it is impossible to provide sufficient lengths of straight pipe to remove swirl and to restore an acceptable reference

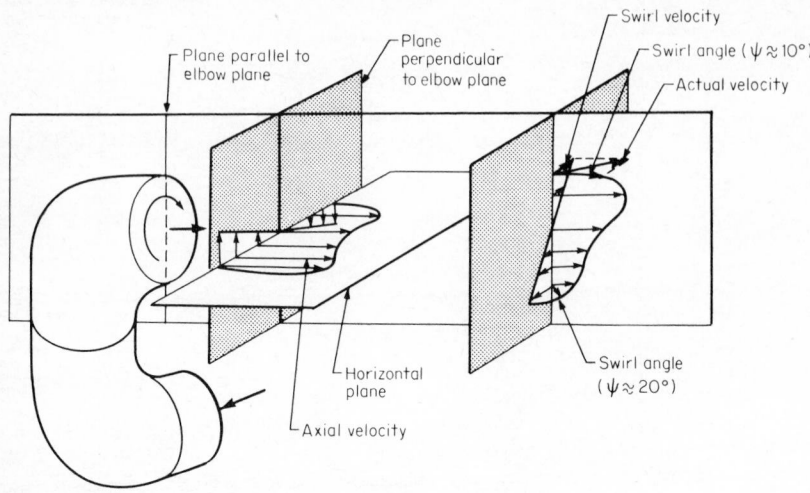

Figure 5.30 Velocity distribution following close-coupled elbows in different planes.

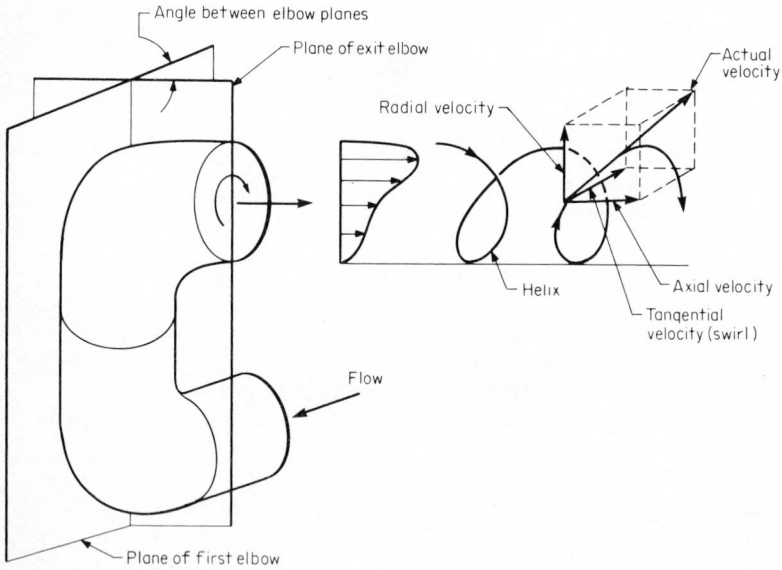

Figure 5.31 Velocity components in swirling flow following close-coupled elbows in different planes.

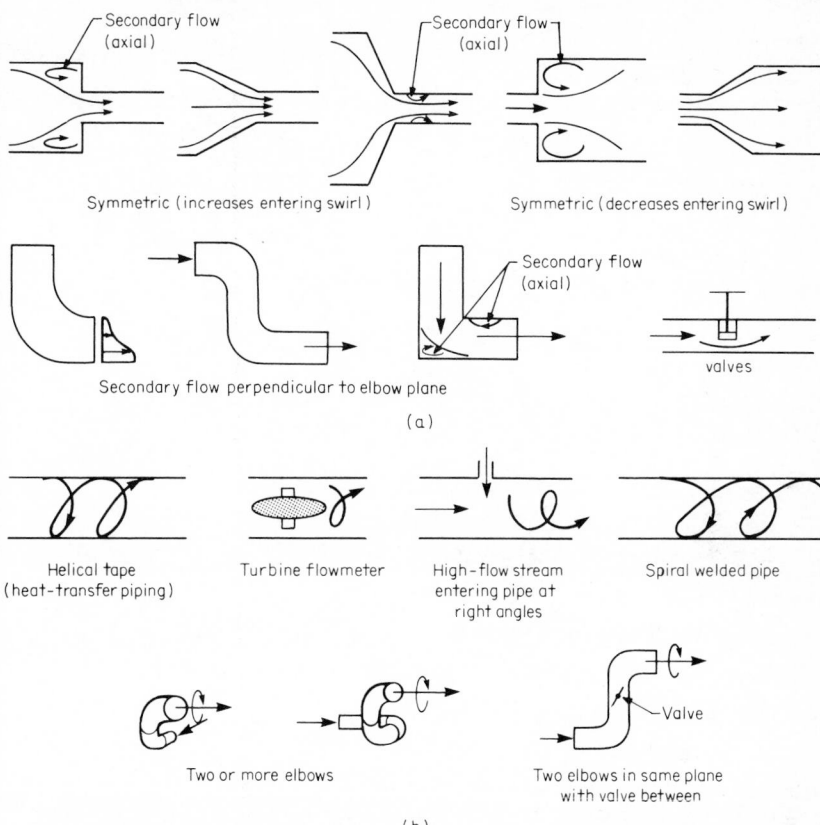

Figure 5.32 Fittings categorized by disturbance type. (*a*) Category 1: profile distortion and secondary flow. (*b*) Category 2: profile distortion, secondary flows, and swirl.

profile geometry. For this reason, flow conditioners (Fig. 5.33) are used in combination with specified pipe lengths.

Shown in Fig. 5.34 are velocity and swirl angles measured at 1 pipe diameter upstream and 7.5 pipe diameters downstream of three flow conditioners. Distortion and swirl were produced by two close-coupled elbows in different planes, with flow conditioners located 1.5 pipe diameters downstream of the last elbow. The swirl is effectively removed by a four-tube-bundle conditioner, but the profile is not brought into symmetry. The swirl is eliminated by both the Zanker and Mitsubishi conditioners, and the profile is made more symmetric.

AGA-ASME tube-bundle conditioners (Fig. 5.35) are widely used because of their low cost, ease of fabrication, and low maintenance. The number of tubes is not specified, but, in general, four to eight tubes effectively remove swirl and secondary flows, and nineteen or more remove moderate distortion. Further increasing the number of tubes does not improve the profile-restoring capability.

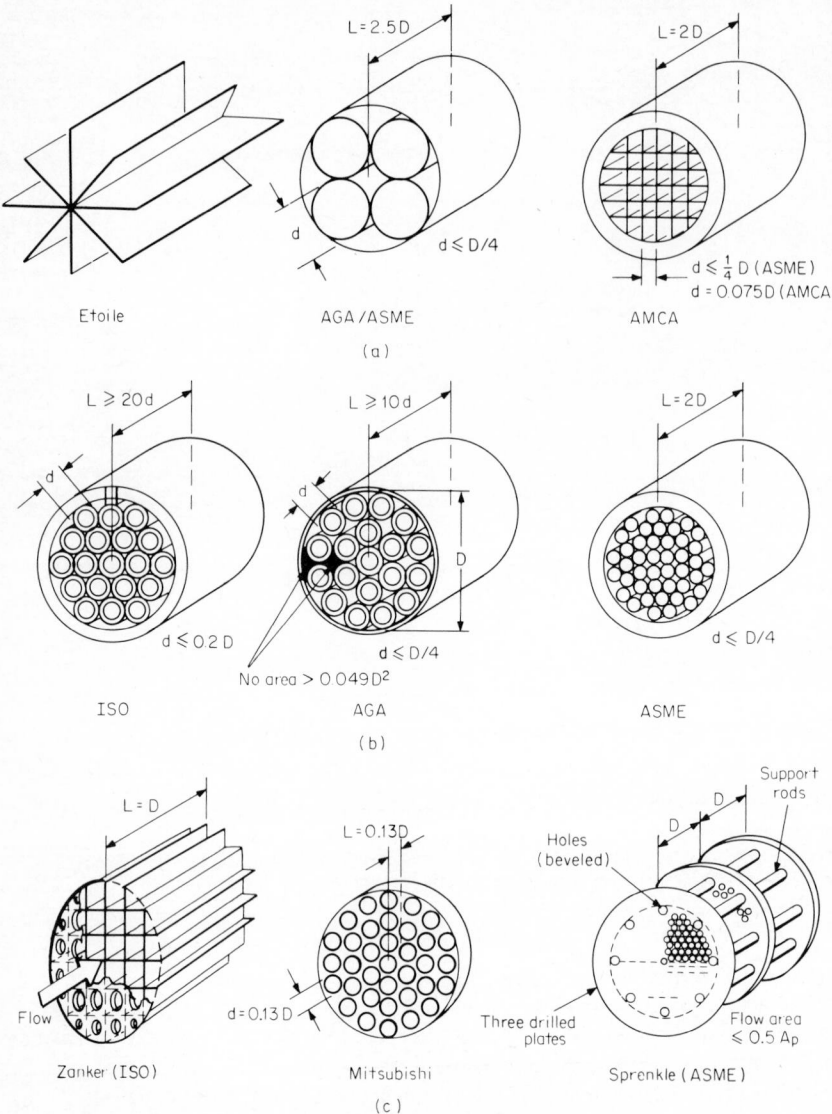

Figure 5.33 Flow conditioners. (*a*) Swirl. (*b*) Swirl and moderate distortion. (*c*) Swirl and distortion.

The hole size, spacing, and dimensions for the downstream box construction of the Zanker (1969) conditioner are shown in Fig. 5.36. The construction details for the single-plate Mitsubishi conditioner (Akashi et al., 1979) are shown in Fig. 5.37. This conditioner has a lower pressure loss and a fixed hole diameter, and it is easier to fabricate than the Zanker or Sprenkle conditioner. The permanent

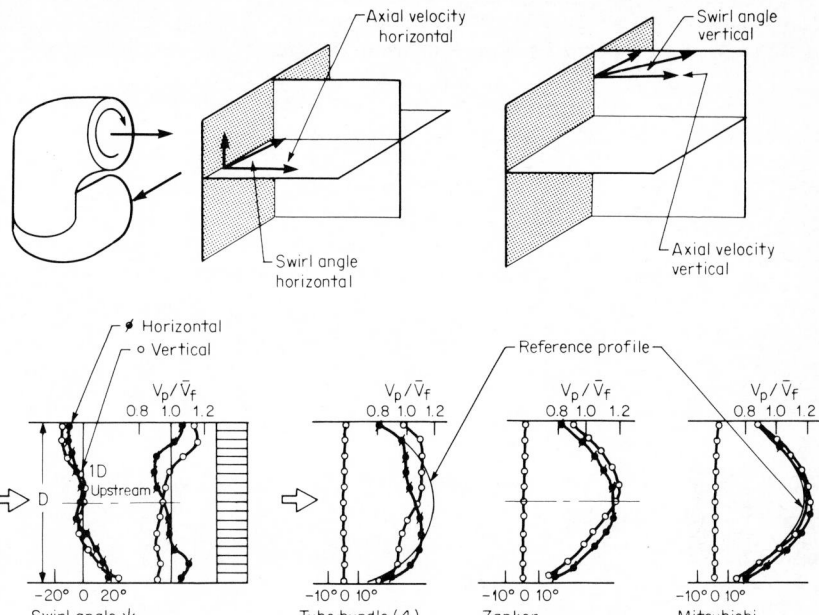

Figure 5.34 Velocity and swirl angle upstream and downstream of three flow conditioners *(from Akashi et al., 1979).*

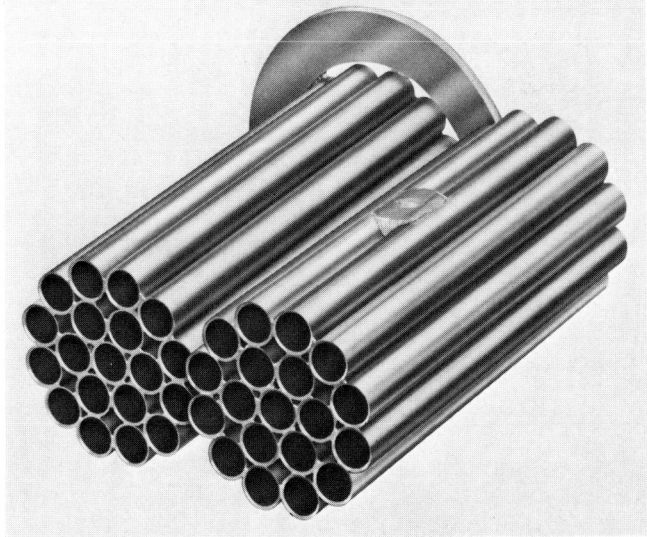

Figure 5.35 **AGA-ASME tube-bundle conditioners** *(courtesy Daniel Industries, Inc.).*

pressure loss increases with the distortion that must be removed. Swirl conditioners (Fig. 5.33a) have the lowest pressure loss but require more straight pipe lengths than the higher-pressure-loss conditioners shown in Fig. 5.33c. The unrecoverable pressure loss for the conditioners shown in Fig. 5.33, expressed in terms of pipeline velocity, is

$$h_l = K_{\text{FC}} \rho_f \overline{V}_f^2 \tag{5.51}$$

in inches of water,

$$H_L = K_{\text{FC,H}} \frac{\overline{V}_f^2}{2g_c} \tag{5.52}$$

in feet of flowing fluid, and

$$\Delta P = K_{\text{FC},\Delta P} \frac{\rho_f \overline{V}_f^2}{2g_c} \tag{5.53}$$

in pounds-force per square foot. Pressure-loss coefficients for these equations are given in Table 5.4.

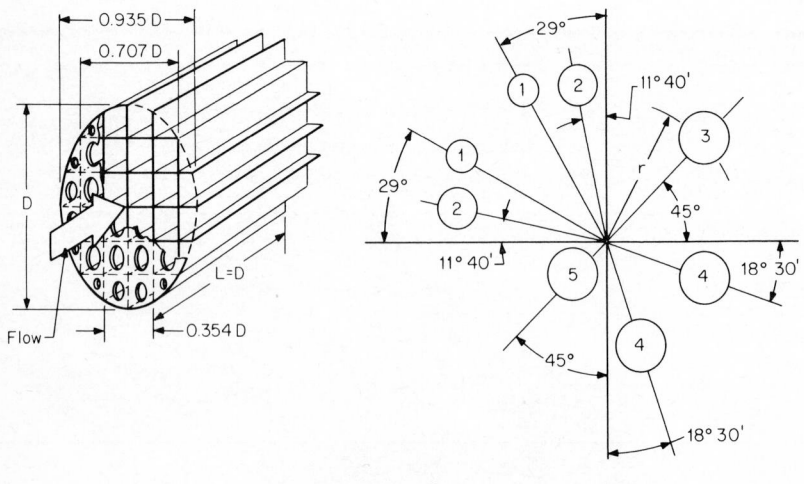

	Hole number				
	1	2	3	4	5
Hole diameter	0.077D	0.110D	0.1365D	0.139D	0.141D
Radius to hole (r)	0.45D	0.425D	0.375D	0.28D	0.125D
Number of holes	8	8	4	8	4

Figure 5.36 **Zanker flow conditioner.**

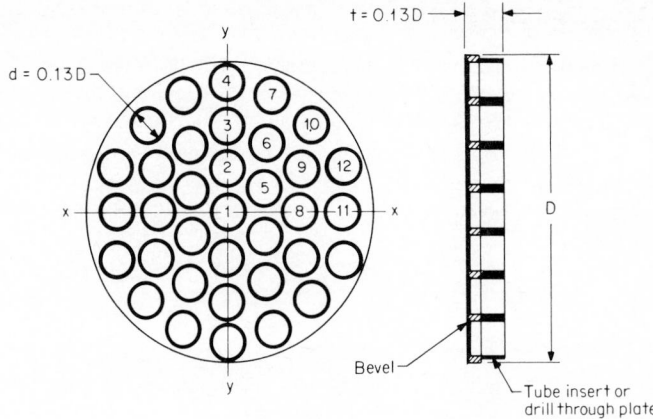

						Hole number						
Coordinate	1	2	3	4	5	6	7	8	9	10	11	12
x	0	0	0	0	0.129D	0.134D	0.156D	0.252D	0.255D	0.288D	0.396D	0.400D
y	0	0.142D	0.283D	0.423D	0.078D	0.225D	0.381D	0	0.146D	0.288D	0	0.151D
Number of holes	1	2	2	2	4	4	4	2	4	4	2	4

Figure 5.37 Mitsubishi flow conditioner (patented in Japan, U.S.A., U.S.S.R., West Germany, U.K., Switzerland, and France).

Table 5.4 Unrecoverable Pressure-Loss Coefficients for Flow Conditioners

Flow conditioner	Unrecoverable pressure-loss coefficient	
	K_{FC}	$K_{FC,H}$ and $K_{FC,\Delta P}$
Tube bundle (4) (Etoile)	0.003	1
Mitsubishi	0.006	2
Tube bundle (20) (Zanker)	0.015	5
Tube bundle (41) (AMCA†)	0.024	8
Sprenkle (50% flow area):		
Beveled holes	0.033	11
Square-edged holes	0.042	14

†Air Moving and Conditioning Association Standard 210-67 (1967).

NONHOMOGENEOUS FLOWS

The theoretical and working flow equations for all flowmeters are based on homogeneous flows. Densities at specific cross sections within the flowmeter must be known from either calculations, direct measurements, or previously determined tabulated values. Depending on the flowmeter's principle of operation, the effect of a nonhomogeneous flow can be slight or quite significant.

It is convenient to group nonhomogeneous fluids into the following three broad categories:

1. Two-phase fluids

2. Fluids with two or more components

3. Two-phase fluids or fluids with two or more components that can be considered homogeneous

Two-phase fluids include *quality* steam and ammonia vapor flows, where the fluid is a single substance partially in a liquid state and partially in a gaseous state. Fluids with two or more components would be air in water, coal mixed with oil, pulp stock in water, or any combination of liquids, gases, and solids of differing substances in varying proportions. In many cases, if the second phase or one of two mixture components is finely dispersed and approximately of the same density as the other, the nonhomogeneous mixture is considered homogenous, and the flow equations are used without additional secondary equipment to measure mixture percentage.

There are two significant differences between nonhomogeneous and homogeneous flows. First, the density is not easily determined; second, one phase or one or more of the components may not be moving at the same velocity as the main flow, and, in some cases, may actually be flowing along the bottom of the pipe. This is called *holdup* or *slip*. Slip is a complex function of viscosity, particle size, density differences, surface tension, and the *superficial* velocity of each component. The effects of gravity also alter flow patterns—horizontal, vertical, and inclined pipe cause differing relative velocities for the components.

While there is a large body of information on nonhomogeneous pressure loss, the added complexity of making a flow measurement has not been well addressed, either experimentally or theoretically. The reader is referred to the texts by Govier and Aziz (1977), Hewitt (1978), and Wasp et al. (1979) for detailed analysis of these complex flows.

PULSATING FLOW

Reference-condition accuracy is based on steady flow, which is defined by ISO Technical Report 3313 (1974) as "flow condition in a measuring section in which the flow does not vary in time." ISO 5167 (1980) notes that "This International Standard applies only to pressure differential devices in which the flow remains subsonic throughout the measurement section, is steady or varies only slowly with time." Although an ISO working group (1975) was established to characterize steady flow, and several draft documents were written, no international or national standards exist that define steady flow. Without a clear distinction between steady and unsteady flow, the user must decide whether a flow measurement standard does or does not apply.

It is important to recognize that flow is a dynamic process, and the dynamics of the pulsating fluid, turbulence intensity, and primary and secondary devices are interrelated in flow measurement. Fluctuating pressure within lead lines, the response to velocity-profile changes, vortex formation, and turbine-rotor inertia are but some of the influences arising from interrelationships between flow and flowmeter dynamics.

In the *ideal* flowmeter, variations in flow rate, pressure, and density are instantaneously sensed, and no error results. But in real flowmeters, the inertia, damping, and fluid-pressure forcing function, as well as profile alterations and pipeline resonant frequencies, affect secondary measuring devices. The response characteristics of each affect the resolution of the overall flow.

Pulsating flow has been the subject of numerous experimental and analytical investigations. This work covers four major investigation areas:

1. Pulsating and unsteady pipeline flow investigation

2. Development of correction factors to be applied to specific flowmeters

3. Define a *threshold pulsation index* below which a certain bias error is assumed

4. Design of passive filters, such as volume tanks and in-line restrictors, that reduce pressure fluctuations to acceptable limits for flow measurement

The following paragraphs citing Muller (1970) appeared in the proceedings of the symposium, *The Measurement of Pulsating Flow:*

The aim of the meeting was to make people aware of the influence of pulsations on the accuracy of measurement. He [Muller] observed that the problem had not really been solved since Hodgson's work. However, delegates will carry away with them important ideas now being developed in relation to the subject.

There are several glimmers of light although the practical solution of the problem at present is still to damp out fluctuations, as a suitable flowmeter for all flow conditions is awaited. Of necessity, measurement must come before control can be attempted and the limiting factor is in the former and not the latter; hence only one paper on pulsating flow control was presented.

Little has changed since this 1970 symposium, and although there is more insight into the problems of pulsating-flow measurement, the practical solution still is to *filter* pulsations prior to making a measurement.

In the absence of standards, the threshold pulsation index I_P defined by Head (1956) is recommended. For index values below the threshold value, reference-condition accuracy applies. For greater index values, passive-type filters (such as restrictors), increased volume, or a combination of both should be used to bring flow variations below the threshold value.

Flow Characteristics

Typical flowmeter signals produce three distinct types of traces on a recorder. The pen may draw a random pattern (Fig. 5.38a), a defined oscillation period with a superimposed noise pattern (Fig. 5.38b), or a complex waveform (Fig. 5.38c).

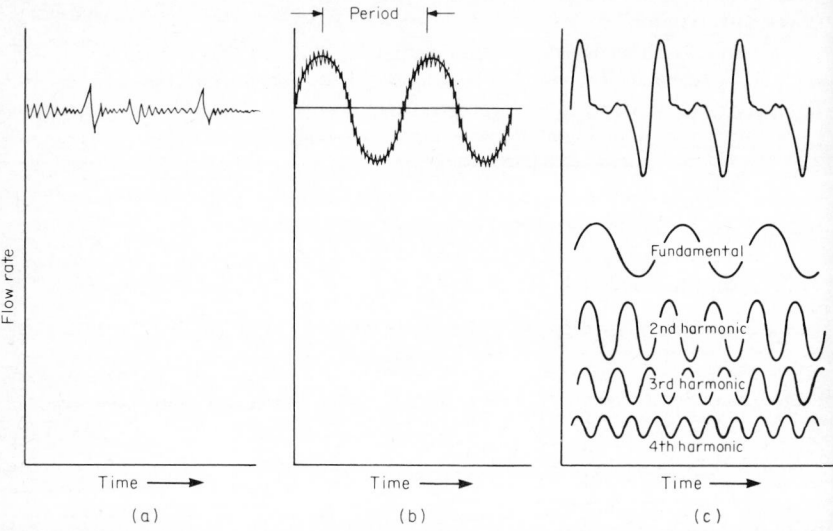

Figure 5.38 Typical recording of a flowmeter signal. (a) Noise. (b) Defined waveform. (c) Complex waveform.

The small-amplitude, rapid pen drawing is caused by the action of the fluid's turbulent intensity on the flowmeter and is referred to as *noise*. Noise is always present and is produced by throttling valves, fittings, slightly protruding gaskets, flow separation, and cavitation within the installation. Noise patterns are usually of smaller amplitude in laminar flow than in turbulent flow. Because they are usually of small amplitude, random, and similar to noise patterns in the laboratory in which the flowmeter accuracy was determined, they lead to no additional error.

A cycling flow rate results from fixed-speed pumps, compressors, and fans, oscillating control valves, and sometimes fixed-pipeline vibration frequencies. The time to complete one cycle may be hours or seconds, but the period is defined by a single value. For long periods and small flow-rate changes, all flowmeters usually follow these changes while remaining within their reference accuracy.

When reciprocating compressors are used to transport large quantities of high-pressure gas in reciprocating steam-engine flows, pressure and flow fluctuations are large and contain multiple frequencies (Fig. 5.38c). This results in observable, usually fixed waveforms with several harmonic frequencies of differing amplitudes.

Edwards and Wilkinson (1971) investigated the effect of changing frequency on laminar flow profiles (Fig. 5.39). At low amplitudes and frequencies the profile is unaffected, but with increasing frequency two symmetrical peaks develop that progressively move outward toward the wall. For turbulent flow (Fig. 5.40), Mizushina et al. (1975) show that a high-amplitude, low-frequency source distorts the profile throughout the pulsating period. The effects diminish with increasing frequency.

Shown in Fig. 5.41 are curves of typical integrated-velocity-profile flow rate, centerline velocity, and longitudinal pressure gradient in pulsating turbulent

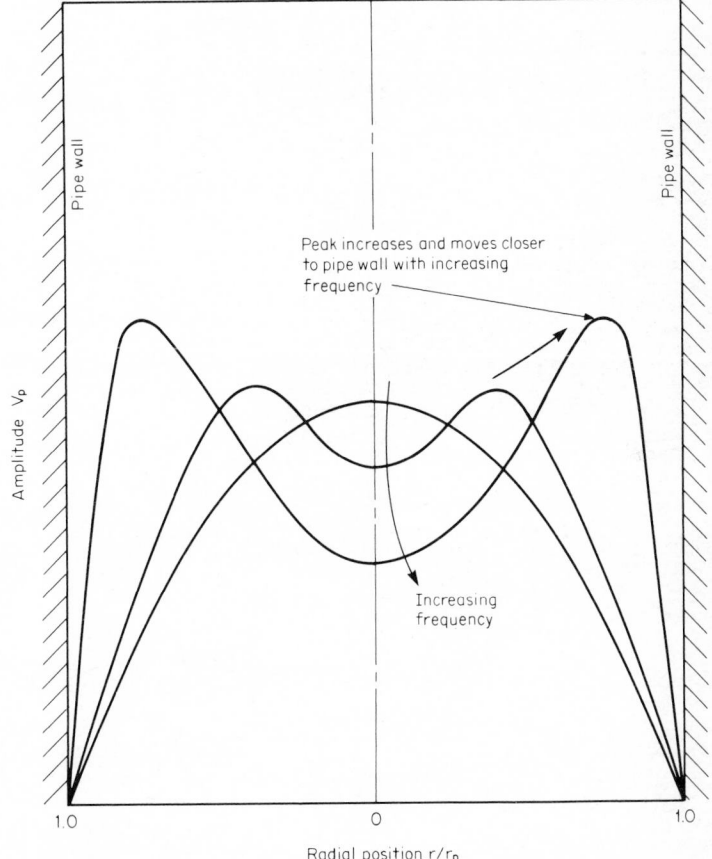

Figure 5.39 Effect of frequency on velocity amplitude in laminar flow.

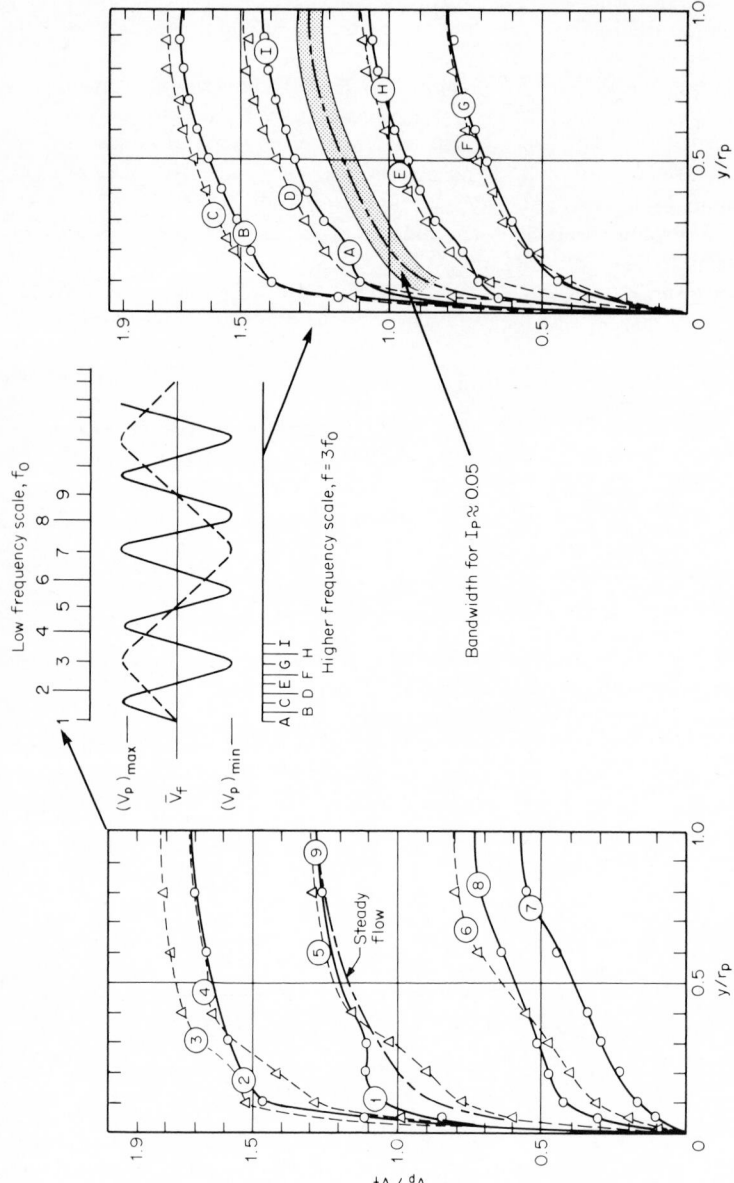

Figure 5.40 Effect of frequency on profile in turbulent flow.

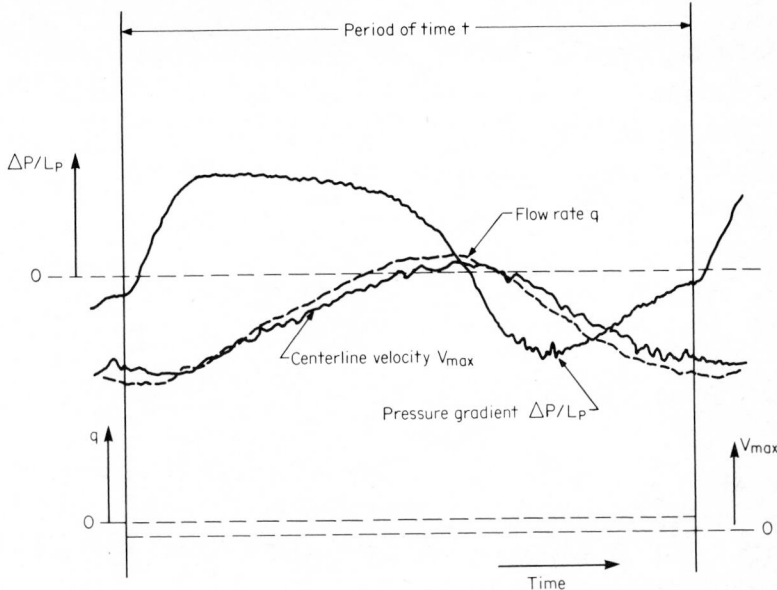

Figure 5.41 Typical fluctuations of discharge pressure gradient and centerline velocity for pulsating turbulent flow ($R_D = 10^5$).

flow, as measured by Kirmse (1979). The integrated velocity profile (flow rate) is not in phase with either pressure gradient (pressure loss) or centerline velocity. In general, velocity and pressure frequencies are combinations of harmonic frequencies of varying amplitude. To fully identify their characteristics requires high-frequency-response velocity and pressure transducers, and a wave analyzer. Additionally, a restrictive device in the line, such as an orifice or vortex-element bluff body, alters the frequency spectrum in the critical measuring planes. In many applications, waveforms in gas pipelines are extremely complex and, as a result, have not been analyzed satisfactorily.

Threshold Pulsation Index

Head (1956) has shown that a pulsation intensity, defined as the ratio of the difference between maximum and minimum flow rate to the average flow rate (Fig. 5.42) can be used to estimate bias errors for a given class of flowmeters. The pulsation index is defined as

$$I_P = \frac{(\overline{V}_f)_{\max} - (\overline{V}_f)_{\min}}{2(\overline{V}_f)_{\text{av}}} = \frac{(q_v)_{\max} - (q_v)_{\min}}{2(q_v)_{\text{av}}} \tag{5.54}$$

The flow-rate correction factor based on this index is

$$F_P = \frac{\text{indicated flow}}{\text{average flow}} = (1 + 4\alpha b_P I_P^2)^n \tag{5.55}$$

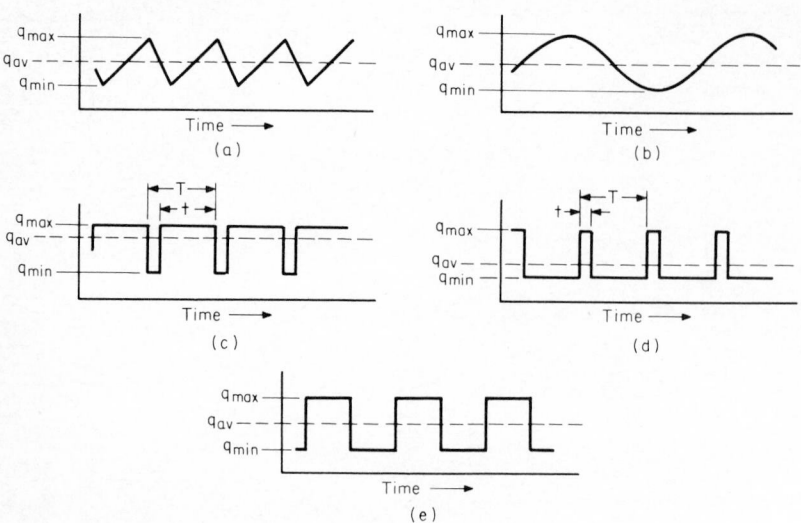

Figure 5.42 Waveform coefficients for some geometrical waveforms. (a) Sawtooth, $\alpha = 0.0833$. (b) Sinusoidal, $\alpha = 0.125$. (c) Dip, $\alpha = (t/T)(1 - t/T)$. (d) Spike, $\alpha = (t/T)(1 - t/T)$. (e) Rectangular, $\alpha = 0.25$, $t/T = 0.5$.

In Eq. (5.55), α is a wave-shape coefficient determined by Fourier analysis of the harmonics contained in the various wave shapes shown in Fig. 5.42. The maximum known value of α is 0.25 for a rectangular flow fluctuation. The frequency coefficient b_P ranges from 0 to 1 and includes flowmeter-system damping and the effect of the pulsation frequency on the primary element. For a given wave shape, b_P is close to 0 (no error) over frequency ranges that flowmeters can follow. These ranges extend from 1 cycle/h to 10 Hz, depending on flowmeter and auxiliary equipment. At greater frequencies, flowmeters cannot follow rapid changes, and b_P approaches 1.0, resulting in maximum bias error.

In turbulent incompressible flow, n in Eq. (5.55) is 0.5 for differential producers and 1.0 for linear meters; it is 0 for all flowmeters in laminar flow. In the laminar-to-turbulent transition regime, n may range from 0 to 0.5 for differential producers and from 0 to 1.0 for linear flowmeters. The maximum bias error then occurs when $n = 0.5$ for differential producers, when $n = 1.0$ for linear meters, and when $b_P = 1.0$.

The pulsation index is maximum for all flowmeters when $n = 1$, $\alpha = 0.25$, and $b_P = 1.0$. If $F_P = 1.001$ (bias $= +0.1$ percent) is selected as an acceptable correction-factor limit that includes normal variations among laboratories, then with these values for n, α, and b_P, Eq. (5.55) becomes

$$1.001 = 1 + I_{PT}^2 \tag{5.56}$$

where I_{PT} is the *threshold pulsation index* that defines steady flow at reference conditions, and

$$I_{PT} = 0.03 \tag{5.57}$$

This index allows for a ± 3 percent flow-rate change for a linear flowmeter (vortex, turbine, ultrasonic, etc.) and a ± 6 percent variation in differential pressure for differentia producers.

Keyser (1981) empirically derived a maximum-bias-error equation that fits 90 percent of available unsteady-flow data for orifice flowmeters. This equation and the data are shown in Fig. 5.43. The bias error is calculated in terms of the threshold pulsation index as

$$B = 20 \exp\left[-(2.7 - 7.2I_{PT})^2\right] = 0.04 \text{ percent} \tag{5.58}$$

In a pulsation study on gas-turbine flowmeters, Lee et al. (1975) presented the data shown in Fig. 5.44. Substituting the threshold pulsation index into the maximum-bias-error equation gives

$$B = 69I_{PT}^2 = 0.06 \text{ percent} \tag{5.59}$$

This recent data and that presented in the original work by Head suggest that a threshold pulsation index of 0.03 is reasonable for the purpose of defining steady flow.

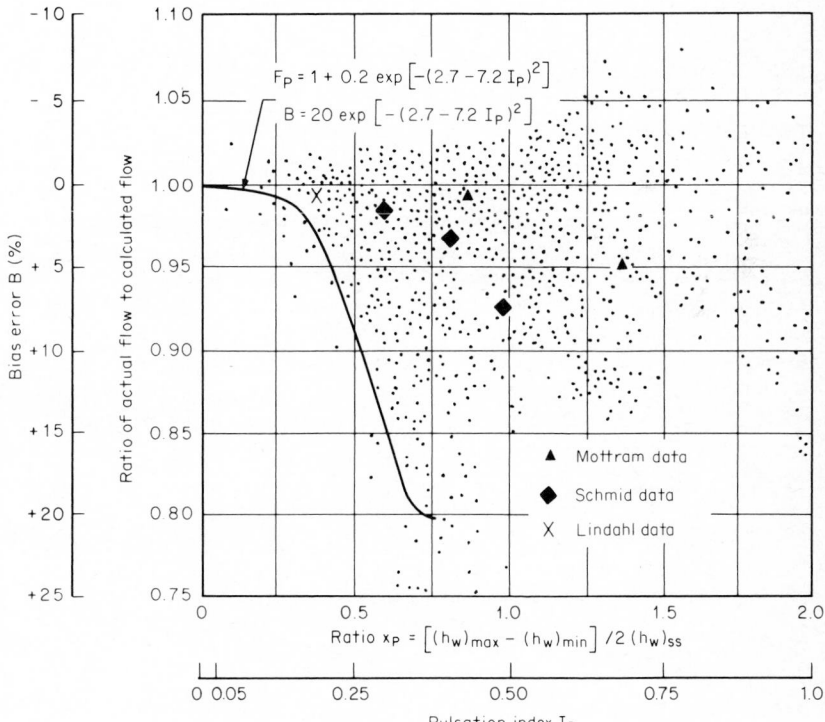

Figure 5.43 Orifice-flowmeter pulsation data.

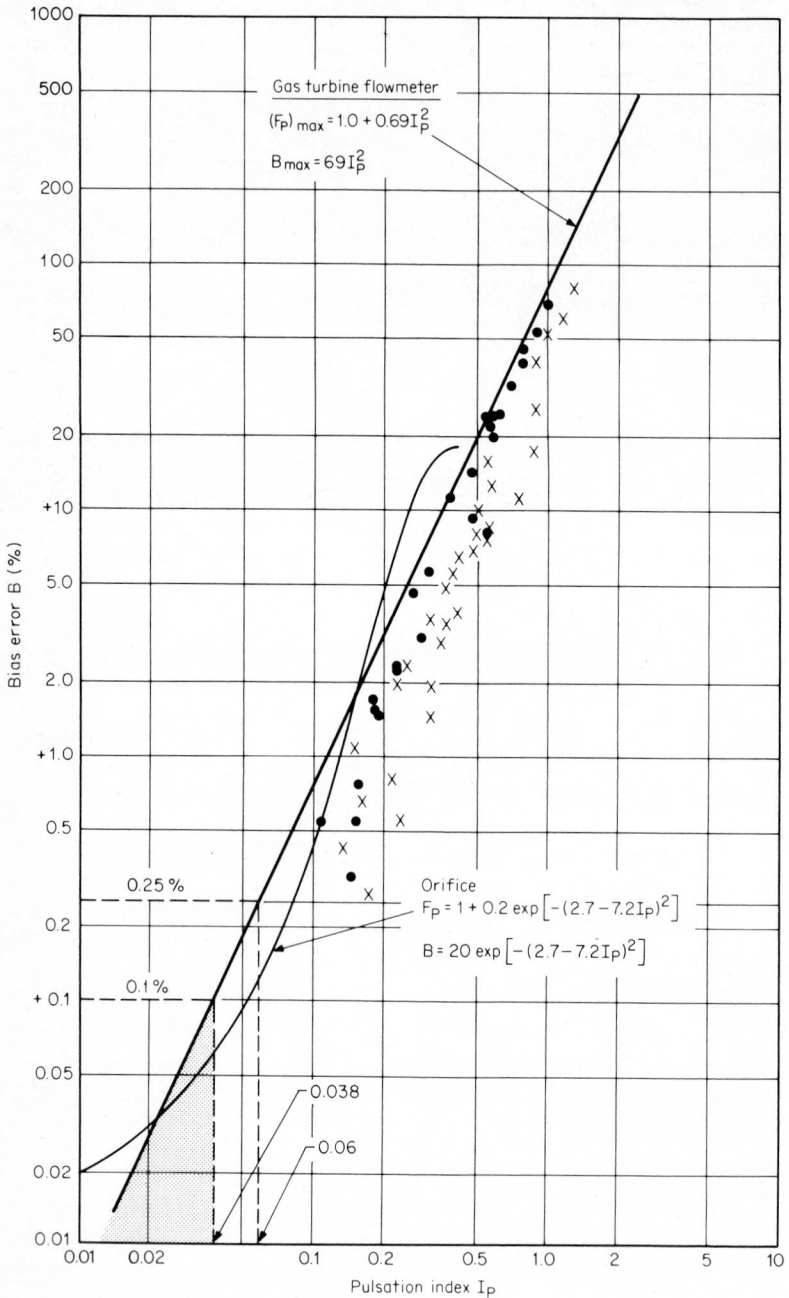

Figure 5.44 Gas turbine-flowmeter pulsation data.

Pulsation Measurement

Severe flow pulsation is usually visually apparent from either a recording of the flowmeter signal, an audible steady rumble, or a sustained hum. For highly damped secondary devices, a visual observation or audible detection may not be sufficient.

If pulsating flow is suspected, a high-frequency-response pressure- or velocity-measuring device, located close to the primary device, should be used to characterize the unsteadiness. A linearized hot-wire anemometer with a light-beam recording oscillograph or an electronic strain-gauge pressure transducer can detect frequencies up to 10 kHz. Harmonics can then be determined using a wave analyzer, Fourier analysis, or spectrum analysis.

Filters

If the threshold pulsation index has been exceeded, corrective action is necessary. Accepted practice is to install a low-pass in-line filter that cuts off all frequencies above the minimum determined value. The design of these filters depends on whether the fluid is a liquid or a gas. Since pulsation not only affects flow measurement but causes unacceptable pipeline vibration as well, pulsation dampers are usually part of the initial piping design. Figure 5.45a shows a typical volume-choke pulsation damper used for gas flows, and Fig. 5.45b a gas-filled surge chamber used for liquids. Industrial *stabilizers* are shown installed on the suction and discharge sides of a four-stage liquid pump in Fig. 5.46.

Numerous passive filter designs have been found effective for various pump types, pump speeds, etc. Location, volume, restrictor size, frequency range, and other parameters for passive filters are the subject of many texts, design brochures, and technical articles. The reader is referred to the works by Campbell (1958), Chilton and Handley (1955), Greer Hydraulics (1981), and Yeaple (1966).

CAVITATION

When decreased line pressure approaches the vapor pressure of a liquid in the line, cavitation begins. In essence, cavitation is boiling of the liquid caused by decreasing pressure rather than by increasing temperature. It is the formation and collapse (implosion) of vapor cavities. This imploding is responsible for the audible noise associated with cavitation, which can range from occasional popping to the sound of moving sand. Extensive cavitation destroys piping, restricts flow, ruins turbine blades, and produces unacceptable noise levels. Cavitation occurs in a system wherever pressure has been reduced sufficiently, by either friction, flow separation, or restrictors such as valves, vortex elements, or differential-producing flowmeters. Even in a well-designed piping system, cavitation can occur if control or relief valves are suddenly opened.

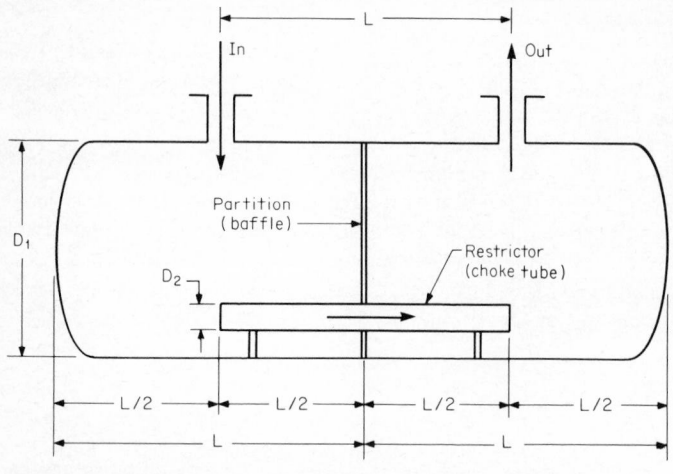

Flow rate $\times 10^3 =$ q_{SCFH}	Tank diameter D_1 (in)	Choke diameter D_2 (in)	Length L (in)
125	16	1.5	38
125 – 450	24	2	34
450 – 700	30	3	40

(a)

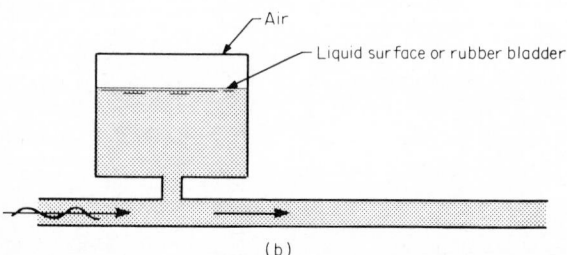

(b)

Figure 5.45 Typical passive filters for gas and liquid flows. (a) Gas. (b) Liquid.

Dissolved gases and gas bubbles in liquids provide nucleative points and assist in the onset of cavitation. With gas concentrations in the range of 40 parts per million, fluids will cavitate at higher static pressures. Generally, cavitation begins at higher static pressures and lower velocities in larger line sizes; once started, it will continue at higher static pressures than the initiating pressure.

Figure 5.46   Suction and discharge stabilizers *(courtesy Greer Hydraulics).*

The number used for correlating cavitation data is the dimensionless *cavitation number,* defined as

$$\sigma_c = \frac{2g_c(P_{f2} - P_{v2})}{\rho_f \overline{V}_{f2}^2} \tag{5.60}$$

where the pressures P_{f2} and P_{v2} are in pounds-force per square foot, and the subscript 2 refers to the location of the maximum velocity within the flowmeter, as shown in Fig. 5.47. In Eq. (5.60) the difference between the static pressure P_{f2} and vapor pressure P_{v2} can be viewed as the force required to *collapse* the vapor bubbles. The velocity-squared term is the dynamic pressure required to *initiate* bubble formation. The cavitation number σ_c is then the ratio between the collapsing and forming forces.

Initially, cavitation begins with very small bubbles isolated in a small section of the flowmeter. As the cavitation number decreases, formation becomes more rapid, with bubble size usually increasing. For cavitation numbers below a certain value $(\sigma_c)_i$, called the *incipient* cavitation number, cavitation becomes increasingly destructive to both piping and flowmeters. The addition of very small amounts of a polymer has been shown by Oba et al. (1978) to reduce both the incipient cavitation number and the audible cavitation noise (Fig. 5.48).

In general, the incipient cavitation number ranges from 1.0 to 2.5 for abrupt obstructions and orifice, vortex, and flow nozzles, where downstream pressure recovery is abrupt. For contoured inlet and exit devices, such as the venturi, ven-

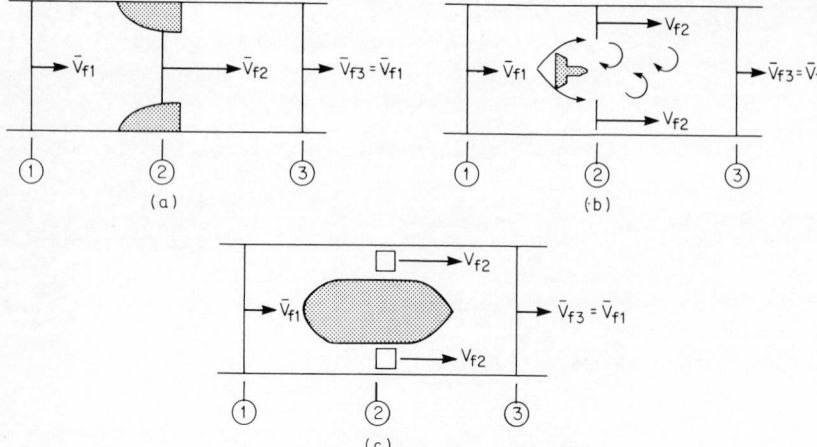

Figure 5.47 Location of maximum velocity and minimum pressure. (*a*) Differential producer. (*b*) Vortex flowmeter. (*c*) Liquid turbine flowmeter.

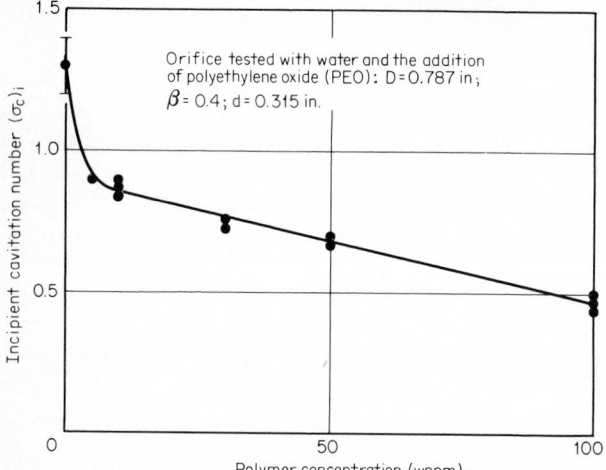

Figure 5.48 Reduction of incipient cavitation number with the addition of polyethylene oxide (**WPPM** = parts per million by weight).

turi nozzle, and Lo-Loss tube, the incipient cavitation number ranges from 0.2 to 0.5.

Three pressure locations are shown in Fig. 5.47. The upstream pressure P_{f1} is reduced to a minimum value P_{f2} at a minimum flow area, while the velocity is increased to a maximum value $\overline{V}_{f2}$ at approximately the same location. At location 2 the cavitation number is minimized, and cavitation will begin if the incipient cavitation number is reached. In some flowmeters, where there is a sudden change in a streamline ahead of a reduced area, separation may cause cavitation to begin slightly upstream of this minimum flow area. The flow will then recover to the full pipe area at location 3. The pressure P_{f3} will be close to P_{f1}, and the average velocity will return to the full-pipe-area value. The difference between the pressures at locations 1 and 3 is the permanent pressure loss; depending on how the flow recovers, it may be a significant percentage of the maximum pressure difference $P_{f1} - P_{f2}$ or it may be negligible.

For differential producers, the required minimum upstream or downstream pressure (back pressure) can be calculated from known incipient cavitation numbers (Table 5.5) by mass flow continuity and Bernoulli's equation. For turbine and vortex flowmeters, the point of minimum pressure depends on design, which obviously varies with line size and manufacturer. For these meters, a minimum back pressure is commonly specified. This minimum is calculated from the pressure loss $P_{f1} - P_{f3}$ and fluid vapor pressure with equations of the general form

$$(P_{f3})_{\min} = A(P_{f1} - P_{f3}) + BP_v \tag{5.61}$$

where P_{f3} is the minimum downstream pressure, and the constants A and B are determined experimentally. Any set of consistent pressure units may be used in Eq. (5.61). For turbine and vortex flowmeters, conservative estimates are 3.0 for A and 1.5 for B; however, the manufacturers should be consulted for actual values.

Fig. 5.49a shows the bias error for a nozzle and a venturi as the cavitation number decreases below the incipient value. Although the bias error occurs at a lower value than the incipient cavitation number $(\sigma_c)_i$, flowmeters should not be used in this range. When the line pressure across a section of the differential producer is the fluid's vapor pressure, the flow is *choked* (see Chap. 13), and increasing the upstream pressure will not increase the flow rate. The effect of cavitation on vortex and turbine flowmeters is shown in Fig. 5.49b. As the back pressure is reduced or the velocity is increased, cavitation results in a positive bias.

Table 5.5   **Incipient Cavitation Numbers for Differential Producers**

Flowmeter	Incipient cavitation number
Venturi	0.33
Nozzle venturi	0.55
Dall tube	1.0
Flow nozzle	1.8
Orifice	3.0

SOURCE: Cousin (1977).

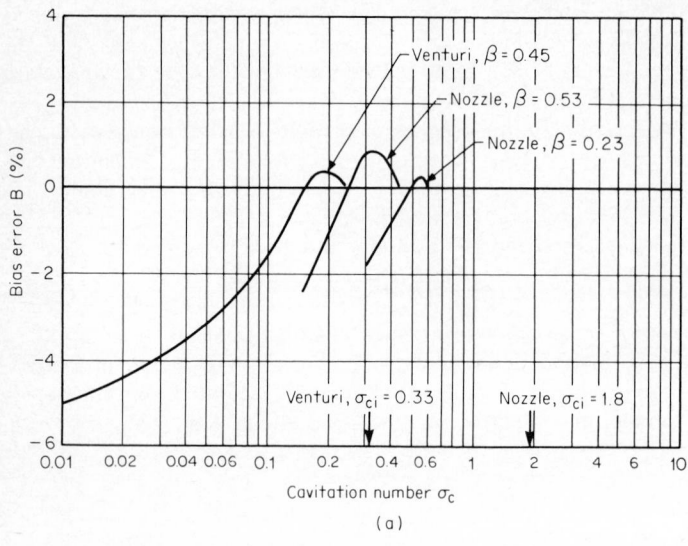

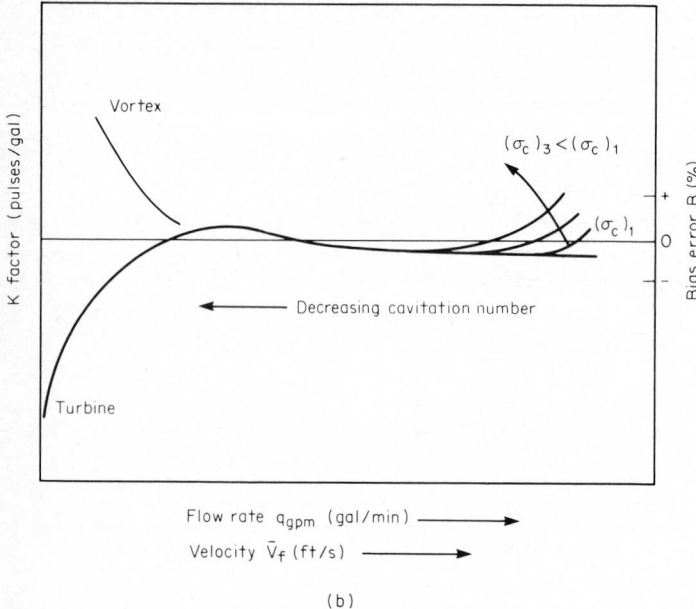

Figure 5.49 Effect of cavitation on flowmeters. (*a*) Nozzle and venturi. (*b*) Vortex and turbine flowmeters.

REFERENCE CONDITIONS

Accuracy

Four methods are commonly used to establish the accuracy of the primary device; the method selected depends on economics, the accuracy required, and whether the primary and secondary devices can be separated. The accuracy statement for all four is based on specified reference conditions.

The four methods for establishing accuracy relative to reference conditions are:

1. The primary device is calibrated, with its associated piping configuration, in a hydraulic laboratory to determine the relationship between the actual and theoretically calculated flow rates. A calibration factor can usually be determined to an accuracy of ± 0.15 percent, and it is used to correct all future on-site flow-rate calculations. The flow-rate accuracy is then the combination of primary-device and secondary-equipment accuracy.

2. If the primary device has dimensional similarity, as does an orifice, flow nozzle, or venturi, the discharge coefficients obtained from many tests are used to establish an equation for calculating the coefficient based on dimensions. Twice the standard deviation of the data with respect to this equation, the *error of estimate,* defines the coefficient accuracy.

3. For some flowmeters it is difficult and costly to control either dimensions or other factors that affect the relationship between output and true flow rate. In these cases, the primary device is water-calibrated to determine an accuracy which is then statistically combined with that of the secondary device to obtain the overall *flowmeter* accuracy (magnetic flowmeter). In some cases, the flowmeter is calibrated as a system to establish a reference accuracy.

4. Based on theory and a fully developed velocity profile, an analog of the flow-meter response to actual flow rate is developed. This analog, combined with suitable dimensional measurements, is used to estimate the accuracy.

Whether theoretically or empirically established, reference flow conditions imply

1. A fully developed laminar or turbulent velocity profile that is nonswirling (rotating) and axisymmetric (symmetric about the pipe centerline)

2. A newtonian fluid

3. A homogeneous single-phase fluid that completely fills the pipe

4. Steady flow

5. Flow-calculation correction for dimensional changes resulting from thermal, pressure, or other known bias errors.

Flowmeter Signature Curve

Under reference conditions, all flowmeters exhibit a characteristic curve that is a function of velocity profile. This *signature curve* has approximately the same shape for all flowmeters of similar design but may vary in level (bias error), depending on how well each individual device is manufactured. For differential producers, the discharge coefficient C defines the relationship between the actual laboratory-determined flow rate and that calculated from the theoretical flow-rate equation. This coefficient is observed to change with pipe Reynolds number (profile). A similar Reynolds-number relationship is observed for other flowmeters. Shown in Fig. 5.50 are typical signature curves for differential producers and turbine and vortex flowmeters. The signature curve is established in a laboratory with long, straight runs of pipe to achieve fully developed reference profiles. Since these are seldom achievable in field installations, tests are conducted with selected upstream flow disturbances, with the primary device incrementally moved closer to the disturbance. The minimum downstream location at which a change occurs in either curve shape or bias defines the reference upstream piping length. If the device is located closer to the disturbance, swirl, secondary flows, and nonaxisymmetric profile result in additional errors.

The various classes of flowmeters tolerate profile variations to differing degrees. Radial flow components from an elbow significantly change a single-path ultrasonic flowmeter output at 5 pipe diameters, but only slightly change a venturi's coefficient. The orientation of an ultrasonic path, the pressure connections for a differential producer, and the electrode axis of a magnetic flowmeter with respect to the distorted profile must be considered in establishing reference lengths.

Wall Roughness

In turbulent flow, relative roughness and Reynolds number establish the flow profile, and some criterion should be available to define *reference-condition piping,* from which profile corrections may be made. Unfortunately, relative roughness and flow profiles are not standardized. However, for a series of hydraulics-laboratory tests conducted with orifice flowmeters, the relative pipe roughness was estimated [in ISO Standard 5167 (1980)] to be 10×10^{-4}. The diameters and materials of pipes that satisfy this "criterion" are given in Table 5.6. Until standards become available, these pipes can be used as reference-condition pipes.

Length of Straight Pipe

ISO Standard 5167 (1980), which treats orifices, venturis, and flow nozzles, defines the general flow condition at the primary device as follows:

Swirl-free conditions at the primary can be taken to exist when the swirl angle over the pipe is less than 2 degrees. Acceptable velocity profile conditions can be presumed to

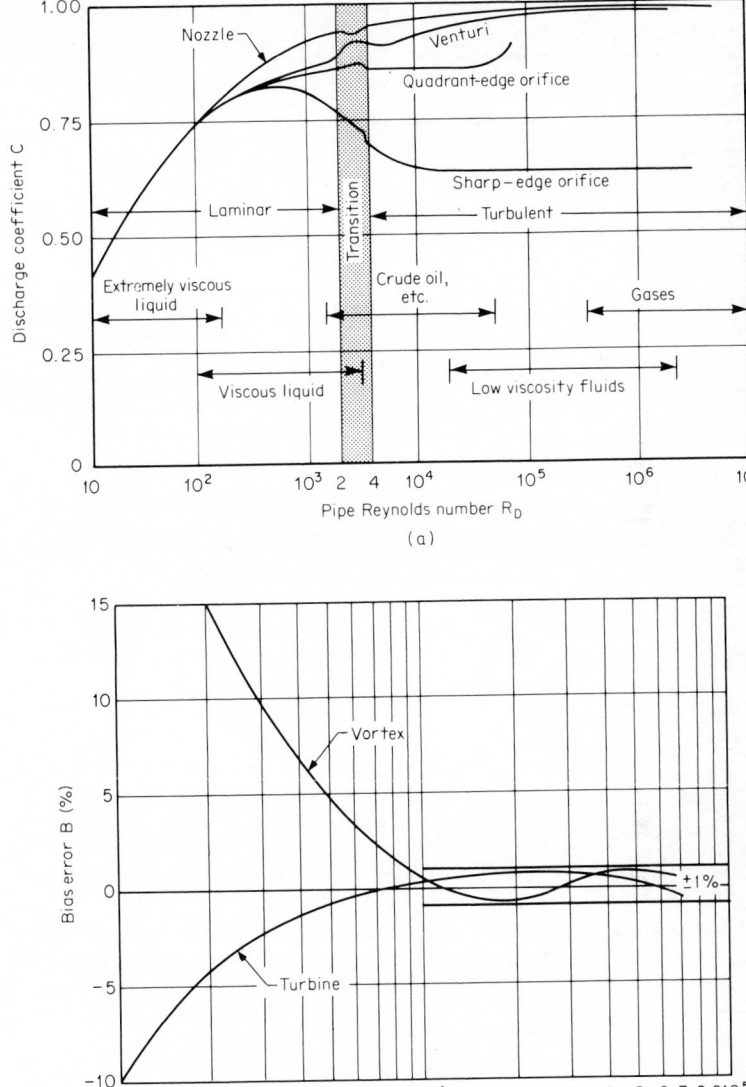

Figure 5.50 Typical flowmeter signature curves. (a) Differential producer. (b) Turbine and vortex flowmeters.

Table 5.6 Pipe Roughness and Acceptable Size Range for Reference Conditions†

Pipe material	Wall roughness ϵ	Pipe size D (reference conditions), in
Aluminum		
Asbestos cement, coated (new)		
Brass		
Copper	≤0.001	≥1
Glass		
Plastic		
Steel, cold-drawn seamless		
Asbestos cement (normal)		
Cast iron, bituminized (new)	≤0.002	≥2
Steel, bituminized (new)		
Steel, rolled seamless		
Steel, spirally welded	≤0.005	≥4
Steel, galvanized		
Cast iron, asphalt dipped		
Steel, bituminized (normal)	≤0.008	≥6
Steel, mildly rusty		
Cast iron (new)		
Galvanized iron	≤0.010	≥8
Steel (rusty)		

†Based on relative roughness of pipes used in orifice flowmeter coefficient test program (flanged D-and-$D/2$ taps), ISO Standard 5167 (1980).

prevail when at each point across the pipe cross-section the ratio of local axial velocity to the maximum axial velocity at the cross-section agree to within 5% with that which would be achieved in swirl-free flow at the same radial position at a cross-section located at the end of a very long straight length (over 100D) of similar pipe.

This definition, although specifically for differential producers, is probably stringent enough to apply to all flowmeters. Without calibration data, a piping length (or the combination of a flow conditioner and piping length) that, by test, has met this definition can be considered a *reference piping length.*

REFERENCES

Akashi, K., H. Watanabe, and K. Kenichi: "Development of New Flow Rectifier for Shortening Upstream Straight Pipe Length of Flowmeter," *Tokyo Flow Symp., IMEKO,* Paper 12b-5, pp. 279–284, Society of Instrumentation and Control, Tokyo, 1979.

Campbell, D. P.: *Process Dynamics,* Wiley, New York, 1958.

Chilton, E. G., and L. R. Handley: "Pulsation Absorbers for Reciprocating Pumps," *Trans. ASME,* vol. 77, pp. 225–230, 1955.

Colebrook, C. F.: "Turbulent Flow in Pipes with Particular Reference to the Transition Region between Smooth and Rough Pipe Laws," *J. Inst. Civ. Eng.,* vol. 11, pp. 133–136, 1938–1939.

Cousin, T.: "The Effect of Cavitation and Air/Water Mixtures on the Discharge Coefficients of Differential Pressure Flowmeters," *IMEKO VII Congr.,* pp. BFL2411–2417, Institute of Measurement and Control, London, 1977.

Darcy, H.: "Sur des recherches experimentales relatives au mouvement des eaux dans les tuyaux," *Comptes Rendu de l'Académie des Sciences de France,* vol. 38, pp. 1109–1121, 1854.

Edwards, M. F., and M. A. Wilkinson: "Review of Potential Applications of Pulsating Flow in Pipes," *Trans. Inst. Chem. Eng.,* vol. 49, pp. 85–94, 1971.

Govier, G. W., and K. Aziz: "The Flow of Complex Mixtures in Pipes," Krieger, Huntington, N.Y., 1977.

Greer Bulletins TR-1, 1500E, 1507, 5005, and 6500, Greer Hydraulics, Chatsworth, Calif., 1979–1981.

Head, V. P.: "A Practical Pulsation Threshold for Flowmeters," *Trans. ASME,* vol. 78, pp. 1471–1473, 1956.

Hewitt, G. F.: *Measurements of Two Phase Flow Parameters,* Academic, London, 1978.

ISO: *Characterization of Steady Flow* (draft), Secretariat ANSI, ISP/TC30/SC2/WG7, doc.(secr.-3)7, 1975.

ISO Standard 5167, *Measurement of Fluid Flow by Means of Orifice Plates, Nozzles, and Venturi Tubes Inserted in Circular Cross-Section Conduits Running Full,* p. 19, ISO 5167-1980(E), Geneva, 1980.

ISO/DIS 7145, *Determination of Flowrate of Fluids in Closed Conduits of Circular Cross-Section—Method of Velocity Measurement at One Point of the Cross-Section,* ISO, UDC 532.542, Geneva, 1981.

ISO Technical Report 3313, *Measurement of Pulsating Fluid Flow in a Pipe by Means of Orifice Plates, Nozzles or Venturi Tubes, in Particular in the Case of Sinusoidal or Square Wave Intermittent Periodic-Type Fluctuations,* ISO/TR 3313(E), 1974.

Keyser, D. R.: "Unsteady Flow Measurement, Its Theory and Observations," in *Flow, Its Measurement and Control in Science and Industry,* vol. 2, Instrument Society of America, Research Triangle Park, N.C., 1981.

Kinghorn, F. C.: "Flow Measurement in Swirling or Asymmetric Flow—A Review," *Flow-con 77 Proc.,* pp. 45–71, Institute of Measurement and Control, Gatton & Kent, U.K., 1977.

Kirmse, R. E.: "Investigations of Pulsating Turbulent Pipe Flows," *J. Fluids Eng.,* vol. 101, pp. 436–442, 1979.

Kreith, F., and O. K. Sonju: "The Decay of a Turbulent Swirl in a Pipe," *J. Fluid Mech.*, vol. 22, pt. 2, pp. 257–271, 1965.

Lee, W. F. Z., M. J. Kirik, and J. A. Bonner: "Gas Turbine Flowmeter Measurement of Pulsating Flow," *J. Eng. Power*, vol. 97, pp. 531–539, October 1975.

Lindahl, E. J.: "Pulsation and Its Effect on Flowmeters," *Trans. ASME*, vol. 68, pp. 893–894, November 1946.

Logan, E. L., and J. B. Jones: "Flow in a Pipe Following an Abrupt Increase in Surface Roughness," ASME Winter Annual Meeting, Paper 62-Hyd-5, 1962.

Metzner, A. B., and J. C. Reed: "Flow of Non-Newtonian Fluids—Correlation of the Laminar Transition and Turbulent-Flow Regions," *AIChE J.*, vol. 1, pp. 434–440, 1955.

Mizushina, T., T. Maruyama, and H. Hirasawa: "Structure of the Turbulence in Pulsating Pipe Flows," *J. Chem. Eng. Jpn.*, vol. 81, no. 3, pp. 210–216, 1975.

Moody, L. F.: "Friction Factors for Pipe Flows," *Trans. ASME*, vol. 66, p. 671, 1944.

Mottram, R. C.: "The Measurement of Pulsating Flow Using Orifice Plate Meters," in *Flow, Its Measurement and Control in Science and Industry*, vol. 1, pp. 197–208, ISA, Pittsburgh, 1974.

Muller, E.: "The Measurement of Pulsating Flow," in *The Measurement of Pulsating Flow* (symp. Proc.), Institute of Measurement and Control, London, 1970.

Nikuradse, J.: *Gestzmassigkeiten der turbulenten Stromung in glatten Rohren*, Forschungsheft 356, VDI Verlag, Berlin, 1932.

————: *Stromungsgesetze in rauhen Rohren*, Forschungsheft 361, VDI Verlag, Berlin, 1933 (translated as NACA TM 1292, 1960).

Oba, R., Y. Ito, and K. Uranishi: "Effect of Polymer Additives on Cavitation Development and Noise in Water Flow through an Orifice," *J. Fluids Eng.*, vol. 100, pp. 493–499, 1978.

Pai, S. I.: "On Turbulent Flow in a Circular Pipe," *J. Franklin Inst.*, vol. 20, no. 4, pp. 337–352, 1953.

Pao, R. H. F.: *Fluid Mechanics*, Wiley, New York, 1961.

Reynolds, O.: "An Experimental Investigation of the Circumstances Which Determine whether the Motion of Water Will Be Direct or Sinuous, and of the Law of Resistance in Parallel Channels," *Philos., Trans. R. Soc. London*, vol. 174, p. 935, 1883.

Robertshaw: Catalog MA-11, Robertshaw Controls Co., Richmond, Va., 1967.

Ryan, N. W., and M. M. Johnson: "Transition from Laminar to Turbulent Flow in Pipes," *AIChE J.*, vol. 5, no. 4, pp. 433–435, 1959.

Teyssandier, R. G.: "Flow Coefficient Variation for Concentric Orifices from an Analysis of Turbulent Velocity Profiles," ASME Winter Annual Meeting, Paper 75 WA/FE-19, 1975.

Townes, H. W., J. L. Gow, R. E. Powe, and N. Weber: "Turbulent Flow in Smooth and Rough Pipes," ASME Winter Annual Meeting, Paper 71-WA/FE-7, 1971.

Van Wazer, J. R., J. W. Lyons, K. Y. Kim, and R. E. Colwell: *Viscosity and Flow Measurement: A Laboratory Handbook of Rheology*, Interscience/Wiley, New York, 1963.

Wasp, E. J., J. P. Kenny, and R. L. Gandhi: *Solid-Liquid Flow Slurry Pipeline Transportation*, Gulf, Houston, Texas, 1979.

Weisbach, J.: *Lehrbuch der Ingenieur und Machinemechanik*, Brunswick, Germany, 1845. Translated by E. B. Cox, Van Nostrand, 1872.

Wilkinson, W. L.: *Non-newtonian Fluids,* Pergamon, New York, 1960.

Yeaple, F.: *Hydraulic and Pneumatic Power and Control: Design, Performance, Application,* McGraw-Hill, New York, 1966.

Zanker, K. J.: "The Development of a Flow Straightener for Use with Orifice Flowmeters in Disturbed Flow," Paper D-2, *Symp. on Flow Measurement,* Scotland, 1969.

FLOWMETER SELECTION

The instrument engineer probably has a wider choice of devices when specifying a flowmeter than for any other process-monitoring application. It is estimated (Hayward, 1975) that at least 100 flowmeter types are commercially available, and new types are being continually introduced. Meters are chosen on the basis of cost, line size, the fluid being metered, its state (gas, vapor, or liquid), meter range, and desired accuracy. Only the more widely used general-purpose flowmeters—those listed in Table 6.1—are covered in this handbook. For these devices, operating principles, selection bases, and equations for the calculation of permanent pressure loss and yearly energy cost are summarized in this chapter.

DIFFERENTIAL PRODUCERS

The differential-producing flowmeters, sometimes called *head-class* flowmeters, are selected most frequently because of their long history of use in many applications. A number of primary elements belong to this class: The concentric orifice, venturi, flow nozzle, Lo-Loss tube, target flowmeter, pitot tube, and Annubar are all differential producers. When some other flowmeter is selected, it is usually because of an obstructionless feature, wider range, a tendency against freezing or condensate buildup in lead lines, or because the fluid is abrasive, dirty, or made up of more than one component (slurry). It is probably true that all new flowmeters must, at least initially, compete in applications where the thin concentric orifice has proved less than satisfactory.

Basic Principle

The underlying principle for all differential-producing flowmeters is Bernoulli's streamline energy equation. When a flow is contracted (Fig. 6.1), either gradually or abruptly, kinetic energy increases at the expense of available potential energy

Table 6.1   **Flowmeter Selection Table**

Flowmeter	Pipe size, in (mm)	Gases (vapors)		Liquids				Slurries	
		Clean	Dirty	Clean	Viscous	Dirty	Corrosive	Fibrous	Abrasive
SQUARE-ROOT SCALE; MAXIMUM SINGLE RANGE 4:1									
Orifice									
Square-edged	>1.5 (40)	■	□	■	□	▨	▨	□	□
Honed meter run	0.5–1.5 (12–40)	■	□	■	▨	□	▨	□	□
Integral	<0.5 (12)	■	□	■	□	□	▨	□	□
Quadrant/conic edge	>1.5 (40)	□	□	■	■	□	▨	□	□
Eccentric	>2 (50)	▨	▨	▨	□	■	▨	□	□
Segmental	>4 (100)	▨	▨	▨	□	■	▨	□	□
Annular	>4 (100)	▨	■	▨	□	■	▨	□	□
Target	>0.5–4 (12–100)	■	■	■	■	■	■	□	□
Venturi	>2 (50)	■	▨	■	▨	▨	▨	▨	▨
Flow nozzle	>2 (50)	■	▨	■	□	▨	▨	□	▨
Lo-Loss	>3 (75)	■	▨	■	□	▨	■	□	□
Pitot	>3 (75)	■	□	■	□	□	▨	□	□
Annubar	>1 (25)	■	▨	■	□	▨	▨	□	□
Elbow	>2 (50)	▨	▨	▨	▨	▨	▨	▨	▨
LINEAR SCALE: TYPICAL RANGE 10:1									
Magnetic	0.1–72 (25–1800)	□	□	■	■	■	■	■	■
Positive-displacement	<12 (300)	■	□	■	■	□	▨	□	□
Turbine	0.25–24 (6–600)	■	□	■	▨	□	▨	□	□
Ultrasonic									
Time-of-flight	>0.5 (12)	□	□	■	▨	□	■	□	□
Doppler	>0.5 (12)	□	□	□	▨	■	■	■	■
Variable-area	≤3 (75)	■	□	■	▨	□	▨	□	□
Vortex	1.5–16 (40–400)	■	▨	■	□	▨	▨	□	□

■ = designed for this application; ▨ = normally applicable; □ = not designed for this application.

Temperature, °F (°C)	Pressure, psig (kPa)	Accuracy, uncalibrated (including transmitter)	Reynolds number
Process temperature to 1000°F (540°C); transmitter limited to −20–250°F (−30–120°C)	To 6000 psig (41,000 kPa)	±1–2% URV*	R_D > 2000
		±1% URV	R_D > 1000
		±2–5% URV	R_D > 100
		±2% URV	R_D > 200
		±2% URV	R_D > 10,000
		±2% URV	R_D > 10,000
		±2% URV	R_D > 10,000
		±1.5–5% URV	R_D > 100
		±1–±2% URV	R_D > 75,000
		±1–±2% URV	R_D > 10,000
		±1.25% URV	R_D > 12,500
		±5% URV	No limit
		±1.25% URV	R_D > 10,000
		±4.25% URV	R_D > 10,000

Temperature, °F (°C)	Pressure, psig (kPa)	Accuracy, uncalibrated (including transmitter)	Reynolds number
360 (180)	≤1500 (10,500)	±0.5% of rate to ±1% URV	No limit
Gases: 250 (120) Liquids: 600 (315)	≤1400 (10,000)	Gases: ±1% URV Liquids: ± 0.5% of rate	≤8000 cSt
−450–500 (−268–260)	≤3000 (21,000)	Gases: ±0.5% of rate Liquids: ±1% of rate	≤2–15 cSt
−450–500 (−268–260)			
−300–500 (180–260)	Pipe rating	±1% of rate to ±5% URV	No limit
−300–250 (−180–120)	Pipe rating	±5% URV	No limit
Glass: ≤400 (200) Metal: ≤1000 (540)	Glass: 350 (2400) Metal: 720 (5000)	±0.5% of rate to ±1% URV	To highly viscous fluids
≤400 (200)	≤1500 (10,500)	±0.75–1.5% of rate	>10,000

*URV = upper range value of the flow rate: formerly full-scale flow rate.

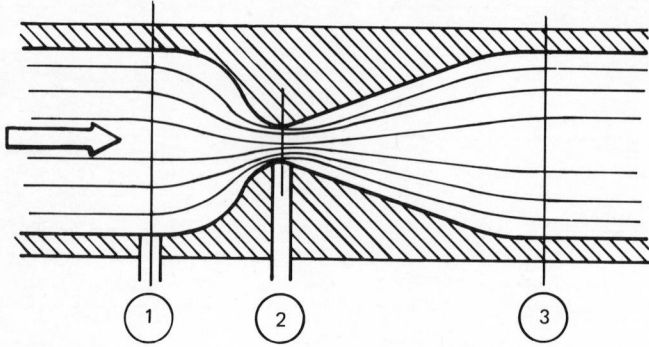

Figure 6.1 Basic principle of the differential producers.

(static pressure). The pressure difference between the taps located at the full pipe section (section 1 in Fig. 6.1) and in the vicinity of the contraction (section 2) is related to the square of the velocity at section 1 less the square of the velocity at section 2, to fluid properties, and to the abruptness of the contraction. Since the velocity times the pipe area is the volumetric flow rate, the basic flow equation may be written as a square-root relationship among measured differential pressure h_w, density ρ_f, and flow rate q as

$$q_{cfs} = F_{MC} \sqrt{\frac{h_w}{\rho_f}} \tag{6.1}$$

The meter constant F_{MC} adjusts for dimensional units and includes a discharge coefficient that corrects for contraction characteristics, pressure-tap location, and velocity profile (Reynolds number). For a gas, the density differences caused by gas expansion between measurement taps requires an expansion-factor correction. This may be based on empirical data, as in the case of the square-edged orifice, or derived from thermodynamic considerations for the more gradually contracting elements (venturi and flow nozzle).

From Eq. (6.1) it is seen that, for example, if the differential pressure h_w decreases by a factor of 9, then the flow rate q decreases by a factor of 3. The accuracy of a fixed-upper-range differential pressure transmitter is therefore degraded. This is a basic reason for considering differential producers to have a maximum flow-rate range of 4:1.

The Orifice

Square Edge In 2-in (50-mm) and larger line sizes the concentric orifice (Fig. 6.2) is the most common restriction for clean liquids, gases, and low-velocity vapor (steam) flows. It is a sharp, square-edged hole bored in a flat, thin plate. The ratio of hole diameter d to pipe diameter D defines the beta ratio β. For most applications this ratio should be between 0.2 and 0.75, depending on desired differential; a high-β orifice produces less differential for the same flow rate than a small-β orifice. Beta ratios greater than 0.75 are sometimes used where accuracy is unimportant.

Depending on upstream and downstream tap location, the flowmeter is referred to as a corner tap, a flange tap, a D-and-$D/2$ tap, a pipe tap ($2\frac{1}{2}D$ and $8D$), or a vena contracta tap-orifice flowmeter. Corner and D-and-$D/2$ taps are widely used in Europe, while flange taps predominate in the United States. Pipe taps ($2\frac{1}{2}D$ and $8D$) are sometimes used as bypass pump restrictors for natural gas, or where the other tapping arrangements require drilling too close to the plate. Vena contracta taps have been replaced by D-and-$D/2$ taps because future changes in orifice bore require no tap relocation.

Figure 6.2 Square-edged concentric orifice.

Overall accuracy ranges from ± 0.8 to ± 5 percent depending on fluid, upstream piping configuration, and whether corrections for Reynolds number, gas expansion factor, and other effects are included in the computation. While standards differ on the minimum pipe Reynolds number that permits this accuracy, a value of 10,000 is considered acceptable by most. The maximum Reynolds number may be as high as 3.3×10^7 (Gorter and Rooij, 1977).

Quadrant and Conic Edge When the pipe Reynolds number is below 10,000, the upstream orifice edge is either rounded (quadrant, Fig. 6.3) or conical (Fig. 6.4); these contours give a more constant and predictable discharge coefficient at lower Reynolds numbers. At low Reynolds numbers the coefficient of a square-edged orifice may change by as much as 30 percent, but for these geometries the effect is only 1 to 2 percent, making a more usable flowmeter for viscous fluids. The quadrant plate is widely used in the United States, whereas the conic plate predominates in European applications.

Small Line Sizes In $\frac{1}{2}$- to $1\frac{1}{2}$-in (12- to 40-mm) line sizes, the effects of pipe roughness, plate eccentricity, and edge sharpness are magnified, resulting in unpredictable coefficients. All indications are that coefficients must be experimentally determined, with selection of a square or quadrant edge depending on Reynolds number.

For these line sizes, corner pressure taps are preferred. Special honed meter-tube assemblies (Fig. 6.5) are available that have predictable coefficients (Filban et al., 1960), and these should be used if accuracy is required.

Integral Orifice When the pipe size is $\frac{1}{2}$ in (12.7 mm) or smaller and the fluid is clean, it is common to select an orifice installed integrally with the differential-pressure transmitter (Fig. 6.6). This provides a compact installation in which overall accuracy, usually ± 2 to ± 5 percent uncalibrated, has been predetermined based on flow-calibrating reproducible fixed-size orifices.

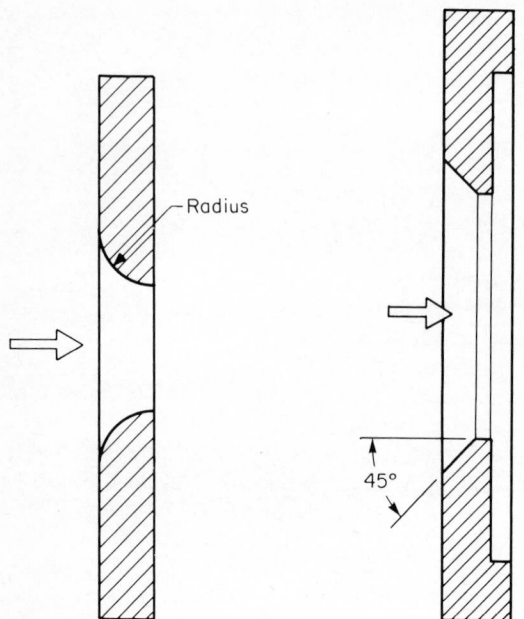

Figure 6.3 Quadrant-edged orifice.

Figure 6.4 Conical orifice.

45°

─Radius

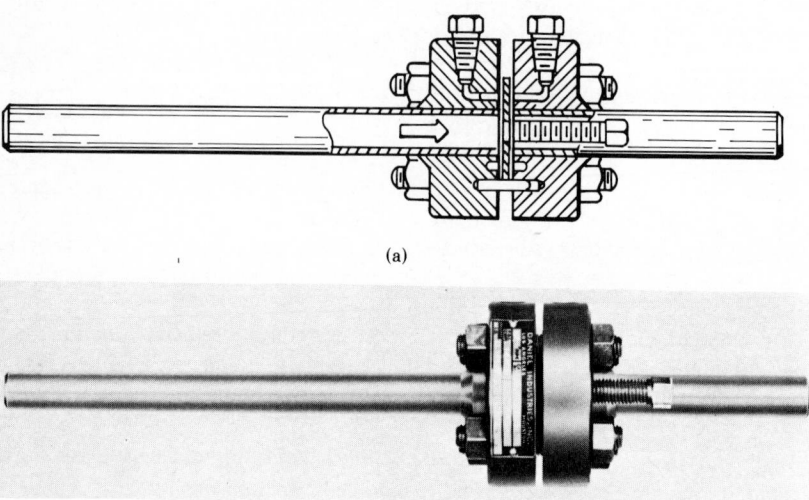

(a)

(b)

Figure 6.5 Honed meter run *(courtesy Daniel Industries).* **(a)** Internal construction. **(b)** External view.

(a)

(b)

Figure 6.6 Integral orifice. (*a*) In-line *(courtesy The Foxboro Co.).* (*b*) U-bend *(courtesy The Foxboro Co.).* (*c*) In-line *(courtesy SYBRON/Taylor).*

6-7

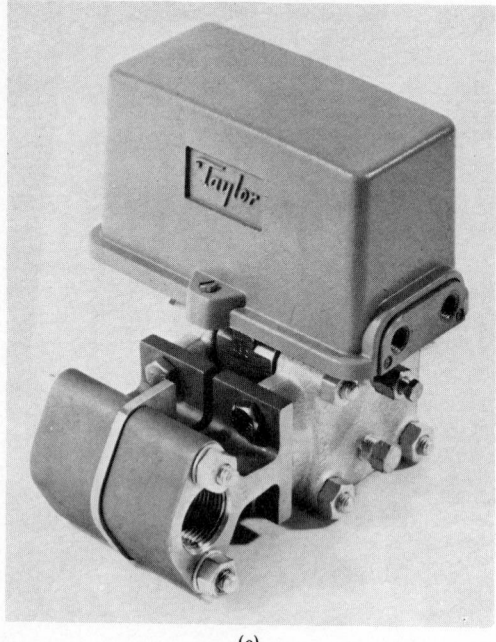

(c)

Figure 6.6 (*Continued*)

Eccentric and Segmental Orifices If the orifice hole is placed at the bottom of the pipe for gases (Fig. 6.7), and at the top for liquids, entrained water or air will flow through the plate rather than build up in front of it. If a segmental opening (Fig. 6.8) is machined in the plate, passage of liquids, air, or particulate matter is also possible. Data for these orifice geometries is limited, but they provide low-cost alternatives for troublesome applications. Their uncalibrated coefficient accuracy is usually estimated as ± 2 percent.

Figure 6.7 Eccentric orifice.

Figure 6.8 Segmental orifice.

Venturi Tube

The tapered inlet and diverging exit of the classical Herschel venturi (Fig. 6.9) substantially reduces permanent pressure loss. Since dirt will not build up as it passes through the contoured sections (as it does in front of an orifice), this differential producer can be used for dirty flow applications. Initially designed for large-line-size [6 in (150 mm) or larger] water and waste applications, the venturi has become more popular in smaller line sizes with the introduction of the proprietary Universal Venturi Tube (Fig. 6.10), a flowmeter of reduced weight and shorter overall length having the advantages of the venturi principle (Halmi, 1974).

Figure 6.9 Classical venturi.

Figure 6.10 Universal Venturi Tube *(courtesy BIF)*.

For pipe Reynolds numbers greater than 100,000, discharge coefficients for venturis are constant and predictable to within ±0.5 to ±2 percent, depending on design. Venturis are not normally used for lower Reynolds numbers without the acceptance of worsened coefficient accuracy. When pumping cost and/or shorter upstream installation length are important, the additional expense for a venturi design is usually warranted.

Flow Nozzle

The flow nozzle (Fig. 6.11) has an elliptical (ASME) or a radius (ISA) entrance and is generally selected for steam (vapor) flows at high pipeline velocities [100 ft/s (30.5 m/s)]. Because of its rigidity it is dimensionally more stable at higher temperatures and velocities than an orifice.

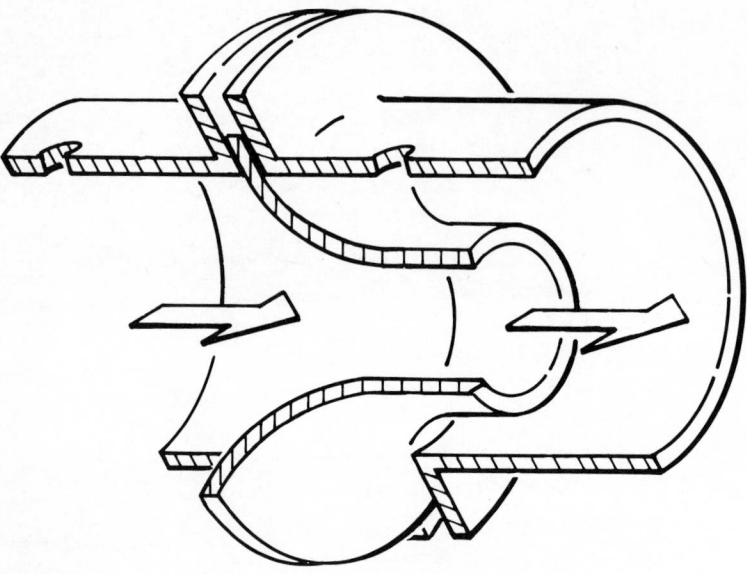

Figure 6.11 Flow nozzle.

The initial cost of a flow nozzle is substantially higher than that of an orifice but lower than that of a venturi; however, its permanent pressure loss is significantly higher than that of a venturi. When both are sized to create the same differential at the same flow rate, the pressure loss of a flow nozzle is approximately the same as for an orifice.

Flow nozzles and critical venturi nozzles (Fig. 6.12) are sometimes operated at critical (choked) flow for flow limiting or as secondary flow standards (Jones,

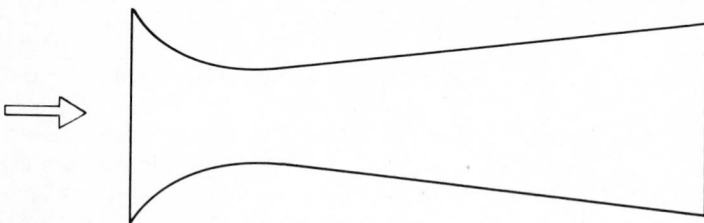

Figure 6.12 Critical-flow venturi nozzle.

1976). The most widely used nozzles are the ASME elliptical-inlet long-radius wall-tap nozzle and the ASME throat-tap nozzle for steam-turbine testing. In Europe the ISA 1932 nozzle is widely used.

Lo-Loss Permanent-Pressure-Loss Devices

Several proprietary designs (Fig. 6.13) have been introduced that produce high differentials with lower permanent pressure loss, lower weight, and shorter overall laying lengths. Coefficient information at high Reynolds numbers is not generally available for these designs. Also, little theoretical work (required for good extrapolation) is available, and the user is cautioned to stay within calibrated ranges. Support information remains proprietary, and the user should contact the manufacturer for the latest information.

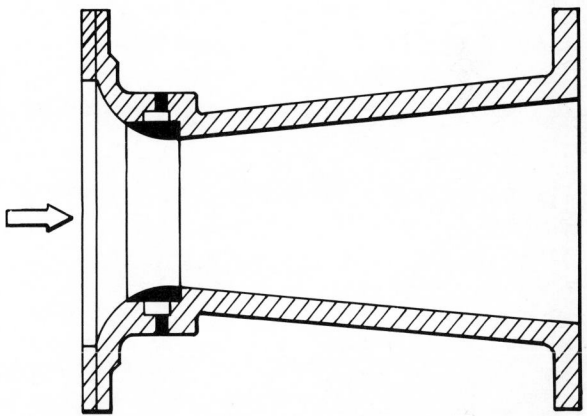

Figure 6.13   Lo-Loss tube *(courtesy Badger Meter, Inc.).*

Elbow Flowmeter

As a fluid passes through a pipe elbow, the pressure at the outside radius of the elbow increases due to centrifugal force (Fig. 6.14). If pressure taps are located at the outside and inside of the elbow at either $22\frac{1}{2}°$ or $45°$, a reproducible measurement can be made. Taps located at angles greater than $45°$ are not recommended, because flow separation may cause erratic readings.

Elbow meters are inexpensive flowmeters, since many installations have elbows. Even when the elbow ID is measured, differences among elbows limit accuracy to ± 4 percent (Murdock et al., 1963), but precision (repeatability) is good (± 0.2 percent). The flow is unobstructed, however, with no additional permanent pressure loss. Several manufacturers offer proprietary machined elbow flowmeters for improved accuracy, but little data has been published on these devices. The major disadvantage of the elbow flowmeter is the very low differential pressure that is created, particularly for gas flows.

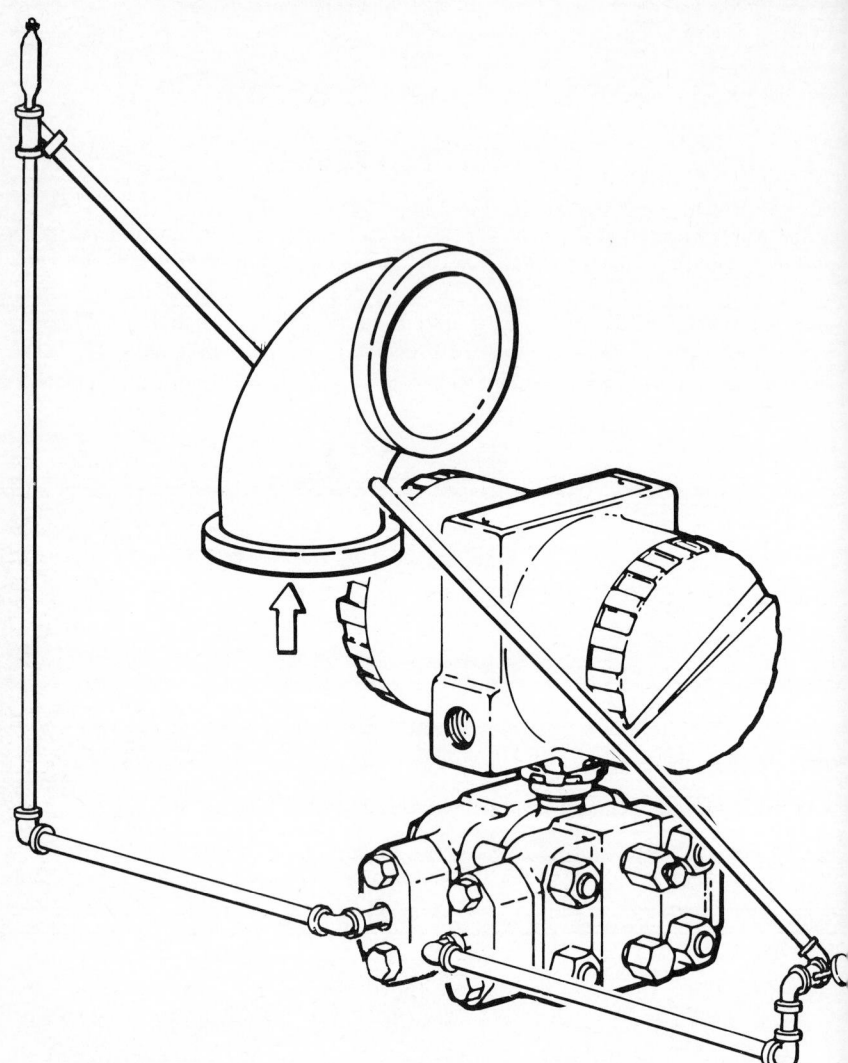

Figure 6.14 Elbow flowmeter.

Pitot Tube and Annubar

The pitot tube (Fig. 6.15) is used for large pipe sizes when the fluid is a clean liquid or gas (vapor) and an inexpensive measurement is required. For this device, the difference between total (stagnation) pressure and static pressure follows the square-root relationship, with velocity being sensed at the insertion depth only. By traversing, an average-velocity point can be located and used to measure flow rate.

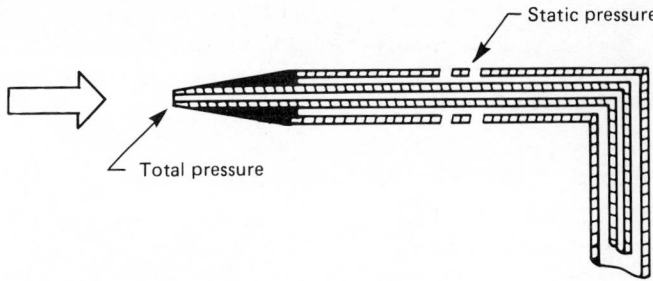

Figure 6.15 Pitot tube.

The Annubar (Fig. 6.16) is a multiple-ported pitot tube that spans the pipe. Pressure taps are located at mathematically defined positions based on published axisymmetric pipeline velocity profiles. These are claimed to average the differential pressure, thereby eliminating the need to locate the average-velocity point.

Ease of installation, low cost, very low permanent pressure loss, and insertability into existing piping make these devices convenient for ducts and large-line-size measurements. The Annubar has essentially replaced the pitot tube for clean liquids, gases, and vapors (steam). Because total-pressure ports face the flow, a purging flow is suggested for dirty-stream applications.

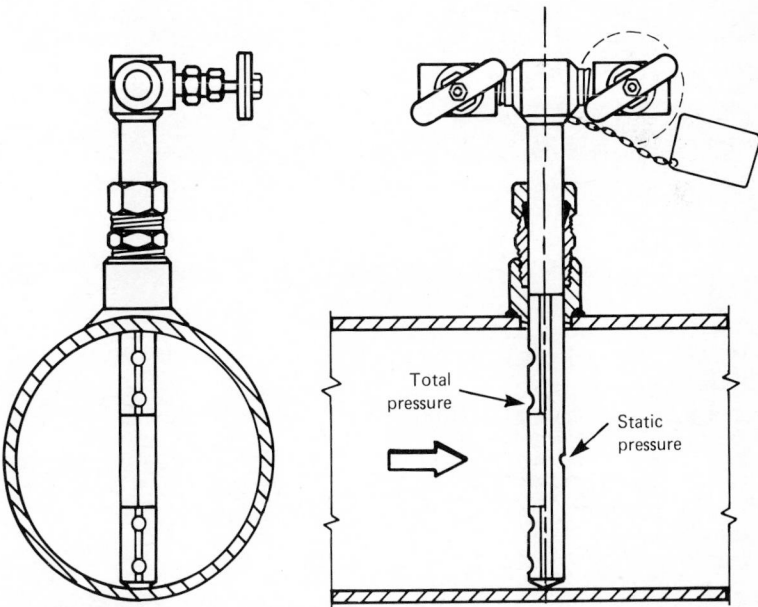

Figure 6.16 Annubar *(courtesy Dietrich Standard Corp.).*

A variety of multiported designs is now available. Users should review individual manufacturers' support data before applying these devices outside ranges for which data is available. Neither ANSI nor ISO has standardized on any of the available designs.

Annular Orifice and Target Flowmeter

The annular orifice (Fig. 6.17) was developed to overcome the problems of dirt buildup in front of an orifice in liquid streams and of liquid buildup in a moist gas stream. Total (stagnation) pressure taps and rearward facing taps produce a high differential for a given β_T here redefined as the ratio of disk diameter to pipe diameter. Little data has been presented for line-size correlation. Only air data is available for the normally used beta ratios. A design that slips between flanges has reportedly been successfully used for air in 24-in (600-mm) and larger line sizes.

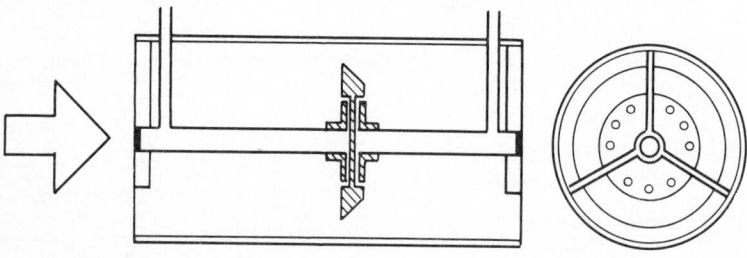

Figure 6.17 Annular orifice.

The target flowmeter (Fig. 6.18) has the features of the annular orifice without the disadvantages of freezing or plugging lead lines. The primary element consists of a sharp-leading-edge disk (target) fastened to a bar. The differential pressure produced by the reduced annular area creates a disk *drag force*. This is transmitted through a bar to a suitable force-measuring secondary device, and the flow rate is calculated as the square root of this output.

Target flowmeters are particularly well suited for dirty and low-Reynolds-number flows, but they are also used with clean fluids and for natural gas (Reinecke et al., 1966). Their uncalibrated accuracy ranges from ±1 to ±5 percent, depending on line size, beta ratio, and Reynolds number.

LINEAR FLOWMETERS

Turbine Flowmeter

The speed of a turbine flowmeter's rotor increases linearly with flow velocity. Blade rotation is thus a measure of velocity, and is detected by noncontacting

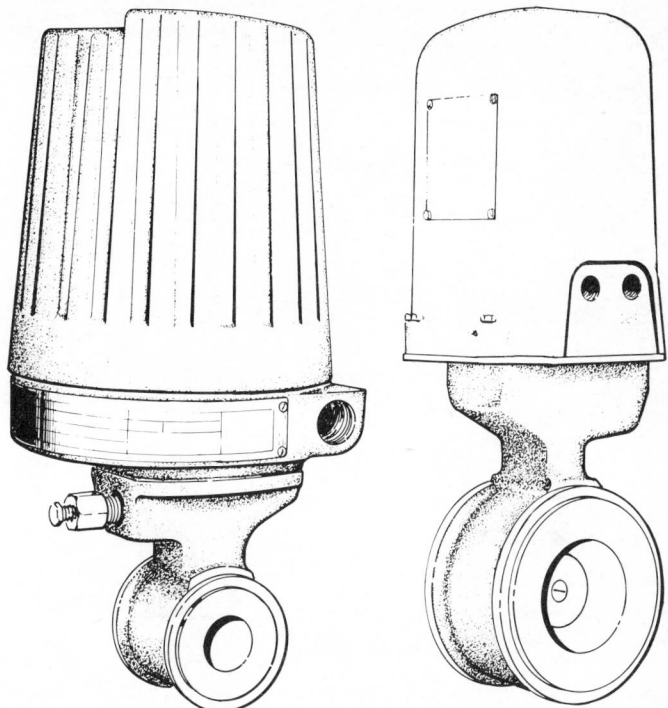

Figure 6.18 Target flowmeter *(courtesy The Foxboro Co.).*

external magnetic or other proximity detectors. The relationship between line velocity and rotor speed is linear (within ±0.5 percent) over a wide flow range of 10:1 to 20:1. Low-velocity performance is affected by velocity profile, tip clearance, friction across the blades, bearing friction, and other retarding torques (Rubin, et al., 1965). The major problems with liquid turbine flowmeters are the detrimental effects of overspeed when the liquid flashes or when slugs of vapor or air enter the line, shifts in calibration with blade wear, bearing friction, and large calibration shifts for liquids containing small amounts of air (Kinghorn and McHugh, 1981).

The turbine flowmeter is used for accurate, wide-range flow measurements of clean gases and liquids, and some manufacturers offer flowmeters designed for steam. Because of the large density differences between liquids and gases, two different turbine flowmeter designs (Figs. 6.19 and 6.20) are required for these two states. Both designs are used in custody-transfer applications requiring accuracy, range, and a pulse-train output.

Turbine flowmeters are calibrated in either a hydraulics or low-pressure-air laboratory. The laboratory-determined meter coefficient is either expressed as a *K factor* in units of pulses per gallon (pulses per cubic meter), or given as a *meter factor* in cubic feet per pulse (cubic meters per pulse). API Publication 2101

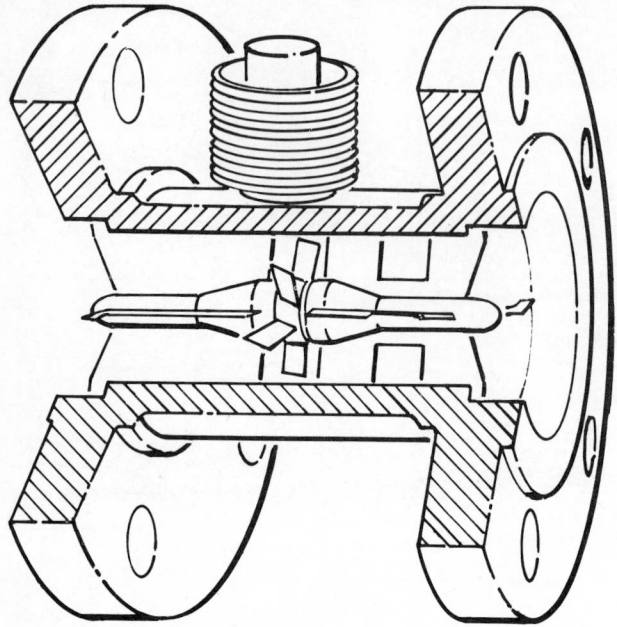

Figure 6.19   Liquid turbine flowmeter.

(1981) defines meter factor as prover volume divided by registered volume. For the metering of liquid hydrocarbons, the requirements of this manual must be strictly adhered to. Accuracy is claimed as ± 0.25 to ± 1 percent of indicated flow rate, depending on flow-rate range, viscosity, and installation conditions.

Vortex Flowmeter

Vortex shedding is a common flow phenomenon that causes bridges to collapse and telephone wires to sing. The instability of the flow field after it splits into two paths around a bluff object causes vortices to shed from alternate sides of the object at a frequency linearly proportional to velocity. If the approximately sinusoidal pressure or velocity changes created by the moving vortices in the fluid are detected, the flow rate can be determined. The relationship between pipeline velocity and shedding frequency is linear and independent of fluid density over flow-rate ranges of 20:1 or 30:1. The K factor, or meter factor, is the same for liquids, gases (vapors), and cryogenic fluids (Brennan et al., 1974), making this device an ideal all-purpose flowmeter.

The barlike construction of the vortex flowmeter (Fig. 6.21) allows for the passage of dirt. The output frequency is lower for the same flow rate than that of a turbine meter, and the vortex device is without overspeed or moving-part problems. Experimental and theoretical information indicates that the calibration factor at moderate Reynolds numbers is not as sensitive to edge sharpness

(a)

(b)

Figure 6.20 Gas turbine flowmeter *(courtesy Rockwell International).* (a) Internal mechanism. (b) Exterior.

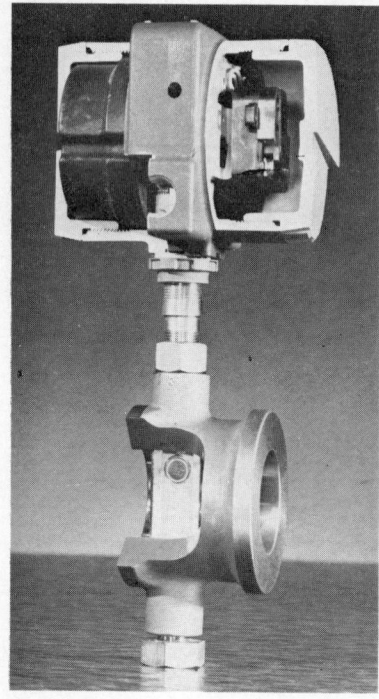

Figure 6.21 Vortex flowmeter
(courtesy The Foxboro Co.).

or dimensional changes as those of square-edged orifices and target flowmeters (Lomas, 1977).

Several manufacturers offer vortex flowmeters for line sizes from 1½ to 16 in (40 to 400 mm). Vortices are shed down to a pipe Reynolds number of 2000, but the shedding rate is nonlinear for Reynolds numbers from 2000 to 10,000; for this reason, a minimum Reynolds number of 10,000 is usually given.

Meter geometry is highly reproducible for a given line size (Miller et al., 1977), and accuracy is normally ±0.5 to ±1 percent over a 15:1 or 20:1 flow-rate range. Vortex flowmeters are increasingly being used because of their accuracy, range, precision, and relative insensitivity to fluid properties.

Magnetic Flowmeter

Based on magnetic induction principles, the magnetic flowmeter (Fig. 6.22) provides an obstructionless flowmeter that essentially averages velocity over the pipe area (Shercliff, 1962). Fluids to be measured must have a conductivity of at least 2 μS/cm to be measurable. A voltage is generated by the flowmeter that is mutually perpendicular to flow direction and magnetic field and is detected by two flush-mounted electrodes on a diameter of a nonconducting pipe wall. The low-level millivolt signal is proportional to the average pipeline velocity, and, for

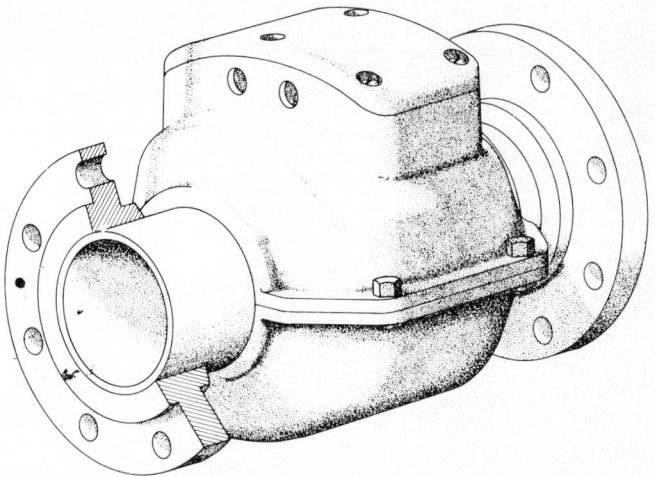

Figure 6.22 Magnetic flowmeter.

this reason, magnetic flowmeters are ideally suited for all conductive fluids that operate in both the laminar and turbulent flow regimes.

Unlike the differential producer, for which fluid density enters into the volumetric flow equation, the magnetic flowmeter's operating principle makes it a *true* volumetric flow device. Entrapped air, foam, two or more component mixtures, aggregates, etc., if uniformly mixed and traveling at the pipeline velocity, are measured on a flowing-volume basis.

Usually a nonmagnetic material (stainless steel) provides the necessary structural integrity, while an inserted nonconducting liner of Teflon or rubber electrically insulates the generated voltage from the supporting structure. Some meters are made wholly of fiberglass, but these have low pressure ratings. The meter housings are made of magnetic materials with coils installed inside and imbedded in a thick insulating liner. With properly selected liners, magnetic flowmeters are ideally suited for slurries, dirty flows, pulp stock, nonnewtonian fluids, corrosive liquids, and other liquids that are difficult to handle. Flowmeters are offered in sizes of from 0.1 to 96 in (2.5 to 2400 mm) with numerous secondary-device output options [4 to 20 mA, 3 to 15 psig (20 to 100 kPa), pulse train, etc.].

Two separate design approaches, based on magnetic-field excitation, are offered. In the first, ac mains directly power the electromagnetic coils; in the second (Fig. 6.23), the field is dc-energized for a short time duration. In the ac system, signal voltage is referenced to excitation voltage, to current, or to a field-measuring flux coil. In the dc system, electrode voltage is measured first with no magnetic field and then with a field. The measured voltage difference is the flow signal.

Magnetic flowmeters (both primary and secondary devices) are linear over a 10:1 range within an accuracy envelope of ±0.5 to ±1 percent of either actual

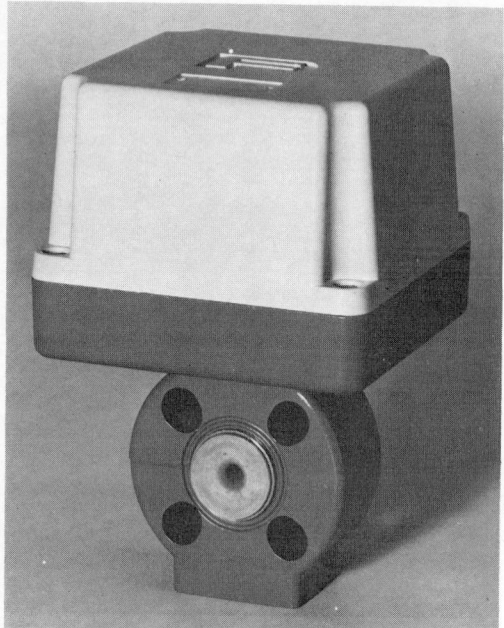

Figure 6.23 DC field electromagnetic flowmeter
(courtesy Fischer & Porter).

or upper-range flow rate. Area averaging makes magnetic flowmeters less sensitive to profile changes than any other type (Haacke, 1974).

Ultrasonic Flowmeters

There are two types of liquid ultrasonic flowmeters. In the first type (*time-of-flight* meters, Fig. 6.24), a high-frequency (approximately 1 MHz) pressure wave is beamed at an acute angle across the pipe. The time required for the wave to reach the opposite wall depends on whether it is moving with or against the flow and on the speed of sound through the liquid. Flow-rate information is obtained from the measured time. In the second type, referred to as the *Doppler* flowmeter, the pressure front does not traverse the pipe but is reflected back to a detector by particulate matter moving with the flow. The difference between reflected frequency and fixed transmitted frequency implies the flow rate.

There are a wide variety of time-of-flight ultrasonic flowmeters. The notable differences are usually in the number of beam paths across the pipe. A single beam averages profile along the beam and not across the pipe area. This makes the single-path measurement dependent on velocity profile. Multipath meters average along several paths, reducing profile dependency. Both single-path and multipath ultrasonic flowmeters are sensitive to swirl.

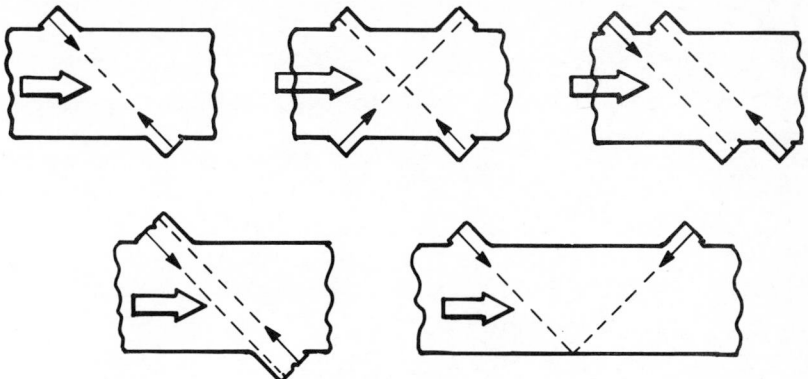

Figure 6.24 Ultrasonic flowmeter paths.

Time-of-flight flowmeters are generally used in clean liquid applications, where the ultrasonic beam is not attenuated or continuously interrupted by fluid particles. Some manufacturers claim success with rather dirty fluids, but then data interrogation and data rejection, which reduce response time, are often necessary. Claimed accuracies range from ± 1 to ± 4 percent, depending on design. Specific, detailed laboratory flow-calibration data, either from manufacturers or independent laboratories, has not been widely published.

The Doppler-type flowmeter relies on small particles or impurities in the flow but has also been used successfully on many almost-clean as well as very dirty streams. Because of the velocity profile, accuracy depends on particle concentration and distribution. Accuracy is also influenced by the relative velocity between particles and fluid.

Positive-Displacement Meters

Positive-displacement (PD) meters segregate the flow into discrete volumes and then sum total volume by counting *unit* volumes passing through the meter. Their major use is for metering household water, low-pressure natural gas, and gasoline. Millions of PD meters are in daily use, in greater quantities than all other flow measurement devices. In recent years, PD meters have been used increasingly in blending and loading applications where accuracy is required. Since PD meters have no time base, they are seldom used as rate flowmeters. Gas and liquid PD meters measure volume at line pressure and temperature; compensation to base volume or to total mass flow is accomplished by calculation or by mechanical ratioing.

Illustrated in Fig. 6.25 are several positive-displacement devices. As fluid enters a chamber, an impeller, a piston, a diaphragm, or a disk rotates or moves to accommodate the entering fluid, and a known volume is discharged. In the bellows type, four measuring compartments operate simultaneously, some filling while others are emptying. As each is filled and discharged, rotation is transmit-

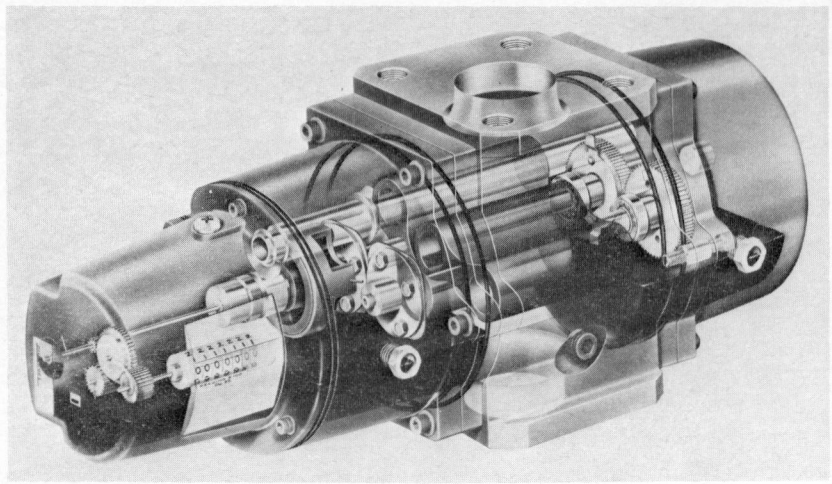

(a)

(b)

Figure 6.25 Positive-displacement meters. (a) ROOTS rotary gas meter *(courtesy ROOTS DRESSER)*. (b) Oscillating-piston and piston-roller water meter *(courtesy Rockwell International)*. (c) Domestic gas meter *(courtesy Rockwell International)*.

Figure 6.25 (*Continued*).

(c)

ted through suitable gearing to a counter that reads total volume. Seals are required to separate the volumes, and the pressure loss across the meter provides the energy to drive the moving parts. Because of the close tolerances required, the fluid temperature and viscosity affect the range as well as the accuracy. Manufacturers should be consulted on temperature, pressure, and viscosity limitations.

Numerous PD-meter designs are available for measuring both liquids and gases. They are ideal for viscous liquids and for batching and blending applications where wide range accuracy is required. Filters are normally installed upstream to prevent dirt from entering the meter, and periodic calibration is usually used to monitor mechanical wear.

Designs are available to handle pressures up to 1500 psig (10,300 kPa) and temperatures to 600°F (320°C) for liquids and 250°F (120°C) for gases. Flow-rate ranges are from 0.1 to 9000 gal/min (0.4 to 34,000 L/min) for liquids and up to 100,000 standard ft³/h (2831 standard m³/h) for gases. Measurement accuracy is 0.5 to 1 percent of the upper range value over typically 10:1 flow-rate ranges.

Variable-Area Meters (Rotameters)

Variable-area flowmeters are a special form of differential producer wherein the area of the flow restriction is varied to maintain constant differential pressure. The rotameter (Fig. 6.26) consists of a vertical tapered tube through which the

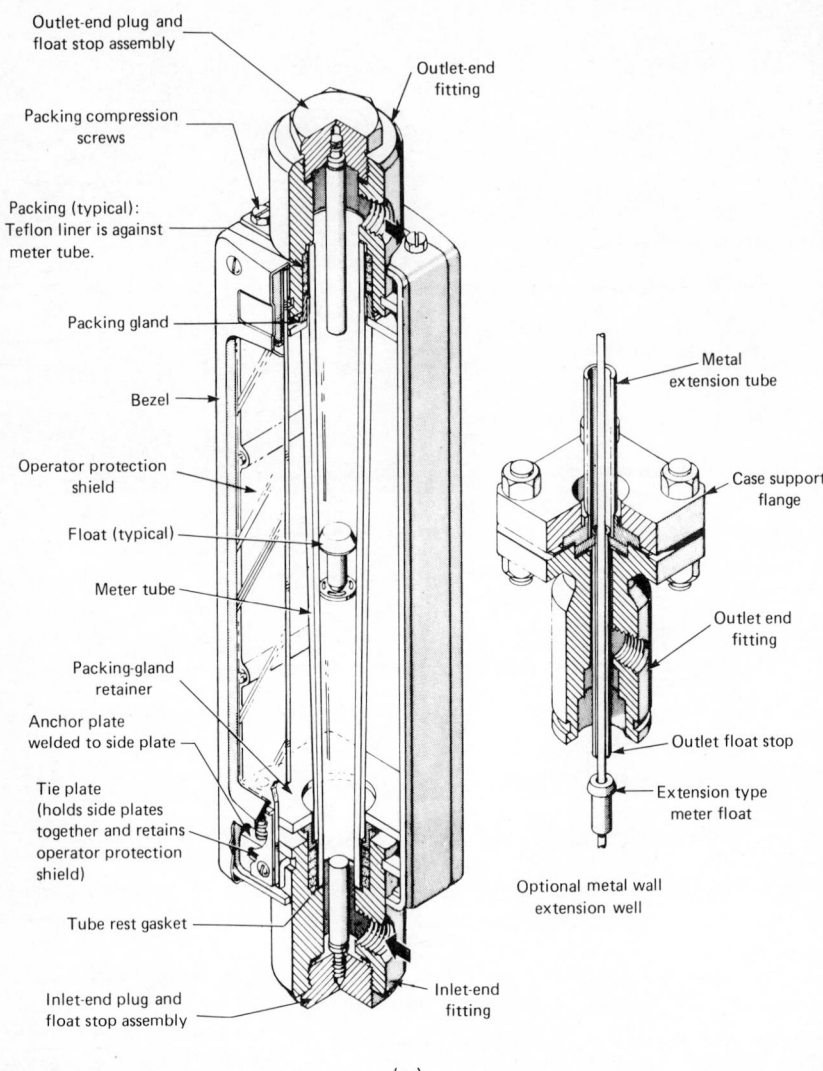

Outlet-end plug and
float stop assembly

Outlet-end
fitting

Packing compression
screws

Packing (typical):
Teflon liner is against
meter tube.

Packing gland

Bezel

Operator protection
shield

Float (typical)

Meter tube

Packing-gland
retainer

Anchor plate
welded to side plate

Tie plate
(holds side plates
together and retains
operator protection
shield)

Tube rest gasket

Inlet-end plug and
float stop assembly

Inlet-end
fitting

Metal
extension tube

Case support
flange

Outlet end
fitting

Outlet float stop

Extension type
meter float

Optional metal wall
extension well

(a)

Figure 6.26 Variable-area meters *(courtesy Fischer & Porter)*. (*a*) Glass (internal). (*b*) Glass (exterior). (*c*) Armored throughflow (internal). (*d*) Armored throughflow (exterior).

fluid moves in an upward direction. A float, either spherical or cone-shaped, with a density higher than that of the fluid, creates an annular passage between its maximum circumference and the wall of the tapered tube. As the flow varies, the float rises or falls to vary the area so that the differential pressure across the float just balances the gravitational force on the float. The differential pressure is then maintained constant, and the float position is a measure of the rate of flow.

A float in a transparent tapered tube makes a very simple flowmeter. The proper taper can be selected to provide a linear flow scale. In applications where the float cannot be seen, as when the flow of an opaque liquid is being measured or when a metal tube must be used because of large flow volume or elevated

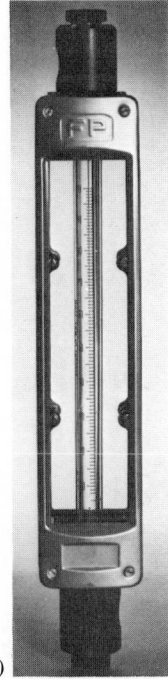

(b)   **Figure 6.26** (*Continued*)

pressure, the float position can be detected electrically or pneumatically for indication and/or control purposes.

A second type of variable-area meter consists of a mechanically variable restriction and a means for direct differential-pressure measurement across it. The area of the restriction is controlled either directly or through a servomechanism actuated by the differential pressure, so as to hold the differential pressure constant. The flow rate is then proportional to the effective area of the restriction. The output can be in the form of a motion that is derived from the motion which varies the restriction, or it may be an electrical or pneumatic signal proportional to the effective area.

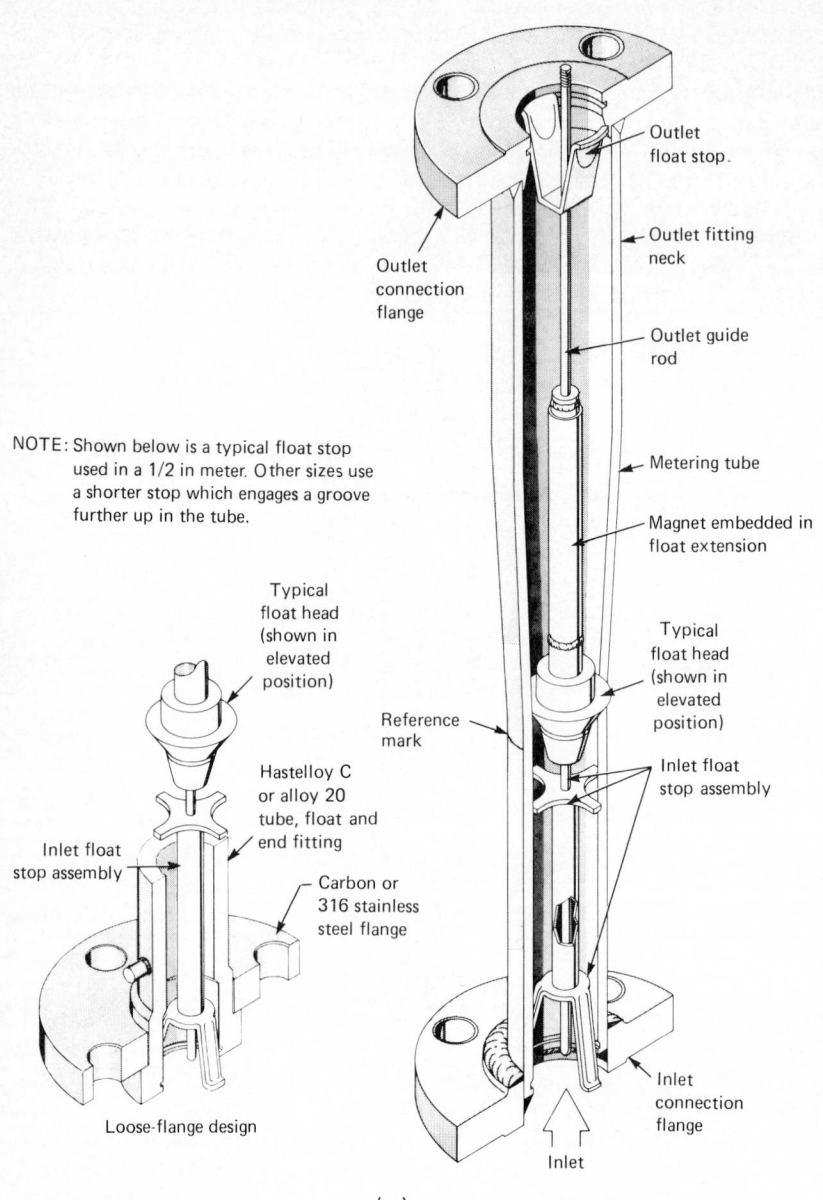

NOTE: Shown below is a typical float stop
used in a 1/2 in meter. Other sizes use
a shorter stop which engages a groove
further up in the tube.

Outlet
float stop.

Outlet fitting
neck

Outlet guide
rod

Metering tube

Magnet embedded in
float extension

Typical
float head
(shown in
elevated
position)

Inlet float
stop assembly

Reference
mark

Outlet
connection
flange

Typical
float head
(shown in
elevated
position)

Hastelloy C
or alloy 20
tube, float and
end fitting

Inlet float
stop assembly

Carbon or
316 stainless
steel flange

Loose-flange design

Inlet
connection
flange

Inlet

(c)

Figure 6.26 (*Continued*).

Variable-area meters are used for liquids, gases, and vapors (steam) with pressures up to 350 psig (2400 kPa) for glass tubes and up to 720 psig (5000 kPa) for metal tubes. Temperature limits range up to 400°F (205°C) for glass tubes and 1000°F (540°C) for metal tubes. Flow rates from 0.01 cm^3/min to 4000 gal/min (15,000 L/min) can be measured over a 5:1 or 10:1 flow-rate range, with an accuracy of ±0.5 percent of flow rate to ±1.0 percent of the upper-range flow rate, depending on type, size, and calibration.

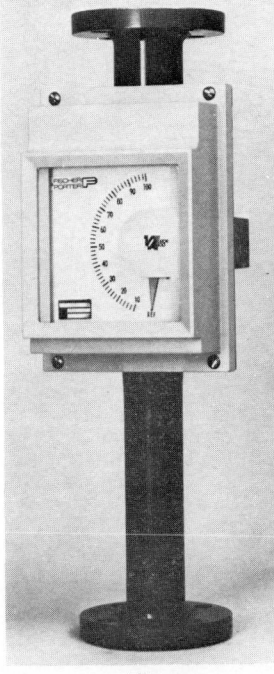

(d)

Figure 6.26 (*Continued*)

METER-SELECTION CONSIDERATIONS

General Guidelines

The choice of flowmeter for a given application depends upon the importance attached to various phases of the measurement problem. Hence, it must remain a matter of individual judgment based on engineering knowledge. The predominance of one class of flowmeter in a particular field with similar operating conditions is usually a good first guide.

It is obvious that an attempt to weigh and balance all qualifications without detailed knowledge of the specific application would be presumptuous. There are, however, five broad questions that are helpful in arriving at a flowmeter decision. There are limited tradeoffs in the first three, since they determine whether or not a particular meter is suitable. The last two are the most difficult because they involve the cost-versus-performance decision. The questions are as follows:

1. What type of fluid is to be handled?

   ■ Is the fluid a liquid, gas, or vapor?

   ■ Is the fluid clean, dirty, or a slurry?

   ■ Is the fluid corrosive?

2. What are the process conditions? The temperature and pressure limits?

3. What are the installation conditions?

   ■ Is an open-channel or closed-conduit measurement planned?

   ■ What is the line size?

   ■ What is the operating pipe Reynolds number?

   ■ Is adequate upstream length available?

   ■ Can flow conditioners be used?

   ■ Is there excessive pipe vibration?

   ■ Will the flow be steady or pulsating?

   ■ What are the ambient or room conditions of temperature and humidity?

4. What are the performance and general flow measurement requirements?

   ■ What overall accuracy is required, and over what range?

   ■ Is the flowmeter normally used at a specific flow rate or over a range of flows?

   ■ Is the flowmeter to be used for control only, and what frequency of response is necessary (0.1 s is usually required for small-line-size control)?

   ■ What is the range of flow rates to be measured, and what rates will occasionally be experienced?

5. What are the economics of installation and operation? Economic considerations include:

   ■ Initial cost of the primary and secondary devices and auxiliary equipment

   ■ Installation cost, including labor and piping

   ■ Energy cost to operate the flowmeter and pumping cost to compensate for the overall pressure loss

   ■ Reliability versus maintenance cost

   ■ Availability of parts and service facilities

■ Possible uses in future applications?

■ The risks in trying a new flowmeter type

Table 6.1 lists information relative to these decision factors for the flowmeters covered in this handbook. In comparing the costs of flowmeters, several factors must be considered other than the actual purchase price. The effects of improved meter accuracy and range on plant efficiency and the quality of the product are important in any decision. Additionally, the costs of installation, long-term maintenance, and operating energy need to be assessed. It is difficult to rank initial purchase price, because material requirements and the need for additional secondary elements vary by meter type. However, as a guide, three relative purchase-price groupings are given in Table 6.2 for the meters listed in Table 6.1.

Table 6.2   Relative Purchase Price by Groups†

Group 1 (low)	Group 2 (medium)	Group 3 (high)
Elbow	Vortex	Venturi
Pitot	Turbine [$D < 6$ in (300 mm)]	Magnetic
Annubar	Variable-area (metal)	Ultrasonic (time of flight)
Variable-area (glass)	Lo-Loss	
Variable-area (metal [$D < 2$ in (50 mm)]	Universal Venturi Tube	Positive-displacement [$D > 4$ in (100 mm)]
Orifice	Flow nozzle	
Target	Ultrasonic (Doppler)	
Positive-displacement [$D < 2$ in (50 mm)]		

†To be used as a guide only. Ranking may be the same within groups, or flowmeters may change groups because of line size, material of construction, auxiliary equipment, etc.

Energy Cost

In many instances, the additional energy cost resulting from permanent pressure loss is a factor in flowmeter selection. Pumping costs are sometimes significant in larger line sizes and may justify the selection of a more expensive flowmeter that either has a lower permanent pressure-loss coefficient or is obstructionless.

Figure 6.27 shows the permanent pressure loss in percent of differential pressure for the differential producers. Equations for calculating permanent pressure loss are given in Table 6.3 for U.S. units and in Table 6.4 for SI units. In Table 6.5 are the power and energy-cost equations for both U.S. and SI units.

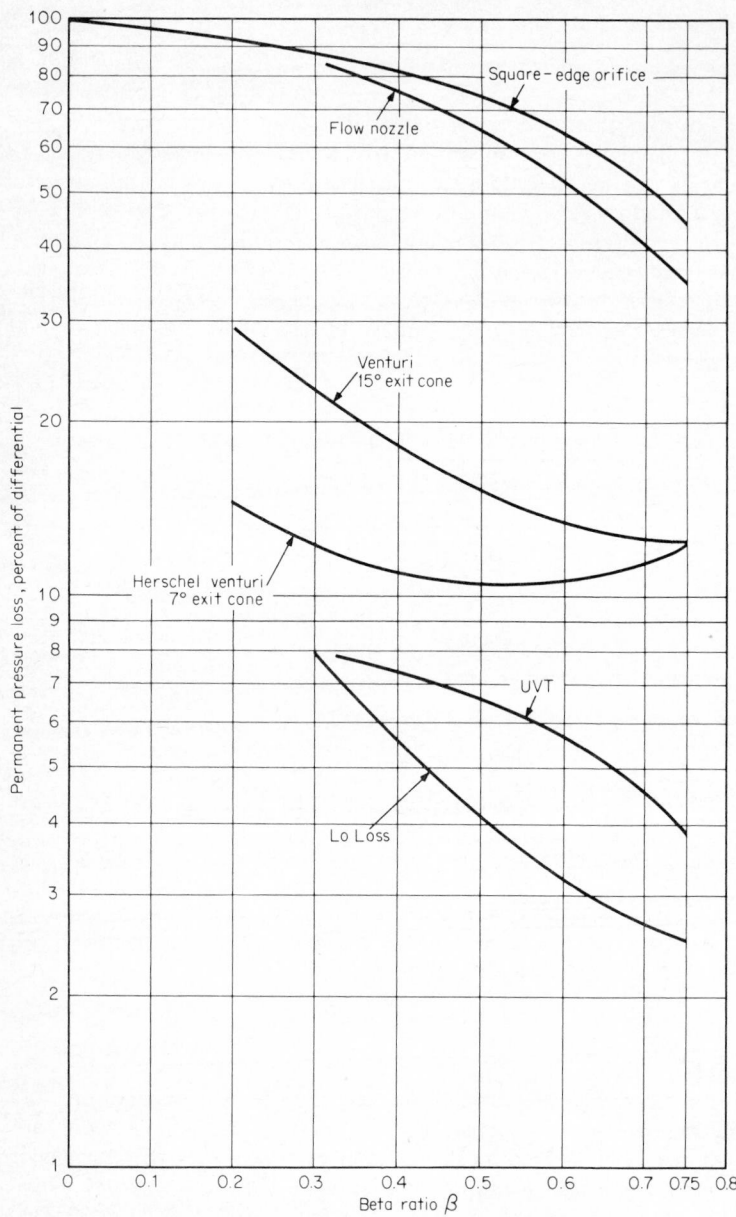

Figure 6.27 Permanent-pressure-loss curves for differential producers.

EXAMPLE 6.1

Calculate the yearly energy cost for an ASME long-radius nozzle and a flange-tap orifice measuring 30,000 lb_m/h (13,610 kg/h) in a 6-in (150-mm) pipe. The nozzle and orifice bore are sized to produce 100 in (25 kPa) at the upper-range flow rate (30,000 lb_m/h). Energy costs \$0.05/kW·h; the motor and pump efficiency is assumed to be 100 percent for a steam flow.

From Example 9.2, β = 0.6085 for the nozzle and β = 0.7432 for the orifice, and ρ_{f1} = 0.3952 lb_m/ft^3.

Energy cost for the nozzle

The permanent pressure loss is, from Eq. (e) of Table 6.3,

$$h_l = (1 + 0.014\beta - 2.06\beta^2 + 1.18\,\beta^3)h_w$$

$$= [1 + (0.014)(0.6085) - (2.06)(0.6085)^2 + (1.18)(0.6085)^3](100) = 51.2 \text{ in}$$

The required horsepower, from Table 6.5, is calculated with Eq. (d) of Table 6.5:

$$\text{hp} = \frac{h_l q_{\text{PPH}}}{3.8 \times 10^5 \eta \rho_{f1}} = \frac{(51.2)(30000)}{(3.8 \times 10^5)(1.0)(0.3952)} = 10.2 \text{ hp}$$

The yearly energy cost is then calculated with Eq. (e) of Table 6.5:

$$\text{Energy cost} = (0.75)(\text{hp})(\text{operating hours/year})(\$/\text{kw}\cdot\text{h})$$
$$= (0.75)(10.2)(8760)(0.05) = \$3357/\text{year}$$

Energy cost for the orifice

The permanent pressure loss is calculated with Eq. (f) of Table 6.3:

$$h_l = (1 - 0.24\beta - 0.52\beta^2 - 0.16\beta^3)h_w$$

$$= [1 - (0.24)(0.7432) - (0.52)(0.7432)^2 - (0.16)(0.7432)^3](100) = 46.9 \text{ in}$$

From Eq. (d) of Table 6.5, the required horsepower is

$$\text{hp} = \frac{h_l q_{\text{PPH}}}{3.8 \times 10^5 \eta \rho_{f1}} = \frac{(46.9)(30000)}{(3.8 \times 10^5)(1.0)(0.3952)} = 9.4 \text{ hp}$$

The energy cost is then, by Eq. (e) of Table 6.5,

$$\text{Energy cost} = (0.75)(\text{hp})(\text{operating hours/year})(\$/\text{kW}\cdot\text{h})$$
$$= (0.75)(9.4)(8760)(0.05) = \$3087/\text{year}$$

EXAMPLE 6.2

Calculate the yearly energy cost for a 200-mm Foxboro vortex flowmeter and for an orifice flowmeter measuring 38°C water flowing at 5.7 m^3/min. For the orifice, β = 0.689, and the differential pressure is 50 kPa. Assume an 80 percent pump and motor efficiency and an energy cost of \$0.05/kW·h.

From Eqs. (2.75) and (2.79), the density and specific gravity of water at 38°C are

$$\rho_F^* = 992.96 \text{ kg/m}^3 \qquad G_F = \frac{992.96}{999.012} = 0.9939$$

Table 6.3 Permanent Pressure-Loss Equations (U.S. Units)

Flowmeter	Liquid	Gas (vapor)		Liquid/gas (vapor); mass flow
Venturi:				
15° exit cone	↑	$h_l = (0.436 - 0.86\,\beta + 0.59\,\beta^2)h_w$	(a)	—
7° exit cone	↑	$h_l = (0.218 - 0.42\,\beta + 0.38\,\beta^2)h_w$	(b)	↓
Universal Venturi Tube†	↑	$h_l = (0.065 + 0.092\,\beta - 0.167\,\beta^2)h_w$	(c)	↓
Lo-Loss tube‡	↑	$h_l = (0.151 - 0.304\,\beta + 0.182\,\beta^2)h_w$	(d)	↓
Nozzle	↑	$h_l = (1 + 0.014\,\beta - 2.06\,\beta^2 + 1.18\,\beta^3)h_w$	(e)	↓
Orifice	↑	$h_l = (1 - 0.24\,\beta - 0.52\,\beta^2 - 0.16\,\beta^3)h_w$	(f)	↓
Annubar: §				
Types 73, 75, 76	↑	$h_l = \dfrac{1.25}{D}\,h_w$	(g)	↓
Types 85, 86	↑	$h_l = \dfrac{3.2}{D}\,h_w$	(h)	↓
Pitot	↑	$h_l = \dfrac{0.6}{D}\,h_w$	(i)	↓
Target¶	↑	$h_l = 0.000467\,\dfrac{\rho_{f1}\overline{V}_f^2}{(1 - \beta_T^2)^{2.75}}$	(j)	↓

$$h_l = \frac{F_p G_F}{(1 - \beta_T)^{2.75}} \left(\frac{q_{\text{gpm}}}{14.3D^2}\right)^2 \quad \text{(k)} \qquad h_l = \frac{Z_{f1}GT_{f1}}{p_{f1}(1 - \beta_T)^{2.75}} \left(\frac{q_{\text{SCFH}}}{19{,}520D^2}\right)^2 \quad \text{(r)} \qquad h_l = \frac{1}{\rho_{f1}(1 - \beta_T)^{2.75}} \left(\frac{q_{\text{PPH}}}{909D^2}\right)^2 \quad \text{(w)}$$

$$h_l = \frac{G_b^2}{F_p G_F (1 - \beta_T)^{2.75}} \left(\frac{q_{\text{GPM}}}{14.3D^2}\right)^2 \quad \text{(l)}$$

$$h_l = 0.00577\,\rho_f \overline{V}_f^2 \quad \text{(m)} \qquad h_l = 0.0129\rho_{f1}\overline{V}_f^2 \quad \text{(s)} \qquad \text{Liquid: } h_l = \frac{1}{\rho_{f1}} \left(\frac{q_{\text{PPH}}}{259D^2}\right)^2 \quad \text{(x)}$$

$$h_l = F_p G_F \left(\frac{q_{\text{gpm}}}{4.08D^2}\right)^2 \quad \text{(n)} \qquad h_l = \frac{Z_{f1}GT_{f1}}{p_{f1}} \left(\frac{q_{\text{SCFH}}}{3714D^2}\right)^2 \quad \text{(t)} \qquad \text{Gas (vapor): } h_l = \frac{1}{\rho_{f1}} \left(\frac{q_{\text{PPH}}}{173D^2}\right)^2 \quad \text{(y)}$$

$$h_l = \frac{G_b^2}{F_p G_F} \left(\frac{q_{\text{GPM}}}{4.08D^2}\right)^2 \quad \text{(o)}$$

Turbine¶

$$\longrightarrow \qquad h_l = \frac{Z_{f1}GT_{f1}}{p_{f1}} \left(\frac{q_{\text{SCFH}}}{5669D^2}\right)^2 \quad \text{(u)}$$

$$h_l = F_p G_F \left(\frac{q_{\text{gpm}}}{4.17D^2}\right)^2 \quad \text{(p)} \qquad h_l = \frac{Z_{f1}GT_{f1}}{p_{f1}} \left(\frac{q_{\text{SCFH}}}{5669D^2}\right)^2 \quad \text{(v)} \qquad h_l = \frac{1}{\rho_{f1}} \left(\frac{q_{\text{PPH}}}{264D^2}\right)^2 \quad \text{(z)}$$

$$h_l = \frac{G_b^2}{F_p G_F} \left(\frac{q_{\text{GPM}}}{4.17D^2}\right)^2 \quad \text{(q)}$$

Vortex¶

$\longrightarrow$

---

†The manufacturer (BIF) should be consulted for exact information.

‡The manufacturer (Badger Meter Inc.) should be consulted for exact information.

§The manufacturer (Dieterich Standard Corp.) should be consulted for exact information.

¶Foxboro flowmeter gas-turbine flowmeter equation based on Rockwell literature; the manufacturer should be consulted for exact information.

6-33

Table 6.4 Permanent Pressure-Loss Equations (SI Units)

Flowmeter	Liquid	Gas (vapor)	Liquid/gas (vapor); mass flow
Venturi:			
15° exit cone	→	$\Delta p_i^* = (0.436 - 0.86\,\beta + 0.59\,\beta^2)\,\Delta p^*$ (a)	↓
7° exit cone	→	$\Delta p_i^* = (0.218 - 0.42\,\beta + 0.38\,\beta^2)\,\Delta p^*$ (b)	↓
Universal Venturi Tube†	→	$\Delta p_i^* = (0.065 + 0.092\,\beta - 0.167\,\beta^2)\,\Delta p^*$ (c)	↓
Lo-Loss tube‡	→	$\Delta p_i^* = (0.151 - 0.304\,\beta + 0.182\,\beta^2)\,\Delta p^*$ (d)	↓
Nozzle	→	$\Delta p_i^* = (1 + 0.014\,\beta - 2.06\,\beta^2 + 1.18\,\beta^3)\,\Delta p^*$ (e)	↓
Orifice	→	$\Delta p_i^* = (1 - 0.24\,\beta - 0.52\,\beta^2 - 0.16\,\beta^3)\,\Delta p^*$ (f)	↓
Annubar: §			
Types 73, 75, 76	→	$\Delta p_i^* = \dfrac{31.8}{D^*}\,\Delta p^*$ (g)	↓
Types 85, 86	→	$\Delta p_i^* = \dfrac{81.3}{D^*}\,\Delta p^*$ (h)	↓
Pitot	→	$\Delta p_i^* = \dfrac{15.2}{D^*}\,\Delta p^*$ (i)	↓
Target¶	→	$\Delta p_i^* = 0.0000779\,\dfrac{\rho_{f1}^* \overline{V}_f^2}{(1 - \beta_T)^{2.75}}$ (j)	↓

$$\Delta p_i^* = \frac{F_p G_F}{(1 - \beta_T)^{2.75}} \left(\frac{5.92 q_{lpm}^*}{D^{*2}} \right)^2 \tag{k}$$

$$\Delta p_i^* = \frac{G_b^2}{F_p G_F (1 - \beta_T)^{2.75}} \left(\frac{5.92 q_{LPM}^*}{D^{*2}} \right)^2 \tag{l}$$

Turbine¶

$$\Delta p_i^* = 0.000964 \rho_f^* \overline{V}_f^2 \tag{m}$$

$$\Delta p_i^* = F_p G_F \left(\frac{20.8 q_{lpm}^*}{D^{*2}} \right)^2 \tag{n}$$

$$\Delta p_i^* = \frac{G_b^2}{F_p G_F} \left(\frac{20.8 q_{LPM}^*}{D^{*2}} \right)^2 \tag{o}$$

Vortex¶ ⟶

$$\Delta p_i^* = F_p G_F \left(\frac{20.4 q_{lpm}^*}{D^{*2}} \right)^2 \tag{p}$$

$$\Delta p_i^* = \frac{G_b^2}{F_p G_F} \left(\frac{20.4 \, q_{LPM}^*}{D^{*2}} \right)^2 \tag{q}$$

$$\Delta p_i^* = \frac{Z_{f1} G T_{K1}}{p_{f1}^* (1 - \beta_T)^{2.75}} \left(\frac{2.05 q_{SCMH}^*}{D^{*2}} \right)^2 \tag{r}$$

$$\Delta p_i^* = 0.00216 \rho_f^* \overline{V}_f^2 \tag{s}$$

$$\Delta p_i^* = \frac{Z_{f1} G T_{K1}}{p_{f1}^*} \left(\frac{10.8 q_{SCMH}^*}{D^{*2}} \right)^2 \tag{t}$$

⟶ (u)

$$\Delta p_i^* = \frac{Z_{f1} G T_{K1}}{p_{f1}^*} \left(\frac{7.06 q_{SCMH}^*}{D^{*2}} \right)^2 \tag{v}$$

$$\Delta p_i^* = \frac{1}{\rho_{f1}^* (1 - \beta_T)^{2.75}} \left(\frac{3.06 q_{KPH}^*}{D^{*2}} \right)^2 \tag{w}$$

Liquid: $\Delta p_i^* = \dfrac{1}{\rho_{f1}^*} \left(\dfrac{11 q_{KPH}^*}{D^{*2}} \right)^2 \tag{x}$

Gas (vapor): $\Delta p_i^* = \dfrac{1}{\rho_{f1}^*} \left(\dfrac{16.4 q_{KPH}^*}{D^{*2}} \right)^2 \tag{y}$

$$\Delta p_i^* = \frac{1}{\rho_{f1}^*} \left(\frac{10.75 \, q_{KPH}^*}{D^{*2}} \right)^2 \tag{z}$$

†The manufacturer (BIF) should be consulted for exact information.

‡The manufacturer (Badger Meter, Inc.) should be consulted for exact information.

§The manufacturer (Dieterich Standard Corp.) should be consulted for exact information.

¶Foxboro flowmeter, gas-turbine flowmeter equation based on Rockwell literature; the manufacturer should be consulted for exact information.

Table 6.5 **Power and Energy-Cost Equations (U.S. and SI Units)**

Liquid	Gas (vapor)	Liquid/gas (vapor); mass flow
	U.S. units (horsepower)	

$$\text{hp} = \frac{h_l q_{\text{gpm}}}{47{,}500\eta} \quad \text{(a)} \qquad \text{hp} = \frac{Z_{f1} T_{f1} h_l q_{\text{SCFH}}}{13.5 \times 10^6 \, \eta p_{f1}} \quad \text{(c)} \qquad \text{hp} = \frac{h_l q_{\text{PPH}}}{3.8 \times 10^5 \eta \rho_{f1}} \quad \text{(d)}$$

$$\text{hp} = \frac{G_b h_l q_{\text{GPM}}}{47{,}500\eta F_p G_F} \quad \text{(b)}$$

Energy cost (\$/year) = (0.75) (hp)(operating hours/year)(\$/kW·h)   (e)

| | SI units (watts) | |

$$W = \frac{\Delta p^* \, q\dagger_{\text{pm}}}{60\eta} \quad \text{(f)} \qquad W = \frac{Z_{f1} T_{K1} \, \Delta p^* q_{\text{SCMH}}^\$}{10.24\eta p_{f1}^*} \quad \text{(h)} \qquad W = \frac{\Delta p^* \, q_{\text{KPH}}^*}{3.6\eta \rho_{f1}^*} \quad \text{(i)}$$

$$W = \frac{G_b \, \Delta p^* \, q_{\text{LPM}}}{60\eta F_p G_F} \quad \text{(g)}$$

Energy cost (\$/year) = (W/1000) (operating hours/year)(\$/kW·h)   (j)

Energy cost for the vortex flowmeter

With negligible liquid compressibility ($F_p = 1.0$), the permanent pressure loss is found with Eq. (p) of Table 6.4:

$$\Delta p\dagger^* = F_p G_F \left(\frac{20.4 q\dagger_{\text{pm}}}{D^{*2}} \right) = (1.0)(0.9939)\left[\frac{(20.4)(5.7)(1000)}{(193.68)^2} \right]^2 = 9.55 \text{ kPa}$$

where, for the Foxboro vortex flowmeter, $D^* = 193.68$ mm. The required energy is then calculated with Eq. (f) of Table 6.5:

$$W = \frac{\Delta p^* \, q\dagger_{\text{pm}}}{60\eta} = \frac{(9.55)(5.7)(1000)}{(60)(0.8)} = 1134 \text{ W}$$

The yearly energy cost is, from Eq. (j) of Table 6.5,

$$\text{Energy cost} = (W/1000)(\text{operating hours/year})(\$/kW\cdot h) = \frac{1134}{1000} (8760)(($$

$$= \$497/\text{year}$$

Energy cost for the orifice

The permanent pressure loss is found with Eq. (f) of Table 6.4:

$$\Delta p\dagger^* = (1 - 0.24\beta - 0.52\beta^2 - 0.16\beta^3)\Delta p^*$$
$$= [1 - (0.24)(0.689) - (0.52)(0.689)^2 - (0.16)(0.689)^3](50) = 26.8 \text{ kPa}$$

The required energy is then, by Eq. (f) of Table 6.5,

$$W = \frac{\Delta p^* \, q\dagger_{\text{pm}}}{60\eta} = \frac{(26.8)(5.7)(1000)}{(60)(0.8)} = 3179 \text{ W}$$

The yearly energy cost is, from Eq. (j) of Table 6.5,

$$\text{Energy cost} = (\text{W}/1000)(\text{operating hours/year})(\$/\text{kW}\cdot\text{h}) = \frac{3179}{1000}(8760)(0.05)$$

$$= \$1393/\text{year}$$

Natural Gas Measurement

In the measurement of natural gas, concentric orifices with either flange taps or $2\frac{1}{2}D$ and $8D$ taps (pipe taps) are used almost exclusively, and the selection of other differential-producing primary devices or other flowmeters may not be acceptable. AGA Report 3 (ANSI 2530, 1978) on natural gas measurement does not include any other primary devices.

Usually, the available pressure is substantially higher than that required, and a pressure regulator is installed near the orifice flowmeter. Hence, a primary device having high pressure-recovery characteristics is seldom needed. In some cases, vortex flowmeters and Annubars are used in *noncustody* transfer applications to reduce the sizeable annual pumping cost required for orifice flowmeters.

Gas-turbine flowmeters are becoming more acceptable in the United States, and AGA Report 7 (1981), which deals with these meters, is available for contractual purposes. In Canada and Europe, turbine flowmeters are well accepted.

REFERENCES

AGA Report 7, *Measurement of Fuel Gas by Turbine Meters,* catalog no. X00580, American Gas Association, Arlington, Va., 1981.

ANSI/API 2530 *Orifice Metering of Natural Gas,* American Gas Association, New York, 1978.

API Publication 2101, chap. 12.2, American Petroleum Institute, Washington, D.C., 1981.

Brennan, J. A., R. W. Stokes, C. H. Kneebone, and D. B. Mann: "An Evaluation of Selected Angular Momentum, Vortex Shedding and Orifice Cryogenic Flowmeters," U.S. Department of Commerce, NBS, Technical Note 650, 1974.

Filban, T. J., and W. A. Griffin: "Small Diameter Orifice Metering," *J. Basic Eng.,* vol. 82, no. 3, p. 735, 1960.

Gorter, J., and D. G. de Rooij: "An Investigation in Widening the Reynolds Number Range for Flow Measurement in Closed Circuits by Means of Orifice Plates," *Fluid Flow Measurements in the Mid-1970's* (conf. proc.), vol. 1, pp. 3–23, Her Majesty's Stationery Office, Edinburgh, 1977.

Haacke, A. C.: "Sensitivity of the Electromagnetic Flowmeter to Flowmeter to Fluid Velocity Profile," *Pittsburg Flow Symposium,* vol. 1, pt. 2, pp. 735–743, ISA, Pittsburgh, 1974.

Halmi, D.: "Metering Performance Investigation and Substantiation of the Universal Venturi Tube," pt. 1, "Hydraulic Shape and Discharge Coefficient," *J. Fluids Eng.,* ser. 1, vol. 96, no. 2, pp. 124–131, 1974.

Halmi, D.: "Metering Performance Investigation and Substantiation of the Universal Venturi Tube," pt. 2, "Installation Effect, Compressible Flow, and Head Loss," *J. Fluids Eng.*, ser. 1, vol. 96, no. 2, pp. 132–138, 1974.

Hayward, A. T. J.: "How to Choose a Flowmeter," *The Chartered Mechanical Engineer*, reprinted from *I. Mech. E.*, 1975.

Jones, J. T.: "Field Experience with Sonic Nozzle," *AGA Operating Section Proceedings*, pp. 315–319, American Gas Association, Arlington, Va., 1976.

Kinghorn, F. C., and A. McHugh: "The Performance of Turbine Meters in Two-Component Gas/Liquid Flow," in *Flow, Its Measurement and Control in Science and Industry*, vol. 2, ISA, Research Triangle Park, N.C., 1981.

Lomas, D.: "Vortex, Turbine, Orifice—Which One Do I Choose," *Flow-Con 77* (symp. proc.), p. 25, Institute of Measurement and Control, Sussex, U.K., 1977.

Miller, R. W., J. P. DeCarlo, and J. T. Cullen: "Vortex Flowmeter Calibration Results and Application Experiences," *Flow-Con 77* (symp. proc.), pp. 341–371, Institute of Measurement and Control, Sussex, U.K., 1977.

Murdock, J. W., C. J. Foltz, and C. Gregory: "Performance Characteristics of Elbow Flowmeters," ASME Winter Annual Meeting, Paper 63-WA-17, 1963.

Reinecke, M. E., W. G. Ragains, and R. W. Miller: "An Experimental Study of the Capability of Measuring Gas Mass Flow," *Flow Measurement Symp.*, pp. 232–252, ASME, New York, 1966.

Rubin, M., R. W. Miller, and W. G. Fox: "Driving Torques in a Theoretical Model of a Turbine Meter," *J. Basic Eng.*, ser. D, vol. 87, no. 2, pp. 413–420, 1965.

Shercliff, J. A.: *The Theory of Electromagnetic Flow-Measurement*, Cambridge University Press, Cambridge, England, 1962.

7

INTRODUCTION TO
THE DIFFERENTIAL
PRODUCER

The differential-producing flowmeters are the most widely used in industrial process-measurement and control applications. The square-edged concentric orifice is selected for 80 percent of all liquid, gas, and vapor (steam) applications. This chapter contains a brief history of the differential producer and a look at the organization of Chaps. 8 through 12, which deal exclusively with differential producers.

HISTORICAL BACKGROUND

There are numerous examples of the early application of the principle of the differential producer. The hourglass and the use of the orifice during Caesar's time to measure the flow of water to householders are but two of many. But the developments which led to the design and widespread use of the various types of primary elements began in the seventeenth century.

At the start of the seventeenth century, Castelli and Torricelli laid the foundation for the theory of differential producers with the concepts that the rate of flow is equal to the velocity times the pipe area, and that the discharge through an orifice varies with the square root of the head. Until recently all differential producers have been called *head-class* flowmeters because of this early work and that of Bernoulli, who, in 1738, developed the hydraulic equation for the calculation of flow rate.

In 1732, Pitot presented his paper on the pitot tube, and in 1797 Venturi published his work on a flowmeter principle that today bears his name. Venturi's work was developed into the first commercial flowmeter in 1887 by Clemens Herschel. Herschel's laboratory work defined the dimensions of the *Herschel venturi* and laid the foundation for future laboratory investigations to determine the relationships between geometry and differential pressure for the other differential producers.

In 1913, E. O. Hickstein (1915) presented early data on orifice flowmeters with pressure taps located $2\frac{1}{2}$ pipe diameters upstream and 8 pipe diameters downstream. This work, and that of others, led to several other pressure-tap locations, such as those for D-and-$D/2$ and vena contracta taps.

In 1916, E. G. Bailey delivered a paper on the measurement of steam with orifice flowmeters, and in 1912 experimental work by Thomas R. Weymouth of the United Natural Gas Company was the basis for the use of the orifice flowmeter for measuring natural gas. For convenience, Weymouth used flange pressure taps located 1 in upstream and 1 in downstream of the faces of a conventional square-edged orifice. In Europe, orifice flowmeters using the corner tap location were being experimentally investigated.

Although the orifice was widely used for many different fluids, it wasn't until 1930 that a joint AGA/ASME/NBS test program obtained sufficient data to develop a coefficient-prediction equation. The Ohio State University (1935) test report and the Buckingham fitting equations for the various tapping arrangements have been used by ASME and AGA since 1935. The ability to predict coefficients from measured dimensions led to the full commercialization of the orifice flowmeter.

In the late 1950s, work in the United States was combined with European practice, and ISO Standards R541 (1967) for orifices and nozzles and R781 (1968) for venturis were issued.

In the mid-1960s, an ASME Fluid Meters Research Committee initiated a study to reevaluate the Ohio State data and to add new coefficient data. The objective of this study was to derive, by regression analysis, a simpler and more accurate coefficient-prediction equation for flange-tapped orifices. The results (Dowdell and Chen, 1970) were not encouraging; the Ohio State data had a ± 1.85 percent deviation (2σ) from the Buckingham equation. By limiting line sizes and beta ratios, ASME *Fluid Meters* (1971) changed the previous ± 0.55 percent coefficient accuracy statement to ± 1 percent.

In 1975, J. Stolz proposed a universal orifice equation to the ISO orifice flowmeter subcommittee. He proposed, based on logical rules, to combine the Ohio State data into a single dimensionless equation suitable for flange, corner, and D-and-$D/2$ taps. This equation was presented in a 1978 paper (Stolz, 1978). The orifice prediction equation appears in ISO Standard 5167 (1980), which combines Standards R541 and R781 into a single differential-producer standard. Based on the Dowdell and Chen (1970) investigation and subsequent papers by Miller et al. (1979), the ASME Fluid Meters Research Committee (1981) adopted the ISO 5167 (1980) equation, used in this handbook for all primary elements.†

For the measurement of natural gas, AGA 3 (ANSI/API 2530, 1978) is usually required for contractual purposes; in it, the Buckingham equations for flange and pipe taps are presented. For this measurement the sizing and flow-rate equations developed in this handbook may be used by substituting the Buckingham flow-coefficient equations given in Sec. A.4.

---

†Subsequently ISO 5167 was developed into ANSI/ASME MFC draft standard 3e (July 1982) to include recognized improvements and U.S. units.

The first investigation of a flow nozzle dates back to the nineteenth century (Froude, 1847). In 1930 Germany standardized on an ISA (1932) nozzle geometry. [The ISA (International Federation of National Standardizing Organizations) was replaced by the present International Organization for Standardization (ISO).] In the United States the long-radius flow nozzle was developed at Ohio State University (Beitler and Bean, 1948) primarily for the measurement of steam flows. The need for improved accuracy when testing steam turbines led to the development of the ASME throat-tap nozzle. The downstream pressure tap in this nozzle is in the cylindrical throat section.

The commercial success of the orifice, venturi, and nozzle led to the development of continually improved secondary measuring elements. This, coupled with test work and user familiarity, led to the further development of primary elements such as the segmental orifice, eccentric orifice, Lo-Loss tube, elbow, and annular orifice. These primary elements, combined with a wide variety of available secondary elements, are used to indicate, record, compute, and control the majority of flow measurements made today.

ORGANIZATION OF DIFFERENTIAL-PRODUCER INFORMATION

Flowmeter Elements

Figure 7.1a shows a differential-producer flowmeter. The meter consists of two separate devices that act in combination to measure flow rate. The first is the primary device (Fig. 7.1b). The geometry of the differential producer, the length and condition of the reference piping, and the pressure-tap location with respect to the primary element are properties of the primary device. Any change in these properties from reference conditions alters the differential pressure–flow rate relationship. Figure 7.1c shows the elements of the secondary device. The differential pressure is measured by transmitting pressure through lead lines to a secondary differential-pressure measuring element. The secondary device in this case consists of three elements: lead lines, valving, and manometer. Since the flow rate can be calculated from visual observation of the differential pressure, no further readout equipment is required. If the flow rate were to be calculated by a central processer or on-site computer or from a chart record, each of these would be an additional element of the secondary device.

As shown in Fig. 7.1a, the differential pressure depends on the pressure-tap locations and on whether the contraction is abrupt or gradual. The relationship between measured differential and flow rate is then a function of tap locations, primary-element design, and associated upstream and downstream piping. These effects are included in the discharge coefficient, which relates the actual flow rate to the theoretically calculated flow rate.

Contents of Chapters 8 to 12

Obviously, with the number of differential producers that have both common and unique features, grouping is necessary to avoid repetition. These meters have, therefore, been separated into two broad groups for discussion. In the first

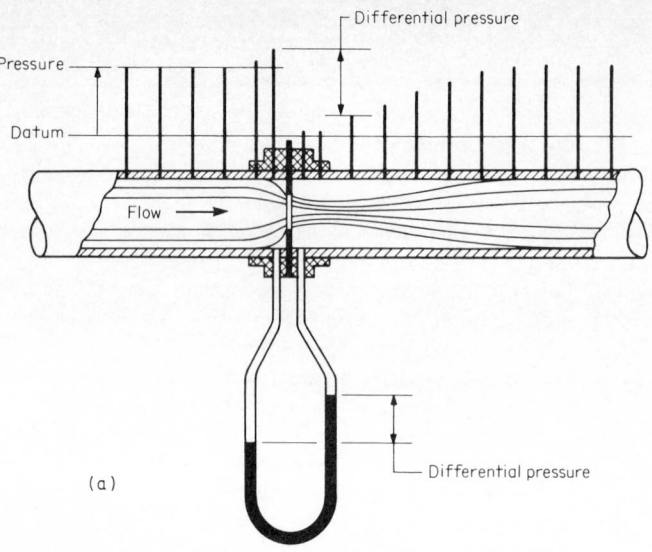

(a)

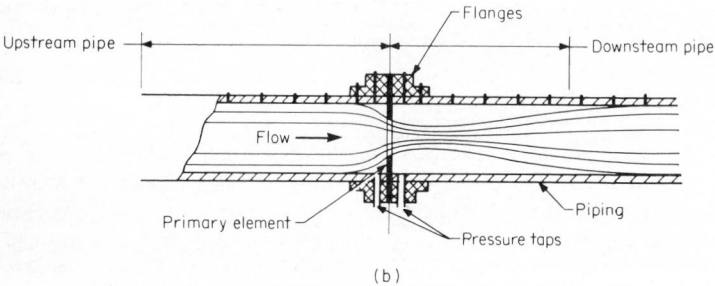

(b)

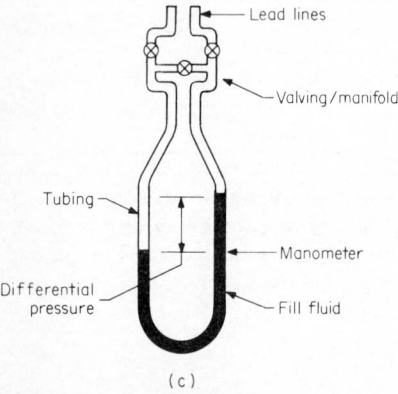

(c)

Figure 7.1 Differential-producer flowmeter. (*a*) Flowmeter. (*b*) Elements of the primary device. (*c*) Elements of the secondary device.

7-4

group are those meters in which the bore is sized to produce a design differential—for example, 100 in (25 kPa) at 1000 gal/min (3800 L/min). In the second group are flowmeters in which geometry is fixed and the differential (or target force) must be calculated for a design flow rate. The content of Chaps. 8 to 12 follows this grouping.

Chapter 8: Installation   In this chapter, pipe circularity, pressure-tap design, the location of thermal wells, lead lines, and installation piping are discussed.

Chapter 9: Engineering Equations   Here the engineering equations for sizing and for calculating flow rates in both U.S. and SI units are derived and presented in tables for primary elements that are sized to produce a design differential at a given flow rate. Procedural tables are presented along with examples that deal with sizing and the calculation of flow rates and accuracy.

Chapter 10: Design Information   Recommended designs for the various primary elements are presented in this chapter. For graphical sizing and flow-rate calculation, discharge-coefficient curves are provided—both based on equations given in Chap. 9 and for meters for which the coefficient has not been reduced to an equation (eccentric and segmental-orifice meters). Thermal-expansion-factor and gas-expansion-factor curves are also given.

Chapter 11: Fixed-Geometry Devices   This chapter presents the geometry-factor equations that are substituted into the engineering equations of Chap. 9 to calculate differential pressure (or target force) for elements with fixed geometry (elbow, target, integral orifice, and arithmetic-progression orifice). This substitution-of variables method allows the retention of the basic form of the equations given in Chap. 9. Tables and graphs are also given for flow-coefficient calculations.

Chapter 12: Computations   This chapter reviews the available computational methods for calculating either flow rate or total flow using chart integration and analog or digital flow computers.

REFERENCES

AGA-ASME: *The Flow of Water Through Orifices,* Ohio State University Studies, Engineering Series, vol. IV, no. 3, Bulletin 89, 1935.

ANSI/API 2530, *Orifice Metering of Natural Gas,* ANSI, New York, 1978.

ANSI (MFC): *Differential Producers Used for the Measurement of Fluid Flow in Pipes (Orifice, Nozzle, Nozzle Venturi, Venturi),* draft 3e, New York, July 1982

ASME: "The ASME-ISO Orifice Equation," *Mech. Eng.,* vol. 103, no. 7, 1981.

ASME: *Fluid Meters,* 6th ed., ASME, New York, 1971.

Bailey, E. G.: "Steam Flow Measurement," *Power,* vol. 38, p. 250, 1916.

Beitler, S. R., and H. S. Bean: *Research on Flow Nozzles,* Ohio State University Engineering Experiment Station Bulletin 131, 1948.

Dowdell, R. B., and Yu Lin Chen: "A Statistical Approach to the Prediction of Discharge Coefficients for Concentric Orifice Plates," *J. Basic Eng.,* vol. 92, no. 4, p. 752, 1970.

Froude, W.: "Discharge of Elastic Fluids under Pressure," *Proc. Inst. Civ. Eng. London,* vol. 6, pp. 356–399, 1847.

Herschel, C.: "The Venturi Water Meter," *Trans. Am. Soc. Civ. Eng.,* vol. 17, p. 228, 1887.

Hickstein, E. O.: "Flow of Air through Thin-Plate Orifices," *Trans. ASME,* vol. 37, p. 765, 1915.

ISA Bulletin 12, International Federation of National Standardizing Organizations, Geneva, 1932.

ISO Standard R541, *Measurement of Fluid Flow by Means of Orifice Plates and Nozzles,* ISO, Geneva, 1967.

ISO Standard R781, *Measurement of Fluid Flow by Means of Venturi,* ISO, Geneva, 1968.

ISO Standard 5167, *Measurement of Fluid Flow by Means of Orifice Plates, Nozzles and Venturi Tubes Inserted in Circular Cross-Section Conduits Running Full,* ISO 5167-1980(E), Geneva, 1980.

Miller, R. W.: "The Stolz and ASME-AGA Orifice Equation Compared to Laboratory Data," *J. Fluids Eng.,* vol. 101, no. 4, pp. 483–490, 1979.

———— and G. A. Koslow: "The Uncertainty Values for the ASME-AGA and ISO 5167 Flange Tap Orifice Equation," ASME Winter Annual Meeting, Paper FM-5, 1979.

Stolz, J.: *An Approach toward a General Correlation of Discharge Coefficients of Orifice Plate Flowmeters,* ISO/TC30/SC2 (France 6) 654, 1975.

————: "A Universal Equation for the Calculation of Orifice Plates," in *FloMeko 1978, Groningen,* 1978, North-Holland, Amsterdam, 1978.

Weymouth, Thomas, R.: "Measurement of Natural Gas," *Trans. ASME,* vol. 34, p. 1091, 1912.

DIFFERENTIAL PRODUCERS: INSTALLATION

It is important that the installation of the primary element approach the standard or reference conditions which prevailed when the flow-coefficient information was obtained. The condition of the pipe, mating of pipe sections, pressure-tap design, straight lengths of pipe preceding and following the primary element, and lead lines that transmit the differential pressure to the secondary measuring element all affect measurement accuracy. While some of these may have a minor effect, others can introduce 5 or 10 percent bias errors. In general, these errors are not predictable, and attempts to adjust coefficients for the effect of nonstandard conditions have not been successful.

PIPING

Reference Piping

ISO Standard 5167† (1980) gives requirements for reference piping concerning the following items:

1. Visual condition of the outside of the pipe as to both straightness and circularity

2. Visual condition of the internal surface of the pipe

3. Reference-condition relative roughness for the internal surface (see Table 5.8)

4. Location of measurement planes and number of measurements for the determination of the average pipe diameter D

5. Circularity of a specified length of pipe preceding the primary element

6. Maximum allowable step height for mating sections of upstream pipe

†Subsequently ISO 5167 was developed into ANSI/ASME MFC draft standard 3e (July 1982) to include recognized improvements and U.S. units.

7. Limits on mating upstream-pipe step-height ratio (for addition of ± 0.2 percent to the discharge-coefficient accuracy)

These requirements are shown in Figs. 8.1 through 8.3.

Pressure Taps

Figure 8.4 illustrates the recommended geometry for wall static-pressure taps. Both pressure taps should be of the same diameter and, where the hole breaks through, should be square with no roughness, burrs, or wire edges. In Fig. 8.5 are shown typical pressure connections used for low- and high-temperature applications.

Protrusions or Pockets

If a pocket or obstruction is present in the upstream or downstream pipe, the velocity profile will be affected. Depending on disturbance diameter, these should be located a sufficient distance from the primary element to ensure fully recovered flow. ISO Standard 5167 (1980) and AGA 3/ANSI 2530 (1978) specifications for the location of these devices are shown in Fig. 8.6. Although ISO allows for an upstream thermal well, it is good practice to locate these downstream to ensure minimum profile distortion.

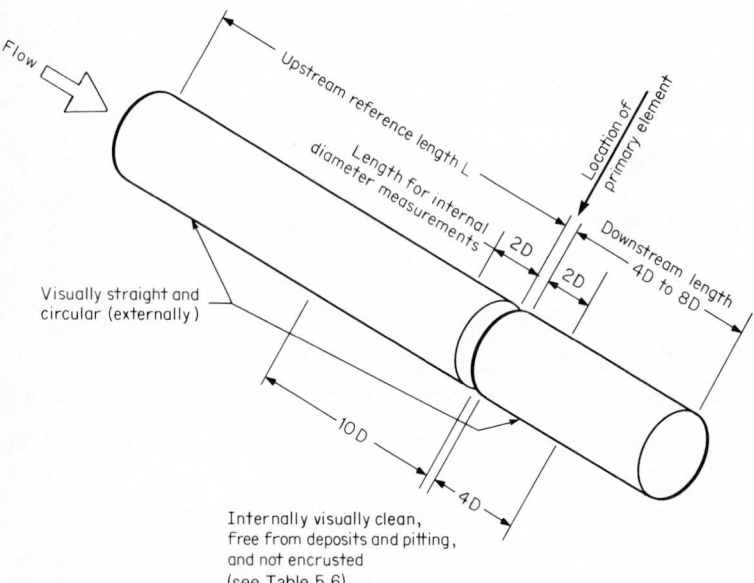

Figure 8.1 ISO visual requirements for pipe condition.

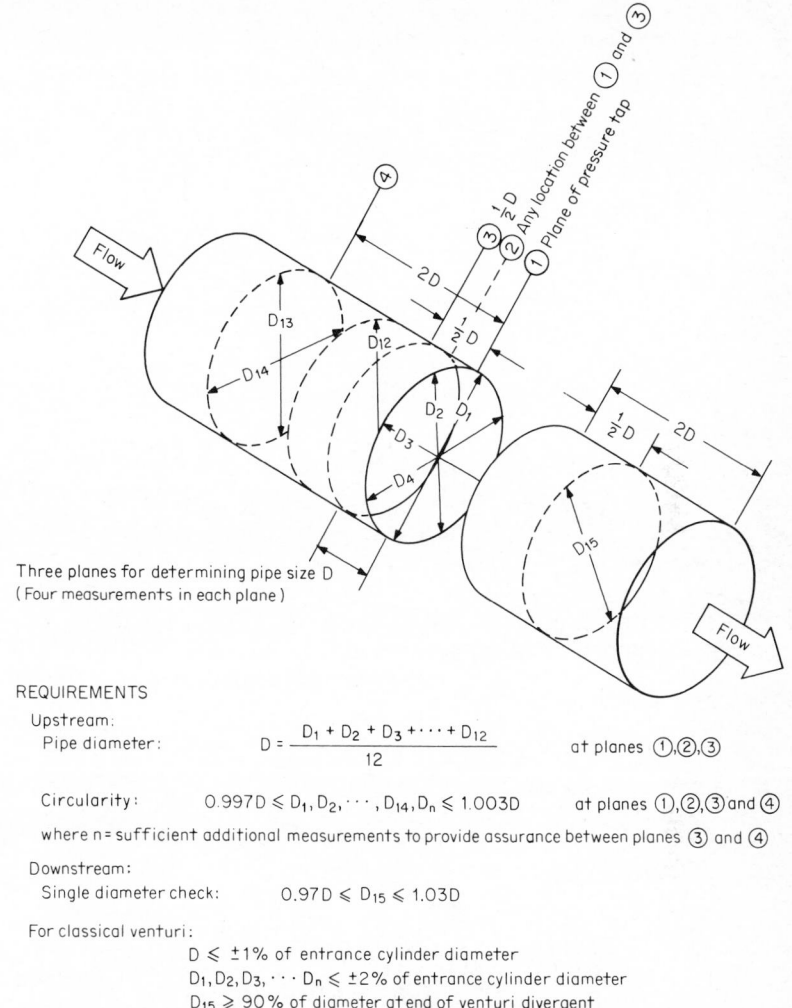

REQUIREMENTS

Upstream:
Pipe diameter:

$$D = \frac{D_1 + D_2 + D_3 + \cdots + D_{12}}{12}$$ at planes ①,②,③

Circularity: $0.997D \leqslant D_1, D_2, \cdots, D_{14}, D_n \leqslant 1.003D$ at planes ①,②,③ and ④

where n = sufficient additional measurements to provide assurance between planes ③ and ④

Downstream:
Single diameter check: $0.97D \leqslant D_{15} \leqslant 1.03D$

For classical venturi:

$D \leqslant \pm 1\%$ of entrance cylinder diameter
$D_1, D_2, D_3, \cdots D_n \leqslant \pm 2\%$ of entrance cylinder diameter
$D_{15} \geqslant 90\%$ of diameter at end of venturi divergent

Figure 8.2   **ISO pipe-diameter measurement and circularity requirements.**

Gaskets or weld beads that extend into the pipe tend to increase fluid turbulence and alter the profile. Welds should be ground smooth, and gaskets trimmed, so that no protrusion is evident on visual inspection.

Straight-Pipe Lengths

The required length of straight pipe preceding and following the primary element depends on the primary element and the type of upstream disturbance.

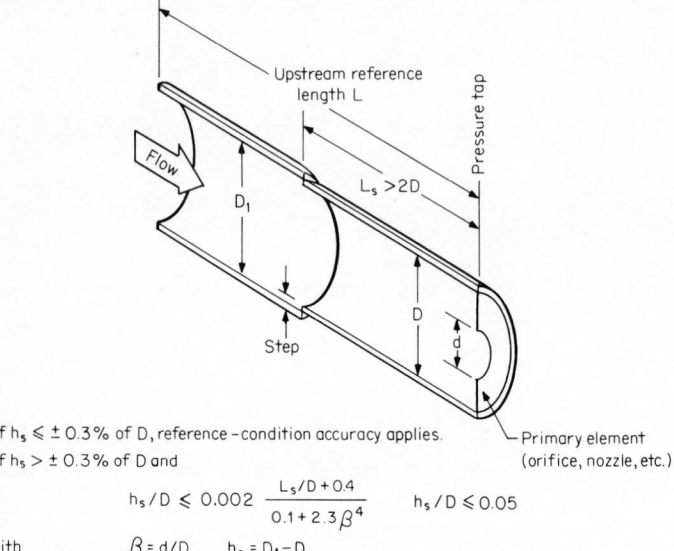

If $h_s \leqslant \pm 0.3\%$ of D, reference-condition accuracy applies.
If $h_s > \pm 0.3\%$ of D and

$$h_s/D \leqslant 0.002 \ \frac{L_s/D + 0.4}{0.1 + 2.3 \beta^4} \qquad h_s/D \leqslant 0.05$$

with $\qquad \beta = d/D \qquad h_s = D_1 - D$

then add 0.2% to discharge-coefficient accuracy.

Figure 8.3 ISO requirements for discharge-coefficient accuracy and for $\pm$ 0.2 percent addition for upstream pipe sections.

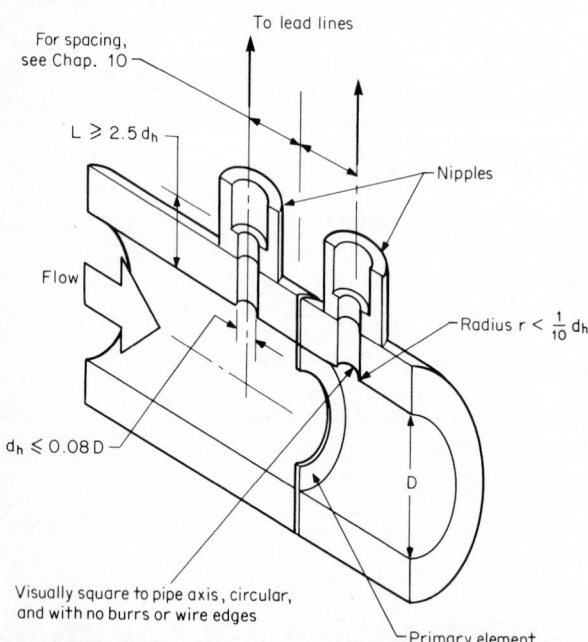

Figure 8.4 ISO pressure-tap design requirements.

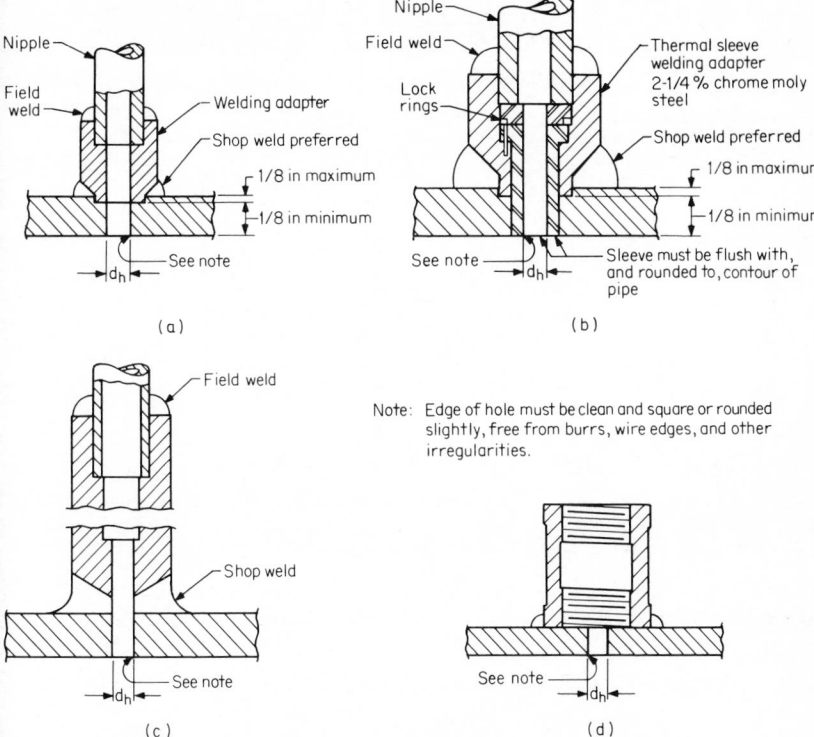

Figure 8.5 Methods of making pressure connections to pipe. (*a*) For temperatures up to 800°F. (*b*) For temperatures above 800°F and a secondary element with appreciable displacement. (*c*) Optional design where full-penetration weld is required. (*d*) For temperatures up to 400°F. (*From ASME Fluid Meters; used with permission.*)

Several U.S. and European test programs have provided data concerning the effects of specific upstream disturbances on orifices, nozzles, venturi nozzles, and venturis. This information has been used to establish *recommended* minimum pipe lengths. However, the recommended minimum lengths differ between U.S. sources (ANSI 2530, ASME *Fluid Meters*) and ISO Standard 5167, with the ISO recommendations being more conservative.

ANSI 2530 and *Fluid Meters*   In Figs. 8.7 through 8.11 are the ANSI 2530 and ASME *Fluid Meters* recommendations for straight pipe with and without a flow conditioner for orifices, nozzles, and venturis. *Fluid Meters* indicates that the diagram which corresponds closest to the actual piping arrangement should be used to determine the required length, and that minimum length is necessary to hold piping bias errors, due to piping influence, to less than ±0.5 percent. Further, it is recommended that for any decrease in length an additional ±0.5 percent be added to the flow-coefficient accuracy value.

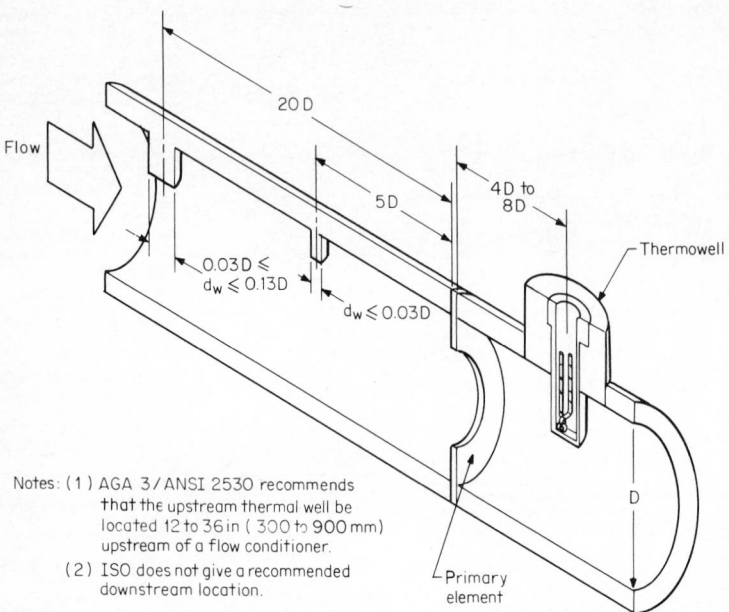

Figure 8.6 ISO- and AGA-recommended locations for thermal wells or pockets.

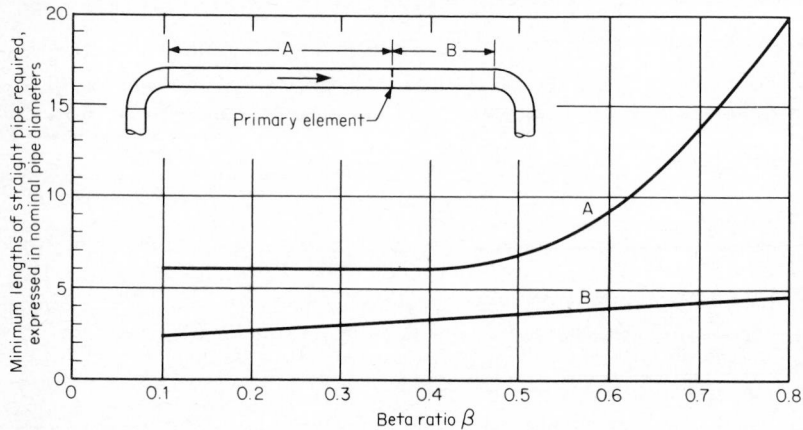

Figure 8.7 Installation lengths for primary elements preceded by an elbow in a single plane.

ISO Standard 5167 ISO straight-length requirements for orifices, nozzles, venturi nozzles, and venturis are listed by beta ratio in Table 8.1. These lengths are generally twice the ANSI 2530 and *Fluid Meters* requirements. For lengths one-half those shown an additional ± 0.5 percent is to be algebraically added to the coefficient accuracy given in Table 9.42. ISO recommends that a flow condi-

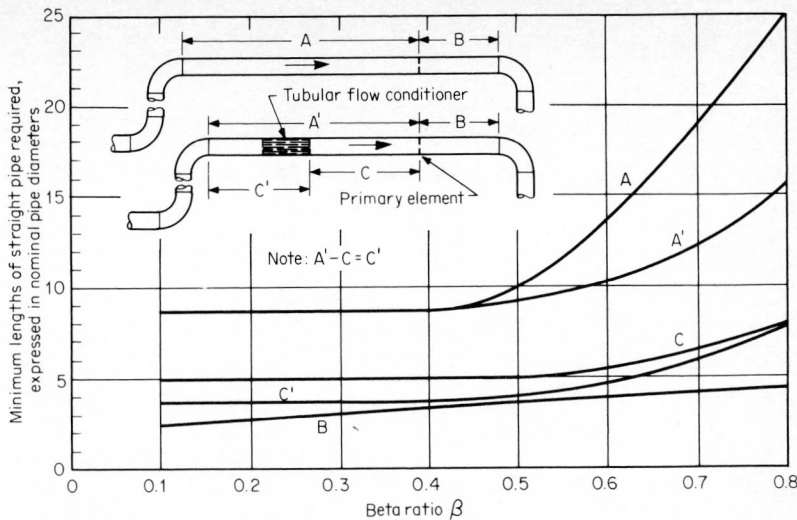

Figure 8.8 Installation lengths for primary element preceded by two elbows in one plane.

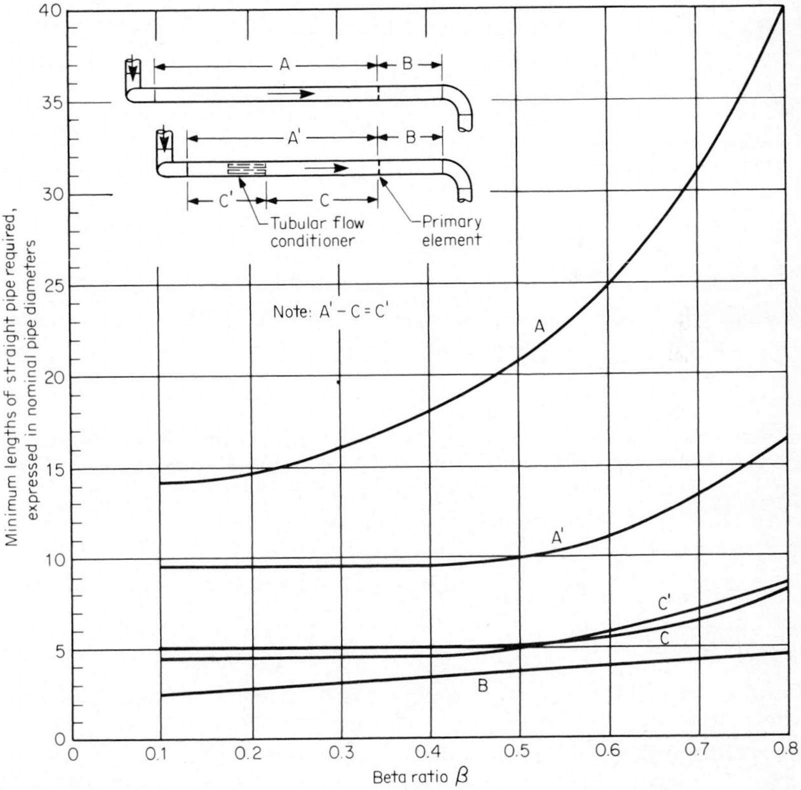

Figure 8.9 Installation lengths for primary element preceded by two elbows in different planes.

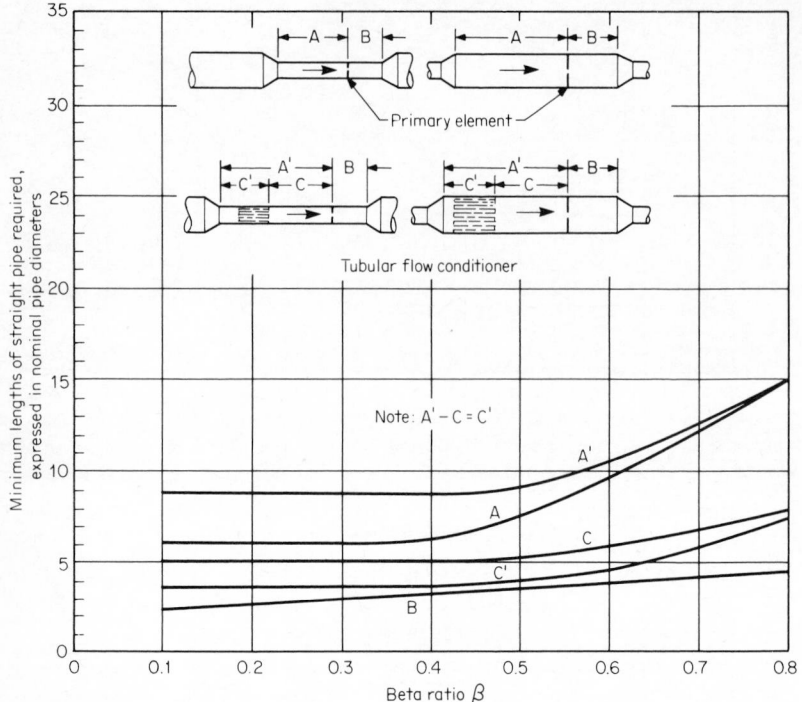

Figure 8.10 Installation lengths for primary element preceded by reducer or expander.

tioner be installed 20 pipe diameters following the disturbance, and that an additional 22 pipe diameters be installed between the conditioner and the primary element.

Proprietary Devices Straight lengths have been standardized only for venturis, nozzles, venturi nozzles, and orifices. For proprietary devices, the user should follow the manufacturer's recommendations. In Table 8.2 are the recommendations for the Annubar, and in Fig. 8.12 curves of bias error versus pipe length for the Lo-Loss tube.

LEAD LINES

Lead lines connect the primary and secondary elements and transmit the process differential pressure to the secondary element. Prior to the introduction of dry-type differential-pressure transmitters, lead-line design was quite complex. Seal pots, condensation chambers, sediment chambers, and purge flows were required

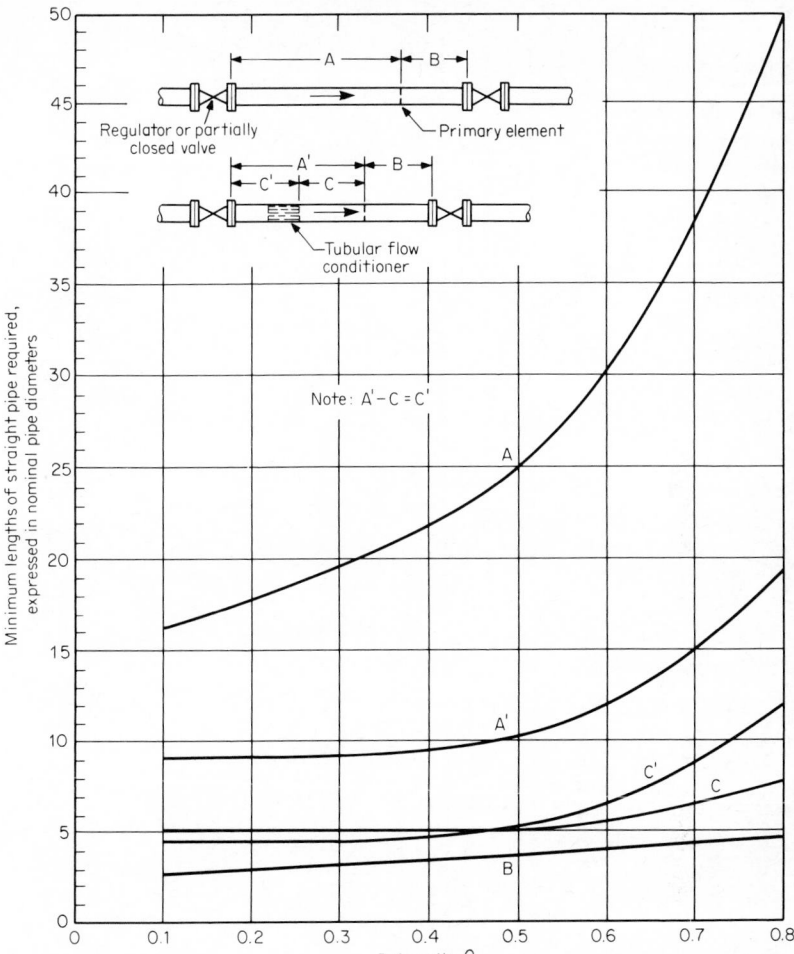

Figure 8.11 Installation lengths for primary element preceded by regulator or valve.

Table 8.1 ISO Standard 5167 Required Straight Lengths for Orifice, Nozzle, ISA Venturi Nozzle, and Venturi in Multiples of Pipe Diameter D

Upstream disturbance	Dimension	Device	β						
			0.2	0.3	0.4	0.5	0.6	0.7	0.75
Single elbow	A	Orifices Nozzles	14	16	18	20	26	28	36
		Venturis		0.5	0.5	1.5	3	4	4.5
Two elbows in same plane	A	Orifices Nozzles	14	16	18	20	26	36	42
		Venturis		1.5	1.5	2.5	3.5	4.5	4.5
Two elbows in different planes	A	Orifices Nozzles	34	34	36	40	48	62	70
		Venturis		0.5	0.5	8.5	17.5	27.5	29.5
Reducer	A	Orifices Nozzles	5	5	5	6	9	14	22
		Venturis		0.5	2.5	5.5	8.5	10.5	11.5

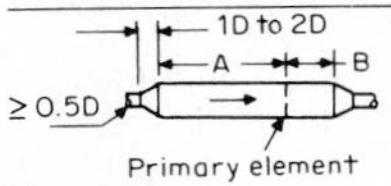 Expander	A	Orifices Nozzles	16	16	16	18	22	30	38
		Venturis		1.5	1.5	2.5	3.5	5.5	6.5
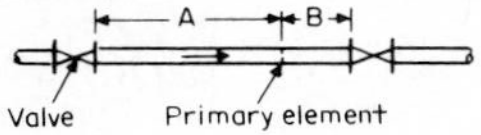 Globe valve, fully open	A	Orifices Nozzles	18	18	20	22	26	32	36
		Venturis							
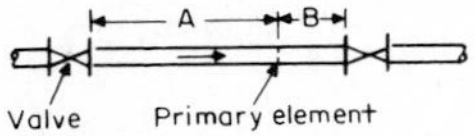 Gate valve, fully open	A	Orifices Nozzles	12	12	12	12	14	20	24
		Venturis		1.5	2.5	3.5	4.5	5.5	5.5
Downstream length for all pictured disturbances	B	Orifices Nozzles	4	5	6	6	7	7	8
		Venturis		$4d$	$4d$	$4d$	$4d$	$4d$	$4d$

NOTES: 1. For upstream and downstream lengths equal to one-half the values shown, add ± 0.5 percent to the accuracy values in Table 9.42.

2. Any flow conditioner shall be installed in the straight length between the primary element and the upstream disturbance, or the fitting closest to the element. The straight length between fitting and conditioner shall be at least $20D$, and the length between conditioner and element shall be at least $22D$.

3. Interpolate pipe diameters for intermediate beta ratios.

Table 8.2 Recommended Lengths for Annubars

Minimum diameters of straight pipe†	Upstream dimension					Downstream dimension B
	Without vanes		With vanes			
	In plane A	Out of plane A	A′	C	C′	
	7	9				3
			6	3	3	
	9	14				3
			8	4	4	
	19	24				4
			9	4	5	
	8	8				3
			8	4	4	
	8	8				3
			8	4	4	
	24	24				4
			9	4	5	

†Values shown are the recommended spacing in terms of internal diameters for normal industrial metering requirements.

SOURCE: Dieterich Standard Corp.

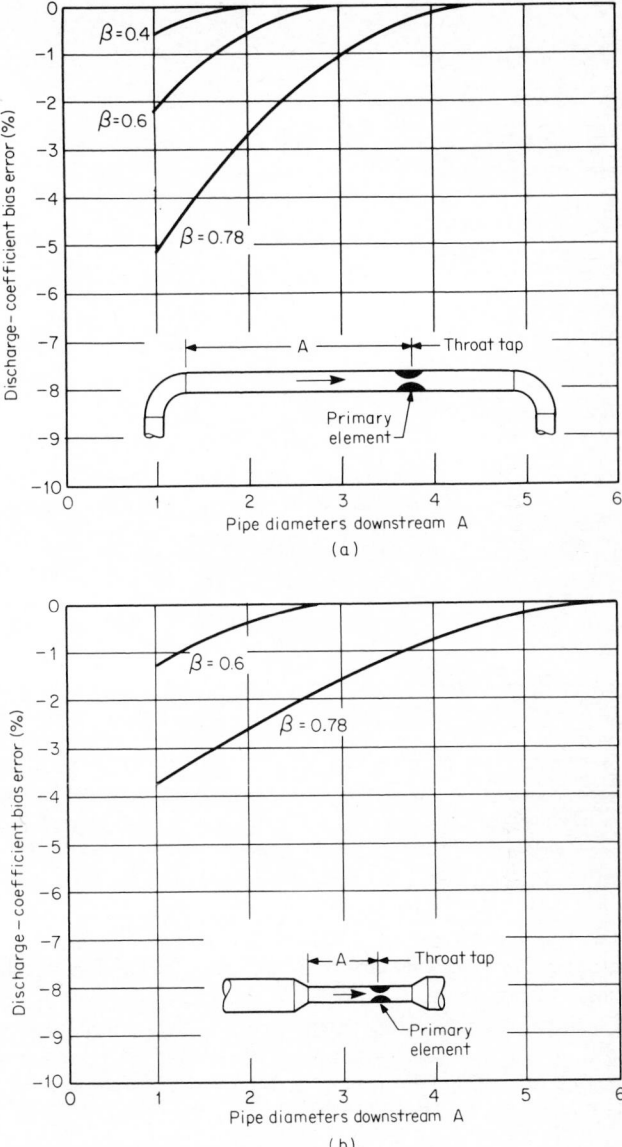

Figure 8.12 Recommended lengths for Lo-Loss tube *(courtesy Badger Meter, Inc.).* (**a**) Single elbow. (**b**) Reducer. (**c**) Tee

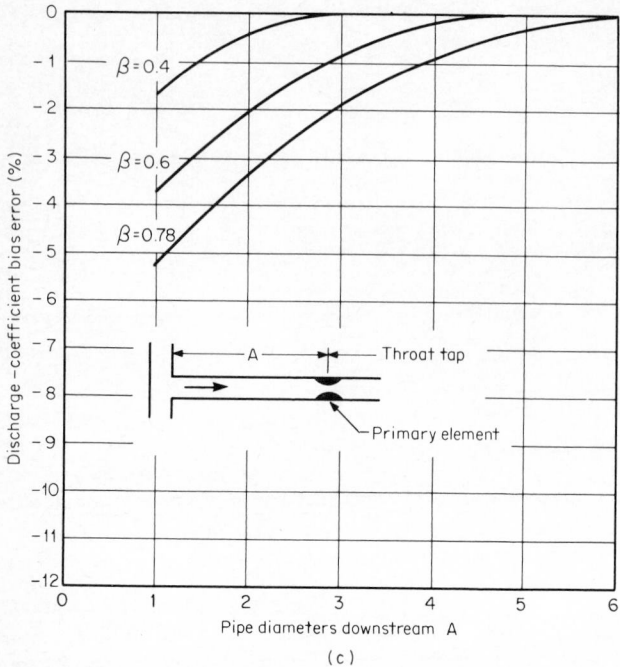

Figure 8.12 (*Continued*).

to ensure that volume changes within the wet-type meters would not affect the measurement or introduce sediment.

Dry-type meters have completely replaced mercury-filled meters in all but a few industries. The recognized hazards of mercury have led to a phasing-out program in the natural gas industry, and special safety precautions are required for laboratory use. Today many instrument manufacturers have discontinued the sale and repair of mercury-filled meters and offer only the dry type. Special high-accuracy mercury-filled manometers are still available, but these are usually used only in laboratories.

The following lead-line requirements, installation diagrams, and notes apply only to the dry-type differential transmitter. For information concerning the installation of wet-type meters, the reader is referred to ISO Standard 2186 (1973).

General Requirements

For liquid flows in a horizontal pipe (Figs. 8.13 and 8.14), the pressure taps should be located at the sides of the pipe, with no more than a ±45° orientation. For horizontal clean gas flows (Fig. 8.15), the pressure taps should be vertical;

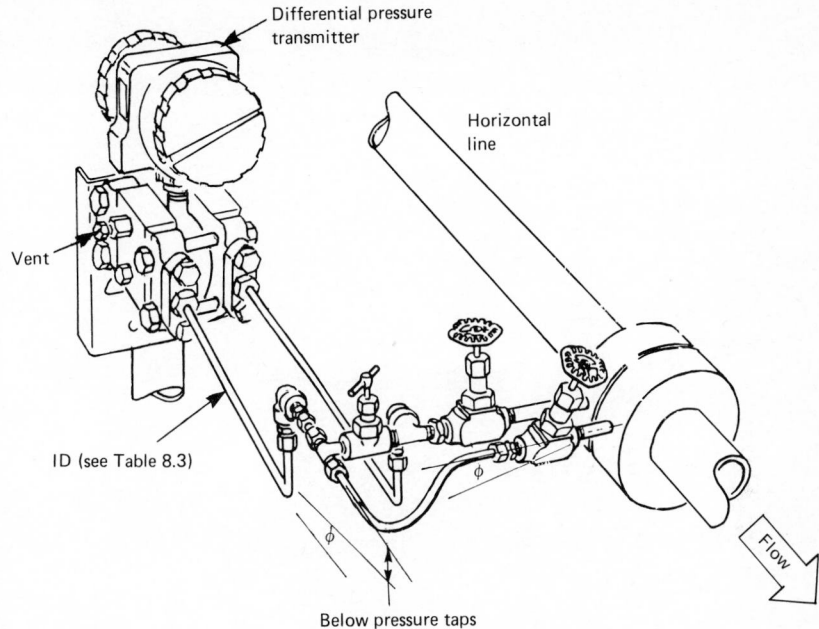

Notes: (1) Slope upward, ϕ = 1 in/ft (80 mm/m) for water;
ϕ = 2 to 4 in/ft (160 to 320 mm/m) for more viscous fluids.
(2) Minimize all lead-line lengths.
(3) For hot liquids, make unlagged lead lines long enough
to minimize density changes.

Figure 8.13 Horizontal installation for clean fluids *(courtesy The Foxboro Co.).*

for vapors (steam, ammonia) and dirty or condensable gases (Fig. 8.16), the taps should be located at the side. For vertical pipe installations, the pressure taps can be at any radial position around the pipe circumference (Figs. 8.17 through 8.20).

The differential-pressure transmitter should be located as close to the taps as possible. This improves the speed of response and reduces the possibility of resonance or attenuation within the leads. Prefabricated manifolds are available for easy close-coupled installation (Fig. 8.21).

Lead lines should be of the same bore, no less than 0.25 in (6 mm) in diameter for clean liquids. For condensable vapor flows, where gas bubbles may be liberated, the diameter should be no less than 0.4 in (10 mm). For long, liquid-filled lines, a gradient of 1 in/ft (80 mm/m) is recommended to allow gas bubbles to rise back into the flow line. If the fluid is of medium to high viscosity (5 to 100 cP), the slope should be increased to 2 to 4 in/ft (160 to 320 mm/m). The tubing bore should be increased to the ISO-recommended values given in Table 8.3 if long lead lines are necessary.

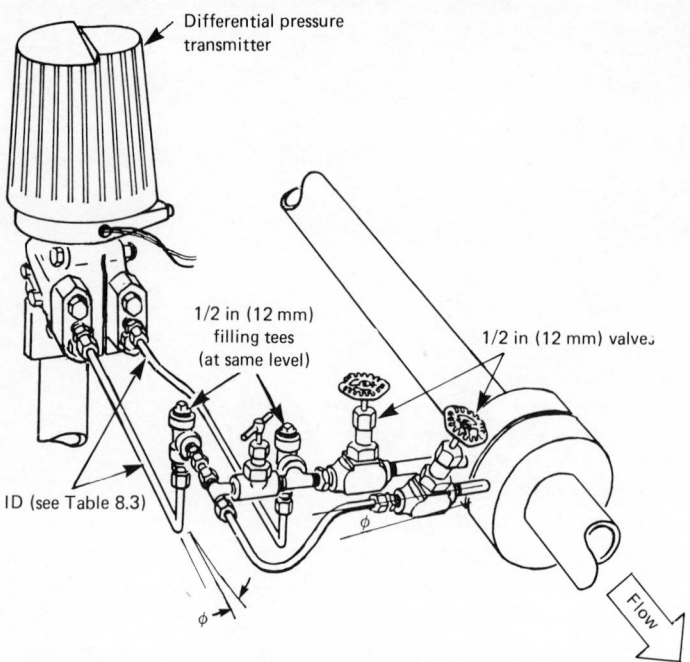

Notes: (1) Slope upward, ϕ = 1 in/ft (80 mm/m) for water ;
           ϕ = 2 to 4 in/ft (160 to 320 mm/m) for more viscous fluids.
       (2) Minimize all lead-line lengths.
       (3) For hot liquids, make unlagged lead lines long enough
           to minimize density changes.

Figure 8.14 Horizontal installation for dirty or corrosive fluids using a seal fluid *(courtesy The Foxboro Co.)*.

Table 8.3   Recommended Minimum Internal Diameters of Lead Lines

Lead-line length, ft	Fluid being metered			
	Water, steam, dry gas	Wet gas	Low- to medium-viscosity fluids	Dirty liquids or gases†
To 50	0.25 in (6 mm)	0.375 in (9 mm)	0.5 in (12.5 mm)	1.0 in (25 mm)
50–135‡	0.25 in (6 mm)	0.375 in (9 mm)	0.75 in (18.8 mm)	1.0 in (25 mm)
135–270	0.50 in (12.5 mm)	0.50 in (12.5 mm)	1.0 in (25 mm)	1.5 in (38 mm)

†Without seal fluid.

‡Lengths longer than 50 ft are not usually recommended and should be used only when absolutely necessary.

SOURCE: ISO 2186 (1973).

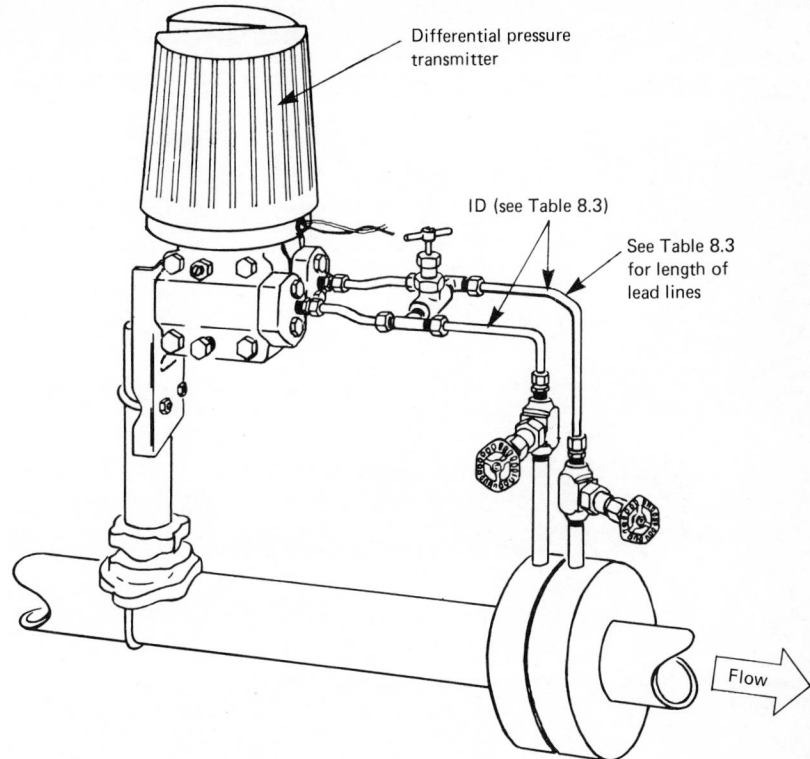

Figure 8.15 Horizontal installation for clean noncondensable gas *(courtesy The Foxboro Co.)*.

The two lead lines should be close together and lagged, if necessary, to reduce density variations due to temperature differences. Any head difference between leads alters the differential, although zeroing under flowing pressure and temperature usually reduces this error.

For clean liquids and dry gases, lead lines are purged through transmitter vents. If the flowing gas or liquid must be isolated from the measuring element because of corrosion, dirt, sediment, or condensation, seal fluids are used. Several slack, flush-mounted diaphragm seals are also available for isolation.

A seal fluid serves two purposes—to isolate the process and to provide protection against freezing. The most common seal liquid for oils is a mixture of 50 percent water and 50 percent ethylene glycol or glycerine, or, for lower-temperature protection, a mixture with 60 percent ethylene glycol. The ethylene glycol mixture has a specific gravity of about 1.07; the glycerine, about 1.13. The 50 percent ethylene glycol mixture freezes at −35°F (−37°C), and the 60 percent mixture at −56°F (−50°C). The 50 percent glycerine mixture freezes at −9.4°F (−23°C) and the 60 percent at −56°F (−49°C).

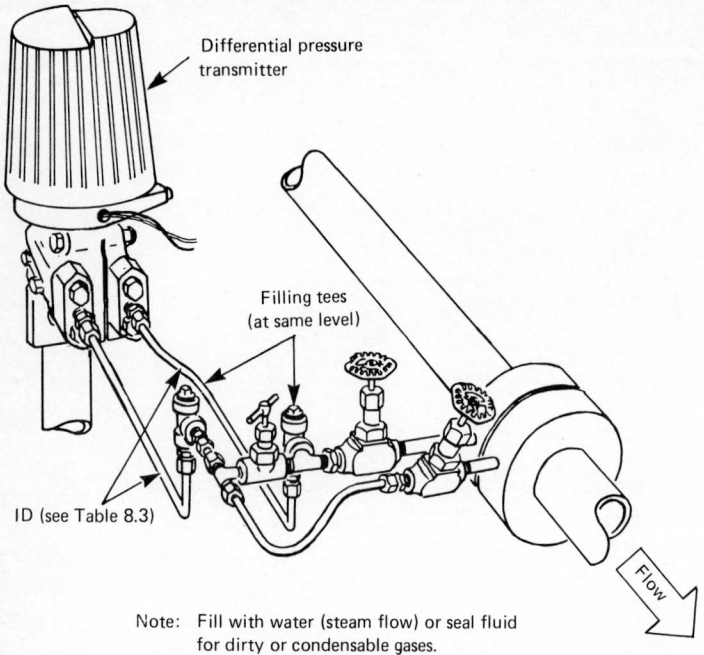

Differential pressure transmitter

Filling tees (at same level)

ID (see Table 8.3)

Flow

Note: Fill with water (steam flow) or seal fluid for dirty or condensable gases.

Figure 8.16   **Horizontal installation for vapor (steam) or dirty or condensable gases** (*courtesy The Foxboro Co.*).

For water and low-gravity aqueous solutions of salts and acids, dibutyl phthalate has been found highly satisfactory. Dibutyl phthalate has a specific gravity of 1.05, freezes at $-31°F$ ($-35°C$), and has a boiling point of $612°F$ ($322°C$). For liquids with a specific gravity greater than 1.01 heavier seal fluids must be found. Chloronaphthalene or Halowax oil, various Arochlors, transformer sealing fluids, Kel-F oil (trifluorochloroethylene polymers), fluorolubes, and acetylene tetrabromide have been used, but many are toxic, highly viscous, or have other disadvantages. No really satisfactory sealing medium for large displacement secondary elements (wet-type meters) has beeen found for materials such as concentrated sulfuric acid, although lighter oils, such as Nujol or other acid and caustic treated oils, have been used. Use of an oil lighter than the flowing fluid introduces serious maintenance problems.

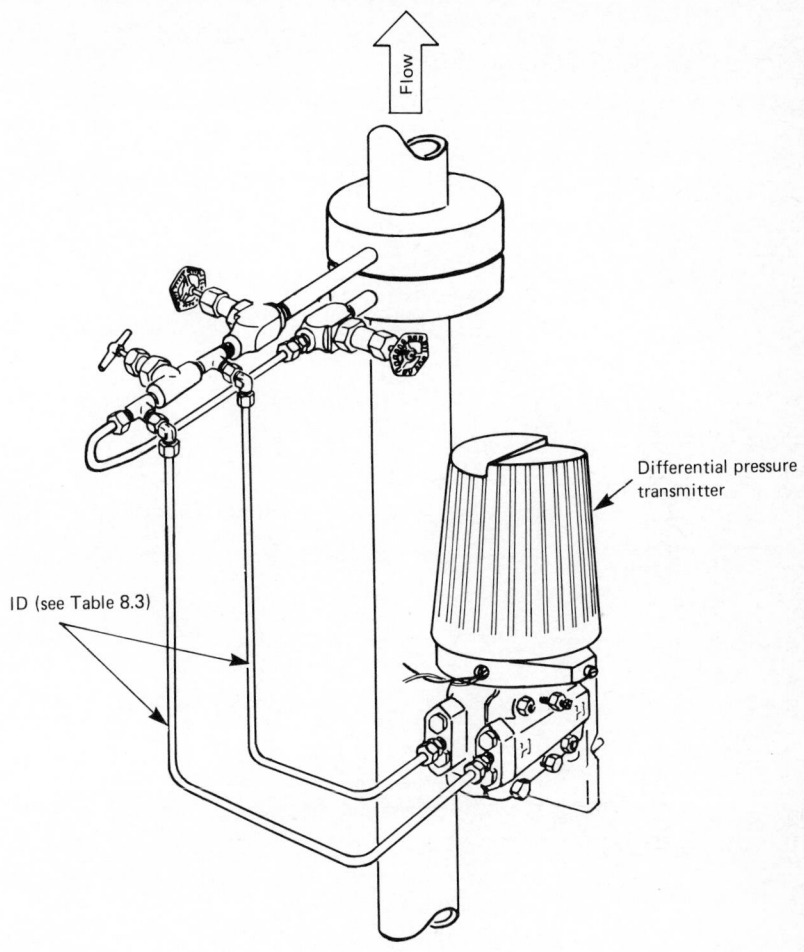

Note: For temperature above 250°F (121°C),
lag leads to pipe.

Figure 8.17 **Vertical installation for clean liquids** *(courtesy The Foxboro Co.).*

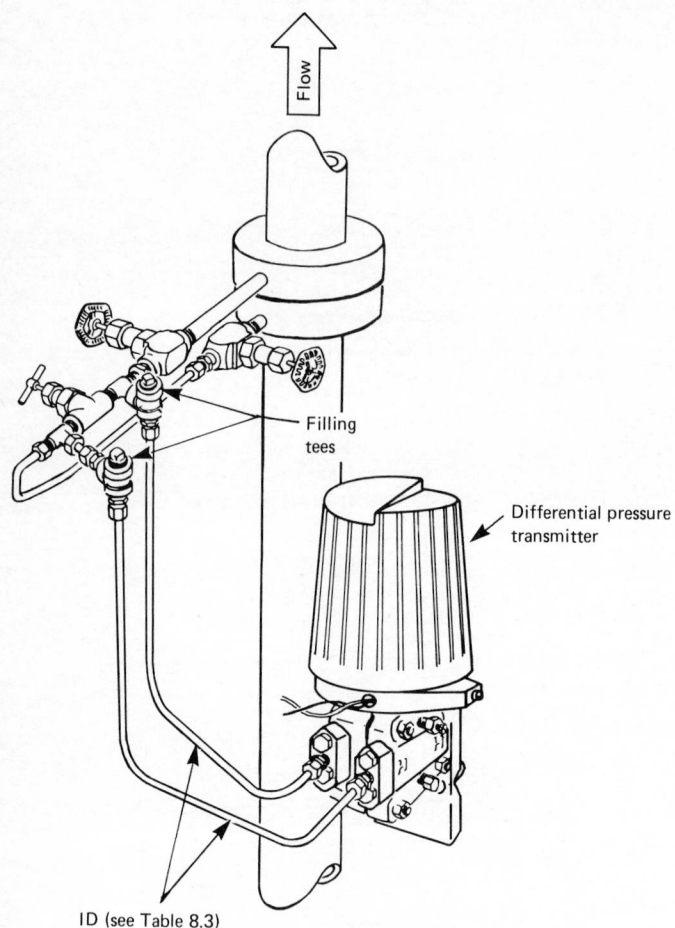

Notes: (1) Two filling tees must be at the same elevation
as the high-pressure tap.
(2) For temperatures above 250°F (121°C), lag
leads to pipe.
(3) Flow upward for liquids containing
appreciable amounts of gas.
(4) Flow downward if liquid contains small
amounts of granular solids.

Figure 8.18 **Vertical installation for dirty or corrosive liquids** (courtesy
The Foxboro Co.).

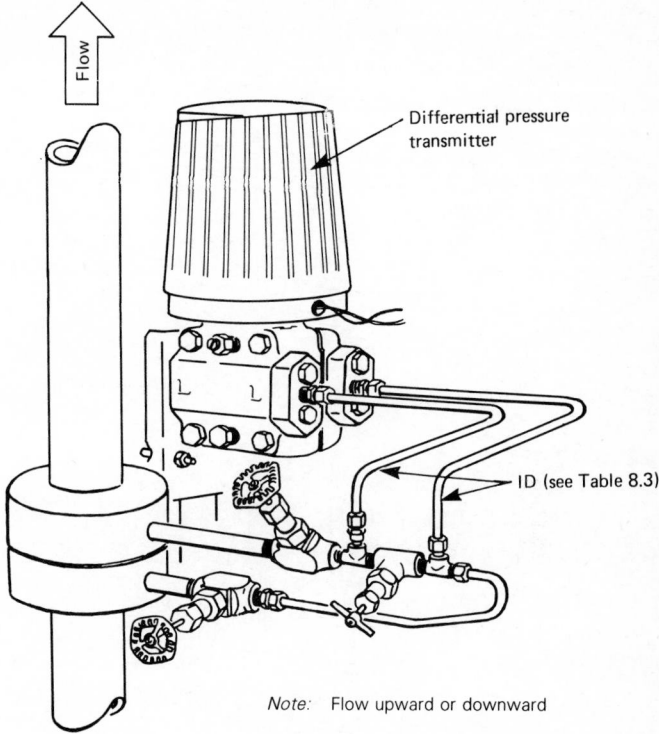

Note: Flow upward or downward

Figure 8.19 **Vertical installation for clean noncondensable gases** *(courtesy The Foxboro Co.).*

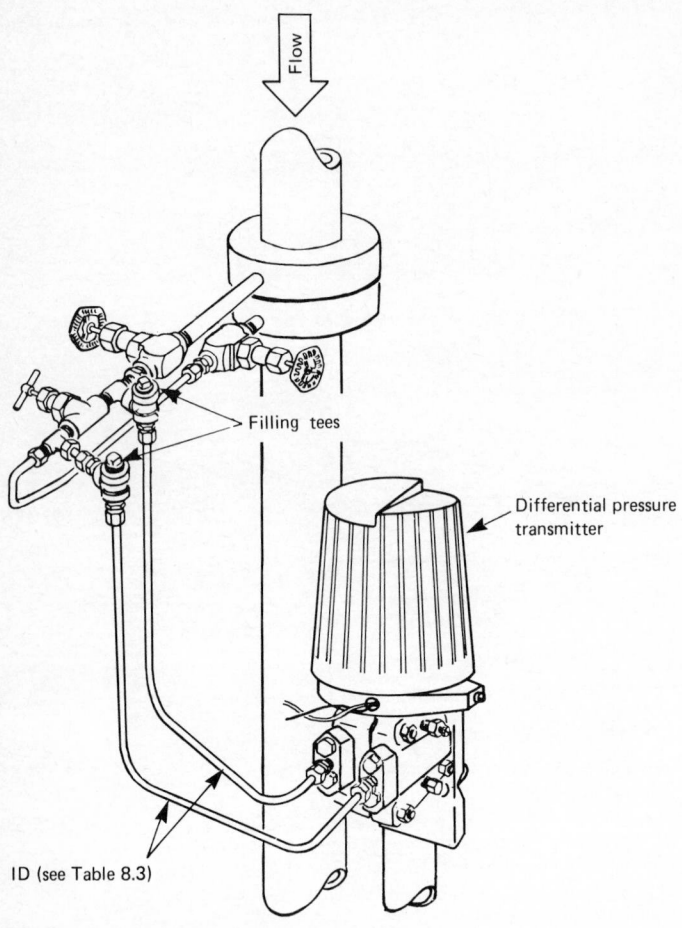

Note: Fill with water (steam flow) or seal fluid
for dirty or condensable gases.

Figure 8.20 **Vertical installation for vapor (steam), condensable gases, or
dirty gases** *(courtesy The Foxboro Co.).*

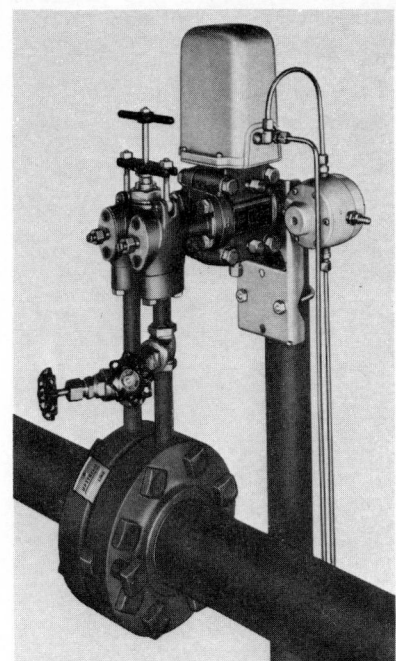

Figure 8.21 Valving manifold
(courtesy The Foxboro Co.).

REFERENCES

ANSI/API 2530, *Orifice Metering of Natural Gas*, catalog no. XQ0178, American Gas Association, Arlington, Va., 1978.

ANSI (MFC): *Differential Producers Used for the Measurement of Fluid Flow in Pipes (Orifice, Nozzle, Nozzle Venturi, Venturi)*, draft 3e, New York, July 1982.

ASME: *Fluid Meters*, 6th ed., ASME, New York, 1971.

ISO Standard 2186, *Fluid Flow in Closed Conduits—Connections for Pressure Signal Transmissions between Primary and Secondary Elements*, ISO 2186-1973, Geneva, 1973.

ISO Standard 5167, *Measurement of Fluid Flow by Means of Plates, Nozzles and Venturi Tubes Inserted in Circular Cross-Section Conduits Running Full*, ISO 5167-1980(E), Geneva, 1980.

DIFFERENTIAL PRODUCERS: ENGINEERING EQUATIONS

The sizing and flow-rate equations for all differential producers are identical. They are developed from theoretical assumptions, modified by correction factors based on empirical evidence, and further altered based on geometric considerations relative to fixed-geometry devices. This chapter develops the engineering equations and presents them in tables for ease in preparing computer programs.

THEORETICAL FLOW-RATE EQUATIONS

Liquid Equation

The dynamic equation for *one-dimensional* flow of incompressible fluids is derived by applying Newton's second law to the fluid element shown in Fig. 9.1a. The sum of the three forces in the direction of flow is equated to the mass of the element times its acceleration.

The external forces acting on the fluid element in the direction of flow are:

1. The net driving force produced by the static pressure acting over the element's upstream and downstream areas

2. The body force (weight) for a nonhorizontal element

3. The viscous shear stress that acts on the circumference of the element

These forces are expressed in differential form, using the English engineering system of units, as

$$- \frac{\partial P_f}{\partial S} \, dS \, dA \tag{9.1}$$

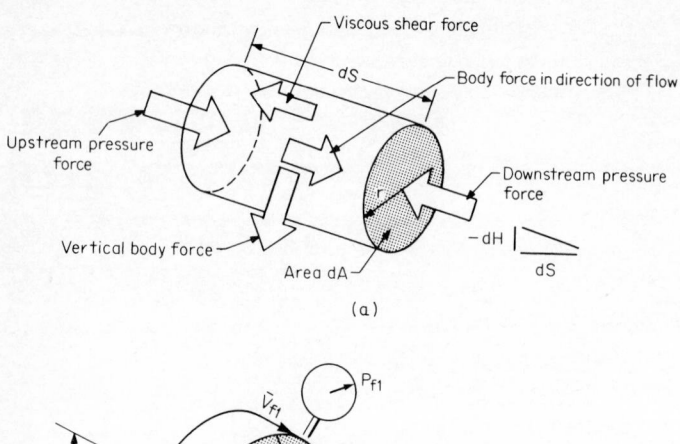

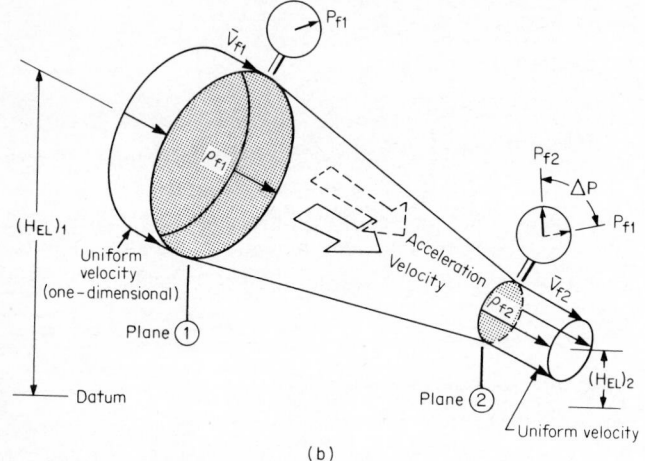

Figure 9.1 Incompressible fluid flow (liquid). (*a*) Streamline element. (*b*) Bernoulli's equation applied to the differential producer.

for the net pressure force,

$$- \rho_f \frac{g_l}{g_c} \, dS \, dA \, \frac{\partial H}{\partial S} \qquad (9.2)$$

for the body force, and

$$- 2\pi r S_s \, dS \qquad (9.3)$$

for the viscous shear force, where the minus signs indicate that pressure is decreasing in the direction of flow, the shear stress opposes the motion, and the body force is decreasing because of the elevation change. Applying Newton's second law of motion to the fluid element yields

$$- \frac{\partial P_l}{\partial S} \, dS \, dA - \rho_f \frac{g_l}{g_c} \, dS \, dA \, \frac{\partial H}{\partial S} - 2\pi r S_s \, dS = \frac{\rho_f}{g_c} \, dS \, dA \, a_S \qquad (9.4)$$

with

$$a_S = \frac{\partial}{\partial S}\left(\frac{\overline{V}_f^2}{2}\right)$$

this becomes, for a nonviscous fluid,

$$-\frac{\partial P_f}{\partial S} - \rho_f \frac{g_l}{g_c}\frac{\partial H}{\partial S} = \frac{\rho_f}{g_c}\frac{\partial}{\partial S}\left(\frac{\overline{V}_f^2}{2}\right) \tag{9.5}$$

The dynamic equation for one-dimensional flow per unit of fluid mass is obtained by dividing Eq. (9.5) by the mass density:

$$-\frac{1}{\rho_f}\frac{\partial P_f}{\partial S} - \frac{g_l}{g_c}\frac{\partial H}{\partial S} = \frac{1}{g_c}\frac{\partial}{\partial S}\left(\frac{\overline{V}_f^2}{2}\right) \tag{9.6}$$

This is known as *Euler's equation.*

Assuming constant density and integrating Eq. (9.6) along the streamline yield *Bernoulli's equation:*

$$\frac{P_f}{\rho_f} + \frac{\overline{V}_f^2}{2g_c} + \frac{g_l}{g_c}H_{EL} = \text{constant} \tag{9.7}$$

$$\frac{\text{ft}\cdot\text{lb}_f}{\text{lb}_m} + \frac{\text{ft}\cdot\text{lb}_f}{\text{lb}_m} + \frac{\text{ft}\cdot\text{lb}_f}{\text{lb}_m} = \text{constant}$$

Equation (9.7) is written in terms of energy per unit mass, and it indicates that *the energy along a streamline remains constant for an incompressible, nonviscous fluid.*

Bernoulli's equation can also be written for *specific weight* instead of mass density. With

$$\gamma_f = \frac{g_l}{g_c}\rho_f \qquad \frac{\text{lb}_f}{\text{ft}^3}$$

Equation (9.7) becomes

$$\frac{P_f}{\gamma_f} + \frac{\overline{V}_f^2}{2g_l} + H_{EL} = \text{constant} \tag{9.8}$$

$$\text{ft} + \text{ft} + \text{ft} = \text{constant}$$

Each term in Eq. (9.8) has the dimensions of feet and, therefore, is a *head.* In some literature, the head equation (9.8) is used to develop the engineering equations, which is the reason for the name *head-class* meters.

The two forms of Bernoulli's equation are equally valid. Note, however, that density values taken from tables or derived from equations must be in units of pounds-mass per cubic foot and, therefore, must be corrected for local gravity.

In the SI system of units, Bernoulli's equation is written as

$$\frac{P_f^*}{\rho_f^*} + \frac{\overline{V}_f^{*2}}{2} + g_l^* H_{EL}^* = \text{constant} \tag{9.9}$$

$$\frac{kg \cdot m^2/s^2}{kg} + \frac{kg \cdot m^2/s^2}{kg} + \frac{kg \cdot m^2/s^2}{kg} = \text{constant}$$

$$\frac{J^*}{kg} + \frac{J^*}{kg} \qquad + \frac{J^*}{kg} \qquad = \text{constant}$$

Here, the unit of pressure is the pascal, the unit of velocity is the meter per second, the unit for local gravity is the meter per second per second, and the unit of elevation is the meter.

Figure 9.1b shows Bernoulli's equation as it applies to a converging stream tube. The density is constant at all sections, the velocity is uniform across each section, and the sum of the pressure energy, kinetic energy, and potential energy is constant. For steady flow, the velocity increases because of the area reduction from plane 1 to plane 2. To maintain an energy balance, the pressure energy is reduced and the kinetic energy is increased. If pressure measurements were made at the two locations, the downstream static pressure would be lower than the upstream pressure. Applying Bernoulli's energy equation between locations 1 and 2 gives

$$\frac{P_{f1}}{\rho_f} = \frac{\overline{V}_{f1}^2}{2g_c} + \frac{g_l}{g_c}(H_{EL})_1 = \frac{P_{f2}}{\rho_f} + \frac{\overline{V}_{f2}^2}{2g_c} + \frac{g_l}{g_c}(H_{EL})_2 \qquad (9.10)$$

In practice, the elevation difference between pressure taps is corrected for by bringing the upstream and downstream lead lines to a differential-pressure measuring device that senses both pressures at a common elevation (see Sec. A.2). For a constant-density fluid, this reduces Eq. (9.10) to the form

$$\frac{P_{f1} - P_{f2}}{\rho_f} = \frac{\overline{V}_{f2}^2 - \overline{V}_{f1}^2}{2g_c} \qquad (9.11)$$

Equation (9.11) is reduced to flow-rate units by applying the mass flow continuity equation between planes 1 and 2. This mass balance simply states that the mass of fluid entering plane 1 equals the mass leaving at plane 2 and stays within the confines of the stream tube; mathematically,

$$q_{PPS} = \rho_{f1}A_1\overline{V}_{f1} = \rho_{f2}A_2\overline{V}_{f2} \qquad (9.12)$$
$$lb_m/s = lb_m/s \quad = lb_m/s$$

and for a constant-density fluid

$$q_{cfs} = A_1\overline{V}_{f1} = A_2\overline{V}_{f2} \qquad (9.13)$$
$$ft^3/s = ft^3/s \quad = ft^3/s$$

Substituting Eq. (9.13) into Eq. (9.11) yields

$$\frac{P_{f1} - P_{f2}}{\rho_f} = \frac{[1 - (A_2/A_1)^2]\overline{V}_{f2}^2}{2g_c} \qquad (9.14)$$

If planes 1 and 2 are circular, with measured pipe and bore diameters at flowing temperature of D_F and d_F in feet, Eq. (9.14) can be rewritten as

$$\frac{P_{f1} - P_{f2}}{\rho_f} = \frac{[1 - (d_F/D_F)^4]\,\overline{V}_f^2}{2g_c} \tag{9.15}$$

Rearranging Eq. (9.12) to solve for the velocity at plane 2 in terms of the mass flow rate yields

$$\overline{V}_{f2} = \frac{q_{\text{PPS}}}{\rho_f A_2} = \frac{q_{\text{PPS}}}{\pi/4\,d_F^2\,\rho_f} \tag{9.16}$$

The theoretical mass flow rate equation for the differential producer is obtained by substituting Eq. (9.16) into Eq. (9.15) and rearranging to obtain

$$(q_{\text{PPS}})_{\text{theo}} = \sqrt{2g_c}\,\frac{\pi}{4}\,d_F^2\,\sqrt{\frac{1}{1 - (d_F/D_F)^4}}\,\sqrt{\Delta P\,\rho_f} \tag{9.17}$$

The units in this equation are pounds-force, pounds-mass, feet, and seconds (English engineering system). With differential pressure in inches of water as defined by Eq. (3.22), and diameters D and d in inches the mass flow equation becomes

$$(q_{\text{PPS}})_{\text{theo}} = 0.09970190\,\frac{d^2}{\sqrt{1 - (d/D)^4}}\,\sqrt{h_w\rho_f} \tag{9.18}$$

where the dimensional constant is exactly

$$\frac{\pi}{4}\left(\frac{1}{12}\right)^2\left[\frac{2g_c(\rho_w)_{68°F}}{12}\right]^{1/2} = 0.09970190\ \text{lb}_m^{1/2}\cdot\text{ft}^{3/2}\cdot\text{in}^{-5/2}\cdot\text{s}^{-1} \tag{9.19}$$

A similar derivation using SI equation (9.9) yields

$$(q_{\text{KPS}}^*)_{\text{theo}} = \frac{\pi}{4}\,\sqrt{2}\,\frac{d_M^2}{\sqrt{1 - (d_M/D_M)^4}}\,\sqrt{\Delta P^*\,\rho_f^*} \tag{9.20}$$

With differential pressure in kilopascals and diameters D^* and d^* in millimeters, Eq. (9.20) becomes

$$(q_{\text{KPS}}^*)_{\text{theo}} = 3.512407 \times 10^{-5}\,\frac{d^{*2}}{\sqrt{1 - (d^*/D^*)^4}}\,\sqrt{\Delta p^*\,\rho_f^*} \tag{9.21}$$

where the constant is exactly

$$\frac{\pi}{4}\,\sqrt{2}\,\frac{\sqrt{1000}}{(1000)^2} = 3.512407 \times 10^{-5}\ \text{kg}^{1/2}\cdot\text{m}^{3/2}\cdot\text{mm}^{-2}\cdot\text{kPa}^{-1/2}\cdot\text{s}^{-1} \tag{9.22}$$

The theoretical volumetric flow rate, in cubic feet per second at flowing conditions, is obtained as

$$(q_{\text{cfs}})_{\text{theo}} = \frac{q_{\text{PPS}}}{\rho_f} = 0.09970190\,\frac{d^2}{\sqrt{1 - (d/D)^4}}\,\sqrt{\frac{h_w}{\rho_f}} \tag{9.23}$$

and, in SI units of cubic meters per second, as

$$(q^*_{\text{cms}})_{\text{theo}} = \frac{q^*_{\text{KPS}}}{\rho^*_f} = 3.512407 \times 10^{-5} \frac{d^{*2}}{\sqrt{1 - (d^*/D^*)^4}} \sqrt{\frac{\Delta p^*}{\rho^*_f}} \quad (9.24)$$

Gas (Vapor) Equation

For gas (vapor) flows, the fluid density is not constant between planes 1 and 2 of Fig. 9.1*b*. As the gas pressure decreases to accommodate the increase in kinetic energy, the density decreases, and the assumption of constant density no longer applies. Only for very low differential pressures will Eq. (9.18) or (9.21) apply with any degree of accuracy.

To correct the equation for gas expansion, a *gas expansion factor* is developed from the thermodynamic steady-flow energy equation. This equation is, for U.S. units,

$$\mathbf{Q} - \mathbf{W} = J(u_2 - u_1) + \frac{P_{f2}}{\rho_{f2}} - \frac{P_{f1}}{\rho_{f1}} + \frac{\overline{V}^2_{f2} - \overline{V}^2_{f1}}{2g_c} + \frac{g_l}{g_c}(H_{EL2} - H_{EL1}) \quad (9.25)$$

This equation expresses the energy balance between planes 1 and 2. For an adiabatic process, with no work **W** entering or leaving the system, and for no elevation difference, Eq. (9.25) reduces to

$$\frac{\overline{V}^2_{f2} - \overline{V}^2_{f1}}{2g_c} = e_1 - e_2 = -\int_{Pf1}^{Pf2} \frac{dP}{\rho} \quad (9.26)$$

where $e_1 - e_2$ is the enthalpy difference between measuring planes. The integral is evaluated by assuming an isentropic expansion from plane 1 to plane 2, expressed by

$$P\left(\frac{1}{\rho}\right)^k = \text{constant} \quad (9.27)$$

Upon integration, Eq. (9.26) yields

$$e_1 - e_2 = \int_{Pf1}^{Pf2} \frac{dP}{\rho} = \frac{P_{f1}}{\rho_{f1}} \frac{k}{k-1}\left[1 - \left(\frac{P_{f2}}{P_{f1}}\right)^{(k-1)/k}\right] \quad (9.28)$$

Substituting this equation into Eq. (9.26) and using mass flow continuity yield the form of the *theoretical adiabatic mass flow equation*. In U.S. units this is

$$(q_{\text{PPS}})_{\text{theo}} = 0.09970190 \frac{Y_1 d^2}{\sqrt{1 - (d/D)^4}} \sqrt{h_w \rho_{f1}} \quad (9.29)$$

where the Y_1 is the *adiabatic gas expansion factor* defined by

$$Y_1 = \left\{\frac{[1 - (d/D)^4][k/(k-1)](p_{f2}/p_{f1})^{2/k}[1 - (p_{f2}/p_{f1})^{(k-1)/k}]}{[1 - (d/D)^4 (p_{f2}/p_{f1})^{2/k}](1 - p_{f2}/p_{f1})}\right\}^{1/2} \quad (9.30)$$

The value of Y_1 depends on the ratio of diameters, the isentropic exponent, and the ratio between the measured downstream and upstream pressures. Equation (9.29) is, therefore, a dimensionless equation that may be written in SI units as

$$(q_{\text{KPS}})_{\text{theo}} = 3.512407 \times 10^{-5} \frac{Y_1 d^{*2}}{\sqrt{1 - (d^*/D^*)^4}} \sqrt{\Delta p^* \, \rho_{f1}^*} \tag{9.31}$$

For volumetric units in actual cubic feet per second, the equation becomes

$$(q_{\text{acfs}})_{\text{theo}} = \frac{q_{\text{PPS}}}{\rho_f} = 0.09970190 \frac{Y_1 d^2}{\sqrt{1 - (d/D)^4}} \sqrt{\frac{h_w}{\rho_{f1}}} \tag{9.32}$$

And in the SI system, in actual cubic meters per second, Eq. (9.29) becomes

$$(q_{\text{acms}}^*)_{\text{theo}} = 3.512407 \times 10^{-5} \frac{Y_1 d^{*2}}{\sqrt{1 - (d^*/D^*)^4}} \sqrt{\frac{\Delta p^*}{\rho_{f1}^*}} \tag{9.33}$$

By suitable algebraic manipulation, the steady-flow energy equation for compressible flow has thus been rearranged into the form of the incompressible mass flow equation. The single difference is that the compressible-flow equation contains the additional factor Y_1. In both U.S. and SI units, the equation requires the determination of the upstream density (at plane 1); hence, when the pvT relationship is used to calculate density, the assumption is that the pressure, temperature, and compressibility are determined there as well. When a densitometer is used, the density and pressure at the upstream tap location are required.

To solve for the adiabatic gas expansion factor, the ratio of downstream to upstream pressure is required. For an upstream tap location, this ratio is calculated from the measured differential as

$$\frac{p_{f2}}{p_{f1}} = \frac{p_{f2}^*}{p_{f1}^*} = 1 - \mathbf{x}_1 \tag{9.34}$$

where

$$\mathbf{x}_1 = \frac{p_{f1} - p_{f2}}{p_{f1}} = \frac{\Delta p^*}{p_{f1}^*} = \frac{h_w}{27.73 p_{f1}} \tag{9.35}$$

When the downstream pressure is measured, this ratio is calculated from the measured downstream tap pressure, and the flow equation must be rewritten for a density determination at that location. With mass flow continuity between locations 1 and 2, this equation is

$$(q_{\text{PPS}})_1 = (q_{\text{PPS}})_2 \propto Y_1 \sqrt{h_w \, \rho_{f1}} = Y_2 \sqrt{h_w \, \rho_{f2}} \tag{9.36}$$

which, for the same differential pressure h_w becomes

$$Y_1^2 \rho_{f1} = Y_2^2 \rho_{f2} \tag{9.37}$$

Substituting the pvT density equation (2.10) into Eq. (9.37) yields

$$Y_2 = Y_1 \left(\frac{p_{f1}}{p_{f2}} \frac{Z_{f2}}{Z_{f1}} \frac{T_{f2}}{T_{f1}} \right)^{1/2} \tag{9.38}$$

The ratio of compressibility factors Z_{f2}/Z_{f1} could be calculated using the Redlich-Kwong extrapolation method presented in Chap. 2 or from available data. However, for most applications, the differential pressure is small compared to the static pressure, and the compressibility change and temperature change during expansion may be assumed to be negligible. This reduces Eq. (9.38) to

$$Y_2 = Y_1 \left(\frac{p_{f1}}{p_{f2}} \right)^{1/2} = Y_1 \left(\frac{p_{f1}^*}{p_{f2}^*} \right)^{1/2} = Y_1 \sqrt{\frac{1}{1 - x_1}} = Y_1 \sqrt{1 + x_2} \tag{9.39}$$

where

$$x_2 = \frac{p_{f1} - p_{f2}}{p_{f2}} = \frac{\Delta p^*}{p_{f2}^*} = \frac{h_w}{27.73 p_{f2}} \tag{9.40}$$

The mass flow equation for downstream tap measurements is then

$$(q_{PPS})_{theo} = 0.09970190 \frac{Y_2 d^2}{\sqrt{1 - (d/D)^4}} \sqrt{h_w \rho_{f2}} \tag{9.41}$$

for U.S. units, and

$$(q_{KPS}^*)_{theo} = 3.512407 \times 10^{-5} \frac{Y_2 d^{*2}}{\sqrt{1 - (d^*/D^*)^4}} \sqrt{\Delta p^* \rho_{f2}^*} \tag{9.42}$$

for SI units, where Y_2 is defined by Eq. (9.39), and the pressure ratio is derived from the downstream pressure as

$$\frac{p_{f2}}{p_{f1}} = \frac{p_{f2}^*}{p_{f1}^*} = \frac{1}{1 + x_2} \tag{9.43}$$

The volumetric flow in actual cubic feet per second is, similarly,

$$(q_{acfs})_{theo} = \frac{q_{PPS}}{\rho_{f2}} = 0.09970190 \frac{Y_2 d^2}{\sqrt{1 - (d/D)^4}} \sqrt{\frac{h_w}{\rho_{f2}}} \tag{9.44}$$

and in SI units it is

$$(q_{acms}^*)_{theo} = \frac{q_{KPS}^*}{\rho_{f2}^*} = 3.512407 \times 10^{-5} \frac{Y_2 d^{*2}}{\sqrt{1 - (d^*/D^*)^4}} \sqrt{\frac{\Delta p^*}{\rho_{f2}^*}} \tag{9.45}$$

CORRECTIONS TO THE THEORETICAL EQUATIONS

The theoretical flow equation calculates the true flow rate only when all the assumptions used to develop it are valid. This is seldom the case, and the true flow rate is almost always less than the theoretically calculated value.

How closely the true flow rate can be calculated depends almost entirely on the geometry of the contraction. For a venturi or flow nozzle, where the area reduction is gradual (Fig. 9.2a), the agreement is within 1 to 3 percent. But for the square-edged orifice (Fig. 9.2b), the abrupt area reduction places the minimum-flow area downstream of the plate at the plane of the vena contracta. Since

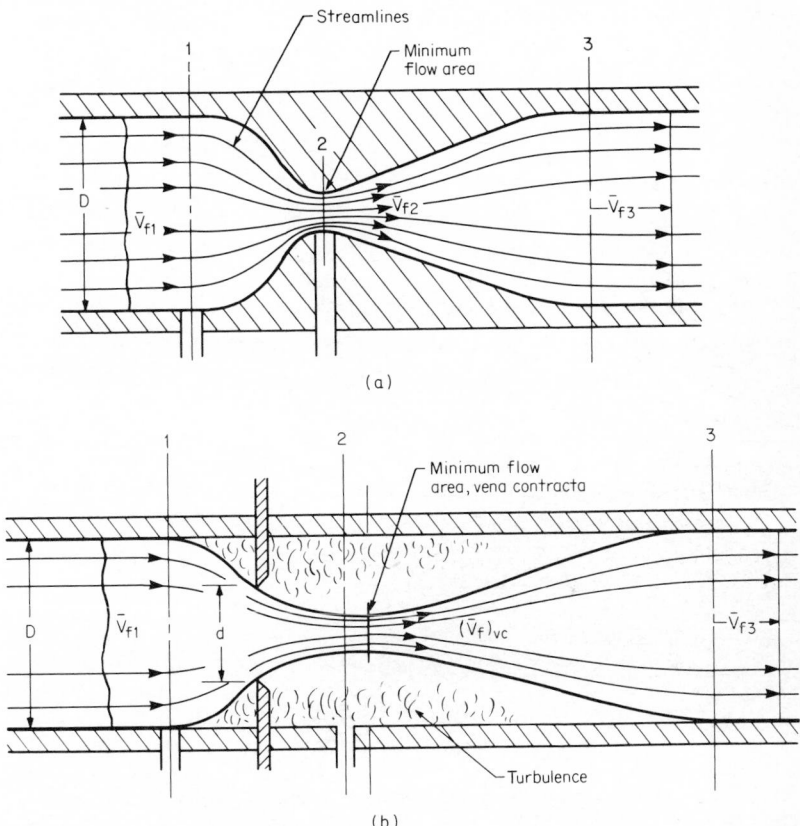

Figure 9.2 Contraction types. (a) Gradual. (b) Abrupt.

the diameter of the vena contracta for an orifice cannot be measured, the theoretical equation includes the measured bore as the correlating diameter. Also, increased downstream turbulence results in an energy loss that is not accounted for by either Bernoulli's equation or the thermodynamic steady flow energy equation. These two factors result in the true flow being approximately 60 percent of the theoretically calculated value. The location of the two measuring taps is also important because it establishes the measured differential.

For both contoured and abrupt reductions, the assumption that the velocity profile is uniform is valid only at very high Reynolds numbers (see Chap. 5). With decreasing velocity or increasing viscosity, the Reynolds number decreases and the profile becomes more peaked, resulting in a further departure from the assumption of a rectilinear or one-dimensional profile. Expansion approaches the adiabatic assumption for contoured elements, but not for the abrupt flow-streamline changes across an orifice, where both radial and axial expansion occur.

The theoretical equation is adjusted for these effects with two empirically determined corrections. The first is the *discharge coefficient C*, which corrects for velocity profile (Reynolds number), tap location, and contraction geometry; the second is an empirically derived *net expansion-factor equation* for orifice flowmeters.

Discharge Coefficient

For a given primary element, the discharge coefficient is derived from laboratory data by ratioing the true and theoretical flow rates. The true flow rate is determined by weighing or volumetric collection of the fluid over a measured time interval, and the theoretical flow rate is calculated with Eq. (9.18). The discharge coefficient is then defined as

$$C = \frac{\text{true flow rate}}{\text{theoretical flow rate}} \tag{9.45}$$

The discharge coefficient corrects the theoretical equation for the influence of velocity profile (Reynolds number), the assumption of no energy loss between taps, and pressure-tap location.

In some flow equations, the discharge coefficient is combined with the velocity of approach and redefined as the *flow coefficient*. For fixed-geometry primary devices, to simplify the equation, or where primary elements are available in a limited range of sizes, the flow coefficient is used in place of the discharge coefficient. The flow coefficient is defined as

$$K = \frac{C}{\sqrt{1 - (d/D)^4}} = \frac{C}{\sqrt{1 - (d^*/D^*)^4}} = \frac{C}{\sqrt{1 - \beta^4}} = EC \tag{9.47}$$

where $\beta = d/D$ and $E = 1/\sqrt{1 - \beta^4}$ is the velocity of approach factor.

Methods of Presenting the Discharge Coefficient

The discharge coefficient may be presented in four different ways.

Method 1   A discharge coefficient–Reynolds number curve, similar to Fig. 9.3*a*, is obtained from laboratory data. This is the most accurate method, since all dimensional effects and other influences on the primary device are included in the data points.

Method 2   A signature curve based on calibration data for many devices of the same geometry is given with an accuracy envelope (Fig. 9.3*b*). The flow coefficient is obtained graphically by reading the coefficient at the pipe Reynolds number.

Method 3 The flow coefficient for a fixed-geometry device is ratioed to a constant K_{ref}, which is usually the flow coefficient at a high Reynolds number (Fig. 9.3c). This ratio,

$$F_K = \frac{K}{K_{ref}} \tag{9.48}$$

is plotted against pipe or bore Reynolds number. The Reynolds-number scale may be either logarithmic or linearized. Curves or tables of F_K versus Reynolds number are used in graphical or tabular computations.

Method 4 For all standardized primary elements, numerous test points have been used to develop an empirical equation that predicts the discharge coefficient from bore and pipe-diameter measurements. The accuracy of these equations is usually acceptable, and a flow calibration is seldom performed. However, for Reynolds number, pipe size, etc., outside the specified range of the equation, a signature curve should be used to obtain the discharge coefficient.

In the turbulent flow regime ($R_D > 4000$), the discharge coefficient for all primary elements can be expressed with an equation of the general form

$$C = C_\infty + \frac{b}{R_D^n} \tag{9.49}$$

and the flow coefficient given by Eq. (9.47) becomes

$$K = K_\infty + \frac{b}{\sqrt{1 - \beta^4}\, R_D^n} = K_\infty + \frac{Eb}{R_D^n} \tag{9.50}$$

in which C_∞ is the discharge coefficient at an infinite Reynolds number, and b is the Reynolds-number correction term. For a graphical solution, the coefficient calculated by Eq. (9.49) is plotted versus Reynolds number. By scaling the Reynolds-number axis as the reciprocal of the Reynolds number raised to the exponent n, the change in discharge coefficient with Reynolds number is linearized, as shown in Fig. 9.3d.

Depending on the primary element, the infinite-Reynolds-number discharge coefficient may be a constant or a function of measured dimensions or tap location. The value of b may also be a function of dimensions, or it may be 0. The Reynolds-number exponent n is constant and depends on the primary element. The equations used for primary elements are presented in Table 9.1; the locations and names of the orifice taps are shown in Fig. 9.4.

Thermal-Expansion Factor

The material of the primary element and pipe expands or contracts with temperature. The pipe and bore diameters are measured at room temperature but

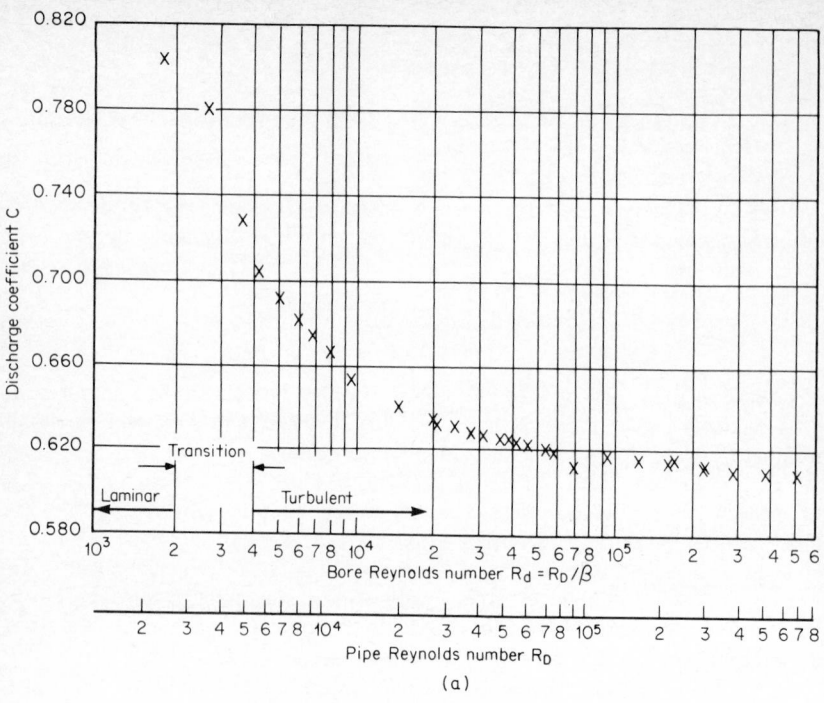

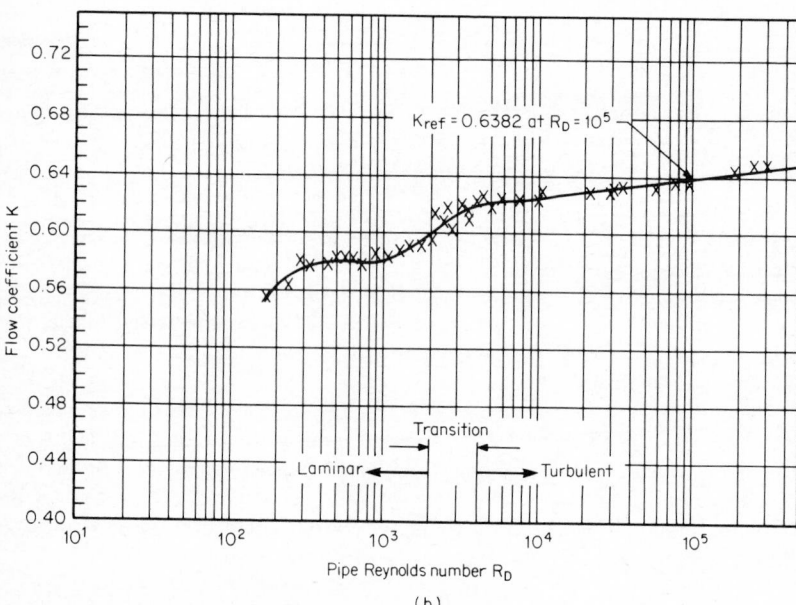

(b)

Figure 9.3 Methods of graphically presenting discharge- or flow-coefficient variations with pipe Reynolds number. (a) Flow-laboratory-determined orifice discharge coefficient $D = 4$ in (100 mm), $\beta = 0.73$. (b) Generalized flow-coefficient curve for a target flowmeter. (c) Reynolds-number correction factor F_K derived from part b. (d) Orifice discharge-coefficient curves.

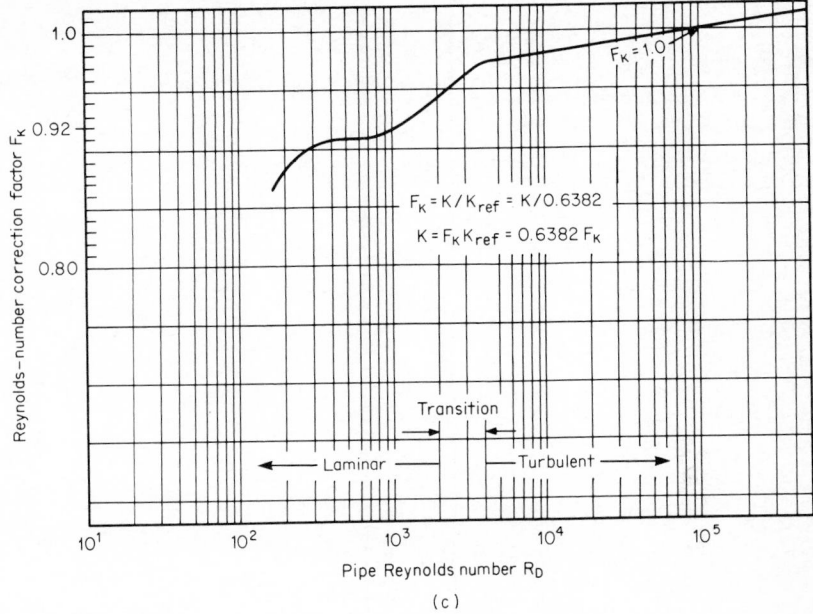

(c)

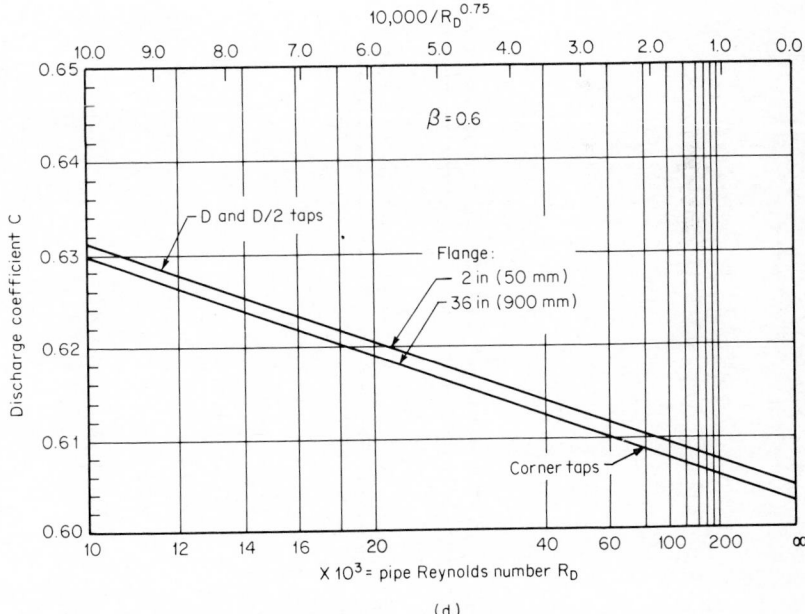

(d)

Figure 9.3 (*Continued*).

Table 9.1 Equations and Values for C_∞, b, and n of Eq. (9.49)[a]

Primary device	Discharge coefficient C_∞ at infinite Reynolds number	Reynolds-number term	
		Coefficient b	Exponent n
Venturi			
Machined inlet	0.995	0	0
Rough cast inlet	0.984	0	0
Rough welded sheet-iron inlet	0.985	0	0
Universal Venturi Tube[b]	0.9797	0	0
Lo-Loss tube[c]	$1.005 - 0.471\beta + 0.564\beta^2 - 0.514\beta^3$	0	0
Nozzle:			
ASME long radius	0.9975	$-6.53\beta^{0.5}$	0.5
ISA	$0.9900 - 0.2262\beta^{4.1}$	$1708 - 8936\beta + 19{,}779\beta^{4.7}$	1.15
Venturi nozzle (ISA inlet)	$0.9858 - 0.196\beta^{4.5}$	0	0

Orifice:

Corner taps	$0.5959 + 0.0312\beta^{2.1} - 0.184\beta^8$	$91.71\beta^{2.5}$	0.75
Flange taps (D in inches) $D \geq 2.3$	$0.5959 + 0.0312\beta^{2.1} - 0.184\beta^8 + 0.09\dfrac{\beta^4}{D(1-\beta^4)} - 0.0337\dfrac{\beta^3}{D}$	$91.71\beta^{2.5}$	0.75
$2 \leq D \leq 2.3^d$	$0.5959 + 0.0312\beta^{2.1} - 0.184\beta^8 + 0.039\dfrac{\beta^4}{1-\beta^4} - 0.0337\dfrac{\beta^3}{D}$	$91.71\beta^{2.5}$	0.75
Flange taps (D^* in millimeters) $D^* \geq 58.4$	$0.5959 + 0.0312\beta^{2.1} - 0.184\beta^8 + 2.286\dfrac{\beta^4}{D^*(1-\beta^4)} - 0.856\dfrac{\beta^3}{D^*}$	$91.71\beta^{2.5}$	0.75
$50.8 \leq D^* \leq 58.4^d$	$0.5959 + 0.0312\beta^{2.1} - 0.184\beta^8 + 0.039\dfrac{\beta^4}{1-\beta^4} - 0.856\dfrac{\beta^3}{D^*}$	$91.71\beta^{2.5}$	0.75
D and $D/2$ taps	$0.5959 + 0.0312\beta^{2.1} - 0.184\beta^8 + 0.039\dfrac{\beta^4}{1-\beta^4} - 0.0158\beta^3$	$91.71\beta^{2.5}$	0.75
$2\frac{1}{2}D$ and $8D$ taps^e	$0.5959 + 0.461\beta^{2.1} + 0.48\beta^8 + 0.039\dfrac{\beta^4}{1-\beta^4}$	$91.71\beta^{2.5}$	0.75

[a] Detailed Reynolds-number, line-size, beta-ratio, and other limitations are given in Table 9.42.

[b] From BIF CALC-440/441; the manufacturer should be consulted for exact coefficient information.

[c] Derived from the Badger Meter, Inc. Lo-Loss tube coefficient curve; the manufacturer should be consulted for exact coefficient information.

[d] For $\frac{1}{2} \leq D \leq 1\frac{1}{2}$ in ($12 \leq D^* \leq 40$ mm) use flow coefficient equation (10.1) or Eq. (10.2) given in Chap. 10, with $C = \sqrt{1-\beta^4}\, K$.

[e] SOURCE: Stolz (1978).

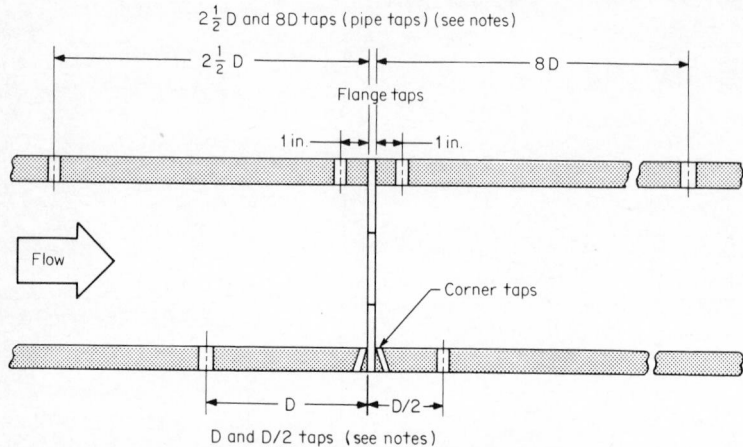

Notes:

(1) 2½D and 8D pipe taps are not recommended in ISO 5167 or ASME Fluid Meters.

(2) D and D/2 are radius taps now used in place of vena contracta taps.

Figure 9.4 Orifice pressure-tap locations and their names.

will be larger or smaller when used at other temperatures. A thermal-expansion factor is introduced to correct for these differences. This multiplying factor is derived by partial differentiation of the theoretical flow equation to determine bore and pipe-diameter sensitivity coefficients.

When the pipe and primary element are of different material, the thermal-expansion-factor correction for measurements at 68°F is

$$F_a = 1 + \frac{2}{1 - \beta^4}(\alpha_{PE} - \beta^4 \alpha_P)(T_F - 68) \qquad (9.51)$$

where α_{PE} is the thermal-expansion coefficient of the primary element and α_p is the thermal-expansion coefficient of the pipe material. When the thermal-expansion coefficients of the primary element and pipe are approximately the same, which is usually the case, Eq. (9.51) reduces to

$$F_a = 1 + 2\alpha_{PE}(T_F - 68) \qquad (9.52)$$

In this equation, the thermal-expansion coefficient of the primary element is used because the flow-rate calculation is more sensitive to bore than to pipe dimensions.

In SI units, Eqs. (9.51) and (9.52) are

$$F_a^* = 1 + \frac{2}{1 - \beta^4}(\alpha_{PE}^* - \beta^4 \alpha_P^*)(T_{°C} - 20) \qquad (9.53)$$

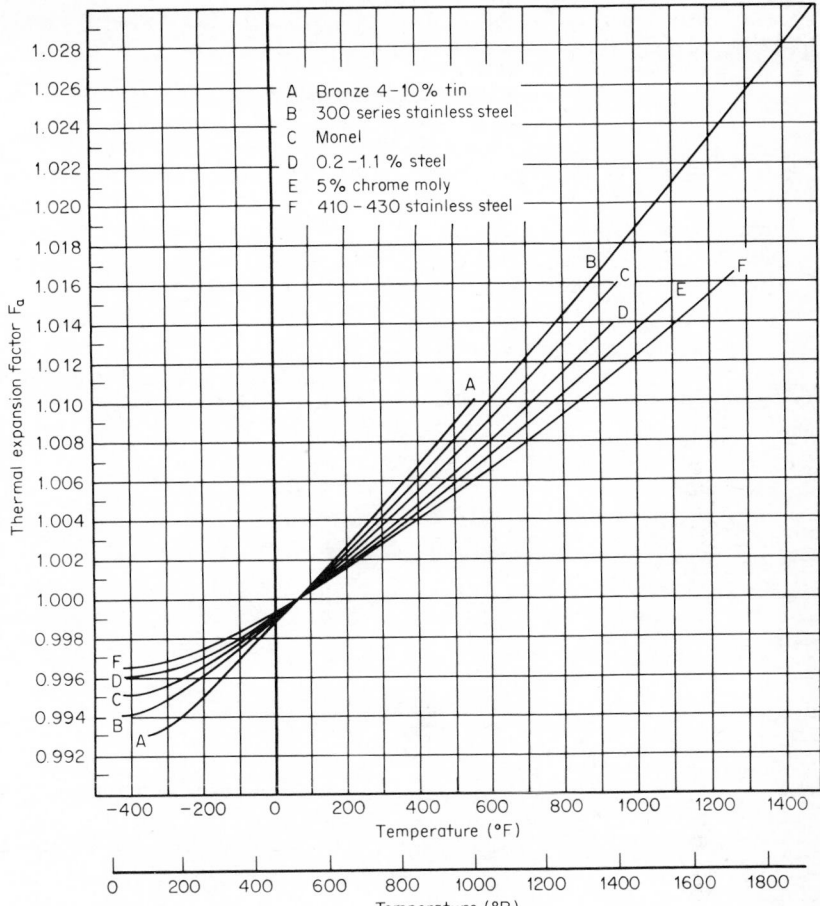

Figure 9.5 **Thermal-expansion factor F_a** *(from ASME Fluid Meters, 1971; used with permission).*

for pipe and primary element of different materials, and

$$F_a^* = 1 + 2\alpha_{PE}^*(T_{°C} - 20) \tag{9.54}$$

for similar pipe and primary-element materials. Table B.4 gives the coefficient of linear expansion for many commonly used materials, and Fig. 9.5 presents F_a graphically.

The Gas Expansion Factor for the Orifice

For convenience, the steady-flow thermodynamic energy equation for gases was rearranged into the form of the liquid equation, which introduced a gas expansion factor to correct for density changes between taps. This factor is defined, based on gas (vapor) test work, as

$$Y = \frac{CY \text{ [from gas (vapor) test]}}{C \text{ (from liquid test)}} \tag{9.55}$$

At small differential pressures, the gas expansion factor approaches 1.

At nozzles, venturis, and other contoured-inlet devices, gas expansion is assumed to be purely axial. For these devices, there is excellent agreement between the adiabatic gas expansion factor [Eq. (9.30)] and laboratory data. Therefore, for contoured devices, the adiabatic gas expansion factor is used to calculate the expansion factor.

At a square-edged orifice, expansion is both axial and radial, and an empirically determined equation is used to calculate the gas expansion factor. Laboratory test work on steam and other compressible fluids was used by Buckingham (1932) to calculate an orifice gas expansion factor. For upstream-tap pressure measurements, Buckingham derived the equation

$$Y_1 = 1 - (0.41 + 0.35\beta^4)\frac{\mathbf{x}_1}{k} \tag{9.56}$$

where k is the isentropic exponent, and the downstream tap location is less than $D/2$ from the upstream face of the orifice. For $2\tfrac{1}{2}D$ and $8D$ taps the equation becomes

$$Y_1 = 1 - [0.333 + 1.145(\beta^2 + 0.7\beta^5 + 12\beta^{13})]\frac{\mathbf{x}}{k} \tag{9.57}$$

These equations may be modified for downstream pressure measurement with Eqs. (9.38), (9.39), and (9.40) to obtain

$$Y_2 = \sqrt{1 + \mathbf{x}_2} - (0.41 + 0.35\beta^4)\frac{\mathbf{x}_2}{k\sqrt{1 + \mathbf{x}_2}} \tag{9.58}$$

for corner, flange, and D and $D/2$ taps, and

$$Y_2 = \sqrt{1 + \mathbf{x}_2} - [0.333 + 1.145(\beta^2 + 0.75\beta^5 + 12\beta^{13})]\frac{\mathbf{x}_2}{k\sqrt{1 + x_2}} \tag{9.59}$$

for $2\tfrac{1}{2}D$ and $8D$ taps.

Fundamental-Unit Liquid and Gas Equation

The *true* mass flow rate equation for both liquids and gases (vapors) is obtained by multiplying the theoretical mass flow equation by the discharge coefficient, the gas expansion factor, and the thermal-expansion factor. The equation for U.S. units is

$$q_{\text{PPS}} = 0.09970190\frac{CY_1F_a d^2}{\sqrt{1 - \beta^4}}\sqrt{h_w\rho_{f1}} \tag{9.60}$$

The equation for SI units is

$$q^*_{\text{KPS}} = 3.512407 \times 10^{-5} \frac{CY_1 F^*_a d^{*2}}{\sqrt{1 - \beta^4}} \sqrt{\Delta p^* \rho^*_{f1}} \qquad (9.61)$$

THE *N* FACTOR

Derived-Unit Conversion-Factor Equation

The relationships between mass and volumetric flow rate, at both flowing and base conditions, are derived in Chap. 3. From these equations and the pvT density equation, the unit constants for the liquid and gas equations are obtained.

As discussed in Chap. 3, there are three commonly used *liquid* flow-rate units. These are:

1. Mass flow rate (pounds-mass per day, pounds-mass per hour, etc.)

2. Volumetric flow rate at flowing conditions (gallons per minute, gallons per hour, barrels per day, etc.)

3. Volumetric flow rate at a 60°F base volume (barrels per day, gallons per minute, etc.)

A flow equation can be written for each unit to calculate the flow rate from density or specific-gravity values obtained from graphs, from tables, or from measurements. This results in six separate equations; which one is selected depends on the desired flow-rate unit and how density is determined.

In Tables 9.2 and 9.3 are equations that relate the derived flow-rate unit and density-determination equation to the fundamental *liquid* flow-rate-unit equation for both U.S. and SI units. The conversion constant N is subscripted sequentially to indicate first the flow-rate unit (such as uppercase M for mass flow), and second the method of density determination (G for specific gravity, ρ for density, and pT for the pvT equation). An uppercase subscript on the flow-rate unit denotes base or fundamental units, such as mass or base volume; a lowercase subscript denotes flowing volume. The conversion constant adjusts the equation both for time base (seconds, minutes, hours) and for the relationship between specific gravity and density.

Four flow-rate units are used in gas (vapor) flows. These are:

1. Mass flow rate (pounds-mass per hour, kilograms per hour, etc.).

2. Volumetric flow rate at flowing conditions (flowing cubic feet per minute, flowing cubic meters per hour, etc.).

3. Volumetric flow rate at a standard base pressure and temperature (standard cubic feet per hour, standard cubic feet per day, etc.). ISO Standard 5024 (1976) standardizes on a base of 14.69595 psia (101.325 kPa) and 59°F (15°C).

Table 9.2 Relationship between Fundamental Constant and Derived Flow-Rate Unit for Liquid Flow: U.S. Units

Flow rate	Letter symbol — Conversion constant	Relationship to fundamental unit equation	Definition
q_M	$N_{M\rho}$	$q_{PPS} = 0.09970190 \dfrac{q_M}{N_{M\rho}}$	Mass flow rate with density determination Example: $q_M = q_{PPD}$ = pounds-mass per day
q_M	N_{MG}	$q_{PPS} = 0.7873692 \dfrac{q_M}{N_{MG}}$	Mass flow rate with a specific-gravity determination Example: $q_M = q_{PPM}$ = pounds-mass per minute
q_v	$N_{v\rho}$	$q_{PPS} = 0.09970190[\rho_f] \dfrac{q_v}{N_{v\rho}}$	Volumetric flow rate at flowing conditions with a density determination Example: $q_v = q_{cfm}$ = cubic feet per minute (flowing)
q_v	N_{vG}	$q_{PPS} = 0.7873692[F_\rho G_F] \dfrac{q_v}{N_{vG}}$	Volumetric flow rate at flowing conditions with a specific-gravity determination Example: $q_v = q_{gpm}$ = gallons per minute (flowing)
q_V	$N_{V\rho}$	$q_{PPS} = 0.09970190[\rho_b] \dfrac{q_V}{N_{V\rho}}$	Volumetric flow rate at base conditions (60°F and 14.696 psia) with a density determination Example: $q_V = q_{GPM}$ = gallons per minute at base conditions
q_V	N_{VG}	$q_{PPS} = 0.7873692[G_b]\left(\dfrac{q_V}{N_{VG}}\right)$	Volumetric flow rate at base conditions (60°F and 14.69 psia) with a specific-gravity determination Example: $q_V = q_{CFM}$ = cubic feet per minute at base conditions

Table 9.3 Relationship between Fundamental Constant and Derived Flow-Rate Unit for Liquid Flows: SI Units

Letter symbol		Relationship to fundamental unit equation	Definition
Flow rate	Conversion constant		
q_M^*	$N_{M\rho}^*$	$q_{KPS}^* = 0.00003512407\,\dfrac{q_M^*}{N_{M\rho}^*}$	Mass flow rate with a density determination Example: $q_M^* = q_{KPD}^*$ = kilograms per day
q_M^*	N_{MG}^*	$q_{KPS}^* = 0.001110172\,\dfrac{q_M^*}{N_{MG}^*}$	Mass flow rate with a specific-gravity determination Example: $q_M^* = q_{KPH}^*$ = kilograms per hour
q_v^*	$N_{v\rho}^*$	$q_{KPS}^* = 0.00003512407[\rho_f^*]\,\dfrac{q_v^*}{N_{v\rho}^*}$	Volumetric flow rate at flowing conditions with a density determination Example: $q_v^* = q_{lpm}^*$ = liters per minute (flowing)
q_v^*	N_{vG}^*	$q_{KPS}^* = 0.001110172[F_p G_F]\,\dfrac{q_v^*}{N_{vG}^*}$	Volumetric flow rate at flowing conditions with a specific-gravity determination Example: $q_v^* = q_{cmm}^*$ = cubic meters per minute (flowing)
q_V^*	$N_{V\rho}^*$	$q_{KPS}^* = 0.00003512407[\rho_b^*]\,\dfrac{q_V^*}{N_{V\rho}^*}$	Volumetric flow rate at base conditions (15.6°C and 101.325 kPa) with a density determination Example: $q_V^* = q_{LPM}^*$ = liters per minute at base conditions
q_V^*	N_{VG}^*	$q_{KPS}^* = 0.001110172[G_b]\,\dfrac{q_V^*}{N_{VG}^*}$	Volumetric flow rate at base conditions (15.6°C and 101.325 kPa) with a specific-gravity determination Example: $q_V^* = q_{CMM}^*$ = cubic meters per minute at base conditions

4. Volumetric flow rate at a selected base pressure and temperature (standard cubic feet per hour at 14.4 psia and 70°F, etc.).

Density may be determined either with a densitometer or by measuring pressure and temperature and calculating the density with the pvT equations developed in Chap. 2. The equations relating these derived units to the fundamental mass flow rate equation are given in Table 9.4 for U.S. units, and in Table 9.5 for SI units.

A second set of gas flow equations is widely used. These are called the *gas-factor equations* because a series of multiplying factors is introduced to aid in graphical or tabular solution. These are well suited for handbook-type lookups but are cumbersome to implement on computers. Also, because a single constant factor is calculated, changes in line temperature, specific gravity, or compressibility from design conditions are easily overlooked, and this introduces significant errors in the final calculated flow rate.

The factors used in the gas-factor equations are, in the U.S. system of units,

$$F_{TB} = \frac{T_b}{518.67} \tag{9.62}$$

for base-temperature correction,

$$F_{PB} = \frac{14.69595}{p_b} \tag{9.63}$$

for base-pressure correction,

$$F_g = \sqrt{\frac{1}{G}} \tag{9.64}$$

for specific-gravity correction,

$$F_{pv1} = \sqrt{\frac{1}{Z_{f1}}} \tag{9.65}$$

(the supercompressibility factor) for compressibility correction, and

$$F_{TF1} = \sqrt{\frac{518.67}{T_{f1}}} \tag{9.66}$$

for flowing-temperature correction.

In the SI system of units, the factors are

$$F_{TB}^* = \frac{T_{Kb}}{288.15} \tag{9.67}$$

for base-temperature correction,

$$F_{PB}^* = \frac{101.325}{p_b^*} \tag{9.68}$$

for base-pressure correction, and

$$F_{TF1}^* = \sqrt{\frac{288.15}{T_{K1}}}$$ (9.69)

for flowing-temperature correction.

The relationships between the fundamental mass flow rate equation and the gas-factor equations are given in Table 9.6 for both U.S. and SI units. The factors may be read from Tables 9.7 through 9.13.

In Tables 9.4 to 9.6, the subscript 1 refers to upstream pressure-tap measurements. For measurements at the downstream tap, these must be changed to the subscript 2, and the expansion-factor equation for a downstream tap location must be used. The compressibility factor and the specific gravity are dimensionless and may be used for both the U.S. and SI system of units. Uppercase subscripts have been given to these factors to indicate that they differ from those used by the natural gas industry, in which similar but lowercase subscripts denote pressure and temperature bases of 14.73 psia and 520°R.

Tables 9.14 through 9.19 contain the N factors for volumetric and mass flow rates in U.S. and SI systems of units.

The Reynolds-Number–N-Factor Relationship

As defined in Chap. 5, the pipe Reynolds number is the basic correlating parameter used to correct for velocity profile. In the English engineering system of units, the Reynolds number is calculated as

$$R_D = \frac{\rho_f \overline{V}_f D_F}{(\mu_m)_e}$$ (5.1)

The mass flow can be calculated as

$$q_{PPS} = \rho_f A_P \overline{V}_f \qquad \text{lb}_m/\text{s}$$ (9.12)

which can be rearranged as

$$\rho_f \overline{V}_f = \frac{q_{PPS}}{A_P} = \frac{q_{PPS}}{(\pi/4)(D/12)^2}$$ (9.70)

Substituting Eq. (9.70) into Eq. (5.1) and converting viscosity to the centipoise unit give the Reynolds number in U.S. units as

$$R_D = 22737.47 \frac{q_{PPS}}{\mu_{cP} D}$$ (9.71)

The Reynolds number for SI units is similarly derived as

$$R_D = 1273239 \frac{q_{KPS}^*}{\mu_{cP} D^*}$$ (9.72)

By substituting Eqs. (9.71) and (9.72) into the relationships given in Tables 9.2 through 9.6, the Reynolds-number equations given in Tables 9.20 through 9.22 are derived in terms of the selected flow-rate unit and the unit conversion factor N.

Table 9.4 Relationship between Fundamental Constant and Derived Flow-Rate Unit for Gas Flow: U.S. Units

Letter symbol		Relationship to fundamental unit equation	Definition
Flow rate	Conversion constant		
q_M	$N_{M\rho}$	$q_{\text{PPS}} = 0.09970190 \; \dfrac{q_M}{N_{M\rho}}$	Mass flow rate with a density determination Example: $q_M = q_{\text{PPD}}$ = pounds-mass per day
q_M	$N_{M\rho T}$	$q_{\text{PPS}} = 0.1637908 \; \dfrac{q_M}{N_{M\rho T}}$	Mass flow rate using the pvT density equation Example: $q_M = q_{\text{PPM}}$ = pounds-mass per minute
q_v	$N_{v\rho}$	$q_{\text{PPS}} = 0.09970190 \; [\rho_{f1}] \; \dfrac{q_v}{N_{v\rho}}$	Volumetric flow rate at flowing conditions with an upstream tap measurement Example: $q_v = q_{\text{acfm}}$ = actual cubic feet per minute
q_v	$N_{v\rho T}$	$q_{\text{PPS}} = 0.1637908 \; \left[\dfrac{G p_{f1}}{Z_{f1} T_{f1}}\right] \dfrac{q_v}{N_{v\rho T}}$	Volumetric flow rate at flowing conditions using the pvT density equation for upstream tap measurements Example: $q_v = q_{\text{acfm}}$ = actual cubic feet per minute

q_V	$N_{V\rho}$	$q_{\mathrm{PPS}} = 0.9970190\,[\rho_b]\,\dfrac{q_V}{N_{V\rho}}$	Volumetric flow rate at standard base conditions with a density determination (ρ_b density at 14.696 and 59°F or other selected base values) Example: $q_V = q_{\mathrm{SCFM}}$ = standard cubic feet per minute
q_V	$N_{V\rho T}$	$q_{\mathrm{PPS}} = 0.1637908\,\left[\dfrac{G}{Z_b}\right]\dfrac{q_V}{N_{V\rho T}}$	Volumetric flow rate at standard base conditions (14.69595 psia and 59°F) using the pvT density equation (recommended standard base volume) Example: $q_v = q_{\mathrm{SCFH}}$ = standard cubic feet per hour
q_{Vb}	$(N_{V\rho T})_b$	$q_{\mathrm{PPS}} = 0.1637908\,\left[\dfrac{G p_b}{Z_b T_b}\right]\dfrac{q_{Vb}}{(N_{V\rho T})_b}$	Volumetric flow rate at a selected base, other than standard, using the pvT equation Example: $q_{Vb} = (q_{\mathrm{SCFD}})_{14.470}$ = standard cubic feet per day at 14.4 psia and 70°F

Table 9.5 Relationship between Fundamental Constant and Derived Flow-Rate Unit for Gas Flow: SI Units

Letter symbol		Relationship to fundamental unit equation	Definition
Flow rate	Conversion constant		
q_M^*	$N_{M_\rho}^*$	$q_{KPS}^* = 0.00003512407 \dfrac{q_M^*}{N_{M_\rho}^*}$	Mass flow rate with a density determination Example: $q_M^* = q_{KPD}^*$ = kilograms per day
q_M^*	$N_{M_{pT}}^*$	$q_{KPS}^* = 0.00006555517 \dfrac{q_M^*}{N_{M_{pT}}^*}$	Mass flow rate using the *pvT* density equation Example: $q_M^* = q_{KPM}^*$ = kilograms per minute
q_v^*	$N_{v_\rho}^*$	$q_{KPS}^* = 0.00003512407\,[\rho_{f1}^*]\,\dfrac{q_v^*}{N_{v_\rho}^*}$	Volumetric flow rate at flowing conditions with an upstream tap density determination Example: $q_v^* = q_{acmm}^*$ = actual cubic meters per minute
q_v^*	$N_{v_{pT}}^*$	$q_{KPS}^* = 0.00006555517 \left[\dfrac{Gp_{f1}^*}{Z_{f1}T_{K1}}\right]\dfrac{q_v^*}{N_{v_{pT}}^*}$	Volumetric flow rate at flowing conditions using the *pvT* density equation for upstream tap measurements Example: $q_v^* = q_{acmm}^*$ = actual cubic meters per minute
q_V^*	$N_{V_\rho}^*$	$q_{KPS}^* = 0.00003512407\,[\rho_b^*]\,\dfrac{q_V^*}{N_{V_\rho}^*}$	Volumetric flow rate at standard or selected base conditions with a density determination Example: $q_V^* = q_{SCMM}^*$ = standard cubic meters per minute
q_V^*	$N_{V_{pT}}^*$	$q_{KPS}^* = 0.00006555517 \left[\dfrac{G}{Z_b}\right]\dfrac{q_V^*}{N_{V_{pT}}^*}$	Volumetric flow rate at standard base conditions (101.325 kPa and 15°C) using the *pvT* density equation (recommended standard) Example: $q_V^* = q_{SCMH}^*$ = standard cubic meters per hour
q_{Vb}^*	$(N_{V_{pT}}^*)_b$	$q_{KPS}^* = 0.00006555517 \left[\dfrac{Gp_b^*}{Z_b T_{Kb}}\right]\dfrac{q_{Vb}^*}{(N_{V_{pT}}^*)_b}$	Volumetric flow rate at a selected base, other than standard, using the *pvT* equation Example: $q_{Vb}^* = (q_{SCMD}^*)_{100,14}$ = standard cubic meters per day at 100 kPa, 14°C

Equation Rearranged in Factor Form (F_{PB}, F_{TB}, T_{TF}, F_{pv})

Flow rate	Letter symbol — Conversion constant	Relationship to fundamental unit equation	Definition
		U.S. UNITS	
q_M	N_{Mhp}	$q_{PPS} = 0.007191927\, \dfrac{q_M}{N_{Mhp}}$	Mass flow rate — Example: $q_M = q_{PPH}$ = pounds-mass per hour
q_v	N_{vhp}	$q_{PPS} = 0.007191927 \left[\dfrac{F_{TF1}^2 F_{pv1}^2 p_1}{F_g^2}\right] \dfrac{q_v}{N_{vhp}}$	Volumetric flow rate at flowing conditions — Example: $q_v = q_{acfs}$ = actual cubic feet per second
q_V	N_{Vhp}	$q_{PPS} = 0.007191927 \left[\dfrac{1}{F_g^2 F_{PB} F_{TB} Z_b}\right] \dfrac{q_{Vb}}{N_{Vhp}}$	Volumetric flow rate at standard base or at selected temperature and pressure base — Example: $q_V = q_{SCFD}$ = standard cubic feet per day at standard base ($p_b = 14.69595$ psia, $T_b = 518.67°R$)
		SI UNITS	
q_M^*	N_{Mhp}^*	$q_{KPS}^* = 0.00000003861870\, \dfrac{q_M^*}{N_{Mhp}^*}$	Mass flow rate — Example: $q_M^* = q_{KPM}^*$ = kilograms per minute
q_v^*	N_{vhp}^*	$q_{KPS}^* = 0.00000003861870 \left[\dfrac{F_{TF1}^{*2} F_{pv1}^2 p_1^*}{F_g^2}\right] \dfrac{q_v^*}{N_{vhp}^*}$	Volume flow rate at flowing conditions — Example: $q_v^* = q_{acmh}^*$ = actual cubic meters per hour
q_V^*	N_{Vhp}^*	$q_{KPS}^* = 0.00000003861870 \left[\dfrac{1}{F_g^2 F_{PB}^* F_{TB}^* Z_b}\right] \dfrac{q_{Vb}}{N_{Vhp}^*}$	Volume flow rate at standard or selected base pressure and temperature — Example: $q_v^* = (q_{SCMH}^*)_{102,16}$ = standard cubic meters per hour at 102 kPa and 16°C

9-27

Table 9.7 **Temperature Base Factor** $F_{TB} = T_b/518.67$: U.S. Units

Temperature			Temperature		
°F	°R	F_{TB}	°F	°R	F_{TB}
33	492.67	0.94987	62	521.67	1.00578
34	493.67	0.95180	63	522.67	1.00771
35	494.67	0.95373	64	523.67	1.00964
36	495.67	0.95566	65	524.67	1.01157
37	496.67	0.95758	66	525.67	1.01350
38	497.67	0.95951	67	526.67	1.01542
39	498.67	0.96144	68	526.67	1.01735
40	499.67	0.96337	69	528.67	1.01928
41	500.67	0.96530	70	529.67	1.02121
42	501.67	0.96722	71	530.67	1.02314
43	502.67	0.96915	72	531.67	1.02506
44	503.67	0.97108	73	532.67	1.02699
45	504.67	0.97301	74	533.67	1.02892
46	505.67	0.97494	75	534.67	1.03085
47	506.67	0.97686	76	535.67	1.03278
48	507.67	0.97879	77	536.67	1.03470
49	508.67	0.98072	78	537.67	1.03663
50	509.67	0.98265	79	538.67	1.03856
51	510.67	0.98458	80	539.67	1.04049
52	511.67	0.98650	81	540.67	1.04242
53	512.67	0.98843	82	541.67	1.04434
54	513.67	0.99036	83	542.67	1.04627
55	514.67	0.99229	84	543.67	1.04820
56	515.67	0.99422	85	544.67	1.05013
57	516.67	0.99614	86	545.67	1.05206
58	517.67	0.99807	87	546.67	1.05393
59	518.67	1.00000	88	547.67	1.05591
60	519.67	1.00193	89	548.67	1.05784
61	520.67	1.00386	90	549.67	1.05977

Table 9.8 Pressure Base Factor $F_{PB} = 14.69595/p_b$: U.S. Units

Pressure base, psia	F_{PB}	Pressure base, psia	F_{PB}
14.40	1.02055	15.60	0.94205
14.50	1.01351	15.70	0.93605
14.60	1.00657	15.80	0.93012
14.70	0.99972	15.90	0.92427
14.80	0.99297	16.00	0.91850
14.90	0.98631	16.10	0.91279
15.00	0.97973	16.20	0.90716
15.10	0.97324	16.30	0.90159
15.20	0.96684	16.40	0.89609
15.30	0.96052	16.50	0.89066
15.40	0.95428	16.60	0.88530
15.50	0.94813		

Table 9.9 **Specific-Gravity Factor $F_g = \sqrt{1/G}$: U.S. and SI Units**

Specific Gravity G	+.000	+.001	+.002	+.003	+.004	+.005	+.006	+.007	+.008	+.009
.06	4.0825	4.0489	4.0161	3.9841	3.9528	3.9223	3.8925	3.8633	3.8348	3.8069
.07	3.7796	3.7529	3.7268	3.7012	3.6761	3.6515	3.6274	3.6038	3.5806	3.5578
.08	3.5355	3.5136	3.4922	3.4711	3.4503	3.4300	3.4100	3.3903	3.3710	3.3520
.09	3.3333	3.3150	3.2969	3.2791	3.2616	3.2444	3.2275	3.2108	3.1944	3.1782
.10	3.1623	3.1466	3.1311	3.1159	3.1009	3.0861	3.0715	3.0571	3.0429	3.0289
.11	3.0151	3.0015	2.9881	2.9748	2.9617	2.9488	2.9361	2.9235	2.9111	2.8989
.12	2.8868	2.8748	2.8630	2.8513	2.8398	2.8284	2.8172	2.8061	2.7951	2.7842
.13	2.7735	2.7629	2.7524	2.7420	2.7318	2.7217	2.7116	2.7017	2.6919	2.6822
.14	2.6726	2.6631	2.6537	2.6444	2.6352	2.6261	2.6171	2.6082	2.5994	2.5906
.15	2.5820	2.5734	2.5649	2.5566	2.5482	2.5400	2.5318	2.5238	2.5158	2.5078
.16	2.5000	2.4922	2.4845	2.4769	2.4693	2.4618	2.4544	2.4470	2.4398	2.4325
.17	2.4254	2.4183	2.4112	2.4042	2.3973	2.3905	2.3837	2.3769	2.3702	2.3636
.18	2.3570	2.3505	2.3440	2.3376	2.3313	2.3250	2.3187	2.3125	2.3063	2.3002
.19	2.2942	2.2881	2.2822	2.2763	2.2704	2.2646	2.2588	2.2530	2.2473	2.2417
.20	2.2361	2.2305	2.2250	2.2195	2.2140	2.2086	2.2033	2.1979	2.1926	2.1874
.21	2.1822	2.1770	2.1719	2.1668	2.1617	2.1567	2.1517	2.1467	2.1418	2.1369
.22	2.1320	2.1272	2.1224	2.1176	2.1129	2.1082	2.1035	2.0989	2.0943	2.0897
.23	2.0851	2.0806	2.0761	2.0717	2.0672	2.0628	2.0585	2.0541	2.0498	2.0455
.24	2.0412	2.0370	2.0328	2.0286	2.0244	2.0203	2.0162	2.0121	2.0080	2.0040
.25	2.0000	1.9960	1.9920	1.9881	1.9842	1.9803	1.9764	1.9726	1.9687	1.9649
.26	1.9612	1.9574	1.9537	1.9499	1.9462	1.9426	1.9389	1.9353	1.9317	1.9281
.27	1.9245	1.9209	1.9174	1.9139	1.9104	1.9069	1.9035	1.9000	1.8966	1.8932
.28	1.8898	1.8865	1.8831	1.8798	1.8765	1.8732	1.8699	1.8666	1.8634	1.8602
.29	1.8570	1.8538	1.8506	1.8474	1.8443	1.8411	1.8380	1.8349	1.8319	1.8288
.30	1.8257	1.8227	1.8197	1.8167	1.8137	1.8107	1.8078	1.8048	1.8019	1.7990
.31	1.7961	1.7932	1.7903	1.7874	1.7846	1.7817	1.7789	1.7761	1.7733	1.7705
.32	1.7678	1.7650	1.7623	1.7595	1.7568	1.7541	1.7514	1.7487	1.7461	1.7434
.33	1.7408	1.7381	1.7355	1.7329	1.7303	1.7277	1.7252	1.7226	1.7201	1.7175
.34	1.7150	1.7125	1.7100	1.7075	1.7050	1.7025	1.7001	1.6976	1.6952	1.6927
.35	1.6903	1.6879	1.6855	1.6831	1.6807	1.6784	1.6760	1.6737	1.6713	1.6690
.36	1.6667	1.6644	1.6621	1.6598	1.6575	1.6552	1.6529	1.6507	1.6485	1.6462
.37	1.6440	1.6418	1.6396	1.6374	1.6352	1.6330	1.6308	1.6287	1.6265	1.6244
.38	1.6222	1.6201	1.6180	1.6158	1.6137	1.6116	1.6096	1.6075	1.6054	1.6033
.39	1.6013	1.5992	1.5972	1.5952	1.5931	1.5911	1.5891	1.5871	1.5851	1.5831
.40	1.5811	1.5792	1.5772	1.5752	1.5733	1.5713	1.5694	1.5675	1.5656	1.5636
.41	1.5617	1.5598	1.5579	1.5561	1.5542	1.5523	1.5504	1.5486	1.5467	1.5449
.42	1.5430	1.5412	1.5394	1.5376	1.5357	1.5339	1.5321	1.5303	1.5285	1.5268
.43	1.5250	1.5232	1.5215	1.5197	1.5179	1.5162	1.5145	1.5127	1.5110	1.5093
.44	1.5076	1.5058	1.5041	1.5024	1.5008	1.4991	1.4974	1.4957	1.4940	1.4924
.45	1.4907	1.4891	1.4874	1.4858	1.4841	1.4825	1.4809	1.4793	1.4776	1.4760
.46	1.4744	1.4728	1.4712	1.4696	1.4681	1.4665	1.4649	1.4633	1.4618	1.4602
.47	1.4586	1.4571	1.4556	1.4540	1.4525	1.4510	1.4494	1.4479	1.4464	1.4449
.48	1.4434	1.4419	1.4404	1.4389	1.4374	1.4359	1.4344	1.4330	1.4315	1.4300
.49	1.4286	1.4271	1.4257	1.4242	1.4228	1.4213	1.4199	1.4185	1.4171	1.4156
.50	1.4142	1.4128	1.4114	1.4100	1.4086	1.4072	1.4058	1.4044	1.4030	1.4017
.51	1.4003	1.3989	1.3975	1.3962	1.3948	1.3935	1.3921	1.3908	1.3894	1.3881
.52	1.3868	1.3854	1.3841	1.3828	1.3814	1.3801	1.3788	1.3775	1.3762	1.3749
.53	1.3736	1.3723	1.3710	1.3697	1.3685	1.3672	1.3659	1.3646	1.3634	1.3621
.54	1.3608	1.3596	1.3583	1.3571	1.3558	1.3546	1.3533	1.3521	1.3509	1.3496
.55	1.3484	1.3472	1.3460	1.3447	1.3435	1.3423	1.3411	1.3399	1.3387	1.3375
.56	1.3363	1.3351	1.3339	1.3327	1.3316	1.3304	1.3292	1.3280	1.3269	1.3257
.57	1.3245	1.3234	1.3222	1.3211	1.3199	1.3188	1.3176	1.3165	1.3153	1.3142
.58	1.3131	1.3119	1.3108	1.3097	1.3086	1.3074	1.3063	1.3052	1.3041	1.3030
.59	1.3019	1.3008	1.2997	1.2986	1.2975	1.2964	1.2953	1.2942	1.2932	1.2921
.60	1.2910	1.2899	1.2888	1.2878	1.2867	1.2856	1.2846	1.2835	1.2825	1.2814
.61	1.2804	1.2793	1.2783	1.2772	1.2762	1.2752	1.2741	1.2731	1.2720	1.2710
.62	1.2700	1.2690	1.2680	1.2669	1.2659	1.2649	1.2639	1.2629	1.2619	1.2609
.63	1.2599	1.2589	1.2579	1.2569	1.2559	1.2549	1.2539	1.2529	1.2520	1.2510
.64	1.2500	1.2490	1.2480	1.2471	1.2461	1.2451	1.2442	1.2432	1.2423	1.2413
.65	1.2403	1.2394	1.2384	1.2375	1.2365	1.2356	1.2347	1.2337	1.2328	1.2318
.66	1.2309	1.2300	1.2290	1.2281	1.2272	1.2263	1.2254	1.2244	1.2235	1.2226
.67	1.2217	1.2208	1.2199	1.2190	1.2181	1.2172	1.2163	1.2154	1.2145	1.2136
.68	1.2127	1.2118	1.2109	1.2100	1.2091	1.2082	1.2074	1.2065	1.2056	1.2047
.69	1.2039	1.2030	1.2021	1.2012	1.2004	1.1995	1.1986	1.1978	1.1969	1.1961

Table 9.9 Specific-Gravity Factor $F_g = \sqrt{1/G}$: U.S. and SI Units (*continued*)

Specific Gravity G	+.000	+.001	+.002	+.003	+.004	+.005	+.006	+.007	+.008	+.009
.70	1.1952	1.1944	1.1935	1.1927	1.1918	1.1910	1.1901	1.1893	1.1884	1.1876
.71	1.1868	1.1859	1.1851	1.1843	1.1834	1.1826	1.1818	1.1810	1.1802	1.1793
.72	1.1785	1.1777	1.1769	1.1761	1.1752	1.1744	1.1736	1.1728	1.1720	1.1712
.73	1.1704	1.1696	1.1688	1.1680	1.1672	1.1664	1.1656	1.1648	1.1640	1.1633
.74	1.1625	1.1617	1.1609	1.1601	1.1593	1.1586	1.1578	1.1570	1.1562	1.1555
.75	1.1547	1.1539	1.1532	1.1524	1.1516	1.1509	1.1501	1.1493	1.1486	1.1478
.76	1.1471	1.1463	1.1456	1.1448	1.1441	1.1433	1.1426	1.1418	1.1411	1.1403
.77	1.1396	1.1389	1.1381	1.1374	1.1366	1.1359	1.1352	1.1345	1.1337	1.1330
.78	1.1323	1.1316	1.1308	1.1301	1.1294	1.1287	1.1279	1.1272	1.1265	1.1258
.79	1.1251	1.1244	1.1237	1.1230	1.1222	1.1215	1.1208	1.1201	1.1194	1.1187
.80	1.1180	1.1173	1.1166	1.1159	1.1152	1.1146	1.1139	1.1132	1.1125	1.1118
.81	1.1111	1.1104	1.1097	1.1090	1.1084	1.1077	1.1070	1.1063	1.1057	1.1050
.82	1.1043	1.1036	1.1030	1.1023	1.1016	1.1010	1.1003	1.0996	1.0990	1.0983
.83	1.0976	1.0970	1.0963	1.0957	1.0950	1.0944	1.0937	1.0930	1.0924	1.0917
.84	1.0911	1.0904	1.0898	1.0891	1.0885	1.0878	1.0872	1.0866	1.0859	1.0853
.85	1.0846	1.0840	1.0834	1.0827	1.0821	1.0815	1.0808	1.0802	1.0796	1.0790
.86	1.0783	1.0777	1.0771	1.0764	1.0758	1.0752	1.0746	1.0740	1.0733	1.0727
.87	1.0721	1.0715	1.0709	1.0703	1.0696	1.0690	1.0684	1.0678	1.0672	1.0666
.88	1.0660	1.0654	1.0648	1.0642	1.0636	1.0630	1.0624	1.0618	1.0612	1.0606
.89	1.0600	1.0594	1.0588	1.0582	1.0576	1.0570	1.0564	1.0558	1.0553	1.0547
.90	1.0541	1.0535	1.0529	1.0523	1.0518	1.0512	1.0506	1.0500	1.0494	1.0489
.91	1.0483	1.0477	1.0471	1.0466	1.0460	1.0454	1.0448	1.0443	1.0437	1.0431
.92	1.0426	1.0420	1.0414	1.0409	1.0403	1.0398	1.0392	1.0386	1.0381	1.0375
.93	1.0370	1.0364	1.0358	1.0353	1.0347	1.0342	1.0336	1.0331	1.0325	1.0320
.94	1.0314	1.0309	1.0303	1.0298	1.0292	1.0287	1.0281	1.0276	1.0270	1.0265
.95	1.0260	1.0254	1.0249	1.0244	1.0238	1.0233	1.0228	1.0222	1.0217	1.0212
.96	1.0206	1.0201	1.0196	1.0190	1.0185	1.0180	1.0174	1.0169	1.0164	1.0159
.97	1.0153	1.0148	1.0143	1.0138	1.0132	1.0127	1.0122	1.0117	1.0112	1.0107
.98	1.0102	1.0096	1.0091	1.0086	1.0081	1.0076	1.0071	1.0066	1.0060	1.0055
.99	1.0050	1.0045	1.0040	1.0035	1.0030	1.0025	1.0020	1.0015	1.0010	1.0005
1.00	1.0000	0.9995	0.9990	0.9985	0.9980	0.9975	0.9970	0.9965	0.9960	0.9955
1.01	0.9950	0.9945	0.9941	0.9936	0.9931	0.9926	0.9921	0.9916	0.9911	0.9906
1.02	0.9901	0.9897	0.9892	0.9887	0.9882	0.9877	0.9872	0.9868	0.9863	0.9858
1.03	0.9853	0.9849	0.9844	0.9839	0.9834	0.9829	0.9825	0.9820	0.9815	0.9811
1.04	0.9806	0.9801	0.9796	0.9792	0.9787	0.9782	0.9778	0.9773	0.9768	0.9764
1.05	0.9759	0.9754	0.9750	0.9745	0.9740	0.9736	0.9731	0.9727	0.9722	0.9717
1.06	0.9713	0.9708	0.9704	0.9699	0.9695	0.9690	0.9685	0.9681	0.9676	0.9672
1.07	0.9667	0.9663	0.9658	0.9654	0.9649	0.9645	0.9640	0.9636	0.9631	0.9627
1.08	0.9623	0.9618	0.9614	0.9609	0.9605	0.9600	0.9596	0.9591	0.9587	0.9583
1.09	0.9578	0.9574	0.9569	0.9565	0.9561	0.9556	0.9552	0.9548	0.9543	0.9539
1.10	0.9535	0.9530	0.9526	0.9522	0.9517	0.9513	0.9509	0.9504	0.9500	0.9496
1.11	0.9492	0.9487	0.9483	0.9479	0.9475	0.9470	0.9466	0.9462	0.9458	0.9453
1.12	0.9449	0.9445	0.9441	0.9436	0.9432	0.9428	0.9424	0.9420	0.9416	0.9411
1.13	0.9407	0.9403	0.9399	0.9395	0.9391	0.9386	0.9382	0.9378	0.9374	0.9370
1.14	0.9366	0.9362	0.9358	0.9354	0.9349	0.9345	0.9341	0.9337	0.9333	0.9329
1.15	0.9325	0.9321	0.9317	0.9313	0.9309	0.9305	0.9301	0.9297	0.9293	0.9289
1.16	0.9285	0.9281	0.9277	0.9273	0.9269	0.9265	0.9261	0.9257	0.9253	0.9249
1.17	0.9245	0.9241	0.9237	0.9233	0.9229	0.9225	0.9221	0.9217	0.9214	0.9210
1.18	0.9206	0.9202	0.9198	0.9194	0.9190	0.9186	0.9182	0.9179	0.9175	0.9171
1.19	0.9167	0.9163	0.9159	0.9155	0.9152	0.9148	0.9144	0.9140	0.9136	0.9133
1.20	0.9129	0.9125	0.9121	0.9117	0.9114	0.9110	0.9106	0.9102	0.9098	0.9095
1.21	0.9091	0.9087	0.9083	0.9080	0.9076	0.9072	0.9068	0.9065	0.9061	0.9057
1.22	0.9054	0.9050	0.9046	0.9042	0.9039	0.9035	0.9031	0.9028	0.9024	0.9020
1.23	0.9017	0.9013	0.9009	0.9006	0.9002	0.8998	0.8995	0.8991	0.8988	0.8984
1.24	0.8980	0.8977	0.8973	0.8969	0.8966	0.8962	0.8959	0.8955	0.8951	0.8948
1.25	0.8944	0.8941	0.8937	0.8934	0.8930	0.8926	0.8923	0.8919	0.8916	0.8912
1.26	0.8909	0.8905	0.8902	0.8898	0.8895	0.8891	0.8888	0.8884	0.8881	0.8877
1.27	0.8874	0.8870	0.8867	0.8863	0.8860	0.8856	0.8853	0.8849	0.8846	0.8842
1.28	0.8839	0.8835	0.8832	0.8828	0.8825	0.8822	0.8818	0.8815	0.8811	0.8808
1.29	0.8805	0.8801	0.8798	0.8794	0.8791	0.8787	0.8784	0.8781	0.8777	0.8774
1.30	0.8771	0.8767	0.8764	0.8760	0.8757	0.8754	0.8750	0.8747	0.8744	0.8740
1.31	0.8737	0.8734	0.8730	0.8727	0.8724	0.8720	0.8717	0.8714	0.8710	0.8707
1.32	0.8704	0.8701	0.8697	0.8694	0.8691	0.8687	0.8684	0.8681	0.8678	0.8674
1.33	0.8671	0.8668	0.8665	0.8661	0.8658	0.8655	0.8652	0.8648	0.8645	0.8642
1.34	0.8639	0.8635	0.8632	0.8629	0.8626	0.8623	0.8619	0.8616	0.8613	0.8610

Table 9.9 Specific-Gravity Factor $F_g = \sqrt{1/G}$: U.S. and SI units (*continued*)

Specific Gravity G	+.000	+.001	+.002	+.003	+.004	+.005	+.006	+.007	+.008	+.009
1.35	0.8607	0.8603	0.8600	0.8597	0.8594	0.8591	0.8588	0.8584	0.8581	0.8578
1.36	0.8575	0.8572	0.8569	0.8565	0.8562	0.8559	0.8556	0.8553	0.8550	0.8547
1.37	0.8544	0.8540	0.8537	0.8534	0.8531	0.8528	0.8525	0.8522	0.8519	0.8516
1.38	0.8513	0.8509	0.8506	0.8503	0.8500	0.8497	0.8494	0.8491	0.8488	0.8485
1.39	0.8482	0.8479	0.8476	0.8473	0.8470	0.8467	0.8464	0.8461	0.8458	0.8455
1.40	0.8452	0.8449	0.8446	0.8443	0.8439	0.8436	0.8433	0.8430	0.8427	0.8425
1.41	0.8422	0.8419	0.8416	0.8413	0.8410	0.8407	0.8404	0.8401	0.8398	0.8395
1.42	0.8392	0.8389	0.8386	0.8383	0.8380	0.8377	0.8374	0.8371	0.8368	0.8365
1.43	0.8362	0.8359	0.8357	0.8354	0.8351	0.8348	0.8345	0.8342	0.8339	0.8336
1.44	0.8333	0.8330	0.8328	0.8325	0.8322	0.8319	0.8316	0.8313	0.8310	0.8307
1.45	0.8305	0.8302	0.8299	0.8296	0.8293	0.8290	0.8287	0.8285	0.8282	0.8279
1.46	0.8276	0.8273	0.8270	0.8268	0.8265	0.8262	0.8259	0.8256	0.8253	0.8251
1.47	0.8248	0.8245	0.8242	0.8239	0.8237	0.8234	0.8231	0.8228	0.8226	0.8223
1.48	0.8220	0.8217	0.8214	0.8212	0.8209	0.8206	0.8203	0.8201	0.8198	0.8195
1.49	0.8192	0.8190	0.8187	0.8184	0.8181	0.8179	0.8176	0.8173	0.8170	0.8168
1.50	0.8165	0.8162	0.8160	0.8157	0.8154	0.8151	0.8149	0.8146	0.8143	0.8141
1.51	0.8138	0.8135	0.8133	0.8130	0.8127	0.8124	0.8122	0.8119	0.8116	0.8114
1.52	0.8111	0.8108	0.8106	0.8103	0.8100	0.8098	0.8095	0.8092	0.8090	0.8087
1.53	0.8085	0.8082	0.8079	0.8077	0.8074	0.8071	0.8069	0.8066	0.8063	0.8061
1.54	0.8058	0.8056	0.8053	0.8050	0.8048	0.8045	0.8043	0.8040	0.8037	0.8035
1.55	0.8032	0.8030	0.8027	0.8024	0.8022	0.8019	0.8017	0.8014	0.8012	0.8009
1.56	0.8006	0.8004	0.8001	0.7999	0.7996	0.7994	0.7991	0.7989	0.7986	0.7983
1.57	0.7981	0.7978	0.7976	0.7973	0.7971	0.7968	0.7966	0.7963	0.7961	0.7958
1.58	0.7956	0.7953	0.7951	0.7948	0.7946	0.7943	0.7941	0.7938	0.7936	0.7933
1.59	0.7931	0.7928	0.7926	0.7923	0.7921	0.7918	0.7916	0.7913	0.7911	0.7908
1.60	0.7906	0.7903	0.7901	0.7898	0.7896	0.7893	0.7891	0.7888	0.7886	0.7884
1.61	0.7881	0.7879	0.7876	0.7874	0.7871	0.7869	0.7866	0.7864	0.7862	0.7859
1.62	0.7857	0.7854	0.7852	0.7849	0.7847	0.7845	0.7842	0.7840	0.7837	0.7835
1.63	0.7833	0.7830	0.7828	0.7825	0.7823	0.7821	0.7818	0.7816	0.7813	0.7811
1.64	0.7809	0.7806	0.7804	0.7802	0.7799	0.7797	0.7794	0.7792	0.7790	0.7787
1.65	0.7785	0.7783	0.7780	0.7778	0.7776	0.7773	0.7771	0.7769	0.7766	0.7764
1.66	0.7762	0.7759	0.7757	0.7755	0.7752	0.7750	0.7748	0.7745	0.7743	0.7741
1.67	0.7738	0.7736	0.7734	0.7731	0.7729	0.7727	0.7724	0.7722	0.7720	0.7717
1.68	0.7715	0.7713	0.7711	0.7708	0.7706	0.7704	0.7701	0.7699	0.7697	0.7695
1.69	0.7692	0.7690	0.7688	0.7685	0.7683	0.7681	0.7679	0.7676	0.7674	0.7672
1.70	0.7670	0.7667	0.7665	0.7663	0.7661	0.7658	0.7656	0.7654	0.7652	0.7649
1.71	0.7647	0.7645	0.7643	0.7640	0.7638	0.7636	0.7634	0.7632	0.7629	0.7627
1.72	0.7625	0.7623	0.7620	0.7618	0.7616	0.7614	0.7612	0.7609	0.7607	0.7605
1.73	0.7603	0.7601	0.7598	0.7596	0.7594	0.7592	0.7590	0.7588	0.7585	0.7583
1.74	0.7581	0.7579	0.7577	0.7574	0.7572	0.7570	0.7568	0.7566	0.7564	0.7561
1.75	0.7559	0.7557	0.7555	0.7553	0.7551	0.7549	0.7546	0.7544	0.7542	0.7540
1.76	0.7538	0.7536	0.7534	0.7531	0.7529	0.7527	0.7525	0.7523	0.7521	0.7519
1.77	0.7516	0.7514	0.7512	0.7510	0.7508	0.7506	0.7504	0.7502	0.7500	0.7497
1.78	0.7495	0.7493	0.7491	0.7489	0.7487	0.7485	0.7483	0.7481	0.7479	0.7476
1.79	0.7474	0.7472	0.7470	0.7468	0.7466	0.7464	0.7462	0.7460	0.7458	0.7456
1.80	0.7454									

Table 9.10 Flowing Temperature Factor $F_{TF} = \sqrt{518.67/T_f}$; U.S. Units

Temperature, °F	F_{TF}	Temperature, °F	F_{TF}	Temperature, °F	F_{TF}	Temperature, °F	F_{TF}
1	1.06109	33	1.02605	65	0.99427	97	0.96527
2	1.05994	34	1.02501	66	0.99332	98	0.96440
3	1.05879	35	1.02397	67	0.99238	99	0.96354
4	1.05765	36	1.02294	68	0.99144	100	0.96267
5	1.05651	37	1.02191	69	0.99050	101	0.96182
6	1.05537	38	1.02088	70	0.98956	102	0.96096
7	1.05424	39	1.01986	71	0.98863	103	0.96010
8	1.05311	40	1.01884	72	0.98770	104	0.95925
9	1.05199	41	1.01782	73	0.98677	105	0.95840
10	1.05087	42	1.01680	74	0.98585	106	0.95756
11	1.04975	43	1.01579	75	0.98492	107	0.95671
12	1.04864	44	1.01478	76	0.98400	108	0.95587
13	1.04753	45	1.01378	77	0.98309	109	0.95503
14	1.04642	46	1.01277	78	0.98217	110	0.95419
15	1.04532	47	1.01177	79	0.98126	111	0.95335
16	1.04422	48	1.01078	80	0.98035	112	0.95252
17	1.04313	49	1.00978	81	0.97944	113	0.95169
18	1.04203	50	1.00879	82	0.97854	114	0.95086
19	1.04094	51	1.00780	83	0.97764	115	0.95003
20	1.03986	52	1.00682	84	0.97674	116	0.94920
21	1.03878	53	1.00583	85	0.97584	117	0.94838
22	1.03770	54	1.00485	86	0.97495	118	0.94756
23	1.03662	55	1.00388	87	0.97405	119	0.94674
24	1.03555	56	1.00290	88	0.97316	120	0.94592
25	1.03448	57	1.00193	89	0.97228	121	0.94511
26	1.03342	58	1.00097	90	0.97139	122	0.94429
27	1.03235	59	1.00000	91	0.97051	123	0.94348
28	1.03129	60	0.99904	92	0.96963	124	0.94267
29	1.03024	61	0.99808	93	0.96875	125	0.94187
30	1.02919	62	0.99712	94	0.96788	126	0.94106
31	1.02814	63	0.99617	95	0.96700	127	0.94026
32	1.02709	64	0.99521	96	0.96613	128	0.93946

Table 9.10 Flowing Temperature Factor $F_{TF} = \sqrt{518.67/T_f}$: U.S. Units (*continued*)

Temperature, °F	F_{TF}	Temperature, °F	F_{TF}	Temperature, °F	F_{TF}	Temperature, °F	F_{TF}
129	0.93866	137	0.93285	145	0.92616	153	0.92009
130	0.93787	138	0.93157	146	0.92540	154	0.91934
131	0.93707	139	0.93079	147	0.92463	155	0.91860
132	0.93628	140	0.93001	148	0.92387	156	0.91785
133	0.93549	141	0.92924	149	0.92311	157	0.91711
134	0.93470	142	0.92847	150	0.92236	158	0.91636
135	0.93392	143	0.92770	151	0.92160	159	0.91562
136	0.93313	144	0.92693	152	0.92085		

Table 9.11 Temperature Base Factor $F_{TB}^* = T_{Kb}/288.15$: SI Units

Temperature			Temperature		
°C	K	F_{TB}^*	°C	K	F_{TB}^*
1	274.15	0.95141	21	294.15	1.02032
2	275.15	0.95488	22	295.15	1.02429
3	276.15	0.95835	23	296.15	1.02776
4	277.15	0.96183	24	297.15	1.03123
5	278.15	0.96530	25	298.15	1.03470
6	279.15	0.96877	26	299.15	1.03817
7	280.15	0.97224	27	300.15	1.04164
8	281.15	0.97571	28	301.15	1.04512
9	282.15	0.97918	29	302.15	1.04859
10	283.15	0.98265	30	303.15	1.05206
11	284.15	0.98612	31	304.15	1.05553
12	285.15	0.98959	32	305.15	1.05900
13	286.15	0.99306	33	306.15	1.06247
14	287.15	0.99653	34	307.15	1.06594
15	288.15	1.00000	35	308.15	1.06941
16	289.15	1.00347	36	309.15	1.07288
17	290.15	1.00694	37	310.15	1.07635
18	291.15	1.01041	38	311.15	1.07982
19	292.15	1.01388	39	312.15	1.08329
20	293.15	1.01735	40	313.15	1.08676

Table 9.12 **Pressure Base Factor $F_{PB}^* = 101.325/p_b^*$: SI Units**

Pressure base, kPa	F_{PB}^*	Pressure base, kPa	F_{PB}^*	Pressure base, kPa	F_{PB}^*
95	1.06658	105	0.96500	115	0.88109
96	1.05547	106	0.95590	116	0.87349
97	1.04459	107	0.94696	117	0.86603
98	1.03393	108	0.93819	118	0.85869
99	1.02348	109	0.92959	119	0.85147
100	1.01325	110	0.92114	120	0.84437
101	1.00322	111	0.91284	121	0.83740
102	0.99338	112	0.90469	122	0.83053
103	0.98374	113	0.89668	123	0.82378
104	0.97428	114	0.88882	124	0.81714

Table 9.13 **Flowing Temperature Factor $F_{TF}^* = \sqrt{288.15/T_K}$: SI Units**

Temperature, °C	F_{TF}^*	Temperature, °C	F_{TF}^*	Temperature, °C	F_{TF}^*	Temperature, °C	F_{TF}^*
1.0	1.02522	6.5	1.01508	12.0	1.00525	17.5	0.99569
1.5	1.02428	7.0	1.01418	12.5	1.00437	18.0	0.99483
2.0	1.02335	7.5	1.01327	13.0	1.00349	19.0	0.99313
2.5	1.02242	8.0	1.01237	13.5	1.00261	19.5	0.99228
3.0	1.02150	8.5	1.01147	14.0	1.00174	20.0	0.99144
3.5	1.02057	9.0	1.01058	14.5	1.00087	20.5	0.99059
4.0	1.01965	9.5	1.00968	15.0	1.00000	21.0	0.98975
4.5	1.01873	10.0	1.00879	15.5	0.99913	21.5	0.98891
5.0	1.01782	10.5	1.00790	16.0	0.99827	22.0	0.98807
5.5	1.01690	11.0	1.00701	16.5	0.99741	22.5	0.98723
6.0	1.01599	11.5	1.00613	17.0	0.99655	23.0	0.98640

Table 9.13 Flowing Temperature Factor $F_{TF}^* = \sqrt{288.15/T_K}$: SI Units (*Continued*)

Temperature, °C	F_{TF}^*	Temperature, °C	F_{TF}^*	Temperature, °C	F_{TF}^*	Temperature, °C	F_{TF}^*
23.5	0.98557	35.5	0.96622	48.0	0.94723	60.5	0.92932
24.0	0.98474	36.0	0.96544	48.5	0.94649	61.0	0.92862
24.5	0.98391	37.0	0.96388	49.0	0.94576	61.5	0.92793
25.0	0.98309	37.5	0.96310	49.5	0.94503	62.0	0.92723
25.5	0.98226	38.0	0.96233	50.0	0.94429	62.5	0.92654
26.0	0.98144	38.5	0.96156	50.5	0.94356	63.0	0.92585
26.5	0.98062	39.0	0.96079	51.0	0.94284	63.5	0.92517
27.0	0.97981	39.5	0.96002	51.5	0.94211	64.0	0.92443
27.5	0.97899	40.0	0.95925	52.0	0.94139	64.5	0.92380
28.0	0.97818	40.5	0.95849	52.5	0.94066	65.0	0.92311
28.5	0.97737	41.0	0.95772	53.0	0.93994	65.5	0.92243
29.0	0.97656	41.5	0.95696	53.5	0.93922	66.0	0.92175
29.5	0.97575	42.0	0.95620	54.0	0.93850	66.5	0.92107
30.0	0.97495	42.5	0.95545	55.0	0.93707	67.0	0.92039
30.5	0.97414	43.0	0.95469	55.5	0.93636	67.5	0.91972
31.0	0.97334	43.5	0.95394	56.0	0.93565	68.0	0.91904
31.5	0.97254	44.0	0.95318	56.5	0.93494	68.5	0.91837
32.0	0.97175	44.5	0.95243	57.0	0.93423	69.0	0.91770
32.5	0.97095	45.0	0.95169	57.5	0.93352	69.5	0.91703
33.0	0.97016	45.5	0.95094	58.0	0.93282	70.0	0.91636
33.5	0.96937	46.0	0.95019	58.5	0.93211	70.5	0.91570
34.0	0.96858	46.5	0.94945	59.0	0.93141	71.0	0.91503
34.5	0.96779	47.0	0.94871	59.5	0.93071	71.5	0.91437
35.0	0.96700	47.5	0.94797	60.0	0.93001	72.0	0.91370

Table 9.14 N Factors for Mass Flow: U.S. Units†

$N_{M\rho}$ Density equation, *liquid and gas* (vapor)

Time	Pound-mass (lb$_m$)	Kilogram (kg)[§]	Gram (g)[§]
s	0.09970190	0.04522402	45.22402
min	5.982114	2.713441	2713.441
h	358.9268	162.8065	162,806.5
24 h	8614.244	3907.36	3,907,356

N_{MG} Specific-gravity‡ equation, *liquid*

Time	Pound-mass (lb$_m$)	Kilogram(kg)[§]	Gram (g)[§]
s	0.7873692	0.3571447	357.1447
min	47.24215	21.42868	21,428.68
h	2834.529	1285.721	1,285,721
24 h	68,028.70	30,857.30	30,857,300

$N_{M\rho T}$ $p\upsilon T$ equation, *gas* (vapor)

Time	Pound-mass (lb$_m$)	Kilogram (kg)[§]	Gram (g)[§]
s	0.1637913	0.07429449	74.29449
min	9.827478	4.4576769	4457.669
h	589.6487	267.4602	267,460.2
24 h	14,151.57	6419.044	6,419,044

†The U.S. units are pressure p_f (psia), differential pressure h_w (inches of water at 68°F, 14.696 psia, and standard gravity, 32.17405 ft/s^2), temperature T_f (°R), dimensions d and D (in), and density ρ (lb$_m$/ft^3).

‡Specific-gravity base: water at 60°F; pressure = 14.69595 psia.

§For sizing and calculating mass flow in SI units, *but* with measurement in the U.S. units defined above.

Table 9.15 N **Factors for Mass Flow: SI Units†**

$N^*_{M\rho}$ **Density equation,** *liquid and gas (vapor)*

Time	Kilogram (kg)	Gram (g)
s	0.00003512407	0.03512407
min	0.0021074444	2.107444
h	0.1264467	126.4467
24 h	3.034720	3034.720

N^*_{MG} **Specific-gravity equation,** *liquid*

Time	Kilogram (kg)	Gram (g)
s	0.001110172	1.110172
min	0.06661032	66.61031
h	3.996619	3996.619
24 h	95.91886	95,918.85

$N^*_{M\rho T}$ **$p\upsilon T$ equation,** *gas* **(vapor)**

Time	Kilogram (kg)	Gram (g)
s	0.00006555517	0.06555517
min	0.003933310	3.933310
h	0.2359986	235.9986
24 h	5.663967	5663.967

†The SI units are pressure p^*_f (kPa), differential pressure Δp^* (kPa), temperature T_K (°K), dimensions d^* and D^* (mm), and density ρ^* (kg/m³). For differential pressure Δp^* in bars, multiply table values by 10. For pressure p^*_f in bars, multiply by 10. For *both* differential pressure and pressure in bars, multiply by 100. For Reynolds-number calculations (equations from Table 9.21) *do not* change tabular values.

Table 9.16 N Factors for Volume Flow: U.S. Units†

N_{vp}, N_{V_ρ} Density equation, *liquid and gas* (vapor)

Time	Cubic foot (ft³)	Cubic meter (m³)‡	Liter (L)‡	U.S. gallon (gal)	U.K. liquid (Imp. gal)	Barrel 42 gal	Barrel 50 gal
s	0.09970190	0.002823244	2.823244	0.7458220	0.6210265	0.01775767	0.01491644
min	5.982114	0.1693946	169.3946	44.74932	37.26159	1.065460	0.8949864
h	358.9268	10.16368	10,163.68	2684.959	2235.696	63.92760	53.69919
24 h	8614.244	243.9283	243,928.3	64,439.02	53,656.69	1534.262	1288.780

N_{vG}, N_{VG} Specific-gravity equation, *liquid*

Time	Cubic foot (ft³)	Cubic meter (m³)	Liter (L)‡	U.S. gallon (gal)	U.K. liquid (Imp. gal)	Barrel 42 gal	Barrel 50 gal
s	0.01262491	0.0003574978	0.3574978	0.09444092	0.07863849	0.002248593	0.001888819
min	0.7574946	0.02144987	21.44987	5.666455	4.718309	0.1349156	0.1133291
h	45.44968	1.286992	1286.992	339.9873	283.0986	8.094936	6.799747
24 h	1090.793	30.88781	30,887.81	8159.696	6794.365	194.2785	163.1939

Table 9.16 *N* Factors for Volume Flow: U.S. Units† (*continued*)

pvT equation, *gas* (vapor)

Time	N_{vpT} (N_{vpT})$_b$			N_{VpT}§		
	Cubic foot (ft³)	Cubic meter (m³)‡	Liter (L)‡	Cubic foot (ft³)	Cubic meter (m³)‡	Liter (L)‡
s	0.0606898	0.001718545	1.718545	2.141951	0.06065330	60.65330
min	3.641391	0.1031127	103.1127	128.5171	3.639198	3639.198
h	218.4834	6.186763	6186.763	7711.023	218.3519	218,351.9
24 h	5243.603	148.4823	148,482.3	185,064.6	5240.445	5,240,445

†The U.S. units are pressure p_f (psia), differential pressure h_w (inches of water at 68°F, 14.696 psia, and standard gravity, 32.17405 ft/s²), temperature T_f (°R), dimensions d and D (in), and density ρ (lb$_m$/ft³).

‡For sizing and calculating volume flow in SI units, *but* with measurements in the U.S. units defined above.

§Standard base volume (ISO 5024): $p_b = 14.69595$ psia; $T_b = 518.67$°R.

Table 9.17 N Factors for Volume Flow: SI Units†

$N_{v\rho}^*$, $N_{V_\rho}^*$	Density equation, *liquid and gas* (vapor)		N_{vG}^*, N_{VG}^*	Specific-gravity equation, *liquid*	
Time	Cubic meter (m^3)	Liter (L)	Time	Cubic meter (m^3)	Liter (L)
s	0.00003512407	0.03512407	s	0.000001111270	0.001111270
min	0.002107444	2.107444	min	0.00006667619	0.06667619
h	0.1264467	126.4467	h	0.004000571	4.000571
24 h	3.034720	3034.720	24 h	0.09601371	96.01371

pvT equation, *gas* (vapor)

	$(N_{vpT}^*, §$ $(N_{VpT}^*)_b$‡			N_{VpT}^*‡	
Time	Cubic meter (m^3)	Liter (L)		Cubic meter (m^3)	Liter (L)
s	0.00001881927	0.01881927		0.00005351861	0.05351861
min	0.001129156	1.129156		0.003211117	3.211117
h	0.06774938	67.74938		0.1926670	192.6670
24 h	1.625985	1625.985		4.624008	4624.008

†The SI units are pressure p_f^* (kPa), differential pressure Δp^* (kPa), temperature T_K (K), dimensions d^* and D^* (mm), and density ρ^* (kg/m^3). For differential pressure Δp^* in bars, multiply table values by 10. For Reynolds-number calculations (equations from Table 9.21), *do not* change tabular values.

‡For pressure p_f^* in bars, multiply by 10. For *both* differential pressure and pressure in bars, multiply by 100.

§For pressure p_f^* in bars, divide by 10. For *both* differential pressure and pressure in bars, there is no change.

¶Standard base volume (ISO 5024): $p_b^* = 101.325$ kPa; $T_{Kb} = 288.15$ K.

Table 9.18 N Factors for Gas-Factor Equations $(F_{PB}, F_{TB}, F_{TF}, F_{pv})$: U.S. Units†

N_{Mhp} **Mass-flow equation**

Time	Pound-mass (lb$_m$)	Kilogram (kg)§	Gram (g)§
s	0.007191927	0.003262203	3.262203
min	0.4315156	0.1957322	195.7322
h	25.89094	11.74393	11,743.93
24 h	621.3825	281.8544	281,854.4

N_{vhp} **Volume flow at flowing conditions**

Time	Cubic foot (ft^3)	Cubic meter (m^3)§	Liter (L)§
s	1.382170	0.03913871	39.13871
min	82.930213	2.348323	2348.323
h	4975.814	140.8994	140,899.4
24 h	119,419.5	3381.585	3,381,585

N_{Vhp} **Volume flow at selected or standard base‡**

Time	Cubic foot (ft^3)	Cubic meter (m^3)§	Liter (L)§
s	0.09405112	0.002663231	2.663231
min	5.643067	0.1597939	159.7939
h	338.5840	9.587633	9587.633
24 h	8126.016	230.1032	230,103.2

†The U.S. units are pressure p_f (psia), differential pressure h_w (inches of water at 68°F, 14.696 psia, and standard gravity, 32.17405 ft/s^2), dimensions d and D (in), and density ρ (lb$_m$/ft^3).

‡Standard base volume (ISO 5024, 1976): p_b = 14.69595 psia; T_b = 518.67°R; F_{PB} = F_{TB} = 1.0.

§For sizing and calculating flow in SI units, *but* with measurements in the U.S. units defined above.

Table 9.19 N Factors for Gas-Factor Equations (F^*_{PB}, F^*_{TB}, F^*_{TF}, F^*_{pv}): SI Units†

N^*_{Mhp} **Mass-flow equation**

Time	Kilogram (kg)	Gram (g)
s	0.000003861870	0.003861870
min	0.0002317122	0.2317122
h	0.01390273	13.90273
24 h	0.3336656	333.6656

N^*_{vhp} **Volume flow at flowing conditions‡**

Time	Cubic meter (m³)	Liter (L)
s	0.0003194568	0.3194568
min	0.01916741	19.16741
h	1.150044	1150.044
24 h	27.60107	27,601.07

N^*_{Vhp} **Volume flow at selected or standard base§**

Time	Cubic meter (m³)	Liter (L)
s	0.000003152793	0.003152793
min	0.0001891676	0.1891676
h	0.01135006	11.35006
24 h	0.2724013	272.4013

†The SI units are pressure p^*_f (kPa), differential pressure Δp^* (kPa), temperature T_K (K), and dimensions d^* and D^* (mm). For differential pressure in bars, multiply table values by 10. For Reynolds-number calculations (equations from Table 9.22) *do not* change tabular values.

‡For pressure in bars, divide by 10. For *both* differential pressure and pressure in bars, there is no change.

§Standard base volume (ISO 5024, 1976): $p^*_b = 101.325$ kPa; $T_{Kb} = 218.15$ K (15°C); $F^*_{PB} = 1.0$. For pressure in bars, multiply by 10. For *both* differential pressure and pressure in bars, multiply by 100.

Table 9.20 Reynolds Number Related to Derived Flow-Rate Unit: U.S. Units

	Liquid	Gas (vapor)
	MASS FLOW RATE	
Density	$R_D = \left[2266.970\, \dfrac{1}{\mu_{cP}DN_{M_p}} \right] q_M$ (a)	$R_D = \left[2266.970\, \dfrac{1}{\mu_{cP}DN_{M_p}} \right] q_M$ (g)
Specific gravity	$R_D = 17{,}902.78\, \dfrac{1}{\mu_{cP}DN_{MG}}\, q_M$ (b)	
pvT equation		$R_D = \left[3724.200\, \dfrac{1}{\mu_{cP}DN_{MpT}} \right] q_M$ (h)
	VOLUMETRIC FLOW RATE AT FLOWING CONDITIONS	
Density	$R_D = \left[2266.970\, \dfrac{\rho_l}{\mu_{cP}DN_{vp}} \right] q_v$ (c)	$R_D = \left[2266.970\, \dfrac{\rho_{f1}}{\mu_{cP}DN_{vp}} \right] q_v$ (i)
Specific gravity	$R_D = \left[17{,}902.78\, \dfrac{F_p G_F}{\mu_{cP}DN_{vG}} \right] q_v$ (d)	
pvT equation		$R_D = \left[3724.200\, \dfrac{G p_{f1}}{Z_{f1}T_{f1}\, \mu_{cP}DN_{vpT}} \right] q_v$ (j)

VOLUMETRIC FLOW RATE AT BASE CONDITIONS

Density	$R_D = \left[2266.970 \dfrac{\rho_b}{\mu_{cP} DN_{V_\rho}} \right] q_V$	(e)	$R_D = \left[2266.970 \rho_b \dfrac{1}{\mu_{cP} DN_{V_\rho}} \right] q_V$	(k)
Specific gravity	$R_D = \left[17{,}902.78 \dfrac{G_b}{\mu_{cP} DN_{VG}} \right] q_V$	(f)		
pvT equation				
Standard base			$R_D = \left[3724.200 \dfrac{G}{Z_b} \dfrac{1}{\mu_{cP} DN_{VPT}} \right] q_V$	(l)
Selected base			$R_D = \left[3724.200 \dfrac{Gp_b}{Z_b T_b} \dfrac{1}{\mu_{cP} D(N_{VPT})_b} \right] q_{vb}$	(m)

Table 9.21   Reynolds Number Related to Derived Flow-Rate Unit: SI Units

	Liquid		Gas (vapor)	
	MASS FLOW RATE			
Density	$R_D = \left[44.72136 \, \dfrac{\rho^*}{\mu_{cP} D^* N^*_{M\rho}} \right] q^*_M$	(a)	$R_D = \left[44.72136 \, \dfrac{1}{\mu_{cP} D^* N^*_{M\rho}} \right] q^*_M$	(g)
Specific gravity	$R_D = \left[1413.515 \, \dfrac{1}{\mu_{cP} D^* N^*_{MG}} \right] q^*_M$	(b)		
pvT equation			$R_D = \left[83.46744 \, \dfrac{1}{\mu_{cP} D^* N^*_{M_{pT}}} \right] q^*_M$	(h)
	VOLUMETRIC FLOW RATE AT FLOWING CONDITIONS			
Density	$R_D = \left[44.72136 \, \dfrac{\rho^*}{\mu_{cP} D^* N^*_{v\rho}} \right] q^*_v$	(c)	$R_D = \left[44.72136 \, \dfrac{\rho^*_1}{\mu_{cP} D^* N^*_{vp}} \right] q^*_v$	(i)
Specific gravity	$R_D = \left[1413.515 \, \dfrac{F_p G_F}{\mu_{cP} D^* N^*_{vG}} \right] q^*_v$	(d)		
pvT equation			$R_D = \left[83.46744 \, \dfrac{G p^*_1}{Z_{f1} T_{K1} \, \mu_{cP} D^* N^*_{v_{pT}}} \right] q^*_v$	(j)

VOLUMETRIC FLOW RATE AT BASE CONDITIONS

Density	$R_D = \left[44.72136\, \dfrac{\rho_b^*}{\mu_{cP} D^* N_{V_\rho}^*} \right] q_V^*$	(e)	$R_D = \left[44.72136\, \rho_b^* \dfrac{1}{\mu_{cP} D^* N_{V_\rho}^*} \right] q_V^*$	(k)
Specific gravity	$R_D = \left[1413.515\, \dfrac{G_b}{\mu_{cP} D^* N_{V_G}^*} \right] q_V^*$	(f)		
pvT equation				
Standard base			$R_D = \left[83.46744\, \dfrac{G}{Z_b} \dfrac{1}{\mu_{cP} D^* N_{V_{pT}}^*} \right] q_V^*$	(l)
Selected base			$R_D = \left[83.46744\, \dfrac{G p_b^*}{Z_b T_{Kb}} \dfrac{1}{\mu_{cP} D^* (N_{V_{pT}}^*)_b} \right] q_{V_b}^*$	(m)

Table 9.22 Reynolds Number Related to Derived Flow-Rate Unit for Gas-Factor Equation: U.S. and SI Units

pvT equation	U.S. units		SI units	
Mass flow rate	$R_D = \left[163.5262 \dfrac{1}{\mu_{cP} D N_{Mhp}} \right] q_M$	(a)	$R_D = \left[4.917086 \dfrac{1}{\mu_{cP} D^* N^*_{Mhp}} \right] q^*_M$	(d)
Volumetric flow rate at flowing conditions	$R_D = \left[163.5262 \dfrac{F^2_{TF1} F^2_{pv1} p_{f1}}{F^2_g \mu_{cP} D N_{vhp}} \right] q_v$	(b)	$R_D = \left[4.917086 \dfrac{F^{*2}_{TF1} F^2_{pv1} p^*_{f1}}{F^2_g \mu_{cP} D^* N^*_{vhp}} \right] q^*_v$	(e)
Volumetric flow rate at standard or selected base conditions	$R_D = \left[163.5262 \dfrac{1}{F^2_g F_{PB} F_{TB} Z_b \mu_{cP} D N_{vhp}} \right] q_{vb}$	(c)	$R_D = \left[4.917086 \dfrac{1}{F^2_g F^*_{PB} F^*_{TB} Z_b \mu_{cP} D^* N^*_{vhp}} \right] q^*_{vb}$	(f)

SIZING VERSUS RATE DETERMINATION

There are two distinctly different problems in the use of a differential producer. First, the designer must establish the bore of the primary element to produce a desirable differential pressure at the design flow rate. Second, after the device has been fabricated, the differential pressure and other selected quantities must be measured and then substituted into the flow equation to calculate the flow rate. The designer has the responsibility of selecting which equation variables should be measured (the measured variables) and which can be grouped as unmeasured variables in the meter coefficient factor.

Depending on the sensitivities of the individual variables, the bias error could be small or significant. For example, a $\pm 10°F$ change in a 90°F (32°C) water flow application causes only a ± 0.1 percent error. But in a 100-psia (700-kPa) flow, a 3:1 flow-rate change results in a 2.5 percent gas-expansion-factor correction. It is important that each variable be investigated and, in particular, that the actual process data be checked against the design information to ensure that no significant differences exist.

To establish the bore, a desired differential pressure is first selected, and the transmitter is then calibrated to that value. Usually the pipe bore is not known initially, and a nominal standard diameter is selected. After the meter run is fabricated, the pipe and primary-element diameters are known, and, although the bore and the range of the differential-pressure transmitter may be exactly as calculated, the beta ratio will be different from the assumed value because the pipe diameter is not the nominal diameter. This means that the dimensional relationships in the flow equation are not the same as those in the sizing equation. This can result in bias errors of 3 to 4 percent in the small line sizes (Fig. 9.6).

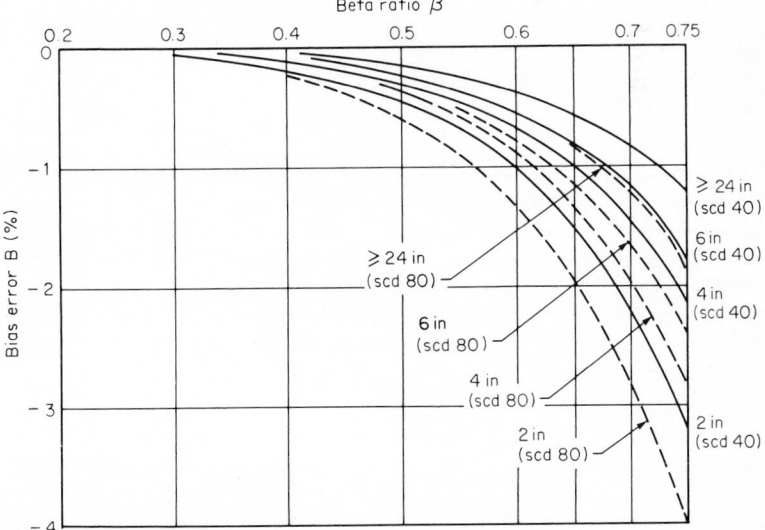

Figure 9.6 Maximum flow-rate bias error for unmeasured pipe diameter.

The decision as to how accurately the flow rate is to be calculated depends entirely on the design objective. For control purposes, when precision (repeatability) is all that is required, the sizing equation can be used to calculate flow. The added cost of correcting for pressure, temperature, specific gravity, or other process variables may not be warranted. But if accuracy is desired, each of these operating variables and dimensional measurements should be included in the flow equation. It should not be assumed that the accuracy at the single design point applies throughout the flow-rate range. The designer should estimate the accuracy over this range, based on expected changes in process conditions.

Sizing or flow-rate computation may be accomplished either graphically or with an iterative process. Graphical solutions are useful for approximate calculations and when the discharge coefficient is not expressed in equation form. Graphical solutions are not as accurate as iterative solutions, and they are not readily accomplished with computer programs.

General Sizing Considerations

The bore is most commonly sized with a chart that directly provides the flow rate. Even when charts are not used, the practice of many companies remains to select the upper-range flow rate and operating flow rate based on prior experience with visually observed circular or strip charts. In Fig. 9.7 are typical square-root flow-rate scale charts. If the maximum design flow rate is selected to be the maximum chart reading except for the decimal point, and the chart is square-root scaled, then a visual chart reading gives the flow rate. If the differential-pressure signal is "square-rooted," then a uniform flow-rate scale is selected (see Fig. 9.8).

The use of a direct-reading chart has the advantage that the rate of flow can be read at a glance. Even if the operating conditions change, this approximate reading is preferred to the use of a multiplying factor. Occasionally, plates are stocked in $\frac{1}{8}$-in orifice-bore increments (or other arithmetic progressions) to permit chart standardization and allow flow capacity to be changed through plate selection. If the operating temperature or specific gravity changes from the design conditions, the chart multiplier factor is changed to compute the new conditions.

When this method is used, an orifice plate with a bore that creates a differential near the maximum chart value is chosen. The square-root scale is not direct-reading but, instead, a multiplying factor is used for correction. In selecting the design-condition flow rate, the expected maximum flow rate is rounded either to coincide (except for the decimal point) with the selected upper-range chart value or to a convenient number. Seldom, if ever, is the maximum value selected as 2100 instead of 2000, for example.

Most plants standardize on a single upper-range value for differential-pressure transmitters, for ease in stocking and for interchangeability. Usually, 0 to 100 inH$_2$O (0 to 25 kPa) is specified for both liquid and gas flows. However, the selection depends on plant experience and preference. It should be noted that as the differential pressure is increased, the overall pressure loss increases, which further increases the pumping cost. But the lower the differential pressure, the

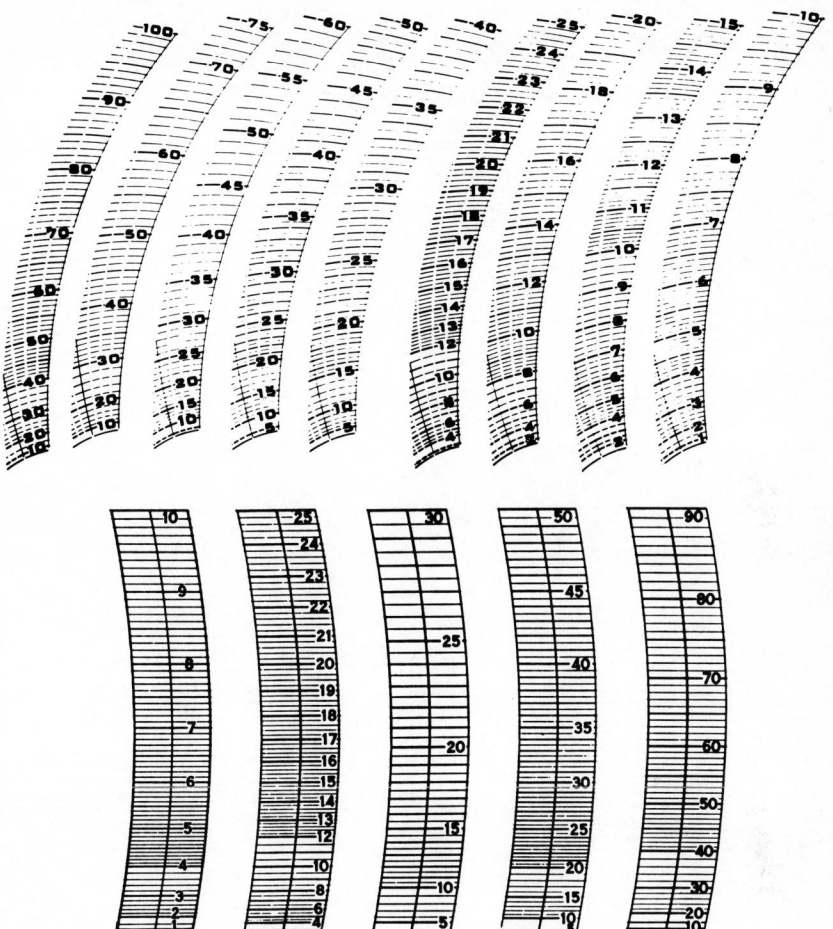

Figure 9.7 **Square-root charts** *(courtesy The Foxboro Co.).*

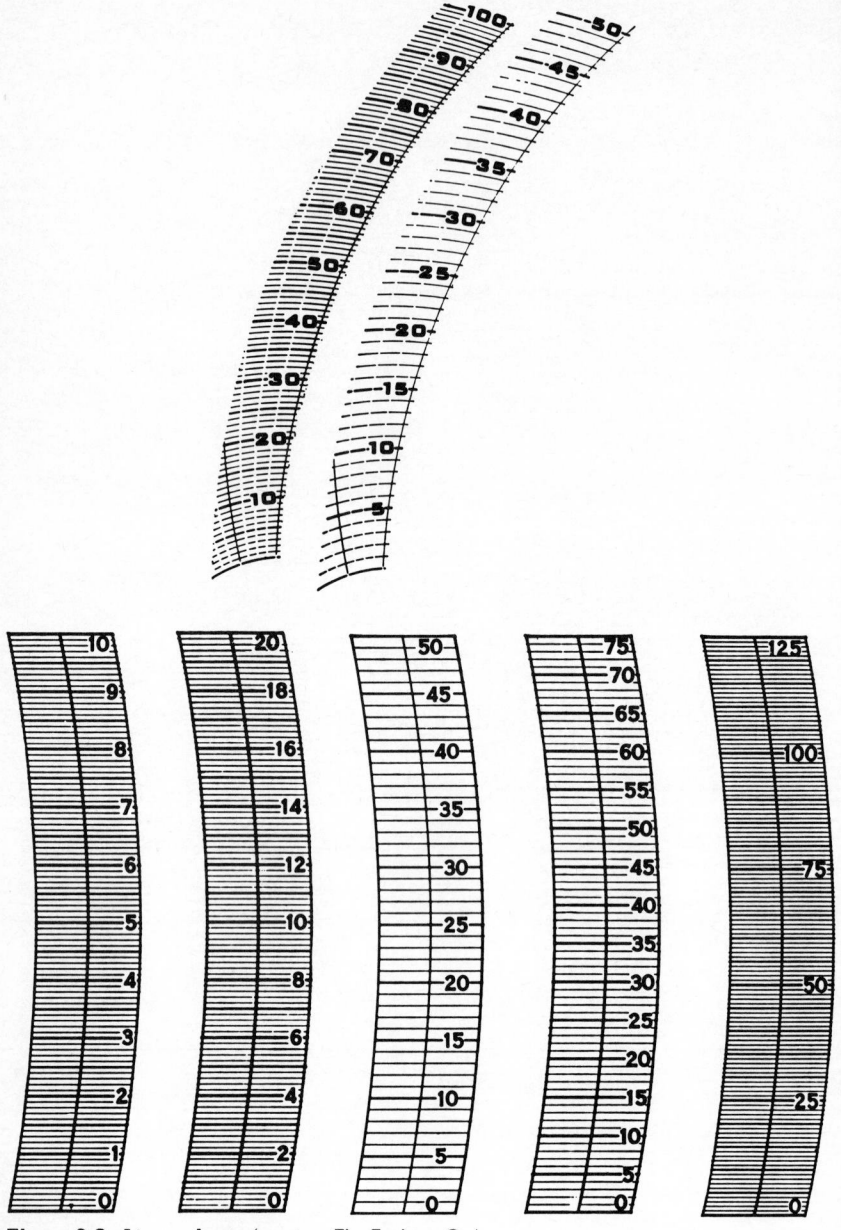

Figure 9.8 Linear charts *(courtesy The Foxboro Co.)*.

9-52

higher the beta ratio becomes, and high-beta-ratio primary elements require more upstream pipe and are inherently less accurate.

The 100-in (25-kPa) range is high enough to minimize differences in temperature if long lead lines are used in lieu of close-coupled manifolds. A few applications may be found in which the available pressure is insufficient to pass the desired flow through an orifice designed for 100 in (25 kPa). In these cases, lower upper-range values or a Lo-Loss tube is selected; or, both a lower-range differential and a Lo-Loss tube may be used to keep the pressure loss within the design objective. The pressure loss can never exceed the difference between the supply and demand pressures at the operating flow rate, and although recovery may be high, the minimum pressure should be kept above the liquid's vapor pressure to avoid cavitation.

Fixed-bore primary elements, such as flow nozzles, Lo-Loss tubes, and venturis, are seldom replaced to account for changes in flow-rate conditions. Rather, the differential-pressure transmitter is recalibrated to a nonstandard value corresponding to a direct-reading flow chart.

An important early decision is whether accuracy or precision (repeatability) is the design objective. The sizing procedure, fabrication of the meter run, flow-meter-system configuration, and final flow-rate equation can be simple or quite complex, depending on how many of the process variables are assumed to be constant (unmeasured variables) and how many are to be measured (measured variables). For all differential producers, if accuracy is required, it is necessary that dimensional measurements and reference conditions be in accordance with the requirements given in Chap. 8. The accuracy of the installation can be estimated with the procedures outlined in Chap. 4.

The S_M-FACTOR EQUATIONS FOR SIZING

The S_M Factor

The fundamental-unit equation [Eq. (9.60) or (9.61)] can be rewritten in a general form for any set of flow-rate units, using the derived-unit conversion factor N, as

$$q = \frac{NCF_a Y d^2}{\sqrt{1 - \beta^4}} f(\rho) \sqrt{h_w} \tag{9.73}$$

where the function of density $f(\rho)$ depends on whether the density is determined by densitometer, or by liquid specific gravity or, for gas flows, inferred from the pvT equation. The functional relationships $f(\rho)$ are shown in brackets in Tables 9.2 through 9.6.

Initially, the bore must be determined so as to satisfy a design operating flow rate, pipe size, and desired differential pressure. For bore sizing, the flow-rate equation is rewritten in terms of these known quantities by introducing a *sizing factor* or S_M *factor* defined as

$$S_M = \frac{CY\beta^2}{\sqrt{1 - \beta^4}}$$

(9.74)

Since $d = \beta D$, Eq. (9.73) can be expressed for both liquids and gases (vapors) as

$$q = NS_M F_a D^2 f(\rho) \sqrt{h_w}$$

(9.75)

Rearranging this equation to solve for S_M gives the relationships

$$S_M = \frac{CY\beta^2}{\sqrt{1 - \beta^4}}$$
$$= \frac{q}{NF_a D^2 f(\rho) \sqrt{h_w}}$$

(9.76)

for U.S. units, and

$$S_M = \frac{CY\beta^2}{\sqrt{1 - \beta^4}}$$
$$= \frac{q^*}{N^* F_a^* D^{*2} f(\rho^*) \sqrt{\Delta p^*}}$$

(9.77)

for SI units.

The value of the S_M factor is *constant* over the flow-rate range. It is calculated by substituting design operating-point data into the right-hand side of Eq. (9.76) or (9.77), in which the β-dependent terms, discharge coefficient, and gas expansion factor are grouped together. Table 9.1 contains the discharge-coefficient equations; Tables 9.23 through 9.25, the S_M factors; and Table 9.26, the gas-expansion-factor equations. Although the thermal-expansion factor F_a is β-dependent for differing pipe and primary-element materials [Eqs. (9.51) and (9.52)], its effect on sizing is negligible.

Table 9.23   Sizing-Factor S_M Equations for U.S. Flow Units

	Liquid		Gas (vapor)	
	MASS FLOW RATE			
Density	$S_M = \dfrac{q_M}{N_{M\rho}F_a D^2 \sqrt{F_p}\sqrt{\rho_F}\sqrt{h_w}}$	(a)	$S_M = \dfrac{q_M}{N_{M\rho}F_a D^2 \sqrt{\rho_{f1}}\sqrt{h_w}}$	(g)
Specific gravity	$S_M = \dfrac{q_M}{N_{MG}F_a D^2 \sqrt{F_p}\sqrt{G_F}\sqrt{h_w}}$	(b)		
$p\upsilon T$ equation			$S_M = \dfrac{\sqrt{Z_{f1}}\sqrt{T_{f1}}q_M}{N_{M\rho T}F_a D^2 \sqrt{G}\;\sqrt{h_w p_{f1}}}$	(h)
	VOLUMETRIC FLOW RATE AT FLOWING CONDITIONS			
Density	$S_M = \dfrac{\sqrt{F_p}\sqrt{\rho_F}\,q_\upsilon}{N_{\upsilon\rho}F_a D^2 \sqrt{h_w}}$	(c)	$S_M = \dfrac{\sqrt{\rho_{f1}}q_\upsilon}{N_{\upsilon\rho}F_a D^2 \sqrt{h_w}}$	(i)
Specific gravity	$S_M = \dfrac{\sqrt{F_p}\sqrt{G_F}\,q_\upsilon}{N_{\upsilon G}F_a D^2 \sqrt{h_w}}$	(d)		
$p\upsilon T$ equation			$S_M = \dfrac{\sqrt{G}\sqrt{p_{f1}}q_\upsilon}{N_{\upsilon\rho T}F_a \sqrt{Z_{f1}}\;\sqrt{T_{f1}}D^2 \sqrt{h_w}}$	(j)
	VOLUMETRIC FLOW RATE AT BASE CONDITIONS			
Density	$S_M = \dfrac{\rho_b\,q_V}{N_{V\rho}F_a D^2 \sqrt{F_p}\sqrt{\rho_F}\sqrt{h_w}}$	(e)	$S_M = \dfrac{\rho_b\,q_V}{N_{V\rho}F_a D^2 \sqrt{\rho_{f1}}\sqrt{h_w}}$	(k)
Specific gravity	$S_M = \dfrac{G_b q_V}{N_{VG}F_a D^2 \sqrt{F_p}\sqrt{G_F}\sqrt{h_w}}$	(f)		
$p\upsilon T$ equation				
Standard base			$S_M = \dfrac{\sqrt{Z_{f1}}\sqrt{T_{f1}}\sqrt{G}\;q_V}{N_{V\rho T}F_a Z_b D^2 \sqrt{h_w p_{f1}}}$	(l)
Selected base			$S_M = \dfrac{\sqrt{Z_{f1}}\sqrt{T_{f1}}\sqrt{G}p_b q_{\upsilon b}}{(N_{V\rho T})_b F_a Z_b T_b D^2 \sqrt{h_w p_{f1}}}$	(m)

Table 9.24 **Sizing-Factor S_M Equations for SI Flow Units**

	Liquid		Gas (vapor)	
	MASS FLOW RATE			
Density	$S_M = \dfrac{q_M^*}{N_{M\rho}^* F_a^* D^{*2} \sqrt{F_p} \sqrt{\rho_F^*} \sqrt{\Delta p^*}}$	(a)	$S_M = \dfrac{q_M^*}{N_{M\rho}^* F_a^* D^{*2} \sqrt{\rho_{f1}^*} \sqrt{\Delta p^*}}$	(g)
Specific gravity	$S_M = \dfrac{q_M^*}{N_{MG}^* F_a^* D^{*2} \sqrt{F_p} \sqrt{G_F} \sqrt{\Delta p^*}}$	(b)		
pvT equation			$S_M = \dfrac{\sqrt{Z_{f1}} \sqrt{T_{K1}}\, q_M^*}{N_{M\rho T}^* F_a^* D^{*2} \sqrt{G} \sqrt{\Delta p^*\, p_{f1}^*}}$	(h)
	VOLUMETRIC FLOW RATE AT FLOWING CONDITIONS			
Density	$S_M = \dfrac{\sqrt{F_p} \sqrt{\rho_F^*}\, q_v^*}{N_{v\rho}^* F_a^* D^{*2} \sqrt{\Delta p^*}}$	(c)	$S_M = \dfrac{\sqrt{\rho_{f1}^*}\, q_v^*}{N_{v\rho}^* F_a^* D^{*2} \sqrt{\Delta p^*}}$	(i)
Specific gravity	$S_M = \dfrac{\sqrt{F_p} \sqrt{G_F}\, q_v^*}{N_{vG}^* F_a^* D^{*2} \sqrt{\Delta p^*}}$	(d)		
pvT equation			$S_M = \dfrac{\sqrt{G} \sqrt{p_{f1}^*}\, q_v^*}{N_{v\rho T}^* F_a^* \sqrt{Z_{f1}} \sqrt{T_{K1}}\, D^{*2} \sqrt{\Delta p^*}}$	(j)
	VOLUMETRIC FLOW RATE AT BASE CONDITIONS			
Density	$S_M = \dfrac{\rho_b^* q_V^*}{N_{V\rho}^* F_a^* D^{*2} \sqrt{F_p} \sqrt{\rho_F^*} \sqrt{\Delta p^*}}$	(e)	$S_M = \dfrac{\rho_b^* q_V^*}{N_{V\rho}^* F_a^* D^{*2} \sqrt{\rho_{f1}^*} \sqrt{\Delta p^*}}$	(k)
Specific gravity	$S_M = \dfrac{G_b\, q_V^*}{N_{VG}^* F_a^* D^{*2} \sqrt{F_p} \sqrt{G_F} \sqrt{\Delta p^*}}$	(f)		
pvT equation				
Standard base			$S_M = \dfrac{\sqrt{Z_{f1}} \sqrt{T_{K1}} \sqrt{G}\, q_V^*}{N_{V\rho T}^* F_a^* Z_b D^{*2} \sqrt{\Delta p^*\, p_{f1}^*}}$	(l)
Selected base			$S_M = \dfrac{\sqrt{Z_{f1}} \sqrt{T_{K1}} \sqrt{G}\, p_b^*\, q_{Vb}^*}{(N_{V\rho T}^*)_b F_a^* Z_b T_{Kb} D^{*2} \sqrt{\Delta p^*\, p_{f1}^*}}$	(m)

Table 9.25 Sizing-Factor S_M Equations for Gas Flow Using Gas Factor Equations (F_{PB}, F_{TB}, F_{TF}, F_{pv}): U.S. and SI Units

	U.S. units		SI units	
Mass flow rate	$S_M = \dfrac{F_g q_M}{N_{Mhp} F_a F_{pv1} F_{TF1} D^2 \sqrt{h_w p_{f1}}}$	(a)	$S_M = \dfrac{F_g q_M^*}{N_{Mhp}^* F_a^* F_{pv1} F_{TF1}^* D^{*2} \sqrt{\Delta p^* p_{f1}^*}}$	(d)
Volumetric flow rate at flowing conditions	$S_M = \dfrac{F_{TF1} F_{pv1} \sqrt{p_{f1}}\, q_v}{N_{vhp} F_a F_g D^2 \sqrt{h_w}}$	(b)	$S_M = \dfrac{F_{TF1}^* F_{pv1} \sqrt{p_{f1}^*}\, q_v^*}{N_{vhp}^* F_a^* F_g D^{*2} \sqrt{\Delta p^*}}$	(e)
Volumetric flow rate at base conditions	$S_M = \dfrac{q_{vb}}{N_{Vhp} F_a F_{PB} F_{TB} F_{TF1} F_{pv1} Z_b F_g D^2 \sqrt{h_w p_{f1}}}$	(c)	$S_M = \dfrac{q_{Vb}^*}{N_{Vhp}^* F_a^* F_{PB}^* F_{TB}^* F_{TF1}^* F_{pv1} Z_b F_g D^{*2} \sqrt{\Delta p^* p_{f1}^*}}$	(f)

Table 9.26 Summary of Gas (Vapor) Expansion-Factor Equations

	Equation	Pressure relationships U.S. units	Pressure relationships SI units
	CONTOURED PRIMARY ELEMENTS (NOZZLE, VENTURI, VENTURI NOZZLE, LO-LOSS,† ETC.)		
Upstream measurements	$Y_1 = \left\{ \dfrac{(1-\beta^4)[k/(k-1)](p_{f2}/p_{f1})^{2/k}[1-(p_{f2}/p_{f1})^{(k-1)/k}]}{[1-\beta^4(p_{f2}/p_{f1})^{2/k}](1-p_{f2}/p_{f1})} \right\}^{1/2}$ (a)‡	$\dfrac{p_{f2}}{p_{f1}} = \dfrac{p_{f2}^*}{p_{f1}^*} = \dfrac{p_{f2}}{(p_{f2}+h_w/27.73)}$	p_{f2} (g)
Downstream measurements	$Y_2 = Y_1\sqrt{1+\mathbf{x}_2}$ (b)‡		
	ORIFICE		
Corner, flange, D and $D/2$ taps Upstream measurements	$Y_1 = 1 - (0.41 + 0.35\beta^4)\,\dfrac{\mathbf{x}_1}{k}$ (c)		
Downstream measurements	$Y_2 = \sqrt{1+\mathbf{x}_2} - (0.41 + 0.35\beta^4)\,\dfrac{\mathbf{x}_2}{k\sqrt{1+\mathbf{x}_2}}$ (d)	$\mathbf{x}_1 = \dfrac{h_w}{27.73 p_{f1}}$	$\mathbf{x}_1 = \dfrac{\Delta p^*}{p_{f1}^*}$ (h)
$2\tfrac{1}{2}D$ and $8D$ Upstream measurements	$Y_1 = 1 - [0.333 + 1.145(\beta^2 + 0.7\beta^5 + 12\beta^{13})]\,\dfrac{\mathbf{x}_1}{k}$ (e)		
Downstream measurements	$Y_2 = \sqrt{1+\mathbf{x}_2} - [0.333 + 1.145(\beta^2 + 0.7\beta^5 + 12\beta^{13})]\,\dfrac{\mathbf{x}_2}{k\sqrt{1+\mathbf{x}_2}}$ (f)	$\mathbf{x}_2 = \dfrac{h_w}{27.73 p_{f2}}$	$\mathbf{x}_2 = \dfrac{\Delta p^*}{p_{f2}^*}$ (i)

†Registered trade mark of Badger Meter, Inc. Manufacturer should be consulted for recommendations.

‡For SI units, substitute $p_{f2}^*/p_{f1}^* = p_{f2}/p_{f1}$ in these equations.

Since Eq. (9.77) is nonlinear, an iterative solution is required to equate the S_M factor to the β-dependent terms. Equation (9.77) can be rearranged to yield

$$\beta = \frac{1}{[1 + (CY/S_M)^2]^{1/4}} = \left[1 + \left(\frac{CY}{S_M}\right)^2\right]^{-1/4} \tag{9.78}$$

It can also be expressed as

$$\beta = \sqrt{\frac{S_M}{KY}} \tag{9.79}$$

where K is the flow coefficient defined by Eq. (9.47).

SIZING THE BORE

The β_0 Approximation

The uncertainty of operating data on viscosity, specific gravity, and temperature often preclude accurate measurement; for many plant *control* operations, close measurement tolerances are unnecessary. When precision (repeatability) is of primary interest, the sizing and flow-rate equations can be simplified by approximating the effect of Reynolds number, assuming that the liquid compressibility is negligible, and, in the case of gas flow, assuming that the expansion factor is negligible. The simplified equations have been referred to by Spink (1967) as *plant equations.*

Over normal operating ranges for liquids and gases (vapors), the discharge coefficient does not significantly vary with Reynolds number, and it can be expressed with reasonable accuracy as

$$C = k_1 + k_2 S_M \tag{9.80}$$

where k_1 and k_2 are constants derived by fitting the discharge coefficient at either the infinite Reynolds number (C_∞) or over a selected Reynolds-number range to the value of S_M calculated from Eq. (9.78).

If the gas expansion factor Y is 1.0, the substitution of Eq. (9.80) into Eq. (9.78) gives β_0, the first approximation to β, as

$$\beta_0 = \left[1 + \left(\frac{k_1 + k_2 S_M}{S_M}\right)^2\right]^{-1/4} = \left[1 + \left(\frac{k_1}{S_M} + k_2\right)^2\right]^{-1/4} \tag{9.81}$$

This simplifying assumption thus allows β_0 to be calculated directly from the S_M factor.

Table 9.27 lists the limits on the variables for approximately 1 to 3 percent accuracy in sizing the bore. In Table 9.28 are the β_0 approximation equations; in Table 9.29, the calculation procedure for liquids; and in Table 9.30, the procedure for gas (vapor) sizing.

Table 9.27 Limits on β_0 Sizing Method for Approximately 2 Percent Accuracy†

	Liquid	Gas (vapor)
Reynolds number		
Orifice	$R_D \geq 10,000$	$R_D \geq 10,000$
Venturi nozzle	$R_D \geq 100,000$	$R_D \geq 10,000$
Lo-Loss	$R_D \geq 100,000$	$R_D \geq 10,000$
Expansion factor‡	$Y_1 = 1.0$	$\dfrac{h_w}{p_{f1}} \leq 0.5; \dfrac{\Delta p^*}{p^*_{f1}} \leq 0.02$
	$Y_2 = 1.0$	$\dfrac{h_w}{p_{f2}} \leq 1.0; \dfrac{\Delta p^*}{p^*_{f2}} \leq 0.04$

†Assumes pipe diameter D or D^* is measured.

‡Ratios are based on upper-range differential pressure; operating flow rate is selected as 0.8 of upper-range flow rate, 0.64 of upper-range differential pressure.

Table 9.28 β_0 Approximate Sizing Equations

Type	Equation	
Venturi		
Machined inlet	$\beta_0 = \left[1 + \left(\dfrac{0.995}{S_M} \right)^2 \right]^{-1/4}$	(a)
Rough-cast inlet	$\beta_0 = \left[1 + \left(\dfrac{0.984}{S_M} \right)^2 \right]^{-1/4}$	(b)
Rough-welded sheet iron	$\beta_0 = \left[1 + \left(\dfrac{0.985}{S_M} \right)^2 \right]^{-1/4}$	(c)
Universal Venturi Tube†	$\beta_0 = \left[1 + \left(\dfrac{0.9797}{S_M} \right)^2 \right]^{-1/4}$	(d)
Lo-Loss tube‡	$\beta_0 = \left[1 + \left(\dfrac{0.92}{S_M} - 0.31 \right)^2 \right]^{-1/4}$	(e)

Table 9.28 β_0 **Approximate Sizing Equations (*continued*)**

Type	Equation
Nozzle	
ASME long radius	$$\beta_0 = \left[1 + \left(\frac{0.9975}{S_M} \right)^2 \right]^{-1/4} \qquad \text{(f)}$$
ISA	$$\beta_0 = \left[1 + \left(\frac{0.9944}{S_M} - 0.118 \right)^2 \right]^{-1/4} \qquad \text{(g)}$$
Venturi nozzle (ISA inlet)	$$\beta_0 = \left[1 + \left(\frac{0.989}{S_M} - 0.09 \right)^2 \right]^{-1/4} \qquad \text{(h)}$$
Orifice	
Corner, flange, D-and-$D/2$ taps	
$\quad R_D < 200{,}000$	$$\beta_0 = \left[1 + \left(\frac{0.6}{S_M} + 0.06 \right)^2 \right]^{-1/4} \qquad \text{(i)}$$
$\quad R_D > 200{,}000$	$$\beta_0 = \left[1 + \left(\frac{0.6}{S_M} \right)^2 \right]^{-1/4} \qquad \text{(j)}$$
$2\frac{1}{2}D$ and $8D$ taps	$$\beta_0 = \left[1 + \left(\frac{0.61}{S_M} + 0.55 \right)^2 \right]^{-1/4} \qquad \text{(k)}$$
Eccentric, all taps	$$\beta_0 = \left[1 + \left(\frac{0.607}{S_M} + 0.088 \right)^2 \right]^{-1/4} \qquad \text{(l)}$$
Segmental, all taps	$$\beta_0 = \left[1 + \left(\frac{0.634}{S_M} - 0.062 \right)^2 \right]^{-1/4} \qquad \text{(m)}$$
Quadrant ($\beta \leq 0.6$)	$$\beta_0 = \left[1 + \left(\frac{0.76}{S_M} + 0.26 \right)^2 \right]^{-1/4} \qquad \text{(n)}$$
Conic, corner ($\beta \leq 0.3$)	$$\beta_0 = \left[1 + \left(\frac{0.734}{S_M} \right)^2 \right]^{-1/4} \qquad \text{(o)}$$

†From BIF CALC 440/441; the manufacturer should be consulted for exact coefficient information.

‡Derived from Badger Meter, Inc. Lo-Loss flow-tube coefficient curve.

Table 9.29 **Procedure for β_0 Sizing: Liquids**

Procedure	Symbol		Reference
	U.S. units	SI units	
1. Select flow rate,† line size, differential pressure, and S_M-factor equation			
a. Select the flow-rate value to be at the upper-range chart value, and the minimum far enough out on the chart for the desired readability (the flow unit may be mass or volume).	$(q)_{URV}$	$(q^*)_{URV}$	Figs. 9.7, 9.8
b. Select the S_M-factor equation for liquid or gas based on the flow-rate unit and the method of density determination (ρ_f, G_F).	S_M	S_M	Tables 9.23–9.24
c. Obtain the N factor for the desired flow-rate unit (q_{PPM}, q_{GPM}, q_{cms}, etc.).	N	N^*	Tables 9.14–9.17
d. Select the upper-range differential pressure for the upper-range flow rate. Use $h_w = 100$ in, $\Delta p^* = 25$ kPa if not otherwise specified.	$(h_w)_{URV}$	$(\Delta p^*)_{URV}$	
e. Obtain the nominal pipe diameter from Table B.1. For SI units, $D^* = 25.4D$.	D	D^*	Table B.1
2. Scale to normal operating flow rate			
a. Multiply the URV flow rate by 0.8 *and use this in all subsequent calculations.*	$(q)_N$	$(q^*)_N$	
b. Multiply the upper-range differential pressure by 0.64 *and use this in all subsequent calculations.*	$(h_w)_N$	$(\Delta p^*)_N$	
3. Obtain fluid properties‡			
a. From the selected S_M-factor equation, tabulate the necessary fluid properties to solve for S_M.	S_M	S_M	Tables 9.23–9.24
b. Obtain the necessary density or specific gravity from measurements, tables, graphs, or prediction equations.	ρ	ρ^*	App. E
c. Obtain the fluid viscosity from measurements, tables, or graphs; convert to centipoises if necessary (Table C.2).	G_F μ_{cP}	G_F μ_{cP}	App. F

Table 9.29   **Procedure for β_0 Sizing: Liquids (*continued*)**

	Symbol		
Procedure	**U.S. units**	**SI units**	**Reference**
d. Calculate the pipe Reynolds number to ensure it is above the minimum Reynolds-number range given in Table 9.27.	R_D	R_D	Tables 9.20–9.21, Tables C.4, C.5
4. Calculate S_M factor			
a. Set the liquid compressibility factor F_p to 1.0 if it is unknown or assumed to be negligible.	F_p	F_p	App. E, Eqs. (2.107), (2.110), (2.111)
b. Calculate or graphically obtain the thermal-expansion factor.	F_a	F_a^*	Eq. (9.52) (U.S. units), Eq. (9.54) (SI units), Fig. 9.5, Table B.4
c. Substitute values into the S_M-factor equation, and solve for S_M.	S_M	S_M	Tables 9.23–9.24
5. Calculate β_0: For the selected primary device, substitute the value of S_M into the β_0 equation.	β_0	β_0	Table 9.28
6. Calculate the bore of the primary element as $d = \beta_0 D$.	d	d^*	

†See Chap. 3.

‡See Chap. 2.

Table 9.30 Procedure for β_0 Sizing: Gases (Vapors)

| | Symbol | | |
Procedure	U.S. units	SI units	Reference
1. Select flow rate† line size, differential pressure, and N factor			
a. Select the upper-range flow-rate value to be at the upper-range chart value, and the minimum far enough out on the chart for the desired readability. The flow unit may be mass or volume; for base volume flow, use 14.69595 psia (101.325 kPa) and 59°F (15°C) as standard, unless contractual or industry standard specifies otherwise.	$(q)_{\mathrm{URV}}$	$(q^*)_{\mathrm{URV}}$	Figs. 9.7, 9.8
b. Select the N factor for the desired flow-rate unit (q_{PPH}, q_{SCFD}, q_{acms}, etc.) and for the method of determining the density (densitometer, tables, pvT equation, or gas-factor equation).	N	N^*	Tables 9.14–9.19
c. Select the upper-range differential pressure for maximum flow rate. Use 100 in and $\Delta p^* = 25$ kPa if not otherwise specified.	$(h_w)_{\mathrm{URV}}$	$(\Delta p^*)_{\mathrm{URV}}$	
d. Limit $h_w/p_f \leq 1.0$ and $\Delta p^*/p_f^* \leq 0.04$; see analytical or graphical-tabular solution procedure for larger ratio calculations.			
e. Obtain the nominal pipe diameter from Table B.1. For SI units, $D^* = 25.4D$.	D	D^*	Table B.1
2. Scale to normal operating flow rate			
a. Multiply the selected maximum flow rate by 0.8 *and use this in all subsequent calculations.*	$(q)_N$	$(q^*)_N$	
b. Multiply the selected upper-range differential pressure by 0.64 *and use this in all subsequent calculations.*	$(h_w)_N$	$(\Delta p^*)_N$	
3 Obtain fluid properties‡			
a. For density determination, obtain the density from tables, graphs, by prediction equation, or from a pvT calculation.	ρ_f	ρ_f^*	Eqs. (2.10), (2.11) Steam (App. G)
b. For pvT equation, select:			
Operating temperature	T_f	T_K	

Table 9.30   Procedure for β_0 Sizing: Gases (Vapors) (*continued*)

Procedure	U.S. units	SI units	Reference
	Symbol		
Specific gravity	G	G	Eq. (2.7) App. D
Compressibility factor	Z_f	Z_f	App. G
c. For the gas-factor equation, calculate or obtain from tables the multiplying factors:			
Temperature base (for standard base, $F_{TB} = F^*_{TB} = 1.0$)	F_{TB}	F^*_{TB}	Eqs. (9.62), (9.67); Tables 9.7, 9.11
Pressure base (for standard base, $F_{PB} = F^*_{PB} = 1.0$)	F_{PB}	F^*_{PB}	Eqs. (9.63), (9.68); Tables 9.8, 9.12
Flowing temperature	F_{TF}	F^*_{TF}	Eqs. (9.66), (9.69); Tables 9.10, 9.13
Specific gravity	F_g	F_g	Eq. (9.64), Table 9.9
Supercompressibility factor	F_{pv}	F_{pv}	Eq. (9.65)
d. Obtain the fluid viscosity from measurements, tables, or graphs; convert to centipoises if necessary (Table C.2).	μ_{cP}	μ_{cP}	App. H, steam (App. G)
e. Calculate the pipe Reynolds number to ensure it is above the minimum Reynolds number given in Table 9.26. For gas (vapor) flows, the Reynolds number is seldom below these limits. *Use the normal flow rate for this calculation.*	R_D	R_D	Tables 9.20–9.22 Tables C.4, C.5
4. Calculate the S_M factor			
a. Set the base compressibility factor Z_b to 1.0 for most practical applications.	Z_b	Z_b	
b. Calculate or graphically determine the thermal-expansion factor	F_a	F^*_a	Fig. 9.5, Eqs. (9.52), (9.54), Table B.4
c. Substitute values into the S_M-factor equation, and solve for S_M.	S_M	S_M	Tables 9.23–9.25
5. Calculate β_0: For the selected primary device, substitute the value of S_M into the β_0 equation.	β_0	β_0	Table 9.28
6. Calculate the bore of the primary element as $d = \beta_0 D$.	d	d^*	

†See Chap. 3.

‡See Chap. 2.

EXAMPLE 9.1

For the gasoline of Example 2.12, determine, by the β_0 method, the bore of a flange-tapped orifice required to measure 100 base gal/min (6.31 L/s) in a 2-in (50-mm) Schedule-80 pipe.

Design information

- Flow rate: $(q_{GPM})_{URV} = 100$ base gal/min
- Differential pressure at URV flow rate: $(h_w)_{URV} = 100$ in
- Flowing pressure: $p_f = 1000$ psia
- Flowing temperature: $T_F = 80.5°F$

Scale to normal operating conditions

- Normal flow rate: $(q_{GPM})_N = 0.8\,(q_{GPM})_{URV} = 80$ base gal/min
- Normal differential pressure: $(h_w)_N = 0.64\,(h_w)_{URV} = 64$ in

Calculate S_M factor

For base volume flow with specific-gravity determination, Eq. (f) of Table 9.23 gives

$$S_M = \frac{G_b q_{GPM}}{N_{VG} F_a D^2 \sqrt{F_p} \sqrt{G_F} \sqrt{h_w}}$$

Data for the right-hand side is obtained as follows:

- From Example 2.12: $G_b = 0.7359$
- From Example 2.12: $G_F = 0.7255$
- From Table 9.16: $N_{VG} = 5.6665$
- From Table B.4: $\alpha_{PE} = 0.0000096$ in/(in·°F)
- From Eq. (9.52) $F_a = 1 + (2)(0.0000096)(80.5 - 68) = 1.0002$
- From Table B.1: $D = 1.939$ in
- Liquid compressibility factor: $F_p = 1.0$

Substitution into the S_M-factor equation gives

$$S_M = \frac{(0.7359)(80)}{(5.6665)(1.0002)(1.939)^2 \sqrt{0.7255}\sqrt{64}} = 0.40546$$

Check Reynolds number

Figure F.13 gives $\nu_{cSt} = 0.575$. Equation (2.120) is used to convert to centipoise:

$$\mu_{cP} = \frac{G_f \nu_{cSt}}{1.001} = \frac{(0.575)(0.7255)}{1.001}$$

Then, from Eq. (f) of Table 9.20,

$$(R_D)_N = 17{,}903 \left[\frac{0.7359}{(0.417)(1.939)(5.6665)} \right] (80) = 230{,}040$$

9-66

Compute β_0

Since $(R_D)_N > 200,000$, Eq. (j) of Table 9.28 is used, and

$$\beta_0 = \left[1 + \left(\frac{0.6}{0.40546}\right)^2\right]^{-1/4} = 0.7483$$

Note: β_0 can be calculated by taking the square root of the bracketed term twice, and then the reciprocal:

$$\beta_0 = \cfrac{1}{\sqrt{\left[1 + \left(\dfrac{0.6}{0.40546}\right)^2\right]^{1/2}}} = \frac{1}{\sqrt{(3.1898)^{1/2}}} = \frac{1}{\sqrt{1.7860}} = \frac{1}{1.3364} = 0.7483$$

Compute bore of orifice

$$d = \beta_0 D = (0.7483)(1.939) = 1.451 \text{ in}$$

EXAMPLE 9.2

Size, by the β_0 method, the bore of an ASME long-radius nozzle and a flange-tapped orifice for a steam flow of 30,000 $\text{lb}_\text{m}/\text{h}$ (13,608 kg/h) in a 6-in (150-mm) Schedule-80S pipe. The primary element is to be of 316 stainless steel.

Design information

- Flow rate: $(q_\text{PPH})_\text{URV} = 30,000 \text{ lb}_\text{m}/\text{h}$
- Differential pressure at URV flow rate: $(h_w)_\text{URV} = 100 \text{ in}$
- Flowing upstream pressure: $p_{f1} = 204 \text{ psia}$
- Flowing upstream temperature: $T_F = 460°\text{F}$

Scale to normal operating flow rate

- Normal flow rate: $(q_\text{PPH})_N = 0.8 \,(q_\text{PPH})_\text{URV} = 24,000 \text{ lb}_\text{m}/\text{h}$
- Normal differential pressure: $(h_w)_N = 0.64(h_w)_\text{URV} = 64 \text{ in}$

Calculate S_M factor

For mass flow with a density determination, Eq. (g) of Table 9.23 is used:

$$S_M = \frac{q_M}{N_{M\rho}F_a D^2 \sqrt{\rho_{f1}}\sqrt{h_w}}$$

Data is as follows:

- From Table 9.14: $N_{M\rho} = 358.93$
- From Table B.4: $\alpha_{PE} = 0.0000096 \text{ in}/(\text{in}\cdot°\text{F})$
- From Eq. (9.52): $F_a = 1 + (2)(0.0000096)(460 - 68) = 1.0075$
- From Table B.1: $D = 5.761 \text{ in}$
- From Table G.1: $\rho_{f1} = 0.39524 \text{ lb}_\text{m}/\text{ft}^3$; $\mu_\text{cP} = 0.01747$

Substitution into the S_M-factor equation gives

$$S_M = \frac{24,000}{(358.93)(1.0075)(5.761)^2 \sqrt{0.39524}\sqrt{64}}$$
$$= 0.39759$$

Bore of ASME long-radius flow nozzle

The Reynolds number at normal flow rate is, from Eq. (g) of Table 9.20,

$$(R_D)_N = 2267 \left[\frac{1}{(0.01747)(5.761)(358.93)}\right](24,000) = 1,506,129$$

Equation (f) of Table 9.28 gives

$$\beta_0 = \left[1 + \left(\frac{0.9975}{0.39759}\right)^2\right]^{-1/4} = 0.6085$$

and the nozzle bore is

$$d = \beta_0 D = (0.6085)(5.761) = 3.506 \text{ in}$$

Bore of flange-tapped orifice

Equation (j) of Table 9.28 is used for $R_D > 200,000$:

$$\beta_0 = \left[1 + \left(\frac{0.6}{0.39759}\right)^2\right]^{-1/4} = 0.7432$$

The orifice bore is then

$$d = \beta_0 D = (0.7432)(5.761) = 4.282 \text{ in}$$

Iterative Solution

The solution for β and, hence, for the bore of the primary element is seen via Eq. (9.78) or (9.79) to require iteration, since both the discharge coefficient C and the gas expansion factor Y are β-dependent. To solve for β, two solution methods are available; both converge rapidly. The first is Newton's method for the solution of numerical equations (Sec. A.1), and the second is simple iteration of Eq. (9.78) or (9.79). Newton's method is the more complex of the two, because of the number of terms contained both in the function and in the derivative of the function. It may be used for computer and programmable-calculator solutions, and it is applied to a sizing example in Sec. A.3. The iteration method is easier to understand, and it rapidly converges with the introduction of β_0 as the initial estimate.

In the iterative method, β_0 is first calculated by following the procedure given in Table 9.29 or 9.30. The substitution of β_0 into the discharge-coefficient equation in the general form of Eq. (9.49) then gives the first estimate of the discharge coefficient as

$$C_0 = (C_\infty)_0 + \frac{b_0}{R_D^n} \tag{9.82}$$

Substituting the isentropic exponent k and β_0 into the proper gas-expansion-factor equation from Table 9.26 gives the first estimate, $(Y_1)_0$. Then substitution of C_0 and $(Y_1)_0$ into Eq. (9.78) yields the second estimate for β as

$$\beta_1 = \left\{ 1 + \left[\frac{C_0(Y_1)_0}{S_M} \right]^2 \right\}^{-1/4} \tag{9.83}$$

Because β_0 is usually within 1 percent of the actual value of β, and the discharge coefficient for most primary elements is not highly sensitive to β, a single iteration is usually all that is required. Successive iterations can be expressed mathematically as follows:

$$\beta_n = \left\{ 1 + \left[\frac{C_{n-1}(Y_1)_{n-1}}{S_M} \right]^2 \right\}^{-1/4} \tag{9.84}$$

The iteration procedure for liquid sizing is given in Table 9.31, and for gas (vapor) sizing in Table 9.32.

Table 9.31 **Procedure for Iterative Solution: Liquids**

Procedure	Symbol		Reference
	U.S. units	SI units	
1. Calculate β_0: Follow the procedure given in Table 9.29	β_0	β_0	Table 9.28
2. Select the discharge-coefficient equation for the primary device. The general form of the equation is $C = C_\infty + b/R_D^n$.	C	C	Table 9.1
3. Iterate to obtain successive values for β.			
$\quad$ a. Set $Y = 1.0$ for liquids.			
$\quad$ b. Substitute β_0 into the discharge-coefficient equation selected in step 2, and solve for C_0.	C_0	C_0	
$\quad$ c. Calculate the second estimate for β as $$\beta_1 = \left[1 + \left(\frac{C_0}{S_M} \right)^2 \right]^{-1/4}$$	β_1	β_1	Eq. (9.83)
$\quad$ d. Repeat from step 3b until $\beta_n - \beta_{n-1} < 0.0001$, where $$\beta_n = \left[1 + \left(\frac{C_{n-1}}{S_M} \right)^2 \right]^{-1/4}$$ Usually a single iteration is all that is required	β_n	β_n	Eq. (9.84)

Table 9.32   Procedure for Iterative Solution: Gases (Vapors)

| | Symbol | | |
Procedure	U.S. units	SI units	Reference
1. Calculate β_0: Follow the procedure given in Table 9.30.	β_0	β_0	Table 9.30
2. Select the discharge-coefficient equation for the primary device. The general form of the equation is $C = C_\infty + b/R_D^n$.	C	C	Table 9.1
3. Iterate to obtain successive values for β.			
a. Substitute β_0 into the discharge-coefficient equation selected in step 2 to obtain an initial estimate.	C_0	C_0	
b. Calculate the initial estimate of the gas expansion factor by substituting β_0 into the expansion-factor equation for the primary device, using the following for the isentropic exponent:			Table 9.26

$$k = k_i = \frac{(C_p)_i}{(C_p)_i - 1.986} \quad \text{for } \frac{h_w}{p_{f1}} \leq 1,\dagger$$

$\quad p_{f1} \leq 0.25 p_c \qquad \dfrac{\Delta p^*}{p_{f1}^*} \leq 0.04$	k_i	k_i	Fig. I.17, Table I.1, Eq. (2.149), Eq. (2.1) Eq. (2.59)

and

$$k = k \text{ (isentropic exponent)}\ddagger \quad \text{for } \frac{h_w}{p_{f1}} > 1,$$

$\quad p_{f1} > 0.25 p_c \qquad \dfrac{\Delta p^*}{p_{f1}^*} > 0.04,$	k	k	Figs. I.1–I.16, Eq. (2.154)
c. Calculate the second estimate for β as			

$$\beta_1 = \left[1 + \left(\frac{C_0(Y_1)_0}{S_M} \right)^2 \right]^{-1/4}$$

(for β_1 equation above)	β_1	β_1	Eq. (9.83)
d. Repeat step 3a until $\beta_n - \beta_{n-1} < 0.0001$, where			

$$\beta_n = \left\{ 1 + \left[\frac{C_{n-1}(Y_1)_{n-1}}{S_M} \right]^2 \right\}^{-1/4}$$

(for β_n equation above)	β_n	β_n	Eq. (9.84)
4. Calculate the bore of the primary element as $d = \beta_n D$.	d	d^*	

†For maximum bias error of ± 0.1% for orifice and ± 0.2% for Nozzle, Venturi, etc.

‡See Chap. 2 for the calculation and Appendix Sec 1.

EXAMPLE 9.3

Correcting for the compressibility of the liquid, determine, by iteration, the orifice bore for the data of Example 9.1.

Recalculate the S_M factor for liquid compressibility

From Example 2.15, $F_p = 1.0088$. From Eq. (f) of Table 9.23,

$$S_M = \frac{(0.7359)(80)}{(5.6665)(1.0002)(1.939)^2 \sqrt{1.0088} \sqrt{0.7255} \sqrt{64}} = 0.40368$$

Iterate for β

From Eq. (j) of Table 9.28,

$$\beta_0 = \left[1 + \left(\frac{0.6}{0.40368} \right)^2 \right]^{-1/4} = 0.74714$$

The discharge-coefficient equation for the initial estimate of beta (β_0) is, from Table 9.1, for $D < 2.3$ in,

$$C_0 = (C_\infty)_0 + \frac{b_0}{R_D^n} = 0.5959 + 0.0312\beta_0^{2.1} - 0.184\beta_0^8$$

$$+ 0.039 \frac{\beta_0^4}{1 - \beta_0^4} - 0.0337 \frac{\beta_0^3}{D} + \frac{91.71\beta_0^{2.5}}{(R_D)_N^{0.75}}$$

Substituting $D = 1.939$, $(R_D)_N = 230{,}040$, and $\beta_0 = 0.74714$ gives

$$C_0 = 0.60957$$

The second estimate for β is then, from Eq. (9.83) with $Y_1 = 1.0$ for liquids,

$$\beta_1 = \left[1 + \left(\frac{0.60957}{0.40368} \right)^2 \right]^{-1/4} = 0.74306$$

Substitution of β_1 into the flange-tap equation gives the second estimate for the discharge coefficient as $C_1 = 0.60965$. The third estimate for β is then, from Eq. (9.83),

$$\beta_2 = \left[1 + \left(\frac{0.60965}{0.40368} \right)^2 \right]^{-1/4} = 0.74303$$

The fourth estimate is similarly obtained as

$$\beta_3 = \left[1 + \left(\frac{0.60965}{0.40368} \right)^2 \right]^{-1/4} = 0.74303$$

The orifice bore is then,

$$d = \beta_3 D = (0.7430)(1.939) = 1.441 \text{ in}$$

EXAMPLE 9.4

Size, by iteration, the flow-nozzle bore for the conditions of Example 9.2.

Data from Example 9.2 is as follows:

$$(q_{PPH})_{URV} = 30,000 \qquad (q_{PPH})_N = 24,000 \qquad \beta_0 = 0.6085$$

$$(h_w)_{URV} = 100 \qquad (h_w)_N = 64 \qquad S_M = 0.39759$$

$$(R_D)_{URV} = 1,882,700 \qquad p_{f1} = 204 \qquad (R_D)_N = 1,506,129$$

The initial estimate of the discharge coefficient, from Table 9.1, is

$$C_0 = 0.9975 - \frac{6.53\beta_0^{0.5}}{(R_D)_n^{0.5}} = 0.99335$$

The expansion factor is found with Eq. (a) of Table 9.26:

$$(Y_1)_0 = \left\{ \frac{(1 - \beta_0^4)[k_i/(k_i - 1)](p_{f2}/p_{f1})^{2/k_i}[1 - (p_{f\,2}/p_{f1})^{(k_i-1)/k_i}]}{[1 - \beta_0^4(p_{f2}/p_{f1})^{2/k_i}](1 - p_{f2}/p_{f1})} \right\}^{1/2}$$

From Eq. (g) of Table 9.26,

$$\frac{p_{f2}}{p_{f1}} = 1 - x_1 = 1 - \frac{h_w}{27.73 p_{f1}} = 1 - \frac{64}{(27.73)(204)} = 0.9887$$

and, using the ideal-gas isentropic exponent from Eq. (2.149), see Fig. I.16 for real-gas isentropic exponent,

$$k_i = \frac{(C_p)_i}{(C_p)_i - 1.986} = \frac{8.44}{8.44 - 1.986} = 1.308$$

where $(C_p)_i = 8.44$ Btu/(lb$_m \cdot$mol$\cdot$°R) is found in Table I.2. Substitution above then gives $(Y_1)_0 = 0.99214$.

The second estimate for β is then, from Eq. (9.83),

$$\beta_1 = \left\{ 1 + \left[\frac{(0.99335)(0.99214)}{0.39759} \right]^2 \right\}^{-1/4}$$

$$= 0.61166$$

Substitution of β_1 into the expansion-factor and discharge-coefficient equations yields, for second estimates,

$$C_1 = 0.99334 \qquad (Y_1)_1 = 0.99210$$

The third estimate for β is then

$$\beta_2 = \left\{ 1 + \left[\frac{(0.99334)(0.99210)}{0.39759} \right]^2 \right\}^{-1/4}$$

$$= 0.61167$$

The nozzle bore is then

$$d = \beta_2 D = (0.6117)(5.761) = 3.524 \text{ in}$$

EXAMPLE 9.5

Size, by iteration, the flange-tap orifice for the data of Example 9.2, assuming (a) pipe and orifice are stainless steel and (b) the pipe material is carbon steel.

a. Stainless-steel orifice and pipe

From Table 9.1, for $D > 2.3$ in,

$$C = 0.5959 + 0.0312\beta^{2.1} - 0.184\beta^8 + 0.09 \frac{\beta^4}{D(1 - \beta^4)} - 0.0337 \frac{\beta^3}{D} + \frac{91.71\beta^{2.5}}{R_D^{0.75}}$$

The expansion factor is, from Eq. (c) of Table 9.26,

$$Y_1 = 1 - (0.41 + 0.35\beta^4) \frac{\mathbf{x}_1}{k}$$

where

$$\mathbf{x}_1 = \frac{h_w}{27.73 p_{f1}} = \frac{64}{(27.73)(204)} = 0.01131$$

and
$$k = k_i = 1.308$$

Thus,

$$Y_1 = 1 - 0.008649(0.41 + 0.35\beta^4)$$

The iteration equation for β is Eq. (9.84):

$$\beta_n = \left\{ 1 + \left[\frac{(C_{n-1})(Y_1)_{n-1}}{S_M} \right]^2 \right\}^{-1/4}$$

where $S_M = 0.39759$ and $\beta_0 = 0.74322$ from Example 9.2. The iteration may be tabulated as follows, using six places for calculation accuracy:

	n		
	1	2	3
β_{n-1}	0.743222	0.743960	0.743989
C_{n-1}	0.600974	0.600907	0.600904
$(Y_1)_{n-1}$	0.995530	0.995530	0.995530
β_n	0.743960	0.743989	0.743990

The orifice bore is then
$$d = \beta_n D = (0.7440)(5.761) = 4.286 \text{ in}$$

b. Stainless-steel orifice plate and carbon-steel pipe

For differing plate and pipe materials, the thermal-expansion factor is given by Eq. (9.51):

$$F_a = 1 + \frac{2}{1 - \beta^4} (\alpha_{PE} - \beta^4 \alpha_P)(T_F - 68)$$

where, from Table B.4, $\alpha_{PE} = 0.0000096$ in/(in $\cdot$ °F) and $\alpha_P = 0.0000067$ in/(in $\cdot$ °F).

Since F_a is a function of β, the S_M-factor equation (9.74) is rewritten as

$$S_M = \left[\frac{F_a C Y \beta^2}{\sqrt{1 - \beta^4}} \right] = \frac{q_M}{N_{M_\rho} D^2 \sqrt{\rho_{f1}} \sqrt{h_w}} = \frac{q_{PPH}}{358.9268 D^2 \sqrt{\rho_{f1}} \sqrt{h_w}} = 0.400580$$

where $S_M = 0.400580$ is constant, and all β-dependent terms are contained within the brackets. The iterative solution for β is, from Eq. (9.84),

$$\beta_n = \left\{ 1 + \left[\frac{(F_a)_{n-1}(C_{n-1})(Y_1)_{n-1}}{S_M} \right]^2 \right\}^{-1/4}$$

The iteration may be tabulated as follows:

	n		
	1	2	3
β_{n-1}	0.745155	0.743778	0.743724
$(F_a)_{n-1}$	1.008540	1.008529	1.008529
C_{n-1}	0.600797	0.600924	0.600929
$(Y_1)_{n-1}$	0.995520	0.995527	0.995527
β_n	0.743778	0.743724	0.743722

The orifice bore is then

$$d = \beta_n D = (0.7437)(5.761) = 4.285 \text{ in}$$

EXAMPLE 9.6

Using the gas-factor equation, size by iteration the bore of a corner-tap orifice for a natural gas with a specific gravity of 0.652, flowing at 340 standard m^3/h at 100 kPa and 14°C in a 52.50-mm pipe. The following design information is selected for an upstream measurement of pressure and temperature.

Design information

- Flow rate: $(q^*_{SCMH})_{URV} = 340$ standard m^3/h
- Differential pressure at maximum flow rate: $(\Delta p^*)_{URV} = 10$ kPa
- Upstream flowing pressure: $p^*_{f1} = 184$ kPa
- Upstream absolute temperature: $T_{K1} = 300$ K (26.85°C)
- Plate material: 304 stainless steel

Scale to normal operating conditions

- Normal operating flow rate: $(q^*_{SCMH})_N = 0.8(q^*_{SCMH})_{URV} = 272$ standard m^3/h
- Normal differential pressure: $(\Delta p^*)_N = 0.64(\Delta p^*)_{URV} = 6.4$ kPa

Calculate S_M factor

Equation (*f*) of Table 9.25,

$$S_M = \frac{q^*_{S\text{CMH}}}{N^*_{Vhp} F^*_a F^*_{PB} F^*_{TB} F^*_{TF1} F_{pv1} Z_b F_g D^{*2} \sqrt{\Delta p^* \, p^*_{f1}}}$$

is used here, with $q^*_{S\text{CMH}}$ in place of q^*_{Vb}, and the denominator factors may be tabulated as follows:

Factor	Reference	Equation	Value
N^*_{Vhp}	Table 9.19		0.011350
F^*_a	Eq. (9.54)	$F^*_a = 1 + 2\alpha^*_{PE}(T_{°C} - 20)$	1.0002
	Table B.4	$\alpha^*_{PE} = 0.0000173 \text{ mm/(mm} \cdot {}^\circ\text{C)}$	
F^*_{PB}	Eq. (9.68)	$F^*_{PB} = \dfrac{101.325}{p^*_b}$	1.0133
F^*_{TB}	Eq. (9.67)	$F^*_{TB} = \dfrac{T_{Kb}}{288.15}$	0.99653
F^*_{TF1}	Eq. (9.69)	$F^*_{TF1} = \sqrt{\dfrac{288.15}{T_{K1}}}$	0.98005
F^*_{pv1}	Eq. (2.37)†	$F_{pv1} = \sqrt{1 + \dfrac{4.66 p_G \, 10^{5+1.6G}}{T^{3.825}_{f1}}}$	1.0011
Z_b	AGA Report 3	Assumed negligible	1.0000
F_g	Eq. (9.64)	$F_g = \sqrt{\dfrac{1}{G}}$	1.2384

†From Table C.1, $T_{f1} = 1.8T_{K1} = 540°\text{R}$, and $p_G = (184 - 101.325)/6.8948 = 11.991$ psig.

Substitution into the S_M-factor equation at normal flow rate gives

$$S_M = \frac{272}{(0.011350)(1.0002)(1.0133)(0.99653)(0.98005)(1.0011)(1.0)(1.2384)(52.50)^2 \sqrt{(6.4)(184)}}$$

$$= 0.20647$$

Calculate β_0

From Eq. (*f*) of Table 9.22,

$$(R_D)_N = 4.9171 \frac{1}{(F^2_g F^*_{PB} F^*_{TB} Z_b \mu_{cP} D^* N^*_{Vhp})} (q^*_{S\text{CMH}})_N$$

From Fig. H.16, $\mu_{cP} = 0.0118$, and substitution gives $(R_D)_N = 122{,}826$. For $R_D < 200{,}000$ Eq. (i) of Table 9.28 is used to compute

$$\beta_0 = \left[1 + \left(\frac{0.6}{0.20647} + 0.06 \right)^2 \right]^{-1/4} = 0.56523$$

Iterate for β

From Eq. (9.84),

$$\beta_n = \left\{ 1 + \left[\frac{C_{n-1}(Y_1)_{n-1}}{S_M} \right]^2 \right\}^{-1/4}$$

where, from Table 9.1 for corner taps,

$$C_{n-1} = 0.5959 + 0.0312\beta_{n-1}^{2.1} - 0.184\beta_{n-1}^8 + \frac{91.71\beta_{n-1}^{2.5}}{(R_D)_N^{0.75}}$$

Also, from Eq. (c) of Table 9.26,

$$(Y_1)_{n-1} = 1 - (0.41 + 0.35\beta_{n-1}^4) \frac{\mathbf{x}_1}{k}$$

where

$$\mathbf{x}_1 = \frac{\Delta p^*}{p_{f1}^*} = \frac{6.4}{184} = 0.03478$$

and, from Fig. I.9, $k = 1.27$. The iteration may be tabulated as follows:

	n	
	1	**2**
β_{n-1}	0.56523	0.57070
C_{n-1}	0.60676	0.60688
$(Y_1)_{n-1}$	0.98780	0.98780
β_n	0.57070	0.57065

The orifice bore is then

$$d = \beta_2 D = (0.5707)(52.50) = 29.96 \text{ mm}$$

Graphical Solution

The procedure for graphical sizing is illustrated in Fig. 9.9, and the actual graphs are presented in Chap. 10. In this procedure, the discharge coefficient and gas

Step 1: Calculate β_0

From Eq. (9.81),

$$\beta_0 = \left[1 + \left(\frac{k_1}{S_M} + k_2 \right)^2 \right]^{-1/4}$$

See Table 9.28 for β_0 equations

For Example 9.5,

$$\beta_0 = \left[1 + \left(\frac{0.6}{0.3976} \right)^2 \right]^{-1/4} = 0.7432$$

Step 2: Find orifice discharge coefficient C_0

For $\beta = 0.74 \approx \beta_0$,

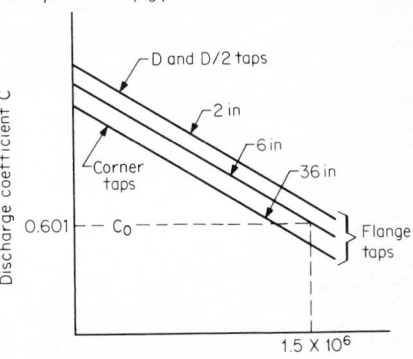

X 1000 = pipe Reynolds number R_D

Step 3: Find expansion factor $(Y_1)_0$

$\left[\text{For liquids}, (Y_1)_0 = 1.0 \right]$

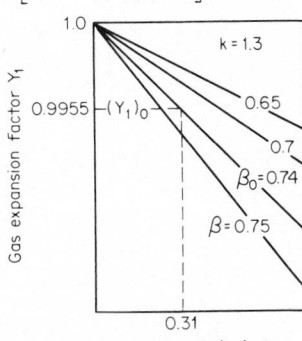

Step 4: Calculate β_1 and d

From Eq. (9.83),

$$\beta_1 = \left[1 + \left(\frac{C_0 (Y_1)_0}{S_M} \right)^2 \right]^{-1/4}$$

For Example 9.5,

$$\beta_1 = \left[1 + \left(\frac{(0.601)(0.9955)}{0.3976} \right)^2 \right]^{-1/4} = 0.7440$$

$$d = \beta_1 D = (0.7440)(5.761) = 4.286 \text{ in}$$

Figure 9.9 Steps in graphical sizing; values shown are from Example 9.5. (Graphs are presented in Chap. 10.)

expansion factor are determined graphically rather than by calculation. The four steps are shown in Fig. 9.9 for Example 9.5 and may be summarized as follows:

- Step 1 (Fig. 9.9): Calculate β_0 and the normal operating Reynolds number by following the procedure for liquids in Table 9.29 or that for gases (vapors) in Table 9.30.

- Step 2 (Fig. 9.9): Enter the discharge coefficient-Reynolds number curves with the value of β_0 and the normal operating Reynolds number and read the discharge coefficient.

- Step 3 (Fig. 9.9): For gases (vapors), enter the appropriate isentropic exponent–primary-element gas expansion factor graph at the normal operating ratio $(h_w)_N/p_f$. Read the gas expansion factor $(Y_1)_0$ from the β line corresponding to β_0.

■ Step 4 (Fig. 9.9*d*): Calculate the second estimate of the beta ratio using Eq. (9.84):

$$\beta_1 = \left\{ 1 + \left[\frac{C_0(Y_1)_0}{S_M} \right]^2 \right\}^{-1/4}$$

(9.84)

where S_M is the value obtained in the β_0 determination, and C_0 and $(Y_1)_0$ are the graphically obtained discharge coefficient and gas expansion factor. Given the accuracy of the graphs, a third estimate is unwarranted.

SPECIAL CONSIDERATIONS

Several additional multiplying factors may be introduced into the flow equation to compensate for particular measurement situations. These factors are used to correct a moist stream to a dry flow-rate basis, adjust for a nonhomogeneous flow, correct for the manometer-elevation bias, and compensate for a small drain hole through the primary element.

Correction for Steam Quality (Gas-Liquid Flows)

Two-phase flows are nonhomogeneous, and the flow equations do not apply. Several compensating equations have been proposed, to permit the use of the basic equations. Among these are the James (1966) effective-density equation, the Smith-Leang (1974) blockage-factor equation, and Murdock's (1961) modified single-phase pseudo-gas flow-rate equation. The Murdock equation is the easiest to apply if certain simplifying assumptions are made. Murdock reports correlation to within ±1.5 percent for orifices, and Smith and Leang suggest that the correlation is valid for venturis and nozzles.

Murdock's correlation is derived from 90 test points for steam-water (wet steam), air-water, natural gas–water, natural gas–salt water, and natural gas–distillate combinations and is applicable to both two-phase and two-component liquid-gas flows. The correlation is developed under the assumption that liquid and gas can be treated as if each were flowing *alone;* the total mass flow is the sum of the mass flows of the individual components, and the total flow area equals the sum of the areas occupied by the individual components. If negligible error is introduced when the gas and liquid flow coefficients are assumed equal, then the total mass flow is

$$(q_M)_{TC} = \frac{N_{M\rho} C_g \beta^2 D^2 F_a Y_1 \sqrt{\rho_{g1}} \sqrt{h_w}}{\sqrt{1 - \beta^4} \, [X + 1.26(1 - X)\sqrt{\rho_{g1}/\rho_l}]}$$

(9.85)

where the subscript TC refers to the two-component flow; g to the gas; and l to the liquid component. The mixture quality X, gas mass flow rate, and liquid mass flow rate are defined by

$$X = \frac{\text{mass of gas}}{\text{mass of mixture}} \qquad (q_M)_g = X(q_M)_{TC} \qquad (q_M)_l = (1 - X)(q_M)_{TC} \quad (9.86)$$

Since the difference between Eq. (9.85) and the mass flow equation (9.60) is the bracketed denominator term in Eq. (9.85), a quality correction factor can be defined as

$$F_X = \frac{1}{X + 1.26(1 - X)\sqrt{\rho_{g1}/\rho_l}} \qquad (9.87)$$

In calculating the total mass flow, the previously presented mass flow equations are multiplied by this correction factor; the discharge coefficient is calculated at the Reynolds number defined by the dry-gas flow rate [in Eq. (9.86)], and the dry-gas density is used in the mass flow equation. The gas expansion factor is calculated on the basis of the differential pressure produced by the dry gas. If the gas and liquid discharge coefficients are assumed to be approximately equal, the equation given by Murdock simplifies to

$$(h_w)_g = \frac{h_w}{\{1 + [1.26(1 - X)/X]\sqrt{\rho_{g1}/\rho_l}\}^2} \qquad (9.88)$$

In the SI system the mass flow equation is

$$(q_M^*)_{TC} = N_{M\rho}^* \frac{C_g\beta^2}{\sqrt{1 - \beta^4}} D^{*2} F_a^* F_X^* Y_1 \sqrt{\rho_{g1}^*} \sqrt{\Delta p^*} \qquad (9.89)$$

where F_X^* is defined by

$$F_X^* = \frac{1}{X + 1.26(1 - X)\sqrt{\rho_{g1}^*/\rho_l^*}} \qquad (9.90)$$

The differential pressure (in kilopascals) of the dry gas is given by

$$(\Delta p^*)_g = \frac{\Delta p^*}{\{1 + [1.26(1 - X)/X]\sqrt{\rho_{g1}^*/\rho_l^*}\}^2} \qquad (9.91)$$

Correction for Drain or Vent Holes

When a wet gas or saturated steam is measured in a horizontal pipe, a pool of liquid may form upstream of the primary element. In liquid flows, small quantities of gas or vapor bubbles tend to collect at the top of the pipe, causing erratic flow measurements. This condition may be alleviated by boring or drilling a small hole nearly flush with the inside diameter of the pipe; the hole is drilled at the bottom (as a drain hole) when condensate is carried in a gas or vapor stream, and at the top (as a vent hole) when a liquid contains entrained gas.

The effect of the hole is to increase the flow by an amount that depends on its area. If the hole diameter is less than one-tenth the bore diameter, the maximum flow through the hole is less than 1 percent of the total flow. The flow equation can be corrected for this added area by introducing a *drain-hole correction factor*. This factor is derived by noting that the total flow is the sum of the main flow and the drain-hole flow, and that the differential pressure is approximately the same for both. The drain-hole (or vent-hole) correction can then be written as

$$F_{DH} = 1 + \sqrt{1 - \beta^2} \frac{C_{DH}}{C} \left(\frac{d_{DH}}{d} \right)^2 \tag{9.92}$$

where C_{DH} is the discharge coefficient of the drain hole, and C is the discharge coefficient of the primary element. If the hole is square-edged, these coefficients are approximately the same, and Eq. (9.92) becomes

$$F_{DH} = 1 + \sqrt{1 - \beta^4} \left(\frac{d_{DH}}{d} \right)^2 \tag{9.93}$$

for orifices, and

$$F_{DH} = 1 + 0.6\sqrt{1 - \beta^4} \left(\frac{d_{DH}}{d} \right)^2 \tag{9.94}$$

for nozzles and venturis.

If the hole is rounded, C_{DH} approaches the discharge coefficient of a nozzle or venturi, and Eq. (9.92) is approximated by

$$F_{DH} = 1 + 1.66\sqrt{1 - \beta^4} \left(\frac{d_{DH}}{d} \right)^2 \tag{9.95}$$

for orifices, and

$$F_{DH} = 1 + \sqrt{1 - \beta^4} \left(\frac{d_{DH}}{d} \right)^2 \tag{9.95}$$

for nozzles and venturis.

When the bore is sized, d is initially unknown, and the drain-hole correction factor must be found by iteration. The hole diameter should be kept as small as possible, and the correction factor limited to a value no greater than 1.02 (and preferably less than 1.01). Table 9.33 gives maximum recommended hole diameters.

In small lines the small drain or vent hole may become plugged with dirt or scale. If this is a possibility, the meter run should be located in a vertical pipe, with upward flow to carry entrained gases in liquid streams through the primary element, and downward flow to purge liquids from gas streams.

Correction for Water Vapor

In the chemical and air-reduction industries (and some others), the volume or mass of the dry gas in a moist gas stream may be of interest. To correct to a dry basis, two separate corrections may be necessary. The first is a correction of the dry-gas specific gravity to that of the wet gas. (This correction is necessary only if the specific gravity was calculated from the molecular weights of the dry components, or if the gas was dried prior to the specific-gravity determination. If the

Table 9.33   Recommended Maximum Drain-Hole (Vent-Hole) Diameters

Primary element bore d, in (mm)	Diameter d_{DH} of vent (drain) hole, in (mm)
1.000–3.500 (25.40–88.90)	$\frac{3}{32}$ (2.38)
3.501–4.125 (88.93–104.78)	$\frac{1}{8}$ (3.18)
4.126–5.000 (104.80–127.00)	$\frac{5}{32}$ (3.96)
5.001–6.000 (127.03–152.40)	$\frac{3}{16}$ (4.76)
6.001–6.750 (152.43–171.45)	$\frac{7}{32}$ (5.56)
6.751–7.500 (171.48–190.50)	$\frac{1}{4}$ (6.35)
7.501–8.375 (190.53–212.73)	$\frac{9}{32}$ (7.14)
8.376–9.250 (212.75–234.95)	$\frac{5}{16}$ (7.94)
9.251–10.000 (234.98–254.00)	$\frac{11}{32}$ (8.73)
10.001–10.875 (254.03–276.23)	$\frac{3}{8}$ (9.53)
10.876–11.625 (276.25–295.28)	$\frac{13}{32}$ (10.32)
11.626–12.500 (295.30–317.50)	$\frac{7}{16}$ (11.11)
12.501–13.250 (317.53–336.55)	$\frac{15}{32}$ (11.91)
> 13.251 (> 336.58)	$\frac{1}{2}$ (12.7)

wet-gas specific gravity was measured, this correction is 1.0.) The second correction is a conversion of the measured wet flow rate to a dry-flow basis, using the water-vapor correction-factor equations developed in Chap. 2. These equations, which are multiplying factors for the flow-rate equations, are presented in Table 9.34.

Correction for Indicated Differential

If the differential-pressure measuring element is located substantially below or above the primary element, its reading will not exactly equal the true differential pressure. Corrections for elevation and for the wet-type manometer are developed in Chap. 3 and are summarized in Table 9.35. The differential pressure h_w in the flow equations is replaced with

$$h_w = F_{EL}F_M\,(h_w)_{\text{ind}} \tag{3.33}$$

for U.S. units, and

$$\Delta p^* = F_{EL}^*F_M^*\,(\Delta p^*)_{\text{ind}} \tag{3.34}$$

for SI units.

Table 9.34   Corrections to Flow Equations for Wet Gases

Use	Correction

SPECIFIC-GRAVITY CORRECTION (MASS AND VOLUMETRIC pvT FLOW-RATE EQUATIONS)

Dry to wet specific-gravity correction	$G = G_{\text{wet}} = \left[1 + \dfrac{p_{wv}}{p_{f1}} \left(\dfrac{0.6220}{G_{\text{dry}}} - 1 \right) \right] G_{\text{dry}}$	(a)
Wet specific gravity measured	$G = G_{\text{wet}} = G_{\text{meas}}$	

WATER-VAPOR CORRECTION FACTOR FOR VOLUMETRIC FLOW-RATE EQUATIONS

Actual wet flow rate at flowing temperature and pressure, at standard, or at a selected base temperature and pressure	$F_{WV,\text{wet}} = 1.0$	(b)
Saturated flow rate at standard, or at a selected base pressure and temperature	$F_{WV,\text{wet}} = \dfrac{1 - p_{wv}/p_{f1}}{1 - p_{wvb}/p_b}$	(c)

$$p_b = 14.696 \text{ for standard conditions}$$

$$p_{wv} = \text{saturated vapor pressure at operating temperature}$$

$$p_{wvb} = \text{saturated vapor pressure at base temperature}$$

Dry flow rate at actual flowing pressure and temperature and at standard or at a selected base temperature and pressure	$F_{WV,\text{dry}} = 1 - \dfrac{p_{wv}}{p_{f1}}$	(d)

MASS FLOW-RATE EQUATIONS

Dry mass flow rate	$F_{WVM,\text{dry}} = 1 - \dfrac{0.6220}{G_{\text{wet}}} \dfrac{Z_{\text{wet}}}{Z_{wv}}$	(e)
	$Z_{\text{wet}} = \left(1 - \dfrac{p_{wv}}{p_{f1}} \right) Z_{\text{dry}} + \dfrac{p_{wv}}{p_{f1}} Z_{wv}$	(f)

Table 9.35 Corrections for Differential-Pressure Measurement

Use	Correction
ELEVATION CORRECTION	

Use	Correction	
Liquid flows and seal fluids for both liquid and gas (vapor) flows (assumes leads at same temperature)	$F_{EL} = 1.0$	(a)
Gas flows†		
U.S. units	$F_{EL} = 1 - 0.01874 \dfrac{F_L G H_{LL}}{Z_{f1} T_{f1}}$	(b)
SI units	$F^*_{EL} = 1 - 0.03416 \dfrac{F_L G H^*_{LL}}{Z_{f1} T_{K1}}$	(c)

MANOMETER FACTOR	

Use	Correction	
Gas and liquid flows for dry-type secondary devices	$F_M = 1.0$	(d)
Gas flow with no seal fluid in leads‡		
U.S. units	$F_M = 1 - \dfrac{2.6988 G p_{f1}}{Z_{f1} T_{f1} \rho_m}$	(e)
SI units	$F^*_M = 1 - \dfrac{3.4834 G p^*_{f1}}{Z_{f1} T_{K1} \rho^*_m}$	(f)
Liquid and gases (vapors) with seal fluid in leads§		
U.S. units	$F_M = 1 - \dfrac{\rho_{f1}}{\rho_m}$	(g)
SI units	$F^*_M = 1 - \dfrac{\rho^*_{f1}}{\rho^*_m}$	(h)

†The sign of H_{LL} is + (plus) when located below primary device, and − (minus) when above.

‡If lead-line temperature is not equal to operating temperature, Z and T are to be calculated at lead-line temperature.

§$\rho_{f1} = \rho_s$ for seal fluid; ρ_m = manometer fluid density.

EXAMPLE 9.7

Size a D and $D/2$ tap orifice to measure wet steam containing 20 percent liquid at 580 psia (4000 kPa) and $T_F = 483°F$ (251°C). The 430-stainless-steel plate is to be installed in a pipe with 2.500-in (63.50-mm) inside diameter. The differential-pressure measuring device has an upper-range value of 500 inH$_2$O (124.3 kPa), and the total mass flow is 10,320 lb$_m$/h (4681 kg/h).

Determine S_M factor

The S_M-factor equation is Eq. (g) of Table 9.23:

$$S_M = \frac{q_M}{N_{M\rho} F_a D^2 F_X \sqrt{\rho_{g1}} \sqrt{h_w}}$$

The quality factor F_X has already been introduced into this equation, and the density has been replaced with the saturated-vapor density.

From Table 9.14, $N_{M\rho} = 358.93$; from Figure 9.5, $F_a = 1.005$; and from Table G.1,

$$\rho_{g1} = 1.254 \text{ lb}_m/\text{ft}^3 \qquad \rho_l = 49.86 \text{ lb}_m/\text{ft}^3$$

The quality correction factor is then, from Eq. (9.87),

$$F_X = \frac{1}{0.8 + (1.26)(1 - 0.8)\sqrt{1.254/49.86}} = 1.1905$$

Substitution into the S_M-factor equation, with the differential pressure and flow rate scaled to normal operating conditions, gives

$$S_M = \frac{8256}{(358.93)(1.005)(2.500)^2(1.1905)\sqrt{1.254}\sqrt{320}}$$

$$= 0.15355$$

Calculate normal operating Reynolds number for the gas phase

From Table G.1, $\mu_{cP} = 0.01760$, and from Eq. (9.86),

$$(q_{PPH})_g = (0.8)(8256) = 6605 \text{ lb}_m/\text{h}$$

Then, from Eq. (g) of Table 9.20 for the gas component (saturated vapor),

$$(R_D)_N = 2266.97 \left[\frac{1}{(0.01760)(2.500)(358.93)} \right] (6605) = 948,104$$

Calculate β_0

From Eq. (j) of Table 9.28 for $R_D > 200,000$,

$$\beta_0 = \left[1 + \left(\frac{0.6}{0.15355} \right)^2 \right]^{-1/4} = 0.49792$$

Size by iteration

The discharge-coefficient equation for D and $D/2$ taps is, from Table 9.1,

$$C = 0.5959 + 0.0312\beta^{2.1} - 0.184\beta^8 + 0.039\frac{\beta^4}{1 - \beta^4} - 0.0158\beta^3 + \frac{91.71\beta^{2.5}}{R_D^{0.75}}$$

The gas-expansion-factor equation is Eq. (c) of Table 9.26:

$$Y_1 = 1 - (0.41 + 0.35\beta^4)\frac{x_1}{k}$$

The real-gas isentropic exponent k is read from Fig. I.16, for a quality of 0.8, as $k = 1.063$.

The differential pressure of the gas component at normal conditions is found with Eq. (9.88):

$$(h_w)_g = \frac{320}{\{1 + [1.26(1 - 0.8)/0.8]\sqrt{1.254/49.86}\}^2} = 290.27 \text{ inH}_2\text{O}$$

Then, from Eq. (h) of Table 9.26,

$$x_1 = \frac{290.27}{(27.73)(580)} = 0.01805$$

and the gas expansion factor is, from Eq. (c) of Table 9.26,

$$Y_1 = 1 - (0.41 + 0.35\beta^4)\left(\frac{0.01805}{1.063}\right)$$

$$= 1 - (0.41 + 0.35\beta^4)(0.01698)$$

Substitution of β_0 into the discharge-coefficient and gas-expansion-factor equations gives

$$C_0 = 0.60355 \qquad (Y_1)_0 = 0.99267$$

The second estimate for β is then, by Eq. (9.83),

$$\beta_1 = \left\{1 + \left[\frac{(0.60355)(0.99267)}{0.15355}\right]^2\right\}^{-1/4} = 0.49826$$

A second iteration gives no change in β, so the orifice bore is

$$d = (0.4983)(2.500) = 1.246 \text{ in}$$

EXAMPLE 9.8

Correcting for differing pipe and plate materials, size a 303 stainless-steel corner-tap orifice in a 3.970-in (100.8-mm) carbon-steel pipe to measure 24,000 flowing gal/h (90.85 m³/h) of water at 100°F (37.8°C) and 1000 psig (6850 kPa). The differential pressure at the upper-range flow rate is to be 200 in (49.73 kPa), and a density measurement is to be made. The water is aerated, and a 0.25-in (6.4-mm) vent hole is to be drilled through the plate.

The S_M-factor equation, Eq. (c) of Table 9.23, can be written as

$$S_M = \left[\frac{F_{DH}F_a C\beta^2}{\sqrt{1 - \beta^4}} \right] = \frac{\sqrt{F_p}\sqrt{\rho_F}\, q_{\text{gph}}}{N_{v\rho}\, D^2 \sqrt{h_w}}$$

where F_{DH} and F_a are β-dependent. The bracketed term contains all beta-dependent functions and can be solved iteratively for β as

$$\beta_n = \left\{ 1 + \left[\frac{(F_{DH})_{n-1}(F_a)_{n-1}C_{n-1}}{S_M} \right]^2 \right\}^{-1/4}$$

where S_M is constant. From Eq. (2.107),

$$F_p = 1 + \frac{p_f}{1000}\, Z_L = 1 + \frac{1000}{1000}\,(0.00262) = 1.00262$$

where $Z_L = 0.00262$ is found in Table E.1.

Now, from Eq. (2.79) or Table E.6, $\rho_F = 61.994$ lb$_m$/ft^3, and from Table 9.16, $N_{v\rho} = 2684.96$. Scaling to normal conditions and substitution give

$$S_M = \frac{\sqrt{1.00262}\,\sqrt{61.994}\,(19,200)}{(2684.96)(3.970)^2\sqrt{128}} = 0.31617$$

The β-dependent terms F_{DH}, F_a, and C are calculated as follows: For a square-edged vent hole, with $d = \beta D$, Eq. (9.93) is written as

$$F_{DH} = 1 + \sqrt{1 - \beta^4}\left(\frac{d_{DH}}{\beta D}\right)^2$$

For differing pipe and orifice plate materials, the thermal-expansion factor is, by Eq. (9.51),

$$F_a = 1 + \frac{2}{1 - \beta^4}\,(\alpha_{PE} - \beta^4\alpha_P)(T_F - 68)$$

where, from Table B.4, $\alpha_{PE} = 0.0000096$ in/(in·°F) and $\alpha_P = 0.0000067$ in/(in·°F). The discharge coefficient for corner taps is, from Table 9.1,

$$C = 0.5959 + 0.0312\beta^{2.1} - 0.184\beta^8 + \frac{91.71\beta^{2.5}}{R_D^{0.75}}$$

The Reynolds number R_D at normal conditions is calculated from Eq. (c) of Table 9.20 as

$$(R_D)_N = 2266.97\left(\frac{F_p\rho_F}{\mu_{cP}\, DN_{vG}}\right)(q_{\text{gph}})_N = 372,600$$

for which Table E.6 gives $\mu_{cP} = 0.6812$, and the substitution $F_p\rho_F = \rho_f$ was used.

With an initial estimate for β from Eq. (j) of Table 9.28,

$$\beta_0 = \left[1 + \left(\frac{0.6}{S_M} \right)^2 \right]^{-1/4} = \left[1 + \left(\frac{0.6}{0.31617} \right)^2 \right]^{-1/4} = 0.68278$$

the iteration may be tabulated as follows:

	n		
	1	2	3
β_{n-1}	0.68278	0.67902	0.67891
$(F_{DH})_{n-1}$	1.00753	1.00763	1.00764
$(F_a)_{n-1}$	1.00067	1.00066	1.00066
C_{n-1}	0.60355	0.60373	0.60374
β_n	0.67902	0.67891	0.67891

The orifice bore is then

$$d = \beta_n D = (0.6789)(3.970) = 2.695 \text{ in}$$

It is important to note that since the drain-hole and thermal-expansion correction factors are now contained in the S_M factor, the flow-rate equation without Reynolds-number correction for a density input is written as

$$q_{\text{gph}} = N_{v\rho} S_M D^2 \frac{\sqrt{h_w}}{\sqrt{F_p}\sqrt{\rho_F}} = (2684.96)(0.31617)(3.970)^2 \frac{\sqrt{h_w}}{\sqrt{\rho_f}}$$

$$= 13{,}379 \sqrt{\frac{h_w}{\rho_f}} = F_{MC} \sqrt{\frac{h_w}{\rho_f}}$$

where the density ρ_f is measured at flowing pressure and temperature.

EXAMPLE 9.9

The natural gas of Example 9.6 is saturated with water vapor. What is the required orifice bore if the flow rate is to be calculated on a dry basis?

The S_M-factor equation, Eq. (f) of Table 9.25, rewritten to include a water-vapor correction, is

$$S_M = \frac{(q^*_{SCMH})_{\text{dry}}}{N^*_{Vhp} F^*_a F^*_{PB} F^*_{TB} F^*_{TF1} F_{pv1} F_{WV,\text{dry}} Z_b F_g D^{*2}} \sqrt{\Delta p^* \, p^*_{f1}}$$

From Table G.1, the water-vapor pressure (saturated) at 184 kPa (26.7 psia) is $p_{wv} = 0.5119$ psia (3.529 kPa). The water-vapor correction factor for conversion to a dry volume is, from Eq. (d) of Table 9.34,

$$F_{WV,\text{dry}} = 1 - \frac{3.529}{184} = 0.98082$$

The specific gravity is based on the *dry* components and is corrected to a *wet* specific gravity with Eq. (a) of Table 9.34:

$$G_{\text{wet}} = \left[1 + \frac{3.529}{184} \left(\frac{0.6220}{0.652} - 1 \right) \right] (0.652) = 0.6514$$

Then, from Eq. (9.64),

$$F_g = \sqrt{\frac{1}{0.6514}} = 1.2390$$

Substitution of values from Example 9.6 as well as $F_{WV,\text{dry}}$ and F_g into the S_M-factor equation gives

$$S_M$$

$$= \frac{272}{(0.011350)(1.0002)(1.0133)(0.99653)(0.98005)(1.0011)(0.98082)(1.2390)(52.50)^2 \sqrt{(6.4)(184)}}$$

$$= 0.21040$$

Now, recalculation of β_0 for the adjusted S_M factor gives

$$\beta_0 = \left[1 + \left(\frac{0.6}{0.21040} + 0.06 \right)^2 \right]^{-1/4} = 0.56993$$

The iteration of Example 9.6, repeated with $S_M = 0.21040$, gives the following:

	n	
	1	2
β_{n-1}	0.56993	0.57532
C_{n-1}	0.60726	0.60739
$(Y_1)_{n-1}$	0.98776	0.98747
β_n	0.57532	0.57534

The orifice bore is then

$$d = (0.5753)(52.50) = 30.20 \text{ mm}$$

EXAMPLE 9.10

A wet-type mercury meter is scaled to 100 in (24.86 kPa) and is to be located 50 ft (15.2 m) below a venturi that is to measure an airflow of 100,000 standard ft³/h (2832 standard m³/h) in a 4-in (100-mm) Schedule-80 pipe. The pressure is 1160 psia (8000 kPa) at a temperature of 80°F (26.7°C). Size the bore if the lead lines are at 60°F (15.6°C). (The venturi is to have a machined inlet.)

The S_M-factor equation [Eq. (l) of Table 9.23], modified for elevation and manometer correction, is

$$S_M = \frac{\sqrt{Z_{f1}}\sqrt{T_{f1}}\sqrt{G}\ q_V}{N_{VpT}F_a Z_b D^2 \sqrt{F_{EL}F_M (h_w)_{ind}P_{f1}}}$$

Data are determined as follows

- From Table G.2: $Z_{f1} = 0.9833$
- From Table 9.16: $N_{VpT} = 7711.02$
- From Table 3.4: $Z_b = 0.99958$
- From Table B.4: $\alpha_{PE} = 0.0000075$ in/(in·°F)
- From Table B.1: $D = 3.826$ in

From Eq. (9.52),

$$F_a = 1 + (2)(0.0000075)(80 - 68) = 1.0002$$

and from Eq. (b) of Table 9.35, assuming $F_L = 1.0$, the elevation correction factor is

$$F_{EL} = 1 - 0.01874 \left[\frac{(1.0)(1.0)(50)}{(0.9833)(459.67 + 60)} \right] = 0.99817$$

Note: The compressibility factor used in this equation is that for the gas at 60°F; it was obtained by the Redlich-Kwong interpolation method presented in Chap. 2.
From Eq. (3.37),

$$\rho_m = \rho_{Hg} = 846.324\,[1 - 0.0000101(60 - 60)] = 846.324\ lb_m/ft^3$$

Then, from Eq. (e) of Table 9.35, the manometer correction is

$$F_M = 1 - \frac{(2.6988)(1.0)(1160)}{(0.9833)(459.67 + 60)(846.324)} = 0.99276$$

The upper-range flow and differential pressure are scaled to normal operating conditions with

$$(q_{SCFH})_N = (0.8)(100,000) = 80,000 \qquad \text{and} \qquad (h_w)_N = (0.64)(100) = 64$$

Substitution into the S_M-factor equation then gives

$$S_M = \frac{\sqrt{0.9833}\sqrt{459.67 + 80}\sqrt{1.0}\ (80,000)}{(7711.02)(1.0002)(0.99958)(3.826)^2\ \sqrt{(0.99817)(0.99276)(64)(1160)}}$$

$$= 0.060207$$

Since $(\mu_{cP})_{air} = 0.021$ from Fig. H.3, the normal operating Reynolds number is, according to Eq. (l) of Table 9.20,

$$(R_D)_N = 3724.20\,\frac{1.0}{(0.9996)} \left[\frac{1}{(0.021)(3.826)(7711)} \right] (80,000)$$

$$= 481,085$$

For a venturi, the discharge coefficent is constant above $R_D = 200{,}000$, and, from Table 9.1, $C = 0.995$.

From Eq. (a) of Table 9.28, for a machined-inlet venturi,

$$\beta_0 = \left[1 + \left(\frac{0.995}{0.060207}\right)^2\right]^{-1/4} = 0.2458$$

From Eq. (h) of Table 9.26,

$$x_1 = \frac{F_{EL}F_M(h_w)_{ind}}{27.73 \, p_{f1}} = \frac{(0.99817)(0.99276)(64)}{(27.73)(1160)} = 0.001972$$

and from Eq. (g) of the same table,

$$\frac{p_{f2}}{p_{f1}} = 1 - 0.001972 = 0.99803$$

The gas expansion factor at normal operating conditions is, by Eq. (a) of Table 9.26,

$$(Y_1)_N = \left\{\frac{[1 - (0.2458)^4]\,[1.5/(1.5-1)](0.99803)^{2/1.5}[1 - (0.99803)^{(1.5-1)/1.5}]}{[1 - (0.2462)^4(0.99803)^{2/1.5}](1 - 0.99803)}\right\}^{1/2}$$

$$= 0.99901$$

where the isentropic exponent $k = 1.5$ is found in Fig. I.4.

The second estimate for beta is then, from Eq. (9.83),

$$\beta_1 = \left\{1 + \left[\frac{(0.995)(0.99901)}{0.060207}\right]^2\right\}^{-1/4} = 0.2459$$

A third estimate gives $\beta_2 = 0.24632$, and the venturi bore is

$$d = (0.2459)(3.826) = 0.9408 \text{ in}$$

FLOW-RATE EQUATIONS

The equations for calculating flow rates may not be exactly the same as the sizing equations. The discharge coefficient, gas expansion factor, pressure, temperature, dimensional measurements, and flow rate may be different from those at design conditions. The number and precision of the corrections made to compensate for these changed variables depends on the accuracy desired. The equations for calculating flow rate are developed in the following sections.

The Basic Flow-Rate Equation

The general form of the flow-rate equation is given by Eq. (9.73), which can be written (by introducing the S_M factor) as

$$q_N = N\left(\frac{CY\beta^2}{\sqrt{1-\beta^4}}\right)_N D^2 f(\rho)\sqrt{h_w} = NS_M D^2 f(\rho)\sqrt{h_w} \tag{9.97}$$

where the subscript N indicates that the discharge coefficient was calculated at the normal operating Reynolds number, and that the gas expansion factor was calculated at normal pressure and differential-pressure values. Because the quantity $NS_M D^2$ is constant, it may conveniently be considered as a *meter coefficient factor* F_{MC}, so that

$$q_N = F_{MC}f(\rho)\sqrt{h_w} \qquad (9.98)$$

If the density function is assumed constant, Eq. (9.98) is further simplified by including that function as an unmeasured variable in the meter coefficient factor; then

$$q_N = F_{MC}\sqrt{h_w} \qquad (9.99)$$

Equation (9.98) or (9.99) may be used only to calculate the true flow rate at the selected normal operating values, since both the discharge coefficient and the gas expansion factor were evaluated at that point. To calculate the *true* flow rate, two correction factors are required. The first is a Reynolds-number factor that corrects for the Reynolds-number influence on the discharge coefficient, and the second is a gas-expansion-factor correction that adjusts the gas expansion factor for the pressure-ratio change. The Reynolds-number correction factor is defined by

$$F_{RD} = \frac{C}{C_N} = \frac{K}{K_N} \qquad (9.100)$$

and the gas-expansion-factor correction by

$$F_Y = \frac{Y}{Y_N} \qquad (9.101)$$

In Eqs. (9.100) and (9.101), the subscript N denotes that C and Y are to be evaluated at the normal operating Reynolds number, differential pressure, and flowing pressure.

The true flow-rate equation can now be written as

$$q = F_{RD}F_Y q_N \qquad (9.102)$$

for the U.S. system of units, and

$$q^* = F_{RD}F_Y q_N^* \qquad (9.103)$$

for the SI system. Presented in Table 9.36 are complete flow equations for U.S. units, and in Table 9.37 those for SI units. In Table 9.38 are flow equations including the gas-expansion-factor correction, for both U.S. and SI units.

Calculating Flow Rate

The Reynolds-number correction factor F_{RD} is dependent on and determined from the actual flow rate and is, therefore, nonlinear. The following methods may be used to determine F_{RD} and the gas-expansion-factor correction F_Y, to calculate the true flow rate based on measured variables.

Table 9.36 Flow-Rate Equations: U.S. Units

	Liquid	Gas (vapor)
MASS FLOW RATE		
Density	$q_M = N_{M\rho} S_M F_a F_{RD} D^2 \sqrt{F_p}\, \sqrt{\rho_F}\, \sqrt{h_w}$ (a)	$q_M = N_{MG} S_M F_a F_{RD} F_Y D^2 \sqrt{\rho_{f1}}\, \sqrt{h_w}$ (g)
Specific gravity	$q_M = N_{MG} S_M F_a F_{RD} D^2 \sqrt{F_p}\, \sqrt{G_F}\, \sqrt{h_w}$ (b)	
pvT equation		$q_M = N_{MPT} S_M F_a F_{RD} F_Y D^2 \sqrt{G}\, \dfrac{\sqrt{h_w p_{f1}}}{\sqrt{Z_{f1}}\, \sqrt{T_{f1}}}$ (h)
VOLUMETRIC FLOW RATE AT FLOWING CONDITIONS		
Density	$q_v = N_{v\rho} S_M F_a F_{RD} D^2 \dfrac{\sqrt{h_w}}{\sqrt{F_p}\, \sqrt{\rho_F}}$ (c)	$q_v = N_{v\rho} S_M F_a F_{RD} F_Y D^2 \dfrac{\sqrt{h_w}}{\sqrt{\rho_{f1}}}$ (i)
Specific gravity	$q_v = N_{vG} S_M F_a F_{RD} D^2 \dfrac{\sqrt{h_w}}{\sqrt{F_p}\, \sqrt{G_F}}$ (d)	
pvT equation		$q_v = N_{v\rho T} S_M F_a F_{RD} F_Y \sqrt{Z_{f1}}\, \sqrt{T_{f1}}\, D^2 \dfrac{\sqrt{h_w}}{\sqrt{G}\, \sqrt{p_{f1}}}$ (j)
VOLUMETRIC FLOW RATE AT BASE CONDITIONS		
Density	$q_V = N_{v\rho} S_M F_a F_{RD} D^2 \sqrt{F_p}\, \sqrt{\rho_F}\, \dfrac{\sqrt{h_w}}{\rho_b}$ (e)	$q_V = N_{v\rho} S_M F_a F_{RD} F_Y D^2 \sqrt{\rho_{f1}}\, \dfrac{\sqrt{h_w}}{\rho_b}$ (k)
Specific gravity	$q_V = N_{VG} S_M F_a F_{RD} D^2 \sqrt{F_p}\, \sqrt{G_F}\, \dfrac{\sqrt{h_w}}{G_b}$ (f)	
pvT equation Standard base		$q_V = N_{v\rho T} S_M F_a F_{RD} F_Y Z_b D^2 \dfrac{\sqrt{h_w p_{f1}}}{\sqrt{Z_{f1}}\, \sqrt{T_{f1}}\, \sqrt{G}}$ (l)
Selected base		$q_{vb} = (N_{v\rho T})_b S_M F_a F_{RD} F_Y Z_b T_b D^2 \dfrac{\sqrt{h_w p_{f1}}}{\sqrt{Z_{f1}}\, \sqrt{T_{f1}}\, \sqrt{G}\, p_b}$ (m)

Table 9.37 Flow–Rate Equations: SI Units

	Liquid	Gas (vapor)
	MASS FLOW RATE	
Density	$q_M^* = N_{Mp}^* S_M F_a^* F_{RD} D^{*2} \sqrt{F_p}\,\sqrt{\rho_F^*}\,\sqrt{\Delta p^*}$ (a)	$q_M^* = N_{Mp}^* S_M F_a^* F_{RD} F_Y D^{*2} \sqrt{\rho_{f1}}\,\sqrt{\Delta p^*}$ (g)
Specific gravity	$q_M^* = N_{MG}^* S_M F_a^* F_{RD} D^{*2} \sqrt{F_p}\,\sqrt{G_F}\,\sqrt{\Delta p^*}$ (b)	
pvT equation		$q_M^* = N_{MpT}^* S_M F_a^* F_{RD} F_Y D^{*2} \sqrt{G}\,\dfrac{\sqrt{\Delta p^*\, p_{f1}^*}}{\sqrt{Z_{f1}}\,\sqrt{T_{K1}}}$ (h)
	VOLUMETRIC FLOW RATE AT FLOWING CONDITIONS	
Density	$q_v^* = N_{vp}^* S_M F_a^* F_{RD} D^{*2}\,\dfrac{\sqrt{\Delta p^*}}{\sqrt{F_p}\,\sqrt{\rho_F^*}}$ (c)	$q_v^* = N_{vp}^* S_M F_a^* F_{RD} F_Y D^{*2}\,\dfrac{\sqrt{\Delta p^*}}{\sqrt{\rho_{f1}^*}}$ (i)
Specific gravity	$q_v^* = N_{vG}^* S_M F_a^* F_{RD} D^{*2}\,\dfrac{\sqrt{\Delta p^*}}{\sqrt{F_p}\,\sqrt{G_F}}$ (d)	
pvT equation		$q_v^* = N_{vpT_m}^* S_m F_a^* F_{RD} F_Y \sqrt{Z_{f1}}\,\sqrt{T_{K1}}\,D^{*2}\,\dfrac{\sqrt{\Delta p^*}}{\sqrt{G}\,\sqrt{p_{f1}^*}}$ (j)

Table 9.37 Flow Rate Equations: SI Units (Continued)

9-94

	Liquid	Gas (vapor)
	VOLUMETRIC FLOW RATE AT BASE CONDITIONS	

Density

$$q_V^* = N_{V\rho}^* S_M F_a^* F_{RD} D^{*2} \sqrt{F_p}\, \sqrt{\rho_F^*}\, \frac{\sqrt{\Delta p^*}}{\rho_b^*} \quad \text{(e)}$$

$$q_V^* = N_{V\rho}^* S_M F_a^* F_{RD} F_Y D^{*2} \frac{\sqrt{\rho_{f1}^*}\,\sqrt{\Delta p^*}}{\rho_b^*} \quad \text{(k)}$$

Specific gravity

$$q_V^* = N_{VG}^* S_M F_a^* F_{RD} D^{*2} \sqrt{F_p}\, \sqrt{G_F}\, \frac{\sqrt{\Delta p^*}}{G_b} \quad \text{(f)}$$

pvT equation

Standard base

$$q_V^* = N_{VpT}^* S_M F_a^* F_{RD} F_Y Z_b D^{*2} \frac{\sqrt{\Delta p^* p_{f1}^*}}{\sqrt{Z_{f1}}\,\sqrt{T_{K1}}\,\sqrt{G}} \quad \text{(l)}$$

Selected base

$$q_{Vb}^* = (N_{VpT}^*)_b S_M F_a^* F_{RD} F_Y Z_b T_{Kb} D^{*2} \frac{\sqrt{\Delta p^* p_{f1}^*}}{\sqrt{Z_{f1}}\,\sqrt{T_{K1}}\,\sqrt{G}\, p_b^*} \quad \text{(m)}$$

Table 9.38 Flow-Rate Equations Using the Gas Factor Equations: U.S. and SI Units

	U.S. units		SI units	
Mass flow rate	$q_M = N_{Mhp} S_M F_a F_{RD} F_Y F_{pv1} F_{TF1} D^2 \dfrac{\sqrt{h_w p_{f1}}}{F_g}$	(a)	$q_M^* = N_{Mhp}^* S_M F_a^* F_{RD} F_Y F_{pv1} F_{TF1}^* D^{*2} \dfrac{\sqrt{\Delta p^* p_{f1}^*}}{F_g}$	(d)
Volumetric flow rate at flowing conditions	$q_v = N_{vhp} S_M F_a F_{RD} F_Y F_g D^2 \dfrac{\sqrt{h_w}}{F_{TF1} F_{pv1} \sqrt{p_{f1}}}$	(b)	$q_v^* = N_{vhp}^* S_M F_a^* F_{RD} F_Y F_g D^{*2} \dfrac{\sqrt{\Delta p^*}}{F_T^* F_{pv1} \sqrt{p_{f1}^*}}$	(e)
Volumetric flow rate at base conditions	$q_{Vb} = N_{Vhp} S_M F_a F_{RD} F_Y F_{PB} F_{TB} F_{TF1} F_{pv1} Z_b F_g D^2 \sqrt{h_w p_{f1}}$	(c)	$q_{Vb}^* = N_{Vhp}^* S_M F_a^* F_{RD} F_Y F_{PB}^* F_{TB}^* F_T^* F_{TF1}^* F_{pv1} Z_b F_g D^{*2} \sqrt{\Delta p^* p_{f1}^*}$	(f)

Graphical Solution The pressure ratio h_w/p_f is calculated from measurements, and the gas expansion factor Y is determined from an appropriate graph (see Fig. 9.9c). The factor F_Y is then determined with Eq. (9.101). For liquids, $F_Y = 1.0$. Under the assumption that $F_{RD} = 1.0$, the flow rate is calculated and the Reynolds number is determined. The discharge coefficient is read from the appropriate discharge coefficient–Reynolds-number curve at this Reynolds number (see Fig. 9.9b). The F_{RD} correction is then calculated with Eq. (9.100), and the flow rate is recalculated. The procedure is repeated until the desired convergence is achieved. Only a single iteration is usually required, unless the discharge coefficient changes significantly with Reynolds number.

Simple Iteration If the discharge coefficient is expressed as an equation (see Table 9.1), then the Reynolds-number correction factor can be written as

$$F_{RD} = \frac{C_\infty}{C_N} + \frac{b}{C_N R_D^n} = \frac{C_\infty}{C_N} + Mq^{-n} \qquad (9.104)$$

where M (called the M *factor*) is defined in terms of the bracketed quantities in Tables 9.20 to 9.22 as

$$M = \frac{b}{C_N}[\ \ ]^{-n} \qquad (9.105)$$

For example, the M factor would be written for a gas mass flow calculation (in U.S. units) using the pvT equation as

$$M = \frac{b}{C_N}\left[3724.200\,\frac{1}{\mu_{cP} D N_{MpT}}\right]^{-n}$$

For gas flows, the expansion factor Y can be calculated from measurements and known beta ratios using the equations given in Table 9.26. This factor is ratioed to the expansion factor at normal conditions to determine the correction factor F_Y, of Eq. (9.101).

To obtain F_Y for the contoured devices (nozzles, venturis, etc.), the form of Eq. (9.30) can be simplified by assuming a linear pressure-ratio relationship defined by

$$(Y_1)_N = 1 - k_1\left(\frac{h_w}{p_{f1}}\right)_N \qquad (9.106)$$

where the constant k_1 is derived from the expansion factor and pressure ratio at normal conditions. The gas-expansion-factor correction is then approximated by

$$F_Y = \frac{1 - k_1 h_w/p_{f1}}{(Y_1)_N} \qquad (9.107)$$

for U.S. units, and

$$F_Y = \frac{1 - k_1 \Delta p^*/p_{f1}^*}{(Y_1)_N} \qquad (9.108)$$

for SI units.

With F_{RD} assumed equal to 1.0 and known gas-expansion-factor correction, the initial estimate of the true flow rate is

$$q_0 = F_Y \, q_N$$

which, when substituted into Eq. (9.104), gives the first estimate for the Reynolds-number correction as

$$(F_{RD})_0 = \frac{C_\infty}{C_N} + M(F_Y \, q_N)_0^{-n} \tag{9.109}$$

The flow rate is then calculated with Eq. (9.102) [or Eq. (9.103) for SI units] as

$$q_1 = (F_{RD})_0 F_Y q_N \tag{9.102}$$

Unless the discharge coefficient changes significantly with Reynolds number, a single estimate yields ± 0.01 to ± 0.05 percent convergence. A second estimate may be computed by substituting the first estimate into Eq. (9.109) and repeating the procedure.

Newton's Method The flow-rate equation can be rearranged for a zero-root solution (see Sec. A.1) by first combining Eqs. (9.102) and (9.104) as

$$q = F_{RD} F_Y q_N = \frac{C_\infty}{C_N} F_Y q_N + M F_Y q_N q^{-n} \tag{9.110}$$

and then subtracting the true flow rate q to obtain the function

$$F = -q + \frac{C_\infty}{C_N} F_Y q_N + M \, F_Y q_N q^{-n} \tag{9.111}$$

The derivative of this function then is

$$F' = -1 - nM F_Y q_N q^{-1-n} \tag{9.112}$$

If the initial estimate of the flow rate q is q_N, then the Reynolds-number correction factor for the first estimate is

$$F_{RD} = 1 + \frac{M(F_Y q_N)^{-n} - (1 - C_\infty/C_N)}{1 + nM(F_Y q_N)^{-n}} \tag{9.113}$$

The flow then is, by Eq. (9.102),

$$q = F_{RD} F_Y q_N \tag{9.102}$$

Further estimates are unnecessary since Eq. (9.113) will estimate F_{RD} to within 0.01 percent over any 4:1 flow-rate range for the equations given in Table 9.1.

EXAMPLE 9.11

For the data of Example 9.3, calculate the flow rate for a measured differential of 10 in by (a) assuming no Reynolds-number correction, (b) correcting by simple interation, and (c) correcting with Newton's method.

a. Assuming no Reynolds-number correction

The flow-rate equation is Eq. (f) from Table 9.36:

$$q_V = N_{VG} S_M F_a F_{RD} D^2 \sqrt{F_p} \sqrt{G_F} \frac{\sqrt{h_w}}{G_b}$$

With no Reynolds-number correction, $F_{RD} = 1.0$, and the flow-rate equation based on the normal operating conditions is

$$(q_{GPM})_N = (5.6665)(0.40368)(1.0002)(1.939)^2 \sqrt{1.0088} \frac{\sqrt{0.7255}}{0.7359} \sqrt{h_w} = 10.00 \sqrt{h_w}$$

and by Eq. (9.99) the flow rate at $h_w = 10$ in is

$$(q_{GPM})_N = F_{MC} \sqrt{h_w} = 10.00 \sqrt{10} = 31.62 \text{ base gal/min}$$

where $F_{MC} = 10.00$ is the meter factor.

b. Correcting for Reynolds number by simple iteration

The flow-rate equation (9.102) becomes, with $F_Y = 1.0$ for liquids,

$$q_{GPM} = F_{RD}(q_{GPM})_N = F_{RD} F_{MC} \sqrt{h_w}$$

where the Reynolds-number factor F_{RD} is defined by Eq. (9.104):

$$F_{RD} = \frac{C_\infty}{C_N} + M q^{-n}$$

From Table 9.1 and Example 9.3, for $\beta = 0.7430$, $C_\infty = 0.6055$, $b = 43.64$, and $n = 0.75$, the discharge coefficient at the normal operating Reynolds number is $C_N = 0.6097$. Substitution into the M-factor equation (9.105) gives

$$M = \frac{b}{C_N} [\ ]^{-n} = \frac{43.64}{0.6097} \left[17,903 \frac{0.7359}{(0.417)(1.939)(5.667)} \right]^{-0.75} = 0.1823$$

where the bracketed term is from Eq. (f) of Table 9.20. The Reynolds-number correction is then

$$F_{RD} = \frac{0.6055}{0.6097} + 0.1823 \, q_{GPM}^{-0.75} = 0.9931 + 0.1823 \, q_{GPM}^{-0.75}$$

For the initial flow-rate estimate, $q_{GPM} = (q_{GPM})_N$ and the Reynolds-number correction becomes, according to Eq. (9.104),

$$F_{RD} = 0.9931 + (0.1823)(31.62)^{-0.75} = 1.0068$$

The second estimate for the flow rate is then, by Eq. (9.102),

$$q_{GPM} = (F_{RD})(q_{GPM})_N = (1.0068)(31.62) = 31.83 \text{ base gal/min}$$

which, when substituted into the Reynolds-number correction-factor equation, gives the second estimate for the Reynolds-number correction factor as

$$F_{RD} = 0.9931 + (0.1823)(31.83)^{-0.75} = 1.0067$$

The flow rate is then

$$q_{GPM} = (1.0067)(31.62) = 31.83 \text{ base gal/min}$$

c. *Correcting with Newton's method*

The Reynolds-number correction factor is given by Eq. (9.113) as

$$F_{RD} = 1 + \frac{M(F_Y q_N)^{-n} - (1 - C_\infty/C_N)}{1 + nM(F_Y q_N)^{-n}}$$

Substituting the values from part b, noting that $F_Y = 1.0$ for liquids, gives

$$F_{RD} = 1 + \frac{0.1823(q_{GPM})_N^{-0.75} - 0.006889}{1 + 0.1367(q_{GPM})_N^{-0.75}}$$

and, with $(q_{GPM})_N = 31.62$,

$$F_{RD} = 1 + \frac{0.1823(31.62)^{-0.75} - 0.006889}{1 + 0.1367(31.62)^{-0.75}} = 1.0067$$

The flow rate is then, by Eq. (9.102),

$$q_{GPM} = F_{RD}(q_{GPM})_N = (1.0067)(31.62) = 31.83 \text{ base gal/min}$$

EXAMPLE 9.12

The internal pipe diameter of Example 9.5 was measured as 5.880 in (149.4 mm), and the bore as 4.289 in (108.9 mm). With other variables remaining constant, set up the calculation procedure for a single iteration to calculate the flow rate for measured differentials of 20, 40, 60, 80, and 100 in (4.97, 9.97, 14.9, 19.9, and 24.9 kPa). Use five significant figures, to show the significance of each correction factor.

Since the pipe and orifice bore are not the same as in the sizing calculation, a recalculation of S_M is required. For $(R_D)_N = 1,475,600$ and $\beta = 4.289/5.880 = 0.72942$, substitution into the discharge-coefficient equation for flange taps given in Table 9.1 yields

$$C_N = 0.60203 \qquad C_\infty = 0.60106 \qquad b = 41.673 \qquad n = 0.75$$

The gas expansion factor for the actual β is then, from Eq. (c) of Table 9.26 and Example 9.5,

$$(Y_1)_N = 1 - 0.008649[0.41 + 0.35(0.72942)^4] = 0.99560$$

and the S_M factor at normal conditions is, by Eq. (9.74),

$$S_M = \frac{(0.60203)(0.99560)(0.72942)^2}{\sqrt{1 - (0.72942)^4}} = 0.37664$$

The flow-rate equation, Eq. (g) of Table 9.36, is

$$q_M = N_{M\rho} S_M F_a F_{RD} F_Y D^2 \sqrt{\rho_{f1}} \sqrt{h_w}$$

Substituting the values given in Example 9.5 provides the flow rate based on normal conditions, for $F_{RD} = 1.0$ and $F_Y = 1.0$:

$$(q_{PPH})_N = (358.93)(0.37664)(1.0075)(5.880)^2 \sqrt{0.39524} \sqrt{h_w}$$

Now, by Eq. (9.99),

$$(q_{\text{PPH}})_N = F_{MC} \sqrt{h_w} = 2960.5 \sqrt{h_w}$$

The M factor is calculated with Eq. (9.105), with the bracketed term from Eq. (g) of Table 9.20:

$$M = \frac{b}{C_N} \left[2267.0 \, \frac{1}{\mu_{\text{cP}} D N_{M\rho}} \right]^{-n} = \frac{41.673}{0.60203} \left[\frac{2267.0}{(0.01747)(5.880)(358.93)} \right]^{-0.75}$$

$$= 3.1525$$

The Reynolds-number correction factor for a single iteration is then, by Eq. (9.109),

$$F_{RD} = \frac{C_\infty}{C_N} + Mq^{-n} = \frac{0.60106}{0.60203} + 3.1525 \, [F_Y(q_{\text{PPH}})_N]^{-0.75}$$

$$= 0.99839 + 3.1525 \, [F_Y(q_{\text{PPH}})_N]^{-0.75}$$

The gas-expansion-factor correction is, from Eq. (9.101),

$$F_Y = \frac{1 - 0.0000688h_w}{0.99560} = 1.00442 - 0.0000691h_w$$

The calculation procedure is then as follows:

- *Step 1:* Calculate $(q_{\text{PPH}})_N = 2960.5 \sqrt{h_w}$.
- *Step 2:* Calculate $F_Y = 1.00442 - 0.0000691h_w$.
- *Step 3:* Calculate $F_Y(q_{\text{PPH}})_N$.
- *Step 4:* Calculate $F_{RD} = 0.99839 + 3.1525[F_Y(q_{\text{PPH}})_N]^{-0.75}$
- *Step 5:* Calculate $q_{\text{PPH}} = F_{RD}F_Y(q_{\text{PPH}})_N$.

The calculation may be tabulated as follows:

h_w	$(q_{\text{PPH}})_N$	F_Y	$F_Y(q_{\text{PPH}})_N$	F_{RD}	q_{PPH}
20	13,240	1.00304	13,280	1.00091	13,292
40	18,724	1.00166	18,755	1.00034	18,761
60	22,932	1.00027	22,938	1.00007	22,940
80	26,480	0.99889	26,450	0.999897	26,448
100	29,605	0.99751	29,531	0.999777	29,525

EXAMPLE 9.13

Assuming normal design conditions apply to the nozzle of Example 9.4, determine, by Newton's method, the actual flow rate for measured differentials of 10, 40, 64, 80, and 100 in (2.49, 9.97, 15.9, 19.9, and 24.9 kPa).

The flow-rate equation, Eq. (g) of Table 9.36, is

$$q_M = N_{M\rho} S_M F_a F_{RD} F_Y D^2 \sqrt{\rho_{f1}} \sqrt{h_w}$$

From Examples 9.2 and 9.4,

$$N_{M_\rho} = 358.93 \qquad S_M = 0.39759 \qquad F_a = 1.0075$$
$$D = 5.761 \qquad \rho_{f1} = 0.39524 \qquad (Y_1)_N = 0.99210$$

The flow rate based on normal conditions is then

$$(q_{\text{PPH}})_N = (358.93)(0.39759)(1.0075)(1.0)(1.0)(5.761)^2 \sqrt{0.39524} \sqrt{h_w}$$

and, by Eq. (9.99),

$$(q_{\text{PPH}})_N = F_{MC}\sqrt{h_w} = 3000.0 \sqrt{h_w}$$

If the gas expansion factor is assumed to be linear for the nozzle, then by Eq. (9.106),

$$(Y_1)_N = 1 - k_1 \left(\frac{h_w}{p_{f1}}\right)_N = 0.99210 = 1 - k_1 \frac{64}{204}$$

$$Y_1 = 1 - 0.02518 \frac{h_w}{p_{f1}}$$

The gas-expansion-factor correction is then, by Eq. (9.101),

$$F_Y = \frac{Y}{Y_N} = \frac{1 - 0.02518 \, (h_w/p_{f1})}{0.99210}$$

and

$$F_Y = 1.00796 - 0.02538 \frac{h_w}{p_{f1}}$$

The Reynolds-number correction factor is found by Newton's method from Eq. (9.113):

$$F_{RD} = 1 + \frac{M[F_Y(q_{\text{PPH}})_N]^{-n} - (1 - C_\infty/C_N)}{1 + nM[F_Y(q_{\text{PPH}})_N]^{-n}}$$

From Example 9.4, $C_\infty = 0.9975$, $C_N = 0.99334$, and $n = 0.5$. In addition,

$$b = -6.53\beta^{0.5} = -6.53(0.61167)^{0.5} = -5.1071$$

The M factor is calculated with Eq. (9.105):

$$M = \frac{b}{C_N}\left[2267.0 \frac{1}{\mu_{cP}DN_{M_\rho}}\right]^{-n} = \frac{-5.1071}{0.99334}\left[\frac{2267.0}{(0.01747)(5.761)(358.93)}\right]^{-0.5}$$

$$= -0.64901$$

where the bracketed term is from Eq. (g) of Table 9.20. Substitution then yields

$$F_{RD} = 1 + \frac{(-0.64901)[F_Y(q_{\text{PPH}})_N]^{-0.5} + 0.004188}{1 - 0.32451[F_Y(q_{\text{PPH}})_N]^{-0.5}}$$

and the actual flow rate is, by Eq. (9.102),

$$q_{\text{PPH}} = F_{RD}F_Y(q_{\text{PPH}})_N$$

The results of the calculation are as follows:

h_w	$(q_{PPH})_N$	F_Y	F_{RD}	q_{PPH}
10	9,486.8	1.00672	0.99754	9,527.0
40	18,974	1.00298	0.99948	19,021
64	24,000	1.00000	1.00000	24,000
80	26,833	0.99801	1.00022	26,786
100	30,000	0.99552	1.00043	29,878

ASME-ISO Flow-Rate Equations

ASME and ISO literature presents the flow-rate equations in a form different from that of Tables 9.36 to 9.38. The form of the ASME-ISO equations was developed earlier in this chapter in Eqs. (9.60) and (9.61). In these equations the flow coefficient K [Eq. (9.47)] replaces S_M and F_{RD}, the gas expansion factor Y replaces F_Y, and the primary-element bore d replaces the pipe diameter D. As an example, the pvT gas flow equation from Table 9.36 is

$$q_V = N_{V_pT} S_M F_a F_{RD} F_Y Z_b D^2 \frac{\sqrt{h_w p_{f1}}}{\sqrt{Z_{f1}} \sqrt{T_{f1}} \sqrt{G}} \tag{9.114}$$

Substitution of variables gives the ASME-ISO equation as

$$q_V = N_{V_pT} K F_a Y_1 Z_b d^2 \frac{\sqrt{h_w p_{f1}}}{\sqrt{Z_{f1}} \sqrt{T_{f1}} \sqrt{G}} \tag{9.115}$$

If the M factor is redefined in terms of the bracketed quantities in Tables 9.20 to 9.22 as

$$M_K = b[\ \ ]^{-n} \tag{9.116}$$

then the flow coefficient, expressed in terms of the flow rate, is

$$K = \frac{1}{\sqrt{1 - \beta^4}} (C_\infty + M_K q^{-n}) \tag{9.117}$$

When this is substituted into Eq. (9.115), that equation is seen to be nonlinear, and, therefore, it requires iterative solution.

Substituting Y_1 for F_Y in Eq. (9.102) gives the first estimate of the flow rate as

$$q_0 = Y_1 q_N = Y_1 F_{MC} \sqrt{h_w} \tag{9.118}$$

where Y_1 is calculated from the measured differential pressure and line static pressure, and F_{MC} is calculated with the flow coefficient at normal conditions.

Successive estimates are then substituted into Eq. (9.117) to obtain the flow coefficient.

Presented in Table 9.39 are the ASME-ISO flow-rate equations for U.S. units; in Table 9.40 are the equations for SI units, and in Table 9.41 the gas-factor equations for both U.S. and SI units. The correction factors F_X, F_{DH}, F_{EL}, and F_M developed in the section on special considerations in this chapter must be included in these equations when necessary.

EXAMPLE 9.14

Set up the calculation procedure and then solve Example 9.12 with a single iteration using the ASME-ISO equation form.

From Table 9.39, the ASME-ISO mass flow-rate equation is

$$q_M = q_{PPH} = N_{M_\rho} K F_a Y_1 d^2 \sqrt{\rho_{f1}} \sqrt{h_w}$$

Data and equations from Example 9.12 are as follows:

$D = 5.880$ in $\qquad\qquad d = 4.289$ in $\qquad\qquad C_N = 0.60203$

$C_\infty = 0.60106$ $\qquad\qquad b = 41.673$ $\qquad\qquad n = 0.75$

$\beta = 0.72942$ $\qquad\qquad N_{M_\rho} = 358.93$ $\qquad\qquad F_a = 1.0075$

$Y_1 = 1 - 0.0000688 h_w$ $\qquad \rho_{f1} = 0.39524 \ \text{lb}_\text{m}/\text{ft}^3$ $\qquad \mu_{cP} = 0.01747$

Substitution gives

$$q_{PPH} = (358.93)(K)(1.00075)(Y_1)(4.289)^2 \sqrt{0.39524} \ \sqrt{h_w} = 4182.1 K Y_1 \ \sqrt{h_w}$$

By Eq. (9.116), with the bracketed term from Eq. (g) of Table 9.20, the M_K factor is

$$M_K = b \left[2267.0 \frac{1}{\mu_{cP} D N_{M_\rho}} \right]^{-n} = 41.673 \left[\frac{2267.0}{(0.01747)(5.880)(358.93)} \right]^{-0.75} = 1.898$$

The flow coefficient K is then calculated with Eq. (9.117):

$$K = \frac{1}{\sqrt{1 - \beta^4}} (C_\infty + M_K q^{-n}) = \frac{1}{\sqrt{1 - (0.72942)^4}} (0.60106 + 1.898 q^{-0.75})$$

$$= 0.70988 + 2.242 q^{-0.75}$$

At normal conditions the flow rate is 23,620 lb_m/h (80 percent URV flow); then

$$K_N = 0.70988 + (2.24)(23,526)^{-0.75} = 0.71106$$

The initial estimate of the flow rate is

$$(q_{PPH})_N = 4154.1 \ K_N Y_1 \sqrt{h_w} = (4182.1)(0.71106)(Y_1) \sqrt{h_w} = 2973.7 Y_1 \sqrt{h_w}$$

and the calculation procedure is as follows:

1. Calculate $Y_1 = 1 - 0.0000688 h_w$.
2. Calculate $(q_{PPH})_N = 2973.7 \ Y_1 \sqrt{h_w}$.
3. Calculate $K = 0.70988 + 2.246 \ (q_{PPH})_N^{-0.75}$.
4. Calculate $q = 4182.1 K Y_1 \sqrt{h_w}$. (*Example 9.14 continues on page 9-109*)

Table 9.39 ASME-ISO Flow-Rate Equations: U.S. Units

	Liquid		Gas (vapor)	
MASS FLOW RATE				
Density	$q_M = N_{M\rho}KF_a d^2 \sqrt{F_p}\,\sqrt{\rho_F}\,\sqrt{h_w}$	(a)	$q_M = N_{M\rho}KY_1 F_a d^2 \sqrt{\rho_{f1}}\,\sqrt{h_w}$	(g)
Specific gravity	$q_M = N_{MG}KF_a d^2 \sqrt{F_p}\,\sqrt{G_F}\,\sqrt{h_w}$	(b)		
pvT equation			$q_M = N_{M\rho T}KY_1 F_a d^2 \sqrt{G}\,\dfrac{\sqrt{h_w p_{f1}}}{\sqrt{Z_{f1}}\,\sqrt{T_{f1}}}$	(h)
VOLUMETRIC FLOW RATE AT FLOWING CONDITIONS				
Density	$q_v = N_{v\rho}KF_a d^2 \dfrac{\sqrt{h_w}}{\sqrt{F_p}\,\sqrt{\rho_F}}$	(c)	$q_v = N_{v\rho}KY_1 F_a d^2 \sqrt{\dfrac{h_w}{\rho_{f1}}}$	(i)
Specific gravity	$q_v = N_{vG}KF_a d^2 \dfrac{\sqrt{h_w}}{\sqrt{F_p}\,\sqrt{G_F}}$	(d)		
pvT equation			$q_v = N_{v\rho T}KY_1 F_a d^2 \sqrt{Z_{f1}}\,\sqrt{T_{f1}}\,\dfrac{\sqrt{h_w}}{\sqrt{G}\,\sqrt{p_{f1}}}$	(j)

VOLUMETRIC FLOW RATE AT BASE CONDITIONS

Density

$$q_V = N_{V\rho}KF_a d^2 \sqrt{F_p}\,\sqrt{\rho_F}\,\frac{\sqrt{h_w}}{\rho_b} \qquad \text{(e)}$$

$$q_V = N_{V\rho}KY_1 F_a d^2 \frac{\sqrt{\rho_{f1}}\,\sqrt{h_w}}{\rho_b} \qquad \text{(k)}$$

Specific gravity

$$q_V = N_{VG}KF_a d^2 \sqrt{F_p}\,\sqrt{G_F}\,\frac{\sqrt{h_w}}{G_b} \qquad \text{(f)}$$

pvT equation

Standard base

$$q_V = N_{V\rho T}KY_1 F_a d^2 Z_b \frac{\sqrt{h_w p_{f1}}}{\sqrt{Z_{f1}}\,\sqrt{T_{f1}}\,\sqrt{G}} \qquad \text{(l)}$$

Selected base

$$q_{Vb} = (N_{V\rho T})_b KY_1 F_a d^2 Z_b T_b \frac{\sqrt{h_w p_{f1}}}{\sqrt{Z_{f1}}\,\sqrt{T_{f1}}\,\sqrt{G}\,p_b} \qquad \text{(m)}$$

Table 9.40 ASME-ISO Flow-Rate Equations: SI Units

	Liquid		Gas (vapor)	

MASS FLOW RATE

	Liquid		Gas (vapor)	
Density	$q_M^* = N_{M\rho}^* K F_a^{*2} d^{*2} \sqrt{F_p}\,\sqrt{\rho_F^*}\,\sqrt{\Delta p^*}$	(a)	$q_M^* = N_{M\rho}^* K Y_1 F_a^{*2} d^{*2} \sqrt{\rho_{f1}^*}\,\sqrt{\Delta p^*}$	(g)
Specific gravity	$q_M^* = N_{MG}^* K F_a^{*2} d^{*2} \sqrt{F_p}\,\sqrt{G_F}\,\sqrt{\Delta p^*}$	(b)		
pvT equation			$q_M^* = N_{M\rho T}^* K Y_1 F_a^{*2} d^{*2} \sqrt{G}\,\dfrac{\sqrt{\Delta p^*}\,p_{f1}^*}{\sqrt{Z_{f1}}\,\sqrt{T_{K1}}}$	(h)

VOLUMETRIC FLOW RATE AT FLOWING CONDITIONS

	Liquid		Gas (vapor)	
Density	$q_v^* = N_{v\rho}^* K F_a^{*2} d^{*2} \dfrac{\sqrt{\Delta p^*}}{\sqrt{F_p}\,\sqrt{\rho_F^*}}$	(c)	$q_v^* = N_{v\rho}^* K Y_1 F_a^{*2} d^{*2} \dfrac{\sqrt{\Delta p^*}}{\sqrt{\rho_{f1}^*}}$	(i)
Specific gravity	$q_v^* = N_{vG}^* K F_a^{*2} d^{*2} \dfrac{\sqrt{\Delta p^*}}{\sqrt{F_p}\,\sqrt{G_F}}$	(d)		
pvT equation			$q_v^* = N_{v\rho T}^* K Y_1 F_a^{*2} d^{*2} \sqrt{Z_{f1}}\,\sqrt{T_{K1}}\,\dfrac{\sqrt{\Delta p^*}}{\sqrt{G}\,\sqrt{p_{f1}^*}}$	(j)

VOLUMETRIC FLOW RATE AT BASE CONDITIONS

Density

$$q_V^* = N_{V_\rho}^* K F_a^* d^{*2} \sqrt{F_p}\, \sqrt{\rho_F^*}\, \frac{\sqrt{\Delta p^*}}{\rho_b^*} \qquad (e)$$

$$q_V^* = N_{V_\rho}^* K Y_1 F_a^* d^{*2}\, \frac{\sqrt{\rho_{f_1}^*}\,\sqrt{\Delta p^*}}{\rho_b^*} \qquad (k)$$

Specific gravity

$$q_V^* = N_{V_G}^* K F_a^* d^{*2} \sqrt{F_p}\, \sqrt{G_F}\, \frac{\sqrt{\Delta p^*}}{G_b} \qquad (f)$$

pvT equation

Standard base

$$q_V^* = N_{V_{pT}}^* K Y_1 F_a^* d^{*2} Z_b\, \frac{\sqrt{\Delta p^*}\,\sqrt{p_{f_1}^*}}{\sqrt{Z_{f1}}\,\sqrt{T_{K1}}\,\sqrt{G}} \qquad (l)$$

Selected base

$$q_{Vb}^* = (N_{VpT}^*)_b K Y_1 F_a^* d^{*2} Z_b T_{Kb}\, \frac{\sqrt{\Delta p^*}\,\sqrt{p_{f_1}^*}}{\sqrt{Z_{f1}}\,\sqrt{T_{K1}}\,\sqrt{G}\, p_b^*} \qquad (m)$$

Table 9.41 ASME-ISO Flow-Rate Equation Using Factor Equations (F_{PB}, F_{TB}, F_{TF}, F_{pv}): U.S. and SI Units

	U.S. units		SI units	
Mass flow rate	$q_M = N_{Mhp}KF_a d^2 Y_1 F_{pv1} F_{TF1} \dfrac{\sqrt{h_w p_{f1}}}{F_g}$	(a)	$q_M^* = N_{Mhp}^* K F_a^{*2} d^{*2} Y_1 F_{pv1}^* F_{TF1}^* \dfrac{\sqrt{\Delta p^* \, p_{f1}^*}}{F_g}$	(d)
Volumetric flow rate at flowing conditions	$q_v = N_{vhp}KF_a d^2 Y_1 F_g \dfrac{\sqrt{h_w}}{F_{TF1} F_{pv1} \sqrt{p_{f1}}}$	(b)	$q_v^* = N_{vhp}^* K F_a^* d^{*2} Y_1 F_g \dfrac{\sqrt{\Delta p^*}}{F_{TF1}^* F_{pv1} \sqrt{p_{f1}^*}}$	(e)
Volumetric flow rate at base conditions	$q_{Vb} = N_{vhp}KF_a d^2 Y_1 F_{PB} F_{TB} F_{TF1} F_{pv1} Z_b F_g \sqrt{h_w p_{f1}}$	(c)	$q_{Vb}^* = N_{Vhp}^* K F_a^* d^{*2} Y_1 F_{PB}^* F_{TB}^* F_{TF1}^* F_{pv1} Z_b F_g \sqrt{\Delta p^* \, p_{f1}^*}$	(f)

The calculation may be tabulated as follows:

h_w	Y_1	$(q_{PPH})_N$	K	q
20	0.99862	13,280	0.71169	13,292
40	0.99725	18,756	0.71128	18,761
60	0.99587	22,939	0.71108	22,940
80	0.99450	26,452	0.71096	26,448
100	0.99312	26,532	0.71087	29,525

ACCURACY

In Chap. 4, accuracy and bias-error equations were developed using sensitivity coefficients, and these can be extended to include each variable in the selected flow-rate equation. Since the accuracy that is finally achieved depends on the selection of the variables to be measured, the decision as to the degree of accuracy can be made only by the designer, who knows the process. Equations can only assist the designer in estimating which variables may be considered negligible and which must be measured. It is obvious that the best accuracy can be achieved only if all variables are measured and accounted for, and that the recommendations of Chap. 8 should be followed.

In Table 9.42 are the recommended coefficient accuracy and restrictions on Reynolds number and line size for the discharge-coefficient equations of Table 9.1; in Table 9.43 are accuracy values for the gas expansion factor; and in Table 9.44 are the sensitivity-coefficient equations. The sensitivity coefficients for both bore and pipe diameters are presented in Fig. 9.10.

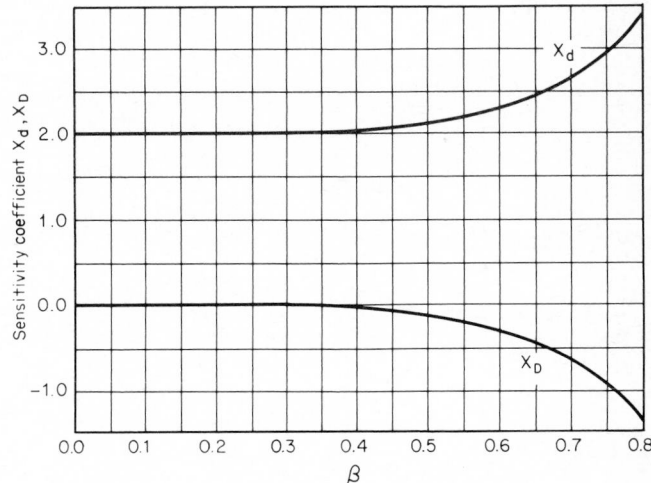

Figure 9.10 Bore and pipe-diameter sensitivity coefficient.

EXAMPLE 9.15

Given the accuracy of each variable (as tabulated below), estimate the flow-rate accuracy for the gasoline flow of Example 9.11 if the measured differential pressure is 72 in (17.9 kPa).

From Eq. (4.27), the accuracy is calculated as

$$(\text{Acc})_{\text{GPM}} = \pm\{[X_{hw}(\text{Acc})_{hw}]^2 + [X_D(\text{Acc})_D]^2 + [X_d(\text{Acc})_d]^2$$
$$+ [X_{Gf}(\text{Acc})_{Gf}]^2 + [X_{Gb}(\text{Acc})_{Gb}]^2 + [X_C(\text{Acc})_C]^2\}^{1/2}$$

The calculation may be tabulated as follows:

Property	Acc, %	Sensitivity coefficient X (Table 9.44)	$[X(\text{Acc})]^2$
Differential pressure h_w			
Transducer, $\pm 0.2\%$ URV	$(\text{Acc})_{hw} = \pm\dfrac{100}{72}\,0.2 = \pm 0.278$	$X_{hw} = \frac{1}{2}$	0.0193
Recorder, $\pm 0.5\%$ URV	$(\text{Acc})_{hw} = \dfrac{100}{72}\,0.5 = \pm 0.694$	$X_{hw} = \frac{1}{2}$	0.1204
Pipe diameter D			
1.939 ± 0.005 in	$(\text{Acc})_D = \pm\dfrac{0.005}{1.939}\,100 = \pm 0.258$	$X_D = -\dfrac{2\beta^4}{1-\beta^4} = -0.88$	0.0515
Orifice bore d			
1.441 ± 0.001 in	$(\text{Acc})_d = \pm\dfrac{0.001}{1.441}\,100 = \pm 0.0694$	$X_d = \dfrac{2}{1-\beta^4} = 2.88$	0.0399
Flowing specific gravity $G_f = F_p\,G_F$			
G_F	$(\text{Acc})_{Gf} = \pm 0.1$	$X_{Gf} = \frac{1}{2}$	0.0025
F_p	$(\text{Acc})_{Gf} = \pm 0.2$	$X_{Gf} = \frac{1}{2}$	0.0100
Base specific gravity G_b	$(\text{Acc})_{Gb} = \pm 0.1$	$X_{Gb} = -1$	0.0100
Discharge coefficient C (Table 9.42)	$(\text{Acc})_C = \pm\beta = \pm 0.743$	$X_C = 1$	$\dfrac{0.5521}{0.8057}$

Then

$$(\text{Acc})_{\text{GPM}} = \pm\{\Sigma[X(\text{Acc})]^2\}^{1/2} = \pm(0.8057)^{1/2} = \pm 0.9 \text{ percent}$$

EXAMPLE 9.16

Estimate the flow-rate accuracy for Example 9.12 if the measured differential pressure is 100 in (24.9 kPa).

By Eq. (4.27), the accuracy equation is written

$$(Acc)_{PPH} = \pm \{[X_{h_w}(Acc)_{h_w}]^2 + [X_d(Acc)_d]^2 + [X_D(Acc)_D]^2$$
$$+ [X_C(Acc)_C]^2 + [X_\rho(Acc)_\rho]^2 + [X_Y(Acc)_Y]^2\}^{1/2}$$

From Example 9.12, $D = 5.880$ in, $d = 4.289$ in, and $\beta = 0.7294$. Then the calculation is tabulated as follows:

Property	Acc	X (Table 9.44)	$[X(Acc)]^2$
h_w	$\pm 0.5\%$	$\frac{1}{2}$	0.0625
d	± 0.002 in; $\pm 0.047\%$	$\frac{2}{1 - \beta^4} = 2.79$	0.0172
D	± 0.004 in; $\pm 0.068\%$	$-\frac{2\beta^4}{1 - \beta^4} = -0.79$	0.0029
C	$\pm \beta = \pm 0.729\%$ (Table 9.42)	1.0	0.5314
Y	$\pm 0.144 \frac{h_w}{p_f} = \pm 0.071\%$ (Table 9.43)	1.0	0.0050
ρ_f	$\pm 0.2\%$	$\frac{1}{2}$	0.0100
			0.6290

The flow-rate accuracy is then

$$(Acc)_{PPH} = \pm \{\Sigma[X(Acc)]^2\}^{1/2} = \pm 0.8 \text{ percent}$$

Table 9.42 Recommended Accuracy and Restrictions for Equations of Table 9.1†

Primary device	Nominal pipe diameter D, in (mm)	Beta ratio β	Pipe Reynolds-number R_D range	Coefficient accuracy, %
Venturi				
Machined inlet	2–10 (50–250)	0.4–0.75	2×10^5 to 10^6	± 1
Rough cast	4–32 (100–800)	0.3–0.75	2×10^5 to 10^6	± 0.7
Rough-welded sheet-iron inlet	8–48 (200–1500)	0.4–0.7	2×10^5 to 10^6	± 1.5
Universal Venturi Tube‡	≥ 3 (≥ 75)	0.2–0.75	$> 7.5 \times 10^4$	± 0.5
Lo-Loss‡	3–120 (75–3000)	0.35–0.85	1.25×10^5 to 3.5×10^6	± 1
Nozzle				
ASME	2–16 (50–400)	0.25–0.75	10^4 to 10^7	± 2.0
ISA	2–20 (50–500)	0.3–0.6	10^5 to 10^6	± 0.8
		0.6–0.75	2×10^5 to 10^7	$2\beta - 0.4$
Venturi nozzle	3–20 (75–500)	0.3–0.75	2×10^5 to 2×10^6	$\pm 1.2 \pm 1.5\beta^4$

	Pipe size, in (mm)	β	R_D	Tolerance, %
Orifice				
Corner, flange, D and $D/2$	2–36 (50–900)¶	0.2–0.6	10^4 to 10^7	±0.6
		0.6–0.75	10^4 to 10^7	±β
		0.2–0.75	2×10^3 to 10^4	±0.6 ± β
$2\frac{1}{2}D$ and $8D$ (Pipetaps)	2–36 (50–900)	0.2–0.5	10^4 to 10^7	±0.8
		0.51–0.7		±1.6
Eccentric§				
Flange and vena contracta	4 (100)	0.3–0.75	10^4 to 10^6	±2
	6–14 (150–350)	0.3–0.75	10^4 to 10^6	±1.5
Segmental§				
Flange and vena contracta	4–14 (150–350)	0.35–0.75	10^4 to 10^6	±2
Quadrant-edged§				
Flange and corner	1–4 (25–100)	0.25–0.6	250 to 6×10^4	±2–±2.5
Conical entrance§				
Corner		0.1–0.3	25 to 2×10^4	±2–±2.5

†ISO 5167 (1980) and ASME *Fluid Meters* (1971) show slightly different values for some devices.

‡The manufacturer should be consulted for recommendations.

§Curves of discharge coefficient versus Reynolds number appear in Chap. 10.

¶For $\frac{1}{2} \leq D \leq 1\frac{1}{2}$ in ($12 \leq D^* \leq 40$ mm) use flow coefficient Eq. (10.1) or Eq. (10.2) given in Chap. 10, with $C = \sqrt{1 - \beta^4} \, K$.

Table 9.43 **Accuracy of Gas-Expansion-Factor Equations in Table 9.26**

Device	Accuracy, %	
	U.S. units	SI units
Contoured		
Venturi, venturi nozzle, Universal Venturi Tube,† Lo-Loss tube†	$\pm(0.144 + 3.61\beta^8)\dfrac{h_w}{p_{f1}}$	$\pm(4 + 100\beta^8)\dfrac{\Delta p^*}{p^*_{f1}}$
Nozzle (ASME, ISA)	$\pm 0.072\dfrac{h_w}{p_{f1}}$	$\pm 2\dfrac{\Delta p^*}{p^*_{f1}}$
Orifice		
Corner, flange, D and $D/2$	$\pm 0.144\dfrac{h_w}{p_{f1}}$	$\pm 4\dfrac{\Delta p^*}{p^*_{f1}}$
$2\tfrac{1}{2}D$ and $8D$, eccentric, segmental‡	$\pm 0.3\dfrac{h_w}{p_{f1}}$	$\pm 8\dfrac{\Delta p^*}{p^*_{f1}}$

†Manufacturers should be consulted for recommendations.

‡Gas-expansion-factor curves are given in Chap. 10.

SOURCE: ISO 5167 (1980).

Wide-Range Metering

The square-root relationship between flow rate and differential pressure limits the practical flow-rate range to 4:1. However, primary elements are capable of operating over wider ranges; they are limited on the high side by permissible pressure loss and structural considerations (plate bending), and on the low side by minimum differential and low-Reynolds-number effects on coefficients. Typically, an orifice which develops 100 in (25 kPa) at 100 percent flow develops only 1 in at 10 percent of the upper-range flow rate. Accurate reading of flow variations at the 10 percent level is impractical on the normal meter scale, which is calibrated uniformly in differential pressure.

The square root of the differential pressure may be taken to obtain a linear response in terms of flow. This increases the readability significantly at flows of less than 50 percent of the upper-range value. Accuracy is, however, not improved, since the same error in differential pressure causes the same error in flow determination.

Several approaches to wide-range flow measurement are available. Where major variations are infrequent and predictable, as with seasonal loads, a simple procedure is to change orifice plates, using a size to match the anticipated upper-range flow rates. Orifice fittings that permit orifice-plate changes without interrupting the flow are available (Fig. 9.11).

For flows that vary frequently and unpredictably, two or more parallel orifice

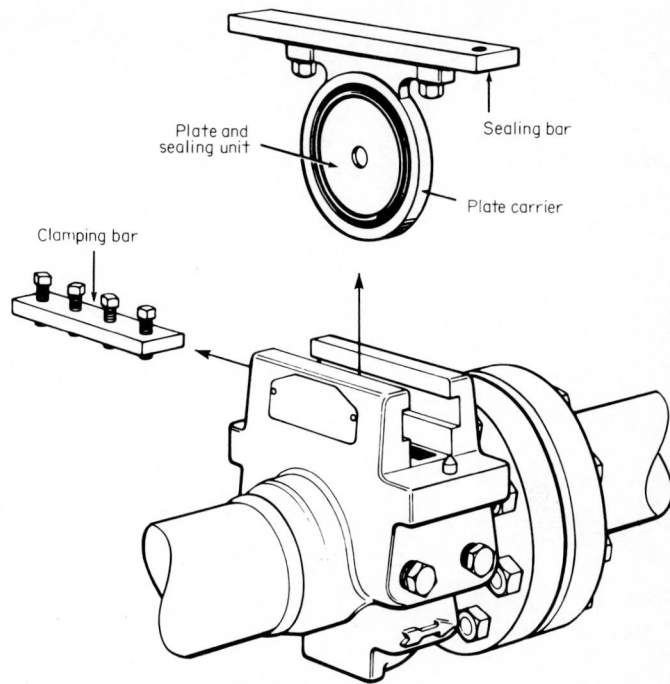

Figure 9.11 Orifice fitting *(courtesy Daniel Industries, Inc.)*. **To change the orifice plate, only five steps are required.**

meter runs are commonly used. Meter runs are sequenced in and out by valves, so that the operating meter or meters are kept in the upper portion of their ranges. A number of versions of this approach are employed, including some that can shift from manual to completely automated operation using computers for sequencing.

An alternative procedure is to connect two differential-pressure transmitters across a single orifice. A typical combination consists of 180- and 20-in (45- and 5-kPa) URV transmitters. For flows about 33 percent of the URV, the measurement is made with the 180-in (45-kPa) transmitter, while the 20-in (5-kPa) transmitter remains inactive. For flows between 10 and 33 percent [differential pressure between 1.8 and 20 in (0.45 and 5 kPa)], the 20-in (5-kPa) transmitter operates. A two-pen recorder can be used to continuously record the output of both transmitters. More complex systems with pneumatic, electronic, and analog computers or with microprocessors are available (see Chap. 11); these have a single output signal for indicating, recording, controlling, and totalizing.

Sensitivity coefficients and Reynolds-number correction factors F_{RD} are used to estimate accuracy over the metering range. Example 9.18 is a sample calculation, with the results shown in Fig. 9.12. In this particular case the low-Reynolds-number influence on the coefficient can result in significant errors (2.7 percent). An analysis of this type is suggested for all wide metering applications.

Table 9.44 Sensitivity Coefficients

	Symbol		Coefficient sign			Sensitivity coefficient	Reference
			Volume		Mass		
	U.S.	SI	Flowing	Base			
Discharge coefficient†	C	C	+	+	+	$X_C = +1.0$	Tables 9.1, 9.42
Reynolds-number influence Venturi (all) Nozzle						$X_{RD} = 1.0$	
ASME long radius						$X_{RD} = +\dfrac{327\beta^{0.5}}{CR_D^{0.5}}$	
ISA						$X_{RD} = \dfrac{-1964 - 10{,}276\beta + 22{,}746\beta^{4.7}}{CR_D^{1.15}}$	
Orifice (all taps)						$X_{RD} = -\dfrac{69\beta^{2.5}}{CR_D^{0.75}}$	
Gas expansion factor‡	Y	Y	+	+	+	$X_Y = 1.0$	Tables 9.26, 9.43
Pipe diameter	D	D^*	−	−	−	$X_D = -\dfrac{2\beta^4}{1-\beta^4}$	Fig. 9.10
Bore diameter	d	d^*	+	+	+	$X_d = +\dfrac{2}{1-\beta^4}$	Fig. 9.10
Differential pressure	h_w	Δp^*	+	+	+	$X_{hw} = \tfrac{1}{2}$	

Density						
Liquids, gases (vapors)	ρ_f	ρ_f^*	−	+	+	$X_\rho = \tfrac{1}{2}$
	G_f	G_f	−	+	+	$X_{G_f} = \tfrac{1}{2}$
Liquids	ρ_b	ρ_b^*	NA	−	NA	$X_{pb} = -1$
	G_b	G_b	NA	−	NA	$X_{Gb} = -1$
pvT equation						
Pressure						
Gauge	p_G	p_G^*	−	+	+	$X_{pG} = \dfrac{1}{2}\dfrac{p_G}{p_G + 14.7}$ for $p_G > 1$ psig (7 kPa)
Absolute	p_{f1}	p_{f1}^*	−	+	+	$X_p = \tfrac{1}{2}$
Temperature						
Fahrenheit (Celsius)	T_F	$T_{°C}$	+	−	−	$X_{TF} = \dfrac{1}{2}\dfrac{|T_F|}{2\,460 + T_F}$ for $T_F > 1°F$ and $T_F < -1°F$
Absolute	T_{f1}	T_{K1}	+	−	−	$X_T = \tfrac{1}{2}$
Specific gravity	G	G	−	−	+	$X_G = \tfrac{1}{2}$
Compressibility	Z_{f1}	Z_{f1}	+	−	−	$X_Z = \tfrac{1}{2}$

†The influences of a change in d, d^* and D, and D^* in the discharge-coefficient equation are considered negligible.

‡The influences of a change in h_w, Δp^*, p_{f1}, p_{f1}^*, k, d, d^*, D, and D^* in the gas-expansion-factor equation are considered negligible.

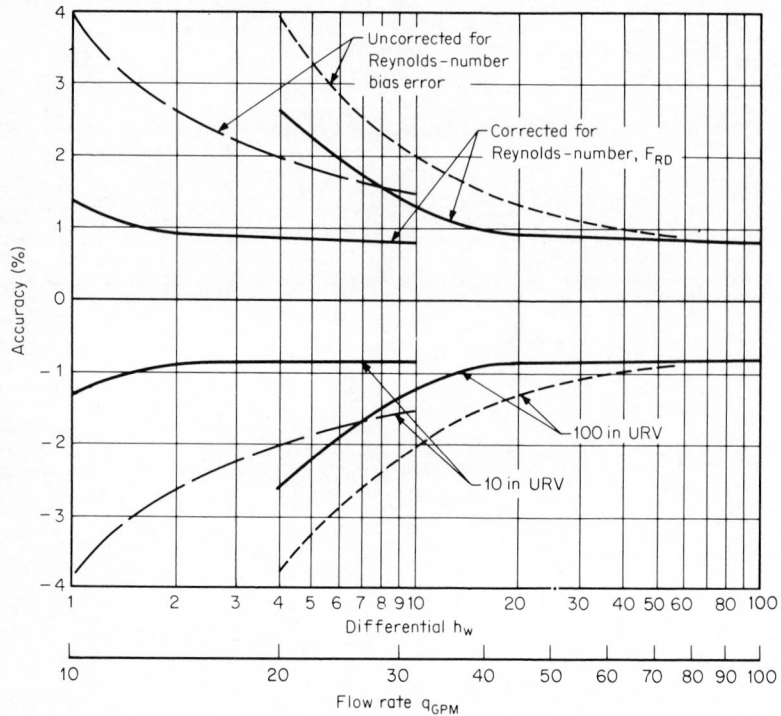

Figure 9.12 Wide-range-metering accuracy envelopes for Example 9.18.

EXAMPLE 9.17

Calculate the accuracy of the natural gas measurement in Example 9.6 if the differential pressure is 3 kPa.

For the flow-rate equation, the accuracy equation is written as

$$(\text{Acc})_{\text{SCMH}} = \pm\{[X_{\Delta p*}(\text{Acc})_{\Delta p*}]^2 + [X_{d*}(\text{Acc})_{d*}]^2 + [X_{D*}(\text{Acc})_{D*}]^2 + [X_C (\text{Acc})_C]^2$$
$$+ [X_p(\text{Acc})_p]^2 + [X_T(\text{Acc})_T]^2 + [X_G(\text{Acc})_G]^2 + [X_Z(\text{Acc})_Z]^2 + [X_Y(\text{Acc})_Y]^2\}^{1/2}$$

In this equation, Z replaces F_{pv} and G replaces F_g of the flow equation.

From Example 9.6, $D^* = 52.50$ mm, $d^* = 29.96$ mm, and $\beta = 0.5707$. The calculation of the accuracy may then be tabulated as

Property	Acc	X (Table 9.44)	$[X(\text{Acc})]^2$
Δp^*	$\pm 0.6\%$	$+\frac{1}{2}$	0.0900
d^*	± 0.02 mm $\pm 0.067\%$	$+\dfrac{2}{1 - \beta^4} = 2.24$	0.0225

Property	Acc	X (Table 9.44)	$[X(\text{Acc})]^2$
D^*	± 0.10 mm $\pm 0.191\%$	$-\dfrac{2\beta^4}{1-\beta^4} = -0.24$	0.0021
C	$\pm 0.6\%$ (Table 9.42)	1.0	0.3600
p^*_{f1}	$\pm 0.5\%$ (Absolute)	$\frac{1}{2}$	0.0625
T_{K1}	$\pm 0.25\%$ (Absolute)	$-\frac{1}{2}$	0.0156
G	$\pm 0.15\%$	$-\frac{1}{2}$	0.0056
Z_f	$\pm 0.2\%$	$-\frac{1}{2}$	0.0100
Y	$\pm 4\,\dfrac{\Delta p^*}{p^*_{f1}} = 0.065\%$ (Table 9.43)	1.0	0.0042
			0.5725

The accuracy is then

$$(\text{Acc})_{\text{SCMH}} = \pm \{\Sigma[X(\text{Acc})]^2\}^{1/2} = \pm 0.76 \text{ percent}$$

EXAMPLE 9.18

In Example 9.15, 10-in (2.5-kPa) and 100-in (25-kPa) URV differential-pressure transmitters are to be used to measure over a 10:1 flow-rate range. Assuming the flow-rate calculation is performed without error, determine the accuracy with (a) no Reynolds-number correction and (b) a Reynolds-number correction. Assume the Reynolds-number bias error can be either positive or negative to obtain a $\pm$ percent accuracy envelope; calculate the accuracy for the low-range transmitter at $h_w = 1, 2.5, 5, 7.5,$ and 10 in, and for the high-range transmitter at $h_w = 11, 25, 50, 75,$ and 100 in.

If the Reynolds-number bias error is treated as a bias range ($\pm$), the accuracy equation becomes, from Eq. (4.9),

$$\text{Acc} = \pm B_q \pm (\text{Acc})_q$$

The Reynolds-number bias error can be written as

$$\pm B_q = \pm \frac{1 - F_{RD}}{F_{RD}} \, 100$$

From Example 9.11, the Reynolds-number correction using Newton's method can be written as

$$F_{RD} = 1 + \frac{0.1823(10\sqrt{h_w})^{-0.75} - 0.006889}{1 + 0.1367(10\sqrt{h_w})^{-0.75}} = 1 + \frac{0.0324 h_w^{-0.375} - 0.006889}{1 + 0.0243 h_w^{-0.375}}$$

If the discharge coefficient is not corrected for Reynolds number in Example 9.15, the accuracy of the flow-rate determination is

$$(Acc)_{GPM} = \pm\{[X_{h_w}(Acc)_{h_w}]^2 + 0.0515 + 0.0399 + 0.0025 + 0.010 + 0.010 + 0.5521\}^{1/2}$$
$$= \pm\{[(X_{h_w})(Acc)_{h_w}]^2 + 0.666\}^{1/2}$$

with

$$X_{h_w}(Acc)_{h_w} = \frac{1}{2}\frac{(h_w)_{URV}}{h_w}0.2 = 0.1\frac{(h_w)_{URV}}{h_w}$$

where $(h_w)_{URV}$ is the upper-range value of the differential-pressure transmitter, and h_w is the measured differential. Then

$$Acc = \pm\frac{1 - F_{RD}}{F_{RD}}100 \pm \left\{\left[0.1\frac{(h_w)_{URV}}{h_w}\right]^2 + 0.666\right\}^{1/2}$$

The calculation is tabulated as follows (see Fig. 9.12 for envelopes):

h_w	F_{RD}	B_q, %	Acc, % (corrected)†	Acc, % (uncorrected)‡
			LOW RANGE	
1	1.0249	±2.43	±1.3	±3.7
2.5	1.0158	±1.56	±0.9	±2.4
5	1.0107	±1.06	±0.8	±1.9
7.5	1.0082	±0.82	±0.8	±1.6
10	1.0067	±0.67	±0.8	±1.5
			HIGH RANGE	
11	1.0062	±0.62	±1.2	±1.8
25	1.0028	±0.28	±0.9	±1.2
50	1.0006	±0.06	±0.8	±0.9
75	0.9995	±0.05	±0.8	±0.9
100	0.9989	±0.11	±0.8	±0.9

†Accuracy corrected for Reynolds number.

‡Accuracy uncorrected for Reynolds number.

EXAMPLE 9.19

For the data given in Example 9.3, use sensitivity coefficients to estimate the bias error and to correct the flow rate for a 10-in (2.5-kPa) differential measurement (a) for Reynolds number and (b) if $G_F = 0.7220$ and $G_b = 0.7300$. (See Example 9.11 for the exact calculation.)

a. Bias error and Reynolds-number correction

The flow rate for a 10-in differential, based on normal conditions, is derived in Example 9.11 as

$$(q_{GPM})_N = 10.00\sqrt{h_w} = 10.00\sqrt{10} = 31.623 \text{ base gal/min}$$

The Reynolds number at flowing conditions is calculated from the equation given in Example 9.2 as

$$R_D = 17{,}903 \frac{G_b}{\mu_{cP}DN_{VG}} q_{GPM} = 17{,}903 \left[\frac{0.7300}{(0.417)(1.939)(5.6665)}\right](31.623) \quad (31.623)$$

$$= 90{,}203$$

The coefficient indicating the sensitivity of the discharge coefficient to Reynolds number is given in Table 9.44 as

$$X_{RD} = -\frac{69\beta^{2.5}}{CR_D^{0.75}} = -\frac{(69)(0.7430)^{2.5}}{(0.6097)(230{,}040)^{0.75}} = -0.00513$$

where $C = C_N = 0.6097$ from Example 9.11, and $(R_D)_N = 230{,}040$ and $\beta = 0.7430$ from Example 9.3.

Since the indicated flow rate is based on normal conditions, the bias error in the Reynolds number is, by Eq. (4.5),

$$B_{RD} = \frac{\bar{I} - I_t}{I_t}(100) = \frac{230{,}040 - 90{,}203}{90{,}203}(100) = +155 \text{ percent}$$

where the true value of the Reynolds number I_t is 90,203. The bias error in the flow rate is then estimated with Eq. (4.26) as

$$B_{GPM} = X_{RD}B_{RD} = (-0.00513)(155) = -0.79 \text{ percent}$$

The Reynolds-number correction, from Eq. (4.6), is then

$$F_{B,RD} = \left(1 + \frac{B_{GPM}}{100}\right)^{-1} = \left(1 + \frac{-0.79}{100}\right)^{-1} = 1.008$$

and the corrected flow rate is

$$q_{GPM} = F_{B,RD}(q_{GPM})_N = (1.008)(31.623) = 31.88 \text{ base gal/min}$$

This compares to 31.83 base gal/min as calculated in Example 9.11. The difference is attributable to the nonlinearity of the sensitivity coefficient with Reynolds number and the change in discharge coefficient with Reynolds number.

b. Bias error and correction for base and flowing specific-gravity differences

The sensitivity coefficient for base specific gravity is, from Table 9.44, $X_{Gb} = -1$; for specific gravity at flowing conditions, it is $X_{GF} = +\frac{1}{2}$. The bias error for a base specific gravity of 0.7300 is, by Eq. (4.5),

$$B_{Gb} = \frac{\bar{I} - I_t}{I_t}(100) = \frac{0.7359 - 0.7300}{0.7300}(100) = +0.81 \text{ percent}$$

The flow-rate bias error is calculated with Eq. (4.26) as

$$B_{GPM} = X_{Gb}B_{Gb} = (-1)(0.81) = -0.81 \text{ percent}$$

and the correction is, from Eq. (4.6),

$$F_{B,Gb} = \left(1 + \frac{B_{GPM}}{100}\right)^{-1} = \left(1 + \frac{-0.81}{100}\right)^{-1} = 1.008$$

The bias error for a flowing specific gravity of 0.7220 is, by Eq. (4.5),

$$B_{GF} = \frac{\bar{I} - I_t}{I_t}(100) = \frac{0.7255 - 0.7220}{0.7220}(100) = +0.49 \text{ percent}$$

The flow-rate bias error is then, by Eq. (4.26),

$$B_{GPM} = X_{GF}B_{GF} = (+\tfrac{1}{2})(+0.49) = +0.24 \text{ percent}$$

and, with Eq. (4.6), the correction becomes

$$F_{B,GF} = \left(1 + \frac{B_{GPM}}{100}\right)^{-1} = \left(1 + \frac{+0.24}{100}\right)^{-1} = 0.9976$$

The flow rate, corrected for Reynolds number and for differences in base specific gravity and flowing specific gravity from normal conditions, is then

$$q_{GPM} = F_{B,RD}F_{B,Gb}F_{B,GF}(q_{GPM})_N = (1.008)(1.008)(0.9976)(31.623)$$
$$= 32.05 \text{ base gal/min}$$

REFERENCES

ANSI (MFC): *Differential Producers Used for the Measurement of Fluid Flow in Pipes (Orifice, Nozzle, Nozzle Venturi, Venturi)*, draft 3e, New York, July 1982.

ASME: *Fluid Meters*, 6th ed., ASME, New York, 1971.

Buckingham, E.: "Notes on the Orifice Meter; the Expansion Factor for Gases," *J. Res. Nat. Bur. Stand.*, vol. 9, p. 61, 1932.

ISO Standard 5024, *Petroleum Liquids and Gases—Measurement—Standard Reference Conditions*, ISO 5024-1976(E), Geneva, 1976.

ISO Standard 5167, *Measurement of Fluid Flow by Means of Orifice Plates, Nozzles, and Venturi Tubes in Circular Cross-Section Conduits Running Full*, ISO 5167-1980(E), 1980.

James, R.: "Metering of Steam/Water Two-Phase Flow by Sharp Edged Orifices," *Proc. Inst. Mech. Eng.*, vol. 180, pt. 1, no. 23, pp. 549–572, 1965–1966.

Murdock, J. W.: "Two-Phase Flow Measurement with Orifice," ASME Paper 61-WA-27, 1961.

Smith, R. V., and J. T. Leang: "Evaluations of Correlations for Two-Phase Flowmeters— Three Current—One New," ASME Paper 74-WA-FM-5, 1974.

Spink, L. K.: *Principles and Practice of Flowmeter Engineering*, 9th ed., The Foxboro Company, Foxboro, Mass., 1967.

Stolz, J.: *OSU 89 Test Analysis*, ISO/TC30/SC2 (France 17) 95E, 1978.

DIFFERENTIAL PRODUCERS: DESIGN INFORMATION

Measured differential pressures depend on both fluid characteristics and primary-element geometry. The proper use of differential producers requires adherence to the installation requirements given in Chap. 8 and the details presented in this chapter.

This chapter is concerned with differential producers that are usually sized to produce a selected differential at a design flow rate. In Chap. 11, design information is presented for fixed-geometry devices, for which the differential (or, for a target flowmeter, the force on the target) must be determined for the design flow rate. The graphs presented for discharge coefficients and gas expansion factors were developed from the equations of Chap. 9 when applicable. Others are based on recommendations given in the technical literature.

ORIFICES

Concentric Square-Edged Orifice

Shown in Fig. 10.1 is the pressure profile along a meter run containing a concentric square-edged orifice. The pressure first increases, beginning at approximately $0.5D$ upstream, and then decreases to a minimum at the vena contracta. From this point, the pressure recovers to the initial upstream pressure (less pressure losses due to friction and energy losses). The discharge-coefficient equation presented in Chap. 9 was developed by Stolz from empirical discharge-coefficient data and this type of pressure-gradient data.

Pressure-tap spacing requirements for flange, D and $D/2$, and $2\frac{1}{2}D$ and $8D$ taps are given in Fig. 10.2. Individual and annular-slot corner-tap design requirements are presented in Fig. 10.3

In the design of the orifice, it is important that the recommendations of Fig. 10.4 and Table 10.1 (for plate thickness) be followed. Edge sharpness, concentricity, plate bending, surface finish, etc., will alter the discharge coefficient and degrade the accuracy of the installation. If the orifice is intended for measuring reverse flows, the plate should not be beveled, and its thickness should be

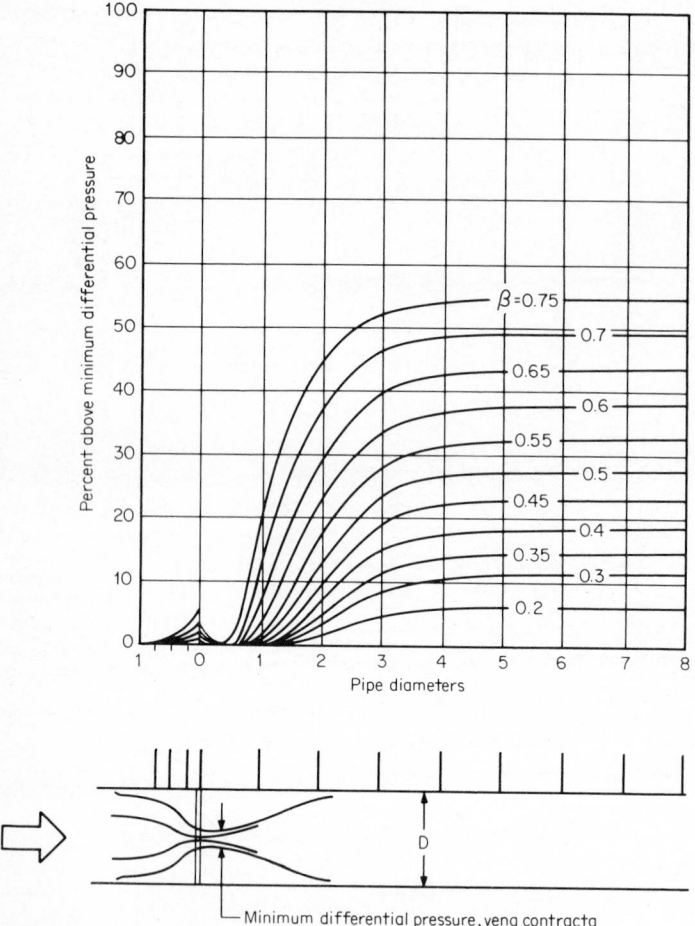

Figure 10.1 Concentric square-edged orifice: pressure profile.

between $0.005D$ and $0.02D$. In these cases, the differential should be as low as practicable consistent with the desired accuracy.

In Figs. 10.5 through 10.22 are graphs of the discharge coefficient developed from equations given in Table 9.1. The minimum Reynolds number is limited to 10,000 for 2-in (50-mm) pipe. For lower values, the coefficient curves of Fig. 10.22 should be used. The equations of Table 9.1 will, however, predict coefficients to within $\pm 0.6 \pm \beta$ percent for Reynolds numbers from 2000 to 10,000, which is the estimated accuracy of the data used to develop the curves in Fig. 10.22. It should be noted that the Reynolds-number axis in Fig. 10.22 is scaled for *bore* Reynolds number, or pipe Reynolds number divided by β.

In Figs. 10.23 through 10.34 are the expansion-factor curves for upstream and downstream pressure measurements. These were developed from equations given in Table 9.26, and the accuracies given in Table 9.43 apply.

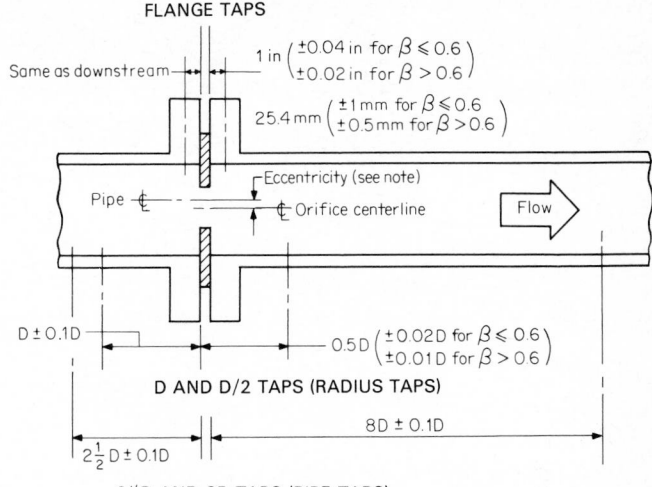

Figure 10.2 Pressure-tap spacing.

Table 10.1 **Minimum Plate Thickness**

Pipe diameter	Minimum plate thickness, in (mm), for differential pressure h_w and Δp^*		
	$h_w = 100$ in $\Delta p^* = 24.9$ kPa	101–200 in 25 to 49.9 kPa	201–1000 in 50.0 to 250 kPa
$D \leq 6$ in $(D^* \leq 150$ mm) $0.2 \leq \beta \leq 0.75$	0.125 (3.18)	0.125 (3.18)	0.125 (3.18)
$8 \leq D \leq 10$ in $(200 \leq D^* \leq 250$ mm) $0.2 \leq \beta \leq 0.75$	0.125 (3.18)	0.125 (3.18)	0.188 (4.78)
$12 \leq D \leq 20$ in $(300 \leq D^* \leq 500$ mm) $\beta \leq 0.5$ $\beta \geq 0.5$	0.250 (6.35) 0.188 (4.76)	0.250 (6.35) 0.188 (4.76)	0.375 (9.53) 0.375 (9.53)
$24 \leq D \leq 36$ in $(600 \leq D^* \leq 900$ mm) $\beta \leq 0.5$ $\beta \geq 0.5$	0.375 (9.53) 0.250 (6.53)	0.375 (9.53) 0.375 (9.53)	0.500 (12.7) 0.500 (12.7)

SOURCE: ASME (1971).

INDIVIDUAL SET OF PRESSURE TAPS

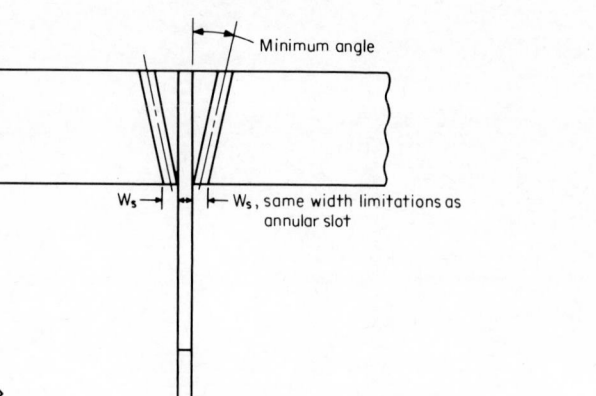

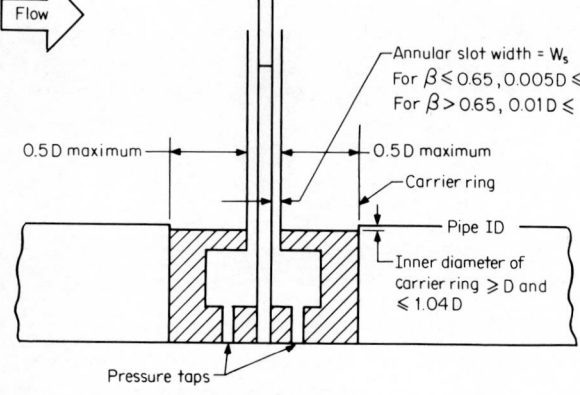

ANNULAR SLOT DESIGN

Figure 10.3 Corner taps.

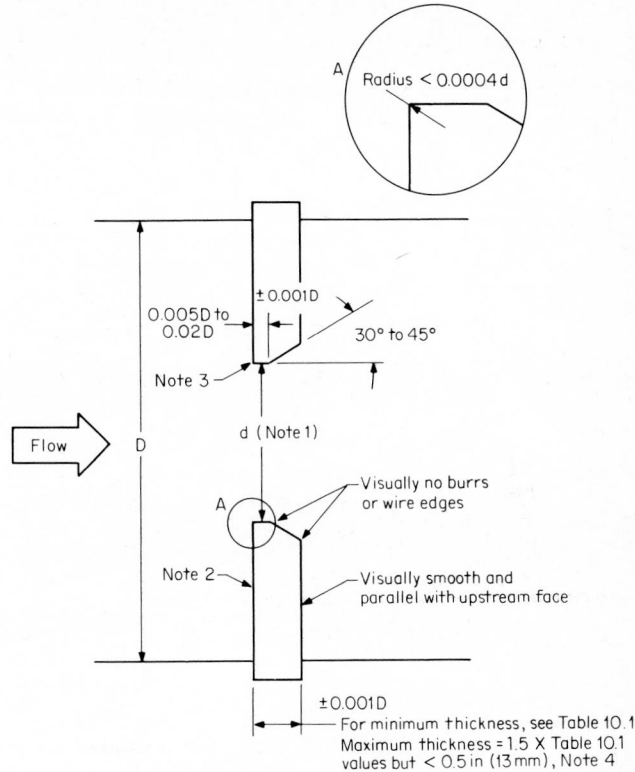

Figure 10.4 Concentric square-edged orifice.

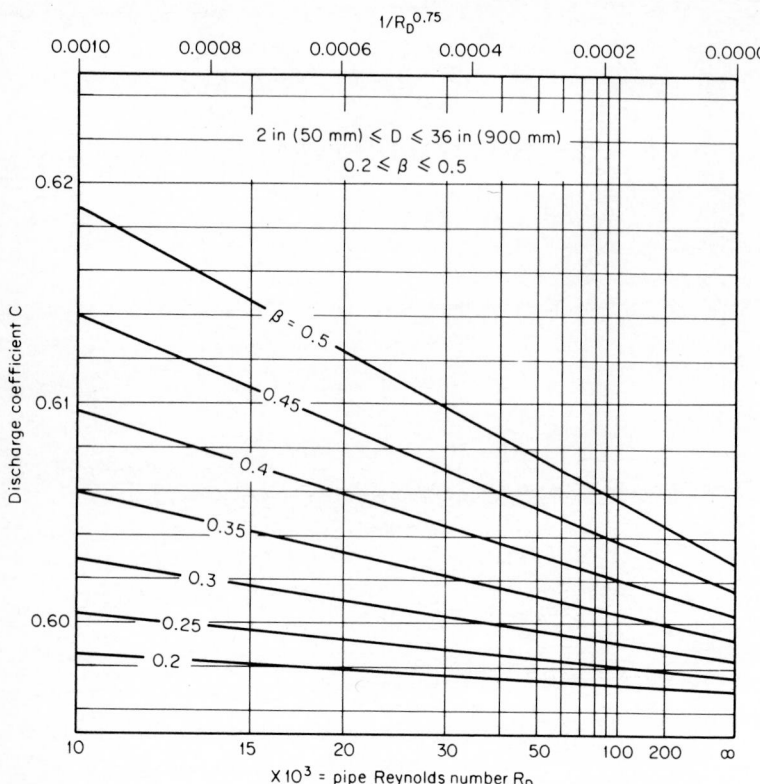

Figure 10.5 *C* versus R_D for concentric square-edged orifice; corner, flange, *D* and *D/2* taps.

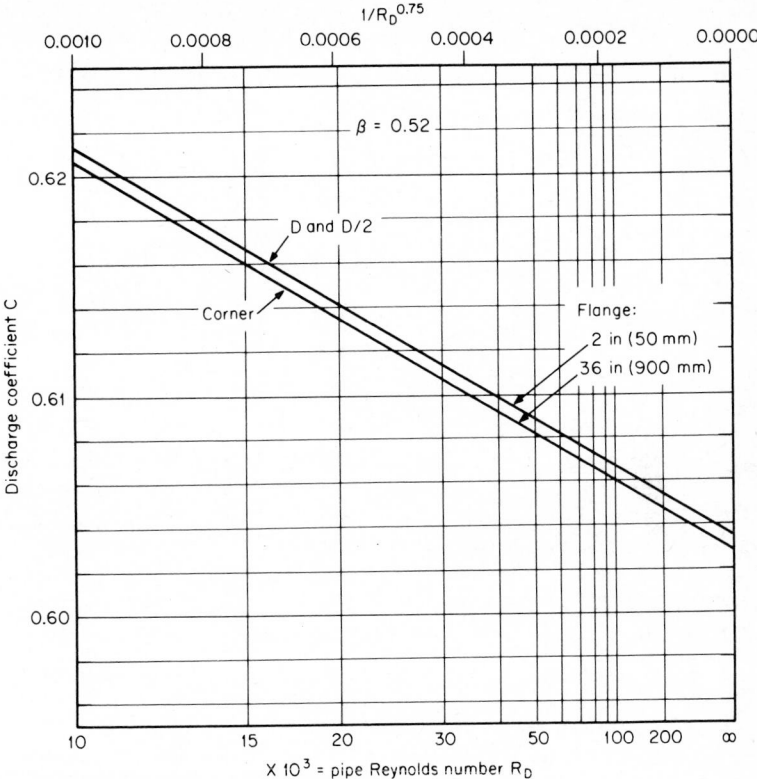

Figure 10.6 C versus R_D for concentric square-edged orifice; corner, flange, D and $D/2$ taps.

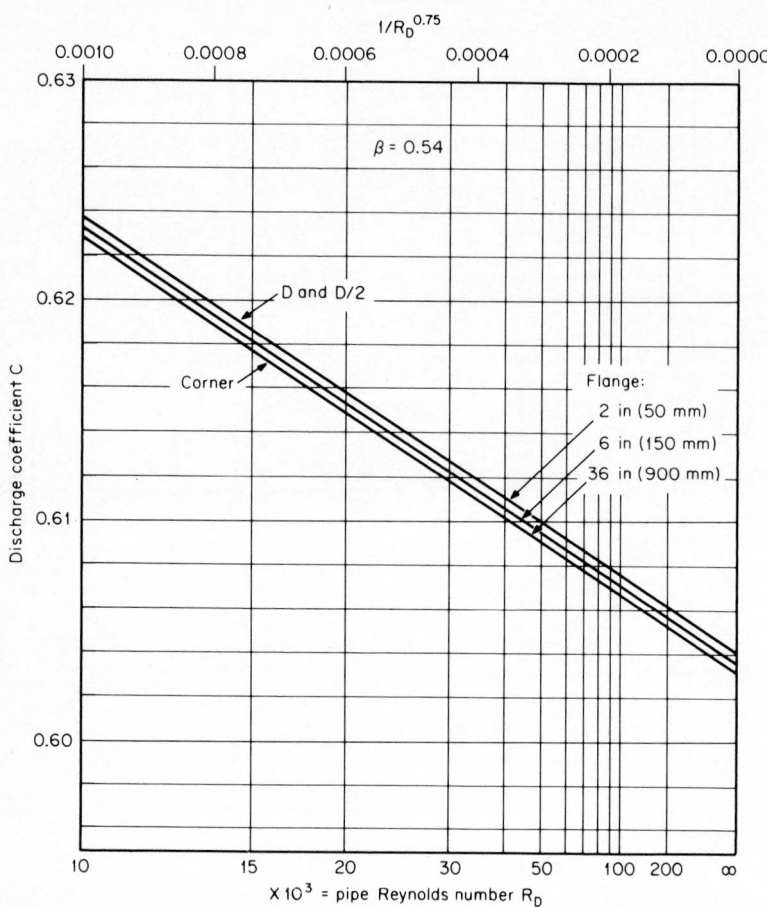

Figure 10.7 *C* versus R_D for concentric square-edged orifice; corner, flange, *D* and *D*/2 taps.

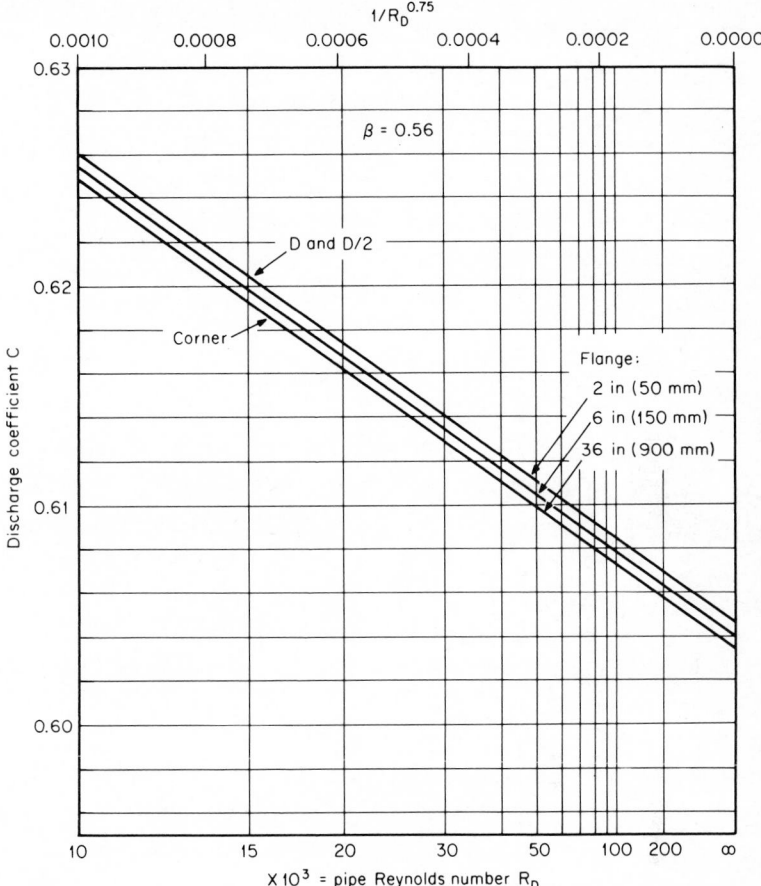

Figure 10.8 C versus R_D for concentric square-edged orifice; corner, flange, D and $D/2$ taps.

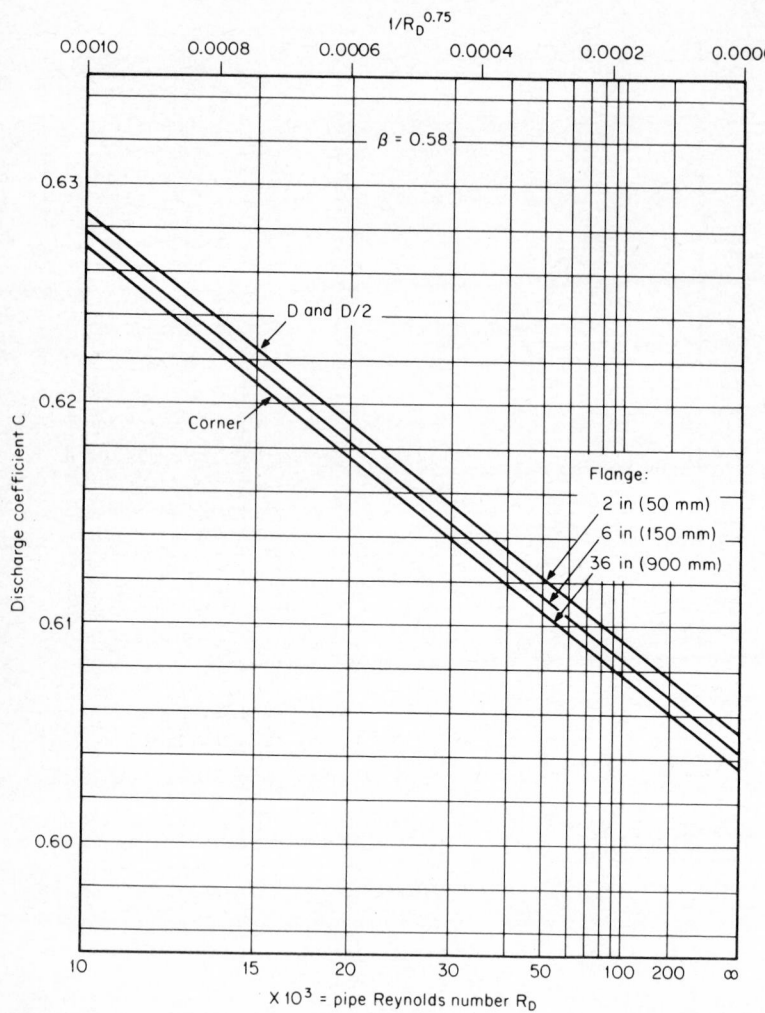

Figure 10.9 C versus R_D for concentric square-edged orifice; corner, flange, D and $D/2$ taps.

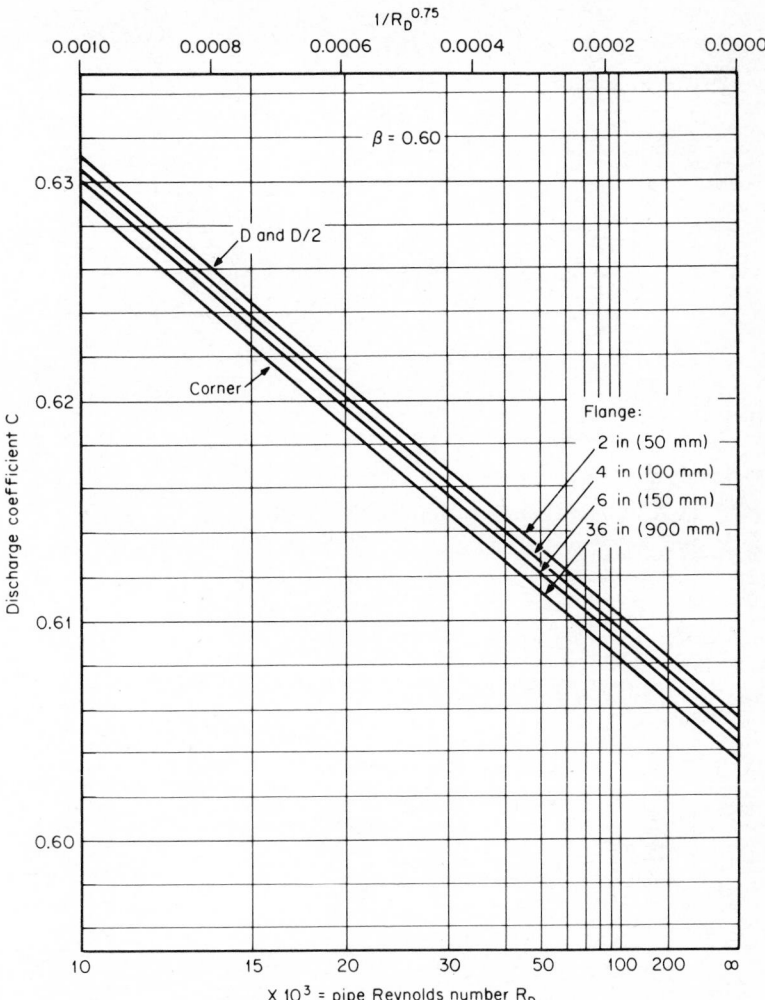

Figure 10.10 C versus R_D for concentric square-edged orifice; corner, flange, D and $D/2$ taps.

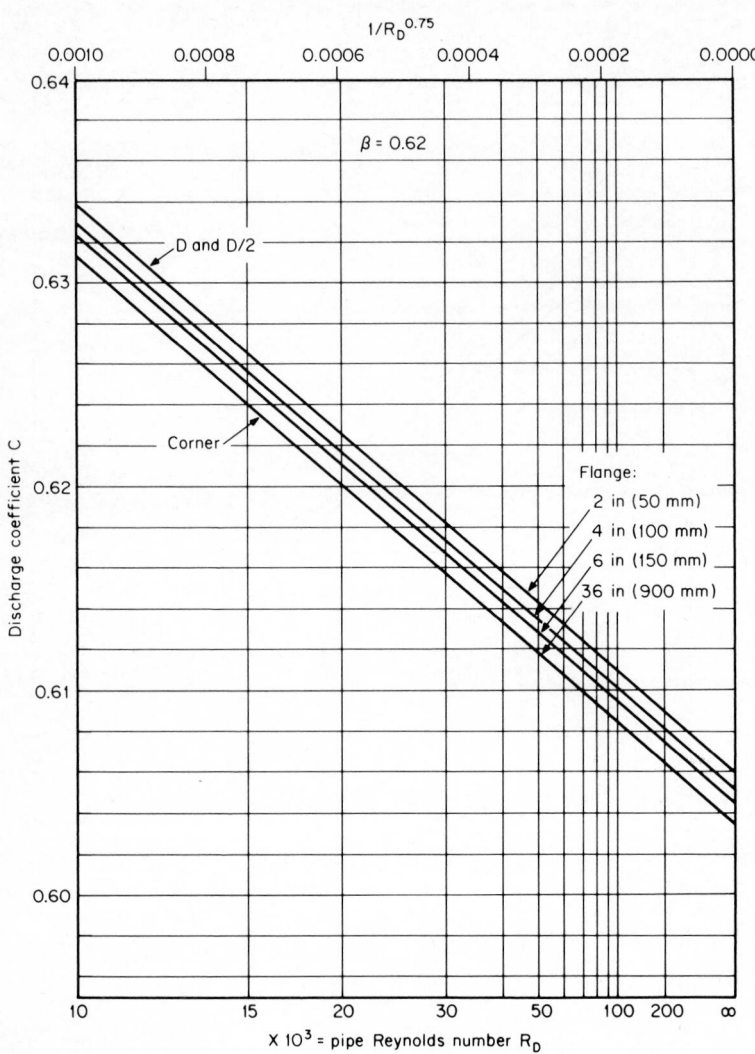

Figure 10.11 C versus R_D for concentric square-edged orifice; corner, flange, D and $D/2$ taps.

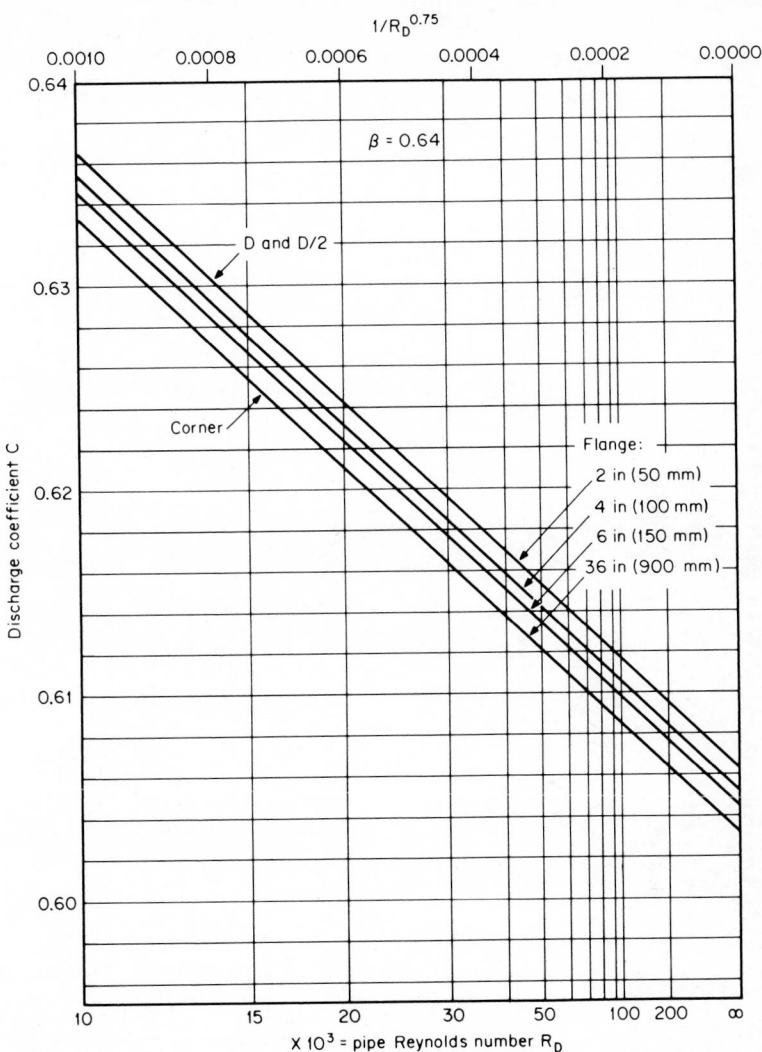

Figure 10.12 C versus R_D for concentric square-edged orifice; corner, flange, D and $D/2$ taps.

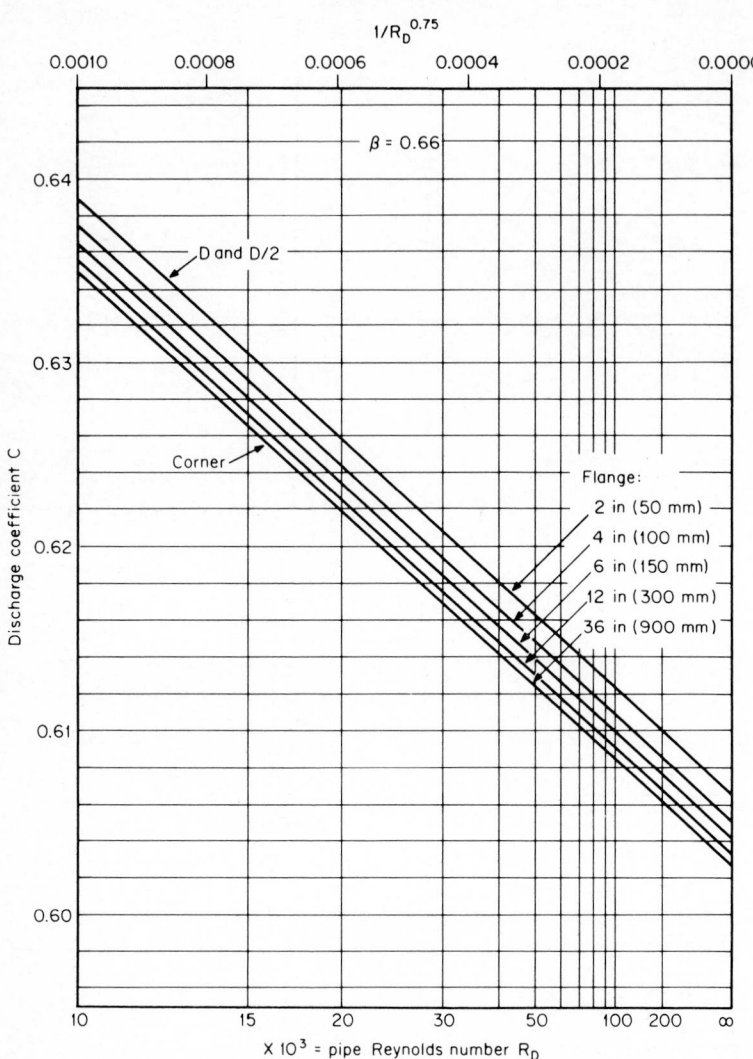

Figure 10.13   *C* versus R_D for concentric square-edged orifice; corner, flange, *D* and *D*/2 taps.

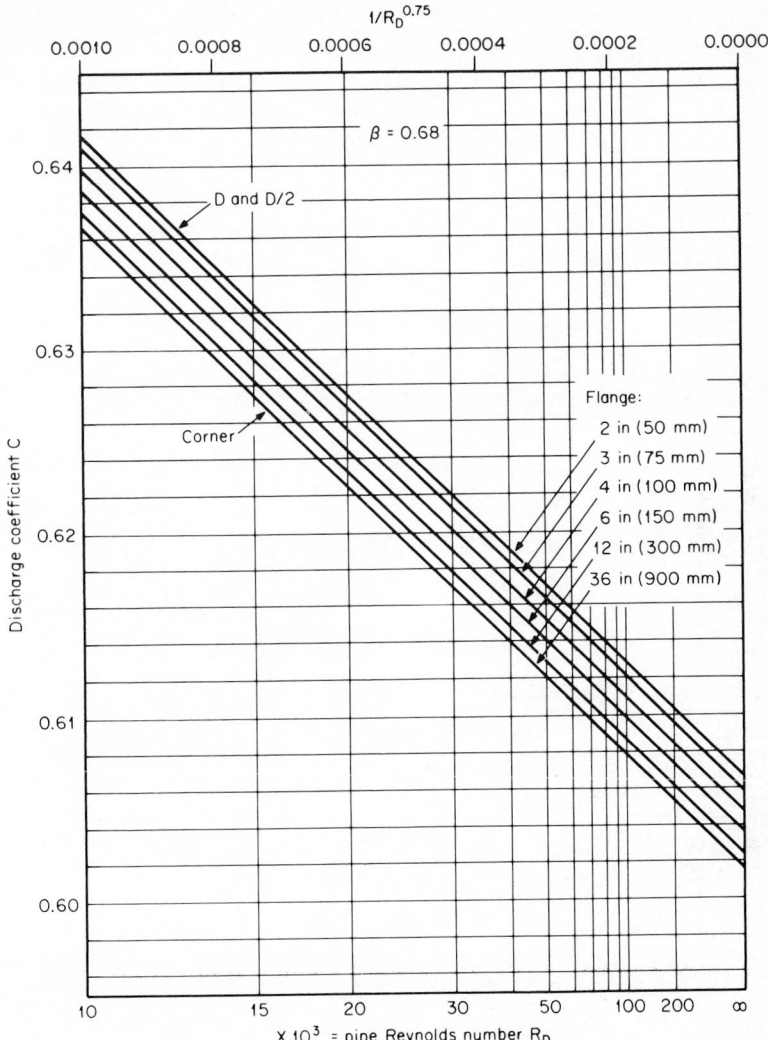

Figure 10.14 *C* versus R_D for concentric square-edged orifice; corner, flange, *D* and *D/2* taps.

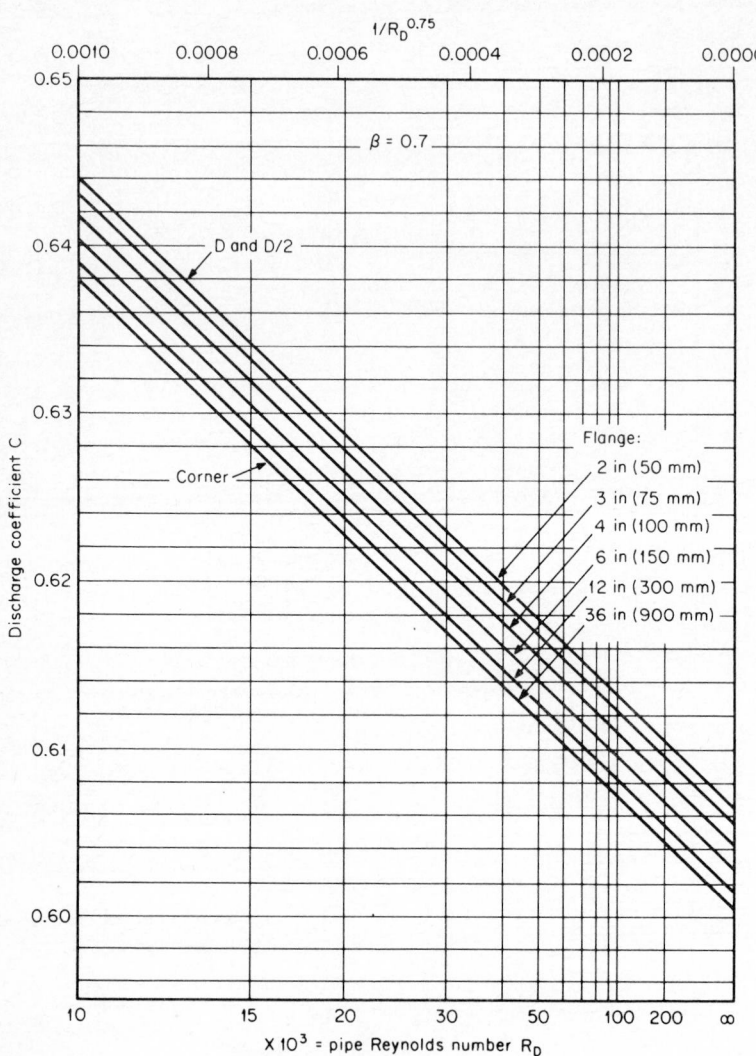

Figure 10.15 C versus R_D for concentric square-edged orifice; corner, flange, D and $D/2$ taps.

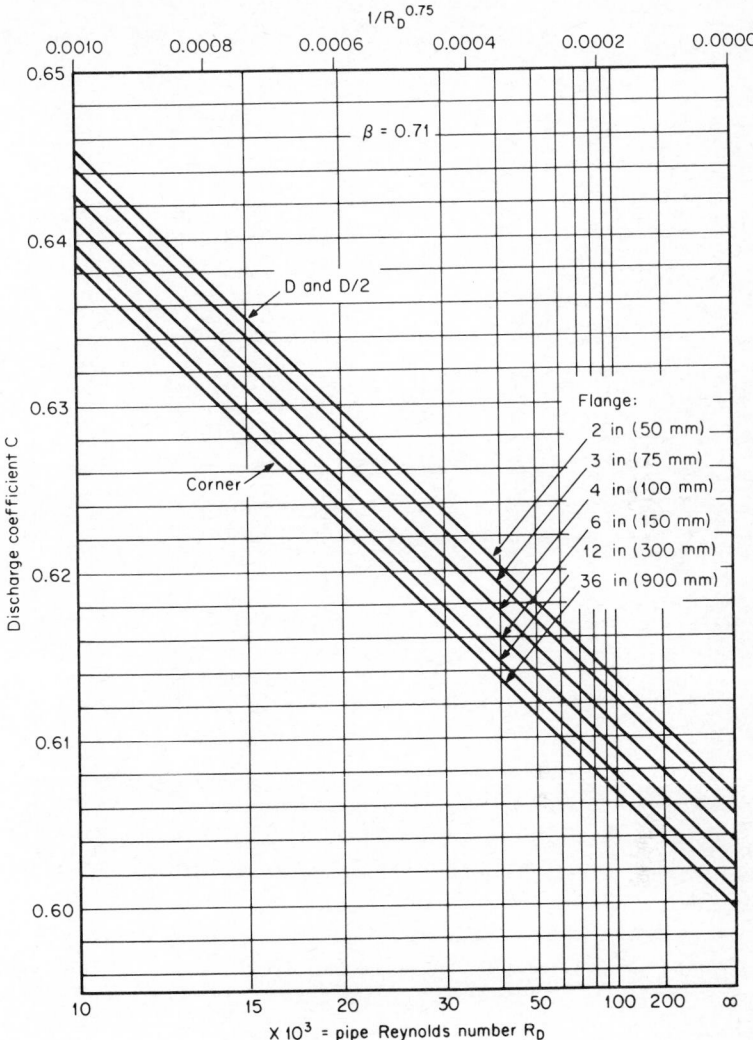

Figure 10.16 C versus R_D for concentric square-edged orifice; corner, flange, D and $D/2$ taps.

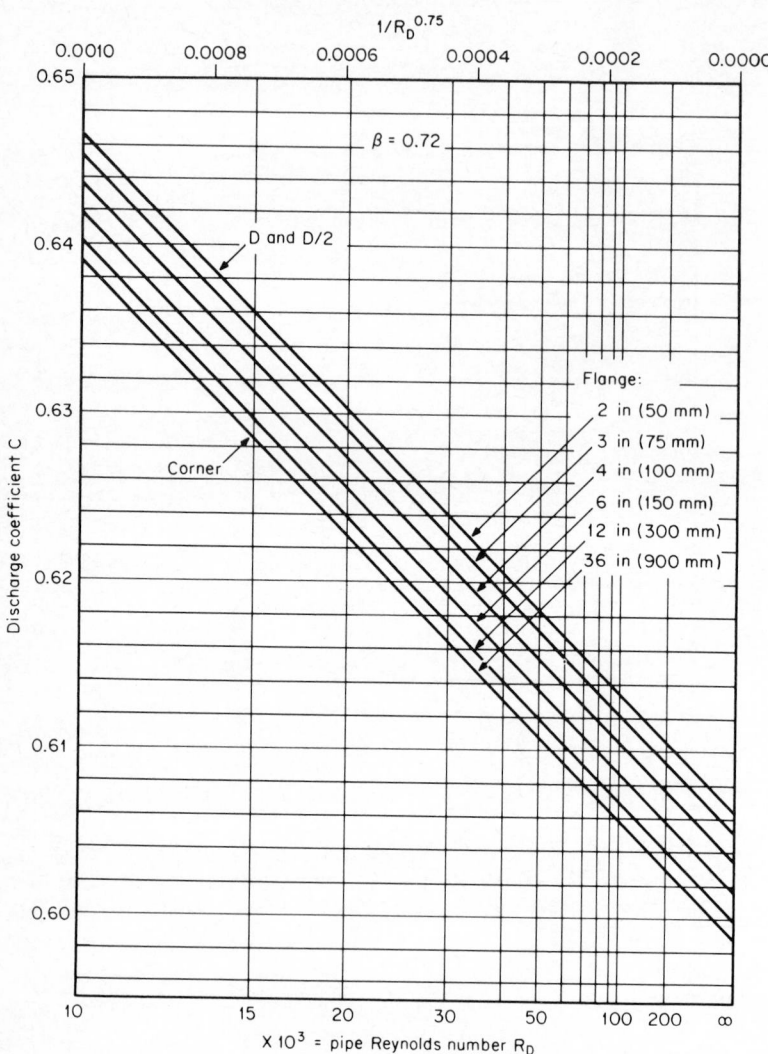

Figure 10.17 C versus R_D for concentric square-edged orifice; corner, flange, D and $D/2$ taps.

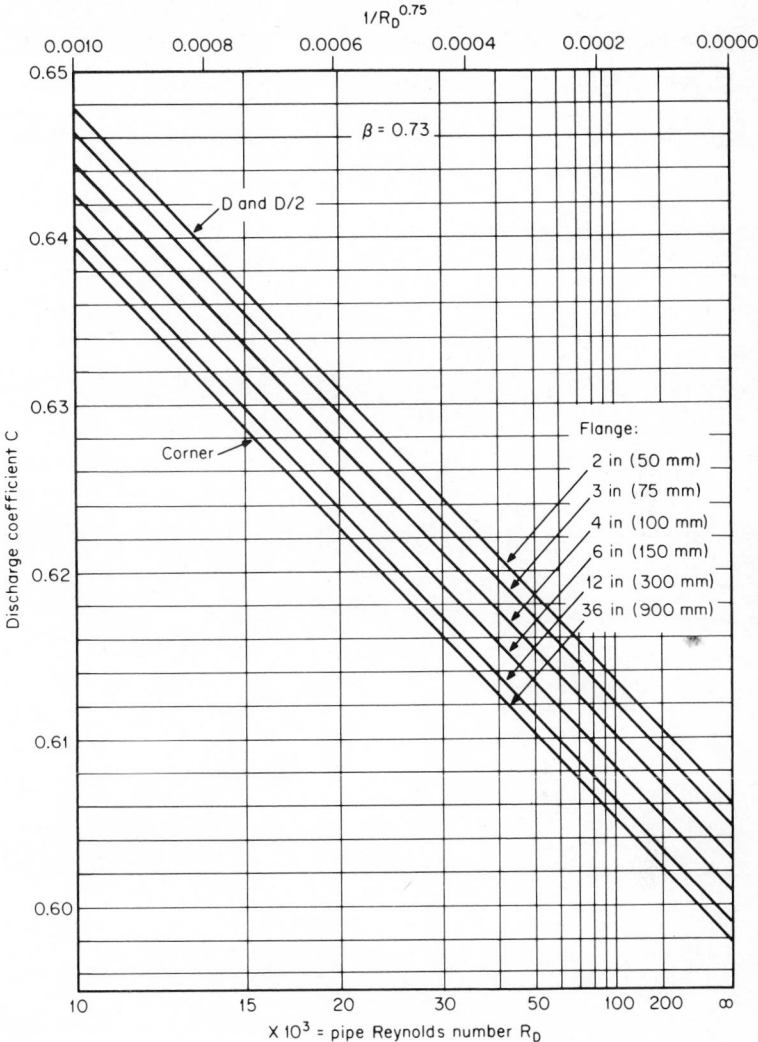

Figure 10.18 C versus R_D for concentric square-edged orifice; corner, flange, D and $D/2$ taps.

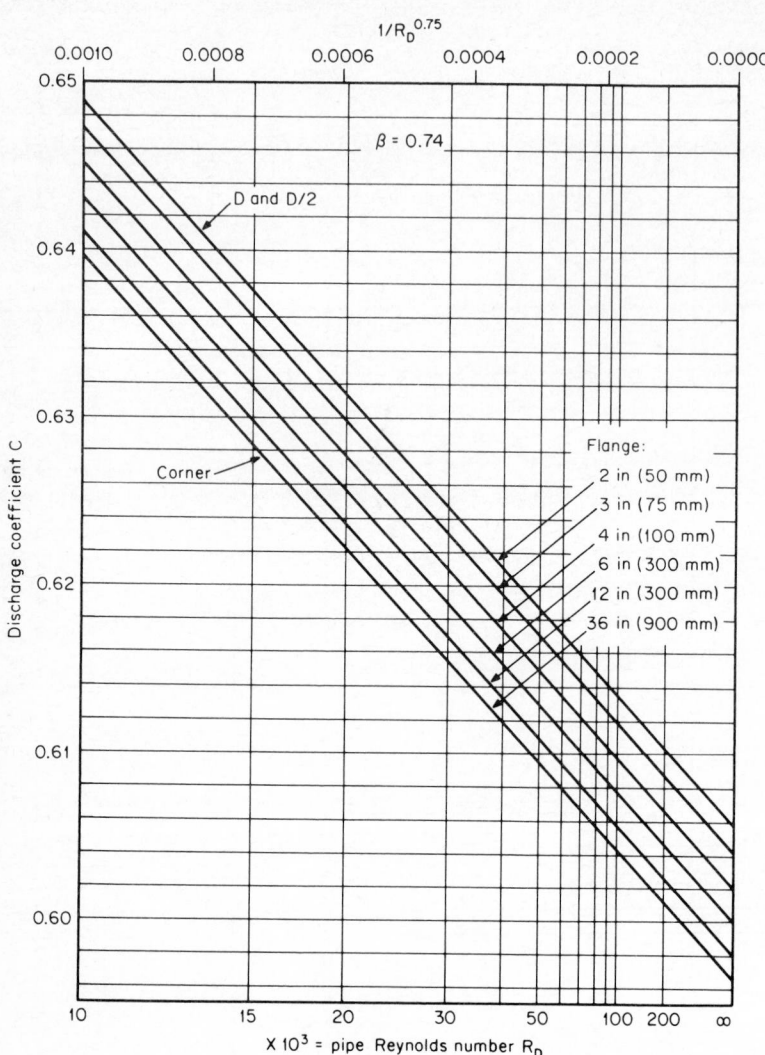

Figure 10.19 C versus R_D for concentric square-edged orifice; corner, flange, D and $D/2$ taps.

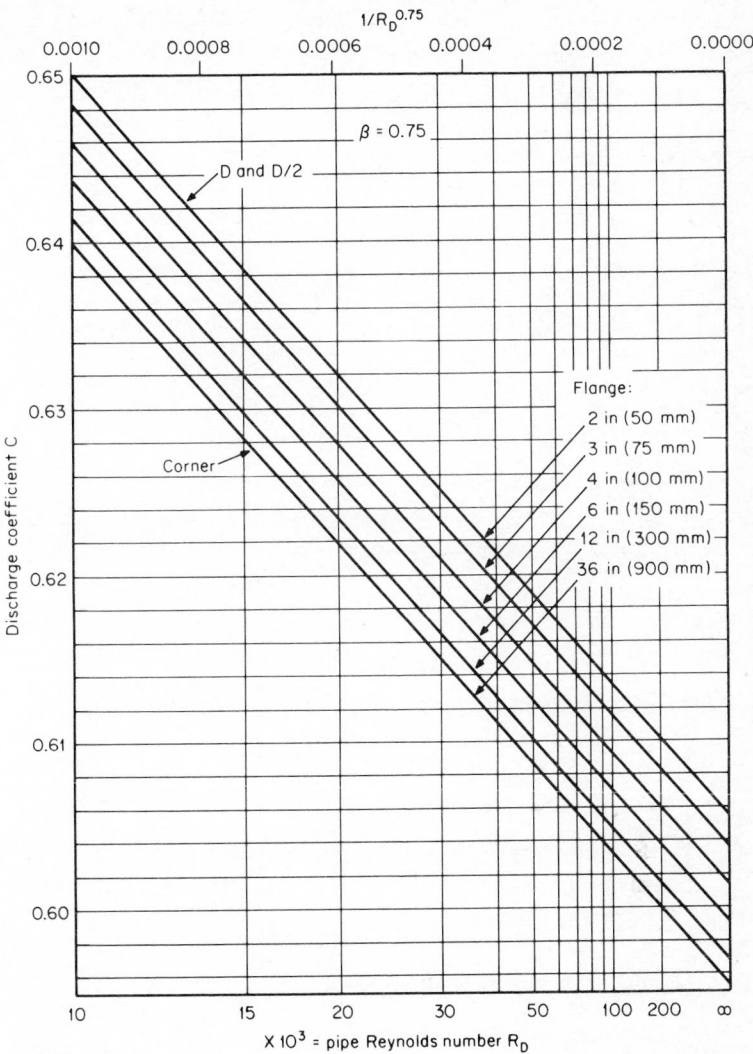

Figure 10.20 C versus R_D for concentric square-edged orifice; corner, flange, D and $D/2$ taps.

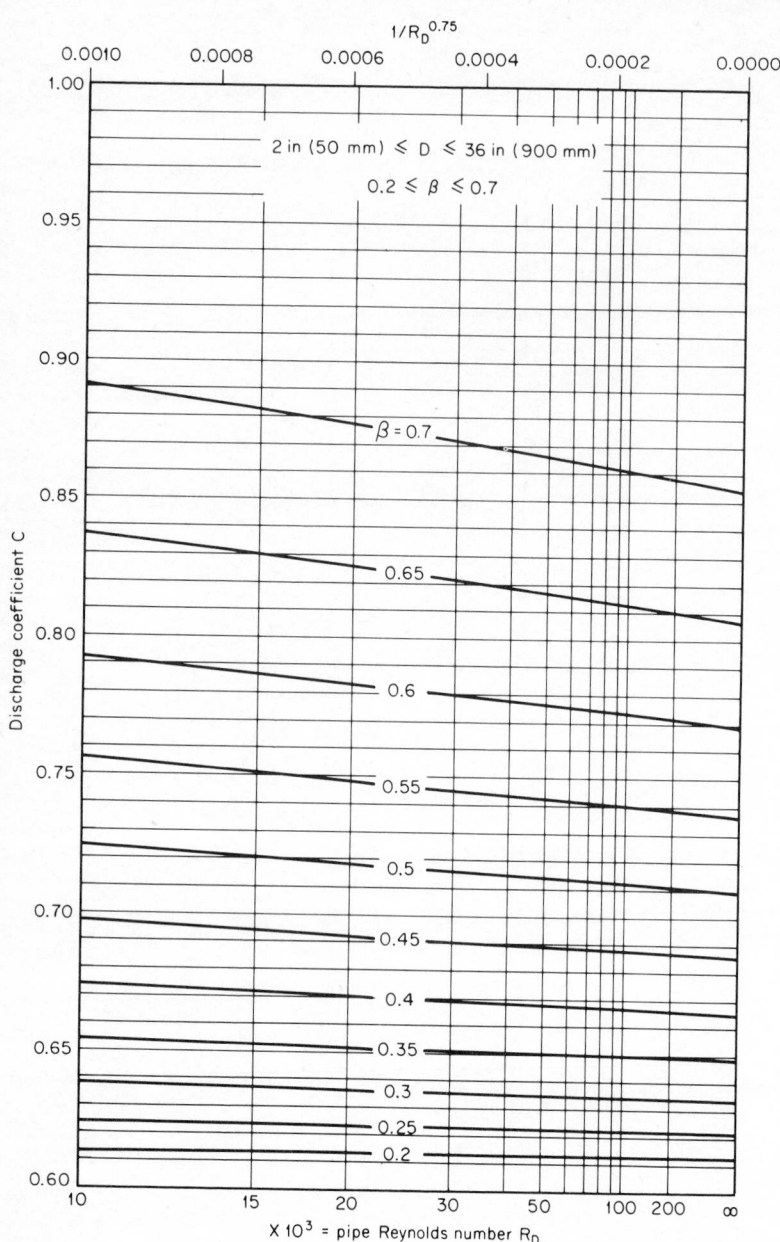

Figure 10.21 C versus R_D for concentric square-edged orifice; $2\frac{1}{2}D$ and $8D$ taps (pipe taps).

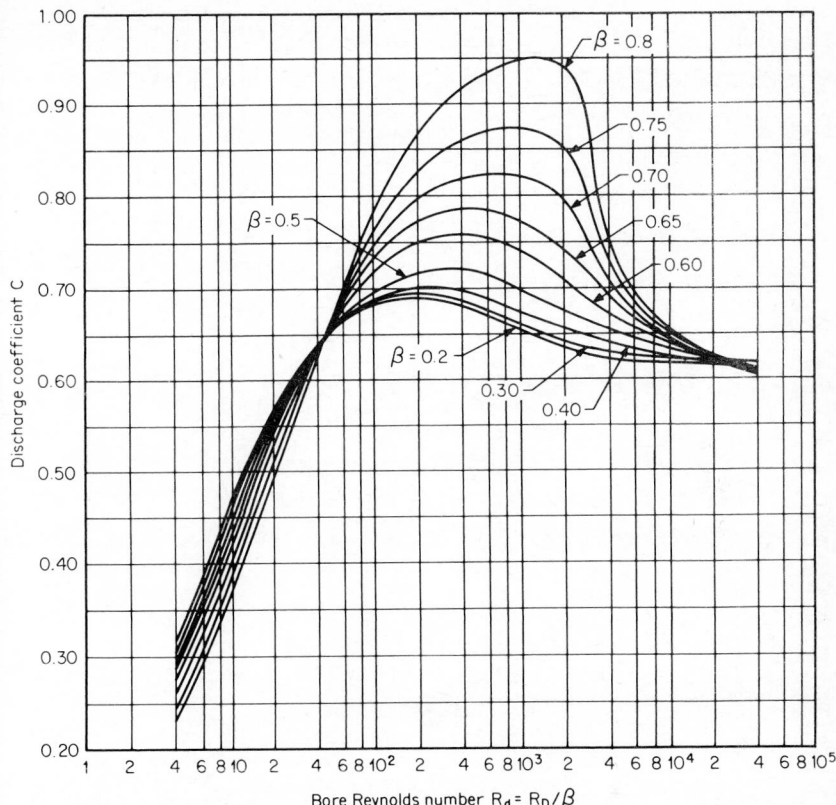

Figure 10.22 C versus R_d for concentric square-edged orifice; corner, flange D and $D/2$ taps.

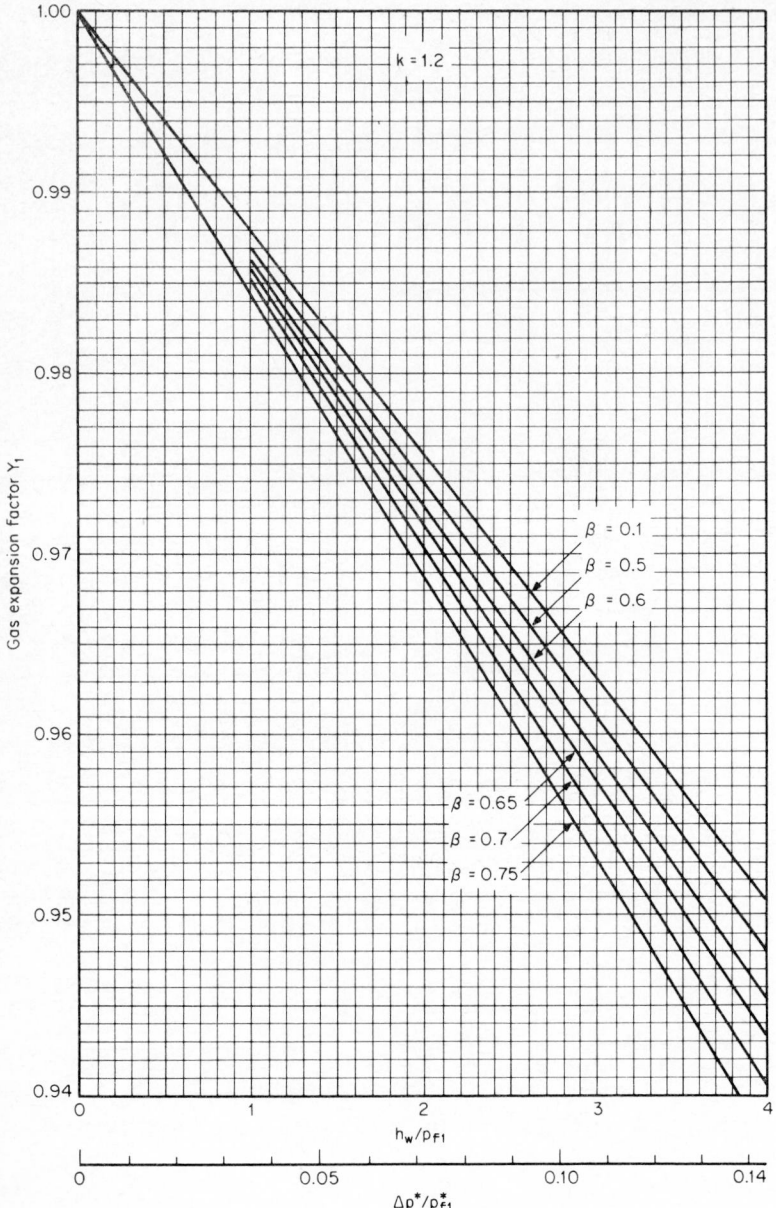

Figure 10.23 Gas expansion factor Y_1 for concentric square-edged orifice; corner, flange, D and $D/2$ taps.

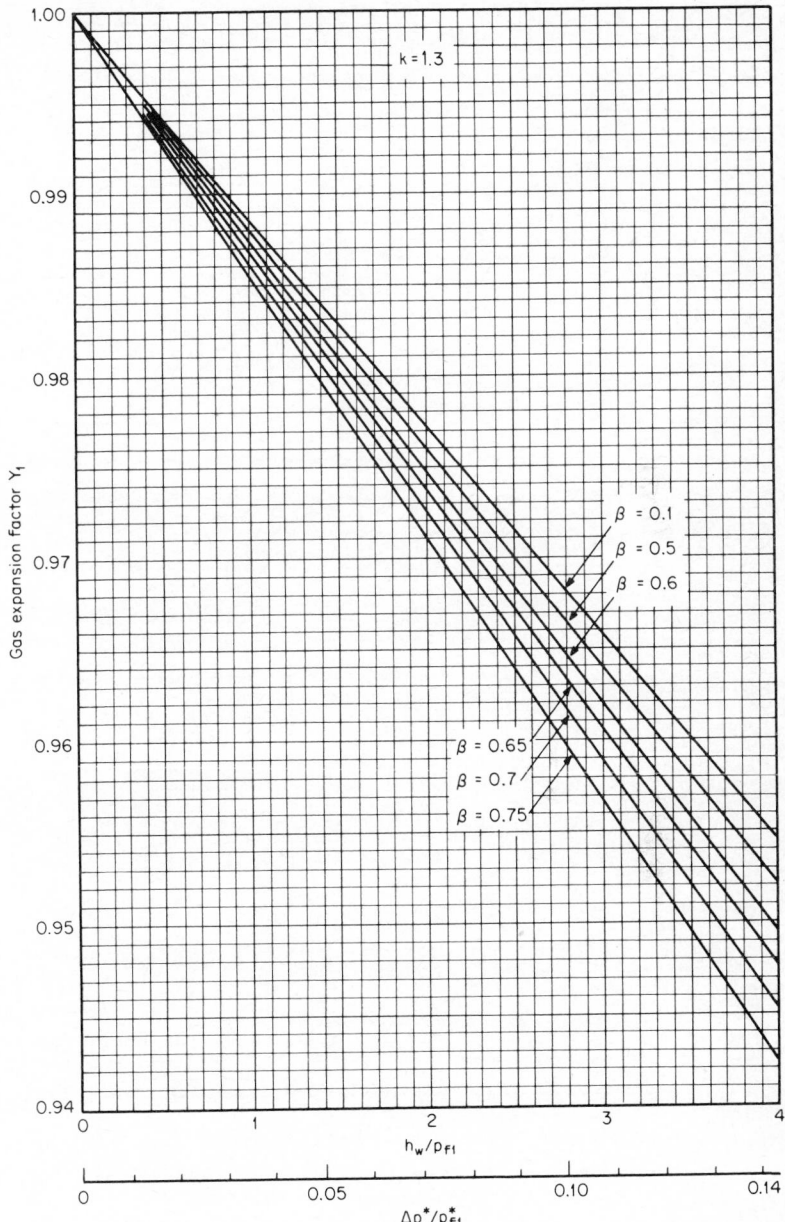

Figure 10.24 Gas expansion factor Y_1 for concentric square-edged orifice; corner, flange, D and $D/2$ taps.

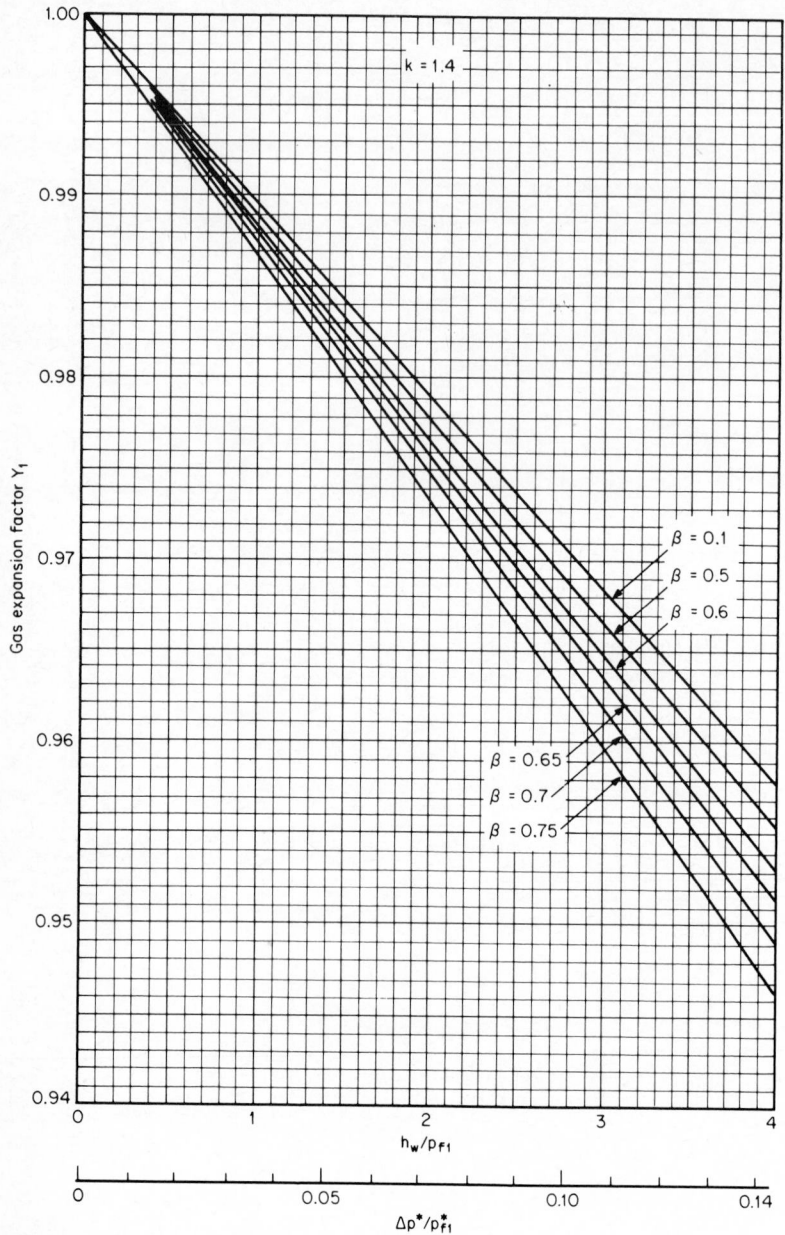

Figure 10.25 Gas expansion factor Y_1 for concentric square-edged orifice; corner, flange, D and $D/2$ taps.

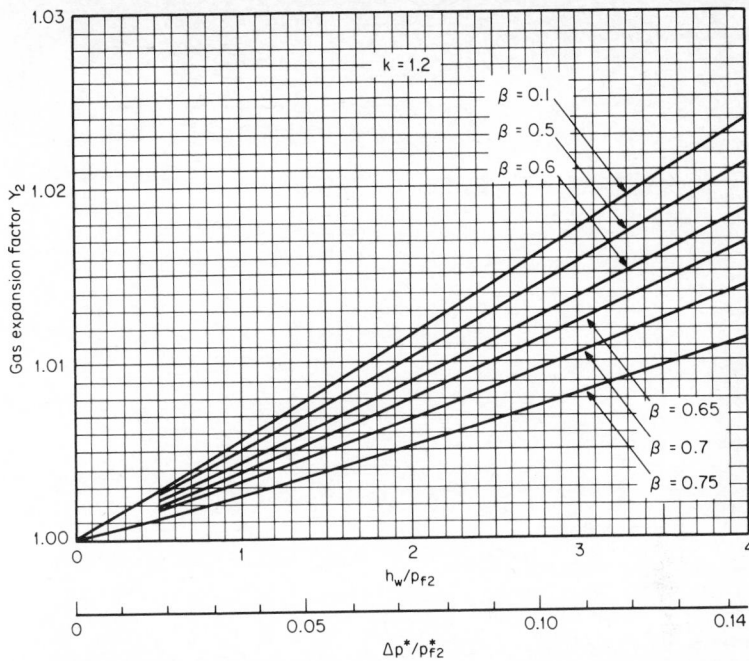

Figure 10.26 Gas expansion factor Y_2 for concentric square-edged orifice; corner, flange, D and $D/2$ taps.

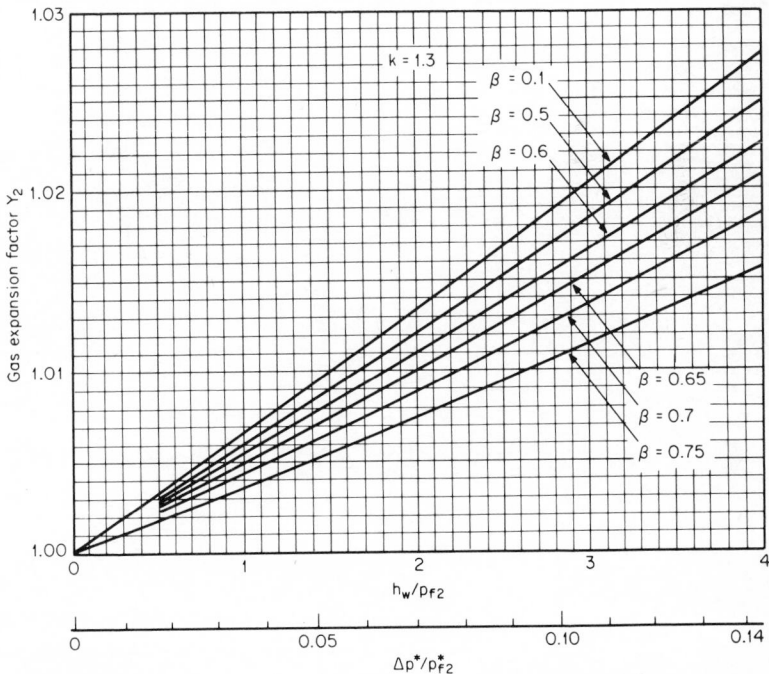

Figure 10.27 Gas expansion factor Y_2 for concentric square-edged orifice; corner, flange, D and $D/2$ taps.

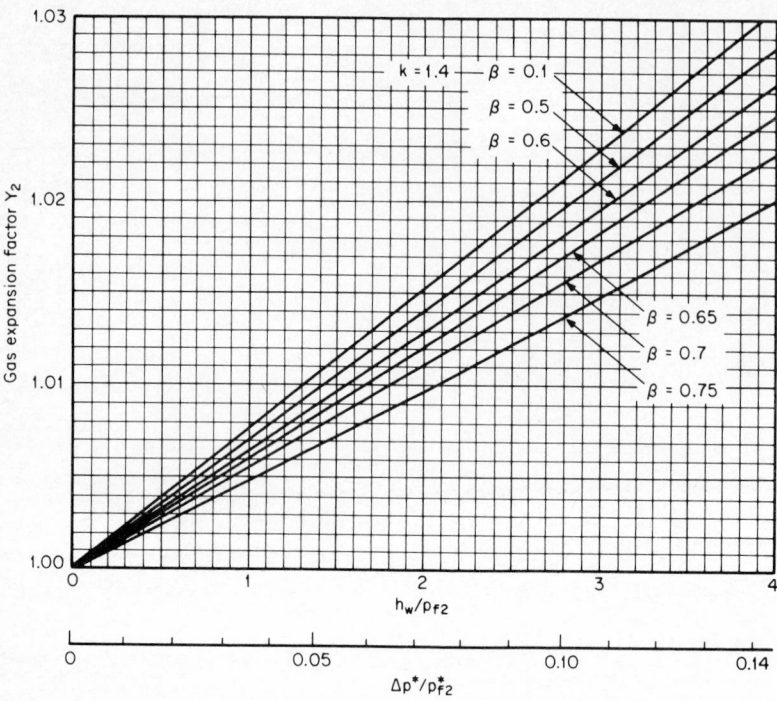

Figure 10.28 Gas expansion factor Y_2 for concentric square-edged orifice; corner, flange, D and $D/2$ taps.

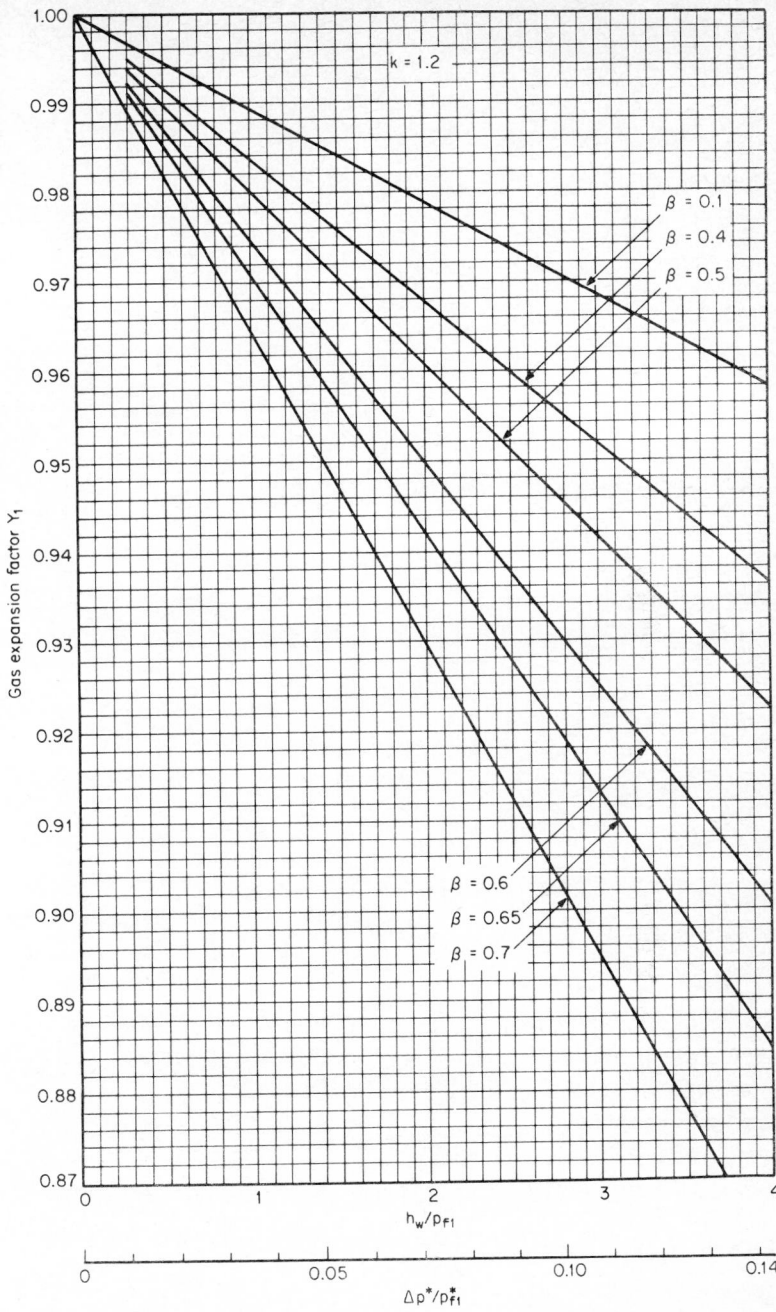

Figure 10.29 Gas expansion factor Y_1 for concentric square-edged orifice; $2\frac{1}{2}$ D and $8D$ taps.

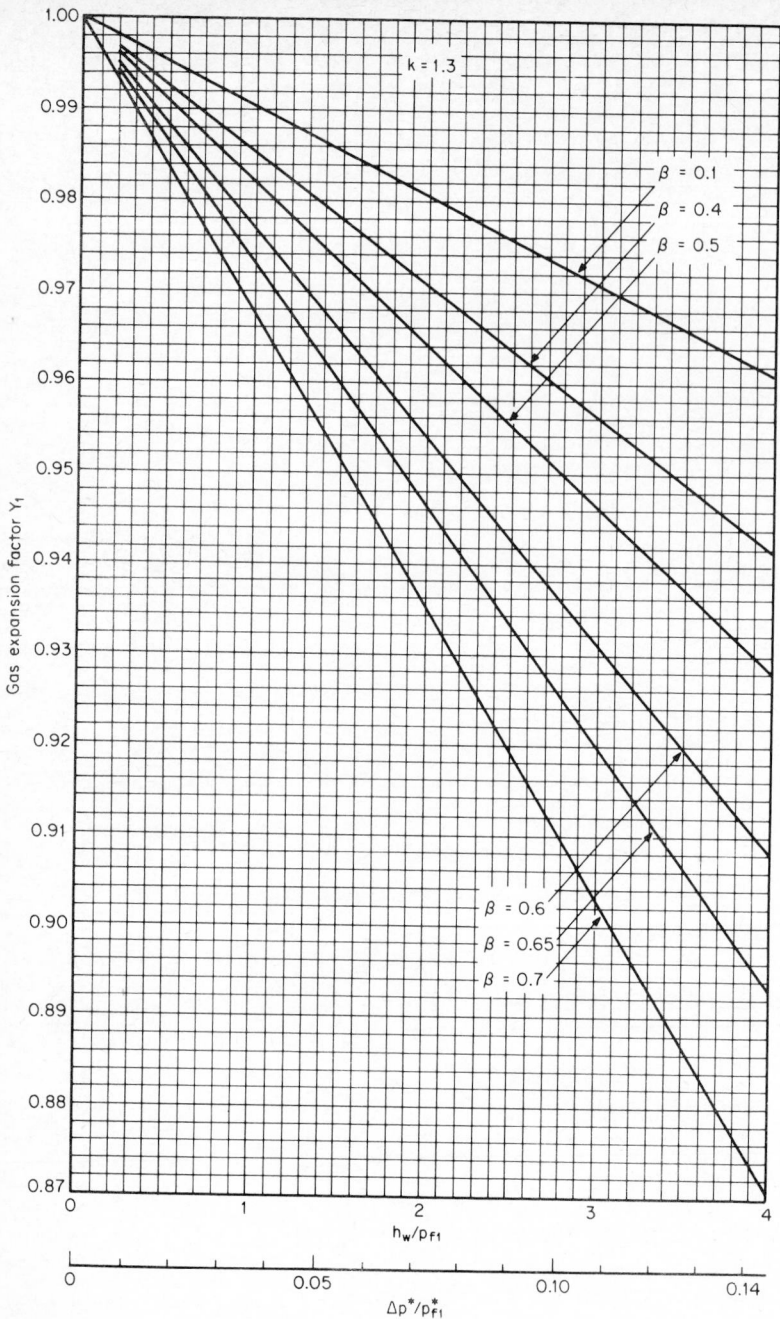

Figure 10.30 Gas expansion factor Y_1 for concentric square-edged orifice; $2\frac{1}{2}\,D$ and $8D$ taps.

10-30

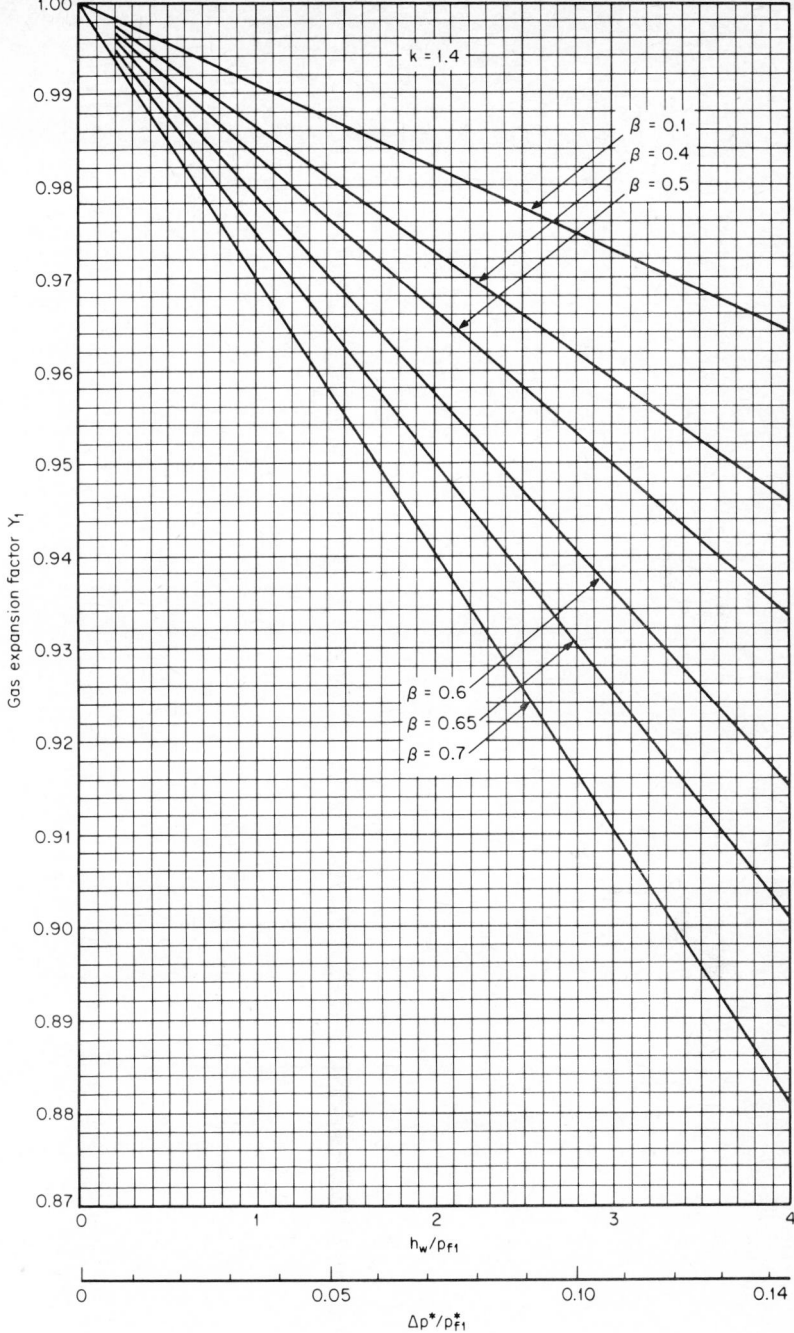

Figure 10.31   Gas expansion factor Y_1 for concentric square-edged orifice; $2\frac{1}{2}$ D and $8D$ taps.

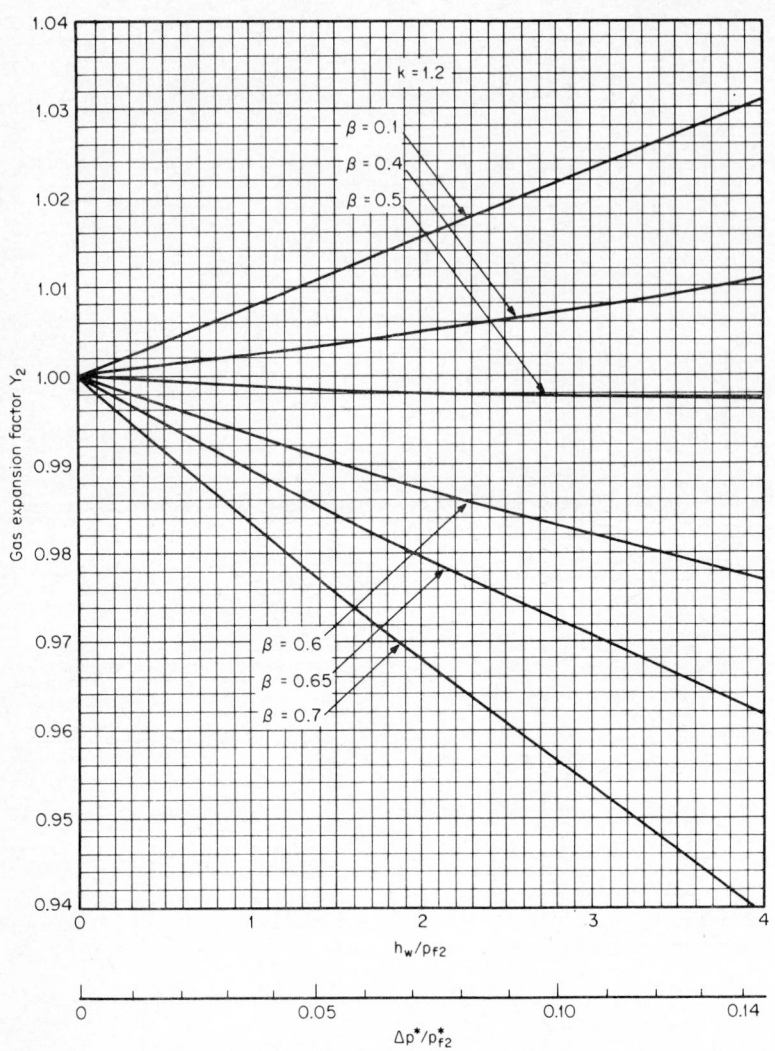

Figure 10.32 Gas expansion factor Y_2 for concentric square-edged orifice; $2\frac{1}{2}$ D and $8D$ taps.

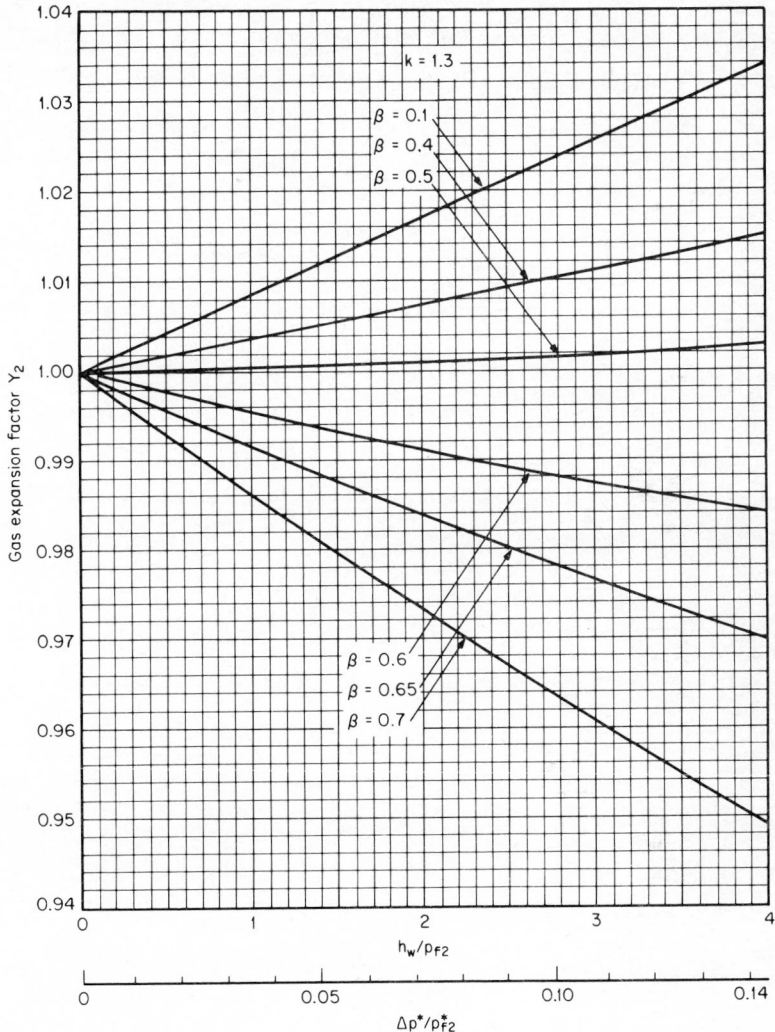

Figure 10.33 Gas expansion factor Y_2 for concentric square-edged orifice; D and $D/2$ taps.

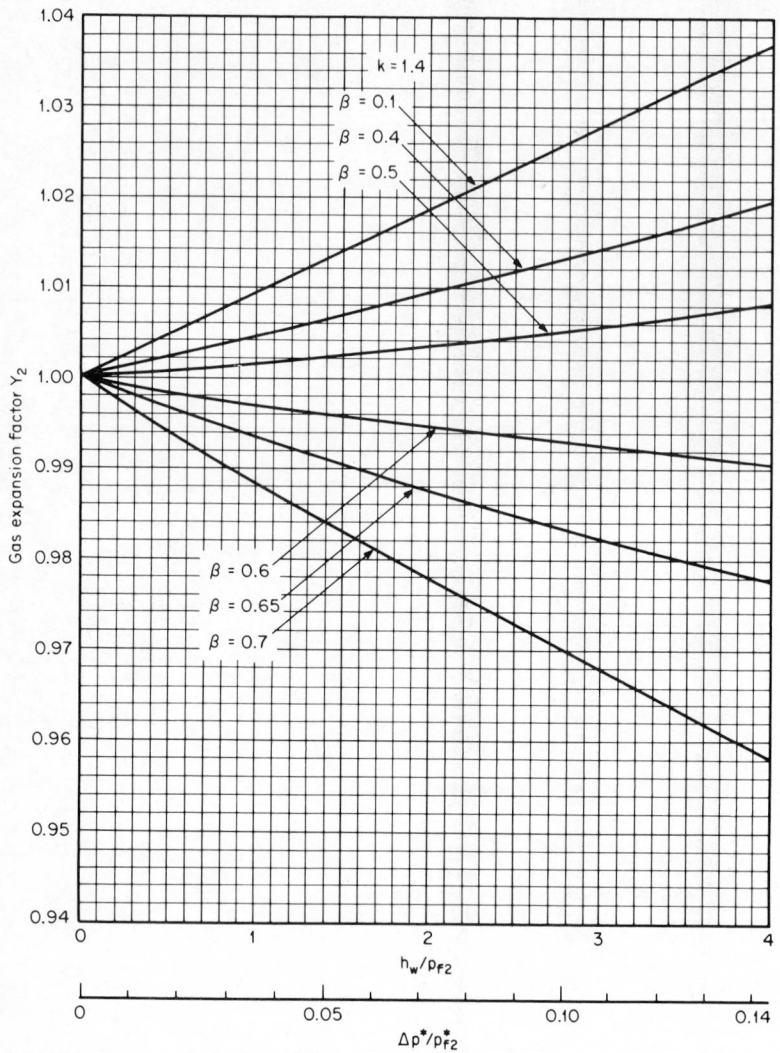

Figure 10.34   Gas expansion factor Y_2 for $2\frac{1}{2}$ D and $8D$ taps.

Small-Bore Honed-Orifice Meter Runs

The effects of plate eccentricity, pipe roughness, and pressure-tap geometry are magnified in smaller line sizes. For this reason, small-line-size orifices are installed in honed meter runs, as shown in Fig. 10.35. As in the corner-tap design, the differential-pressure taps are annular grooves on each side of the plate. The use of grooves ensures manufacturing uniformity.

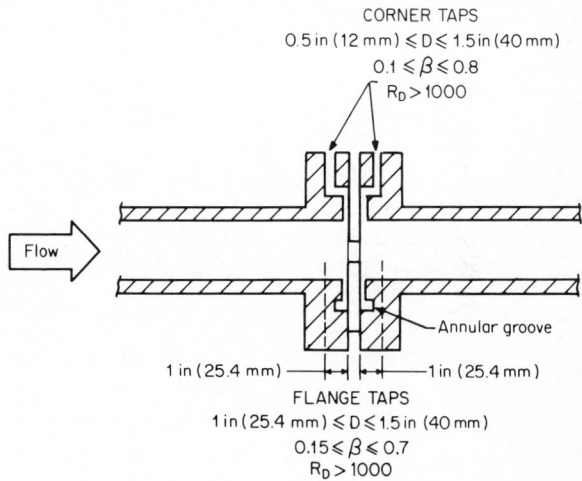

Figure 10.35 gives the labels:

CORNER TAPS
0.5 in (12 mm) $\leqslant D \leqslant$ 1.5 in (40 mm)
$0.1 \leqslant \beta \leqslant 0.8$
$R_D > 1000$

Flow

Annular groove

1 in (25.4 mm) — 1 in (25.4 mm)

FLANGE TAPS
1 in (25.4 mm) $\leqslant D \leqslant$ 1.5 in (40 mm)
$0.15 \leqslant \beta \leqslant 0.7$
$R_D > 1000$

Note: Threaded or flanged tube ends, assembled and then bored and honed

Figure 10.35 Honed-orifice meter run.

The equation for the flow coefficient K for meter runs in line sizes from $\frac{1}{2}$ to $1\frac{1}{2}$ in (12 to 40 mm) is given in *Fluid Meters* (ASME, 1971) as

$$\frac{C}{\sqrt{1 - \beta^4}} = K = \left[0.5991 + \frac{0.0044}{D} + \left(0.3155 + \frac{0.0175}{D} \right) (\beta^4 + 2\beta^{16}) \right]$$
$$+ \left[\frac{0.52}{D} - 0.192 + \left(16.48 - \frac{1.16}{D} \right) (\beta^4 + 4\beta^{16}) \right] R_D^{-0.5} \quad (10.1)$$

For corner taps the accuracy of this equation is ± 0.75 percent for pipe Reynolds numbers greater than 1000 and β ranging from 0.1 to 0.8.

Occasionally, flange taps are used in line sizes between 1 and $1\frac{1}{2}$ in (25 and 40 mm), in which case the flow-coefficient equation is

$$\frac{C}{\sqrt{1 - \beta^4}} = K = 0.5980 + 0.468(\beta^4 + 10\beta^{12}) + (0.87 + 8.1\beta^4)R_D^{-0.5} \quad (10.2)$$

The gas expansion factors shown in Figs. 10.23 through 10.28 are used for corner taps.

Low-Reynolds-Number Orifices

For Reynolds numbers below 10,000, the discharge coefficient of a square-edged orifice changes appreciably with either flow rate or viscosity (see Fig. 10.22). For this reason, either a quadrant- or conical-edged inlet is used. Both geometries have an essentially constant coefficient in the laminar flow regime. These geometries are covered in BSI Standard 1042 (1964), but no U.S. or international standard is available.

Eujen (1977) and others have reported that upstream pipe length alters the coefficient for both designs by from ±1.5 to ±2 percent. The laminar parabolic profile is not fully developed for short pipe lengths, and the coefficient rises when the upstream length is increased. This is practically eliminated by installing upstream diffuser plates, similar to the Sprenkle plates discussed in Chap. 5.

Quadrant-Edged Orifice   The quadrant, or quarter-circle orifice, is shown in Fig. 10.36. It is readily manufactured from standard orifice-plate blanks with a high degree of conformity. It is relatively immune to the effects of corrosion, erosion, and the deposit of solids at the surface of the orifice.

The Reynolds-number range over which the coefficient is constant is shown in Fig. 10.37. In Fig. 10.38, the discharge is plotted versus β. Since the effect of upstream length is greatest for large β, it is recommended that β be sized as small

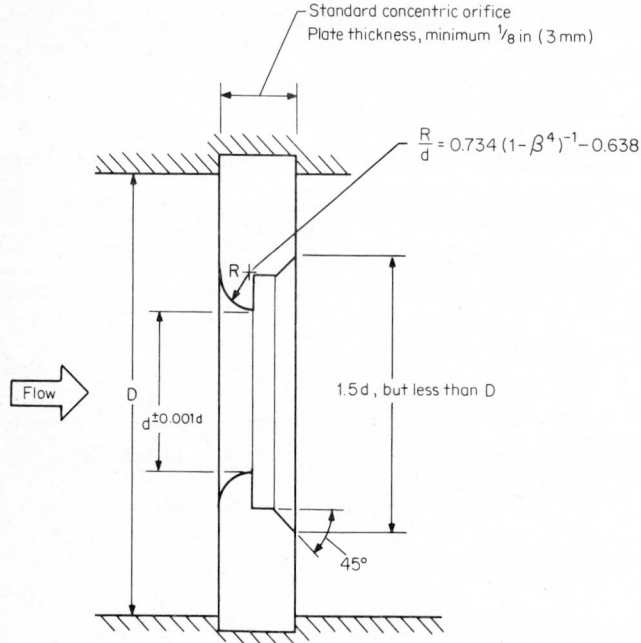

Standard concentric orifice
Plate thickness, minimum $\frac{1}{8}$ in (3 mm)

$$\frac{R}{d} = 0.734\,(1-\beta^4)^{-1} - 0.638$$

Flow

D

$d^{\pm 0.001d}$

R

1.5 d, but less than D

45°

Figure 10.36   Quadrant (quarter-circle) concentric orifice.

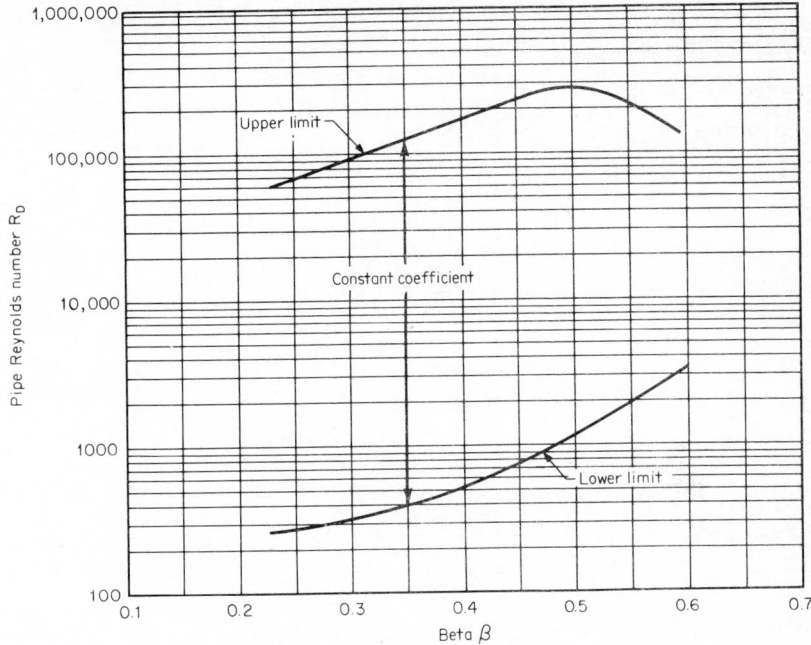

Figure 10.37 Quadrant-orifice Reynolds-number range.

as practicable. The beta ratio is sized graphically by the procedure given in Chap. 9; the β_0 equation is given in Table 9.28.

Conical-Entrance Orifice   In the conical-entrance orifice the quarter-circle of the quadrant is replaced by a conical inlet, as shown in Fig. 10.39. This inlet geometry provides for a lower Reynolds-number limit. The plate is considered by some to be more difficult to manufacture than the quadrant-edged orifice plate, because the bore-section land is difficult to reproduce accurately in some sizes.

Depending on the Reynolds-number range, the discharge coefficient is assigned two different values. These are shown in Fig. 10.40 (and in Fig. 10.38). Substitution of the discharge coefficient for the selected Reynolds-number range into Eq. (9.83) will exactly size the bore, since the coefficient is assumed constant over each range.

Eccentric and Segmental Orifices

These orifices were developed to meter both liquids containing sediment and gases entrained within liquids. Their construction materials, plate thickness, edge sharpness, etc., are identical to those for the square-edged concentric ori-

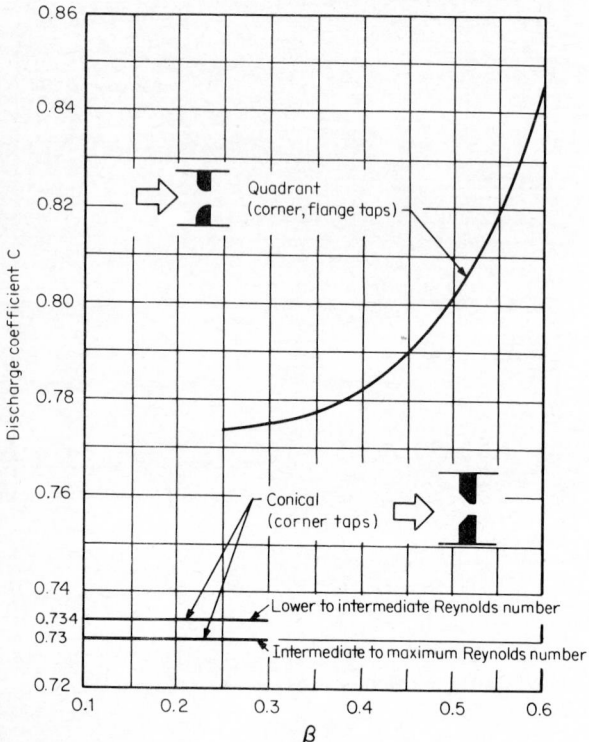

Figure 10.38 **Discharge coefficient for quadrant and conical orifices.**

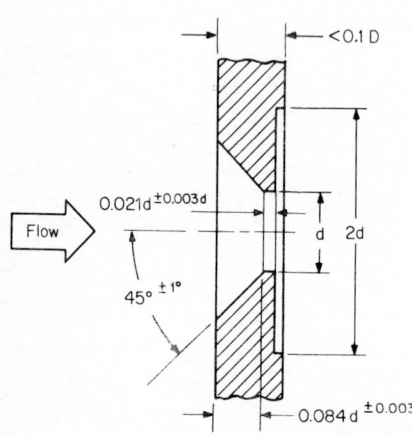

Figure 10.39 **Conical-entrance orifice.**

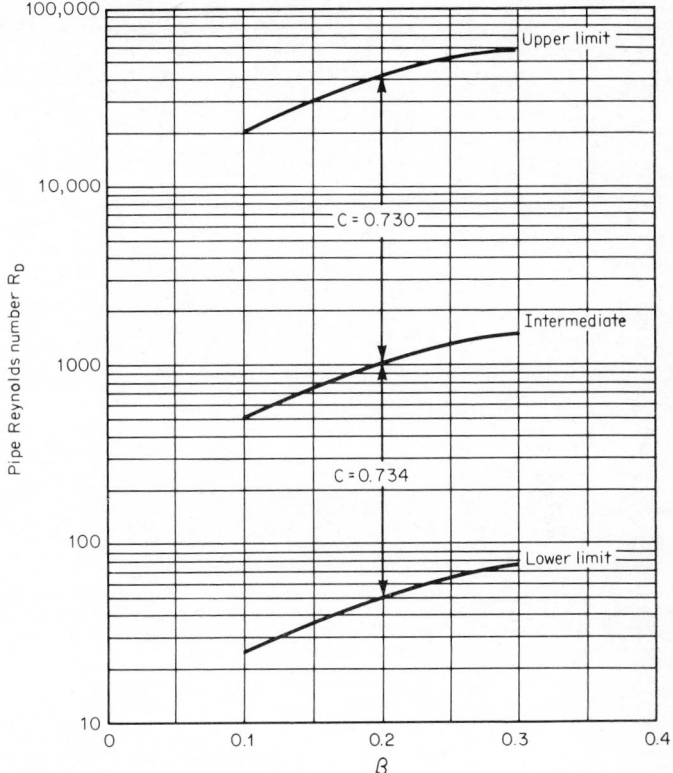

Figure 10.40 Conical-orifice Reynolds-number ranges.

fice. Also, their sizing and flow equations are the same as for other differential producers.

The discharge coefficient curves were developed from data obtained in an ASME test program. The initial 4-in (100-mm) and 6-in (150-mm) calibration results were reported by Beitler and Masson (1949). A later ASME fluid meters committee report extending line sizes to 10 and 14 in (250 and 350 mm) was prepared by Lindahl and Beitler (1954). The test data was for both flange and vena contracta taps, with the flange taps located at the conventional 1-in (25-mm) spacing. The location of the vena contracta, which was determined experimentally, is shown in Fig. 10.41.

Segmental Orifice The segmental orifice is shown in Fig. 10.42. The opening is a segment of a circle having a diameter of 98 percent of the pipe diameter. The circular section of the opening should be concentric with the pipe and accurately centered, to ensure that the opening is not covered by the inlet pipe or gaskets or because of the pipe eccentricity.

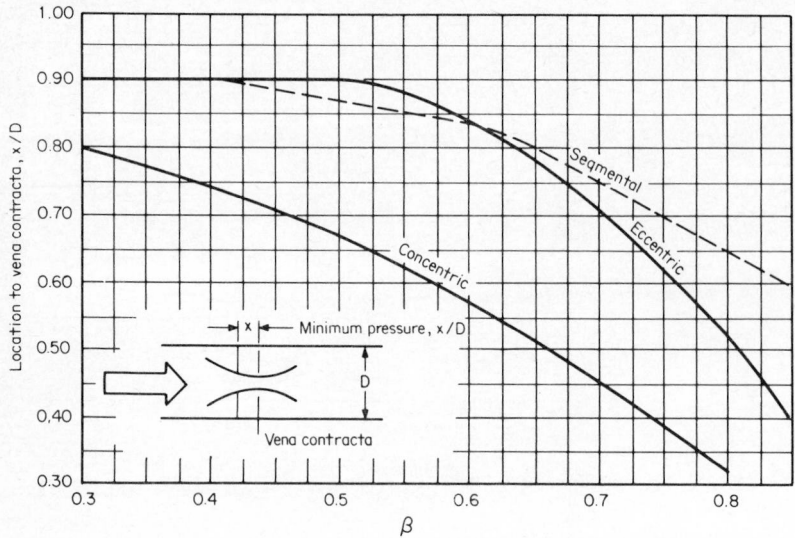

Figure 10.41   Location of minimum pressure (vena contracta).

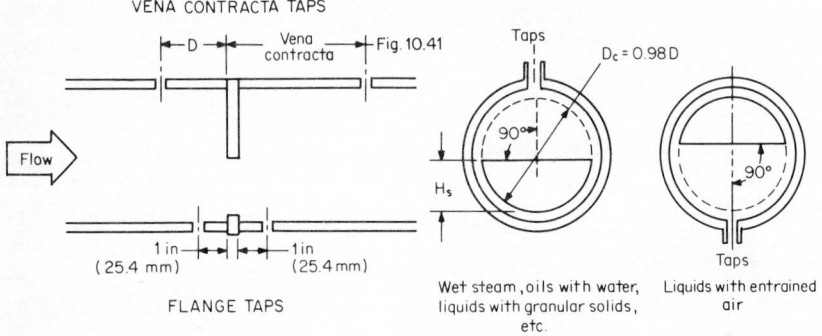

Figure 10.42   Segmental orifice.

For segmental orifices, an equivalent value β_c is used for the diameter ratio; it is equal to the diameter ratio of a circular orifice which has an aperture equal in area to the area of the segment. The relationship between the sized beta ratio and the segment geometry is

$$\beta_c = \frac{d}{D_c} = \frac{\beta}{0.98} = \left(\frac{1}{\pi} \left\{ \arccos\left(1 - \frac{2H_s}{D_c} \right) \right.\right.$$

$$\left.\left. - 2\left(1 - \frac{2H_s}{D_c} \right)\left[\frac{H_s}{D_c} - \left(\frac{H_s}{D_c} \right)^2 \right]^{1/2} \right\} \right)^{1/2} \quad (10.3)$$

An iterative solution is required to obtain the segment height H_s. Table 10.2 presents the dimensional relationship given by Eq. (10.3) for a concentric orifice and a segmental orifice.

The segmental orifice has one characteristic that makes it useful for special service; it eliminates damming of foreign material at the side of the pipe on which the opening is located. This characteristic makes it useful for measuring wet steam, liquids containing granular solids, and oils containing water when the measurement must be made in a horizontal pipe. When the orifice may be located in a vertical run with the flow in a *downward* direction, the concentric orifice is preferred because of its better discharge-coefficient accuracy.

When liquids containing gas or vapor must be measured in a horizontal pipe, the segmental orifice may be used with the opening at the top of the pipe. How-

Table 10.2 **Segmental-Orifice Dimensional Table**

Pipe diameter, D
Segment diameter
$D_c = 0.98D$
H_s

$$\beta_c = \frac{d}{D_c} = \frac{\beta}{0.98} = \left(\frac{1}{\pi} \left\{ \arccos\left(1 - \frac{2H_s}{D_c} \right) - 2\left(1 - \frac{2H_s}{D_c} \right)\left[\frac{H_s}{D_c} - \left(\frac{H_s}{D_c} \right)^2 \right]^{1/2} \right\} \right)^{1/2}$$

d/D_c	H_s/D_c	d/D_c	H_s/D_c	d/D_c	H_s/D_c	d/D_c	H_s/D_c	d/D_c	H_s/D_c
.28255	.1340	.29321	.1410	.30372	.1480	.31405	.1550	.32426	.1620
.28286	.1342	.29352	.1412	.30401	.1482	.31435	.1552	.32455	.1622
.28316	.1344	.29382	.1414	.30431	.1484	.31465	.1554	.32483	.1624
.28347	.1346	.29412	.1416	.30460	.1486	.31493	.1556	.32512	.1626
.28379	.1348	.29442	.1418	.30490	.1488	.31523	.1558	.32541	.1628
.28409	.1350	.29472	.1420	.30520	.1490	.31553	.1560	.32570	.1630
.28440	.1352	.29502	.1422	.30550	.1492	.31582	.1562	.32599	.1632
.28470	.1354	.29533	.1424	.30579	.1494	.31611	.1564	.32628	.1634
.28501	.1356	.29563	.1426	.30609	.1496	.31640	.1566	.32656	.1636
.28531	.1358	.29592	.1428	.30638	.1498	.31670	.1568	.32685	.1638
.28562	.1360	.29623	.1430	.30668	.1500	.31698	.1570	.32714	.1640
.28592	.1362	.29653	.1432	.30698	.1502	.31728	.1572	.32743	.1642
.28622	.1364	.29683	.1434	.30728	.1504	.31757	.1574	.32772	.1644
.28653	.1366	.29713	.1436	.30758	.1506	.31787	.1576	.32801	.1646
.28684	.1368	.29743	.1438	.30787	.1508	.31815	.1578	.32830	.1648
.28714	.1370	.29773	.1440	.30817	.1510	.31844	.1580	.32858	.1650
.28745	.1372	.29803	.1442	.30846	.1512	.31874	.1582	.32887	.1652
.28775	.1374	.29833	.1444	.30876	.1514	.31903	.1584	.32916	.1654
.28805	.1376	.29863	.1446	.30905	.1516	.31932	.1586	.32944	.1656
.28836	.1378	.29893	.1448	.30935	.1518	.31961	.1588	.32973	.1658
.28867	.1380	.29924	.1450	.30965	.1520	.31990	.1590	.33002	.1660
.28897	.1382	.29953	.1452	.30993	.1522	.32019	.1592	.33030	.1662
.28927	.1384	.29983	.1454	.31023	.1524	.32048	.1594	.33059	.1664
.28957	.1386	.30013	.1456	.31053	.1526	.32077	.1596	.33088	.1666
.28988	.1388	.30043	.1458	.31082	.1528	.32107	.1598	.33117	.1668
.29019	.1390	.30073	.1460	.31112	.1530	.32136	.1600	.33145	.1670
.29049	.1392	.30103	.1462	.31141	.1532	.32165	.1602	.33174	.1672
.29079	.1394	.30133	.1464	.31171	.1534	.32194	.1604	.33203	.1674
.29110	.1396	.30162	.1466	.31200	.1536	.32223	.1606	.33232	.1676
.29140	.1398	.30192	.1468	.31229	.1538	.32252	.1608	.33260	.1678
.29170	.1400	.30222	.1470	.31259	.1540	.32281	.1610	.33289	.1680
.29202	.1402	.30252	.1472	.31289	.1542	.32309	.1612	.33317	.1682
.29231	.1404	.30282	.1474	.31317	.1544	.32338	.1614	.33346	.1684
.29261	.1406	.30312	.1476	.31347	.1546	.32367	.1616	.33374	.1686
.29291	.1408	.30341	.1478	.31377	.1548	.32397	.1618	.33403	.1688

Table 10.2   Segmental-Orifice Dimensional Table *(continued)*

d/D_c	H_a/D_c	d/D_c	H_a/D_c	d/D_c	H_a/D_c	d/D_c	H_a/D_c	d/D_c	H_a/D_c
.33431	.1690	.35404	.1830	.37327	.1970	.39204	.2110	.41037	.2250
.33460	.1692	.35431	.1832	.37353	.1972	.39230	.2112	.41064	.2252
.33488	.1694	.35459	.1834	.37380	.1974	.39257	.2114	.41089	.2254
.33517	.1696	.35486	.1836	.37408	.1976	.39283	.2116	.41115	.2256
.33545	.1698	.35515	.1838	.37435	.1978	.39309	.2118	.41141	.2258
.33574	.1700	.35543	.1840	.37462	.1980	.39336	.2120	.41168	.2260
.33602	.1702	.35570	.1842	.37489	.1982	.39362	.2122	.41193	.2262
.33631	.1704	.35598	.1844	.37516	.1984	.39389	.2124	.41219	.2264
.33659	.1706	.35625	.1846	.37543	.1986	.39415	.2126	.41245	.2266
.33688	.1708	.35654	.1848	.37571	.1988	.39441	.2128	.41270	.2268
.33716	.1710	.35681	.1850	.37597	.1990	.39468	.2130	.41296	.2270
.33744	.1712	.35709	.1852	.37624	.1992	.39494	.2132	.41322	.2272
.33773	.1714	.35736	.1854	.37651	.1994	.39521	.2134	.41348	.2274
.33801	.1716	.35764	.1856	.37678	.1996	.39547	.2136	.41374	.2276
.33830	.1718	.35792	.1858	.37705	.1998	.39574	.2138	.41400	.2278
.33858	.1720	.35820	.1860	.37733	.2000	.39600	.2140	.41425	.2280
.33886	.1722	.35847	.1862	.37759	.2002	.39627	.2142	.41451	.2282
.33915	.1724	.35875	.1864	.37786	.2004	.39653	.2144	.41477	.2284
.33943	.1726	.35903	.1866	.37813	.2006	.39679	.2146	.41503	.2286
.33971	.1728	.35930	.1868	.37840	.2008	.39705	.2148	.41528	.2288
.34000	.1730	.35958	.1870	.37867	.2010	.39731	.2150	.41554	.2290
.34028	.1732	.35985	.1872	.37894	.2012	.39758	.2152	.41580	.2292
.34057	.1734	.36013	.1874	.37921	.2014	.39784	.2154	.41606	.2294
.34085	.1736	.36041	.1876	.37948	.2016	.39811	.2156	.41631	.2296
.34114	.1738	.36068	.1878	.37975	.2018	.39837	.2158	.41657	.2298
.34142	.1740	.36096	.1880	.38002	.2020	.39863	.2160	.41683	.2300
.34170	.1742	.36123	.1882	.38029	.2022	.39889	.2162	.41708	.2302
.34198	.1744	.36151	.1884	.38056	.2024	.39915	.2164	.41734	.2304
.34227	.1746	.36179	.1886	.38082	.2026	.39942	.2166	.41760	.2306
.34255	.1748	.36206	.1888	.38110	.2028	.39968	.2168	.41786	.2308
.34283	.1750	.36233	.1890	.38136	.2030	.39995	.2170	.41811	.2310
.34311	.1752	.36261	.1892	.38163	.2032	.40021	.2172	.41837	.2312
.34339	.1754	.36289	.1894	.38190	.2034	.40047	.2174	.41862	.2314
.34367	.1756	.36316	.1896	.38217	.2036	.40073	.2176	.41888	.2316
.34395	.1758	.36343	.1898	.38243	.2038	.40099	.2178	.41914	.2318
.34424	.1760	.36371	.1900	.38270	.2040	.40126	.2180	.41940	.2320
.34452	.1762	.36399	.1902	.38297	.2042	.40151	.2182	.41965	.2322
.34480	.1764	.36426	.1904	.38324	.2044	.40178	.2184	.41991	.2324
.34509	.1766	.36453	.1906	.38351	.2046	.40204	.2186	.42016	.2326
.34537	.1768	.36481	.1908	.38377	.2048	.40231	.2188	.42042	.2328
.34565	.1770	.36508	.1910	.38404	.2050	.40256	.2190	.42067	.2330
.34593	.1772	.36535	.1912	.38431	.2052	.40283	.2192	.42093	.2332
.34621	.1774	.36563	.1914	.38457	.2054	.40309	.2194	.42119	.2334
.34649	.1776	.36590	.1916	.38485	.2056	.40335	.2196	.42145	.2336
.34677	.1778	.36618	.1918	.38511	.2058	.40361	.2198	.42170	.2338
.34705	.1780	.36645	.1920	.38538	.2060	.40387	.2200	.42196	.2340
.34733	.1782	.36672	.1922	.38565	.2062	.40413	.2202	.42221	.2342
.34761	.1784	.36700	.1924	.38591	.2064	.40439	.2204	.42247	.2344
.34789	.1786	.36727	.1926	.38618	.2066	.40466	.2206	.42272	.2346
.34817	.1788	.36754	.1928	.38645	.2068	.40491	.2208	.42298	.2348
.34845	.1790	.36782	.1930	.38672	.2070	.40518	.2210	.42323	.2350
.34873	.1792	.36809	.1932	.38698	.2072	.40544	.2212	.42349	.2352
.34901	.1794	.36836	.1934	.38725	.2074	.40570	.2214	.42374	.2354
.34929	.1796	.36863	.1936	.38752	.2076	.40596	.2216	.42400	.2356
.34957	.1798	.36891	.1938	.38778	.2078	.40622	.2218	.42425	.2358
.34986	.1800	.36918	.1940	.38805	.2080	.40648	.2220	.42450	.2360
.35013	.1802	.36945	.1942	.38831	.2082	.40674	.2222	.42476	.2362
.35041	.1804	.36973	.1944	.38858	.2084	.40700	.2224	.42501	.2364
.35069	.1806	.37000	.1946	.38885	.2086	.40726	.2226	.42527	.2366
.35097	.1808	.37027	.1948	.38911	.2088	.40752	.2228	.42552	.2368
.35125	.1810	.37054	.1950	.38938	.2090	.40778	.2230	.42578	.2370
.35152	.1812	.37082	.1952	.38964	.2092	.40804	.2232	.42603	.2372
.35181	.1814	.37109	.1954	.38991	.2094	.40830	.2234	.42629	.2374
.35209	.1816	.37136	.1956	.39018	.2096	.40856	.2236	.42654	.2376
.35237	.1818	.37164	.1958	.39044	.2098	.40882	.2238	.42680	.2378
.35264	.1820	.37190	.1960	.39071	.2100	.40908	.2240	.42705	.2380
.35292	.1822	.37217	.1962	.39097	.2102	.40934	.2242	.42730	.2382
.35320	.1824	.37245	.1964	.39124	.2104	.40960	.2244	.42755	.2384
.35348	.1826	.37272	.1966	.39150	.2106	.40986	.2246	.42781	.2386
.35376	.1828	.37299	.1968	.39176	.2108	.41012	.2248	.42806	.2388

Table 10.2 **Segmental-Orifice Dimensional Table (*continued*)**

d/D_c	H_s/D_c	d/D_c	H_s/D_c	d/D_c	H_s/D_c	d/D_c	H_s/D_c	d/D_c	H_s/D_c
.42832	.2390	.44588	.2530	.46309	.2670	.47994	.2810	.49648	.2950
.42857	.2392	.44613	.2532	.46332	.2672	.48018	.2812	.49671	.2952
.42882	.2394	.44637	.2534	.46357	.2674	.48042	.2814	.49694	.2954
.42908	.2396	.44663	.2536	.46381	.2676	.48066	.2816	.49718	.2956
.42933	.2398	.44687	.2538	.46405	.2678	.48090	.2818	.49741	.2958
.42958	.2400	.44712	.2540	.46430	.2680	.48114	.2820	.49764	.2960
.42984	.2402	.44737	.2542	.46454	.2682	.48137	.2822	.49788	.2962
.43009	.2404	.44762	.2544	.46478	.2684	.48161	.2824	.49811	.2964
.43035	.2406	.44786	.2546	.46503	.2686	.48185	.2826	.49835	.2966
.43060	.2408	.44811	.2548	.46527	.2688	.48208	.2828	.49858	.2968
.43085	.2410	.44836	.2550	.46551	.2690	.48232	.2830	.49881	.2970
.43110	.2412	.44861	.2552	.46575	.2692	.48256	.2832	.49905	.2972
.43136	.2414	.44885	.2554	.46600	.2694	.48280	.2834	.49928	.2974
.43161	.2416	.44910	.2556	.46624	.2696	.48304	.2836	.49951	.2976
.43186	.2418	.44935	.2558	.46648	.2698	.48328	.2838	.49974	.2978
.43211	.2420	.44959	.2560	.46672	.2700	.48351	.2840	.49998	.2980
.43236	.2422	.44985	.2562	.46696	.2702	.48375	.2842	.50021	.2982
.43262	.2424	.45009	.2564	.46721	.2704	.48399	.2844	.50044	.2984
.43287	.2426	.45034	.2566	.46745	.2706	.48422	.2846	.50067	.2986
.43312	.2428	.45058	.2568	.46769	.2708	.48446	.2848	.50091	.2988
.43337	.2430	.45083	.2570	.46793	.2710	.48470	.2850	.50114	.2990
.43363	.2432	.45108	.2572	.46818	.2712	.48493	.2852	.50137	.2992
.43388	.2434	.45132	.2574	.46842	.2714	.48517	.2854	.50160	.2994
.43413	.2436	.45157	.2576	.46866	.2716	.48541	.2856	.50184	.2996
.43438	.2438	.45181	.2578	.46890	.2718	.48564	.2858	.50207	.2998
.43463	.2440	.45207	.2580	.46914	.2720	.48589	.2860	.50230	.3000
.43488	.2442	.45231	.2582	.46938	.2722	.48612	.2862	.50253	.3002
.43514	.2444	.45256	.2584	.46963	.2724	.48636	.2864	.50277	.3004
.43538	.2446	.45280	.2586	.46986	.2726	.48660	.2866	.50300	.3006
.43564	.2448	.45305	.2588	.47011	.2728	.48683	.2868	.50323	.3008
.43589	.2450	.45329	.2590	.47035	.2730	.48707	.2870	.50346	.3010
.43614	.2452	.45354	.2592	.47059	.2732	.48730	.2872	.50369	.3012
.43639	.2454	.45379	.2594	.47083	.2734	.48754	.2874	.50393	.3014
.43665	.2456	.45404	.2596	.47107	.2736	.48777	.2876	.50416	.3016
.43689	.2458	.45428	.2598	.47132	.2738	.48801	.2878	.50439	.3018
.43715	.2460	.45452	.2600	.47155	.2740	.48825	.2880	.50462	.3020
.43739	.2462	.45477	.2602	.47179	.2742	.48848	.2882	.50485	.3022
.43765	.2464	.45502	.2604	.47204	.2744	.48872	.2884	.50508	.3024
.43790	.2466	.45526	.2606	.47227	.2746	.48896	.2886	.50532	.3026
.43815	.2468	.45551	.2608	.47252	.2748	.48919	.2888	.50555	.3028
.43840	.2470	.45575	.2610	.47276	.2750	.48943	.2890	.50578	.3030
.43865	.2472	.45600	.2612	.47300	.2752	.48967	.2892	.50601	.3032
.43890	.2474	.45624	.2614	.47324	.2754	.48990	.2894	.50625	.3034
.43915	.2476	.45649	.2616	.47348	.2756	.49013	.2896	.50648	.3036
.43940	.2478	.45674	.2618	.47372	.2758	.49037	.2898	.50671	.3038
.43965	.2480	.45698	.2620	.47396	.2760	.49061	.2900	.50693	.3040
.43990	.2482	.45722	.2622	.47420	.2762	.49084	.2902	.50717	.3042
.44015	.2484	.45747	.2624	.47444	.2764	.49108	.2904	.50740	.3044
.44040	.2486	.45771	.2626	.47468	.2766	.49132	.2906	.50763	.3046
.44065	.2488	.45796	.2628	.47492	.2768	.49155	.2908	.50786	.3048
.44090	.2490	.45820	.2630	.47516	.2770	.49178	.2910	.50809	.3050
.44115	.2492	.45845	.2632	.47540	.2772	.49202	.2912	.50832	.3052
.44140	.2494	.45869	.2634	.47564	.2774	.49226	.2914	.50855	.3054
.44165	.2496	.45894	.2636	.47588	.2776	.49249	.2916	.50878	.3056
.44190	.2498	.45918	.2638	.47612	.2778	.49272	.2918	.50902	.3058
.44215	.2500	.45942	.2640	.47636	.2780	.49296	.2920	.50924	.3060
.44240	.2502	.45967	.2642	.47660	.2782	.49319	.2922	.50947	.3062
.44264	.2504	.45991	.2644	.47684	.2784	.49343	.2924	.50970	.3064
.44289	.2506	.46016	.2646	.47708	.2786	.49367	.2926	.50994	.3066
.44314	.2508	.46040	.2648	.47731	.2788	.49390	.2928	.51017	.3068
.44340	.2510	.46064	.2650	.47756	.2790	.49413	.2930	.51040	.3070
.44364	.2512	.46089	.2652	.47779	.2792	.49437	.2932	.51063	.3072
.44389	.2514	.46113	.2654	.47803	.2794	.49460	.2934	.51085	.3074
.44414	.2516	.46138	.2656	.47827	.2796	.49484	.2936	.51108	.3076
.44439	.2518	.46162	.2658	.47851	.2798	.49507	.2938	.51132	.3078
.44464	.2520	.46187	.2660	.47875	.2800	.49531	.2940	.51155	.3080
.44189	.2522	.46211	.2662	.47899	.2802	.49554	.2942	.51178	.3082
.44514	.2524	.46235	.2664	.47923	.2804	.49577	.2944	.51200	.3084
.44538	.2526	.46260	.2666	.47947	.2806	.49600	.2946	.51223	.3086
.44563	.2528	.46284	.2668	.47971	.2808	.49624	.2948	.51246	.3088

Table 10.2 Segmental-Orifice Dimensional Table (*continued*)

d/D_c	H_s/D_c	d/D_c	H_s/D_c	d/D_c	H_s/D_c	d/D_c	H_s/D_c	d/D_c	H_s/D_c
.51269	.3090	.52861	.3230	.54423	.3370	.55958	.3510	.57464	.3650
.51292	.3092	.52884	.3232	.54446	.3372	.55979	.3512	.57486	.3652
.51315	.3094	.52906	.3234	.54468	.3374	.56001	.3514	.57507	.3654
.51338	.3096	.52929	.3236	.54490	.3376	.56023	.3516	.57529	.3656
.51361	.3098	.52951	.3238	.54512	.3378	.56045	.3518	.57550	.3658
.51384	.3100	.52974	.3240	.54534	.3380	.56066	.3520	.57571	.3660
.51407	.3102	.52996	.3242	.54556	.3382	.56088	.3522	.57592	.3662
.51430	.3104	.53019	.3244	.54578	.3384	.56110	.3524	.57614	.3664
.51453	.3106	.53041	.3246	.54600	.3386	.56131	.3526	.57635	.3666
.51476	.3108	.53064	.3248	.54622	.3388	.56153	.3528	.57656	.3668
.51499	.3110	.53086	.3250	.54645	.3390	.56175	.3530	.57678	.3670
.51521	.3112	.53109	.3252	.54666	.3392	.56196	.3532	.57699	.3672
.51544	.3114	.53131	.3254	.54688	.3394	.56218	.3534	.57720	.3674
.51567	.3116	.53154	.3256	.54710	.3396	.56240	.3536	.57741	.3676
.51590	.3118	.53176	.3258	.54732	.3398	.56261	.3538	.57762	.3678
.51613	.3120	.53198	.3260	.54755	.3400	.56283	.3540	.57784	.3680
.51636	.3122	.53221	.3262	.54777	.3402	.56305	.3542	.57805	.3682
.51659	.3124	.53243	.3264	.54799	.3404	.56326	.3544	.57826	.3684
.51682	.3126	.53266	.3266	.54821	.3406	.56348	.3546	.57848	.3686
.51704	.3128	.53288	.3268	.54843	.3408	.56370	.3548	.57869	.3688
.51727	.3130	.53311	.3270	.54864	.3410	.56391	.3550	.57890	.3690
.51750	.3132	.53333	.3272	.54887	.3412	.56412	.3552	.57911	.3692
.51773	.3134	.53355	.3274	.54909	.3414	.56434	.3554	.57932	.3694
.51796	.3136	.53378	.3276	.54931	.3416	.56455	.3556	.57954	.3696
.51819	.3138	.53400	.3278	.54952	.3418	.56477	.3558	.57975	.3698
.51840	.3140	.53422	.3280	.54974	.3420	.56499	.3560	.57996	.3700
.51864	.3142	.53445	.3282	.54996	.3422	.56520	.3562	.58017	.3702
.51887	.3144	.53467	.3284	.55019	.3424	.56542	.3564	.58039	.3704
.51910	.3146	.53489	.3286	.55041	.3426	.56564	.3566	.58060	.3706
.51932	.3148	.53512	.3288	.55063	.3428	.56585	.3568	.58081	.3708
.51955	.3150	.53534	.3290	.55084	.3430	.56607	.3570	.58102	.3710
.51978	.3152	.53556	.3292	.55106	.3432	.56628	.3572	.58123	.3712
.52001	.3154	.53579	.3294	.55128	.3434	.56650	.3574	.58145	.3714
.52023	.3156	.53601	.3296	.55150	.3436	.56672	.3576	.58165	.3716
.52046	.3158	.53623	.3298	.55172	.3438	.56693	.3578	.58187	.3718
.52069	.3160	.53646	.3300	.55194	.3440	.56714	.3580	.58208	.3720
.52092	.3162	.53668	.3302	.55216	.3442	.56736	.3582	.58229	.3722
.52114	.3164	.53690	.3304	.55238	.3444	.56757	.3584	.58250	.3724
.52137	.3166	.53713	.3306	.55260	.3446	.56779	.3586	.58271	.3726
.52160	.3168	.53735	.3308	.55282	.3448	.56800	.3588	.58292	.3728
.52182	.3170	.53757	.3310	.55303	.3450	.56822	.3590	.58313	.3730
.52205	.3172	.53780	.3312	.55325	.3452	.56844	.3592	.58335	.3732
.52228	.3174	.53802	.3314	.55347	.3454	.56865	.3594	.58356	.3734
.52250	.3176	.53824	.3316	.55369	.3456	.56887	.3596	.58377	.3736
.52273	.3178	.53847	.3318	.55391	.3458	.56908	.3598	.58398	.3738
.52296	.3180	.53869	.3320	.55413	.3460	.56929	.3600	.58419	.3740
.52319	.3182	.53891	.3322	.55435	.3462	.56951	.3602	.58440	.3742
.52342	.3184	.53913	.3324	.55457	.3464	.56972	.3604	.58461	.3744
.52364	.3186	.53935	.3326	.55478	.3466	.56994	.3606	.58482	.3746
.52387	.3188	.53957	.3328	.55500	.3468	.57015	.3608	.58503	.3748
.52409	.3190	.53980	.3330	.55522	.3470	.57037	.3610	.58524	.3750
.52432	.3192	.54002	.3332	.55544	.3472	.57058	.3612	.58546	.3752
.52455	.3194	.54024	.3334	.55566	.3474	.57079	.3614	.58566	.3754
.52477	.3196	.54047	.3336	.55587	.3476	.57101	.3616	.58588	.3756
.52500	.3198	.54069	.3338	.55609	.3478	.57122	.3618	.58609	.3758
.52522	.3200	.54091	.3340	.55631	.3480	.57144	.3620	.58630	.3760
.52545	.3202	.54114	.3342	.55653	.3482	.57165	.3622	.58651	.3762
.52567	.3204	.54136	.3344	.55675	.3484	.57187	.3624	.58672	.3764
.52590	.3206	.54157	.3346	.55697	.3486	.57208	.3626	.58693	.3766
.52613	.3208	.54180	.3348	.55718	.3488	.57229	.3628	.58714	.3768
.52635	.3210	.54202	.3350	.55740	.3490	.57251	.3630	.58735	.3770
.52658	.3212	.54224	.3352	.55762	.3492	.57272	.3632	.58756	.3772
.52680	.3214	.54246	.3354	.55784	.3494	.57294	.3634	.58777	.3774
.52703	.3216	.54269	.3356	.55806	.3496	.57315	.3636	.58798	.3776
.52726	.3218	.54291	.3358	.55827	.3498	.57336	.3638	.58819	.3778
.52748	.3220	.54313	.3360	.55849	.3500	.57358	.3640	.58840	.3780
.52771	.3222	.54335	.3362	.55871	.3502	.57379	.3642	.58861	.3782
.52794	.3224	.54357	.3364	.55892	.3504	.57401	.3644	.58882	.3784
.52816	.3226	.54379	.3366	.55914	.3506	.57422	.3646	.58903	.3786
.52839	.3228	.54401	.3368	.55936	.3508	.57443	.3648	.58924	.3788

Table 10.2 Segmental-Orifice Dimensional Table (*continued*)

d/D_c	H_s/D_c	d/D_c	H_s/D_c	d/D_c	H_s/D_c	d/D_c	H_s/D_c	d/D_c	H_s/D_c
.58945	.3790	.60399	.3930	.61828	.4070	.63232	.4210	.64611	.4350
.58966	.3792	.60419	.3932	.61848	.4072	.63252	.4212	.64631	.4352
.58987	.3794	.60440	.3934	.61868	.4074	.63271	.4214	.64651	.4354
.59008	.3796	.60461	.3936	.61888	.4076	.63291	.4216	.64670	.4356
.59028	.3798	.60481	.3938	.61908	.4078	.63312	.4218	.64689	.4358
.59049	.3800	.60502	.3940	.61929	.4080	.63331	.4220	.64709	.4360
.59071	.3802	.60522	.3942	.61949	.4082	.63351	.4222	.64729	.4362
.59091	.3804	.60543	.3944	.61969	.4084	.63371	.4224	.64748	.4364
.59112	.3806	.60563	.3946	.61990	.4086	.63391	.4226	.64768	.4366
.59133	.3808	.60584	.3948	.62010	.4088	.63411	.4228	.64787	.4368
.59154	.3810	.60604	.3950	.62030	.4090	.63430	.4230	.64807	.4370
.59175	.3812	.60625	.3952	.62050	.4092	.63450	.4232	.64826	.4372
.59196	.3814	.60646	.3954	.62070	.4094	.63470	.4234	.64846	.4374
.59217	.3816	.60666	.3956	.62090	.4096	.63490	.4236	.64865	.4376
.59237	.3818	.60687	.3958	.62110	.4098	.63509	.4238	.64884	.4378
.59258	.3820	.60707	.3960	.62131	.4100	.63530	.4240	.64904	.4380
.59280	.3822	.60728	.3962	.62151	.4102	.63549	.4242	.64924	.4382
.59300	.3824	.60748	.3964	.62171	.4104	.63569	.4244	.64943	.4384
.59321	.3826	.60769	.3966	.62191	.4106	.63589	.4246	.64963	.4386
.59342	.3828	.60789	.3968	.62211	.4108	.63608	.4248	.64982	.4388
.59363	.3830	.60809	.3970	.62231	.4110	.63629	.4250	.65001	.4390
.59383	.3832	.60830	.3972	.62252	.4112	.63648	.4252	.65021	.4392
.59404	.3834	.60851	.3974	.62272	.4114	.63668	.4254	.65040	.4394
.59426	.3836	.60871	.3976	.62292	.4116	.63688	.4256	.65059	.4396
.59446	.3838	.60891	.3978	.62312	.4118	.63707	.4258	.65079	.4398
.59467	.3840	.60912	.3980	.62332	.4120	.63727	.4260	.65098	.4400
.59488	.3842	.60933	.3982	.62352	.4122	.63747	.4262	.65118	.4402
.59509	.3844	.60953	.3984	.62372	.4124	.63767	.4264	.65137	.4404
.59529	.3846	.60973	.3986	.62392	.4126	.63787	.4266	.65157	.4406
.59550	.3848	.60994	.3988	.62412	.4128	.63806	.4268	.65176	.4408
.59571	.3850	.61015	.3990	.62432	.4130	.63826	.4270	.65196	.4410
.59592	.3852	.61035	.3992	.62452	.4132	.63846	.4272	.65215	.4412
.59613	.3854	.61055	.3994	.62473	.4134	.63866	.4274	.65234	.4414
.59633	.3856	.61076	.3996	.62493	.4136	.63885	.4276	.65254	.4416
.59654	.3858	.61096	.3998	.62513	.4138	.63905	.4278	.65273	.4418
.59675	.3860	.61117	.4000	.62533	.4140	.63925	.4280	.65292	.4420
.59696	.3862	.61137	.4002	.62553	.4142	.63944	.4282	.65312	.4422
.59716	.3864	.61157	.4004	.62573	.4144	.63964	.4284	.65331	.4424
.59737	.3866	.61178	.4006	.62593	.4146	.63984	.4286	.65350	.4426
.59758	.3868	.61198	.4008	.62613	.4148	.64004	.4288	.65370	.4428
.59779	.3870	.61218	.4010	.62633	.4150	.64023	.4290	.65389	.4430
.59799	.3872	.61239	.4012	.62653	.4152	.64043	.4292	.65409	.4432
.59820	.3874	.61259	.4014	.62673	.4154	.64062	.4294	.65428	.4434
.59841	.3876	.61280	.4016	.62693	.4156	.64082	.4296	.65447	.4436
.59862	.3878	.61300	.4018	.62713	.4158	.64102	.4298	.65467	.4438
.59882	.3880	.61320	.4020	.62733	.4160	.64122	.4300	.65486	.4440
.59903	.3882	.61341	.4022	.62753	.4162	.64141	.4302	.65505	.4442
.59924	.3884	.61361	.4024	.62773	.4164	.64161	.4304	.65524	.4444
.59945	.3886	.61381	.4026	.62793	.4166	.64180	.4306	.65544	.4446
.59965	.3888	.61402	.4028	.62813	.4168	.64200	.4308	.65563	.4448
.59986	.3890	.61422	.4030	.62833	.4170	.64220	.4310	.65582	.4450
.60006	.3892	.61443	.4032	.62853	.4172	.64240	.4312	.65601	.4452
.60027	.3894	.61463	.4034	.62873	.4174	.64259	.4314	.65621	.4454
.60048	.3896	.61483	.4036	.62893	.4176	.64279	.4316	.65640	.4456
.60069	.3898	.61503	.4038	.62913	.4178	.64298	.4318	.65659	.4458
.60089	.3900	.61524	.4040	.62933	.4180	.64318	.4320	.65679	.4460
.60110	.3902	.61544	.4042	.62953	.4182	.64337	.4322	.65698	.4462
.60131	.3904	.61564	.4044	.62973	.4184	.64357	.4324	.65717	.4464
.60151	.3906	.61584	.4046	.62993	.4186	.64376	.4326	.65737	.4466
.60172	.3908	.61605	.4048	.63013	.4188	.64396	.4328	.65756	.4468
.60193	.3910	.61625	.4050	.63032	.4190	.64416	.4330	.65775	.4470
.60213	.3912	.61645	.4052	.63053	.4192	.64436	.4332	.65794	.4472
.60234	.3914	.61665	.4054	.63072	.4194	.64455	.4334	.65814	.4474
.60254	.3916	.61686	.4056	.63092	.4196	.64475	.4336	.65833	.4476
.60275	.3918	.61706	.4058	.63112	.4198	.64494	.4338	.65852	.4478
.60296	.3920	.61726	.4060	.63132	.4200	.64514	.4340	.65871	.4480
.60316	.3922	.61747	.4062	.63152	.4202	.64533	.4342	.65890	.4482
.60337	.3924	.61767	.4064	.63172	.4204	.64553	.4344	.65910	.4484
.60358	.3926	.61787	.4066	.63192	.4206	.64572	.4346	.65929	.4486
.60378	.3928	.61808	.4068	.63212	.4208	.64592	.4348	.65948	.4488

Table 10.2 Segmental-Orifice Dimensional Table (*continued*)

d/D_c	H_s/D_c	d/D_c	H_s/D_c	d/D_c	H_s/D_c	d/D_c	H_s/D_c	d/D_c	H_s/D_c
.65967	.4490	.67300	.4630	.68609	.4770	.69895	.4910	.71160	.5050
.65986	.4492	.67319	.4632	.68627	.4772	.69913	.4912	.71178	.5052
.66006	.4494	.67337	.4634	.68646	.4774	.69932	.4914	.71196	.5054
.66025	.4496	.67356	.4636	.68664	.4776	.69950	.4916	.71214	.5056
.66044	.4498	.67375	.4638	.68683	.4778	.69968	.4918	.71232	.5058
.66063	.4500	.67394	.4640	.68702	.4780	.69986	.4920	.71250	.5060
.66082	.4502	.67413	.4642	.68720	.4782	.70004	.4922	.71267	.5062
.66101	.4504	.67432	.4644	.68739	.4784	.70023	.4924	.71285	.5064
.66121	.4506	.67451	.4646	.68757	.4786	.70040	.4926	.71303	.5066
.66140	.4508	.67469	.4648	.68775	.4788	.70059	.4928	.71321	.5068
.66159	.4510	.67488	.4650	.68794	.4790	.70077	.4930	.71339	.5070
.66178	.4512	.67507	.4652	.68812	.4792	.70095	.4932	.71357	.5072
.66198	.4514	.67525	.4654	.68831	.4794	.70113	.4934	.71375	.5074
.66217	.4516	.67544	.4656	.68849	.4796	.70131	.4936	.71392	.5076
.66236	.4518	.67563	.4658	.68868	.4798	.70150	.4938	.71410	.5078
.66255	.4520	.67582	.4660	.68886	.4800	.70168	.4940	.71428	.5080
.66274	.4522	.67601	.4662	.68905	.4802	.70186	.4942	.71446	.5082
.66293	.4524	.67620	.4664	.68923	.4804	.70204	.4944	.71464	.5084
.66312	.4526	.67638	.4666	.68942	.4806	.70222	.4946	.71481	.5086
.66331	.4528	.67657	.4668	.68960	.4808	.70240	.4948	.71499	.5088
.66350	.4530	.67676	.4670	.68979	.4810	.70259	.4950	.71517	.5090
.66369	.4532	.67695	.4672	.68997	.4812	.70276	.4952	.71535	.5092
.66389	.4534	.67713	.4674	.69016	.4814	.70294	.4954	.71553	.5094
.66408	.4536	.67732	.4676	.69034	.4816	.70313	.4956	.71571	.5096
.66427	.4538	.67751	.4678	.69052	.4818	.70331	.4958	.71588	.5098
.66446	.4540	.67770	.4680	.69071	.4820	.70349	.4960	.71606	.5100
.66465	.4542	.67789	.4682	.69089	.4822	.70367	.4962	.71624	.5102
.66484	.4544	.67807	.4684	.69107	.4824	.70385	.4964	.71641	.5104
.66503	.4546	.67826	.4686	.69126	.4826	.70403	.4966	.71659	.5106
.66522	.4548	.67845	.4688	.69144	.4828	.70421	.4968	.71677	.5108
.66541	.4550	.67864	.4690	.69163	.4830	.70439	.4970	.71695	.5110
.66560	.4552	.67882	.4692	.69181	.4832	.70458	.4972	.71713	.5112
.66579	.4554	.67901	.4694	.69200	.4834	.70476	.4974	.71730	.5114
.66598	.4556	.67920	.4696	.69218	.4836	.70493	.4976	.71748	.5116
.66617	.4558	.67938	.4698	.69236	.4838	.70511	.4978	.71766	.5118
.66636	.4560	.67957	.4700	.69255	.4840	.70530	.4980	.71784	.5120
.66655	.4562	.67976	.4702	.69273	.4842	.70548	.4982	.71801	.5122
.66674	.4564	.67995	.4704	.69292	.4844	.70566	.4984	.71819	.5124
.66693	.4566	.68013	.4706	.69310	.4846	.70584	.4986	.71837	.5126
.66713	.4568	.68032	.4708	.69328	.4848	.70602	.4988	.71854	.5128
.66732	.4570	.68050	.4710	.69346	.4850	.70620	.4990	.71872	.5130
.66750	.4572	.68069	.4712	.69365	.4852	.70638	.4992	.71890	.5132
.66769	.4574	.68088	.4714	.69383	.4854	.70656	.4994	.71908	.5134
.66789	.4576	.68106	.4716	.69402	.4856	.70674	.4996	.71925	.5136
.66808	.4578	.68125	.4718	.69420	.4858	.70692	.4998	.71943	.5138
.66826	.4580	.68144	.4720	.69438	.4860	.70711	.5000	.71961	.5140
.66845	.4582	.68162	.4722	.69457	.4862	.70729	.5002	.71978	.5142
.66864	.4584	.68181	.4724	.69475	.4864	.70747	.5004	.71996	.5144
.66883	.4586	.68200	.4726	.69493	.4866	.70765	.5006	.72014	.5146
.66902	.4588	.68218	.4728	.69512	.4868	.70783	.5008	.72032	.5148
.66921	.4590	.68237	.4730	.69530	.4870	.70801	.5010	.72049	.5150
.66940	.4592	.68256	.4732	.69548	.4872	.70819	.5012	.72067	.5152
.66959	.4594	.68274	.4734	.69566	.4874	.70837	.5014	.72084	.5154
.66978	.4596	.68293	.4736	.69585	.4876	.70855	.5016	.72102	.5156
.66997	.4598	.68312	.4738	.69603	.4878	.70873	.5018	.72120	.5158
.67016	.4600	.68330	.4740	.69621	.4880	.70891	.5020	.72137	.5160
.67035	.4602	.68349	.4742	.69640	.4882	.70909	.5022	.72155	.5162
.67054	.4604	.68368	.4744	.69658	.4884	.70927	.5024	.72173	.5164
.67073	.4606	.68386	.4746	.69676	.4886	.70945	.5026	.72190	.5166
.67092	.4608	.68405	.4748	.69694	.4888	.70963	.5028	.72208	.5168
.67111	.4610	.68423	.4750	.69713	.4890	.70981	.5030	.72225	.5170
.67130	.4612	.68442	.4752	.69731	.4892	.70999	.5032	.72243	.5172
.67149	.4614	.68460	.4754	.69749	.4894	.71017	.5034	.72261	.5174
.67168	.4616	.68479	.4756	.69768	.4896	.71035	.5036	.72278	.5176
.67186	.4618	.68497	.4758	.69786	.4898	.71053	.5038	.72296	.5178
.67205	.4620	.68516	.4760	.69804	.4900	.71071	.5040	.72314	.5180
.67224	.4622	.68535	.4762	.69822	.4902	.71089	.5042	.72331	.5182
.67243	.4624	.68553	.4764	.69840	.4904	.71107	.5044	.72349	.5184
.67262	.4626	.68572	.4766	.69859	.4906	.71125	.5046	.72366	.5186
.67281	.4628	.68590	.4768	.69877	.4908	.71142	.5048	.72384	.5188

Table 10.2 Segmental-Orifice Dimensional Table (*continued*)

d/D_c	H_s/D_c	d/D_c	H_s/D_c	d/D_c	H_s/D_c	d/D_c	H_s/D_c	d/D_c	H_s/D_c
.72401	.5190	.72944	.5252	.7348	.5314	.74016	.5376	.74546	.5438
.72419	.5192	.72961	.5254	.73500	.5316	.74033	.5378	.74563	.5440
.72437	.5194	.72979	.5256	.73517	.5318	.74050	.5380	.74580	.5442
.72454	.5196	.72996	.5258	.73534	.5320	.74068	.5382	.74597	.5444
.72472	.5198	.73014	.5260	.73551	.5322	.74085	.5384	.74614	.5446
.72489	.5200	.73031	.5262	.73569	.5324	.74102	.5386	.74631	.5448
.72507	.5202	.73048	.5264	.73586	.5326	.74119	.5388	.74648	.5450
.72524	.5204	.73066	.5266	.73603	.5328	.74136	.5390	.74665	.5452
.72542	.5206	.73083	.5268	.73621	.5330	.74153	.5392	.74682	.5454
.72559	.5208	.73101	.5270	.73638	.5332	.74270	.5394	.74699	.5456
.72577	.5210	.73118	.5272	.73655	.5334	.74187	.5396	.74716	.5458
.72595	.5212	.73136	.5274	.73672	.5336	.74205	.5398	.74733	.5460
.72612	.5214	.73153	.5276	.73690	.5338	.74222	.5400	.74750	.5462
.72629	.5216	.73170	.5278	.73707	.5340	.74239	.5402	.74767	.5464
.72647	.5218	.73188	.5280	.73724	.5342	.74256	.5404	.74783	.5466
.72664	.5220	.73205	.5282	.73741	.5344	.74273	.5406	.74800	.5468
.72682	.5222	.73222	.5284	.73759	.5346	.74290	.5408	.74817	.5470
.72699	.5224	.73240	.5286	.73776	.5348	.74307	.5410	.74834	.5472
.72717	.5226	.73257	.5288	.73793	.5350	.74324	.5412	.74851	.5474
.72734	.5228	.73275	.5290	.73810	.5352	.74341	.5414	.74868	.5476
.72752	.5230	.73292	.5292	.73827	.5354	.74358	.5416	.74885	.5478
.72769	.5232	.73309	.5294	.73844	.5356	.74375	.5418	.74902	.5480
.72787	.5234	.73326	.5296	.73861	.5358	.74392	.5420	.74919	.5482
.72804	.5236	.73344	.5298	.73879	.5360	.74409	.5422	.74936	.5484
.72822	.5238	.73361	.5300	.73896	.5362	.74427	.5424	.74953	.5486
.72839	.5240	.73378	.5302	.73913	.5364	.74444	.5426	.74970	.5488
.72857	.5242	.73396	.5304	.73930	.5366	.74461	.5428	.74987	.5490
.72874	.5244	.73413	.5306	.73947	.5368	.74478	.5430	.75004	.5492
.72892	.5246	.73430	.5308	.73965	.5370	.74495	.5432	.75021	.5494
.72909	.5248	.73448	.5310	.73982	.5372	.74512	.5434	.75037	.5496
.72926	.5250	.73465	.5312	.73999	.5374	.74529	.5436	.75054	.5498

ever, for these fluids, the concentric orifice in a vertical pipe, with the flow in an *upward* direction, is preferred for greater accuracy—if a suitable location is available. Segmental orifice plates should not be used to measure liquids containing sticky solids or solids having a density near that of the flowing liquid. It is affected in the same way as the concentric orifice by deposits on the face or edge of the orifice.

Coefficient data are presented in Fig. 10.43 for both flange and vena contracta taps, and in Figs. 10.44 through 10.46 are graphs of the gas expansion factor. These apply only where taps are located at points in line with the midpoint of the solid segment of the orifice plate. This restriction often makes application to liquids containing both vapors and solids undesirable, because it necessitates bringing the taps out above the horizontal line at an angle such that excessive maintenance results, due to the accumulation of air or vapor. Data for eccentric orifices with taps at 90° from the point of tangency are available; hence eccentric orifices are preferable for such applications.

The spread of segmental-orifice data points was so great that any effect of Reynolds number on the discharge coefficient was obscured. Twice the standard deviation of the test data is ±2.2 percent for vena contracta taps, and ±1.9 percent for flange taps.

Eccentric Orifice The eccentric orifice is shown in Fig. 10.47, with two tap arrangements. The orifice bore is tangent to a circle with a diameter equal to 98

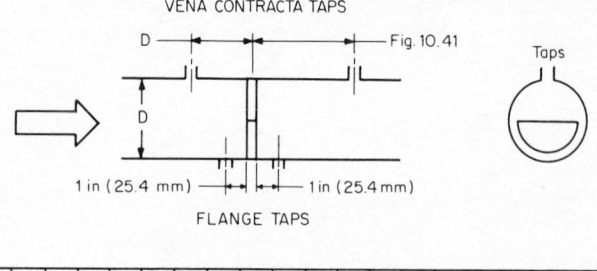

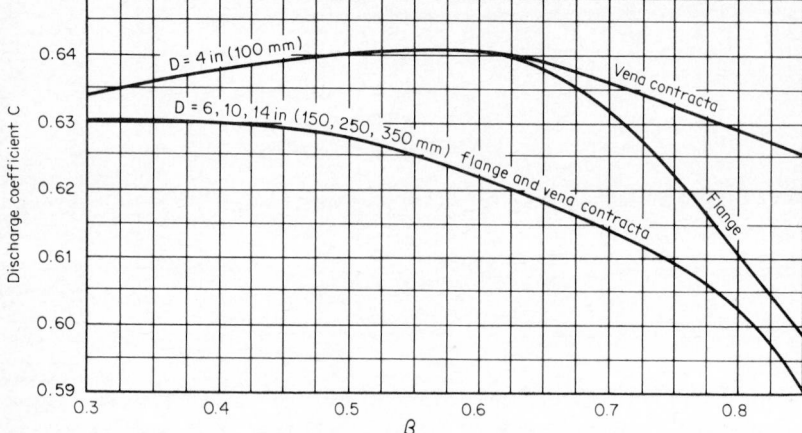

Figure 10.43 Discharge coefficient for segmental orifice; flange and vena contracta taps.

percent of that of the pipe and concentric with the pipe. When the plate is installed, no portion of the hole must be covered by the flange or gasket. The orifice bore is calculated in exactly the same way as the bore of a square-edged orifice.

The eccentric orifice has the advantages of the segmental orifice, but it does not allow free drainage over as much of the pipe circumference. In all other respects, the eccentric orifice is superior to the segmental.

Discharge coefficients are presented in Fig. 10.48 for diametrically opposite flange taps, and in Fig. 10.49 for side, or 90°, taps. In Figs. 10.50 and 10.51 are discharge coefficients for diametrically opposite and 90° vena contracta taps. Figures 10.52 through 10.54 contain the gas-expansion-factor curves.

The discharge-coefficient curves are presented at pipe Reynolds numbers of 10,000 and 1,000,000. The general form of the equation for coefficient interpolation is

$$C = k_1 + \frac{k_2}{R_D^{0.5}} \qquad (10.4)$$

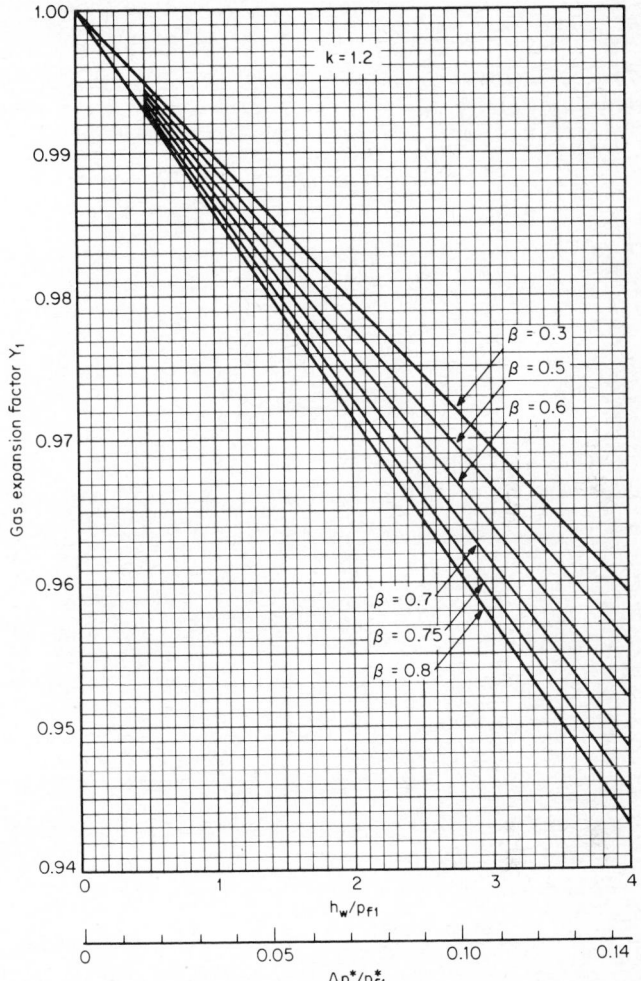

Figure 10.44 Gas expansion factor Y_1 for segmental orifice.

where the constants k_1 and k_2 are derived from the graphically determined coefficients at the value of β_0 as

$$k_1 = C_{max} - 1.11111(C_{max} - C_{min}) \qquad (10.5)$$

and

$$k_2 = 111.111(C_{max} - C_{min}) \qquad (10.6)$$

The subscripts max and min refer to the maximum and minimum discharge coefficients at pipe Reynolds numbers of 10,000 and 1,000,000.

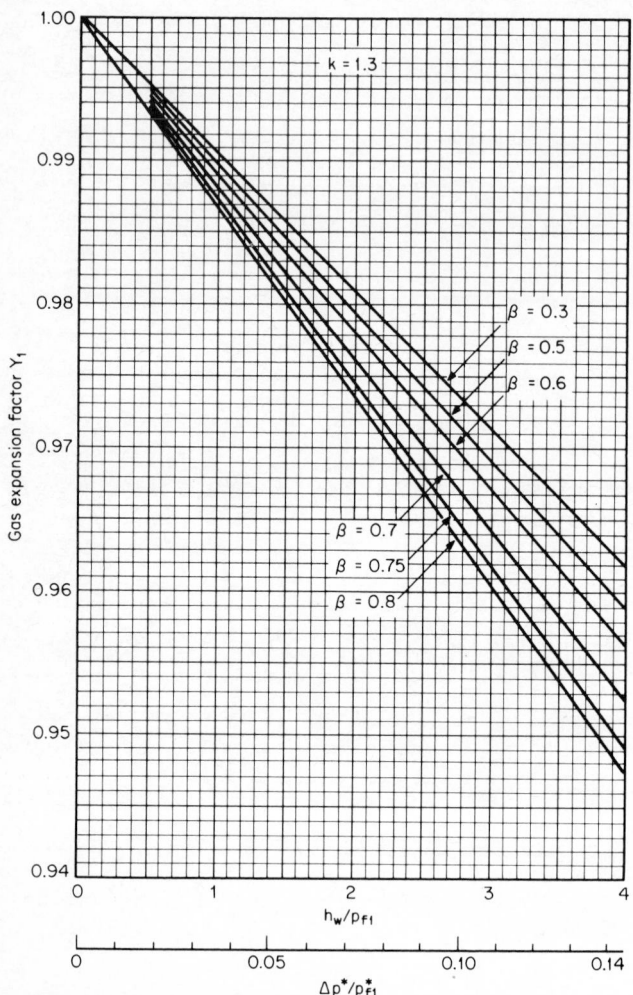

Figure 10.45   Gas expansion factor Y_1 for segmental orifice.

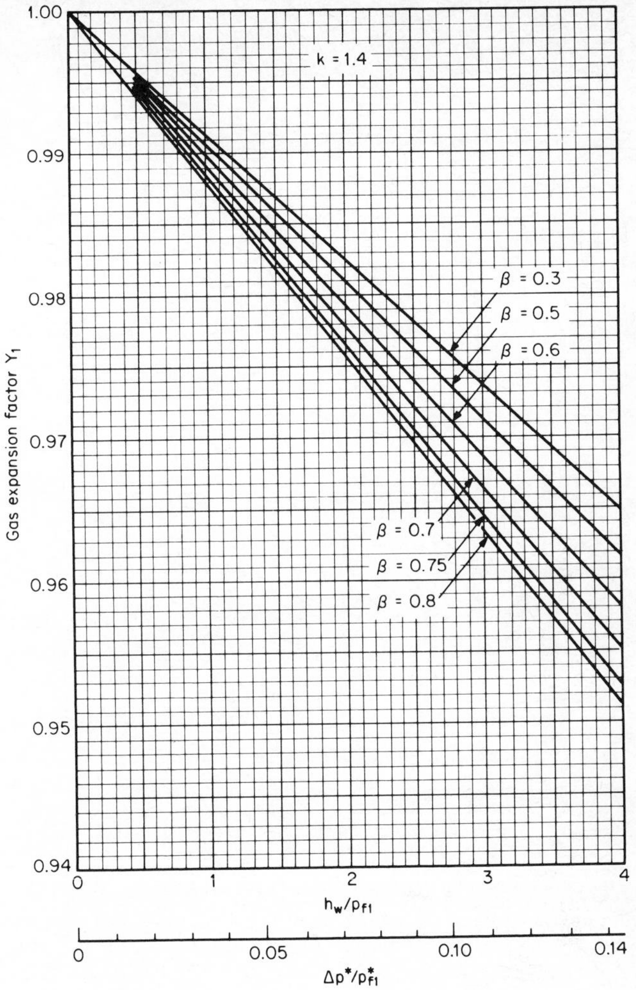

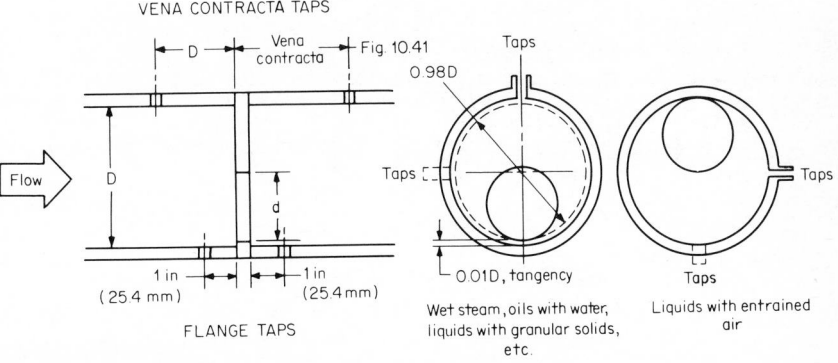

Figure 10.46 Gas expansion factor Y_1 for segmental orifice.

The chart shows Gas expansion factor Y_1 (vertical axis, 0.94 to 1.00) versus h_w/p_{f1} (horizontal axis, 0 to 4) and $\Delta p^*/p_{f1}^*$ (0 to 0.14), with $k = 1.4$ and curves for $\beta = 0.3$, $\beta = 0.5$, $\beta = 0.6$, $\beta = 0.7$, $\beta = 0.75$, $\beta = 0.8$.

VENA CONTRACTA TAPS

Flow

D

D   Vena contracta   Fig. 10.41

d

1 in
(25.4 mm)

1 in
(25.4 mm)

FLANGE TAPS

Taps

0.98D

Taps

0.01D, tangency

Wet steam, oils with water,
liquids with granular solids,
etc.

Taps

Taps

Liquids with entrained
air

Figure 10.47 Eccentric orifice.

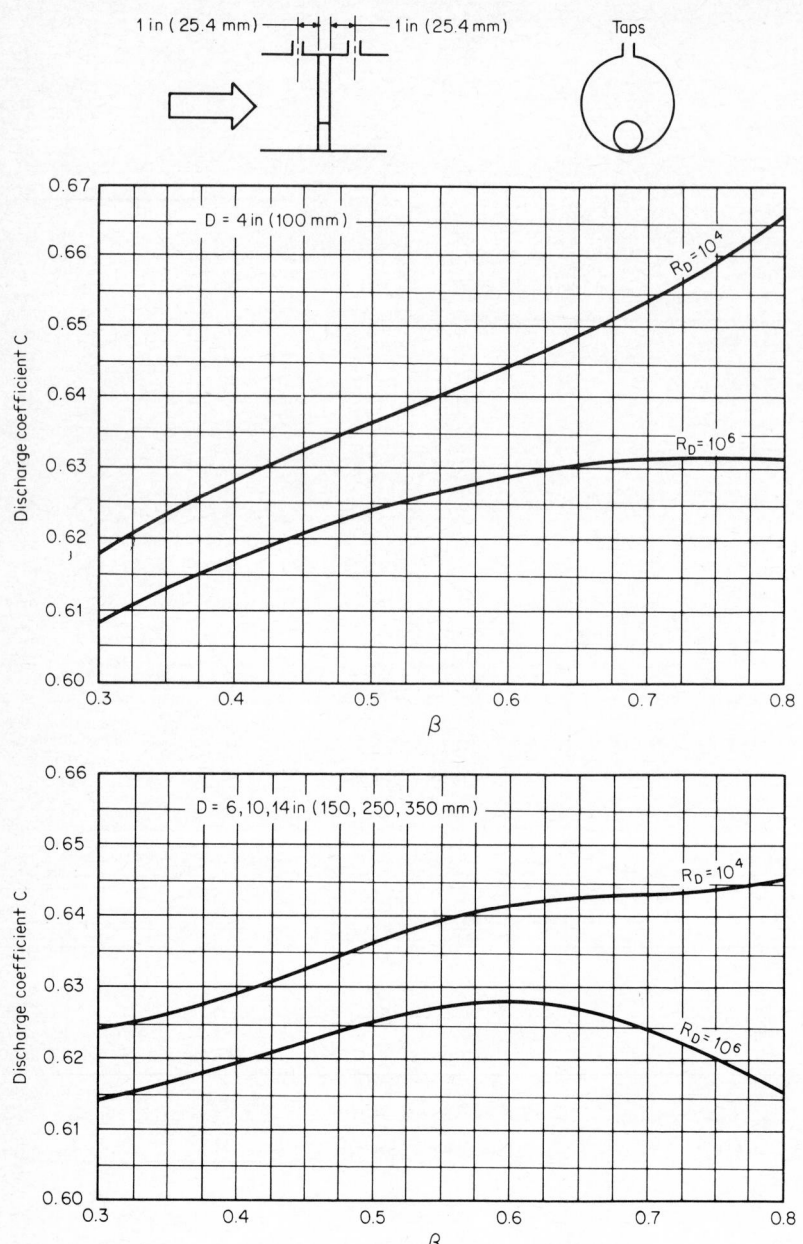

Figure 10.48 Discharge coefficient for eccentric orifice; flange taps.

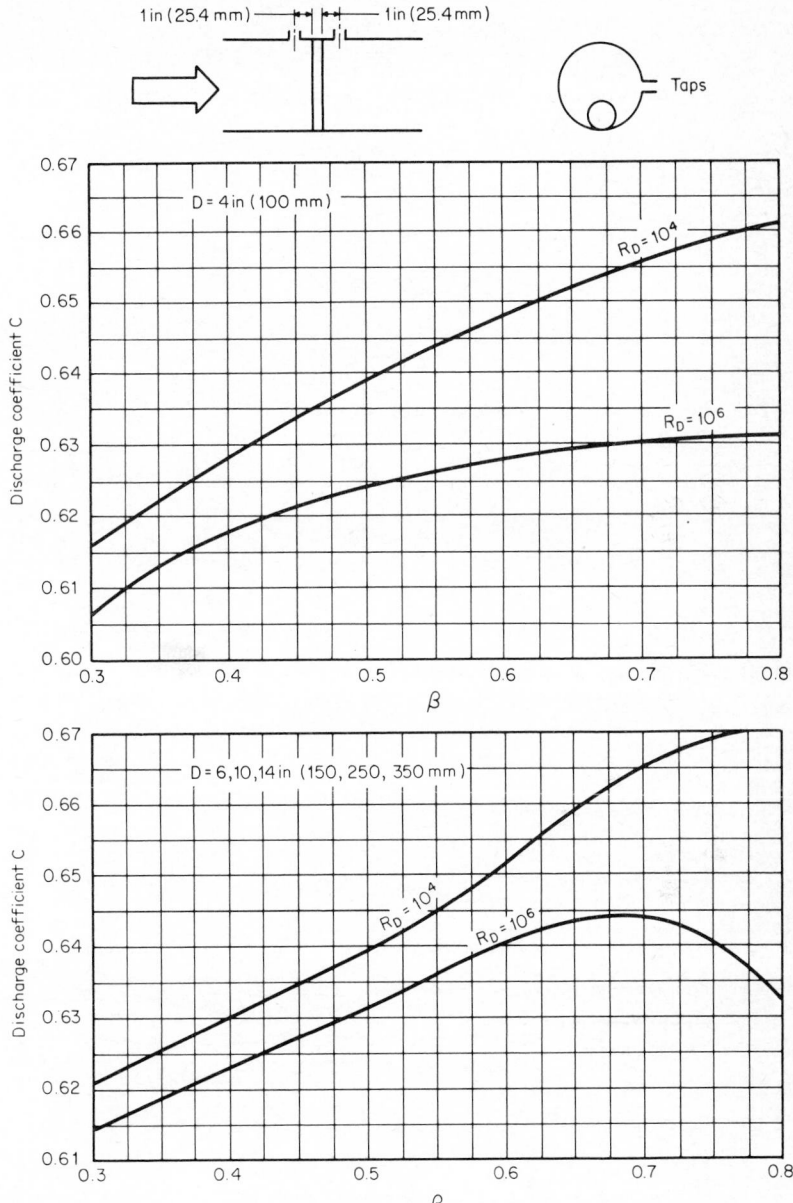

Figure 10.49 Discharge coefficient for eccentric orifice; flange taps.

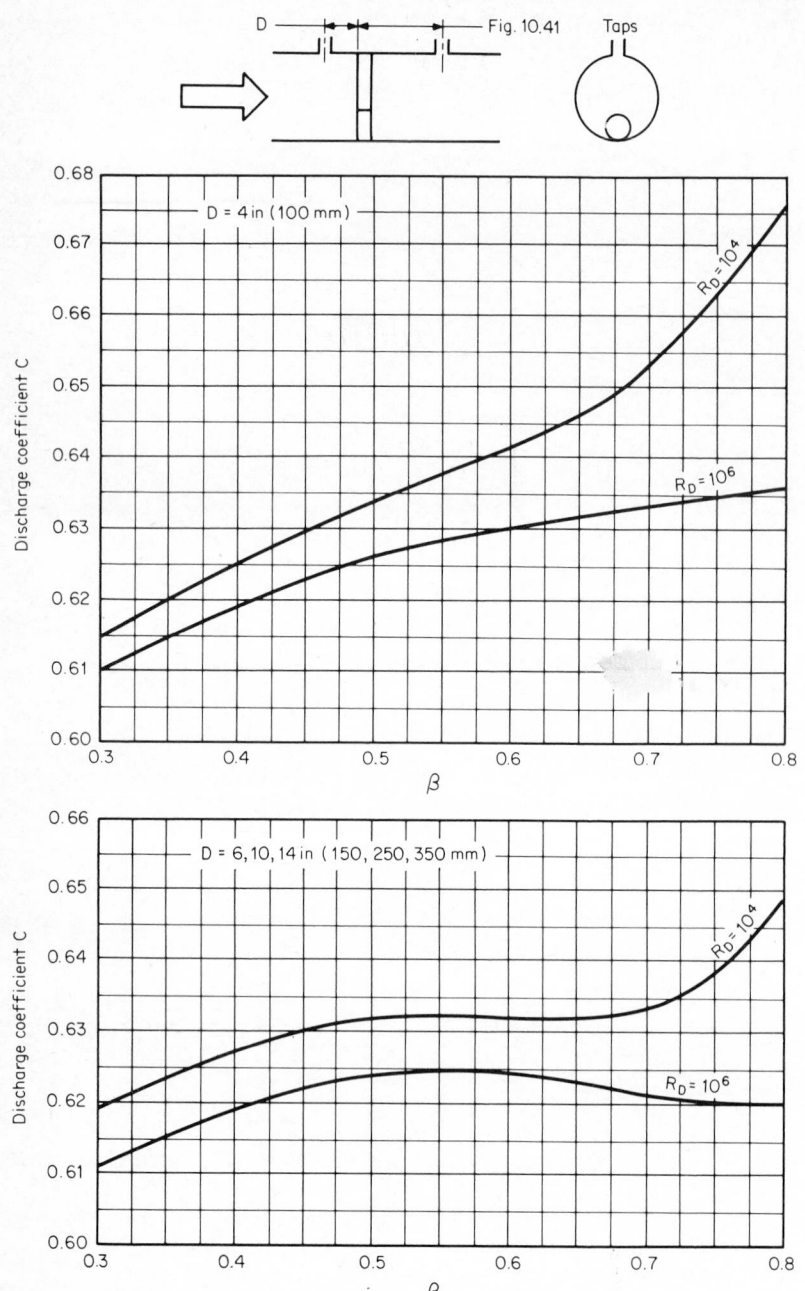

Figure 10.50 Discharge coefficient for eccentric orifice; vena contracta taps.

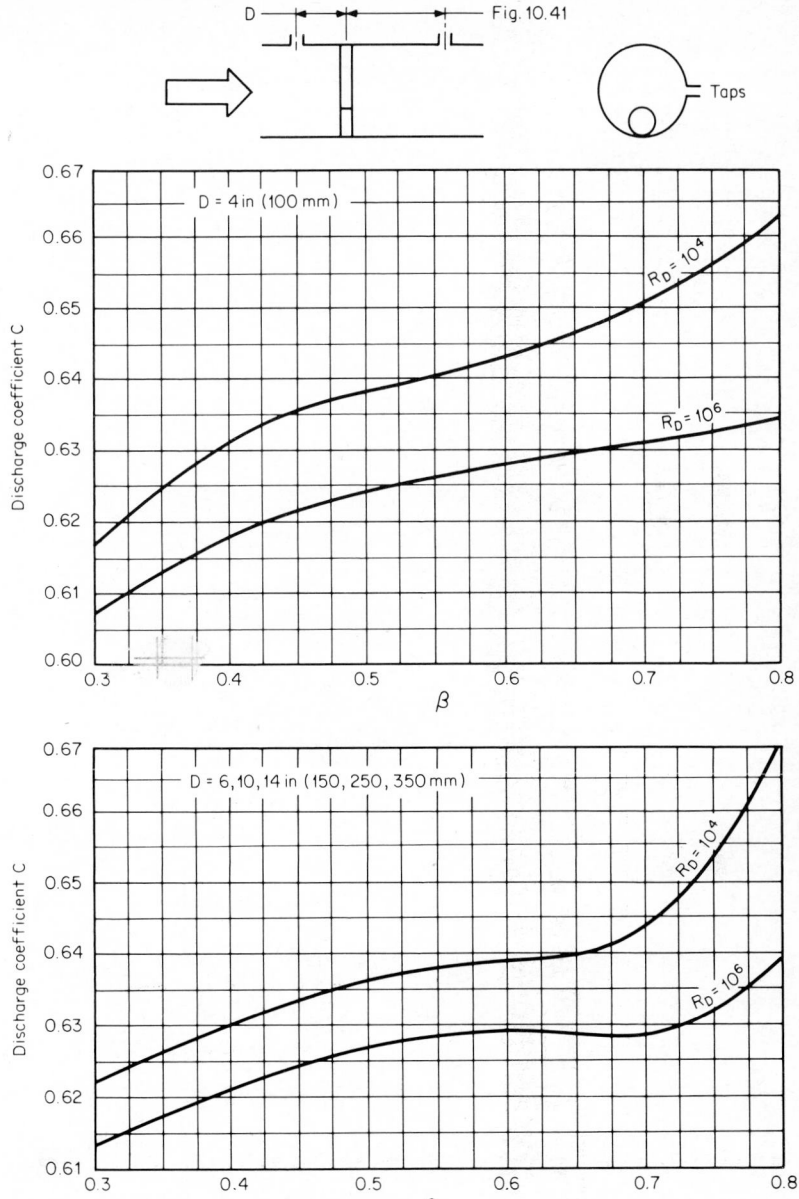

Figure 10.51 Discharge coefficient for eccentric orifice; vena contracta taps.

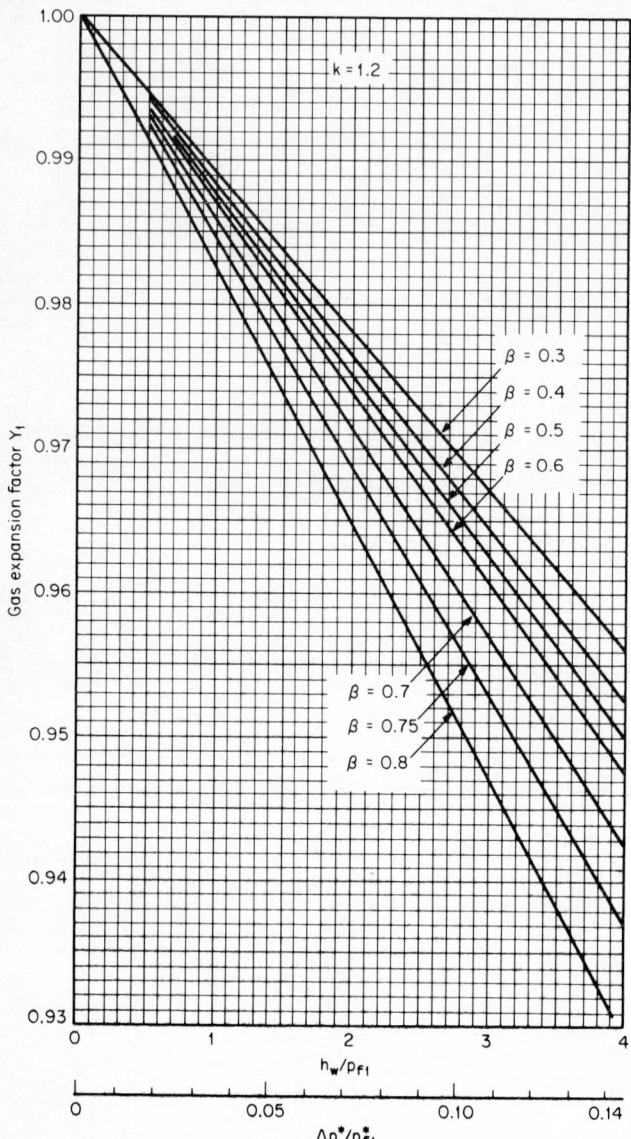

Figure 10.52 Gas expansion factor for eccentric orifice.

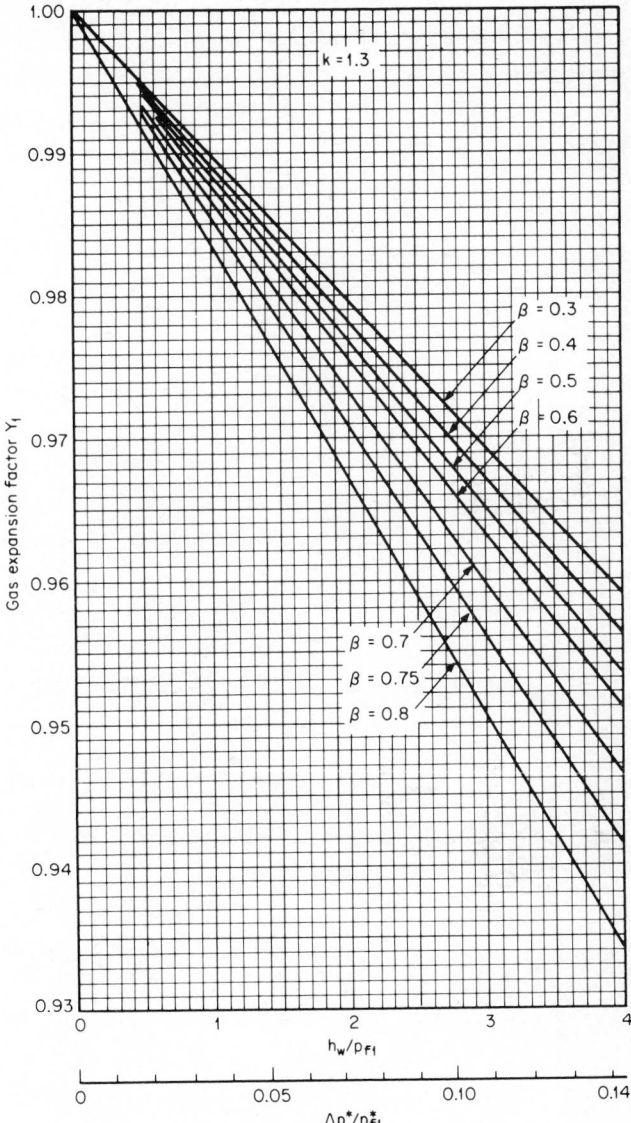

Figure 10.53 Gas expansion factor for eccentric orifice.

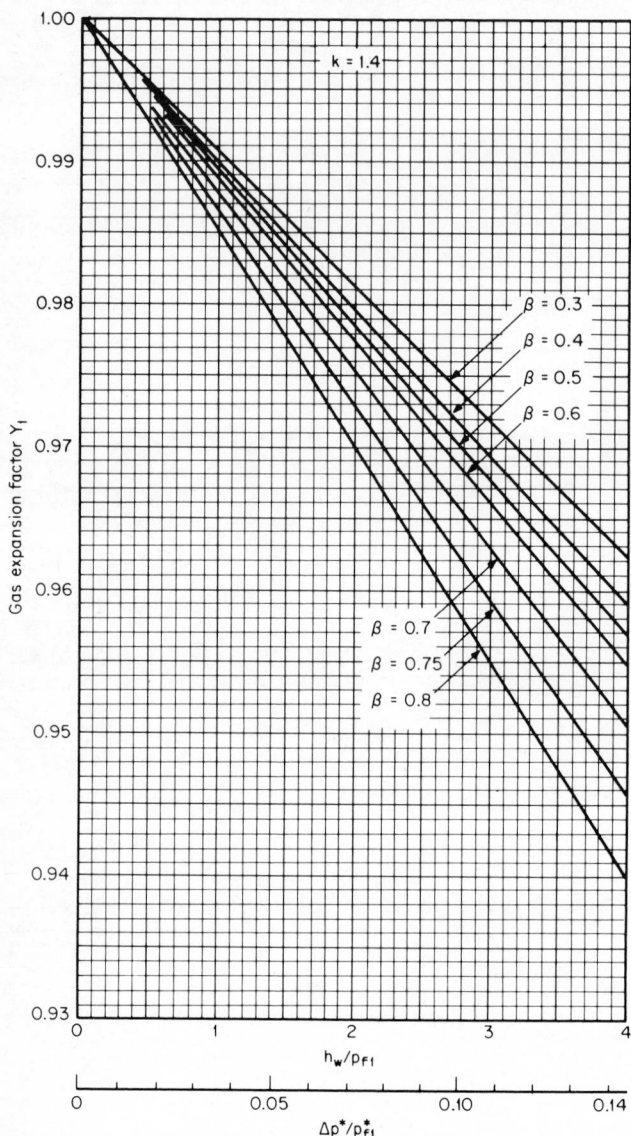

Figure 10.54 Gas expansion factor for eccentric orifice.

EXAMPLE 10.1

For liquid flow in a 4.000-in (101.6-mm) pipe, the S_M factor is calculated as 0.2500. Size a segmental and an eccentric orifice for a pipe Reynolds number of 250,000. Both are to have diametrically opposite flange taps.

Segmental orifice

From Eq. (m) of Table 9.28, the initial estimate for β is

$$\beta_0 = \left[1 + \left(\frac{0.634}{S_M} - 0.062\right)^2\right]^{-1/4} = \left[1 + \left(\frac{0.634}{0.2500} - 0.062\right)^2\right]^{-1/4} = 0.6122$$

By entering Fig. 10.43 at this value, the discharge coefficient is read as $C = 0.641$. Substitution into Eq. (9.83), with $(Y_1)_0 = 1.0$ for liquids, gives

$$\beta_1 = \left\{1 + \left[\frac{C_0(Y_1)_0}{S_M}\right]^2\right\}^{-1/4} = \left\{1 + \left[\frac{(0.641)(1.0)}{0.2500}\right]^2\right\}^{-1/4} = 0.6028$$

A third estimate is unnecessary, since the discharge coefficient remains the same.
The diameter of the segment is

$$D_c = 0.980D = (0.98)(4.00) = 3.92 \text{ in}$$

and
$$\beta_c = \frac{\beta}{0.98} = \frac{0.6028}{0.98} = 0.6151$$

Either from Eq. (10.3) or by interpolation in Table 10.2,

$$\frac{H_s}{D_c} = 0.4038 + \left(\frac{0.4040 - 0.4038}{0.61524 - 0.61503}\right)(0.6151 - 0.61503) = 0.4039$$

and
$$H_s = 0.4039D_c = (0.4039)(3.92) = 1.583 \text{ in}$$

Eccentric orifice

From Eq. (l) of Table 9.28, the initial estimate for β is

$$\beta_0 = \left[1 + \left(\frac{0.607}{S_M} + 0.088\right)^2\right]^{-1/4} = \left[1 + \left(\frac{0.607}{0.2500} + 0.088\right)^2\right]^{-1/4} = 0.6077$$

From Figure 10.48 for flange taps and with this estimate of β,

$$C_{\max} = 0.6442 \qquad R_D = 10,000$$
$$C_{\min} = 0.6285 \qquad R_D = 1,000,000$$

The equation constants are then, by Eqs. (10.5) and (10.6),

$$k_1 = 0.6442 - 1.11111(0.6442 - 0.6285) = 0.6268$$
$$k_2 = 111.111(0.6442 - 0.6285) = 1.744$$

The discharge coefficient at $R_D = 250,000$ is then, by Eq. (10.4),

$$C = k_1 + \frac{k_2}{R_D^{0.5}} = 0.6268 + \frac{1.744}{(250,000)^{0.5}} = 0.6303$$

The second estimate for β is, by Eq. (9.83) with $Y_1 = 1.0$ for liquids,

$$\beta_1 = \left\{ 1 + \left[\frac{C_0(Y_1)_0}{S_M} \right]^2 \right\}^{-1/4} = \left\{ 1 + \left[\frac{(0.6303)(1.0)}{0.2500} \right]^2 \right\}^{-1/4} = 0.6072$$

A third estimate is unwarranted, and the orifice bore is

$$d = \beta D = (0.6072)(4.000) = 2.429 \text{ in}$$

CONTOURED DEVICES

Contoured inlet devices have the advantage of sweeping solids through the throat, and, with an attached divergent cone, of reducing permanent pressure loss. Usually these devices are of more rugged construction than orifices and, therefore, are more suitable for steam or other high-velocity applications.

Flow Nozzles

Several flow-nozzle contours are available. The ASME wall-tap flow nozzle (either low- or high-β series) is preferred in the United States, while in Europe the ISA 1932 design is widely used. In the United States, the ASME low-β-series nozzle is sometimes modified to have throat taps, and this has become the accepted standard for testing steam-turbine performance.

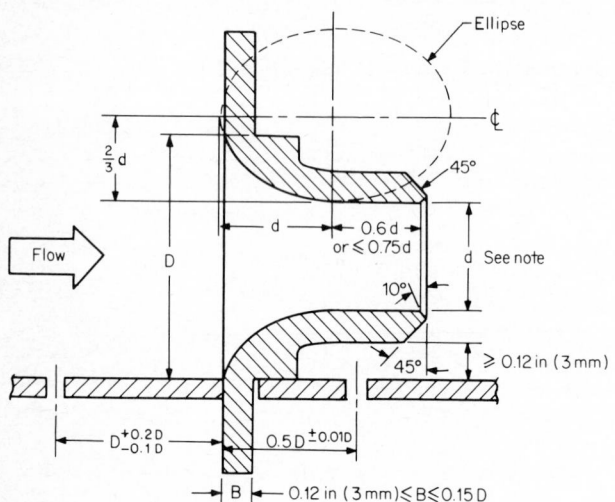

Note: For throat diameter, no measurement more than 0.05% from average of at least four measurements.

Figure 10.55 ASME long-radius nozzle, $0.2 \leq \beta \leq 0.5$ (low-beta).

ASME Long-Radius Wall-Tap Nozzle The ASME flow nozzle has a contoured elliptical inlet in which the curvature is the quadrant of an ellipse. For beta ratios from 0.2 to 0.5 (the low-β series), the design is as shown in Fig. 10.55. For beta ratios between 0.45 and 0.8 (the high-β series), the design is as shown in Fig. 10.56.

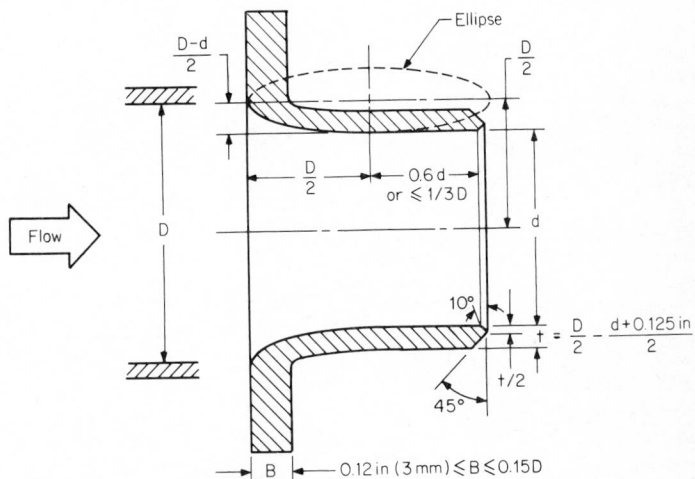

Figure 10.56 ASME long-radius nozzle, $0.25 \leq \beta \leq 0.8$ (high-beta).

The nozzle wall and flange thickness should be sufficient to prevent distortion caused by pipeline pressure, temperature, or bolting strains. The nozzle should be well centered in the pipe, and the installation requirements of Chap. 8 relative to upstream piping conditions, thermal wells, etc., should be followed for good measurements. It is important that the downstream pressure tap be inside the exit. Also, outward widening of the cylindrical throat should be avoided, since this causes a rising discharge coefficient at high Reynolds numbers. The throat taper, if any, should always decrease the throat diameter toward the exit, with no bell mouth or diameter increase.

Discharge-coefficient data has been reduced to an empirical equation for pipe Reynolds numbers greater than 10,000. This equation is given in Table 9.1 and graphed in Fig. 10.57. For low Reynolds numbers, Benedict's (1966) equation, which was empirically fitted to the ASME low-Reynolds-number data, may be used. This equation is

$$C = 0.19436 + 0.152884 \ln R_d - 0.0097785(\ln R_d)^2$$
$$+ 0.00020903(\ln R_d)^3 \quad (10.7)$$

Figure 10.58 is a plot of this equation.

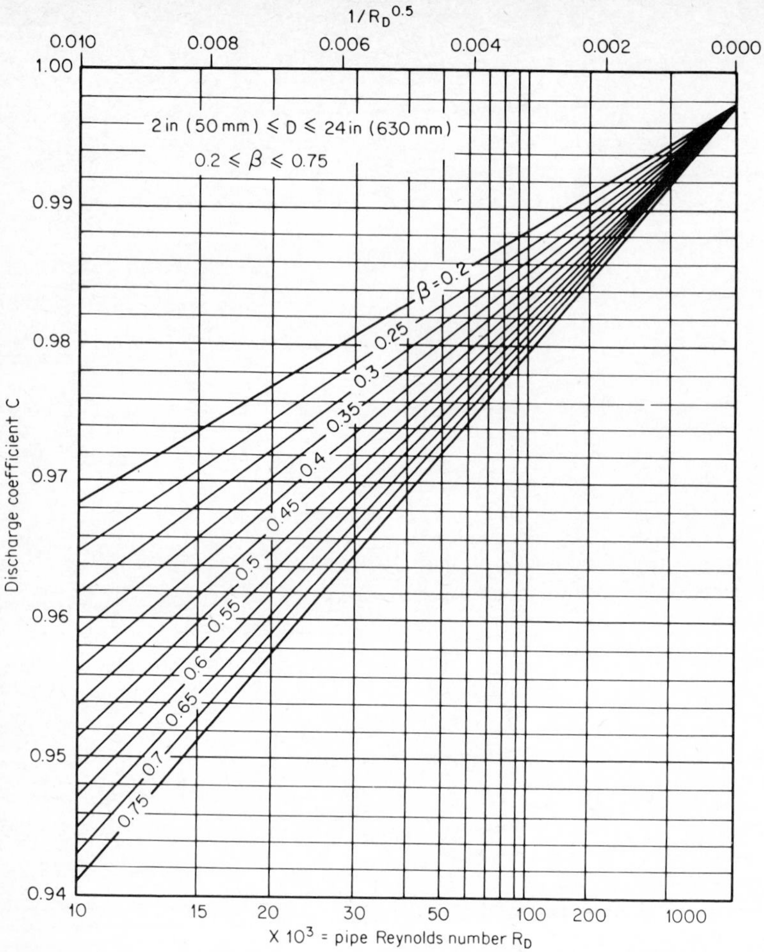

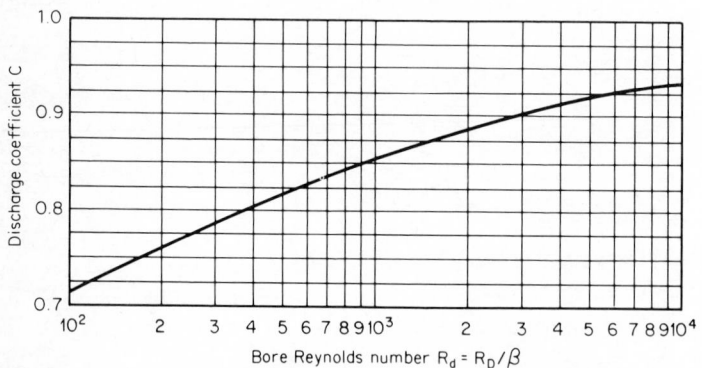

Figure 10.57 Discharge coefficient for ASME flow nozzle.

Figure 10.58 Discharge coefficient for ASME long-radius nozzle.

10-62

ASME Throat-Tap Nozzle (PTC-6 Nozzle) In the testing of steam turbines, a high degree of accuracy is required. Extensive test work on an ASME long-radius nozzle with diametric throat taps has led to standardization on the ANSI/ASME PTC-6 (1976) nozzle for these performance tests. The nozzle design is shown in Fig. 10.59.

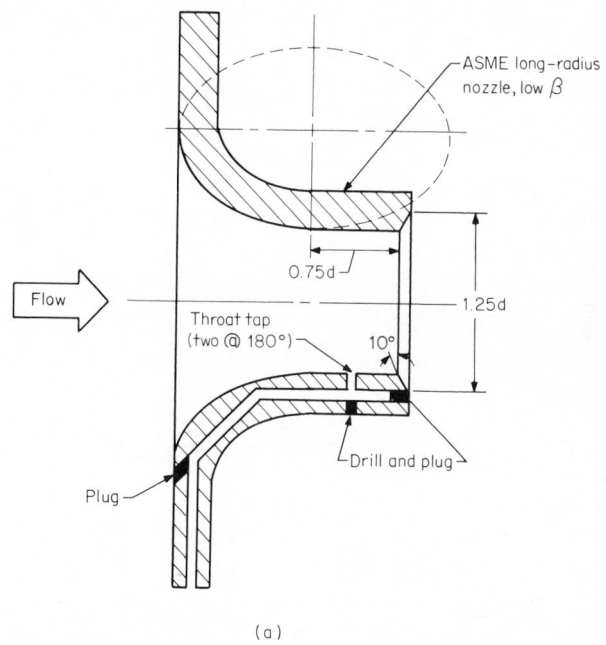

(a)

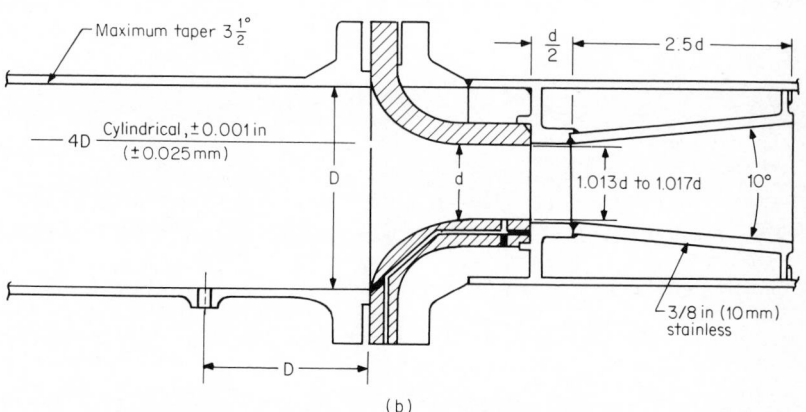

(b)

Figure 10.59 ASME PTC-6 throat-tap nozzle. (*a*) Throat-tap nozzle. (*b*) With diffuser cone.

It is required that the nozzle be water-calibrated to the highest obtainable Reynolds number, and that the discharge-coefficient curve be within ±0.25 percent of the curve shown in Fig. 10.60. The beta ratio is limited to 0.25 to 0.5. For beta ratios other than 0.43, the discharge coefficient–versus–bore Reynolds number curve must fit, within ±0.25 percent, the equation shown in Fig. 10.60.

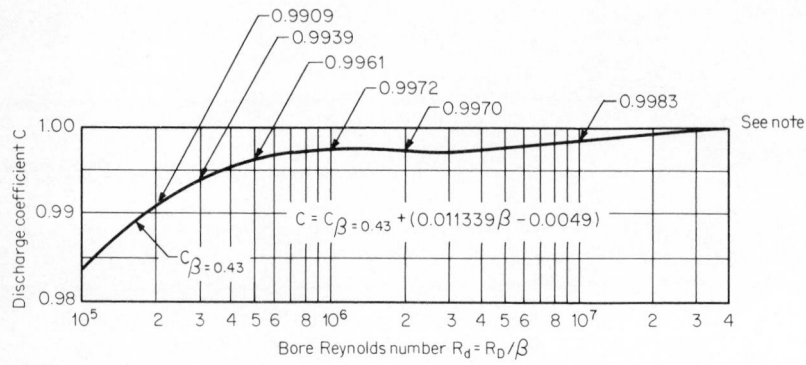

Note: Flow nozzle calibration curve to be within ±0.25% of curve shown, or ±0.25% of parallel curve based on equation shown for other βs.

Figure 10.60   Discharge coefficient for ASME PTC-6 throat-tap nozzle, $0.25 \leq \beta \leq 0.5$.

For general applications, where performance testing is not required, the PTC-6 nozzle has a ±0.5 percent uncalibrated accuracy. This compares to ±2 percent for the wall-tap nozzle, and the added expense of tap drilling may be warranted if accuracy is desired.

ISA 1932 Nozzle   The ISA 1932 nozzle shown in Fig. 10.61 has a double-radius-contoured inlet with corner tapping. It is used extensively in Europe but seldom in the United States. The nozzle's discharge-coefficient equation is shown in Table 9.1 and graphed in Fig. 10.62.

Venturi Nozzle   The ISA 1932 nozzle, fitted with a venturi divergent downstream section, reduces the overall pressure loss. This nozzle, too, is widely used in Europe but seldom in the United States. The included angle may be from 10° to 30°, as shown in Fig. 10.63. The length of the divergent section has practically no influence on the discharge coefficient, but the truncated version, which has the maximum included angle, has a slightly higher pressure loss.

Classical Venturi Tube

The design of the classical, or Herschel, venturis is shown in Fig. 10.64. Depending on the manufacturing method, these are conveniently grouped as:

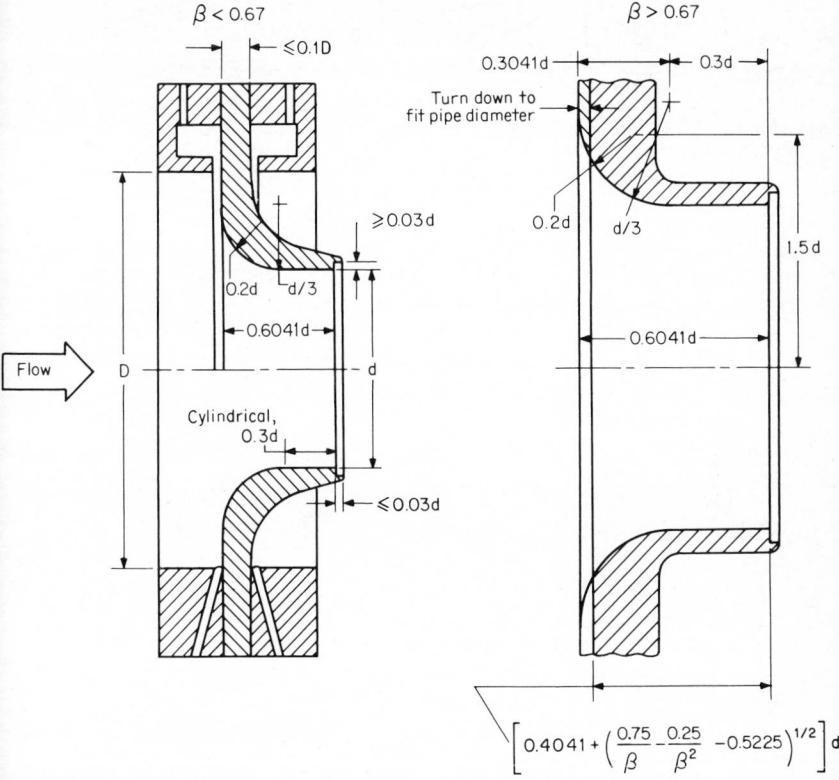

Figure 10.61 ISA nozzle.

1. Classical venturi with rough-cast inlet, recommended for line sizes of 4 to 32 in (100 to 800 mm)

2. Classical venturi with machined inlet, recommended for line sizes of 2 to 10 in (50 to 250 mm)

3. Classical venturi with rough-welded sheet-iron inlet, for line sizes of 8 to 48 in (200 to 1200 mm)

Venturis in the first two groups are cast in a sand mold, and the inlet may be left as cast (group 1) or machined (group 2). Although standards limit the maximum line size to 48 in (1200 mm), classical venturis have been fabricated for use in 120-in (3000-mm) lines. In the United States, the rough-cast inlet is almost always used, while in Europe all three groups are commonly used.

The exit cone (recovery cone) may have an included angle between 7° and 15°, with 7° being preferred for minimum permanent pressure loss. The 7° cone may be shortened to reduce lay-in length without significantly altering recovery.

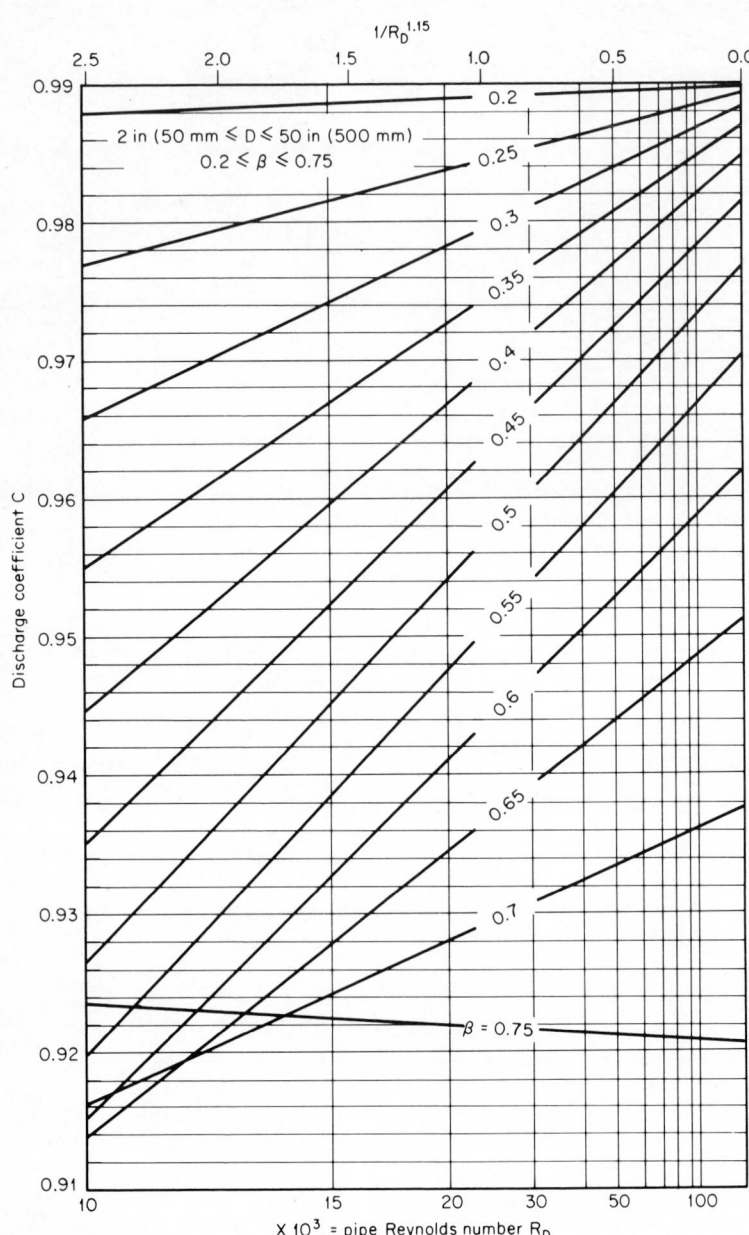

Figure 10.62 Discharge coefficient for ISA flow nozzle.

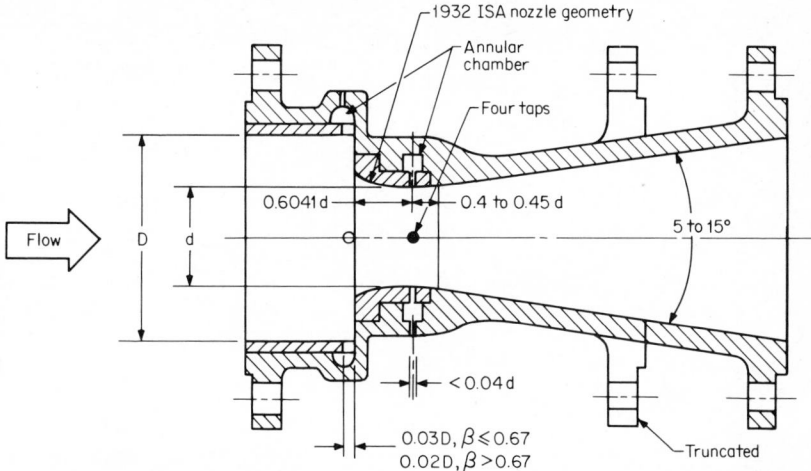

Figure 10.63 ISA venturi nozzle.

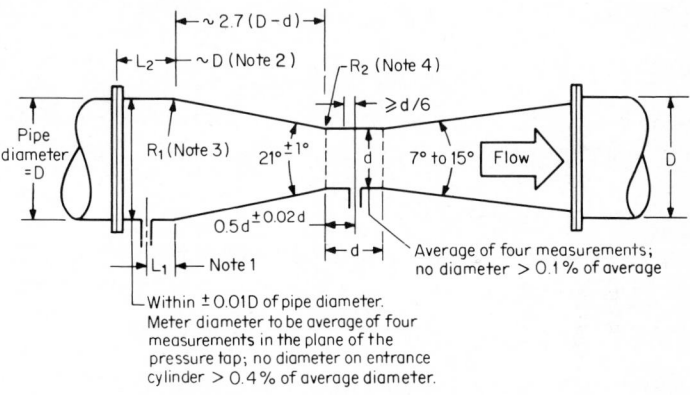

Note	Rough cast inlet	Machined inlet	Rough sheet-weld inlet
1	$L_1 = 0.5D \begin{array}{l} \pm 0.25D \;\; 4\text{ in} \leqslant D \leqslant 6\text{ in} \\ +0 \\ -0.25D \end{array} \;\; 6\text{ in} \leqslant D \leqslant 30\text{ in}$	$L_1 = 0.5D \pm 0.05D$	$L_1 = 0.5D \pm 0.05D$
2	$L_2 =$ smallest value of either 1D or 0.25D + 9.84 in	$L_2 \geqslant D$ (inlet diameter)	$L_2 \geqslant D$ (inlet diameter)
3	$R_1 = 1.375D + 20\%$	$R_1 < 0.25D$	$R_1 = 0$, other than welding
4	$R_2 = 3.625d$ to 3.8d	$R_2 < 0.25d$	$R_2 = 0$, other than welding

Figure 10.64 Venturi designs.

The discharge-coefficient accuracy and allowable β range are different for each group and are given in Tables 9.1 and 9.42. The discharge coefficient is constant for pipe Reynolds numbers greater than 200,000. For small-line-size venturis with machined inlets, the discharge-coefficient curve given in Fig. 10.65 may be used for lower-Reynolds-number applications.

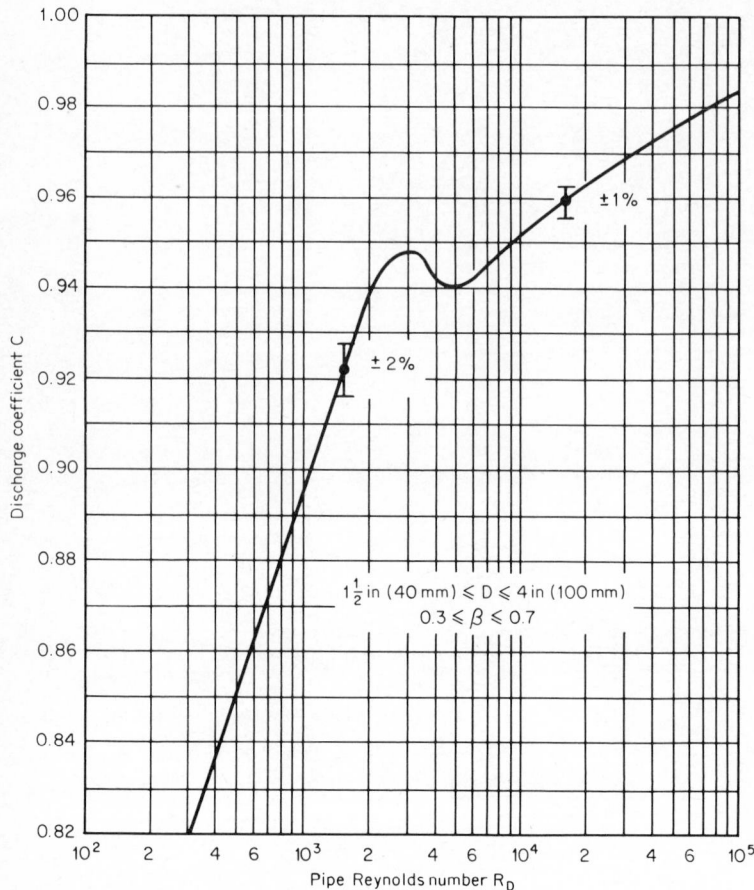

Figure 10.65 Discharge coefficient for venturi *(Fluidic Techniques, Inc., data, obtained in Foxboro Co. hydraulic laboratory; used with permission).*

Universal Venturi Tube

This proprietary device was designed to reduce overall lay-in length but retain the pressure recovery and coefficient constancy of the classical venturi. The *hydraulic shape* is shown in Fig. 10.66.

The manufacturer claims ±0.5 percent discharge-coefficient accuracy with uncalibrated tubes for pipe Reynolds numbers greater than 75,000. Figure 10.67 gives the percent variation in discharge coefficient for Reynolds numbers down to approximately 7000. Several ASME papers by Halmi (1974) detail the hydraulic shape and discharge coefficient as well as the installation requirements, head loss, and compressible-flow characteristics. The manufacturer should be consulted for exact sizing information, accuracy specifications, and installation recommendations.

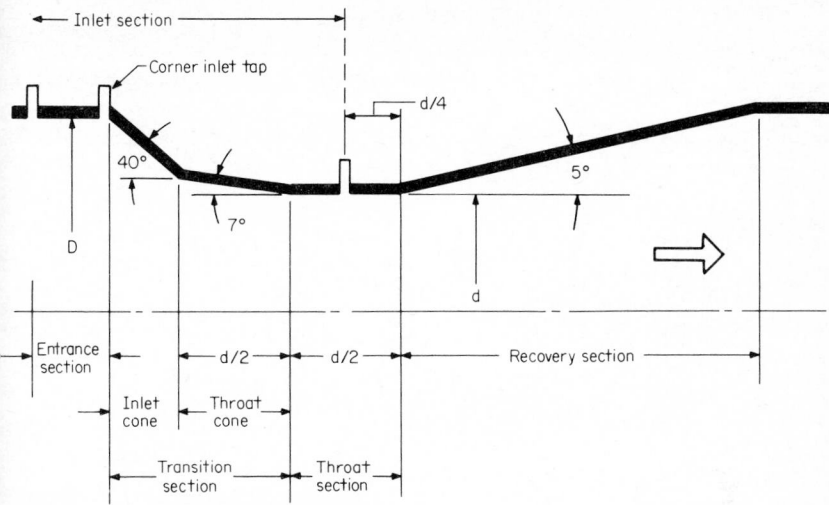

Figure 10.66 Hydraulic shape of the UVT *(courtesy BIF).*

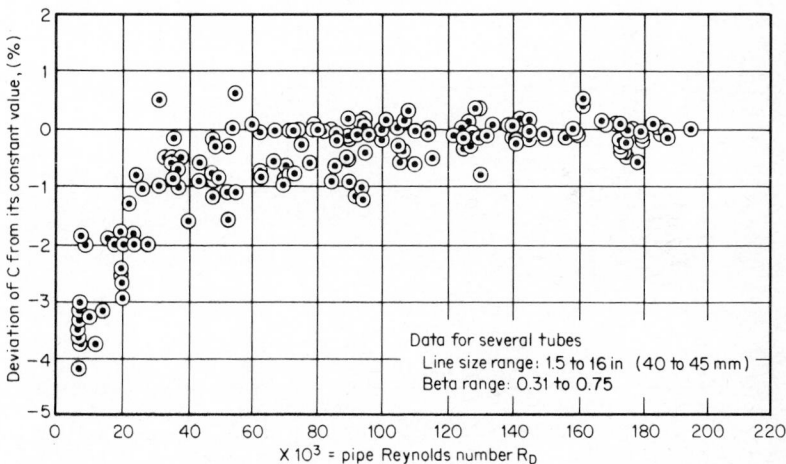

Figure 10.67 Discharge coefficient for Universal Venturi Tube *(compiled by BIF).*

Lo-Loss Flow Tube

Several proprietary primary devices are available which have low overall pressure loss relative to the produced differential. These include the Lo-Loss flow tube, the Dall Tube, the Foster flow tube (Gentile Tube), and the Twin Throat venturi tube.

In the Lo-Loss flow tube (Fig. 10.68), the upstream tap is located in the corner, and the downstream tap at or very close to the minimum throat diameter. This gives a higher differential than other tap locations, and the exit recovery cone reduces overall pressure loss.

The discharge coefficient is a function of β and is usually constant for pipe Reynolds numbers above 100,000. Its Reynolds-number range depends, however,

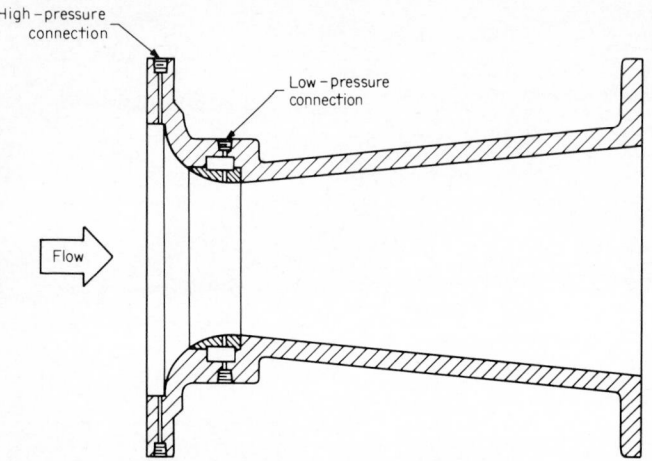

Figure 10.68   Lo-Loss flow tube *(courtesy Badger Meter, Inc.).*

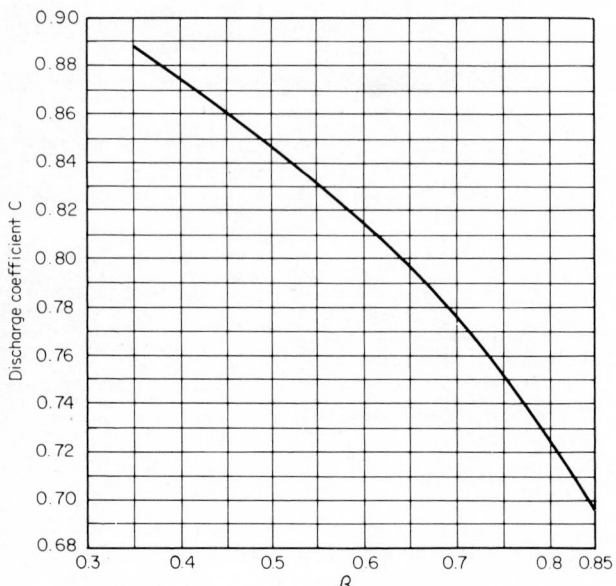

Figure 10.69   Discharge coefficient for Lo-Loss flow tube *(courtesy Badger Meter, Inc.).*

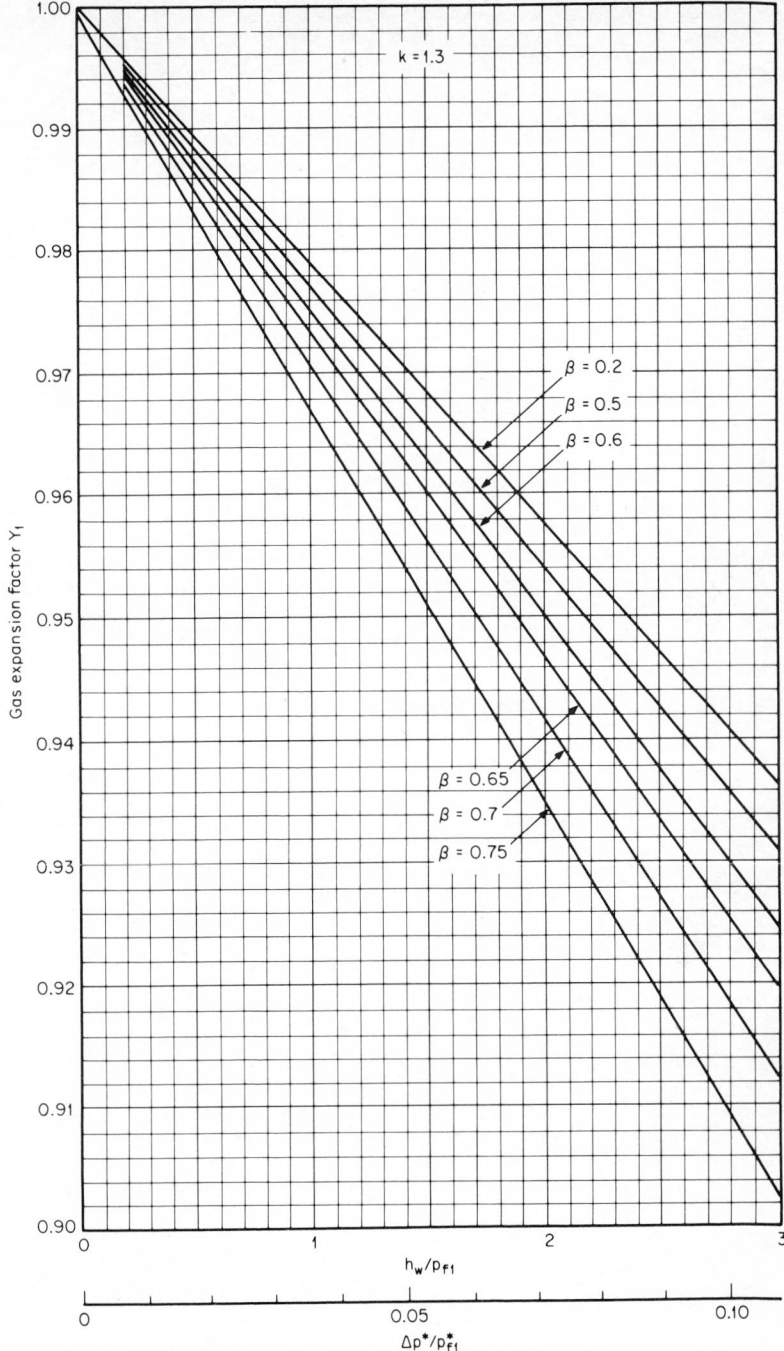

Figure 10.70 Gas expansion factor Y_1 for contoured-inlet primaries.

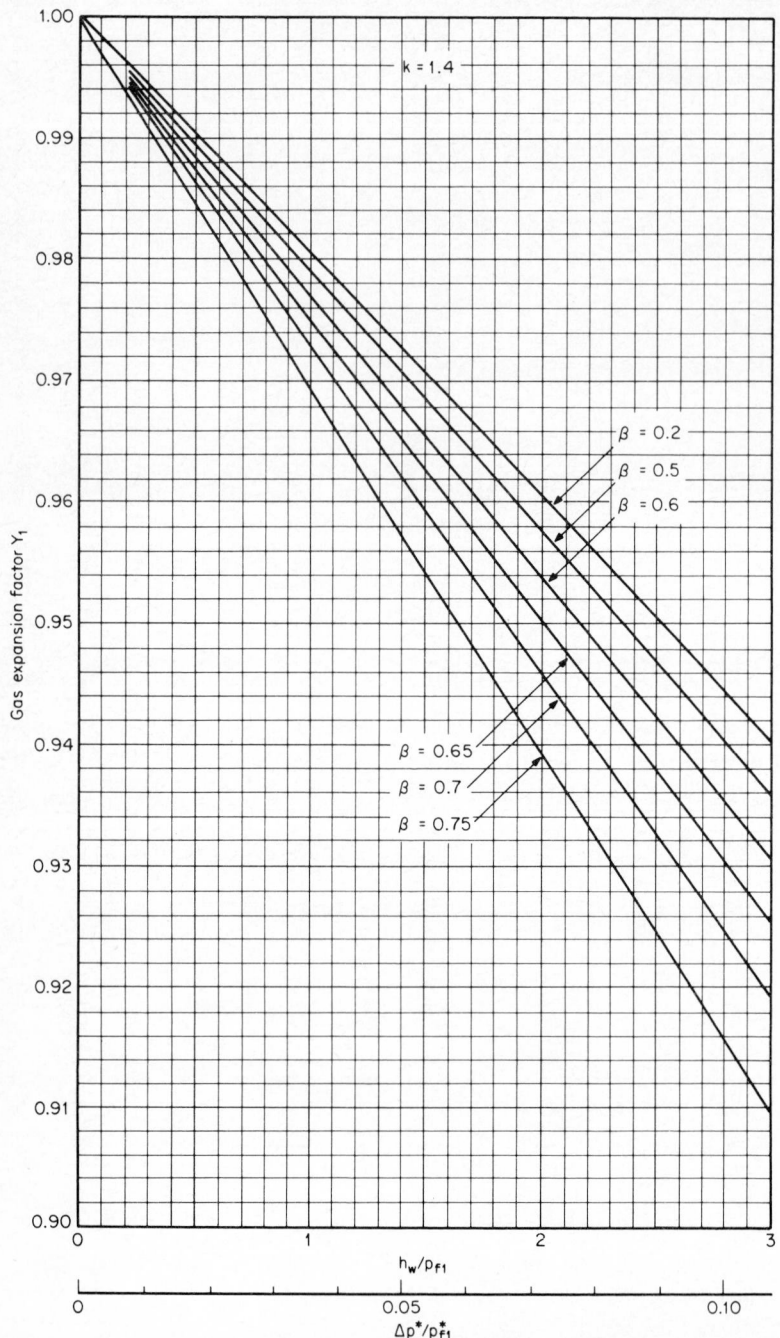

Figure 10.71   Gas expansion factor Y_1 for contoured-inlet primaries.

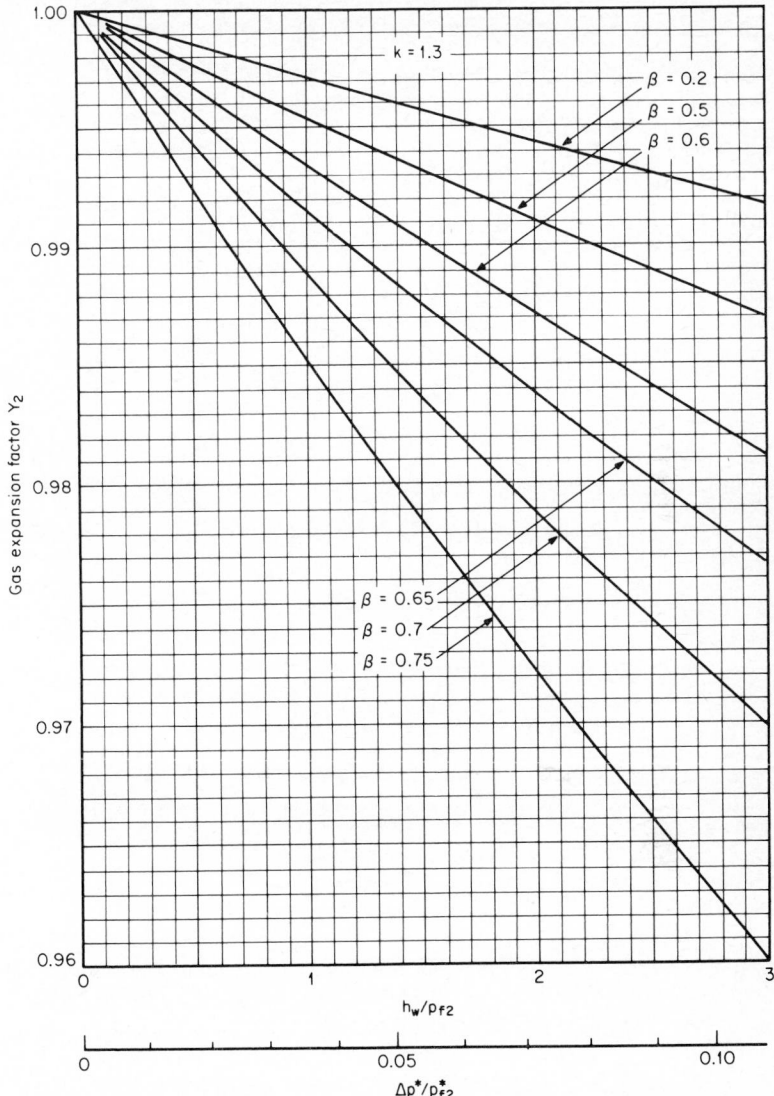

Figure 10.72 Gas expansion factor Y_2 for contoured inlet primaries.

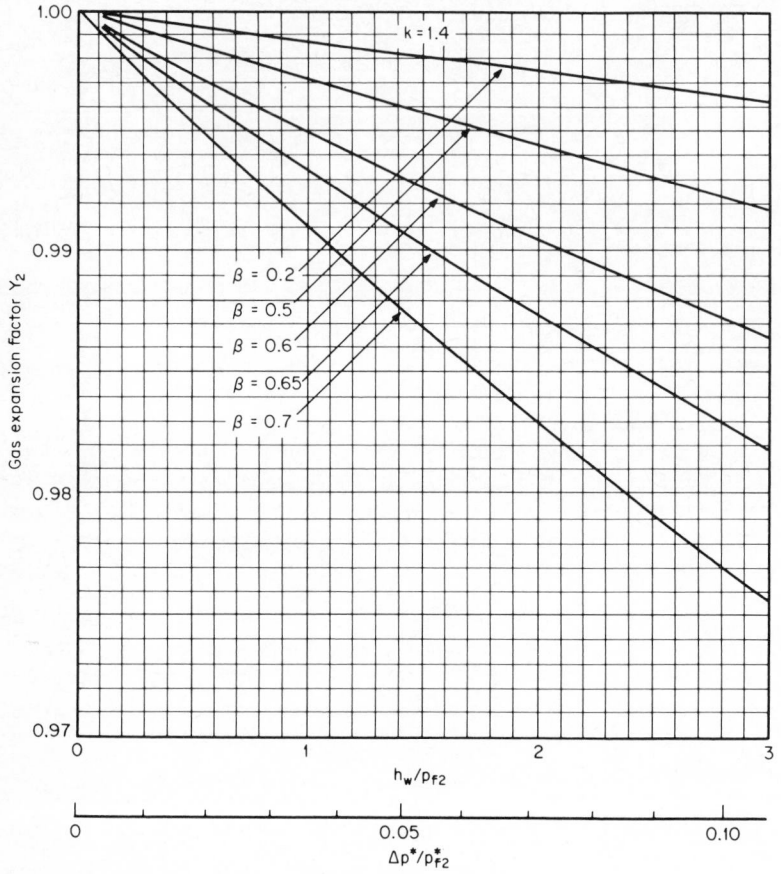

Figure 10.73 Gas expansion factor Y_2 for contoured inlet primaries.

on the manufacturer's recommendations and available test data. A graph of discharge coefficient versus β for a Lo-Loss flow tube is shown in Fig. 10.69. Similar curves are available from the manufacturers of the other designs.

GAS-EXPANSION-FACTOR CURVES

The gas expansion factor for contoured-inlet devices was derived in Chap. 9 and is expressed by Eq. (9.30). Curves of this equation are given in Figs. 10.70 through 10.73 for two different isentropic exponents and for upstream and downstream pressure measurements.

REFERENCES

ANSI/ASME MFC: *Differential Producers Used for the Measurement of Fluid Flow in Pipes (Orifice, Nozzle, Nozzle Venturi, Venturi),* Draft 3e, ASME, New York, 1982.

ANSI/ASME PTC-6, *Steam Turbines* (performance test code), ASME, New York, 1976.

ASME: *Fluid Meters,* pp. 214–215, ASME, New York, 1971.

Beitler, S. R., and D. J. Masson: "Calibration of Eccentric and Segmental Orifices in 4- and 6-in. Pipe Lines," *Trans. ASME,* vol. 71, p. 751, 1949.

Benedict, R. P.: "Most Probable Discharge Coefficients for ASME Flow Nozzles," *J. Basic Eng.,* vol. 88, p. 734, 1966.

BSI Standard 1042, *Methods for the Measurement of Fluid Flow in Pipes,* pt. 1, *Orifice Plates, Nozzles and Venturi Tubes,* UDC 53254.08, British Standards Institution, London, 1964.

Eujen, E.: *Comparison Measurements between Two Different Primary Elements in the Range of Laminar Flow,* ISO/TC/30 (Germany 5) 83, International Standards Organization, Paris, 1977.

Halmi, D.: "Metering Performance Investigation and Substantiation of the Universal Venturi Tube," pts. 1 and 2, *Fluids Eng.,* vol. 96, no. 2, pp. 124–138, 1974.

Lindahl, E. J., and S. R. Beitler: *Coefficients of Discharge for Eccentric and Segmental Orifices in 4-in., 6-in., 10-in., and 14-in. Pipe,* ASME Fluid Meters Research Report, 1954.

Stolz, J.: "A Universal Equation for the Calculation of Discharge Coefficients of Orifice Plates," ISO 5167, *Flomeko 1978, IMEKO Conference on Flow Measurement of Fluids, Groningen,* North-Holland, Amsterdam, 1978.

DIFFERENTIAL PRODUCERS: FIXED-GEOMETRY DEVICES

Chapter 9 covers differential producers that are sized by determining primary-element dimensions that will produce a chosen differential at a design flow rate. An alternative is to select a fixed-geometry primary device. These have limited dimensional selectivity; therefore, the differential pressure or target force, rather than the flowmeter dimensions, must be calculated to match the design flow rate.

Arithmetic-progression orifices, annular orifices, target flowmeters, integral orifices, Annubars, and elbow flowmeters are covered in this chapter. The flow-rate equations developed in Chap. 9 (Tables 9.36 through 9.38) apply to these devices. However, several of the symbols may be changed, grouped, or set equal to 1, depending on the device, how the geometry affects the differential pressure or target force, and whether an expansion factor is required. Table 11.1 presents the necessary modifications to these equations. The necessary graphs and equations and examples of the calculation procedure are given in the remainder of this chapter.

ARITHMETIC-PROGRESSION ORIFICES (EVEN-SIZED ORIFICES)

To change flow capacity, many plants stock a series of orifice plates with fixed-increment (arithmetic-progression) bore increases. Measurement equipment, pipe diameter, and fluid properties remain constant, and it becomes necessary to determine flow rates for fixed-range differential-pressure transmitters (50 in, 100 in, etc.). The general form of the flow-rate equation is given by Eq. (9.97) as

$$q = N \frac{CY\beta^2}{\sqrt{1 - \beta^4}} D^2 f(\rho) \sqrt{h_w} \qquad (9.97)$$

With constant fluid properties, design URV differential, and pipe size, the variables are conveniently grouped as

$$q = \frac{CY\beta^2}{\sqrt{1 - \beta^4}} [ND^2 f(\rho) \sqrt{h_w}] \qquad (11.1)$$

Table 11.1 Substitution of Variables in the Flow-Rate Equations of Tables 9.36 through 9.38

Primary element	S_M	F_{RD}	F_Y	F_a	Geometry factor D^2		Differential pressure	
					U.S. units	SI units	U.S. (h_w)	SI (Δp^*)
Arithmetic: progression orifice (even-sized)	$\dfrac{C\beta^2 Y}{\sqrt{1-\beta^4}}$	1.0	1.0	F_a, F_a^* [Eqs. (9.52), (9.54)]	D^2	D^{*2}	h_w	Δp^*
Annular orifice	K (Fig. 11.2)	1.0	1.0	F_a, F_a^* [Eqs. (11.6), (11.7)]	$(1 - \beta_T^2)D^2$	$(1 - \beta_T^2)D^{*2}$	h_w	Δp^*
Target flowmeter (Foxboro)	K (Figs. 11.4–11.11, Table 11.3)	1.0	1.0	F_a, F_a^* [Eqs. (11.6), (11.7)]	$5.941939\,\dfrac{1 - \beta_T^2}{\beta_T}\,D$ (D = meter bore)†	$35.68248\left(\dfrac{1 - \beta_T^2}{\beta_T}\right)D^*$ (D^* = meter bore)†	F_T	F_T^*
Integral orifice	K (Foxboro, Table 11.4, Figs. 11.17–11.22; Taylor, Table 11.4)	1.0	1.0	F_a, F_a^* [Eqs. (9.52), (9.54)]	d^2	d^{*2}	h_w	Δp^*
Annubar (Dieterich Standard Corp.)	K_{ref} (Table 11.5)	F_K (Table 11.6)	Y_1 (Table 11.7)	F_a, F_a^* [Eqs. (9.52), (9.54)]	D^2	D^{*2}	h_w	Δp^*
Elbow	K [Table 11.8, Eq. (11.8)]	1.0	1.0	F_a, F_a^* [Eqs. (9.52), (9.54)]	D^2	D^{*2}	h_w	Δp^*

†See Table 11.2.

where the bracketed term remains constant for a given differential, and the β-dependent quantities change with bore increment and Reynolds number. Equation (11.1) is nonlinear and therefore must be solved by iteration to calculate the flow rate.

The solution rapidly converges if an initial discharge coefficient of 0.6 is assumed. The first estimate for the flow rate then is

$$q_0 = \frac{0.6 Y \beta^2}{\sqrt{1 - \beta^4}} ND^2 f(\rho) \sqrt{h_w} \qquad (11.2)$$

With this estimate, the Reynolds number is calculated by using the equations of Tables 9.20 through 9.22, and the discharge coefficient is either calculated with the equations of Table 9.1 or obtained from the curves presented in Chap. 10. The second flow-rate estimate is then

$$q_1 = \frac{C_1 Y \beta^2}{\sqrt{1 - \beta^4}} ND^2 f(\rho) \sqrt{h_w} \qquad (11.3)$$

Depending on the desired accuracy, the procedure may be repeated by recalculating the Reynolds number, recalculating the discharge coefficient, and then substituting into Eq. (11.3).

EXAMPLE 11.1

Orifices with bores of 1.250, 1.350, and 1.450 in (31.75, 34.29, and 36.83 mm) are to be used to measure the gasoline flow of Example 9.3. Determine the flow rate at differential pressures of 64 and 100 in (15.9 and 24.9 kPa), using a single iteration. [The arithmetic-progression difference is 0.100 in (2.54 mm).]

From Eq. (f) of Table 9.36, the flow-rate equation is

$$q_{\text{GPM}} = N_{VG} S_M F_a F_{RD} D^2 \sqrt{F_p} \sqrt{G_F} \frac{\sqrt{h_w}}{G_b}$$

Rearranging into the form of Eq. (11.1), with $S_M = CY_1 \beta^2 / \sqrt{1 - \beta^4}$ and $F_{RD} = 1.0$, gives

$$q_{\text{GPM}} = \frac{CY_1 \beta^2}{\sqrt{1 - \beta^4}} N_{VG} F_a D^2 \sqrt{F_p} \sqrt{G_F} \frac{\sqrt{h_w}}{G_b}$$

From Example 9.3,

$$N_{VG} = 5.6665 \qquad F_a = 1.0002 \qquad D = 1.939 \text{ in} \qquad F_p = 1.0088$$

$$G_b = 0.7359 \qquad G_F = 0.7255 \qquad Y_1 = 1.0 \text{ for liquids}$$

Substitution then gives

$$q_{\text{GPM}} = 24.772 \frac{C\beta^2}{\sqrt{1 - \beta^4}} \sqrt{h_w}$$

The discharge-coefficient equation from Table 9.1 is

$$C = C_\infty + \frac{b}{R_D^n}$$

where, with $D = 1.939$ in,

$$C_\infty = 0.5959 + 0.0312\beta^{2.1} - 0.184\beta^8 + 0.039\frac{\beta^4}{1 - \beta^4} - \frac{0.0337}{1.939}\beta^3$$

and from Eq. (f) of Table 9.20 with $\mu_{cP} = 0.417$,

$$R_D = \left[17{,}903\left(\frac{G_b}{\mu_{cP}DN_{VG}}\right)\right] q_{GPM} = 2875.5\, q_{GPM}$$

then

$$\frac{b}{R_D^n} = \frac{91.71\beta^{2.5}}{R_D^{0.75}} = 0.2336\beta^{2.5}q_{GPM}^{-0.75}$$

For a differential of 64 in, the initial estimate for the flow rate is, by Eq. (11.2),

$$(q_{GPM})_0 = 24.772\frac{0.6\beta^2}{\sqrt{1 - \beta^4}}(1.0)\sqrt{64} = 118.905\frac{\beta^2}{\sqrt{1 - \beta^4}}$$

At 100 in differential the initial estimate is

$$(q_{GPM})_0 = 24.772\frac{0.6\beta^2}{\sqrt{1 - \beta^4}}(1.0)\sqrt{100} = 148.632\frac{\beta^2}{\sqrt{1 - \beta^4}}$$

Tabulating the calculation at $h_w = 64$ in gives

	Bore d, in		
	1.250	**1.350**	**1.450**
β	0.64466	0.69624	0.74781
$(q_{GPM})_0$	54.330	65.900	80.207
C_∞	0.60631	0.60644	0.60533
$b/R_D^{0.75}$	0.00390	0.00409	0.00421
C	0.61021	0.61053	0.60954
q_{GPM}†	55.25	67.06	81.48

†$q_{GPM} = 24.772(C\beta^2/\sqrt{1 - \beta^4})\sqrt{h_w}$.

and the calculation at $h_w = 100$ in is as follows:

	Bore d, in		
	1.250	**1.350**	**1.450**
β	0.64466	0.69624	0.74781
$(q_{\mathrm{GPM}})_0$	67.912	82.375	100.26
C_∞	0.60631	0.60644	0.60533
$b/R_D^{0.75}$	0.00329	0.00346	0.00357
C	0.60960	0.60990	0.60890
q_{GPM}†	69.00	83.73	101.7

†$q_{\mathrm{GPM}} = 24.772(C\beta^2/\sqrt{1-\beta^4})\sqrt{h_w}$.

ANNULAR ORIFICE

The annular orifice is shown in Fig. 11.1. It consists of a disk supported concentrically in a pipe section by supporting spiders. Upstream and downstream pressures are transmitted through the central shaft to the differential-pressure transmitter. The earliest experimental test work on this device was done by Howell (1939), using air and diameter ratios from 0.7 to 0.9. Bell and Bergelin (1957, 1962) experimentally tested 21 annular orifices with beta ratios between 0.95 and 0.996, and they reported good agreement with other published results. Although limited low-Reynolds-number data are available for β_T in the range of 0.7 to 0.9, the flow coefficient appears to be constant above a pipe Reynolds number of 10,000.

The annular orifice has the advantages of providing free drainage for heavy materials at the bottom of the pipe while, at the same time, allowing gas or vapor to pass along the top of the pipe. Its major disadvantages are the lack of exten-

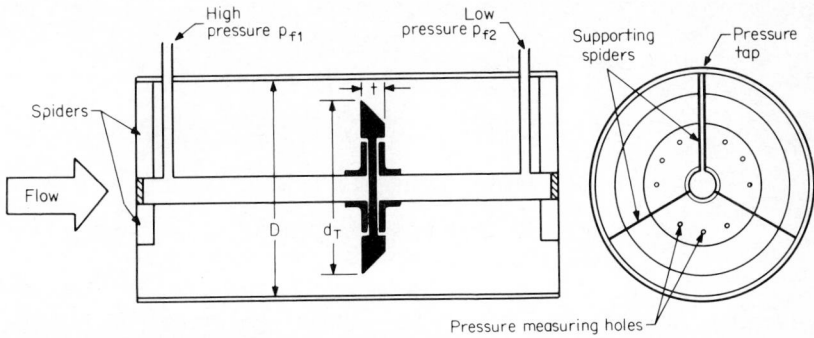

Figure 11.1 Annular orifice.

sive data and the dependence on pipe dimension to define flow area; the latter becomes more serious as β_T approaches 0.9.

The flow coefficient K as determined by Howell (1939) is presented in Fig. 11.2. For gas (vapor) flows, no gas expansion factor has been determined. The thermal-expansion-factor correction is the same as for other differential producers. The overall pressure loss is approximately the same as that of the target flowmeter for the same beta ratio.

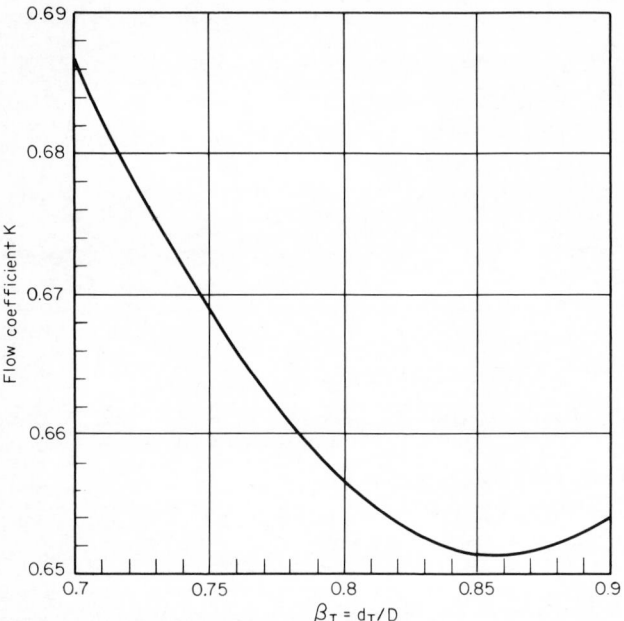

Figure 11.2 Annular-orifice flow coefficient.

EXAMPLE 11.2

An annular orifice with $\beta_T = 0.75$ is to measure the steam flow of Example 9.2. Calculate the upper-range value for the differential-pressure transmitter and the flow rate if the differential pressure is 100 in (24.9 kPa).

The flow-rate equation is, from Eq. (g) of Table 9.36,

$$q_{\text{PPH}} = N_{M\rho}S_MF_aF_{RD}D^2\sqrt{\rho_{f1}}\sqrt{h_w}$$

Substituting variables from Table 11.1 gives the equation for an annular orifice as

$$q_{\text{PPH}} = N_{M\rho}KF_a(1 - \beta_T^2)D^2\sqrt{\rho_{f1}}\sqrt{h_w}$$

From Example 9.2,

$$(q_{\text{PPH}})_{\text{URV}} = 30{,}000 \text{ lb}_m/\text{h} \qquad N_{M\rho} = 358.93 \qquad F_a = 1.0075$$
$$\beta_T = 0.75 \qquad D = 5.761 \text{ in} \qquad \rho_{f1} = 0.39524 \text{ lb}_m/\text{ft}^3$$

and from Fig. 11.2, $K = 0.669$. Substitution and rearrangement then give

$$(h_w)_{\text{URV}} = \left\{ \frac{30,000}{(358.93)(0.669)(1.0075)[1 - (0.75)^2](5.761)^2 \sqrt{0.39524}} \right\}^2 = 184.53 \text{ in}$$

At $h_w = 100$, the flow rate is calculated as

$$q_{\text{PPH}} = (358.93)(0.669)(1.0075)[1 - (0.75)^2](5.761)^2 \sqrt{0.39524} \sqrt{h_w}$$
$$= F_{MC} \sqrt{h_w} = 2208.4 \sqrt{h_w} = 2208.4 \sqrt{100} = 22,084 \text{ lb}_m/\text{h}$$

TARGET FLOWMETER

The acceleration of fluid into the annular space around a concentric target (disk) creates a reduced pressure at the target's rear surface (Fig. 11.3a). The force on

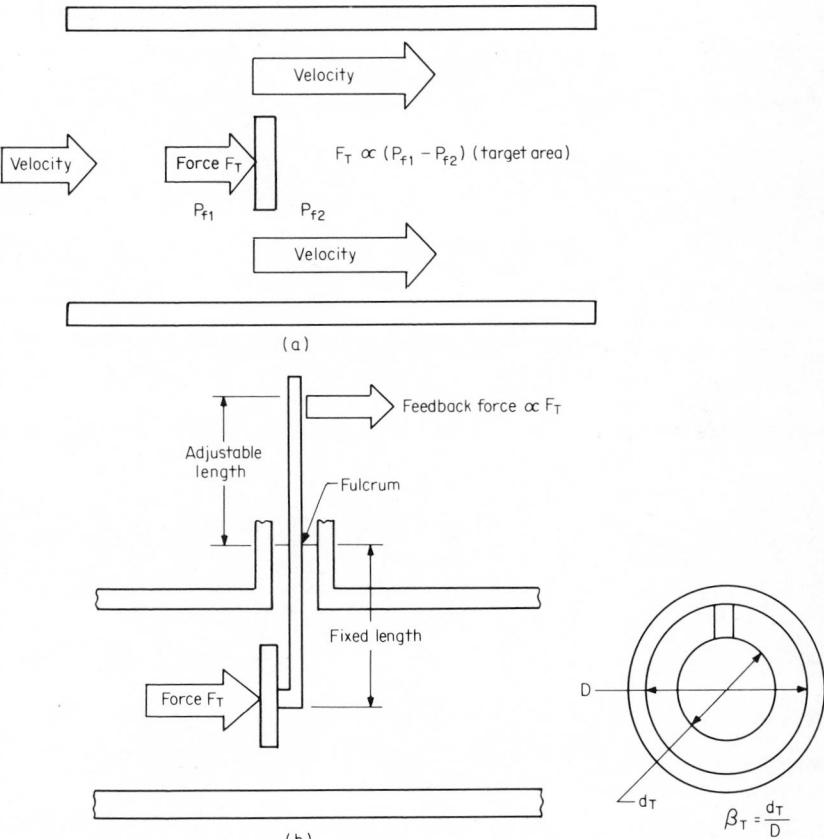

Figure 11.3 Target flowmeter *(courtesy The Foxboro Co.)*. (a) Basic principle. (b) Foxboro force-balance target flowmeter.

the target is the difference between the upstream and downstream surface pressures, integrated over the target area. By Bernoulli's equation, the square root of this force is proportional to the flow rate.

Foxboro's target flowmeter balances the target force with a feedback force applied above a fulcrum (Fig. 11.3b). The moment arm to the target force is constant, and a variable range is achieved by adjusting the position of the fulcrum. The output is a standardized 3- to 15-lb/in² (20- to 100-kPa) or 4- to 20-mA signal proportional to target force. Flowmeters are calibrated by suspending weights from the target centerline and adjusting the variable moment arm to the desired upper-range value.

The flow-rate equations given in Tables 9.36 through 9.38 are modified by substituting the geometry factor for D^2 and the flow coefficient K for the S_M factor to obtain the flow-rate equation for the target flowmeter. The Reynolds-number correction factor F_{RD} and gas-expansion-factor correction F_Y are set equal to 1.0. Geometry factors and allowable force limits for available fixed-geometry flowmeters are given in Table 11.2. Recommended installation lengths are based on equivalent orifice beta ratios; these are also given in Table 11.2.

Curran (1981) gives the flow-coefficient equations and pipe-Reynolds-number curves for Foxboro's fixed-geometry targets in $\frac{1}{2}$ (15-mm) through 4-in (100-mm) meter sizes. These are presented in Table 11.3 and Figs. 11.4 through 11.11. Test work on natural gas by Reinecke et al. (1966) with similar calibrated flowmeters showed no gas-expansion-factor error and flow performance equivalent to that of a series-installed orifice.

The permanent pressure loss in pounds-force per square inch can be calculated as

$$\Delta p_L = \frac{0.199K^2(1 - \beta_T^2)^2}{D^2\beta_T^2(1 - \beta_T)^{2.75}} F_T = K_{LT}F_T \tag{11.4}$$

and in kilopascals as

$$\Delta p_L^* = \frac{198.2K^2(1 - \beta_T^2)^2}{D^{*2}\beta_T^2(1 - \beta_T)^{2.75}} F_T^* = K_{LT}^*F_T^* \tag{11.5}$$

Values of the loss coefficients K_{LT} and K_{LT}^* at $R_D = 100{,}000$ for fixed-geometry targets are given in Table 11.2.

The thermal-expansion-factor correction for target flowmeters is

$$F_a = 1 + \alpha_{PE}(T_F - 68) \tag{11.6}$$

for U.S. units, and

$$F_a = 1 + \alpha_{PE}^*(T_{\circ C} - 20) \tag{11.7}$$

for SI units.

Table 11.2 Geometry Constants for Target Flowmeters (Foxboro)

Nominal line size, in (mm)	Meter bore D, in (mm)	Target diameter d_T in (mm)	β_T $\left(\dfrac{d_T}{D}\right)$	Pressure-loss coefficient		Force limits		Geometry factor		Installation (equivalent orifice beta ratio)
				U.S. units (K_{LT}†)	SI units (K'_{LT}‡)	U.S. units (lbf)	SI units (N)	U.S. units§	SI units¶	
0.5 (15)	0.546 (13.87)	0.4368 (11.09)	0.8	5.41	8.35	$1.5 \leq F_T \leq 15.3$	$6.67 \leq F_T^* \leq 68.1$	1.459934	222.6965	0.6
1 (25)	0.957 (24.31)	0.7656 (19.45)	0.8	1.82	2.81	$1.3 \leq F_T \leq 13.1$	$5.78 \leq F_T^* \leq 58.3$	2.558896	390.2132	0.6
2 (50)	2.067 (52.50)	1.3610 (34.57)	0.6584	0.322	0.497	$1.9 \leq F_T \leq 19.2$	$8.5 \leq F_T^* \leq 85.4$	10.56612	1611.670	0.75
2 (50)	2.067 (52.50)	1.6679 (42.37)	0.8069	0.330	0.509	$1.9 \leq F_T \leq 19.2$	$8.5 \leq F_T^* \leq 85.4$	5.310298	809.9895	0.6
3 (75)	3.068 (77.93)	1.604 (40.74)	0.5228	0.126	0.195	$1.6 \leq F_T \leq 16.2$	$7.12 \leq F_T^* \leq 72.1$	25.33773	3864.811	0.8
3 (75)	3.068 (77.93)	2.301 (58.45)	0.75	0.133	0.205	$1.6 \leq F_T \leq 16.2$	$7.12 \leq F_T^* \leq 72.1$	10.63409	1622.038	0.65
4 (100)	4.026 (102.26)	1.6679 (42.37)	0.4143	0.093	0.144	$1.4 \leq F_T \leq 13.7$	$6.23 \leq F_T^* \leq 60.9$	47.83329	7296.099	0.75
4 (100)	4.026 (102.26)	2.678 (68.02)	0.6652	0.072	0.111	$1.4 \leq F_T \leq 13.7$	$6.23 \leq F_T^* \leq 60.9$	20.05125	3058.455	0.75

†$K_{LT} = (0.199/D^2)K^2[(1 - \beta_T^2)/\beta_T]^2[1/(1 - \beta_T)^{2.75}]$.

‡$K'_{LT} = (198.2/D^{*2})K^2[(1 - \beta_T^2)/\beta_T]^2[1/(1 - \beta_T)^{2.75}]$, where K is the flow coefficient at R_D = 100,000.

§Equal to $5.941939[(1 - \beta_T^2)/\beta_T]D$.

¶Equal to $35.68248[(1 - \beta_T^2)/\beta_T]D^*$.

11-9

Table 11.3 Target-Flowmeter Flow-Coefficient Equations (Foxboro)

Nominal line size, in (mm)	β_T	Reynolds-number range	Accuracy (uncalibrated), %	Flow coefficient K
0.5 (15)	0.8	$111 \leq R_D \leq 4000$	±2	$K = -15.8447766 + 23.0693751 \log R_D - 11.958139(\log R_D)^2 + 2.7183724(\log R_D)^3 - 0.2283791(\log R_D)^4$
0.5 (15)	0.8	$R_D > 4000$	±2.7	$K = 0.69163$
1 (25)	0.8	$44 \leq R_D \leq 4000$	±2.5	$K = 27.7292586 - 57.2484112 \log R_D + 46.3134298(\log R_D)^2 - 18.1122003(\log R_D)^3 + 3.4484014(\log R_D)^4 - 0.2568661(\log R_D)^5$
1 (25)	0.8	$R_D > 4000$	±1.8	$K = 0.70365$
2 (50)	0.658	$167 \leq R_D \leq 4000$	±1.9	$K = -9.3752781 + 13.5112009 \log R_D - 6.8267566(\log R_D)^2 + 1.5191852(\log R_D)^3 - 0.1253178(\log R_D)^4$
2 (50)	0.658	$R_D > 4000$	±1.0	$K = 0.57553 + 0.012744 \log R_D$
2 (50)	0.807	$129 \leq R_D \leq 4000$	±2.7	$K = -10.8240856 + 14.7439321 \log R_D - 6.9467876(\log R_D)^2 + 1.4199689(\log R_D)^3 - 0.1061497(\log R_D)^4$
2 (50)	0.807	$R_D > 4000$	±1.1	$K = 0.61852 + 0.0045082 \log R_D$
3 (80)	0.523	$319 \leq R_D \leq 4000$	±6.3	$K = -18.0866104 + 24.5119444 \log R_D - 11.9261559(\log R_D)^2 + 2.5315746(\log R_D)^3 - 0.1969396(\log R_D)^4$
3 (80)	0.523	$R_D > 4000$	±2.0	$K = 0.57810 + 0.011170 \log R_D$
3 (80)	0.750	$178 \leq R_D \leq 4000$	±2.3	$K = -2.5116530 + 3.3406322 \log R_D - 1.1874923(\log R_D)^2 + 0.1393851(\log R_D)^3$
3 (80)	0.750	$R_D > 4000$	±1.7	$K = 0.59294 + 0.0090958 \log R_D$
4 (100)	0.414	$530 \leq R_D \leq 4000$	±1.4	$K = -66.9527450 + 86.3212859 \log R_D - 41.0847774(\log R_D)^2 + 8.6108367(\log R_D)^3 - 0.6694608(\log R_D)^4$
4 (100)	0.414	$R_D > 4000$	±0.7	$K = 0.587071 + 0.014652 \log R_D$
4 (100)	0.665	$240 \leq R_D \leq 4000$	±2.1	$K = -18.7909023 + 26.1845236 \log R_D - 13.0980936(\log R_D)^2 + 2.8696266(\log R_D)^3 - 0.2319703(\log R_D)^4$
4 (100)	0.665	$R_D > 4000$	±1.1	$K = 0.613920 + 0.0057003 \log R_D$

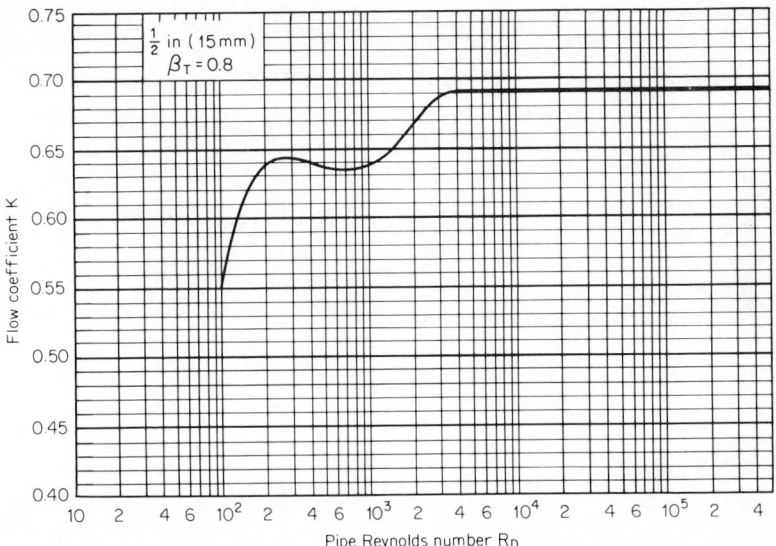

Figure 11.4   Target-flowmeter flow coefficient: $\frac{1}{2}$-in (15-mm) bore, $\beta_T = 0.8$.

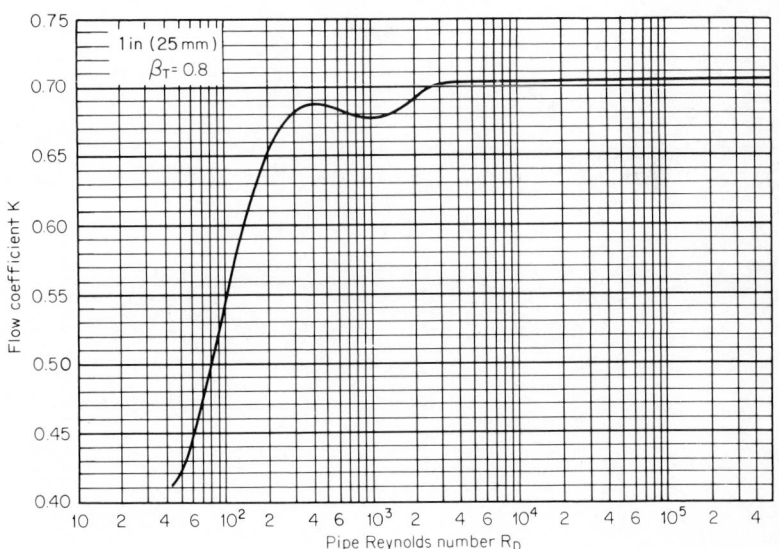

Figure 11.5   Target-flowmeter flow coefficient, 1-in (25-mm) bore, $\beta_T = 0.8$.

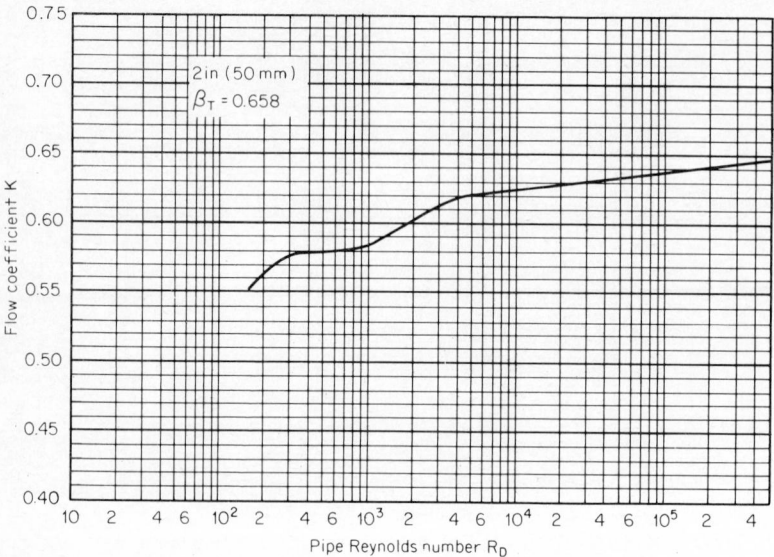

Figure 11.6   Target-flowmeter flow coefficient, 2-in (50-mm) bore, $\beta_T = 0.658$.

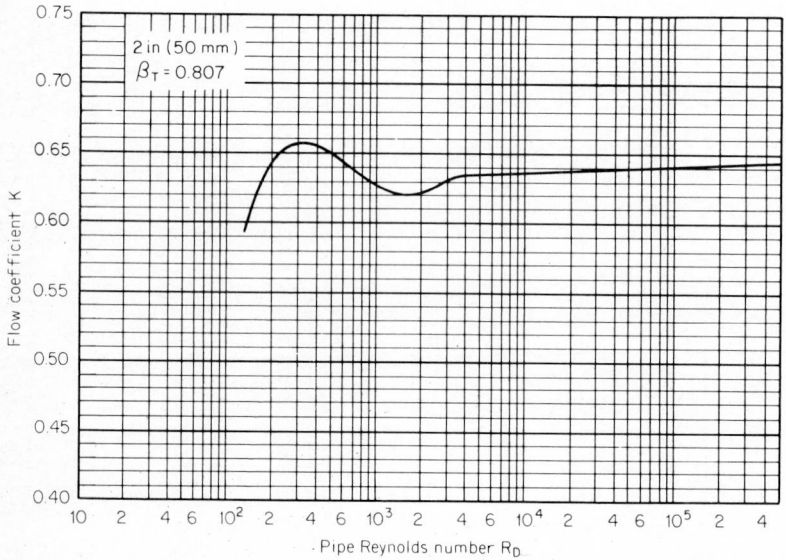

Figure 11.7   Target-flowmeter flow coefficient, 2-in (50-mm) bore, $\beta_T = 0.807$.

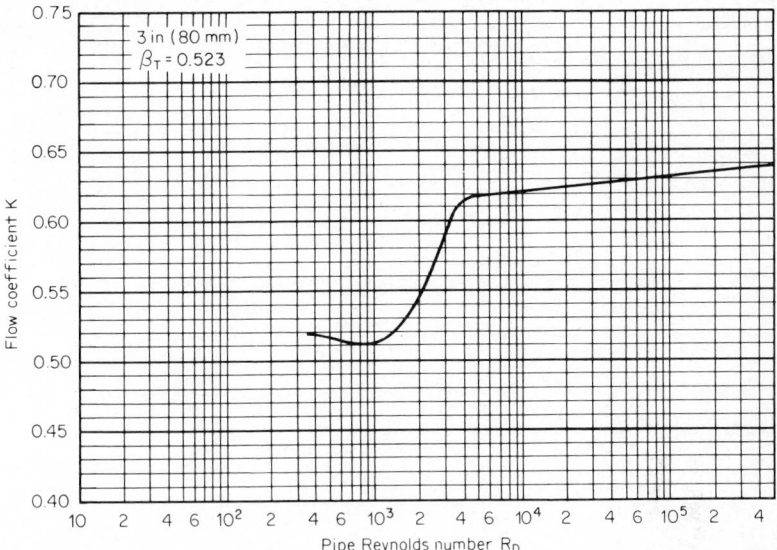

Figure 11.8 Target-flowmeter flow coefficient, 3-in (75-mm) bore, $\beta_T = 0.523$.

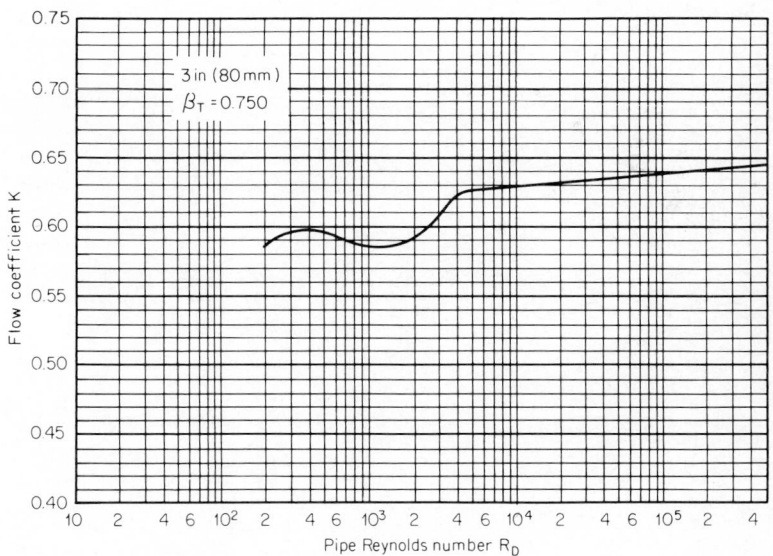

Figure 11.9 Target-flowmeter flow coefficient, 3-in (75-mm) bore, $\beta_T = 0.750$.

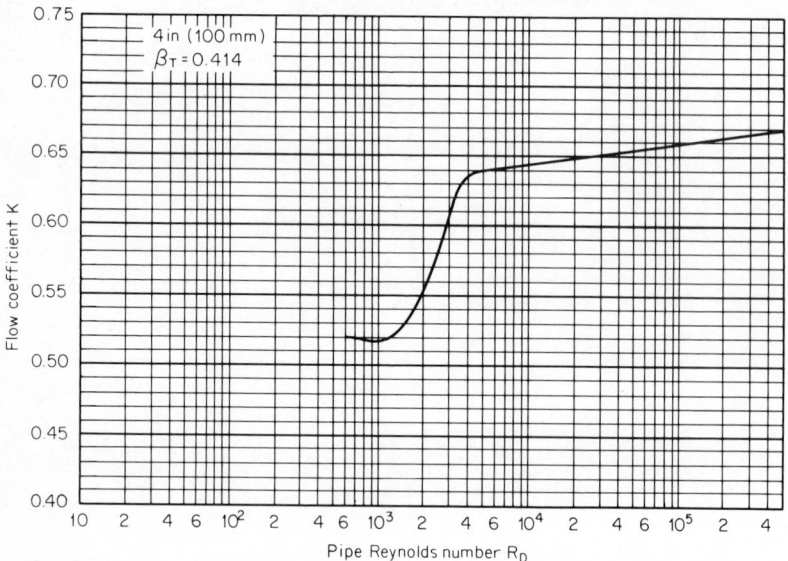

Figure 11.10   Target-flowmeter flow coefficient, 4-in (100-mm) bore, $\beta_T = 0.414$.

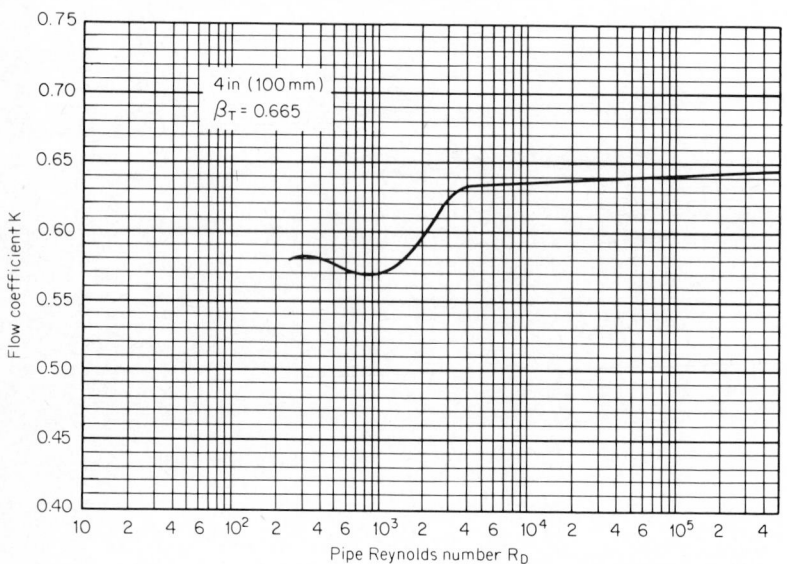

Figure 11.11   Target-flowmeter flow coefficient, 4-in (100-mm) bore, $\beta_T = 0.665$.

EXAMPLE 11.3

A 15-mm target flowmeter is to measure the flow rate of the gasoline of Example 2.12. The maximum flow rate is 50 L/min. Determine (a) the upper-range force value F_{URV}, (b) the flow rate when the reading is 20 percent of the URV, and (c) the overall pressure loss at maximum flow. (The flow rate is to be measured in base volume units.)

The flow-rate equation [Eq. (f) of Table 9.37] is

$$q_{\text{LPM}}^* = N_{VG}^* S_M F_a^* F_{RD} D^{*2} \sqrt{F_p} \sqrt{G_F} \frac{\sqrt{\Delta p^*}}{G_b}$$

Substitution of variables from Table 11.1 gives the flow-rate equation for a target flowmeter as

$$q_{\text{LPM}}^* = N_{VG}^* K F_a^* (35.68248) \frac{1 - \beta_T^2}{\beta_T} D^* \sqrt{F_p} \sqrt{G_F} \frac{\sqrt{F_T^*}}{G_b}$$

Data is determined as follows: From Table 9.17, $N_{VG}^* = 0.0666762$. From Eq. (11.7),

$$F_a^* = 1 + \alpha_{PE}^*(T_{^\circ C} - 20) = 1 + 0.0000173(26.9 - 20) = 1.00012$$

where, from Table B.4, $\alpha_{PE}^* = 0.0000173$ mm/(mm·°C). The geometry factor is, from Table 11.2,

$$35.68248 \frac{1 - \beta_T^2}{\beta_T} D^* = 222.6825$$

From Example 2.15, $F_P = 1.0088$; from Example 2.12, $G_F = 0.7255$ and $G_b = 0.7359$; and, from Example 9.1, $\mu_{cP} = 0.417$.

Substitution now gives

$$q_{\text{LPM}}^* = (0.0666762)(K)(1.00012)(222.6865)\sqrt{1.0088}\sqrt{0.7255}\,\frac{\sqrt{F_T^*}}{0.7359}$$

$$= 17.263 K \sqrt{F_T^*}$$

The pipe Reynolds number is, by Eq. (f) of Table 9.21,

$$R_D = 1413.5 \frac{G_b}{\mu_{cP} D^* N_{VG}^*} q_V^* = 1413.5 \frac{0.7359}{(0.417)(13.87)(0.06667)} q_{\text{LPM}}^*$$

$$= 2698 q_{\text{LPM}}^*$$

Scaling to normal operating conditions gives

$$(q_{\text{LPM}}^*)_N = 0.8(q_{\text{LPM}}^*)_{\text{URV}} = 40$$

The Reynolds number is then

$$(R_D)_N = (2698)(40) = 107{,}920$$

From Table 11.3 or Fig. 11.4, the flow coefficient is $K = 0.69163$. Rearranging the flow-rate equation to solve for target force yields

$$(F_T^*)_N = \left(\frac{q_{\text{LPM}}^*}{17.263 K}\right)_N^2 = \left[\frac{40}{(17.263)(0.69163)}\right]^2 = 11.224 \text{ N}$$

a. The upper-range force value is then

$$(F_T^*)_{\text{URV}} = 11.224 \left(\frac{50}{40}\right)^2 = 17.54 \text{ N}$$

b. The flow rate at 20 percent of URV is

$$q_{\text{LPM}}^* = 17.263K\sqrt{F_T^*} = (17.263)(0.69163)\sqrt{F_T^*} = F_{MC}\sqrt{F_T^*} = 11.940\sqrt{(0.2)(17.54)}$$

$$= 22.36 \text{ base L/min}$$

(*Note:* The flow coefficient is constant for $R_D > 4000$.)

c. The overall pressure loss at URV flow is, from Eq. (11.5) and Table 11.2 for the loss coefficient,

$$\Delta p_L^* = K_{LT}^* F_T^* = (8.35)(17.54) = 146 \text{ kPa}$$

EXAMPLE 11.4

A 3-in (75-mm) target flowmeter with $\beta_T = 0.523$ is to measure the flow rate of ammonia vapor. The URV flow rate is 12,500 lb$_m$/h (5670 kg/h). Determine (*a*) the upper-range force value $(F_{T,})_{\text{URV}}$, (*b*) the maximum permanent pressure loss, and (*c*) the flow rate at 25 percent of URV target force.

The process conditions are

$$T_F = 100°\text{F} \qquad p_G = 165 \text{ psig} \qquad \mu_{\text{cP}} = 0.0105 \qquad \rho_f = 0.5797 \text{ lb}_m/\text{ft}^3$$

From Eq. (g) of Table 9.36, the flow-rate equation is

$$q_{\text{PPH}} = N_{M\rho}S_M F_a F_{RD} F_Y D^2 \sqrt{\rho_{f1}} \sqrt{h_w}$$

Substitution of variables from Table 11.1 gives the target-flowmeter flow-rate equation as

$$q_{\text{PPH}} = N_{M\rho}KF_a(5.941939)\frac{1 - \beta_T^2}{\beta_T} D\sqrt{\rho_{f1}}\sqrt{F_T}$$

From Table 9.14, $N_{M\rho} = 358.9268$, and from Eq. (11.6),

$$F_a = 1 + \alpha_{PE}(T_F - 68) = 1 + (0.0000096)(100 - 68) = 1.00031$$

where, from Table B.4, $\alpha_{PE} = 0.0000096$ in/(in·°F). The geometry factor, from Table 11.2 is,

$$5.941939\frac{1 - \beta_T^2}{\beta_T} D = 25.33773$$

Substitution now gives

$$q_{\text{PPH}} = (358.9268)(K)(1.00031)(25.33773)\sqrt{0.5797}\sqrt{F_T}$$

$$= 6926.4K\sqrt{F_T}$$

The pipe Reynolds number is, from Eq. (g) of Table 9.20,

$$R_D = 2266.97\frac{1}{\mu_{\text{cP}}DN_{M\rho}}q_{\text{PPH}} = 2266.97\frac{1}{(0.0105)(3.068)(358.93)}q_{\text{PPH}}$$

$$= 196.1q_{\text{PPH}}$$

Scaled to normal operating conditions, the flow rate is

$$(q_{PPH})_N = 0.8(q_{PPH})_{URV} = (0.8)(12,500) = 10,000$$

The Reynolds number at normal conditions is then

$$(R_D)_N = (196.1)(10,000) = 1,961,000$$

From Table 11.3, for a target flowmeter with $\beta = 0.523$,

$$K = 0.57810 + 0.011170 \log R_D = 0.57810 + 0.011170 \log 1,961,000$$
$$= 0.64839$$

The target force at normal conditions is then

$$(F_T)_N = \left(\frac{q_{PPH}}{6926.4K} \right)^2 = \left[\frac{10,000}{(6926.4)(0.64839)} \right]^2 = 4.958 \text{ lb}_f$$

a. The upper-range force is then

$$(F_T)_{URV} = 4.958 \left(\frac{12,500}{10,000} \right)^2 = 7.747 \text{ lb}_f$$

b. The maximum permanent pressure loss is, from Eq. (11.4), with the loss coefficient from Table 11.2,

$$\Delta p_L = K_{LT}F_T = (0.126)(7.747) = 0.98 \text{ lb/in}^2$$

c. The flow rate at 25 percent URV target force is

$$q_{PPH} = 6926.4K \sqrt{F_T} = 6926.4K \sqrt{(0.25)(7.747)} = 9639.3K$$

For an initial estimate $K_0 = 0.64839$, the Reynolds number is

$$R_D = 196.1q_{PPH} = (196.1)(9639.3)(0.64839) = 1,225,600$$

which, when substituted into the flow-coefficient equation, gives the second estimate for K as

$$K_1 = 0.57810 + 0.011170 \log R_D = 0.64611$$

A third estimate gives

$$(q_{PPH})_2 = 9639.3K_1 = 6228.0 \qquad R_D = 1,221,300$$

and

$$K_2 = 0.64609$$

The flow rate is then

$$q_{PPH} = 9639.3K_2 = (9639.3)(0.64609) = 6227.9 \text{ lb}_m/\text{h}$$

Without iteration the flow rate would be calculated as

$$q_{PPH} = F_{MC}\sqrt{F_T} = (6926.4)(0.64839)\sqrt{F_T} = 4491\sqrt{F_T} = 6250 \text{ lb}_m/\text{h}$$

INTEGRAL ORIFICE

Several manufacturers install fixed-bore orifices in manifolds that are an *integral* part of the differential-pressure transmitter. These not only provide a compact installation but also extend differential producers to very low flow-rate capabilities—for liquids down to 0.04 gal/min (0.015 L/min), and for gases as low as 0.9 standard ft³/h (25 standard L/h).

The wide range of today's differential-pressure transmitters [5 to 850 in (1.3 to 210 kPa)], combined with available bore sizes, provides a wide flow-rate capability. Accuracies of $1\frac{1}{2}$ to 3 percent are claimed for uncalibrated units over selected Reynolds-number ranges. Felton (1972) presents calibration data collected over a 4-year period that demonstrates a precision of ±0.25 percent.

Foxboro offers two integral-orifice flow assemblies, the in-line type and the U-bend type. The in-line type, illustrated in Fig. 11.12a, is designed for the process fluid to pass through the high-pressure side of the transmitter and the orifice only. A cross-port in the manifold, just downstream of the orifice, allows the lower downstream pressure to be sensed by the low-pressure side of the transmitter. In the 180° U-bend type, shown in Fig. 11.12b, process fluid enters the

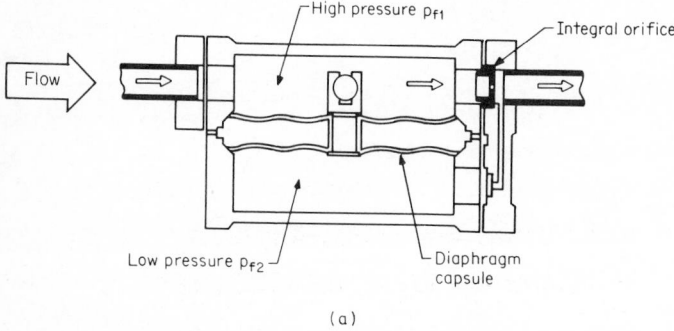

(a)

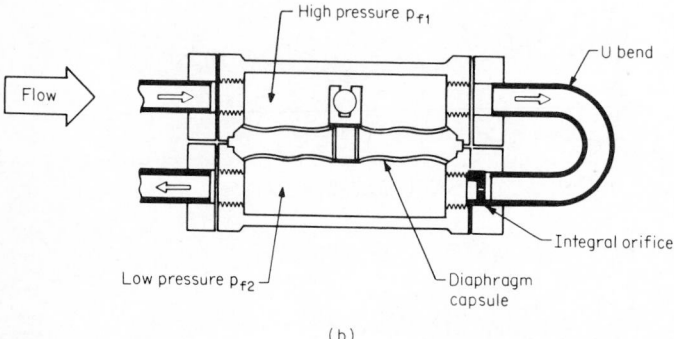

(b)

Figure 11.12 Foxboro in-line and U-bend integral orifices. (*a*) In-line. (*b*) U-bend.

transmitter's high-pressure side, passes through a U bend which contains the orifice, and then exits through the low-pressure side. This design virtually eliminates cavities and dead-end passageways, making it ideal for fluids which may precipitate solids, congeal, or polymerize and for vapors that tend to condense. Both process connections are on the same end of the transmitter body.

The Taylor integral orifice is an in-line flow-element assembly (Fig. 11.13*a*), close-coupled to the differential-pressure transmitter (Fig. 11.13*b*). Inlet piping is either 1-in NPT (25 mm) or 1½-in NPT (40 mm). Liquids, gases, and steam can be measured.

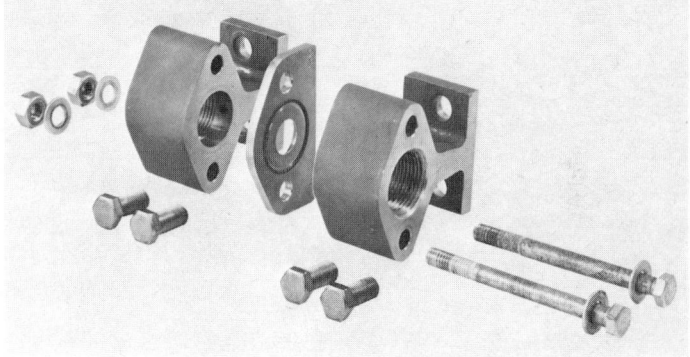

Figure 11.13 Taylor in-line integral orifice. (*a*) Orifice assembly. (*b*) Installed with transmitter.

Table 11.4 Integral-Orifice Bores and Flow Coefficients (Foxboro, Taylor)

Orifice bore d, in (mm)	Accuracy (uncalibrated flow coefficient), %	Reynolds-number limits†	Flow coefficients K	
			In line	U bend
FOXBORO				
0.0200 (0.508)	±3	$R_d \geq 3,000$	0.868	0.921
		$R_d < 3,000$	See Fig. 11.14	
0.0340 (0.8636)	±3	$R_d \geq 10,000$	0.906	0.930
		$R_d < 10,000$	See Fig. 11.15	
0.0595 (1.5113)	±3	$R_d \geq 2,000$	0.882	0.890
		$R_d < 2,000$	See Fig. 11.16	
0.0995 (2.5273)	±3	$R_d \geq 1,500$	0.806	0.850
		$R_d < 1,500$	See Fig. 11.17	
0.1590 (4.0386)	±3	$R_d \geq 1,000$	0.772	0.821
		$R_d < 1,000$	See Fig. 11.18	
0.250 (6.350)	±3	$R_d \geq 2,000$	0.855	0.911
		$R_d < 2,000$	See Fig. 11.19	
TAYLOR				
0.020 (0.508)	±5		0.645	
0.035 (0.888)	±5		0.635	
0.065 (1.651)	±4		0.620	
0.113 (2.870)	±2	$R_D \geq 5,000$	0.605	
0.196 (4.9784)	±2	$D = 1$ in	0.603	
0.340 (8.636)	±2	$D^* = 25$ mm	0.605	
0.500 (12.700)	±2		0.630	
0.735 (18.669)	±2		0.715	
0.500 (12.700)	±1.5		0.611	
0.612 (15.545)	±1.5	$R_D \geq 5,000$	0.614	
0.750 (19.05)	±1.5	$D = 1.5$ in	0.623	
0.917 (23.292)	±1.5	$D^* = 40$ mm	0.650	
1.127 (28.626)	±1.5		0.714	

†R_d = bore Reynolds number; R_D = pipe Reynolds number.

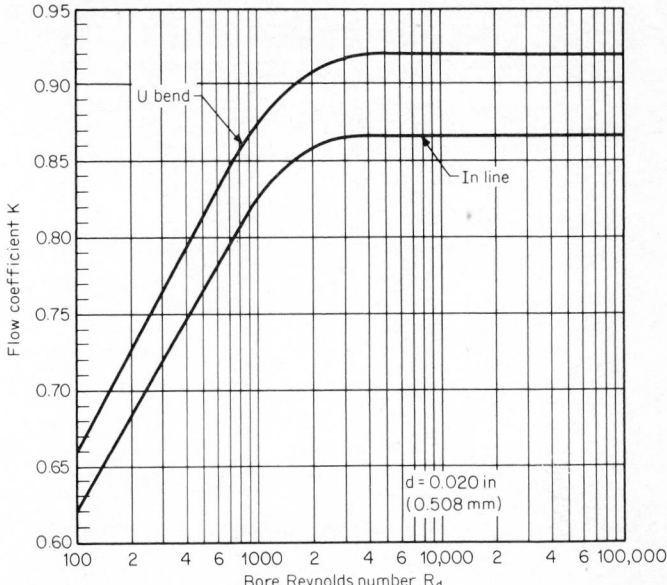

Figure 11.14 Integral-orifice flow coefficient K: d = 0.020 in (0.508 mm) *(courtesy The Foxboro Co.).*

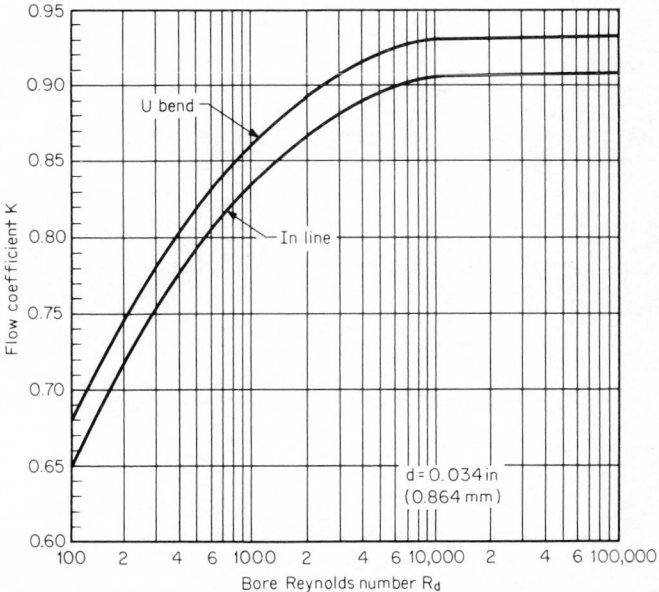

Figure 11.15 Integral-orifice flow coefficient K: d = 0.034 in (0.864 mm) *(courtesy The Foxboro Co.).*

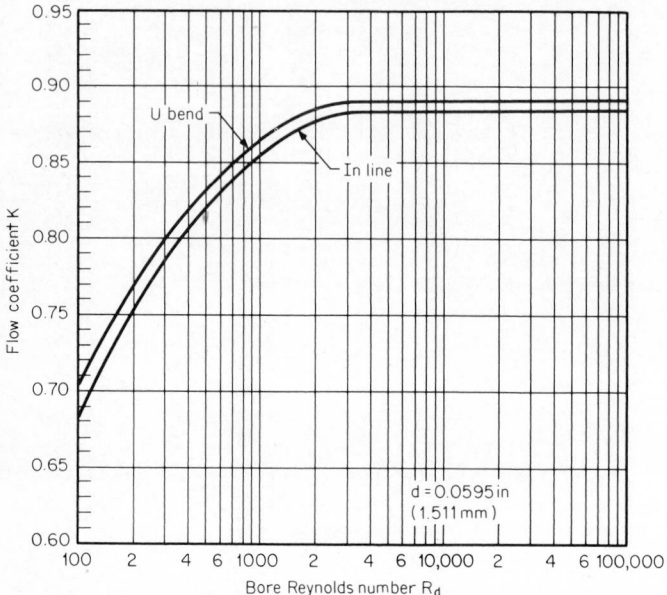

Figure 11.16 Integral-orifice flow coefficient K: d = 0.0595 in (1.511 mm) *(courtesy The Foxboro Co.).*

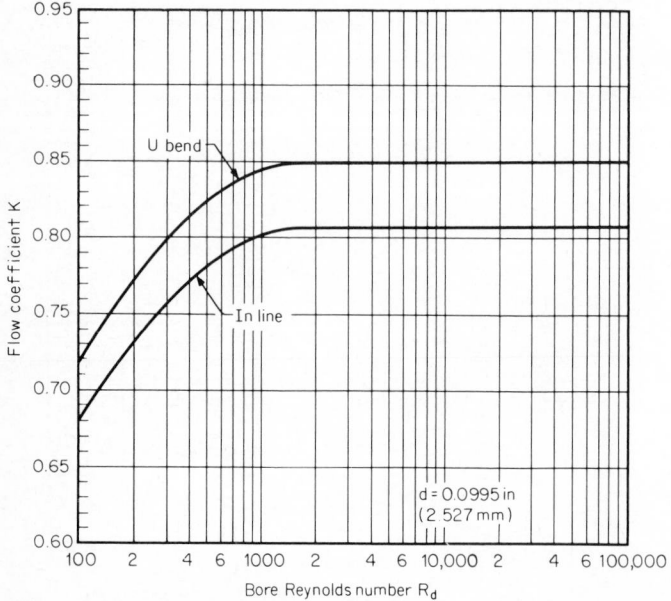

Figure 11.17 Integral-orifice flow coefficient K: d = 0.0995 in (2.527 mm) *(courtesy The Foxboro Co.).*

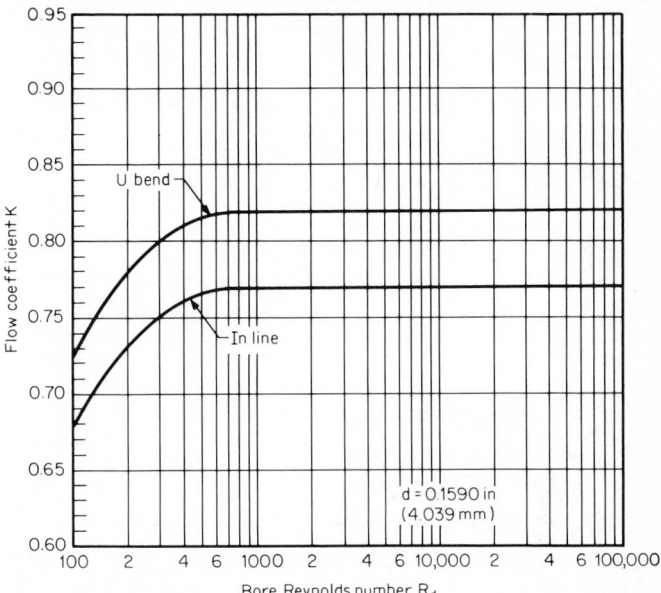

Figure 11.18 Integral-orifice flow coefficient K: d = 0.1590 in (4.039 mm) (*courtesy The Foxboro Co.*).

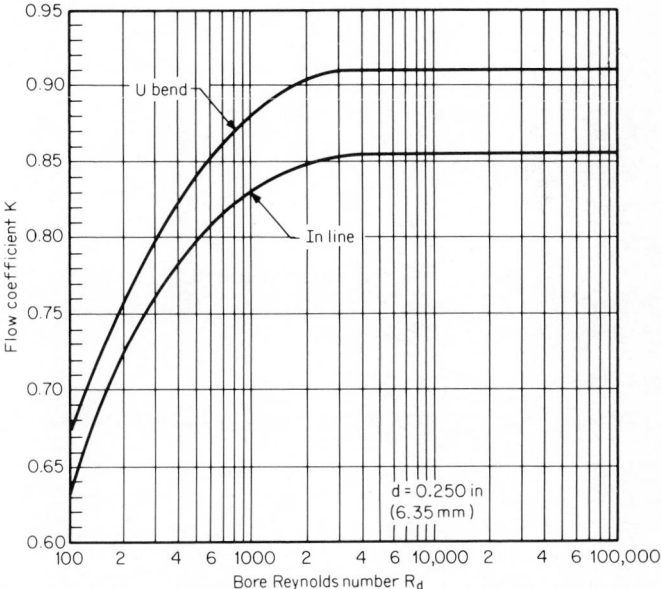

Figure 11.19 Integral-orifice flow coefficient K: d = 0.250 in (6.35 mm) (*courtesy The Foxboro Co.*).

For both Foxboro and Taylor integral orifices, the flow equation is modified by replacing S_M with the flow coefficient K, F_{RD} with 1.0, F_Y with 1.0, and the pipe diameter D with the orifice bore d.

Available orifice bores, uncalibrated accuracy, Reynolds-number ranges, and flow coefficients are given in Table 11.4. In Figs. 11.14 through 11.19 are curves of flow coefficient versus bore Reynolds number for the Foxboro orifices. Because of the tap locations, the permanent pressure loss equals the operating differential pressure. For gas (vapor) flows, no gas expansion factor has been determined. It is recommended that pressure ratios be limited to

$$\frac{h_w}{p_{f1}} \leq 1.5 \quad \text{and} \quad \frac{\Delta p^*}{p^*_{f1}} \leq 0.054$$

EXAMPLE 11.5

A 0.020-in (0.508-mm) Foxboro stainless-steel integral orifice is to measure the flow rate of the gasoline of Example 2.12. The maximum flow rate is 0.025 flowing gal/min (0.095 flowing L/min). Determine (a) the upper-range value for the differential-pressure transmitter and (b) the flow rate when the differential pressure is 20 percent of the URV.

The flow-rate equation for flowing gallons per minute is, from Eq. (d) of Table 9.36,

$$q_{\text{gpm}} = N_{vG} S_M F_a F_{RD} D^2 \frac{\sqrt{h_w}}{\sqrt{F_p} \sqrt{G_F}}$$

Substitution of variables from Table 11.1 gives the integral-orifice flow-rate equation as

$$q_{\text{gpm}} = N_{vG} K F_a d^2 \frac{\sqrt{h_w}}{\sqrt{F_p} \sqrt{G_F}}$$

The design information is as follows:

■ From Table 9.16: $N_{vG} = 5.6665$

■ From Example 9.1: $F_a = 1 + (2)(0.0000096)(80.5 - 68) = 1.0002$

■ From Example 2.15: $F_p = 1.0088$

■ From Example 2.12: $G_F = 0.7255$

Scaling the flow rate to normal operating conditions and rearranging the flow-rate equation to solve for the differential give

$$h_w = \left(\frac{\sqrt{F_p}\sqrt{G_F}\, q_{\text{gpm}}}{N_{vG} K F_a d^2} \right)^2 = \left[\frac{\sqrt{1.0088}\sqrt{0.7255}\,(0.025)(0.8)}{(5.6665)(K)(1.0002)(0.020)^2} \right]^2$$

$$= \frac{56.96}{K^2}$$

The bore Reynolds number at normal conditions is calculated from Eq. (d) of Table 9.20 with bore diameter substituted for pipe diameter:

$$R_d = 17{,}903 \frac{F_p G_F}{\mu_{cP} d N_{vG}} q_{\text{gpm}} = 17{,}903 \frac{(1.0088)(0.7255)}{(0.417)(0.020)(5.6665)} q_{\text{gpm}} = 277{,}260 q_{\text{gpm}}$$

$$(R_d)_N = (277{,}260)(0.025)(0.8) = 5{,}545$$

where $\mu_{cP} = 0.417$ from Example 9.3. The flow coefficient $K = 0.868$ is constant for $R_d > 3000$.

a. Substitution gives the differential at normal operating conditions as

$$(h_w)_N = \frac{56.96}{K^2} = \frac{56.96}{(0.868)^2} = 75.60 \text{ in}$$

The differential at URV flow is then

$$(h_w)_{URV} = 75.60 \frac{1}{0.64} = 118.1 \text{ in}$$

b. Assuming the flow coefficient is constant, the flow-rate equation gives

$$q_{gpm} = N_{vG}KF_a d^2 \frac{\sqrt{h_w}}{\sqrt{F_p}\sqrt{G_F}} = (5.6665)(0.868)(1.0002)(0.020)^2 \frac{\sqrt{h_w}}{\sqrt{1.0088}\sqrt{0.7255}}$$

$$= F_{MC}\sqrt{h_w} = 0.00230\sqrt{h_w}$$

where the meter-coefficient factor $F_{MC} = 0.00230$. The flow rate at 20 percent URV is then

$$q_{gpm} = F_{MC}\sqrt{h_w} = 0.00230\sqrt{(0.20)(118.1)} = 0.0112 \text{ gal/min}$$

EXAMPLE 11.6

For a maximum differential pressure of 50 kPa, select the bore and determine the URV for a differential-pressure transmitter for (a) Foxboro and (b) Taylor stainless-steel integral orifices to measure an airflow of 50 standard L/min (ISO Standard 5024 base). The air is flowing at a temperature of 27°C and an absolute pressure of 1000 kPa.

The flow-rate equation, from Eq. (1) of Table 9.37, is

$$q^*_{SLPM} = N^*_{V_pT}S_M F^*_a F_{RD}F_Y Z_b D^{*2} \frac{\sqrt{\Delta p^* p^*_{f1}}}{\sqrt{Z_{f1}}\sqrt{T_{K1}}\sqrt{G}}$$

Substitution of variables from Table 11.1 gives the integral-orifice flow-rate equation as

$$q^*_{SLPM} = N^*_{V_pT}KF^*_a Z_b d^{*2} \frac{\sqrt{\Delta p^* p^*_{f1}}}{\sqrt{Z_{f1}}\sqrt{T_{K1}}\sqrt{G}}$$

From Table 9.17, $N^*_{V_pT} = 3.2111$, and from Eq. (9.54) and Table B.4,

$$F^*_a = 1 + 2\alpha^*_{PE}(T_{°C} - 20) = 1 + 2(0.0000173)(27 - 20) = 1.00024$$

From Table G.2, $Z_{f1} = 0.9974$, and from Eq. (3.39),

$$T_{K1} = 273.17 + 27 = 300.15 \text{ K}$$

For air, $G = 1.0$, and Z_b is assumed equal to 1.0. Then, rearranging the flow-rate equation to solve for the differential pressure gives

$$\Delta p^* = \left(\frac{\sqrt{Z_{f1}}\sqrt{T_{K1}}\sqrt{G}q^*_{SLPM}}{N^*_{V_pT}KF^*_a Z_b d^{*2}\sqrt{p^*_{f1}}}\right)^2 = \left[\frac{\sqrt{0.9974}\sqrt{300.15}\sqrt{1.0}\,(50)}{(3.2111)(1.00024)Kd^{*2}\sqrt{1000}}\right]^2$$

$$= \frac{72.549}{K^2 d^{*4}}$$

and

$$d^* = \left(\frac{72.549}{K^2 \, \Delta p^*} \right)^{1/4}$$

a. Foxboro integral orifice

Assuming $K = 0.9$, the bore diameter for $\Delta p^* = 50$ kPa is

$$d^* = \left[\frac{72.549}{(0.9)^2(50)} \right]^{1/4} = 1.157 \text{ mm}$$

A bore equal to or greater than 1.157 mm must be selected so as *not* to produce a differential greater than 50 kPa.

Now, from Table 11.4, $d^* = 1.5113$ mm. The bore Reynolds number is calculated from Eq. (1) of Table 9.21 by replacing D^* with d^*:

$$R_d = 83.467 \, \frac{G}{Z_b} \, \frac{1}{\mu_{cP} d^* N_{V_{pT}}^*} \, q_{\text{SLPM}}^*$$

$$= 83.467 \, \frac{1.0}{1.0} \, \frac{1}{(0.018)(1.5113)(3.2111)} \, q_{\text{SLPM}}^* = 955.5 q_{\text{SLPM}}^*$$

where $\mu_{cP} = 0.018$ from Fig. H.1. At normal operating conditions the flow rate is $(q_{\text{SLPM}}^*)_N = (0.8)(50) = 40$, and the bore Reynolds number is $(R_d)_N = (955.5)(40) = 38{,}220$. From Table 11.4 or Fig. 11.16, the flow coefficient is $K = 0.882$. Since K is constant for $R_d > 2000$, the URV of the differential-pressure transmitter is

$$\Delta p^* = \frac{72.549}{K^2 d^{*4}} = \frac{72.549}{(0.882)^2(1.5113)^4} = 17.88 \text{ kPa}$$

b. Taylor integral orifice

Assuming $K = 0.6$ (see Table 11.4), the bore for $\Delta p^* = 50$ kPa is

$$d^* = \left[\frac{72.549}{(0.6)(50)} \right]^{1/4} = 1.247 \text{ mm}$$

The next largest bore from Table 11.4 is $d^* = 1.651$ mm, and $K = 0.620$. The URV of the differential-pressure transmitter is then

$$\Delta p^* = \frac{72.549}{(0.620)^2(1.651)^4} = 25.401 \text{ kPa}$$

ANNUBAR

Pitot tubes are seldom used in industrial applications because of vibration, lack of ruggedness, and the need for a velocity traverse to obtain an accurate measurement. The reader who requires information regarding pitot measurements is

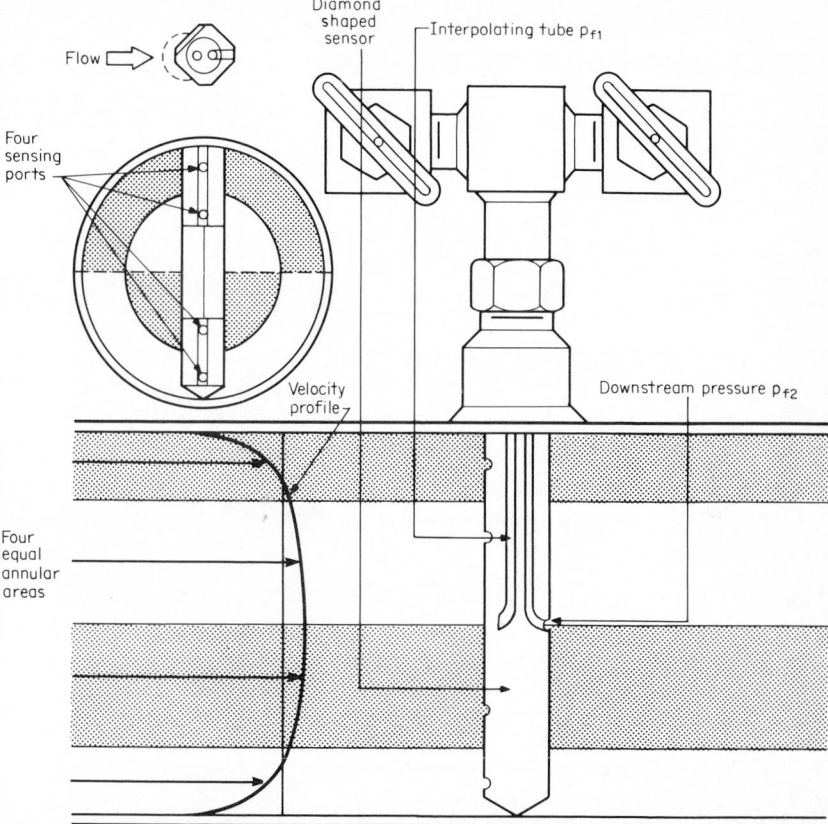

Figure 11.20 Annubar flow sensor *(courtesy Dieterich Standard Corp.)*.

referred to ISO/DIS 7145 (1981). Multiport elements based on the pitot principle are, however, widely used in many applications.

The Annubar is a proprietary differential producer designed for easy pipe insertion for liquid, gas, and vapor (steam) flows. A flow-coefficient accuracy of ± 1 percent is claimed, based on testing in many different laboratories. Shown in Fig. 11.20 is the basic Annubar. It is a diamond-shaped strut that spans the pipe. Four upstream-facing pressure ports are used to *interpolate* the velocity profile for averaging purposes. A single downstream rear-facing pressure port is used to sense the downstream pressure. Four basic Annubar types are available (Fig. 11.21), with selection depending on line size and application.

The flow-rate equation is modified by substitution of the flow coefficient K_{ref} for S_M, F_K for F_{RD}, and Y_1 for F_Y. The coefficient K_{ref} for the four available Annubar types is given in Table 11.5, the Reynolds-number correction factor F_K in

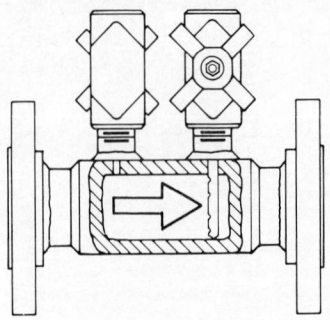

Type 61, 1 to 1½ in
(25 to 40 mm)

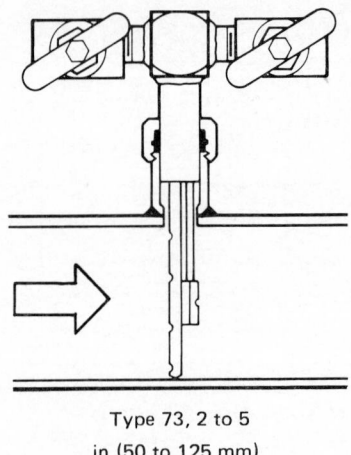

Type 73, 2 to 5
in (50 to 125 mm)

Figure 11.21 **Annubar types** *(courtesy Dieterich Standard Corp.).*

Table 11.6, and the gas expansion factor Y_1 in Table 11.7 for two isentropic exponents.

Recommended upstream straight pipe lengths are shown in Table 8.2. Differential pressure is measured with conventional industrial differential-pressure transmitters.

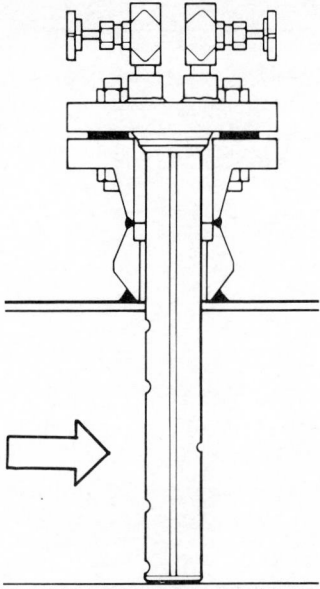

Type 75 and 76, 6 to 72 in
(150 to 1800 mm)

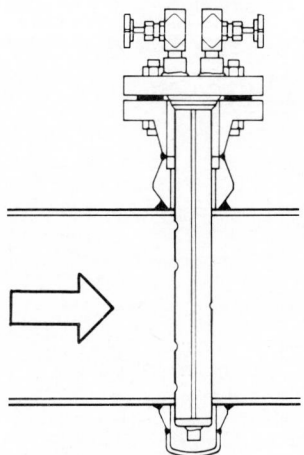

Type 86, end support

Figure 11.21 (*continued*)

Table 11.5   Annubar Flow Coefficient K_{ref}: U.S. Pipe Sizes†

Nominal pipe size, in	Schedule	Pipe ID, in	K_{ref}	Nominal pipe size, in	Schedule	Pipe ID, in	K_{ref}
	Type 61				**Type 75/76**		
1	80	0.957	0.9506	4	XX-STG	3.152	0.4570
	40	1.049	0.9561		X-STG	3.826	0.5348
					40	4.026	0.5528
1¼	80	1.278	0.9696				
	40	1.380	0.9754	6	XX-STG	4.987	0.6194
					160	5.189	0.6303
1½	80	1.500	0.9779		80	5.761	0.6569
	40	1.610	0.9802		....	6.000	0.6612
					40	6.065	0.6622
	Type 73			8	160	6.813	0.6718
2	XX-STG	1.503	0.5718		XX-STG	6.875	0.6725
	160	1.689	0.6197		80	7.625	0.6800
	80	1.939	0.6677		40	7.981	0.6830
	....	2.000	0.6773		....	8.000	0.6831
	40	2.067	0.6871	10	160	8.500	0.6869
2½	XX-STG	1.771	0.6372		80	9.564	0.6936
	160	2.125	0.6950		X-STG	9.750	0.6947
	80	2.323	0.7062		....	10.000	0.6960
	40	2.469	0.7122		40	10.020	0.6961
	....	2.500	0.7134	12	160	10.126	0.6966
3	XX-STG	2.300	0.7052		80	11.376	0.7021
	160	2.624	0.7177		X-STG	11.750	0.7035
	80	2.900	0.7257		40	11.938	0.7042
	....	3.000	0.7282		STD	12.000	0.7044
	40	3.068	0.7298	14	80	12.500	0.7060
3½	XX-STG	2.728	0.7209		X-STG	13.000	0.7069
	80	3.364	0.7359		40	13.124	0.7070
	....	3.500	0.7383		STD	13.250	0.7071
	40	3.548	0.7391		....	14.000	0.7074
4	XX-STG	3.152	0.7317	16	80	14.314	0.7076
	160	3.438	0.7372		40	15.000	0.7079
	80	3.826	0.7432		STD	15.250	0.7080
	....	4.000	0.7455		....	16.000	0.7082
	40	4.026	0.7459	18	80	16.126	0.7083
5	XX-STG	4.063	0.7463		X-STG	17.000	0.7086
	160	4.313	0.7491		STD	17.250	0.7087
	80	4.813	0.7503		....	18.000	0.7089
	....	5.000	0.7506				
	40	5.047	0.7506				

†Linear interpolation may be used.

SOURCE: Dieterich Standard Corp.

Nominal pipe size, in	Schedule	Pipe ID, in	K_{ref}	Nominal pipe size, in	Schedule	Pipe ID, in	K_{ref}
Type 75/76 (*continued*)				**Type 85/86 (*continued*)**			
20	80	17.938	0.7089	20	80	17.938	0.6264
	X-STG	19.000	0.7091		X-STG	19.000	0.6300
	STD	19.250	0.7092		STD	19.250	0.6307
	....	20.000	0.7094		....	20.000	0.6329
24	80	21.564	0.7097	24	80	21.564	0.6370
	X-STG	23.000	0.7100		X-STG	23.000	0.6402
	STD	23.250	0.7100		STD	23.250	0.6408
	....	24.000	0.7101		....	24.000	0.6423
30	X-STG	29.000	0.7108	30	X-STG	29.000	0.6502
	STD	29.250	0.7108		STD	29.250	0.6505
	...	30.000	0.7109		....	30.000	0.6515
36	X-STG	35.000	0.7113	36	X-STG	35.000	0.6526
	STD	35.250	0.7113		STD	35.250	0.6526
	....	36.000	0.7114		....	36.000	0.6527
42	X-STG	41.000	0.7117	42	X-STG	41.000	0.6534
	STD	41.250	0.7117		STD	41.250	0.6534
	....	42.000	0.7117		....	42.000	0.6535
48 and larger		48.000	0.7120	48	....	48.000	0.6541
Type 85/86				60	....	60.000	0.6549
12	160	10.126	0.5208	72	....	72.000	0.6555
	80	11.376	0.5567				
	X-STG	11.750	0.5659	84	....	84.000	0.6559
	40	11.938	0.5703				
	STD	12.000	0.5718	96	....	96.000	0.6562
14	80	12.500	0.5828	108 and larger		108.000	0.6564
	X-STG	13.000	0.5930				
	40	13.124	0.5954				
	STD	13.250	0.5978				
	....	14.000	0.6083				
16	80	14.314	0.6101				
	40	15.000	0.6139				
	STD	15.250	0.6151				
	....	16.000	0.6187				
18	80	16.126	0.6192				
	X-STG	17.000	0.6229				
	STD	17.250	0.6239				
	....	18.000	0.6266				

K_{ref} for pipe ID, mm continues on next page.

†Linear interpolation maybe used.

SOURCE: Dieterich Standard Corp.

Table 11.5 Annubar Flow Coefficient K_{ref}; U.S. Pipe Sizes† (*Continued*)

Pipe ID, mm	K_{ref}	Pipe ID, mm	K_{ref}	Pipe ID, mm	K_{ref}
Type 61		Type 75/76 (*continued*)		Type 85/86	
20	0.9406			300	0.5674
25	0.9522	200	0.6820	350	0.6070
32	0.9685	225	0.6894	400	0.6175
40	0.9795	250	0.6953	450	0.6256
50	0.9877	300	0.7036	500	0.6320
		350	0.7074	550	0.6372
Type 73		400	0.7081	600	0.6415
Pipe ID, mm	K_{ref}	450	0.7088	650	0.6451
40	0.5919	500	0.7093	700	0.6482
50	0.6726	550	0.7097	750	0.6509
65	0.7155	600	0.7101	800	0.6520
80	0.7316	650	0.7104	850	0.6523
90	0.7390	700	0.7106	900	0.6527
100	0.7449	750	0.7108	1000	0.6532
125	0.7504	800	0.7110	1100	0.6537
150	0.7516	850	0.7112	1200	0.6540
		900	0.7113	1350	0.6545
Type 75/76		1000	0.7116	1500	0.6549
90	0.5057	1100	0.7118	1750	0.6553
100	0.5450	1200	0.7120	2000	0.6557
125	0.6157	1500	0.7123	2250	0.6560
150	0.6597	1800	0.7126	2500	0.6562
175	0.6726			2750 and larger	0.6564

†Linear interpolation maybe used.

SOURCE: Dieterich Standard Corp.

Table 11.6 Annubar Reynolds-Number Correction Factor F_K

Annubar sensor type 61

Nominal		Pipe Reynolds number R_D†														
D, in	*D**, mm	1.0 E4	1.5 E4	2.0 E4	3.0 E4	5.0 E4	7.0 E4	1.0 E5	1.5 E5	2.0 E5	3.0 E5	5.0 E5	7.0 E5	1.0 E6	1.5 E6	2.0 E6
1	25	0.9551	0.9682	0.9761	0.9854	0.9947	0.9997	1.0041	1.0083	1.0107	1.0137	1.0166	1.0182	1.0196	1.0209	
1.25	32			0.9424	0.9619	0.9815	0.9919	1.0012	1.0100	1.0152	1.0213	1.0275	1.0308	1.0338	1.0365	
1.50	40			0.9477	0.9644	0.9811	0.9900	0.9980	1.0054	1.0099	1.0151	1.0204	1.0232	1.0257	1.0281	1.0295

Annubar sensor type 73

Nominal		Pipe Reynolds number R_D†													
D, in	*D**, mm	2.0 E4	2.5 E4	3.0 E4	5.0 E4	7.0 E4	1.0 E5	2.0 E5	3.0 E5	5.0 E5	7.0 E5	1.0 E6	2.0 E6	3.0 E6	5.0	
2	50	0.9515	0.9597	0.9657	0.9800	0.9876	0.9944	1.0044	1.0090	1.0136	1.0160	1.0181	1.0213	1.0227		
2.5	65		0.9569	0.9630	0.9774	0.9850	0.9918	1.0021	1.0066	1.0112	1.0136	1.0158	1.0190	1.0204	1.0219	
3	75		0.9632	0.9681	0.9796	0.9858	0.9912	0.9995	1.0031	1.0068	1.0087	1.0104	1.0130	1.0142	1.0153	
3.5	80			0.9679	0.9790	0.9850	0.9902	0.9982	1.0017	1.0052	1.0071	1.0088	1.0113	1.0124	1.0135	
4	100				0.9737	0.9808	0.9871	0.9966	1.0007	1.0050	1.0072	1.0092	1.0122	1.0135	1.0148	1.0155
5	125				0.9722	0.9791	0.9853	0.9946	0.9987	1.0028	1.0050	1.0070	1.0099	1.0112	1.0125	1.0132

†Pipe Reynolds numbers are expressed in computer notation. For example, R_D = 5.0 E5 means $R_D = 5.0 \times 10^5 = 500{,}000$.

SOURCE: Dietrich Standard Corp.

Table 11.6 Annubar Reynolds-Number Correction Factor F_K (*Continued*)

Annubar sensor type 75/76

Nominal		Pipe Reynolds number R_D†														
D, in	*D**, mm	5.0 E4	7.0 E4	1.0 E5	1.5 E5	2.0 E5	3.0 E5	5.0 E5	7.0 E5	1.0 E6	2.0 E6	3.0 E6	5.0 E6	7.0 E6	1.0 E7	2.0 E7
4	100	0.9779	0.9839	0.9892	0.9941	0.9971	1.0006	1.0042	1.0060	1.0077	1.0102	1.0113	1.0125			
6	150	0.9692	0.9765	0.9830	0.9891	0.9928	0.9971	1.0014	1.0037	1.0058	1.0088	1.0102	1.0116	1.0123		
8	200		0.9751	0.9814	0.9874	0.9909	0.9951	0.9993	1.0016	1.0036	1.0066	1.0079	1.0092	1.0099	1.0106	
10	250			0.9735	0.9814	0.9860	0.9916	0.9972	1.0001	1.0028	1.0067	1.0085	1.0102	1.0112	1.0120	
12	300			0.9637	0.9739	0.9800	0.9872	0.9945	0.9983	1.0018	1.0070	1.0093	1.0115	1.0128	1.0139	1.0155
14	350				0.9719	0.9783	0.9858	0.9933	0.9973	1.0009	1.0062	1.0086	1.0110	1.0122	1.0134	1.0150

D, in	*D**, mm	Pipe Reynolds number R_D†														
		1.5 E5	2.0 E5	3.0 E5	4.0 E5	5.0 E5	6.0 E5	8.0 E5	1.0 E6	1.5 E6	2.0 E6	3.0 E6	5.0 E6	7.0 E6	1.0 E7	2.0 E7
16	400	0.9697	0.9763	0.9840	0.9886	0.9917	0.9941	0.9973	0.9995	1.0030	1.0051	1.0075	1.0100	1.0113	1.0124	1.0142
18	450	0.9675	0.9743	0.9823	0.9870	0.9903	0.9927	0.9961	0.9984	1.0020	1.0041	1.0066	1.0092	1.0105	1.0117	1.0135
20	500		0.9737	0.9816	0.9863	0.9895	0.9919	0.9952	0.9975	1.0010	1.0031	1.0056	1.0081	1.0094	1.0106	1.0124
24	600		0.9688	0.9776	0.9829	0.9865	0.9892	0.9929	0.9955	1.9994	1.0018	1.0046	1.0074	1.0089	1.0102	1.0122
30	750			0.9746	0.9801	0.9839	0.9867	0.9906	0.9933	0.9975	0.9999	1.0029	1.0058	1.0074	1.0088	1.0109
36	900				0.9750	0.9795	0.9828	0.9873	0.9905	0.9954	0.9983	1.0017	1.0052	1.0070	1.0087	1.0111
42	1000				0.9672	0.9728	0.9769	0.9826	0.9865	0.9926	0.9962	1.0005	1.0048	1.0071	1.0092	1.0123
>48	>1200				0.9599	0.9664	0.9712	0.9779	0.9825	0.9896	0.9939	0.9989	1.0040	1.0067	1.0091	1.0127

Annubar sensor type 85/86

Nominal		Pipe Reynolds number R_D†														
D, in	D*, mm	1.5 E5	2.0 E5	3.0 E5	4.0 E5	5.0 E5	6.0 E5	8.0 E5	1.0 E6	2.0 E6	3.0 E6	5.0 E6	7.0 E6	1.0 E7	2.0 E7	3.0 E7
12	300	0.9851	0.9886	0.9927	0.9952	0.9969	0.9981	0.9998	1.0010	1.0040	1.0053	1.0066	1.0073	1.0079		
14	350	0.9845	0.9880	0.9921	0.9946	0.9963	0.9975	0.9993	1.0005	1.0034	1.0047	1.0060	1.0067	1.0074		
16	400	0.9801	0.9844	0.9895	0.9925	0.9946	0.9961	0.9982	0.9997	1.0033	1.0049	1.0066	1.0074	1.0082	1.0093	
18	450		0.9836	0.9887	0.9918	0.9938	0.9954	0.9975	0.9990	1.0026	1.0042	1.0058	1.0067	1.0075	1.0086	
20	500		0.9821	0.9874	0.9907	0.9928	0.9945	0.9967	0.9983	1.0021	1.0038	1.0055	1.0064	1.0072	1.0085	
24	600		0.9811	0.9864	0.9896	0.9918	0.9934	0.9957	0.9973	1.0011	1.0028	1.0045	1.0054	1.0062	1.0074	1.0080
30	750			0.9801	0.9845	0.9874	0.9896	0.9927	0.9948	1.0000	1.0023	1.0046	1.0058	1.0069	1.0085	1.0093

	Pipe Reynolds number R_D†														
D, in	3.0 E5	4.0 E5	5.0 E5	6.0 E5	7.0 E5	8.0 E5	9.0 E5	1.0 E6	1.5 E6	2.0 E6	3.0 E6	5.0 E6	7.0 E6	1.0 E7	3.0 E7
36	0.9790	0.9834	0.9863	0.9885	0.9902	0.9916	0.9927	0.9937	0.9969	0.9988	1.0011	1.0034	1.0047	1.0058	1.0081
42		0.9761	0.9801	0.9831	0.9854	0.9873	0.9888	0.9901	0.9946	0.9972	1.0004	1.0035	1.0052	1.0067	1.0100
48			0.9730	0.9769	0.9799	0.9823	0.9843	0.9860	0.9917	0.9951	0.9991	1.0032	1.0054	1.0073	1.0115
60				0.9751	0.9781	0.9805	0.9825	0.9842	0.9893	0.9933	0.9973	1.0014	1.0036	1.0055	1.0096
72					0.9768	0.9792	0.9812	0.9829	0.9886	0.9920	0.9960	1.0001	1.0022	1.0041	1.0083

†Pipe Reynolds numbers are expressed in computer notation. For example, R_D = 5.0 E5 means R_D = 5.0 × 10^5 = 500,000.

SOURCE: Dieterich Standard Corp.

Table 11.7 Gas Expansion Factor Y_1 for Annubars

Open area — D — Annubar area

$$\text{Area ratio} = \frac{\text{pipe area} - \text{Annubar area}}{\text{pipe area}}$$

Type 61: $\text{Area ratio} = 1 - \dfrac{0.2387}{D} = 1 - \dfrac{6.063}{D^*}$

Type 73: $\text{Area ratio} = 1 - \dfrac{0.1592 + 0.3979D}{D^2} = 1 - \dfrac{102.7 + 10.107D^*}{D^{*2}}$

Type 75/76: $\text{Area ratio} = 1 - \dfrac{1.273}{D} = 1 - \dfrac{32.33}{D^*}$

Type 85/86: $\text{Area ratio} = 1 - \dfrac{3.024}{D} = 1 - \dfrac{76.81}{D^*}$

$k = 1.4$

$\dfrac{h_w}{p_{f1}}$	Area ratio						
	0.70	0.75	0.80	0.85	0.90	0.95	1.00
0.0	1.0000	1.0000	1.0000	1.0000	1.0000	1.0000	1.0000
0.2	0.9987	0.9986	0.9986	0.9985	0.9984	0.9984	0.9983
0.4	0.9974	0.9972	0.9971	0.9970	0.9969	0.9967	0.9966
0.6	0.9960	0.9958	0.9957	0.9955	0.9953	0.9950	0.9948
0.8	0.9947	0.9944	0.9942	0.9939	0.9937	0.9934	0.9931
1.0	0.9934	0.9931	0.9928	0.9924	0.9921	0.9917	0.9914
1.2	0.9921	0.9917	0.9914	0.9910	0.9906	0.9901	0.9897
1.4	0.9908	0.9903	0.9899	0.9895	0.9890	0.9885	0.9879
1.6	0.9894	0.9890	0.9885	0.9880	0.9874	0.9868	0.9862
1.8	0.9881	0.9876	0.981	0.9864	0.9858	0.9851	0.9845
2.0	0.9868	0.9862	0.9856	0.9850	0.9843	0.9835	0.9828

$k = 1.4$

$\dfrac{\Delta p^*}{p^*_{f1}}$	Area ratio						
	0.70	0.75	0.80	0.85	0.90	0.95	1.00
0.00	1.0000	1.0000	1.0000	1.0000	1.0000	1.0000	1.0000
0.01	0.9982	0.9981	0.9980	0.9979	0.9978	0.9977	0.9976
0.02	0.9963	0.9962	0.9960	0.9958	0.9956	0.9954	0.9952
0.03	0.9945	0.9943	0.9940	0.9937	0.9935	0.9931	0.9928
0.04	0.9927	0.9924	0.9920	0.9917	0.9913	0.9909	0.9904
0.05	0.9909	0.9905	0.9900	0.9896	0.9891	0.9886	0.9880
0.06	0.9890	0.9885	0.9880	0.9875	0.9869	0.9863	0.9857
0.07	0.9872	0.9866	0.9860	0.9854	0.9847	0.9840	0.9833
0.08	0.9854	0.9847	0.9840	0.9833	0.9825	0.9817	0.9809
0.09	0.9835	0.9828	0.9820	0.9812	0.9804	0.9794	0.9785
0.10	0.9817	0.9809	0.9801	0.9791	0.9782	0.9772	0.9761

$$k = 1.3$$

$\dfrac{h_w}{p_{f1}}$	Area ratio						
	0.70	0.75	0.80	0.85	0.90	0.95	1.00
0.0	1.0000	1.0000	1.0000	1.0000	1.0000	1.0000	1.0000
0.2	0.9986	0.9985	0.9985	0.9984	0.9983	0.9982	0.9981
0.4	0.9972	0.9970	0.9969	0.9968	0.9966	0.9964	0.9963
0.6	0.9957	0.9956	0.9954	0.9952	0.9949	0.9946	0.9944
0.8	0.9943	0.9940	0.9938	0.9936	0.9932	0.9929	0.9926
1.0	0.9929	0.9926	0.9923	0.9919	0.9915	0.9911	0.9907
1.2	0.9915	0.9911	0.9907	0.9903	0.9898	0.9894	0.9889
1.4	0.9901	0.9896	0.9892	0.9887	0.9881	0.9876	0.9870
1.6	0.9886	0.9881	0.9876	0.9870	0.9864	0.9858	0.9851
1.8	0.9872	0.9866	0.9861	0.9854	0.9847	0.9840	0.9833
2.0	0.9858	0.9852	0.9845	0.9838	0.9831	0.9822	0.9814

$$k = 1.3$$

$\dfrac{\Delta p^*}{p^*_{f1}}$	Area ratio						
	0.70	0.75	0.80	0.85	0.90	0.95	1.00
0.00	1.0000	1.0000	1.0000	1.0000	1.0000	1.0000	1.0000
0.01	0.9980	0.9979	0.9979	0.9978	0.9976	0.9975	0.9974
0.02	0.9961	0.9959	0.9957	0.9955	0.9953	0.9951	0.9948
0.03	0.9941	0.9938	0.9936	0.9933	0.9929	0.9926	0.9923
0.04	0.9921	0.9918	0.9914	0.9910	0.9906	0.9902	0.9897
0.05	0.9901	0.9897	0.9893	0.9888	0.9882	0.9877	0.9871
0.06	0.9882	0.9877	0.9871	0.9865	0.9859	0.9852	0.9845
0.07	0.9862	0.9856	0.9850	0.9843	0.9835	0.9828	0.9820
0.08	0.9842	0.9835	0.9828	0.9820	0.9812	0.9803	0.9794
0.09	0.9823	0.9815	0.9807	0.9798	0.9788	0.9779	0.9768
0.10	0.9803	0.9794	0.9785	0.9775	0.9765	0.9754	0.9742

EXAMPLE 11.7

Determine the upper-range value of a differential-pressure transmitter for use with a type-75 Annubar measuring a 0.63-specific-gravity natural gas in a carbon-steel 300-mm pipe. The maximum flow rate is 170,000 standard m³/h at 15°C and 103 kPa (not the ISO Standard 5024 base); the normal operating pressure is 8700 kPa, and the temperature is 50°C.

For a selected base, the flow-rate equation, from Eq. (m) of Table 9.37, is

$$q^*_{S\text{CMH}} = (N^*_{\dot{V}pT})_b S_M F^*_a F_{RD} F_Y Z_b T_{Kb} D^{*2} \frac{\sqrt{\Delta p^*}\, p^*_{f1}}{\sqrt{Z_{f1}}\sqrt{T_{K1}}\sqrt{G}p^*_b}$$

Substituting the variables from Table 11.1 gives the Annubar flow-rate equation as

$$q^*_{S\text{CMH}} = (N^*_{\dot{V}pT})_b K_{\text{ref}} F^*_a F_K Y_1 Z_b D^{*2} T_{Kb} \frac{\sqrt{\Delta p^*}\, p_{f1}}{\sqrt{Z_{f1}}\sqrt{T_{K1}}\sqrt{G}p^*_b}$$

Data is as follows:

- From Table 9.17: $(N^*_{\dot{V}pT})_b = 0.06775$
- From Table 11.5: $K_{\text{ref}} = 0.7036$
- From Eq. (9.54): $F^*_a = 1 + 2\alpha^*_{PE}(T^\circ_C - 20) = 1 + (2)(0.0000120)(50 - 20) = 1.0007$
- From Table B.4: $\alpha^*_{PE} = 0.0000120$ mm/(mm · °C)

Then, from Eq. (m) of Table 9.21,

$$R_D = 83.467 \frac{G p^*_b}{Z_b T_{Kb}} \frac{1}{\mu_{cP} D^* (N^*_{\dot{V}pT})_b} q^*_{S\text{CMH}}$$

From Fig. H.6, $\mu_{cP} = 0.0145$, and, assuming $Z_b = 1.0$,

$$R_D = 83.467 \frac{(0.63)(103)}{(1.0)(288.15)} \frac{1}{(0.0145)(300)(0.06775)} q_{S\text{CMH}}$$

$$= 63.78 q^*_{S\text{CMH}}$$

At normal flow rate, $(R_D)_N = (63.780)(0.8)(170,000) = 8,674,000$. From Table 11.6, the Reynolds-number factor is interpolated to be $F_K = 1.0133$. From Table G.20, the super-compressibility factor F_{pv} is

$$F_{pv} = \sqrt{\frac{1}{Z_{f1}}} = 1.0636 \quad \text{so that} \quad Z_{f1} = 0.8840$$

Scaling of the flow rate to normal operating conditions and substitution gives

$$(0.8)(170,000)$$

$$= (0.06775)(0.7036)(1.0007)(1.0133)(1.0)(300)^2(288.15)\, Y_1 \frac{\sqrt{\Delta p^*}\sqrt{8700}}{\sqrt{0.8840}\sqrt{323.15}\sqrt{0.63}\,(103)}$$

and

$$Y_1\sqrt{\Delta p^*} = 1.6072$$

Initially assuming $Y_1 = 1.0$ gives

$$\Delta p^* = \left(\frac{1.6072}{Y_1}\right)^2 = 2.583 \text{ kPa}$$

The pressure ratio is then

$$\frac{\Delta p^*}{p_{f1}^*} = \frac{2.583}{8700} = 0.0003$$

Table 11.7 for $k = 1.4$ (for air) gives

$$\text{Area ratio} = 1 - \frac{32.33}{D^*} = 1 - \frac{32.33}{300} = 0.8922$$

Then, by interpolation, $Y_1 = 0.9999$, and the URV of the differential-pressure transmitter is

$$\Delta p^* = \left(\frac{1.6072}{0.9999}\right)^2 \frac{1}{0.64} = 4.037 \text{ kPa}$$

ELBOW FLOWMETER

When fluid moves around the curved path of an elbow, it is subjected to an angular acceleration. The resulting centrifugal force creates a differential pressure between the inner and outer radii. The square root of this differential is proportional to the flow rate and is the basis of the operation of the elbow flowmeter.

The elbow flowmeter is shown in Fig. 11.22. Pressure taps are located at 45° and are of the same construction as orifice taps. Special care must be used to align the tap holes in both planes; the alignment should be checked by sliding a rod through the taps. The inside of the elbow should be commercially smooth, and the mean diameter D must be taken as the average of four measurements

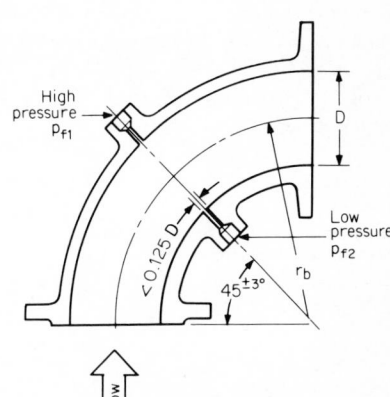

Figure 11.22 Elbow flowmeter

made at the entrance and exit elbow plane. The bend radius r_b is the mean of the inside and outside radii.

Murdock (1963) correlated available elbow-tap data with the Lansford (1936) flow-coefficient equation to within ± 4 percent, provided normal upstream straight-pipe requirements for high-β orifices are followed. The Lansford flow-coefficient equation is, with the addition of the recommended ASME (1971) Reynolds-number correction term,

$$K = \sqrt{\frac{r_b}{2D} + \frac{6.5\sqrt{r_b/2D}}{R_D^{0.5}}} \tag{11.8}$$

where $R_D > 10^4$ and $r_b/D > 1.25$. For these limits, Eq. (11.8) provides the flow coefficient to within ± 4 percent. Values of $\sqrt{r_b/2D}$ for standard elbows are listed in Table 11.8. Precision is comparable with that of an orifice with a calibrated elbow.

Table 11.8   **Values of $\sqrt{r_b/2D}$ for Welding-End Elbows**

Nominal line size D, in (mm)	$\sqrt{r_b/2D}$			
	Short radius		Long radius	
	Schedule 40	Schedule 80	Schedule 40	Schedule 80
2(50)	0.696	0.718	0.852	0.880
3(75)	0.699	0.719	0.856	0.881
4(100)	0.705	0.723	0.863	0.885
6(150)	0.703	0.721	0.861	0.884
8(200)	0.708	0.724	0.867	0.887
10(250)	0.706	0.716	0.865	0.877
12(300)	0.707	0.715	0.866	0.875

The gas expansion factor has not been experimentally determined but is assumed equal to 1 for the low differentials produced. In the flow equation, the flow coefficient K replaces S_M; the thermal-expansion-factor equation is the same as for other differential producers. Upstream piping should be in accordance with an orifice for which $\beta = 0.75$, with a minimum of 10 pipe diameters of straight pipe.

EXAMPLE 11.8

For the gas flow of Example 11.7, calculate the upper-range value for a differential-pressure transmitter to be used with a Schedule-80 long-radius elbow; also calculate the flow rate when the differential is 50 percent of the URV.

For a selected base, the flow-rate equation [Eq. (m) of Table 9.37] is

$$q^*_{\text{SCMH}} = (N^*_{V_pT})_b S_M F^*_a F_{RD} F_Y Z_b T_{Kb} D^{*2} \frac{\sqrt{\Delta p^* \, p^*_{f1}}}{\sqrt{Z_{f1}} \sqrt{T_{K1}} \sqrt{G} \, p^*_b}$$

Substituting the variables from Table 11.1 gives the elbow-flowmeter flow-rate equation as

$$q^*_{SCMH} = (N^*_{VpT})_b KF^*_a Z_b T_{Kb} D^{*2} \frac{\sqrt{\Delta p^*}\, p^*_{f1}}{\sqrt{Z_{f1}}\sqrt{T_{K1}}\sqrt{G}\, p^*_b}$$

Data from Example 11.7 is:

$$(N^*_{VpT})_b = 0.067749 \qquad D^* = 300 \text{ mm} \qquad G = 0.63$$
$$Z_b = 1.0 \qquad p^*_{f1} = 8700 \text{ kPa} \qquad p^*_b = 103 \text{ kPa}$$
$$T_{Kb} = 288.15 \text{ K} \qquad Z_{f1} = 0.8840 \qquad F^*_a = 1.0007$$
$$T_{K1} = 323.15 \text{ K}$$

Scaling to normal operating conditions and substituting gives

$$(0.8)(170,000) = (0.067749)(1.0007)(1.0)(288.15)(300)^2 K_N \frac{\sqrt{(\Delta p^*)_N}\sqrt{8700}}{\sqrt{0.8840}\sqrt{323.15}\sqrt{0.63}\,(103)}$$

$$(0.8)(170,000) = 118,684 K_N \sqrt{(\Delta p^*)_N}$$

$$K_N \sqrt{(\Delta p^*)_N} = 1.1459$$

The flow coefficient K is given by Eq. (11.8) in SI units as

$$K = \sqrt{\frac{r^*_b}{2D^*} + \frac{6.5\sqrt{r^*_b/2D^*}}{R_D^{0.5}}}$$

For a long-radius bend, $r^*_b/D^* = r_b/D$, and, from Table 11.8, $\sqrt{r_b/2D} = 0.875$. At normal operating conditions, the Reynolds number from Example 11.7 is $(R_D)_N = 8,674,000$. Substitution into Eq. (11.8) gives

$$K_N = 0.875 + \frac{(6.5)(0.875)}{(8,674,000)^{0.5}} = 0.8769$$

The differential pressure at normal conditions is then

$$(\Delta p^*)_N = \left(\frac{1.1459}{K_N}\right)^2 = \left(\frac{1.1459}{0.8769}\right)^2 = 1.708 \text{ kPa}$$

and the differential at the URV is

$$(\Delta p^*)_{URV} = 1.708\,\frac{1}{0.64} = 2.668 \text{ kPa}$$

The initial estimate of the flow rate at 20 percent of the URV is

$$q^*_{SCMH} = 118,684 K \sqrt{\Delta p^*} = (118,684)(0.8769)\sqrt{(0.2)(2.668)} = 76,024$$

From Example 11.7, the Reynolds number is calculated as

$$R_D = 63.780 q^*_{SCMH} = (63.780)(76,024) = 4,849,000$$

and the flow coefficient K becomes

$$K = 0.875 + \frac{(6.5)(0.875)}{(4,849,000)^{0.5}} = 0.8776$$

The flow rate is then

$$q^*_{\text{SCMH}} = (118,684)(0.8776)\sqrt{(0.2)(2.668)} = 76,085 \text{ standard m}^3/\text{h}$$

Without a Reynolds-number correction, the flow-rate equation is written as

$$q^*_{\text{SCMH}} = 118,684 K_N \sqrt{\Delta p^*} = 104,074\sqrt{\Delta p^*} = F_{MC}\sqrt{\Delta p^*}$$

where F_{MC}, the meter-coefficient factor, is 104,074. The uncorrected flow rate at 20 percent of the URV is then

$$q^*_{\text{SCMH}} = F_{MC}\sqrt{\Delta p^*} = 104,074\sqrt{(0.2)(2.668)} = 76,024$$

REFERENCES

ASME: *Fluid Meters,* 6th ed., ASME, New York, 1971.

Bell, K. J., and O. D. Bergelin: "Flow through Annular Orifices," *Trans. ASME,* vol. 79, pp. 593–601, 1957.

——— and ———: "Laminar and Transition Flow in Annular Orifices of Fine Clearance," *Flow Measurement in Closed Conduits* (symp. proc.), Her Majesty's Stationery Office, Edinburgh, 1962.

Curran, D. E.: "Laboratory Determination of Flow Coefficient Values for the Target Type Flowmeter at Low Reynolds Number Flows," in *Flow, Its Measurement and Control in Science and Industry,* vol. 2, ISA, Research Triangle Park, N.C., 1981.

Felton, G. L.: "Low Flow Measurement with the Integral Orifice," *Chem. Eng. Proc.,* vol. 68, no. 1, 1972.

Howell, A. R.: *Annular Airflow Orifice,* Report 1934, Ministry of Aircraft, London, 1939.

ISO 7145, *Determination of Flowrate of Fluids in Closed Conduits or Circular Cross Section—Method of Velocity Measurement at One Point of the Cross Section,* ISO UDC 532.57.082.25:532.542, 1st ed., Geneva, 1982.

Lansford, W. M.: *The Use of an Elbow in a Pipe Line for Determining the Rate of Flow in a Pipe,* University of Illinois Engineering Experimental Station Bulletin 289, 1936.

Murdock, J. W., C. J. Foltz, and C. Gregory: "Performance Characteristics of Elbow Flowmeters," ASME WAM Paper 63-WA-17, 1963.

Reinecke, M. E., W. G. Ragains, W. G. Fox, and R. W. Miller: "An Experimental Study of the Capability of Measuring Gas Mass Flow," *Pittsburgh Flow Measurement Symposium,* vol. 1, ASME, New York, 1966.

12

DIFFERENTIAL PRODUCERS: COMPUTATIONS

Depending on the desired accuracy, flow-rate determination may require only a simple visual observation of differential pressure on a square-root chart, or it may involve the use of a dedicated microprocessor that receives several measurement signals and calculates the flow rate. Compensation for pressure and/or temperature variations on chart indications may mean using pneumatic or electronic analog computers. Total flow, rather than flow rate, can be computed or determined by chart integration. The choice of measurement equipment, calculation procedure, computation means, and data-transmission means is extensive. This chapter presents some of the commonly used equipment and calculations for chart integration.

GENERAL PRINCIPLES

Measured and Unmeasured Variables

The flow-rate calculation can be viewed as the product of three terms: an unmeasured-variable term, a measured-variable term, and differential pressure. Differential pressure is always measured. The unmeasured-variables term includes a unit conversion factor and all factors assumed to be constant; the measured variables are quantities that must be measured for the desired accuracy (see Chap. 4). The unmeasured variables are combined into a meter-coefficient factor F_{MC} which commonly includes pipe and primary-element bore dimensions and the discharge coefficient. Measured variables are usually density-related (such as pressure and temperature) or are derived from other measurements (such as the Reynolds-number correction, which is derived from the flow rate, and the gas expansion factor, which is derived from differential- and absolute-pressure measurements). Depending on process variations, the designer determines which variables must be measured and which can be assumed constant.

As an example, the mass flow equation for liquids may be written as

$$q_{PPH} = [N_{M\rho}S_M F_a F_{RD} D^2 \sqrt{F_p}]_{UMV}[\sqrt{G_F}]_{MV} \sqrt{h_w} \qquad (12.1)$$

The first bracketed term contains the unmeasured variables; that is, after the pipe and bore diameter are measured and the thermal-expansion factor, liquid-compressibility factor, and discharge coefficient are calculated, the designer considers their product constant. The Reynolds-number correction F_{RD} appears within the brackets, which implies that this factor is also assumed constant. The second bracketed term is the measured variable, specific gravity. The equation thus reduces to

$$q_{PPH} = [F_{MC}]_{UMV}[\sqrt{G_F}]_{MV} \sqrt{h_w} \qquad (12.2)$$

If the specific gravity were assumed constant, it would be included as an unmeasured variable, and the equation would be simplified to

$$q_{PPH} = [F_{MC}]_{UMV} \sqrt{h_w} \qquad (12.3)$$

Similar simplified equations can be written for the equations given in Tables 9.36 through 9.38, and for the fixed-geometry devices discussed in Chap. 11. For many flow measurements, particularly when differential producers have been sized at normal flow-rate conditions, Eq. (12.3) is acceptable. However, for range and when temperature and pressure vary significantly, the designer must decide whether the added cost of including additional *measured* variables is warranted. *It is worth noting that initially selected design conditions may be different from process conditions. Significant errors can result if the values of the unmeasured variables are not checked. Using computers for more accurate calculation when unmeasured variables are inaccurate would be of little value.*

Analog Computers

Many manufacturers supply analog computers that receive two or three standardized 3- to 15-lb/in² (20- to 100-kPa), 4- to 20-mA, or 0- to 5-V input signals proportional to measured variables—for example, differential pressure, temperature, and absolute pressure. Pneumatic computers receive scaled inputs and *mechanically* perform the calculation, whereas electronic computers use passive circuits to ratio and scale. Both output a standardized signal proportional to

1. The product of two or more inputs

2. The quotient of two or more inputs

3. The square root of a single input or the square root of the product of two inputs

Figure 12.1 shows an analog computer that receives three 4- to 20-mA signals, and in Fig. 12.2 is a pneumatic analog computer that receives scaled 3- to 15-lb/in² (20- to 100-kPa) inputs. Provided the inputs are properly scaled, the com-

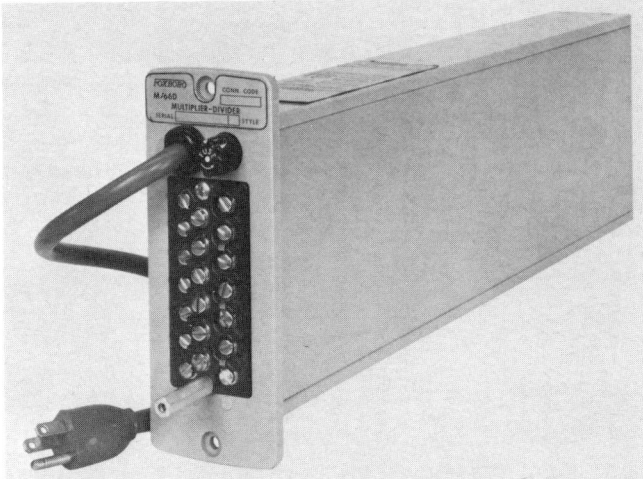

Figure 12.1 Three-input electronic analog computer *(courtesy The Fox-boro Co.).*

puter will correct for pressure, temperature, and specific-gravity changes for a gas-flow application, or for any combination of selected variables. The output is then a corrected standardized signal suitable for use with chart recorders, indicators, controllers, etc. Figure 12.3 shows a typical pneumatic-computer arrangement for pressure and temperature computation for a gas flow.

Modern analog circuits have excellent computational accuracy, along with stability. Analog units are small in size, have a good price-to-performance ratio, and are widely used when process variables change significantly.

Digital Flow Computers

In the natural gas industry, the need for improved calculation accuracy led to the development of dedicated on-site digital computers. Receiving inputs from transmitters (Fig. 12.4), these computers can be programmed to correct for Reynolds number, gas expansion factor, temperature, pressure, and gas compressibility, and to display either flow rate or total volume (Figs. 12.5 and 12.6). Also, meter runs and differential-pressure transmitters with different upper-range values may be computer-selected for wide-range metering.

Figure 12.2 Pneumatic analog computer *(courtesy The Fox-boro Co.).*

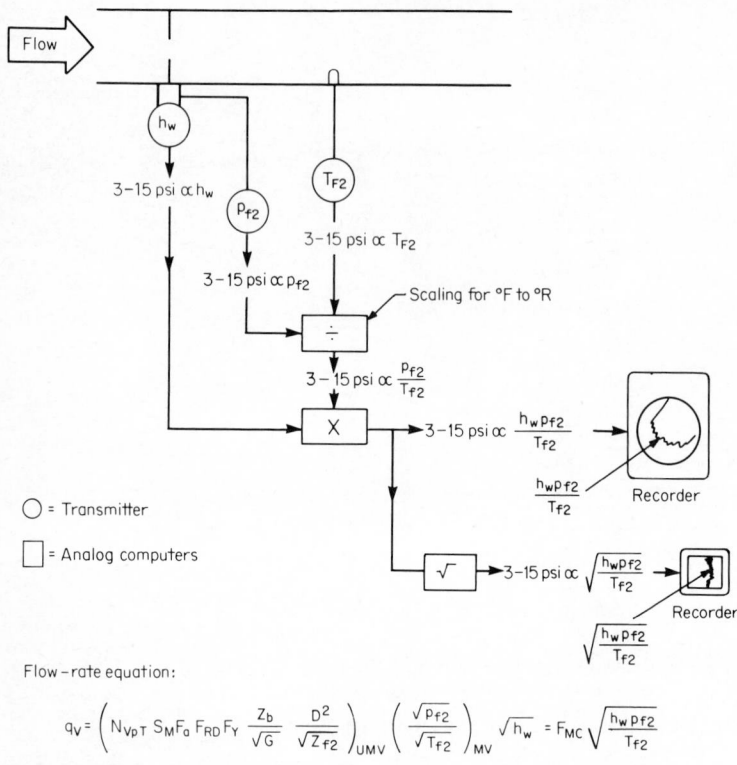

Figure 12.3 Computing gas flow with pneumatic analog computers.

In the process industries, digital instrumentation and microprocessors are causing a fundamental change in orifice flow measurement methods. The use of either central or dedicated on-site computers to perform calculations is becoming commonplace. Many of the process variables previously grouped as unmeasured are now measured and accounted for in calculations. In addition, instrument bias error is now being continuously corrected by computers, via stored calibration data. Transmitter zeroing and changes of differential-pressure transmitters for wide-range meters are now also computer-controlled.

Integrated Flow

When the flow-rate equation is used, it is written for a particular item unit, as in gallons per minute or gallons per day. However, the differential pressure and measured variables change continually within any unit of time. The total quantity to be measured over a selected time interval then must be the *integrated* total over this interval for these changing conditions. The total flow over any

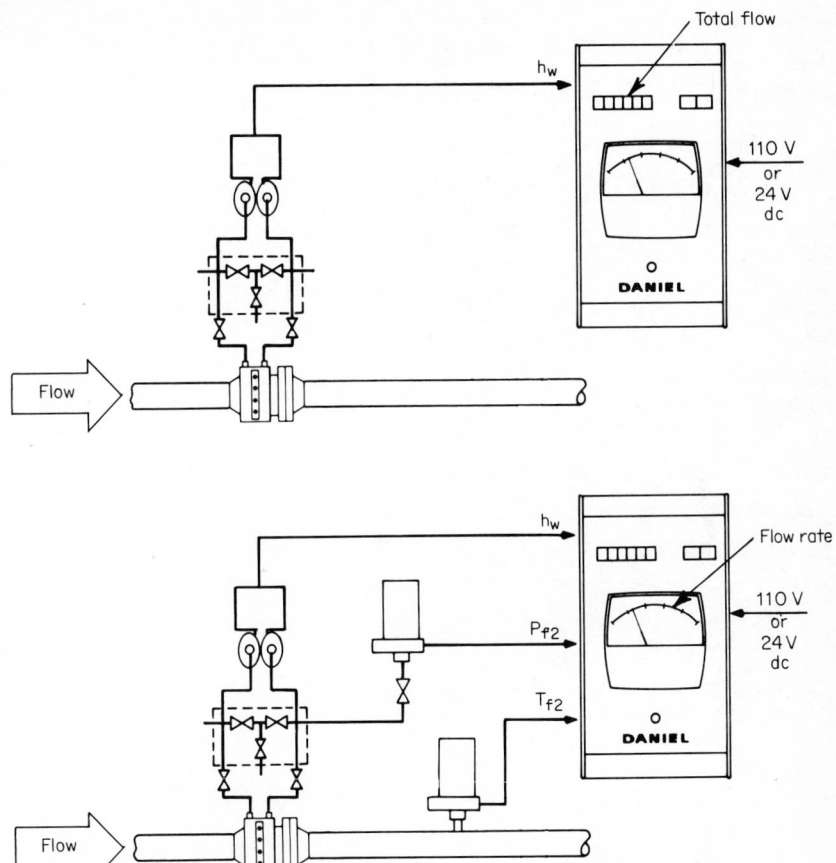

Figure 12.4 Digital on-site flow computer *(courtesy Daniel Industries, Inc.).*

time interval is then

$$Q = \int_0^t q \, dt \tag{12.4}$$

For example, for Eq. (12.3) the integrated flow on a daily basis would be written as

$$Q_{\text{PPD}} = \int q_{\text{PPH}} \, dt = F_{MC} \int_0^t \sqrt{h_w} \, dt \tag{12.5}$$

where the integral is evaluated by chart integration, computer summation, or visual observation of chart records. If the flow-rate unit is gallons per minute

Figure 12.5 Solar-powered on-site computer *(courtesy Daniel Industries, Inc.).*

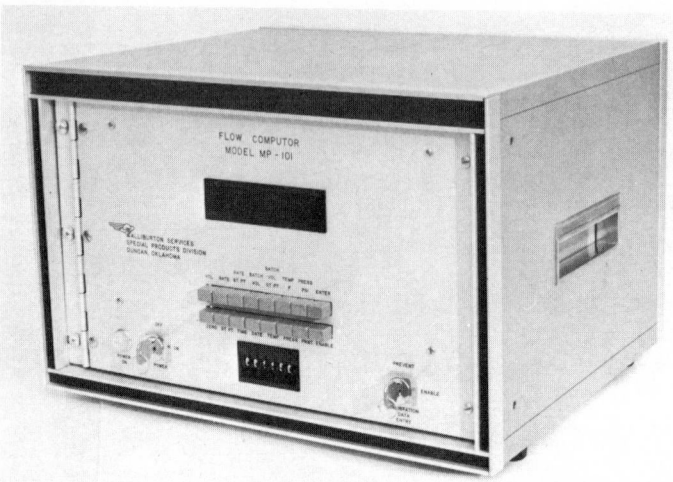

Figure 12.6 Digital flow computer *(courtesy Halliburton).*

and total gallons per day are required, Eq. (12.4) would be written as

$$Q_{GPD} = \int q_{GPM} \, dt = F_{MC} \int_0^t \sqrt{h_w} \, dt \tag{12.6}$$

where the integration limits in Eqs. (12.5) and (12.6) define the period of interest. The general expression for total flow is then, by Eq. (12.1),

$$Q = \int_0^t q \, dt = [\quad]_{UMV} \int_0^t [\quad]_{MV} \sqrt{h_w} \, dt \tag{12.7}$$

PLANIMETERING

Many industries retain charts as permanent records for accounting purposes. It is also common practice to use chart recorders (Fig. 12.7) for visual observation

Figure 12.7 Recorder *(courtesy The Foxboro Co.).*

of flow rate and flow-rate trends, and for daily or weekly accounting of feedstocks and steam usage. Total flow is obtained by *planimetering* either a circular- or strip-chart recording. A single-pen recording of differential pressure or a three-pen recording of pressure, temperature, and differential pressure is available, with selection depending on the number of measured variables.

The integration device can be a manually operated desktop integrator for circular charts (Fig. 12.8), a strip-chart integrator (Fig. 12.9), or a computer-based optical-scanner system (Fig. 12.10).

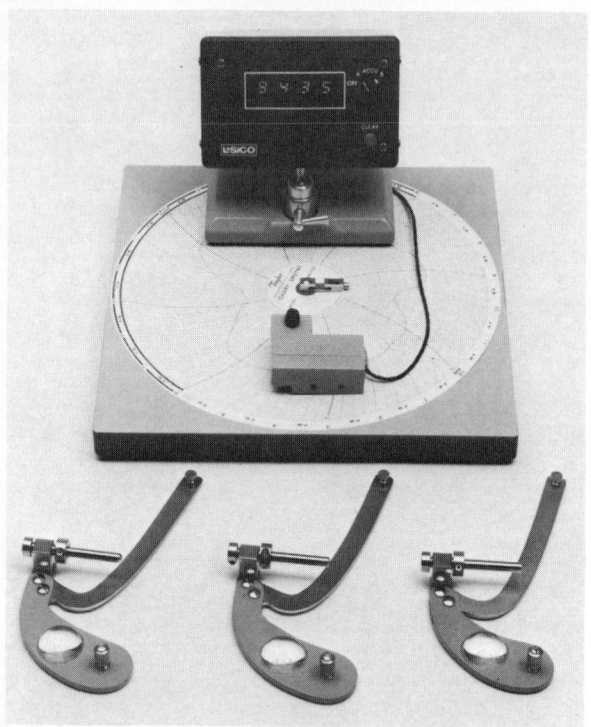

Figure 12.8 Square-root planimeter *(courtesy Lasico).*

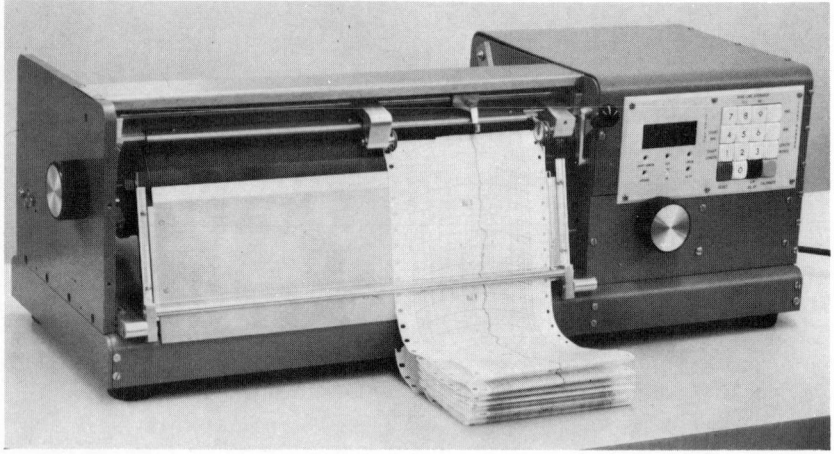

Figure 12.9 Strip-chart planimeter *(courtesy Flow Measurement Inc., Tulsa, OK).*

Figure 12.10 Computer-based optical-scanner integrator *(courtesy UGC Industries Inc.).*

Uniform Circular Charts

Uniform charts are divided into equal segments between an inner radius r_i and an outer radius r_o. Along an arc over which the pen travels, charts may be marked in either percentage of span (URV to LRV) or in units of the measured variables (°F, psia, q).

Shown in Fig. 12.11a is the nomenclature for a uniform circular chart. For a planimeter that integrates radially, the average value of the measured variable is calculated by substituting the planimeter reading $(P_{rdg})_{MV}$ into the equation

$$(MV)_{av} = [(MV)_{URV} - (MV)_{LRV}] \left(\frac{r_i}{r_o - r_i} \right) \left[\frac{F_{ARC} N_t (P_{rdg})_{MV}}{(P_{rdg})_0} - 1 \right] + (MV)_{URV} \quad (12.8)$$

where N_t is calculated as

$$N_t = \frac{\text{number of hours represented by } 360° \text{ chart rotation}}{\text{number of hours planimetered}} \quad (12.9)$$

The total quantity of the measured variable over the integration period is then

$$(MV)_t = t(MV)_{av} \quad (12.10)$$

where t is the number of hours planimetered.

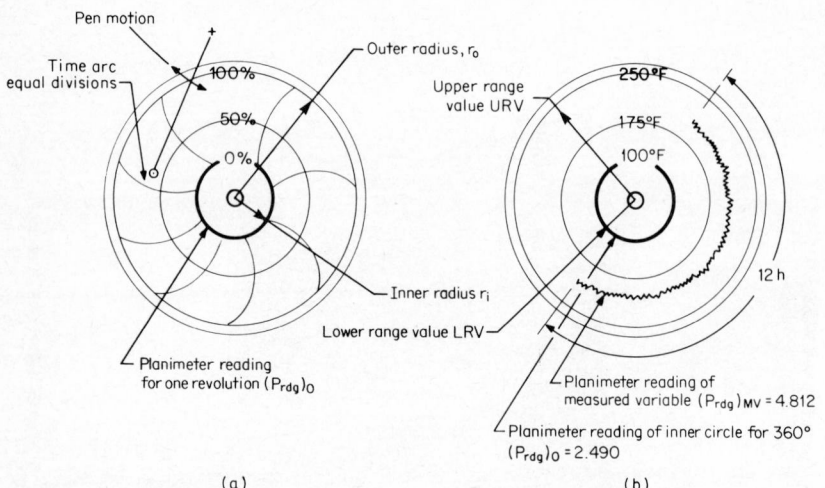

Figure 12.11   Uniform chart. (*a*) Chart nomenclature. (*b*) Radial-planimeter nomenclature (values from Example 12.1).

The arc correction factor F_{ARC} is read from Fig. 12.12 for the product of $N_t(P_{\text{rdg}})_{\text{MV}}$. This correction is required because radial planimeters average radial distances and, hence, are accurate only with a chart employing equal radial divisions, whereas uniform charts employ equal increments along an arc (see Fig. 12.11).

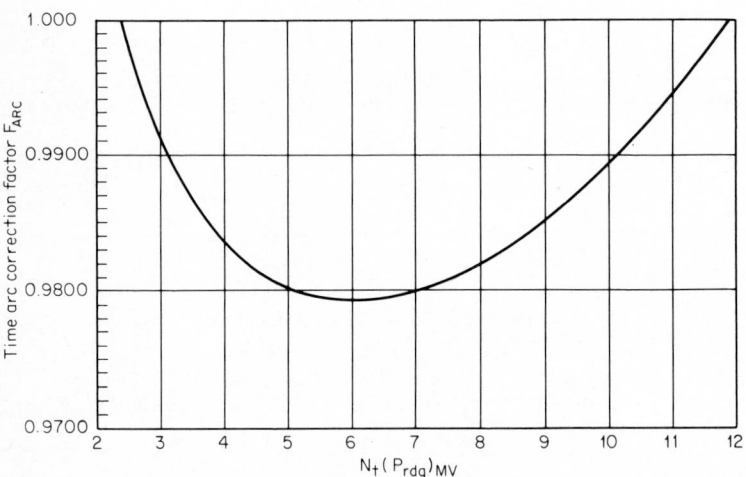

Figure 12.12   Radial-planimeter time-arc correction.

Unless the differential and static pressures remain essentially constant (Fig. 12.13) or a square-root signal inputs the recorder, radial planimeters should *not* be used to average differential-pressure or static-pressure charts. For example, in Fig. 12.14, the average for the high period is approximately 75, and for the low period approximately 35.6. If the entire record were averaged at one operation

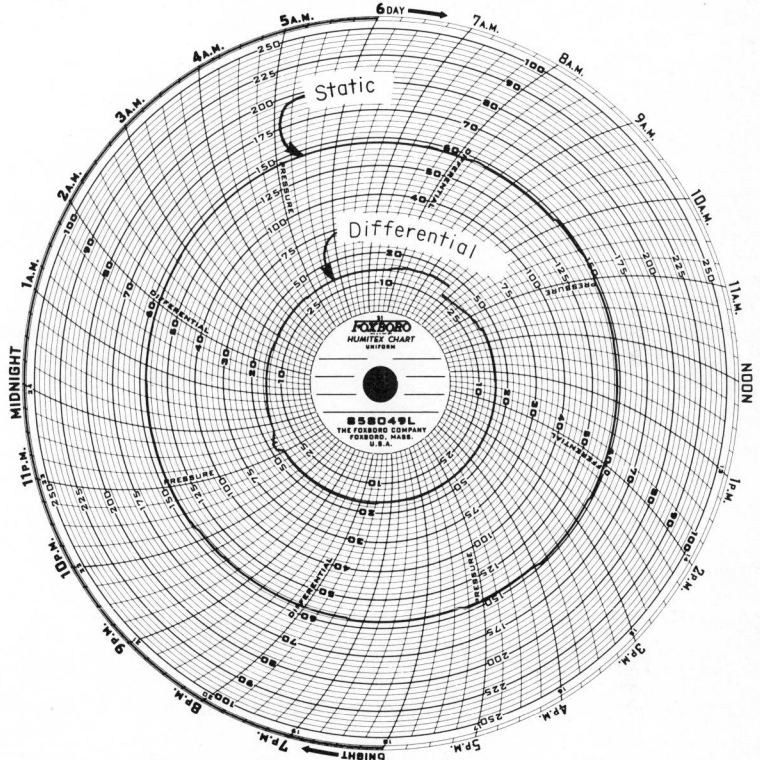

Figure 12.13 Chart that can be averaged by observation or by a radial planimeter.

by a radial planimeter, the result would be 55.3, and the square root of 55.3 is 7.4364. If, however, the periods were averaged separately and the average square root determined, the result would be one-half the sum of the square root of 75 (8.6603) and the square root of 35.6 (5.9666), or 7.3135. The error in this particular case is not very large, but it would increase rapidly as the difference between the averages for the high and low periods increased. For example, if the high period averaged approximately 90.6 and the low period 20, the average for the entire record would still be 55.3, and its square root 7.4364. But the average of the square root of 90.6 (9.5184) and the square root of 20 (4.4721) would be 6.9955. It is obvious that the desired result is the average square root, and not the square root of the average.

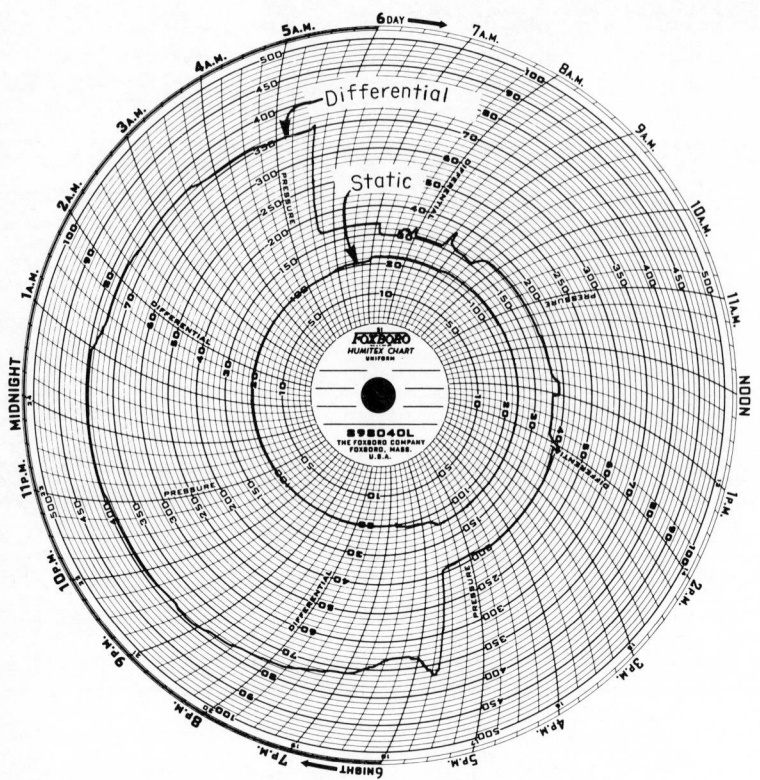

Figure 12.14 Chart which should not be averaged by a radial planimeter.

EXAMPLE 12.1

A temperature measurement is recorded on a 24-h, 12-in (300-mm) uniform Foxboro circular chart (Fig. 12.11b). The chart range is 100 to 250°F (37.8 to 121°C). Planimetering the chart for 12 h with a radial planimeter gives a reading of 4.812. Determine the average temperature.

For a Foxboro chart,

$$r_i = 0.996 \text{ in} \qquad r_o = 4.96 \text{ in}$$

Planimetering of the inner circle for 360° gives

$$(P_{\text{rdg}})_0 = 2.490$$

The time factor is, by Eq. (12.9),

$$N_t = \frac{24}{12} = 2$$

From Fig. 12.12, the time-arc correction factor for

$$N_t(P_{\text{rdg}})_{\text{MV}} = (2)(4.812) = 9.624$$

is $F_{ARC} = 0.9876$. From Eq. (12.8), the average temperature is

$$(MV)_{av} = [(MV)_{URV} - (MV)_{LRV}] \frac{r_i}{r_o - r_i} \left[\frac{F_{ARC} N_t (P_{rdg})_{MV}}{(P_{rdg})_0} - 1 \right] + (MV)_{LRV}$$

Substitution gives

$$(T_F)_{av} = (250 - 100)(0.2513) \left[\frac{(0.9876)(2)(4.812)}{2.490} - 1 \right] + 100 = 206.2°F$$

EXAMPLE 12.2

For Example 9.11, an analog computer transmits the square root of the differential-pressure-transmitter signal to a chart recorder. A planimeter reading of 4.812 is obtained for a 12-h integration of a 12-in, 24-h Foxboro 0- to 100-in (24.9-kPa) uniform circular chart. What is the total flow, in gallons, for the 12-h period?

From Example 9.11, the flow rate is calculated as

$$q_{GPM} = F_{MC} \sqrt{h_w} = 10.00 \sqrt{h_w}$$

The upper-range value of the flow rate on an hourly basis is

$$(q_{GPH})_{URV} = (10.00)(60) \sqrt{(h_w)_{URV}} = (10.00)(60) \sqrt{100} = 6000 \text{ base gal/h}$$

where $(h_w)_{URV} = 100$, and the lower-range value $(h_w)_{LRV}$ is 0. For the values of Example 12.1, the average of the measured variable is, by Eq. (12.8),

$$(MV)_{av} = (q_{GPH})_{av} = 0.2513[(MV)_{URV} - (MV)_{LRV}] \left[\frac{F_{ARC} N_t (P_{rdg})_{MV}}{2.490} - 1 \right] + (MV)_{LRV}$$

$$= (0.2513)(6000) \left[\frac{(0.9876)(2)(4.812)}{2.490} - 1 \right] + 0 = 4248$$

Then, from Eq. (12.10),

$$Q_{GAL} = t(MV)_{av} = (12)(4248) = 50,970 \text{ base gal}$$

Strip Charts

A strip-chart planimeter is shown in Fig. 12.9. Standard strip charts are 4 in (100 mm) wide, with a recorder speed of 0.75 in/h (19 mm/h). Chart widths of 1 to 12 in (25 to 300 mm) and varying chart speeds are available. The planimeter is calibrated to an upper-range reading corresponding to the upper-range value of the measured variable; at a zero chart reading, the planimeter registers zero. Total integrated flow is then calculated as

$$Q_t = P_{rdg} \left[\frac{q}{P_{rdg}} \right]_{URV} \tag{12.11}$$

where the bracketed values are to be taken at the upper-range value for the flow rate and planimetered over the time period specified for the planimeter.

EXAMPLE 12.3

For Example 9.11, a pneumatic analog computer transmits the square root of the differential-pressure-transmitter signal to a strip-chart recorder. A linear strip-chart planimeter records 860 when the record is traced for 24 h. If the planimeter's upper-range value is 1000 counts for a 24-h trace, what is the total flow?

The upper-range value of the flow rate for 24 h is, from Example 9.11,

$$q_{GPD} = (24)(60)q_{GPM} = (24)(60)(10.00)\sqrt{(h_w)_{URV}} = 14,400\sqrt{100} = 144,000 \text{ base gal/24 h}$$

The total flow for 24 h is then

$$Q_{GAL} = (P_{rdg})_{MV}\left(\frac{q}{P_{rdg}}\right)_{URV} = 860\,\frac{144,000}{1000} = 123,840 \text{ base gal}$$

EXAMPLE 12.4

For the gasoline flow of Example 9.11, the differential pressure is recorded on a 12-in (300-mm), 24-h circular chart. The record is planimetered over a 24-h period, and a reading of 4.720 is obtained. Determine the total flow over the planimetered period if the planimeter reads 5.00 when tracing the 100-in (25-kPa) upper-range differential-pressure value for 24 h.

The upper-range value of the flow rate over the chart's time period (24 h) is, from Example 9.11,

$$q_{GPD} = (24)(60)(q_{GPM})_{URV} = (24)(60)(10)\sqrt{100} = 144,000 \text{ base gal/24 h}$$

The total flow for 24 h is then, from Eq. (12.12),

$$Q_{GAL} = (P_{rdg})_{hw}\left(\frac{q}{P_{rdg}}\right)_{URV} = (4.720)\,\frac{144,000}{5} = 136,000 \text{ base gal}$$

Square-Root Planimeters

Square-root planimeters instantaneously average the square root of the measured variable as the chart record (differential- or static-pressure recording) is traced. Integration is along the arc, and the record may be mechanically averaged or computer calculated (Fig. 12.10). For gas flows, the square root of the product of the differential and absolute pressures is required; this may be calculated from two planimeter readings or by a computer-based integrator that displays a single value.

When only the differential pressure is recorded, the total flow over a specified time period is calculated as

$$Q_t = (P_{rdg})_{hw}\left[\frac{q}{(P_{rdg})_{hw}}\right]_{URV} \qquad (12.12)$$

where the quantities within the brackets are to be taken at the upper-range value for both flow rate and planimeter for 360° chart rotation. In gas (vapor) flows, both the differential and static pressures are usually planimetered. In this case, the total flow over a specified time period is calculated as

$$Q_t = N_t (P_{rdg})_{hw} (P_{rdg})_{pf1} \left[\frac{q}{(P_{rdg})_{hw} (P_{rdg})_{pf1}} \right]_{URV} \tag{12.13}$$

EXAMPLE 12.5

For the natural gas flow of Example 9.6, the differential and *absolute* static pressure are recorded for 12 h on a 12-in (300-mm), 24-h circular chart. The planimetered reading of the differential-pressure record is 2.415, and that of the static-pressure record is 2.250. Determine the total flow in standard cubic meters over the 12-h planimetered period. (The planimeter reads 5.00 when tracing the upper-range values.)

From Eq. (f) of Table 9.38, the flow-rate equation can be written in terms of measured and unmeasured variables as

$$q_{Vb}^* = (N_{Vhp}^* S_M F_a^* F_{RD} F_Y F_{PB}^* F_{TB}^* F_{TF1}^* F_{pv1} Z_b F_g D^{*2})_{UMV} (\sqrt{p_{f1}^*})_{MV} \sqrt{\Delta p^*}$$

Substituting the values given in Example 9.6 yields

$$q_{SCMH}^* = F_{MC} \sqrt{p_{f1}^*} \sqrt{\Delta p_{f1}^*} = 7.9264 \sqrt{p_{f1}^*} \sqrt{\Delta p_{f1}^*}$$

The upper-range value for 24 h is

$$q_{SCMD}^* = 24 F_{MC} \sqrt{(p_{f1}^*)_{URV}} \sqrt{(\Delta p^*)_{URV}} = (24)(7.9264) \sqrt{184} \sqrt{10} = 8160 \text{ standard m}^3/24 \text{ h}$$

The planimetered time factor is, by Eq. (12.9), $N_t = 24/12 = 2$. The total flow for 12 h is then, by Eq. (12.13),

$$Q_{SCM}^* = N_t (P_{rdg})_{\Delta p^*} (P_{rdg})_{pf1}^* \left[\frac{q^*}{(P_{rdg})_{\Delta p^*} (P_{rdg})_{pf1}^*} \right]_{URV}$$

$$= (2)(2.415)(2.250) \frac{8160}{(5)(5)} = 3547 \text{ standard m}^3$$

13

CRITICAL
FLOW

When a gas accelerates through a restriction, its density decreases and its velocity increases. Since the mass flow per unit area (mass flux) is a function of both density and velocity, a critical area exists at which the mass flux is at a maximum. In this area, the velocity is sonic, and further decreasing the downstream pressure will not increase the mass flow. This is referred to as *choked* or *critical flow*. For liquids, if the pressure at the minimum area is reduced to the liquid's vapor pressure, a cavitation zone is formed which restricts the flow. Further decreases in pressure will not increase the flow rate. In both cases, mass flow can only be increased by increasing the upstream pressure.

Critical flow nozzles are widely used as secondary standards to test air compressors, steam generators, and natural gas flowmeters. Over the last 20 years the aerospace industry has developed a critical nozzle with a downstream diffuser (venturi) recovery section that gives minimum overall pressure loss to maintain critical flow. Cavitating venturis or restrictive orifices are used as flow limiters in the event of a downstream system failure.

GASES

Basic Principles

Figure 13.1 shows the pressure-velocity relationship for a convergent-divergent passage through which a compressible fluid accelerates. As the downstream pressure p_{f3} decreases, the throat velocity $\overline{V}_t$ increases until a *critical pressure ratio*† is reached at which the throat velocity is sonic. Further decreases in the downstream pressure will not increase the mass flow rate. The flow is referred to as *subsonic* down to the critical pressure ratio, and *critical* below this ratio. In critical flow the throat velocity is always sonic, but the velocity increases in the dif-

†The critical (or choking) pressure ratio is discussed in detail later in this chapter.

13-1

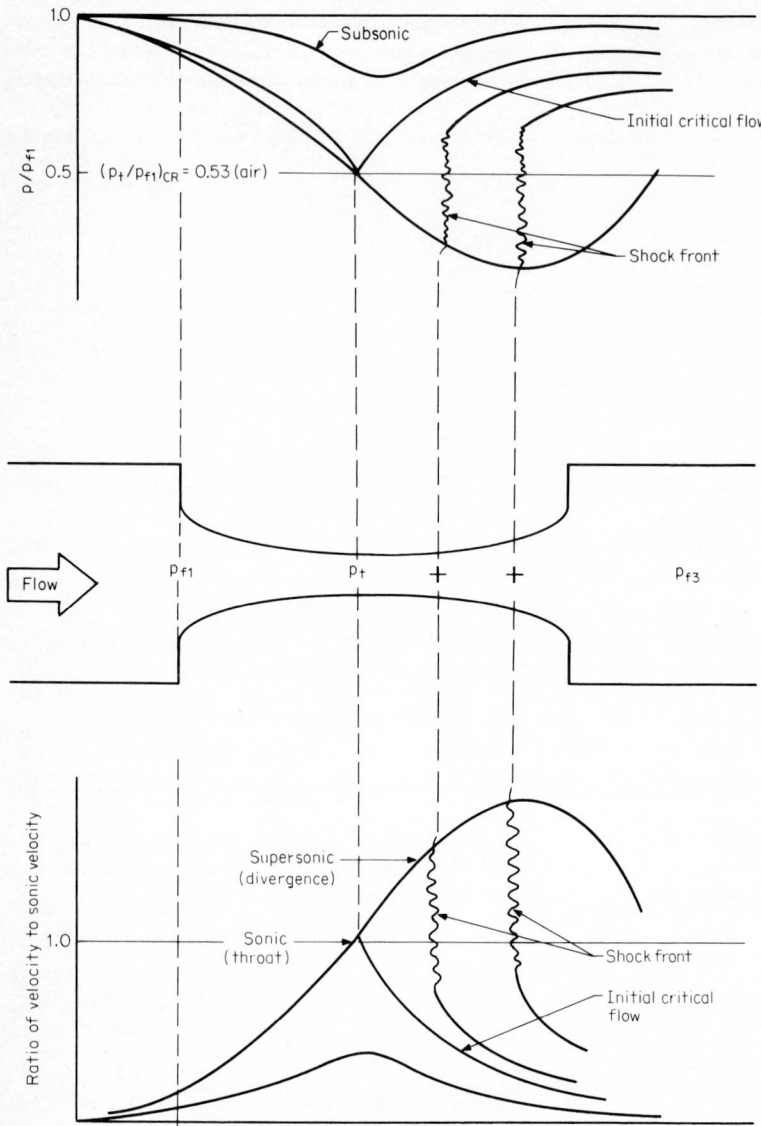

Figure 13.1 Pressure and velocity relationships for a venturi nozzle.

fuser section, where a normal shock front occurs. Depending on the downstream pressure, four flow conditions are possible:

1. For pressure ratios greater than critical, the flow remains subsonic and may be calculated with the relationships given in Chap. 9.

2. When p_{f3} is reduced to the value at which sonic throat velocity first occurred, the flow decelerates in the divergent section to a subsonic velocity; the gas expands isentropically to throat pressure and then returns to the higher downstream pressure.

3. Reducing the downstream pressure further will not alter the critical pressure ratio between the upstream and throat pressures, but the velocity increases up to a normal shock wave in the divergent section. The flow across the shock is not isentropic, and the velocity abruptly changes from a supersonic speed to a subsonic speed.

4. The shock-front location moves progressively downstream with further pressure reductions, until it no longer occurs within the divergent section. The flow then accelerates steadily throughout the nozzle and is subsonic in the convergent passage, sonic at the throat, and supersonic within the divergent section.

The minimum downstream pressure at which critical flow begins is dependent on the geometry and the isentropic exponent of the gas. With a divergent section the choking pressure ratio is approximately 5 to 10 percent of the upstream pressure, but with no divergent section it is approximately 50 percent of the upstream pressure.

Engineering Equations

The mass flow-rate equation is derived from mass flow continuity between upstream and throat sections as

$$(q_{PPS})_1 = A_t \rho_{ft} \overline{V}_t \tag{13.1}$$

where the subscript t refers to throat conditions. Sonic velocity is the speed at which small pressure changes propagate as waves through a compressible fluid. A one-dimensional isentropic model gives the sonic velocity as the square root of the ratio of pressure to density change; that is,

$$\overline{V}_t = \overline{V}_{son} = \sqrt{\left(\frac{dP}{d\rho}\right)_s} \tag{13.2}$$

where the subscript s refers to an isentropic expansion from inlet to throat. Substitution of Eq. (9.27) into Eq. (13.2) yields the sonic throat velocity as

$$\overline{V}_t = \overline{V}_{son} = \sqrt{\frac{kg_c P_t}{\rho_{ft}}} \tag{13.3}$$

where k is the isentropic exponent (see Chap. 2). Substituting Eq. (13.3) into Eq. (13.1) gives the mass flow equation in U.S. units as

$$q_{PPS} = CA_t \sqrt{kg_c P_t \rho_{ft}} \tag{13.4}$$

where C is the *coefficient of discharge* relating the actual flow rate to a one-dimensional nonviscous (ideal) flow rate.

In practice, static pressures and temperatures are measured at an upstream location, and equations are developed using total (stagnation) pressure and temperature (see Chap. 3). For isentropic expansion, the mass flow-rate equation for a density determination can be developed from the isentropic relationships given by

$$\left(\frac{P_t}{P_{f1}}\right)_T = \left(\frac{2}{k+1}\right)^{k/(k-1)} \qquad \left(\frac{\rho_t}{\rho_{f1}}\right)_T = \left(\frac{2}{k+1}\right)^{1/(k-1)}$$

$$\left(\frac{T_t}{T_{f1}}\right)_T = \frac{2}{k+1} \tag{13.5}$$

where the subscript T refers to total, or stagnation, upstream conditions. Substituting Eq. (13.5) into Eq. (13.4) yields the fundamental mass flow-rate equation

$$q_{\text{PPS}} = CA_t \left[k \left(\frac{2}{k+1}\right)^{(k+1)/(k-1)} \right]^{1/2} \sqrt{g_c (P_{f1}\rho_{f1})_T} \tag{13.6}$$

for U.S. units (pounds-force, pounds-mass, feet, seconds), and

$$q_{\text{KPS}}^* = CA_t^* \left[k \left(\frac{2}{k+1}\right)^{(k+1)/(k-1)} \right]^{1/2} \sqrt{(P_{f1}^*\rho_{f1}^*)_T} \tag{13.7}$$

for SI units (kilograms, newtons, seconds).

In practice, wall pressure taps are used to measure pressure; the total pressure is seldom measured with impact or simple pitot tubes that might disrupt the flow. The relationship between the total and static pressures for any value of β is

$$\left[\frac{P_{f1}}{(P_{f1})_T}\right]^{2/k} - \left[\frac{P_{f1}}{(P_{f1})_T}\right]^{(k+1)/k} = \left(\frac{k-1}{2}\frac{2}{k+1}\right)^{(k+1)/(k-1)} \beta^4 \tag{13.8}$$

For $\beta < 0.5$ the equation relating total to measured static pressure is well approximated by

$$(p_{f1})_T = \left[1 + \frac{k}{2}\left(\frac{2}{k+1}\right)^{(k+1)/(k-1)} \beta^4 \right] p_{f1} = F_{TP} p_{f1} \tag{13.9}$$

As noted in Chap. 3, a thermal well needs no correction for stagnation conditions since, in general, the total temperature is measured. The upstream density ρ_{f1} may be calculated from the pvT equations rather than found by measurement. Substituting Eqs. (2.10) and (2.11) into Eqs. (13.6) and (13.7) and using U.S. and SI engineering pressure units (pounds per square inch absolute, kilopascals) give the engineering equations for density and for pvT density determination for a wall pressure tap:

$$q_{\text{PPS}} = 0.3712458 C d^2 \sqrt{Z_{f1}}\ Y_{\text{CR}} \sqrt{\rho_{f1}}\ \sqrt{F_{TP}}\ \sqrt{p_{f1}} \tag{13.10}$$

$$q_{PPS} = 0.6098863 Cd^2 Y_{CR} \sqrt{\frac{G}{T_{f1}}} F_{TP} p_{f1} \qquad (13.11)$$

for U.S. units, and

$$q^*_{KPS} = 0.00002483647 Cd^{*2} \sqrt{Z_{f1}} \; Y_{CR} \sqrt{\rho^*_{f1}} \; \sqrt{F_{TP} p^*_{f1}} \qquad (13.12)$$

$$q^*_{KPS} = 0.00004635451 Cd^{*2} Y_{CR} \sqrt{\frac{G}{T_{K1}}} F_{TP} p^*_{f1} \qquad (13.13)$$

for SI units. In these equations Y_{CR} is the critical flow function, defined as

$$Y_{CR} = \left[\frac{k}{Z_{f1}} \left(\frac{2}{k+1} \right)^{(k+1)/(k-1)} \right]^{1/2} \qquad (13.14)$$

Table 13.1 presents the U.S.- and SI-unit equations for mass and volumetric flow-rate calculations on an hourly basis.

Critical Flow Function

The critical flow function in Eq. (13.14) may be calculated with the isentropic-exponent curves presented in App. I, found by the graphical method presented in Chap. 2, or obtained from tables.

Sullivan Tables   Presented in Tables 13.2 through 13.7 are the critical flow functions for air, argon, carbon dioxide, methane, nitrogen, and oxygen provided by D. A. Sullivan (1981). These tables are incorporated in an ISO critical-flow-meter standard (1982) and are based on the Bender (1970) thermal equations of state.

Murdock-Bauman Tables (Superheated Steam)   Table 13.8 lists the critical flow function for superheated steam, as developed by Murdock and Bauman (1963). The original table was developed using Arnberg's critical flow function (1962), which included both the specific gas constant R and the gravitational constant g_c. The relationship between Arnberg's critical flow function and Eq. (13.14) is

$$Y_{CR} = \sqrt{\frac{144 R_0}{(M_w)_{H_2O} g_c}} \; Y_{Arn} = 1.632819 Y_{Arn} \qquad (13.15)$$

In Table 13.8 the Murdock-Bauman tabular values are modified by Eq. (13.15).

Johnson Tables (Methane and Natural Gas)   In the natural gas industry, the Johnson (1970, 1971) methane and natural gas critical-flow-function tables are the standard. These were calculated using the state equation of Benedict et al. (1940) and are based on 19 natural gas mixtures in the pressure range 0 to 1000 psia (0 to 7000 kPa) and temperature range 450 to 700°R (250 to 390 K).

Table 13.1 Mass and Volumetric Flow-Rate Equations for Critical Flowmeters: U.S. and SI Units

	U.S. units	SI units
MASS FLOW RATE		
Density	$q_{PPH} = 1336.485 Cd^2 \sqrt{Z_{f1}}\, Y_{CR} \sqrt{\rho_{f1}}\, \sqrt{F_{TP}}\, \sqrt{p_{f1}}$ (a)	$q^*_{KPH} = 0.08941130 Cd^{*2} \sqrt{Z_{f1}}\, Y_{CR} \sqrt{\rho^*_{f1}}\, \sqrt{F_{TP}}\, \sqrt{p^*_{f1}}$ (h)
pvT equation	$q_{PPH} = 2195.591 Cd^2 Y_{CR} \sqrt{\dfrac{G}{T_{f1}}}\, F_{TP} p_{f1}$ (b)	$q^*_{KPH} = 0.166876\,2 Cd^{*2} Y_{CR} \sqrt{\dfrac{G}{T_{K1}}}\, F_{TP} p^*_{f1}$ (i)
VOLUMETRIC FLOW RATE AT FLOWING CONDITIONS		
Density	$q_{acfh} = 1336.485 Cd^2 \sqrt{Z_{f1}}\, Y_{CR} \dfrac{\sqrt{F_{TP}}}{\sqrt{\rho_{f1}}}\, \sqrt{p_{f1}}$ (c)	$q^*_{acmh} = 0.08941130 Cd^{*2} \sqrt{Z_{f1}}\, Y_{CR} \dfrac{\sqrt{F_{TP}}}{\sqrt{\rho^*_{f1}}}\, \sqrt{p^*_{f1}}$ (j)
pvT equation	$q_{acfh} = 813.5358 Cd^2 Z_{f1} Y_{CR} \sqrt{\dfrac{T_{f1}}{G}}$ (d)	$q^*_{acmh} = 0.04790604 Cd^{*2} Z_{f1} Y_{CR} \sqrt{\dfrac{T_{K1}}{G_1}}$ (k)
VOLUMETRIC FLOW RATE AT BASE CONDITIONS		
Density	$(q_{SCFH})_b = 1336.485 Cd^2 \sqrt{Z_{f1}}\, Y_{CR} \dfrac{\sqrt{\rho_{f1}}}{\rho_b}\, \sqrt{F_{TP}}\, \sqrt{p_{f1}}$ (e)	$(q^*_{CMH})_b = 0.08941130 Cd^{*2} \sqrt{Z_{f1}}\, Y_{CR} \dfrac{\sqrt{\rho^*_{f1}}}{\rho^*_b}\, \sqrt{F_{TP}}\, \sqrt{p^*_{f1}}$ (l)
pvT equation (standard or selected base)	$q_{SCFH} = 28712.44 Cd^2 Z_b Y_{CR} \sqrt{\dfrac{1}{T_{f1}G}}\, F_{TP} p_{f1}$ (f)	$q^*_{SCMH} = 0.1362361 Cd^{*2} Z_b Y_{CR} \sqrt{\dfrac{1}{T_{K1}G}}\, F_{TP} p^*_{f1}$ (m)
	$(q_{SCFH})_b = 813.5358 Cd^2 \dfrac{Z_b T_b}{p_b} Y_{CR} \sqrt{\dfrac{1}{T_{f1}G}}\, F_{TP} p_{f1}$ (g)	$(q^*_{SCMH})_b = 0.04790604 Cd^{*2} \dfrac{Z_b T_{Kb}}{p^*_b} Y_{CR} \sqrt{\dfrac{1}{T_{K1}G}}\, F_{TP} p^*_{f1}$ (n)

Table 13.2 Critical Flow Function Y_{CR} for Air

Temperature		Pressure $F_{TP}p_{f1}$, psia (kPa)										
°F	°C	0 (0)	145 (1000)	290 (2000)	435 (3000)	580 (4000)	725 (5000)	870 (6000)	1015 (7000)	1160 (8000)	1305 (9000)	1450 (10,000)
−58	−50	0.6850	0.6945	0.6994	0.7069	0.7148	0.7230	0.7316	0.7407	0.7497	0.7588	0.7675
−13	−25	0.6850	0.6900	0.6952	0.7004	0.7059	0.7113	0.7168	0.7224	0.7281	0.7336	0.7391
32	0	0.6887	0.6887	0.6925	0.6962	0.7000	0.7038	0.7076	0.7113	0.7150	0.7187	0.7224
77	25	0.6850	0.6876	0.6904	0.6927	0.6958	0.6982	0.7011	0.7037	0.7062	0.7087	0.7111
122	50	0.6849	0.6870	0.6889	0.6910	0.6930	0.6949	0.6968	0.6987	0.7005	0.7022	0.7039
167	75	0.6848	0.6863	0.6878	0.6893	0.6907	0.6921	0.6935	0.6949	0.6961	0.6974	0.6985
212	100	0.6847	0.6858	0.6869	0.6880	0.6890	0.6900	0.6910	0.6920	0.6928	0.6937	0.6946

SOURCE: Sullivan (1981).

Table 13.3 Critical Flow Function Y_{CR} for Argon

Temperature		Total pressure F_{TPPf1}, psia (kPa)											
°F	°C	0 (0)	72.5 (500)	145 (1000)	290 (2000)	435 (3000)	580 (4000)	725 (5000)	870 (6000)	1015 (7000)	1160 (8000)	1305 (9000)	1450 (10,000)
−58	−50	0.7262	0.7310	0.7358	0.7460	0.7567	0.7679	0.7797	0.7922	0.8053	0.8191	0.8335	0.8484
−13	−25	0.7262	0.7297	0.7333	0.7407	0.7482	0.7560	0.7639	0.7720	0.7803	0.7888	0.7975	0.8062
32	0	0.7262	0.7289	0.7316	0.7372	0.7427	0.7484	0.7540	0.7598	0.7655	0.7713	0.7772	0.7830
77	25	0.7262	0.7283	0.7304	0.7347	0.7389	0.7432	0.7474	0.7517	0.7559	0.7601	0.7643	0.7684
122	50	0.7262	0.7279	0.7295	0.7329	0.7362	0.7395	0.7427	0.7460	0.7492	0.7523	0.7555	0.7585
167	75	0.7262	0.7275	0.7289	0.7315	0.7342	0.7367	0.7393	0.7418	0.7443	0.7467	0.7491	0.7515
212	100	0.7262	0.7273	0.7284	0.7305	0.7326	0.7347	0.7367	0.7387	0.7406	0.7426	0.7444	0.7463

SOURCE: Sullivan (1981).

Table 13.4 Critical Flow Function Y_{CR} for Carbon Dioxide

| Temperature | | Total pressure F_{TPPh}, psia (kPa) | | | | | | | | | | | |
°F	°C	0 (0)	72.5 (500)	145 (1000)	290 (2000)	435 (3000)	580 (4000)	725 (5000)	870 (6000)	1015 (7000)	1160 (8000)	1305 (9000)	1450 (10,000)
−58	−50	0.6739	—	—	—	—	—	—	—	—	—	—	—
−13	−25	7.6713	0.6864	—	—	—	—	—	—	—	—	—	—
32	0	0.6689	0.6797	0.6918	—	—	—	—	—	—	—	—	—
77	25	0.6668	0.6748	0.6834	0.7032	0.7277	—	—	—	—	—	—	—
122	50	0.6649	0.6709	0.6774	0.6915	0.7077	0.7267	0.7497	0.7783	0.8162	—	—	—
167	75	0.6632	0.6679	0.6728	0.6833	0.6949	0.7078	0.7222	0.7386	0.7575	0.7795	0.8056	0.8369
212	100	0.6616	0.6653	0.6692	0.6772	0.6859	0.6952	0.7053	0.7163	0.7282	0.7412	0.7555	0.7713

SOURCE: Sullivan (1981).

13-9

Table 13.5 Critical Flow Function Y_{CR} for Methane

| Temperature | | Total pressure $F_{TPP\eta}$, psia (kPa) | | | | | | | | | | | |
°F	°C	0 (0)	72.5 (500)	145 (1000)	290 (2000)	435 (3000)	580 (4000)	725 (5000)	870 (6000)	1015 (7000)	1160 (8000)	1305 (9000)	1450 (10,000)
−58	−50	0.6726	0.6798	0.6875	0.7048	0.7254	0.7506	0.7827	0.8249	0.8824			
−13	−25	0.6719	0.6771	0.6825	0.6943	0.7075	0.7223	0.7390	0.7581	0.7799	0.8047	0.8324	0.8623
32	0	0.6708	0.6747	0.6787	0.6872	0.6963	0.7061	0.7167	0.7281	0.7404	0.7536	0.7676	0.7823
77	25	0.6694	0.6724	0.6754	0.6817	0.6884	0.6953	0.7026	0.7102	0.7182	0.7266	0.7352	0.7441
122	50	0.6678	0.6701	0.6724	0.6772	0.6822	0.6873	0.6925	0.6980	0.7036	0.7093	0.7151	0.7211
167	75	0.6659	0.6677	0.6695	0.6732	0.6770	0.6808	0.6848	0.6888	0.6929	0.6970	0.7012	0.7054
212	100	0.6639	0.6653	0.6667	0.6696	0.6725	0.6755	0.6784	0.6815	0.6845	0.6876	0.6907	0.6938

SOURCE: Sullivan (1981).

Table 13.6 Critical Flow Function Y_{CR} for Nitrogen

Temperature		Total pressure F_{TPPf1}, psia (kPa)											
°F	°C	0 (0)	72.5 (500)	145 (1000)	290 (2000)	435 (3000)	580 (4000)	725 (5000)	870 (6000)	1015 (7000)	1160 (8000)	1305 (9000)	1450 (10,000)
−58	−50	0.6848	0.6878	0.6908	0.6970	0.7035	0.7102	0.7171	0.7243	0.7315	0.7389	0.7462	0.7536
−13	−25	0.6848	0.6869	0.6891	0.6934	0.6978	0.7023	0.7069	0.7115	0.7161	0.7208	0.7254	0.7299
32	0	0.6848	0.6863	0.6879	0.6910	0.6941	0.6972	0.7004	0.7035	0.7067	0.7097	0.7128	0.7158
77	25	0.6848	0.6859	0.6870	0.6893	0.6915	0.6938	0.6960	0.6982	0.7004	0.7025	0.7046	0.7066
122	50	0.6847	0.6855	0.6864	0.6880	0.6896	0.6913	0.6928	0.6944	0.6959	0.6974	0.6989	0.7003
167	75	0.6846	0.6853	0.6859	0.6871	0.6882	0.6894	0.6905	0.6916	0.6927	0.6938	0.6948	0.6958
212	100	0.6845	0.6850	0.6854	0.6863	0.6871	0.6880	0.6888	0.6895	0.6903	0.6910	0.6917	0.6924

SOURCE: Sullivan (1981).

Table 13.7 Critical Flow Function Y_{CR} for Oxygen

Temperature		Total pressure $F_{TP}P_{fl}$, psia (kPa)											
°F	°C	0 (0)	72.5 (500)	145 (1000)	290 (2000)	435 (3000)	580 (4000)	725 (5000)	870 (6000)	1015 (7000)	1160 (8000)	1305 (9000)	1450 (10,000)
−58	−50	0.6846	0.6886	0.6927	0.7013	0.7104	0.7201	0.7304	0.7413	0.7528	0.7650	0.7779	0.7914
−13	−25	0.6845	0.6875	0.6905	0.6966	0.7030	0.7096	0.7164	0.7234	0.7307	0.7381	0.7457	0.7535
32	0	0.6844	0.6866	0.6889	0.6934	0.6981	0.7028	0.7076	0.7125	0.7175	0.7225	0.7276	0.7326
77	25	0.6842	0.6859	0.6876	0.6911	0.6946	0.6981	0.7016	0.7052	0.7087	0.7123	0.7159	0.7194
122	50	0.6839	0.6852	0.6865	0.6892	0.6919	0.6945	0.6972	0.6999	0.7025	0.7051	0.7078	0.7103
167	75	0.6835	0.6845	0.6855	0.6876	0.6897	0.6917	0.6938	0.6958	0.6978	0.6998	0.7017	0.7037
212	100	0.6829	0.6837	0.6845	0.6861	0.6877	0.6893	0.6909	0.6925	0.6940	0.6955	0.6970	0.6984

SOURCE: Sullivan (1981).

Table 13.8 Critical Flow Function Y_{CR} for Superheated Steam

Temperature						Total pressure $F_{TP}P_1$, psia (kPa)				
°C	°F	0 (0)	1 (7)	100 (700)	500 (3450)	1000 (6900)	2000 (13,800)	3000 (20,700)	4000 (27,600)	5000 (34,500)
93	200	0.6724	0.6727							
104	220	0.6722	0.6726							
116	240	0.6719	0.6724							
127	260	0.6719	0.6722							
138	280	0.6716	0.6721							
149	300	0.6714	0.6717							
160	320	0.6713	0.6716							
171	340	0.6711	0.6714							
182	360	0.6711	0.6711							
193	380	0.6708	0.6709							
204	400	0.6706	0.6708							
216	420	0.6703	0.6704	0.6802						
227	440	0.6701	0.6703	0.6791						
238	460	0.6699	0.6699	0.6781						
249	480	0.6698	0.6698	0.6673						
260	500	0.6695	0.6696	0.6765						
271	520	0.6693	0.6695	0.6757						

Table 13.8 Critical Flow Function Y_{CR} for Superheated Steam (Continued)

| Temperature | | Total pressure $F_{TP}Pf_1$, psia (kPa) | | | | | | | | |
°C	°F	0 (0)	1 (7)	100 (700)	500 (3450)	1000 (6900)	2000 (13,800)	3000 (20,700)	4000 (27,600)	5000 (34,500)
282	540	0.6690	0.6693	0.6750	0.7041					
293	560	0.6688	0.6690	0.6744	0.7011					
304	580	0.6686	0.6688	0.6739	0.6985					
316	600	0.6683	0.6685	0.6732	0.6957					
327	620	0.6680	0.6683	0.6727	0.6935	0.7274				
338	640	0.6678	0.6680	0.6722	0.6912	0.7201				
349	660	0.6675	0.6680	0.6717	0.6890	0.7150				
360	680	0.6673	0.6677	0.6713	0.6871	0.7111				
371	700	0.6670	0.6675	0.6708	0.6855	0.7075	0.7772			
382	720	0.6667	0.6672	0.6703	0.6838	0.7044	0.7632			
393	740	0.6665	0.6670	0.6698	0.6824	0.7011	0.7534			
404	760	0.6662	0.6668	0.6693	0.6812	0.6984	0.7455	0.8262		
416	780	0.6660	0.6665	0.6688	0.6801	0.6961	0.7385	0.8071		
427	800	0.6657	0.6664	0.6685	0.6789	0.6938	0.7316	0.8803	0.8803	
438	820	0.6654	0.6662	0.6680	0.6780	0.6918	0.7261	0.7746	0.8461	0.9879
449	840	0.6652	0.6659	0.6673	0.6770	0.6899	0.7212	0.7638	0.8252	0.9418
460	860	0.6650	0.6657	0.6670	0.6760	0.6981	0.7171	0.7549	0.8060	0.8853
471	880	0.6649	0.6655	0.6667	0.6750	0.6864	0.7132	0.7473	0.7900	0.8562

482	900	0.6646	0.6652	0.6664	0.6742	0.6848	0.7098	0.7406	0.7778	0.8839
493	920	0.6642	0.6650	0.6660	0.6734	0.6833	0.7065	0.7349	0.7679	0.8159
504	940	0.6641	0.6647	0.6657	0.6727	0.6820	0.7037	0.7300	0.7589	0.8014
516	960	0.6637	0.6646	0.6654	0.6721	0.6807	0.7011	0.7255	0.7513	0.7885
527	980	0.6636	0.6644	0.6650	0.6714	0.6780	0.6988	0.7212	0.7452	0.7775
538	1000	0.6633	0.6641	0.6649	0.6708	0.6788	0.6962	0.7168	0.7406	0.7691
566	1050	0.6626	0.6636	0.6640	0.6695	0.6765	0.6915	0.7088	0.7304	0.7503
593	1100	0.6621	0.6629	0.6639	0.6685	0.6745	0.6876	0.7023	0.7180	0.7379
621	1150	0.6615	0.6624	0.6633	0.6675	0.6729	0.6843	0.6971	0.7126	0.7266
649	1200	0.6608	0.6618	0.6628	0.6665	0.6714	0.6815	0.6925	0.7047	0.7171
677	1250	0.6601	0.6611	0.6623	0.6657	0.6698	0.6789	0.6839	0.6990	0.7098
704	1300	0.6597	0.6605	0.6616	0.6647	0.6683	0.6765	0.6855	0.6946	0.7041
732	1350	0.6592	0.6598	0.6608	0.6637	0.6668	0.6742	0.6820	0.6904	0.6990
760	1400	0.6587	0.6592	0.6600	0.6626	0.6655	0.6719	0.6791	0.6866	0.6944
788	1450	0.6582	0.6585	0.6592	0.6616	0.6641	0.6698	0.6760	0.6832	0.6904
816	1500	0.6577	0.6579	0.6582	0.6605	0.6628	0.6677	0.6745	0.6797	0.6868
843	1550	0.6572	0.6570	0.6572	0.6593	0.6613	0.6657	0.6732	0.6765	0.6830
871	1600	0.6567	0.6564	0.6561	0.6582	0.6600	0.6637	0.6678	0.6734	0.6794

SOURCE: Murdock and Bauman (1963).

Johnson reduced the critical flow function to two empirical linear equations for components of no more than four carbon atoms. First a composition factor is determined as

$$f = x_{C2H6} + x_{CO2} - \tfrac{1}{2}x_{N2} + 2x_{C3H8} + 3x_{C4H10} \tag{13.16}$$

[In Table 13.9 are the mole-fraction composition limits for Eq. (13.16). Sullivan (1981) suggests that Eq. (13.16) *not* be extrapolated outside these values and that f be limited to the range 0 to 0.2.] Next, from the composition factor f, the isentropic-exponent relationship of Eq. (13.14) is calculated as

$$\sqrt{Z_{f1}} \; Y_{CR} = \left[k \left(\frac{2}{k+1} \right)^{(k+1)/(k-1)} \right]^{1/2} = a_c f + b_c \tag{13.17}$$

where a_c and b_c are functions of total pressure and temperature. Values are listed in Table 13.10. [In Eq. (13.17), b_c represents the contribution of methane, and $a_c f$ the contribution of other components.] The square root of the compressibility factor in Eq. (13.17) is determined as

$$\sqrt{Z_{f1}} = a_z f + b_z \tag{13.18}$$

where a_z and b_z are functions of total pressure and temperature. These are listed in Table 13.11. Finally, substitution of Eq. (13.18) into Eq. (13.17) and rearranging give the critical flow function as

$$Y_{CR} = \frac{a_c f + b_c}{a_z f + b_z} \tag{13.19}$$

Sullivan, in a private communication, proposed the following accuracy formula for Eq. (13.19):

$$(ACC)_{YCR} = 0.04 + 0.00021 F_{TP} p_{f1} + f \tag{13.20}$$

Table 13.9 **Permissible Mole Fractions for Johnson's Methane–Natural Gas Mixtures**

Substance	Mole fraction x
Methane (CH$_4$)	0.840–1.000
Ethane (C$_2$H$_6$)	0–0.11
Propane (C$_3$H$_8$)	0–0.020
2-Methyl propane (C$_3$H$_8$)	0–0.004
Butane (C$_4$H$_{10}$)	0–0.004
Nitrogen (N$_2$)	0–0.023
Carbon dioxide (CO$_2$)	0–0.017

SOURCE: Sullivan (1981).

Table 13.10   Values of a_c and b_c for Methane and Natural Gas

Temperature		— Total pressure F_{TPPf1}, psia (kPa)										
K	°R	0 (0)	100 (690)	200 (1380)	300 (2070)	400 (2760)	500 (3450)	600 (4140)	700 (4825)	800 (5120)	900 (6200)	1000 (6900)
						VALUES OF a_c						
250	450	−0.0265	−0.0297	−0.0330	−0.0365	−0.0399	−0.0430	−0.0452	−0.0458	−0.0435	−0.0361	−0.0206
256	460	−0.0272	−0.0302	−0.0334	−0.0366	−0.0398	−0.0426	−0.0448	−0.0457	−0.0443	−0.0394	−0.0290
261	470	−0.0279	−0.0308	−0.0338	−0.0368	−0.0398	−0.0424	−0.0445	−0.0455	−0.0448	−0.0414	−0.0343
267	480	−0.0285	−0.0313	−0.0342	−0.0371	−0.0398	−0.0423	−0.0442	−0.0453	−0.0451	−0.0428	−0.0376
272	490	−0.0292	−0.0318	−0.0346	−0.0373	−0.0399	−0.0422	−0.0441	−0.0452	−0.0452	−0.0437	−0.0399
278	500	−0.0298	−0.0324	−0.0350	−0.0375	−0.0400	−0.0422	−0.0440	−0.0451	−0.0454	−0.0443	−0.0416
283	510	−0.0304	−0.0329	−0.0353	−0.0378	−0.0401	−0.0422	−0.0439	−0.0451	−0.0455	−0.0448	−0.0427
289	520	−0.0310	−0.0333	−0.0357	−0.0380	−0.0402	−0.0422	−0.0439	−0.0450	−0.0455	−0.0451	−0.0436
294	530	−0.0315	−0.0338	−0.0361	−0.0383	−0.0404	−0.0423	−0.0439	−0.0450	−0.0456	−0.0454	−0.0443
300	540	−0.0321	−0.0343	−0.0365	−0.0386	−0.0406	−0.0424	−0.0439	−0.0450	−0.0456	−0.0456	−0.0448
306	550	−0.0326	−0.0347	−0.0368	−0.0388	−0.0407	−0.0425	−0.0439	−0.0450	−0.0457	−0.0458	−0.0452
311	560	−0.0331	−0.0351	−0.0371	−0.0391	−0.0409	−0.0426	−0.0440	−0.0451	−0.0458	−0.0459	−0.0455
317	570	−0.0335	−0.0355	−0.0375	−0.0393	−0.0411	−0.0427	−0.0440	−0.0451	−0.0458	−0.0461	−0.0458
322	580	−0.0340	−0.0359	−0.0378	−0.0396	−0.0412	−0.0428	−0.0441	−0.0451	−0.0458	−0.0462	−0.0460
328	590	−0.0344	−0.0362	−0.0380	−0.0398	−0.0414	−0.0428	−0.0441	−0.0451	−0.0459	−0.0462	−0.0462
333	600	−0.0348	−0.0366	−0.0383	−0.0400	−0.0415	−0.0429	−0.0441	−0.0451	−0.0459	−0.0463	−0.0463
338	610	−0.0351	−0.0368	−0.0385	−0.0401	−0.0416	−0.0430	−0.0442	−0.0451	−0.0459	−0.0463	−0.0464

Table 13.10 Values of a_c and b_c for Methane and Natural Gas (Continued)

| Temperature | | Total pressure $F_{TP}p_{t_1}$, psia (kPa) | | | | | | | | | | |
K	°R	0 (0)	100 (690)	200 (1380)	300 (2070)	400 (2760)	500 (3450)	600 (4140)	700 (4825)	800 (5120)	900 (6200)	1000 (6900)
344	620	−0.0354	−0.0371	−0.0387	−0.0403	−0.0417	−0.0430	−0.0442	−0.0451	−0.0458	−0.0463	−0.0464
350	630	−0.0357	−0.0373	−0.0389	−0.0404	−0.0418	−0.0430	−0.0442	−0.0451	−0.0458	−0.0462	−0.0464
356	640	−0.0360	−0.0375	−0.0390	−0.0405	−0.0418	−0.0430	−0.0441	−0.0450	−0.0457	−0.0462	−0.0464
361	650	−0.0362	−0.0377	−0.0392	−0.0406	−0.0418	−0.0430	−0.0441	−0.0449	−0.0456	−0.0461	−0.0463
367	660	−0.0363	−0.0378	−0.0392	−0.0406	−0.0418	−0.0430	−0.0440	−0.0448	−0.0455	−0.0460	−0.0462
372	670	−0.0365	−0.0379	−0.0393	−0.0406	−0.0418	−0.0429	−0.0439	−0.0447	−0.0453	−0.0458	−0.0461
378	680	−0.0366	−0.0380	−0.0393	−0.0406	−0.0417	−0.0428	−0.0437	−0.0445	−0.0452	−0.0456	−0.0459
383	690	−0.0367	−0.0380	−0.0393	−0.0405	−0.0416	−0.0427	−0.0436	−0.0443	−0.0450	−0.0454	−0.0457
389	700	−0.0367	−0.0380	−0.0392	−0.0404	−0.0415	−0.0425	−0.0434	−0.0441	−0.0447	−0.0452	−0.0455

VALUES OF b_c

K	°R	0 (0)	100 (690)	200 (1380)	300 (2070)	400 (2760)	500 (3450)	600 (4140)	700 (4825)	800 (5120)	900 (6200)	1000 (6900)
250	450	0.6719	0.6715	0.6713	0.6712	0.6713	0.6717	0.6723	0.6733	0.6747	0.6767	0.6791
256	460	0.6717	0.6714	0.6712	0.6712	0.6713	0.6717	0.6724	0.6734	0.6748	0.6766	0.6789
261	470	0.6714	0.6712	0.6711	0.6711	0.6714	0.6718	0.6725	0.6734	0.6747	0.6764	0.6786
267	480	0.6712	0.6710	0.6710	0.6711	0.6713	0.6718	0.6725	0.6734	0.6747	0.6763	0.6783
272	490	0.6709	0.6708	0.6708	0.6709	0.6712	0.6717	0.6724	0.6734	0.6746	0.6761	0.6780

278	500	0.6707	0.6706	0.6706	0.6708	0.6711	0.6716	0.6723	0.6733	0.6745	0.6759	0.6777
283	510	0.6704	0.6703	0.6704	0.6706	0.6710	0.6715	0.6722	0.6731	0.6743	0.6757	0.6774
289	520	0.6701	0.6701	0.6702	0.6704	0.6708	0.6714	0.6721	0.6730	0.6741	0.6755	0.6771
294	530	0.6698	0.6698	0.6699	0.6702	0.6706	0.6712	0.6719	0.6728	0.6739	0.6752	0.6767
300	540	0.6694	0.6695	0.6697	0.6700	0.6704	0.6709	0.6717	0.6726	0.6736	0.6749	0.6764
306	550	0.6691	0.6692	0.6694	0.6697	0.6701	0.6707	0.6714	0.6723	0.6734	0.6746	0.6760
311	560	0.6687	0.6689	0.6691	0.6694	0.6699	0.6704	0.6712	0.6720	0.6731	0.6743	0.6756
317	570	0.6684	0.6685	0.6687	0.6691	0.6696	0.6701	0.6709	0.6717	0.6727	0.6739	0.6753
322	580	0.6680	0.6681	0.6684	0.6688	0.6692	0.6698	0.6706	0.6714	0.6724	0.6735	0.6748
328	590	0.6676	0.6678	0.6680	0.6684	0.6689	0.6695	0.6702	0.6711	0.6720	0.6732	0.6744
333	600	0.6672	0.6674	0.6677	0.6681	0.6686	0.6692	0.6699	0.6707	0.6717	0.6728	0.6740
338	610	0.6668	0.6670	0.6673	0.6677	0.6682	0.6688	0.6695	0.6703	0.6713	0.6723	0.6735
344	620	0.6663	0.6666	0.6669	0.6673	0.6678	0.6684	0.6691	0.6699	0.6709	0.6719	0.6731
350	630	0.6659	0.6662	0.6665	0.6669	0.6674	0.6680	0.6687	0.6695	0.6705	0.6715	0.6726
356	640	0.6655	0.6657	0.6661	0.6665	0.6670	0.6676	0.6683	0.6691	0.6700	0.6710	0.6721
361	650	0.6650	0.6653	0.6656	0.6661	0.6666	0.6672	0.6679	0.6687	0.6696	0.6706	0.6717
367	660	0.6646	0.6649	0.6652	0.6657	0.6662	0.6668	0.6675	0.6683	0.6691	0.6701	0.6712
372	670	0.6641	0.6644	0.6648	0.6652	0.6658	0.6664	0.6671	0.6678	0.6687	0.6696	0.6707
378	680	0.6637	0.6640	0.6643	0.6648	0.6653	0.6659	0.6666	0.6674	0.6682	0.6692	0.6702
383	690	0.6632	0.6635	0.6639	0.6644	0.6649	0.6655	0.6662	0.6669	0.6678	0.6687	0.6697
389	700	0.6627	0.6631	0.6635	0.6639	0.6644	0.6651	0.6657	0.6665	0.6673	0.6682	0.6692

13-20

Table 13.11   Values of a_z and b_z for Methane and Natural Gas

| Temperature | | Total pressure F_{TPP}, psia (kPa) | | | | | | | | | | |
K	°R	0 (0)	100 (690)	200 (1380)	300 (2070)	400 (2760)	500 (3450)	600 (4140)	700 (4825)	800 (5120)	900 (6200)	1000 (6900)
						VALUES OF a_z						
250	450	0	−0.0252	−0.0530	−0.0837	−0.1179	−0.1561	−0.1988	−0.2464	−0.2991	−0.3564	−0.4162
256	460	0	−0.0234	−0.0490	−0.0770	−0.1078	−0.1417	−0.1790	−0.2199	−0.2644	−0.3118	−0.3610
261	470	0	−0.0218	−0.0454	−0.0710	−0.0989	−0.1293	−0.1622	−0.1978	−0.2360	−0.2762	−0.3175
267	480	0	−0.0203	−0.0422	−0.0657	−0.0911	−0.1184	−0.1478	0.1791	−0.2123	−0.2469	−0.2823
272	490	0	−0.0190	−0.0393	−0.0610	−0.0842	−0.1089	−0.1353	−0.1631	−0.1923	−0.2225	−0.2532
278	500	0	−0.0178	−0.0366	−0.0567	−0.0780	−0.1005	−0.1243	−0.1493	0.1752	−0.2018	−0.2288
283	510	0	−0.0166	−0.0342	−0.0528	−0.0724	−0.0931	−0.1147	−0.1371	−0.1604	−0.1841	−0.2079
289	520	0	−0.0156	−0.0321	−0.0493	−0.0675	−0.0864	−0.1061	−0.1265	−0.1474	−0.1687	−0.1900
294	530	0	−0.0147	−0.0301	−0.0462	−0.0630	−0.0804	−0.0984	−0.1170	−0.1360	−0.1552	−0.1743
300	540	0	−0.0138	−0.0283	−0.0433	−0.0589	−0.0750	−0.0916	−0.1086	−0.1259	−0.1433	−0.1606
306	550	0	−0.0130	−0.0266	−0.0406	−0.0551	−0.0701	−0.0854	−0.1010	−0.1169	−0.1327	−0.1485
311	560	0	−0.0123	−0.0251	−0.0382	−0.0518	−0.0656	−0.0798	−0.0942	−0.1088	−0.1233	−0.1377
317	570	0	−0.0116	−0.0236	−0.0360	−0.0487	−0.0616	−0.0748	−0.0881	−0.1015	−0.1149	−0.1281
322	580	0	−0.0110	−0.0223	−0.0339	−0.0458	−0.0579	−0.0701	−0.0825	−0.0949	−0.1072	−0.1194
328	590	0	−0.0104	−0.0211	−0.0320	−0.0432	−0.0545	−0.0659	−0.0774	−0.0889	−0.1003	−0.1116

VALUES OF b_z

333	−0.1045	−0.0941	−0.0835	−0.0728	−0.0621	−0.0514	−0.0408	−0.0303	−0.0200	−0.0099	0	600
338	−0.0980	−0.0883	−0.0785	−0.0685	−0.0585	−0.0485	−0.0385	−0.0287	−0.0190	−0.0094	0	610
344	−0.0921	−0.0831	−0.0739	−0.0646	−0.0552	−0.0458	−0.0365	−0.0272	−0.0180	−0.0089	0	620
350	−0.0867	−0.0783	−0.0697	−0.0610	−0.0522	−0.0434	−0.0346	−0.0258	−0.0171	−0.0085	0	630
356	−0.0818	−0.0739	−0.0658	−0.0577	−0.0494	−0.0411	−0.0328	−0.0245	−0.0162	−0.0081	0	640
361	−0.0772	−0.0698	−0.0623	−0.0546	−0.0468	−0.0390	−0.0311	−0.0233	−0.0154	−0.0077	0	650
367	−0.0730	−0.0660	−0.0589	−0.0517	−0.0444	−0.0370	−0.0296	−0.0221	−0.0147	−0.0073	0	660
372	−0.0691	−0.0625	−0.0559	−0.0491	−0.0422	−0.0352	−0.0281	−0.0211	−0.0140	−0.0070	0	670
378	−0.0654	−0.0593	−0.0530	−0.0466	−0.0401	−0.0334	−0.0268	−0.0201	−0.0134	−0.0067	0	680
383	−0.0621	−0.0563	−0.0503	−0.0443	−0.0381	−0.0318	−0.0255	−0.0192	−0.0128	−0.0064	0	690
389	−0.0589	−0.0535	−0.0479	−0.0421	−0.0363	−0.0303	−0.0243	−0.0183	−0.0122	−0.0061	0	700

250	0.8837	0.8955	0.9075	0.9195	0.9315	0.9434	0.9552	0.9667	0.9880	0.9891	1.0000	450
256	0.8938	0.9044	0.9152	0.9261	0.9370	0.9478	0.9585	0.9692	0.9796	0.9899	1.0000	460
261	0.9028	0.9124	0.9221	0.9320	0.9419	0.9518	0.9616	0.9714	0.9810	0.9906	1.0000	470
267	0.9109	0.9195	0.9283	0.9373	0.9463	0.9553	0.9644	0.9734	0.9824	0.9912	1.0000	480
272	0.9181	0.9259	0.9339	0.9420	0.9503	0.9586	0.9659	0.9753	0.9836	0.9918	1.0000	490
278	0.9245	0.9316	0.9389	0.9464	0.9539	0.9616	0.9693	0.9770	0.9847	0.9924	1.0000	500
283	0.9304	0.9369	0.9435	0.9503	0.9573	0.9643	0.9714	0.9785	0.9857	0.9929	1.0000	510
289	0.9357	0.9416	0.9477	0.9539	0.9603	0.9668	0.9734	0.9800	0.9866	0.9933	1.0000	520
294	0.9406	0.9459	0.9515	0.9572	0.9631	0.9691	0.9752	0.9813	0.9875	0.9937	1.0000	530
300	0.9450	0.9499	0.9550	0.9603	0.9657	0.9712	0.9768	0.9826	0.9883	0.9941	1.0000	540

Table 13.11 Values of a_z and b_z for Methane and Natural Gas (Continued)

Temperature		Total pressure F_{TPpf}, psia (kPa)										
K	°R	0 (0)	100 (690)	200 (1380)	300 (2070)	400 (2760)	500 (3450)	600 (4140)	700 (4825)	800 (5120)	900 (6200)	1000 (6900)
306	550	1.0000	0.9945	0.9891	0.9837	0.9784	0.9732	0.9681	0.9631	0.9582	0.9535	0.9490
311	560	1.0000	0.9949	0.9898	0.9848	0.9798	0.9750	0.9702	0.9656	0.9612	0.9569	0.9527
317	570	1.0000	0.9952	0.9904	0.9857	0.9811	0.9766	0.9723	0.9680	0.9639	0.9599	0.9562
322	580	1.0000	0.9955	0.9910	0.9867	0.9824	0.9782	0.9741	0.9702	0.9664	0.9628	0.9593
328	590	1.0000	0.9958	0.9916	0.9875	0.9835	0.9797	0.9759	0.9723	0.9688	0.9654	0.9622
333	600	1.0000	0.9960	0.9921	0.9883	0.9846	0.9810	0.9775	0.9741	0.9709	0.9678	0.9649
338	610	1.0000	0.9963	0.9926	0.9891	0.9856	0.9823	0.9790	0.9759	0.9729	0.9701	0.9674
344	620	1.0000	0.9965	0.9931	0.9898	0.9865	0.9834	0.9804	0.9776	0.9748	0.9722	0.9698
350	630	1.0000	0.9967	0.9935	0.9904	0.9874	0.9845	0.9817	0.9791	0.9766	0.9742	0.9719
356	640	1.0000	0.9969	0.9939	0.9910	0.9882	0.9856	0.9830	0.9805	0.9782	0.9760	0.9740
361	650	1.0000	0.9971	0.9943	0.9916	0.9890	0.9865	0.9841	0.9819	0.9797	0.9777	0.9758
367	660	1.0000	0.9973	0.9947	0.9922	0.9897	0.9874	0.9852	0.9831	0.9812	0.9793	0.9776
372	670	1.0000	0.9975	0.9950	0.9927	0.9904	0.9883	0.9862	0.9843	0.9825	0.9808	0.9793
378	680	1.0000	0.9976	0.9953	0.9931	0.9911	0.9891	0.9872	0.9854	0.9838	0.9822	0.9808
383	690	1.0000	0.9978	0.9956	0.9936	0.9917	0.9898	0.9881	0.9865	0.9849	0.9835	0.9823
389	700	1.0000	0.9979	0.9959	0.9940	0.9922	0.9905	0.9889	0.9874	0.9861	0.9848	0.9836

for U.S. units, and

$$(ACC)_{YCR} = 0.04 + 0.00003 F_{TP} p_{f1}^* + f \qquad (13.21)$$

for SI units. Equations (13.20) and (13.21) yield an accuracy of ± 0.25 percent at 1000 psia (7000 kPa) for methane ($f = 0$) and, for a maximum composition factor of 0.2, yield the value ± 0.45 percent.

Discharge Coefficient

The venturi nozzle, ASME long-radius nozzle, and square-edged orifice are commonly used critical flowmeters; the accepted standard is the venturi nozzle. The discharge coefficient depends on the inlet contour (sharp or contoured), throat length, and bore Reynolds number (Fig. 13.2).

ISO Critical Venturi Nozzle Toroidal and cylindrical-throat ISO (1982) critical venturi nozzles are shown in Fig. 13.3. In Fig. 13.4 are the available data on these nozzles (Hillbrath, 1981), plotted against throat Reynolds number R_d. This data is from the work of Smith and Matz (1973), Arnberg (1962), and Brain and Macdonald (1977). The general form of the prediction equation is

$$C = C_\infty - \frac{b}{R_d^n} \qquad (13.22)$$

Constants for this equation and uncalibrated accuracy values are given in Table 13.12.

Table 13.12 **Discharge-Coefficient-Equation Constants for Critical Flowmeters**

Device	Constants†			Uncalibrated accuracy, %
	a	b	n	
ISO toroidal throat ($\beta < 0.25$)				
$\quad 10^5 < R_d < 10^7$	0.99354	1.525	0.5	± 0.5
ISO cylindrical throat and ASME long-radius nozzle‡ ($\beta < 0.25$)				
$\quad 10^4 < R_d < 4 \times 10^5$	1.0000	7.21	0.5	± 0.5
$\quad 4 \times 10^5 < R_d < 2.8 \times 10^6$	0.9886	0	0	± 0.5
$\quad 2.8 \times 10^6 < R_d < 2 \times 10^7$	1.000	0.222	0.2	± 0.5
Square-edged orifice ($\beta < 0.5$; plate thickness, d to $6d$)				
$\quad R_d > 10^5$	0.83932	0	0	± 1.6

†$C = a - b/R_d^n$.

‡Add ± 0.25 percent to uncalibrated accuracy.

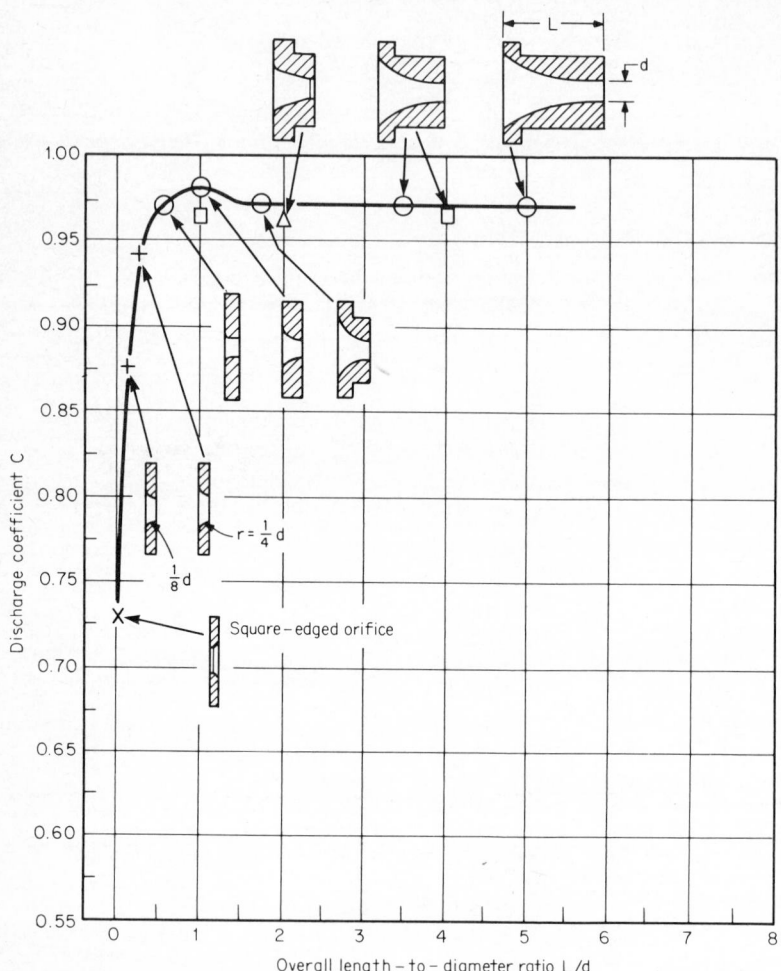

Figure 13.2 Discharge coefficient versus nozzle length-to-bore ratio.

ASME Long-Radius Nozzle The discharge-coefficient data for the ASME long-radius nozzle presented by Szaniszlo (1975) is within the accuracy envelope for the ISO cylindrical-throat venturi nozzle. Although only a single value of β (0.18) was tested, other investigators have found agreement within the accuracy specified for the subsonic flow-coefficient equation given in Table 9.1. While no standard exists for this nozzle, it is suggested that the ISO prediction equation be used with the accuracy increased to ± 0.75 percent.

Square-Edged Orifice Cunningham (1951) first drew attention to the fact that choked flow will not occur across a standard, thin, square-edged orifice. Recently Ward-Smith (1979) showed theoretically that, for choking to occur, the

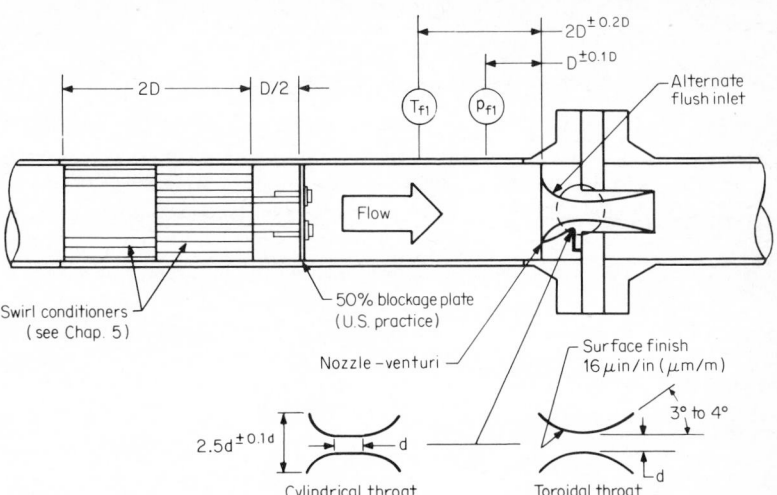

Figure 13.3   ISO critical venturi nozzle.

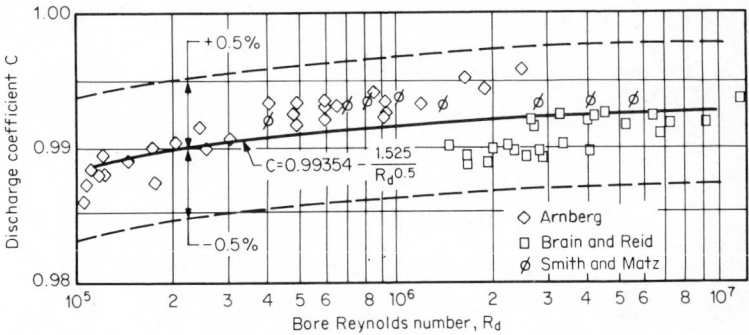

Figure 13.4   Discharge coefficient for venturi nozzle.

discharge coefficient must approach unity. His highest measured value was approximately 0.9, and during his work the orifice remained unchoked down to a choking pressure ratio of 0.2. However, choked flow is achieved by increasing the plate thickness. For a sharp edge and for ratios of plate thickness to bore diameter between 1 and 6, the discharge coefficient is

$$C = 0.83932 \qquad (13.23)$$

This value, the average of the data given in the Ward-Smith paper, has an estimated accuracy of ± 1.6 percent.

Choking Pressure Ratio

The downstream pressure p_{f3} at which choked flow first occurs depends on the exit conditions. For flow without a diffuser, the choking pressure ratio is developed on the assumption of steady isentropic flow as

$$\frac{p_{f3}}{F_{TP}p_{f1}} \leq \left(\frac{2}{k_i + 1}\right)^{k_i/(k_i-1)} \tag{13.24}$$

where k_i is the isentropic exponent of an ideal gas (see Chap. 2). This equation applies to ASME long-radius nozzles and to *thick* square-edged orifices.

For a venturi nozzle, Hillbrath et al. (1975) have shown that the choking pressure ratio may vary from 0.8 to 0.95, depending on the diffuser area ratio. A diffuser that is four throat diameters in length with a divergence half-angle of 4° (included angle of 8°) produces a choking pressure ratio of approximately 0.9; the ratio is 0.8 for a one-diameter length. For longer diffuser lengths, critical flow can be produced at a ratio of 0.95.

EXAMPLE 13.1

A toroidal-throat critical venturi nozzle is to be used to measure natural gas flowing at 7,500,000 standard ft^3/h (212,380 standard m^3/h) in a 12-in [D = 11.370 in (290 mm)] line. For a design pressure of 740 psia (5100 kPa) and temperature of 75°F (24°C), what is the nozzle bore if pressure and temperature measurements are made?

The gas composition is as follows:

Substance	Mole fraction x_i	Molecular weight M_{wi}†	$x_i M_{wi}$
Methane	0.9150	16.043	14.6793
Ethane	0.0370	30.070	1.1126
Propane	0.0090	44.097	0.3969
Butane	0.0010	58.124	0.0581
Nitrogen	0.0110	28.013	0.3081
Carbon dioxide	0.0270	44.010	1.1883
			17.7433

†From Table D.1.

From Eq. (2.44),

$$(M_w)_{\text{mix}} = \Sigma x_i M_{wi} = 17.7433 \text{ lb}_m/(\text{lb}_m \cdot \text{mole})$$

The mixture's specific gravity is, by Eq. (2.7),

$$G = \frac{(M_w)_{\text{mix}}}{(M_w)_{\text{air}}} = \frac{17.7433}{28.9625} = 0.61263$$

The flow-rate equation for pressure and temperature measurements (pvT equation) is, from Eq. (f) of Table 13.1,

$$q_{\text{SCFH}} = 28{,}712.44 C d^2 Z_b Y_{\text{CR}} \sqrt{\frac{1}{T_{f1}G}} \, F_{TP}p_{f1}$$

Rearrangement to solve for bore yields

$$d = \left(\frac{q_{\text{SCFH}} \sqrt{T_{f1}G}}{28{,}712.44 C Z_b Y_{\text{CR}} F_{TP} p_{f1}} \right)^{1/2}$$

From Eq. (13.18), $Z_b = (a_z f + b_z)^2$. At base conditions (Table 13.11), for $T_b = 518.67°R$ and $p_b = 14.69595$ psia,

$$a_z = -0.00231 \qquad b_z = 0.99901$$

and, by Eq. (13.16),

$$f = x_{\text{C2H6}} + x_{\text{CO2}} - \tfrac{1}{2} x_{\text{N2}} + 2 x_{\text{C3H8}} + 3 x_{\text{C4H10}}$$
$$= 0.0370 + 0.0270 - (\tfrac{1}{2})(0.0110) + (2)(0.0090) + (3)(0.0010) = 0.0795$$

Then

$$Z_b = [(-0.00231)(0.0795) + 0.99901]^2 = 0.99765$$

From Tables 13.10 and 13.11, for $p_{f1} = 740$ psia and $T_{f1} = 459.67 + 75 = 534.67°R$

$$a_c = -0.0452 \qquad b_c = 0.6731$$
$$a_z = -0.1201 \qquad b_z = 0.95655$$

The critical flow function Y_{CR} is then, by Eq. (13.19),

$$Y_{\text{CR}} = \frac{a_c f + b_c}{a_z f + b_z} = \frac{(-0.0452)(0.0795) + 0.6731}{(-0.1201)(0.0795) + 0.95655} = 0.70697$$

Substitution into the diameter equation yields

$$d = \left[\frac{7{,}500{,}000 \sqrt{(534.67)(0.61263)}}{(28{,}712.44)(0.99765)(0.70697)(740)} \right]^{1/2} \left(\frac{1}{CF_{TP}} \right)^{1/2} = 3.0096 \left(\frac{1}{CF_{TP}} \right)^{1/2}$$

The initial assumptions for the values of the discharge coefficient and the total-pressure correction are $C_0 = 0.99354$ and $F_{PT} = 1.0$. Then

$$d = 3.0096 \left[\frac{1}{(0.99354)(1.0)} \right]^{1/2} = 3.0194 \text{ in}$$

The discharge coefficient is given by Eq. (13.22) and Table 13.12 as

$$C = 0.99354 - \frac{1.525}{R_d^{0.5}}$$

From Table C.4 with $d = D$, the bore Reynolds number is

$$R_d = \frac{0.4830 G q_{\text{SCFH}}}{Z_b d \mu_{\text{cP}}} = \frac{(0.4830)(0.6126)(7{,}500{,}000)}{(0.9977)(3.019)(0.0142)} = 51{,}884{,}000$$

where, from Fig. H.16, $\mu_{\text{cP}} = 0.0142$. Substitution then gives

$$C = 0.99354 - \frac{1.525}{(51{,}884{,}000)^{0.5}} = 0.99333$$

The total-pressure correction for $\beta = d/D = 3.0194/11.370 = 0.2655$ is, by Eq. (13.9),

$$F_{TP} = 1 + \frac{k_i}{2}\left(\frac{2}{k_i + 1}\right)^{(k_i+1)/(k_i-1)}\beta^4$$

$$= 1 + \frac{1.32}{2}\left(\frac{2}{1.32 + 1}\right)^{(1.32+1)/(1.32-1)}(0.2655)^4 = 1.0011$$

where, from Fig. I.9, $k = 1.32$.

Substitution into the bore equation now gives

$$d = 3.0096\left(\frac{1}{CF_{TP}}\right)^{1/2} = 3.0096\left[\frac{1}{(0.99333)(1.0011)}\right]^{1/2} = 3.018 \text{ in}$$

Further iterations are unwarranted, since neither F_{TP} nor C is highly sensitive to bore changes.

EXAMPLE 13.2

Superheated steam at 500 psia (3450 kPa) and 660°F (350°C) is flowing in a 4-in (100-mm) Schedule-80 pipe through a 2-in (50-mm) thick square-edged orifice that has a 1.000-in (25-mm) bore. If the downstream pressure is 200 psia (1380 kPa), what is the mass flow?

For a thick orifice the choking pressure ratio is given by Eq. (13.24) as

$$\frac{p_{f3}}{F_{TP}p_{f1}} \le \left(\frac{2}{k_i + 1}\right)^{k_i/(k_i-1)}$$

From Eq. (2.149), with $(C_p)_i = 8.73$ from Table I.2,

$$k_i = \frac{(C_p)_i}{(C_p)_i - 1.986} = \frac{8.73}{8.73 - 1.986} = 1.295$$

Then, by Eq. (13.9), with $\beta = d/D = 1.000/3.826 = 0.2614$ (where D was found from Table B.1),

$$F_{TP} = 1 + \frac{k_i}{2}\left(\frac{2}{k_i + 1}\right)^{(k_i+1)/(k_i-1)}\beta^4$$

$$= 1 + \frac{1.295}{2}\left(\frac{2}{1.295 + 1}\right)^{(1.295+1)/(1.295-1)}(0.2614)^4 = 1.0010$$

and

$$p_{f3} \le \left(\frac{2}{k_i + 1}\right)^{k_i/(k_i-1)}F_{TP}p_{f1} = \left(\frac{2}{1.295 + 1}\right)^{1.295/(1.295-1)}(1.0010)(500) = 274 \text{ psia}$$

Since $p_{f3} = 200 \le 274$, the orifice is choked.

The mass flow-rate equation, from Eq. (b) of Table 13.1, is

$$q_{PPH} = 2195.591 C d^2 Y_{CR} \sqrt{\frac{G}{T_{f1}}} F_{TP} p_{f1}$$

For a thick square-edged orifice, the discharge coefficient is, by Eq. (13.23), $C = 0.83932$. Other required data are as follows: From Table 13.8, $Y_{CR} = 0.6890$. From Eq. (2.7),

$$G = \frac{(M_w)_{H2O}}{M_{w,air}} = \frac{18.0153}{28.9625} = 0.6220$$

where, from Table D.1, $(M_w)_{H2O} = 18.0153$ and $M_{w,air} = 28.9625$. Finally, from Eq. (3.38),

$$T_{f1} = T_F + 459.67 = 660 + 459.67 = 1119.67°R$$

and substitution into the mass flow equation yields

$$q_{PPH} = (2195.591)(0.83932)(1.0)^2(0.6890) \sqrt{\frac{0.6220}{1119.67}} (1.0010)(500)$$

$$= 14,978 \text{ lb}_m/h$$

LIQUIDS

In some applications it is important to limit the flow rate to a maximum value if a valve, pipe, or other system component fails downstream. Liquid *choked* flow occurs if a cavitation barrier exists within the flowmeter. Downstream pressure changes cannot be transmitted through the barrier, and only an upstream pressure change can increase the flow rate.

As shown in Chap. 5, the cavitation zone is more clearly defined in a venturi, whereas for an orifice the location may change. For these reasons, *cavitating* choked venturis are usually selected for flow-limiting use. Choked, thin, sharp-edged orifices may be used, but they are less accurate and not as structurally suited should cavitation persist for long periods. A square-edged or slightly rounded, *thick* inlet orifice is suggested as an inexpensive alternative.

Engineering Equations

The sizing and flow-rate equations developed in Chap. 9 may be used after substitution, for the differential pressure, of

$$h_w = 27.72976(p_{f1} - p_v) \tag{13.25}$$

for U.S. units, and

$$\Delta p^* = p_{f1}^* - p_v^* \tag{13.26}$$

for SI units, where p_{f1}, p_{f1}^*, p_v and p_v^* are in absolute pressure units. With these substitutions, sizing and flow-rate calculations are performed in the manner out-

Table 13.13 **Recommended Designs and Coefficients for Liquid Choked Flowmeters**

Device		Discharge coefficient C	Source
Square-edged orifice (refer to Fig. 10.4 for additional details)	 Minimum = 0.125 in (3mm)	0.6	Numachi et al. (1960)
Thick square-edged orifice (refer to Fig. 10.4 for additional details)	 6d to 20 d	0.899	Nurick (1976)
Venturi (refer to Fig. 10.64 for additional details)	3.5d < radius < 4d 	0.985	Bonnington (1961)

lined in Chap. 9. Table 13.13 gives the recommended design and discharge coefficients to be used in the equations.

EXAMPLE 13.3

Water at 200 psig (1380 kPa) is flowing at 100°F (37.8°C) in a 2-in (50-mm) Schedule-80 pipe. Size a machined-inlet stainless-steel venturi to limit the flow rate to 100 gal/min (378.5 L/min). (Assume the barometric pressure is 14.47 psia.)

From Eq. (c) of Table 9.23, the sizing equation for gallons per minute at flowing conditions is

$$S_M = \frac{\sqrt{F_p}\,\sqrt{\rho_F}\,q_v}{N_{v\rho}F_a D^2 \sqrt{h_w}}$$

Substitution of variables from Eq. (13.25),

$$h_w = 27.72976(p_{f1} - p_v)$$

yields

$$S_M = \frac{\sqrt{F_p}\,\sqrt{\rho_F}\,q_v}{N_{v\rho}F_a D^2 \sqrt{27.72976}\,\sqrt{p_{f1} - p_v}} = \frac{\sqrt{F_p}\,\sqrt{\rho_F}\,q_{\text{gpm}}}{5.2659 N_{v\rho}F_a D^2 \sqrt{p_{f1} - p_v}}$$

From Table 9.16, $N_{vp} = 44.749$, and from Eq. (9.52),

$$F_a = 1 + 2\alpha_{PE}(T_F - 68) = 1 + 2(0.0000096)(100 - 68) = 1.0006$$

where, from Table B.4, $\alpha_{PE} = 0.0000096 \text{in}/(\text{in}\cdot°\text{F})$. In addition,

- From Table B.1: $D = 1.939$ in
- By Eq. (3.15): $p_{f1} = p_G + p_B = 200 + 14.47 = 214.47$ psia
- From Fig. D.5: $p_v = 0.95$ psia at $100°\text{F}$
- From Eq. (2.75) or Table E.6: $\rho_F = 61.994 \text{ lb}_\text{m}/\text{ft}^3$

For negligible liquid compressibility, $F_p = 1.0$, and substitution gives

$$S_M = \frac{\sqrt{1.0} \ \sqrt{61.994} \ 100}{(5.2659)(44.759)(1.0006)(1.939)^2 \sqrt{214.47 - 0.95}} = 0.06077$$

The beta ratio is calculated from Eq. (9.78), with $Y_1 = 1.0$ for liquids and (from Table 9.1) $C = 0.995$ for a venturi, as

$$\beta = \left[1 + \left(\frac{CY_1}{S_M}\right)^2\right]^{-1/4} = \left\{1 + \left[\frac{(0.995)(1.0)}{0.06077}\right]^2\right\}^{-1/4} = 0.2469$$

The venturi bore is then

$$d = \beta D = (0.2469)(1.939) = 0.4788 \text{ in}$$

EXAMPLE 13.4

In Example 13.3 the pressure increases to 220 psig (1520 kPa). Determine the flow rate if the temperature remains at $100°\text{F}$ ($37.8°\text{C}$). From Eq. (c) of Table 9.36, the flow-rate equation is

$$q_\text{gpm} = N_{vp}S_M F_a F_{RD} D^2 \frac{\sqrt{h_w}}{\sqrt{F_p} \ \sqrt{\rho_F}}$$

Substitution of the values given in Example 13.3 with $F_{RD} = 1.0$ and $F_p = 1.0$ yields

$$q_\text{gpm} = (5.2659)(44.749)(0.06078)(1.0006)(1.0)(1.939)^2 \frac{\sqrt{p_{f1} - 0.95}}{\sqrt{61.995}}$$

$$= 6.843 \ \sqrt{p_{f1} - 0.95} = 6.843 \ \sqrt{(220 + 14.47) - 0.95} = 104.6 \text{ gal/min}$$

REFERENCES

Arnberg, B. T.: "Review of Critical Flow Meters for Gas Flow Measurements," *J. Basic Eng.*, ser. D, vol. 84, pp. 447–460, 1962.

Bender, E.: "Equations of State Exactly Representing the Phase Behavior of Pure Substances," *Proc. of the 5th Symp. on Thermodynamic Properties*, pp. 227–235, ASME, New York, 1970.

Benedict, M., G. B. Webb, and L. C. Rubin: "An Empirical Equation for Thermodynamic Properties of Light Hydrocarbons and Their Mixtures," *J. Chem. Phys.*, vol. 8, no. 4, pp. 334–345, 1940.

Bonnington, S. T.: *The Influence of Cavitation on the Performance of Standard Flowmeters*, BHRA Report RR701, Bedford, U.K., 1961.

Brain, T. J. S., and L. M. Macdonald: "Evaluation of the Performance of Small-Scale Critical Flow Venturis Using the NEL Gravimetric Gas Flow Standard Test Facility," *Fluid Flow Measurement in the Mid 1970's*, pp. 103–125, Her Majesty's Stationery Office, Edinburgh, 1977.

Cunningham, R. G.: "Orifice Meters with Supercritical Flow," *Trans. ASME*, vol. 73, pp. 625–630, 1951.

Hillbrath, H. S.: "The Critical Flow Venturi—An Update," in *Flow, Its Measurement and Control in Science and Industry*, vol. 2, pp. 407–420, ISA, Research Triangle Park, N.C., 1981.

―――, W. P. Dill, and W. A. Wacker: "The Choking Pressure Ratio of a Critical Flow Venturi," *J. Eng. Ind.*, ser. B, vol. 97, no. 4, pp. 1251–1256, 1975.

ISO: *Measurement of Gas Flow by Means of Critical Flow Venturi Nozzles* (draft), ISO/TC30/SC2/WG5, ASME, New York, July, 1982.

Johnson, R. C.: "Calculations of the Flow of Natural Gas through Critical Flowmeters," *J. Basic Eng.*, ser. D, vol. 92, no. 3, pp. 580–589, 1970.

―――: *A Set of Fortran IV Routines Used to Calculate the Mass Flow Rate of Natural Gas through Nozzles*, NASA TM X-2240, 1971.

―――: *Tables of Critical Flow Functions and Thermodynamic Properties for Methane and Computational Procedures for both Methane and Natural Gas*, NASA SP-3074, 1972.

Murdock, J. W., and J. M. Bauman: "The Critical Flow Function for Superheated Steam," *ASME Winter Annual Meeting*, Paper 63-WA-19, New York, 1963.

Numachi, F., M. Yamabe, and R. Oba: "Cavitation Effect on the Discharge Coefficient of the Sharp-Edged Orifice Plate," *J. Basic Eng.*, vol. 82, pp. 1–11, 1960.

Nurick, W. H.: "Orifice Cavitation and Its Effect on Spray Mixing," *J. Fluids Eng.*, vol. 98, pp. 681–687, 1976.

Smith, R. E., and R. J. Matz: "Performance Characteristics of an 8-Inch-Diameter ASME Nozzle Operating at Compressible and Incompressible Conditions," *J. Basic Eng.*, vol. 95, no. 4, 1973.

Sullivan, D. A.: "Historical Review of Real-Fluid Isentropic Flow Models," *ASME Winter Annual Meeting*, Paper 79-WA/FM-1, New York, 1979.

―――: private communication concerning real-fluid isentropic models, 1981.

Szaniszlo, F. C.: "Experimental and Analytical Sonic Nozzle Discharge Coefficients for Reynolds Numbers up to 8×10^6," *J. Eng. Power*, vol. 97, pp. 521–526, 1975.

Ward-Smith, A. J.: "Critical Flowmetering: The Characteristics of Cylindrical Nozzles with Sharp Upstream Edges," *Int. J. Heat Fluid Flow*, vol. 1, no. 3, pp. 123–132, 1979.

14

LINEAR FLOWMETERS

In general, volumetric flowmeters whose output is not proportional to the square of the flow rate divided by the fluid density are *linear* flowmeters. Either the operating principle yields a direct linear output or, through electronics, the output is linearized to volumetric units. These meters can be grouped into two classes: pulse-frequency types and linear-scale flowmeters. Both are discussed in this chapter.

PULSE-FREQUENCY TYPE

Turbine and vortex flowmeters produce a frequency (pulse train) proportional to the pipeline velocity, and positive-displacement meters produce one pulse per unit volume. Although based on different operating principles, these *pulse-type* meters respond to flowing conditions and, therefore, the pertinent engineering equations for flowing and base volumes and for mass flow are the same. With turbine and vortex flowmeters, flow rate is commonly measured by frequency or by frequency-to-analog conversion, but this is seldom the case for the low-resolution positive-displacement meters.

The signature curves for vortex and turbine flowmeters, although different in shape, are linear over 20:1 or 30:1 flow-rate ranges, and, hence, a mean meter coefficient (K factor) is given. Positive-displacement flowmeters are usually calibrated in the desired volumetric units and, through suitable internal gearing, directly display the total volume. For turbine and vortex flowmeters, the integrated count is electronically scaled, using the K factor, to display the total volume. Through suitable electronics and computer or mechanical computations, base volume or mass flow is also displayed.

Engineering Equations

K Factor   The K factor defines the relationship between flow rate and frequency for vortex and turbine flowmeters. For liquid turbine meters, this factor is obtained by water calibration; for gas meters, by a low-pressure bell prover test. The K factor (in any volume units) is defined as

$$K_{F,v} = \frac{f_{Hz}}{q_v} = \frac{\text{pulses}}{\text{unit volume}} \tag{14.1}$$

where q_v denotes the volume flow rate (in gallons, cubic meters, etc., per second), and f_{Hz} is the frequency in pulses per second. The K factor then has the units of pulses per unit volume. For both liquids and gases (vapors), the K factor is obtained over a flow-rate range and plotted versus flow rate. Shown in Fig. 14.1a is a typical signature curve for a turbine flowmeter.

Depending on the flow-rate range, which is usually specified as 20:1, 25:1, or 30:1, a *linearity* envelope is specified, such as ± 0.5 percent, ± 0.75 percent, or ± 1 percent. Since signature curves vary with design, manufacturers' designated linear ranges are different. Usually, the percentage assigned to the linearity envelope is acceptable, and the *true K* factor [Eq. (14.1)] is seldom corrected. A *mean K* factor, defined as the mean of the maximum and minimum K factors over the designated range,

$$\overline{K}_{F,v} = \frac{(K_{F,v})_{\max} + (K_{F,v})_{\min}}{2} \tag{14.2}$$

is also used.

Several influence quantities bias or alter the signature curve. Viscosity, meter-body expansion with pressure and temperature, and (for gas turbine flowmeters) the influence of flowing pressure (density) are sometimes considered. Manufacturers specify, based on test data, a limiting viscosity or Reynolds number to maintain linearity. For liquids, the effects of both pressure and temperature are small (≤ 0.01 percent) and usually considered negligible. However, for high-accuracy custody-transfer measurements API 2101 (1981), corrections are applied and sometimes calibration curves are programmed for computer corrections.

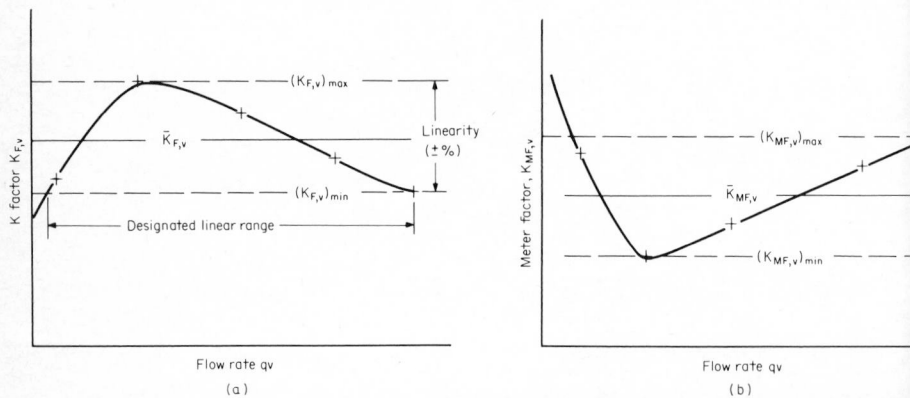

Figure 14.1 *K*-factor and meter-factor signature curves. (*a*) *K* factor. (*b*) Meter factor.

Table 14.1 Liquid Flow Rate Equations: U.S. and SI Units

	U.S. units $(\overline{K}_{F,v},\ \text{pulses/gal})$†		SI units $(\overline{K}^*_{F,v},\ \text{pulses/L})$††	
	MASS FLOW RATE			
Density	$q_{\mathrm{PPH}} = 481.2500\, \dfrac{\rho_f f_{\mathrm{Hz}}}{\overline{K}_{F,\mathrm{gal}}}$	(a)	$q^*_{\mathrm{KPH}} = 3.6\, \dfrac{\rho^*_f f_{\mathrm{Hz}}}{\overline{K}^*_{F,l}}$	(f)
Specific gravity	$q_{\mathrm{PPH}} = 30{,}013.780\, \dfrac{F_p G_f f_{\mathrm{Hz}}}{\overline{K}_{F,\mathrm{gal}}}$	(b)	$q^*_{\mathrm{KPH}} = 3596.443\, \dfrac{F_p G_f f_{\mathrm{Hz}}}{\overline{K}^*_{F,l}}$	(g)
	VOLUMETRIC FLOW RATE AT FLOWING CONDITIONS			
	$q_{\mathrm{gpm}} = 60\, \dfrac{f_{\mathrm{Hz}}}{\overline{K}_{F,\mathrm{gal}}}$	(c)	$q^*_{\mathrm{l\,pm}} = 60\, \dfrac{f_{\mathrm{Hz}}}{\overline{K}^*_{F,l}}$	(h)
	VOLUMETRIC FLOW RATE AT BASE CONDITONS			
Density	$q_{\mathrm{GPM}} = 60\, \dfrac{\rho_f}{\rho_b}\, \dfrac{f_{\mathrm{Hz}}}{K_{F,\mathrm{gal}}}$	(d)	$q^*_{\mathrm{LPM}} = 60\, \dfrac{\rho^*_f f_{\mathrm{Hz}}}{\rho^*_b \overline{K}^*_{F,l}}$	(i)
Specific gravity	$q_{\mathrm{GPM}} = 60\, \dfrac{F_p G_f f_{\mathrm{Hz}}}{G_b \overline{K}_{F,\mathrm{gal}}}$	(e)	$q^*_{\mathrm{LPM}} = 60\, \dfrac{F_p G_f f_{\mathrm{Hz}}}{G_b \overline{K}^*_{F,l}}$	(j)

†Substitute $K_{F,v}$ for $\overline{K}_{F,v}$ if signature curve or an equation is used.

Table 14.2 Liquid Total-Flow Equations: U.S. and SI Units†

	U.S. units $(\overline{K}_{F,v},\ \text{pulses/gal})$†		SI units $(\overline{K}^*_{F,v},\ \text{pulses/L})$†‡	
	MASS FLOW			
Density	$Q_{\mathrm{lbm}} = 0.1336806 \rho_f\, \dfrac{\text{pulses}}{\overline{K}_{F,\mathrm{gal}}}$	(a)	$Q^*_{\mathrm{kg}} = \dfrac{\rho^*_f}{1000}\, \dfrac{\text{pulses}}{\overline{K}^*_{F,l}}$	(f)
Specific gravity	$Q_{\mathrm{lbm}} = 8.337161 F_p G_F\, \dfrac{\text{pulses}}{\overline{K}_{F,\mathrm{gal}}}$	(b)	$Q^*_{\mathrm{kg}} = 0.999012 F_p G_F\, \dfrac{\text{pulses}}{\overline{K}^*_{F,l}}$	(g)
	VOLUMETRIC FLOW AT FLOWING CONDITONS			
	$Q_{\mathrm{gal}} = \dfrac{\text{pulses}}{\overline{K}_{F,\mathrm{gal}}}$	(c)	$Q^*_l = \dfrac{\text{pulses}}{\overline{K}^*_{F,l}}$	(h)
	VOLUMETRIC FLOW AT BASE CONDITIONS			
Density	$Q_{\mathrm{GAL}} = \dfrac{\rho_f}{\rho_b}\, \dfrac{\text{pulses}}{\overline{K}_{F,\mathrm{gal}}}$	(d)	$Q^*_L = \dfrac{\rho^*_f}{\rho^*_b}\, \dfrac{\text{pulses}}{\overline{K}^*_{F,l}}$	(i)
Specific gravity	$Q_{\mathrm{GAL}} = \dfrac{F_p G_F}{G_b}\, \dfrac{\text{pulses}}{\overline{K}_{F,\mathrm{gal}}}$	(e)	$Q^*_L = \dfrac{F_p G_F}{G_b}\, \dfrac{\text{pulses}}{\overline{K}^*_{F,l}}$	(j)

†Substitute $K_{F,v}$ for $\overline{K}_{F,v}$ if signature curve or an equation is used.

‡Subscript l refers to flowing liters, and L to base liters.

Table 14.3 Gas (Vapor) Flow-Rate Equations: U.S. and SI Units

	U.S. units ($\underline{K}_{F,v}$, pulses/ft³)†		SI units ($\overline{K}_{F,v}^*$, pulses/m³)†
	MASS FLOW RATE		
Density	$q_{PPH} = 3600\rho_f \dfrac{f_{Hz}}{\underline{K}_{F,ft3}}$	(a)	$q_{KPH}^* = 3600\rho_f^* \dfrac{f_{Hz}}{\overline{K}_{F,m3}^*}$ (g)
pvT equation	$q_{PPH} = 9715.77\,\dfrac{Gp_f}{Z_f T_f}\,\dfrac{f_{Hz}}{\underline{K}_{F,ft3}}$	(b)	$q_{KPH}^* = 12{,}540.27\,\dfrac{Gp_f^*}{Z_f T_K}\,\dfrac{f_{Hz}}{\overline{K}_{F,m3}^*}$ (h)
	VOLUMETRIC FLOW RATE AT FLOWING CONDITONS		
	$q_{acfh} = 3600\,\dfrac{f_{Hz}}{\underline{K}_{F,ft3}}$	(c)	$q_{acmh}^* = 3600\,\dfrac{f_{Hz}}{\overline{K}_{F,m3}^*}$ (i)
	VOLUMETRIC FLOW RATE AT BASE CONDITIONS		
Density	$q_{SCFH} = 3600\,\dfrac{\rho_f}{\rho_b}\,\dfrac{f_{Hz}}{\underline{K}_{F,ft3}}$	(d)	$q_{SCMH}^* = 3600\,\dfrac{\rho_f^*}{\rho_b^*}\,\dfrac{f_{Hz}}{\overline{K}_{F,m3}^*}$ (j)
pvT equation Standard base	$q_{SCFH} = 127{,}056.24\,\dfrac{Z_b}{Z_f}\,\dfrac{p_f}{T_f}\,\dfrac{f_{Hz}}{\underline{K}_{F,ft3}}$	(e)	$q_{SCMH}^* = 10{,}237.75\,\dfrac{Z_b}{Z_f}\,\dfrac{p_f^*}{T_K}\,\dfrac{f_{Hz}}{\overline{K}_{F,m3}^*}$ (k)
Selected base	$q_{SCFH} = 3600\,\dfrac{Z_b}{Z_f}\,\dfrac{T_b}{T_f}\,\dfrac{p_f}{p_b}\,\dfrac{f_{Hz}}{\underline{K}_{F,ft3}}$	(f)	$q_{SCMH}^* = 3600\,\dfrac{Z_b}{Z_f}\,\dfrac{T_{Kb}}{T_K}\,\dfrac{p_f^*}{p_b^*}\,\dfrac{f_{Hz}}{\overline{K}_{F,m3}^*}$ (l)

†Substitute $K_{F,v}$ for $\overline{K}_{F,v}$ if signature curve or an equation is used.

Table 14.4 Gas (Vapor) Total-Flow Equations: U.S. and SI Units

	U.S. units ($\overline{K}_{F,v}$, pulses/ft³)†		SI units ($\overline{K}^*_{Fv} = $ pulses/m³)ᵈ‡	
	MASS FLOW			
Density	$Q_{lbm} = \rho_f \dfrac{\text{pulses}}{\overline{K}_{F,ft3}}$	(a)	$Q^*_{kg} = \rho^*_f \dfrac{\text{pulses}}{\overline{K}^*_{F,m3}}$	(g)
pvT equation	$Q_{lbm} = 2.698825 \dfrac{p_f G}{Z_f T_f} \dfrac{\text{pulses}}{\overline{K}_{F,ft3}}$	(b)	$Q^*_{kg} = 3.483407 \dfrac{p^*_f G}{Z_f T_K} \dfrac{\text{pulses}}{\overline{K}^*_{F,m3}}$	(h)
	VOLUMETRIC FLOW AT FLOWING CONDITIONS			
	$Q_{acf} = \dfrac{\text{pulses}}{\overline{K}_{F,ft3}}$	(c)	$Q^*_{a\,cm} = \dfrac{\text{pulses}}{\overline{K}^*_{F,m3}}$	(i)
	VOLUMETRIC FLOW AT BASE CONDITIONS			
Density pvT equation	$Q_{SCF} = \dfrac{\rho_f}{\rho_b} \dfrac{\text{pulses}}{\overline{K}_{F,ft3}}$	(d)	$Q^*_{SCM} = \dfrac{\rho^*_f}{\rho^*_b} \dfrac{\text{pulses}}{\overline{K}^*_{F,m3}}$	(j)
Standard base	$Q_{SCF} = 35.29340 \dfrac{Z_b}{Z_f} \dfrac{p_f}{T_f} \dfrac{\text{pulses}}{\overline{K}_{F,ft3}}$	(e)	$Q^*_{SCM} = 2.843819 \dfrac{Z_b}{Z_f} \dfrac{p^*_f}{T_K} \dfrac{\text{pulses}}{\overline{K}^*_{F,m3}}$	(k)
Selected base	$Q_{SCF} = \dfrac{Z_b}{Z_f} \dfrac{T_b}{T_f} \dfrac{p_f}{p_b} \dfrac{\text{pulses}}{\overline{K}_{F,ft3}}$	(f)	$Q^*_{SCM} = \dfrac{Z_b}{Z_f} \dfrac{T_{Kb}}{T_K} \dfrac{p^*_f}{p^*_b} \dfrac{\text{pulses}}{\overline{K}^*_{F,m3}}$	(l)

†Substitute $K_{F,v}$ for $\overline{K}_{F,v}$ if signature curve or an equation is used.

‡Subscripts acm and SCM refer to flowing and base or standard cubic meters, respectively.

Meter Factor In some applications, the reciprocal of the K factor is used. This is referred to as the *meter factor* and defined as

$$K_{MF,v} = \frac{1}{K_{F,v}} = \frac{q_v}{f_{Hz}} \qquad (14.3)$$

The meter factor has the units of unit volume per pulse.† The *mean* meter factor is defined as

$$\overline{K}_{MF,v} = \frac{1}{\overline{K}_{F,v}} = \frac{(K_{MF,v})_{max} + (K_{MF,v})_{min}}{2} \qquad (14.4)$$

When the meter factor is given, flow rate or total volume is calculated by multiplication rather than division. As shown in Fig. 14.1b, the meter-factor signature curve is the mirror image of the K-factor curve.

Flow Rate and Total Volume The K factor has the units of pulses per unit volume—that is, pulses per gallon, per barrel, per cubic meter, etc. Since the frequency is directly proportional to the volumetric flow rate, conversion to base volume or to mass is by the equations developed in Chap. 3. Tables 14.1 through 14.4 list the commonly used U.S.- and SI-unit flow-rate and total-volume equations for mass flow and volumetric flow rate at both flowing and base conditions.

Two-Phase Flows Vortex and specially designed turbine flowmeters are used to meter quality (wet) steam and other two-phase fluids, such as ammonia, where both liquid and gas (vapor) are present. In these nonhomogeneous applications, mass flow is calculated, and an *effective density* must be used in the equation. Some manufacturers recommend dividing the gas-phase density by the mixture's quality to yield a higher density value, under the assumption that the liquid is a mist homogeneously mixed with the gas. This approach was suggested by Spink (1967) for differential producers and has been widely used. However, the work of Smith and Leang (1974) suggests that the effective density proposed by James (1966) better correlates available orifice data over wider quality ranges. The James equation,

$$\rho_{TP} = \frac{\rho_g}{\mathbf{X}^{1.53} + (1 - \mathbf{X}^{1.53})\rho_g/\rho_l} \qquad (14.5)$$

is suggested for these two-phase applications until sufficient calibration data becomes available. It is important that the density of either a single- or two-phase vapor mixture be calculated or measured at the location recommended by the manufacturer.

Accuracy For an estimate of measurement accuracy, the accuracy of the K factor (or meter factor) must be estimated and then combined with the accuracy of

†API Publication 2101 (1981) defines meter factor as prover volume divided by registered volume. In the metering of liquid hydrocarbons, the requirements of this manual must be strictly adhered to.

each measured quantity (see Chap. 4) by the root-sum-squared method. The sensitivity coefficients for the variables in Tables 14.1 to 14.4 are all 1.0, with the sign either positive for a numerator value or negative for a denominator value.

The accuracy of the K factor (or meter factor) should be obtained from the manufacturer, but it may be conservatively estimated as

$$(\text{Acc})_{KF,v} = \pm (\text{Acc})_{\text{LAB}} \pm \sqrt{\overline{L}^2 + 4\sigma_P^2} \qquad (14.6)$$

where $(\text{Acc})_{\text{LAB}}$ is the laboratory accuracy, and σ_P is the precision of the two data points used to define the linearity envelope.

Turbine Flowmeter

Principle of Operation   Figure 14.2a shows the velocity vector diagram for an *ideal* helical-bladed turbine flowmeter. The velocity of the entering fluid displaces the blades, and the vector sum of the fluid velocity with respect to the blade $\mathbf{V}_{f/B}$ and the blade rotational velocity $\mathbf{V}_B$ equals the entering axial velocity vector $\mathbf{V}_f$. In this ideal case, the velocity leaving the blades is, therefore, not displaced with respect to the entering velocity vector. In the real case, retarding torques from blade fluid friction, bearing friction, tip clearing (windage), etc., are present and cause the exit velocity to be displaced from the entering vector; the blades therefore rotate at a speed below the theoretically predicted speed. This decrease in rotational velocity, referred to as *slip*, results in an exit swirl velocity component that changes with retarding torque. This velocity component provides the kinetic energy to balance the retarding torques.

Assuming a flat (rectilinear) entering velocity profile, Rubin et al. (1965) considered both momentum and airfoil approaches to explain the relationship between driving and retarding torques. The data was presented in terms of a slip parameter, which is the ratio of the tangent of the effective angle of attack to the blade angle. Thompson and Grey (1970) extended this analysis to include a variable lift coefficient and annular velocity profile. These analyses led to a better understanding of the relationships among the numerous parameters affecting performance.

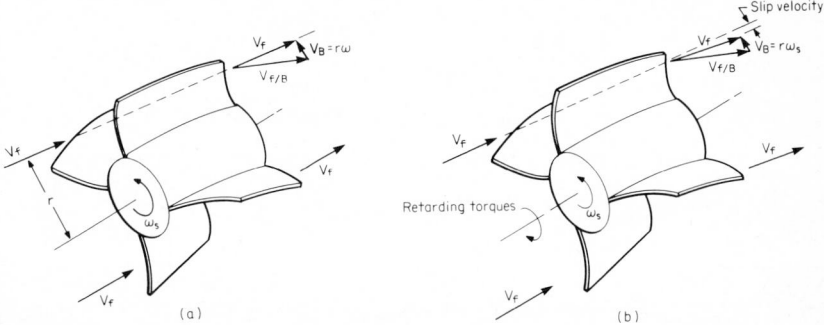

Figure 14.2   Fluid-velocity vector diagram for a turbine flowmeter. (*a*) Ideal. (*b*) With retarding torque.

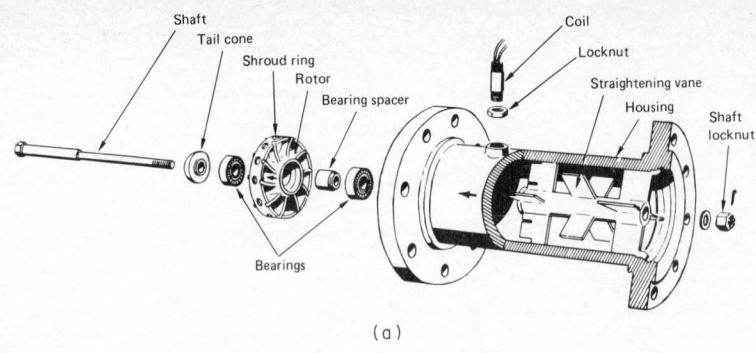

(a)

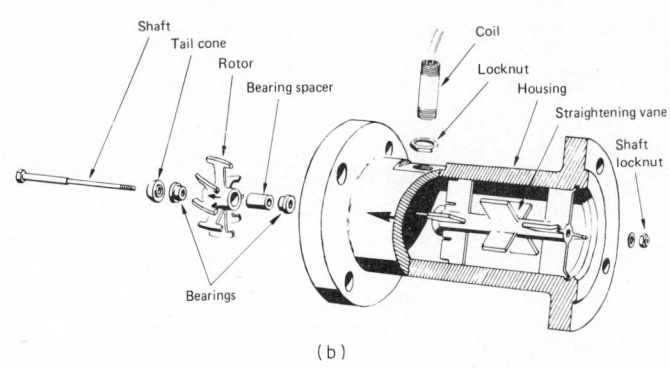

(b)

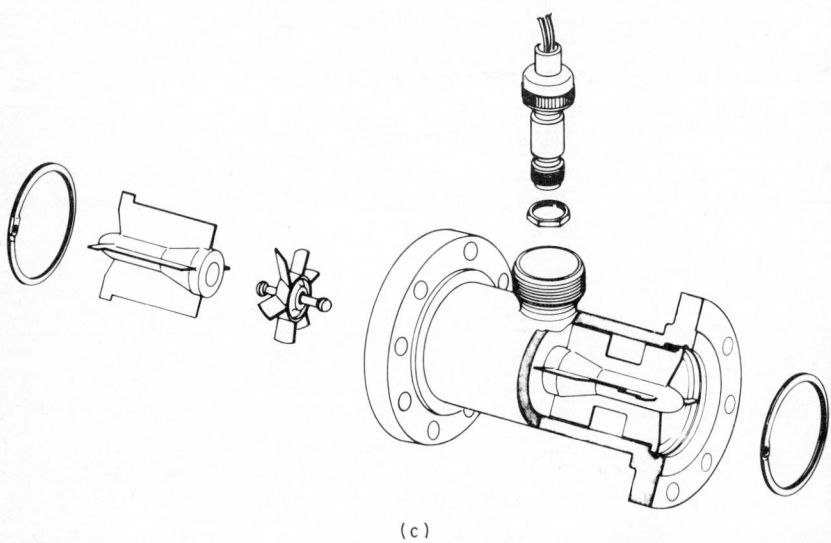

(c)

Figure 14.3 Turbine-flowmeter blade and bearing designs *(courtesy The Foxboro Co.)*. **(a)** Helical blades with ball bearings. (*b*) Helical T blades with journal bearing. (*c*) Straight blades with journal bearing and thrust ball.

14-8

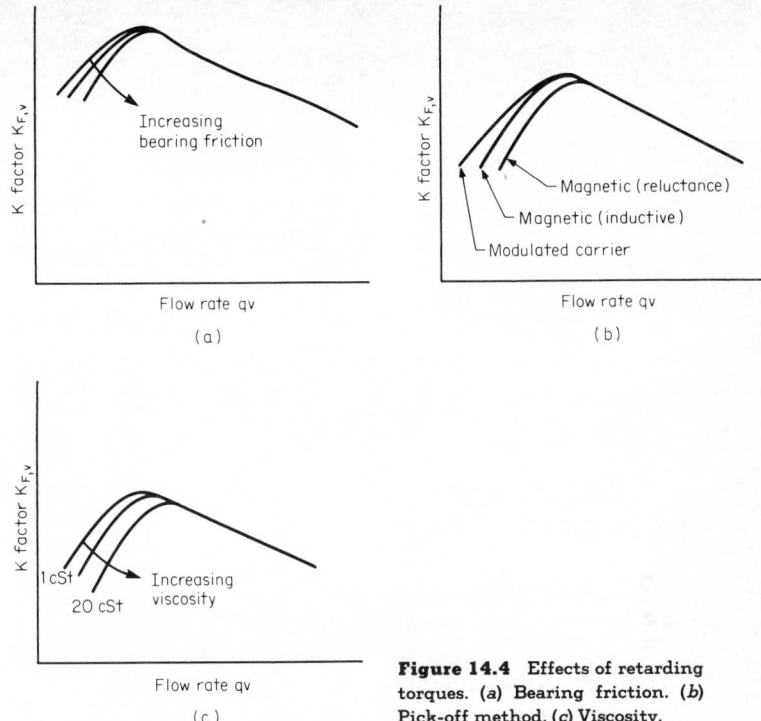

Figure 14.4 Effects of retarding torques. (*a*) Bearing friction. (*b*) Pick-off method. (*c*) Viscosity.

Liquid Flowmeters Liquid turbine flowmeters are available with straight, helical, and T-shaped blades (Fig. 14.3). A multiholed shroud ring (Fig. 14.3*a*) is also widely used to increase the frequency for the same number of blades. Depending on line size, flow-rate range, and metered fluid, either ball or journal bearings (Fig. 14.3*b*) are selected. Except for very clean lubricating-type liquids, journal bearings with a thrust ball (Fig. 14.3*c*) are usually chosen. Bearing wear (in both ball and journal types) increases the retarding torques, which, in turn, alter the K factor at lower flow rates.

Two basic types of detection devices are available: magnetic and modulated-carrier. In the magnetic type, the magnetic pick-off may be in the external coil assembly (reluctance pick-off) or magnets may be inserted into the rotor assembly (inductive pick-off). Less drag is associated with the inductive pick-off, and, due to the reduced retarding torques, linearity is slightly improved (Fig. 14.4*b*). The modulated-carrier pick-off detector operates through the modulation of a 45-kHz carrier frequency by the passage of the turbine blade. Since retarding torques are minimal, this detector can improve low-velocity (or low-density-flow) performance. The magnetic-reluctance pick-off is the most widely used for reasons of economics, but the other detection methods may be selected for better low-range performance.

The effect of fluid viscosity is illustrated in Fig. 14.4*c*. In general, blade friction increases with viscosity, resulting in a decrease in speed for the same flow rate. The effect of viscosity is, however, difficult to predict, because bearing friction, tip clearance, drag, velocity profile, and blade friction are also viscosity-dependent.

Turbine meters are sometimes checked by a *spin* test in which the rotor is spun by a jet of air. The time for the rotor to stop after removal of the jet is checked against prior spin-test times to detect increased torques. Although no specific value for the change in K factor can be determined, a decreased time usually indicates that recalibration or new bearings are required. Additionally, a spin test is an excellent check on overall system circuitry. Extreme care should

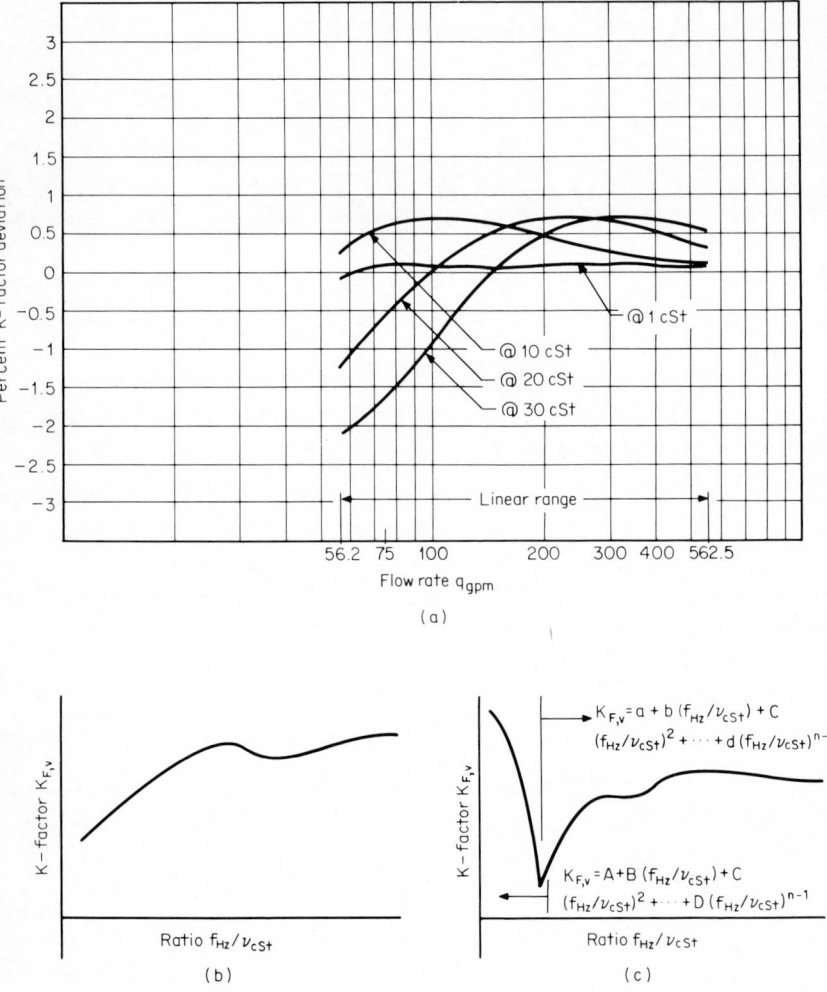

Figure 14.5 Viscosity corrections. (*a*) Type testing. (*b*) Universal viscosity curve. (*c*) Polynomial fitted curves for a single fluid.

be exercised, however, so as not to overspeed the rotor during the spin test. Figure 14.4a illustrates the change in K factor with increasing bearing friction.

Industrial liquid turbine flowmeters are water-calibrated to obtain the K-factor signature curve. For fluids with other viscosities, any of five correction methods may be used; the method selected depends on the manufacturer and the desired accuracy. The five methods are as follows:

1. A type test is conducted for each meter size, and a bias error curve is given (Fig. 14.5a).

2. The flowmeter is calibrated using the fluid to be metered, and a calibration curve is used for correction; alternatively, a fluid with similar viscosity and density may be used.

3. The flow rate or frequency is divided by the kinematic viscosity, usually in centistokes, and a universal K-factor curve is presented (Jennings, 1977). Shown in Fig. 14.5b is a typical universal viscosity curve that some manufacturers use to represent the viscosity effect on a single plot. This curve is then used to calculate flow rates by iteration from known frequency and viscosity values.

4. Numerous data points are obtained at different viscosities, for the fluid to be metered. This data is then fitted to two polynomial series-expansion equations (Fig. 14.4c). The true flow rate is then computer-calculated from viscosity and frequency values (Ball, 1977).

5. The meter is calibrated in situ with a ball prover. This method is used for the custody transfer of liquid hydrocarbons; API 2101 should be consulted for calculation procedures.

Turbine flowmeters are usually calibrated at moderate pressures [<100 psia (700 kPa)] and with ambient-temperature fluids. Except for the custody transfer of petroleum liquids, small geometry changes with pressure are not normally considered. API 2101 should be consulted for the combined correction for pressure and temperature. ANSI (ISA RP) 31.1 (1977) recommends a temperature correction for accurate measurement. This correction is

$$(K_{F,v})_{\text{flow}} = \frac{(K_{F,v})_{\text{calib}}}{1 + 3\alpha_{\text{HO}}(T_F - T_{F,\text{calib}})} \tag{14.7}$$

for U.S. units, and

$$(K_{F,v})_{\text{flow}} = \frac{(K_{F,v})_{\text{calib}}}{1 + 3\alpha_{\text{HO}}^*(T_{\text{°C}} - T_{\text{°C,calib}})} \tag{14.8}$$

for SI units, where α_{HO} and α_{HO}^* are the thermal-expansion coefficients of the meter housing, and the subscripts flow and calib refer to flowing and calibration conditions.

Since blade and bearing designs vary widely among manufacturers, it is important that each meter be installed in accordance with the manufacturer's recommendations. Precautions such as overspeeding protection and the installation of upstream strainers must be strictly followed for reliable performance. Typically, 10 diameters of upstream pipe and 5 diameters of downstream pipe are required. The strainer mesh numbers shown in Table 14.5 are usually suggested.

Table 14.5 Typical Mesh Numbers for Upstream Strainers

Flowmeter size, in (mm)	Mesh number	Hole size, in (mm)
$\frac{1}{2}, \frac{3}{4}$ (12, 20)	100	0.0059 (0.15)
1, 2, 3 (25, 50, 75)	80	0.0070 (0.18)
4, 6 (100, 150)	60	0.0098 (0.25)
8, 10, 12 (200, 250, 300)	40	0.0165 (0.42)

Gas Flowmeters   Figure 14.6*a* illustrates a single-rotor gas turbine flowmeter. As the gas enters, it is constricted into an annular space that is approximately one-third the area of the housing. The area reduction increases the gas velocity, providing more driving torque and thereby improving low-flow-rate performance. The rotor is usually of high-strength plastic, molded to the desired angle of attack. The rotor is supported by ball bearings that are either permanently lubricated or lubricated through a wet-wicking system. For protection, the bearings are enclosed in the central housing. Integral gearing and electronic pick-off are available.

In the Rockwell Auto-Adjust Turbo-Meter (Fig. 14.6*b*), a sensor rotor, placed downstream from the main rotor, responds to changes in the exit angle of the fluid leaving the main rotor (Lee, 1982). The pulse output of this sensor rotor is electronically subtracted from the pulse output of the main rotor to produce an adjusted output which automatically and continuously corrects to the original meter calibration accuracy. This adjustment occurs despite differences in retarding torques, bearing wear and contamination, or upstream flow between field conditions and calibration conditions. The ratio of sensor-rotor output to main-rotor output at flowing conditions is also automatically and continuously compared with that at calibration conditions. This provides a self-check of the device's own mechanical and electrical operation as well as the upstream flow conditions; it also gives an indication of the deviation from original calibration conditions that is being corrected for by the sensor rotor.

In the mechanical gear-train type pick-off, blade rotation is transmitted through a magnetic coupling to an index counter. Through suitable gear ratios, the index reads in flowing volume units. The mechanical gearing increases the retarding torque, and, therefore, its low-flow-rate performance is not as good as that of the magnetic–pick-off types.

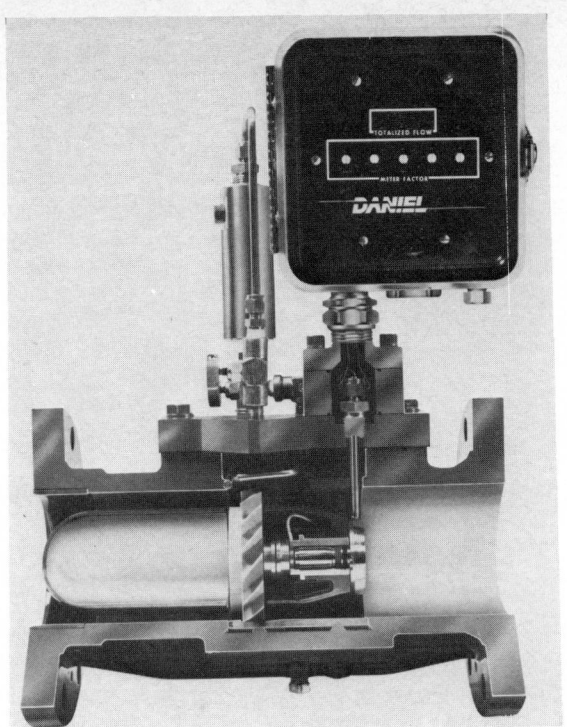

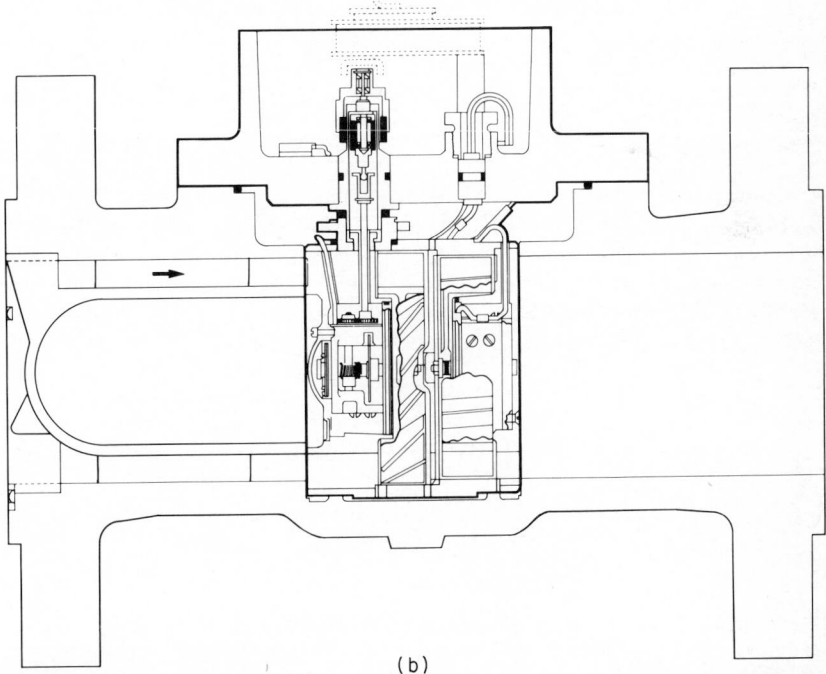

(b)

Figure 14.6 Gas turbine flowmeter. (*a*) **Single rotor** *(courtesy Daniel Industries)*. (*b*) **Auto-Adjust** *(courtesy Rockwell International)*.

14-13

The measuring unit, consisting of rotor, shaft, and bearing assembly, is removable through the side or top of the meter housing. For electrical pulse meters, the detector system and associated electronics are outside the body.

AGA Report 7 (1981) gives the performance of gas turbine metering as ± 1 percent of flowing volume, with a precision of ± 0.1 percent. The designated linearity range is a function of velocity profile, fluid drag, and nonfluid drag effects. The effects of these quantities on rotor speed are summarized in Fig. 14.7 for pipe Reynolds numbers below 100,000.

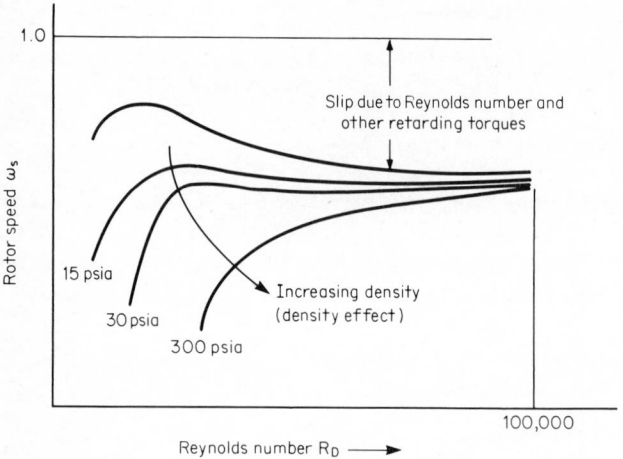

Figure 14.7 Gas-turbine-flowmeter performance at low Reynolds numbers.

To achieve the desired accuracy, the flowmeter should be installed according to the manufacturer's recommendations. Shown in Fig. 14.8 are the straight-length requirements given in AGA Report 7. Rotor overspeeding resulting from pressuring, purging, venting, or operating the meter in reverse can damage the meter. Usually the manufacturer limits overspeed to 150 percent of the rated capacity, and that only for short time periods. AGA Report 7 suggests that blow-

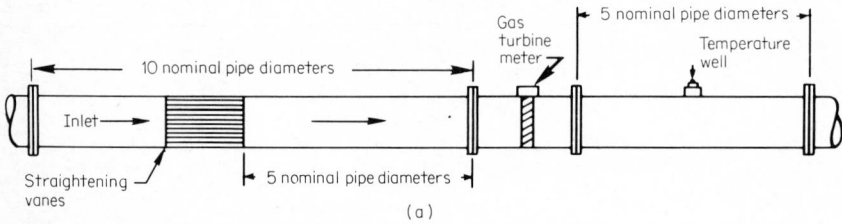

Figure 14.8 Recommended minimum lengths for gas turbine flowmeter *(from AGA Report No. 7).* **(a)** Straight-through recommended design. **(b)** Optional design for close coupling. **(c)** Optional design for close coupling with integral flow conditioners. **(d)** Recommended for angle-body design.

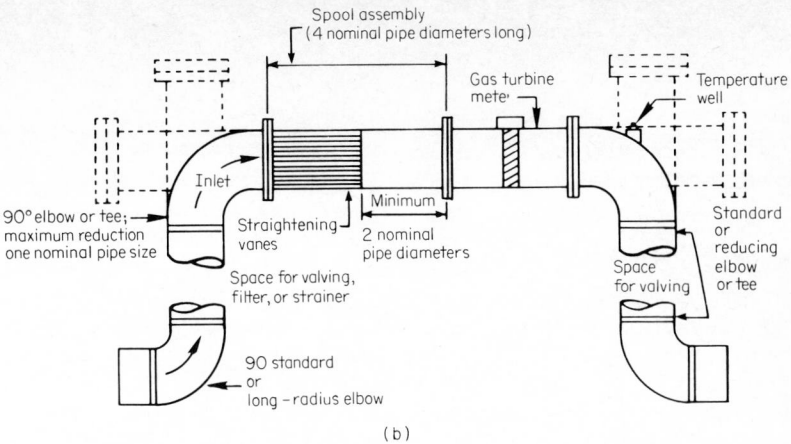

Spool assembly
(4 nominal pipe diameters long)

Gas turbine
meter

Temperature
well

Inlet

90° elbow or tee;
maximum reduction
one nominal pipe size

Straightening
vanes

Minimum

2 nominal
pipe diameters

Standard
or
reducing
elbow
or tee

Space
for valving

Space for valving,
filter, or strainer

90 standard
or
long–radius elbow

(b)

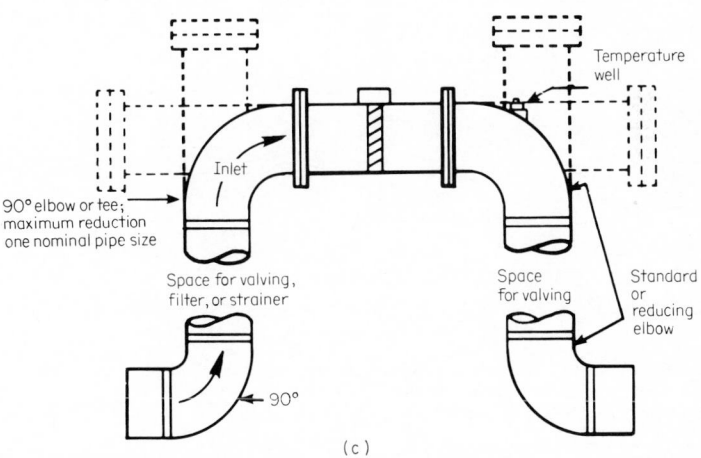

Temperature
well

Inlet

90° elbow or tee;
maximum reduction
one nominal pipe size

Space for valving,
filter, or strainer

Space
for valving

Standard
or
reducing
elbow

90°

(c)

Inlet piping (10 nominal pipe diameters long;
5 nominal pipe diameters with straightening vane)

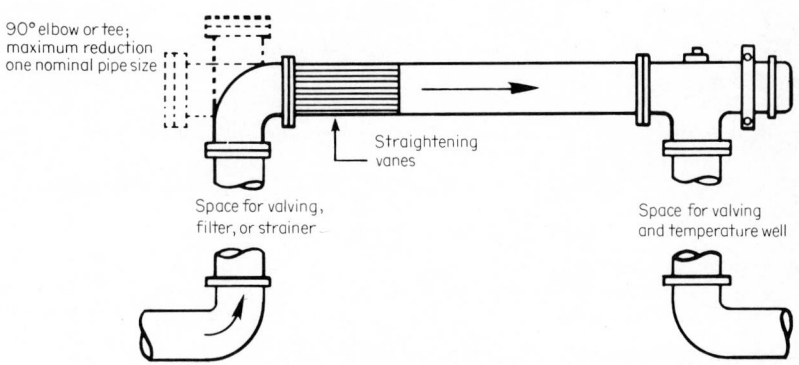

90° elbow or tee;
maximum reduction
one nominal pipe size

Straightening
vanes

Space for valving,
filter, or strainer

Space for valving
and temperature well

Horizontal installation (inlet in norizontal plane, outlet down)

(d)

14-15

down valves be less than one-sixth the meter size, or, for high-pressure installations, that critical flow nozzles (see Chap. 13) be installed to limit the flow to 120 percent of rated capacity.

Vortex Flowmeters

Principle of Operation Vortex shedding is a common flow phenomenon in which rotating zones of fluid are shed downstream of a barrier placed in a moving stream. This *roll-up* of fluid, first from one side of the barrier and then the other, is responsible for the waving of flags and the singing (aeolian tones) of telephone wires. The subject of vortex shedding is treated in most fluid dynamics texts and in numerous technical papers. To date, however, no complete theoretical explanation of the fundamental nature of vortex formation and the resulting alternate shedding pattern has been advanced.

Figure 14.9 shows the basic vortex-shedding principle, and in Fig. 14.10 are sequential photographs of vortex shedding. As the flow is split into two streams, the instability of the shear layer causes the fluid to roll up into a well-defined

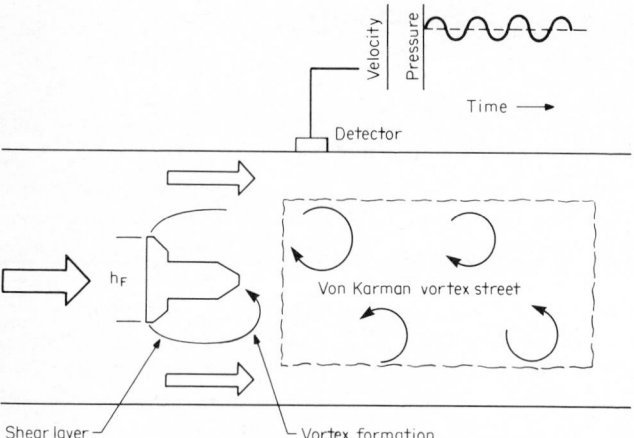

Figure 14.9 Vortex shedding from a bluff body.

vortex, the time for complete vortex formation depending on vortex-element (barrier) geometry. After it is formed, the vortex sheds, and a second vortex begins to form on the opposite side of the element. If the vortex shedding is stable, the time for complete formation of the second vortex is the same as that for the first vortex, with the formation time proportional to velocity. The pattern is then repeated alternately from side to side, resulting in the familiar downstream *Von Karman* (1912) *vortex street*. The vortex shedding results in pressure and velocity changes around and downstream of the vortex element. By placing pressure, thermal, or ultrasonic detectors in a location where the signal is high, the vortex-shedding frequency may be measured.

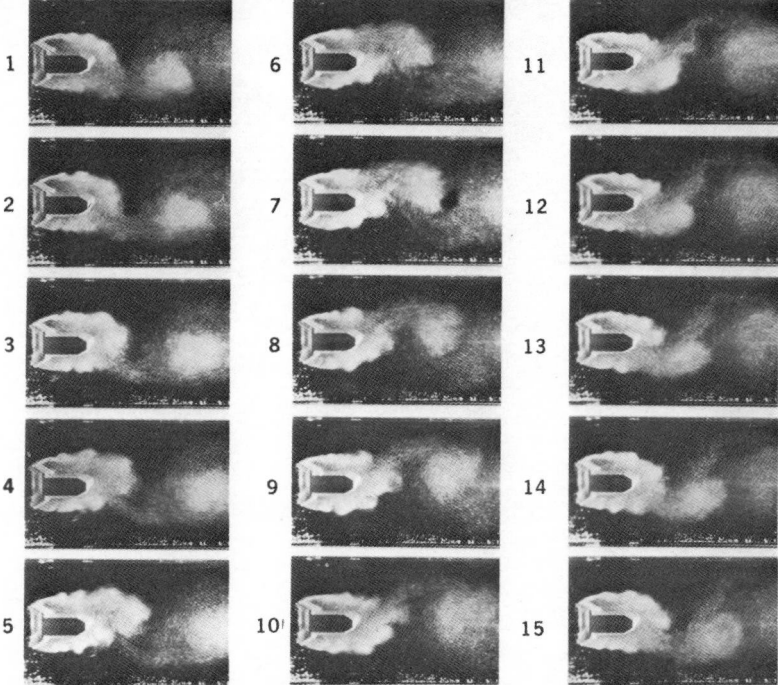

Figure 14.10 Sequential photographs of vortex shedding, showing one cycle of shedding from a T-cross-section cylinder in a pipe at 0.04 s/frame *(courtesy The Foxboro Co.)*.

Strouhal (1878) made the first experimental observation of the shedding phenomenon. He showed that the shedding frequency of a vibrating wire in the wind was related to the wind velocity and wire diameter. The *Strouhal number* is still widely used as the basic vortex-shedding correlation; it is computed as

$$S = \frac{f_{\text{Hz}} h_F}{V_{\text{free}}} \tag{14.9}$$

where S is the Strouhal number, h_F the barrier width in feet, and V_{free} the free-stream velocity.

Experimental evidence indicates that the Strouhal number is essentially constant over wide velocity ranges and is independent of fluid density. Data obtained with industrial meters on liquids (Inkley et al., 1980), steam, natural gas, and air (White et al., 1974) substantiates this density independence. The pipeline velocity profile does, however, influence the constancy of the Strouhal number.

Industrial Flowmeters   Generally, vortex flowmeters are made up of three parts: a vortex-generating element fully spanning the pipe, a sensor to convert shedding energy into an electrical pulse signal, and a transmitter for amplification and conditioning. The significant differences among vortex-flowmeter designs are in the element shape and the sensor. Shown in Fig. 14.11 are the geometries of some vortex-generating elements.

Figure 14.11   Vortex-element geometries.

The shedding frequency is a function of barrier width, barrier length, and the ratio of barrier width to meter bore. The relationship between frequency and geometry is empirically determined, since even for free-stream flows the theoretical relationship between near and far wake has not been developed.

The signature curves for vortex flowmeters generally have two distinct shapes, depending on element design, as shown in Fig. 14.12. Linearity ranges from 0.5 percent of rate to ±1 percent of the upper-range flow rate over a 20:1 range. The lower-limit Reynolds number is usually specified as 10,000; no upper limit has been experimentally determined.

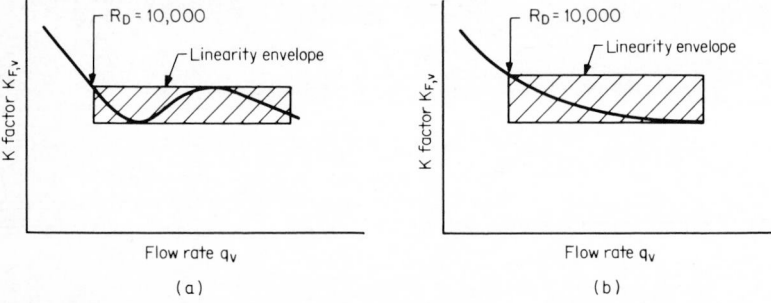

Figure 14.12   Typical vortex-flowmeter signature curves. (*a*) S shape. (*b*) C shape.

The minimum velocity that can be measured depends on the detector type. For liquids, where density variations are small, the lower limit is generally 1 to 2 ft/s (0.3 to 0.6 m/s). For gases (vapors), the minimum velocity may depend on the fluid density, which is a function of process pressure, temperature, and specific gravity. Manufacturers should be consulted for lower-limit values.

Presently, vortex flowmeters are individually water-calibrated to determine the mean K factor. Miller et al. (1977) present data on 167 meters that indicate a ±0.3 percent agreement in mean K factors. This suggests that, in the future, meters may be assigned K factors and not individually calibrated.

Positive-Displacement Flowmeters

Principle of Operation Positive-displacement meters measure total flowing volumes by repeatedly filling and discharging fixed volumes. Many different designs are available for both liquid and gas flows, ranging from household water meters to large-volume natural gas meters (see Chap. 6). *Fluid Meters* (ASME, 1971) divides positive-displacement meters into three broad classes. In the first are meters in which one wall is of a flexible material that moves to displace the volume with no leakage into another chamber. In the second are meters in which a mechanical seal is used between movable and stationary walls. The third class of meters employs a capillary or film seal between the moving and stationary elements.

The permanent pressure loss (pressure drop) provides the energy to drive the mechanical system and moving compartments. This is usually lowest for the film-seal class and highest for meters with reciprocating pistons.

Positive-displacement meters are widely used as *cash* (cash transaction) register meters, and for this reason are invariably calibrated and adjusted to state or national regulatory standards. In the United States, for gases (vapors), standards prepared by AGA (ANSI B109.2, 1980; ANSI B109.3, 1980) and AGA Report 6 (1975) are followed; for liquid petroleum products, ANSI Z11.170 (1960) is followed.

These standards specify a ± 1 percent accuracy between the register reading and the *true* flow as defined by certified volume cans, weigh-tank readings, bell prover tests, or a critical-flow prover (critical nozzle). Mechanical adjustments within the meter are usually provided to correct register readings to the referenced volume.

Because of the many available designs, no single signature curve can be established. However, the curve shown in Fig. 14.13 can be used qualitatively to rep-

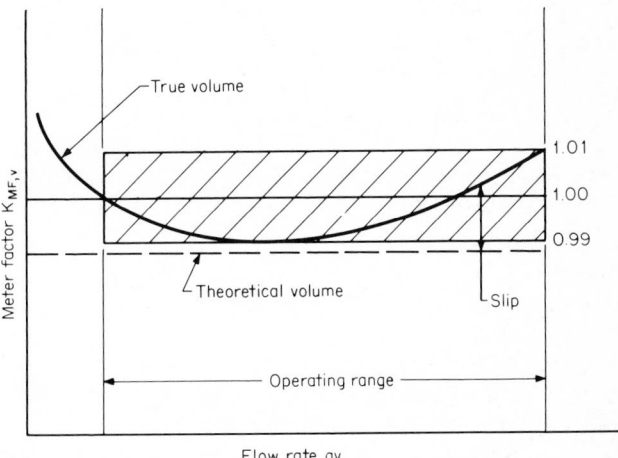

Figure 14.13 Representative positive-displacement signature curve.

resent the characteristics of positive-displacement meters. The horizontal line represents the calculated dimensional volume displacements, which are theoretically unaffected by flow rate. The C-shaped curve is the true meter factor (unit volume per pulse count); because of leakage (slip), the curve is above the calculated volume line. The meter is calibrated, or *proofed,* by adjusting the register reading to within ±1 percent of the true volume over the rated flow-rate range.

Liquid Flowmeters Liquid meters range from household water meters to those used to measure liquefied petroleum gas (LPG). Corrections for liquid compressibility and temperature may be required, depending on the desired accuracy. Regarding meters (or auxiliary equipment) for high-accuracy applications, ANSI Z11.170 suggests the following considerations:

1. Range of operating flow rate and whether the flow is continuous or intermittent

2. Maximum flowing pressure and allowable permanent pressure loss

3. Type of liquid (or liquids) the meter will measure, including viscosity and corrosivity

4. Temperature range and whether automatic temperature compensation is applicable

5. Type of register or ticket printer required

6. Required accuracy and ease of meter-registration adjustment

7. Type and method of meter proving (volume cans, weighing, etc.)

8. Maintenance cost

9. Quantity and size of foreign matter

10. Space availability

Certain precautions are necessary to prevent the passage of vapor or air, including the installation of air eliminators and eliminating leaky valves and pump seals. Further, it is important that system design ensure that vapor from liquids is not released, and that strainers are installed to remove sediment.

Gas Flowmeters The rated capacity of a natural gas positive-displacement meter is the maximum flow rate at which the meter may be operated; it is specified in flowing cubic feet (flowing cubic meters), or standard cubic feet (standard cubic meters). In the United States, standard volumes are calculated at 14.73 psia (101.56 kPa) and 60°F (15.6°C) (ANSI/API 2562, 1969) or at other selected (contract) base pressures and temperatures.

In many designs, mechanical temperature and pressure compensations are available to adjust the register reading to standard volume units. ANSI/API 2562 (1969) requires that the index dial of a temperature-compensated meter with pressure compensation indicate in contrasting colors that the meter is temperature-compensated to a given base. For example, such a dial indication is "Cubic feet, pressure-compensated to 10 psig (69 kPa) nominal and temperature-compensated to 60°F (15.6°C)."

The measuring compartment of the rotary-type gas meter is located between a stationary case and rotating elements. Rotation is usually transmitted mechanically to an index which registers the flowing volume. Because of the close clearances between mating elements, wear can degrade performance. ANSI B109.3 (1980) requires that meters be given a 4000-h, 50-psig (344.7-kPa) life test during which the meter factor must remain within 1 percent of rated capacity over a 10:1 range. The permanent pressure loss is used as an indication of meter condition, to detect wear. A maximum differential pressure of 0.1 in (0.025 kPa) is specified when air is used to start rotation. When the differential exceeds 50 percent of the original value, the meter should be opened and inspected for wear.

Diaphragm-type gas displacement meters are classified via the flowing volume of a 0.6-specific-gravity gas at 0.5 in (0.125 kPa) differential pressure across the

Table 14.6 Classes of Diaphragm-Type Gas Displacement Meters

Class		Range, ft³/h (m³/h)
U.S.	OIML†	
500	16	500–899 (14.2–25.5)
900	25	900–1399 (25.5–39.6)
1400	40	1400–2299 (39.6–65.1)
2300	65	2300–3499 (65.1–99.1)
3500	100	3500–5599 (99.1–158.5)

†International Organization of Legal Metrology, approximate ratings.

SOURCE: ANSI B109.2 (1980).

meter (ANSI B109.2, 1980). Meter classes are shown in Table 14.6. These classes are determined with a bell prover, using air, with test data referred to standard conditions of 60°F (15.6°C) and 14.73 psia (101.6 kPa).

EXAMPLE 14.1

A turbine meter with a stainless-steel body and a mean K factor of 50.00 pulses/gal, determined with 68°F (20°C) water, is metering the gasoline of Example 2.15. Assuming no correction for viscosity, what is the base flow rate if the pulse frequency is 200 Hz?

The flow rate in base gallons per minute is calculated from Eq. (e) of Table 14.1:

$$q_{\text{GPM}} = 60 \frac{F_p G_F f_{\text{Hz}}}{G_b \overline{K}_{F,\text{gal}}}$$

Data is as follows:

■ From Example 2.15: $F_p = 1.0088$

■ From Example 2.12: $G_F = 0.7255$ and $G_b = 0.7359$ at $T_F = 80.5°F$

Assuming minimum and maximum data points are equally biased by temperature, the mean K factor is, from Eq. (14.7),

$$(\overline{K}_{F,\text{gal}})_{\text{flow}} = \frac{(\overline{K}_{F,\text{gal}})_{\text{calib}}}{1 + 3\alpha_{\text{HO}}(T_F - T_{F,\text{calib}})} = \frac{50.00}{1 + (3)(0.0000096)(80.5 - 68)}$$

$$= 49.982$$

where, from Table B.4, $\alpha_{\text{HO}} = 0.0000096$ in/(in·°F). Substitution then gives the base volumetric flow rate as

$$q_{\text{GPM}} = 60\,\frac{(1.0088)(0.7255)(200)}{(0.7359)(49.982)} = 238.8 \text{ base gal/min}$$

EXAMPLE 14.2

Estimate the measurement accuracy for the gasoline flow of Example 14.1 given the following accuracy estimates:

Manufacturer's data

■ Linearity: $\overline{L} = \pm0.5$ percent

■ Laboratory accuracy: $(\text{ACC})_{\text{LAB}} = \pm0.15$ percent

■ Data-point precision: $\sigma_P = \pm0.05$ percent

Fluid-property estimates

■ Flowing specific gravity: $(\text{ACC})_{GF} = \pm0.1$ percent

■ Base specific gravity: $(\text{ACC})_{Gb} = \pm0.1$ percent

■ Compressibility factor: $(\text{ACC})_{F_p} = \pm0.2$ percent

Since *all* sensitivity coefficients are 1.0, the flow-rate accuracy is estimated with Eq. (4.24):

$$(\text{ACC})_{\text{GPM}} = \pm[(\text{ACC})^2_{\overline{K}_{F,\text{gal}}} + (\text{ACC})^2_{GF} + (\text{ACC})^2_{F_p} + (\text{ACC})^2_{Gb}]^{1/2}$$

The estimated accuracy of the mean K factor is, by Eq. (14.6),

$$(\text{ACC})_{\overline{K}_{F,\text{gal}}} = \pm(\text{ACC})_{\text{LAB}} \pm \sqrt{\overline{L}^2 + 4\sigma_p^2} = \pm0.15 \pm \sqrt{(0.5)^2 + 4(0.05)^2} = \pm0.66 \text{ percent}$$

The accuracy is then estimated as

$$(\text{ACC})_{\text{GPM}} = \pm[(0.66)^2 + (0.1)^2 + (0.1)^2 + (0.2)^2]^{1/2} = \pm0.7 \text{ percent}$$

EXAMPLE 14.3

A vortex flowmeter is metering 140-psia (965-kPa) 95 percent quality steam. If the mean K factor is 64 pulses/ft³ (2295 pulses/m³), what are the (*a*) total mass flow if the counter registers 100,000, and (*b*) the mass of saturated vapor (steam)?

For a two-phase flow, total flow is calculated from Eq. (a) of Table 14.4 by substituting $\rho_{TP} = \rho_f$:

$$Q_{lbm} = \rho_{TP} \frac{\text{pulses}}{\overline{K}_{F,ft3}}$$

The effective two-phase density is, by Eq. (14.5),

$$\rho_{TP} = \frac{\rho_g}{X^{1.53} + (1 - X^{1.53})\rho_g/\rho_l} = \frac{0.31066}{(0.95)^{1.53} + [1 - (0.95)^{1.53}](0.31066/55.476)}$$

$$= 0.33587 \; lb_m/ft^3$$

where, from Table G.1, $\rho_g = 0.31066 \; lb_m/ft^3$ and $\rho_l = 55.476 \; lb_m/ft^3$.

a. The total mass flow is then

$$Q_{lbm} = 0.33587 \frac{100{,}000}{65} = 516.7 \; lb_m$$

b. The mass of saturated vapor is, by Eq. (9.86),

$$(Q_{lbm})_g = X(Q_{lbm})_{TC} = (0.95)(516.7) = 490.9 \; lb_m$$

EXAMPLE 14.4

A vortex flowmeter is measuring the water-vapor-saturated natural gas of Example 9.9. The mean K factor is 6600 pulses/m³. For 100,000 pulses, what is the total dry volume (ISO Standard 5024 base), and what are the wet and dry flow rates if the pulse frequency is 285 Hz?

Since pulse-type meters measure at flowing conditions, the standard volume of wet natural gas is calculated from Eq. (k) of Table 14.4 as

$$(Q^*_{SCM})_{wet} = 2.843819 \frac{Z_b}{Z_f} \frac{p_f^*}{T_K} \frac{\text{pulses}}{\overline{K}_{F,m3}}$$

Data from Examples 9.9 and 9.6 is

$$Z_b = 1.0 \qquad F_{pv} = 1.0011 \qquad p_f^* = 184 \; kPa \qquad T_K = 300 \; K$$

and, from Eq. (9.65),

$$Z_f = \frac{1}{F_{pv}^2} = \frac{1}{(1.0011)^2} = 0.9978$$

Substitution gives

$$(Q^*_{SCM})_{wet} = 2.843819 \frac{1.0}{0.9978} \frac{184}{300} \frac{100{,}000}{6600} = 26.486 \; \text{standard m}^3 \; (\text{wet})$$

The dry volume is calculated, using Eq. (2.66), as

$$F_{WV,dry} = \frac{V_{dry}}{V_{wet}} = \frac{(Q_{SCM})_{dry}}{(Q_{SCM})_{wet}}$$

Then, with $F_{WV,dry} = 0.98082$ from Example 9.9,

$$(Q^*_{SCM})_{dry} = (0.98082)(26.486) = 25.978 \; \text{standard m}^3 \; (\text{dry})$$

The wet flow rate is calculated with Eq. (k) of Table 14.3:

$$(q^*_{\text{SCMH}})_{\text{wet}} = 10{,}237.75 \frac{Z_b}{Z_f} \frac{p^*_f}{T_K} \frac{f_{\text{Hz}}}{\overline{K}_{F,\text{m3}}} = 10{,}237.75 \frac{1.0}{0.9978} \frac{184}{300} \frac{285}{6600}$$

$$= 272 \text{ standard m}^3/\text{h (wet)}$$

The dry flow rate is

$$(q^*_{\text{SCMH}})_{\text{dry}} = (F_{WV,\text{dry}})(q^*_{\text{SCMH}})_{\text{wet}} = (0.98082)(272)$$

$$= 266.5 \text{ standard m}^3/\text{h (dry)}$$

LINEAR-OUTPUT FLOWMETERS

Magnetic Flowmeter

Principle of Operation   The electromagnetic flowmeter is based on Faraday's (1832) law of magnetic induction. When a conductive fluid passes through a magnetic field (Fig. 14.14), a voltage is generated at right angles to the velocity and magnetic-field vectors. This signal voltage e_s is a summation of individual volt-

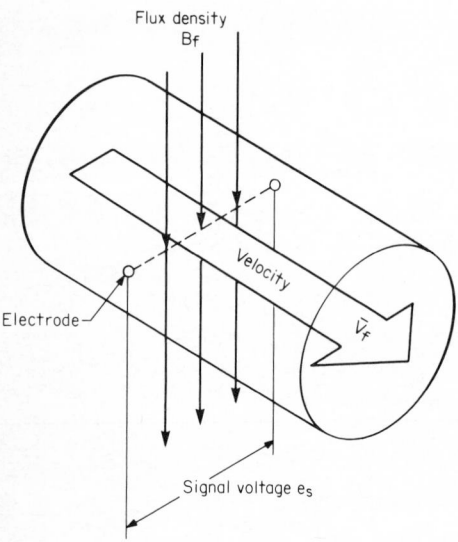

Figure 14.14 Principle of operation of the magnetic flowmeter.

ages generated by differential volumes moving at differing velocities across the pipe section. Shercliff (1961) mathematically demonstrated, by introducing a *weighing function,* that the signal voltage represents the average velocity for axisymmetric profiles. The signal voltage can then be simply expressed as

$$e_s = kDB_f \overline{V}_f \tag{14.10}$$

where k is constant, D is the distance between electrodes, B_f is the magnetic flux density, and $\overline{V}_f$ is the average velocity. For a circular pipe the volumetric flow rate then is

$$q_{ve} = \frac{\pi}{4} k_1 D \frac{e_s}{B_f} \tag{14.11}$$

where k_1 is a constant determined by calibration.

The three principal methods of producing flux density B_f are

1. Permanent-magnet or dc excitation

2. Normal 50- or 60-Hz ac power-line sinusoidal excitation

3. Pulsed excitation of any frequency

The dc magnetic flowmeter is well suited to pulsating flows and for measuring very high conductivity fluids, such as molten metals and mercury. However, for lower-conductivity, ion-conducting fluids, time-variant polarization occurs at the electrode, and for this reason dc excitation is not used. Since polarization potentials are eliminated by sinusoidal excitation, ac or short-duration pulsed-dc systems are most commonly used for these fluids.

Figure 14.15 presents the relationship between signal voltage, excitation voltage, and current for ac and dc systems. In ac systems (Fig. 14.15a), the signal voltage is in phase with the supply current and approximately 90° out of phase with the supply voltage. Since a change in voltage or current results in a signal-voltage change, a reference signal is used, and the ratio of e_s/B_f is measured by the secondary [Eq. (14.11)]. This reference signal may be derived as a constant signal by regulating the coil current, derived from the supply voltage or current, or derived by measuring flux density in the laminated core or within the air gap.

In pulsed-dc systems, a low-frequency on-off (or plus-minus) square-wave current powers the electromagnet, producing the signal voltage shown in Fig. 14.15b. When the field is *on*, the secondary device measures and stores this signal voltage, which is a composite of the flow signal plus non-flow-related voltages. During the period when the field is turned *off*, the secondary measures the non-flow-related voltages. The flow-rate signal is obtained by subtraction.

Whether ac- or dc-powered, the flow signal varies with the field intensity, is in the millivolt range, and contains non-flow-related voltages. These are compensated for by continuously forming a ratio of the signal to a reference signal. It is this ratio that is scaled by water calibration to obtain an overall meter factor in the desired flowing volumetric flow-rate unit.

It is important to recognize that magnetic flowmeters are designed as total flowmetering systems consisting of primary and secondary devices. The function of the primary device is to produce the signal; the secondary device corrects for changes in phase relationships and in the supply voltage (or current) and minimizes non-flow-related voltages associated with the primary and the connecting cables.

Industrial Flowmeters   ISO 6817 (1980) defines the electromagnetic flowmeter as follows:

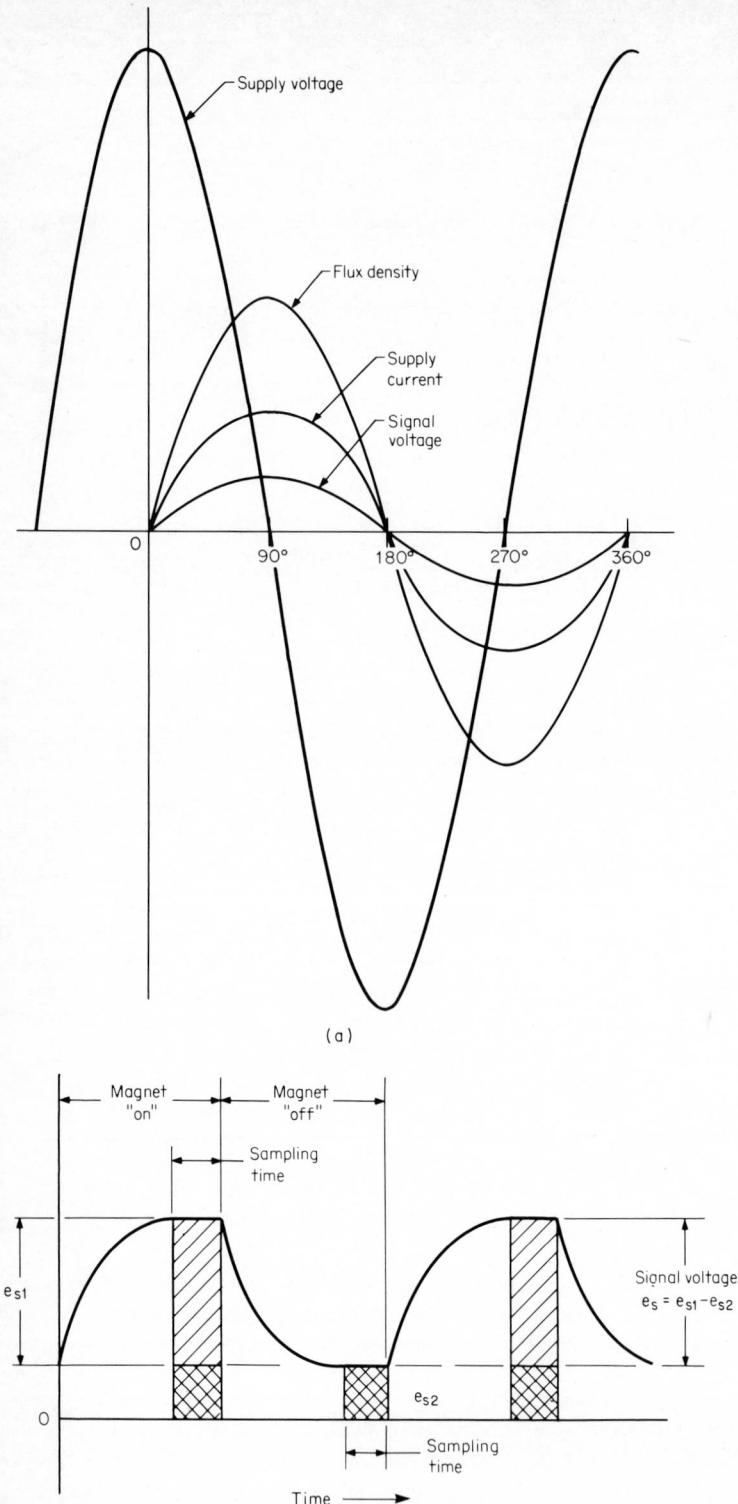

Supply voltage

Flux density

Supply
current

Signal
voltage

0   90°   180°   270°   360°

(a)

Magnet
"on"

Magnet
"off"

Sampling
time

e_{s1}

e_{s2}

Signal voltage
$e_s = e_{s1} - e_{s2}$

Sampling
time

0

Time ⟶

(b)

The electromagnetic flowmeter is an assembly of a primary device through which the process fluid flows, and of a secondary device which converts the low level signal generated by the primary device into a suitable standardized signal for acceptance by industrial instrumentation. The system produces an output signal proportional to volumetric flowrate (or average velocity). Its application is generally limited only by the requirement that the metered fluid shall be conductive and nonmagnetic.

Shown in Fig. 14.16 is an exploded view of one design of primary device. The primary consists of (1) a tube, (2) two coils, (3) electrodes, (4) a laminated iron core, (5) a cover, and (6) end connections. The tube provides support for the assembly and may be of nonmagnetic stainless steel or fiberglass-reinforced plastic. Stainless-steel tubes are lined with a suitable insulating material, such as ptfe, polyurethane, or rubber, to prevent short-circuiting of the generated voltage. The saddle-shaped coils of copper wire produce the magnetic field, which is focused at right angles into the process fluid by the laminated iron core. The generated voltage is detected by two small, diametrically opposed electrodes.

The meter size and liner and electrode materials depend largely on the fluid to be metered. Table 14.7 lists the upper-range-value flow rates given by one manufacturer, and Table 14.8 is a material selection table for electrodes and liners. In general, five basic tube types are used:

1. *Unlined resin tubes.* These are unlined glass-fiber-reinforced tubes with epoxy or polyester resin as the glass bonding agent. The nonconductive, nonpermeable tubes of this type require no liner and are free from the eddy-current problems associated with lined metal tubes. They are used for applications with limited temperature, pressure, and corrosive severity. Reinforced resin flow tubes are lower in price than corresponding sizes of lined metal units but are suitable for a wide variety of applications.

2. *ptfe-lined tubes.* Because of its high temperature rating and its inertness to a wide variety of acids and bases, ptfe is the most commonly used flow-tube lining material. A preformed "tube" of ptfe is mechanically fitted into an AISI type-304 nonmagnetic stainless steel tube, and the ptfe ends are flared out over the tube end connections to effectively isolate the process liquid from the steel. The recommended maximum process temperature for this type of tube is 350°F (175°C).

3. *Polyurethane-lined tubes.* For some applications, a ptfe flow-tube lining does not have adequate abrasion resistance. Where extreme resistance to wear or erosion by solid particles in the process stream is required, polyurethane lining is often the best choice. The polyurethane liner extends over the raised face of the flange. Polyurethane is resilient and abrasion resistant, but it cannot be used at very high temperatures or with strong acids or bases.

4. *Neoprene-lined tubes.* Neoprene-lined flow tubes have greater abrasion resistance than ptfe-lined tubes and are moderately resistant to chemical attack.

Figure 14.15 (opposite) Magnetic-flowmeter systems. (*a*) AC system. (*b*) Pulsed-dc system *(courtesy Fischer & Porter).*

Table 14.7 Flow-Rate Ranges for Selected Liner Materials

Nominal upper-range value flow rates, L/s (gal/min)

Nominal flow-tube size		Power-supply frequency, Hz	ptfe-lined sanitary flow tube		ptfe-lined flow tube		Polyurethane-lined flow tube		Neoprene-lined flow tube		Unlined glass-fiber-reinforced resin flow tube			
											Epoxy resin		Polyester resin	
mm	in		Minimum	Maximum	Minimum	Maximum	Minimum	Maximum	Minimum	Maximum	Minimum	Maximum	Minimum	Maximum
15	½	50 or 60	0.145 (2.3)	0.75 (12.0)	0.145 (2.3)	0.75 (12.0)								
25	1	50 or 60	0.505 (8.0)	3.03 (48.0)	0.505 (8.0)	3.03 (48.0)								
40	1½	50 or 60	1.26 (20.0)	75.7 (120)	1.26 (20.0)	75.7 (120)								
50	2	50 or 60	2.14 (34.0)	12.8 (203)	2.14 (34.0)	12.8 (203)	1.45 (23.0)	7.70 (122)			1.8 (29)	9.7 (154)		
80	3	50 or 60	4.67 (74.0)	29.6 (469)	4.67 (74.0)	29.6 (469)	3.91 (62.0)	24.4 (387)			4.6 (72)	25 (400)		
100	4	50 or 60			8.45 (134)	51.7 (820)	6.75 (107)	41.1 (652)			8.1 (129)	47 (752)		
150	6	50 or 60			18.9 (300)	114 (1815)	16.5 (262)	111 (1760)			19 (295)	111 (1765)		
200	8	50 or 60			33.1 (524)	198 (3130)	30.2 (478)	208 (3300)			34 (527)	195 (3121)		
250	10	50 or 60			52.1 (826)	315 (5000)	48.6 (770)	343 (5430)					54 (834)	309 (4895)
300	12	50 or 60			74.4 (1180)	451 (7150)	70.4 (1116)	500 (7920)					75 (1177)	445 (7050)

350	14	50	96 (1516)	676 (10,744)	96 (1516)	553 (8758)		96 (1516)	532 (8434)
350	14	60	115 (1819)	676 (10,744)	115 (1819)	553 (8758)		113 (1786)	532 (8434)
400	16	50	127 (2000)	895 (14,232)	127 (2000)	730 (11,573)		124 (1961)	695 (11,015)
400	16	60	151 (2381)	895 (14,232)	151 (2381)	730 (11,573)		147 (2326)	695 (11,015)
450	18	50	158 (2500)	1145 (18,210)	158 (2500)	933 (14,780)		158 (2500)	880 (13,942)
450	18	60	191 (3031)	1145 (18,210)	191 (3031)	933 (14,780)		186 (2942)	880 (13,942)
500	20	50	198 (3125)	1427 (22,678)	198 (3125)	1160 (18,378)		192 (3031)	1093 (17,326)
500	20	60	234 (3704)	1427 (22,678)	234 (3704)	1160 (18,378)		235 (3704)	1093 (17,326)
600	24	50	287 (4546)	2037 (32,374)	287 (4546)	1689 (26,751)		281 (4445)	1582 (25,061)
600	24	60	351 (5556)	2037 (32,374)	351 (5556)	1689 (26,751)		334 (5291)	1582 (25,061)
750	30	50	454 (7195)	3239 (51,468)	454 (7195)	2673 (42,246)		445 (7043)	2577 (40,820)
750	30	60	544 (8621)	3239 (51,468)	544 (8621)	2673 (42,246)		535 (8475)	2577 (40,820)
900	36	50	658 (10,417)	4718 (74,969)	658 (10,417)	3868 (61,267)		646 (10,205)	3732 (59,121)
900	36	60	789 (12,500)	4718 (74,969)	789 (12,500)	3868 (61,267)		776 (12,196)	3732 (59,121)

SOURCE: The Foxboro Co.

Table 14.8 Liner and Electrode Selection Table

Process liquid	Flow-tube lining material						Electrode material			
	ptfe	Sanitary	Polyurethane	Epoxy glass fiber	Polyester glass fiber	Neoprene	Platinum–10% iridium	Tantalum	316 stainless steel	Hastelloy C
Acetaldehyde	A†		X	X	X	X	A	A	B	A
Acetic acid, 10%	A		X	65	90	A	A	A	B	A
Acetic acid, 50%	A		X	35	70		A	A	X	A
Acetone	A		X			B	A	A	A	A
Alcohol and glycerin	A	X	X			B	A	A	A	A
Alum	A			90	90	B	A	A	X	X
Alumina	A	X	X	X	X	X	A	A	A	A
Aluminum chloride	A			90	90	A	A	A	X	X
Aluminum chlorohydrate	A				90		A	A	X	X
Aluminum fluoride	A			65	25	B	A	X	X	X
Aluminum hydroxide	A			65			A	A	B	X
Aluminum nitrate	A			90	70	A	A	X	X	X
Aluminum sulfate	A		B	90	90	A	A	A	B	B
Ammonium bicarbonate	A			90	70		A	A	X	X
Ammonium bifluoride	A						A	A	X	X
Ammonium bisulfate	A	X		90	70	A	A	A		X
Ammonium carbonate	A			90	25	A	A	A	X	X
Ammonium chloride	A			90	90	A	A	A	X	B
Ammonium fluoride	A			65			A	X	X	B
Ammonium hydroxide, 10%	A		A	90	60	A	A	B	B	B
Ammonium hydroxide, 20%	A		A	65	60	A	A	B	B	B
Ammonium hydroxide, 30%	A		A	35	35	A	B	A	B	B
Ammonium nitrate	A		X	90	90	C	A	A	X	B
Ammonium persulfate	A		X	X	80	B	A	A	A	X
Ammonium phosphate	A			65	25	B	A	A	X	X
Ammonium sulfate	A		A	90	90		A	A	X	X
Antimony pentachloride	A				20		A	A	X	X
Antimony trichloride	A			65	90		A	A	X	X
Arsenic acid	A	X			20	B	A	A	X	X
Arsenious acid	A				80		A	A	X	X
Barium acetate	A				60		A	A	X	X
Barium carbonate	A			90	90		A	A	X	X
Barium chloride	A		A	90	90	A	A	A	X	B
Barium hydroxide	A		A	90	20	A	A	X	B	X
Barium sulfate	A		A	90	20	B	A	A	X	X

Table 14.8 Liner and Electrode Selection Table (*Continued*)

Process liquid	ptfe	Sanitary	Polyurethane	Epoxy glass fiber	Polyester glass fiber	Neoprene	Platinum- 10% iridium	Tantalum	316 stainless steel	Hastelloy C
			Flow-tube lining material					Electrode material		
Barium sulfide	A		A	90	60	B	A	A	X	X
Bauxite slurry	B	X	A			B	A	A	A	A
Beer	A	X					A	A	A	A
Black liquor (strong)	A	X	X	X	X	X	A	B		
Black liquor (weak)	A	X	X	X	X	X	A	A		
Borax	A		A	90	20	B	A	A	X	X
Brine	A		X	X	X	A	A	A	B	A
Cadium chloride	A			90	20		B	A	X	X
Calcium bisulfite	A		A	90	80	B	A	A	X	X
Calcium carbonate	A			90	90		A	A	B	B
Calcium chlorate	A			90	90		A	A	X	X
Calcium chloride	A		B	90	90	A	A	A	X	X
Calcium hydroxide, 25%	A		A	90	60	A	A	A	X	B
Calcium hypochlorite	A			X	30	X	A	A	X	B
Calcium nitrate	A		A	90	90	A	A	A	X	X
Calcium sulfate	A			90	90		A	A	X	X
Cheese	A	A	X	X	X	X	A	A	A	A
Chlorine dioxide	A	X	X	X	X	X	X	A	X	X
Chloroacetic acid, 25%	A			35	90		A	A	X	B
Chloroacetic acid, 50%	A				60		A	A	X	B
Chloroacetic acid, 100%	A						A	A	X	B
Chromic acid, 30%	A		X	25	X	X	A	A	X	X
Chromic acid, 100%	A		X	X	X	X	A	A	X	X
Chromium sulfate	A		X		60	X	A	A	X	B
Clay slurry	A	X	B	90	60	A	A	A	X	B
Coal and water slurry	B	X	A	X	X	A	A	A	A	A
Copper chloride	A		A	90	90	A	X	A	X	X
Copper cyanide	A		A		90	A	A	A	B	B
Copper fluoride	A			90			A	X	X	X
Copper nitrate	A			90	90		A	A	X	B
Copper oxychloride	A						A	X	X	X
Copper sulfate	A		A	90	90	A	A	A	B	B
Dairy products	B	A	X	X	X	X	A	A	A	A
Dyes	A	X	X	X	X	X	A	A	A	A
Ferric chloride	A			90	90	B	X	A	X	B

Table 14.8 Liner and Electrode Selection Table (*Continued*)

Process liquid	ptfe	Sanitary	Polyurethane	Epoxy glass fiber	Polyester glass fiber	Neoprene	Platinum–10% iridium	Tantalum	316 stainless steel	Hastelloy C
Ferric nitrate	A			90	90	B	A	A	X	B
Ferric sulfate	A			90	90	A	A	A	X	B
Ferrous chloride	A			90	90		A	A	X	X
Ferrous nitrate	A				90		A	A	X	B
Ferrous sulfate	A			90	90		A	A	X	X
Fluosilicic acid, 10%	A			90	65	B	A	X	X	X
Fluosilicic acid, 40%	A					B	A	X	X	X
Formaldehyde	A		X	65		C	A	A	X	B
Formic acid, 10%	A		X		65	B	A	A	X	B
Formic acid, 50%	A		X		20	X	A	A	X	B
Formic acid, 100%	A		X			X	A	A	X	B
Glucose syrup	A						A	A	A	A
Glycerin	A		A	90	90	A	A	A	A	A
Green liquor	A	X	X	X	X	X	A	A	A	A
Hydrobromic acid, 50%	A		X	65	70		X	A	X	X
Hydrochloric (cone)	A		X	90	70	C	B	A	X	X
Hydrocyanic (all)	A			X	90	C	A	A	B	B
Hydrofluoric acid, 10%	A		X	X	90	C	A	X	X	X
Hydrofluoric acid, 20%	A		X	X	50	C	A	X	X	X
Hydrofluosilicic, 35%	A				20	C	A	X	X	B
Hydrogen peroxide, 5%	A		X	65	65		A	A	B	B
Hydrogen peroxide, 30%	A		X	25	20		A	A	B	B
Hypochlorous acid, 10%	A			90			X	A	X	B
Hypochlorous acid, 20%	A				70		X	A	X	B
Latex	A	X	X	65	65	X	A	A	A	A
Lead acetate	A			90	90	C	X	A	X	X
Lime slurry	A	X	A	90	60	A	A	A	X	B
Limestone slurry	A	X	A	90	60	A	A	A	X	B
Lithium chloride	A						A	A	X	B
Magnesium bisulfite	A			90	80				B	B
Magnesium carbonate	A			90	80		A	A	B	B
Magnesium chloride	A		B	90	90	B	A	A	X	B
Magnesium hydroxide	A		B	90		A	A	X	X	X
Magnesium nitrate	A			90			A	A	X	X
Magnesium sulfate	A			90	90	A	A	A	B	X

Table 14.8 Liner and Electrode Selection Table (*Continued*)

Process liquid	ptfe	Sanitary	Polyurethane	Epoxy glass fiber	Polyester glass fiber	Neoprene	Platinum– 10% iridium	Tantalum	316 stainless steel	Hastelloy C
Mercuric chloride	A				90	B	A	A	X	X
Mercurous chloride	A				90					
Molasses	A	A	X			X	A	A	A	A
Mud drilling	X	X	A	X	X	A	A	A	A	A
Neuphor	A		X	X	X	X	A	A	A	A
Nickel chloride	A			90	90	B	A	A	X	B
Nickel nitrate	A			90	90		A	A	X	X
Nickel sulfate	A		A		90	A	A	A	X	X
Nitric acid, 40%	A		X	X	20	X	A	A	X	X
Nitric acid, 70%	A		X	X	X	X	A	A	X	X
Oleum	A	X	X	X	X	X	A	X	X	X
Oxalic acid (saturated)	A		X	X	X	C				
Paper stock	A	X		90	90	X	A	A	A	A
Perchloric acid, 70%	A			25		A	A	A	X	X
Phosphate slurry	A	X		X	X		A	A	X	X
Phosphoric, 75%	A			90	90		A	A	X	X
Phosphoric, 85%	A			X	90		A	A	X	X
Potassium aluminum sulfate	A				90		A	A	X	X
Potassium bicarbonate	A			90	20		A	A	B	B
Potassium carbonate	A			90	20		A	A	B	B
Potassium chloride	A		X	90	90	A	A	A	X	B
Potassium dichromate	A		X	90	90	A	A	A	B	X
Potassium hydroxide, 10%	A		C	90	65	B	A	X	X	X
Potassium hydroxide, 20%	A		C		20	B	A	X	X	X
Potassium hydroxide, 45%	A		C			B	A	X	X	X
Potassium nitrate	A		X	90	90	A	A	A	X	X
Potassium permanganate	A			65	90		A	A	X	X
Potassium persulfate	A				90		A	A	B	B
Potassium sulfate	A		X	65	90	A	A	A	B	X
Sewage, raw	A	X	A			A	A	A	A	A
Silver nitrate	A		A	90	90	X	A	A	X	B
Sludge, activated	A	X	A	90	90	A	A	A	A	A
Sludge, primary	A	X	B			B	A	A	A	A
Sludge, thickened	X	X	A	X	B	A	A	A	A	A
Sludge, waste	A	X	B			B	A	A	A	A

Table 14.8 **Liner and Electrode Selection Table (*Continued*)**

Process liquid	ptfe	Sanitary	Polyurethane	Epoxy glass fiber	Polyester glass fiber	Neoprene	Platinum– 10% iridium	Tantalum	316 stainless steel	Hastelloy C
				Flow-tube lining material				Electrode material		
Sodium acetate	A		X	90	90	C	A	A	X	X
Sodium bicarbonate	A			90	60	A	A	A	B	B
Sodium bisulfate	A			90	90		A	A	X	X
Sodium bisulfide	A						A	A	B	X
Sodium bisulfite	A				90	A	A	A	X	B
Sodium borate	A					B	A	A	X	X
Sodium bromide	A			90	90		A	A	X	X
Sodium carbonate	A			90	70		A	A	B	B
Sodium chlorate	A				90		A	A	X	X
Sodium chloride	A		C	90	90	B	A	A	X	X
Sodium chlorite	A				65		B	A	X	X
Sodium chromate	A				90		A	A	X	X
Sodium cyanide	A			90	90	B	X	A	B	
Sodium dichromate	A			90	90					
Sodium ferricyanide	A				90		X	B	X	
Sodium ferrocyanide	A			90			X	B	X	
Sodium fluoride	A			90			A	X	X	X
Sodium hexametaphos	A									
Sodium hydrosulfide	A						A	A	B	B
Sodium hydroxide, 5%	A		C	90	90	X	A	X	B	X
Sodium hydroxide, 25%	A		C	90	60	X	A	X	B	X
Sodium hydroxide, 50%	A		C	90	20	X	A	X	B	X
Sodium hypochlorite, 20%	A		X	X		B	A	A	X	B
Sodium nitrate	A			90	90	X	A	A	B	X
Sodium nitrite	A				90		A	A	X	X
Sodium silicate	A			65	90	X	A	A	X	X
Sodium sulfate	A		C	90	90	A	A	A	X	X
Sodium sulfide	A				90	X	A	A	X	X
Sodium sulfite	A			90	90	X	A	A	B	X
Sodium tetraborate	A						A	A	B	B
Sodium thiosulfate	A		X	65		X	X	X	X	A
Sulfuric acid, 10%	A		C	90	90	C	A	A	X	X
Sulfuric acid, 25%	A		X	65	90	B	A	A	X	X
Sulfuric acid, 50%	A		X	35	90	C	A	A	X	X
Sulfuric acid, 100%	A		X	X		X	A	X	X	X

Table 14.8 Liner and Electrode Selection Table (*Continued*)

Process liquid	Flow-tube lining material						Electrode material			
	ptfe	Sanitary	Polyurethane	Epoxy glass fiber	Polyester glass fiber	Neoprene	Platinum–10% iridium	Tantalum	316 stainless steel	Hastelloy C
Sulfurous acid, 10%	A		X	90		B	A	A	X	X
Titanium dioxide	A	X	X			X	A	A	A	A
Trisodium phosphate	A			65	20		A	A	B	B
Urea, 50%	A	X	X	X	X	X	A		A	A
Water, fresh	A		A	90	90	A	A	A	A	A
White liquor	A	X				B	A	A	B	A
Zinc chloride	A			90	90	X	A	A	X	X
Zinc sulfate	A	X				A	A	A	B	B

†Key: A = preferred material—virtually unlimited life; B = satisfactory material—reasonable life under most conditions; C = fair—some wear effect; X = not recommended—definitely troublesome, will not survive, or unacceptable; a *blank* indicates no information is available; the *numbers* are maximum recommended temperatures (°C) of process liquids in flow tubes.

SOURCE: The Foxboro Co.

They can be used at higher temperatures than polyurethane-lined tubes and are considered a good general-purpose flow tube.

5. *Sanitary lined tube.* A ptfe-lined flow tube with quick-disconnect sanitary end connections is used in consumable-product processes. These sanitary flow tubes are ideally suited for use with dairy products, beer, soft drinks, coffee, molasses, and corn syrup. Processed products, such as catsup and other viscous, sticky, or otherwise difficult-to-measure liquids, are also easily measured.

The electrodes are made of a nonmagnetic material such as AISI type-316L stainless steel, platinum–10 percent iridium, hastelloy C, titanium, or tantalum. Conical electrodes, electrode boil-off, and ultrasonic electrode cleaning are sometimes available as optional features for removing contamination from the electrodes.

Application Magnetic flowmeters are suitable for use with liquids that have electrical conductivities of 1 μS/cm or larger. For clean liquids, the minimum upper-range velocity is approximately 1 ft/s (0.3 m/s), and a 10:1 range is usually given. When liquid-bearing solids, abrasive slurries, sludges, and immiscible liq-

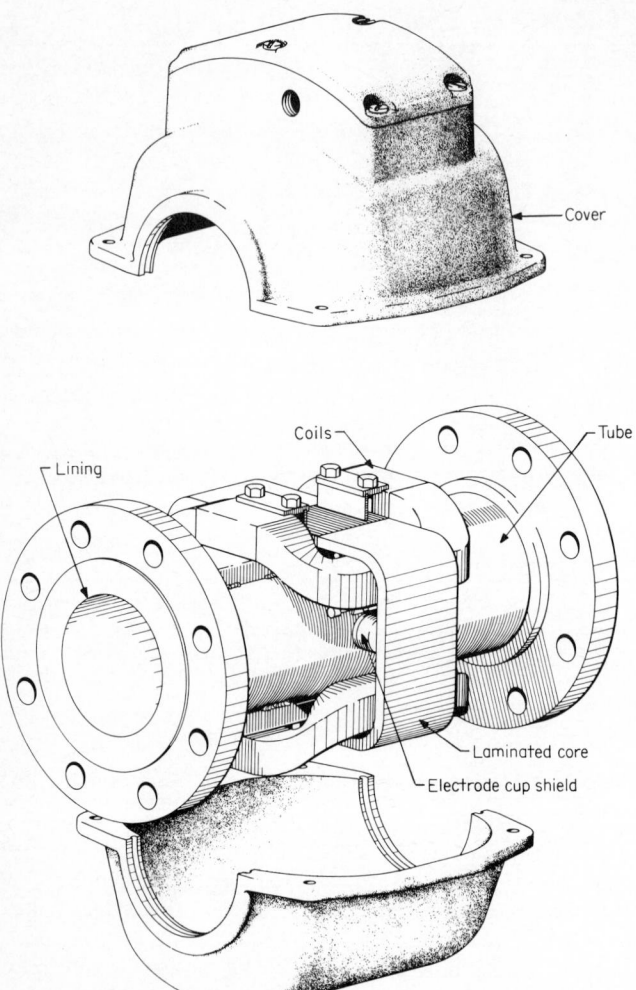

Figure 14.16 Basic construction of the primary device of a magnetic flowmeter *(courtesy The Foxboro Co.).*

uids are metered, a minimum or maximum velocity is suggested by manufacturers to reduce coating, liner wear, and maintenance cost.

The energy dissipated by the coils causes local heating in the flow tube. At no-flow condition, with the coils energized, the temperature rise is approximately 30°F for an unfilled tube and 20°F for a filled tube. Depending on whether the coils are series- or parallel-connected, the temperature rise can be as high as 100°F (parallel) for an unfilled tube. For this reason, manufacturers suggest that the primary always be kept filled with liquid, to prevent liner damage and to minimize baking of solids into the liner and/or electrodes.

In general, settleable solids require a minimum velocity of 5 ft/s (1.5 m/s) to minimize coating. The primary should be installed vertically, with the flow upward, and at no-flow condition the tube should remain full to prevent excessive temperature rise. For abrasive slurries, rubber or some other abrasion-resistant material should be selected for the liner. An upstream protection ring, similar to a high-beta-ratio orifice, is sometimes installed to protect the leading edge of the flowmeter. The velocity should not exceed 10 ft/s (3 m/s).

Performance   Magnetic flowmeters are water-calibrated, and a standardized output signal is scaled to suitable volumetric flow-rate units. Performance specifications are given at reference conditions based on type testing for ranges of influence quantities, such as supply voltage, frequency, temperature, and fluid conductivity. These are specified by the manufacturer.

Reference accuracy envelopes are specified, within which bias error and data-point precision are included. Figure 14.17 presents typical reference-condition accuracy envelopes, and Table 14.9 contains reference conditions specified in ISO 6817 (1980) for flow conditions and for the primary and secondary device.

Table 14.9   Magnetic-Flowmeter Reference Conditions

	Flow (water)	Primary device	Secondary device
Temperature	40–100°F (4–40°C)	32–86°F (0–30°C)	50–86°F (10–30°C)
Pressure	<75 psig (<500 kPa)	<75 psig (<500 kPa)	
Conductivity	50–5000 μS/cm†		
Power supply			
Voltage		Rated voltage ± 2%	
Frequency		Nominal frequency ± 0.5%	
Distortion		≤ 0.05% (IEC Publication 359)	
Piping	Minimum: 5 pipe diameters		
Signal output			See IEC publication 381A

†Microsiemens per centimeter.

SOURCE: ISO 6817 (1980).

The magnetic flowmeter's signature curve shows little, if any, change over the laminar-to-turbulent profile transition regime, and Reynolds-number corrections are not applied. Except for a minimum conductivity level, the effects of fluid properties on performance are minimal. However, magnetic-flowmeter system designs vary widely among manufacturers, and it is important that the manufacturer's recommendations as to conductivity limits, cable lengths, installation procedure, etc., be followed for reliable performance.

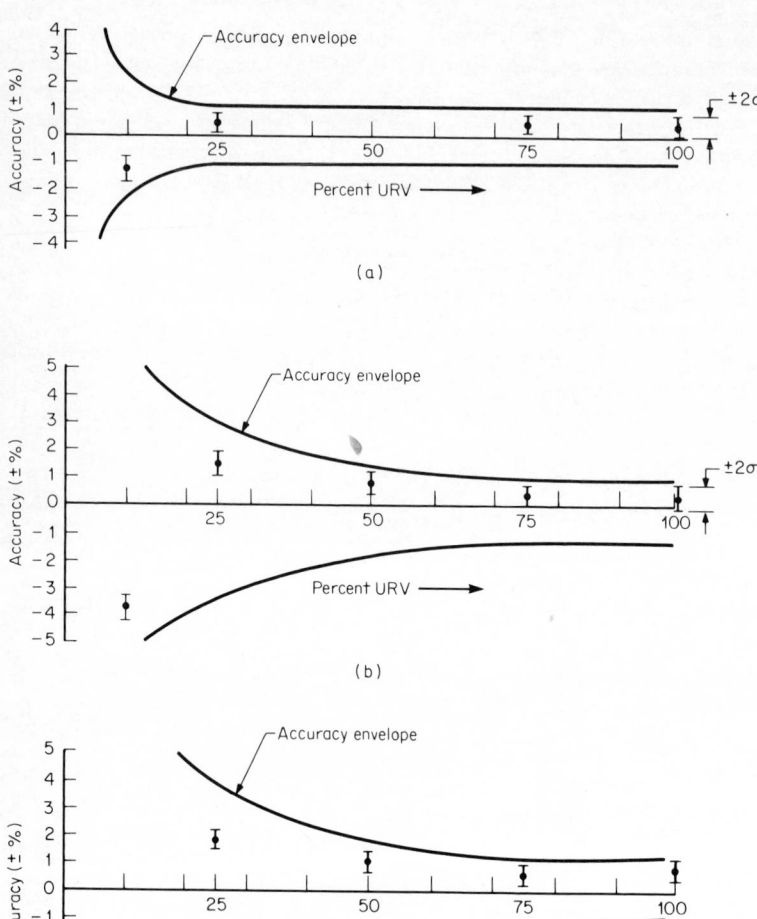

Figure 14.17   Reference-condition accuracy envelopes. (*a*) 1 percent of rate or ±0.2 percent of URV, whichever is greatest. (*b*) ±1 percent of URV flow rate. (*c*) ±0.5 percent URV ±0.5 percent of rate.

Ultrasonic Flowmeters

Lynnworth (1979) lists eight different types of ultrasonic flowmeters, but two basic types are most commonly used in industry: counterpropagating (time-of-flight) and reflection (Doppler). Both are used for liquids—counterpropagating meters for clean liquids, and the Doppler type for flows that contain the particulate matter required for signal reflection.

Counterpropagating Flowmeter The operation of the counterpropagating flowmeter is based on the fact that the speed of an acoustic pressure wave increases in the direction of flow and decreases when directed against the flow, which causes differing transit times. By measuring these two times, or by measuring frequency differences, the pipeline velocity is inferred. In practice, up to four acoustic paths are used to better estimate the average velocity and, hence, the average volumetric flow rate.

Figure 14.18 shows the velocity vector diagrams that describe the relationships used to develop the flow equations for counterpropagating flowmeters. An upstream piezoelectric transmitter provides a pulse that travels at the speed of

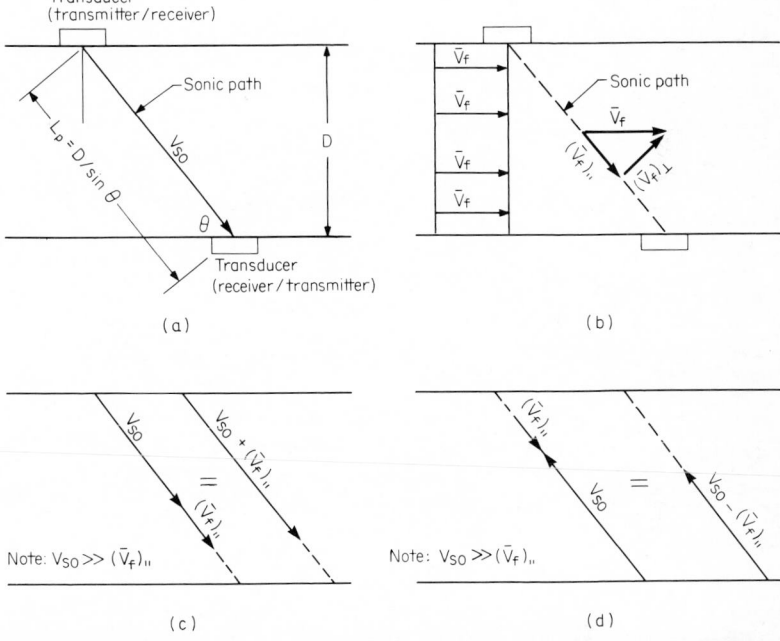

Figure 14.18 Vector diagrams for counterpropagating ultrasonic flowmeters. (*a*) No flow. (*b*) Velocity vector parallel to sonic path. (*c*) Upstream apparent velocity. (*d*) Downstream apparent velocity.

sound through the liquid, along path length L_p and at path angle θ. This pulse is detected by the downstream piezoelectric receiver. The upstream transmitter and downstream receiver (transducers) are actually used alternately as transmitter and receiver, first to direct a pulse downstream, parallel to the velocity vector (Fig. 14.18*b*), and then upstream, in opposition to the velocity vector (Fig. 14.18*c*). The apparent velocity of the downstream pulses is higher than sonic velocity, and that of the upstream pulses is lower.

The downstream transit time is

$$t_{\text{DN}} = \frac{L_p}{V_{\text{SO}} + \overline{V}_f \cos \theta} \tag{14.12}$$

and the upstream time is

$$t_{\text{UP}} = \frac{L_p}{V_{\text{SO}} - \overline{V}_f \cos \theta} \tag{14.13}$$

From these time equations the following relationships are derived:

$$\frac{1}{t_{\text{DN}}} - \frac{1}{t_{\text{UP}}} = \frac{t_{\text{UP}} - t_{\text{DN}}}{t_{\text{DN}} t_{\text{UP}}} = \frac{2 \cos \theta}{L_p} \overline{V}_f \tag{14.14}$$

The average volumetric flow rate for a circular pipe then is

$$q_{\text{cfs}} = K_{UF} \left[\frac{\pi D_F^3}{4 \sin 2\theta} \right] \frac{t_{\text{UP}} - t_{\text{DN}}}{t_{\text{UP}} t_{\text{DN}}} \tag{14.15}$$

for U.S. units, and

$$q_{\text{c ms}}^* = K_{UF} \left[\frac{\pi D_M^3}{4 \sin 2\theta} \right] \frac{t_{\text{UP}} - t_{\text{DN}}}{t_{\text{UP}} t_{\text{DN}}} \tag{14.16}$$

for SI units, where the flow coefficient K_{UF} is introduced as a Reynolds-number correction for velocity profile. The bracketed quantities are usually assumed constant, and the flow coefficient and time term are computer-calculated to provide the flow rate.

In another approach, called the *frequency-difference* or (sing-around) approach, reception of the acoustical signal at the receiver initiates transmission of the subsequent signal. Upstream and downstream frequencies are generated proportional to their respective transit times, so that

$$(f_{\text{Hz}})_{\text{UP}} = \frac{1}{t_{\text{UP}}} \qquad (f_{\text{Hz}})_{\text{DN}} = \frac{1}{t_{\text{DN}}} \tag{14.17}$$

Substitution into Eq. (14.15) and (14.16) yields

$$q_{\text{cfs}} = K_{UF} \left[\frac{\pi D_F^3}{4 \sin 2\theta} \right] \{ (f_{\text{Hz}})_{\text{DN}} - (f_{\text{Hz}})_{\text{UP}} \} \tag{14.18}$$

for U.S. units, and

$$q_{\text{c ms}}^* = K_{UF} \left[\frac{\pi D_M^3}{4 \sin 2\theta} \right] \{ (f_{\text{Hz}})_{\text{DN}} - (f_{\text{Hz}})_{\text{UP}} \} \tag{14.19}$$

for SI units.

Equations (14.12) through (14.19) are based on the assumption that the transducers are in direct contact with the fluid. If a protecting intervening material or a clamp-on transducer is used, the transit times and entering path angle are altered. These effects are mathematically modeled and corrected for by the secondary device.

The flow coefficient K_{UF} for a multipath ultrasonic flowmeter is defined in ANSI/ASME MFC-YY (1982) as

$$K_{UF} = \frac{q_v}{A_P \sum_{i=1}^{n} W_i \overline{V}_i} \qquad (14.20)$$

where A_P is the pipe area (in square feet or square meters), W_i is a weighting function for each path, $\overline{V}_i$ is the axial velocity along each path, and n is the number of paths. The flow coefficient–pipe Reynolds number signature curve shown in Fig. 14.19 was derived by Vignos (1981) in the turbulent flow regime for a

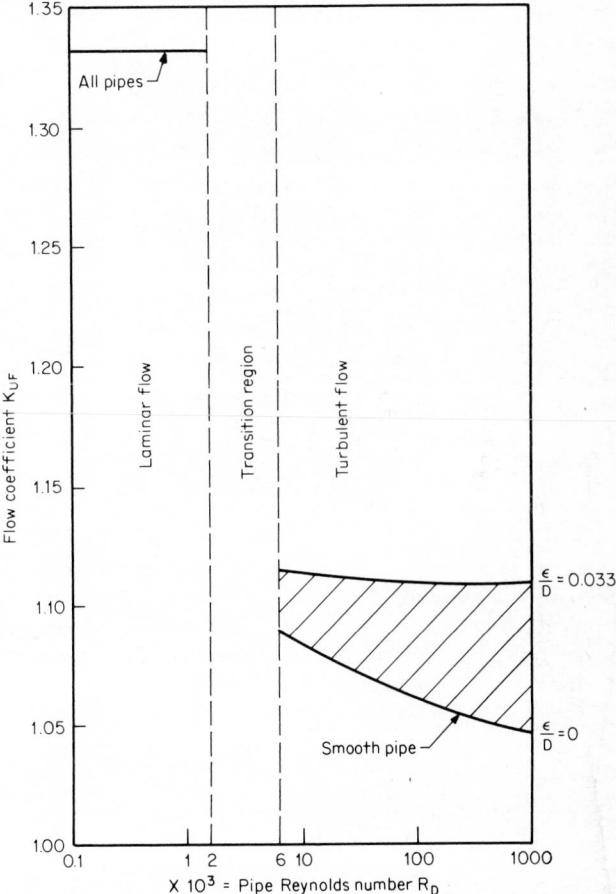

Figure 14.19 Ultrasonic-flowmeter signature curve for single path along pipe axis.

single-path flowmeter with transducers mounted along the pipe axis. The Vignos equation for the flowmeter with a single path along the pipe axis is

$$K_{UF} = 1 + \frac{5}{4} \sqrt{\frac{f}{8}} \qquad (14.21)$$

where f is the pipe friction factor defined by Eq. (5.9).

The flow-coefficient variation from laminar to turbulent flow (approximately 30 percent) and variations within the turbulent regime can be reduced by locating the transducers off the pipe axis or by increasing the number of paths.

Secondary flows caused by upstream fittings (see Chap. 5) can significantly alter the flow coefficient. The magnitude of this effect depends on the number of transducers and their orientation with respect to the nonaxial velocity components. Long, straight runs of pipe, the use of flow conditioners, and multipath flowmeters are suggested (ANSI/ASME MFC-YY, 1982) to obtain desired accuracy.

Doppler Flowmeter   Doppler ultrasonic flowmeters operate by the reflection of sonic energy from scatterers (particulate matter) in the fluid, back to a receiver. If the scatterers are moving at the velocity of the fluid, the Doppler frequency shift is proportional to the volumetric flow rate. The receiver may be the same transducer as the transmitter or a separate transducer. A typical meter consists of two piezoelectric transducers (Fig. 14.20) clamped to the pipe or con-

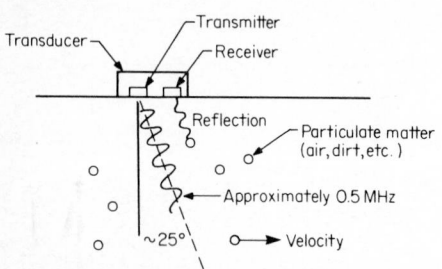

Figure 14.20   Doppler flowmeter.

tained within a housing. The transmitter projects a continuous beam into the fluid, at a frequency of from 0.5 to 10 MHz. The receiver measures the frequency of the reflections from particles in the fluid. This frequency is Doppler-shifted by the velocity of the particles. The volumetric flow rate is then calculated as

$$q_{cfs} = \left[\frac{\pi}{4} D_F^2 \frac{V_{SO}}{\sin \theta} \right] \frac{\Delta f_{Hz}}{(f_{Hz})_{CW}} \qquad (14.22)$$

for U.S. units, and

$$q_{cms}^* = \left[\frac{\pi}{4} D_M^2 \frac{V_{SO}^*}{\sin \theta} \right] \frac{\Delta f_{Hz}}{(f_{Hz})_{CW}} \qquad (14.23)$$

for SI units, where Δf_{Hz} is the Doppler frequency shift, θ the beam angle, and $(f_{Hz})_{CW}$ the transmitter's fixed frequency.

Industrial Flowmeters   Industrial ultrasonic flowmeters are designed as total systems. The primary device consists of two or more transducers which are either permanently attached to a spool piece (Fig. 14.21a) or clamped onto existing pipe (Fig. 14.21b). However, transducer designs vary widely among manufacturers, ranging from clamp-on types to those in recessed pockets in the pipe wall (Fig.

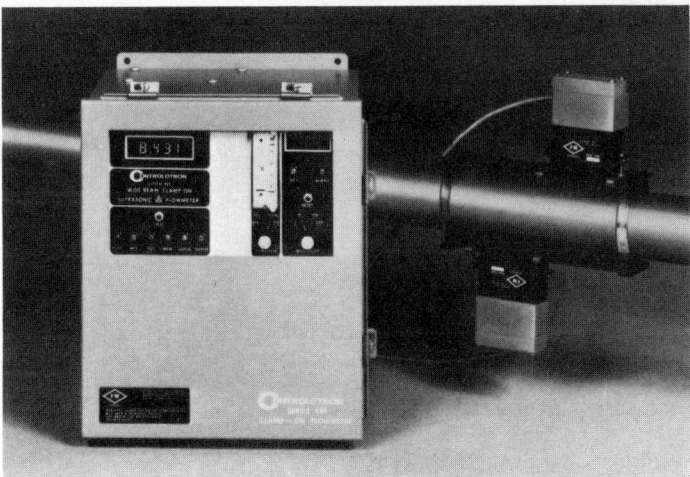

Figure 14.21   Industrial ultrasonic flowmeters. (a) **Spool-piece meter** *(courtesy Panametrics).* (b) **Clamp-on meter** *(courtesy Controlotron).*

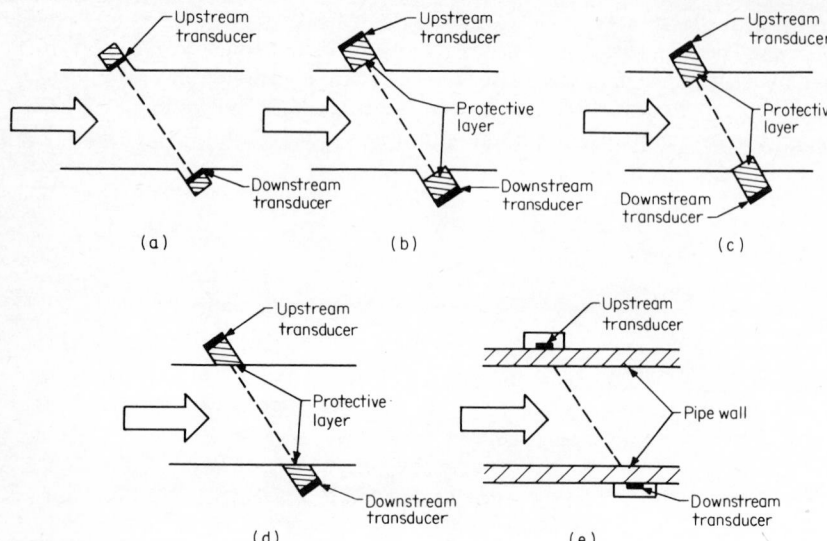

Figure 14.22 Transducer designs. (*a*) Recessed process exposed. (*b*) Recessed process protected. (*c*) Protruding process protected. (*d*) Flush process protected. (*e*) Clamp-on.

14.22). Single paths or multiacoustic paths may be used, and they may traverse the pipe along a diameter, along a chord, or along the pipe axis (Fig. 14.23). The function of the secondary device is to make the measurements (of time or frequency), process the data, perform computations for Reynolds-number and other corrections, and output a standardized signal.

Ultrasonic flowmeters have been successfully used for many fluids, ranging from crude oil to sludge flows. However, the applicability of a particular type of meter is very much a function of the experience of its manufacturer. For this reason, the performance and application specifications of the various manufacturers can differ widely. Schmidt (1980), reporting on Doppler and counterpropagating industrial flowmeters, notes that meters *appear* to be ±0.5 percent repeatable, and ±1 percent repeatable among different but related fluids. However, he indicates that the as-received calibration factor may vary considerably and suggests that meters be field calibrated.

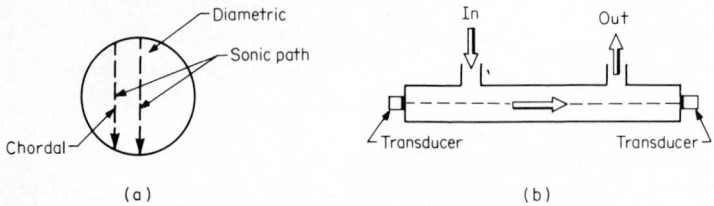

Figure 14.23 Acoustic paths. (*a*) Chordal and diametric. (*b*) Axial.

Presently, no single design can be considered a multipurpose flowmeter; each manufacturer tends to choose certain desirable features for a particular design. Many proprietary (and unpublished) laboratory flow tests have been conducted on the various designs, but little independent data is available to indicate performance under laboratory or industrial conditions. Usually, the manufacturer and user work closely together to verify applicability.

Variable-Area Flowmeters

Principle of Operation In differential producers the restricted flow area is fixed, and the differential pressure varies and is used to infer flow rate. In the most common variable-area flowmeters (Fig. 14.24a), a float, moving vertically within a linearly tapered tube, exposes a variable area to the flow. At design conditions the differential pressure remains constant across the float, because buoyant and weight forces are constant and the float height changes to expose the area that satisfies Bernoulli's equation. The flow area between tube and float is then an indication of the flow rate. Since the float height is a linear function of

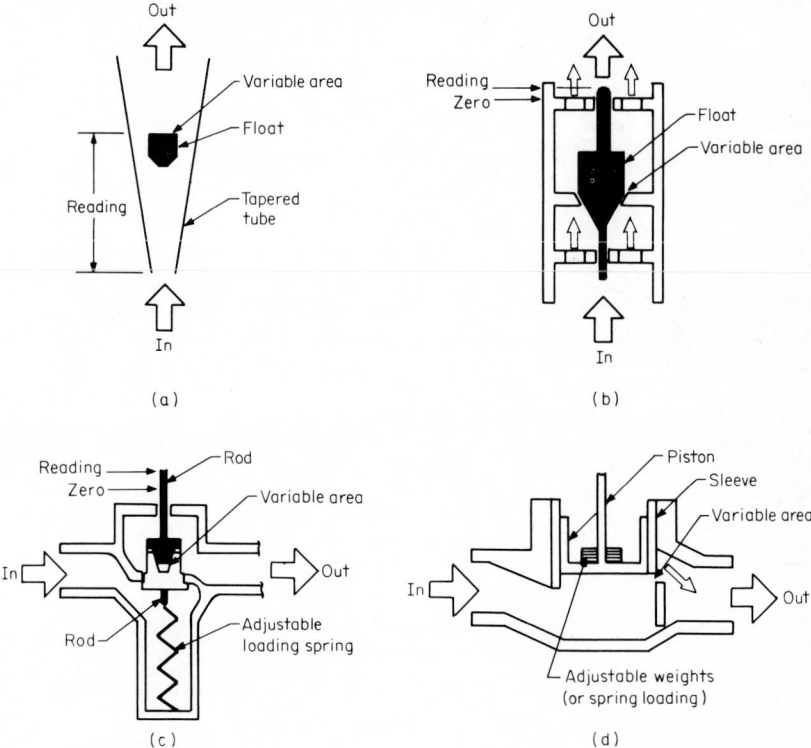

Figure 14.24 Types of variable flowmeters. (a) Tapered tube. (b) Orifice and plug. (c) Slotted cylinder. (d) Piston.

the flow area, visual observation or some other detection means allows the flow rate to be observed.

Several design variations are used to provide an area that varies with flow rate. These include an orifice and plug (Fig. 14.24b), a slotted cylinder and piston (Fig. 14.24c), and a piston (Fig. 14.24d). The basic operating principle remains the same in all cases: to alter the flow area by maintaining a constant pressure drop across a variable area, and then to infer the flow rate from position.

The basic geometry of a tapered-tube variable-area meter is shown in Fig. 14.25. The fluid enters at the bottom, flows upward through the tapered tube,

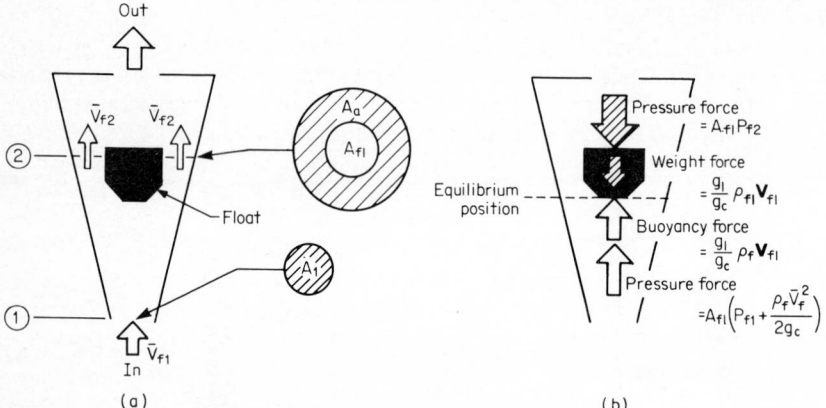

Figure 14.25 Variable-area flowmeter. (a) Area and velocity relationships. (b) Force diagram.

accelerates to a higher velocity through the annular area around the float, and leaves at the top. Bernoulli's general energy equation (9.10) for a liquid, written between entering area and annular area (Fig. 14.25a), yields, in the English engineering system of units (pounds-force, pounds-mass, feet, seconds),

$$\frac{P_{f1}}{\rho_f} + \frac{\overline{V}_{f1}^2}{2g_c} + H_{EL1} = \frac{P_{f2}}{\rho_f} + \frac{\overline{V}_{f2}^2}{2g_c} + H_{EL2} \qquad (14.24)$$

Neglecting elevation difference gives

$$P_{f1} - P_{f2} = \frac{\rho_f}{2g_c}(\overline{V}_{f2}^2 - \overline{V}_{f1}^2) \qquad (14.25)$$

If it is assumed for simplicity that the downward pressure acting on the float is the static pressure at section 1, and that the upward pressure is the total pressure [Eq. (9.7)], then for static equilibrium (Fig. 14.25b) at any float position the sum of the forces becomes

$$A_{fl}\left(P_{f1} + \frac{\rho_f \overline{V}_f^2}{2g_c}\right) + \frac{g_l}{g_c}\rho_f \mathbf{V}_{fl} = \frac{g_l}{g_c}\rho_{fl}\mathbf{V}_{fl} + A_{fl}P_{f2} \qquad (14.26)$$

where A_{fl} is the effective area of the float, $\mathbf{V}_{fl}$ the float volume, P_{f1} and P_{f2} the upstream and downstream presssures at sections 1 and 2, and $\overline{V}_{f1}$ and $\overline{V}_{f2}$ the upstream and downstream velocities at sections 1 and 2. Mass flow continuity between sections 1 and 2 for a constant-density fluid (liquid) gives

$$A_a \overline{V}_{f2} = A_1 \overline{V}_{f1} \tag{14.27}$$

where A_a is the annular area between float and tube wall. Introducing a flow coefficient K_{VA} to correct for factors not included in the analysis and combining Eqs. (14.25) to (14.27) give the basic liquid equation for variable-area meters in U.S. units as

$$q_{\text{cfs}} = \left[K_{VA} \sqrt{\frac{2g_l \mathbf{V}_{fl}}{A_{fl}}} \right] \sqrt{\frac{\rho_{fl}}{\rho_f} - 1} \; A_a \tag{14.28}$$

By a similar derivation the SI-unit equation becomes

$$q^*_{\text{c ms}} = \left[K_{VA} \sqrt{\frac{2g^*_l \mathbf{V}_{fl}}{A^*_{fl}}} \right] \sqrt{\frac{\rho^*_{fl}}{\rho^*_f} - 1} \; A^*_a \tag{14.29}$$

For a given flowmeter the quantity in brackets is assumed constant, and the annular area A_a is directly proportional to the float height. For compressible fluids, *Fluid Meters* (ASME, 1971) introduces an expansion factor to correct for density variation across the float. This correction is usually negligible and, by calibration, is included in the flow coefficient. Equations (14.28) and (14.29) are used as the basic correlation equations for liquids and gases, for selecting float materials and predicting changes in fluid density.

Engineering Equations   Variable-area flowmeters are usually calibrated for a specific fluid, and a correction factor is applied if a meter is to be used for a fluid of different density or viscosity. Viscosity corrections are, however, difficult to calculate, and for this reason special float designs (Fig. 14.26) are used or the meter is calibrated with the new fluid. A *viscosity immunity ceiling*, below which no viscosity correction is required, is sometimes given for a float design. Viscosity

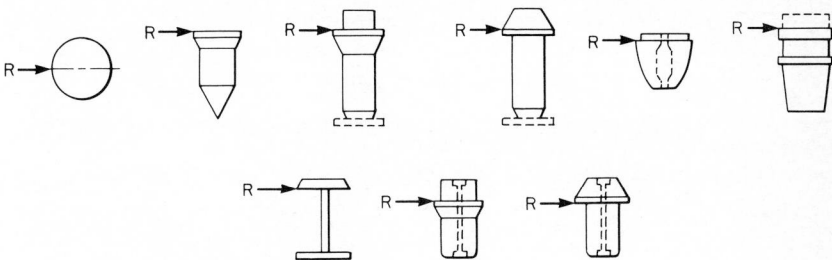

Figure 14.26   Float designs *(courtesy Fischer and Porter)*. R = reading edge of float. All floats are shown in the correct metering position.

effects are also reduced by using cylinder-and-plug type variable-area meters for measuring high-viscosity fluids (bunker C, tar, liquid chemicals). The manufacturer should be consulted for the viscosity limit for a given design.

When the fluid is at design (or calibration) conditions, the meter reads correctly in the desired volume or mass flow-rate units. However, if conditions change or the meter is used for some other fluid, a correction factor is necessary to obtain the true flow rate. This factor is derived as the ratio of Eq. (14.28) or Eq. (14.29) at flowing conditions to the same equation at design conditions. The corrected flow rate is then calculated with

$$F_{VA} = \frac{q_v}{(q_v)_{\text{des}}} \qquad F_{VA} = \frac{q_V}{(q_V)_{\text{des}}} \qquad F_{VA} = \frac{q_M}{(q_M)_{\text{des}}} \qquad (14.30)$$

where F_{VA} is the correction factor for the variable-area flowmeter reading.

Table 14.10 gives the specific gravities of some float materials, and Table 14.11 presents the U.S.-unit liquid and gas correction factors for mass, flowing-volume, and base-volume flow rates. Table 14.12 presents the correction factors for S.I. units, and Table 14.13 presents typical ratings for glass variable-area flowmeters. For gas (vapor) flows the buoyancy of the float is normally assumed negligible. The manufacturer should be consulted for the engineering equations for spring-loaded variable flowmeters.

Table 14.10 **Specific Gravities G_{fl}† of Float Materials**

Float material	Specific gravity G_{fl}	Float material	Specific gravity G_{fl}
Dowmetal[a]	1.80	Everdur[c]	8.54
Porcelain	2.41	Monel	8.80
Glass	2.54	Bronze	8.80
Aluminum	2.79	Nickel	8.85
Titanium	4.50	Hastelloy C[d]	8.94
Steel (carbon)	7.80	Hastelloy B[e]	9.24
Durimet[b]	8.02	Lead (10% antimony)	10.67
Stainless steel (316)	8.02		
Brass	8.46	Tantalum	16.6

†$G_{fl} = \rho_{fl}/62.37$.

[a]Trademark of Dow Chemical Co.

[b]Trademark of Duriron Co.

[c]Trademark of Anaconda American Brass Co.

[d]Trademark of Union Carbide Corp.

[e]Trademark of Huntington Alloy Products Division, International Nickel Co.

Table 14.11 Correction Factor F_{VA} for Variable-Area Flowmeters: U.S. Units

	Liquid	Gas (vapor)
	MASS FLOW RATE	
Density	(a) $F_{VA} = \sqrt{\dfrac{(\rho_{fl} - \rho_f)\rho_f}{[(\rho_{fl} - \rho_f)\rho_f]_{des}}}$	(g) $F_{VA} = \sqrt{\dfrac{\rho_f}{(\rho_f)_{des}}}$
Specific gravity	(b) $F_{VA} = \sqrt{\dfrac{(G_{fl} - F_p G_F)F_p G_F}{[(G_{fl} - F_p G_F)F_p G_F]_{des}}}$	
pvT equation		(h) $F_{VA} = \sqrt{\left(\dfrac{Z_f T_f}{G p_f}\right)_{des}\dfrac{G p_f}{Z_f T_f}}$
	VOLUMETRIC FLOW RATE AT FLOWING CONDITIONS	
Density	(c) $F_{VA} = \sqrt{\left(\dfrac{\rho_f}{\rho_{fl} - \rho_f}\right)_{des}\dfrac{\rho_{fl} - \rho_f}{\rho_f}}$	(i) $F_{VA} = \sqrt{\dfrac{(\rho_f)_{des}}{\rho_f}}$
Specific gravity	(d) $F_{VA} = \sqrt{\left(\dfrac{F_p G_F}{G_{fl} - F_p G_F}\right)_{des}\dfrac{G_{fl} - F_p G_F}{F_p G_F}}$	
pvT equation		(j) $F_{VA} = \sqrt{\left(\dfrac{G p_f}{Z_f T_f}\right)_{des}\dfrac{Z_f T_f}{G p_f}}$

Table 14.11 Correction Factor F_{VA} for Variable-Area Flowmeters: U.S. Units (*Continued*)

	Liquid	Gas (vapor)

VOLUMETRIC FLOW RATE AT BASE CONDITIONS

Density

(e)
$$F_{VA} = \frac{(\rho_b)_{des}}{\rho_b} \sqrt{\frac{(\rho_{fl} - \rho_f)\rho_f}{[(\rho_{fl} - \rho_f)\rho_f]_{des}}}$$

(k)
$$F_{VA} = \frac{(\rho_b)_{des}}{\rho_b} \sqrt{\frac{\rho_f}{(\rho_f)_{des}}}$$

Specific gravity

(f)
$$F_{VA} = \frac{(G_b)_{des}}{G_b} \sqrt{\frac{(G_{fl} - F_p G_b)F_p G_F}{[(G_{fl} - F_p G_F)F_p G_F]_{des}}}$$

pvT equation

Standard base

(l)
$$F_{VA} = \left(\frac{G}{Z_b}\right)_{des} \frac{Z_b}{G} \sqrt{\left(\frac{Z_f T_f}{G p_f}\right)_{des} \frac{G p_f}{Z_f T_f}}$$

Selected base

(m)
$$F_{VA} = \left(\frac{G p_b}{Z_b T_b}\right)_{des} \frac{Z_b T_b}{G p_b} \sqrt{\left(\frac{Z_f T_f}{G p_f}\right)_{des} \frac{G p_f}{Z_f T_f}}$$

Table 14.12 Correction Factor F_{VA} for Variable-Area Flowmeters: SI Units (*Continued*)

	Liquid	Gas (vapor)
MASS FLOW RATE		
Density	(a) $F_{VA} = \sqrt{\dfrac{(\rho_{fl}^* - \rho_f^*)\rho_f}{[(\rho_{fl} - \rho_f^*)\rho_f]_{des}}}$	(g) $F_{VA} = \sqrt{\dfrac{\rho_f^*}{(\rho_f^*)_{des}}}$
Specific gravity	(b) $F_{VA} = \sqrt{\dfrac{(G_{fl} - F_p G_F)F_p G_F}{[(G_{fl} - F_p G_F)F_p G_F]_{des}}}$	
$\rho v T$ equation		(h) $F_{VA} = \sqrt{\left(\dfrac{Z_f T_K}{G p_f^*}\right)_{des} \dfrac{G p_f^*}{Z_f T_K}}$
VOLUMETRIC FLOW RATE AT FLOWING CONDITIONS		
Density	(c) $F_{VA} = \sqrt{\left(\dfrac{\rho_f^*}{\rho_{fl} - \rho_f^*}\right)_{des} \dfrac{\rho_{fl} - \rho_f^*}{\rho_f^*}}$	(i) $F_{VA} = \sqrt{\dfrac{(\rho_f^*)_{des}}{\rho_f^*}}$
Specific gravity	(d) $F_{VA} = \sqrt{\left(\dfrac{F_p G_F}{G_{fl} - F_p G_F}\right)_{des} \dfrac{G_{fl} - F_p G_F}{F_p G_F}}$	
$\rho v T$ equation		(j) $F_{VA} = \sqrt{\left(\dfrac{G p_f^*}{Z_f T_K}\right)_{des} \dfrac{Z_f T_K}{G p_f^*}}$

Table 14.12 Correction Factor F_{VA} for Variable-Area Flowmeters: SI Units (*Continued*)

VOLUMETRIC FLOW RATE AT BASE CONDITIONS

	Liquid	Gas (vapor)
Density	$F_{VA} = \dfrac{(\rho_b^*)_{des}}{\rho_b} \sqrt{\dfrac{(\rho_{fl}^* - \rho_f^*\rho_f^*)}{[(\rho_f^* - \rho_f^*)\rho_{fl}^*]_{des}}}$ (e)	$F_{VA} = \dfrac{(\rho_b^*)_{des}}{\rho_b} \sqrt{\dfrac{\rho_f^*}{(\rho_f^*)_{des}}}$ (k)
Specific gravity	$F_{VA} = \dfrac{(G_b)_{des}}{G_b} \sqrt{\dfrac{(G_{fl} - F_p G_F)F_p G_F}{[(G_{fl} - F_p G_F)F_p G_F]_{des}}}$ (f)	
pvT equation Standard base		$F_{VA} = \left(\dfrac{G}{Z_b}\right)_{des} \dfrac{Z_b}{G} \sqrt{\left(\dfrac{Z_f T_f}{G p_f^*}\right)_{des} \left(\dfrac{G p_f^*}{Z_f T_K}\right)}$ (l)
Selected base		$F_{VA} = \left(\dfrac{G p_b^*}{Z_b T_{Kb}}\right)_{des} \left(\dfrac{Z_b T_{Kb}}{G p_b^*}\right) \sqrt{\left(\dfrac{Z_f T_K}{G p_f^*}\right)_{des} \left(\dfrac{G p_f^*}{Z_f T_K}\right)_{calib} \left(\dfrac{G p_f^*}{Z_f T_K}\right)}$ (m)

Table 14.13 Typical Ratings for Glass Variable-Area Flowmeters

Tube diameter, in (mm)	Safe working pressure, psig (kPa)	Temperature, °F (°C)	Capacity		Overall pressure loss (water), psid (kPa)
			Air†, SCFM (L/s)	Water‡, gal/min (L/s)	
¼ (6.3)	450 (3100)	200 (90)	0.15–1.8 (0.07–0.85)	0.025–0.5 (0.002–0.03)	0.04–0.37 (0.3–2.5)
½ (12.5)	300 (2070)	180 (80)	1–20 (0.5–10)	0.25–5 (0.02–0.3)	0.04–1.2 (0.3–8)
1 (25)	180 (1240)	160 (70)	18–122 (8.5–58)	4–30 (0.25–1.9)	0.3–0.4 (2–3)
1½ (40)	130 (900)	150 (65)	48–226 (23–107)	13–54 (0.8–3.4)	0.3–3.6 (2–25)
2 (50)	100 (700)	140 (60)	78–384 (37–180)	20–90 (1.3–5.7)	0.4–4 (3–28)
3 (75)	60 (414)	130 (55)	205–500 (100–236)	60–120 (3.8–7.6)	0.7–1.4 (5–10)

†Air at standard conditions: p_b = 14.696 psia (p_b^* = 101.325 kPa); T_b = 518.67°R (T_{Kb} = 288.15 K).

‡Water at 60°F (15.6°C) and 14.696 psia (101.325 kPa).

SOURCE: Adapted from Fees (1974).

Industrial Flowmeters   Industrial variable-area flowmeters range from a simple ball floating in a glass tube to high-pressure piston-type meters in 600-lb (4000-kPa) valve bodies. Several flowmeters are illustrated in Chap. 6. Readings may be obtained visually or through a magnetic coupling, a dial reading, or some standardized output. Variable-area meters have advantages in that the overall pressure loss is nearly constant, the meters are small in size, and almost any corrosive fluid can be metered. Their accuracy ranges from ±0.5 to ±2 percent of the URV flow rate, depending on design.

The simple direct-viewing glass-tube meter is the most popular for reasons of economics, and armored-tube meters extend applications to high-pressure service. Orifice and tapered-plug meters (Fig. 14.25b) are low-cost, compact variable-area flowmeters. Because of their small size and low cost, these meters are frequently selected for purge-water flows where accuracy is not required.

Piston-type meters are made of steel, cast iron, and other metals and are installed in a valve body in line sizes up to 4 in (100 mm). These meters are suitable for high-pressure liquids and gas (vapor) applications. Their range is adjustable by varying a spring force or changing weights. Since these meters are relatively less sensitive to vertical installation (at least within ±5° of vertical), they are preferred for mobile or marine installations. They are quite rugged and are commonly used in gasoline, kerosene, and tar applications. However, the fluid must be clean, and a strainer may be necessary.

Periodic cleaning and an upstream strainer are usually recommended for dirty flows. A leak-type bypass manifold (Fig. 14.27) is suggested for applications where shutdown is undesirable. The upstream and downstream piping configurations have little effect on meter performance, and no straight lengths are given.

EXAMPLE 14.5

A variable-area flowmeter with a stainless-steel float was calibrated to measure the gasoline of Example 2.12 in base gallons per minute at a line pressure of 100 psig (690 kPa). If the flowing temperature is 100°F (37.8°C) and the pressure is 300 psig (2070 kPa), what is the flow rate for a meter reading of 10 gal/min (0.631 L/s)?

The flow rate is calculated with Eq. (14.30)

$$q_{\text{GPM}} = F_{VA}(q_{\text{GPM}})_{\text{des}}$$

where the correction factor F_{VA} is, from Eq. (f) of Table 14.11,

$$F_{VA} = \frac{(G_b)_{\text{des}}}{G_b} \sqrt{\frac{(G_{fl} - F_p G_F)F_p G_F}{[(G_{fl} - F_p G_F)F_p G_F]_{\text{des}}}}$$

Required data at design (calibrated) conditions is:

■ From Example 2.12: $(G_b)_{\text{des}} = 0.7359$

■ From Example 2.12: $(G_F)_{\text{des}} = 0.7255$

■ From Table 14.10: $G_{fl} = 8.02$

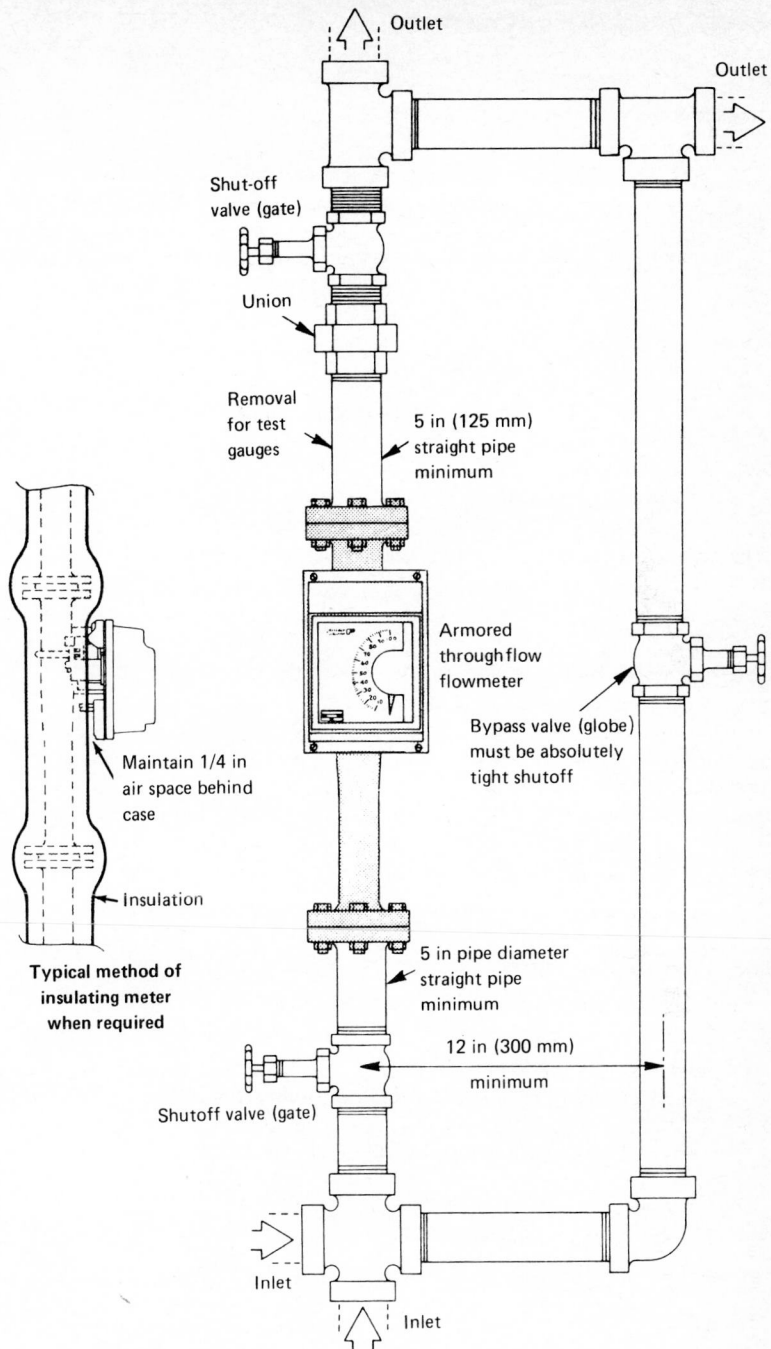

Outlet

Outlet

Shut-off
valve (gate)

Union

Removal
for test
gauges

5 in (125 mm)
straight pipe
minimum

Armored
through flow
flowmeter

Bypass valve (globe)
must be absolutely
tight shutoff

Maintain 1/4 in
air space behind
case

Insulation

**Typical method of
insulating meter
when required**

5 in pipe diameter
straight pipe
minimum

12 in (300 mm)

minimum

Shutoff valve (gate)

Inlet

Inlet

Figure 14.27 Typical variable-area-flowmeter installation with bypass manifold
(courtesy Fischer and Porter).

14-55

The liquid compressibility factor is, from Eq. (2.107),

$$(F_p)_{des} = 1 + Z_L \frac{p_f}{1000} = 1 + 0.0088 \frac{114.7}{1000} = 1.0010$$

where $Z_L = 0.0088$ from Fig. E.22.

The density at flowing conditions (uncorrected for pressure) is, from Eq. (2.97),

$$\rho_F^* = \rho_b^* \exp\left[-\alpha_b \, \Delta T_F \, (1 + 0.8 \, \alpha_b \, \Delta T_F)\right]$$

where, from Example 2.12, $\rho_b^* = 735.14$ kg/m^3, and, from Eq. (2.96),

$$\alpha_b = \frac{K_0}{\rho_b^{*2}} + \frac{K_1}{\rho_b^*} = \frac{192.4511}{(735.14)^2} + \frac{0.2438}{735.14} = 0.0006878$$

Since $\Delta T_F = 100 - 60 = 40°$F,

$$\rho_F^* = 735.14 \exp\{(-0.0006878)(40)[1 + (0.8)(0.0006878)(40)]\} = 714.76 \text{ kg/m}^3$$

The specific gravity at flowing conditions is then, by Eq. (2.78),

$$G_F = \frac{\rho_F^*}{999.012} = \frac{714.76}{999.012} = 0.7155$$

The liquid compressibility factor at flowing conditions is, from Eq. (2.107),

$$F_p = 1 + Z_L \frac{p_f}{1000} = 1 + 0.009 \frac{314.7}{1000} = 1.0028$$

where $Z_L = 0.009$ from Fig. E.22. Because $(G_b)_{des} = G_b$ and $(G_{fl})_{des} = G_{fl}$,

$$F_{VA} = \frac{0.7359}{0.7359} \sqrt{\frac{[(8.02 - (1.0028)(0.7155)](1.0028)(0.7155)}{[(8.02 - (1.0010)(0.7255)](1.0010)(0.7255)}} = 0.9946$$

The flow rate is then

$$q_{GPM} = (0.9946)(10) = 9.946 \text{ base gal/min}$$

EXAMPLE 14.6

The variable-area flowmeter of Example 14.5 is being used to meter water at 100°F (37.8°C) and 14.7 psia (101 kPa). If the reading is 10 gal/min (37.9 L/min), what is the flow rate?

The water flow rate is calculated by Eq. (14.30):

$$(q_{GPM})_{water} = F_{VA} \, (q_{GPM})_{gasoline}$$

The correction factor is, from Eq. (f) of Table 14.11,

$$F_{VA} = \frac{(G_b)_{des}}{G_b} \sqrt{\frac{(G_{fl} - F_p G_F)F_p G_F}{[(G_{fl} - F_p G_F)F_p G_F]_{des}}}$$

Design data, from Example 14.5, is:

$$(G_{fl})_{des} = 8.02 \qquad (F_p)_{des} = 1.0010 \qquad (G_F)_{des} = 0.7255 \qquad (G_b)_{des} = 0.7359$$

For flowing conditions, $G_{fl} = 8.02$. From Table E.6, $\rho_F = 61.994$ lb$_m$/ft^3, and, by Eq. (2.79), $G_F = 61.994/62.3663 = 0.99402$. It is assumed that $F_p = 1.0$, and, for water, $G_b = 1.0$. Then substitution gives

$$F_{VA} = \frac{0.7359}{1.0}\sqrt{\frac{[(8.02 - (1.0)(0.99402)](1.0)(0.99402)}{[8.02 - (1.0010)(0.7255)](1.0010)(0.7255)}} = 0.8450$$

The flow rate is then

$$q_{GPM} = (0.8450)(10) = 8.45 \text{ base gal/min}$$

EXAMPLE 14.7

A variable-area flowmeter was calibrated on air at 101.325 kPa and 15°C to read standard cubic meters per hour. The meter is being used to measure the water-vapor-saturated natural gas of Example 9.9. For a reading of 10 standard m^3/h, what is the dry natural gas flow rate?

The dry-gas flow rate is calculated by combining Eqs. (14.30) and (2.66) as

$$(q^*_{SCMH})_{dry} = F_{WV,dry}(q^*_{SCMH})_{wet} = F_{WV,dry}F_{VA}(q^*_{SCMH})_{air}$$

From Example 9.9, $F_{WV,dry} = 0.98082$. From Eq. (l) of Table 14.12, the correction factor F_{VA} is

$$F_{VA} = \left(\frac{G}{Z_b}\right)_{des}\left(\frac{Z_b}{G_{wet}}\right)\sqrt{\left(\frac{Z_f T_K}{Gp_f^*}\right)_{des}\frac{G_{wet}p_f^*}{Z_f T_K}}$$

where G_{wet} replaces G because the meter is measuring wet gas.

At design (calibrated) conditions,

$$G_{des} = 1.0 \quad \text{for air}$$
$$(Z_b)_{des} = (Z_f)_{des} = 1.0 \quad \text{(assumed)}$$
$$(T_K)_{des} = 288.15 \text{ K}$$
$$(p_b^*)_{des} = (p_f^*)_{des} = 101.325 \text{ kPa}$$

and at flowing conditions,

$$G_{wet} = 0.6514 \qquad T_K = 300 \text{ K}$$

from Example 9.9;

$$Z_f = \frac{1}{F_{pv}^2} = \frac{1}{(1.0011)^2} = 0.9978$$

from Eq. (9.65); and

$$Z_b = 1.0 \qquad p_f^* = 184 \text{ kPa}$$

Substitution then gives

$$F_{VA} = \frac{1.0}{1.0}\frac{1.0}{0.6514}\sqrt{\frac{(1.0)(288.15)}{(1.0)(101.325)}\frac{(0.6514)(184)}{(0.9978)(300)}} = 1.6381$$

and the dry-volume flow rate is

$$(q^*_{\text{SCMH}})_{\text{dry}} = (0.98082)(1.6381)(10) = 16.07 \text{ standard m}^3/\text{h}$$

REFERENCES

AGA Report 6, *Methods of Testing Large Capacity Displacement Meters,* American Gas Association, Arlington, Va., 1975.

AGA Report 7, *Measurement of Fuel Gas by Turbine Meters,* catalog no. XQ0580, American Gas Association, Arlington, Va., 1981.

ANSI (ISA RP) 31.1, *Specifications, Installation, and Calibration of Turbine Flowmeters,* ANSI, New York, 1977.

ANSI B109.2, *Gas Capacity at 0.5 Inches of Water (0.125 kPa) Differential for a 0.6 Specific Gravity Gas Referred to 60°F (15.6°C) and 14.73 psia (101.6 kPa),* ANSI, New York, 1980.

ANSI B109.3, *Rotary Type Gas Displacement Meters,* ANSI, New York, 1980.

ANSI Z11.170, *Measurement of Petroleum Liquid Hydrocarbon by Positive Displacement Meter* (API Standard 1101), ANSI, New York, 1960.

ANSI/API 2562, *Natural Gas,* ANSI, New York, 1969.

ANSI/ASME MFC-YY, *Measurement of Liquid Flow in Closed Conduits Using Transit-Time Ultrasonic Flowmeters* (draft standard), ASME, New York, 1982.

API Publication 2101, *Calculation of Liquid Petroleum Quantities by Turbine or Displacement Meters,* chap. 12.2, American Petroleum Institute Document 852-30302, Washington, D.C., 1981.

ASME: *Fluid Meters, Their Theory and Application,* 6th ed., ASME, New York, 1971.

Ball, J. M.: "Viscosity Effects of the Turbine Flowmeter," *Proc. of the Symp. on Flow,* NBS publication 484, pp. 847–869, U.S. Government Printing Office, Washington, D.C., 1977.

Faraday, M.: *Philos. Trans. R. Soc. London Ser. A,* p. 125, 1892.

Fees, Charles E.: *Process Instruments and Control Handbook,* chap. 4, McGraw-Hill, New York, 1974.

IEC Publication 381A, International Electrical Commission, Geneva.

Inkley, F. A., D. C. Walden, and D. J. Scott: "Flow Characteristics of Vortex Flowmeters," *Meas. Control,* vol. 13, pp. 166–169, 1980.

ISO 6817, *Measurement of Conductive Fluid Fluorate in Closed Conduits—Methods Using Electromagnetic Flowmeters,* ISO/TR6817-1980(E), 1980.

James, R.: "Metering of Steam/Water Two-Phase Flow by Sharp-Edged Orifices," *Proc. Inst. Mech. Eng.,* vol. 180, pt. 1, no. 23, 1965–1966.

Jennings, R.: "Digital Compensation Techniques for Positive Displacement and Turbine Flowmeters," *Proc. of the Symp. on Flow,* NBS publication 484, pp. 821–846, 1977.

Lee, W. F. Z., D. C. Blakeslee, and R. V. White: "A Self-Correcting and Self-Checking Gas Turbine Meter," *J. Fluid Eng.,* vol. 104, no2, pp. 143–149, 1982.

Lynnworth, L. C.: "Ultrasonic Flowmeters," pp. 407–525 in W. P. Mason and R. N. Thurston (eds.), *Physical Acoustics,* vol. 14, Academic, New York, 1979.

Miller, R. W., J. P. DeCarlo, and J. T. Cullen: "A Vortex Flowmeter—Calibration Results and Application Experiences," *Proc. of the Symp. on Flow,* NBS publication 484, pp. 549–570, 1977.

Rubin, M., R. W. Miller, and W. G. Fox: "Driving Torques in a Theoretical Model of a Turbine Meter," *J. Basic Eng.,* ser. D, vol. 87, no. 2, pp. 413–420, 1965.

Schmidt, T. R.: "Clamp-on Ultrasonic Flowmeters," *Trans. ISA,* vol. 4, pp. 111–126, 1980.

Shercliff, J. A.: *The Theory of Electromagnetic Flow Measurement,* Cambridge University Press, Cambridge, England, 1961.

Smith, R. V., and J. T. Leang: "Evaluations of Correlations for Two-Phase Flowmeters—Three current—One New," ASME Paper 74-FM-5, New York, 1974.

Spink, L. K.: *Principles and Practice of Flowmeter Engineering,* 9th ed., The Foxboro Company, Foxboro, Mass., 1967.

Strouhal, F.: "Über eine besondere Art der Tonerregung," *Ann. Phys. Chem.,* vol. 5, p. 216, 1878.

Thompson, R. E., and J. Grey: "Turbine Flowmeter Performance Model," *J. Basic Eng.,* ser. D, vol. 92, no. 4, pp. 712–723, 1970.

von Karman, T., and H. Rubach: "Über den Mechanismus des Flüssigkeits und Luftwiderstanders" ("On the Mechanism of Fluid Resistance"), *Phys. Z.,* vol. 13, p. 49, 1912.

Vignos, J. H.: "Effect of Velocity Profile on Flow Measurement in the Turbulent Regime," private communication, 1981.

White, D. F., A. E. Rodely, and C. L. McMurtie: "The Vortex Shedding Flowmeter," *Flow, Its Measurement and Control in Science and Industry,* vol. 1, pt. 2, pp. 967–974, ISA, Research Triangle Park, N.C., 1974.

DISCUSSIONS AND PROOFS

A.1 NEWTON'S METHOD FOR THE APPROXIMATE SOLUTION OF NUMERICAL EQUATIONS

Many of the equations used in flow measurement require an iterative solution for the flow rate, compressibility factor, or orifice bore. Newton's method for the approximate solution of numerical equations is a convenient trial-and-error technique that requires fewer estimates than other methods. In many cases the initial solution is sufficiently accurate, and a single calculation can be used. The calculations are readily programmable on hand calculators, dedicated microprocessors, or central computers.

As an example of the use of Newton's method, consider a 2-in (50-mm) orifice flowmeter operating at a Reynolds number of 10,000, for which the flow equation reduces to

$$q = 4.019 + \frac{0.8884}{q^{0.75}} \tag{A.1}$$

In this equation, the first constant (4.019) is the calculated flow rate at an infinite Reynolds number for the measured differential and fluid density. The second constant includes the coefficient correction for Reynolds number, a dimensional term, and any necessary unit conversion.

Equation (A.1) is nonlinear, and to solve it estimates of the flow rate q must be successively substituted until the relationship is satisfied. Instead, Eq. (A.1) can be rearranged into a *function* equation as

$$F = 4.019 + \frac{0.8884}{q^{0.75}} - q \tag{A.2}$$

Then, to solve Eq. (A.2) for the flow rate q, successive estimates of the flow rate are substituted into Eq. (A.2) until F is calculated to be zero.

The values of F for several flow rates are given in Table A.1, beginning with the infinite flow rate. These pairs of values are shown plotted in Fig. A.1. The zero crossing provides the zero root of Eq. (A.2), which is the desired flow rate. Its value can be read as 4.316; when substituted into Eq. (A.1) or (A.2), this value satisfies the equality.

A-1

Table A.1 Values of the Function
F of Eq. (A.2) for Various Flow
Rates

Flow rate q_{PPS}	F
4.019	$+0.313$
4.119	$+0.207$
4.229	$+0.091$
4.329	-0.014
4.429	-0.119

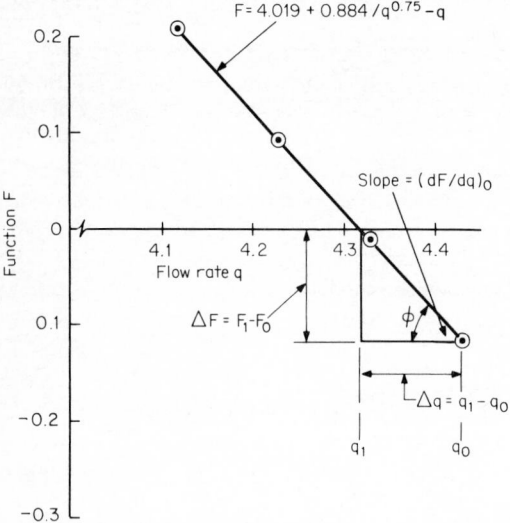

Figure A.1 Plot of the function *F*.

The number of iterations (or estimates or guesses) is reduced if the equation of the tangent to the curve at the initial estimate q_0 is used to calculate the second flow-rate estimate q_1. For a tangent to the curve shown in Fig. A.1 at the point q_0, the angle ϕ is approximately given by

$$\tan \phi = \frac{\Delta F}{\Delta q} = \frac{F_1 - F_0}{q_1 - q_0} \tag{A.3}$$

Under the assumption that the second estimate q_1 is the desired zero root, F_1 equals zero. Equation (A.3) then becomes

$$q_1 = q_0 - \frac{F_0}{\tan \phi} \tag{A.4}$$

where F_0 is the value of the function *F* for the initial flow-rate estimate.

The slope of the function F at any point is calculated by substituting the flow rate at that point into the derivative of the function. In general,

$$\frac{dF}{dq} = F' = -(0.6663q^{-1.75} + 1) \tag{A.5}$$

and at the point where $q = q_0$, the slope is

$$\tan \phi = F'_0 = -(0.6663q_0^{-1.75} + 1) \tag{A.6}$$

Substituting Eq. (A.6) into Eq. (A.4) yields the equation for calculating the second estimate as

$$q_1 = q_0 - \frac{F_0}{F'_0} = q_0 + \frac{4.019 + 0.8884q_0^{-0.75} - q_0}{0.6663q_0^{-1.75} + 1} \tag{A.7}$$

Equation (A.7) can be solved directly to obtain the zero root q_1, the desired flow rate. In most cases this single equation gives results to within 0.01 percent of the true value, which is sufficient for most flow applications. Subsequent estimates for improved calculation accuracy are obtained as

$$q_2 = q_1 - \frac{F_1}{F'_1} \qquad q_3 = q_2 - \frac{F_2}{F'_2} \qquad \cdots \tag{A.8}$$

and, in general,

$$q_n = q_{n-1} - \frac{F_{n-1}}{F'_{n-1}} \tag{A.9}$$

where n is the estimate number.

Table A.2 **Sample Iterations for Calculating the Flow Rate q**

Trial n	Flow rate q_{n-1}	F, Eq. (A.2)	F', Eq. (A.6)	Flow rate q_n, Eq. (A.9)	Deviation, %
		FOR INITIAL ESTIMATE CLOSE TO ZERO ROOT			
1	4.019	0.3130	−1.0584	4.3147	−0.02
2	4.3147	0.0010	−1.0516	4.3157	0.00
3	4.3157	0.0000	−1.0516	4.3157	0.00
		FOR INITIAL GUESS SUBSTANTIALLY DIFFERENT FROM ZERO ROOT			
1	1.000	3.9074	−1.6663	3.3450	−22.49
2	3.3450	1.0332	−1.0805	4.3012	−0.34
3	4.3012	0.0153	−1.0519	4.3157	0.00
1	100	−95.9529	−1.0002	4.0673	−5.76
2	4.0673	0.2619	−1.0572	4.3150	−0.02
3	4.3150	0.0007	−1.0516	4.3157	0.00

Table A.2 lists values of the function F, the derivative F', and the calculated flow rate for three initial estimates. In the first calculation, the initial assumption of a flow rate equal to the constant 4.019 in Eq. (A.1) gives a calculation accuracy of 0.02 percent; for an initial guess substantially different from the zero root, two or three iterations are required for that accuracy.

Newton's method can be used to solve any nonlinear equation that is first rearranged into the general function form of Eq. (A.2), provided the function F is continuous.

A.2 PROOF THAT THE MEASURED DIFFERENTIAL INCLUDES THE POTENTIAL-ENERGY TERM FOR INCLINED INSTALLATIONS

Shown in Fig. A.2a is a differential producer inclined with respect to the horizontal. Bernoulli's equation along a stream tube can be written as

$$\frac{P_{f1}}{\rho_f} + \frac{\overline{V}_{f1}^2}{2g_c} + \frac{g_l}{g_c} H_{EL,1} = \frac{P_{f2}}{\rho_f} + \frac{\overline{V}_{f2}^2}{2g_c} + \frac{g_l}{g_c} H_{EL,2} \qquad (A.10)$$

Rearranging gives

$$\frac{P_{f1} - P_{f2}}{\rho_f} + \frac{g_l}{g_c} (H_{EL,1} - H_{EL,2}) = \frac{\overline{V}_{f2}^2 - \overline{V}_{f1}^2}{2g_c} \qquad (A.11)$$

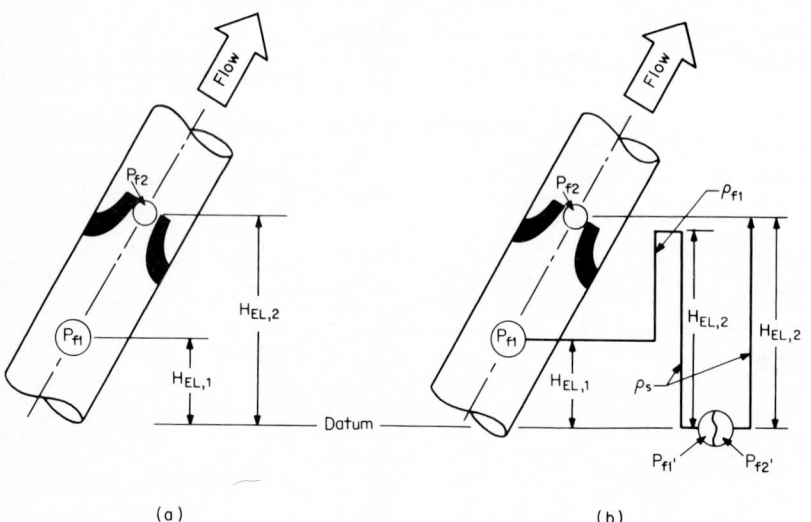

(a)   (b)

Figure A.2   Bernoulli's equation applied to an inclined differential producer. (a) Stream tube. (b) Lead lines at same elevation.

Mass flow continuity requires that

$$q_{PPS} = \rho_{f1} A_1 \overline{V}_{f1} = \rho_{f2} A_2 \overline{V}_{f2} \tag{A.12}$$

Then, for a constant-density fluid,

$$\overline{V}_{f1}^2 = \left(\frac{\rho_f A_2}{\rho_f A_1} \right)^2 \overline{V}_{f2}^2 = \beta^4 \overline{V}_{f2}^2 \tag{A.13}$$

Substitution of Eq. (A.13) into Eq. (A.11) yields

$$\frac{P_{f1} - P_{f2}}{\rho_f} + \frac{g_l}{g_c} (H_{EL,1} - H_{EL,2}) = \frac{1 - \beta^4}{2g_c} \overline{V}_{f2}^2 \tag{A.14}$$

which, when rewritten for mass flow rate using Eq. (A.12), yields

$$\frac{P_{f1} - P_{f2}}{\rho_f} + \frac{g_l}{g_c} (H_{EL,1} - H_{EL,2}) = \frac{1 - \beta^4}{2g_c} \frac{q_{PPS}^2}{\rho_f^2 A_2^2} \tag{A.15}$$

The mass flow equation for an inclined installation thus is

$$q_{PPS} = \frac{\pi}{4} \sqrt{2g_c} \frac{d_F^2}{\sqrt{1 - \beta^4}} \sqrt{\left[P_{f1} - P_{f2} + \frac{g_l}{g_c} (H_{EL,1} - H_{EL,2}) \right] \rho_f} \tag{A.16}$$

For a horizontal installation ($H_{EL,1} = H_{EL,2}$), this reduces to the fundamental mass flow equation (9.17):

$$q_{PPS} = \frac{\pi}{4} \sqrt{2g_c} \frac{d_F^2}{\sqrt{1 - \beta^4}} \sqrt{(P_{f1} - P_{f2}) \rho_f} \tag{A.17}$$

The difference between the equations for a horizontal and an inclined installation is the potential-energy term

$$\frac{g_l}{g_c} (H_{EL,1} - H_{EL,2}) \rho_f \tag{A.18}$$

within the radical.

Shown in Fig. A.2b is a differential-pressure measuring device located at a datum. Upstream lead lines have been brought to the same elevation as the downstream lead lines, so that $H_1 = H_2$. The upstream pressure measured at the datum can be expressed as

$$P_{f1'} = P_{f1} - \frac{g_l}{g_c} (H_{EL,2} - H_{EL,1}) \rho_{f1} + \frac{g_l}{g_c} H_{EL,2} \rho_s \tag{A.19}$$

and the downstream pressure at the datum is

$$P_{f2'} = P_{f2} + \frac{g_l}{g_c} H_{EL,2} \rho_s \tag{A.20}$$

where ρ_s is the density of the seal fluid in the lines.

The differential pressure transmitted to the differential-pressure measuring device, for the same seal-fluid density and provided the lines are at the same elevation, is

$$P_{f1'} - P_{f2'} = P_{f1} - P_{f2} + \frac{g_l}{g_c}(H_{EL,1} - H_{EL,2})\rho_{f1} \tag{A.21}$$

Equation (A.21) is the measured differential pressure; when it is substituted into Eq. (A.16), the result is

$$q_{PPS} = \frac{\pi}{4}\sqrt{2g_c}\,\frac{d_F^2}{\sqrt{1 - \beta^4}}\,\sqrt{(P_{f1'} - P_{f2'})\rho_f} \tag{A.22}$$

Equation (A.22) thus relates the measured differential to the flow rate, and the measured differential adjusts for the potential-energy term in Eq. (A.18).

A.3 USE OF NEWTON'S METHOD FOR SIZING THE BORE OF A DIFFERENTIAL PRODUCER

The relationship between the S_M factor, the beta ratio, the discharge coefficient, and the known sizing information is given by Eq. (9.77). This equation can be written for solution via Newton's method (see Sec. A.1) as

$$F = S_M - \frac{CY\beta^2}{\sqrt{1 - \beta^4}} \tag{A.23}$$

The gas expansion factor Y is not highly sensitive to β and negligible error is introduced by simplifying the derivative of the function to

$$F' = -Y\frac{d}{d\beta}\left(\frac{C\beta^2}{\sqrt{1 - \beta^4}}\right) \tag{A.24}$$

With the substitution of the general form of the discharge-coefficient equation [Eq. (9.49)], the derivative of the function can be written as

$$F' = \frac{-\beta Y}{(1 - \beta^4)^{1.5}}\left[2C_\infty + (\beta - \beta^5)K_1 + \frac{2b + (\beta - \beta^5)K_2}{R_D^n}\right] \tag{A.25}$$

Succeeding estimates for β are then

$$\beta_1 = \beta_0 - \frac{F_0}{F_0'} \qquad \beta_2 = \beta_1 - \frac{F_1}{F_1'} \qquad \cdots \qquad \beta_n = \beta_{n-1} - \frac{F_{n-1}}{F_{n-1}'}$$

A single iteration for β_1 generally yields a calculation accuracy of 0.001 percent or better, and a second iteration is unnecessary.

The function F_0 and its derivative F_0' are evaluated by substituting β_0 for β into the discharge-coefficient equations given in Table 9.1 to determine C. For liquids, the gas expansion factor Y is 1.0, and for gases (vapors), the expansion factor is calculated by substitution of β_0 for β into the equations given in Table 9.26. With C and Y known, F_0 may be calculated from Eq. (A.23); F_0' is found using Table A.3.

Table A.3   Values of K_1 and K_2 in the Equation $F' = \dfrac{-\beta Y}{(1-\beta^4)^{1.5}}\left[2C_\infty + (\beta - \beta^5)K_1 + \dfrac{2b + (\beta - \beta^5)K_2}{R_D^n}\right]^\dagger$

Primary device	K_1	K_2
Venturi		
Machined inlet	0	0
Rough-cast inlet	0	0
Rough-welded sheet-iron inlet	0	0
Universal Venturi tube ‡	0	0
Lo-Loss Tube§	$-0.471 + 1.128\beta - 1.542\beta^2$	0
Nozzle		
ASME long radius	0	$-3.265\beta^{-0.5}$
ISA	$-0.9274\beta^{3.1}$	$-8936 + 92{,}961\beta^{3.7}$
Venturi nozzle (ISA inlet)	$-0.882\beta^{3.5}$	0
Orifice		
Corner taps	$0.0655\beta^{1.1} - 1.472\beta^7$	$229.3\beta^{1.5}$
Flange taps		
$D \geq 2.3$ in	$0.0655\beta^{1.1} - 1.472\beta^7 + \dfrac{0.36}{D}\dfrac{\beta^3}{(1-\beta^4)^2} - \dfrac{0.101}{D}\beta^2$	$229.3\beta^{1.5}$
$2 \leq D \leq 2.3$ in	$0.0655\beta^{1.1} - 1.472\beta^7 + 0.156\dfrac{\beta^3}{(1-\beta^4)^2} - \dfrac{0.101\beta^2}{D}$	$229.3\beta^{1.5}$
$D^* > 58.4$ mm	$0.0655\beta^{1.1} - 1.472\beta^7 + \dfrac{9.144}{D^*}\dfrac{\beta^3}{(1-\beta^4)^2} - 2.568\dfrac{\beta^2}{D^*}$	$229.3\beta^{1.5}$
$50.8 \leq D^* \leq 58.4$ mm	$0.0655\beta^{1.1} - 1.472\beta^7 + 0.156\dfrac{\beta^3}{(1-\beta^4)^2} - 2.568\dfrac{\beta^2}{D^*}$	$229.3\beta^{1.5}$
D and $D/2$ taps	$0.0655\beta^{1.1} - 1.472\beta^7 + 0.156\dfrac{\beta^3}{(1-\beta^4)^2} - 0.0474\beta^2$	$229.3\beta^{1.5}$
$2\frac{1}{2}D$ and $8D$ taps¶	$0.9681\beta^{1.1} + 3.84\beta^7 + 0.156\dfrac{\beta^3}{(1-\beta^4)^2}$	$229.3\beta^{1.5}$

†F' is the derivative of the function $F = S_M - CY\beta^2/\sqrt{1-\beta^4}$.

‡ From BIF CALC-440/441; the manufacturer should be consulted for exact coefficient information.

§Derived from Badger Meter, Inc., Lo-Loss tube coefficient curve; the manufacturer should be consulted for exact coefficient information.

¶From Stolz (1978).

EXAMPLE A.1

Use Newton's method to determine the bore for the flange-tapped orifice of Example 9.2.
    From Example 9.2,

$$(q_{\text{PPH}})_{\text{URV}} = 30{,}000 \ \text{lb}_m/\text{h} \qquad (q_{\text{PPH}})_N = 0.8(q_{\text{PPH}})_{\text{URV}} = 24{,}000 \ \text{lb}_m/\text{h}$$

$$(h_w)_{\text{URV}} = 100 \ \text{in} \qquad (h_w)_N = 0.64(h_w)_{\text{URV}} = 64 \ \text{in}$$

$$(R_D)_{\text{URV}} = 1,882,700 \qquad (R_D)_N = 0.8(R_D)_{\text{URV}} = 1,506,200$$

$$S_M = 0.39759 \qquad \beta_0 = 0.74322 \qquad D = 5.761 \text{ in}$$

and from Example 9.4,

$$k_i = 1.308$$

The function equation (for the initial estimate) is, by Eq. (A.1),

$$F_0 = S_M - \frac{C_0(Y_1)_0 \beta_0^2}{\sqrt{1 - \beta_0^4}}$$

and, from Table 9.1,

$$C_0 = 0.5959 + 0.0312\beta_0^{2.1} - 0.184\beta_0^8 + 0.09 \frac{\beta_0^4}{D(1 - \beta_0^4)} - 0.0337 \frac{\beta_0^3}{D} + 91.71 \frac{\beta_0^{2.5}}{(R_D)_N^{0.75}}$$

Substituting the initial estimate $\beta_0 = 0.74322$ into this equation gives $C_0 = 0.60097$. From Eqs. (c) and (h) of Table 9.26,

$$\mathbf{x}_1 = \frac{(h_w)_N}{27.73 p_{f1}} = \frac{64}{(27.73)(204)} = 0.01131$$

and

$$(Y_1)_0 = 1 - (0.41 + 0.35\beta_0^4) \frac{\mathbf{x}_1}{k_i} = 0.99553$$

Substitution into the function equation gives

$$F_0 = 0.39759 - \frac{(0.60097)(0.99553)(0.74322)^2}{\sqrt{1 - (0.74322)^4}} = 0.001142$$

The derivative of the function equation (for the initial estimate) is

$$F_0' = \frac{-\beta_0(Y_1)_0}{(1 - \beta_0^4)^{1.5}} \left[2C_\infty + (\beta_0 - \beta_0^5)K_1 + \frac{2b + (\beta_0 - \beta_0^5)K_2}{R_D^{0.75}} \right]$$

From Table 9.1, for $\beta_0 = 0.74322$,

$$C_\infty = 0.5959 + 0.0312\beta_0^{2.1} - 0.184\beta_0^8 + 0.09 \frac{\beta_0^4}{D(1 - \beta_0^4)} - 0.0337 \frac{\beta_0^3}{D} = 0.59997$$

$$b = 91.71\beta_0^{2.5} = 43.6555$$

and from Table A.3,

$$K_1 = 0.0655\beta_0^{1.1} - 1.472\beta_0^7 + \frac{0.36}{D} \frac{\beta_0^3}{(1 - \beta_0^4)^2} - \frac{0.101}{D} \beta_0^2 = -0.093536$$

$$K_2 = 229.3\beta_0^{1.5} = 146.92$$

Substitution then gives $F_0' = -1.4758$, and the second estimate for β is, from Eq. (A.4),

$$\beta_1 = \beta_0 - \frac{F_0}{F_0'} = 0.74322 - \frac{0.001142}{-1.4758} = 0.74399$$

Tabulated below are successive iterations for β, carried to seven places to show the rapid convergence:

	n		
	1	**2**	**3**
β_{n-1}	0.7432227	0.7439916	0.7439904
C_{n-1}	0.6009739	0.6009042	0.6009042
$(Y_1)_{n-1}$	0.9955299	0.9955299	0.9955299
F_{n-1}	0.00113475	-0.00000185	1.6×10^{-12}
F'_{n-1}	-1.475	-1.480580	-1.480572
β_n	0.7439916	0.7439904	0.7439904

The orifice bore is then

$$d = \beta_n D = (0.7440)(5.761) = 4.286 \text{ in}$$

Note: There is no justification for seven-place calculations in determining β, other than to maintain calculational accuracy, since the discharge coefficients for primary devices are known, at best, to within ± 0.1 percent under laboratory conditions.

A.4 AGA ORIFICE-COEFFICIENT EQUATIONS

ISO (1980), ASME *Fluid Meters* (1981), and ANSI/ASME MFC (1982) recommend the use of the Stolz orifice-coefficient equation as presented in Table 9.1. The American Gas Association (AGA) *does not* recommend this equation in its Report 3. The Buckingham flange and $2\frac{1}{2}D$ and $8D$ equations are given in this report. The following flow-coefficient equations are to be used when natural gas is to be metered according to AGA Report 3.

For Flange Taps

$$K = K_0 \left(1 + \frac{E_f}{R_d} \right) = K_0 \left(1 + \frac{E_f \beta}{R_D} \right)$$

$$K_0 = \left[0.5993 + \frac{0.007}{D} + \left(0.364 + \frac{0.076}{\sqrt{D}} \right) \beta^4 \right.$$

$$+ 0.4 \left(1.6 - \frac{1}{D} \right)^5 \left(0.07 + \frac{0.5}{D} - \beta \right)^{5/2}$$

$$- \left(0.009 + \frac{0.034}{D} \right) (0.5 - \beta)^{3/2} + \left(\frac{65}{D^2} + 3 \right) (\beta - 0.7)^{5/2} \right]$$

$$\div \left[1 + 0.000015 \left(830 - 5000\beta + 9000\beta^2 - 4200\beta^3 + \frac{530}{\sqrt{D}} \right) \right]$$

$$E_f = d \left(830 - 5000\beta + 9000\beta^2 - 4200\beta^3 + \frac{530}{\sqrt{D}} \right)$$

$$C = \sqrt{1 - \beta^4} K$$

For $2\frac{1}{2}D$ and $8D$ Taps

$$K = K_0 \left(1 + \frac{E_p}{R_d} \right) = K_0 \left(1 + \frac{E_p\beta}{R_D} \right)$$

$$K_0 = \left[0.5925 + \frac{0.0182}{D} + \left(0.440 - \frac{0.06}{D} \right) \beta^2 + \left(0.935 + \frac{0.225}{D} \right) \beta^5 \right.$$

$$\left. + 1.35\beta^{14} + \frac{1.43}{\sqrt{D}} (0.25 - \beta)^{5/2} \right]$$

$$\div \left[1 + 0.000015 \left(905 - 5000\beta + 9000\beta^2 - 4200\beta^3 + \frac{875}{D} \right) \right]$$

$$E_p = d \left(905 - 5000\beta + 9000\beta^2 - 4200\beta^3 + \frac{875}{D} \right)$$

$$C = \sqrt{1 - \beta^4} K$$

Note: Any negative factor, such as $(\beta - 0.7)^{5/2}$ when β is less than 0.7, should be taken as equal to zero, so that the whole term $[(65/D^2) + 3](\beta - 0.7)^{5/2}$ drops out of the equation.

REFERENCES FOR APPENDIXES A TO I

ANSI B36.10, *Welded and Seamless Wrought Steel Pipe,* New York, 1979.

ANSI/API 2530, *Orifice Metering of Natural Gas,* AGA, Arlington, Va., 1978.

ANSI/ASME (MFC): *Differential Producers Used for the Measurement of Fluid Flow in Pipes (Orifice, Nozzle, Nozzle Venturi, Venturi),* draft 3e, New York, July 1982.

ASME: "Fluid Meters: The ASME-ISO Orifice Equation," *Mech. Eng.,* vol. 103, no. 7, 1981.

ASME: *Steam Tables,* 3d ed., American Society of Mechanical Engineers, New York, 1977.

ASTM E380-79, *Annual Book of ASTM Standards,* pt. 41, American Society for Testing and Materials, Philadelphia, 1980.

Baumeister, T.: *Marks' Standard Handbook for Mechanical Engineers,* 8th ed., McGraw-Hill, New York, 1978.

Dieterich Standard Corporation: *Annubar Flow Handbook,* Dieterich publication DS-7300M (1/79), Boulder, Colo., 1979.

Edmister, Wayne: *Applied Hydrocarbon Thermodynamics,* vol. 1, Gulf, Houston, 1961.

———: *Applied Hydrocarbon Thermodynamics,* vol. 2, Gulf, Houston, 1974.

Fischer & Porter Company, *Catalog 10-A-54,* Fischer & Porter publication 10355, Warminster, Pa., 1950.

Fluidic Techniques, Inc.: *Technical Data, Orifice Flange, Pipe,* Mansfield, Texas, 1982.

Gambill, W. R.: "How P and T Change Liquid Viscosity," *Chem. Eng.,* p. 126, McGraw-Hill, New York, 1959.

Gas Processors Supplier Association: *Engineering Data Book,* 9th ed., 4th rev., Tulsa, 1979.

Giles, R. W.: *Fluid Mechanics and Hydraulics,* 2d ed., McGraw-Hill, New York, 1962.

Hardy, R. C., and R. L. Cottington: *J. Res. Nat. Bur. Stand.,* vol. 42, p. 573, 1949.

PEA: *Bulletin T.S. 622-77,* Pacific Energy Association, Los Angeles, 1977.

Perry, R. H., and C. H. Chilton: *Chemical Engineers' Handbook,* 5th ed., McGraw-Hill, New York, 1973.

Reid, R. C., J. M. Prausnitz, and T. K. Sherwood: *The Properties of Gases and Liquids,* 3d ed., McGraw-Hill, New York, 1977.

Schoenthaler, J. L.: Standard Oil of California Technical Reprint, San Francisco, 1966.

Spink, L. K., *Principles and Practice of Flowmeter Engineering,* 9th ed., The Foxboro Company, Foxboro, Mass., 1967.

Stolz, J: *OSU 89 Test Analysis: Interim report on Pipe Taps Discharge Coefficient,* ISO/TC30/SC2 (France 17) 95E, AFNOR, Paris, 1978.

Swindells, J. F.: Unpublished results (National Bureau of Standards).

VDI: Publication 2040, Verein Deutscher Ingenieure, Düsseldorf, 1970.

PIPE DATA AND REYNOLDS-NUMBER CURVES

Table B.1 Pipe Schedules and Nominal Pipe Diameters

Nominal OD, in	ASA B36.10 steel pipe	ASA B36.10 steel pipe, nominal wall thickness	ASA B36.10 stainless pipe	Wall thickness, in	D, in
		Schedule number			
$\frac{1}{8}$ 0.405	. . .	. . . .	10S	0.049	0.307
	40	Std	40S	0.068	0.269
	80	XS	80S	0.095	0.215
$\frac{1}{4}$ 0.540	. . .	. . . .	10S	0.065	0.410
	40	Std	40S	0.088	0.364
	80	XS	80S	0.119	0.302
$\frac{3}{8}$ 0.675	. . .	. . . .	10S	0.065	0.545
	40	Std	40S	0.091	0.493
	80	XS	80S	0.126	0.423
$\frac{1}{2}$ 0.840	. . .	. . . .	10S	0.083	0.674
	40	Std	40S	0.109	0.622
	80	XS	80S	0.147	0.546
	160	. . . .	. . .	0.188	0.464
	. . .	XXS	. . .	0.294	0.252
$\frac{3}{4}$ 1.050	. . .	. . . .	5S	0.065	0.920
	. . .	. . . .	10S	0.083	0.884
	40	Std	40S	0.113	0.824
	80	XS	80S	0.154	0.742
	160	. . . .	. . .	0.219	0.612
	. . .	XXS	. . .	0.308	0.434

Table B.1 **Pipe Schedules and Nominal Pipe Diameters** (*Continued*)

Nominal OD, in	ASA B36.10 steel pipe	ASAB36.10 steel pipe, nominal wall thickness	ASA B36.10 stainless pipe	Wall thickness, in	D, in
				Schedule number	
	. . .	. . . .	5S	0.065	1.185
	. . .	. . . .	10S	0.109	1.097
1	40	Std	40S	0.133	1.049
1.315	80	XS	80S	0.179	0.957
	160	. . . .	. . .	0.250	0.815
	. . .	XXS	. . .	0.358	0.599
	. . .	. . . .	5S	0.065	1.530
	. . .	. . . .	10S	0.109	1.442
$1\frac{1}{4}$	40	Std	40S	0.140	1.380
1.660	80	XS	80S	0.191	1.278
	160	. . . .	. . .	0.250	1.160
	. . .	XXS	. . .	0.382	0.896
	. . .	. . . .	5S	0.065	1.770
	. . .	. . . .	10S	0.109	1.682
$1\frac{1}{2}$	40	Std	40S	0.145	1.610
1.900	80	XS	80S	0.200	1.500
	160	. . . .	. . .	0.281	1.338
	. . .	XXS	. . .	0.400	1.100
	. . .	. . . .	5S	0.065	2.245
	. . .	. . . .	10S	0.109	2.157
2	40	Std	40S	0.154	2.067
2.375	80	XS	80S	0.218	1.939
	160	. . . .	. . .	0.344	1.687
	. . .	XXS	. . .	0.436	1.503
	. . .	. . . .	5S	0.083	2.709
	. . .	. . . .	10S	0.120	2,635
$2\frac{1}{2}$	40	Std	40S	0.203	2.469
2.875	80	XS	80S	0.276	2.323
	160	. . . .	. . .	0.375	2.125
	. . .	XXS	. . .	0.552	1.771
	. . .	. . . .	5S	0.083	3.334
	. . .	. . . .	10S	0.120	3.260
3	40	Std	40S	0.216	3.068
3.500	80	XS	80S	0.300	2.900
	160	. . . .	. . .	0.438	2.624
	. . .	XXS	. . .	0.600	2.300

Table B.1 Pipe Schedules and Nominal Pipe Diameters
(*Continued*)

| Nominal OD, in | Schedule number | | | Wall thickness, in | D, in |
	ASA B36.10 steel pipe	ASA B36.10 steel pipe, nominal wall thickness	ASA B36.10 stainless pipe		
$3\frac{1}{2}$ 4.000	. . .	. . . .	5S	0.083	3.834
	. . .	. . . .	10S	0.120	3.760
	40	Std	40S	0.226	3.548
	80	XS	80S	0.318	3.364
4 4.500	. . .	. . . .	5S	0.083	4.334
	. . .	. . . .	10S	0.120	4.260
	40	Std	40S	0.237	4.026
	80	XS	80S	0.337	3.826
	120	. . . .	. . .	0.438	3.624
	160	. . . .	. . .	0.531	3.438
	. . .	XXS	. . .	0.674	3.152
5 5.563	. . .	. . . .	5S	0.109	5.345
	. . .	. . . .	10S	0.134	5.295
	40	Std	40S	0.258	5.047
	80	XS	80S	0.375	4.813
	120	. . . .	. . .	0.500	4.563
	160	. . . .	. . .	0.625	4.313
	. . .	XXS	. . .	0.750	4.063
6 6.625	. . .	. . . .	5S	0.109	6.407
	. . .	. . . .	10S	0.134	6.357
	40	Std	40S	0.280	6.065
	80	XS	80S	0.432	5.761
	120	. . . .	. . .	0.562	5.501
	160	. . . .	. . .	0.719	5.187
	. . .	XXS	. . .	0.864	4.897
8 8.625	. . .	. . . .	5S	0.109	8.407
	. . .	. . . .	10S	0.148	8.329
	20	. . . .	. . .	0.250	8.125
	30	. . . .	. . .	0.277	8.071
	40	Std	40S	0.322	7.981
	60	. . . .	. . .	0.406	7.813
	80	XS	80S	0.500	7.625
	100	. . . .	. . .	0.594	7.437
	120	. . . .	. . .	0.719	7.187
	140	. . . .	. . .	0.812	7.001
	. . .	XXS	. . .	0.875	6.875
	160	. . . .	. . .	0.906	6.813

Table B.1 **Pipe Schedules and Nominal Pipe Diameters** (*Continued*)

| Nominal OD, in | Schedule number | | | Wall thickness, in | D, in |
	ASA B36.10 steel pipe	ASA B36.10 steel pipe, nominal wall thickness	ASA B36.10 stainless pipe		
	...	....	5S	0.134	10.482
	...	....	10S	0.165	10.420
	20	....	...	0.250	10.250
	...	....	...	0.279	10.192
	30	....	...	0.307	10.136
10	40	Std	40S	0.365	10.020
10.750	60	XS	80S	0.500	9.750
	80	....	...	0.594	9.562
	100	....	...	0.719	9.312
	120	....	...	0.844	9.062
	140	....	...	1.000	8.750
	160	....	...	1.125	8.500
	...	....	5S	0.165	12.420
	...	....	10S	0.180	12.390
	20	....	...	0.250	12.250
	30	....	...	0.330	12.090
	...	Std	40S	0.375	12.000
12	40	....	...	0.406	11.938
12.750	...	XS	80S	0.500	11.750
	60	....	...	0.562	11.626
	80	....	...	0.688	11.374
	100	....	...	0.844	11.062
	120	....	...	1.000	10.750
	140	....	...	1.125	10.500
	60	....	...	1.312	10.126
	10	....	...	0.250	13.500
	20	....	...	0.312	13.376
	30	Std	...	0.375	13.250
	40	....	...	0.438	13.124
	...	XS	...	0.500	13.000
	...	....	...	0.562	12.876
14	60	....	...	0.594	12.812
14.000	...	....	...	0.625	12.750
	...	....	...	0.688	12.624
	80	....	...	0.750	12.500
	...	....	...	0.875	12.250
	100	....	...	0.938	12.124
	120	....	...	1.094	11.812

Table B.1 Pipe Schedules and Nominal Pipe Diameters (*Continued*)

Nominal OD, in	Schedule number			Wall thickness, in	D, in
	ASA B36.10 steel pipe	ASA B36.10 steel pipe, nominal wall thickness	ASA B36.10 stainless pipe		
	140	....	...	1.250	11.500
	160	....	...	1.406	11.188
	10	....	...	0.250	15.500
	20	....	...	0.312	15.376
	30	Std	...	0.375	15.250
	...	....	...	0.438	15.124
	40	XS	...	0.500	15.000
	...	....	...	0.562	14.876
	...	....	...	0.625	14.750
16	60	....	...	0.656	14.688
16.000	...	....	...	0.688	14.624
	...	....	...	0.750	14.500
	80	....	...	0.844	14.312
	...	....	...	0.875	14.250
	100	....	...	1.031	13.938
	120	....	...	1.219	13.562
	140	....	...	1.438	13.124
	160	....	...	1.594	12.812
	10	....	...	0.250	17.500
	20	....	...	0.312	17.376
	...	Std	...	0.375	17.250
	30	....	...	0.438	17.124
	...	XS	...	0.500	17.000
	40	....	...	0.562	16.876
	...	....	...	0.625	16.750
18	...	....	...	0.688	16.624
18.000	60	....	...	0.750	16.500
	...	....	...	0.875	16.250
	80	....	...	0.938	16.124
	100	....	...	1.156	15.688
	120	....	...	1.375	15.250
	140	....	...	1.562	14.876
	160	....	...	1.781	14.438
	10	....	...	0.250	19.500
	...	....	...	0.312	19.376
	20	Std	...	0.375	19.250
	...	....	...	0.438	19.124

Table B.1 **Pipe Schedules and Nominal Pipe Diameters**
(*Continued*)

| Nominal OD, in | Schedule number | | | Wall thickness, in | D, in |
	ASA B36.10 steel pipe	ASAB36.10 steel pipe, nominal wall thickness	ASA B36.10 stainless pipe		
	30	XS	. . .	0.500	19.000
	. . .	. . . .	. . .	0.562	18.876
	40	. . . .	. . .	0.594	18.812
	. . .	. . . .	. . .	0.625	18.750
20	. . .	. . . .	. . .	0.688	18.624
20.000	. . .	. . . .	. . .	0.750	18.500
	60	. . . .	. . .	0.812	18.376
	. . .	. . . .	. . .	0.875	18.250
	80	. . . .	. . .	1.031	17.938
	100	. . . .	. . .	1.281	17.438
	120	. . . .	. . .	1.500	17.000
	140	. . . .	. . .	1.750	16.500
	160	. . . .	. . .	1.969	16.062
	10	. . . .	. . .	0.250	23.500
	. . .	. . . .	. . .	0.312	23.376
	20	Std	. . .	0.375	23.250
	. . .	. . . .	. . .	0.438	23.124
	. . .	XS	. . .	0.500	23.000
	30	. . . .	. . .	0.562	22.876
24	. . .	. . . .	. . .	0.625	22.750
24.000	40	. . . .	. . .	0.688	22.624
	. . .	. . . .	. . .	0.750	22.500
	60	. . . .	. . .	0.969	22.062
	80	. . . .	. . .	1.219	21.562
	100	. . . .	. . .	1.531	20.938
	120	. . . .	. . .	1.812	20.376
	140	. . . .	. . .	2.062	19.876
	160	. . . .	. . .	2.344	19.312
30	10	. . . .	. . .	0.312	29.376
30.000	20	. . . .	. . .	0.500	29.000
	30	. . . .	. . .	0.625	28.750

SOURCE: Modified from Baumeister (1978), with data from ANSI B36.10 (1979).

Table B.2 Maximum Permissible Inside Diameters and Minimum Wall Thicknesses for ASTM A106 Pipe

Nominal pipe size, in	Maximum outside diameter, in	Wall, ID, in	Schedule 10	Schedule 20	Schedule 30	Standard weight	Schedule 40	Schedule 60	Extra strong	Schedule 80	Schedule 100	Schedule 120	Schedule 140	Schedule 160	Double extra strong
$\frac{1}{8}$	0.421	Wall				0.060	0.060		0.083	0.083					
		ID				0.302	0.302		0.254	0.254					
$\frac{1}{4}$	0.556	Wall				0.077	0.077		0.110	0.110					
		ID				0.402	0.402		0.335	0.335					
$\frac{3}{8}$	0.691	Wall				0.080	0.080		0.110	0.110					
		ID				0.531	0.531		0.470	0.470					
$\frac{1}{2}$	0.856	Wall				0.095	0.095		0.129	0.129				0.164	0.257
		ID				0.665	0.665		0.598	0.598				0.528	0.341
$\frac{3}{4}$	1.066	Wall				0.099	0.099		0.135	0.135				0.191	0.270
		ID				0.868	0.868		0.796	0.796				0.684	0.527
1	1.331	Wall				0.116	0.116		0.157	0.157				0.219	0.313
		ID				1.098	1.098		1.017	1.017				0.893	0.704

Table B.2 Maximum Permissible Inside Diameters and Minimum Wall Thicknesses for ASTM A106 Pipe (*Continued*)

Nominal pipe size, in	Maximum outside diameter, in	Wall, ID, in	Schedule 10	Schedule 20	Schedule 30	Standard weight	Schedule 40	Schedule 60	Extra strong	Schedule 80	Schedule 100	Schedule 120	Schedule 140	Schedule 160	Double extra strong
1¼	1.676	Wall				0.123	0.123		0.167	0.167				0.219	0.334
		ID				1.431	1.431		1.341	1.341				1.238	1.007
1½	1.916	Wall				0.127	0.127		0.175	0.175				0.246	0.350
		ID				1.662	1.662		1.566	1.566				1.424	1.216
2	2.406	Wall				0.135	0.135		0.191	0.191				0.300	0.382
		ID				2.138	2.137		2.025	2.025				1.806	1.643
2½	2.906	Wall				0.178	0.178		0.242	0.242				0.328	0.483
		ID				2.551	2.551		2.423	2.423				2.250	1.940
3	3.531	Wall				0.189	0.189		0.263	0.263				0.383	0.525
		ID				3.153	3.153		3.006	3.006				2.765	2.481
3½	4.031	Wall				0.198	0.198		0.278	0.278		0.383			0.557
		ID				3.636	3.636		3.475	3.475					2.918
4	4.531	Wall				0.207	0.207		0.295	0.295				0.465	0.590

Pipe dimension table (nominal size, outside diameter, and schedule data). Dimensions for each nominal size are given as Wall thickness and inside diameter (ID). Schedule columns numbered 1–13 (heaviest wall at 1).

Nom.	OD	Dim	1	2	3	4	5	6	7	8	9	10	11	12	13
		ID	3.352	3.602	3.765		3.942	3.942			4.117	4.117			
5	5.626	Wall	0.656	0.547	0.438		0.328	0.328			0.226	0.226			
		ID	4.313	4.532	4.751		4.969	4.969			5.174	5.174			
6	6.688	Wall	0.756	0.628	0.492		0.378	0.378			0.245	0.245			
		ID	5.176	5.431	5.704		5.932	5.932			6.198	6.198			
8	8.719	Wall	0.766	0.793	0.628	0.711	0.519	0.438	0.438	0.355	0.282	0.282	0.242	0.219	
		ID	7.156	7.102	7.431	7.267	7.650	7.813	7.813	7.977	8.124	8.124	8.203	8.250	
10	10.844	Wall		0.984	0.738	0.875	0.628	0.519	0.438	0.438	0.319	0.319	0.269	0.219	
		ID		8.875	9.369	9.094	9.587	9.806	9.969	9.969	10.205	10.205	10.307	10.406	
12	12.844	Wall		1.148	0.875	0.984	0.738	0.601	0.438	0.492	0.355	0.328	0.289	0.219	
		ID		10.548	11.094	10.875	11.369	11.642	11.969	11.860	12.133	12.188	12.266	12.406	
14	14.094	Wall		1.230	0.956	1.094	0.820	0.656	0.438	0.519	0.383	0.328	0.328	0.273	0.219
		ID		11.633	12.181	11.906	12.454	12.781	13.219	13.056	13.327	13.438	13.438	13.548	13.656
16	16.094	Wall		1.394	1.066	1.258	0.902	0.738	0.438	0.574	0.438	0.328	0.328	0.273	0.219
		ID		13.306	13.962	13.577	14.290	14.619	15.219	14.946	15.219	15.438	15.438	15.548	15.656
18	18.094	Wall		1.558	1.203	1.367	1.012	0.820	0.438	0.656	0.492	0.328	0.383	0.273	0.219
		ID		14.977	15.688	15.360	16.071	16.454	17.219	16.781	17.110	17.438	17.327	17.548	17.656

Table B.2 Maximum Permissible Inside Diameters and Minimum Wall Thicknesses for ASTM A106 Pipe (*Continued*)

Nominal pipe size, in	Maximum outside diameter, in	Wall, ID, in	Nominal wall thickness and inside diameter, in												
			Schedule 10	Schedule 20	Schedule 30	Standard weight	Schedule 40	Schedule 60	Extra strong	Schedule 80	Schedule 100	Schedule 120	Schedule 140	Schedule 160	Double extra strong
20	20.125	Wall	0.219	0.328	0.438	0.328	0.519	0.711	0.438	0.902	1.121	1.313	1.531	1.722	
		ID	19.688	19.469	19.250	19.469	19.087	18.704	19.250	18.321	17.883	17.500	17.063	16.681	
24	24.125	Wall	0.219	0.328	0.492	0.328	0.601	0.847	0.438	1.066	1.340	1.586	1.804	2.050	
		ID	23.688	23.469	23.142	23.469	22.923	22.431	23.250	21.994	21.446	20.954	20.517	20.025	
30	30.125	Wall	0.273	0.438	0.547	0.328			0.438						
		ID	29.579	29.250	29.031	29.469			29.250						

SOURCE: Fluidic Techniques, Inc. (1982); used with permission.

Table B.3   Standard Honed Internal Diameters for Flow-Nozzle Sections

Nominal pipe size D, in	Schedule no.	Bored diameter, in	Tolerance, in	Nominal pipe size D, in	Schedule no.	Bored diameter, in	Tolerance, in
4	40	4.065	± 0.005	10	20	10.291	+0.005 −0.010
	80	3.881	± 0.005		30	10.186	+0.005 −0.010
	120	3.697	± 0.005		40	10.081	+0.005 −0.010
	160	3.524	± 0.005		60	9.831	+0.005 −0.010
					80	9.660	+0.005 −0.010
5	40	5.089	± 0.005		100	9.431	+0.005 −0.010
	80	4.874	± 0.005		120	9.201	+0.005 −0.010
	120	4.611	± 0.005		140	8.913	+0.005 −0.010
	160	4.415	± 0.005		160	8.683	+0.005 −0.010
6	40	6.111	± 0.005	12	20	12.291	+0.005 −0.015
	80	5.831	± 0.005		30	12.144	+0.005 −0.015
	120	5.592	± 0.005		40	12.004	+0.005 −0.015
	160	5.306	± 0.005		60	11.717	+0.005 −0.015
					80	11.488	+0.005 −0.015
8	20	8.166	+0.005 −0.010		100	11.201	+0.005 −0.015
	30	8.116	+0.005 −0.010		120	10.913	+0.005 −0.015
	40	8.033	+0.005 −0.010		140	10.683	+0.005 −0.015
	60	7.879	+0.005 −0.010		160	10.339	+0.005 −0.015
	80	7.706	+0.005 −0.010				
	100	7.535	+0.005 −0.010				
	120	7.306	+0.005 −0.010				
	140	7.133	+0.005 −0.010				
	160	6.960	+0.005 −0.010				

Table B.3 Standard Honed Internal Diameters for Flow-Nozzle Sections (Continued)

Nominal pipe size D, in	Schedule no.	Bored diameter, in	Tolerance, in		Nominal pipe size D, in	Schedule no.	Bored diameter, in	Tolerance, in	
14	10	13.511	+0.005	−0.020	18	10	17.541	+0.005	−0.030
	20	13.427	+0.005	−0.020		20	17.427	+0.005	−0.030
	30	13.311	+0.005	−0.020		30	17.197	+0.005	−0.030
	40	13.197	+0.005	−0.020		40	16.967	+0.005	−0.030
	60	12.910	+0.005	−0.020		60	16.622	+0.005	−0.030
	80	12.622	+0.005	−0.020		80	16.278	+0.005	−0.030
	100	12.278	+0.005	−0.020		100	15.875	+0.005	−0.030
	120	11.992	+0.005	−0.020		120	15.474	+0.005	−0.030
	140	11.703	+0.005	−0.020		140	15.130	+0.005	−0.030
	160	11.416	+0.005	−0.020		160	14.728	+0.005	−0.030
16	10	15.541	+0.005	−0.020	20	10	19.541	+0.005	−0.030
	20	15.427	+0.005	−0.020		20	19.311	+0.005	−0.030
	30	15.311	+0.005	−0.020		30	19.081	+0.005	−0.030
	40	15.081	+0.005	−0.020		40	18.910	+0.005	−0.030
	60	14.795	+0.005	−0.020		60	18.508	+0.005	−0.030
	80	14.451	+0.005	−0.020		80	18.106	+0.005	−0.030
	100	14.106	+0.005	−0.020		100	17.616	+0.005	−0.030
	120	13.762	+0.005	−0.020		120	17.244	+0.005	−0.030
	140	13.360	+0.005	−0.020		140	16.784	+0.005	−0.030
	160	13.073	+0.005	−0.020		160	16.384	+0.005	−0.030

SOURCE: Fluidic Techniques, Inc. (1982); used with permission.

Table B.4 Coefficient of Thermal Expansion†

Material	Coefficient of thermal expansion	
	α, in/(in · °F)	α', mm/(mm · °C)
Plain carbon steel (SAE 1020)	0.0000067	0.0000120
Stainless steels		
301 70–600°F (21–315°C)	0.0000097	0.0000175
−300–70°F (−185–21°C)	0.0000076	0.0000137
304 70–600°F (21–315°C)	0.0000095	0.0000171
−300–70°F (−185–21°C)	0.0000074	0.0000133
310 70–600°F (21–315°C)	0.0000090	0.0000162
−300–70°F (−185–21°C)	0.0000070	0.0000126
316 70–600°F (21–315°C)	0.0000096	0.0000173
−300–70°F (−185–21°C)	0.0000071	0.0000128
330 70–600°F (21–315°C)	0.0000089	0.0000160
−300–70°F (−185–21°C)	0.0000058	0.0000104
347 70–600°F (21–315°C)	0.0000097	0.0000175
−300–70°F (−185–21°C)	0.0000075	0.0000135
Hastelloy B 32–212°F (0–100°C)	0.0000056	0.0000101
Hastelloy C	0.0000063	0.0000113
Inconel X, annealed	0.0000067	0.0000120
Haynes Stellite 25 (L605)	0.0000076	0.0000137
Copper (ASTM B152, B124, B133)	0.0000093	0.0000167
Yellow brass (ASTM B36, B134, B135)	0.0000105	0.0000189
Aluminum bronze (ASTM B169 Alloy A)	0.0000092	0.0000166
Beryllium copper 25 (ASTM B194)	0.0000093	0.0000167
Cupronickel 30%	0.0000085	0.0000154
K-Monel	0.0000074	0.0000133
Nickel	0.0000083	0.0000149
Pyrex glass 32–580°F (0–300°C)	0.0000002	0.0000004
Titanium 70–212°F (20–100°C)	0.0000047	0.0000085
Tantalum 70–212°F (20–100°C)	0.0000036	0.0000065

†Ranges of 70 to 600°F (21–315°C) unless otherwise noted.

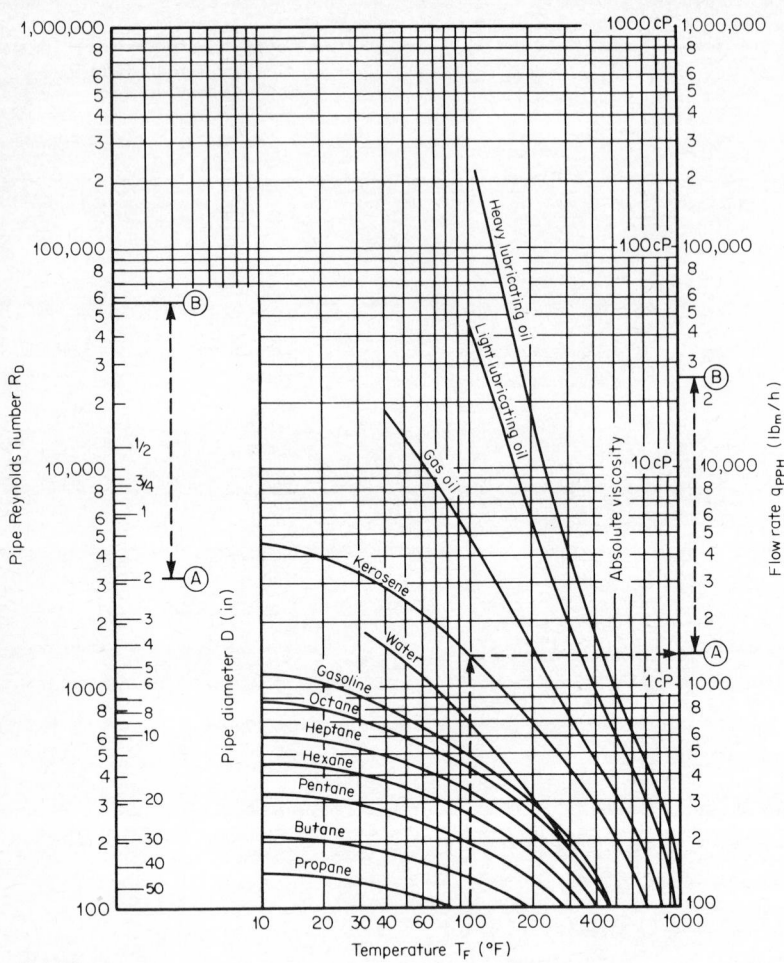

Figure B.1   Reynolds-number alignment curves for liquids. The example is for kerosene (q_{PPH} = 25,000, T_F = 100°F, D = 2 in): (1) Locate point A using T_F. (2) Determine distance AB to flow-rate value. (3) Lay out distance AB from pipe size. (4) Read Reynolds number R_D = 57,000 *(from Spink, 1967)*.

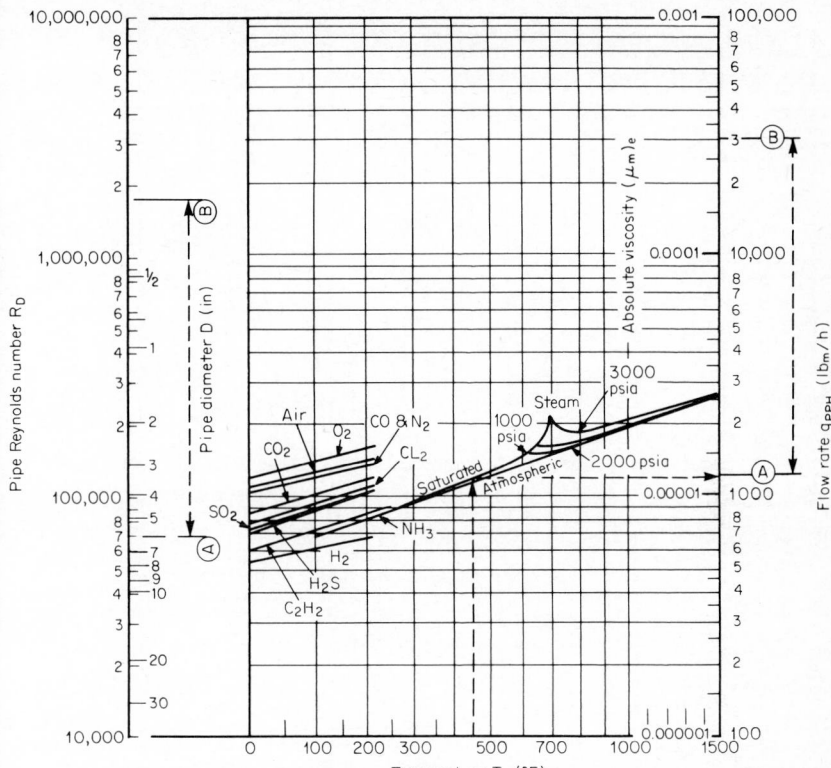

Figure B.2 Reynolds-number alignment curves for gas (vapor). The example is for saturated steam ($q_{PPH} = 30,000$, $T_F = 450°F$, $D = 6$ in): (1) Locate point A using T_F. (2) Determine distance AB to flow-rate value. (3) Lay out distance AB from pipe size. (4) Read Reynolds number $R_D = 1,800,000$ *(from Spink, 1967).*

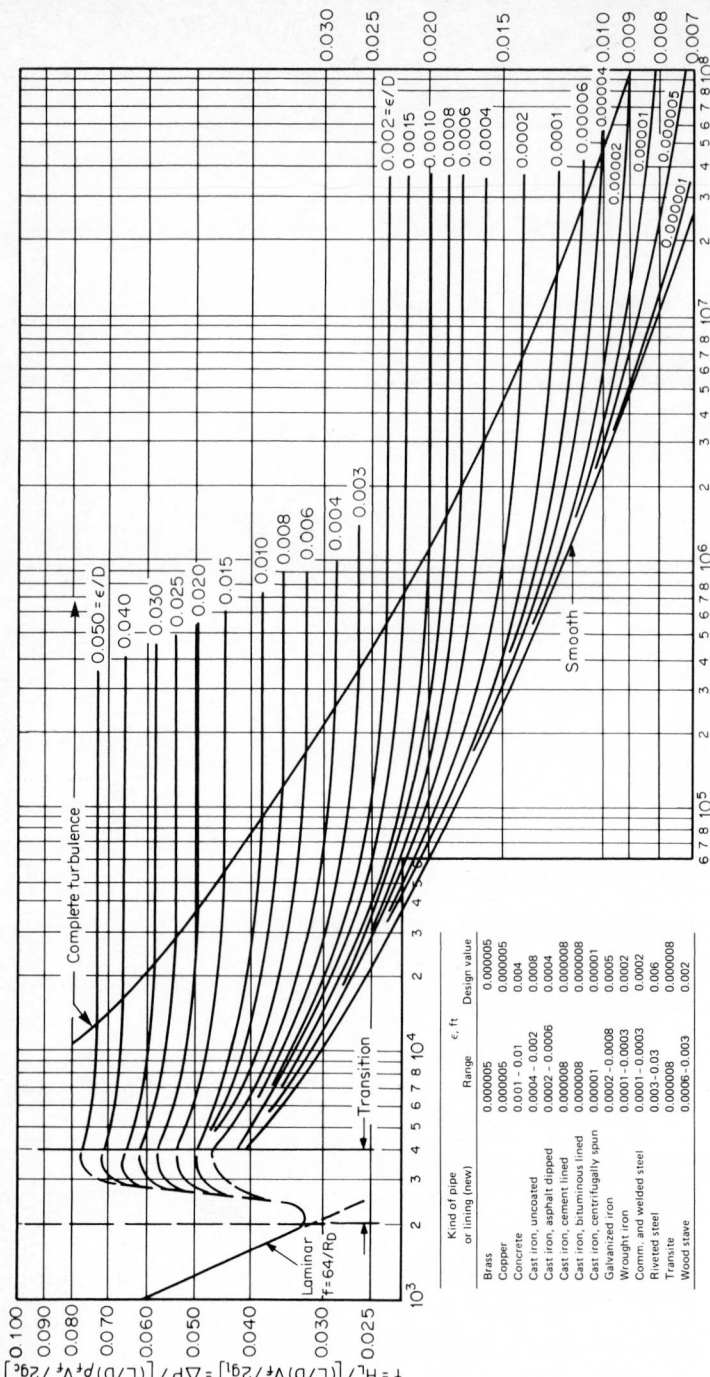

Figure B.3 Moody friction-factor diagram. Curves for relative roughness ϵ/D from 0.000001 to 0.050, where ϵ = size of surface imperfections in inches, D = actual inside diameter in inches (*from Giles, 1962*).

REYNOLDS-NUMBER EQUATIONS AND UNIT CONVERSION TABLES

Table C.1 SI-Unit Conversion Factors†

To convert from	To	Multiply by
ACCELERATION		
ft/s^2	meter per second^2 (m/s^2)	3.048 000*E − 01
free fall, standard (g)	meter per second^2 (m/s^2)	9.806 650*E + 00
in/s^2	meter per second^2 (m/s^2)	2.540 000*E − 02
ANGLE		
degree (angle)	radian (rad)	1.745 329 E − 02
minute (angle)	radian (rad)	2.908 882 E − 04
second (angle)	radian (rad)	4.848 137 E − 06
AREA		
ft^2	meter^2 (m^2)	9.290 304*E − 02
in^2	meter^2 (m^2)	6.451 600*E − 04
mi^2 (international)	meter^2 (m^2)	2.589 988 E + 06
mi^2 (U.S. survey)	meter^2 (m^2)	2.589 998 E + 06
BENDING MOMENT OR TORQUE		
dyne·cm	newton meter (N·m)	1.000 000*E − 07
kf_f·m	newton meter (N·m)	9.806 650*E + 00
oz_f·in	newton meter (N·m)	7.061 552 E − 03
lb_f·in	newton meter (N·m)	1.129 848 E − 01
lb_f·ft	newton meter (N·m)	1.355 818 E + 00
BENDING MOMENT OR TORQUE PER UNIT LENGTH		
lb_f·ft/in	newton meter per meter (N·m/m)	5.337 866 E + 01
lb_f·in/in	newton meter per meter (N·m/m)	4.448 222 E + 00
ENERGY (INCLUDES WORK)		
British thermal unit (International Table)	joule (J)	1.055 056 E + 03
British thermal unit (mean)	joule (J)	1.055 87 E + 03
British thermal unit (thermochemical)	joule (J)	1.054 350 E + 03

†Factors with an asterisk are exact.

Table C.1 SI-Unit Conversion Factors† (*Continued*)

To convert from	To	Multiply by
British thermal unit (39°F)	joule (J)	1.059 67 E + 03
British thermal unit (59°F)	joule (J)	1.054 80 E + 03
British thermal unit (60°F)	joule (J)	1.054 68 E + 03
calorie (International Table)	joule (J)	4.186 800*E + 00
calorie (mean)	joule (J)	4.190 02 E + 00
calorie (thermochemical)	joule (J)	4.184 000*E + 00
calorie (15°C)	joule (J)	4.185 80 E + 00
calorie (20°C)	joule (J)	4.181 90 E + 00
calorie (kilogram, International Table)	joule (J)	4.186 800*E + 03
calorie (kilogram, mean)	joule (J)	4.190 02 E + 03
calorie (kilogram, thermochemical)	joule (J)	4.184 000*E + 03
electronvolt	joule (J)	1.602 19 E − 19
erg	joule (J)	1.000 000*E − 07
ft·lb$_f$	joule (J)	1.355 818 E + 00
ft·poundal	joule (J)	4.214 011 E − 02
kilocalorie (International Table)	joule (J)	4.186 800*E + 03
kilocalorie (mean)	joule (J)	4.190 02 E + 03
kilocalorie (thermochemical)	joule (J)	4.184 000*E + 03
kW·h	joule (j)	3.600 000*E + 06
therm	joule (J)	1.055 056 E + 08
ton (nuclear equivalent of TNT)	joule (J)	4.184 E + 09
W·h	joule (J)	3.600 000*E + 03
W·s	joule (J)	1.000 000*E + 00

	MASS PER UNIT VOLUME	
grain (lb avoirdupois/7000)/gal (U.S. liquid)	kilogram per meter^3 (kg/m^3)	1.711 806 E − 02
g/cm^3	kilogram per meter^3 (kg/m^3)	1.000 000*E + 03
oz (avoirdupois)/gal (U.K. liquid)	kilogram per meter^3 (kg/m^3)	6.236 021 E + 00
oz (avoirdupois)/gal (U.S. liquid)	kilogram per meter^3 (kg/m^3)	7.489 152 E + 00
oz (avoirdupois)/in^3	kilogram per meter^3 (kg/m^3)	1.729 994 E + 03
lb/ft^3	kilogram per meter^3 (kg/m^3)	1.601 846 E + 01
lb/in^3	kilogram per meter^3 (kg/m^3)	2.767 990 E + 04
lb/gal (U.K. liquid)	kilogram per meter^3 (kg/m^3)	9.977 633 E + 01
lb/gal (U.S. liquid)	kilogram per meter^3 (kg/m^3)	1.198 264 E + 02
lb/yd^3	kilogram per meter^3 (kg/m^3)	5.932 764 E − 01
slug/ft^3	kilogram per meter^3 (kg/m^3)	5.153 788 E + 02
ton (long)/yd^3	kilogram per meter^3 (kg/m^3)	1.328 939 E + 03
ton (short)/yd^3	kilogram per meter^3 (kg/m^3)	1.186 553 E + 03

	POWER	
Btu (International Table)/h	watt (W)	2.930 711 E − 01
Btu (International Table)/s	watt (W)	1.055 056 E + 03
Btu (thermochemical)/h	watt (W)	2.928 751 E − 01
Btu (thermochemical)/min	watt (W)	1.757 250 E + 01
Btu (thermochemical)/s	watt (W)	1.054 350 E + 03
cal (thermochemical)/min	watt (W)	6.973 333 E − 02
cal (thermochemical)/s	watt (W)	4.184 000*E + 00
erg/s	watt (W)	1.000 000*E − 07
ft·lb$_f$/h	watt (W)	3.766 161 E − 04
ft·lb$_f$/min	watt (W)	2.259 697 E − 02

†Factors with an asterisk are exact.

Table C.1 **SI-Unit Conversion Factors†** (*Continued*)

To convert from	To	Multiply by
ft·lb$_f$/s	watt (W)	1.355 818 E + 00
horsepower (550 ft·lb$_f$/s)	watt (W)	7.456 999 E + 02
horsepower (boiler)	watt (W)	9.809 50 E + 03
horsepower (electric)	watt (W)	7.460 000*E + 02
horsepower (metric)	watt (W)	7.354 99 E + 02
horsepower (water)	watt (W)	7.460 43 E + 02
horsepower (U.K.)	watt (W)	7.457 0 E + 02
kilocalorie (thermochemical)/min	watt(W)	6.973 333 E + 01
kilocalorie (thermochemical)/s	watt (W)	4.184 000*E + 03
ton (refrigeration)	watt (W)	3.516 800 E + 03

PRESSURE OR STRESS (FORCE PER UNIT AREA)

atmosphere (standard)	pascal (Pa)	1.013 250*E + 05
atmosphere (technical = 1 kg$_f$/cm^2)	pascal (Pa)	9.806 650*E + 04
bar	pascal (Pa)	1.000 000*E + 05
centimeter of mercury (0°C)	pascal (Pa)	1.333 22 E + 03
centimeter of water (4°C)	pascal (Pa)	9.806 38 E + 01
dyne/cm^2	pascal (Pa)	1.000 000*E − 01
foot of water (39.2°F)	pascal (Pa)	2.988 98 E + 03
gram-force/cm^2	pascal (Pa)	9.806 650*E + 01
inch of mercury (32°F)	pascal (Pa)	3.386 38 E + 03
inch of mercury (60°F)	pascal (Pa)	3.376 85 E + 03
inch of water (60°F)	pascal (Pa)	2.488 429 E + 02
inch of water (68°F)	pascal (Pa)	2.4864107 E + 02
kg$_f$/cm^2	pascal (Pa)	9.806 650*E + 04
kg$_f$/m^2	pascal (Pa)	9,806 650*E + 00
kg$_f$/mm^2	pascal (Pa)	9,806 650*E + 06
kip/in^2 (ksi)	pascal (Pa)	6.894 757 E + 06
millibar	pascal (Pa)	1.000 000*E + 02
millimeter of mercury (0°C)	pascal (Pa)	1.333 22 E + 02
poundal/ft^2	pascal (Pa)	1.488 164 E + 00
lb$_f$/ft^2	pascal (Pa)	4.788 026 E + 01
lb$_f$/in^2 (psi)	pascal (Pa)	6.894 757 E + 03
psi	pascal (Pa)	6.894 757 E + 03
torr (mm Hg, 0°C)	pascal (Pa)	1.333 22 E + 02

TEMPERATURE

degree Celsius	kelvin (K)	$T_K = T_{°C} + 273.15$
degree Fahrenheit	degree Celsius	$T_{°C} = (T_F - 32)/1.8$
degree Fahrenheit	kelvin (K)	$T_K = (T_F + 459.67)/1.8$
degree Rankine	kelvin (K)	$T_{°R} = T_F/1.8$
kelvin	degree Celsius	$T_{°C} = T_K - 273.15$

VELOCITY

ft/h	meter per second (m/s)	8.466 667 E − 05
ft/min	meter per second (m/s)	5.080 000*E − 03
ft/s	meter per second (m/s)	3.048 000*E − 01
in/s	meter per second (m/s)	2.540 000*E − 02

†Factors with an asterisk are exact.

Table C.1 **SI-Unit Conversion Factors† (*Continued*)**

To convert from	To	Multiply by
	VISCOSITY	
centipoise	pascal second (Pa·s)	1.000 000*E − 03
centistoke	meter^2 per second (m^2/s)	1.000 000*E − 06
ft^2/s	meter^2 per second (m^2/s)	9.290 304*E − 02
poise	pascal second (Pa·s)	1.000 000*E − 01
poundal·s/ft^2	pascal second (Pa·s)	1.488 164 E + 00
lb/(ft·h)	pascal second (Pa·s)	4.133 789 E − 04
lb/(ft·s)	pascal second (Pa·s)	1.488 164 E + 00
lb$_f$·s/ft^2	pascal second (Pa·s)	4.788 026 E + 01
lb$_f$·s/in^2	pascal second (Pa·s)	6.894 757 E + 03
rhe	1 per pascal second [1/(Pa·s)]	1.000 000*E + 01
slug/(ft·s)	pascal second (Pa·s)	4.788 026 E + 01
stokes	meter^2 per second (m^2/s)	1.000 000*E − 04
	VOLUME	
fluid ounce (U.S.)	meter^3 (m^3)	2.957 353 E − 05
ft^3	meter^3 (m^3)	2.831 685 E − 02
gallon (Canadian liquid)	meter^3 (m^3)	4.546 090 E − 03
gallon (U.K. liquid)	meter^3 (m^3)	4.546 092 E − 03
gallon (U.S. dry)	meter^3 (m^3)	4.404 884 E − 03
gallon (U.S. liquid)	meter^3 (m^3)	3.785 412 E − 03
pint (U.S. dry)	meter^3 (m^3)	5.506 105 E − 04
pint (U.S. liquid)	meter^3 (m^3)	4.731 765 E − 04
quart (U.S. dry)	meter^3 (m^3)	1.101 221 E − 03
quart (U.S. liquid)	meter^3 (m^3)	9.463 529 E − 04
	VOLUME PER UNIT TIME	
ft^3/min	meter^3 per second (m^3/s)	4.719 474 E − 04
ft^3/s	meter^3 per second (m^3/s)	2.831 685 E − 02
gallon (U.S. liquid)/(hp·h) (SFC, specific fuel consumption)	meter^3 per joule (m^3/J)	1.410 089 E − 09
in^3/min	meter^3 per second (m^3/s)	2.731 177 E − 07
gallon (U.S. liquid) per day	meter^3 per second (m^3/s)	4.381 264 E − 08
gallon (U.S. liquid) per minute	meter^3 per second (m^3/s)	6.309 020 E − 05

†Factors with an asterisk are exact.

SOURCE: Reprinted/adapted, with permission, from ASTM (1980). Copyright, American Society for Testing and Materials, 1916 Race Street, Philadelphia, Pa. 19103.

Table C.2 **Viscosity Conversion Factors**

To convert from	To		Multiply by
	Units	Symbol	
ABSOLUTE VISCOSITY μ			
poise μ_P	$lb_f \cdot s/ft^2$	$(\mu_f)_e$	2.088543 E − 03
centipoise μ_{cP}	$lb_f \cdot s/ft^2$	$(\mu_f)_e$	2.088543 E − 05
poise μ_P	centipoise	μ_{cP}	1.0000 E + 02
poise μ_P	$lb_m/(ft \cdot s)$	$(\mu_m)_e$	6.719689 E − 02
centipoise μ_{cP}	$lb_m/(ft \cdot s)$	$(\mu_m)_e$	6.719689 E − 04
$lb_f \cdot s/ft^2$ $(\mu_f)_e$	centipoise	μ_{cP}	4.788026 E + 04
$lb_f \cdot s/ft^2$ $(\mu_f)_e$	poise	μ_P	4.788026 E + 02
$lb_f \cdot s/ft^2$ $(\mu_f)_e$	$lb_m/(ft \cdot s)$	$(\mu_m)_e$	3.217405 E + 01
$lb_m/(ft \cdot s)$ $(\mu_m)_e$	$lb_f \cdot s/ft^2$	$(\mu_f)_e$	3.108095 E − 02
$lb_m/(ft \cdot s)$ $(\mu_m)_e$	centipoise	μ_{cP}	1.488164 E + 03
$lb_m/(ft \cdot s)$ $(\mu_m)_e$	poise	μ_P	1.488164 E + 01
KINEMATIC VISCOSITY (ν)			
stokes ν_{St}	ft^2/s	ν_e	1.076391 E − 03
centistokes ν_{cSt}	ft^2/s	ν_e	1.076391 E − 05
ft^2/s ν_e	stokes	ν_{St}	9.290304 E + 02
ft^2/s ν_e	centistokes	ν_{cSt}	9.290304 E + 04

Table C.3 **Conversion Equations for U.S. Flow-Rate Units**

	Liquid	Gas	Vapor (steam)
	$\longrightarrow$	$\overline{V}_f = 183.3465 \dfrac{q_{cfs}}{D^2}$	$\longleftarrow$
	$\longrightarrow$	$q_{cfs} = 5.454154 \times 10^{-3} D^2 \overline{V}_f$	$\longleftarrow$
Mass flow rate, lb_m/h	$\longrightarrow$	$q_{PPH} = 19.63495 \, \rho_f D^2 \overline{V}_f$	$\longleftarrow$
	$\longrightarrow$	$q_{PPH} = 3600 \, \rho_f \, q_{cfs}$	$\longleftarrow$
	$q_{PPH} = 500.230 G_f q_{gpm}$	$q_{PPH} = 9715.775 \dfrac{\rho_f G}{T_f Z_f} q_{cfs}$†	$\longleftarrow$
Volumetric flow rate at flowing conditions	$q_{gpm} = 2.447994 \, D^2 \overline{V}_f$		
	$q_{gpm} = 1.999082 \times 10^{-3} \dfrac{q_{PPH}}{G_f}$		
	$q_{gpm} = 448.8312 \, q_{cfs}$		
	$q_{gpm} = \dfrac{\rho_b}{\rho_f} q_{GPM}$		
	$q_{gpm} = \dfrac{G_b}{G_f} q_{GPM}$		
Volumetric flow rate at base conditions‡	$q_{GPM} = \dfrac{\rho_f}{\rho_b} q_{gpm}$	$q_{SCFH} = 35.29340 \dfrac{\rho_f}{T_f} \dfrac{Z_b}{Z_f} q_{acfh}$	
	$q_{GPM} = \dfrac{G_f}{G_b} q_{gpm}$		
	$q_{GPM} = 448.8312 \dfrac{G_f}{G_b} q_{cfs}$	$q_{SCFH} = 692.9843 \dfrac{\rho_f}{T_f} \dfrac{Z_b}{Z_f} D^2 \overline{V}_f$	
	$q_{GPM} = 2.447994 \dfrac{G_f}{G_b} D^2 \overline{V}_f$	$q_{SCFH} = 13.07732 \dfrac{Z_b}{G} q_{PPH}$	
	$q_{GPM} = 1.999082 \times 10^{-3} \dfrac{q_{PPH}}{G_b}$	$q_{SCFH} = \dfrac{1}{\rho_b} q_{PPH}$	

†For liquids, q_{cfs}; for gases (vapors), q_{acfs}.

‡Liquid base volumes at 60°F and 14.69595 psia; gas base volumes at 59°F and 14.69595 psia.

Table C.4 Reynolds-Number Equations for U.S. Flow Units

Flow rate unit	Absolute viscosity				Kinematic viscosity		
	English		cgs Metric		English	cgs Metric	
	$(\mu_f)_e$	$(\mu_m)_e$	Poise μP	Centipoise μcP	ν_e	Stokes νSt	Centistokes νcSt
	$\text{lb}_f\cdot\text{s/ft}^2$	$\text{lb}_m/(\text{ft}\cdot\text{s})$	$\text{g}/(\text{cm}\cdot\text{s})$	$\text{g}/(\text{cm}\cdot\text{s})$	ft^2/St	cm^2/St	cm^2/St
Average velocity $\overline{V}_f$	$\dfrac{\rho_f \overline{V}_f D}{386.0886(\mu_f)_e}$	$\dfrac{\rho_f \overline{V}_f D}{12(\mu_m)_e}$	$\dfrac{1.24013T\rho_f \overline{V}_f D}{\mu P}$	$\dfrac{124.0137\rho_f \overline{V}_f D}{\mu cP}$	$\dfrac{\overline{V}_f D}{12\nu_e}$	$\dfrac{77.41920\overline{V}_f D}{\nu St}$	$\dfrac{7741.920\overline{V}_f D}{\nu cSt}$
Mass flow $q\text{PPS}$	$\dfrac{0.4748820q\text{PPS}}{D(\mu_f)_e}$	$\dfrac{15.27888q\text{PPS}}{D(\mu_m)_e}$	$\dfrac{227.374q\text{PPS}}{D\mu P}$	$\dfrac{22{,}737.47q\text{PPS}}{D\mu cP}$	$\dfrac{15.27888q\text{PPS}}{\rho_f D\nu_e}$	$\dfrac{14{,}194.54q\text{PPS}}{\rho_f D\nu St}$	$\dfrac{1{,}419{,}454q\text{PPS}}{\rho_f D\nu cSt}$
$q\text{PPH}$	$\dfrac{q\text{PPH}}{7580.830D(\mu_f)_e}$	$\dfrac{q\text{PPH}}{235.6194D(\mu_m)_e}$	$\dfrac{q\text{PPH}}{15.83288D\mu P}$	$\dfrac{6.315964q\text{PPH}}{D\mu cP}$	$\dfrac{q\text{PPH}}{235.6194\rho_f D\nu_e}$	$\dfrac{3.942929q\text{PPH}}{\rho_f D\nu St}$	$\dfrac{394.2929q\text{PPH}}{\rho_f D\nu cSt}$
Volumetric at flowing conditions $q\text{cfs}$	$\dfrac{0.4748820\rho_f q\text{cfs}}{D(\mu_f)_e}$	$\dfrac{15.27888\rho_f q\text{cfs}}{D(\mu_m)_e}$	$\dfrac{227.374\rho_f q\text{cfs}}{D\mu P}$	$\dfrac{22{,}737.47\rho_f q\text{cfs}}{D\mu cP}$	$\dfrac{15.27888q\text{cfs}}{D\nu_e}$	$\dfrac{14{,}194.54q\text{cfs}}{D\nu St}$	$\dfrac{1{,}419{,}454q\text{cfs}}{D\nu cSt}$
$q\text{gpm}$	$\dfrac{\rho_f q\text{gpm}}{945.1426D(\mu_f)_e}$	$\dfrac{\rho_f q\text{gpm}}{29.37593D(\mu_m)_e}$	$\dfrac{0.5065929\rho_f q\text{gpm}}{D\mu P}$	$\dfrac{50.65929\rho_f q\text{gpm}}{D\mu cP}$	$\dfrac{q\text{gpm}}{29.37593D\nu_e}$	$\dfrac{31.62557q\text{gpm}}{D\nu St}$	$\dfrac{3162.557q\text{gpm}}{D\nu cSt}$
Volumetric at base conditions $q\text{GPM}$	$\dfrac{G_b q\text{GPM}}{15.15470D(\mu_f)_e}$	$\dfrac{2.123041G_b q\text{GPM}}{D(\mu_m)_e}$	$\dfrac{31.59433G_b q\text{GPM}}{D\mu P}$	$\dfrac{3159.433G_b q\text{GPM}}{D\mu cP}$	$\dfrac{G_b q\text{GPM}}{29.37593G_f D\nu_e}$	$\dfrac{31.62557G_b q\text{GPM}}{G_f D\nu St}$	$\dfrac{3159.434G_b q\text{GPM}}{G_{Fe}\nu cSt}$
$q\text{SCFH}$	$\dfrac{Gq\text{SCFH}}{99136.68 Z_b D(\mu_f)_e}$	$\dfrac{Gq\text{SCFH}}{3081.269 Z_b D(\mu_m)_e}$	$\dfrac{Gq\text{SCFH}}{207.0515 Z_b D\mu P}$	$\dfrac{0.4829716Gq\text{SCFH}}{Z_b D\mu cP}$	$\dfrac{TZ_f q\text{SCFH}}{8315.808\,\rho_f Z_b D\nu_e}$	$\dfrac{0.1117186TZ_f q\text{SCFH}}{\rho_f Z_b D\nu St}$	$\dfrac{11.17186TZ_f q\text{SCFH}}{\rho_f Z_b D\nu cSt}$

NOTE: D = pipe size (in); $\overline{V}_f$ = fluid velocity (ft/s); ρ_f = fluid density (lb_m/ft^3).

Table C.5 **Reynolds-Number Equations for SI Units**

Flow-rate unit	Absolute viscosity		Kinematic viscosity	
	μ_P	μ_{cP}	ν_{St}	ν_{cSt}
Average velocity $\overline{V}_f^*$	$\dfrac{\rho_f^* \overline{V}_f^* D^*}{100\mu_P}$	$\dfrac{\rho_f \overline{V}_f^* D^*}{\mu_{cP}}$	$\dfrac{10\overline{V}_f^* D^*}{\nu_{St}}$	$\dfrac{1000\overline{V}_f^* D^*}{\nu_{cSt}}$
Mass flow q_{KPH}	$\dfrac{3.536777 q_{KPH}}{D^* \mu_P}$	$\dfrac{353.6777 q_{KPH}}{D^* \mu_{cP}}$	$\dfrac{3536.777 q_{KPH}}{\rho_f^* D^* \nu_{St}}$	$\dfrac{353,677.7 q_{KPH}}{\rho_f^* D^* \nu_{cSt}}$
Volumetric at flowing conditions	$\dfrac{0.2122066 \rho_f^* q_{lpm}}{D^* \mu_P}$	$\dfrac{21.22066 \rho_f^* q_{lpm}}{D^* \mu_{cP}}$	$\dfrac{212.2066 q_{lpm}}{D^* \nu_{St}}$	$\dfrac{21,220.66 q_{lpm}}{D^* \nu_{cSt}}$
Volumetric at base conditions (liquid)† q_{LPM}	$\dfrac{211.9970 G_b q_{LPM}}{D^* \mu_P}$	$\dfrac{21199.70 G_b q_{LPM}}{D^* \mu_{cP}}$	$\dfrac{212.2070 G_b q_{LPM}}{G_f D^* \nu_{St}}$	$\dfrac{212,207.0 G_b q_{LPM}}{G_f D^* \nu_{cSt}}$
Volumetric at base conditions‡ (gas) q_{SCMH}	$\dfrac{4.332211 G q_{SCMH}}{Z_b D^* \mu_P}$	$\dfrac{433.2211 G q_{SCMH}}{Z_b D^* \mu_{cP}}$	$\dfrac{1243.671 T_K Z_f q_{SCMH}}{p_f^* Z_b D^* \nu_{St}}$	$\dfrac{124,367.2 T_K Z_f q_{SCMH}}{p_f^* Z_b d^* \nu_{cSt}}$

†Base temperature = 15.56°C; base pressure = 101.325 kPa.
‡ISO 5024 base: $p_f^* = 101.325$ kPa, $T_{Kb} = 288.15$ K.
NOTE: The poise (P) and the stokes (St) are cgs metric units, not SI metric units; 1 P = 0.1 Pa.s; 1 St = 0.0001 m²/s.

GENERALIZED FLUID PROPERTIES

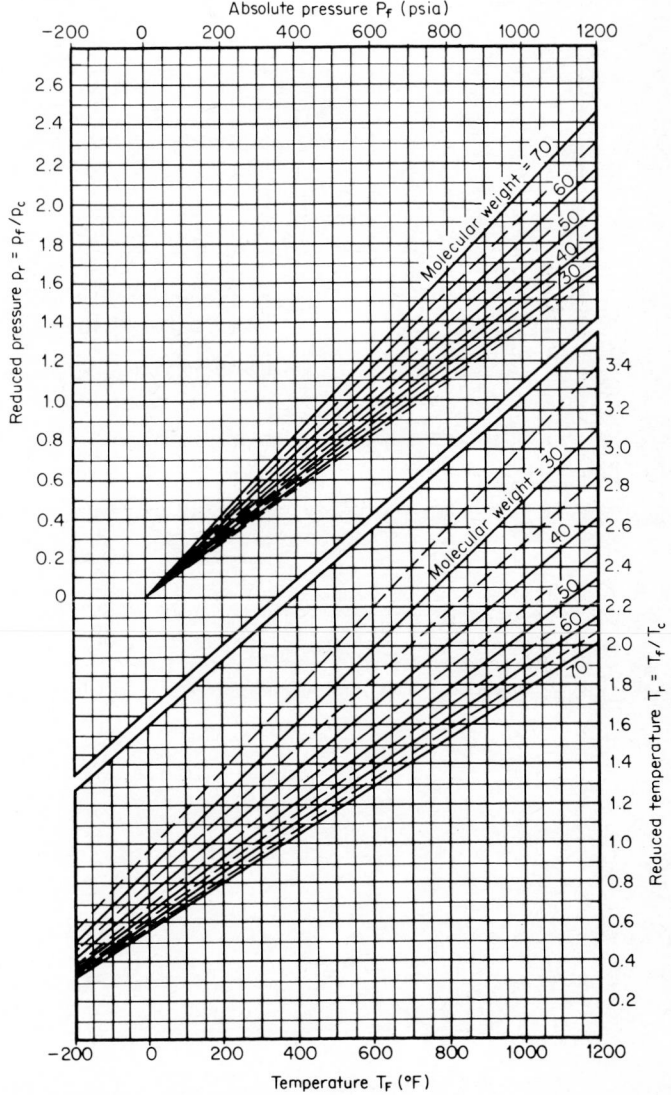

Figure D.1 Estimated reduced pressure and temperature from molecular weight *(from GPSA, 1979; used with permission).*

D-1

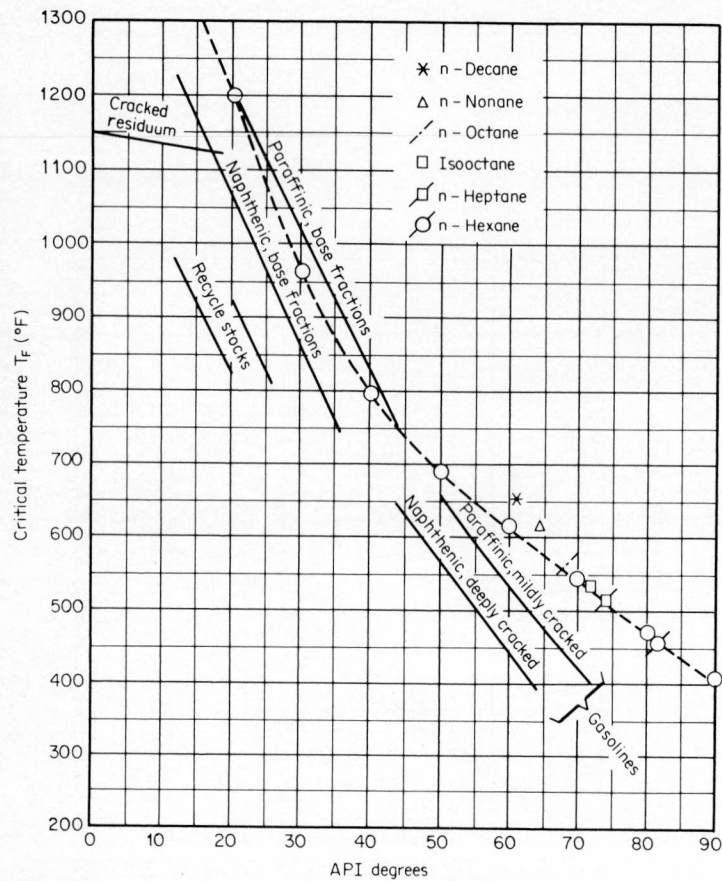

Figure D.2 **Estimated critical temperature for typical petroleum oils** *(from Spink, 1967).*

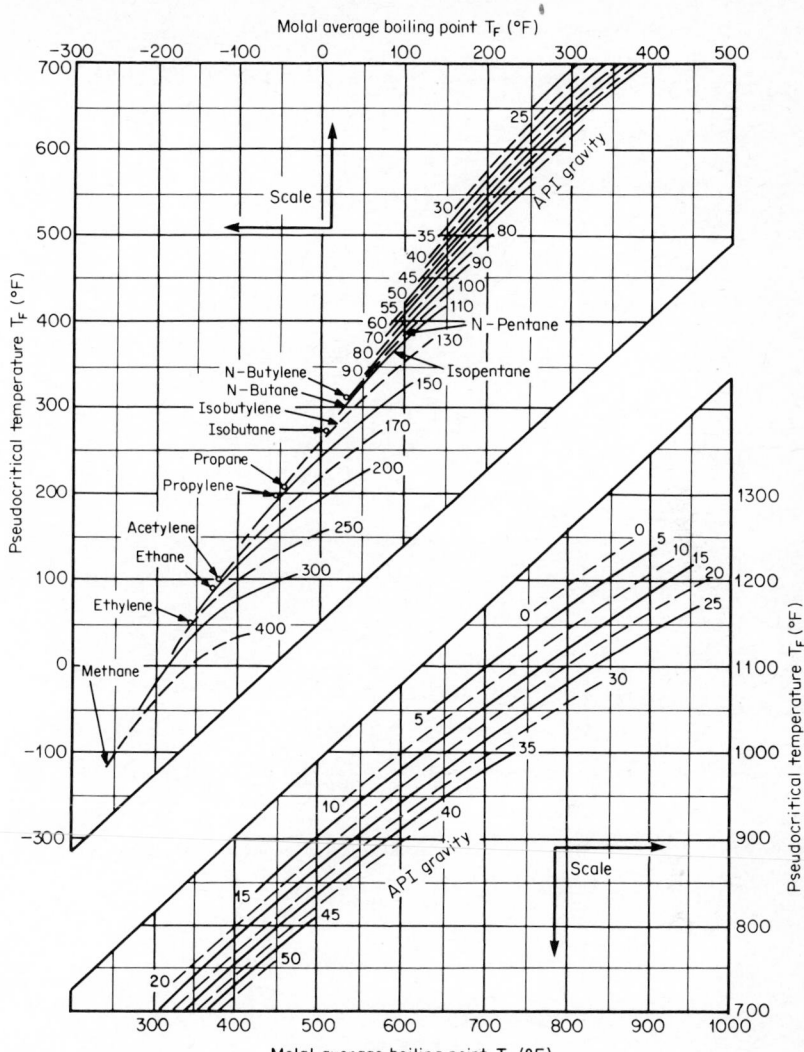

Figure D.3 Chart for estimating pseudocritical temperature from molal average boiling point *(from Spink, 1967).*

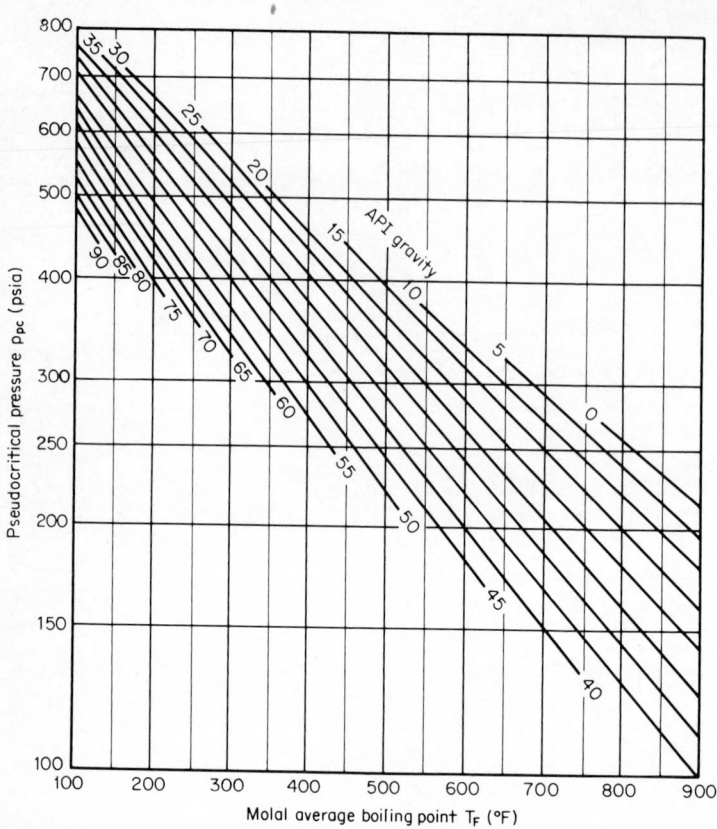

Figure D.4 **Chart for estimating critical pressure from molal average boiling point** *(from Spink, 1967).*

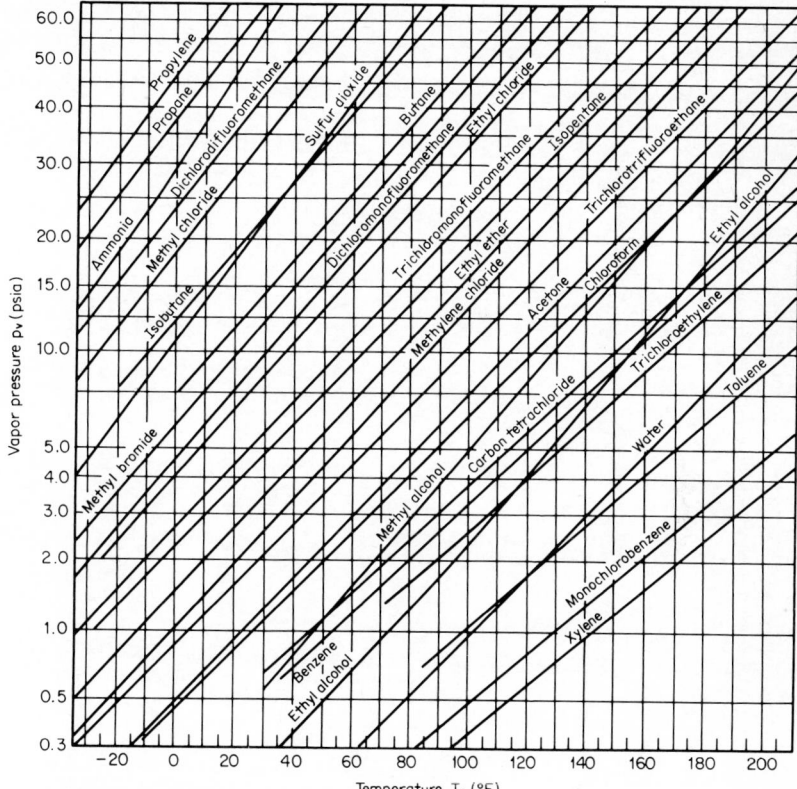

Figure D.5   Vapor-pressure curves for some fluids.

Table D.1 Physical Properties

Substance	Molecular weight M_w	Normal boiling point °F	Normal boiling point °R	Critical properties Temperature T_c °F	Critical properties Temperature T_c °R	Pressure p_c psia	Volume V_c ft³/(lb$_m$·mol)	Volume v_c ft³/lb	Compressibility Z_c dimensionless	Acentric factor ω	Isentropic exponent (ideal gas) at 60°F, $k_p = \dfrac{c_p}{c_v}$
Acetylene	26.038	−119.1	340.6	95.3	554.9	890.6	1.812	0.0696	0.271	0.184	1.26
Air	28.9625	−317.6	142.1	−220.8	238.9	546.7	1.463	0.0505	0.312	−0.012†	1.400
Ammonia	17.031	−28.2	431.5	270.4	730.1	1635.7	1.159	0.0680	0.242	0.250	1.310
Argon	39.9480	−302.5	157.1	−188.2	271.4	706.9	1.199	0.0300	0.291	−0.004	1.668
Benzene	70.114	176.3	635.9	552.1	1011.8	709.8	4.145	0.0530	0.271	0.212	1.08
1,2-Butadiene	54.092	51.5	511.2	339.0	798.7	652.5	3.507	0.0648	0.267	0.255	1.120
1,3-Butadiene	54.092	24.0	483.7	305.3	765.0	627.5	3.532	0.0653	0.270	0.195	1.120
Isobutane	58.1243	10.67	470.3	274.9	734.6	529.1	4.217	0.0726	0.283	0.176	1.11
n-Butane	58.1243	31.2	490.9	305.7	765.4	551.1	4.0836	0.0703	0.274	0.193	1.10
Isobutylene	56.108	19.7	479.3	292.6	752.2	580.5	3.824	0.0682	0.275	0.190	1.12
1-Butene	56.108	20.8	480.4	295.6	755.3	583.4	3.848	0.0686	0.277	0.187	1.11
Carbon tetrachloride	153.82	169.8	629.5	541.9	1001.5	661.3	4.421	0.0287	0.272	0.194	1.067
Carbon dioxide	44.010	−109.2	350.5	87.9	547.6	1069.9	1.505	0.0342	0.274	0.225	1.295
Carbon monoxide	28.01055	−312.6	147.1	−220.5	239.2	507.0	1.494	0.0533	0.295	0.049	1.395
cis-2-Butene	56.108	38.8	498.4	324.4	784.1	609.9	4.514	0.0805	0.272	0.202	1.121
Chlorine	70.906	−30.0	429.7	290.9	750.6	1116.9	1.983	0.0280	0.275	0.073	1.355

Cyclohexane	84.162	177.4	637.0	536.5	996.1	590.8	4.940	0.0587	0.273	0.213	1.08
Cyclopentane	70.135	120.7	580.3	461.2	920.9	654.0	4.171	0.0595	0.276	0.192	1.117
n-Decane	142.286	345.5	805.1	652.0	1111.7	305.7	9.640	0.0678	0.247	0.490	1.038
2,3-Dimethyl butane	86.178	136.5	596.2	440.2	899.8	454.1	5.742	0.0666	0.270	0.247	1.065
2,2-Dimethyl pentane	100.205	174.7	634.3	477.1	936.7	402.7	6.666	0.0665	0.267	0.289	1.053
2,4-Dimethyl pentane	100.205	177.0	636.7	475.8	935.5	396.8	6.705	0.0669	0.265	0.306	1.053
3,3-Dimethyl pentane	100.205	186.9	646.6	505.7	965.3	427.7	6.637	0.0662	0.274	0.270	1.073
Ethane	30.0701	-127.6	332.1	90.1	549.7	708.4	2.374	0.0789	0.285	0.098	1.18
Ethyl alcohol	46.069	173.0	632.7	469.5	929.2	925.9	2.671	0.0580	0.248	0.635	1.13
Ethylene	28.054	-154.8	304.9	48.7	508.3	730.4	2.061	0.0028	0.276	0.085	1.22
Ethylbenzene	106.168	277.1	736.7	651.1	1110.8	523.2	5.992	0.0564	0.263	0.301	1.072
3-Ethylpentane	100.205	200.2	659.9	513.4	973.1	418.8	6.657	0.0664	0.267	0.310	1.054
Helium-4	4.003	-452.1	7.6	-450.3	9.34	32.9	0.917	0.2290	0.301	0.387	1.660
n-Heptane	100.205	209.2	668.9	512.7	972.4	396.8	6.916	0.0690	0.263	0.351	1.052
n-Hexane	86.178	155.8	615.4	453.7	913.3	430.6	5.918	0.0687	0.260	0.296	1.062
Hydrogen	2.016	-423.0	36.7	-399.9	59.8	188.1	1.040	0.5158	0.305	0.22	1.412
Hydrogen chloride	34.461	-121.1	338.6	124.6	584.3	1205.7	1.295	0.0376	0.249	0.12	1.41
Hydrogen sulfide	34.08	-76.6	383.1	212.1	671.8	1269.2	1.613	0.0473	0.284	0.100	1.32
Methane	16.043	-258.6	201.1	-116.6	343.1	667.2	1.589	0.0991	0.288	0.008	1.315
Methyl alcohol	32.042	148.4	608.4	463.0	922.7	1174.2	1.889	0.0589	0.224	0.559	1.203
Neon	20.183	-411.1	48.6	-379.8	79.9	399.7	0.667	0.0330	0.311	0.000	1.667

Table D.1 Physical Properties (Continued)

Substance	Molecular weight M_w	Normal boiling point		Critical properties							Acentric factor ω	Isentropic exponent (ideal gas) at 60°F, $k_p = \dfrac{c_p}{c_v}$
				Temperature T_c		Pressure p_c psia	Volume		Compressibility Z_c dimensionless			
		°F	°R	°F	°R		V_c ft³/(lbm·mol)	v_c ft³/lb				
Nitrogen	28.0134	−320.4	139.3	−232.5	227.2	492.3	1.436	0.0513	0.290		0.040	1.400
n-Nonane	128.259	303.5	763.2	610.6	1070.3	335.1	8.913	0.0695	0.26		0.444	1.042
Octane (average)	114.2327	258.2	717.9	564.8	1024.5	364.5	7.873	0.0689	0.261		0.399	1.046
n-Octane	114.232	256.7	716.4	564.2	1023.8	360.1	7.904	0.0692	0.259		0.394	1.046
2,2,4-Trimethylpentane‡	114.232	210.7	670.3	519.4	979.0	371.8	7.517	0.0658	0.266		0.303	1.046
Oxygen	31.9988	−297.3	162.4	−181.4	278.3	731.9	1.175	0.0367	0.288		0.021	1.397
2-Methyl butane (isopentane)‡	72.1514	82.1	541.8	369.1	828.7	490.9	4.910	0.0680	0.271		0.227	1.076
2,2-Dimethyl propane, (neopentane)‡	72.151	49.0	508.7	321.2	780.8	464.4	4.854	0.0673	0.269		0.197	1.076
n-Pentane	72.1514	96.9	556.6	385.6	845.3	489.4	4.856	0.0673	0.262		0.251	1.07

Propane	44.0972	−43.7	416	206.0	665.6	615.8	3.260	0.0739	0.281	0.152	1.13
Propylene	42.081	−54.0	405.7	197.3	657.0	670.1	2.893	0.0688	0.275	0.148	1.154
Sulfur dioxide	64.063	13.7	473.4	315.8	775.4	1143.4	1.951	0.0304	0.268	0.251	1.29
Styrene	104.152	293.3	752.9	704.9	1164.6	579.0	5.634	0.0541	0.261	0.257†	1.076
Toluene	92.141	231.2	690.8	605.4	1065.1	596.7	5.057	0.0548	0.264	0.257	1.06
Triptane (2,2,3-trimethylbutane)	100.205	177.5	637.2	496.3	956.0	429.1	6.383	0.0637	0.267	0.251	1.052
Water	18.01534	212.1	671.8	705.5	1165.1	3197.9	0.895	0.0497	0.229	0.344	1.335
m-Xylene	106.168	282.5	742.1	650.9	1110.6	514.4	6.025	0.0567	0.260	0.331	1.072
O-Xylene	106.168	292.0	751.7	674.7	1134.4	540.8	5.920	0.0558	0.263	0.314	1.049
p-Xylene	106.168	281.0	740.7	649.5	1109.2	510.0	6.069	0.0572	0.260	0.324	1.073
Xenon	131.30	−162.7	297.0	61.8	521.5	846.5	1.891	0.0144	0.286	0.002	1.665

†This acentric factor is calculated as

$$\omega = \frac{3}{7} \frac{\log(p_c/14.7)}{T_c/T_B - 1.0} - 1.0$$

‡Source: Gas Processors Suppliers Association, "Engineering Data Book", Ninth edition, fourth revision, 1979

SOURCE: Reid, et al. (1977).

LIQUID DENSITY AND SPECIFIC GRAVITY

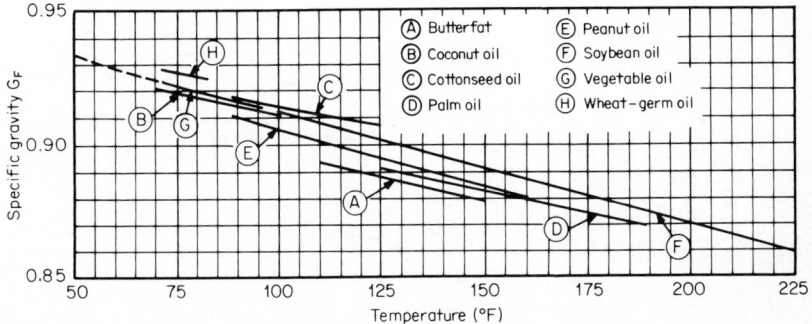

Figure E.1 **Specific gravities of fats and oils** *(from Fischer & Porter Catalog 10-A-54; used with permission).*

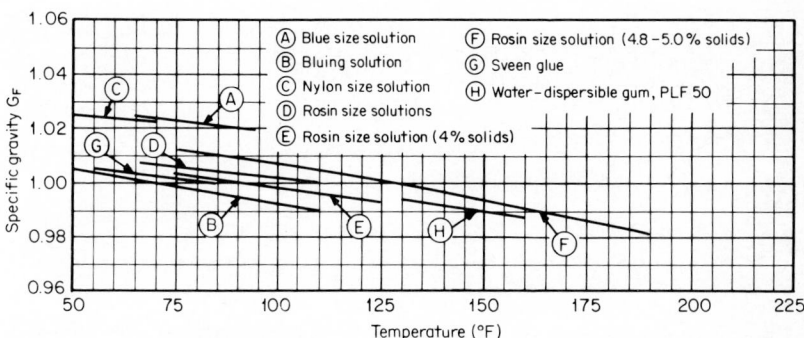

Figure E.2 **Specific gravities of liquids related to the paper, leather, and textile industries** *(from Fischer & Porter Catalog 10-A54; used with permission).*

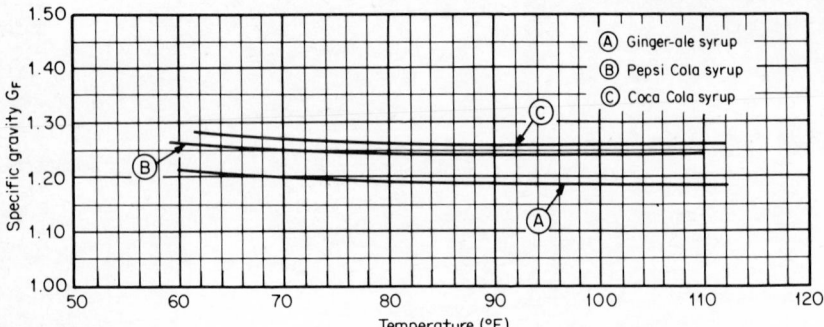

Figure E.3 **Specific gravities of syrups** *(from Fischer & Porter Catalog 10-A-54; used with permission).*

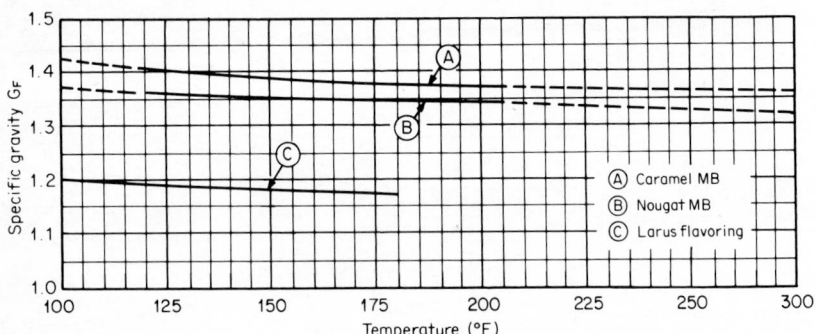

Figure E.4 **Specific gravities of liquids related to the candy industry** *(from Fischer & Porter Catalog 10-A-54; used with permission).*

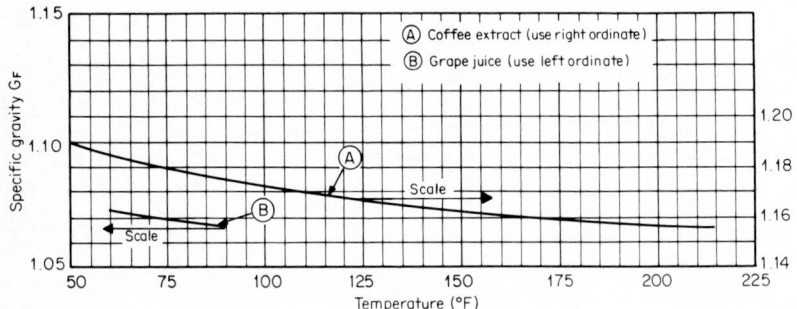

Figure E.5 **Specific gravities of liquids related to the food industry** *(from Fischer & Porter Catalog 10-A-54; used with permission).*

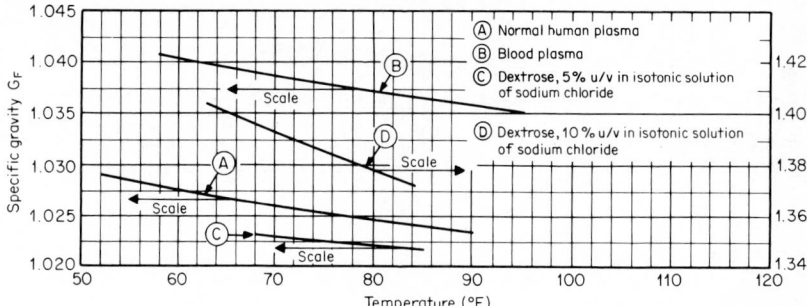

Figure E.6 **Specific gravities of intravenous solutions** *(from Fischer & Porter Catalog 10-A-54; used with permission).*

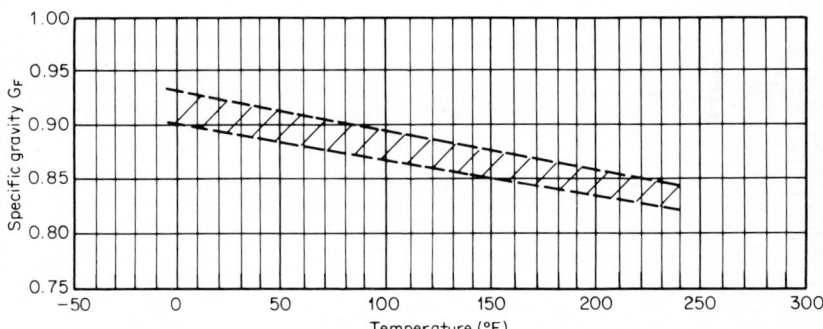

Figure E.7 **Specific gravities of Garboyle D.T.E. oils** *(from Fischer & Porter Catalog 10-A-54; used with permission).*

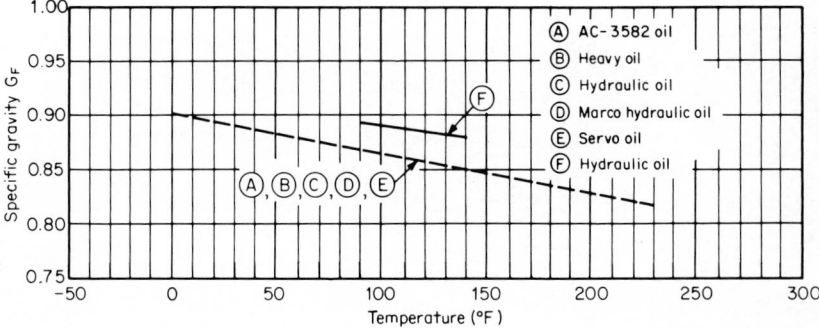

Figure E.8 **Specific gravities of hydraulic oils** *(from Fischer & Porter Catalog 10-A-54; used with permission).*

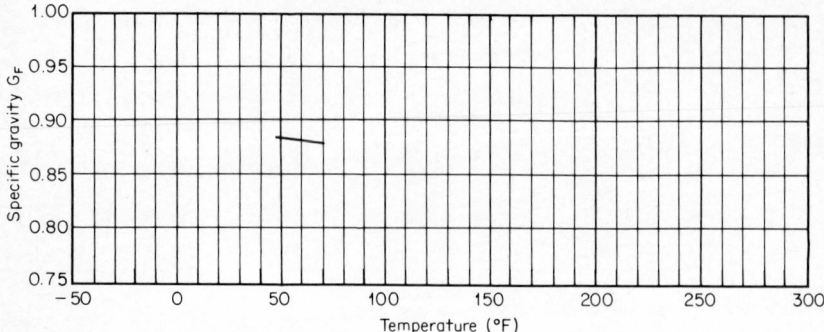

Figure E.9 **Specific gravities of Delco lube oil** *(from Fischer & Porter Catalog 10-A-54; used with permission).*

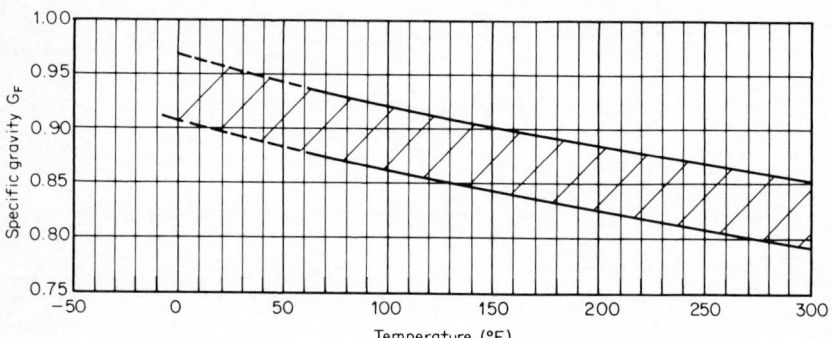

Figure E.10 **Specific gravities of S.A.E. (Pennsylvania base) oils** *(from Fischer & Porter Catalog 10-A-54; used with permission).*

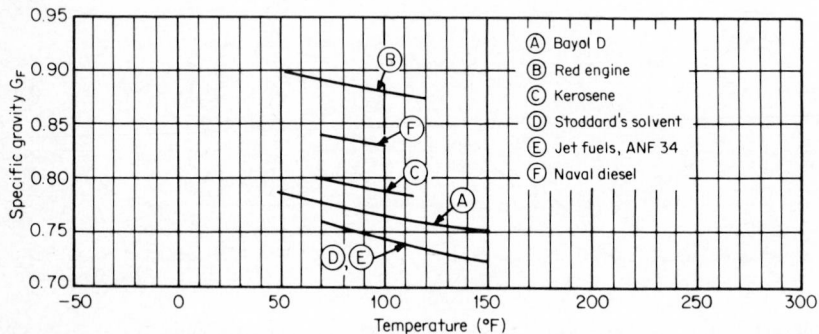

Figure E.11 **Specific gravities of miscellaneous fuel oils** *(from Fischer & Porter Catalog 10-A-54; used with permission).*

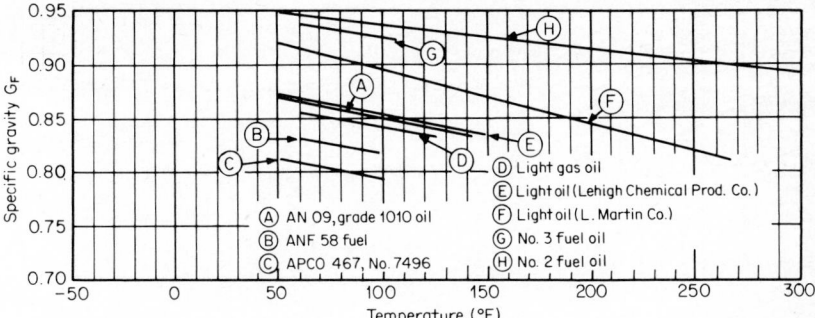

Figure E.12 **Specific gravities of miscellaneous fuel oils** *(from Fischer & Porter Catalog 10-A-54; used with permission).*

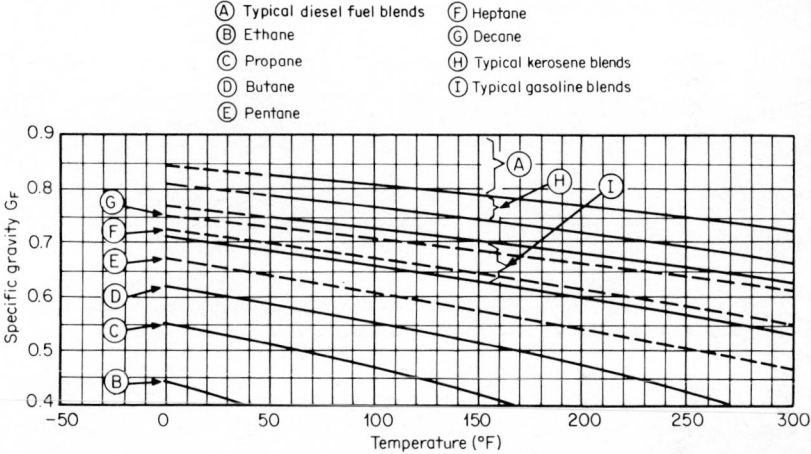

Figure E.13 **Specific gravities of typical fuel blends** *(from Fischer & Porter Catalog 10-A-54; used with permission).*

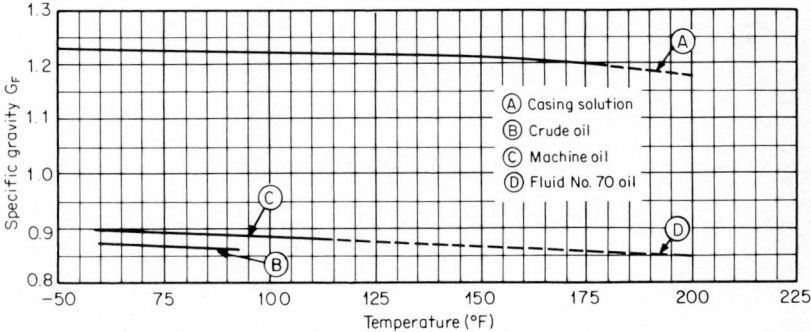

Figure E.14 **Specific gravities of miscellaneous oils** *(from Fischer & Porter Catalog 10-A-54; used with permission).*

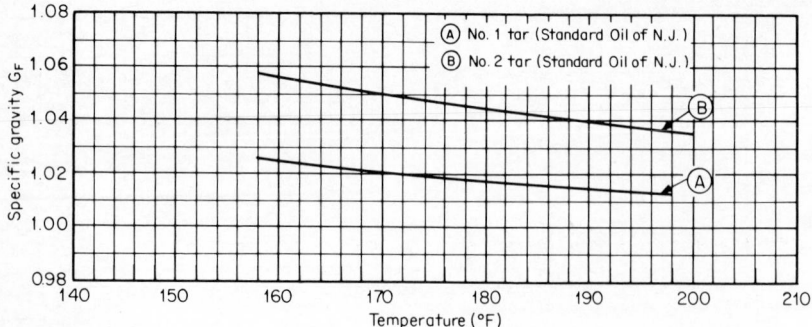

Figure E.15 Specific gravities of high boiling fractions *(from Fischer & Porter Catalog 10-A-54; used with permission).*

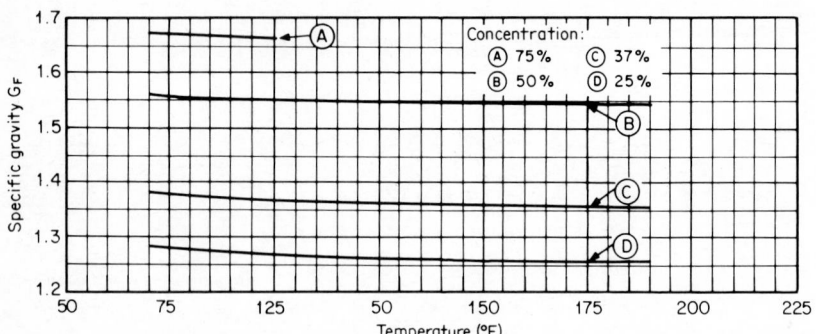

Figure E.16 Specific gravities of phosphoric acid solutions *(from Fischer & Porter Catalog 10-A-54; used with permission).*

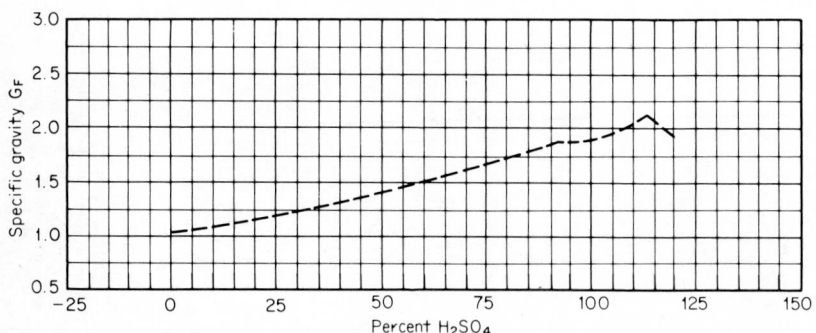

Figure E.17 Specific gravities of sulfuric acid solutions at a temperature T_F **of 60°F (15.6°C)** *(from Fischer & Porter Catalog 10-A-54; used with permission).*

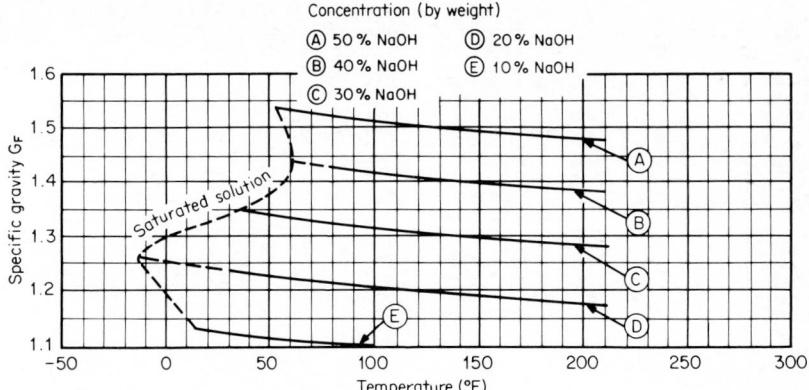

Figure E.18 **Specific gravities of sodium hydroxide solutions** *(from Fischer & Porter Catalog 10-A-54; used with permission).*

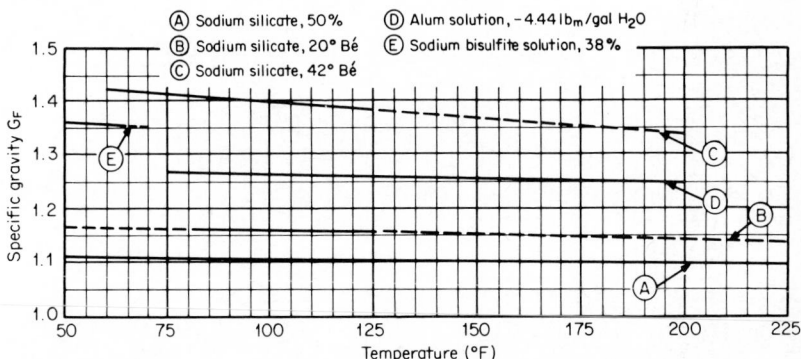

Figure E.19 **Specific gravities of inorganic solutions** *(from Fischer & Porter Catalog 10-A-54; used with permission).*

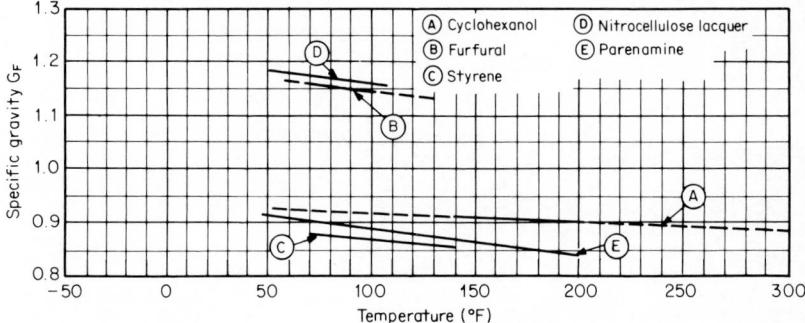

Figure E.20 **Specific gravities of miscellaneous organic compounds** *(from Fischer & Porter Catalog 10-A-54; used with permission).*

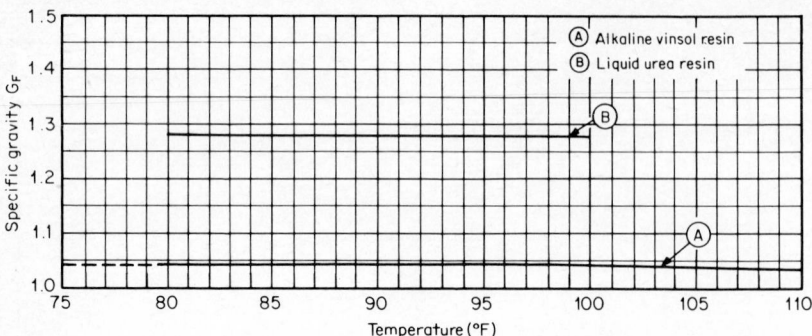

Figure E.21 **Specific gravities of resins** *(from Fischer & Porter Catalog 10-A-54; used with permission).*

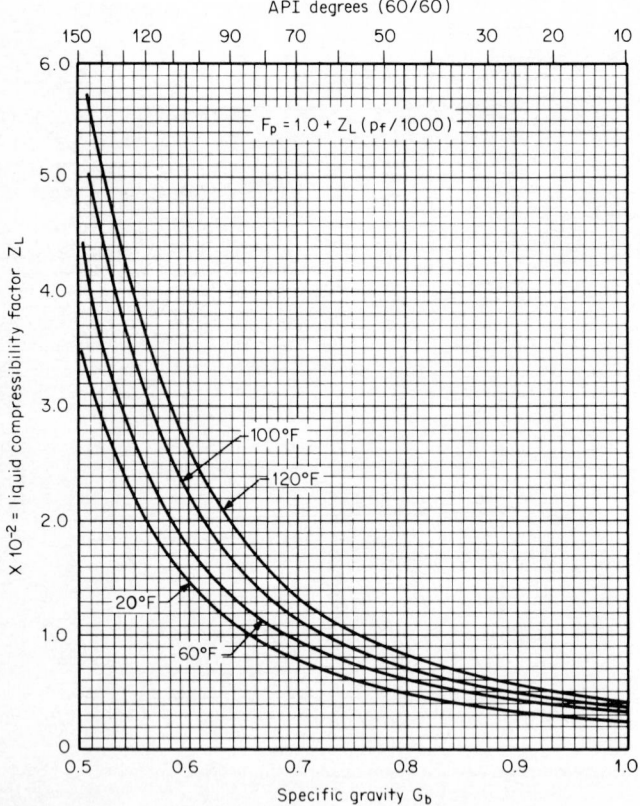

Figure E.22 Average liquid-hydrocarbon compressibility factor.

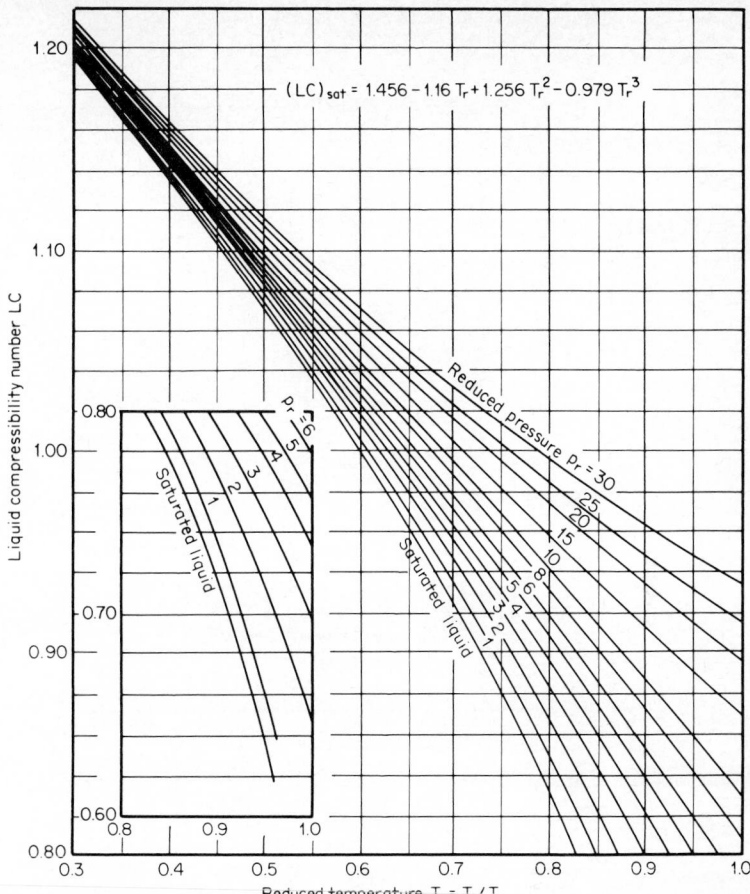

Figure E.23 Lu's generalized liquid compressibility diagram *(from Perry and Chilton, 1973).*

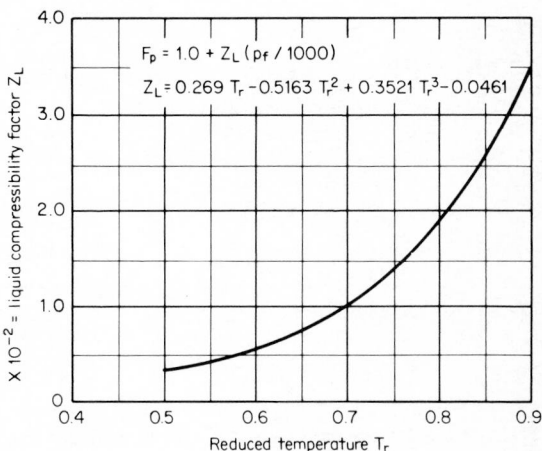

Figure E.24 Generalized liquid compressibility factor.

E-9

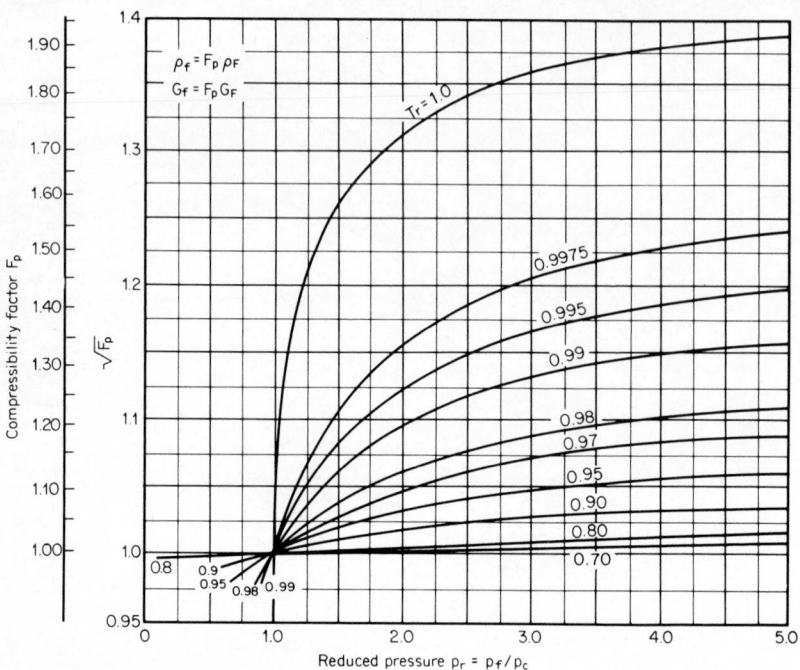

Figure E.25 Generalized petroleum-oil compressibility factor *(from Spink, 1967).*

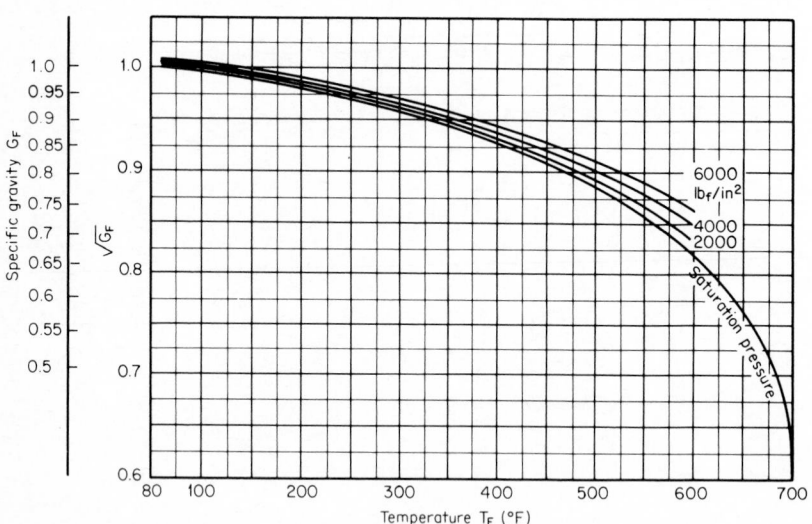

Figure E.26 Specific gravity and the square root of specific gravity for water *(from Spink, 1967).*

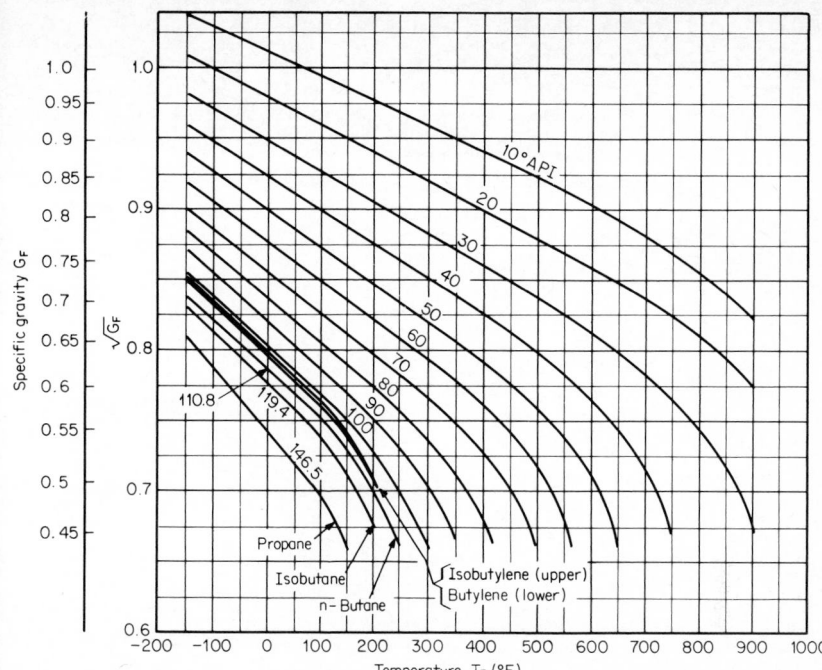

Figure E.27 Specific gravity and the square root of specific gravity for hydrocarbons *(from Spink, 1967).*

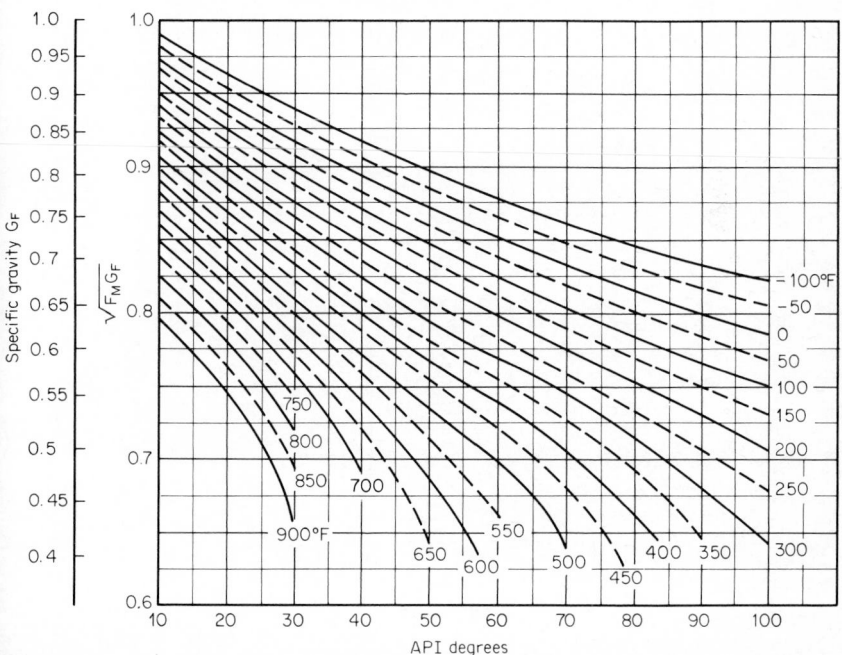

Figure E.28 Specific gravity and the square root of specific gravity for hydrocarbons, based on API-degrees reading *(from Spink, 1967).*

E-11

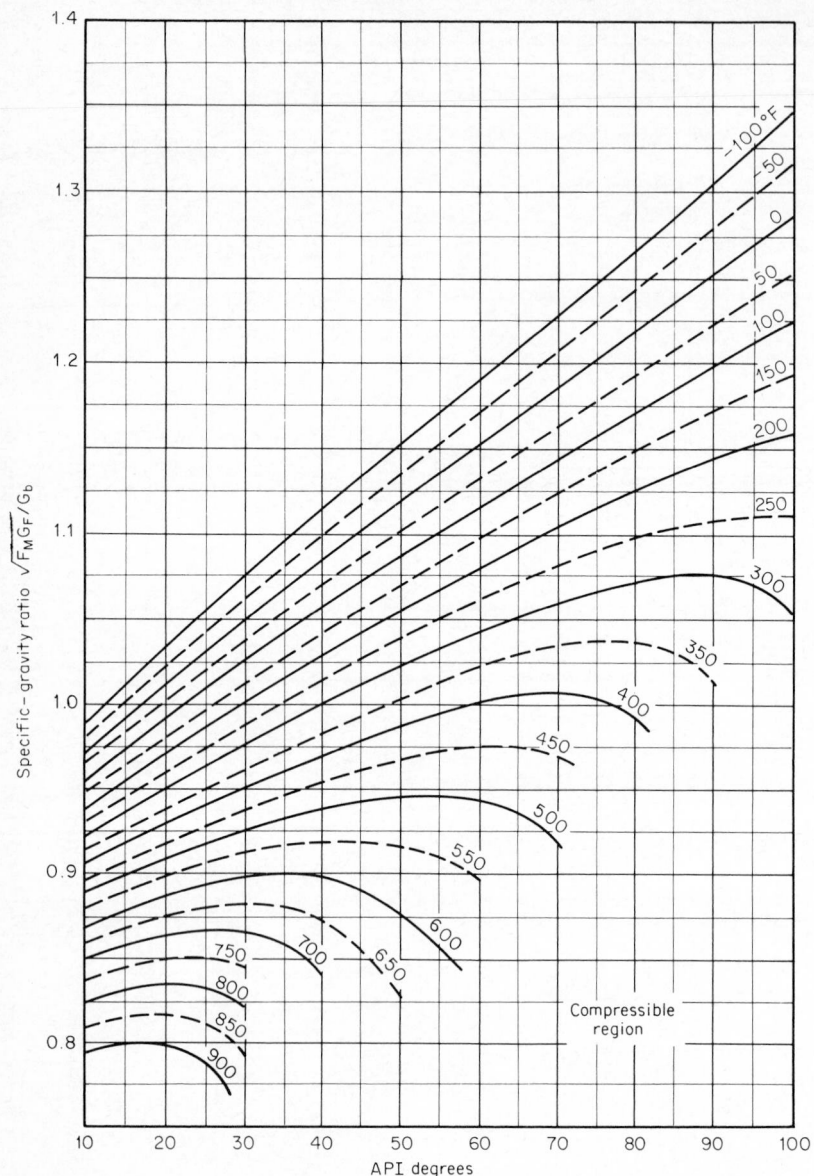

Figure E.29   **Square root of specific gravity divided by base specific gravity for petroleum oils** *(from Spink, 1967).*

Table E.1 Liquid Compressibility Data

Substance	Temperature, °F	Bulk-modulus zero intercept K_{BM0}	Slope b_c	Liquid compressibility factor Z_L
Acetone	68	240,000	5.7	0.00417
Amyl alcohol (iso)	68	172,500	5.95	0.00580
Amyl alcohol (n)	68	240,000	10.7	0.00417
Benzene	68	188,000	9.7	0.00532
Bromine	68	259,000	8.8	0.00386
Butyl alcohol (iso)	68	179,000	16.4	0.00559
Carbon bisulfide	68	254,000	6.6	0.00394
Carbon tetrachloride	68	168,500	10.4	0.00594
Chloroform	68	174,500	11.2	0.00573
Dichloroethyl sulfide	68	426,000	12.3	0.00235
Ethyl acetate	68	161,000	11.1	0.00621
Ethyl alcohol	68	230,000	5.3	0.00435
Ethyl bromide	68	145,000	11.0	0.00690
Ethyl chloride	68	142,000	10.7	0.00704
Ethyl ether	68	172,000	9.0	0.00581
Ethyl iodide	68	179,000	10.7	0.00559
Hexane	68	123,500	29.7	0.00810
Kerosene	68	263,500	8.1	0.00380
Mercury	71.5	3,650,000	7.7	0.000274
Methyl alcohol	68	152,500	9.9	0.00656
Phosphorus trichloride	68	230,000	10.8	0.00435
Propyl alcohol	68	188,000	9.7	0.00532
Toluene	68	196,000	10.6	0.00510
Water	68	337,000	7.8	0.00297
Water	104	381,500	8.0	0.00262

Table E.2 Specific Gravities at 60°F/60°F Corresponding to API Degrees and Weights per U.S. Gallon at 60°F†

API degrees	Specific gravity	Pounds per U.S. gallon	API degrees	Specific gravity	Pounds per U.S. gallon	API degrees	Specific gravity	Pounds per U.S. gallon	API degrees	Specific gravity	Pounds per U.S. gallon
10	1.0000	8.328	33	0.8602	7.163	56	0.7547	6.283	79	0.6722	5.595
11	0.9930	8.270	34	0.8550	7.119	57	0.7507	6.249	80	0.6690	5.568
12	0.9861	8.212	35	0.8498	7.076	58	0.7467	6.216	81	0.6659	5.542
13	0.9792	8.155	36	0.8448	7.034	59	0.7428	6.184	82	0.6628	5.516
14	0.9725	8.099	37	0.8398	6.993	60	0.7389	6.151	83	0.6597	5.491
15	0.9659	8.044	38	0.8348	6.951	61	0.7351	6.119	84	0.6566	5.465
16	0.9593	7.989	39	0.8299	6.910	62	0.7313	6.087	85	0.6536	5.440
17	0.9529	7.935	40	0.8251	6.870	63	0.7275	6.056	86	0.6506	5.415
18	0.9465	7.882	41	0.8203	6.830	64	0.7238	6.025	87	0.6476	5.390
19	0.9402	7.830	42	0.8155	6.790	65	0.7201	5.994	88	0.6446	5.365
20	0.9340	7.778	43	0.8109	6.752	66	0.7165	5.964	89	0.6417	5.341
21	0.9279	7.727	44	0.8063	6.713	67	0.7128	5.934	90	0.6388	5.316
22	0.9218	7.676	45	0.8017	6.675	68	0.7093	5.904	91	0.6360	5.293
23	0.9159	7.627	46	0.7972	6.637	69	0.7057	5.874	92	0.6331	5.269
24	0.9100	7.578	47	0.7927	6.600	70	0.7022	5.845	93	0.6303	5.246
25	0.9042	7.529	48	0.7883	6.563	71	0.6988	5.817	94	0.6275	5.222
26	0.8984	7.481	49	0.7839	6.526	72	0.6953	5.788	95	0.6247	5.199
27	0.8927	7.434	50	0.7796	6.490	73	0.6919	5.759	96	0.6220	5.176
28	0.8871	7.387	51	0.7753	6.455	74	0.6886	5.731	97	0.6193	5.154
29	0.8816	7.341	52	0.7711	6.420	75	0.6852	5.703	98	0.6166	5.131
30	0.8762	7.296	53	0.7669	6.385	76	0.6819	5.676	99	0.6139	5.109
31	0.8708	7.251	54	0.7628	6.350	77	0.6787	5.649	100	0.6112	5.086
32	0.8654	7.206	55	0.7587	6.316	78	0.6754	5.622			

†Calculated from the formula specific gravity = $141.5/(131.5 + $ API degrees). The weights in this table are weights in air at 60°F with 50 percent humidity and 760 mm pressure.

SOURCE: Baumeister (1978).

Table E.3 Specific Gravities at 60°F/60°F Corresponding to Degrees Baumé for Liquids Lighter than Water and Weights per U.S. Gallon at 60°F†

Degrees Baumé	Specific gravity	Pounds per gallon	Degrees Baumé	Specific gravity	Pounds per gallon	Degrees Baumé	Specific gravity	Pounds per gallon	Degrees Baumé	Specific gravity	Pounds per gallon
10.0	1.0000	8.328	33.0	0.8589	7.152	55.0	0.7568	6.300	78.0	0.6731	5.602
11.0	0.9929	8.269	34.0	0.8537	7.108	56.0	0.7527	6.266	79.0	0.6699	5.576
12.0	0.9859	8.211	35.0	0.8485	7.065	57.0	0.7487	6.233	80.0	0.6667	5.549
13.0	0.9790	8.153	36.0	0.8434	7.022	58.0	0.7447	6.199	81.0	0.6635	5.522
14.0	0.9722	8.096	37.0	0.8383	6.980	59.0	0.7407	6.166	82.0	0.6604	5.497
15.0	0.9655	8.041	38.0	0.8333	6.939	60.0	0.7368	6.134	83.0	0.6573	5.471
16.0	0.9589	7.986	39.0	0.8284	6.898	61.0	0.7330	6.102	84.0	0.6542	5.445
17.0	0.9524	7.931	40.0	0.8235	6.857	62.0	0.7292	6.070	85.0	0.6512	5.420
18.0	0.9459	7.877	41.0	0.8187	6.817	63.0	0.7254	6.038	86.0	0.6482	5.395
19.0	0.9396	7.825	42.0	0.8140	6.777	64.0	0.7216	6.007	87.0	0.6452	5.370
20.0	0.9333	7.772	43.0	0.8092	6.738	65.0	0.7179	5.976	88.0	0.6422	5.345
21.0	0.9272	7.721	44.0	0.8046	6.699	66.0	0.7143	5.946	89.0	0.6393	5.320
22.0	0.9211	7.670	45.0	0.8000	6.661	67.0	0.7107	5.916	90.0	0.6364	5.296
23.0	0.9150	7.620	46.0	0.7955	6.623	68.0	0.7071	5.886	91.0	0.6335	5.272
24.0	0.9091	7.570	47.0	0.7910	6.586	69.0	0.7035	5.856	92.0	0.6306	5.248
25.0	0.9032	7.522	48.0	0.7865	6.548	70.0	0.7000	5.827	93.0	0.6278	5.225
26.0	0.8974	7.473	49.0	0.7821	6.511	71.0	0.6965	5.798	94.0	0.6250	5.201
27.0	0.8917	7.425	50.0	0.7778	6.476	72.0	0.6931	5.769	95.0	0.6222	5.178
28.0	0.8861	7.378	51.0	0.7735	6.440	73.0	0.6897	5.741	96.0	0.6195	5.155
29.0	0.8805	7.332	52.0	0.7692	6.404	74.0	0.6863	5.712	97.0	0.6167	5.132
30.0	0.8750	7.286	53.0	0.7650	6.369	75.0	0.6829	5.685	98.0	0.6140	5.110
31.0	0.8696	7.241	54.0	0.7609	6.334	76.0	0.6796	5.657	99.0	0.6114	5.088
32.0	0.8642	7.196				77.0	0.6763	5.629	100.0	0.6087	5.066

†Calculated from the formula specific gravity $\dfrac{60°}{60°}$ F $= 140/(130 - $ Baumé degrees).

Table E.4 Specific Gravities at 60°F/60°F Corresponding to Degrees Baumé for Liquids Heavier than Water†

Degrees Baumé	Specific gravity	Degrees Baumé	Specific gravity	Degrees Baumé	Specific gravity	Degrees Baumé	Specific gravity	Degrees Baumé	Specific gravity	Degrees Baumé	Specific gravity
0	1.0000	12	1.0902	24	1.1983	36	1.3303	48	1.4948	60	1.7059
1	1.0069	13	1.0985	25	1.2083	37	1.3426	49	1.5104	61	1.7262
2	1.0140	14	1.1069	26	1.2185	38	1.3551	50	1.5263	62	1.7470
3	1.0211	15	1.1154	27	1.2288	39	1.3679	51	1.5426	63	1.7683
4	1.0284	16	1.1240	28	1.2393	40	1.3810	52	1.5591	64	1.7901
5	1.0357	17	1.1328	29	1.2500	41	1.3942	53	1.5761	65	1.8125
6	1.0432	18	1.1417	30	1.2609	42	1.4078	54	1.5934	66	1.8354
7	1.0507	19	1.1508	31	1.2719	43	1.4216	55	1.6111	67	1.8590
8	1.0584	20	1.1600	32	1.2832	44	1.4356	56	1.6292	68	1.8831
9	1.0662	21	1.1694	33	1.2946	45	1.4500	57	1.6477	69	1.9079
10	1.0741	22	1.1789	34	1.3063	46	1.4646	58	1.6667	70	1.9333
11	1.0821	23	1.1885	35	1.3182	47	1.4796	59	1.6860		

†Calculated from the formula specific gravity $\frac{60°}{60°}$ F $= 145/(145 -$ Baumé degrees).

SOURCE: Baumeister (1978).

Table E.5 Density of Mercury: U.S. and SI Units†

Temperature				Temperature			
°F	°C	lb$_m$/ft³	kg/m³	°F	°C	lb$_m$/ft³	kg/m³
32	0.00	848.72	13595.2	56	13.33	846.67	13562.3
33	0.56	848.63	13593.8	57	13.89	846.58	13560.9
34	1.11	848.55	13592.4	58	14.44	846.50	13559.5
35	1.67	848.46	13591.0	59	15.00	846.41	13558.2
36	2.22	848.38	13589.7	60	15.56	846.32	13556.8
37	2.78	848.29	13588.3	61	16.11	846.24	13555.4
38	3.33	848.20	13586.9	62	16.67	846.15	13554.1
39	3.89	848.12	13585.6	63	17.22	846.07	13552.7
40	4.44	848.03	13584.2	64	17.78	845.98	13551.3
41	5.00	847.95	13582.8	65	18.33	845.90	13550.0
42	5.56	847.86	13581.5	66	18.89	845.81	13548.6
43	6.11	847.78	13580.1	67	19.44	845.73	13547.2
44	6.67	847.69	13578.7	68	20.00	845.64	13545.9
45	7.22	847.61	13577.3	69	20.56	845.55	13544.5
46	7.78	847.52	13576.0	70	21.11	845.47	13543.1
47	8.33	847.44	13574.6	71	21.67	845.38	13541.8
48	8.89	847.35	13573.2	72	22.22	845.30	13540.4
49	9.44	847.26	13571.9	73	22.78	845.21	13539.0
50	10.00	847.18	13570.5	74	23.33	845.13	13537.6
51	10.56	847.09	13569.1	75	23.89	845.04	13536.3
52	11.11	847.01	13567.8	76	24.44	844.96	13534.9
53	11.67	846.92	13566.4	77	25.00	844.87	13533.5
54	12.22	846.84	13565.0	78	25.56	844.79	13532.2
55	12.78	846.75	13563.7	79	26.11	844.70	13530.8

Table E.5 Density of Mercury: U.S. and SI Units† (Continued)

Temperature				Temperature			
°F	°C	lb_m/ft³	kg/m³	°F	°C	lb_m/ft³	kg/m³
80	26.67	844.61	13529.4	91	32.78	843.67	13514.4
81	27.22	844.53	13528.1	92	33.33	843.59	13513.0
82	27.78	844.44	13526.7	93	33.89	843.50	13511.6
83	28.33	844.36	13525.3	94	34.44	843.42	13510.3
84	28.89	844.27	13523.9	95	35.00	843.33	13508.9
85	29.44	844.19	13522.6	96	35.56	843.25	13507.5
86	30.00	844.10	13521.2	97	36.11	843.16	13506.1
87	30.56	844.02	13519.8	98	36.67	843.08	13504.8
88	31.11	843.93	13518.5	99	37.22	842.99	13503.4
89	31.67	843.85	13517.1	100	37.78	842.91	13502.0
90	32.22	843.76	13515.7	101	38.33	842.82	13500.7

†$\rho_{Hg} = 846.324[1 - 0.000101(T_F - 60)]$. Equation from AGA Report 3 (ANSI/API 2530, 1978).

Table E.6 Density and Viscosity of Water: U.S. and SI Units

Temperature		Density†		
°F	°C	lb$_m$/ft^3	kg/m^3	Viscosity‡ μ_{cP}
32	0.00	62.41796	999.839	1.7869
33	0.56	62.42014	999.874	1.7536
34	1.11	62.42198	999.904	1.7212
35	1.67	62.42348	999.928	1.6898
36	2.22	62.42465	999.947	1.6592
37	2.78	62.42549	999.960	1.6295
38	3.33	62.42601	999.968	1.6005
39	3.89	62.42622	999.972	1.5724
40	4.44	62.42611	999.970	1.5450
41	5.00	62.42571	999.964	1.5183
42	5.56	62.42500	999.952	1.4923
43	6.11	62.42400	999.936	1.4670
44	6.67	62.42271	999.916	1.4424
45	7.22	62.42114	999.890	1.4184
46	7.78	62.41928	999.861	1.3950
47	8.33	62.41715	999.827	1.3722
48	8.89	62.41476	999.788	1.3499
49	9.44	62.41209	999.746	1.3283
50	10.00	62.40916	999.699	1.3071
51	10.56	62.40598	999.648	1.2865
52	11.11	62.40254	999.592	1.2664
53	11.67	62.39885	999.533	1.2468
54	12.22	62.39491	999.470	1.2276
55	12.78	62.39073	999.403	1.2089
56	13.33	62.38631	999.333	1.1907
57	13.89	62.38165	999.258	1.1729
58	14.44	62.37676	999.180	1.1555
59	15.00	62.37164	999.098	1.1385
60	15.56	62.36630	999.012	1.1219
61	16.11	62.36073	998.923	1.1057
62	16.67	62.35494	998.830	1.0898
63	17.22	62.34893	998.734	1.0743
64	17.78	62.34271	998.634	1.0592
65	18.33	62.33627	998.531	1.0444
66	18.89	62.32963	998.425	1.0299
67	19.44	62.32277	998.315	1.0158
68	20.00	62.31572	998.202	1.0019
69	20.56	62.30846	998.085	0.9884
70	21.11	62.30100	997.966	0.9753
71	21.67	62.29334	997.843	0.9624
72	22.22	62.28549	997.718	0.9498
73	22.78	62.27744	997.589	0.9374
74	23.33	62.26920	997.457	0.9253
75	23.89	62.26078	997.322	0.9134
76	24.44	62.25216	997.184	0.9018
77	25.00	62.24336	997.043	0.8904
78	25.56	62.23438	996.899	0.8793
79	26.11	62.22521	996.752	0.8683

Temperature		Density†		
°F	°C	lb$_m$/ft^3	kg/m^3	Viscosity‡ μ_{cP}
80	26.67	62.21587	996.602	0.8576
81	27.22	62.20635	996.450	0.8471
82	27.78	62.19665	996.294	0.8368
83	28.33	62.18678	996.136	0.8267
84	28.89	62.17673	995.975	0.8168
85	29.44	62.16651	995.812	0.8071
86	30.00	62.15612	995.645	0.7975
87	30.56	62.14557	995.476	0.7882
88	31.11	62.13484	995.304	0.7790
89	31.67	62.12396	995.130	0.7700
90	32.22	62.11291	994.953	0.7612
91	32.78	62.10169	994.773	0.7525
92	33.33	62.09032	994.591	0.7440
93	33.89	62.07879	994.407	0.7356
94	34.44	62.06710	994.219	0.7274
95	35.00	62.05525	994.030	0.7194
96	35.56	62.04325	993.837	0.7115
97	36.11	62.03110	993.643	0.7037
98	36.67	62.01880	993.446	0.6960
99	37.22	62.00634	993.246	0.6885
100	37.78	61.99374	993.044	0.6812
101	38.33	61.98099	992.840	0.6739
102	38.89	61.96810	992.633	0.6668
103	39.44	61.95506	992.425	0.6598
104	40.00	61.94188	992.214	0.6529
105	40.56	61.92856	992.000	0.6462
106	41.11	61.91510	991.785	0.6395
107	41.67	61.90151	991.567	0.6330
108	42.22	61.88778	991.347	0.6266
109	42.78	61.87391	991.125	0.6202
110	43.33	61.85992	990.901	0.6140
111	43.89	61.84579	990.674	0.6079
112	44.44	61.83154	990.446	0.6019
113	45.00	61.81716	990.216	0.5960
114	45.56	61.80266	989.983	0.5902
115	46.11	61.78803	989.749	0.5844
116	46.67	61.77329	989.513	0.5788
117	47.22	61.75843	989.275	0.5733
118	47.78	61.74345	989.035	0.5678
119	48.33	61.72836	988.793	0.5624
120	48.89	61.71315	988.550	0.5571
121	49.44	61.69784	988.304	0.5519
122	50.00	61.68243	988.057	0.5468
123	50.56	61.66691	987.809	0.5418
124	51.11	61.65128	987.559	0.5368
125	51.67	61.63556	987.307	0.5319

Table E.6 Density and Viscosity of Water: U.S. and SI Units (*Continued*)

Temperature		Density†		
°F	°C	lb$_m$/ft^3	kg/m^3	Viscosity‡ μ_{cP}
126	52.22	61.61975	987.053	0.5271
127	52.78	61.60384	986.799	0.5223
128	53.33	61.58784	986.542	0.5176
129	53.89	61.57176	986.285	0.5130
130	54.44	61.55559	986.026	0.5085
131	55.00	61.53934	985.765	0.5040
132	55.56	61.52302	985.504	0.4996
133	56.11	61.50662	985.241	0.4953
134	56.67	61.49015	984.977	0.4910
135	57.22	61.47361	984.713	0.4867
136	57.78	61.45701	984.447	0.4826
137	58.33	61.44035	984.180	0.4785
138	58.89	61.42364	983.912	0.4744
139	59.44	61.40687	983.644	0.4705
140	60.00	61.39006	983.374	0.4665
141	60.56	61.37321	983.104	0.4626
142	61.11	61.35632	982.834	0.4588
143	61.67	61.33939	982.562	0.4550
144	62.22	61.32243	982.291	0.4513
145	62.78	61.30545	982.019	0.4477
146	63.33	61.28845	981.747	0.4440
147	63.89	61.27144	981.474	0.4405
148	64.44	61.25441	981.201	0.4370
149	65.00	61.23739	980.929	0.4335
150	65.56	61.22036	980.656	0.4300
151	66.11	61.20334	980.383	0.4267
152	66.67	61.18633	980.111	0.4233
153	67.22	61.16935	979.839	0.4200
154	67.78	61.15238	979.567	0.4168
155	68.33	61.13545	979.296	0.4136
156	68.89	61.11856	979.025	0.4104
157	69.44	61.10170	978.755	0.4073
158	70.00	61.08490	978.486	0.4042
159	70.56	61.06816	978.218	0.4011
160	71.11	61.05148	977.951	0.3981
161	71.67	61.03488	977.685	0.3952
162	72.22	61.01835	977.420	0.3922
163	72.78	61.00191	977.157	0.3893

†ρ(kg/m^3) = 999.8395639 + 6.798299989 × 10^{-2} $T_{°C}$ − 9.10602556 × 10^{-3} ($T_{°C}$)^2 + 1.005272999 × 10^{-4} ($T_{°C}$)^4 − 1.126713526 × 10^{-6} ($T_{°C}$)^4 + 6.591795606 × 10^{-9} ($T_{°C}$)^5

‡If $0 \leq T_{°C} \leq 20$°C, use $A = 1301/[998.333 + 8.1855(T_{°C} − 20) + 0.00585 (T_{°C} − 20)^2] − 3.30233$ and $\mu_{cP} = (100)(10^4)$. If $20 \leq T_{°C} \leq 100$°C, use $A = [1.3272 (20 − T_{°C}) − 0.001053 (T_{°C} − 20)^2]/(T_{°C} + 105)$ and $\mu_{cP} = (\mu_{cP}$ at 20°C)(10^4). The equations for viscosity are from Hardy and Cottington (1949) and Swindells (NBS). The density of water is calculated from Eq. (2.75).

VISCOSITIES OF LIQUIDS

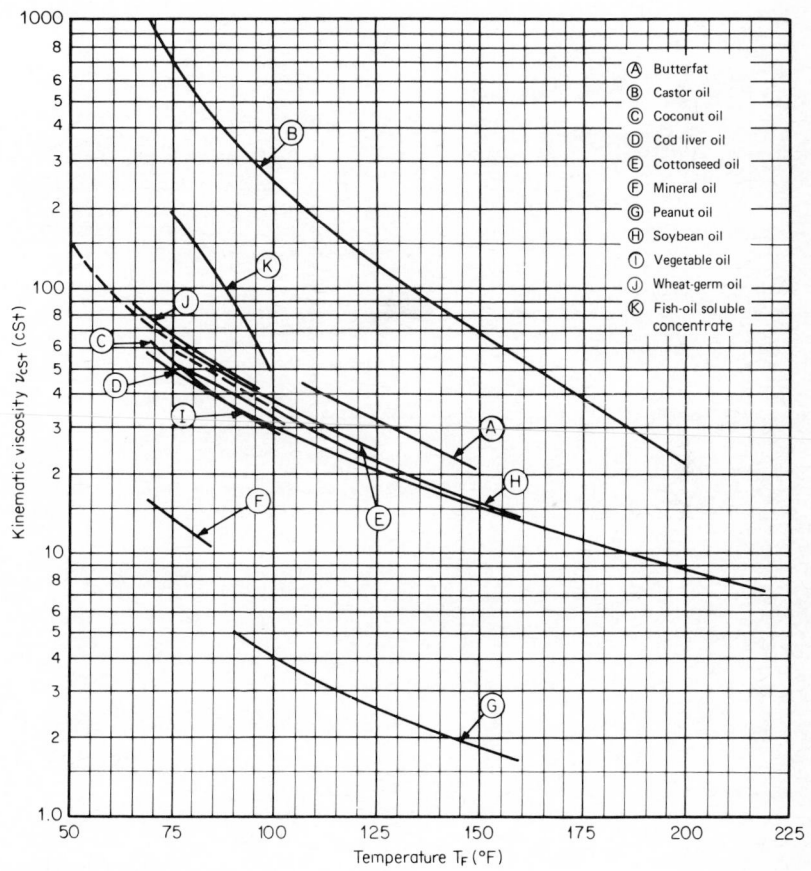

Figure F.1 Kinematic viscosities of fats and oils *(from Fischer & Porter Catalog 10-A-54; used with permission).*

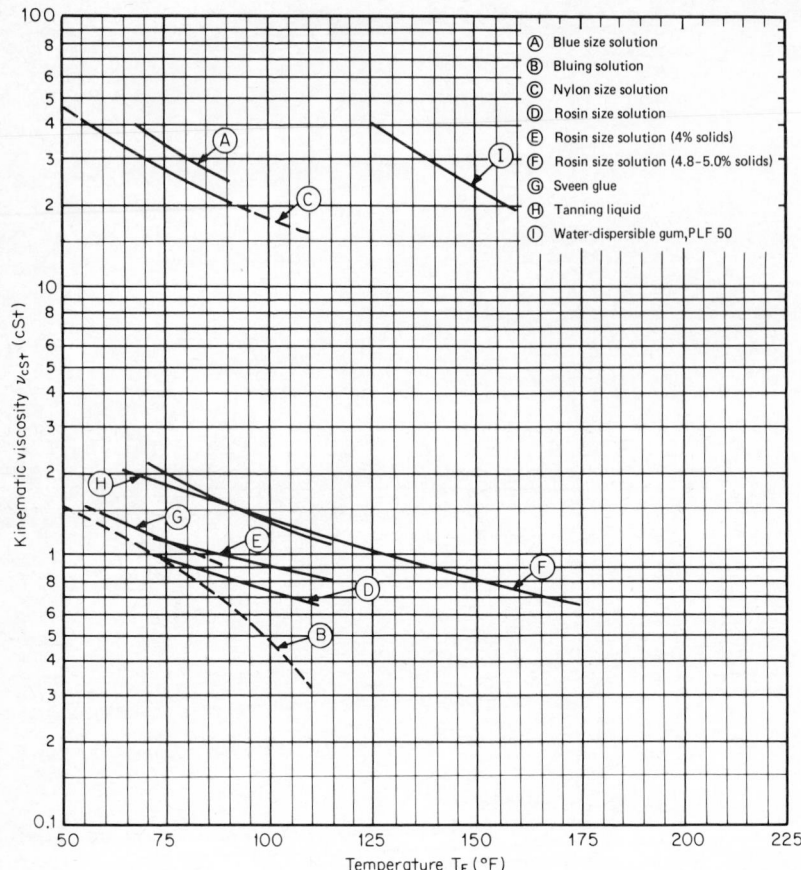

Figure F.2 **Kinematic viscosities of liquids related to the paper, leather, and textile industries** *(from Fischer & Porter Catalog 10-A-54; used with permission).*

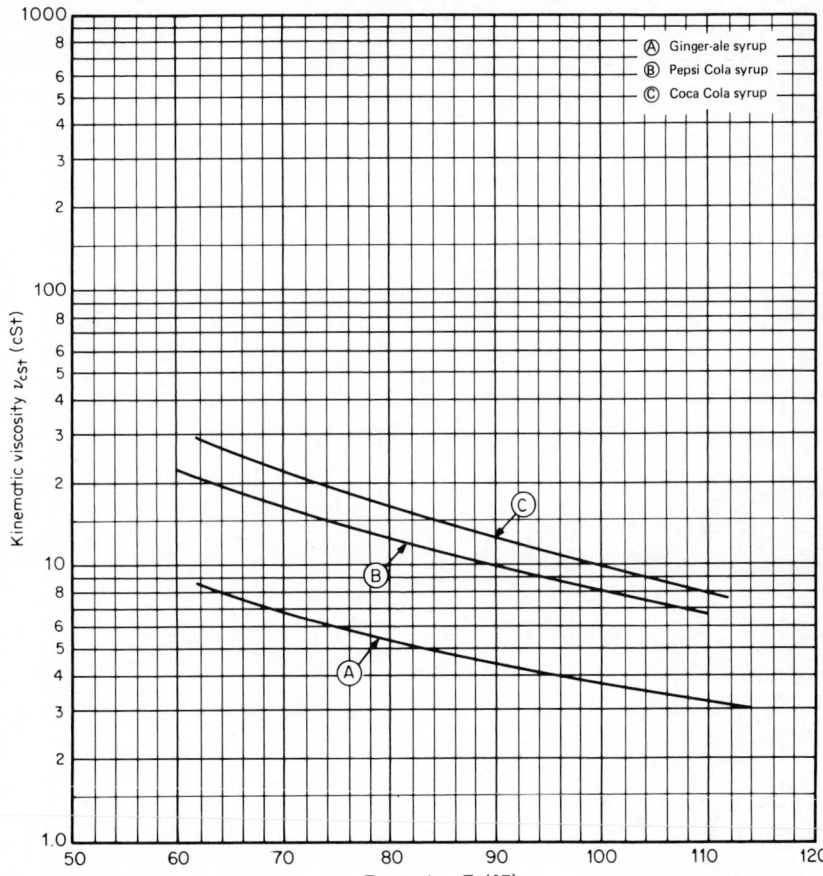

Figure F.3 **Kinematic viscosities of syrups** *(from Fischer & Porter Catalog 10-A-54; used with permission).*

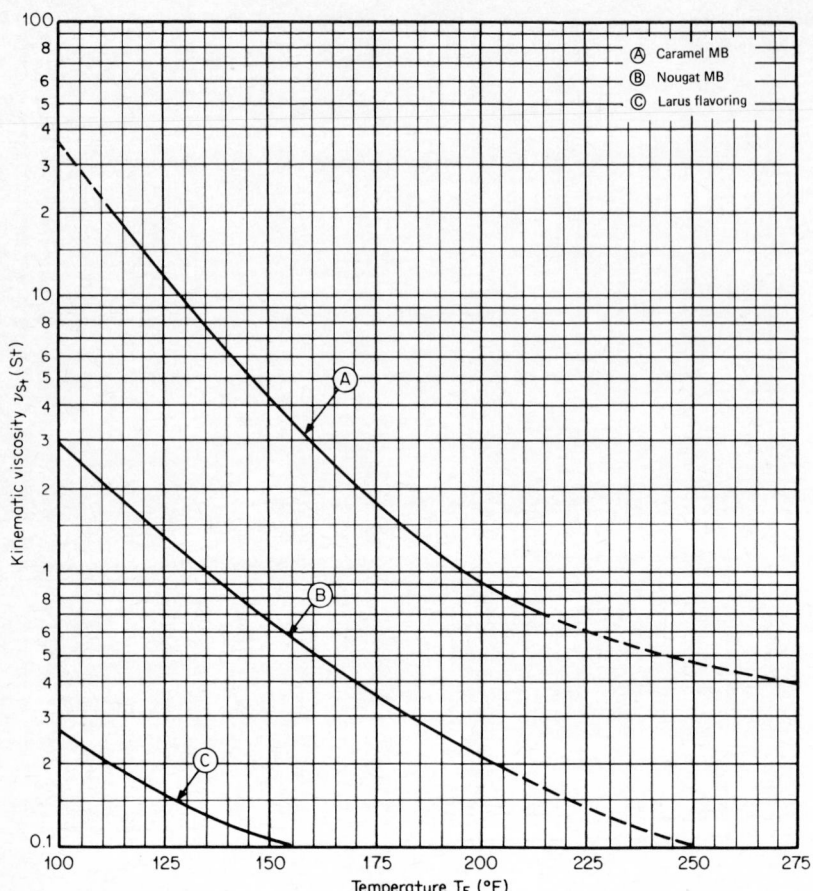

Figure F.4 **Kinematic viscosities of liquids related to the candy industry** *(from Fischer & Porter Catalog 10-A-54; used with permission).*

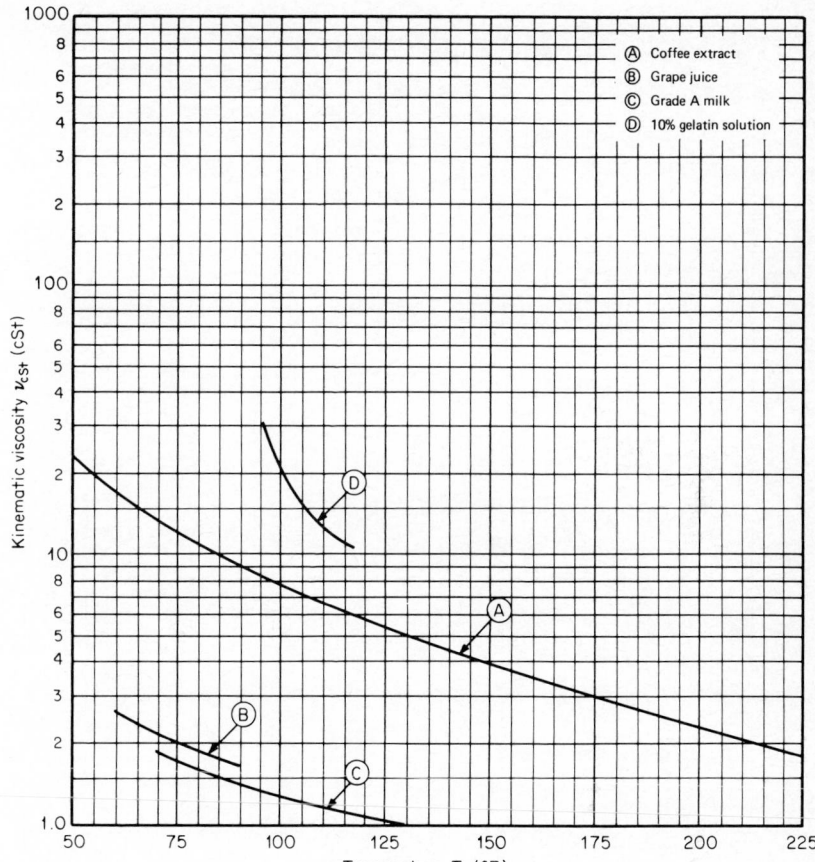

Figure F.5 **Kinematic viscosities of miscellaneous liquids related to the food industry**
(from Fischer & Porter Catalog 10-A-54; used with permission).

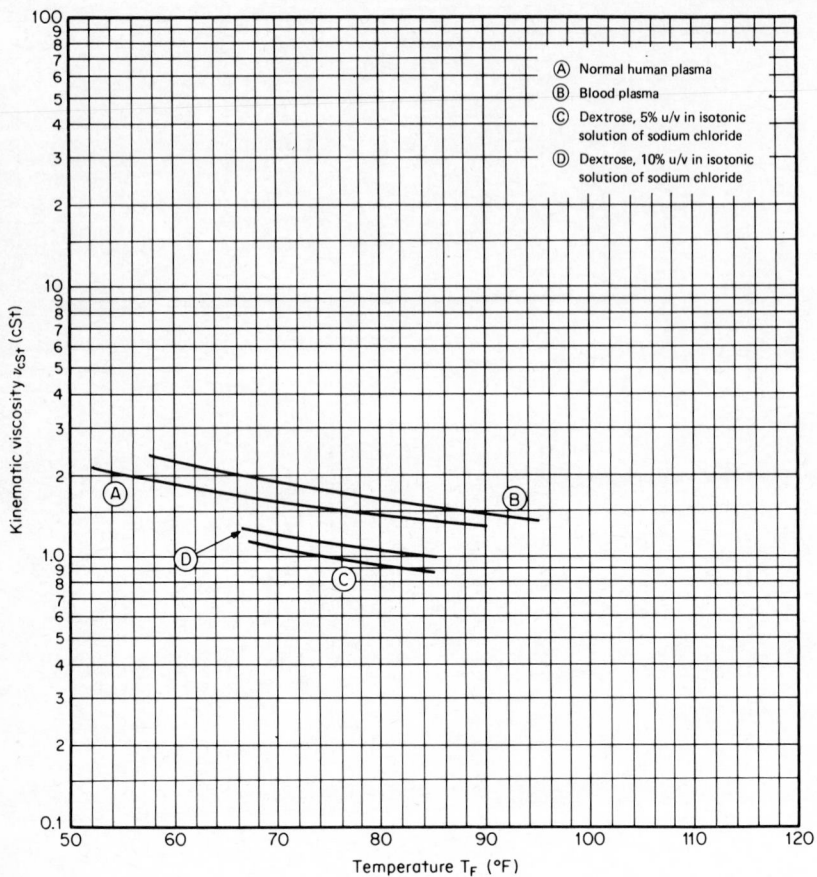

Figure F.6   **Kinematic viscosities of intravenous solutions** *(from Fischer & Porter Catalog 10-A-54; used with permission).*

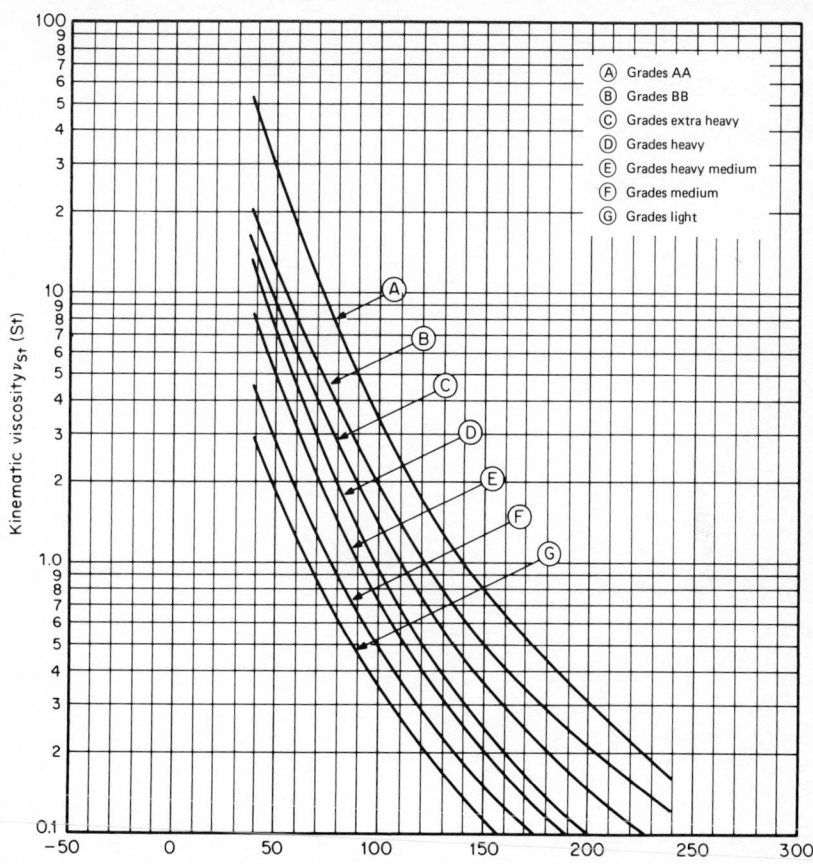

Figure F.7 **Kinematic viscosities of Gargoyle D.T.E. oils** *(from Fischer & Porter Catalog 10-A-54; used with permission).*

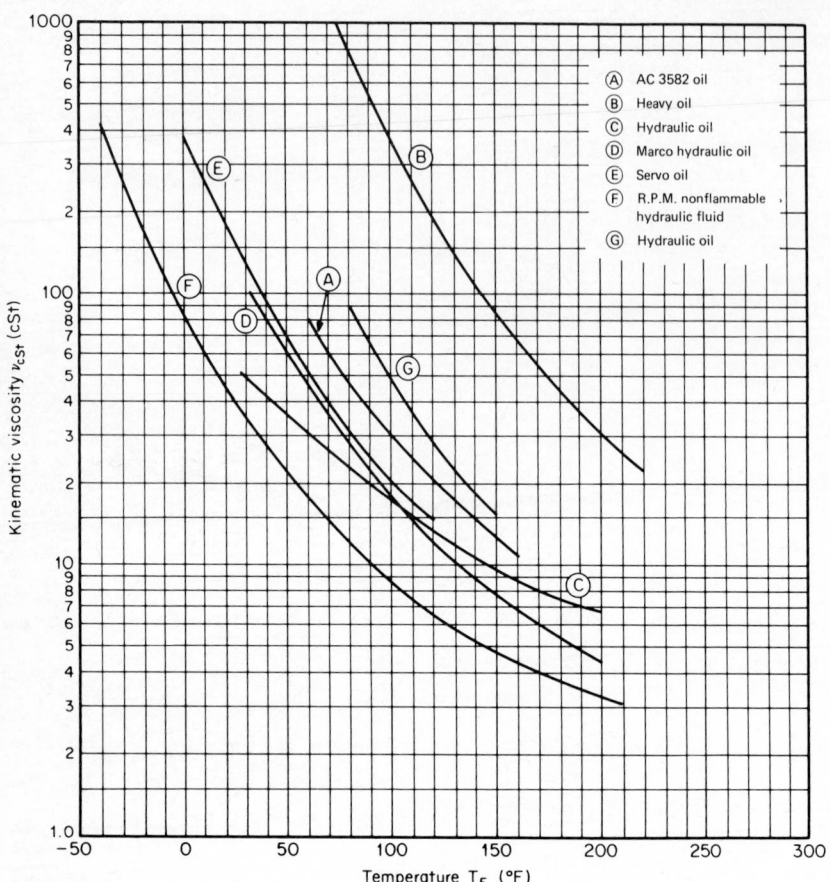

Figure F.8 **Kinematic viscosities of hydraulic oils** *(from Fischer & Porter Catalog 10-A-54; used with permission).*

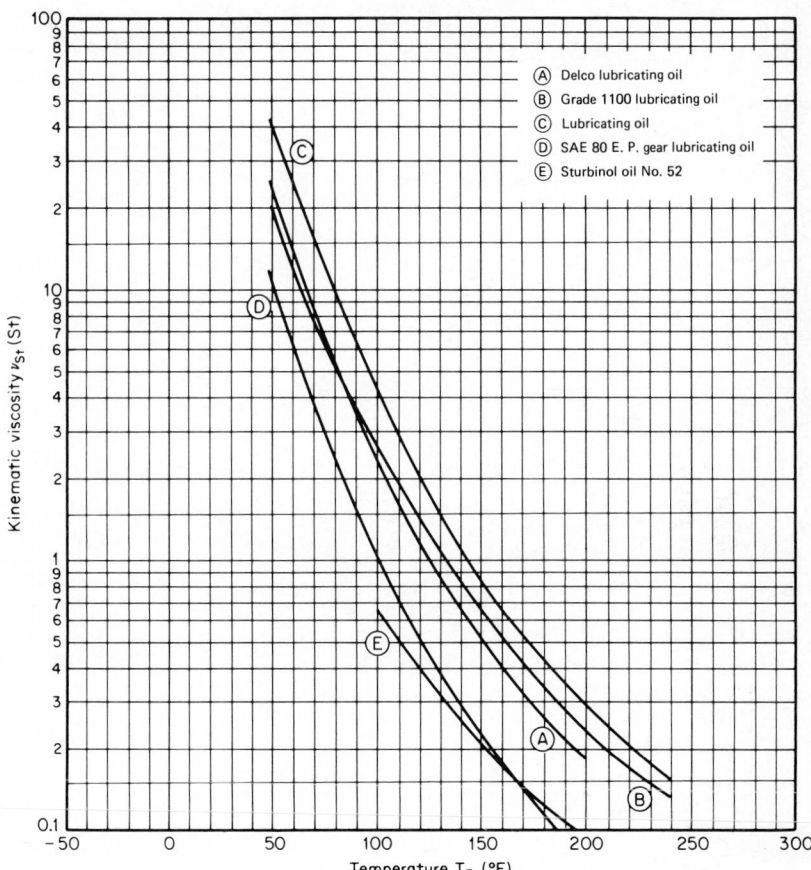

Figure F.9 **Kinematic viscosities of lube oils** *(from Fischer & Porter Catalog 10-A-54; used with permission).*

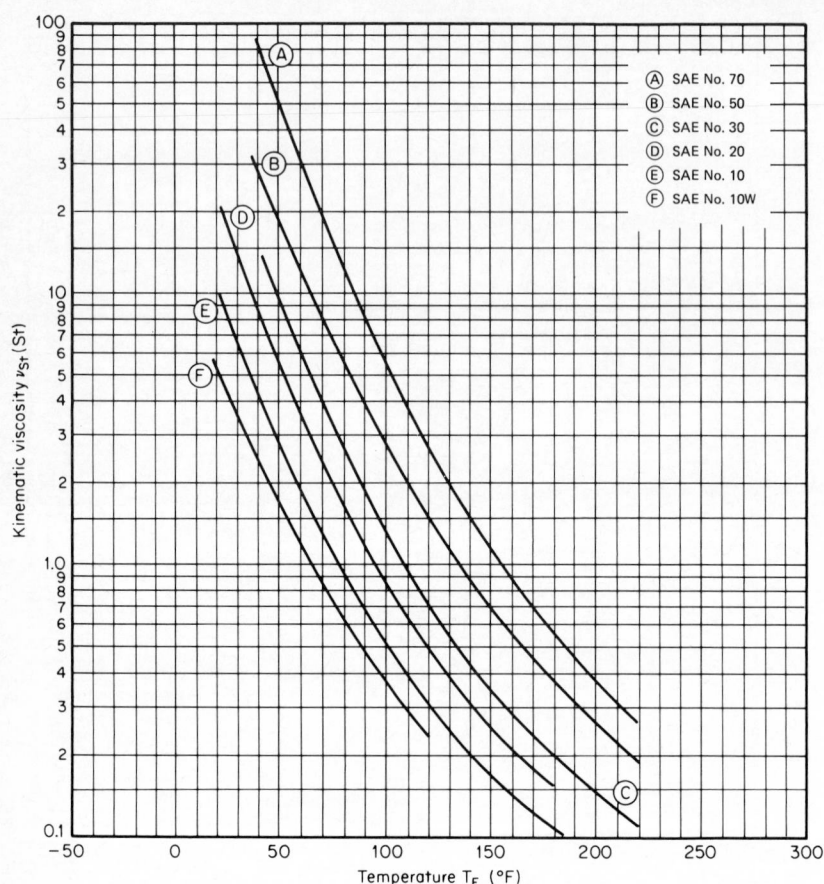

Figure F.10 **Kinematic viscosities of SAE (Pennsylvania base) oils** *(from Fischer & Porter Catalog 10-A-54; used with permission).*

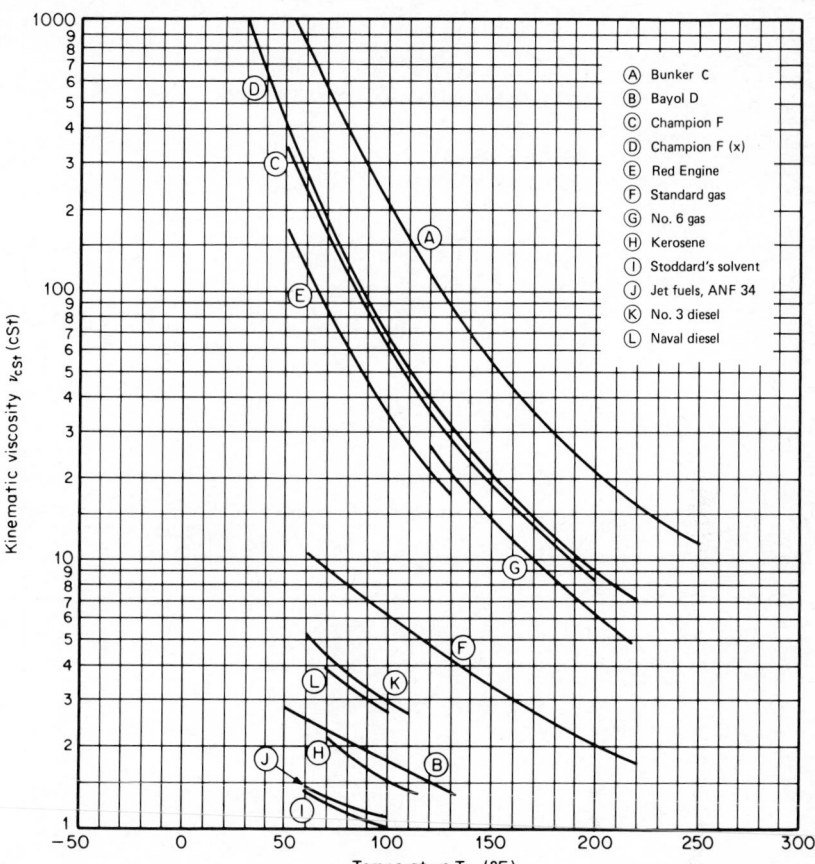

Figure F.11 **Kinematic viscosities of miscellaneous fuel oils** *(from Fischer & Porter Catalog 10-A-54; used with permission).*

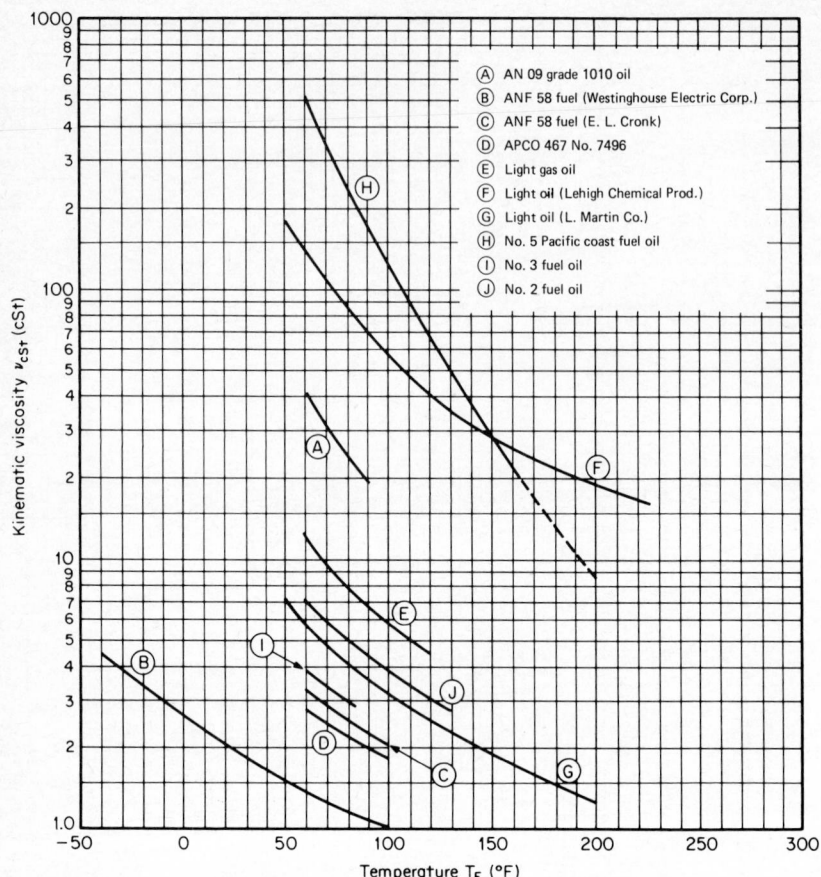

Figure F.12 Kinematic viscosities of miscellaneous fuel oils *(from Fischer & Porter Catalog 10-A-54; used with permission).*

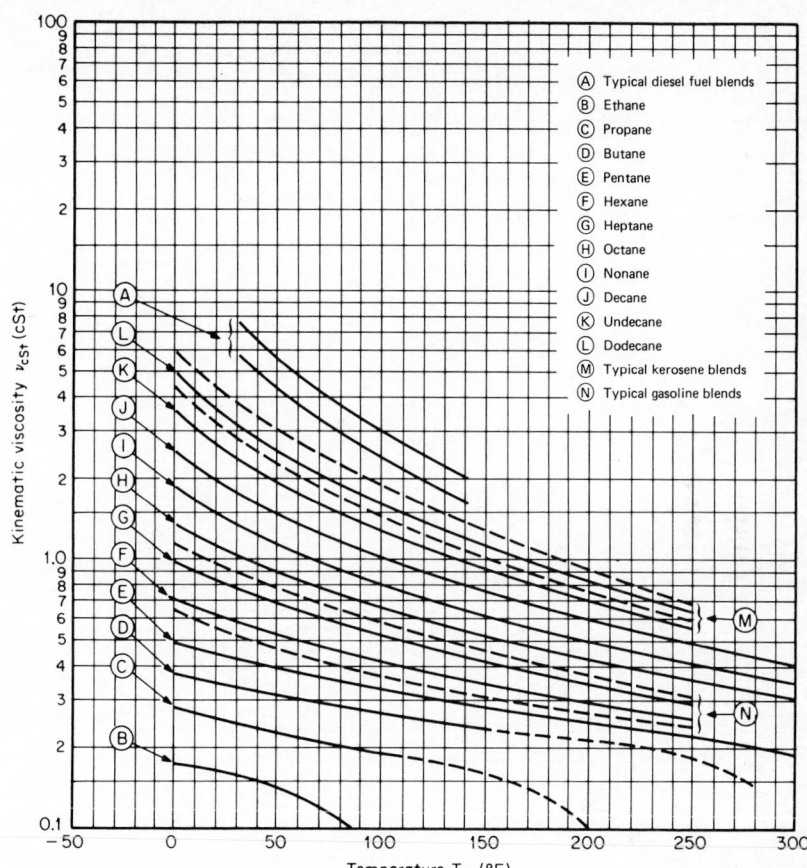

Figure F.13 **Kinematic viscosities of typical fuel blends** *(from Fischer & Porter Catalog 10-A-54; used with permission).*

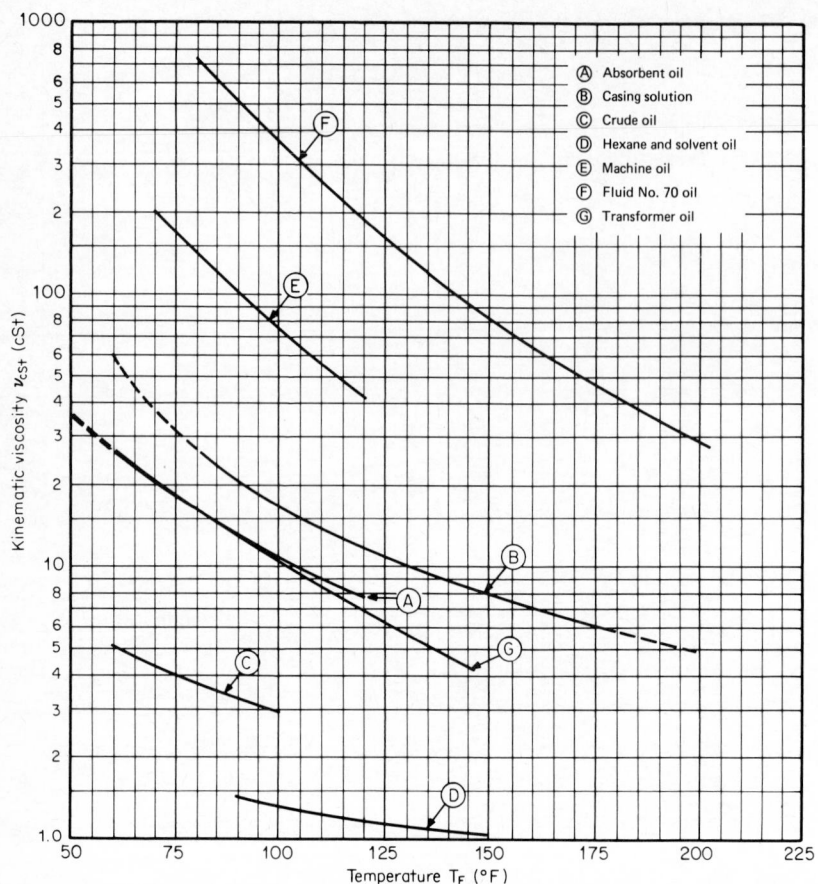

Figure F.14 **Kinematic viscosities of miscellaneous oils** *(from Fischer & Porter Catalog 10-A-54; used with permission).*

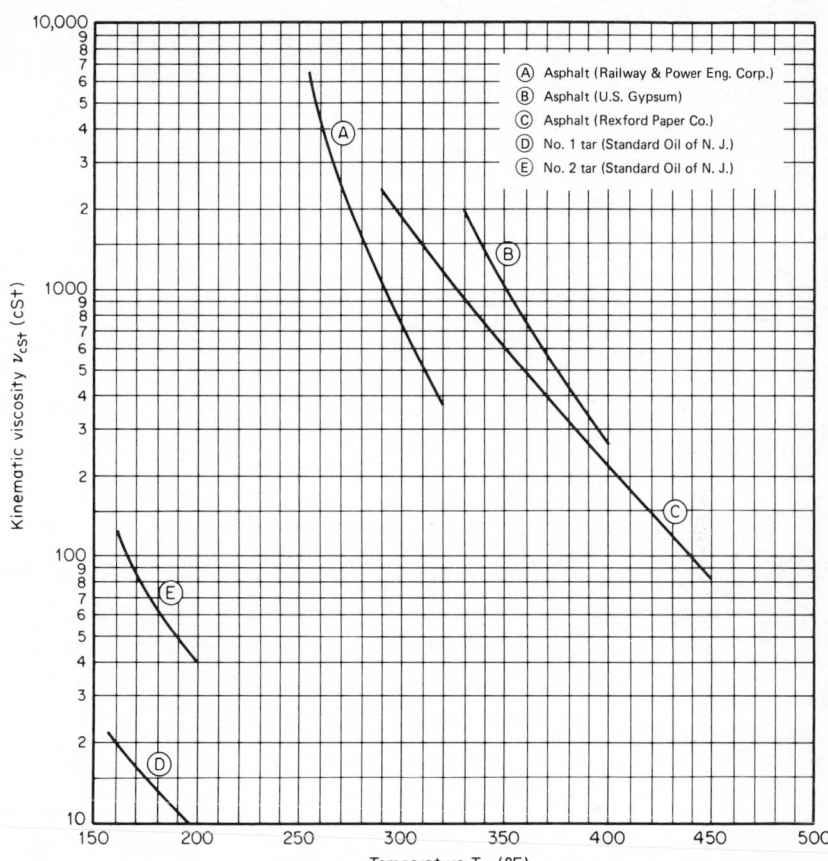

Figure F.15 **Kinematic viscosities of high boiling fractions** *(from Fischer & Porter Catalog 10-A-54; used with permission).*

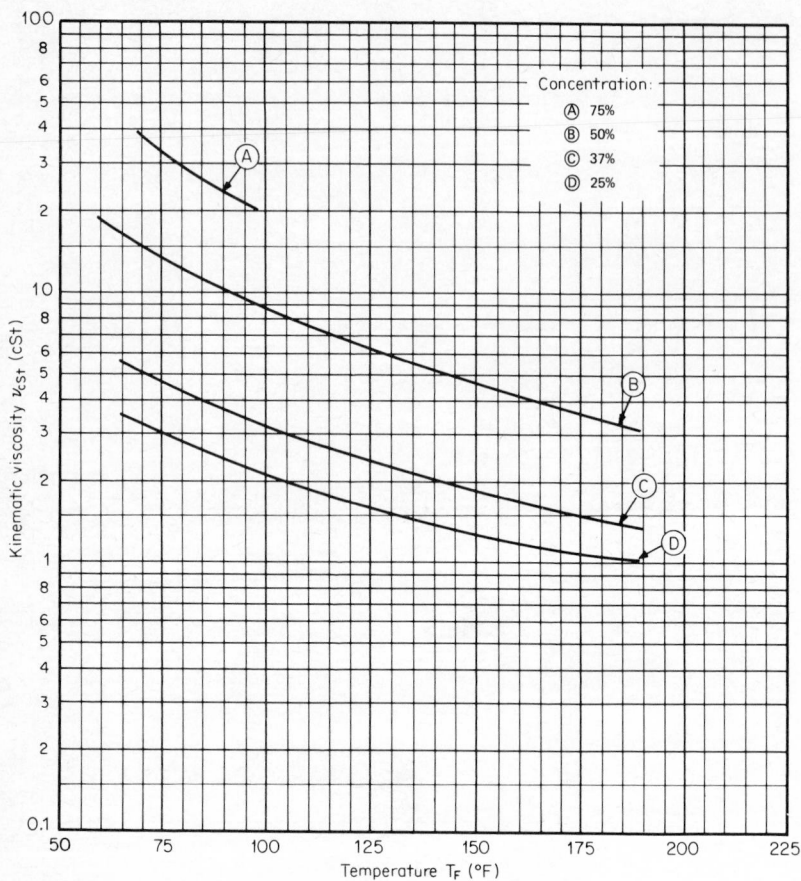

Figure F.16 **Kinematic viscosities of phosphoric acid solutions** *(from Fischer & Porter Catalog 10-A-54; used with permission).*

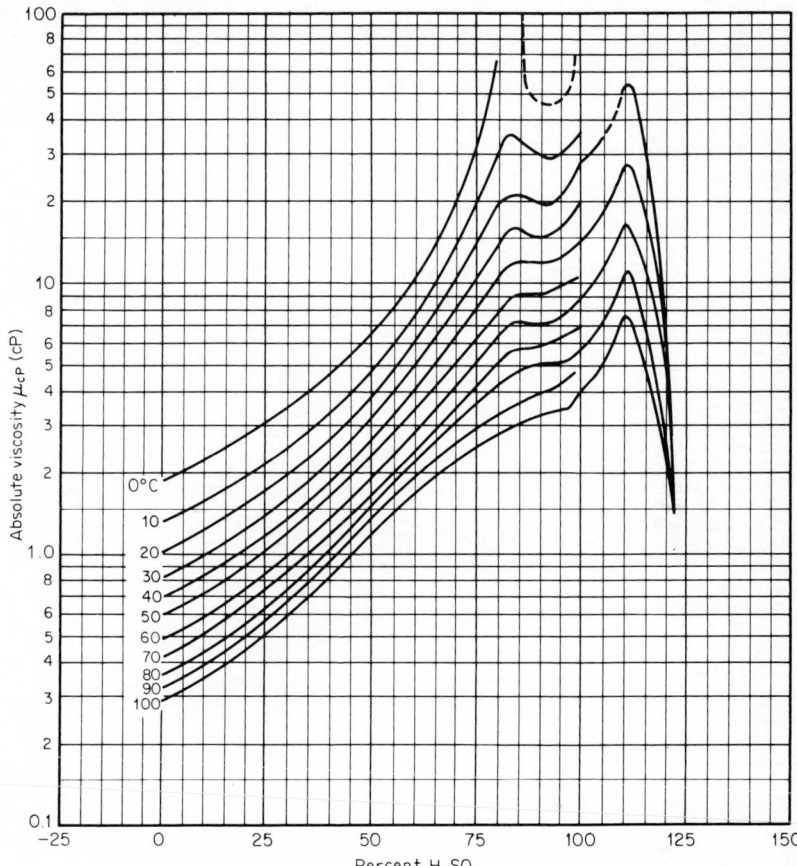

Figure F.17 **Absolute viscosities of sulfuric acid solutions** *(from Fischer & Porter Catalog 10-A-54; used with permission).*

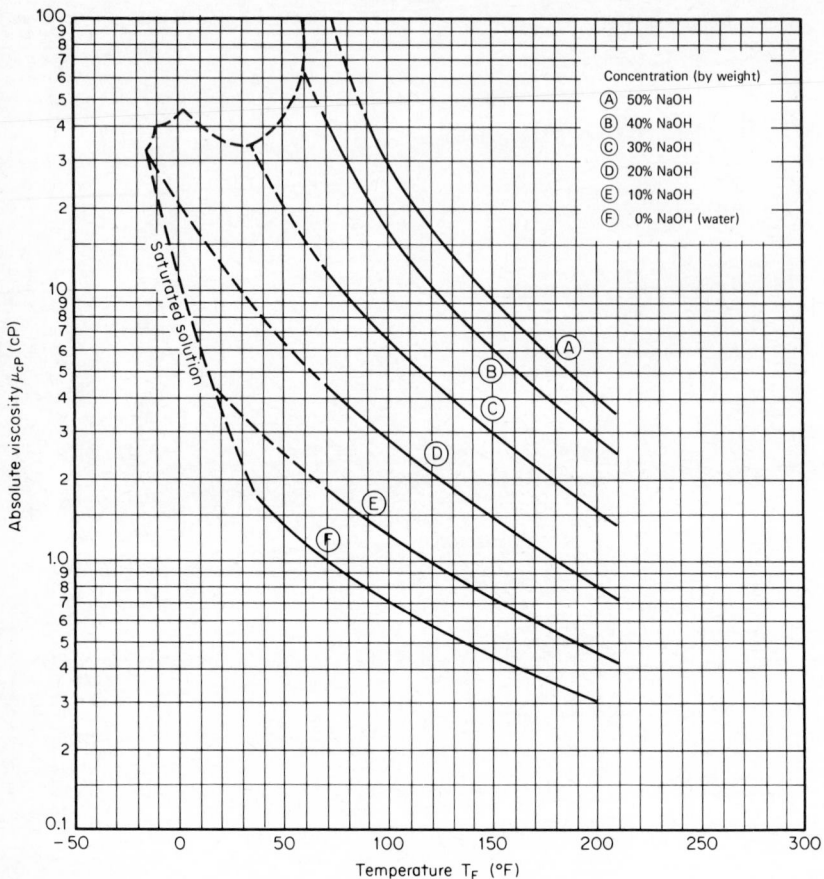

Figure F.18   **Absolute viscosities of sodium hydroxide solutions** *(from Fischer & Porter Catalog 10-A-54; used with permission).*

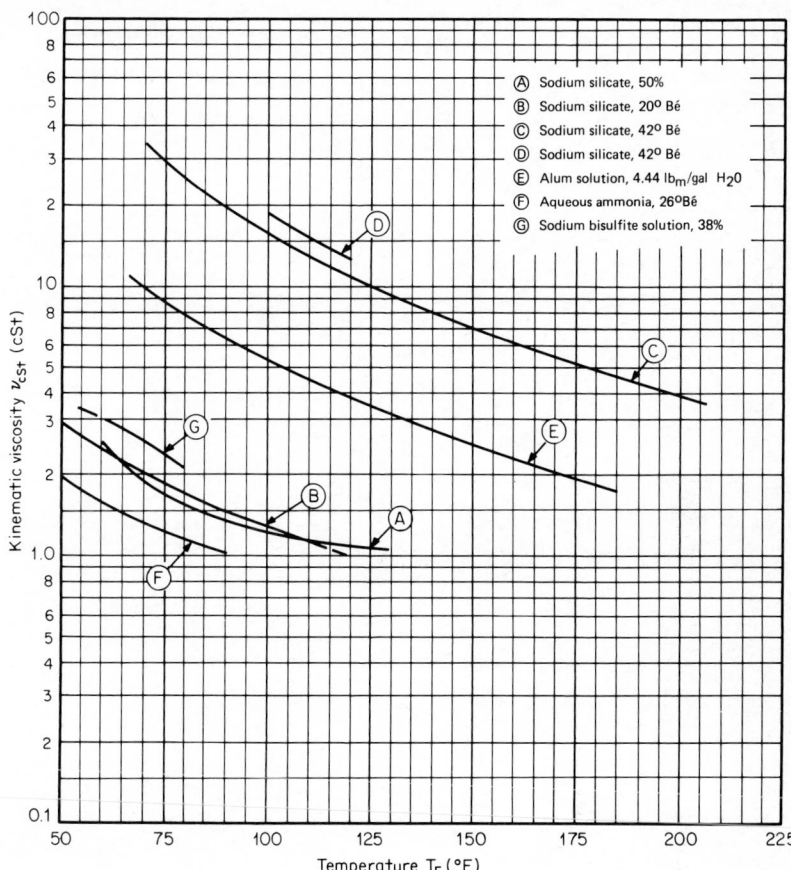

Figure F.19 **Kinematic viscosities of inorganic solutions** *(from Fischer & Porter Catalog 10-A-54; used with permission).*

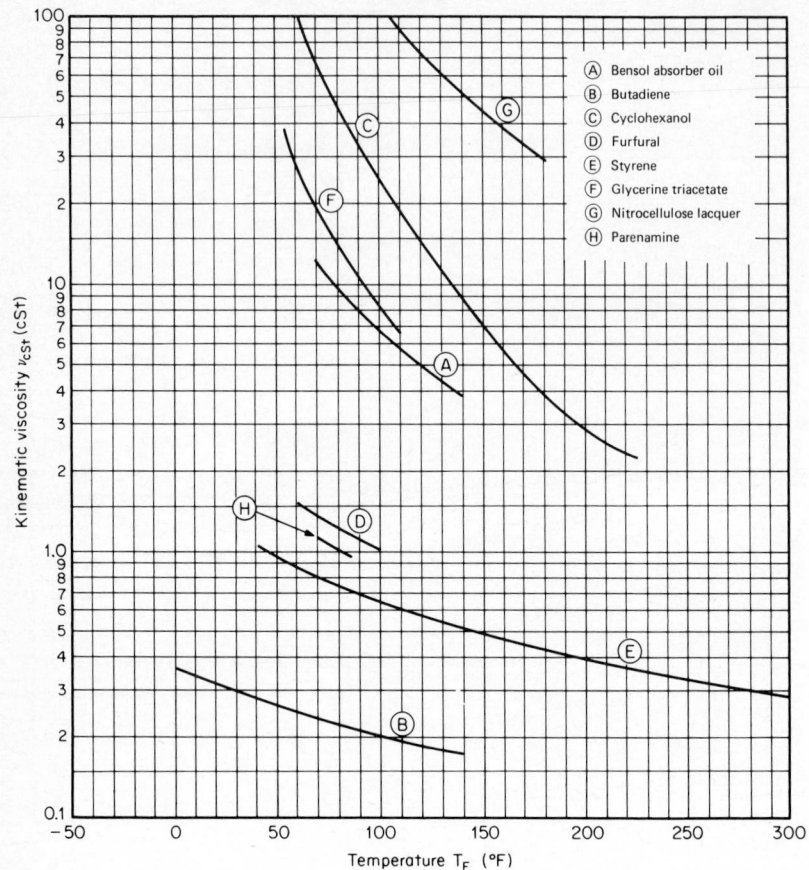

Figure F.20 **Absolute viscosities of miscellaneous organic compounds** *(from Fischer & Porter Catalog 10-A-54; used with permission).*

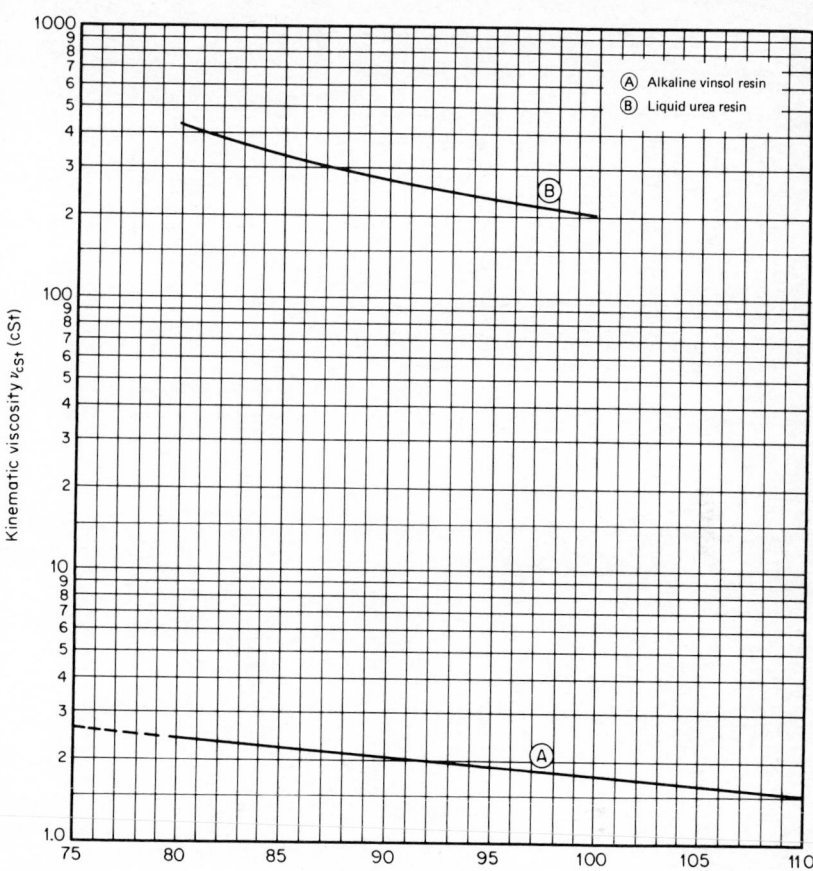

Figure F.21   **Kinematic viscosities of resins** *(from Fischer & Porter Catalog 10-A-54; used with permission).*

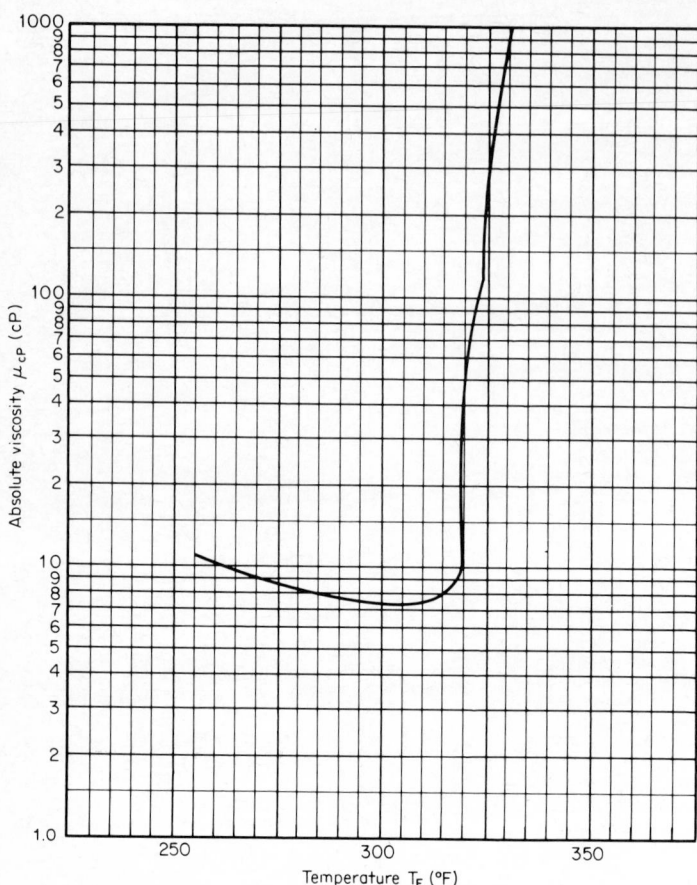

Figure F.22 **Absolute viscosities of liquid sulfur** *(from Fischer & Porter Catalog 10-A-54; used with permission).*

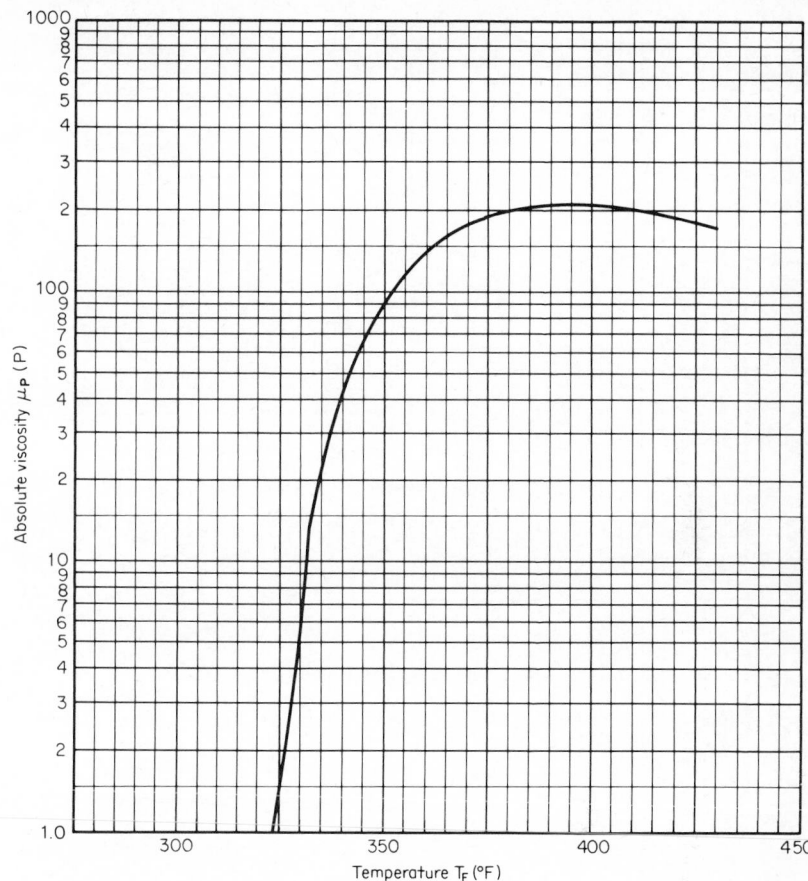

Figure F.23 **Absolute viscosities of liquid sulfur** *(continued) (from Fischer & Porter Catalog 10-A-54; used with permission).*

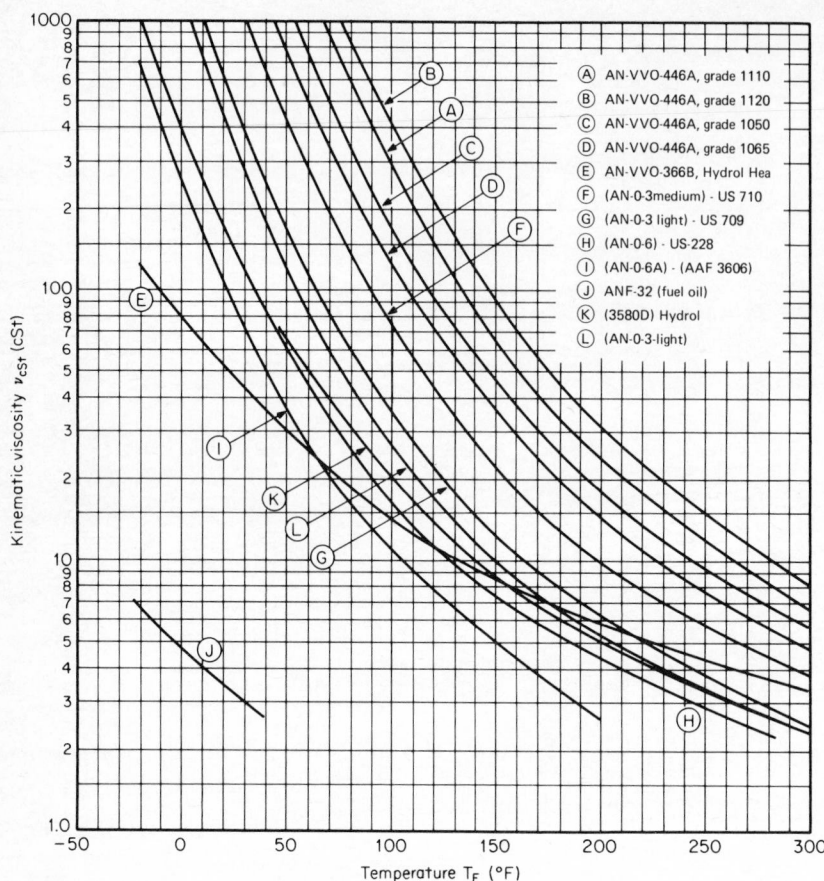

Figure F.24 **Kinematic viscosities of Army and Navy lube and hydraulic oils** *(from Fischer & Porter Catalog 10-A-54; used with permission).*

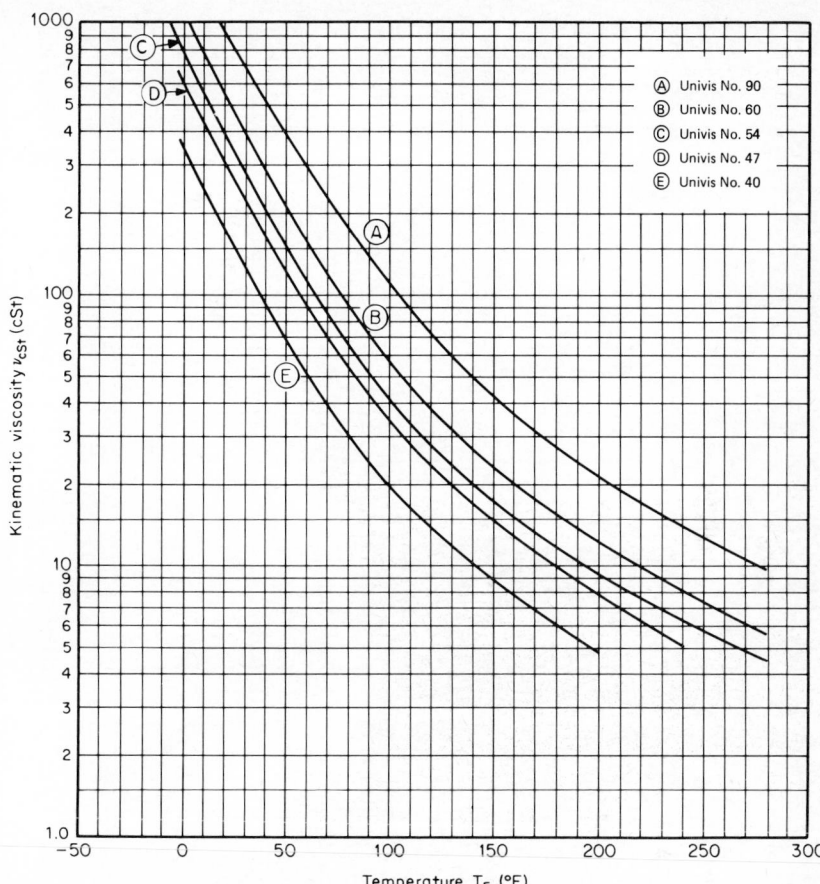

Figure F.25 Kinematic viscosities of Univis oils (Standard Oil of New Jersey) *(from Fischer & Porter Catalog 10-A-54; used with permission).*

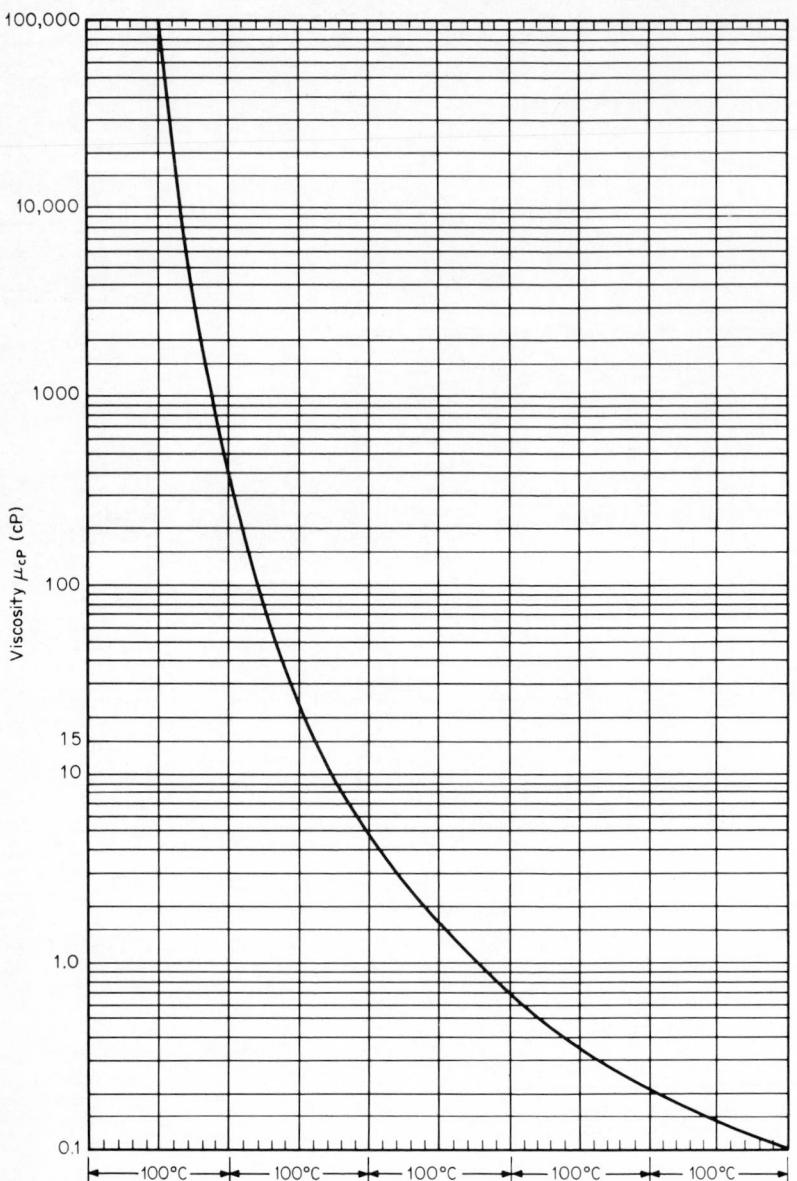

Figure F.26 Curve for estimating viscosity from a single measurement value *(from Gambill, 1959).*

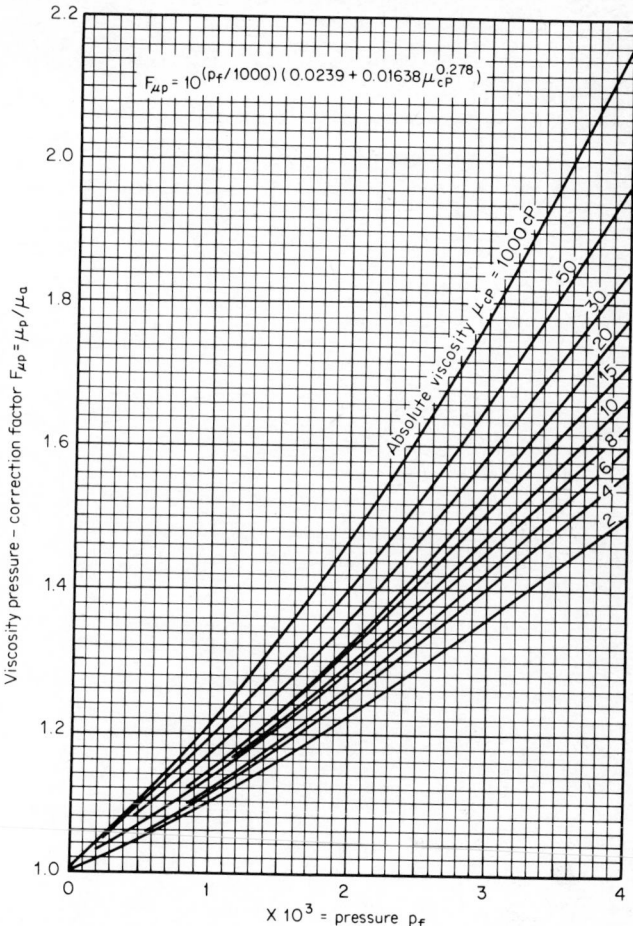

Figure F.27 Viscosity pressure-correction factor for oils.

GAS COMPRESSIBILITY FACTORS AND STEAM DENSITY TABLES

Table G.1 Density (in Pounds-Mass per Cubic Foot) and Absolute Viscosity μ_{cP} (in Centipoises) of Steam and Compressed Water†

	Pressure, psia					
	12.	14.696	16.	18.	20.	22.
Temp., °F	201.96	212.00	216.32	222.41	227.96	233.07
Sat. vap.	0.03087	0.03732	0.04040	0.04511	0.04978	0.05443
	(0.01208)	(0.01228)	(0.01236)	(0.01248)	(0.01258)	(0.01268)
Sat. liq.	60.06059	59.81363	59.70482	59.54873	59.40363	59.26779
	(0.29966)	(0.28214)	(0.27515)	(0.26580)	(0.25775)	(0.25071)
Temp., °F						
220.	0.03000	0.03684	0.04017	59.61096	59.61138	59.61180
	(0.01245)	(0.01244)	(0.01243)	(0.26943)	(0.26943)	(0.26943)
240.	0.02910	0.03572	0.03894	0.04388	0.04885	0.05384
	(0.01287)	(0.01286)	(0.01285)	(0.01284)	(0.01284)	(0.01283)
260.	0.02825	0.03467	0.03779	0.04258	0.04738	0.05221
	(0.01329)	(0.01328)	(0.01328)	(0.01327)	(0.01326)	(0.01325)
280.	0.02746	0.03369	0.03671	0.04136	0.04601	0.05069
	(0.01372)	(0.01371)	(0.01371)	(0.01370)	(0.01369)	(0.01369)
300.	0.02671	0.03276	0.03570	0.04021	0.04473	0.04927
	(0.01415)	(0.01414)	(0.01414)	(0.01413)	(0.01413)	(0.01412)
320.	0.02600	0.03189	0.03474	0.03913	0.04352	0.04793
	(0.01459)	(0.01458)	(0.01458)	(0.01457)	(0.01457)	(0.01456)
340.	0.02533	0.03107	0.03384	0.03811	0.04239	0.04667
	(0.01503)	(0.01502)	(0.01502)	(0.01501)	(0.01501)	(0.01500)
360.	0.02470	0.03029	0.03299	0.03715	0.04131	0.04548
	(0.01547)	(0.01546)	(0.01546)	(0.01546)	(0.01545)	(0.01545)
380.	0.02410	0.02955	0.03218	0.03624	0.04029	0.04436
	(0.01592)	(0.01591)	(0.01591)	(0.01590)	(0.01590)	(0.01590)
400.	0.02353	0.02884	0.03142	0.03537	0.03933	0.04329
	(0.01636)	(0.01636)	(0.01636)	(0.01635)	(0.01635)	(0.01635)
420.	0.02299	0.02817	0.03069	0.03455	0.03841	0.04228
	(0.01682)	(0.01681)	(0.01681)	(0.01681)	(0.01680)	(0.01680)
440.	0.02247	0.02754	0.02999	0.03376	0.03753	0.04131
	(0.01727)	(0.01726)	(0.01726)	(0.01726)	(0.01726)	(0.01725)
460.	0.02197	0.02693	0.02933	0.03301	0.03670	0.04039
	(0.01772)	(0.01772)	(0.01772)	(0.01771)	(0.01771)	(0.01771)
480.	0.02150	0.02635	0.02869	0.03230	0.03590	0.03951
	(0.01818)	(0.01818)	(0.01817)	(0.01817)	(0.01817)	(0.01817)
500.	0.02105	0.02579	0.02809	0.03161	0.03514	0.03867
	(0.01864)	(0.01863)	(0.01863)	(0.01863)	(0.01863)	(0.01863)
520.	0.02061	0.02526	0.02751	0.03096	0.03441	0.03787
	(0.01910)	(0.01909)	(0.01909)	(0.01909)	(0.01909)	(0.01909)
540.	0.02020	0.02475	0.02695	0.03033	0.03371	0.03710
	(0.01956)	(0.01955)	(0.01955)	(0.01955)	(0.01955)	(0.01955)
560.	0.01980	0.02426	0.02642	0.02973	0.03304	0.03636
	(0.02002)	(0.02001)	(0.02001)	(0.02001)	(0.02001)	(0.02001)

	Pressure, psia					
	12.	14.696	16.	18.	20.	22.
Temp., °F						
600.	0.01905	0.02333	0.02541	0.02859	0.03178	0.03497
	(0.02094)	(0.02094)	(0.02094)	(0.02093)	(0.02093)	(0.02093)
650.	0.01818	0.02227	0.02426	0.02729	0.03033	0.03338
	(0.02209)	(0.02209)	(0.02209)	(0.02209)	(0.02209)	(0.02209)
700.	0.01739	0.02131	0.02320	0.02611	0.02902	0.03192
	(0.02325)	(0.02325)	(0.02325)	(0.02325)	(0.02325)	(0.02325)
750.	0.01667	0.02042	0.02224	0.02502	0.02781	0.03059
	(0.02440)	(0.02440)	(0.02440)	(0.02440)	(0.02440)	(0.02440)
800.	0.01601	0.01961	0.02135	0.02402	0.02670	0.02937
	(0.02556)	(0.02556)	(0.02555)	(0.02555)	(0.02555)	(0.02555)
850.	0.01539	0.01886	0.02053	0.02310	0.02567	0.02824
	(0.02670)	(0.02670)	(0.02670)	(0.02670)	(0.02670)	(0.02670)
900.	0.01483	0.01816	0.01977	0.02225	0.02472	0.02720
	(0.02784)	(0.02784)	(0.02784)	(0.02784)	(0.02785)	(0.02785)
950.	0.01430	0.01751	0.01907	0.02146	0.02384	0.02623
	(0.02898)	(0.02898)	(0.02898)	(0.02898)	(0.02898)	(0.02898)
1000.	0.01381	0.01691	0.01842	0.02072	0.02302	0.02533
	(0.03011)	(0.03011)	(0.03011)	(0.03011)	(0.03011)	(0.03011)
1050.	0.01335	0.01635	0.01780	0.02003	0.02226	0.02449
	(0.03123)	(0.03123)	(0.03123)	(0.03123)	(0.03123)	(0.03124)
1100.	0.01292	0.01583	0.01723	0.01939	0.02154	0.02370
	(0.03235)	(0.03235)	(0.03235)	(0.03235)	(0.03235)	(0.03235)
1150.	0.01252	0.01533	0.01670	0.01878	0.02087	0.02296
	(0.03346)	(0.03346)	(0.03346)	(0.03346)	(0.03346)	(0.03346)
1200.	0.01214	0.01487	0.01619	0.01822	0.02024	0.02227
	(0.03455)	(0.03455)	(0.03455)	(0.03456)	(0.03456)	(0.03456)
1250.	0.01179	0.01444	0.01572	0.01768	0.01965	0.02161
	(0.03564)	(0.03564)	(0.03564)	(0.03565)	(0.03565)	(0.03565)
1300.	0.01145	0.01402	0.01527	0.01718	0.01909	0.02100
	(0.03672)	(0.03673)	(0.03673)	(0.03673)	(0.03673)	(0.03673)
1350.	0.01113	0.01364	0.01485	0.01670	0.01856	0.02042
	(0.03780)	(0.03780)	(0.03780)	(0.03780)	(0.03780)	(0.03780)
1400.	0.01083	0.01327	0.01445	0.01625	0.01806	0.01987
	(0.03886)	(0.03886)	(0.03886)	(0.03886)	(0.03886)	(0.03886)
1450.	0.01055	0.01292	0.01407	0.01583	0.01759	0.01935
	(0.03991)	(0.03991)	(0.03991)	(0.03992)	(0.03992)	(0.03992)
1500.	0.01028	0.01259	0.01371	0.01542	0.01714	0.01885
	(0.04096)	(0.04096)	(0.04096)	(0.04096)	(0.04096)	(0.04096)

†Absolute viscosity μ_{cP} is shown in parentheses.

SOURCE: Derived from ASME, 1977.

Table G.1 Density (in Pounds-Mass per Cubic Foot) and Absolute Viscosity μ_{cP} (in Centipoises) of Steam and Compressed Water† (*continued*)

	Pressure, psia					
	24.	26.	28.	30.	32.	34.
Temp. °F	237.82	242.25	246.41	250.34	254.05	257.58
Sat. vap.	0.05904	0.06364	0.06821	0.07276	0.07729	0.08181
	(0.01277)	(0.01286)	(0.01294)	(0.01301)	(0.01309)	(0.01315)
Sat. liq.	59.13985	59.01871	58.90354	58.79364	58.68843	58.58742
	(0.24449)	(0.23892)	(0.23390)	(0.22934)	(0.22517)	(0.22133)
Temp. °F						
220.	59.61222	59.61264	59.61306	59.61349	59.61391	59.61433
	(0.26944)	(0.26944)	(0.26945)	(0.26945)	(0.26945)	(0.26946)
240.	0.05884	59.08068	59.08111	59.08155	59.08199	59.08243
	(0.01282)	(0.24172)	(0.24173)	(0.24173)	(0.24173)	(0.24174)
260.	0.05705	0.06190	0.06678	0.07167	0.07657	0.08150
	(0.01325)	(0.01324)	(0.01323)	(0.01322)	(0.01321)	(0.01321)
280.	0.05538	0.06008	0.06479	0.06952	0.07426	0.07902
	(0.01368)	(0.01367)	(0.01366)	(0.01366)	(0.01365)	(0.01364)
300.	0.05381	0.05837	0.06294	0.06752	0.07211	0.07672
	(0.01412)	(0.01411)	(0.01410)	(0.01410)	(0.01409)	(0.01408)
320.	0.05234	0.05677	0.06120	0.06565	0.07011	0.07457
	(0.01455)	(0.01455)	(0.01454)	(0.01454)	(0.01453)	(0.01453)
340.	0.05096	0.05526	0.05957	0.06389	0.06822	0.07256
	(0.01500)	(0.01499)	(0.01499)	(0.01498)	(0.01498)	(0.01497)
360.	0.04966	0.05385	0.05804	0.06224	0.06645	0.07066
	(0.01544)	(0.01544)	(0.01543)	(0.01543)	(0.01543)	(0.01542)
380.	0.04843	0.05250	0.05659	0.06068	0.06477	0.06888
	(0.01589)	(0.01589)	(0.01588)	(0.01588)	(0.01588)	(0.01587)
400.	0.04726	0.05123	0.05521	0.05920	0.06319	0.06719
	(0.01634)	(0.01634)	(0.01634)	(0.01633)	(0.01633)	(0.01632)
420.	0.04615	0.05003	0.05391	0.05780	0.06169	0.06559
	(0.01680)	(0.01679)	(0.01679)	(0.01679)	(0.01678)	(0.01678)
440.	0.04509	0.04888	0.05267	0.05646	0.06026	0.06407
	(0.01725)	(0.01725)	(0.01724)	(0.01724)	(0.01724)	(0.01724)
460.	0.04409	0.04778	0.05149	0.05519	0.05891	0.06262
	(0.01771)	(0.01770)	(0.01770)	(0.01770)	(0.01770)	(0.01769)
480.	0.04312	0.04674	0.05036	0.05398	0.05761	0.06124
	(0.01816)	(0.01816)	(0.01816)	(0.01816)	(0.01816)	(0.01815)
500.	0.04221	0.04574	0.04928	0.05283	0.05638	0.05993
	(0.01862)	(0.01862)	(0.01862)	(0.01862)	(0.01862)	(0.01861)
520.	0.04133	0.04479	0.04826	0.05172	0.05519	0.05867
	(0.01908)	(0.01908)	(0.01908)	(0.01908)	(0.01908)	(0.01907)
540.	0.04049	0.04388	0.04727	0.05067	0.05406	0.05746
	(0.01954)	(0.01954)	(0.01954)	(0.01954)	(0.01954)	(0.01954)
560.	0.03968	0.04300	0.04633	0.04965	0.05298	0.05631
	(0.02001)	(0.02000)	(0.02000)	(0.02000)	(0.02000)	(0.02000)

Temp., °F	Pressure, psia 24.	26.	28.	30.	32.	34.
600.	0.03816 (0.02093)	0.04135 (0.02093)	0.04455 (0.02093)	0.04774 (0.02093)	0.05094 (0.02093)	0.05414 (0.02093)
650.	0.03642 (0.02209)	0.03946 (0.02209)	0.04251 (0.02209)	0.04556 (0.02209)	0.04861 (0.02209)	0.05166 (0.02208)
700.	0.03483 (0.02325)	0.03774 (0.02325)	0.04066 (0.02324)	0.04357 (0.02324)	0.04649 (0.02324)	0.04940 (0.02324)
750.	0.03338 (0.02440)	0.03617 (0.02440)	0.03896 (0.02440)	0.04175 (0.02440)	0.04454 (0.02440)	0.04733 (0.02440)
800.	0.03205 (0.02555)	0.03472 (0.02555)	0.03740 (0.02555)	0.04008 (0.02555)	0.04276 (0.02555)	0.04544 (0.02555)
850.	0.03081 (0.02670)	0.03339 (0.02670)	0.03596 (0.02670)	0.03853 (0.02670)	0.04111 (0.02670)	0.04368 (0.02670)
900.	0.02968 (0.02785)	0.03215 (0.02785)	0.03463 (0.02785)	0.03711 (0.02785)	0.03959 (0.02785)	0.04207 (0.02785)
950.	0.02862 (0.02898)	0.03101 (0.02898)	0.03339 (0.02898)	0.03578 (0.02898)	0.03817 (0.02898)	0.04056 (0.02898)
1000.	0.02763 (0.03011)	0.02994 (0.03011)	0.03224 (0.03011)	0.03455 (0.03011)	0.03686 (0.03011)	0.03916 (0.03011)
1050.	0.02671 (0.03124)	0.02894 (0.03124)	0.03117 (0.03124)	0.03340 (0.03124)	0.03563 (0.03124)	0.03786 (0.03124)
1100.	0.02585 (0.03235)	0.02801 (0.03235)	0.03017 (0.03235)	0.03232 (0.03235)	0.03448 (0.03235)	0.03664 (0.03235)
1150.	0.02505 (0.03346)	0.02714 (0.03346)	0.02923 (0.03346)	0.03132 (0.03346)	0.03341 (0.03346)	0.03550 (0.03346)
1200.	0.02429 (0.03456)	0.02632 (0.03456)	0.02834 (0.03456)	0.03037 (0.03456)	0.03240 (0.03456)	0.03442 (0.03456)
1250.	0.02358 (0.03565)	0.02555 (0.03565)	0.02751 (0.03565)	0.02948 (0.03565)	0.03145 (0.03565)	0.03341 (0.03565)
1300.	0.02291 (0.03673)	0.02482 (0.03673)	0.02673 (0.03673)	0.02864 (0.03673)	0.03055 (0.03673)	0.03246 (0.03673)
1350.	0.02227 (0.03780)	0.02413 (0.03780)	0.02599 (0.03780)	0.02785 (0.03780)	0.02970 (0.03780)	0.03156 (0.03780)
1400.	0.02167 (0.03886)	0.02348 (0.03886)	0.02529 (0.03886)	0.02709 (0.03887)	0.02890 (0.03887)	0.03071 (0.03887)
1450.	0.02111 (0.03992)	0.02286 (0.03992)	0.02462 (0.03992)	0.02638 (0.03992)	0.02814 (0.03992)	0.02990 (0.03992)
1500.	0.02057 (0.04096)	0.02228 (0.04096)	0.02399 (0.04096)	0.02571 (0.04096)	0.02742 (0.04096)	0.02914 (0.04097)

†Absolute viscosity μ_{cP} is shown in parentheses.

SOURCE: Derived from ASME, 1977.

Table G.1 Density (in Pounds-Mass per Cubic Foot) and
Absolute Viscosity μ_{cP} (in Centipoises) of Steam and
Compressed Water† (*continued*)

	Pressure, psia					
	36.	**38.**	**40.**	**42.**	**44.**	**46.**
Temp., °F	260.95	264.17	267.25	270.21	273.06	275.80
Sat. vap.	0.08631	0.09080	0.09527	0.09973	0.10418	0.10861
	(0.01322)	(0.01328)	(0.01334)	(0.01340)	(0.01345)	(0.01351)
Sat. liq.	58.49021	58.39647	58.30586	58.21816	58.13312	58.05053
	(0.21779)	(0.21450)	(0.21143)	(0.20856)	(0.20587)	(0.20333)
Temp., °F						
220.	59.61475	59.61517	59.61559	59.61602	59.61644	59.61686
	(0.26946)	(0.26946)	(0.26947)	(0.26947)	(0.26948)	(0.26948)
240.	59.08287	59.08330	59.08374	59.08418	59.08462	59.08505
	(0.24174)	(0.24174)	(0.24175)	(0.24175)	(0.24176)	(0.24176)
260.	58.51786	58.51832	58.51878	58.51923	58.51969	58.52015
	(0.21878)	(0.21878)	(0.21879)	(0.21879)	(0.21879)	(0.21880)
280.	0.08380	0.08858	0.09339	0.09821	0.10304	0.10789
	(0.01364)	(0.01363)	(0.01362)	(0.01362)	(0.01361)	(0.01360)
300.	0.08134	0.08597	0.09062	0.09527	0.09994	0.10463
	(0.01408)	(0.01407)	(0.01407)	(0.01406)	(0.01405)	(0.01405)
320.	0.07905	0.08354	0.08804	0.09254	0.09706	0.10160
	(0.01452)	(0.01452)	(0.01451)	(0.01450)	(0.01450)	(0.01449)
340.	0.07690	0.08126	0.08562	0.09000	0.09438	0.09877
	(0.01497)	(0.01496)	(0.01496)	(0.01495)	(0.01495)	(0.01494)
360.	0.07489	0.07912	0.08336	0.08761	0.09186	0.09613
	(0.01542)	(0.01541)	(0.01541)	(0.01540)	(0.01540)	(0.01539)
380.	0.07299	0.07710	0.08123	0.08536	0.08950	0.09364
	(0.01587)	(0.01586)	(0.01586)	(0.01586)	(0.01585)	(0.01585)
400.	0.07119	0.07520	0.07922	0.08324	0.08727	0.09130
	(0.01632)	(0.01632)	(0.01631)	(0.01631)	(0.01631)	(0.01630)
420.	0.06949	0.07340	0.07731	0.08123	0.08516	0.08909
	(0.01678)	(0.01677)	(0.01677)	(0.01677)	(0.01676)	(0.01676)
440.	0.06788	0.07169	0.07551	0.07933	0.08316	0.08699
	(0.01723)	(0.01723)	(0.01723)	(0.01722)	(0.01722)	(0.01722)
460.	0.06634	0.07006	0.07379	0.07752	0.08126	0.08500
	(0.01769)	(0.01769)	(0.01769)	(0.01768)	(0.01768)	(0.01768)
480.	0.06488	0.06851	0.07216	0.07580	0.07945	0.08310
	(0.01815)	(0.01815)	(0.01815)	(0.01814)	(0.01814)	(0.01814)
500.	0.06348	0.06704	0.07060	0.07416	0.07773	0.08130
	(0.01861)	(0.01861)	(0.01861)	(0.01860)	(0.01860)	(0.01860)
520.	0.06215	0.06563	0.06911	0.07259	0.07608	0.07957
	(0.01907)	(0.01907)	(0.01907)	(0.01907)	(0.01906)	(0.01906)
540.	0.06087	0.06427	0.06768	0.07109	0.07451	0.07793
	(0.01953)	(0.01953)	(0.01953)	(0.01953)	(0.01953)	(0.01953)
560.	0.05964	0.06298	0.06632	0.06966	0.07300	0.07635
	(0.02000)	(0.02000)	(0.01999)	(0.01999)	(0.01999)	(0.01999)

	Pressure, psia					
	36.	38.	40.	42.	44.	46.
Temp., °F						
600.	0.05734 (0.02092)	0.06055 (0.02092)	0.06376 (0.02092)	0.06696 (0.02092)	0.07017 (0.02092)	0.07339 (0.02092)
650.	0.05471 (0.02208)	0.05777 (0.02208)	0.06082 (0.02208)	0.06388 (0.02208)	0.06694 (0.02208)	0.07000 (0.02208)
700.	0.05232 (0.02324)	0.05524 (0.02324)	0.05816 (0.02324)	0.06108 (0.02324)	0.06400 (0.02324)	0.06692 (0.02324)
750.	0.05013 (0.02440)	0.05292 (0.02440)	0.05572 (0.02440)	0.05852 (0.02440)	0.06131 (0.02440)	0.06411 (0.02440)
800.	0.04812 (0.02555)	0.05080 (0.02555)	0.05348 (0.02555)	0.05616 (0.02555)	0.05885 (0.02555)	0.06153 (0.02555)
850.	0.04626 (0.02670)	0.04884 (0.02670)	0.05142 (0.02670)	0.05399 (0.02670)	0.05657 (0.02670)	0.05915 (0.02670)
900.	0.04455 (0.02785)	0.04703 (0.02785)	0.04951 (0.02785)	0.05199 (0.02785)	0.05447 (0.02785)	0.05695 (0.02785)
950.	0.04295 (0.02898)	0.04534 (0.02898)	0.04774 (0.02898)	0.05013 (0.02898)	0.05252 (0.02898)	0.05491 (0.02898)
1000.	0.04147 (0.03011)	0.04378 (0.03011)	0.04609 (0.03011)	0.04840 (0.03011)	0.05071 (0.03012)	0.05302 (0.03012)
1050.	0.04009 (0.03124)	0.04232 (0.03124)	0.04455 (0.03124)	0.04678 (0.03124)	0.04901 (0.03124)	0.05125 (0.03124)
1100.	0.03880 (0.03235)	0.04096 (0.03235)	0.04311 (0.03235)	0.04527 (0.03235)	0.04743 (0.03235)	0.04959 (0.03236)
1150.	0.03759 (0.03346)	0.03968 (0.03346)	0.04177 (0.03346)	0.04386 (0.03346)	0.04595 (0.03346)	0.04804 (0.03346)
1200.	0.03645 (0.03456)	0.03848 (0.03456)	0.04050 (0.03456)	0.04253 (0.03456)	0.04456 (0.03456)	0.04659 (0.03456)
1250.	0.03538 (0.03565)	0.03735 (0.03565)	0.03931 (0.03565)	0.04128 (0.03565)	0.04325 (0.03565)	0.04522 (0.03565)
1300.	0.03437 (0.03673)	0.03628 (0.03673)	0.03819 (0.03673)	0.04010 (0.03673)	0.04202 (0.03673)	0.04393 (0.03673)
1350.	0.03342 (0.03780)	0.03528 (0.03780)	0.03713 (0.03781)	0.03899 (0.03781)	0.04085 (0.03781)	0.04271 (0.03781)
1400.	0.03252 (0.03887)	0.03432 (0.03887)	0.03613 (0.03887)	0.03794 (0.03887)	0.03975 (0.03887)	0.04156 (0.03887)
1450.	0.03166 (0.03992)	0.03342 (0.03992)	0.03518 (0.03992)	0.03694 (0.03992)	0.03870 (0.03992)	0.04046 (0.03992)
1500.	0.03085 (0.04097)	0.03257 (0.04097)	0.03428 (0.04097)	0.03600 (0.04097)	0.03771 (0.04097)	0.03943 (0.04097)

†Absolute viscosity μ_{cP} is shown in parentheses.

SOURCE: Derived from ASME, 1977.

Table G.1 Density (in Pounds-Mass per Cubic Foot) and Absolute Viscosity μ_{cP} (in Centipoises) of Steam and Compressed Water† (*continued*)

	Pressure, psia					
	48.	**50.**	**52.**	**54.**	**56.**	**58.**
Temp., °F	278.45	281.02	283.50	285.90	288.24	290.51
Sat. vap.	0.11304	0.11745	0.12186	0.12626	0.13065	0.13503
	(0.01356)	(0.01361)	(0.01366)	(0.01370)	(0.01375)	(0.01379)
Sat. liq.	57.97024	57.89207	57.81589	57.74157	57.66899	57.59805
	(0.20094)	(0.19868)	(0.19654)	(0.19451)	(0.19257)	(0.19072)
Temp., °F						
220.	59.61728	59.61770	59.61812	59.61855	59.61897	59.61939
	(0.26948)	(0.26949)	(0.26949)	(0.26949)	(0.26950)	(0.26950)
240.	59.08549	59.08593	59.08637	59.08681	59.08724	59.08768
	(0.24176)	(0.24177)	(0.24177)	(0.24177)	(0.24178)	(0.24178)
260.	58.52060	58.52106	58.52152	58.52197	58.52243	58.52288
	(0.21880)	(0.21880)	(0.21881)	(0.21881)	(0.21882)	(0.21882)
280.	0.11276	57.92332	57.92380	57.92428	57.92476	57.92523
	(0.01359)	(0.19958)	(0.19958)	(0.19958)	(0.19959)	(0.19959)
300.	0.10932	0.11403	0.11876	0.12350	0.12825	0.13302
	(0.01404)	(0.01403)	(0.01403)	(0.01402)	(0.01401)	(0.01401)
320.	0.10614	0.11069	0.11526	0.11984	0.12443	0.12903
	(0.01449)	(0.01448)	(0.01448)	(0.01447)	(0.01447)	(0.01446)
340.	0.10317	0.10758	0.11200	0.11644	0.12088	0.12533
	(0.01494)	(0.01493)	(0.01493)	(0.01492)	(0.01492)	(0.01491)
360.	0.10040	0.10468	0.10897	0.11326	0.11757	0.12188
	(0.01539)	(0.01539)	(0.01538)	(0.01538)	(0.01537)	(0.01537)
380.	0.09779	0.10195	0.10612	0.11029	0.11447	0.11866
	(0.01584)	(0.01584)	(0.01584)	(0.01583)	(0.01583)	(0.01582)
400.	0.09534	0.09938	0.10344	0.10749	0.11156	0.11563
	(0.01630)	(0.01630)	(0.01629)	(0.01629)	(0.01629)	(0.01628)
420.	0.09302	0.09696	0.10091	0.10486	0.10881	0.11277
	(0.01676)	(0.01675)	(0.01675)	(0.01675)	(0.01674)	(0.01674)
440.	0.09082	0.09467	0.09851	0.10236	0.10622	0.11008
	(0.01722)	(0.01721)	(0.01721)	(0.01721)	(0.01720)	(0.01720)
460.	0.08874	0.09249	0.09624	0.10000	0.10376	0.10752
	(0.01768)	(0.01767)	(0.01767)	(0.01767)	(0.01767)	(0.01766)
480.	0.08676	0.09042	0.09408	0.09775	0.10142	0.10510
	(0.01814)	(0.01813)	(0.01813)	(0.01813)	(0.01813)	(0.01812)
500.	0.08487	0.08845	0.09203	0.09561	0.09920	0.10279
	(0.01860)	(0.01860)	(0.01859)	(0.01859)	(0.01859)	(0.01859)
520.	0.08307	0.08657	0.09007	0.09357	0.09707	0.10058
	(0.01906)	(0.01906)	(0.01906)	(0.01906)	(0.01905)	(0.01905)
540.	0.08134	0.08477	0.08819	0.09162	0.09505	0.09848
	(0.01952)	(0.01952)	(0.01952)	(0.01952)	(0.01952)	(0.01952)
560.	0.07970	0.08305	0.08640	0.08975	0.09311	0.09647
	(0.01999)	(0.01999)	(0.01999)	(0.01998)	(0.01998)	(0.01998)

			Pressure, psia			
	48.	**50.**	**52.**	**54.**	**56.**	**58.**
Temp., °F						
600.	0.07660 (0.02092)	0.07982 (0.02092)	0.08303 (0.02091)	0.08625 (0.02091)	0.08948 (0.02091)	0.09270 (0.02091)
650.	0.07306 (0.02208)	0.07613 (0.02208)	0.07919 (0.02208)	0.08226 (0.02208)	0.08533 (0.02207)	0.08840 (0.02207)
700.	0.06985 (0.02324)	0.07278 (0.02324)	0.07570 (0.02324)	0.07863 (0.02324)	0.08156 (0.02324)	0.08449 (0.02324)
750.	0.06691 (0.02440)	0.06971 (0.02440)	0.07252 (0.02440)	0.07532 (0.02440)	0.07812 (0.02440)	0.08093 (0.02439)
800.	0.06422 (0.02555)	0.06690 (0.02555)	0.06959 (0.02555)	0.07228 (0.02555)	0.07497 (0.02555)	0.07766 (0.02555)
850.	0.06173 (0.02670)	0.06431 (0.02670)	0.06690 (0.02670)	0.06948 (0.02670)	0.07206 (0.02670)	0.07465 (0.02670)
900.	0.05944 (0.02785)	0.06192 (0.02785)	0.06441 (0.02785)	0.06689 (0.02785)	0.06938 (0.02785)	0.07186 (0.02785)
950.	0.05731 (0.02898)	0.05970 (0.02898)	0.06210 (0.02898)	0.06449 (0.02898)	0.06689 (0.02899)	0.06928 (0.02899)
1000.	0.05533 (0.03012)	0.05764 (0.03012)	0.05995 (0.03012)	0.06226 (0.03012)	0.06457 (0.03012)	0.06688 (0.03012)
1050.	0.05348 (0.03124)	0.05571 (0.03124)	0.05794 (0.03124)	0.06018 (0.03124)	0.06241 (0.03124)	0.06465 (0.03124)
1100.	0.05175 (0.03236)	0.05391 (0.03236)	0.05607 (0.03236)	0.05823 (0.03236)	0.06039 (0.03236)	0.06256 (0.03236)
1150.	0.05013 (0.03346)	0.05223 (0.03346)	0.05432 (0.03346)	0.05641 (0.03347)	0.05850 (0.03347)	0.06060 (0.03347)
1200.	0.04862 (0.03456)	0.05064 (0.03456)	0.05267 (0.03456)	0.05470 (0.03456)	0.05673 (0.03457)	0.05876 (0.03457)
1250.	0.04719 (0.03565)	0.04915 (0.03565)	0.05112 (0.03565)	0.05309 (0.03566)	0.05506 (0.03566)	0.05703 (0.03566)
1300.	0.04584 (0.03674)	0.04775 (0.03674)	0.04966 (0.03674)	0.05158 (0.03674)	0.05349 (0.03674)	0.05540 (0.03674)
1350.	0.04457 (0.03781)	0.04643 (0.03781)	0.04828 (0.03781)	0.05014 (0.03781)	0.05200 (0.03781)	0.05386 (0.03781)
1400.	0.04336 (0.03887)	0.04517 (0.03887)	0.04698 (0.03887)	0.04879 (0.03887)	0.05060 (0.03887)	0.05241 (0.03887)
1450.	0.04223 (0.03993)	0.04399 (0.03993)	0.04575 (0.03993)	0.04751 (0.03993)	0.04927 (0.03993)	0.05103 (0.03993)
1500.	0.04114 (0.04097)	0.04286 (0.04097)	0.04458 (0.04097)	0.04629 (0.04097)	0.04801 (0.04097)	0.04972 (0.04097)

†Absolute viscosity μ_{cP} is shown in parentheses.

SOURCE: Derived from ASME, 1977.

	Pressure, psia					
	60.	**62.**	**64.**	**66.**	**68.**	**70.**
Temp., °F	292.71	294.86	296.95	298.99	300.99	302.93
Sat. vap.	0.13940	0.14377	0.14812	0.15248	0.15682	0.16116
	(0.01384)	(0.01388)	(0.01392)	(0.01396)	(0.01400)	(0.01404)
Sat. liq.	57.52867	57.46073	57.39418	57.32895	57.26494	57.20212
	(0.18896)	(0.18727)	(0.18566)	(0.18411)	(0.18262)	(0.18119)
Temp., °F						
220.	59.61981	59.62023	59.62065	59.62107	59.62150	59.62192
	(0.26951)	(0.26951)	(0.26951)	(0.26952)	(0.26952)	(0.26952)
240.	59.08812	59.08856	59.08899	59.08943	59.08987	59.09031
	(0.24179)	(0.24179)	(0.24179)	(0.24180)	(0.24180)	(0.24180)
260.	58.52334	58.52380	58.52425	58.52471	58.52517	58.52562
	(0.21882)	(0.21883)	(0.21883)	(0.21883)	(0.21884)	(0.21884)
280.	57.92571	57.92619	57.92667	57.92715	57.92762	57.92810
	(0.19959)	(0.19960)	(0.19960)	(0.19961)	(0.19961)	(0.19961)
300.	0.13780	0.14259	0.14740	0.15223	57.29690	57.29740
	(0.01400)	(0.01400)	(0.01399)	(0.01398)	(0.18336)	(0.18336)
320.	0.13364	0.13826	0.14290	0.14755	0.15221	0.15689
	(0.01445)	(0.01445)	(0.01444)	(0.01444)	(0.01443)	(0.01443)
340.	0.12979	0.13426	0.13874	0.14323	0.14774	0.15225
	(0.01491)	(0.01490)	(0.01490)	(0.01489)	(0.01489)	(0.01488)
360.	0.12620	0.13053	0.13487	0.13922	0.14358	0.14795
	(0.01536)	(0.01536)	(0.01535)	(0.01535)	(0.01535)	(0.01534)
380.	0.12285	0.12705	0.13126	0.13548	0.13971	0.14394
	(0.01582)	(0.01582)	(0.01581)	(0.01581)	(0.01580)	(0.01580)
400.	0.11971	0.12379	0.12788	0.13197	0.13608	0.14019
	(0.01628)	(0.01627)	(0.01627)	(0.01627)	(0.01626)	(0.01626)
420.	0.11674	0.12071	0.12469	0.12868	0.13267	0.13666
	(0.01674)	(0.01673)	(0.01673)	(0.01673)	(0.01672)	(0.01672)
440.	0.11394	0.11781	0.12169	0.12557	0.12945	0.13334
	(0.01720)	(0.01720)	(0.01719)	(0.01719)	(0.01719)	(0.01718)
460.	0.11129	0.11506	0.11884	0.12262	0.12641	0.13020
	(0.01766)	(0.01766)	(0.01765)	(0.01765)	(0.01765)	(0.01765)
480.	0.10877	0.11246	0.11614	0.11983	0.12353	0.12722
	(0.01812)	(0.01812)	(0.01812)	(0.01812)	(0.01811)	(0.01811)
500.	0.10638	0.10998	0.11358	0.11718	0.12079	0.12440
	(0.01859)	(0.01858)	(0.01858)	(0.01858)	(0.01858)	(0.01858)
520.	0.10410	0.10761	0.11113	0.11465	0.11818	0.12170
	(0.01905)	(0.01905)	(0.01905)	(0.01904)	(0.01904)	(0.01904)
540.	0.10192	0.10536	0.10880	0.11224	0.11569	0.11914
	(0.01951)	(0.01951)	(0.01951)	(0.01951)	(0.01951)	(0.01951)
560.	0.09983	0.10320	0.10657	0.10994	0.11331	0.11668
	(0.01998)	(0.01998)	(0.01998)	(0.01997)	(0.01997)	(0.01997)

	Pressure, psia					
	60.	**62.**	**64.**	**66.**	**68.**	**70.**
Temp., °F						
600.	0.09593 (0.02091)	0.09915 (0.02091)	0.10239 (0.02091)	0.10562 (0.02091)	0.10885 (0.02090)	0.11209 (0.02090)
650.	0.09147 (0.02207)	0.09455 (0.02207)	0.09762 (0.02207)	0.10070 (0.02207)	0.10378 (0.02207)	0.10686 (0.02207)
700.	0.08743 (0.02323)	0.09036 (0.02323)	0.09330 (0.02323)	0.09623 (0.02323)	0.09917 (0.02323)	0.10211 (0.02323)
750.	0.08373 (0.02439)	0.08654 (0.02439)	0.08935 (0.02439)	0.09216 (0.02439)	0.09497 (0.02439)	0.09778 (0.02439)
800.	0.08035 (0.02555)	0.08304 (0.02555)	0.08573 (0.02555)	0.08843 (0.02555)	0.09112 (0.02555)	0.09382 (0.02555)
850.	0.07723 (0.02670)	0.07982 (0.02670)	0.08240 (0.02670)	0.08499 (0.02670)	0.08758 (0.02670)	0.09017 (0.02670)
900.	0.07435 (0.02785)	0.07684 (0.02785)	0.07933 (0.02785)	0.08181 (0.02785)	0.08430 (0.02785)	0.08679 (0.02785)
950.	0.07168 (0.02899)	0.07408 (0.02899)	0.07647 (0.02899)	0.07887 (0.02899)	0.08127 (0.02899)	0.08367 (0.02899)
1000.	0.06920 (0.03012)	0.07151 (0.03012)	0.07382 (0.03012)	0.07614 (0.03012)	0.07845 (0.03012)	0.08077 (0.03012)
1050.	0.06688 (0.03124)	0.06912 (0.03124)	0.07135 (0.03124)	0.07359 (0.03124)	0.07582 (0.03124)	0.07806 (0.03124)
1100.	0.06472 (0.03236)	0.06688 (0.03236)	0.06904 (0.03236)	0.07120 (0.03236)	0.07337 (0.03236)	0.07553 (0.03236)
1150.	0.06269 (0.03347)	0.06478 (0.03347)	0.06688 (0.03347)	0.06897 (0.03347)	0.07107 (0.03347)	0.07316 (0.03347)
1200.	0.06079 (0.03457)	0.06282 (0.03457)	0.06485 (0.03457)	0.06688 (0.03457)	0.06891 (0.03457)	0.07094 (0.03457)
1250.	0.05900 (0.03566)	0.06097 (0.03566)	0.06294 (0.03566)	0.06491 (0.03566)	0.06688 (0.03566)	0.06885 (0.03566)
1300.	0.05731 (0.03674)	0.05923 (0.03674)	0.06114 (0.03674)	0.06305 (0.03674)	0.06497 (0.03674)	0.06688 (0.03674)
1350.	0.05572 (0.03781)	0.05758 (0.03781)	0.05944 (0.03781)	0.06130 (0.03781)	0.06316 (0.03781)	0.06502 (0.03781)
1400.	0.05422 (0.03887)	0.05603 (0.03888)	0.05783 (0.03888)	0.05964 (0.03888)	0.06145 (0.03888)	0.06326 (0.03888)
1450.	0.05279 (0.03993)	0.05455 (0.03993)	0.05631 (0.03993)	0.05807 (0.03993)	0.05984 (0.03993)	0.06160 (0.03993)
1500.	0.05144 (0.04097)	0.05315 (0.04097)	0.05487 (0.04098)	0.05659 (0.04098)	0.05830 (0.04098)	0.06002 (0.04098)

†Absolute viscosity μ_{cP} is shown in parentheses.

SOURCE: Derived from ASME, 1977.

	Pressure, psia					
	72.	**74.**	**76.**	**78.**	**80.**	**82.**
Temp., °F						
Sat. vap.	304.83	306.69	308.51	310.29	312.04	313.75
	0.16549	0.16982	0.17415	0.17846	0.18278	0.18709
	(0.01407)	(0.01411)	(0.01415)	(0.01418)	(0.01421)	(0.01425)
Sat. liq.	57.14044	57.07982	57.02023	56.96161	56.90393	56.84715
	(0.17981)	(0.17849)	(0.17721)	(0.17597)	(0.17478)	(0.17362)
Temp., °F						
220.	59.62234	59.62276	59.62318	59.62360	59.62402	59.62444
	(0.26953)	(0.26953)	(0.26954)	(0.26954)	(0.26954)	(0.26955)
240.	59.09074	59.09118	59.09162	59.09206	59.09249	59.09293
	(0.24181)	(0.24181)	(0.24182)	(0.24182)	(0.24182)	(0.24183)
260.	58.52608	58.52653	58.52699	58.52745	58.52790	58.52836
	(0.21884)	(0.21885)	(0.21885)	(0.21886)	(0.21886)	(0.21886)
280.	57.92858	57.92906	57.92953	57.93001	57.93049	57.93097
	(0.19962)	(0.19962)	(0.19962)	(0.19963)	(0.19963)	(0.19964)
300.	57.29790	57.29840	57.29891	57.29941	57.29991	57.30041
	(0.18336)	(0.18337)	(0.18337)	(0.18337)	(0.18338)	(0.18338)
320.	0.16158	0.16628	0.17099	0.17572	0.18046	0.18522
	(0.01442)	(0.01441)	(0.01441)	(0.01440)	(0.01440)	(0.01439)
340.	0.15677	0.16131	0.16585	0.17041	0.17498	0.17956
	(0.01488)	(0.01487)	(0.01487)	(0.01486)	(0.01486)	(0.01485)
360.	0.15232	0.15671	0.16110	0.16550	0.16991	0.17434
	(0.01534)	(0.01533)	(0.01533)	(0.01532)	(0.01532)	(0.01531)
380.	0.14818	0.15243	0.15668	0.16094	0.16522	0.16950
	(0.01580)	(0.01579)	(0.01579)	(0.01578)	(0.01578)	(0.01578)
400.	0.14430	0.14842	0.15255	0.15669	0.16083	0.16498
	(0.01626)	(0.01625)	(0.01625)	(0.01625)	(0.01624)	(0.01624)
420.	0.14066	0.14467	0.14868	0.15270	0.15673	0.16076
	(0.01672)	(0.01671)	(0.01671)	(0.01671)	(0.01671)	(0.01670)
440.	0.13723	0.14113	0.14504	0.14895	0.15286	0.15678
	(0.01718)	(0.01718)	(0.01718)	(0.01717)	(0.01717)	(0.01717)
460.	0.13399	0.13779	0.14160	0.14541	0.14922	0.15304
	(0.01764)	(0.01764)	(0.01764)	(0.01764)	(0.01763)	(0.01763)
480.	0.13092	0.13463	0.13834	0.14205	0.14577	0.14949
	(0.01811)	(0.01811)	(0.01810)	(0.01810)	(0.01810)	(0.01810)
500.	0.12801	0.13163	0.13525	0.13887	0.14250	0.14613
	(0.01857)	(0.01857)	(0.01857)	(0.01857)	(0.01856)	(0.01856)
520.	0.12523	0.12877	0.13230	0.13584	0.13939	0.14293
	(0.01904)	(0.01904)	(0.01903)	(0.01903)	(0.01903)	(0.01903)
540.	0.12259	0.12604	0.12950	0.13296	0.13642	0.13989
	(0.01950)	(0.01950)	(0.01950)	(0.01950)	(0.01950)	(0.01950)
560.	0.12006	0.12344	0.12682	0.13021	0.13359	0.13698
	(0.01997)	(0.01997)	(0.01997)	(0.01997)	(0.01996)	(0.01996)

			Pressure, psia			
	72.	74.	76.	78.	80.	82.
Temp., °F						
600.	0.11533	0.11857	0.12181	0.12505	0.12830	0.13155
	(0.02090)	(0.02090)	(0.02090)	(0.02090)	(0.02090)	(0.02090)
650.	0.10994	0.11302	0.11611	0.11919	0.12228	0.12537
	(0.02207)	(0.02207)	(0.02207)	(0.02206)	(0.02206)	(0.02206)
700.	0.10505	0.10799	0.11094	0.11388	0.11683	0.11977
	(0.02323)	(0.02323)	(0.02323)	(0.02323)	(0.02323)	(0.02323)
750.	0.10059	0.10341	0.10622	0.10904	0.11186	0.11467
	(0.02439)	(0.02439)	(0.02439)	(0.02439)	(0.02439)	(0.02439)
800.	0.09651	0.09921	0.10191	0.10460	0.10730	0.11000
	(0.02555)	(0.02555)	(0.02555)	(0.02555)	(0.02555)	(0.02555)
850.	0.09275	0.09534	0.09793	0.10053	0.10312	0.10571
	(0.02670)	(0.02670)	(0.02670)	(0.02670)	(0.02670)	(0.02670)
900.	0.08928	0.09178	0.09427	0.09676	0.09925	0.10175
	(0.02785)	(0.02785)	(0.02785)	(0.02785)	(0.02785)	(0.02785)
950.	0.08607	0.08847	0.09087	0.09327	0.09567	0.09807
	(0.02899)	(0.02899)	(0.02899)	(0.02899)	(0.02899)	(0.02899)
1000.	0.08308	0.08540	0.08771	0.09003	0.09235	0.09466
	(0.03012)	(0.03012)	(0.03012)	(0.03012)	(0.03012)	(0.03012)
1050.	0.08030	0.08253	0.08477	0.08701	0.08925	0.09148
	(0.03124)	(0.03124)	(0.03124)	(0.03125)	(0.03125)	(0.03125)
1100.	0.07769	0.07986	0.08202	0.08419	0.08635	0.08852
	(0.03236)	(0.03236)	(0.03236)	(0.03236)	(0.03236)	(0.03236)
1150.	0.07526	0.07735	0.07945	0.08154	0.08364	0.08574
	(0.03347)	(0.03347)	(0.03347)	(0.03347)	(0.03347)	(0.03347)
1200.	0.07297	0.07500	0.07703	0.07906	0.08110	0.08313
	(0.03457)	(0.03457)	(0.03457)	(0.03457)	(0.03457)	(0.03457)
1250.	0.07082	0.07279	0.07476	0.07673	0.07870	0.08067
	(0.03566)	(0.03566)	(0.03566)	(0.03566)	(0.03566)	(0.03566)
1300.	0.06879	0.07071	0.07262	0.07454	0.07645	0.07836
	(0.03674)	(0.03674)	(0.03674)	(0.03674)	(0.03674)	(0.03675)
1350.	0.06688	0.06874	0.07060	0.07246	0.07432	0.07618
	(0.03782)	(0.03782)	(0.03782)	(0.03782)	(0.03782)	(0.03782)
1400.	0.06507	0.06688	0.06869	0.07050	0.07231	0.07412
	(0.03888)	(0.03888)	(0.03888)	(0.03888)	(0.03888)	(0.03888)
1450.	0.06336	0.06512	0.06688	0.06865	0.07041	0.07217
	(0.03993)	(0.03993)	(0.03993)	(0.03994)	(0.03994)	(0.03994)
1500.	0.06174	0.06345	0.06517	0.06689	0.06860	0.07032
	(0.04098)	(0.04098)	(0.04098)	(0.04098)	(0.04098)	(0.04098)

†Absolute viscosity μ_{cP} is shown in parentheses.

SOURCE: Derived from ASME, 1977.

Table G.1 Density (in Pounds-Mass per Cubic Foot) and Absolute Viscosity μ_{cP} (in Centipoises) of Steam and Compressed Water† (*continued*)

	Pressure, psia					
	84.	86.	88.	90.	92.	94.
Temp., °F	315.43	317.08	318.69	320.28	321.84	323.37
Sat. vap.	0.19139	0.19569	0.19998	0.20428	0.20856	0.21285
	(0.01428)	(0.01431)	(0.01434)	(0.01438)	(0.01441)	(0.01444)
Sat. liq.	56.79123	56.73613	56.68184	56.62830	56.57550	56.52342
	(0.17251)	(0.17142)	(0.17038)	(0.16936)	(0.16837)	(0.16741)
Temp., °F						
220.	59.62487	59.62529	59.62571	59.62613	59.62655	59.62697
	(0.26955)	(0.26955)	(0.26956)	(0.26956)	(0.26957)	(0.26957)
240.	59.09337	59.09380	59.09424	59.09468	59.09512	59.09555
	(0.24183)	(0.24183)	(0.24184)	(0.24184)	(0.24185)	(0.24185)
260.	58.52881	58.52927	58.52973	58.53018	58.53064	58.53109
	(0.21887)	(0.21887)	(0.21887)	(0.21888)	(0.21888)	(0.21889)
280.	57.93144	57.93192	57.93240	57.93288	57.93335	57.93383
	(0.19964)	(0.19964)	(0.19965)	(0.19965)	(0.19965)	(0.19966)
300.	57.30091	57.30142	57.30192	57.30242	57.30292	57.30343
	(0.18339)	(0.18339)	(0.18339)	(0.18340)	(0.18340)	(0.18340)
320.	0.18998	0.19476	0.19956	56.63792	56.63845	56.63898
	(0.01439)	(0.01438)	(0.01437)	(0.16954)	(0.16954)	(0.16955)
340.	0.18415	0.18875	0.19336	0.19799	0.20262	0.20727
	(0.01485)	(0.01484)	(0.01484)	(0.01483)	(0.01483)	(0.01482)
360.	0.17877	0.18321	0.18766	0.19212	0.19659	0.20107
	(0.01531)	(0.01530)	(0.01530)	(0.01530)	(0.01529)	(0.01529)
380.	0.17378	0.17808	0.18238	0.18669	0.19101	0.19534
	(0.01577)	(0.01577)	(0.01576)	(0.01576)	(0.01576)	(0.01575)
400.	0.16914	0.17330	0.17747	0.18165	0.18583	0.19003
	(0.01623)	(0.01623)	(0.01623)	(0.01622)	(0.01622)	(0.01622)
420.	0.16479	0.16883	0.17288	0.17694	0.18100	0.18506
	(0.01670)	(0.01670)	(0.01669)	(0.01669)	(0.01669)	(0.01668)
440.	0.16071	0.16464	0.16857	0.17251	0.17646	0.18041
	(0.01716)	(0.01716)	(0.01716)	(0.01715)	(0.01715)	(0.01715)
460.	0.15686	0.16068	0.16451	0.16835	0.17219	0.17603
	(0.01763)	(0.01763)	(0.01762)	(0.01762)	(0.01762)	(0.01762)
480.	0.15321	0.15694	0.16068	0.16441	0.16815	0.17190
	(0.01809)	(0.01809)	(0.01809)	(0.01809)	(0.01809)	(0.01808)
500.	0.14976	0.15340	0.15704	0.16069	0.16433	0.16799
	(0.01856)	(0.01856)	(0.01856)	(0.01855)	(0.01855)	(0.01855)
520.	0.14648	0.15003	0.15359	0.15715	0.16071	0.16427
	(0.01903)	(0.01903)	(0.01902)	(0.01902)	(0.01902)	(0.01902)
540.	0.14336	0.14683	0.15030	0.15378	0.15726	0.16074
	(0.01949)	(0.01949)	(0.01949)	(0.01949)	(0.01949)	(0.01949)
560.	0.14037	0.14377	0.14717	0.15057	0.15397	0.15737
	(0.01996)	(0.01996)	(0.01996)	(0.01996)	(0.01996)	(0.01995)

			Pressure, psia			
	84.	**86.**	**88.**	**90.**	**92.**	**94.**
Temp., °F						
600.	0.13480 (0.02090)	0.13805 (0.02089)	0.14131 (0.02089)	0.14456 (0.02089)	0.14782 (0.02089)	0.15108 (0.02089)
650.	0.12846 (0.02206)	0.13155 (0.02206)	0.13465 (0.02206)	0.13774 (0.02206)	0.14084 (0.02206)	0.14394 (0.02206)
700.	0.12272 (0.02323)	0.12567 (0.02323)	0.12862 (0.02323)	0.13157 (0.02323)	0.13453 (0.02323)	0.13748 (0.02322)
750.	0.11749 (0.02439)	0.12031 (0.02439)	0.12313 (0.02439)	0.12596 (0.02439)	0.12878 (0.02439)	0.13160 (0.02439)
800.	0.11271 (0.02555)	0.11541 (0.02555)	0.11811 (0.02555)	0.12081 (0.02555)	0.12352 (0.02555)	0.12623 (0.02555)
850.	0.10830 (0.02670)	0.11090 (0.02670)	0.11349 (0.02670)	0.11609 (0.02670)	0.11868 (0.02670)	0.12128 (0.02670)
900.	0.10424 (0.02785)	0.10674 (0.02785)	0.10923 (0.02785)	0.11173 (0.02785)	0.11422 (0.02785)	0.11672 (0.02785)
950.	0.10048 (0.02899)	0.10288 (0.02899)	0.10528 (0.02899)	0.10769 (0.02899)	0.11009 (0.02899)	0.11250 (0.02899)
1000.	0.09698 (0.03012)	0.09930 (0.03012)	0.10162 (0.03012)	0.10394 (0.03012)	0.10626 (0.03012)	0.10858 (0.03012)
1050.	0.09372 (0.03125)	0.09596 (0.03125)	0.09820 (0.03125)	0.10044 (0.03125)	0.10268 (0.03125)	0.10492 (0.03125)
1100.	0.09068 (0.03236)	0.09285 (0.03236)	0.09501 (0.03236)	0.09718 (0.03236)	0.09935 (0.03237)	0.10151 (0.03237)
1150.	0.08783 (0.03347)	0.08993 (0.03347)	0.09203 (0.03347)	0.09412 (0.03347)	0.09622 (0.03347)	0.09832 (0.03347)
1200.	0.08516 (0.03457)	0.08719 (0.03457)	0.08922 (0.03457)	0.09126 (0.03457)	0.09329 (0.03457)	0.09532 (0.03458)
1250.	0.08265 (0.03566)	0.08462 (0.03566)	0.08659 (0.03567)	0.08856 (0.03567)	0.09053 (0.03567)	0.09251 (0.03567)
1300.	0.08028 (0.03675)	0.08219 (0.03675)	0.08411 (0.03675)	0.08602 (0.03675)	0.08794 (0.03675)	0.08985 (0.03675)
1350.	0.07804 (0.03782)	0.07991 (0.03782)	0.08177 (0.03782)	0.08363 (0.03782)	0.08549 (0.03782)	0.08735 (0.03782)
1400.	0.07593 (0.03888)	0.07774 (0.03888)	0.07955 (0.03888)	0.08136 (0.03888)	0.08317 (0.03889)	0.08499 (0.03889)
1450.	0.07393 (0.03994)	0.07569 (0.03994)	0.07746 (0.03994)	0.07922 (0.03994)	0.08098 (0.03994)	0.08275 (0.03994)
1500.	0.07204 (0.04098)	0.07375 (0.04098)	0.07547 (0.04098)	0.07719 (0.04098)	0.07890 (0.04099)	0.08062 (0.04099)

†Absolute viscosity μ_{cP} is shown in parentheses.

SOURCE: Derived from ASME, 1977.

	Pressure, psia					
	96.	98.	100.	102.	104.	106.
Temp., °F	324.88	326.36	327.82	329.26	330.67	332.06
Sat. vap.	0.21713	0.22141	0.22568	0.22995	0.23422	0.23849
	(0.01446)	(0.01449)	(0.01452)	(0.01455)	(0.01458)	(0.01461)
Sat. liq.	56.47200	56.42126	56.37115	56.32164	56.27275	56.22442
	(0.16648)	(0.16557)	(0.16468)	(0.16382)	(0.16299)	(0.16217)
Temp., °F						
220.	59.62739	59.62781	59.62823	59.62865	59.62908	59.62950
	(0.26957)	(0.26958)	(0.26958)	(0.26958)	(0.26959)	(0.26959)
240.	59.09599	59.09643	59.09687	59.09730	59.09774	59.09818
	(0.24185)	(0.24186)	(0.24186)	(0.24186)	(0.24187)	(0.24187)
260.	58.53155	58.53201	58.53246	58.53292	58.53337	58.53383
	(0.21889)	(0.21889)	(0.21890)	(0.21890)	(0.21890)	(0.21891)
280.	57.93431	57.93479	57.93526	57.93574	57.93622	57.93669
	(0.19966)	(0.19966)	(0.19967)	(0.19967)	(0.19968)	(0.19968)
300.	57.30393	57.30443	57.30493	57.30543	57.30593	57.30644
	(0.18341)	(0.18341)	(0.18341)	(0.18342)	(0.18342)	(0.18343)
320.	56.63951	56.64004	56.64057	56.64111	56.64164	56.64217
	(0.16955)	(0.16955)	(0.16956)	(0.16956)	(0.16956)	(0.16957)
340.	0.21193	0.21660	0.22129	0.22599	0.23070	0.23542
	(0.01482)	(0.01481)	(0.01481)	(0.01480)	(0.01480)	(0.01479)
360.	0.20556	0.21006	0.21457	0.21909	0.22362	0.22816
	(0.01528)	(0.01528)	(0.01527)	(0.01527)	(0.01526)	(0.01526)
380.	0.19968	0.20403	0.20838	0.21274	0.21711	0.22149
	(0.01575)	(0.01574)	(0.01574)	(0.01574)	(0.01573)	(0.01573)
400.	0.19422	0.19843	0.20264	0.20686	0.21109	0.21533
	(0.01621)	(0.01621)	(0.01621)	(0.01620)	(0.01620)	(0.01620)
420.	0.18913	0.19321	0.19730	0.20139	0.20549	0.20959
	(0.01668)	(0.01668)	(0.01667)	(0.01667)	(0.01667)	(0.01666)
440.	0.18437	0.18833	0.19229	0.19627	0.20024	0.20423
	(0.01715)	(0.01714)	(0.01714)	(0.01714)	(0.01713)	(0.01713)
460.	0.17988	0.18373	0.18759	0.19146	0.19532	0.19920
	(0.01761)	(0.01761)	(0.01761)	(0.01761)	(0.01760)	(0.01760)
480.	0.17565	0.17940	0.18316	0.18692	0.19068	0.19445
	(0.01808)	(0.01808)	(0.01808)	(0.01807)	(0.01807)	(0.01807)
500.	0.17164	0.17530	0.17896	0.18263	0.18630	0.18997
	(0.01855)	(0.01855)	(0.01854)	(0.01854)	(0.01854)	(0.01854)
520.	0.16784	0.17141	0.17499	0.17856	0.18214	0.18573
	(0.01902)	(0.01901)	(0.01901)	(0.01901)	(0.01901)	(0.01901)
540.	0.16423	0.16771	0.17120	0.17470	0.17819	0.18169
	(0.01948)	(0.01948)	(0.01948)	(0.01948)	(0.01948)	(0.01948)
560.	0.16078	0.16419	0.16760	0.17102	0.17443	0.17785
	(0.01995)	(0.01995)	(0.01995)	(0.01995)	(0.01995)	(0.01994)

		Pressure, psia				
	96.	98.	100.	102.	104.	106.
Temp., °F						
600.	0.15434	0.15761	0.16088	0.16415	0.16742	0.17069
	(0.02089)	(0.02089)	(0.02089)	(0.02088)	(0.02088)	(0.02088)
650.	0.14704	0.15014	0.15325	0.15636	0.15946	0.16257
	(0.02206)	(0.02206)	(0.02206)	(0.02205)	(0.02205)	(0.02205)
700.	0.14044	0.14340	0.14636	0.14932	0.15228	0.15524
	(0.02322)	(0.02322)	(0.02322)	(0.02322)	(0.02322)	(0.02322)
750.	0.13443	0.13726	0.14008	0.14291	0.14574	0.14857
	(0.02439)	(0.02439)	(0.02439)	(0.02439)	(0.02439)	(0.02439)
800.	0.12893	0.13164	0.13435	0.13706	0.13977	0.14248
	(0.02555)	(0.02555)	(0.02555)	(0.02555)	(0.02555)	(0.02555)
850.	0.12388	0.12648	0.12908	0.13168	0.13428	0.13688
	(0.02670)	(0.02670)	(0.02670)	(0.02670)	(0.02670)	(0.02670)
900.	0.11922	0.12172	0.12422	0.12672	0.12922	0.13172
	(0.02785)	(0.02785)	(0.02785)	(0.02785)	(0.02785)	(0.02785)
950.	0.11490	0.11731	0.11972	0.12212	0.12453	0.12694
	(0.02899)	(0.02899)	(0.02899)	(0.02899)	(0.02899)	(0.02899)
1000.	0.11090	0.11322	0.11554	0.11786	0.12018	0.12250
	(0.03012)	(0.03012)	(0.03012)	(0.03012)	(0.03012)	(0.03012)
1050.	0.10716	0.10941	0.11165	0.11389	0.11613	0.11837
	(0.03125)	(0.03125)	(0.03125)	(0.03125)	(0.03125)	(0.03125)
1100.	0.10368	0.10585	0.10801	0.11018	0.11235	0.11452
	(0.03237)	(0.03237)	(0.03237)	(0.03237)	(0.03237)	(0.03237)
1150.	0.10042	0.10252	0.10461	0.10671	0.10881	0.11091
	(0.03348)	(0.03348)	(0.03348)	(0.03348)	(0.03348)	(0.03348)
1200.	0.09736	0.09939	0.10142	0.10346	0.10549	0.10753
	(0.03458)	(0.03458)	(0.03458)	(0.03458)	(0.03458)	(0.03458)
1250.	0.09448	0.09645	0.09843	0.10040	0.10237	0.10435
	(0.03567)	(0.03567)	(0.03567)	(0.03567)	(0.03567)	(0.03567)
1300.	0.09177	0.09369	0.09560	0.09752	0.09943	0.10135
	(0.03675)	(0.03675)	(0.03675)	(0.03675)	(0.03675)	(0.03675)
1350.	0.08921	0.09108	0.09294	0.09480	0.09666	0.09852
	(0.03782)	(0.03782)	(0.03782)	(0.03782)	(0.03783)	(0.03783)
1400.	0.08680	0.08861	0.09042	0.09223	0.09404	0.09585
	(0.03889)	(0.03889)	(0.03889)	(0.03889)	(0.03889)	(0.03889)
1450.	0.08451	0.08627	0.08803	0.08980	0.09156	0.09332
	(0.03994)	(0.03994)	(0.03994)	(0.03994)	(0.03994)	(0.03994)
1500.	0.08234	0.08406	0.08577	0.08749	0.08921	0.09093
	(0.04099)	(0.04099)	(0.04099)	(0.04099)	(0.04099)	(0.04099)

†Absolute viscosity μ_{cP} is shown in parentheses.

SOURCE: Derived from ASME, 1977.

Table G.1 Density (in Pounds-Mass per Cubic Foot) and
Absolute Viscosity μ_{cP} (in Centipoises) of Steam and
Compressed Water† (*continued*)

	Pressure, psia					
	108.	**112.**	**116.**	**120.**	**124.**	**128.**
Temp., °F	333.44	336.12	338.73	341.27	343.74	346.15
Sat. vap.	0.24275	0.25127	0.25977	0.26827	0.27676	0.28525
	(0.01463)	(0.01468)	(0.01474)	(0.01478)	(0.01483)	(0.01488)
Sat. liq.	56.17665	56.08272	55.99081	55.90083	55.81264	55.72618
	(0.16137)	(0.15984)	(0.15837)	(0.15698)	(0.15564)	(0.15436)
Temp., °F						
220.	59.62992	59.63076	59.63160	59.63244	59.63328	59.63413
	(0.26960)	(0.26960)	(0.26961)	(0.26962)	(0.26963)	(0.26963)
240.	59.09861	59.09949	59.10036	59.10124	59.10211	59.10298
	(0.24188)	(0.24188)	(0.24189)	(0.24190)	(0.24191)	(0.24191)
260.	58.53428	58.53520	58.53611	58.53702	58.53793	58.53884
	(0.21891)	(0.21892)	(0.21893)	(0.21893)	(0.21894)	(0.21895)
280.	57.93717	57.93813	57.93908	57.94003	57.94099	57.94194
	(0.19968)	(0.19969)	(0.19970)	(0.19971)	(0.19971)	(0.19972)
300.	57.30694	57.30794	57.30894	57.30995	57.31095	57.31195
	(0.18343)	(0.18344)	(0.18344)	(0.18345)	(0.18346)	(0.18347)
320.	56.64270	56.64376	56.64482	56.64588	56.64693	56.64799
	(0.16957)	(0.16958)	(0.16958)	(0.16959)	(0.16960)	(0.16961)
340.	0.24015	0.24966	0.25923	55.94636	55.94749	55.94861
	(0.01479)	(0.01478)	(0.01477)	(0.15767)	(0.15768)	(0.15769)
360.	0.23271	0.24185	0.25103	0.26025	0.26952	0.27883
	(0.01525)	(0.01525)	(0.01524)	(0.01523)	(0.01522)	(0.01521)
380.	0.22588	0.23469	0.24353	0.25241	0.26132	0.27027
	(0.01572)	(0.01571)	(0.01571)	(0.01570)	(0.01569)	(0.01568)
400.	0.21957	0.22808	0.23661	0.24518	0.25378	0.26242
	(0.01619)	(0.01618)	(0.01618)	(0.01617)	(0.01616)	(0.01616)
420.	0.21370	0.22194	0.23020	0.23849	0.24681	0.25516
	(0.01666)	(0.01665)	(0.01665)	(0.01664)	(0.01663)	(0.01663)
440.	0.20822	0.21621	0.22423	0.23227	0.24033	0.24841
	(0.01713)	(0.01712)	(0.01712)	(0.01711)	(0.01711)	(0.01710)
460.	0.20307	0.21084	0.21863	0.22644	0.23426	0.24211
	(0.01760)	(0.01759)	(0.01759)	(0.01758)	(0.01758)	(0.01757)
480.	0.19823	0.20579	0.21336	0.22096	0.22857	0.23620
	(0.01807)	(0.01806)	(0.01806)	(0.01805)	(0.01805)	(0.01804)
500.	0.19365	0.20102	0.20840	0.21579	0.22320	0.23063
	(0.01854)	(0.01853)	(0.01853)	(0.01852)	(0.01852)	(0.01851)
520.	0.18932	0.19650	0.20370	0.21091	0.21813	0.22537
	(0.01900)	(0.01900)	(0.01900)	(0.01899)	(0.01899)	(0.01899)
540.	0.18520	0.19221	0.19924	0.20627	0.21332	0.22038
	(0.01947)	(0.01947)	(0.01947)	(0.01946)	(0.01946)	(0.01946)
560.	0.18128	0.18813	0.19499	0.20187	0.20875	0.21565
	(0.01994)	(0.01994)	(0.01994)	(0.01993)	(0.01993)	(0.01993)

	Pressure, psia					
	108.	**112.**	**116.**	**120.**	**124.**	**128.**
Temp., °F						
600.	0.17396	0.18052	0.18709	0.19366	0.20024	0.20683
	(0.02088)	(0.02088)	(0.02088)	(0.02087)	(0.02087)	(0.02087)
650.	0.16568	0.17191	0.17814	0.18438	0.19063	0.19688
	(0.02205)	(0.02205)	(0.02205)	(0.02205)	(0.02205)	(0.02204)
700.	0.15820	0.16414	0.17007	0.17602	0.18196	0.18792
	(0.02322)	(0.02322)	(0.02322)	(0.02322)	(0.02322)	(0.02321)
750.	0.15140	0.15707	0.16274	0.16842	0.17410	0.17978
	(0.02439)	(0.02438)	(0.02438)	(0.02438)	(0.02438)	(0.02438)
800.	0.14519	0.15062	0.15605	0.16148	0.16692	0.17236
	(0.02555)	(0.02555)	(0.02554)	(0.02554)	(0.02554)	(0.02554)
850.	0.13948	0.14469	0.14990	0.15511	0.16033	0.16554
	(0.02670)	(0.02670)	(0.02670)	(0.02670)	(0.02670)	(0.02670)
900.	0.13422	0.13922	0.14423	0.14924	0.15425	0.15927
	(0.02785)	(0.02785)	(0.02785)	(0.02785)	(0.02785)	(0.02785)
950.	0.12935	0.13417	0.13899	0.14381	0.14864	0.15346
	(0.02899)	(0.02899)	(0.02899)	(0.02899)	(0.02899)	(0.02899)
1000.	0.12483	0.12947	0.13412	0.13877	0.14343	0.14808
	(0.03012)	(0.03013)	(0.03013)	(0.03013)	(0.03013)	(0.03013)
1050.	0.12062	0.12511	0.12959	0.13408	0.13858	0.14307
	(0.03125)	(0.03125)	(0.03125)	(0.03125)	(0.03125)	(0.03125)
1100.	0.11669	0.12103	0.12537	0.12971	0.13405	0.13839
	(0.03237)	(0.03237)	(0.03237)	(0.03237)	(0.03237)	(0.03237)
1150.	0.11301	0.11721	0.12141	0.12561	0.12982	0.13402
	(0.03348)	(0.03348)	(0.03348)	(0.03348)	(0.03348)	(0.03348)
1200.	0.10956	0.11363	0.11770	0.12178	0.12585	0.12992
	(0.03458)	(0.03458)	(0.03458)	(0.03458)	(0.03458)	(0.03458)
1250.	0.10632	0.11027	0.11422	0.11817	0.12212	0.12607
	(0.03567)	(0.03567)	(0.03567)	(0.03567)	(0.03568)	(0.03568)
1300.	0.10327	0.10710	0.11094	0.11477	0.11861	0.12244
	(0.03675)	(0.03675)	(0.03676)	(0.03676)	(0.03676)	(0.03676)
1350.	0.10039	0.10411	0.10784	0.11157	0.11529	0.11902
	(0.03783)	(0.03783)	(0.03783)	(0.03783)	(0.03783)	(0.03783)
1400.	0.09767	0.10129	0.10491	0.10854	0.11216	0.11579
	(0.03889)	(0.03889)	(0.03889)	(0.03889)	(0.03890)	(0.03890)
1450.	0.09509	0.09862	0.10214	0.10567	0.10920	0.11273
	(0.03995)	(0.03995)	(0.03995)	(0.03995)	(0.03995)	(0.03995)
1500.	0.09265	0.09608	0.09952	0.10295	0.10639	0.10983
	(0.04099)	(0.04099)	(0.04099)	(0.04099)	(0.04100)	(0.04100)

†Absolute viscosity μ_{cP} is shown in parentheses.

SOURCE: Derived from ASME, 1977.

Table G.1 Density (in Pounds-Mass per Cubic Foot) and Absolute Viscosity μ_{cP} (in Centipoises) of Steam and Compressed Water† (*continued*)

	Pressure, psia					
	132.	**136.**	**140.**	**144.**	**148.**	**152.**
Temp., °F	348.50	350.80	353.04	355.23	357.38	359.48
Sat. vap.	0.29372	0.30219	0.31066	0.31911	0.32757	0.33602
	(0.01493)	(0.01497)	(0.01501)	(0.01506)	(0.01510)	(0.01514)
Sat. liq.	55.64133	55.55800	55.47616	55.39568	55.31655	55.23868
	(0.15313)	(0.15195)	(0.15081)	(0.14972)	(0.14867)	(0.14765)
Temp., °F						
220.	59.63497	59.63581	59.63665	59.63749	59.63833	59.63917
	(0.26964)	(0.26965)	(0.26966)	(0.26966)	(0.26967)	(0.26968)
240.	59.10386	59.10473	59.10560	59.10648	59.10735	59.10822
	(0.24192)	(0.24193)	(0.24194)	(0.24194)	(0.24195)	(0.24196)
260.	58.53975	58.54066	58.54157	58.54248	58.54339	58.54430
	(0.21896)	(0.21896)	(0.21897)	(0.21898)	(0.21899)	(0.21899)
280.	57.94289	57.94385	57.94480	57.94576	57.94671	57.94766
	(0.19973)	(0.19973)	(0.19974)	(0.19975)	(0.19976)	(0.19976)
300.	57.31296	57.31396	57.31496	57.31596	57.31697	57.31797
	(0.18347)	(0.18348)	(0.18349)	(0.18349)	(0.18350)	(0.18351)
320.	56.64905	56.65011	56.65117	56.65223	56.65329	56.65435
	(0.16961)	(0.16962)	(0.16963)	(0.16963)	(0.16964)	(0.16965)
340.	55.94973	55.95086	55.95198	55.95311	55.95423	55.95535
	(0.15770)	(0.15770)	(0.15771)	(0.15772)	(0.15772)	(0.15773)
360.	0.28819	0.29760	0.30706	0.31656	0.32611	0.33572
	(0.01520)	(0.01519)	(0.01518)	(0.01517)	(0.01516)	(0.01515)
380.	0.27926	0.28829	0.29736	0.30647	0.31562	0.32482
	(0.01567)	(0.01567)	(0.01566)	(0.01565)	(0.01564)	(0.01563)
400.	0.27108	0.27978	0.28851	0.29727	0.30607	0.31490
	(0.01615)	(0.01614)	(0.01613)	(0.01613)	(0.01612)	(0.01611)
420.	0.26353	0.27193	0.28036	0.28881	0.29730	0.30581
	(0.01662)	(0.01661)	(0.01661)	(0.01660)	(0.01660)	(0.01659)
440.	0.25652	0.26465	0.27280	0.28098	0.28919	0.29742
	(0.01709)	(0.01709)	(0.01708)	(0.01708)	(0.01707)	(0.01706)
460.	0.24998	0.25786	0.26577	0.27370	0.28165	0.28962
	(0.01757)	(0.01756)	(0.01756)	(0.01755)	(0.01755)	(0.01754)
480.	0.24384	0.25150	0.25919	0.26688	0.27460	0.28234
	(0.01804)	(0.01803)	(0.01803)	(0.01802)	(0.01802)	(0.01802)
500.	0.23807	0.24553	0.25300	0.26048	0.26799	0.27551
	(0.01851)	(0.01851)	(0.01850)	(0.01850)	(0.01849)	(0.01849)
520.	0.23262	0.23988	0.24716	0.25445	0.26176	0.26907
	(0.01898)	(0.01898)	(0.01898)	(0.01897)	(0.01897)	(0.01896)
540.	0.22745	0.23454	0.24164	0.24874	0.25587	0.26300
	(0.01945)	(0.01945)	(0.01945)	(0.01944)	(0.01944)	(0.01944)
560.	0.22255	0.22947	0.23639	0.24333	0.25028	0.25724
	(0.01993)	(0.01992)	(0.01992)	(0.01992)	(0.01991)	(0.01991)

	Pressure, psia					
	132.	136.	140.	144.	148.	152.
Temp., °F						
600.	0.21343 (0.02087)	0.22004 (0.02086)	0.22666 (0.02086)	0.23329 (0.02086)	0.23992 (0.02086)	0.24656 (0.02086)
650.	0.20314 (0.02204)	0.20941 (0.02204)	0.21568 (0.02204)	0.22196 (0.02204)	0.22825 (0.02204)	0.23455 (0.02203)
700.	0.19388 (0.02321)	0.19984 (0.02321)	0.20581 (0.02321)	0.21179 (0.02321)	0.21777 (0.02321)	0.22375 (0.02321)
750.	0.18547 (0.02438)	0.19116 (0.02438)	0.19686 (0.02438)	0.20256 (0.02438)	0.20827 (0.02438)	0.21398 (0.02438)
800.	0.17780 (0.02554)	0.18325 (0.02554)	0.18870 (0.02554)	0.19415 (0.02554)	0.19961 (0.02554)	0.20507 (0.02554)
850.	0.17076 (0.02670)	0.17599 (0.02670)	0.18122 (0.02670)	0.18645 (0.02670)	0.19168 (0.02670)	0.19691 (0.02670)
900.	0.16428 (0.02785)	0.16930 (0.02785)	0.17433 (0.02785)	0.17935 (0.02785)	0.18438 (0.02785)	0.18941 (0.02785)
950.	0.15829 (0.02899)	0.16312 (0.02899)	0.16796 (0.02899)	0.17279 (0.02899)	0.17763 (0.02899)	0.18247 (0.02899)
1000.	0.15274 (0.03013)	0.15739 (0.03013)	0.16205 (0.03013)	0.16671 (0.03013)	0.17138 (0.03013)	0.17604 (0.03013)
1050.	0.14756 (0.03126)	0.15206 (0.03126)	0.15656 (0.03126)	0.16106 (0.03126)	0.16556 (0.03126)	0.17006 (0.03126)
1100.	0.14274 (0.03237)	0.14709 (0.03238)	0.15143 (0.03238)	0.15578 (0.03238)	0.16013 (0.03238)	0.16448 (0.03238)
1150.	0.13823 (0.03348)	0.14243 (0.03349)	0.14664 (0.03349)	0.15085 (0.03349)	0.15506 (0.03349)	0.15927 (0.03349)
1200.	0.13400 (0.03459)	0.13807 (0.03459)	0.14215 (0.03459)	0.14623 (0.03459)	0.15030 (0.03459)	0.15438 (0.03459)
1250.	0.13002 (0.03568)	0.13397 (0.03568)	0.13793 (0.03568)	0.14188 (0.03568)	0.14584 (0.03568)	0.14979 (0.03568)
1300.	0.12628 (0.03676)	0.13012 (0.03676)	0.13395 (0.03676)	0.13779 (0.03676)	0.14163 (0.03677)	0.14547 (0.03677)
1350.	0.12275 (0.03783)	0.12648 (0.03784)	0.13021 (0.03784)	0.13394 (0.03784)	0.13767 (0.03784)	0.14140 (0.03784)
1400.	0.11941 (0.03890)	0.12304 (0.03890)	0.12667 (0.03890)	0.13030 (0.03890)	0.13392 (0.03890)	0.13755 (0.03891)
1450.	0.11626 (0.03995)	0.11979 (0.03995)	0.12332 (0.03996)	0.12685 (0.03996)	0.13038 (0.03996)	0.13391 (0.03996)
1500.	0.11327 (0.04100)	0.11671 (0.04100)	0.12014 (0.04100)	0.12358 (0.04100)	0.12702 (0.04100)	0.13046 (0.04101)

†Absolute viscosity μ_{cP} is shown in parentheses.

SOURCE: Derived from ASME, 1977.

	Pressure, psia					
	156.	**160.**	**164.**	**168.**	**172.**	**176.**
Temp., °F	361.53	363.55	365.53	367.47	369.37	371.24
Sat. vap.	0.34446	0.35290	0.36134	0.36978	0.37821	0.38664
	(0.01518)	(0.01522)	(0.01526)	(0.01530)	(0.01533)	(0.01537)
Sat. liq.	55.16200	55.08649	55.01210	54.93876	54.86643	54.79509
	(0.14667)	(0.14572)	(0.14480)	(0.14391)	(0.14305)	(0.14222)
Temp., °F						
220.	59.64001	59.64085	59.64169	59.64254	59.64338	59.64422
	(0.26969)	(0.26969)	(0.26970)	(0.26971)	(0.26972)	(0.26972)
240.	59.10910	59.10997	59.11084	59.11171	59.11259	59.11346
	(0.24197)	(0.24197)	(0.24198)	(0.24199)	(0.24200)	(0.24200)
260.	58.54521	58.54612	58.54703	58.54794	58.54885	58.54976
	(0.21900)	(0.21901)	(0.21902)	(0.21902)	(0.21903)	(0.21904)
280.	57.94861	57.94957	57.95052	57.95147	57.95242	57.95338
	(0.19977)	(0.19978)	(0.19979)	(0.19979)	(0.19980)	(0.19981)
300.	57.31897	57.31997	57.32097	57.32198	57.32298	57.32398
	(0.18352)	(0.18352)	(0.18353)	(0.18354)	(0.18355)	(0.18355)
320.	56.65541	56.65646	56.65752	56.65858	56.65964	56.66070
	(0.16966)	(0.16966)	(0.16967)	(0.16968)	(0.16969)	(0.16969)
340.	55.95648	55.95760	55.95872	55.95985	55.96097	55.96209
	(0.15774)	(0.15775)	(0.15775)	(0.15776)	(0.15777)	(0.15777)
360.	55.22012	55.22132	55.22252	55.22372	55.22492	55.22612
	(0.14741)	(0.14741)	(0.14742)	(0.14743)	(0.14743)	(0.14744)
380.	0.33405	0.34333	0.35265	0.36201	0.37142	0.38088
	(0.01562)	(0.01562)	(0.01561)	(0.01560)	(0.01559)	(0.01558)
400.	0.32377	0.33267	0.34161	0.35058	0.35959	0.36864
	(0.01610)	(0.01610)	(0.01609)	(0.01608)	(0.01607)	(0.01607)
420.	0.31435	0.32292	0.33153	0.34016	0.34882	0.35752
	(0.01658)	(0.01658)	(0.01657)	(0.01656)	(0.01656)	(0.01655)
440.	0.30567	0.31395	0.32225	0.33058	0.33893	0.34731
	(0.01706)	(0.01705)	(0.01705)	(0.01704)	(0.01704)	(0.01703)
460.	0.29761	0.30562	0.31366	0.32171	0.32979	0.33789
	(0.01754)	(0.01753)	(0.01752)	(0.01752)	(0.01751)	(0.01751)
480.	0.29009	0.29786	0.30565	0.31346	0.32129	0.32914
	(0.01801)	(0.01801)	(0.01800)	(0.01800)	(0.01799)	(0.01799)
500.	0.28304	0.29059	0.29816	0.30574	0.31334	0.32096
	(0.01849)	(0.01848)	(0.01848)	(0.01847)	(0.01847)	(0.01847)
520.	0.27641	0.28376	0.29112	0.29849	0.30588	0.31329
	(0.01896)	(0.01896)	(0.01895)	(0.01895)	(0.01895)	(0.01894)
540.	0.27014	0.27730	0.28447	0.29166	0.29885	0.30606
	(0.01943)	(0.01943)	(0.01943)	(0.01942)	(0.01942)	(0.01942)
560.	0.26421	0.27119	0.27819	0.28519	0.29221	0.29923
	(0.01991)	(0.01990)	(0.01990)	(0.01990)	(0.01990)	(0.01989)

			Pressure, psia			
	156.	**160.**	**164.**	**168.**	**172.**	**176.**
Temp., °F						
600.	0.25321 (0.02085)	0.25988 (0.02085)	0.26654 (0.02085)	0.27322 (0.02085)	0.27991 (0.02084)	0.28661 (0.02084)
650.	0.24085 (0.02203)	0.24715 (0.02203)	0.25347 (0.02203)	0.25979 (0.02203)	0.26612 (0.02203)	0.27245 (0.02202)
700.	0.22974 (0.02321)	0.23574 (0.02321)	0.24174 (0.02320)	0.24775 (0.02320)	0.25376 (0.02320)	0.25978 (0.02320)
750.	0.21969 (0.02438)	0.22541 (0.02438)	0.23114 (0.02438)	0.23686 (0.02437)	0.24260 (0.02437)	0.24833 (0.02437)
800.	0.21054 (0.02554)	0.21601 (0.02554)	0.22148 (0.02554)	0.22696 (0.02554)	0.23244 (0.02554)	0.23792 (0.02554)
850.	0.20215 (0.02670)	0.20740 (0.02670)	0.21264 (0.02670)	0.21789 (0.02670)	0.22314 (0.02670)	0.22839 (0.02670)
900.	0.19444 (0.02785)	0.19947 (0.02785)	0.20451 (0.02785)	0.20955 (0.02785)	0.21459 (0.02785)	0.21963 (0.02785)
950.	0.18731 (0.02900)	0.19215 (0.02900)	0.19700 (0.02900)	0.20185 (0.02900)	0.20670 (0.02900)	0.21155 (0.02900)
1000.	0.18071 (0.03013)	0.18537 (0.03013)	0.19004 (0.03013)	0.19471 (0.03013)	0.19939 (0.03013)	0.20406 (0.03014)
1050.	0.17456 (0.03126)	0.17907 (0.03126)	0.18357 (0.03126)	0.18808 (0.03126)	0.19259 (0.03126)	0.19710 (0.03126)
1100.	0.16883 (0.03238)	0.17319 (0.03238)	0.17754 (0.03238)	0.18190 (0.03238)	0.18626 (0.03238)	0.19061 (0.03238)
1150.	0.16348 (0.03349)	0.16769 (0.03349)	0.17191 (0.03349)	0.17612 (0.03349)	0.18034 (0.03349)	0.18455 (0.03350)
1200.	0.15846 (0.03459)	0.16254 (0.03459)	0.16662 (0.03459)	0.17071 (0.03460)	0.17479 (0.03460)	0.17887 (0.03460)
1250.	0.15375 (0.03569)	0.15771 (0.03569)	0.16166 (0.03569)	0.16562 (0.03569)	0.16958 (0.03569)	0.17354 (0.03569)
1300.	0.14931 (0.03677)	0.15316 (0.03677)	0.15700 (0.03677)	0.16084 (0.03677)	0.16468 (0.03677)	0.16853 (0.03677)
1350.	0.14513 (0.03784)	0.14886 (0.03784)	0.15260 (0.03784)	0.15633 (0.03785)	0.16006 (0.03785)	0.16380 (0.03785)
1400.	0.14118 (0.03891)	0.14481 (0.03891)	0.14844 (0.03891)	0.15207 (0.03891)	0.15570 (0.03891)	0.15933 (0.03891)
1450.	0.13744 (0.03996)	0.14098 (0.03996)	0.14451 (0.03996)	0.14804 (0.03997)	0.15158 (0.03997)	0.15511 (0.03997)
1500.	0.13390 (0.04101)	0.13734 (0.04101)	0.14078 (0.04101)	0.14422 (0.04101)	0.14767 (0.04101)	0.15111 (0.04101)

†Absolute viscosity μ_{cP} is shown in parentheses.

SOURCE: Derived from ASME, 1977.

Table G.1 Density (in Pounds-Mass per Cubic Foot) and Absolute Viscosity μ_{cP} (in Centipoises) of Steam and Compressed Water† (*continued*)

	Pressure, psia					
	180.	184.	188.	192.	196.	200.
Temp., °F	373.08	374.88	376.65	378.40	380.12	381.80
Sat. vap.	0.39507	0.40350	0.41193	0.42035	0.42878	0.43720
	(0.01541)	(0.01544)	(0.01548)	(0.01551)	(0.01554)	(0.01558)
Sat. liq.	54.72470	54.65521	54.58657	54.51879	54.45183	54.38565
	(0.14141)	(0.14062)	(0.13986)	(0.13911)	(0.13839)	(0.13768)
Temp., °F						
220.	59.64506	59.64590	59.64674	59.64758	59.64842	59.64926
	(0.26973)	(0.26974)	(0.26975)	(0.26975)	(0.26976)	(0.26977)
240.	59.11433	59.11520	59.11608	59.11695	59.11782	59.11869
	(0.24201)	(0.24202)	(0.24203)	(0.24203)	(0.24204)	(0.24205)
260.	58.55067	58.55158	58.55249	58.55340	58.55431	58.55522
	(0.21905)	(0.21905)	(0.21906)	(0.21907)	(0.21908)	(0.21908)
280.	57.95433	57.95528	57.95623	57.95719	57.95814	57.95909
	(0.19982)	(0.19982)	(0.19983)	(0.19984)	(0.19984)	(0.19985)
300.	57.32498	57.32598	57.32698	57.32798	57.32898	57.32998
	(0.18356)	(0.18357)	(0.18357)	(0.18358)	(0.18359)	(0.18360)
320.	56.66175	56.66281	56.66387	56.66493	56.66598	56.66704
	(0.16970)	(0.16971)	(0.16971)	(0.16972)	(0.16973)	(0.16974)
340.	55.96322	55.96434	55.96546	55.96658	55.96770	55.96883
	(0.15778)	(0.15779)	(0.15780)	(0.15780)	(0.15781)	(0.15782)
360.	55.22731	55.22851	55.22971	55.23091	55.23210	55.23330
	(0.14745)	(0.14745)	(0.14746)	(0.14747)	(0.14748)	(0.14748)
380.	0.39038	0.39993	0.40952	0.41917	54.45643	54.45772
	(0.01558)	(0.01557)	(0.01556)	(0.01555)	(0.13844)	(0.13844)
400.	0.37773	0.38686	0.39602	0.40523	0.41447	0.42376
	(0.01606)	(0.01605)	(0.01605)	(0.01604)	(0.01603)	(0.01602)
420.	0.36624	0.37500	0.38379	0.39262	0.40147	0.41036
	(0.01654)	(0.01654)	(0.01653)	(0.01652)	(0.01652)	(0.01651)
440.	0.35572	0.36416	0.37262	0.38111	0.38963	0.39817
	(0.01702)	(0.01702)	(0.01701)	(0.01701)	(0.01700)	(0.01700)
460.	0.34602	0.35416	0.36233	0.37052	0.37874	0.38698
	(0.01750)	(0.01750)	(0.01749)	(0.01749)	(0.01748)	(0.01748)
480.	0.33700	0.34489	0.35280	0.36072	0.36867	0.37663
	(0.01798)	(0.01798)	(0.01797)	(0.01797)	(0.01796)	(0.01796)
500.	0.32859	0.33624	0.34391	0.35159	0.35930	0.36702
	(0.01846)	(0.01846)	(0.01845)	(0.01845)	(0.01844)	(0.01844)
520.	0.32071	0.32814	0.33559	0.34305	0.35053	0.35803
	(0.01894)	(0.01893)	(0.01893)	(0.01893)	(0.01892)	(0.01892)
540.	0.31328	0.32052	0.32776	0.33503	0.34230	0.34959
	(0.01941)	(0.01941)	(0.01941)	(0.01940)	(0.01940)	(0.01940)
560.	0.30627	0.31332	0.32038	0.32745	0.33454	0.34163
	(0.01989)	(0.01989)	(0.01988)	(0.01988)	(0.01988)	(0.01988)

	Pressure, psia					
	180.	184.	188.	192.	196.	200.
Temp., °F						
600.	0.29331 (0.02084)	0.30003 (0.02084)	0.30675 (0.02084)	0.31348 (0.02083)	0.32023 (0.02083)	0.32698 (0.02083)
650.	0.27880 (0.02202)	0.28515 (0.02202)	0.29150 (0.02202)	0.29787 (0.02202)	0.30424 (0.02202)	0.31061 (0.02201)
700.	0.26581 (0.02320)	0.27184 (0.02320)	0.27787 (0.02320)	0.28391 (0.02320)	0.28996 (0.02320)	0.29601 (0.02320)
750.	0.25407 (0.02437)	0.25982 (0.02437)	0.26557 (0.02437)	0.27133 (0.02437)	0.27708 (0.02437)	0.28285 (0.02437)
800.	0.24341 (0.02554)	0.24890 (0.02554)	0.25439 (0.02554)	0.25989 (0.02554)	0.26539 (0.02554)	0.27090 (0.02554)
850.	0.23365 (0.02670)	0.23891 (0.02670)	0.24417 (0.02670)	0.24944 (0.02670)	0.25470 (0.02670)	0.25998 (0.02670)
900.	0.22468 (0.02785)	0.22973 (0.02785)	0.23478 (0.02785)	0.23983 (0.02785)	0.24489 (0.02785)	0.24995 (0.02785)
950.	0.21640 (0.02900)	0.22126 (0.02900)	0.22612 (0.02900)	0.23098 (0.02900)	0.23584 (0.02900)	0.24070 (0.02900)
1000.	0.20874 (0.03014)	0.21341 (0.03014)	0.21809 (0.03014)	0.22278 (0.03014)	0.22746 (0.03014)	0.23214 (0.03014)
1050.	0.20161 (0.03126)	0.20613 (0.03127)	0.21064 (0.03127)	0.21516 (0.03127)	0.21968 (0.03127)	0.22419 (0.03127)
1100.	0.19497 (0.03239)	0.19933 (0.03239)	0.20370 (0.03239)	0.20806 (0.03239)	0.21242 (0.03239)	0.21679 (0.03239)
1150.	0.18877 (0.03350)	0.19299 (0.03350)	0.19721 (0.03350)	0.20143 (0.03350)	0.20565 (0.03350)	0.20987 (0.03350)
1200.	0.18296 (0.03460)	0.18705 (0.03460)	0.19113 (0.03460)	0.19522 (0.03460)	0.19931 (0.03460)	0.20340 (0.03460)
1250.	0.17750 (0.03569)	0.18147 (0.03569)	0.18543 (0.03569)	0.18939 (0.03570)	0.19336 (0.03570)	0.19732 (0.03570)
1300.	0.17237 (0.03678)	0.17622 (0.03678)	0.18006 (0.03678)	0.18391 (0.03678)	0.18776 (0.03678)	0.19160 (0.03678)
1350.	0.16753 (0.03785)	0.17127 (0.03785)	0.17501 (0.03785)	0.17874 (0.03785)	0.18248 (0.03786)	0.18622 (0.03786)
1400.	0.16297 (0.03891)	0.16660 (0.03892)	0.17023 (0.03892)	0.17386 (0.03892)	0.17750 (0.03892)	0.18113 (0.03892)
1450.	0.15864 (0.03997)	0.16218 (0.03997)	0.16571 (0.03997)	0.16925 (0.03997)	0.17279 (0.03998)	0.17632 (0.03998)
1500.	0.15455 (0.04102)	0.15799 (0.04102)	0.16143 (0.04102)	0.16488 (0.04102)	0.16832 (0.04102)	0.17176 (0.04102)

†Absolute viscosity μ_{cP} is shown in parentheses.

SOURCE: Derived from ASME, 1977.

	Pressure, psia					
	204.	**208.**	**212.**	**216.**	**220.**	**224.**
Temp., °F	383.47	385.11	386.72	388.31	389.88	391.42
Sat. vap.	0.44562	0.45405	0.46247	0.47090	0.47932	0.48775
	(0.01561)	(0.01564)	(0.01567)	(0.01571)	(0.01574)	(0.01577)
Sat. liq.	54.32020	54.25551	54.19153	54.12823	54.06556	54.00357
	(0.13700)	(0.13633)	(0.13568)	(0.13504)	(0.13442)	(0.13381)
Temp., °F						
220.	59.65010	59.65094	59.65178	59.65262	59.65345	59.65429
	(0.26978)	(0.26978)	(0.26979)	(0.26980)	(0.26981)	(0.26981)
240.	59.11956	59.12044	59.12131	59.12218	59.12305	59.12392
	(0.24206)	(0.24206)	(0.24207)	(0.24208)	(0.24209)	(0.24209)
260.	58.55613	58.55704	58.55794	58.55885	58.55976	58.56067
	(0.21909)	(0.21910)	(0.21911)	(0.21911)	(0.21912)	(0.21913)
280.	57.96004	57.96099	57.96194	57.96289	57.96384	57.96480
	(0.19986)	(0.19987)	(0.19987)	(0.19988)	(0.19989)	(0.19990)
300.	57.33098	57.33198	57.33298	57.33398	57.33498	57.33598
	(0.18360)	(0.18361)	(0.18362)	(0.18363)	(0.18363)	(0.18364)
320.	56.66810	56.66915	56.67021	56.67126	56.67232	56.67338
	(0.16974)	(0.16975)	(0.16976)	(0.16976)	(0.16977)	(0.16978)
340.	55.96995	55.97107	55.97219	55.97331	55.97443	55.97555
	(0.15782)	(0.15783)	(0.15784)	(0.15785)	(0.15785)	(0.15786)
360.	55.23450	55.23569	55.23689	55.23809	55.23928	55.24048
	(0.14749)	(0.14750)	(0.14750)	(0.14751)	(0.14752)	(0.14753)
380.	54.45900	54.46029	54.46157	54.46285	54.46414	54.46542
	(0.13845)	(0.13846)	(0.13847)	(0.13847)	(0.13848)	(0.13849)
400.	0.43308	0.44245	0.45186	0.46132	0.47082	0.48036
	(0.01602)	(0.01601)	(0.01600)	(0.01599)	(0.01599)	(0.01598)
420.	0.41929	0.42825	0.43724	0.44627	0.45534	0.46444
	(0.01650)	(0.01650)	(0.01649)	(0.01648)	(0.01648)	(0.01647)
440.	0.40674	0.41535	0.42398	0.43264	0.44133	0.45005
	(0.01699)	(0.01698)	(0.01698)	(0.01697)	(0.01697)	(0.01696)
460.	0.39524	0.40353	0.41184	0.42018	0.42854	0.43693
	(0.01747)	(0.01747)	(0.01746)	(0.01746)	(0.01745)	(0.01745)
480.	0.38462	0.39263	0.40066	0.40871	0.41678	0.42487
	(0.01796)	(0.01795)	(0.01795)	(0.01794)	(0.01794)	(0.01793)
500.	0.37475	0.38251	0.39028	0.39807	0.40588	0.41371
	(0.01844)	(0.01843)	(0.01843)	(0.01842)	(0.01842)	(0.01842)
520.	0.36554	0.37306	0.38060	0.38816	0.39573	0.40331
	(0.01892)	(0.01891)	(0.01891)	(0.01891)	(0.01890)	(0.01890)
540.	0.35689	0.36420	0.37153	0.37887	0.38622	0.39359
	(0.01940)	(0.01939)	(0.01939)	(0.01939)	(0.01938)	(0.01938)
560.	0.34874	0.35586	0.36299	0.37013	0.37729	0.38446
	(0.01987)	(0.01987)	(0.01987)	(0.01986)	(0.01986)	(0.01986)

	Pressure, psia					
	204.	208.	212.	216.	220.	224.
Temp., °F						
600.	0.33374 (0.02083)	0.34051 (0.02082)	0.34729 (0.02082)	0.35408 (0.02082)	0.36088 (0.02082)	0.36768 (0.02082)
650.	0.31700 (0.02201)	0.32339 (0.02201)	0.32979 (0.02201)	0.33620 (0.02201)	0.34261 (0.02201)	0.34903 (0.02201)
700.	0.30207 (0.02319)	0.30813 (0.02319)	0.31420 (0.02319)	0.32027 (0.02319)	0.32635 (0.02319)	0.33244 (0.02319)
750.	0.28862 (0.02437)	0.29439 (0.02437)	0.30017 (0.02437)	0.30595 (0.02437)	0.31173 (0.02437)	0.31752 (0.02437)
800.	0.27640 (0.02554)	0.28192 (0.02554)	0.28743 (0.02554)	0.29295 (0.02554)	0.29848 (0.02554)	0.30400 (0.02554)
850.	0.26525 (0.02670)	0.27053 (0.02670)	0.27581 (0.02670)	0.28109 (0.02670)	0.28638 (0.02670)	0.29167 (0.02670)
900.	0.25501 (0.02785)	0.26007 (0.02785)	0.26514 (0.02785)	0.27021 (0.02786)	0.27528 (0.02786)	0.28035 (0.02786)
950.	0.24557 (0.02900)	0.25044 (0.02900)	0.25531 (0.02900)	0.26018 (0.02900)	0.26505 (0.02900)	0.26993 (0.02900)
1000.	0.23683 (0.03014)	0.24152 (0.03014)	0.24621 (0.03014)	0.25090 (0.03014)	0.25559 (0.03014)	0.26029 (0.03014)
1050.	0.22872 (0.03127)	0.23324 (0.03127)	0.23776 (0.03127)	0.24229 (0.03127)	0.24681 (0.03127)	0.25134 (0.03127)
1100.	0.22116 (0.03239)	0.22553 (0.03239)	0.22990 (0.03239)	0.23427 (0.03239)	0.23864 (0.03239)	0.24301 (0.03240)
1150.	0.21410 (0.03350)	0.21832 (0.03350)	0.22255 (0.03351)	0.22678 (0.03351)	0.23100 (0.03351)	0.23523 (0.03351)
1200.	0.20749 (0.03461)	0.21158 (0.03461)	0.21567 (0.03461)	0.21977 (0.03461)	0.22386 (0.03461)	0.22795 (0.03461)
1250.	0.20129 (0.03570)	0.20525 (0.03570)	0.20922 (0.03570)	0.21319 (0.03570)	0.21716 (0.03570)	0.22113 (0.03571)
1300.	0.19545 (0.03678)	0.19930 (0.03678)	0.20315 (0.03679)	0.20700 (0.03679)	0.21085 (0.03679)	0.21470 (0.03679)
1350.	0.18996 (0.03786)	0.19369 (0.03786)	0.19743 (0.03786)	0.20117 (0.03786)	0.20491 (0.03786)	0.20866 (0.03786)
1400.	0.18477 (0.03892)	0.18840 (0.03892)	0.19204 (0.03893)	0.19567 (0.03893)	0.19931 (0.03893)	0.20295 (0.03893)
1450.	0.17986 (0.03998)	0.18340 (0.03998)	0.18693 (0.03998)	0.19047 (0.03998)	0.19401 (0.03998)	0.19755 (0.03999)
1500.	0.17521 (0.04102)	0.17865 (0.04103)	0.18210 (0.04103)	0.18554 (0.04103)	0.18899 (0.04103)	0.19243 (0.04103)

†Absolute viscosity μ_{cP} is shown in parentheses.

source: Derived from ASME, 1977.

Table G.1 Density (in Pounds-Mass per Cubic Foot) and Absolute Viscosity μ_{cP} (in Centipoises) of Steam and Compressed Water† (*continued*)

	Pressure, psia					
	228.	**232.**	**236.**	**240.**	**244.**	**248.**
Temp., °F	392.94	394.45	395.93	397.39	398.84	400.26
Sat. vap.	0.49617	0.50460	0.51303	0.52146	0.52989	0.53833
	(0.01580)	(0.01583)	(0.01586)	(0.01588)	(0.01591)	(0.01594)
Sat. liq.	53.94220	53.88144	53.82126	53.76162	53.70257	53.64403
	(0.13322)	(0.13264)	(0.13208)	(0.13152)	(0.13098)	(0.13045)
Temp., °F						
220.	59.65513	59.65597	59.65681	59.65765	59.65849	59.65933
	(0.26982)	(0.26983)	(0.26984)	(0.26984)	(0.26985)	(0.26986)
240.	59.12479	59.12566	59.12653	59.12741	59.12828	59.12915
	(0.24210)	(0.24211)	(0.24212)	(0.24212)	(0.24213)	(0.24214)
260.	58.56158	58.56249	58.56339	58.56430	58.56521	58.56612
	(0.21913)	(0.21914)	(0.21915)	(0.21916)	(0.21916)	(0.21917)
280.	57.96575	57.96670	57.96765	57.96860	57.96955	57.97050
	(0.19990)	(0.19991)	(0.19992)	(0.19993)	(0.19993)	(0.19994)
300.	57.33698	57.33798	57.33898	57.33998	57.34098	57.34198
	(0.18365)	(0.18365)	(0.18366)	(0.18367)	(0.18368)	(0.18368)
320.	56.67443	56.67549	56.67654	56.67760	56.67865	56.67971
	(0.16979)	(0.16979)	(0.16980)	(0.16981)	(0.16981)	(0.16982)
340.	55.97667	55.97779	55.97891	55.98003	55.98115	55.98227
	(0.15787)	(0.15787)	(0.15788)	(0.15789)	(0.15790)	(0.15790)
360.	55.24167	55.24287	55.24406	55.24526	55.24645	55.24765
	(0.14753)	(0.14754)	(0.14755)	(0.14755)	(0.14756)	(0.14757)
380.	54.46670	54.46799	54.46927	54.47055	54.47183	54.47312
	(0.13849)	(0.13850)	(0.13851)	(0.13852)	(0.13852)	(0.13853)
400.	0.48995	0.49958	0.50926	0.51899	0.52876	53.65511
	(0.01597)	(0.01596)	(0.01596)	(0.01595)	(0.01594)	(0.13055)
420.	0.47358	0.48275	0.49196	0.50122	0.51051	0.51984
	(0.01646)	(0.01646)	(0.01645)	(0.01644)	(0.01644)	(0.01643)
440.	0.45880	0.46758	0.47640	0.48524	0.49412	0.50303
	(0.01695)	(0.01695)	(0.01694)	(0.01694)	(0.01693)	(0.01693)
460.	0.44534	0.45378	0.46225	0.47074	0.47925	0.48780
	(0.01744)	(0.01744)	(0.01743)	(0.01743)	(0.01742)	(0.01742)
480.	0.43298	0.44112	0.44927	0.45745	0.46566	0.47388
	(0.01793)	(0.01792)	(0.01792)	(0.01791)	(0.01791)	(0.01791)
500.	0.42155	0.42942	0.43730	0.44520	0.45312	0.46106
	(0.01841)	(0.01841)	(0.01840)	(0.01840)	(0.01840)	(0.01839)
520.	0.41092	0.41854	0.42617	0.43382	0.44149	0.44918
	(0.01889)	(0.01889)	(0.01889)	(0.01888)	(0.01888)	(0.01888)
540.	0.40097	0.40837	0.41578	0.42321	0.43064	0.43810
	(0.01938)	(0.01937)	(0.01937)	(0.01937)	(0.01936)	(0.01936)
560.	0.39163	0.39883	0.40603	0.41325	0.42048	0.42772
	(0.01986)	(0.01985)	(0.01985)	(0.01985)	(0.01984)	(0.01984)

Temp., °F	Pressure, psia					
	228.	232.	236.	240.	244.	248.
600.	0.37450 (0.02081)	0.38133 (0.02081)	0.38816 (0.02081)	0.39501 (0.02081)	0.40187 (0.02080)	0.40873 (0.02080)
650.	0.35546 (0.02200)	0.36189 (0.02200)	0.36834 (0.02200)	0.37479 (0.02200)	0.38125 (0.02200)	0.38771 (0.02200)
700.	0.33853 (0.02319)	0.34463 (0.02319)	0.35073 (0.02319)	0.35684 (0.02318)	0.36296 (0.02318)	0.36908 (0.02318)
750.	0.32332 (0.02437)	0.32912 (0.02436)	0.33492 (0.02436)	0.34073 (0.02436)	0.34655 (0.02436)	0.35236 (0.02436)
800.	0.30953 (0.02554)	0.31507 (0.02554)	0.32061 (0.02554)	0.32615 (0.02554)	0.33169 (0.02554)	0.33724 (0.02554)
850.	0.29696 (0.02670)	0.30226 (0.02670)	0.30755 (0.02670)	0.31286 (0.02670)	0.31816 (0.02670)	0.32347 (0.02670)
900.	0.28543 (0.02786)	0.29051 (0.02786)	0.29559 (0.02786)	0.30067 (0.02786)	0.30576 (0.02786)	0.31085 (0.02786)
950.	0.27481 (0.02900)	0.27969 (0.02900)	0.28457 (0.02901)	0.28946 (0.02901)	0.29434 (0.02901)	0.29923 (0.02901)
1000.	0.26499 (0.03014)	0.26969 (0.03014)	0.27439 (0.03015)	0.27909 (0.03015)	0.28379 (0.03015)	0.28850 (0.03015)
1050.	0.25587 (0.03127)	0.26040 (0.03128)	0.26493 (0.03128)	0.26947 (0.03128)	0.27400 (0.03128)	0.27854 (0.03128)
1100.	0.24739 (0.03240)	0.25176 (0.03240)	0.25614 (0.03240)	0.26052 (0.03240)	0.26489 (0.03240)	0.26928 (0.03240)
1150.	0.23946 (0.03351)	0.24369 (0.03351)	0.24793 (0.03351)	0.25216 (0.03351)	0.25639 (0.03351)	0.26063 (0.03351)
1200.	0.23205 (0.03461)	0.23615 (0.03461)	0.24024 (0.03462)	0.24434 (0.03462)	0.24844 (0.03462)	0.25254 (0.03462)
1250.	0.22510 (0.03571)	0.22907 (0.03571)	0.23304 (0.03571)	0.23701 (0.03571)	0.24098 (0.03571)	0.24496 (0.03571)
1300.	0.21856 (0.03679)	0.22241 (0.03679)	0.22626 (0.03679)	0.23012 (0.03680)	0.23397 (0.03680)	0.23783 (0.03680)
1350.	0.21240 (0.03787)	0.21614 (0.03787)	0.21988 (0.03787)	0.22363 (0.03787)	0.22737 (0.03787)	0.23111 (0.03787)
1400.	0.20658 (0.03893)	0.21022 (0.03893)	0.21386 (0.03893)	0.21750 (0.03894)	0.22114 (0.03894)	0.22478 (0.03894)
1450.	0.20109 (0.03999)	0.20463 (0.03999)	0.20817 (0.03999)	0.21171 (0.03999)	0.21525 (0.03999)	0.21879 (0.03999)
1500.	0.19588 (0.04103)	0.19933 (0.04103)	0.20277 (0.04104)	0.20622 (0.04104)	0.20967 (0.04104)	0.21312 (0.04104)

†Absolute viscosity μ_{cP} is shown in parentheses.

SOURCE: Derived from ASME, 1977.

	Pressure, psia					
	250.	260.	270.	280.	290.	300.
Temp., °F	400.97	404.44	407.80	411.07	414.25	417.35
Sat. vap.	0.54254	0.56364	0.58475	0.60588	0.62703	0.64820
	(0.01595)	(0.01602)	(0.01609)	(0.01615)	(0.01622)	(0.01628)
Sat. liq.	53.61499	53.47158	53.33117	53.19358	53.05867	52.92622
	(0.13019)	(0.12893)	(0.12772)	(0.12657)	(0.12548)	(0.12443)
Temp., °F						
220.	59.65975	59.66184	59.66394	59.66604	59.66813	59.67023
	(0.26986)	(0.26988)	(0.26990)	(0.26992)	(0.26994)	(0.26996)
240.	59.12958	59.13176	59.13394	59.13611	59.13829	59.14046
	(0.24214)	(0.24216)	(0.24218)	(0.24220)	(0.24222)	(0.24224)
260.	58.56657	58.56884	58.57111	58.57338	58.57564	58.57791
	(0.21918)	(0.21919)	(0.21921)	(0.21923)	(0.21925)	(0.21927)
280.	57.97097	57.97335	57.97572	57.97810	57.98047	57.98284
	(0.19994)	(0.19996)	(0.19998)	(0.20000)	(0.20002)	(0.20004)
300.	57.34248	57.34498	57.34747	57.34997	57.35246	57.35495
	(0.18369)	(0.18371)	(0.18372)	(0.18374)	(0.18376)	(0.18378)
320.	56.68024	56.68287	56.68551	56.68814	56.69078	56.69341
	(0.16982)	(0.16984)	(0.16986)	(0.16988)	(0.16990)	(0.16991)
340.	55.98283	55.98563	55.98842	55.99122	55.99401	55.99681
	(0.15791)	(0.15792)	(0.15794)	(0.15796)	(0.15798)	(0.15799)
360.	55.24824	55.25123	55.25421	55.25719	55.26018	55.26315
	(0.14757)	(0.14759)	(0.14761)	(0.14762)	(0.14764)	(0.14766)
380.	54.47376	54.47696	54.48016	54.48336	54.48656	54.48976
	(0.13853)	(0.13855)	(0.13857)	(0.13859)	(0.13860)	(0.13862)
400.	53.65580	53.65927	53.66272	53.66618	53.66964	53.67309
	(0.13055)	(0.13057)	(0.13059)	(0.13061)	(0.13063)	(0.13064)
420.	0.52451	0.54806	0.57187	0.59594	0.62029	0.64493
	(0.01643)	(0.01641)	(0.01640)	(0.01638)	(0.01636)	(0.01635)
440.	0.50749	0.52995	0.55261	0.57550	0.59860	0.62194
	(0.01692)	(0.01691)	(0.01689)	(0.01688)	(0.01686)	(0.01685)
460.	0.49208	0.51359	0.53528	0.55715	0.57920	0.60144
	(0.01741)	(0.01740)	(0.01739)	(0.01738)	(0.01736)	(0.01735)
480.	0.47800	0.49869	0.51953	0.54051	0.56165	0.58295
	(0.01790)	(0.01789)	(0.01788)	(0.01787)	(0.01786)	(0.01785)
500.	0.46504	0.48500	0.50508	0.52529	0.54563	0.56611
	(0.01839)	(0.01838)	(0.01837)	(0.01836)	(0.01835)	(0.01834)
520.	0.45303	0.47233	0.49174	0.51126	0.53090	0.55064
	(0.01888)	(0.01887)	(0.01886)	(0.01885)	(0.01884)	(0.01883)
540.	0.44183	0.46054	0.47935	0.49825	0.51725	0.53634
	(0.01936)	(0.01935)	(0.01934)	(0.01933)	(0.01933)	(0.01932)
560.	0.43134	0.44952	0.46777	0.48611	0.50453	0.52303
	(0.01984)	(0.01983)	(0.01983)	(0.01982)	(0.01981)	(0.01981)

Temp., °F	Pressure, psia					
	250.	260.	270.	280.	290.	300.
600.	0.41217 (0.02080)	0.42939 (0.02080)	0.44667 (0.02079)	0.46402 (0.02079)	0.48143 (0.02078)	0.49890 (0.02077)
650.	0.39095 (0.02200)	0.40715 (0.02199)	0.42340 (0.02199)	0.43970 (0.02198)	0.45605 (0.02198)	0.47244 (0.02198)
700.	0.37214 (0.02318)	0.38747 (0.02318)	0.40284 (0.02318)	0.41825 (0.02317)	0.43369 (0.02317)	0.44917 (0.02317)
750.	0.35527 (0.02436)	0.36984 (0.02436)	0.38444 (0.02436)	0.39907 (0.02436)	0.41373 (0.02436)	0.42842 (0.02436)
800.	0.34002 (0.02554)	0.35391 (0.02553)	0.36783 (0.02553)	0.38177 (0.02553)	0.39574 (0.02553)	0.40972 (0.02553)
850.	0.32612 (0.02670)	0.33941 (0.02670)	0.35272 (0.02670)	0.36604 (0.02670)	0.37938 (0.02670)	0.39275 (0.02670)
900.	0.31339 (0.02786)	0.32613 (0.02786)	0.33888 (0.02786)	0.35165 (0.02786)	0.36443 (0.02786)	0.37723 (0.02786)
950.	0.30168 (0.02901)	0.31391 (0.02901)	0.32616 (0.02901)	0.33842 (0.02901)	0.35069 (0.02901)	0.36298 (0.02901)
1000.	0.29085 (0.03015)	0.30263 (0.03015)	0.31441 (0.03015)	0.32621 (0.03015)	0.33801 (0.03015)	0.34983 (0.03016)
1050.	0.28081 (0.03128)	0.29216 (0.03128)	0.30352 (0.03128)	0.31489 (0.03129)	0.32627 (0.03129)	0.33765 (0.03129)
1100.	0.27147 (0.03240)	0.28242 (0.03240)	0.29339 (0.03241)	0.30436 (0.03241)	0.31535 (0.03241)	0.32633 (0.03241)
1150.	0.26275 (0.03352)	0.27334 (0.03352)	0.28394 (0.03352)	0.29455 (0.03352)	0.30516 (0.03353)	0.31579 (0.03353)
1200.	0.25459 (0.03462)	0.26485 (0.03462)	0.27511 (0.03463)	0.28538 (0.03463)	0.29565 (0.03463)	0.30592 (0.03463)
1250.	0.24694 (0.03571)	0.25688 (0.03572)	0.26683 (0.03572)	0.27677 (0.03572)	0.28673 (0.03573)	0.29669 (0.03573)
1300.	0.23976 (0.03680)	0.24940 (0.03680)	0.25904 (0.03680)	0.26869 (0.03681)	0.27835 (0.03681)	0.28801 (0.03681)
1350.	0.23299 (0.03787)	0.24235 (0.03788)	0.25172 (0.03788)	0.26109 (0.03788)	0.27046 (0.03789)	0.27984 (0.03789)
1400.	0.22660 (0.03894)	0.23570 (0.03894)	0.24480 (0.03895)	0.25391 (0.03895)	0.26302 (0.03895)	0.27214 (0.03896)
1450.	0.22056 (0.03999)	0.22941 (0.04000)	0.23827 (0.04000)	0.24713 (0.04000)	0.25599 (0.04001)	0.26486 (0.04001)
1500.	0.21484 (0.04104)	0.22346 (0.04104)	0.23209 (0.04105)	0.24071 (0.04105)	0.24934 (0.04105)	0.25797 (0.04106)

†Absolute viscosity μ_{cP} is shown in parentheses.

SOURCE: Derived from ASME, 1977.

	Pressure, psia					
	310.	320.	330.	340.	350.	360.
Temp., °F	420.36	423.31	426.18	428.98	431.73	434.41
Sat. vap.	0.66939	0.69060	0.71184	0.73311	0.75441	0.77574
	(0.01634)	(0.01640)	(0.01645)	(0.01651)	(0.01656)	(0.01662)
Sat. liq.	52.79612	52.66825	52.54246	52.41873	52.29684	52.17674
	(0.12342)	(0.12246)	(0.12153)	(0.12064)	(0.11978)	(0.11895)
Temp., °F						
220.	59.67232	59.67441	59.67651	59.67860	59.68069	59.68278
	(0.26997)	(0.26999)	(0.27001)	(0.27003)	(0.27005)	(0.27007)
240.	59.14263	59.14481	59.14698	59.14915	59.15132	59.15349
	(0.24225)	(0.24227)	(0.24229)	(0.24231)	(0.24233)	(0.24235)
260.	58.58018	58.58244	58.58470	58.58697	58.58923	58.59149
	(0.21929)	(0.21931)	(0.21932)	(0.21934)	(0.21936)	(0.21938)
280.	57.98521	57.98758	57.98995	57.99232	57.99469	57.99706
	(0.20005)	(0.20007)	(0.20009)	(0.20011)	(0.20013)	(0.20015)
300.	57.35745	57.35994	57.36243	57.36492	57.36741	57.36990
	(0.18380)	(0.18381)	(0.18383)	(0.18385)	(0.18387)	(0.18389)
320.	56.69604	56.69867	56.70130	56.70393	56.70656	56.70919
	(0.16993)	(0.16995)	(0.16997)	(0.16999)	(0.17000)	(0.17002)
340.	55.99960	56.00239	56.00518	56.00797	56.01076	56.01355
	(0.15801)	(0.15803)	(0.15805)	(0.15807)	(0.15808)	(0.15810)
360.	55.26613	55.26911	55.27209	55.27506	55.27803	55.28101
	(0.14768)	(0.14769)	(0.14771)	(0.14773)	(0.14775)	(0.14777)
380.	54.49295	54.49615	54.49934	54.50253	54.50572	54.50891
	(0.13864)	(0.13866)	(0.13867)	(0.13869)	(0.13871)	(0.13873)
400.	53.67654	53.67999	53.68344	53.68689	53.69033	53.69378
	(0.13066)	(0.13068)	(0.13070)	(0.13071)	(0.13073)	(0.13075)
420.	52.81233	52.81608	52.81984	52.82359	52.82734	52.83108
	(0.12355)	(0.12356)	(0.12358)	(0.12360)	(0.12362)	(0.12364)
440.	0.64552	0.66933	0.69340	0.71774	0.74233	0.76721
	(0.01684)	(0.01682)	(0.01681)	(0.01679)	(0.01678)	(0.01676)
460.	0.62387	0.64651	0.66934	0.69239	0.71565	0.73913
	(0.01734)	(0.01732)	(0.01731)	(0.01730)	(0.01729)	(0.01727)
480.	0.60441	0.62603	0.64782	0.66978	0.69192	0.71423
	(0.01784)	(0.01782)	(0.01781)	(0.01780)	(0.01779)	(0.01778)
500.	0.58671	0.60746	0.62834	0.64937	0.67055	0.69187
	(0.01833)	(0.01832)	(0.01831)	(0.01830)	(0.01829)	(0.01828)
520.	0.57050	0.59047	0.61057	0.63078	0.65112	0.67158
	(0.01882)	(0.01881)	(0.01880)	(0.01880)	(0.01879)	(0.01878)
540.	0.55553	0.57482	0.59421	0.61370	0.63330	0.65301
	(0.01931)	(0.01930)	(0.01930)	(0.01929)	(0.01928)	(0.01927)
560.	0.54162	0.56029	0.57906	0.59791	0.61685	0.63588
	(0.01980)	(0.01979)	(0.01979)	(0.01978)	(0.01977)	(0.01977)

	Pressure, psia					
	310.	320.	330.	340.	350.	360.
Temp., °F						
600.	0.51644 (0.02077)	0.53404 (0.02076)	0.55172 (0.02076)	0.56946 (0.02075)	0.58727 (0.02075)	0.60514 (0.02074)
650.	0.48888 (0.02197)	0.50538 (0.02197)	0.52192 (0.02197)	0.53851 (0.02196)	0.55515 (0.02196)	0.57185 (0.02196)
700.	0.46469 (0.02317)	0.48024 (0.02317)	0.49584 (0.02316)	0.51147 (0.02316)	0.52714 (0.02316)	0.54285 (0.02316)
750.	0.44314 (0.02435)	0.45788 (0.02435)	0.47266 (0.02435)	0.48747 (0.02435)	0.50230 (0.02435)	0.51717 (0.02435)
800.	0.42373 (0.02553)	0.43777 (0.02553)	0.45183 (0.02553)	0.46591 (0.02553)	0.48002 (0.02553)	0.49415 (0.02553)
850.	0.40613 (0.02670)	0.41953 (0.02670)	0.43295 (0.02670)	0.44639 (0.02670)	0.45985 (0.02670)	0.47333 (0.02670)
900.	0.39005 (0.02786)	0.40288 (0.02786)	0.41572 (0.02786)	0.42859 (0.02787)	0.44147 (0.02787)	0.45436 (0.02787)
950.	0.37528 (0.02902)	0.38759 (0.02902)	0.39992 (0.02902)	0.41226 (0.02902)	0.42461 (0.02902)	0.43698 (0.02902)
1000.	0.36166 (0.03016)	0.37350 (0.03016)	0.38535 (0.03016)	0.39721 (0.03016)	0.40908 (0.03017)	0.42097 (0.03017)
1050.	0.34905 (0.03129)	0.36045 (0.03129)	0.37187 (0.03130)	0.38329 (0.03130)	0.39473 (0.03130)	0.40617 (0.03130)
1100.	0.33733 (0.03242)	0.34834 (0.03242)	0.35935 (0.03242)	0.37037 (0.03242)	0.38140 (0.03243)	0.39243 (0.03243)
1150.	0.32641 (0.03353)	0.33705 (0.03353)	0.34769 (0.03354)	0.35833 (0.03354)	0.36899 (0.03354)	0.37965 (0.03355)
1200.	0.31621 (0.03464)	0.32650 (0.03464)	0.33679 (0.03464)	0.34709 (0.03465)	0.35740 (0.03465)	0.36771 (0.03465)
1250.	0.30665 (0.03573)	0.31661 (0.03574)	0.32659 (0.03574)	0.33656 (0.03574)	0.34654 (0.03575)	0.35653 (0.03575)
1300.	0.29767 (0.03682)	0.30734 (0.03682)	0.31701 (0.03682)	0.32668 (0.03683)	0.33636 (0.03683)	0.34604 (0.03683)
1350.	0.28922 (0.03789)	0.29861 (0.03790)	0.30799 (0.03790)	0.31738 (0.03790)	0.32678 (0.03791)	0.33618 (0.03791)
1400.	0.28125 (0.03896)	0.29037 (0.03896)	0.29950 (0.03897)	0.30862 (0.03897)	0.31775 (0.03897)	0.32688 (0.03898)
1450.	0.27373 (0.04002)	0.28260 (0.04002)	0.29147 (0.04002)	0.30034 (0.04003)	0.30922 (0.04003)	0.31810 (0.04003)
1500.	0.26660 (0.04106)	0.27524 (0.04107)	0.28388 (0.04107)	0.29251 (0.04107)	0.30116 (0.04108)	0.30980 (0.04108)

†Absolute viscosity μ_{cP} is shown in parentheses.

SOURCE: Derived from ASME, 1977.

Table G.1 Density (in Pounds-Mass per Cubic Foot) and
Absolute Viscosity μ_{cP} (in Centipoises) of Steam and
Compressed Water† (*continued*)

	Pressure, psia					
	370.	**380.**	**390.**	**400.**	**410.**	**420.**
Temp., °F	437.04	439.61	442.13	444.60	447.02	449.40
Sat. vap.	0.79709	0.81848	0.83991	0.86136	0.88285	0.90438
	(0.01667)	(0.01672)	(0.01677)	(0.01682)	(0.01687)	(0.01692)
Sat. liq.	52.05840	51.94167	51.82650	51.71276	51.60048	51.48955
	(0.11815)	(0.11738)	(0.11663)	(0.11590)	(0.11520)	(0.11452)
Temp., °F						
220.	59.68487	59.68697	59.68906	59.69114	59.69323	59.69532
	(0.27009)	(0.27011)	(0.27012)	(0.27014)	(0.27016)	(0.27018)
240.	59.15566	59.15783	59.16000	59.16217	59.16434	59.16650
	(0.24237)	(0.24239)	(0.24240)	(0.24242)	(0.24244)	(0.24246)
260.	58.59376	58.59602	58.59828	58.60054	58.60280	58.60505
	(0.21940)	(0.21942)	(0.21944)	(0.21945)	(0.21947)	(0.21949)
280.	57.99943	58.00179	58.00416	58.00652	58.00889	58.01125
	(0.20016)	(0.20018)	(0.20020)	(0.20022)	(0.20024)	(0.20026)
300.	57.37239	57.37487	57.37736	57.37984	57.38233	57.38481
	(0.18390)	(0.18392)	(0.18394)	(0.18396)	(0.18398)	(0.18400)
320.	56.71181	56.71444	56.71706	56.71968	56.72231	56.72493
	(0.17004)	(0.17006)	(0.17008)	(0.17009)	(0.17011)	(0.17013)
340.	56.01633	56.01912	56.02190	56.02468	56.02746	56.03024
	(0.15812)	(0.15814)	(0.15815)	(0.15817)	(0.15819)	(0.15821)
360.	55.28398	55.28695	55.28992	55.29288	55.29585	55.29881
	(0.14778)	(0.14780)	(0.14782)	(0.14784)	(0.14785)	(0.14787)
380.	54.51210	54.51528	54.51847	54.52165	54.52483	54.52801
	(0.13874)	(0.13876)	(0.13878)	(0.13880)	(0.13881)	(0.13883)
400.	53.69722	53.70066	53.70410	53.70753	53.71097	53.71440
	(0.13077)	(0.13078)	(0.13080)	(0.13082)	(0.13084)	(0.13086)
420.	52.83483	52.83857	52.84231	52.84605	52.84979	52.85352
	(0.12365)	(0.12367)	(0.12369)	(0.12371)	(0.12373)	(0.12374)
440.	0.79237	0.81783	51.92727	51.93137	51.93547	51.93957
	(0.01675)	(0.01673)	(0.11728)	(0.11729)	(0.11731)	(0.11733)
460.	0.76285	0.78679	0.81098	0.83541	0.86011	0.88506
	(0.01726)	(0.01725)	(0.01723)	(0.01722)	(0.01721)	(0.01720)
480.	0.73674	0.75943	0.78232	0.80541	0.82870	0.85220
	(0.01777)	(0.01776)	(0.01775)	(0.01773)	(0.01772)	(0.01771)
500.	0.71335	0.73499	0.75678	0.77874	0.80086	0.82316
	(0.01827)	(0.01826)	(0.01825)	(0.01824)	(0.01823)	(0.01822)
520.	0.69218	0.71290	0.73376	0.75475	0.77588	0.79715
	(0.01877)	(0.01876)	(0.01875)	(0.01874)	(0.01874)	(0.01873)
540.	0.67283	0.69276	0.71279	0.73295	0.75322	0.77361
	(0.01927)	(0.01926)	(0.01925)	(0.01924)	(0.01924)	(0.01923)
560.	0.65501	0.67423	0.69355	0.71297	0.73248	0.75210
	(0.01976)	(0.01975)	(0.01975)	(0.01974)	(0.01973)	(0.01973)

Temp., °F	Pressure, psia					
	370.	380.	390.	400.	410.	420.
600.	0.62309 (0.02074)	0.64111 (0.02073)	0.65920 (0.02073)	0.67737 (0.02072)	0.69561 (0.02072)	0.71392 (0.02071)
650.	0.58859 (0.02195)	0.60539 (0.02195)	0.62224 (0.02195)	0.63914 (0.02194)	0.65610 (0.02194)	0.67311 (0.02194)
700.	0.55860 (0.02315)	0.57439 (0.02315)	0.59021 (0.02315)	0.60608 (0.02315)	0.62199 (0.02315)	0.63794 (0.02314)
750.	0.53207 (0.02435)	0.54700 (0.02435)	0.56196 (0.02435)	0.57695 (0.02434)	0.59197 (0.02434)	0.60702 (0.02434)
800.	0.50831 (0.02553)	0.52249 (0.02553)	0.53669 (0.02553)	0.55092 (0.02553)	0.56518 (0.02553)	0.57946 (0.02553)
850.	0.48683 (0.02670)	0.50035 (0.02670)	0.51389 (0.02671)	0.52745 (0.02671)	0.54103 (0.02671)	0.55462 (0.02671)
900.	0.46727 (0.02787)	0.48020 (0.02787)	0.49314 (0.02787)	0.50610 (0.02787)	0.51908 (0.02787)	0.53207 (0.02787)
950.	0.44936 (0.02902)	0.46175 (0.02903)	0.47415 (0.02903)	0.48657 (0.02903)	0.49901 (0.02903)	0.51145 (0.02903)
1000.	0.43286 (0.03017)	0.44477 (0.03017)	0.45669 (0.03017)	0.46862 (0.03018)	0.48055 (0.03018)	0.49251 (0.03018)
1050.	0.41762 (0.03131)	0.42908 (0.03131)	0.44055 (0.03131)	0.45203 (0.03131)	0.46352 (0.03132)	0.47502 (0.03132)
1100.	0.40348 (0.03243)	0.41453 (0.03244)	0.42559 (0.03244)	0.43665 (0.03244)	0.44773 (0.03244)	0.45881 (0.03245)
1150.	0.39031 (0.03355)	0.40099 (0.03355)	0.41166 (0.03355)	0.42235 (0.03356)	0.43304 (0.03356)	0.44374 (0.03356)
1200.	0.37802 (0.03466)	0.38835 (0.03466)	0.39867 (0.03466)	0.40901 (0.03466)	0.41934 (0.03467)	0.42969 (0.03467)
1250.	0.36652 (0.03575)	0.37652 (0.03575)	0.38652 (0.03576)	0.39652 (0.03576)	0.40653 (0.03576)	0.41655 (0.03577)
1300.	0.35573 (0.03684)	0.36542 (0.03684)	0.37511 (0.03684)	0.38481 (0.03685)	0.39451 (0.03685)	0.40422 (0.03685)
1350.	0.34558 (0.03791)	0.35498 (0.03792)	0.36439 (0.03792)	0.37380 (0.03792)	0.38322 (0.03793)	0.39264 (0.03793)
1400.	0.33602 (0.03898)	0.34515 (0.03898)	0.35429 (0.03899)	0.36344 (0.03899)	0.37258 (0.03899)	0.38173 (0.03900)
1450.	0.32699 (0.04004)	0.33587 (0.04004)	0.34476 (0.04004)	0.35365 (0.04005)	0.36254 (0.04005)	0.37144 (0.04006)
1500.	0.31844 (0.04108)	0.32709 (0.04109)	0.33574 (0.04109)	0.34439 (0.04109)	0.35305 (0.04110)	0.36170 (0.04110)

†Absolute viscosity μ_{cP} is shown in parentheses.

SOURCE: Derived from ASME, 1977.

	Pressure, psia					
	430.	**440.**	**450.**	**460.**	**470.**	**480.**
Temp., °F	451.74	454.03	456.28	458.50	460.68	462.82
Sat. vap.	0.92595	0.94755	0.96919	0.99088	1.01260	1.03437
	(0.01697)	(0.01701)	(0.01706)	(0.01711)	(0.01715)	(0.01719)
Sat. liq.	51.37991	51.27152	51.16440	51.05838	50.95347	50.84962
	(0.11385)	(0.11321)	(0.11258)	(0.11197)	(0.11138)	(0.11080)
220.	59.69741	59.69950	59.70158	59.70367	59.70576	59.70784
	(0.27020)	(0.27022)	(0.27024)	(0.27026)	(0.27027)	(0.27029)
Temp., °F						
240.	59.16867	59.17084	59.17300	59.17517	59.17733	59.17950
	(0.24248)	(0.24250)	(0.24252)	(0.24254)	(0.24255)	(0.24257)
260.	58.60731	58.60957	58.61183	58.61408	58.61634	58.61859
	(0.21951)	(0.21953)	(0.21955)	(0.21957)	(0.21958)	(0.21960)
280.	58.01361	58.01597	58.01834	58.02070	58.02306	58.02541
	(0.20027)	(0.20029)	(0.20031)	(0.20033)	(0.20035)	(0.20037)
300.	57.38729	57.38978	57.39226	57.39474	57.39722	57.39970
	(0.18401)	(0.18403)	(0.18405)	(0.18407)	(0.18409)	(0.18410)
320.	56.72755	56.73017	56.73279	56.73541	56.73802	56.74064
	(0.17015)	(0.17016)	(0.17018)	(0.17020)	(0.17022)	(0.17024)
340.	56.03302	56.03580	56.03858	56.04136	56.04413	56.04691
	(0.15822)	(0.15824)	(0.15826)	(0.15828)	(0.15830)	(0.15831)
360.	55.30178	55.30474	55.30770	55.31066	55.31362	55.31657
	(0.14789)	(0.14791)	(0.14792)	(0.14794)	(0.14796)	(0.14798)
380.	54.53119	54.53436	54.53754	54.54071	54.54388	54.54706
	(0.13885)	(0.13887)	(0.13888)	(0.13890)	(0.13892)	(0.13894)
400.	53.71783	53.72126	53.72469	53.72811	53.73154	53.73496
	(0.13087)	(0.13089)	(0.13091)	(0.13093)	(0.13094)	(0.13096)
420.	52.85725	52.86098	52.86471	52.86844	52.87216	52.87588
	(0.12376)	(0.12378)	(0.12380)	(0.12382)	(0.12383)	(0.12385)
440.	51.94367	51.94776	51.95185	51.95594	51.96002	51.96411
	(0.11735)	(0.11737)	(0.11739)	(0.11740)	(0.11742)	(0.11744)
460.	0.91029	0.93580	0.96159	0.98769	50.98760	50.99213
	(0.01718)	(0.01717)	(0.01716)	(0.01714)	(0.11157)	(0.11159)
480.	0.87592	0.89986	0.92403	0.94843	0.97307	0.99795
	(0.01770)	(0.01769)	(0.01768)	(0.01767)	(0.01766)	(0.01765)
500.	0.84563	0.86828	0.89111	0.91413	0.93734	0.96074
	(0.01821)	(0.01820)	(0.01819)	(0.01818)	(0.01817)	(0.01816)
520.	0.81857	0.84013	0.86184	0.88371	0.90573	0.92791
	(0.01872)	(0.01871)	(0.01870)	(0.01869)	(0.01869)	(0.01868)
540.	0.79412	0.81475	0.83551	0.85639	0.87740	0.89855
	(0.01922)	(0.01921)	(0.01921)	(0.01920)	(0.01919)	(0.01918)
560.	0.77181	0.79163	0.81156	0.83159	0.85174	0.87199
	(0.01972)	(0.01971)	(0.01971)	(0.01970)	(0.01969)	(0.01969)

Temp., °F	Pressure, psia					
	430.	440.	450.	460.	470.	480.
600.	0.73231 (0.02071)	0.75078 (0.02070)	0.76932 (0.02070)	0.78795 (0.02070)	0.80665 (0.02069)	0.82543 (0.02069)
650.	0.69018 (0.02193)	0.70730 (0.02193)	0.72447 (0.02193)	0.74171 (0.02192)	0.75900 (0.02192)	0.77634 (0.02192)
700.	0.65393 (0.02314)	0.66996 (0.02314)	0.68603 (0.02314)	0.70215 (0.02314)	0.71830 (0.02313)	0.73450 (0.02313)
750.	0.62211 (0.02434)	0.63722 (0.02434)	0.65237 (0.02434)	0.66755 (0.02434)	0.68276 (0.02434)	0.69801 (0.02434)
800.	0.59376 (0.02553)	0.60809 (0.02553)	0.62245 (0.02553)	0.63682 (0.02553)	0.65123 (0.02553)	0.66566 (0.02553)
850.	0.56824 (0.02671)	0.58188 (0.02671)	0.59554 (0.02671)	0.60922 (0.02671)	0.62292 (0.02671)	0.63664 (0.02671)
900.	0.54508 (0.02788)	0.55810 (0.02788)	0.57114 (0.02788)	0.58420 (0.02788)	0.59728 (0.02788)	0.61037 (0.02788)
950.	0.52391 (0.02903)	0.53638 (0.02904)	0.54887 (0.02904)	0.56137 (0.02904)	0.57389 (0.02904)	0.58641 (0.02904)
1000.	0.50447 (0.03018)	0.51644 (0.03018)	0.52843 (0.03019)	0.54042 (0.03019)	0.55243 (0.03019)	0.56445 (0.03019)
1050.	0.48652 (0.03132)	0.49804 (0.03132)	0.50957 (0.03133)	0.52110 (0.03133)	0.53265 (0.03133)	0.54420 (0.03133)
1100.	0.46990 (0.03245)	0.48100 (0.03245)	0.49211 (0.03245)	0.50322 (0.03246)	0.51434 (0.03246)	0.52547 (0.03246)
1150.	0.45445 (0.03357)	0.46516 (0.03357)	0.47588 (0.03357)	0.48661 (0.03358)	0.49734 (0.03358)	0.50808 (0.03358)
1200.	0.44004 (0.03467)	0.45039 (0.03468)	0.46076 (0.03468)	0.47112 (0.03468)	0.48150 (0.03469)	0.49187 (0.03469)
1250.	0.42656 (0.03577)	0.43659 (0.03577)	0.44662 (0.03578)	0.45665 (0.03578)	0.46669 (0.03578)	0.47673 (0.03579)
1300.	0.41393 (0.03686)	0.42365 (0.03686)	0.43337 (0.03687)	0.44309 (0.03687)	0.45281 (0.03687)	0.46255 (0.03688)
1350.	0.40206 (0.03794)	0.41149 (0.03794)	0.42092 (0.03794)	0.43035 (0.03795)	0.43978 (0.03795)	0.44922 (0.03795)
1400.	0.39088 (0.03900)	0.40004 (0.03901)	0.40919 (0.03901)	0.41836 (0.03901)	0.42752 (0.03902)	0.43669 (0.03902)
1450.	0.38033 (0.04006)	0.38923 (0.04006)	0.39814 (0.04007)	0.40704 (0.04007)	0.41595 (0.04007)	0.42486 (0.04008)
1500.	0.37036 (0.04111)	0.37902 (0.04111)	0.38768 (0.04111)	0.39635 (0.04112)	0.40502 (0.04112)	0.41369 (0.04112)

†Absolute viscosity μ_{cP} is shown in parentheses.

source: Derived from ASME, 1977.

			Pressure, psia			
	490.	500.	510.	520.	530.	540.
Temp., °F	464.93	467.01	469.05	471.07	473.06	475.01
Sat. vap.	1.05618	1.07803	1.09993	1.12187	1.14386	1.16589
	(0.01724)	(0.01728)	(0.01732)	(0.01736)	(0.01741)	(0.01745)
Sat. liq.	50.74672	50.64486	50.54395	50.44395	50.34483	50.24664
	(0.11024)	(0.10969)	(0.10915)	(0.10862)	(0.10811)	(0.10760)
Temp., °F						
220.	59.70993	59.71201	59.71409	59.71618	59.71826	59.72034
	(0.27031)	(0.27033)	(0.27035)	(0.27037)	(0.27039)	(0.27041)
240.	59.18166	59.18382	59.18598	59.18815	59.19031	59.19247
	(0.24259)	(0.24261)	(0.24263)	(0.24265)	(0.24267)	(0.24268)
260.	58.62085	58.62310	58.62535	58.62760	58.62986	58.63211
	(0.21962)	(0.21964)	(0.21966)	(0.21968)	(0.21970)	(0.21971)
280.	58.02777	58.03013	58.03249	58.03484	58.03720	58.03955
	(0.20038)	(0.20040)	(0.20042)	(0.20044)	(0.20046)	(0.20048)
300.	57.40217	57.40465	57.40713	57.40960	57.41208	57.41455
	(0.18412)	(0.18414)	(0.18416)	(0.18418)	(0.18419)	(0.18421)
320.	56.74325	56.74587	56.74848	56.75109	56.75370	56.75632
	(0.17025)	(0.17027)	(0.17029)	(0.17031)	(0.17033)	(0.17034)
340.	56.04968	56.05245	56.05522	56.05799	56.06076	56.06353
	(0.15833)	(0.15835)	(0.15837)	(0.15838)	(0.15840)	(0.15842)
360.	55.31953	55.32248	55.32544	55.32839	55.33134	55.33429
	(0.14799)	(0.14801)	(0.14803)	(0.14805)	(0.14806)	(0.14808)
380.	54.55022	54.55339	54.55656	54.55972	54.56289	54.56605
	(0.13896)	(0.13897)	(0.13899)	(0.13901)	(0.13903)	(0.13904)
400.	53.73838	53.74180	53.74521	53.74863	53.75204	53.75545
	(0.13098)	(0.13100)	(0.13101)	(0.13103)	(0.13105)	(0.13107)
420.	52.87960	52.88332	52.88703	52.89074	52.89446	52.89816
	(0.12387)	(0.12389)	(0.12390)	(0.12392)	(0.12394)	(0.12396)
440.	51.96819	51.97226	51.97634	51.98041	51.98448	51.98855
	(0.11746)	(0.11748)	(0.11750)	(0.11751)	(0.11753)	(0.11755)
460.	50.99665	51.00117	51.00568	51.01020	51.01471	51.01921
	(0.11161)	(0.11163)	(0.11165)	(0.11167)	(0.11168)	(0.11170)
480.	1.02309	1.04850	1.07417	1.10012	1.12636	1.15289
	(0.01763)	(0.01762)	(0.01761)	(0.01760)	(0.01759)	(0.01758)
500.	0.98435	1.00816	1.03219	1.05643	1.08089	1.10558
	(0.01815)	(0.01814)	(0.01814)	(0.01813)	(0.01812)	(0.01811)
520.	0.95026	0.97277	0.99545	1.01830	1.04133	1.06454
	(0.01867)	(0.01866)	(0.01865)	(0.01864)	(0.01864)	(0.01863)
540.	0.91983	0.94124	0.96280	0.98450	1.00634	1.02832
	(0.01918)	(0.01917)	(0.01916)	(0.01916)	(0.01915)	(0.01914)
560.	0.89236	0.91284	0.93343	0.95415	0.97498	0.99594
	(0.01968)	(0.01968)	(0.01967)	(0.01966)	(0.01966)	(0.01965)

	Pressure, psia					
	490.	**500.**	**510.**	**520.**	**530.**	**540.**
Temp., °F						
600.	0.84430	0.86325	0.88228	0.90140	0.92060	0.93989
	(0.02068)	(0.02068)	(0.02067)	(0.02067)	(0.02066)	(0.02066)
650.	0.79375	0.81121	0.82873	0.84632	0.86396	0.88166
	(0.02191)	(0.02191)	(0.02191)	(0.02190)	(0.02190)	(0.02190)
700.	0.75074	0.76703	0.78336	0.79973	0.81615	0.83261
	(0.02313)	(0.02313)	(0.02313)	(0.02313)	(0.02312)	(0.02312)
750.	0.71329	0.72860	0.74394	0.75932	0.77473	0.79018
	(0.02434)	(0.02434)	(0.02433)	(0.02433)	(0.02433)	(0.02433)
800.	0.68012	0.69460	0.70911	0.72364	0.73820	0.75279
	(0.02553)	(0.02553)	(0.02553)	(0.02553)	(0.02553)	(0.02553)
850.	0.65038	0.66414	0.67793	0.69173	0.70555	0.71940
	(0.02671)	(0.02671)	(0.02671)	(0.02671)	(0.02672)	(0.02672)
900.	0.62348	0.63660	0.64974	0.66290	0.67607	0.68927
	(0.02788)	(0.02788)	(0.02789)	(0.02789)	(0.02789)	(0.02789)
950.	0.59895	0.61151	0.62408	0.63666	0.64926	0.66187
	(0.02904)	(0.02905)	(0.02905)	(0.02905)	(0.02905)	(0.02905)
1000.	0.57648	0.58852	0.60057	0.61263	0.62471	0.63679
	(0.03020)	(0.03020)	(0.03020)	(0.03020)	(0.03020)	(0.03021)
1050.	0.55577	0.56734	0.57892	0.59052	0.60212	0.61373
	(0.03134)	(0.03134)	(0.03134)	(0.03134)	(0.03135)	(0.03135)
1100.	0.53661	0.54776	0.55891	0.57007	0.58124	0.59242
	(0.03247)	(0.03247)	(0.03247)	(0.03247)	(0.03248)	(0.03248)
1150.	0.51882	0.52958	0.54034	0.55110	0.56187	0.57265
	(0.03358)	(0.03359)	(0.03359)	(0.03359)	(0.03360)	(0.03360)
1200.	0.50226	0.51265	0.52304	0.53344	0.54385	0.55426
	(0.03469)	(0.03470)	(0.03470)	(0.03470)	(0.03471)	(0.03471)
1250.	0.48678	0.49683	0.50689	0.51695	0.52702	0.53709
	(0.03579)	(0.03579)	(0.03580)	(0.03580)	(0.03581)	(0.03581)
1300.	0.47228	0.48202	0.49176	0.50151	0.51126	0.52102
	(0.03688)	(0.03688)	(0.03689)	(0.03689)	(0.03689)	(0.03690)
1350.	0.45867	0.46811	0.47756	0.48702	0.49647	0.50593
	(0.03796)	(0.03796)	(0.03796)	(0.03797)	(0.03797)	(0.03797)
1400.	0.44585	0.45503	0.46420	0.47338	0.48256	0.49175
	(0.03902)	(0.03903)	(0.03903)	(0.03904)	(0.03904)	(0.03904)
1450.	0.43377	0.44269	0.45161	0.46053	0.46945	0.47837
	(0.04008)	(0.04008)	(0.04009)	(0.04009)	(0.04010)	(0.04010)
1500.	0.42236	0.43103	0.43971	0.44838	0.45706	0.46574
	(0.04113)	(0.04113)	(0.04114)	(0.04114)	(0.04114)	(0.04115)

†Absolute viscosity μ_{cP} is shown in parentheses.

SOURCE: Derived from ASME, 1977.

	Pressure, psia					
	550.	**560.**	**570.**	**580.**	**590.**	**600.**
Temp., °F	476.94	478.84	480.72	482.57	484.40	486.20
Sat. vap.	1.18798	1.21011	1.23229	1.25452	1.27680	1.29913
	(0.01749)	(0.01753)	(0.01757)	(0.01760)	(0.01764)	(0.01768)
Sat. liq.	50.14921	50.05258	49.95671	49.86160	49.76720	49.67341
	(0.10711)	(0.10663)	(0.10616)	(0.10570)	(0.10524)	(0.10480)
Temp., °F						
220.	59.72242	59.72451	59.72659	59.72867	59.73075	59.73282
	(0.27042)	(0.27044)	(0.27046)	(0.27048)	(0.27050)	(0.27052)
240.	59.19463	59.19679	59.19895	59.20110	59.20326	59.20542
	(0.24270)	(0.24272)	(0.24274)	(0.24276)	(0.24278)	(0.24280)
260.	58.63436	58.63661	58.63886	58.64110	58.64335	58.64560
	(0.21973)	(0.21975)	(0.21977)	(0.21979)	(0.21981)	(0.21983)
280.	58.04191	58.04426	58.04661	58.04897	58.05132	58.05367
	(0.20049)	(0.20051)	(0.20053)	(0.20055)	(0.20057)	(0.20059)
300.	57.41702	57.41950	57.42197	57.42444	57.42691	57.42938
	(0.18423)	(0.18425)	(0.18427)	(0.18428)	(0.18430)	(0.18432)
320.	56.75892	56.76153	56.76414	56.76675	56.76935	56.77196
	(0.17036)	(0.17038)	(0.17040)	(0.17041)	(0.17043)	(0.17045)
340.	56.06629	56.06906	56.07182	56.07459	56.07735	56.08011
	(0.15844)	(0.15845)	(0.15847)	(0.15849)	(0.15851)	(0.15852)
360.	55.33724	55.34019	55.34313	55.34608	55.34902	55.35196
	(0.14810)	(0.14812)	(0.14813)	(0.14815)	(0.14817)	(0.14819)
380.	54.56921	54.57237	54.57552	54.57868	54.58184	54.58499
	(0.13906)	(0.13908)	(0.13910)	(0.13911)	(0.13913)	(0.13915)
400.	53.75886	53.76227	53.76568	53.76908	53.77248	53.77588
	(0.13108)	(0.13110)	(0.13112)	(0.13114)	(0.13116)	(0.13117)
420.	52.90187	52.90557	52.90928	52.91298	52.91667	52.92037
	(0.12398)	(0.12399)	(0.12401)	(0.12403)	(0.12405)	(0.12407)
440.	51.99261	51.99667	52.00073	52.00479	52.00884	52.01289
	(0.11757)	(0.11759)	(0.11761)	(0.11762)	(0.11764)	(0.11766)
460.	51.02371	51.02821	51.03271	51.03720	51.04169	51.04617
	(0.11172)	(0.11174)	(0.11176)	(0.11178)	(0.11180)	(0.11182)
480.	1.17973	1.20688	49.99543	50.00046	50.00550	50.01052
	(0.01757)	(0.01756)	(0.10635)	(0.10637)	(0.10639)	(0.10641)
500.	1.13050	1.15567	1.18108	1.20675	1.23267	1.25887
	(0.01810)	(0.01809)	(0.01808)	(0.01807)	(0.01806)	(0.01805)
520.	1.08793	1.11152	1.13530	1.15927	1.18345	1.20784
	(0.01862)	(0.01861)	(0.01860)	(0.01860)	(0.01859)	(0.01858)
540.	1.05046	1.07275	1.09520	1.11781	1.14057	1.16351
	(0.01914)	(0.01913)	(0.01912)	(0.01912)	(0.01911)	(0.01910)
560.	1.01702	1.03822	1.05955	1.08102	1.10261	1.12434
	(0.01965)	(0.01964)	(0.01963)	(0.01963)	(0.01962)	(0.01962)

			Pressure, psia			
	550.	560.	570.	580.	590.	600.
Temp., °F						
600.	0.95927	0.97874	0.99830	1.01795	1.03769	1.05753
	(0.02065)	(0.02065)	(0.02065)	(0.02064)	(0.02064)	(0.02063)
650.	0.89943	0.91725	0.93514	0.95310	0.97112	0.98920
	(0.02190)	(0.02189)	(0.02189)	(0.02189)	(0.02189)	(0.02188)
700.	0.84911	0.86567	0.88226	0.89891	0.91560	0.93233
	(0.02312)	(0.02312)	(0.02312)	(0.02312)	(0.02312)	(0.02311)
750.	0.80566	0.82117	0.83672	0.85230	0.86792	0.88357
	(0.02433)	(0.02433)	(0.02433)	(0.02433)	(0.02433)	(0.02433)
800.	0.76740	0.78204	0.79671	0.81140	0.82612	0.84087
	(0.02553)	(0.02553)	(0.02553)	(0.02553)	(0.02553)	(0.02553)
850.	0.73326	0.74715	0.76106	0.77499	0.78894	0.80291
	(0.02672)	(0.02672)	(0.02672)	(0.02672)	(0.02672)	(0.02672)
900.	0.70248	0.71570	0.72894	0.74220	0.75548	0.76878
	(0.02789)	(0.02789)	(0.02790)	(0.02790)	(0.02790)	(0.02790)
950.	0.67449	0.68713	0.69978	0.71245	0.72513	0.73782
	(0.02906)	(0.02906)	(0.02906)	(0.02906)	(0.02906)	(0.02907)
1000.	0.64889	0.66100	0.67312	0.68525	0.69740	0.70955
	(0.03021)	(0.03021)	(0.03021)	(0.03022)	(0.03022)	(0.03022)
1050.	0.62535	0.63698	0.64862	0.66027	0.67193	0.68360
	(0.03135)	(0.03135)	(0.03136)	(0.03136)	(0.03136)	(0.03136)
1100.	0.60360	0.61480	0.62600	0.63721	0.64842	0.65965
	(0.03248)	(0.03249)	(0.03249)	(0.03249)	(0.03249)	(0.03250)
1150.	0.58344	0.59423	0.60503	0.61584	0.62665	0.63747
	(0.03360)	(0.03361)	(0.03361)	(0.03361)	(0.03362)	(0.03362)
1200.	0.56468	0.57510	0.58553	0.59596	0.60640	0.61685
	(0.03471)	(0.03472)	(0.03472)	(0.03472)	(0.03473)	(0.03473)
1250.	0.54716	0.55725	0.56733	0.57742	0.58752	0.59762
	(0.03581)	(0.03582)	(0.03582)	(0.03582)	(0.03583)	(0.03583)
1300.	0.53077	0.54054	0.55030	0.56008	0.56985	0.57963
	(0.03690)	(0.03690)	(0.03691)	(0.03691)	(0.03691)	(0.03692)
1350.	0.51540	0.52486	0.53433	0.54381	0.55328	0.56276
	(0.03798)	(0.03798)	(0.03799)	(0.03799)	(0.03799)	(0.03800)
1400.	0.50093	0.51012	0.51932	0.52851	0.53771	0.54691
	(0.03905)	(0.03905)	(0.03905)	(0.03906)	(0.03906)	(0.03906)
1450.	0.48730	0.49623	0.50516	0.51410	0.52304	0.53198
	(0.04010)	(0.04011)	(0.04011)	(0.04011)	(0.04012)	(0.04012)
1500.	0.47443	0.48311	0.49180	0.50049	0.50919	0.51788
	(0.04115)	(0.04115)	(0.04116)	(0.04116)	(0.04117)	(0.04117)

†Absolute viscosity μ_{cP} is shown in parentheses.

SOURCE: Derived from ASME, 1977.

			Pressure, psia			
	610.	620.	630.	640.	650.	660.
Temp., °F	487.98	489.74	491.48	493.20	494.89	496.57
Sat. vap.	1.32151	1.34395	1.36643	1.38898	1.41158	1.43423
	(0.01772)	(0.01776)	(0.01779)	(0.01783)	(0.01787)	(0.01790)
Sat. liq.	49.58036	49.48797	49.39621	49.30506	49.21459	49.12463
	(0.10436)	(0.10393)	(0.10351)	(0.10310)	(0.10270)	(0.10230)
Temp., °F						
220.	59.73490	59.73698	59.73906	59.74114	59.74321	59.74529
	(0.27054)	(0.27055)	(0.27057)	(0.27059)	(0.27061)	(0.27063)
240.	59.20758	59.20973	59.21189	59.21404	59.21620	59.21835
	(0.24282)	(0.24283)	(0.24285)	(0.24287)	(0.24289)	(0.24291)
260.	58.64785	58.65009	58.65234	58.65458	58.65683	58.65907
	(0.21984)	(0.21986)	(0.21988)	(0.21990)	(0.21992)	(0.21994)
280.	58.05602	58.05837	58.06072	58.06306	58.06541	58.06776
	(0.20060)	(0.20062)	(0.20064)	(0.20066)	(0.20068)	(0.20070)
300.	57.43185	57.43431	57.43678	57.43925	57.44171	57.44418
	(0.18434)	(0.18436)	(0.18437)	(0.18439)	(0.18441)	(0.18443)
320.	56.77456	56.77717	56.77977	56.78237	56.78497	56.78757
	(0.17047)	(0.17049)	(0.17050)	(0.17052)	(0.17054)	(0.17056)
340.	56.08287	56.08563	56.08839	56.09115	56.09390	56.09666
	(0.15854)	(0.15856)	(0.15858)	(0.15860)	(0.15861)	(0.15863)
360.	55.35490	55.35784	55.36078	55.36372	55.36666	55.36959
	(0.14820)	(0.14822)	(0.14824)	(0.14826)	(0.14827)	(0.14829)
380.	54.58814	54.59129	54.59444	54.59759	54.60073	54.60388
	(0.13917)	(0.13918)	(0.13920)	(0.13922)	(0.13924)	(0.13925)
400.	53.77928	53.78268	53.78608	53.78947	53.79286	53.79625
	(0.13119)	(0.13121)	(0.13123)	(0.13124)	(0.13126)	(0.13128)
420.	52.92406	52.92775	52.93144	52.93513	52.93882	52.94250
	(0.12408)	(0.12410)	(0.12412)	(0.12414)	(0.12415)	(0.12417)
440.	52.01694	52.02098	52.02503	52.02907	52.03310	52.03714
	(0.11768)	(0.11770)	(0.11772)	(0.11773)	(0.11775)	(0.11777)
460.	51.05066	51.05513	51.05961	51.06408	51.06855	51.07302
	(0.11183)	(0.11185)	(0.11187)	(0.11189)	(0.11191)	(0.11193)
480.	50.01555	50.02057	50.02558	50.03059	50.03560	50.04060
	(0.10643)	(0.10645)	(0.10646)	(0.10648)	(0.10650)	(0.10652)
500.	1.28534	1.31210	1.33916	1.36652	1.39419	1.42218
	(0.01804)	(0.01803)	(0.01802)	(0.01801)	(0.01801)	(0.01800)
520.	1.23244	1.25726	1.28230	1.30757	1.33308	1.35883
	(0.01857)	(0.01856)	(0.01856)	(0.01855)	(0.01854)	(0.01853)
540.	1.18661	1.20988	1.23333	1.25696	1.28078	1.30478
	(0.01909)	(0.01909)	(0.01908)	(0.01908)	(0.01907)	(0.01906)
560.	1.14621	1.16822	1.19037	1.21266	1.23511	1.25770
	(0.01961)	(0.01961)	(0.01960)	(0.01959)	(0.01959)	(0.01958)

Temp., °F	Pressure, psia					
	610.	620.	630.	640.	650.	660.
600.	1.07746 (0.02063)	1.09750 (0.02063)	1.11762 (0.02062)	1.13785 (0.02062)	1.15818 (0.02061)	1.17862 (0.02061)
650.	1.00734 (0.02188)	1.02556 (0.02188)	1.04384 (0.02188)	1.06219 (0.02187)	1.08060 (0.02187)	1.09909 (0.02187)
700.	0.94912 (0.02311)	0.96595 (0.02311)	0.98282 (0.02311)	0.99975 (0.02311)	1.01673 (0.02311)	1.03375 (0.02311)
750.	0.89926 (0.02433)	0.91498 (0.02433)	0.93074 (0.02433)	0.94654 (0.02433)	0.96237 (0.02433)	0.97823 (0.02433)
800.	0.85565 (0.02553)	0.87045 (0.02553)	0.88528 (0.02553)	0.90013 (0.02553)	0.91502 (0.02553)	0.92993 (0.02554)
850.	0.81690 (0.02672)	0.83092 (0.02672)	0.84495 (0.02673)	0.85901 (0.02673)	0.87309 (0.02673)	0.88719 (0.02673)
900.	0.78209 (0.02790)	0.79542 (0.02790)	0.80876 (0.02790)	0.82213 (0.02791)	0.83551 (0.02791)	0.84891 (0.02791)
950.	0.75053 (0.02907)	0.76325 (0.02907)	0.77598 (0.02907)	0.78873 (0.02907)	0.80150 (0.02908)	0.81428 (0.02908)
1000.	0.72172 (0.03022)	0.73390 (0.03023)	0.74609 (0.03023)	0.75829 (0.03023)	0.77050 (0.03023)	0.78273 (0.03024)
1050.	0.69527 (0.03137)	0.70696 (0.03137)	0.71866 (0.03137)	0.73037 (0.03138)	0.74208 (0.03138)	0.75381 (0.03138)
1100.	0.67088 (0.03250)	0.68212 (0.03250)	0.69337 (0.03251)	0.70463 (0.03251)	0.71590 (0.03251)	0.72717 (0.03252)
1150.	0.64829 (0.03362)	0.65913 (0.03363)	0.66997 (0.03363)	0.68081 (0.03363)	0.69166 (0.03363)	0.70252 (0.03364)
1200.	0.62730 (0.03473)	0.63775 (0.03474)	0.64822 (0.03474)	0.65869 (0.03474)	0.66916 (0.03475)	0.67964 (0.03475)
1250.	0.60772 (0.03583)	0.61783 (0.03584)	0.62794 (0.03584)	0.63806 (0.03584)	0.64819 (0.03585)	0.65832 (0.03585)
1300.	0.58941 (0.03692)	0.59920 (0.03693)	0.60899 (0.03693)	0.61879 (0.03693)	0.62859 (0.03694)	0.63839 (0.03694)
1350.	0.57225 (0.03800)	0.58174 (0.03800)	0.59123 (0.03801)	0.60072 (0.03801)	0.61022 (0.03802)	0.61972 (0.03802)
1400.	0.55612 (0.03907)	0.56532 (0.03907)	0.57453 (0.03908)	0.58375 (0.03908)	0.59296 (0.03908)	0.60218 (0.03909)
1450.	0.54092 (0.04013)	0.54987 (0.04013)	0.55881 (0.04013)	0.56776 (0.04014)	0.57672 (0.04014)	0.58567 (0.04015)
1500.	0.52658 (0.04117)	0.53528 (0.04118)	0.54398 (0.04118)	0.55268 (0.04118)	0.56139 (0.04119)	0.57009 (0.04119)

†Absolute viscosity μ_{cP} is shown in parentheses.

SOURCE: Derived from ASME, 1977.

G-43

Table G.1 Density (in Pounds-Mass per Cubic Foot) and Absolute Viscosity μ_{cP} (in Centipoises) of Steam and Compressed Water† (*continued*)

	Pressure, psia					
	670.	680.	690.	700.	710.	720.
Temp., °F	498.22	499.86	501.48	503.08	504.67	506.23
Sat. vap.	1.45694	1.47971	1.50254	1.52542	1.54836	1.57136
	(0.01794)	(0.01798)	(0.01801)	(0.01805)	(0.01808)	(0.01812)
Sat. liq.	49.03523	48.94637	48.85805	48.77023	48.68291	48.59598
	(0.10190)	(0.10152)	(0.10114)	(0.10077)	(0.10040)	(0.10003)
Temp., °F						
220.	59.74736	59.74944	59.75151	59.75359	59.75566	59.75773
	(0.27065)	(0.27067)	(0.27069)	(0.27070)	(0.27072)	(0.27074)
240.	59.22050	59.22265	59.22481	59.22696	59.22911	59.23126
	(0.24293)	(0.24295)	(0.24297)	(0.24298)	(0.24300)	(0.24302)
260.	58.66131	58.66355	58.66580	58.66804	58.67028	58.67252
	(0.21995)	(0.21997)	(0.21999)	(0.22001)	(0.22003)	(0.22005)
280.	58.07010	58.07245	58.07479	58.07714	58.07948	58.08182
	(0.20071)	(0.20073)	(0.20075)	(0.20077)	(0.20079)	(0.20081)
300.	57.44664	57.44910	57.45156	57.45403	57.45649	57.45895
	(0.18445)	(0.18446)	(0.18448)	(0.18450)	(0.18452)	(0.18454)
320.	56.79017	56.79277	56.79536	56.79796	56.80055	56.80315
	(0.17057)	(0.17059)	(0.17061)	(0.17063)	(0.17065)	(0.17066)
340.	56.09941	56.10216	56.10492	56.10767	56.11042	56.11317
	(0.15865)	(0.15867)	(0.15868)	(0.15870)	(0.15872)	(0.15874)
360.	55.37252	55.37546	55.37839	55.38132	55.38425	55.38718
	(0.14831)	(0.14833)	(0.14834)	(0.14836)	(0.14838)	(0.14840)
380.	54.60702	54.61016	54.61330	54.61644	54.61958	54.62271
	(0.13927)	(0.13929)	(0.13931)	(0.13932)	(0.13934)	(0.13936)
400.	53.79964	53.80303	53.80642	53.80980	53.81318	53.81656
	(0.13130)	(0.13131)	(0.13133)	(0.13135)	(0.13137)	(0.13138)
420.	52.94618	52.94986	52.95354	52.95721	52.96088	52.96455
	(0.12419)	(0.12421)	(0.12423)	(0.12424)	(0.12426)	(0.12428)
440.	52.04117	52.04520	52.04923	52.05326	52.05728	52.06130
	(0.11779)	(0.11781)	(0.11782)	(0.11784)	(0.11786)	(0.11788)
460.	51.07748	51.08194	51.08639	51.09085	51.09530	51.09974
	(0.11195)	(0.11197)	(0.11198)	(0.11200)	(0.11202)	(0.11204)
480.	50.04560	50.05060	50.05559	50.06057	50.06555	50.07053
	(0.10654)	(0.10656)	(0.10658)	(0.10660)	(0.10662)	(0.10664)
500.	1.45051	1.47919	48.94394	48.94960	48.95526	48.96092
	(0.01799)	(0.01798)	(0.10150)	(0.10153)	(0.10155)	(0.10157)
520.	1.38482	1.41108	1.43759	1.46438	1.49144	1.51879
	(0.01853)	(0.01852)	(0.01851)	(0.01850)	(0.01850)	(0.01849)
540.	1.32897	1.35336	1.37795	1.40274	1.42775	1.45297
	(0.01906)	(0.01905)	(0.01904)	(0.01904)	(0.01903)	(0.01902)
560.	1.28044	1.30334	1.32640	1.34962	1.37301	1.39656
	(0.01958)	(0.01957)	(0.01957)	(0.01956)	(0.01956)	(0.01955)

	Pressure, psia					
	670.	**680.**	**690.**	**700.**	**710.**	**720.**
Temp., °F						
600.	1.19915	1.21979	1.24054	1.26139	1.28236	1.30344
	(0.02061)	(0.02060)	(0.02060)	(0.02060)	(0.02059)	(0.02059)
650.	1.11764	1.13626	1.15496	1.17373	1.19256	1.21148
	(0.02187)	(0.02186)	(0.02186)	(0.02186)	(0.02186)	(0.02186)
700.	1.05083	1.06795	1.08512	1.10235	1.11963	1.13695
	(0.02311)	(0.02311)	(0.02310)	(0.02310)	(0.02310)	(0.02310)
750.	0.99414	1.01008	1.02606	1.04207	1.05813	1.07422
	(0.02433)	(0.02433)	(0.02433)	(0.02433)	(0.02433)	(0.02433)
800.	0.94487	0.95984	0.97484	0.98987	1.00492	1.02000
	(0.02554)	(0.02554)	(0.02554)	(0.02554)	(0.02554)	(0.02554)
850.	0.90132	0.91546	0.92963	0.94382	0.95804	0.97227
	(0.02673)	(0.02673)	(0.02673)	(0.02673)	(0.02674)	(0.02674)
900.	0.86232	0.87576	0.88921	0.90268	0.91616	0.92967
	(0.02791)	(0.02791)	(0.02792)	(0.02792)	(0.02792)	(0.02792)
950.	0.82707	0.83988	0.85270	0.86553	0.87838	0.89125
	(0.02908)	(0.02908)	(0.02909)	(0.02909)	(0.02909)	(0.02909)
1000.	0.79496	0.80721	0.81947	0.83174	0.84403	0.85632
	(0.03024)	(0.03024)	(0.03024)	(0.03025)	(0.03025)	(0.03025)
1050.	0.76554	0.77729	0.78905	0.80081	0.81258	0.82437
	(0.03138)	(0.03139)	(0.03139)	(0.03139)	(0.03140)	(0.03140)
1100.	0.73845	0.74974	0.76104	0.77234	0.78366	0.79498
	(0.03252)	(0.03252)	(0.03252)	(0.03253)	(0.03253)	(0.03253)
1150.	0.71339	0.72426	0.73514	0.74603	0.75692	0.76782
	(0.03364)	(0.03364)	(0.03365)	(0.03365)	(0.03365)	(0.03366)
1200.	0.69012	0.70061	0.71111	0.72161	0.73212	0.74263
	(0.03475)	(0.03476)	(0.03476)	(0.03476)	(0.03477)	(0.03477)
1250.	0.66845	0.67859	0.68873	0.69888	0.70903	0.71919
	(0.03585)	(0.03586)	(0.03586)	(0.03587)	(0.03587)	(0.03587)
1300.	0.64820	0.65801	0.66783	0.67765	0.68747	0.69730
	(0.03694)	(0.03695)	(0.03695)	(0.03696)	(0.03696)	(0.03696)
1350.	0.62923	0.63873	0.64825	0.65776	0.66728	0.67680
	(0.03802)	(0.03803)	(0.03803)	(0.03803)	(0.03804)	(0.03804)
1400.	0.61140	0.62063	0.62986	0.63909	0.64832	0.65756
	(0.03909)	(0.03910)	(0.03910)	(0.03910)	(0.03911)	(0.03911)
1450.	0.59463	0.60359	0.61255	0.62152	0.63048	0.63945
	(0.04015)	(0.04015)	(0.04016)	(0.04016)	(0.04016)	(0.04017)
1500.	0.57880	0.58751	0.59623	0.60494	0.61366	0.62238
	(0.04120)	(0.04120)	(0.04120)	(0.04121)	(0.04121)	(0.04122)

†Absolute viscosity μ_{cP} is shown in parentheses.

SOURCE: Derived from ASME, 1977.

804 of 980

Table G.1 **Density (in Pounds-Mass per Cubic Foot) and Absolute Viscosity μ_{cP} (in Centipoises) of Steam and Compressed Water† (*continued*)**

	Pressure, psia					
	740.	**750.**	**760.**	**770.**	**780.**	**790.**
Temp., °F	509.32	510.84	512.34	518.83	515.30	516.76
Sat. vap.	1.61754	1.64073	1.66397	1.68728	1.71065	1.73409
	(0.01818)	(0.01822)	(0.01825)	(0.01828)	(0.01832)	(0.01835)
Sat. liq.	48.42369	48.33821	48.25316	48.16854	48.08432	48.00049
	(0.09933)	(0.09898)	(0.09864)	(0.09830)	(0.09797)	(0.09764)
Temp., °F						
220.	59.76188	59.76395	59.76602	59.76809	59.77016	59.77223
	(0.27078)	(0.27080)	(0.27082)	(0.27084)	(0.27085)	(0.27087)
240.	59.23556	59.23771	59.23985	59.24200	59.24415	59.24630
	(0.24306)	(0.24308)	(0.24310)	(0.24311)	(0.24313)	(0.24315)
260.	58.67699	58.67923	58.68147	58.68371	58.68594	58.68818
	(0.22008)	(0.22010)	(0.22012)	(0.22014)	(0.22016)	(0.22018)
280.	58.08650	58.08884	58.09118	58.09352	58.09586	58.09820
	(0.20084)	(0.20086)	(0.20088)	(0.20090)	(0.20091)	(0.20093)
300.	57.46386	57.46632	57.46878	57.47123	57.47369	57.47614
	(0.18457)	(0.18459)	(0.18461)	(0.18463)	(0.18464)	(0.18466)
320.	56.80833	56.81093	56.81352	56.81611	56.81869	56.82128
	(0.17070)	(0.17072)	(0.17073)	(0.17075)	(0.17077)	(0.17079)
340.	56.11866	56.12141	56.12415	56.12690	56.12964	56.13238
	(0.15877)	(0.15879)	(0.15881)	(0.15882)	(0.15884)	(0.15886)
360.	55.39303	55.39595	55.39887	55.40180	55.40472	55.40764
	(0.14843)	(0.14845)	(0.14847)	(0.14848)	(0.14850)	(0.14852)
380.	54.62898	54.63211	54.63524	54.63837	54.64150	54.64463
	(0.13939)	(0.13941)	(0.13943)	(0.13944)	(0.13946)	(0.13948)
400.	53.82332	53.82669	53.83007	53.83344	53.83681	53.84018
	(0.13142)	(0.13144)	(0.13145)	(0.13147)	(0.13149)	(0.13151)
420.	52.97189	52.97555	52.97922	52.98288	52.98654	52.99019
	(0.12431)	(0.12433)	(0.12435)	(0.12437)	(0.12438)	(0.12440)
440.	52.06933	52.07334	52.07735	52.08136	52.08536	52.08936
	(0.11791)	(0.11793)	(0.11795)	(0.11797)	(0.11799)	(0.11800)
460.	51.10863	51.11306	51.11750	51.12193	51.12635	51.13078
	(0.11208)	(0.11210)	(0.11212)	(0.11213)	(0.11215)	(0.11217)
480.	50.08047	50.08544	50.09040	50.09536	50.10031	50.10526
	(0.10668)	(0.10670)	(0.10672)	(0.10674)	(0.10676)	(0.10678)
500.	48.97221	48.97785	48.98348	48.98911	48.99473	49.00034
	(0.10161)	(0.10163)	(0.10165)	(0.10167)	(0.10169)	(0.10171)
520.	1.57439	1.60265	1.63123	1.66015	1.68942	1.71904
	(0.01847)	(0.01847)	(0.01846)	(0.01845)	(0.01845)	(0.01844)
540.	1.50408	1.52998	1.55611	1.58250	1.60913	1.63602
	(0.01901)	(0.01901)	(0.01900)	(0.01900)	(0.01899)	(0.01898)
560.	1.44419	1.46826	1.49253	1.51697	1.54161	1.56644
	(0.01954)	(0.01954)	(0.01953)	(0.01953)	(0.01952)	(0.01952)

	Pressure, psia					
	740.	750.	760.	770.	780.	790.
Temp., °F						
600.	1.34593 (0.02058)	1.36735 (0.02058)	1.38889 (0.02057)	1.41055 (0.02057)	1.43233 (0.02057)	1.45423 (0.02056)
650.	1.24952 (0.02185)	1.26866 (0.02185)	1.28787 (0.02185)	1.30716 (0.02185)	1.32653 (0.02184)	1.34598 (0.02184)
700.	1.17176 (0.02310)	1.18925 (0.02310)	1.20678 (0.02310)	1.22438 (0.02310)	1.24202 (0.02310)	1.25972 (0.02310)
750.	1.10651 (0.02433)	1.12272 (0.02433)	1.13896 (0.02433)	1.15525 (0.02433)	1.17157 (0.02433)	1.18793 (0.02433)
800.	1.05026 (0.02554)	1.06543 (0.02554)	1.08063 (0.02554)	1.09586 (0.02554)	1.11111 (0.02554)	1.12640 (0.02555)
850.	1.00081 (0.02674)	1.01511 (0.02674)	1.02944 (0.02674)	1.04379 (0.02674)	1.05816 (0.02675)	1.07255 (0.02675)
900.	0.95673 (0.02792)	0.97029 (0.02793)	0.98387 (0.02793)	0.99746 (0.02793)	1.01108 (0.02793)	1.02471 (0.02793)
950.	0.91702 (0.02910)	0.92993 (0.02910)	0.94285 (0.02910)	0.95579 (0.02910)	0.96874 (0.02911)	0.98171 (0.02911)
1000.	0.88095 (0.03026)	0.89328 (0.03026)	0.90562 (0.03026)	0.91798 (0.03026)	0.93034 (0.03027)	0.94272 (0.03027)
1050.	0.84797 (0.03140)	0.85978 (0.03141)	0.87160 (0.03141)	0.88344 (0.03141)	0.89528 (0.03142)	0.90713 (0.03142)
1100.	0.81765 (0.03254)	0.82899 (0.03254)	0.84035 (0.03255)	0.85171 (0.03255)	0.86308 (0.03255)	0.87446 (0.03256)
1150.	0.78964 (0.03367)	0.80056 (0.03367)	0.81149 (0.03367)	0.82243 (0.03368)	0.83337 (0.03368)	0.84431 (0.03368)
1200.	0.76368 (0.03478)	0.77421 (0.03478)	0.78475 (0.03479)	0.79529 (0.03479)	0.80584 (0.03479)	0.81639 (0.03480)
1250.	0.73952 (0.03588)	0.74969 (0.03588)	0.75987 (0.03589)	0.77005 (0.03589)	0.78023 (0.03589)	0.79043 (0.03590)
1300.	0.71697 (0.03697)	0.72681 (0.03697)	0.73665 (0.03698)	0.74650 (0.03698)	0.75635 (0.03699)	0.76621 (0.03699)
1350.	0.69585 (0.03805)	0.70539 (0.03805)	0.71492 (0.03806)	0.72446 (0.03806)	0.73400 (0.03807)	0.74355 (0.03807)
1400.	0.67604 (0.03912)	0.68529 (0.03912)	0.69454 (0.03913)	0.70379 (0.03913)	0.71304 (0.03913)	0.72230 (0.03914)
1450.	0.65740 (0.04018)	0.66638 (0.04018)	0.67536 (0.04018)	0.68434 (0.04019)	0.69333 (0.04019)	0.70231 (0.04020)
1500.	0.63983 (0.04122)	0.64855 (0.04123)	0.65728 (0.04123)	0.66601 (0.04124)	0.67475 (0.04124)	0.68348 (0.04124)

†Absolute viscosity μ_{cP} is shown in parentheses.

SOURCE: Derived from ASME, 1977.

Table G.1 Density (in Pounds-Mass per Cubic Foot) and
Absolute Viscosity μ_{cP} (in Centipoises) of Steam and
Compressed Water† (*continued*)

	Pressure, psia					
	800.	**810.**	**820.**	**830.**	**840.**	**850.**
Temp., °F	518.21	519.64	521.06	522.46	523.86	525.24
Sat. vap.	1.75759	1.78117	1.80481	1.82851	1.85228	1.87612
	(0.01838)	(0.01842)	(0.01845)	(0.01848)	(0.01851)	(0.01855)
Sat. liq.	47.91706	47.83411	47.75142	47.66909	47.58711	47.50547
	(0.09732)	(0.09700)	(0.09668)	(0.09637)	(0.09606)	(0.09576)
Temp., °F						
220.	59.77430	59.77637	59.77844	59.78050	59.78257	59.78464
	(0.27089)	(0.27091)	(0.27093)	(0.27095)	(0.27097)	(0.27098)
240.	59.24844	59.25059	59.25273	59.25488	59.25702	59.25916
	(0.24317)	(0.24319)	(0.24321)	(0.24323)	(0.24325)	(0.24326)
260.	58.69041	58.69265	58.69488	58.69711	58.69934	58.70158
	(0.22020)	(0.22021)	(0.22023)	(0.22025)	(0.22027)	(0.22029)
280.	58.10053	58.10287	58.10520	58.10754	58.10987	58.11221
	(0.20095)	(0.20097)	(0.20099)	(0.20101)	(0.20102)	(0.20104)
300.	57.47859	57.48105	57.48350	57.48595	57.48840	57.49085
	(0.18468)	(0.18470)	(0.18472)	(0.18473)	(0.18475)	(0.18477)
320.	56.82387	56.82646	56.82904	56.83163	56.83421	56.83679
	(0.17081)	(0.17082)	(0.17084)	(0.17086)	(0.17088)	(0.17089)
340.	56.13512	56.13786	56.14060	56.14334	56.14607	56.14881
	(0.15888)	(0.15889)	´(0.15891)	(0.15893)	(0.15895)	(0.15896)
360.	55.41055	55.41347	55.41639	55.41930	55.42221	55.42513
	(0.14854)	(0.14855)	(0.14857)	(0.14859)	(0.14861)	(0.14862)
380.	54.64775	54.65087	54.65399	54.65711	54.66023	54.66335
	(0.13950)	(0.13951)	(0.13953)	(0.13955)	(0.13957)	(0.13958)
400.	53.84354	53.84691	53.85027	53.85363	53.85699	53.86035
	(0.13152)	(0.13154)	(0.13156)	(0.13158)	(0.13159)	(0.13161)
420.	52.99385	52.99750	53.00115	53.00480	53.00844	53.01209
	(0.12442)	(0.12444)	(0.12446)	(0.12447)	(0.12449)	(0.12451)
440.	52.09336	52.09736	52.10136	52.10535	52.10934	52.11332
	(0.11802)	(0.11804)	(0.11806)	(0.11808)	(0.11810)	(0.11811)
460.	51.13520	51.13962	51.14403	51.14844	51.15285	51.15725
	(0.11219)	(0.11221)	(0.11223)	(0.11225)	(0.11226)	(0.11228)
480.	50.11020	50.11514	50.12008	50.12501	50.12994	50.13486
	(0.10680)	(0.10682)	(0.10684)	(0.10686)	(0.10687)	(0.10689)
500.	49.00595	49.01156	49.01716	49.02275	49.02834	49.03392
	(0.10173)	(0.10175)	(0.10177)	(0.10179)	(0.10181)	(0.10183)
520.	1.74903	1.77940	47.81798	47.82444	47.83090	47.83734
	(0.01843)	(0.01843)	(0.09693)	(0.09695)	(0.09698)	(0.09700)
540.	1.66317	1.69059	1.71830	1.74629	1.77457	1.80316
	(0.01898)	(0.01897)	(0.01897)	(0.01896)	(0.01896)	(0.01895)
560.	1.59148	1.61671	1.64215	1.66781	1.69368	1.71978
	(0.01951)	(0.01951)	(0.01950)	(0.01950)	(0.01950)	(0.01949)

Temp., °F	Pressure, psia					
	800.	810.	820.	830.	840.	850.
600.	1.47626 (0.02056)	1.49841 (0.02056)	1.52070 (0.02056)	1.54311 (0.02055)	1.56566 (0.02055)	1.58834 (0.02055)
650.	1.36551 (0.02184)	1.38511 (0.02184)	1.40480 (0.02184)	1.42457 (0.02184)	1.44443 (0.02183)	1.46436 (0.02183)
700.	1.27748 (0.02310)	1.29528 (0.02310)	1.31315 (0.02309)	1.33107 (0.02309)	1.34905 (0.02309)	1.36709 (0.02309)
750.	1.20434 (0.02433)	1.22078 (0.02433)	1.23726 (0.02433)	1.25379 (0.02433)	1.27035 (0.02433)	1.28696 (0.02433)
800.	1.14172 (0.02555)	1.15707 (0.02555)	1.17245 (0.02555)	1.18786 (0.02555)	1.20330 (0.02555)	1.21877 (0.02555)
850.	1.08697 (0.02675)	1.10141 (0.02675)	1.11588 (0.02675)	1.13037 (0.02675)	1.14488 (0.02676)	1.15941 (0.02676)
900.	1.03836 (0.02794)	1.05203 (0.02794)	1.06571 (0.02794)	1.07942 (0.02794)	1.09314 (0.02795)	1.10688 (0.02795)
950.	0.99469 (0.02911)	1.00768 (0.02911)	1.02070 (0.02912)	1.03372 (0.02912)	1.04676 (0.02912)	1.05982 (0.02912)
1000.	0.95511 (0.03027)	0.96752 (0.03028)	0.97993 (0.03028)	0.99236 (0.03028)	1.00480 (0.03028)	1.01725 (0.03029)
1050.	0.91900 (0.03142)	0.93087 (0.03143)	0.94275 (0.03143)	0.95465 (0.03143)	0.96655 (0.03144)	0.97846 (0.03144)
1100.	0.88585 (0.03256)	0.89724 (0.03256)	0.90864 (0.03257)	0.92006 (0.03257)	0.93148 (0.03257)	0.94290 (0.03258)
1150.	0.85527 (0.03369)	0.86623 (0.03369)	0.87720 (0.03369)	0.88817 (0.03370)	0.89915 (0.03370)	0.91014 (0.03370)
1200.	0.82695 (0.03480)	0.83751 (0.03480)	0.84808 (0.03481)	0.85866 (0.03481)	0.86924 (0.03481)	0.87983 (0.03482)
1250.	0.80062 (0.03590)	0.81082 (0.03591)	0.82103 (0.03591)	0.83124 (0.03591)	0.84145 (0.03592)	0.85167 (0.03592)
1300.	0.77607 (0.03699)	0.78593 (0.03700)	0.79580 (0.03700)	0.80567 (0.03701)	0.81555 (0.03701)	0.82543 (0.03701)
1350.	0.75310 (0.03807)	0.76265 (0.03808)	0.77221 (0.03808)	0.78177 (0.03809)	0.79133 (0.03809)	0.80090 (0.03809)
1400.	0.73156 (0.03914)	0.74082 (0.03915)	0.75009 (0.03915)	0.75936 (0.03915)	0.76863 (0.03916)	0.77790 (0.03916)
1450.	0.71130 (0.04020)	0.72030 (0.04020)	0.72929 (0.04021)	0.73829 (0.04021)	0.74729 (0.04022)	0.75629 (0.04022)
1500.	0.69222 (0.04125)	0.70096 (0.04125)	0.70970 (0.04126)	0.71844 (0.04126)	0.72719 (0.04126)	0.73594 (0.04127)

†Absolute viscosity μ_{cP} is shown in parentheses.

SOURCE: Derived from ASME, 1977.

Table G.1 Density (in Pounds-Mass per Cubic Foot) and Absolute Viscosity μ_{cP} (in Centipoises) of Steam and Compressed Water† (*continued*)

	Pressure, psia					
	875.	**900.**	**925.**	**950.**	**975.**	**1000.**
Temp., °F	528.63	531.95	535.21	538.39	541.52	544.58
Sat. vap.	1.93602	1.99636	2.05716	2.11842	2.18015	2.24237
	(0.01862)	(0.01870)	(0.01878)	(0.01886)	(0.01893)	(0.01901)
Sat. liq.	47.30277	47.10200	46.90302	46.70571	46.50985	46.31558
	(0.09501)	(0.09429)	(0.09358)	(0.09290)	(0.09223)	(0.09157)
Temp., °F						
220.	59.78980	59.79496	59.80012	59.80527	59.81043	59.81557
	(0.27103)	(0.27108)	(0.27112)	(0.27117)	(0.27122)	(0.27126)
240.	59.26452	59.26987	59.27522	59.28056	59.28590	59.29124
	(0.24331)	(0.24336)	(0.24340)	(0.24345)	(0.24350)	(0.24354)
260.	58.70715	58.71272	58.71829	58.72386	58.72942	58.73498
	(0.22033)	(0.22038)	(0.22043)	(0.22047)	(0.22052)	(0.22056)
280.	58.11804	58.12386	58.12968	58.13550	58.14131	58.14712
	(0.20109)	(0.20113)	(0.20118)	(0.20122)	(0.20127)	(0.20132)
300.	57.49697	57.50308	57.50919	57.51530	57.52140	57.52750
	(0.18482)	(0.18486)	(0.18491)	(0.18495)	(0.18500)	(0.18504)
320.	56.84324	56.84969	56.85613	56.86257	56.86900	56.87542
	(0.17094)	(0.17098)	(0.17103)	(0.17107)	(0.17112)	(0.17116)
340.	56.15564	56.16247	56.16929	56.17611	56.18292	56.18972
	(0.15901)	(0.15905)	(0.15910)	(0.15914)	(0.15918)	(0.15923)
360.	55.43240	55.43967	55.44693	55.45418	55.46143	55.46867
	(0.14867)	(0.14871)	(0.14875)	(0.14880)	(0.14884)	(0.14888)
380.	54.67114	54.67892	54.68669	54.69445	54.70220	54.70995
	(0.13963)	(0.13967)	(0.13971)	(0.13976)	(0.13980)	(0.13984)
400.	53.86874	53.87712	53.88549	53.89385	53.90219	53.91053
	(0.13165)	(0.13170)	(0.13174)	(0.13178)	(0.13183)	(0.13187)
420.	53.02119	53.03028	53.03935	53.04841	53.05747	53.06650
	(0.12455)	(0.12460)	(0.12464)	(0.12468)	(0.12473)	(0.12477)
440.	52.12328	52.13322	52.14315	52.15305	52.16295	52.17283
	(0.11816)	(0.11820)	(0.11825)	(0.11829)	(0.11834)	(0.11838)
460.	51.16825	51.17923	51.19019	51.20113	51.21205	51.22295
	(0.11233)	(0.11238)	(0.11242)	(0.11247)	(0.11251)	(0.11256)
480.	50.14715	50.15942	50.17166	50.18387	50.19606	50.20823
	(0.10694)	(0.10699)	(0.10704)	(0.10709)	(0.10714)	(0.10718)
500.	49.04785	49.06175	49.07562	49.08945	49.10325	49.11702
	(0.10189)	(0.10194)	(0.10199)	(0.10204)	(0.10209)	(0.10214)
520.	47.85342	47.86945	47.88544	47.90138	47.91728	47.93313
	(0.09705)	(0.09711)	(0.09716)	(0.09722)	(0.09727)	(0.09733)
540.	1.87603	1.95102	2.02831	2.10809	46.61444	46.63310
	(0.01894)	(0.01893)	(0.01891)	(0.01890)	(0.09258)	(0.09264)
560.	1.78603	1.85380	1.92319	1.99431	2.06728	2.14223
	(0.01948)	(0.01947)	(0.01946)	(0.01945)	(0.01944)	(0.01944)

	Pressure, psia					
	875.	**900.**	**925.**	**950.**	**975.**	**1000.**
Temp., °F						
600.	1.64566	1.70388	1.76303	1.82316	1.88431	1.94653
	(0.02054)	(0.02053)	(0.02053)	(0.02052)	(0.02052)	(0.02051)
650.	1.51458	1.56535	1.61668	1.66859	1.72110	1.77422
	(0.02183)	(0.02183)	(0.02182)	(0.02182)	(0.02182)	(0.02182)
700.	1.41243	1.45815	1.50423	1.55071	1.59758	1.64484
	(0.02309)	(0.02309)	(0.02309)	(0.02309)	(0.02309)	(0.02309)
750.	1.32865	1.37061	1.41284	1.45534	1.49812	1.54118
	(0.02433)	(0.02433)	(0.02434)	(0.02434)	(0.02434)	(0.02434)
800.	1.25758	1.29659	1.33580	1.37521	1.41482	1.45464
	(0.02556)	(0.02556)	(0.02556)	(0.02557)	(0.02557)	(0.02557)
850.	1.19586	1.23245	1.26919	1.30609	1.34315	1.38035
	(0.02676)	(0.02677)	(0.02677)	(0.02678)	(0.02678)	(0.02679)
900.	1.14132	1.17588	1.21055	1.24534	1.28026	1.31529
	(0.02795)	(0.02796)	(0.02797)	(0.02797)	(0.02798)	(0.02798)
950.	1.09252	1.12532	1.15821	1.19120	1.22428	1.25746
	(0.02913)	(0.02914)	(0.02914)	(0.02915)	(0.02916)	(0.02917)
1000.	1.04843	1.07969	1.11102	1.14243	1.17392	1.20548
	(0.03030)	(0.03030)	(0.03031)	(0.03032)	(0.03033)	(0.03033)
1050.	1.00828	1.03817	1.06812	1.09813	1.12821	1.15835
	(0.03145)	(0.03145)	(0.03146)	(0.03147)	(0.03148)	(0.03149)
1100.	0.97151	1.00017	1.02888	1.05764	1.08645	1.11532
	(0.03259)	(0.03259)	(0.03260)	(0.03261)	(0.03262)	(0.03263)
1150.	0.93765	0.96519	0.99278	1.02041	1.04808	1.07580
	(0.03371)	(0.03372)	(0.03373)	(0.03374)	(0.03375)	(0.03376)
1200.	0.90632	0.93285	0.95942	0.98602	1.01266	1.03934
	(0.03483)	(0.03484)	(0.03485)	(0.03486)	(0.03487)	(0.03488)
1250.	0.87724	0.90284	0.92848	0.95414	0.97983	1.00555
	(0.03593)	(0.03594)	(0.03595)	(0.03596)	(0.03597)	(0.03598)
1300.	0.85015	0.87489	0.89967	0.92446	0.94928	0.97413
	(0.03702)	(0.03703)	(0.03704)	(0.03705)	(0.03706)	(0.03707)
1350.	0.82483	0.84878	0.87276	0.89675	0.92077	0.94481
	(0.03810)	(0.03811)	(0.03812)	(0.03813)	(0.03814)	(0.03815)
1400.	0.80110	0.82432	0.84756	0.87081	0.89408	0.91737
	(0.03917)	(0.03918)	(0.03919)	(0.03920)	(0.03921)	(0.03922)
1450.	0.77881	0.80134	0.82389	0.84645	0.86903	0.89163
	(0.04023)	(0.04024)	(0.04025)	(0.04026)	(0.04027)	(0.04028)
1500.	0.75782	0.77971	0.80161	0.82353	0.84546	0.86741
	(0.04128)	(0.04129)	(0.04130)	(0.04131)	(0.04132)	(0.04133)

†Absolute viscosity μ_{cP} is shown in parentheses.

SOURCE: Derived from ASME, 1977.

Table G.1 Density (in Pounds-Mass per Cubic Foot) and Absolute Viscosity μ_{cP} (in Centipoises) of Steam and Compressed Water† (*continued*)

	Pressure, psia					
	1025.	1050.	1075.	1100.	1125.	1150.
Temp., °F	547.59	550.54	553.43	556.28	559.07	561.82
Sat. vap.	2.30509	2.36832	2.43207	2.49635	2.56118	2.62656
	(0.01909)	(0.01916)	(0.01923)	(0.01931)	(0.01938)	(0.01946)
Sat. liq.	46.12268	45.93107	45.74065	45.55135	45.36322	45.17592
	(0.09093)	(0.09031)	(0.08970)	(0.08910)	(0.08851)	(0.08794)
Temp., °F						
220.	59.82072	59.82586	59.83100	59.83614	59.84127	59.84640
	(0.27131)	(0.27136)	(0.27140)	(0.27145)	(0.27150)	(0.27154)
240.	59.29658	59.30191	59.30724	59.31256	59.31788	59.32320
	(0.24359)	(0.24364)	(0.24368)	(0.24373)	(0.24378)	(0.24382)
260.	58.74053	58.74608	58.75162	58.75717	58.76271	58.76824
	(0.22061)	(0.22066)	(0.22070)	(0.22075)	(0.22080)	(0.22084)
280.	58.15293	58.15873	58.16452	58.17031	58.17610	58.18188
	(0.20136)	(0.20141)	(0.20145)	(0.20150)	(0.20154)	(0.20159)
300.	57.53359	57.53967	57.54576	57.55183	57.55790	57.56397
	(0.18508)	(0.18513)	(0.18517)	(0.18522)	(0.18526)	(0.18531)
320.	56.88184	56.88826	56.89467	56.90107	56.90747	56.91386
	(0.17120)	(0.17125)	(0.17129)	(0.17134)	(0.17138)	(0.17143)
340.	56.19651	56.20330	56.21009	56.21687	56.22364	56.23040
	(0.15927)	(0.15931)	(0.15936)	(0.15940)	(0.15945)	(0.15949)
360.	55.47590	55.48312	55.49034	55.49754	55.50475	55.51194
	(0.14893)	(0.14897)	(0.14901)	(0.14906)	(0.14910)	(0.14914)
380.	54.71768	54.72541	54.73313	54.74084	54.74854	54.75623
	(0.13989)	(0.13993)	(0.13997)	(0.14002)	(0.14006)	(0.14010)
400.	53.91886	53.92717	53.93548	53.94377	53.95206	53.96034
	(0.13192)	(0.13196)	(0.13200)	(0.13205)	(0.13209)	(0.13213)
420.	53.07553	53.08454	53.09354	53.10253	53.11151	53.12047
	(0.12482)	(0.12486)	(0.12490)	(0.12495)	(0.12499)	(0.12504)
440.	52.18269	52.19254	52.20237	52.21219	52.22199	52.23178
	(0.11843)	(0.11847)	(0.11852)	(0.11856)	(0.11861)	(0.11865)
460.	51.23383	51.24469	51.25554	51.26636	51.27717	51.28795
	(0.11261)	(0.11265)	(0.11270)	(0.11274)	(0.11279)	(0.11284)
480.	50.22037	50.23249	50.24458	50.25665	50.26869	50.28071
	(0.10723)	(0.10728)	(0.10733)	(0.10738)	(0.10742)	(0.10747)
500.	49.13075	49.14446	49.15813	49.17176	49.18537	49.19894
	(0.10219)	(0.10224)	(0.10229)	(0.10234)	(0.10239)	(0.10244)
520.	47.94893	47.96469	47.98041	47.99608	48.01171	48.02730
	(0.09738)	(0.09744)	(0.09749)	(0.09755)	(0.09760)	(0.09765)
540.	46.65170	46.67024	46.68871	46.70712	46.72547	46.74375
	(0.09270)	(0.09276)	(0.09282)	(0.09288)	(0.09294)	(0.09300)
560.	2.21933	2.29875	2.38067	2.46531	2.55294	45.31490
	(0.01943)	(0.01942)	(0.01942)	(0.01941)	(0.01941)	(0.08836)

	Pressure, psia					
	1025.	**1050.**	**1075.**	**1100.**	**1125.**	**1150.**
Temp., °F						
600.	2.00988 (0.02051)	2.07440 (0.02050)	2.14017 (0.02050)	2.20725 (0.02050)	2.27570 (0.02050)	2.34562 (0.02050)
650.	1.82797 (0.02182)	1.88238 (0.02182)	1.93746 (0.02182)	1.99324 (0.02182)	2.04974 (0.02182)	2.10698 (0.02182)
700.	1.69252 (0.02309)	1.74062 (0.02310)	1.78915 (0.02310)	1.83811 (0.02310)	1.88753 (0.02310)	1.93740 (0.02310)
750.	1.58452 (0.02435)	1.62816 (0.02435)	1.67209 (0.02435)	1.71632 (0.02436)	1.76086 (0.02436)	1.80571 (0.02437)
800.	1.49467 (0.02558)	1.53491 (0.02558)	1.57536 (0.02559)	1.61604 (0.02559)	1.65693 (0.02560)	1.69805 (0.02560)
850.	1.41772 (0.02679)	1.45525 (0.02680)	1.49294 (0.02680)	1.53079 (0.02681)	1.56881 (0.02682)	1.60699 (0.02682)
900.	1.35045 (0.02799)	1.38574 (0.02800)	1.42114 (0.02800)	1.45668 (0.02801)	1.49234 (0.02802)	1.52812 (0.02803)
950.	1.29074 (0.02917)	1.32412 (0.02918)	1.35759 (0.02919)	1.39116 (0.02920)	1.42484 (0.02920)	1.45861 (0.02921)
1000.	1.23713 (0.03034)	1.26885 (0.03035)	1.30065 (0.03036)	1.33253 (0.03037)	1.36449 (0.03038)	1.39653 (0.03038)
1050.	1.18855 (0.03150)	1.21882 (0.03151)	1.24915 (0.03152)	1.27955 (0.03152)	1.31001 (0.03153)	1.34053 (0.03154)
1100.	1.14423 (0.03264)	1.17321 (0.03265)	1.20223 (0.03266)	1.23131 (0.03267)	1.26044 (0.03268)	1.28962 (0.03269)
1150.	1.10356 (0.03377)	1.13137 (0.03378)	1.15921 (0.03379)	1.18711 (0.03380)	1.21504 (0.03381)	1.24302 (0.03382)
1200.	1.06605 (0.03489)	1.09280 (0.03490)	1.11958 (0.03491)	1.14640 (0.03492)	1.17326 (0.03493)	1.20015 (0.03494)
1250.	1.03130 (0.03599)	1.05709 (0.03600)	1.08290 (0.03601)	1.10875 (0.03602)	1.13462 (0.03603)	1.16053 (0.03604)
1300.	0.99900 (0.03708)	1.02390 (0.03709)	1.04883 (0.03710)	1.07378 (0.03711)	1.09875 (0.03712)	1.12375 (0.03714)
1350.	0.96887 (0.03816)	0.99296 (0.03817)	1.01706 (0.03819)	1.04119 (0.03820)	1.06533 (0.03821)	1.08950 (0.03822)
1400.	0.94068 (0.03923)	0.96401 (0.03924)	0.98736 (0.03925)	1.01072 (0.03927)	1.03410 (0.03928)	1.05750 (0.03929)
1450.	0.91424 (0.04029)	0.93686 (0.04030)	0.95950 (0.04031)	0.98216 (0.04032)	1.00483 (0.04033)	1.02752 (0.04035)
1500.	0.88936 (0.04134)	0.91134 (0.04135)	0.93332 (0.04136)	0.95532 (0.04137)	0.97733 (0.04138)	0.99935 (0.04139)

†Absolute viscosity μ_{cP} is shown in parentheses.

SOURCE: Derived from ASME, 1977.

	Pressure, psia					
	1800.	1900.	2000.	2100.	2200.	2300.
Temp., °F	621.02	628.56	635.80	642.76	649.45	655.89
Sat. vap.	4.57423	4.93149	5.31041	5.71404	6.14558	6.60778
	(0.02140)	(0.02173)	(0.02208)	(0.02245)	(0.02285)	(0.02328)
Sat. liq.	40.45756	39.72799	38.98833	38.23518	37.46419	36.66831
	(0.07546)	(0.07379)	(0.07215)	(0.07052)	(0.06891)	(0.06728)
Temp., °F						
220.	59.97871	59.99889	60.01902	60.03911	60.05915	60.07914
	(0.27276)	(0.27294)	(0.27313)	(0.27331)	(0.27350)	(0.27368)
240.	59.46029	59.48118	59.50203	59.52282	59.54356	59.56426
	(0.24503)	(0.24521)	(0.24540)	(0.24558)	(0.24577)	(0.24595)
260.	58.91084	58.93256	58.95423	58.97583	58.99739	59.01889
	(0.22204)	(0.22222)	(0.22240)	(0.22258)	(0.22277)	(0.22295)
280.	58.33079	58.35346	58.37607	58.39861	58.42108	58.44350
	(0.20277)	(0.20295)	(0.20313)	(0.20331)	(0.20349)	(0.20367)
300.	57.72009	57.74383	57.76751	57.79111	57.81464	57.83811
	(0.18647)	(0.18665)	(0.18682)	(0.18700)	(0.18718)	(0.18735)
320.	57.07820	57.10317	57.12806	57.15288	57.17761	57.20227
	(0.17257)	(0.17274)	(0.17292)	(0.17309)	(0.17327)	(0.17344)
340.	56.40416	56.43053	56.45682	56.48301	56.50911	56.53513
	(0.16062)	(0.16079)	(0.16096)	(0.16113)	(0.16131)	(0.16148)
360.	55.69653	55.72451	55.75239	55.78016	55.80783	55.83540
	(0.15026)	(0.15043)	(0.15060)	(0.15077)	(0.15094)	(0.15111)
380.	54.95334	54.98319	55.01291	55.04250	55.07198	55.10133
	(0.14122)	(0.14139)	(0.14156)	(0.14173)	(0.14189)	(0.14206)
400.	54.17206	54.20406	54.23592	54.26762	54.29919	54.33062
	(0.13325)	(0.13342)	(0.13359)	(0.13376)	(0.13392)	(0.13409)
420.	53.34940	53.38393	53.41829	53.45247	53.48648	53.52033
	(0.12616)	(0.12633)	(0.12650)	(0.12667)	(0.12684)	(0.12701)
440.	52.48118	52.51872	52.55604	52.59315	52.63005	52.66674
	(0.11979)	(0.11997)	(0.12014)	(0.12031)	(0.12049)	(0.12066)
460.	51.56207	51.60321	51.64408	51.68468	51.72504	51.76514
	(0.11401)	(0.11419)	(0.11437)	(0.11454)	(0.11472)	(0.11489)
480.	50.58514	50.63066	50.67584	50.72070	50.76524	50.80946
	(0.10869)	(0.10887)	(0.10906)	(0.10924)	(0.10942)	(0.10960)
500.	49.54127	49.59223	49.64275	49.69285	49.74254	49.79182
	(0.10373)	(0.10392)	(0.10411)	(0.10430)	(0.10449)	(0.10468)
520.	48.41819	48.47602	48.53329	48.59000	48.64616	48.70180
	(0.09902)	(0.09923)	(0.09943)	(0.09963)	(0.09983)	(0.10003)
540.	47.19883	47.26563	47.33165	47.39690	47.46140	47.52519
	(0.09448)	(0.09470)	(0.09492)	(0.09514)	(0.09536)	(0.09557)
560.	45.85856	45.93746	46.01523	46.09190	46.16750	46.24208
	(0.09002)	(0.09026)	(0.09050)	(0.09074)	(0.09098)	(0.09121)

	Pressure, psia					
	1800.	**1900.**	**2000.**	**2100.**	**2200.**	**2300.**
Temp., °F						
600.	42.63936	42.76192	42.88143	42.99809	43.11204	43.22344
	(0.08084)	(0.08117)	(0.08149)	(0.08180)	(0.08211)	(0.08241)
650.	3.99269	4.39810	4.86296	5.41385	6.11282	37.66215
	(0.02208)	(0.02219)	(0.02234)	(0.02254)	(0.02285)	(0.06935)
700.	3.44152	3.72114	4.01939	4.33957	4.68593	5.06401
	(0.02333)	(0.02340)	(0.02349)	(0.02359)	(0.02371)	(0.02386)
750.	3.10227	3.32903	3.56498	3.81104	4.06829	4.33800
	(0.02459)	(0.02465)	(0.02472)	(0.02480)	(0.02489)	(0.02498)
800.	2.85692	3.05302	3.25472	3.46240	3.67645	3.89732
	(0.02584)	(0.02590)	(0.02596)	(0.02602)	(0.02610)	(0.02618)
850.	2.66490	2.84015	3.01919	3.20219	3.38935	3.58086
	(0.02707)	(0.02713)	(0.02718)	(0.02725)	(0.02731)	(0.02739)
900.	2.50752	2.66736	2.82991	2.99525	3.16350	3.33477
	(0.02829)	(0.02834)	(0.02839)	(0.02845)	(0.02852)	(0.02859)
950.	2.37444	2.52227	2.67211	2.82401	2.97803	3.13423
	(0.02948)	(0.02954)	(0.02959)	(0.02965)	(0.02971)	(0.02977)
1000.	2.25931	2.39742	2.53707	2.67828	2.82108	2.96553
	(0.03066)	(0.03072)	(0.03077)	(0.03083)	(0.03089)	(0.03095)
1050.	2.15800	2.28801	2.41923	2.55165	2.68532	2.82024
	(0.03183)	(0.03188)	(0.03193)	(0.03199)	(0.03205)	(0.03211)
1100.	2.06767	2.19079	2.31486	2.43988	2.56588	2.69287
	(0.03298)	(0.03303)	(0.03308)	(0.03314)	(0.03320)	(0.03326)
1150.	1.98632	2.10345	2.22135	2.34001	2.45945	2.57968
	(0.03411)	(0.03417)	(0.03422)	(0.03427)	(0.03433)	(0.03439)
1200.	1.91245	2.02431	2.13680	2.24991	2.36365	2.47802
	(0.03524)	(0.03529)	(0.03534)	(0.03539)	(0.03545)	(0.03551)
1250.	1.84492	1.95210	2.05978	2.16798	2.27669	2.38592
	(0.03634)	(0.03639)	(0.03645)	(0.03650)	(0.03655)	(0.03661)
1300.	1.78285	1.88581	1.98920	2.09300	2.19723	2.30189
	(0.03744)	(0.03749)	(0.03754)	(0.03759)	(0.03765)	(0.03770)
1350.	1.72551	1.82466	1.92416	2.02402	2.12422	2.22477
	(0.03852)	(0.03857)	(0.03862)	(0.03867)	(0.03872)	(0.03878)
1400.	1.67232	1.76800	1.86397	1.96023	2.05679	2.15364
	(0.03959)	(0.03964)	(0.03969)	(0.03974)	(0.03979)	(0.03984)
1450.	1.62279	1.71529	1.80804	1.90102	1.99425	2.08772
	(0.04064)	(0.04069)	(0.04074)	(0.04079)	(0.04084)	(0.04090)
1500.	1.57653	1.66610	1.75587	1.84585	1.93603	2.02641
	(0.04169)	(0.04174)	(0.04178)	(0.04183)	(0.04188)	(0.04194)

†Absolute viscosity μ_{cP} is shown in parentheses.

SOURCE: Derived from ASME, 1977.

Table G.1 Density (in Pounds-Mass per Cubic Foot) and
Absolute Viscosity μ_{cP} (in Centipoises) of Steam and
Compressed Water† (*continued*)

	Pressure, psia					
	2400.	**2500.**	**2600.**	**2700.**	**2800.**	**2900.**
Temp., °F	662.11	668.11	673.91	679.53	684.96	690.22
Sat. vap.	7.10411	7.65209	8.25779	8.93348	9.70407	10.61518
	(0.02374)	(0.02426)	(0.02484)	(0.02550)	(0.02627)	(0.02721)
Sat. liq.	35.82105	34.97454	34.07850	33.10424	32.01994	30.77498
	(0.06559)	(0.06395)	(0.06225)	(0.06044)	(0.05847)	(0.05626)
Temp., °F						
220.	60.09909	60.11900	60.13886	60.15867	60.17845	60.19817
	(0.27387)	(0.27406)	(0.27424)	(0.27443)	(0.27461)	(0.27480)
240.	59.58490	59.60549	59.62604	59.64653	59.66698	59.68738
	(0.24614)	(0.24632)	(0.24651)	(0.24669)	(0.24688)	(0.24706)
260.	59.04033	59.06172	59.08305	59.10433	59.12556	59.14673
	(0.22313)	(0.22331)	(0.22350)	(0.22368)	(0.22386)	(0.22404)
280.	58.46586	58.48815	58.51039	58.53256	58.55468	58.57674
	(0.20385)	(0.20403)	(0.20421)	(0.20438)	(0.20456)	(0.20474)
300.	57.86150	57.88483	57.90809	57.93128	57.95441	57.97747
	(0.18753)	(0.18771)	(0.18788)	(0.18806)	(0.18824)	(0.18841)
320.	57.22684	57.25134	57.27577	57.30012	57.32439	57.34859
	(0.17361)	(0.17379)	(0.17396)	(0.17413)	(0.17431)	(0.17448)
340.	56.56105	56.58689	56.61264	56.63830	56.66388	56.68937
	(0.16165)	(0.16182)	(0.16199)	(0.16216)	(0.16233)	(0.16250)
360.	55.86287	55.89023	55.91749	55.94465	55.97172	55.99869
	(0.15128)	(0.15145)	(0.15162)	(0.15179)	(0.15196)	(0.15212)
380.	55.13056	55.15968	55.18868	55.21757	55.24634	55.27500
	(0.14223)	(0.14240)	(0.14257)	(0.14273)	(0.14290)	(0.14307)
400.	54.36190	54.39305	54.42405	54.45493	54.48567	54.51628
	(0.13426)	(0.13443)	(0.13460)	(0.13476)	(0.13493)	(0.13509)
420.	53.55400	53.58751	53.62086	53.65405	53.68708	53.71995
	(0.12718)	(0.12735)	(0.12752)	(0.12768)	(0.12785)	(0.12802)
440.	52.70324	52.73953	52.77563	52.81154	52.84725	52.88278
	(0.12083)	(0.12100)	(0.12117)	(0.12134)	(0.12151)	(0.12167)
460.	51.80499	51.84460	51.88397	51.92310	51.96200	52.00067
	(0.11507)	(0.11524)	(0.11541)	(0.11559)	(0.11576)	(0.11593)
480.	50.85337	50.89697	50.94028	50.98329	51.02601	51.06844
	(0.10978)	(0.10996)	(0.11014)	(0.11032)	(0.11049)	(0.11067)
500.	49.84071	49.88921	49.93733	49.98507	50.03244	50.07946
	(0.10487)	(0.10505)	(0.10524)	(0.10542)	(0.10561)	(0.10579)
520.	48.75691	48.81151	48.86561	48.91923	48.97237	49.02504
	(0.10023)	(0.10042)	(0.10062)	(0.10081)	(0.10101)	(0.10120)
540.	47.58826	47.65065	47.71237	47.77344	47.83387	47.89368
	(0.09578)	(0.09599)	(0.09620)	(0.09641)	(0.09661)	(0.09682)
560.	46.31566	46.38827	46.45995	46.53072	46.60061	46.66964
	(0.09144)	(0.09167)	(0.09190)	(0.09212)	(0.09234)	(0.09257)

Temp., °F	Pressure, psia					
	2400.	2500.	2600.	2700.	2800.	2900.
600.	43.33241 (0.08271)	43.43909 (0.08300)	43.54357 (0.08329)	43.64597 (0.08357)	43.74639 (0.08385)	43.84490 (0.08413)
650.	37.93040 (0.06995)	38.18086 (0.07051)	38.41622 (0.07105)	38.63854 (0.07155)	38.84947 (0.07203)	39.05037 (0.07250)
700.	5.48112 (0.02404)	5.94721 (0.02426)	6.47631 (0.02454)	7.08906 (0.02489)	7.82501 (0.02536)	8.78607 (0.02603)
750.	4.62164 (0.02510)	4.92096 (0.02522)	5.23806 (0.02537)	5.57540 (0.02553)	5.93597 (0.02572)	6.32337 (0.02593)
800.	4.12551 (0.02627)	4.36155 (0.02637)	4.60607 (0.02647)	4.85975 (0.02659)	5.12334 (0.02672)	5.39770 (0.02686)
850.	3.77695 (0.02746)	3.97785 (0.02755)	4.18379 (0.02764)	4.39505 (0.02774)	4.61191 (0.02784)	4.83467 (0.02795)
900.	3.50915 (0.02866)	3.68678 (0.02873)	3.86777 (0.02881)	4.05224 (0.02890)	4.24032 (0.02899)	4.43216 (0.02909)
950.	3.29268 (0.02984)	3.45344 (0.02991)	3.61658 (0.02999)	3.78215 (0.03007)	3.95024 (0.03015)	4.12091 (0.03023)
1000.	3.11165 (0.03101)	3.25947 (0.03108)	3.40905 (0.03115)	3.56041 (0.03123)	3.71359 (0.03130)	3.86864 (0.03138)
1050.	2.95644 (0.03217)	3.09394 (0.03224)	3.23277 (0.03231)	3.37294 (0.03238)	3.51449 (0.03245)	3.65742 (0.03252)
1100.	2.82086 (0.03332)	2.94986 (0.03338)	3.07989 (0.03345)	3.21096 (0.03351)	3.34308 (0.03358)	3.47627 (0.03366)
1150.	2.70071 (0.03445)	2.82254 (0.03451)	2.94518 (0.03457)	3.06864 (0.03464)	3.19292 (0.03471)	3.31804 (0.03477)
1200.	2.59303 (0.03557)	2.70869 (0.03563)	2.82500 (0.03569)	2.94196 (0.03575)	3.05959 (0.03582)	3.17787 (0.03588)
1250.	2.49567 (0.03667)	2.60594 (0.03673)	2.71674 (0.03679)	2.82806 (0.03685)	2.93992 (0.03691)	3.05231 (0.03698)
1300.	2.40697 (0.03776)	2.51248 (0.03782)	2.61842 (0.03787)	2.72479 (0.03793)	2.83159 (0.03800)	2.93882 (0.03806)
1350.	2.32568 (0.03883)	2.42693 (0.03889)	2.52854 (0.03895)	2.63050 (0.03901)	2.73282 (0.03907)	2.83548 (0.03913)
1400.	2.25077 (0.03990)	2.34820 (0.03995)	2.44592 (0.04001)	2.54393 (0.04007)	2.64222 (0.04012)	2.74081 (0.04018)
1450.	2.18144 (0.04095)	2.27539 (0.04100)	2.36959 (0.04106)	2.46402 (0.04111)	2.55870 (0.04117)	2.65361 (0.04123)
1500.	2.11700 (0.04199)	2.20778 (0.04204)	2.29877 (0.04209)	2.38996 (0.04215)	2.48134 (0.04220)	2.57292 (0.04226)

†Absolute viscosity μ_{cP} is shown in parentheses.

SOURCE: Derived from ASME, 1977.

	Pressure, psia					
	3000.	**3100.**	**3200.**	**3300.**	**3400.**	**3500.**
Temp., °F						
220.	60.21786	60.23750	60.25710	60.27666	60.29617	60.31565
	(0.27498)	(0.27517)	(0.27535)	(0.27554)	(0.27572)	(0.27591)
240.	59.70773	59.72804	59.74830	59.76851	59.78868	59.80880
	(0.24724)	(0.24743)	(0.24761)	(0.24780)	(0.24798)	(0.24816)
260.	59.16785	59.18893	59.20994	59.23091	59.25183	59.27269
	(0.22422)	(0.22441)	(0.22459)	(0.22477)	(0.22495)	(0.22513)
280.	58.59874	58.62068	58.64256	58.66439	58.68616	58.70788
	(0.20492)	(0.20510)	(0.20528)	(0.20546)	(0.20563)	(0.20581)
300.	58.00046	58.02340	58.04626	58.06906	58.09180	58.11448
	(0.18859)	(0.18876)	(0.18894)	(0.18911)	(0.18929)	(0.18946)
320.	57.37271	57.39677	57.42075	57.44465	57.46849	57.49226
	(0.17465)	(0.17482)	(0.17500)	(0.17517)	(0.17534)	(0.17551)
340.	56.71478	56.74010	56.76535	56.79051	56.81559	56.84059
	(0.16267)	(0.16284)	(0.16301)	(0.16318)	(0.16335)	(0.16352)
360.	56.02556	56.05234	56.07902	56.10561	56.13210	56.15851
	(0.15229)	(0.15246)	(0.15263)	(0.15279)	(0.15296)	(0.15313)
380.	55.30354	55.33198	55.36031	55.38853	55.41664	55.44465
	(0.14323)	(0.14340)	(0.14356)	(0.14373)	(0.14389)	(0.14406)
400.	54.54675	54.57710	54.60732	54.63741	54.66738	54.69723
	(0.13526)	(0.13543)	(0.13559)	(0.13575)	(0.13592)	(0.13608)
420.	53.75267	53.78523	53.81764	53.84991	53.88202	53.91399
	(0.12818)	(0.12835)	(0.12851)	(0.12868)	(0.12884)	(0.12901)
440.	52.91811	52.95327	52.98824	53.02304	53.05766	53.09211
	(0.12184)	(0.12201)	(0.12218)	(0.12234)	(0.12251)	(0.12267)
460.	52.03911	52.07732	52.11532	52.15310	52.19067	52.22803
	(0.11610)	(0.11627)	(0.11644)	(0.11661)	(0.11678)	(0.11694)
480.	51.11060	51.15247	51.19408	51.23542	51.27650	51.31731
	(0.11084)	(0.11102)	(0.11119)	(0.11136)	(0.11154)	(0.11171)
500.	50.12612	50.17243	50.21840	50.26403	50.30934	50.35432
	(0.10597)	(0.10615)	(0.10633)	(0.10651)	(0.10669)	(0.10686)
520.	49.07726	49.12903	49.18036	49.23126	49.28174	49.33181
	(0.10139)	(0.10158)	(0.10177)	(0.10195)	(0.10214)	(0.10232)
540.	47.95288	48.01149	48.06953	48.12700	48.18392	48.24030
	(0.09702)	(0.09722)	(0.09742)	(0.09762)	(0.09781)	(0.09801)
560.	46.73784	46.80524	46.87185	46.93770	47.00280	47.06718
	(0.09278)	(0.09300)	(0.09321)	(0.09343)	(0.09364)	(0.09385)

Temp., °F	Pressure, psia					
	3000.	3100.	3200.	3300.	3400.	3500.
600.	43.94161 (0.08440)	44.03657 (0.08466)	44.12987 (0.08493)	44.22158 (0.08519)	44.31175 (0.08544)	44.40045 (0.08570)
650.	39.24231 (0.07294)	39.42623 (0.07337)	39.60290 (0.07379)	39.77298 (0.07419)	39.93703 (0.07458)	40.09555 (0.07496)
700.	10.18665 (0.02715)	27.60501 (0.05083)	29.88794 (0.05488)	31.06777 (0.05704)	31.91140 (0.05862)	32.58142 (0.05989)
750.	6.74195 (0.02618)	7.19706 (0.02646)	7.69526 (0.02680)	8.24476 (0.02718)	8.85586 (0.02765)	9.54311 (0.02820)
800.	5.68378 (0.02701)	5.98267 (0.02718)	6.29556 (0.02736)	6.62380 (0.02757)	6.96889 (0.02779)	7.33252 (0.02803)
850.	5.06366 (0.02807)	5.29922 (0.02820)	5.54174 (0.02833)	5.79162 (0.02848)	6.04929 (0.02863)	6.31522 (0.02880)
900.	4.62789 (0.02919)	4.82766 (0.02929)	5.03162 (0.02941)	5.23994 (0.02952)	5.45277 (0.02965)	5.67030 (0.02978)
950.	4.29422 (0.03033)	4.47026 (0.03042)	4.64910 (0.03052)	4.83081 (0.03062)	5.01547 (0.03073)	5.20315 (0.03084)
1000.	4.02559 (0.03147)	4.18448 (0.03155)	4.34534 (0.03164)	4.50823 (0.03173)	4.67318 (0.03183)	4.84022 (0.03193)
1050.	3.80177 (0.03260)	3.94756 (0.03268)	4.09480 (0.03277)	4.24352 (0.03285)	4.39375 (0.03294)	4.54549 (0.03303)
1100.	3.61054 (0.03373)	3.74590 (0.03381)	3.88237 (0.03388)	4.01996 (0.03396)	4.15868 (0.03405)	4.29854 (0.03413)
1150.	3.44400 (0.03485)	3.57082 (0.03492)	3.69849 (0.03499)	3.82702 (0.03507)	3.95642 (0.03515)	4.08670 (0.03523)
1200.	3.29682 (0.03595)	3.41644 (0.03602)	3.53674 (0.03609)	3.65772 (0.03616)	3.77938 (0.03624)	3.90173 (0.03631)
1250.	3.16523 (0.03704)	3.27869 (0.03711)	3.39269 (0.03718)	3.50723 (0.03725)	3.62232 (0.03732)	3.73795 (0.03739)
1300.	3.04648 (0.03812)	3.15458 (0.03819)	3.26311 (0.03825)	3.37208 (0.03832)	3.48149 (0.03839)	3.59133 (0.03846)
1350.	2.93850 (0.03919)	3.04188 (0.03925)	3.14561 (0.03932)	3.24969 (0.03938)	3.35412 (0.03945)	3.45891 (0.03951)
1400.	2.83969 (0.04024)	2.93886 (0.04030)	3.03831 (0.04037)	3.13806 (0.04043)	3.23809 (0.04049)	3.33841 (0.04056)
1450.	2.74877 (0.04129)	2.84415 (0.04134)	2.93978 (0.04140)	3.03564 (0.04147)	3.13174 (0.04153)	3.22807 (0.04159)
1500.	2.66470 (0.04232)	2.75667 (0.04237)	2.84884 (0.04243)	2.94120 (0.04249)	3.03375 (0.04255)	3.12650 (0.04261)

†Absolute viscosity μ_{cP} is shown in parentheses.

SOURCE: Derived from ASME, 1977.

Table G.1 Density (in Pounds-Mass per Cubic Foot) and Absolute Viscosity μ_{cP} (in Centipoises) of Steam and Compressed Water† (*continued*)

| Temp., °F | Pressure, psia | | | | | |
	3600.	3700.	3800.	3900.	4000.	4100.
220.	60.33508	60.35447	60.37382	60.39313	60.41240	60.43162
	(0.27609)	(0.27628)	(0.27646)	(0.27665)	(0.27683)	(0.27702)
240.	59.82887	59.84891	59.86889	59.88883	59.90873	59.92858
	(0.24835)	(0.24853)	(0.24871)	(0.24890)	(0.24908)	(0.24926)
260.	59.29351	59.31427	59.33499	59.35566	59.37628	59.39685
	(0.22531)	(0.22549)	(0.22567)	(0.22585)	(0.22603)	(0.22621)
280.	58.72954	58.75114	58.77269	58.79418	58.81562	58.83701
	(0.20599)	(0.20617)	(0.20634)	(0.20652)	(0.20670)	(0.20688)
300.	58.13710	58.15965	58.18214	58.20457	58.22694	58.24926
	(0.18964)	(0.18981)	(0.18998)	(0.19016)	(0.19033)	(0.19050)
320.	57.51595	57.53958	57.56314	57.58663	57.61005	57.63341
	(0.17568)	(0.17585)	(0.17602)	(0.17619)	(0.17636)	(0.17653)
340.	56.86551	56.89036	56.91512	56.93981	56.96442	56.98896
	(0.16368)	(0.16385)	(0.16402)	(0.16419)	(0.16435)	(0.16452)
360.	56.18482	56.21105	56.23718	56.26323	56.28919	56.31506
	(0.15329)	(0.15346)	(0.15362)	(0.15379)	(0.15395)	(0.15412)
380.	55.47255	55.50035	55.52804	55.55563	55.58313	55.61052
	(0.14422)	(0.14439)	(0.14455)	(0.14471)	(0.14488)	(0.14504)
400.	54.72695	54.75655	54.78603	54.81540	54.84464	54.87377
	(0.13625)	(0.13641)	(0.13657)	(0.13673)	(0.13690)	(0.13706)
420.	53.94582	53.97750	54.00904	54.04045	54.07171	54.10284
	(0.12917)	(0.12933)	(0.12950)	(0.12966)	(0.12982)	(0.12998)
440.	53.12638	53.16048	53.19442	53.22819	53.26180	53.29524
	(0.12284)	(0.12300)	(0.12316)	(0.12333)	(0.12349)	(0.12365)
460.	52.26518	52.30212	52.33886	52.37541	52.41175	52.44791
	(0.11711)	(0.11728)	(0.11744)	(0.11761)	(0.11777)	(0.11794)
480.	51.35788	51.39819	51.43825	51.47808	51.51766	51.55700
	(0.11188)	(0.11205)	(0.11222)	(0.11239)	(0.11256)	(0.11272)
500.	50.39899	50.44334	50.48739	50.53113	50.57458	50.61773
	(0.10704)	(0.10722)	(0.10739)	(0.10757)	(0.10774)	(0.10791)
520.	49.38148	49.43074	49.47962	49.52812	49.57624	49.62399
	(0.10251)	(0.10269)	(0.10287)	(0.10305)	(0.10324)	(0.10341)
540.	48.29616	48.35150	48.40633	48.46068	48.51454	48.56793
	(0.09820)	(0.09840)	(0.09859)	(0.09878)	(0.09897)	(0.09916)
560.	47.13085	47.19384	47.25615	47.31782	47.37884	47.43925
	(0.09406)	(0.09426)	(0.09447)	(0.09467)	(0.09487)	(0.09507)

	Pressure, psia					
	3600.	**3700.**	**3800.**	**3900.**	**4000.**	**4100.**
Temp., °F						
600.	44.48773 (0.08595)	44.57364 (0.08619)	44.65823 (0.08644)	44.74156 (0.08668)	44.82366 (0.08692)	44.90457 (0.08715)
650.	40.24897 (0.07532)	40.39767 (0.07568)	40.54199 (0.07603)	40.68222 (0.07637)	40.81863 (0.07671)	40.95147 (0.07703)
700.	33.14320 (0.06097)	33.63021 (0.06192)	34.06209 (0.06277)	34.45142 (0.06355)	34.80679 (0.06426)	35.13436 (0.06492)
750.	10.35615 (0.02890)	11.32553 (0.02979)	12.51317 (0.03098)	13.99956 (0.03258)	15.83815 (0.03475)	17.94110 (0.03745)
800.	7.71657 (0.02830)	8.12312 (0.02860)	8.55451 (0.02893)	9.01328 (0.02930)	9.50227 (0.02971)	10.02459 (0.03016)
850.	6.58990 (0.02898)	6.87387 (0.02917)	7.16769 (0.02937)	7.47195 (0.02959)	7.78727 (0.02982)	8.11432 (0.03007)
900.	5.89271 (0.02992)	6.12018 (0.03007)	6.35292 (0.03022)	6.59111 (0.03038)	6.83497 (0.03055)	7.08472 (0.03073)
950.	5.39395 (0.03096)	5.58792 (0.03108)	5.78517 (0.03121)	5.98577 (0.03134)	6.18980 (0.03148)	6.39736 (0.03162)
1000.	5.00940 (0.03203)	5.18075 (0.03214)	5.35431 (0.03225)	5.53013 (0.03237)	5.70824 (0.03249)	5.88867 (0.03261)
1050.	4.69878 (0.03312)	4.85363 (0.03322)	5.01006 (0.03332)	5.16810 (0.03342)	5.32775 (0.03353)	5.48904 (0.03364)
1100.	4.43955 (0.03422)	4.58172 (0.03431)	4.72507 (0.03440)	4.86960 (0.03449)	5.01533 (0.03459)	5.16226 (0.03469)
1150.	4.21787 (0.03531)	4.34992 (0.03539)	4.48287 (0.03548)	4.61672 (0.03557)	4.75148 (0.03566)	4.88714 (0.03575)
1200.	4.02477 (0.03639)	4.14850 (0.03647)	4.27293 (0.03655)	4.39806 (0.03663)	4.52388 (0.03672)	4.65041 (0.03680)
1250.	3.85413 (0.03746)	3.97085 (0.03754)	4.08813 (0.03762)	4.20596 (0.03769)	4.32433 (0.03777)	4.44326 (0.03786)
1300.	3.70162 (0.03853)	3.81234 (0.03860)	3.92349 (0.03867)	4.03509 (0.03875)	4.14713 (0.03882)	4.25960 (0.03890)
1350.	3.56405 (0.03958)	3.66954 (0.03965)	3.77539 (0.03972)	3.88159 (0.03979)	3.98814 (0.03986)	4.09504 (0.03994)
1400.	3.43901 (0.04062)	3.53991 (0.04069)	3.64108 (0.04076)	3.74255 (0.04082)	3.84429 (0.04089)	3.94632 (0.04096)

†Absolute viscosity μ_{cP} is shown in parentheses.

SOURCE: Derived from ASME, 1977.

Table G.1 Density (in Kilograms per Cubic Meter) and Absolute Viscosity μ_{cP} (in Centipoises) of Steam and Compressed Water†

	Absolute pressure, kPa					
	80.	**100.**	**120.**	**140.**	**160.**	**180.**
Temp., °C	93.51	99.63	104.81	109.32	113.32	116.93
Sat. vap.	0.47917	0.59041	0.70022	0.80885	0.91650	1.02330
	(0.01205)	(0.01226)	(0.01244)	(0.01260)	(0.01274)	(0.01286)
Sat. liq.	962.70856	958.38788	954.60342	951.21361	948.12905	945.28822
	(0.30269)	(0.28324)	(0.26843)	(0.25661)	(0.24687)	(0.23864)
Temp., °C						
100.	0.47031	0.58979	958.13142	958.14108	958.15075	958.16041
	(0.01229)	(0.01228)	(0.28214)	(0.28215)	(0.28215)	(0.28216)
110.	0.45736	0.57331	0.68994	0.80726	950.69955	950.70953
	(0.01266)	(0.01265)	(0.01264)	(0.01262)	(0.25491)	(0.25491)
120.	0.44516	0.55783	0.67108	0.78492	0.89936	1.01443
	(0.01304)	(0.01303)	(0.01302)	(0.01300)	(0.01299)	(0.01298)
130.	0.43365	0.54325	0.65335	0.76395	0.87507	0.98671
	(0.01342)	(0.01341)	(0.01340)	(0.01339)	(0.01338)	(0.01337)
140.	0.42276	0.52948	0.63663	0.74421	0.85224	0.96071
	(0.01381)	(0.01380)	(0.01379)	(0.01378)	(0.01377)	(0.01376)
150.	0.41244	0.51645	0.62083	0.72558	0.83072	0.93624
	(0.01420)	(0.01419)	(0.01418)	(0.01417)	(0.01416)	(0.01415)
160.	0.40265	0.50409	0.60586	0.70796	0.81038	0.91315
	(0.01459)	(0.01458)	(0.01457)	(0.01457)	(0.01456)	(0.01455)
170.	0.39333	0.49235	0.59165	0.69124	0.79112	0.89130
	(0.01498)	(0.01498)	(0.01497)	(0.01496)	(0.01496)	(0.01495)
180.	0.38445	0.48117	0.57814	0.67536	0.77284	0.87057
	(0.01538)	(0.01538)	(0.01537)	(0.01536)	(0.01536)	(0.01535)
190.	0.37598	0.47051	0.56527	0.66024	0.75545	0.85088
	(0.01578)	(0.01578)	(0.01577)	(0.01576)	(0.01576)	(0.01575)
200.	0.36788	0.46034	0.55298	0.64583	0.73888	0.83213
	(0.01619)	(0.01618)	(0.01617)	(0.01617)	(0.01616)	(0.01616)
210.	0.36014	0.45061	0.54125	0.63207	0.72307	0.81425
	(0.01659)	(0.01659)	(0.01658)	(0.01658)	(0.01657)	(0.01657)
220.	0.35273	0.44130	0.53003	0.61891	0.70796	0.79717
	(0.01700)	(0.01699)	(0.01699)	(0.01698)	(0.01698)	(0.01697)
230.	0.34563	0.43238	0.51928	0.60632	0.69350	0.78083
	(0.01740)	(0.01740)	(0.01740)	(0.01739)	(0.01739)	(0.01738)
240.	0.33881	0.42382	0.50897	0.59425	0.67965	0.76518
	(0.01781)	(0.01781)	(0.01781)	(0.01780)	(0.01780)	(0.01780)
250.	0.33226	0.41561	0.49908	0.58266	0.66636	0.75018
	(0.01823)	(0.01822)	(0.01822)	(0.01821)	(0.01821)	(0.01821)
260.	0.32596	0.40771	0.48957	0.57153	0.65360	0.73578
	(0.01864)	(0.01863)	(0.01863)	(0.01863)	(0.01862)	(0.01862)
270.	0.31990	0.40012	0.48043	0.56084	0.64134	0.72194
	(0.01905)	(0.01905)	(0.01904)	(0.01904)	(0.01904)	(0.01904)

	Absolute pressure, kPa					
	80.	**100.**	**120.**	**140.**	**160.**	**180.**
Temp., °C						
325.	0.29028 (0.02133)	0.36300 (0.02133)	0.43578 (0.02133)	0.50862 (0.02133)	0.58152 (0.02133)	0.65448 (0.02132)
350.	0.27857 (0.02237)	0.34834 (0.02237)	0.41815 (0.02237)	0.48802 (0.02237)	0.55793 (0.02237)	0.62790 (0.02237)
375.	0.26777 (0.02341)	0.33482 (0.02341)	0.40191 (0.02341)	0.46904 (0.02341)	0.53621 (0.02341)	0.60342 (0.02341)
400.	0.25779 (0.02445)	0.32232 (0.02445)	0.38689 (0.02445)	0.45149 (0.02445)	0.51613 (0.02445)	0.58080 (0.02445)
425.	0.24852 (0.02549)	0.31073 (0.02549)	0.37296 (0.02549)	0.43522 (0.02549)	0.49751 (0.02549)	0.55983 (0.02548)
450.	0.23990 (0.02652)	0.29994 (0.02652)	0.36000 (0.02652)	0.42009 (0.02652)	0.48020 (0.02652)	0.54034 (0.02652)
475.	0.23186 (0.02755)	0.28988 (0.02755)	0.34792 (0.02755)	0.40598 (0.02755)	0.46407 (0.02755)	0.52217 (0.02755)
500.	0.22435 (0.02857)	0.28048 (0.02857)	0.33663 (0.02857)	0.39280 (0.02857)	0.44898 (0.02857)	0.50519 (0.02857)
525.	0.21731 (0.02959)	0.27167 (0.02959)	0.32605 (0.02959)	0.38045 (0.02959)	0.43486 (0.02959)	0.48928 (0.02959)
550.	0.21069 (0.03061)	0.26340 (0.03061)	0.31612 (0.03061)	0.36885 (0.03061)	0.42159 (0.03061)	0.47435 (0.03061)
575.	0.20447 (0.03161)	0.25562 (0.03161)	0.30677 (0.03161)	0.35794 (0.03161)	0.40912 (0.03162)	0.46031 (0.03162)
600.	0.19861 (0.03261)	0.24828 (0.03262)	0.29797 (0.03262)	0.34766 (0.03262)	0.39737 (0.03262)	0.44708 (0.03262)
625.	0.19307 (0.03361)	0.24136 (0.03361)	0.28965 (0.03361)	0.33796 (0.03361)	0.38627 (0.03361)	0.43460 (0.03361)
650.	0.18783 (0.03460)	0.23481 (0.03460)	0.28179 (0.03460)	0.32878 (0.03460)	0.37578 (0.03460)	0.42279 (0.03460)
675.	0.18287 (0.03558)	0.22861 (0.03558)	0.27435 (0.03558)	0.32010 (0.03558)	0.36585 (0.03558)	0.41161 (0.03558)
700.	0.17817 (0.03655)	0.22273 (0.03655)	0.26729 (0.03655)	0.31186 (0.03655)	0.35643 (0.03656)	0.40101 (0.03656)
725.	0.17370 (0.03752)	0.21714 (0.03752)	0.26058 (0.03752)	0.30403 (0.03752)	0.34748 (0.03752)	0.39094 (0.03752)
750.	0.16946 (0.03848)	0.21183 (0.03848)	0.25421 (0.03848)	0.29659 (0.03848)	0.33898 (0.03848)	0.38137 (0.03848)
775.	0.16541 (0.03943)	0.20677 (0.03943)	0.24814 (0.03943)	0.28951 (0.03943)	0.33088 (0.03943)	0.37225 (0.03943)

†Absolute viscosity μ_{cP} is shown in parentheses.

SOURCE: Derived from ASME, 1977.

	Absolute pressure, kPa					
	200.	**220.**	**240.**	**260.**	**280.**	**300.**
Temp., °C	120.23	123.27	126.09	128.73	131.20	133.54
Sat. vap.	1.12938	1.23481	1.33967	1.44402	1.54790	1.65136
	(0.01298)	(0.01308)	(0.01318)	(0.01327)	(0.01336)	(0.01344)
Sat. liq.	942.64737	940.17407	937.84347	935.63621	933.53667	931.53221
	(0.23155)	(0.22535)	(0.21987)	(0.21496)	(0.21053)	(0.20651)
Temp., °C						
100.	958.17008	958.17974	958.18940	958.19906	958.20873	958.21839
	(0.28216)	(0.28217)	(0.28217)	(0.28218)	(0.28219)	(0.28219)
110.	950.71951	950.72949	950.73947	950.74945	950.75942	950.76940
	(0.25492)	(0.25492)	(0.25493)	(0.25493)	(0.25494)	(0.25494)
120.	942.83477	942.84511	942.85545	942.86579	942.87613	942.88647
	(0.23204)	(0.23204)	(0.23205)	(0.23205)	(0.23206)	(0.23206)
130.	1.09890	1.21163	1.32492	1.43879	934.56549	934.57624
	(0.01336)	(0.01335)	(0.01333)	(0.01332)	(0.21266)	(0.21267)
140.	1.06965	1.17906	1.28894	1.39931	1.51017	1.62154
	(0.01375)	(0.01374)	(0.01373)	(0.01372)	(0.01371)	(0.01370)
150.	1.04217	1.14849	1.25522	1.36237	1.46994	1.57794
	(0.01414)	(0.01413)	(0.01413)	(0.01412)	(0.01411)	(0.01410)
160.	1.01626	1.11971	1.22352	1.32769	1.43222	1.53712
	(0.01454)	(0.01453)	(0.01452)	(0.01452)	(0.01451)	(0.01450)
170.	0.99177	1.09255	1.19363	1.29502	1.39673	1.49875
	(0.01494)	(0.01493)	(0.01493)	(0.01492)	(0.01491)	(0.01490)
180.	0.96857	1.06683	1.16536	1.26416	1.36323	1.46258
	(0.01534)	(0.01534)	(0.01533)	(0.01532)	(0.01532)	(0.01531)
190.	0.94654	1.04243	1.13856	1.23493	1.33153	1.42838
	(0.01575)	(0.01574)	(0.01573)	(0.01573)	(0.01572)	(0.01572)
200.	0.92558	1.01924	1.11310	1.20718	1.30146	1.39596
	(0.01615)	(0.01615)	(0.01614)	(0.01614)	(0.01613)	(0.01613)
210.	0.90560	0.99714	1.08887	1.18078	1.27288	1.36516
	(0.01656)	(0.01656)	(0.01655)	(0.01655)	(0.01654)	(0.01654)
220.	0.88653	0.97606	1.06576	1.15562	1.24565	1.33584
	(0.01697)	(0.01697)	(0.01696)	(0.01696)	(0.01695)	(0.01695)
230.	0.86830	0.95592	1.04369	1.13160	1.21966	1.30787
	(0.01738)	(0.01738)	(0.01737)	(0.01737)	(0.01736)	(0.01736)
240.	0.85085	0.93664	1.02257	1.10863	1.19483	1.28116
	(0.01779)	(0.01779)	(0.01778)	(0.01778)	(0.01778)	(0.01777)
250.	0.83412	0.91817	1.00235	1.08664	1.17106	1.25559
	(0.01820)	(0.01820)	(0.01820)	(0.01819)	(0.01819)	(0.01819)
260.	0.81806	0.90045	0.98295	1.06556	1.14828	1.23110
	(0.01862)	(0.01862)	(0.01861)	(0.01861)	(0.01861)	(0.01860)
270.	0.80264	0.88344	0.96433	1.04532	1.12642	1.20761
	(0.01903)	(0.01903)	(0.01903)	(0.01902)	(0.01902)	(0.01902)

	Absolute pressure, kPa					
	200.	**220.**	**240.**	**260.**	**280.**	**300.**
Temp., °C						
325.	0.72751 (0.02132)	0.80059 (0.02132)	0.87374 (0.02132)	0.94694 (0.02132)	1.02021 (0.02132)	1.09354 (0.02131)
350.	0.69791 (0.02236)	0.76798 (0.02236)	0.83810 (0.02236)	0.90826 (0.02236)	0.97848 (0.02236)	1.04875 (0.02236)
375.	0.67067 (0.02341)	0.73797 (0.02341)	0.80531 (0.02340)	0.87269 (0.02340)	0.94011 (0.02340)	1.00757 (0.02340)
400.	0.64551 (0.02445)	0.71025 (0.02445)	0.77503 (0.02445)	0.83984 (0.02444)	0.90469 (0.02444)	0.96957 (0.02444)
425.	0.62218 (0.02548)	0.68456 (0.02548)	0.74697 (0.02548)	0.80941 (0.02548)	0.87188 (0.02548)	0.93438 (0.02548)
450.	0.60050 (0.02652)	0.66069 (0.02652)	0.72090 (0.02652)	0.78114 (0.02652)	0.84140 (0.02652)	0.90168 (0.02652)
475.	0.58029 (0.02755)	0.63844 (0.02755)	0.69660 (0.02755)	0.75479 (0.02755)	0.81300 (0.02755)	0.87123 (0.02755)
500.	0.56141 (0.02857)	0.61765 (0.02857)	0.67391 (0.02857)	0.73018 (0.02857)	0.78648 (0.02857)	0.84279 (0.02857)
525.	0.54372 (0.02959)	0.59818 (0.02959)	0.65265 (0.02959)	0.70714 (0.02959)	0.76164 (0.02960)	0.81616 (0.02960)
550.	0.52713 (0.03061)	0.57991 (0.03061)	0.63271 (0.03061)	0.68552 (0.03061)	0.73835 (0.03061)	0.79118 (0.03061)
575.	0.51152 (0.03162)	0.56273 (0.03162)	0.61395 (0.03162)	0.66519 (0.03162)	0.71644 (0.03162)	0.76770 (0.03162)
600.	0.49681 (0.03262)	0.54654 (0.03262)	0.59629 (0.03262)	0.64604 (0.03262)	0.69581 (0.03262)	0.74558 (0.03262)
625.	0.48293 (0.03361)	0.53127 (0.03361)	0.57961 (0.03361)	0.62797 (0.03362)	0.67633 (0.03362)	0.72471 (0.03362)
650.	0.46980 (0.03460)	0.51682 (0.03460)	0.56385 (0.03460)	0.61089 (0.03460)	0.65793 (0.03460)	0.70498 (0.03461)
675.	0.45738 (0.03558)	0.50315 (0.03558)	0.54893 (0.03558)	0.59471 (0.03559)	0.64050 (0.03559)	0.68630 (0.03559)
700.	0.44559 (0.03656)	0.49018 (0.03656)	0.53478 (0.03656)	0.57938 (0.03656)	0.62398 (0.03656)	0.66860 (0.03656)
725.	0.43440 (0.03752)	0.47787 (0.03752)	0.52134 (0.03753)	0.56482 (0.03753)	0.60830 (0.03753)	0.65179 (0.03753)
750.	0.42376 (0.03848)	0.46616 (0.03848)	0.50857 (0.03849)	0.55097 (0.03849)	0.59339 (0.03849)	0.63580 (0.03849)
775.	0.41363 (0.03944)	0.45502 (0.03944)	0.49641 (0.03944)	0.53780 (0.03944)	0.57919 (0.03944)	0.62059 (0.03944)

†Absolute viscosity μ_{cP} is shown in parentheses.

SOURCE: Derived from ASME, 1977.

	Absolute pressure, kPa					
	320.	340.	360.	380.	400.	420.
Temp., °C	135.75	137.86	139.87	141.78	143.62	145.39
Sat. vap.	1.75443	1.85715	1.95954	2.06164	2.16345	2.26500
	(0.01352)	(0.01359)	(0.01366)	(0.01373)	(0.01379)	(0.01386)
Sat. liq.	929.61239	927.76856	925.99319	924.28012	922.62376	921.01959
	(0.20283)	(0.19945)	(0.19632)	(0.19342)	(0.19071)	(0.18818)
Temp., °C						
100.	958.22805	958.23770	958.24736	958.25702	958.26668	958.27633
	(0.28220)	(0.28220)	(0.28221)	(0.28221)	(0.28222)	(0.28222)
110.	950.77937	950.78935	950.79932	950.80930	950.81927	950.82924
	(0.25495)	(0.25495)	(0.25496)	(0.25497)	(0.25497)	(0.25498)
120.	942.89681	942.90714	942.91748	942.92781	942.93815	942.94848
	(0.23207)	(0.23207)	(0.23208)	(0.23208)	(0.23209)	(0.23210)
130.	934.58699	934.59774	934.60849	934.61923	934.62998	934.64072
	(0.21268)	(0.21268)	(0.21269)	(0.21269)	(0.21270)	(0.21270)
140.	1.73343	1.84584	1.95879	925.88385	925.89506	925.90627
	(0.01369)	(0.01368)	(0.01367)	(0.19612)	(0.19612)	(0.19613)
150.	1.68638	1.79527	1.90461	2.01441	2.12468	2.23543
	(0.01409)	(0.01408)	(0.01407)	(0.01406)	(0.01405)	(0.01404)
160.	1.64239	1.74805	1.85409	1.96053	2.06736	2.17460
	(0.01449)	(0.01448)	(0.01448)	(0.01447)	(0.01446)	(0.01445)
170.	1.60110	1.70378	1.80679	1.91014	2.01383	2.11786
	(0.01490)	(0.01489)	(0.01488)	(0.01487)	(0.01487)	(0.01486)
180.	1.56221	1.66213	1.76233	1.86283	1.96362	2.06471
	(0.01530)	(0.01530)	(0.01529)	(0.01528)	(0.01528)	(0.01527)
190.	1.52548	1.62282	1.72041	1.81826	1.91636	2.01472
	(0.01571)	(0.01570)	(0.01570)	(0.01569)	(0.01569)	(0.01568)
200.	1.49068	1.58561	1.68076	1.77614	1.87174	1.96756
	(0.01612)	(0.01611)	(0.01611)	(0.01610)	(0.01610)	(0.01609)
210.	1.45764	1.55031	1.64317	1.73623	1.82949	1.92294
	(0.01653)	(0.01653)	(0.01652)	(0.01652)	(0.01651)	(0.01651)
220.	1.42620	1.51674	1.60745	1.69833	1.78938	1.88061
	(0.01694)	(0.01694)	(0.01693)	(0.01693)	(0.01692)	(0.01692)
230.	1.39624	1.48475	1.57342	1.66225	1.75122	1.84036
	(0.01736)	(0.01735)	(0.01735)	(0.01734)	(0.01734)	(0.01734)
240.	1.36762	1.45422	1.54096	1.62784	1.71485	1.80201
	(0.01777)	(0.01777)	(0.01776)	(0.01776)	(0.01775)	(0.01775)
250.	1.34025	1.42504	1.50994	1.59497	1.68012	1.76540
	(0.01818)	(0.01818)	(0.01818)	(0.01817)	(0.01817)	(0.01817)
260.	1.31404	1.39709	1.48025	1.56352	1.64690	1.73040
	(0.01860)	(0.01860)	(0.01859)	(0.01859)	(0.01859)	(0.01858)
270.	1.28890	1.37029	1.45179	1.53338	1.61508	1.69688
	(0.01902)	(0.01901)	(0.01901)	(0.01901)	(0.01901)	(0.01900)

	Absolute pressure, kPa					
	320.	340.	360.	380.	400.	420.
Temp., °C						
325.	1.16693 (0.02131)	1.24038 (0.02131)	1.31390 (0.02131)	1.38747 (0.02131)	1.46111 (0.02131)	1.53481 (0.02130)
350.	1.11907 (0.02236)	1.18943 (0.02236)	1.25985 (0.02235)	1.33032 (0.02235)	1.40084 (0.02235)	1.47142 (0.02235)
375.	1.07508 (0.02340)	1.14262 (0.02340)	1.21021 (0.02340)	1.27784 (0.02340)	1.34551 (0.02340)	1.41323 (0.02340)
400.	1.03448 (0.02444)	1.09944 (0.02444)	1.16442 (0.02444)	1.22945 (0.02444)	1.29450 (0.02444)	1.35960 (0.02444)
425.	0.99690 (0.02548)	1.05946 (0.02548)	1.12204 (0.02548)	1.18466 (0.02548)	1.24730 (0.02548)	1.30998 (0.02548)
450.	0.96199 (0.02652)	1.02233 (0.02652)	1.08269 (0.02652)	1.14308 (0.02652)	1.20349 (0.02652)	1.26392 (0.02652)
475.	0.92948 (0.02755)	0.98775 (0.02755)	1.04605 (0.02755)	1.10436 (0.02755)	1.16270 (0.02755)	1.22105 (0.02755)
500.	0.89912 (0.02858)	0.95547 (0.02858)	1.01183 (0.02858)	1.06822 (0.02858)	1.12462 (0.02858)	1.18104 (0.02858)
525.	0.87070 (0.02960)	0.92525 (0.02960)	0.97981 (0.02960)	1.03439 (0.02960)	1.08899 (0.02960)	1.14360 (0.02960)
550.	0.84403 (0.03061)	0.89690 (0.03061)	0.94978 (0.03061)	1.00267 (0.03061)	1.05557 (0.03061)	1.10849 (0.03061)
575.	0.81897 (0.03162)	0.87025 (0.03162)	0.92155 (0.03162)	0.97285 (0.03162)	1.02417 (0.03162)	1.07550 (0.03162)
600.	0.79536 (0.03262)	0.84516 (0.03262)	0.89496 (0.03262)	0.94477 (0.03262)	0.99460 (0.03262)	1.04443 (0.03263)
625.	0.77309 (0.03362)	0.82148 (0.03362)	0.86988 (0.03362)	0.91828 (0.03362)	0.96670 (0.03362)	1.01512 (0.03362)
650.	0.75204 (0.03461)	0.79910 (0.03461)	0.84617 (0.03461)	0.89325 (0.03461)	0.94034 (0.03461)	0.98743 (0.03461)
675.	0.73211 (0.03559)	0.77792 (0.03559)	0.82374 (0.03559)	0.86956 (0.03559)	0.91539 (0.03559)	0.96122 (0.03559)
700.	0.71321 (0.03656)	0.75784 (0.03656)	0.80246 (0.03656)	0.84710 (0.03656)	0.89174 (0.03657)	0.93638 (0.03657)
725.	0.69528 (0.03753)	0.73877 (0.03753)	0.78227 (0.03753)	0.82578 (0.03753)	0.86929 (0.03753)	0.91280 (0.03753)
750.	0.67822 (0.03849)	0.72065 (0.03849)	0.76308 (0.03849)	0.80551 (0.03849)	0.84795 (0.03849)	0.89039 (0.03849)
775.	0.66199 (0.03944)	0.70340 (0.03944)	0.74480 (0.03944)	0.78622 (0.03944)	0.82763 (0.03944)	0.86905 (0.03945)

†Absolute viscosity μ_{cP} is shown in parentheses.

SOURCE: Derived from ASME, 1977.

	Absolute pressure, kPa					
	440.	**460.**	**480.**	**500.**	**520.**	**540.**
Temp., °C	147.09	148.73	150.31	151.84	153.33	154.77
Sat. vap.	2.36632	2.46741	2.56829	2.66897	2.76947	2.86979
	(0.01392)	(0.01397)	(0.01403)	(0.01408)	(0.01414)	(0.01419)
Sat. liq.	919.46338	917.95150	916.48075	915.04841	913.65166	912.28834
	(0.18580)	(0.18357)	(0.18146)	(0.17947)	(0.17758)	(0.17578)
Temp., °C						
100.	958.28599	958.29564	958.30530	958.31495	958.32460	958.33425
	(0.28223)	(0.28223)	(0.28224)	(0.28225)	(0.28225)	(0.28226)
110.	950.83921	950.84918	950.85915	950.86912	950.87908	950.88905
	(0.25498)	(0.25499)	(0.25499)	(0.25500)	(0.25500)	(0.25501)
120.	942.95881	942.96914	942.97947	942.98980	943.00013	943.01046
	(0.23210)	(0.23211)	(0.23211)	(0.23212)	(0.23212)	(0.23213)
130.	934.65147	934.66221	934.67295	934.68369	934.69443	934.70517
	(0.21271)	(0.21271)	(0.21272)	(0.21272)	(0.21273)	(0.21273)
140.	925.91748	925.92869	925.93989	925.95110	925.96231	925.97351
	(0.19613)	(0.19614)	(0.19614)	(0.19615)	(0.19616)	(0.19616)
150.	2.34666	2.45839	916.77467	916.78641	916.79814	916.80987
	(0.01404)	(0.01403)	(0.18187)	(0.18188)	(0.18188)	(0.18189)
160.	2.28225	2.39032	2.49882	2.60774	2.71711	2.82692
	(0.01444)	(0.01444)	(0.01443)	(0.01442)	(0.01441)	(0.01440)
170.	2.22225	2.32699	2.43210	2.53757	2.64341	2.74964
	(0.01485)	(0.01485)	(0.01484)	(0.01483)	(0.01482)	(0.01482)
180.	2.16610	2.26779	2.36980	2.47212	2.57476	2.67773
	(0.01526)	(0.01526)	(0.01525)	(0.01524)	(0.01524)	(0.01523)
190.	2.11334	2.21223	2.31139	2.41082	2.51052	2.61050
	(0.01567)	(0.01567)	(0.01566)	(0.01566)	(0.01565)	(0.01564)
200.	2.06362	2.15990	2.25642	2.35317	2.45017	2.54740
	(0.01609)	(0.01608)	(0.01608)	(0.01607)	(0.01607)	(0.01606)
210.	2.01660	2.11045	2.20452	2.29879	2.39326	2.48795
	(0.01650)	(0.01650)	(0.01649)	(0.01649)	(0.01648)	(0.01648)
220.	1.97202	2.06360	2.15537	2.24732	2.33945	2.43177
	(0.01692)	(0.01691)	(0.01691)	(0.01690)	(0.01690)	(0.01689)
230.	1.92965	2.01910	2.10871	2.19849	2.28842	2.37852
	(0.01733)	(0.01733)	(0.01732)	(0.01732)	(0.01731)	(0.01731)
240.	1.88931	1.97674	2.06432	2.15205	2.23991	2.32793
	(0.01775)	(0.01774)	(0.01774)	(0.01774)	(0.01773)	(0.01773)
250.	1.85081	1.93634	2.02200	2.10779	2.19370	2.27975
	(0.01816)	(0.01816)	(0.01816)	(0.01815)	(0.01815)	(0.01815)
260.	1.81401	1.89773	1.98157	2.06553	2.14960	2.23378
	(0.01858)	(0.01858)	(0.01858)	(0.01857)	(0.01857)	(0.01857)
270.	1.77878	1.86079	1.94290	2.02511	2.10743	2.18985
	(0.01900)	(0.01900)	(0.01899)	(0.01899)	(0.01899)	(0.01899)

	Absolute pressure, kPa					
	440.	460.	480.	500.	520.	540.
Temp., °C						
325.	1.60857 (0.02130)	1.68240 (0.02130)	1.75629 (0.02130)	1.83024 (0.02130)	1.90426 (0.02130)	1.97833 (0.02130)
350.	1.54204 (0.02235)	1.61271 (0.02235)	1.68344 (0.02235)	1.75422 (0.02235)	1.82505 (0.02235)	1.89593 (0.02234)
375.	1.48099 (0.02340)	1.54879 (0.02340)	1.61663 (0.02339)	1.68451 (0.02339)	1.75244 (0.02339)	1.82041 (0.02339)
400.	1.42472 (0.02444)	1.48989 (0.02444)	1.55508 (0.02444)	1.62032 (0.02444)	1.68559 (0.02444)	1.75089 (0.02444)
425.	1.37268 (0.02548)	1.43541 (0.02548)	1.49818 (0.02548)	1.56097 (0.02548)	1.62379 (0.02548)	1.68664 (0.02548)
450.	1.32438 (0.02652)	1.38487 (0.02652)	1.44538 (0.02652)	1.50592 (0.02652)	1.56648 (0.02652)	1.62706 (0.02652)
475.	1.27943 (0.02755)	1.33783 (0.02755)	1.39625 (0.02755)	1.45469 (0.02755)	1.51316 (0.02755)	1.57164 (0.02755)
500.	1.23748 (0.02858)	1.29393 (0.02858)	1.35041 (0.02858)	1.40690 (0.02858)	1.46341 (0.02858)	1.51994 (0.02858)
525.	1.19823 (0.02960)	1.25287 (0.02960)	1.30753 (0.02960)	1.36220 (0.02960)	1.41689 (0.02960)	1.47159 (0.02960)
550.	1.16142 (0.03061)	1.21437 (0.03061)	1.26732 (0.03061)	1.32029 (0.03062)	1.37328 (0.03062)	1.42628 (0.03062)
575.	1.12684 (0.03162)	1.17819 (0.03162)	1.22955 (0.03162)	1.28092 (0.03162)	1.33231 (0.03163)	1.38370 (0.03163)
600.	1.09427 (0.03263)	1.14412 (0.03263)	1.19399 (0.03263)	1.24386 (0.03263)	1.29374 (0.03263)	1.34363 (0.03263)
625.	1.06356 (0.03362)	1.11200 (0.03362)	1.16045 (0.03362)	1.20891 (0.03362)	1.25737 (0.03362)	1.30585 (0.03363)
650.	1.03453 (0.03461)	1.08164 (0.03461)	1.12876 (0.03461)	1.17588 (0.03461)	1.22301 (0.03461)	1.27015 (0.03461)
675.	1.00707 (0.03559)	1.05292 (0.03559)	1.09877 (0.03559)	1.14463 (0.03560)	1.19050 (0.03560)	1.23638 (0.03560)
700.	0.98103 (0.03657)	1.02569 (0.03657)	1.07035 (0.03657)	1.11502 (0.03657)	1.15969 (0.03657)	1.20437 (0.03657)
725.	0.95632 (0.03753)	0.99985 (0.03754)	1.04337 (0.03754)	1.08691 (0.03754)	1.13045 (0.03754)	1.17399 (0.03754)
750.	0.93283 (0.03849)	0.97528 (0.03850)	1.01774 (0.03850)	1.06019 (0.03850)	1.10266 (0.03850)	1.14512 (0.03850)
775.	0.91048 (0.03945)	0.95190 (0.03945)	0.99334 (0.03945)	1.03477 (0.03945)	1.07621 (0.03945)	1.11765 (0.03945)

†Absolute viscosity μ_{cP} is shown in parentheses.

SOURCE: Derived from ASME, 1977.

	Absolute pressure, kPa					
	560.	580.	600.	620.	640.	660.
Temp., °C	156.16	157.52	158.84	160.12	161.38	162.60
Sat. vap.	2.96995	3.06997	3.16983	3.26956	3.36917	3.46865
	(0.01423)	(0.01428)	(0.01433)	(0.01437)	(0.01442)	(0.01446)
Sat. liq.	910.95636	909.65403	908.37933	907.13084	905.90736	904.70721
	(0.17407)	(0.17244)	(0.17088)	(0.16940)	(0.16797)	(0.16660)
Temp., °C						
100.	958.34391	958.35356	958.36321	958.37286	958.38250	958.39215
	(0.28226)	(0.28227)	(0.28227)	(0.28228)	(0.28228)	(0.28229)
110.	950.89902	950.90898	950.91895	950.92891	950.93887	950.94884
	(0.25501)	(0.25502)	(0.25503)	(0.25503)	(0.25504)	(0.25504)
120.	943.02079	943.03111	943.04144	943.05176	943.06209	943.07241
	(0.23213)	(0.23214)	(0.23214)	(0.23215)	(0.23216)	(0.23216)
130.	934.71591	934.72664	934.73738	934.74812	934.75885	934.76958
	(0.21274)	(0.21275)	(0.21275)	(0.21276)	(0.21276)	(0.21277)
140.	925.98471	925.99592	926.00712	926.01832	926.02952	926.04071
	(0.19617)	(0.19617)	(0.19618)	(0.19618)	(0.19619)	(0.19619)
150.	916.82160	916.83333	916.84506	916.85679	916.86851	916.88024
	(0.18190)	(0.18190)	(0.18191)	(0.18191)	(0.18192)	(0.18192)
160.	2.93718	3.04790	3.15909	907.25211	907.26443	907.27675
	(0.01439)	(0.01439)	(0.01438)	(0.16954)	(0.16954)	(0.16955)
170.	2.85624	2.96323	3.07062	3.17840	3.28659	3.39519
	(0.01481)	(0.01480)	(0.01479)	(0.01479)	(0.01478)	(0.01477)
180.	2.78102	2.88464	2.98859	3.09288	3.19752	3.30251
	(0.01522)	(0.01522)	(0.01521)	(0.01520)	(0.01520)	(0.01519)
190.	2.71077	2.81131	2.91214	3.01327	3.11468	3.21640
	(0.01564)	(0.01563)	(0.01563)	(0.01562)	(0.01561)	(0.01561)
200.	2.64488	2.74260	2.84057	2.93879	3.03727	3.13600
	(0.01605)	(0.01605)	(0.01604)	(0.01604)	(0.01603)	(0.01603)
210.	2.58285	2.67797	2.77330	2.86885	2.96461	3.06060
	(0.01647)	(0.01647)	(0.01646)	(0.01646)	(0.01645)	(0.01645)
220.	2.52427	2.61696	2.70984	2.80291	2.89618	2.98963
	(0.01689)	(0.01688)	(0.01688)	(0.01688)	(0.01687)	(0.01687)
230.	2.46878	2.55921	2.64980	2.74057	2.83150	2.92260
	(0.01731)	(0.01730)	(0.01730)	(0.01729)	(0.01729)	(0.01729)
240.	2.41608	2.50439	2.59284	2.68145	2.77020	2.85911
	(0.01773)	(0.01772)	(0.01772)	(0.01771)	(0.01771)	(0.01771)
250.	2.36593	2.45223	2.53867	2.62525	2.71196	2.79880
	(0.01814)	(0.01814)	(0.01814)	(0.01813)	(0.01813)	(0.01813)
260.	2.31809	2.40251	2.48705	2.57171	2.65649	2.74138
	(0.01856)	(0.01856)	(0.01856)	(0.01855)	(0.01855)	(0.01855)
270.	2.27238	2.35501	2.43775	2.52060	2.60355	2.68662
	(0.01898)	(0.01898)	(0.01898)	(0.01897)	(0.01897)	(0.01897)

			Absolute pressure, kPa			
	560.	**580.**	**600.**	**620.**	**640.**	**660.**
Temp., °C						
325.	2.05248	2.12668	2.20095	2.27528	2.34968	2.42414
	(0.02129)	(0.02129)	(0.02129)	(0.02129)	(0.02129)	(0.02129)
350.	1.96686	2.03784	2.10888	2.17997	2.25111	2.32230
	(0.02234)	(0.02234)	(0.02234)	(0.02234)	(0.02234)	(0.02234)
375.	1.88842	1.95648	2.02457	2.09271	2.16090	2.22912
	(0.02339)	(0.02339)	(0.02339)	(0.02339)	(0.02339)	(0.02339)
400.	1.81623	1.88161	1.94702	2.01246	2.07795	2.14347
	(0.02444)	(0.02444)	(0.02444)	(0.02444)	(0.02443)	(0.02443)
425.	1.74953	1.81244	1.87538	1.93835	2.00135	2.06438
	(0.02548)	(0.02548)	(0.02548)	(0.02548)	(0.02548)	(0.02548)
450.	1.68768	1.74831	1.80897	1.86966	1.93037	1.99111
	(0.02652)	(0.02652)	(0.02652)	(0.02652)	(0.02652)	(0.02652)
475.	1.63015	1.68867	1.74722	1.80579	1.86438	1.92300
	(0.02755)	(0.02755)	(0.02755)	(0.02755)	(0.02755)	(0.02755)
500.	1.57649	1.63305	1.68964	1.74624	1.80286	1.85950
	(0.02858)	(0.02858)	(0.02858)	(0.02858)	(0.02858)	(0.02858)
525.	1.52631	1.58105	1.63580	1.69057	1.74535	1.80015
	(0.02960)	(0.02960)	(0.02960)	(0.02960)	(0.02960)	(0.02960)
550.	1.47929	1.53231	1.58535	1.63840	1.69146	1.74454
	(0.03062)	(0.03062)	(0.03062)	(0.03062)	(0.03062)	(0.03062)
575.	1.43511	1.48653	1.53796	1.58940	1.64086	1.69232
	(0.03163)	(0.03163)	(0.03163)	(0.03163)	(0.03163)	(0.03163)
600.	1.39354	1.44345	1.49337	1.54330	1.59324	1.64319
	(0.03263)	(0.03263)	(0.03263)	(0.03263)	(0.03263)	(0.03263)
625.	1.35433	1.40282	1.45132	1.49983	1.54835	1.59688
	(0.03363)	(0.03363)	(0.03363)	(0.03363)	(0.03363)	(0.03363)
650.	1.31730	1.36445	1.41161	1.45878	1.50595	1.55313
	(0.03462)	(0.03462)	(0.03462)	(0.03462)	(0.03462)	(0.03462)
675.	1.28226	1.32814	1.37404	1.41994	1.46584	1.51176
	(0.03560)	(0.03560)	(0.03560)	(0.03560)	(0.03560)	(0.03560)
700.	1.24905	1.29374	1.33844	1.38314	1.42784	1.47255
	(0.03657)	(0.03657)	(0.03657)	(0.03658)	(0.03658)	(0.03658)
725.	1.21754	1.26109	1.30465	1.34821	1.39178	1.43535
	(0.03754)	(0.03754)	(0.03754)	(0.03754)	(0.03754)	(0.03754)
750.	1.18759	1.23007	1.27255	1.31503	1.35752	1.40001
	(0.03850)	(0.03850)	(0.03850)	(0.03850)	(0.03850)	(0.03850)
775.	1.15910	1.20054	1.24200	1.28345	1.32491	1.36638
	(0.03945)	(0.03945)	(0.03945)	(0.03946)	(0.03946)	(0.03946)

†Absolute viscosity μ_{cP} is shown in parentheses.

SOURCE: Derived from ASME, 1977.

	Absolute pressure, kPa					
	680.	**700.**	**720.**	**740.**	**760.**	**780.**
Temp., °C	163.79	164.96	166.10	167.21	168.30	169.37
Sat. vap.	3.56802	3.66728	3.76645	3.86552	3.96449	4.06339
	(0.01450)	(0.01454)	(0.01458)	(0.01462)	(0.01466)	(0.01470)
Sat. liq.	903.52958	902.37297	901.23677	900.11982	899.02107	897.94009
	(0.16529)	(0.16403)	(0.16281)	(0.16164)	(0.16051)	(0.15942)
Temp., °C						
100.	958.40180	958.41145	958.42109	958.43074	958.44038	958.45002
	(0.28229)	(0.28230)	(0.28231)	(0.28231)	(0.28232)	(0.28232)
110.	950.95880	950.96876	950.97872	950.98868	950.99864	951.00860
	(0.25505)	(0.25505)	(0.25506)	(0.25506)	(0.25507)	(0.25507)
120.	943.08273	943.09306	943.10338	943.11370	943.12401	943.13433
	(0.23217)	(0.23217)	(0.23218)	(0.23218)	(0.23219)	(0.23219)
130.	934.78032	934.79105	934.80178	934.81251	934.82324	934.83396
	(0.21277)	(0.21278)	(0.21278)	(0.21279)	(0.21279)	(0.21280)
140.	926.05191	926.06311	926.07430	926.08550	926.09669	926.10788
	(0.19620)	(0.19620)	(0.19621)	(0.19621)	(0.19622)	(0.19622)
150.	916.89196	916.90368	916.91540	916.92712	916.93884	916.95056
	(0.18193)	(0.18193)	(0.18194)	(0.18194)	(0.18195)	(0.18195)
160.	907.28907	907.30139	907.31371	907.32603	907.33834	907.35066
	(0.16955)	(0.16956)	(0.16956)	(0.16957)	(0.16957)	(0.16958)
170.	3.50421	3.61364	3.72351	3.83381	3.94456	4.05575
	(0.01476)	(0.01476)	(0.01475)	(0.01474)	(0.01473)	(0.01473)
180.	3.40784	3.51354	3.61959	3.72601	3.83281	3.93997
	(0.01518)	(0.01518)	(0.01517)	(0.01516)	(0.01516)	(0.01515)
190.	3.31841	3.42073	3.52336	3.62629	3.72955	3.83312
	(0.01560)	(0.01560)	(0.01559)	(0.01558)	(0.01558)	(0.01557)
200.	3.23499	3.33424	3.43375	3.53354	3.63359	3.73391
	(0.01602)	(0.01602)	(0.01601)	(0.01601)	(0.01600)	(0.01599)
210.	3.15682	3.25326	3.34993	3.44683	3.54397	3.64133
	(0.01644)	(0.01644)	(0.01643)	(0.01643)	(0.01642)	(0.01642)
220.	3.08329	3.17714	3.27119	3.36544	3.45989	3.55455
	(0.01686)	(0.01686)	(0.01685)	(0.01685)	(0.01684)	(0.01684)
230.	3.01388	3.10533	3.19695	3.28875	3.38073	3.47288
	(0.01728)	(0.01728)	(0.01727)	(0.01727)	(0.01727)	(0.01726)
240.	2.94816	3.03737	3.12674	3.21625	3.30593	3.39576
	(0.01770)	(0.01770)	(0.01770)	(0.01769)	(0.01769)	(0.01768)
250.	2.88577	2.97289	3.06014	3.14752	3.23505	3.32271
	(0.01812)	(0.01812)	(0.01812)	(0.01811)	(0.01811)	(0.01811)
260.	2.82640	2.91154	2.99681	3.08219	3.16770	3.25334
	(0.01855)	(0.01854)	(0.01854)	(0.01854)	(0.01853)	(0.01853)
270.	2.76979	2.85307	2.93646	3.01996	3.10357	3.18729
	(0.01897)	(0.01896)	(0.01896)	(0.01896)	(0.01896)	(0.01895)

	Absolute Pressure, kPa					
	680.	700.	720.	740.	760.	780.
Temp., °C						
325.	2.49867	2.57326	2.64791	2.72263	2.79742	2.87227
	(0.02128)	(0.02128)	(0.02128)	(0.02128)	(0.02128)	(0.02128)
350.	2.39354	2.46484	2.53619	2.60759	2.67904	2.75055
	(0.02234)	(0.02234)	(0.02233)	(0.02233)	(0.02233)	(0.02233)
375.	2.29739	2.36571	2.43406	2.50246	2.57090	2.63939
	(0.02339)	(0.02339)	(0.02338)	(0.02338)	(0.02338)	(0.02338)
400.	2.20902	2.27461	2.34023	2.40590	2.47159	2.53733
	(0.02443)	(0.02443)	(0.02443)	(0.02443)	(0.02443)	(0.02443)
425.	2.12744	2.19054	2.25366	2.31681	2.37999	2.44320
	(0.02548)	(0.02548)	(0.02548)	(0.02548)	(0.02548)	(0.02548)
450.	2.05187	2.11266	2.17347	2.23431	2.29517	2.35606
	(0.02652)	(0.02652)	(0.02652)	(0.02652)	(0.02652)	(0.02652)
475.	1.98163	2.04029	2.09896	2.15766	2.21638	2.27512
	(0.02755)	(0.02755)	(0.02755)	(0.02755)	(0.02755)	(0.02755)
500.	1.91615	1.97283	2.02952	2.08623	2.14296	2.19971
	(0.02858)	(0.02858)	(0.02858)	(0.02858)	(0.02858)	(0.02858)
525.	1.85496	1.90979	1.96463	2.01949	2.07437	2.12926
	(0.02960)	(0.02960)	(0.02960)	(0.02960)	(0.02960)	(0.02960)
550.	1.79763	1.85073	1.90385	1.95698	2.01013	2.06328
	(0.03062)	(0.03062)	(0.03062)	(0.03062)	(0.03062)	(0.03062)
575.	1.74380	1.79529	1.84679	1.89830	1.94982	2.00135
	(0.03163)	(0.03163)	(0.03163)	(0.03163)	(0.03163)	(0.03163)
600.	1.69315	1.74312	1.79310	1.84310	1.89310	1.94311
	(0.03263)	(0.03263)	(0.03264)	(0.03264)	(0.03264)	(0.03264)
625.	1.64541	1.69395	1.74250	1.79107	1.83963	1.88821
	(0.03363)	(0.03363)	(0.03363)	(0.03363)	(0.03363)	(0.03363)
650.	1.60032	1.64752	1.69473	1.74194	1.78916	1.83638
	(0.03462)	(0.03462)	(0.03462)	(0.03462)	(0.03462)	(0.03462)
675.	1.55768	1.60360	1.64953	1.69547	1.74142	1.78737
	(0.03560)	(0.03560)	(0.03560)	(0.03561)	(0.03561)	(0.03561)
700.	1.51727	1.56199	1.60672	1.65145	1.69619	1.74094
	(0.03658)	(0.03658)	(0.03658)	(0.03658)	(0.03658)	(0.03658)
725.	1.47893	1.52251	1.56610	1.60969	1.65329	1.69689
	(0.03755)	(0.03755)	(0.03755)	(0.03755)	(0.03755)	(0.03755)
750.	1.44250	1.48500	1.52751	1.57001	1.61253	1.65504
	(0.03851)	(0.03851)	(0.03851)	(0.03851)	(0.03851)	(0.03851)
775.	1.40784	1.44932	1.49079	1.53227	1.57375	1.61524
	(0.03946)	(0.03946)	(0.03946)	(0.03946)	(0.03946)	(0.03946)

†Absolute viscosity μ_{cP} is shown in parentheses.

SOURCE: Derived from ASME, 1977.

	Absolute pressure, kPa					
	800.	850.	900.	950.	1000.	1050.
Temp., °C	170.42	172.94	175.36	177.67	179.88	182.02
Sat. vap.	4.16220	4.40892	4.65523	4.90120	5.14686	5.39228
	(0.01474)	(0.01482)	(0.01491)	(0.01499)	(0.01507)	(0.01514)
Sat. liq.	896.87571	894.28436	891.78374	889.36561	887.02271	884.74908
	(0.15836)	(0.15588)	(0.15357)	(0.15144)	(0.14944)	(0.14758)
Temp., °C						
100.	958.45967	958.48377	958.50787	958.53197	958.55606	958.58015
	(0.28233)	(0.28234)	(0.28235)	(0.28237)	(0.28238)	(0.28239)
110.	951.01855	951.04344	951.06833	951.09321	951.11808	951.14295
	(0.25508)	(0.25509)	(0.25511)	(0.25512)	(0.25513)	(0.25515)
120.	943.14465	943.17044	943.19622	943.22200	943.24778	943.27355
	(0.23220)	(0.23221)	(0.23223)	(0.23224)	(0.23225)	(0.23227)
130.	934.84469	934.87150	934.89831	934.92511	934.95191	934.97870
	(0.21280)	(0.21282)	(0.21283)	(0.21284)	(0.21286)	(0.21287)
140.	926.11907	926.14705	926.17501	926.20297	926.23093	926.25887
	(0.19623)	(0.19624)	(0.19626)	(0.19627)	(0.19628)	(0.19630)
150.	916.96228	916.99156	917.02084	917.05011	917.07938	917.10864
	(0.18196)	(0.18197)	(0.18198)	(0.18200)	(0.18201)	(0.18202)
160.	907.36297	907.39374	907.42451	907.45527	907.48602	907.51676
	(0.16958)	(0.16960)	(0.16961)	(0.16962)	(0.16964)	(0.16965)
170.	897.30381	897.33628	897.36874	897.40119	897.43363	897.46607
	(0.15878)	(0.15880)	(0.15881)	(0.15882)	(0.15883)	(0.15885)
180.	4.04752	4.31809	4.59115	4.86677	5.14505	886.93287
	(0.01514)	(0.01513)	(0.01511)	(0.01509)	(0.01507)	(0.14935)
190.	3.93701	4.19818	4.46145	4.72687	4.99452	5.26446
	(0.01557)	(0.01555)	(0.01553)	(0.01552)	(0.01550)	(0.01549)
200.	3.83451	4.08723	4.34174	4.59808	4.85631	5.11646
	(0.01599)	(0.01598)	(0.01596)	(0.01595)	(0.01593)	(0.01592)
210.	3.73894	3.98401	4.23061	4.47877	4.72855	4.97997
	(0.01641)	(0.01640)	(0.01639)	(0.01637)	(0.01636)	(0.01635)
220.	3.64942	3.88749	4.12688	4.36764	4.60977	4.85332
	(0.01683)	(0.01682)	(0.01681)	(0.01680)	(0.01679)	(0.01678)
230.	3.56522	3.79685	4.02965	4.26362	4.49879	4.73518
	(0.01726)	(0.01725)	(0.01724)	(0.01723)	(0.01722)	(0.01721)
240.	3.48575	3.71143	3.93812	4.16585	4.39463	4.62447
	(0.01768)	(0.01767)	(0.01766)	(0.01765)	(0.01764)	(0.01764)
250.	3.41052	3.63065	3.85168	4.07362	4.29649	4.52029
	(0.01810)	(0.01810)	(0.01809)	(0.01808)	(0.01807)	(0.01806)
260.	3.33910	3.55405	3.76979	3.98634	4.20371	4.42191
	(0.01853)	(0.01852)	(0.01851)	(0.01850)	(0.01850)	(0.01849)
270.	3.27113	3.48121	3.69199	3.90350	4.11574	4.32872
	(0.01895)	(0.01894)	(0.01894)	(0.01893)	(0.01892)	(0.01892)

	Absolute pressure, kpa					
	800.	**850.**	**900.**	**950.**	**1000.**	**1050.**
Temp., °C						
325.	2.94718 (0.02128)	3.13476 (0.02127)	3.32275 (0.02127)	3.51116 (0.02126)	3.69998 (0.02126)	3.88923 (0.02126)
350.	2.82211 (0.02233)	3.00125 (0.02233)	3.18071 (0.02232)	3.36051 (0.02232)	3.54065 (0.02232)	3.72113 (0.02232)
375.	2.70792 (0.02338)	2.87943 (0.02338)	3.05122 (0.02338)	3.22328 (0.02338)	3.39561 (0.02337)	3.56822 (0.02337)
400.	2.60310 (0.02443)	2.76768 (0.02443)	2.93248 (0.02443)	3.09752 (0.02443)	3.26278 (0.02443)	3.42827 (0.02442)
425.	2.50644 (0.02548)	2.66468 (0.02547)	2.82310 (0.02547)	2.98172 (0.02547)	3.14052 (0.02547)	3.29952 (0.02547)
450.	2.41698 (0.02652)	2.56937 (0.02652)	2.72193 (0.02652)	2.87465 (0.02652)	3.02752 (0.02652)	3.18055 (0.02651)
475.	2.33388 (0.02755)	2.48088 (0.02755)	2.62802 (0.02755)	2.77529 (0.02755)	2.92269 (0.02755)	3.07023 (0.02755)
500.	2.25647 (0.02858)	2.39847 (0.02858)	2.54058 (0.02858)	2.68280 (0.02858)	2.82514 (0.02858)	2.96759 (0.02858)
525.	2.18417 (0.02960)	2.32150 (0.02961)	2.45893 (0.02961)	2.59646 (0.02961)	2.73409 (0.02961)	2.87182 (0.02961)
550.	2.11645 (0.03062)	2.24944 (0.03062)	2.38251 (0.03062)	2.51566 (0.03063)	2.64889 (0.03063)	2.78221 (0.03063)
575.	2.05290 (0.03163)	2.18181 (0.03163)	2.31080 (0.03164)	2.43986 (0.03164)	2.56899 (0.03164)	2.69819 (0.03164)
600.	1.99313 (0.03264)	2.11822 (0.03264)	2.24337 (0.03264)	2.36859 (0.03264)	2.49387 (0.03264)	2.61921 (0.03265)
625.	1.93680 (0.03364)	2.05830 (0.03364)	2.17985 (0.03364)	2.30146 (0.03364)	2.42312 (0.03364)	2.54484 (0.03364)
650.	1.88362 (0.03463)	2.00173 (0.03463)	2.11990 (0.03463)	2.23810 (0.03463)	2.35636 (0.03463)	2.47466 (0.03464)
675.	1.83332 (0.03561)	1.94825 (0.03561)	2.06321 (0.03561)	2.17821 (0.03561)	2.29325 (0.03562)	2.40833 (0.03562)
700.	1.78569 (0.03658)	1.89759 (0.03659)	2.00952 (0.03659)	2.12149 (0.03659)	2.23349 (0.03659)	2.34553 (0.03659)
725.	1.74050 (0.03755)	1.84953 (0.03755)	1.95860 (0.03756)	2.06769 (0.03756)	2.17682 (0.03756)	2.28598 (0.03756)
750.	1.69756 (0.03851)	1.80388 (0.03851)	1.91023 (0.03852)	2.01660 (0.03852)	2.12300 (0.03852)	2.22942 (0.03852)
775.	1.65673 (0.03946)	1.76046 (0.03947)	1.86423 (0.03947)	1.96801 (0.03947)	2.07182 (0.03947)	2.17565 (0.03948)

†Absolute viscosity μ_{cP} is shown in parentheses.

SOURCE: Derived from ASME, 1977.

	Absolute pressure, kPa					
	1100.	**1150.**	**1200.**	**1250.**	**1300.**	**1350.**
Temp., °C	184.07	186.05	187.96	189.81	191.61	193.35
Sat. vap.	5.63749	5.88253	6.12743	6.37223	6.61694	6.86161
	(0.01522)	(0.01529)	(0.01535)	(0.01542)	(0.01548)	(0.01554)
Sat. liq.	882.53862	880.38707	878.28958	876.24292	874.24308	872.28766
	(0.14583)	(0.14418)	(0.14262)	(0.14115)	(0.13975)	(0.13842)
Temp., °C						
100.	958.60423	958.62831	958.65238	958.67645	958.70052	958.72458
	(0.28241)	(0.28242)	(0.28244)	(0.28245)	(0.28246)	(0.28248)
110.	951.16782	951.19268	951.21754	951.24239	951.26724	951.29208
	(0.25516)	(0.25518)	(0.25519)	(0.25520)	(0.25522)	(0.25523)
120.	943.29931	943.32507	943.35082	943.37657	943.40231	943.42805
	(0.23228)	(0.23229)	(0.23231)	(0.23232)	(0.23233)	(0.23235)
130.	935.00548	935.03226	935.05904	935.08580	935.11257	935.13932
	(0.21289)	(0.21290)	(0.21291)	(0.21293)	(0.21294)	(0.21295)
140.	926.28682	926.31475	926.34268	926.37060	926.39852	926.42643
	(0.19631)	(0.19632)	(0.19634)	(0.19635)	(0.19636)	(0.19638)
150.	917.13789	917.16713	917.19637	917.22560	917.25482	917.28403
	(0.18204)	(0.18205)	(0.18206)	(0.18208)	(0.18209)	(0.18210)
160.	907.54750	907.57823	907.60895	907.63966	907.67036	907.70106
	(0.16966)	(0.16968)	(0.16969)	(0.16970)	(0.16971)	(0.16973)
170.	897.49849	897.53091	897.56331	897.59571	897.62809	897.66047
	(0.15886)	(0.15887)	(0.15889)	(0.15890)	(0.15891)	(0.15892)
180.	886.96722	887.00157	887.03590	887.07022	887.10453	887.13883
	(0.14937)	(0.14938)	(0.14939)	(0.14940)	(0.14942)	(0.14943)
190.	5.53676	5.81148	6.08870	6.36849	876.06887	876.10538
	(0.01547)	(0.01546)	(0.01544)	(0.01543)	(0.14101)	(0.14103)
200.	5.37860	5.64277	5.90904	6.17745	6.44807	6.72095
	(0.01591)	(0.01589)	(0.01588)	(0.01586)	(0.01585)	(0.01584)
210.	5.23308	5.48791	5.74451	6.00292	6.26318	6.52534
	(0.01634)	(0.01632)	(0.01631)	(0.01630)	(0.01629)	(0.01627)
220.	5.09832	5.34479	5.59277	5.84229	6.09339	6.34609
	(0.01677)	(0.01676)	(0.01674)	(0.01673)	(0.01672)	(0.01671)
230.	4.97283	5.21175	5.45196	5.69350	5.93640	6.18066
	(0.01720)	(0.01719)	(0.01718)	(0.01717)	(0.01716)	(0.01715)
240.	4.85540	5.08744	5.32061	5.55492	5.79040	6.02706
	(0.01763)	(0.01762)	(0.01761)	(0.01760)	(0.01759)	(0.01758)
250.	4.74505	4.97077	5.19749	5.42520	5.65393	5.88369
	(0.01805)	(0.01805)	(0.01804)	(0.01803)	(0.01802)	(0.01801)
260.	4.64095	4.86085	5.08162	5.30326	5.52579	5.74923
	(0.01848)	(0.01847)	(0.01847)	(0.01846)	(0.01845)	(0.01844)
270.	4.54244	4.75693	4.97218	5.18821	5.40502	5.62264
	(0.01891)	(0.01890)	(0.01890)	(0.01889)	(0.01888)	(0.01888)

			Absolute pressure, kPa			
	1100.	1150.	1200.	1250.	1300.	1350.
Temp., °C						
325.	4.07890	4.26900	4.45952	4.65049	4.84189	5.03373
	(0.02125)	(0.02125)	(0.02125)	(0.02124)	(0.02124)	(0.02123)
350.	3.90195	4.08311	4.26461	4.44646	4.62866	4.81121
	(0.02231)	(0.02231)	(0.02231)	(0.02231)	(0.02230)	(0.02230)
375.	3.74111	3.91428	4.08773	4.26146	4.43547	4.60976
	(0.02337)	(0.02337)	(0.02337)	(0.02336)	(0.02336)	(0.02336)
400.	3.59399	3.75994	3.92612	4.09253	4.25917	4.42605
	(0.02442)	(0.02442)	(0.02442)	(0.02442)	(0.02442)	(0.02442)
425.	3.45870	3.61808	3.77765	3.93741	4.09736	4.25751
	(0.02547)	(0.02547)	(0.02547)	(0.02547)	(0.02547)	(0.02547)
450.	3.33375	3.48711	3.64062	3.79430	3.94814	4.10214
	(0.02651)	(0.02651)	(0.02651)	(0.02651)	(0.02651)	(0.02651)
475.	3.21791	3.36572	3.51367	3.66175	3.80997	3.95833
	(0.02755)	(0.02755)	(0.02755)	(0.02755)	(0.02755)	(0.02755)
500.	3.11016	3.25284	3.39564	3.53855	3.68158	3.82473
	(0.02858)	(0.02859)	(0.02859)	(0.02859)	(0.02859)	(0.02859)
525.	3.00964	3.14756	3.28558	3.42370	3.56192	3.70023
	(0.02961)	(0.02961)	(0.02961)	(0.02961)	(0.02961)	(0.02962)
550.	2.91562	3.04911	3.18268	3.31633	3.45007	3.58390
	(0.03063)	(0.03063)	(0.03063)	(0.03063)	(0.03064)	(0.03064)
575.	2.82746	2.95680	3.08622	3.21571	3.34527	3.47490
	(0.03164)	(0.03164)	(0.03165)	(0.03165)	(0.03165)	(0.03165)
600.	2.74461	2.87008	2.99561	3.12119	3.24685	3.37256
	(0.03265)	(0.03265)	(0.03265)	(0.03265)	(0.03265)	(0.03266)
625.	2.66660	2.78843	2.91030	3.03223	3.15421	3.27625
	(0.03365)	(0.03365)	(0.03365)	(0.03365)	(0.03365)	(0.03366)
650.	2.59301	2.71140	2.82984	2.94833	3.06686	3.18544
	(0.03464)	(0.03464)	(0.03464)	(0.03464)	(0.03465)	(0.03465)
675.	2.52345	2.63861	2.75381	2.86905	2.98433	3.09965
	(0.03562)	(0.03562)	(0.03563)	(0.03563)	(0.03563)	(0.03563)
700.	2.45760	2.56971	2.68185	2.79402	2.90623	3.01848
	(0.03660)	(0.03660)	(0.03660)	(0.03660)	(0.03661)	(0.03661)
725.	2.39516	2.50438	2.61363	2.72290	2.83221	2.94155
	(0.03757)	(0.03757)	(0.03757)	(0.03757)	(0.03757)	(0.03758)
750.	2.33588	2.44235	2.54885	2.65538	2.76194	2.86852
	(0.03853)	(0.03853)	(0.03853)	(0.03853)	(0.03854)	(0.03854)
775.	2.27950	2.38337	2.48727	2.59119	2.69514	2.79910
	(0.03948)	(0.03948)	(0.03948)	(0.03949)	(0.03949)	(0.03949)

†Absolute viscosity μ_{cP} is shown in parentheses.

SOURCE: Derived from ASME, 1977.

Table G.1 Density (in Kilograms per Cubic Meter) and Absolute Viscosity μ_{cP} (in Centipoises) of Steam and Compressed Water† (*continued*)

	Absolute pressure, kPa					
	1400.	1450.	1500.	1550.	1600.	1650.
Temp., °C	195.04	196.69	198.29	199.85	201.37	202.86
Sat. vap.	7.10625	7.35088	7.59553	7.84021	8.08495	8.32975
	(0.01560)	(0.01566)	(0.01572)	(0.01577)	(0.01583)	(0.01588)
Sat. liq.	870.37316	868.49795	866.65923	864.85468	863.08313	861.34205
	(0.13716)	(0.13595)	(0.13480)	(0.13369)	(0.13263)	(0.13162)
Temp., °C						
100.	958.74864	958.77269	958.79674	958.82079	958.84483	958.86886
	(0.28249)	(0.28250)	(0.28252)	(0.28253)	(0.28254)	(0.28256)
110.	951.31692	951.34175	951.36658	951.39141	951.41622	951.44104
	(0.25524)	(0.25526)	(0.25527)	(0.25528)	(0.25530)	(0.25531)
120.	943.45379	943.47951	943.50524	943.53095	943.55667	943.58237
	(0.23236)	(0.23238)	(0.23239)	(0.23240)	(0.23242)	(0.23243)
130.	935.16607	935.19282	935.21956	935.24629	935.27302	935.29974
	(0.21297)	(0.21298)	(0.21299)	(0.21301)	(0.21302)	(0.21303)
140.	926.45433	926.48223	926.51012	926.53800	926.56588	926.59376
	(0.19639)	(0.19640)	(0.19642)	(0.19643)	(0.19644)	(0.19646)
150.	917.31324	917.34244	917.37164	917.40083	917.43001	917.45918
	(0.18212)	(0.18213)	(0.18214)	(0.18216)	(0.18217)	(0.18218)
160.	907.73174	907.76242	907.79309	907.82376	907.85441	907.88506
	(0.16974)	(0.16975)	(0.16977)	(0.16978)	(0.16979)	(0.16981)
170.	897.69284	897.72520	897.75755	897.78990	897.82223	897.85455
	(0.15894)	(0.15895)	(0.15896)	(0.15898)	(0.15899)	(0.15900)
180.	887.17312	887.20740	887.24167	887.27593	887.31018	887.34442
	(0.14944)	(0.14946)	(0.14947)	(0.14948)	(0.14949)	(0.14951)
190.	876.14188	876.17836	876.21484	876.25130	876.28775	876.32419
	(0.14104)	(0.14105)	(0.14107)	(0.14108)	(0.14109)	(0.14110)
200.	6.99615	7.27375	7.55381	7.83639	864.71587	864.75486
	(0.01582)	(0.01581)	(0.01579)	(0.01578)	(0.13360)	(0.13361)
210.	6.78945	7.05556	7.32370	7.59394	7.86632	8.14090
	(0.01626)	(0.01625)	(0.01624)	(0.01622)	(0.01621)	(0.01620)
220.	6.60044	6.85647	7.11422	7.37373	7.63503	7.89817
	(0.01670)	(0.01669)	(0.01668)	(0.01667)	(0.01665)	(0.01664)
230.	6.42633	6.67344	6.92200	7.17206	7.42363	7.67676
	(0.01714)	(0.01713)	(0.01712)	(0.01711)	(0.01710)	(0.01709)
240.	6.26494	6.50405	6.74441	6.98605	7.22898	7.47324
	(0.01757)	(0.01756)	(0.01755)	(0.01754)	(0.01753)	(0.01753)
250.	6.11450	6.34638	6.57934	6.81341	7.04859	7.28492
	(0.01800)	(0.01800)	(0.01799)	(0.01798)	(0.01797)	(0.01796)
260.	5.97359	6.19888	6.42512	6.65231	6.88048	7.10964
	(0.01844)	(0.01843)	(0.01842)	(0.01842)	(0.01841)	(0.01840)
270.	5.84106	6.06030	6.28037	6.50129	6.72305	6.94568
	(0.01887)	(0.01886)	(0.01886)	(0.01885)	(0.01884)	(0.01884)

	Absolute pressure, kPa					
	1400.	**1450.**	**1500.**	**1550.**	**1600.**	**1650.**
Temp., °C						
325.	5.22601 (0.02123)	5.41874 (0.02123)	5.61192 (0.02122)	5.80556 (0.02122)	5.99965 (0.02122)	6.19421 (0.02121)
350.	4.99411 (0.02230)	5.17737 (0.02229)	5.36098 (0.02229)	5.54495 (0.02229)	5.72928 (0.02229)	5.91397 (0.02228)
375.	4.78434 (0.02336)	4.95921 (0.02336)	5.13436 (0.02336)	5.30980 (0.02335)	5.48554 (0.02335)	5.66156 (0.02335)
400.	4.59316 (0.02442)	4.76050 (0.02442)	4.92808 (0.02441)	5.09590 (0.02441)	5.26395 (0.02441)	5.43224 (0.02441)
425.	4.41785 (0.02547)	4.57839 (0.02547)	4.73912 (0.02547)	4.90004 (0.02547)	5.06117 (0.02547)	5.22248 (0.02547)
450.	4.25631 (0.02651)	4.41063 (0.02651)	4.56512 (0.02651)	4.71978 (0.02651)	4.87459 (0.02651)	5.02957 (0.02651)
475.	4.10682 (0.02755)	4.25545 (0.02756)	4.40422 (0.02756)	4.55312 (0.02756)	4.70216 (0.02756)	4.85134 (0.02756)
500.	3.96799 (0.02859)	4.11137 (0.02859)	4.25486 (0.02859)	4.39847 (0.02859)	4.54220 (0.02859)	4.68604 (0.02859)
525.	3.83865 (0.02962)	3.97716 (0.02962)	4.11577 (0.02962)	4.25449 (0.02962)	4.39330 (0.02962)	4.53221 (0.02962)
550.	3.71781 (0.03064)	3.85180 (0.03064)	3.98588 (0.03064)	4.12004 (0.03064)	4.25428 (0.03064)	4.38862 (0.03064)
575.	3.60461 (0.03165)	3.73439 (0.03165)	3.86424 (0.03165)	3.99416 (0.03166)	4.12416 (0.03166)	4.25423 (0.03166)
600.	3.49833 (0.03266)	3.62417 (0.03266)	3.75007 (0.03266)	3.87603 (0.03266)	4.00206 (0.03267)	4.12814 (0.03267)
625.	3.39834 (0.03366)	3.52048 (0.03366)	3.64267 (0.03366)	3.76492 (0.03366)	3.88723 (0.03367)	4.00958 (0.03367)
650.	3.30406 (0.03465)	3.42273 (0.03465)	3.54145 (0.03465)	3.66021 (0.03466)	3.77902 (0.03466)	3.89787 (0.03466)
675.	3.21501 (0.03563)	3.33041 (0.03564)	3.44585 (0.03564)	3.56133 (0.03564)	3.67685 (0.03564)	3.79241 (0.03564)
700.	3.13076 (0.03661)	3.24307 (0.03661)	3.35542 (0.03662)	3.46781 (0.03662)	3.58022 (0.03662)	3.69268 (0.03662)
725.	3.05091 (0.03758)	3.16031 (0.03758)	3.26974 (0.03758)	3.37919 (0.03759)	3.48868 (0.03759)	3.59820 (0.03759)
750.	2.97513 (0.03854)	3.08176 (0.03854)	3.18842 (0.03855)	3.29510 (0.03855)	3.40182 (0.03855)	3.50855 (0.03855)
775.	2.90309 (0.03949)	3.00710 (0.03950)	3.11114 (0.03950)	3.21520 (0.03950)	3.31928 (0.03950)	3.42338 (0.03951)

†Absolute viscosity μ_{cP} is shown in parentheses.

SOURCE: Derived from ASME, 1977.

	Absolute pressure, kPa					
	1700.	**1750.**	**1800.**	**1850.**	**1900.**	**1950.**
Temp., °C	204.31	205.72	207.11	208.47	209.80	211.10
Sat. vap.	8.57465	8.81966	9.06476	9.31000	9.55538	9.80091
	(0.01593)	(0.01598)	(0.01603)	(0.01608)	(0.01613)	(0.01617)
Sat. liq.	859.63078	857.94746	856.28997	854.65792	853.04989	851.46481
	(0.13064)	(0.12970)	(0.12879)	(0.12792)	(0.12708)	(0.12626)
Temp., °C						
100.	958.89289	958.91692	958.94094	958.96496	958.98898	959.01299
	(0.28257)	(0.28258)	(0.28260)	(0.28261)	(0.28263)	(0.28264)
110.	951.46585	951.49066	951.51546	951.54025	951.56504	951.58983
	(0.25532)	(0.25534)	(0.25535)	(0.25537)	(0.25538)	(0.25539)
120.	943.60808	943.63377	943.65947	943.68515	943.71084	943.73651
	(0.23244)	(0.23246)	(0.23247)	(0.23248)	(0.23250)	(0.23251)
130.	935.32646	935.35317	935.37988	935.40658	935.43327	935.45996
	(0.21305)	(0.21306)	(0.21307)	(0.21309)	(0.21310)	(0.21311)
140.	926.62162	926.64948	926.67733	926.70518	926.73302	926.76086
	(0.19647)	(0.19648)	(0.19650)	(0.19651)	(0.19652)	(0.19654)
150.	917.48835	917.51750	917.54666	917.57580	917.60494	917.63407
	(0.18220)	(0.18221)	(0.18222)	(0.18223)	(0.18225)	(0.18226)
160.	907.91570	907.94633	907.97696	908.00757	908.03818	908.06878
	(0.16982)	(0.16983)	(0.16984)	(0.16986)	(0.16987)	(0.16988)
170.	897.88687	897.91917	897.95147	897.98376	898.01604	898.04831
	(0.15901)	(0.15903)	(0.15904)	(0.15905)	(0.15907)	(0.15908)
180.	887.37864	887.41286	887.44707	887.48127	887.51545	887.54963
	(0.14952)	(0.14953)	(0.14955)	(0.14956)	(0.14957)	(0.14958)
190.	876.36061	876.39703	876.43343	876.46982	876.50620	876.54256
	(0.14112)	(0.14113)	(0.14114)	(0.14116)	(0.14117)	(0.14118)
200.	864.79383	864.83278	864.87173	864.91065	864.94957	864.98847
	(0.13363)	(0.13364)	(0.13365)	(0.13366)	(0.13368)	(0.13369)
210.	8.41774	8.69689	8.97841	9.26236	9.54880	852.83837
	(0.01619)	(0.01617)	(0.01616)	(0.01615)	(0.01613)	(0.12696)
220.	8.16318	8.43012	8.69901	8.96992	9.24288	9.51794
	(0.01663)	(0.01662)	(0.01661)	(0.01660)	(0.01659)	(0.01657)
230.	7.93146	8.18778	8.44575	8.70539	8.96675	9.22986
	(0.01708)	(0.01707)	(0.01705)	(0.01704)	(0.01703)	(0.01702)
240.	7.71884	7.96582	8.21419	8.46398	8.71522	8.96794
	(0.01752)	(0.01751)	(0.01750)	(0.01749)	(0.01748)	(0.01747)
250.	7.52240	7.76105	8.00091	8.24198	8.48428	8.72784
	(0.01796)	(0.01795)	(0.01794)	(0.01793)	(0.01792)	(0.01791)
260.	7.33980	7.57098	7.80319	8.03645	8.27077	8.50618
	(0.01839)	(0.01839)	(0.01838)	(0.01837)	(0.01836)	(0.01836)
270.	7.16919	7.39359	7.61888	7.84509	8.07222	8.30028
	(0.01883)	(0.01882)	(0.01882)	(0.01881)	(0.01880)	(0.01880)

	Absolute pressure, kPa					
	1700.	1750.	1800.	1850.	1900.	1950.
Temp., °C						
325.	6.38922	6.58471	6.78067	6.97710	7.17402	7.37141
	(0.02121)	(0.02121)	(0.02120)	(0.02120)	(0.02120)	(0.02119)
350.	6.09903	6.28446	6.47026	6.65642	6.84297	7.02988
	(0.02228)	(0.02228)	(0.02228)	(0.02227)	(0.02227)	(0.02227)
375.	5.83788	6.01449	6.19140	6.36860	6.54610	6.72391
	(0.02335)	(0.02335)	(0.02335)	(0.02334)	(0.02334)	(0.02334)
400.	5.60077	5.76954	5.93855	6.10780	6.27730	6.44703
	(0.02441)	(0.02441)	(0.02441)	(0.02441)	(0.02441)	(0.02440)
425.	5.38400	5.54572	5.70763	5.86974	6.03205	6.19456
	(0.02547)	(0.02546)	(0.02546)	(0.02546)	(0.02546)	(0.02546)
450.	5.18472	5.34003	5.49551	5.65115	5.80696	5.96293
	(0.02651)	(0.02651)	(0.02651)	(0.02651)	(0.02651)	(0.02651)
475.	5.00066	5.15012	5.29972	5.44945	5.59933	5.74934
	(0.02756)	(0.02756)	(0.02756)	(0.02756)	(0.02756)	(0.02756)
500.	4.83000	4.97408	5.11827	5.26259	5.40702	5.55156
	(0.02859)	(0.02859)	(0.02860)	(0.02860)	(0.02860)	(0.02860)
525.	4.67122	4.81032	4.94953	5.08884	5.22825	5.36776
	(0.02962)	(0.02962)	(0.02963)	(0.02963)	(0.02963)	(0.02963)
550.	4.52303	4.65753	4.79212	4.92679	5.06155	5.19639
	(0.03065)	(0.03065)	(0.03065)	(0.03065)	(0.03065)	(0.03065)
575.	4.38437	4.51458	4.64487	4.77523	4.90566	5.03616
	(0.03166)	(0.03166)	(0.03166)	(0.03167)	(0.03167)	(0.03167)
600.	4.25429	4.38050	4.50677	4.63311	4.75951	4.88597
	(0.03267)	(0.03267)	(0.03267)	(0.03267)	(0.03268)	(0.03268)
625.	4.13199	4.25446	4.37698	4.49955	4.62217	4.74485
	(0.03367)	(0.03367)	(0.03367)	(0.03368)	(0.03368)	(0.03368)
650.	4.01678	4.13572	4.25472	4.37376	4.49285	4.61198
	(0.03466)	(0.03466)	(0.03467)	(0.03467)	(0.03467)	(0.03467)
675.	3.90801	4.02365	4.13933	4.25506	4.37082	4.48662
	(0.03565)	(0.03565)	(0.03565)	(0.03565)	(0.03566)	(0.03566)
700.	3.80516	3.91769	4.03024	4.14284	4.25546	4.36812
	(0.03662)	(0.03663)	(0.03663)	(0.03663)	(0.03663)	(0.03664)
725.	3.70774	3.81732	3.92693	4.03657	4.14623	4.25593
	(0.03759)	(0.03760)	(0.03760)	(0.03760)	(0.03760)	(0.03761)

†Absolute viscosity μ_{cP} is shown in parentheses.

SOURCE: Derived from ASME, 1977.

	Absolute pressure, kPa					
	2000.	**2100.**	**2200.**	**2300.**	**2400.**	**2500.**
Temp., °C	212.37	214.85	217.24	219.55	221.78	223.94
Sat. vap.	10.04662	10.53854	11.03124	11.52480	12.01931	12.51478
	(0.01622)	(0.01631)	(0.01639)	(0.01647)	(0.01655)	(0.01663)
Sat. liq.	849.90217	846.83775	843.85079	840.93468	838.08547	835.29644
	(0.12547)	(0.12396)	(0.12255)	(0.12121)	(0.11994)	(0.11874)
Temp., °C						
100.	959.03700	959.08500	959.13298	959.18094	959.22889	959.27682
	(0.28265)	(0.28268)	(0.28271)	(0.28273)	(0.28276)	(0.28279)
110.	951.61461	951.66416	951.71370	951.76321	951.81271	951.86218
	(0.25541)	(0.25543)	(0.25546)	(0.25549)	(0.25552)	(0.25554)
120.	943.76219	943.81351	943.86482	943.91611	943.96738	944.01862
	(0.23252)	(0.23255)	(0.23258)	(0.23261)	(0.23263)	(0.23266)
130.	935.48664	935.53999	935.59332	935.64662	935.69991	935.75317
	(0.21313)	(0.21315)	(0.21318)	(0.21321)	(0.21323)	(0.21326)
140.	926.78868	926.84432	926.89994	926.95552	927.01109	927.06663
	(0.19655)	(0.19658)	(0.19660)	(0.19663)	(0.19666)	(0.19668)
150.	917.66319	917.72142	917.77962	917.83779	917.89594	917.95405
	(0.18227)	(0.18230)	(0.18233)	(0.18235)	(0.18238)	(0.18241)
160.	908.09937	908.16053	908.22166	908.28276	908.34383	908.40486
	(0.16990)	(0.16992)	(0.16995)	(0.16997)	(0.17000)	(0.17003)
170.	898.08057	898.14506	898.20952	898.27394	898.33833	898.40268
	(0.15909)	(0.15912)	(0.15914)	(0.15917)	(0.15919)	(0.15922)
180.	887.58379	887.65209	887.72035	887.78857	887.85675	887.92488
	(0.14960)	(0.14962)	(0.14965)	(0.14967)	(0.14970)	(0.14972)
190.	876.57892	876.65159	876.72421	876.79679	876.86931	876.94179
	(0.14119)	(0.14122)	(0.14125)	(0.14127)	(0.14130)	(0.14132)
200.	865.02735	865.10508	865.18275	865.26037	865.33793	865.41543
	(0.13370)	(0.13373)	(0.13375)	(0.13378)	(0.13381)	(0.13383)
210.	852.88021	852.96384	853.04740	853.13090	853.21433	853.29769
	(0.12697)	(0.12700)	(0.12702)	(0.12705)	(0.12707)	(0.12710)
220.	9.79515	10.35624	10.92656	11.50659	840.43745	840.52772
	(0.01656)	(0.01654)	(0.01652)	(0.01649)	(0.12097)	(0.12100)
230.	9.49476	10.03006	10.57297	11.12381	11.68295	12.25074
	(0.01701)	(0.01699)	(0.01697)	(0.01695)	(0.01693)	(0.01691)
240.	9.22216	9.73521	10.25462	10.78062	11.31347	11.85345
	(0.01746)	(0.01744)	(0.01743)	(0.01741)	(0.01739)	(0.01737)
250.	8.97267	9.46626	9.96522	10.46974	10.98000	11.49620
	(0.01791)	(0.01789)	(0.01787)	(0.01786)	(0.01784)	(0.01783)
260.	8.74268	9.21904	9.69998	10.18565	10.67620	11.17178
	(0.01835)	(0.01834)	(0.01832)	(0.01831)	(0.01829)	(0.01828)
270.	8.52930	8.99024	9.45513	9.92409	10.39722	10.87466
	(0.01879)	(0.01878)	(0.01876)	(0.01875)	(0.01874)	(0.01873)

		Absolute pressure, kPa				
	2000.	**2100.**	**2200.**	**2300.**	**2400.**	**2500.**
Temp., °C						
325.	7.56930	7.96654	8.36578	8.76704	9.17037	9.57580
	(0.02119)	(0.02118)	(0.02118)	(0.02117)	(0.02116)	(0.02116)
350.	7.21718	7.59292	7.97020	8.34905	8.72948	9.11151
	(0.02227)	(0.02226)	(0.02226)	(0.02225)	(0.02225)	(0.02224)
375.	6.90201	7.25913	7.61747	7.97705	8.33787	8.69996
	(0.02334)	(0.02334)	(0.02333)	(0.02333)	(0.02333)	(0.02332)
400.	6.61702	6.95772	7.29941	7.64209	7.98578	8.33049
	(0.02440)	(0.02440)	(0.02440)	(0.02440)	(0.02440)	(0.02440)
425.	6.35728	6.68331	7.01015	7.33781	7.66628	7.99559
	(0.02546)	(0.02546)	(0.02546)	(0.02546)	(0.02546)	(0.02546)
450.	6.11907	6.43186	6.74532	7.05945	7.37426	7.68976
	(0.02651)	(0.02652)	(0.02652)	(0.02652)	(0.02652)	(0.02652)
475.	5.89950	6.20023	6.50152	6.80338	7.10580	7.40880
	(0.02756)	(0.02756)	(0.02756)	(0.02756)	(0.02757)	(0.02757)
500.	5.69623	5.98592	6.27608	6.56672	6.85783	7.14943
	(0.02860)	(0.02860)	(0.02860)	(0.02860)	(0.02861)	(0.02861)
525.	5.50737	5.78688	6.06680	6.34713	6.62785	6.90898
	(0.02963)	(0.02963)	(0.02964)	(0.02964)	(0.02964)	(0.02964)
550.	5.33132	5.60143	5.87188	6.14267	6.41381	6.68530
	(0.03065)	(0.03066)	(0.03066)	(0.03066)	(0.03067)	(0.03067)
575.	5.16674	5.42812	5.68978	5.95174	6.21400	6.47655
	(0.03167)	(0.03167)	(0.03168)	(0.03168)	(0.03168)	(0.03169)
600.	5.01249	5.26573	5.51922	5.77295	6.02694	6.28118
	(0.03268)	(0.03268)	(0.03269)	(0.03269)	(0.03270)	(0.03270)
625.	4.86759	5.11322	5.35906	5.60512	5.85139	6.09789
	(0.03368)	(0.03369)	(0.03369)	(0.03369)	(0.03370)	(0.03370)
650.	4.73116	4.96966	5.20834	5.44722	5.68627	5.92552
	(0.03468)	(0.03468)	(0.03468)	(0.03469)	(0.03469)	(0.03470)
675.	4.60246	4.83426	5.06622	5.29835	5.53063	5.76307
	(0.03566)	(0.03567)	(0.03567)	(0.03567)	(0.03568)	(0.03568)
700.	4.48082	4.70632	4.93195	5.15772	5.38364	5.60969
	(0.03664)	(0.03664)	(0.03665)	(0.03665)	(0.03666)	(0.03666)
725.	4.36566	4.58520	4.80487	5.02465	5.24456	5.46459
	(0.03761)	(0.03761)	(0.03762)	(0.03762)	(0.03763)	(0.03763)
750.	4.25645	4.47037	4.68439	4.89852	5.11276	5.32709
	(0.03857)	(0.03858)	(0.03858)	(0.03859)	(0.03859)	(0.03860)
775.	4.15273	4.36132	4.57001	4.77878	4.98764	5.19660
	(0.03952)	(0.03953)	(0.03953)	(0.03954)	(0.03954)	(0.03955)

†Absolute viscosity μ_{cP} is shown in parentheses.

source: Derived from ASME, 1977.

	Absolute pressure, kPa					
	3000.	**3100.**	**3200.**	**3300.**	**3400.**	**3500.**
Temp., °C	233.84	235.67	237.45	239.18	240.88	242.54
Sat. vap.	15.00909	15.51171	16.01567	16.52102	17.02776	17.53600
	(0.01699)	(0.01706)	(0.01712)	(0.01719)	(0.01725)	(0.01731)
Sat. liq.	822.13694	819.64046	817.18211	814.76050	812.37238	810.01778
	(0.11352)	(0.11261)	(0.11173)	(0.11088)	(0.11006)	(0.10927)
Temp., °C						
100.	959.51623	959.56406	959.61188	959.65967	959.70745	959.75522
	(0.28292)	(0.28295)	(0.28298)	(0.28300)	(0.28303)	(0.28306)
110.	952.10929	952.15866	952.20801	952.25734	952.30665	952.35594
	(0.25568)	(0.25571)	(0.25573)	(0.25576)	(0.25579)	(0.25581)
120.	944.27456	944.32569	944.37679	944.42788	944.47894	944.52999
	(0.23279)	(0.23282)	(0.23285)	(0.23288)	(0.23290)	(0.23293)
130.	936.01913	936.07226	936.12536	936.17844	936.23150	936.28454
	(0.21340)	(0.21342)	(0.21345)	(0.21348)	(0.21350)	(0.21353)
140.	927.34395	927.39934	927.45470	927.51004	927.56536	927.62065
	(0.19681)	(0.19684)	(0.19687)	(0.19689)	(0.19692)	(0.19695)
150.	918.24421	918.30216	918.36008	918.41798	918.47585	918.53369
	(0.18254)	(0.18256)	(0.18259)	(0.18261)	(0.18264)	(0.18267)
160.	908.70957	908.77041	908.83123	908.89201	908.95277	909.01349
	(0.17016)	(0.17018)	(0.17021)	(0.17023)	(0.17026)	(0.17029)
170.	898.72390	898.78804	898.85214	898.91621	898.98024	899.04424
	(0.15935)	(0.15937)	(0.15940)	(0.15943)	(0.15945)	(0.15948)
180.	888.26495	888.33284	888.40069	888.46850	888.53627	888.60400
	(0.14985)	(0.14988)	(0.14990)	(0.14993)	(0.14995)	(0.14998)
190.	877.30348	877.37568	877.44783	877.51993	877.59199	877.66399
	(0.14145)	(0.14147)	(0.14150)	(0.14153)	(0.14155)	(0.14158)
200.	865.80212	865.87929	865.95640	866.03346	866.11047	866.18742
	(0.13396)	(0.13398)	(0.13401)	(0.13404)	(0.13406)	(0.13409)
210.	853.71352	853.79649	853.87939	853.96223	854.04500	854.12771
	(0.12723)	(0.12726)	(0.12728)	(0.12731)	(0.12733)	(0.12736)
220.	840.97787	841.06767	841.15738	841.24702	841.33657	841.42605
	(0.12113)	(0.12116)	(0.12118)	(0.12121)	(0.12124)	(0.12126)
230.	827.51913	827.61705	827.71488	827.81261	827.91025	828.00779
	(0.11555)	(0.11558)	(0.11560)	(0.11563)	(0.11566)	(0.11568)
240.	14.67086	15.26027	15.85924	16.46823	813.66978	813.77705
	(0.01728)	(0.01726)	(0.01724)	(0.01723)	(0.11050)	(0.11053)
250.	14.17418	14.73089	15.29528	15.86767	16.44838	17.03776
	(0.01775)	(0.01773)	(0.01771)	(0.01770)	(0.01768)	(0.01767)
260.	13.73084	14.26015	14.79576	15.33789	15.88676	16.44262
	(0.01821)	(0.01819)	(0.01818)	(0.01817)	(0.01815)	(0.01814)
270.	13.33068	13.83660	14.34777	14.86435	15.38650	15.91440
	(0.01866)	(0.01865)	(0.01864)	(0.01863)	(0.01861)	(0.01860)

	Absolute pressure, kPa					
	3000.	**3100.**	**3200.**	**3300.**	**3400.**	**3500.**
Temp., °C						
325.	11.63558	12.05433	12.47543	12.89893	13.32485	13.75326
	(0.02112)	(0.02112)	(0.02111)	(0.02111)	(0.02110)	(0.02110)
350.	11.04650	11.43862	11.83250	12.22816	12.62562	13.02493
	(0.02222)	(0.02222)	(0.02221)	(0.02221)	(0.02221)	(0.02220)
375.	10.52979	10.89973	11.27102	11.64369	12.01775	12.39321
	(0.02331)	(0.02331)	(0.02330)	(0.02330)	(0.02330)	(0.02330)
400.	10.06949	10.42045	10.77248	11.12559	11.47980	11.83511
	(0.02439)	(0.02439)	(0.02439)	(0.02439)	(0.02438)	(0.02438)
425.	9.65469	9.98907	10.32431	10.66043	10.99742	11.33529
	(0.02546)	(0.02546)	(0.02546)	(0.02546)	(0.02546)	(0.02546)
450.	9.27760	9.59727	9.91764	10.23873	10.56053	10.88305
	(0.02652)	(0.02652)	(0.02652)	(0.02652)	(0.02652)	(0.02652)
475.	8.93240	9.23886	9.54591	9.85355	10.16178	10.47061
	(0.02757)	(0.02758)	(0.02758)	(0.02758)	(0.02758)	(0.02758)
500.	8.61464	8.90914	9.20413	9.49961	9.79559	10.09206
	(0.02862)	(0.02862)	(0.02862)	(0.02863)	(0.02863)	(0.02863)
525.	8.32076	8.60435	8.88835	9.17276	9.45759	9.74284
	(0.02966)	(0.02966)	(0.02966)	(0.02966)	(0.02967)	(0.02967)
550.	8.04791	8.32148	8.59540	8.86967	9.14428	9.41926
	(0.03069)	(0.03069)	(0.03069)	(0.03070)	(0.03070)	(0.03070)
575.	7.79372	8.05804	8.32266	8.58758	8.85280	9.11832
	(0.03171)	(0.03171)	(0.03171)	(0.03172)	(0.03172)	(0.03173)
600.	7.55618	7.81194	8.06795	8.32422	8.58074	8.83752
	(0.03272)	(0.03272)	(0.03273)	(0.03273)	(0.03274)	(0.03274)
625.	7.33360	7.58140	7.82941	8.07764	8.32609	8.57476
	(0.03372)	(0.03373)	(0.03373)	(0.03374)	(0.03374)	(0.03375)
650.	7.12454	7.36490	7.60545	7.84619	8.08712	8.32823
	(0.03472)	(0.03472)	(0.03473)	(0.03473)	(0.03474)	(0.03474)
675.	6.92771	7.16112	7.39470	7.62843	7.86233	8.09638
	(0.03571)	(0.03571)	(0.03572)	(0.03572)	(0.03573)	(0.03573)
700.	6.74203	6.96892	7.19595	7.42311	7.65042	7.87786
	(0.03669)	(0.03669)	(0.03670)	(0.03670)	(0.03671)	(0.03671)
725.	6.56653	6.78728	7.00815	7.22914	7.45025	7.67148
	(0.03766)	(0.03766)	(0.03767)	(0.03767)	(0.03768)	(0.03768)
750.	6.40035	6.61531	6.83038	7.04555	7.26083	7.47621
	(0.03862)	(0.03863)	(0.03863)	(0.03864)	(0.03864)	(0.03865)
775.	6.24273	6.45223	6.66182	6.87150	7.08127	7.29113
	(0.03958)	(0.03958)	(0.03959)	(0.03959)	(0.03960)	(0.03960)

†Absolute viscosity μ_{cP} is shown in parentheses.

SOURCE: Derived from ASME, 1977.

G-85

			Absolute pressure, kPa			
	4000.	4100.	4200.	4300.	4400.	4500.
Temp., °C	250.33	251.80	253.24	254.66	256.05	257.41
Sat. vap.	20.10084	20.61870	21.13835	21.65976	22.18299	22.70814
	(0.01761)	(0.01766)	(0.01772)	(0.01777)	(0.01782)	(0.01788)
Sat. liq.	798.68317	796.49264	794.32617	792.18173	790.05834	787.95650
	(0.10569)	(0.10504)	(0.10440)	(0.10378)	(0.10318)	(0.10259)
Temp., °C						
100.	959.99379	960.04145	960.08910	960.13673	960.18434	960.23194
	(0.28319)	(0.28322)	(0.28325)	(0.28328)	(0.28330)	(0.28333)
110.	952.60214	952.65132	952.70049	952.74963	952.79877	952.84788
	(0.25595)	(0.25598)	(0.25600)	(0.25603)	(0.25606)	(0.25609)
120.	944.78492	944.83585	944.88676	944.93764	944.98851	945.03936
	(0.23307)	(0.23309)	(0.23312)	(0.23315)	(0.23317)	(0.23320)
130.	936.54940	936.60231	936.65519	936.70805	936.76089	936.81371
	(0.21366)	(0.21369)	(0.21372)	(0.21374)	(0.21377)	(0.21380)
140.	927.89674	927.95188	928.00701	928.06210	928.11717	928.17222
	(0.19708)	(0.19711)	(0.19713)	(0.19716)	(0.19718)	(0.19721)
150.	918.82247	918.88015	918.93780	918.99542	919.05301	919.11058
	(0.18280)	(0.18282)	(0.18285)	(0.18288)	(0.18290)	(0.18293)
160.	909.31664	909.37718	909.43769	909.49817	909.55861	909.61903
	(0.17041)	(0.17044)	(0.17047)	(0.17049)	(0.17052)	(0.17054)
170.	899.36369	899.42748	899.49123	899.55494	899.61863	899.68227
	(0.15961)	(0.15963)	(0.15966)	(0.15968)	(0.15971)	(0.15973)
180.	888.94203	889.00952	889.07697	889.14438	889.21174	889.27907
	(0.15011)	(0.15013)	(0.15016)	(0.15018)	(0.15021)	(0.15023)
190.	878.02333	878.09506	878.16675	878.23838	878.30997	878.38152
	(0.14170)	(0.14173)	(0.14175)	(0.14178)	(0.14180)	(0.14183)
200.	866.57135	866.64797	866.72454	866.80106	866.87752	866.95392
	(0.13421)	(0.13424)	(0.13426)	(0.13429)	(0.13432)	(0.13434)
210.	854.54027	854.62259	854.70484	854.78703	854.86916	854.95122
	(0.12749)	(0.12751)	(0.12754)	(0.12756)	(0.12759)	(0.12761)
220.	841.87227	841.96129	842.05022	842.13908	842.22786	842.31657
	(0.12139)	(0.12142)	(0.12144)	(0.12147)	(0.12150)	(0.12152)
230.	828.49406	828.59103	828.68791	828.78469	828.88138	828.97798
	(0.11582)	(0.11584)	(0.11587)	(0.11590)	(0.11592)	(0.11595)
240.	814.31165	814.41822	814.52467	814.63100	814.73722	814.84332
	(0.11067)	(0.11069)	(0.11072)	(0.11075)	(0.11077)	(0.11080)
250.	799.20259	799.32092	799.43910	799.55713	799.67502	799.79275
	(0.10585)	(0.10587)	(0.10590)	(0.10593)	(0.10596)	(0.10599)
260.	19.33647	19.94033	20.55338	21.17604	21.80874	22.45194
	(0.01807)	(0.01806)	(0.01804)	(0.01803)	(0.01802)	(0.01800)
270.	18.64690	19.21350	19.78736	20.36875	20.95795	21.55527
	(0.01854)	(0.01853)	(0.01852)	(0.01851)	(0.01850)	(0.01849)

	Absolute pressure, kPa					
	4000.	**4100.**	**4200.**	**4300.**	**4400.**	**4500.**
Temp., °C						
325.	15.93393 (0.02107)	16.37811 (0.02106)	16.82509 (0.02106)	17.27490 (0.02105)	17.72761 (0.02105)	18.18326 (0.02104)
350.	15.04981 (0.02218)	15.46065 (0.02218)	15.87350 (0.02218)	16.28840 (0.02218)	16.70537 (0.02217)	17.12445 (0.02217)
375.	14.29209 (0.02329)	14.67629 (0.02329)	15.06201 (0.02328)	15.44926 (0.02328)	15.83806 (0.02328)	16.22843 (0.02328)
400.	13.62851 (0.02438)	13.99064 (0.02438)	14.35393 (0.02438)	14.71841 (0.02438)	15.08409 (0.02438)	15.45097 (0.02438)
425.	13.03813 (0.02546)	13.38143 (0.02546)	13.72567 (0.02546)	14.07084 (0.02546)	14.41695 (0.02546)	14.76401 (0.02546)
450.	12.50661 (0.02653)	12.83354 (0.02653)	13.16122 (0.02653)	13.48965 (0.02653)	13.81884 (0.02654)	14.14880 (0.02654)
475.	12.02374 (0.02759)	12.33619 (0.02759)	12.64925 (0.02760)	12.96293 (0.02760)	13.27723 (0.02760)	13.59214 (0.02760)
500.	11.58193 (0.02864)	11.88141 (0.02865)	12.18141 (0.02865)	12.48191 (0.02865)	12.78293 (0.02865)	13.08446 (0.02866)
525.	11.17536 (0.02969)	11.46313 (0.02969)	11.75132 (0.02969)	12.03994 (0.02970)	12.32899 (0.02970)	12.61847 (0.02970)
550.	10.79942 (0.03072)	11.07653 (0.03072)	11.35399 (0.03073)	11.63180 (0.03073)	11.90998 (0.03074)	12.18852 (0.03074)
575.	10.45042 (0.03175)	10.71774 (0.03175)	10.98537 (0.03175)	11.25330 (0.03176)	11.52154 (0.03176)	11.79008 (0.03177)
600.	10.12527 (0.03276)	10.38359 (0.03277)	10.64216 (0.03277)	10.90100 (0.03277)	11.16010 (0.03278)	11.41945 (0.03278)
625.	9.82140 (0.03377)	10.07139 (0.03377)	10.32159 (0.03378)	10.57202 (0.03378)	10.82267 (0.03379)	11.07354 (0.03379)
650.	9.53664 (0.03477)	9.77888 (0.03477)	10.02132 (0.03478)	10.26394 (0.03478)	10.50675 (0.03479)	10.74976 (0.03479)
675.	9.26910 (0.03576)	9.50413 (0.03576)	9.73932 (0.03577)	9.97468 (0.03577)	10.21020 (0.03578)	10.44588 (0.03578)
700.	9.01718 (0.03674)	9.24546 (0.03674)	9.47388 (0.03675)	9.70245 (0.03675)	9.93115 (0.03676)	10.15999 (0.03676)
725.	8.77945 (0.03771)	9.00141 (0.03772)	9.22349 (0.03772)	9.44569 (0.03773)	9.66800 (0.03773)	9.89044 (0.03774)
750.	8.55469 (0.03867)	8.77070 (0.03868)	8.98681 (0.03869)	9.20303 (0.03869)	9.41935 (0.03870)	9.63578 (0.03870)
775.	8.34179 (0.03963)	8.55220 (0.03964)	8.76269 (0.03964)	8.97328 (0.03965)	9.18395 (0.03965)	9.39472 (0.03966)

†Absolute viscosity μ_{cP} is shown in parentheses.

SOURCE: Derived from ASME, 1977.

Table G.1 Density (in Kilograms per Cubic Meter) and Absolute Viscosity μ_{cP} (in Centipoises) of Steam and Compressed Water† (*continued*)

	Absolute pressure, kPa					
	5000.	**5100.**	**5200.**	**5300.**	**5400.**	**5500.**
Temp., °C	263.91	265.15	266.37	267.58	268.76	269.93
Sat. vap.	25.36229	25.89912	26.43803	26.97904	27.52221	28.06754
	(0.01813)	(0.01818)	(0.01823)	(0.01828)	(0.01833)	(0.01838)
Sat. liq.	777.71406	775.71519	773.73127	771.76172	769.80595	767.86344
	(0.09985)	(0.09934)	(0.09883)	(0.09834)	(0.09786)	(0.09739)
Temp., °C						
100.	960.46968	960.51718	960.56466	960.61213	960.65958	960.70701
	(0.28346)	(0.28349)	(0.28352)	(0.28355)	(0.28357)	(0.28360)
110.	953.09317	953.14217	953.19116	953.24013	953.28908	953.33801
	(0.25622)	(0.25625)	(0.25628)	(0.25630)	(0.25633)	(0.25636)
120.	945.29330	945.34403	945.39474	945.44543	945.49610	945.54675
	(0.23334)	(0.23336)	(0.23339)	(0.23342)	(0.23344)	(0.23347)
130.	937.07748	937.13017	937.18283	937.23548	937.28810	937.34070
	(0.21393)	(0.21396)	(0.21398)	(0.21401)	(0.21404)	(0.21406)
140.	928.44710	928.50200	928.55688	928.61174	928.66657	928.72138
	(0.19734)	(0.19737)	(0.19740)	(0.19742)	(0.19745)	(0.19748)
150.	919.39801	919.45541	919.51279	919.57014	919.62746	919.68476
	(0.18306)	(0.18309)	(0.18311)	(0.18314)	(0.18316)	(0.18319)
160.	909.92065	909.98088	910.04109	910.10126	910.16140	910.22151
	(0.17067)	(0.17070)	(0.17072)	(0.17075)	(0.17078)	(0.17080)
170.	899.99999	900.06343	900.12684	900.19021	900.25354	900.31685
	(0.15986)	(0.15989)	(0.15991)	(0.15994)	(0.15996)	(0.15999)
180.	889.61512	889.68221	889.74926	889.81627	889.88324	889.95018
	(0.15036)	(0.15039)	(0.15041)	(0.15044)	(0.15046)	(0.15049)
190.	878.73855	878.80982	878.88104	878.95222	879.02335	879.09444
	(0.14196)	(0.14198)	(0.14201)	(0.14203)	(0.14206)	(0.14208)
200.	867.33515	867.41124	867.48727	867.56325	867.63918	867.71505
	(0.13447)	(0.13449)	(0.13452)	(0.13454)	(0.13457)	(0.13459)
210.	855.36058	855.44226	855.52388	855.60544	855.68693	855.76836
	(0.12774)	(0.12777)	(0.12779)	(0.12782)	(0.12784)	(0.12787)
220.	842.75895	842.84719	842.93537	843.02346	843.11149	843.19943
	(0.12165)	(0.12168)	(0.12170)	(0.12173)	(0.12176)	(0.12178)
230.	829.45956	829.55560	829.65155	829.74741	829.84317	829.93885
	(0.11608)	(0.11611)	(0.11614)	(0.11616)	(0.11619)	(0.11622)
240.	815.37209	815.47750	815.58280	815.68798	815.79305	815.89801
	(0.11094)	(0.11097)	(0.11099)	(0.11102)	(0.11105)	(0.11108)
250.	800.37925	800.49611	800.61283	800.72941	800.84584	800.96213
	(0.10613)	(0.10616)	(0.10619)	(0.10622)	(0.10624)	(0.10627)
260.	784.32558	784.45672	784.58767	784.71843	784.84901	784.97941
	(0.10158)	(0.10161)	(0.10164)	(0.10167)	(0.10170)	(0.10173)
270.	24.67569	25.32932	25.99384	26.66980	27.35774	28.05828
	(0.01843)	(0.01842)	(0.01841)	(0.01840)	(0.01839)	(0.01838)

Temp., °C	Absolute pressure, kPa					
	5000.	5100.	5200.	5300.	5400.	5500.
325.	20.50759 (0.02102)	20.98207 (0.02102)	21.45989 (0.02101)	21.94112 (0.02101)	22.42583 (0.02100)	22.91409 (0.02100)
350.	19.25244 (0.02216)	19.68478 (0.02215)	20.11944 (0.02215)	20.55645 (0.02215)	20.99586 (0.02215)	21.43769 (0.02214)
375.	18.20440 (0.02327)	18.60455 (0.02327)	19.00639 (0.02327)	19.40996 (0.02327)	19.81526 (0.02327)	20.22231 (0.02327)
400.	17.30381 (0.02438)	17.67815 (0.02438)	18.05378 (0.02438)	18.43070 (0.02438)	18.80894 (0.02438)	19.18849 (0.02438)
425.	16.51381 (0.02547)	16.86671 (0.02547)	17.22062 (0.02547)	17.57553 (0.02547)	17.93146 (0.02547)	18.28841 (0.02547)
450.	15.81016 (0.02655)	16.14478 (0.02655)	16.48020 (0.02655)	16.81642 (0.02655)	17.15345 (0.02656)	17.49129 (0.02656)
475.	15.17618 (0.02762)	15.49490 (0.02762)	15.81426 (0.02762)	16.13427 (0.02762)	16.45493 (0.02763)	16.77624 (0.02763)
500.	14.59989 (0.02867)	14.90455 (0.02868)	15.20973 (0.02868)	15.51545 (0.02868)	15.82170 (0.02869)	16.12848 (0.02869)
525.	14.07234 (0.02972)	14.36442 (0.02972)	14.65694 (0.02973)	14.94990 (0.02973)	15.24330 (0.02974)	15.53714 (0.02974)
550.	13.58666 (0.03076)	13.86738 (0.03076)	14.14847 (0.03077)	14.42993 (0.03077)	14.71175 (0.03078)	14.99395 (0.03078)
575.	13.13739 (0.03179)	13.40778 (0.03179)	13.67847 (0.03180)	13.94948 (0.03180)	14.22079 (0.03181)	14.49242 (0.03181)
600.	12.72013 (0.03281)	12.98105 (0.03281)	13.24223 (0.03282)	13.50367 (0.03282)	13.76538 (0.03283)	14.02735 (0.03283)
625.	12.33122 (0.03382)	12.58343 (0.03382)	12.83585 (0.03383)	13.08851 (0.03383)	13.34138 (0.03384)	13.59448 (0.03384)
650.	11.96762 (0.03482)	12.21176 (0.03482)	12.45610 (0.03483)	12.70062 (0.03483)	12.94534 (0.03484)	13.19025 (0.03484)
675.	11.62673 (0.03581)	11.86339 (0.03581)	12.10021 (0.03582)	12.33719 (0.03582)	12.57434 (0.03583)	12.81166 (0.03584)
700.	11.30631 (0.03679)	11.53600 (0.03680)	11.76582 (0.03680)	11.99579 (0.03681)	12.22590 (0.03681)	12.45614 (0.03682)
725.	11.00446 (0.03776)	11.22762 (0.03777)	11.45091 (0.03778)	11.67431 (0.03778)	11.89784 (0.03779)	12.12149 (0.03779)
750.	10.71948 (0.03873)	10.93653 (0.03874)	11.15369 (0.03874)	11.37095 (0.03875)	11.58831 (0.03875)	11.80579 (0.03876)
775.	10.44990 (0.03969)	10.66121 (0.03969)	10.87261 (0.03970)	11.08410 (0.03970)	11.29567 (0.03971)	11.50734 (0.03971)

†Absolute viscosity μ_{cP} is shown in parentheses.

SOURCE: Derived from ASME, 1977.

	Absolute pressure, kPa					
	6000.	**6200.**	**6400.**	**6600.**	**6800.**	**7000.**
Temp., °C	275.55	277.70	279.79	281.84	283.84	285.79
Sat. vap.	30.82829	31.94895	33.07939	34.21969	35.37046	36.53183
	(0.01861)	(0.01870)	(0.01879)	(0.01888)	(0.01897)	(0.01906)
Sat. liq.	758.33519	754.59965	750.90291	747.24014	743.61233	740.01526
	(0.09515)	(0.09431)	(0.09349)	(0.09270)	(0.09193)	(0.09118)
Temp., °C						
100.	960.94393	961.03859	961.13318	961.22770	961.32216	961.41655
	(0.28374)	(0.28379)	(0.28384)	(0.28390)	(0.28395)	(0.28401)
110.	953.58241	953.68005	953.77761	953.87510	953.97253	954.06988
	(0.25649)	(0.25655)	(0.25660)	(0.25666)	(0.25671)	(0.25676)
120.	945.79972	945.90076	946.00174	946.10263	946.20345	946.30419
	(0.23361)	(0.23366)	(0.23371)	(0.23377)	(0.23382)	(0.23387)
130.	937.60339	937.70831	937.81315	937.91791	938.02258	938.12716
	(0.21420)	(0.21425)	(0.21430)	(0.21436)	(0.21441)	(0.21446)
140.	928.99505	929.10436	929.21357	929.32268	929.43171	929.54063
	(0.19761)	(0.19766)	(0.19771)	(0.19777)	(0.19782)	(0.19787)
150.	919.97085	920.08510	920.19924	920.31328	920.42721	920.54104
	(0.18332)	(0.18337)	(0.18342)	(0.18348)	(0.18353)	(0.18358)
160.	910.52163	910.64146	910.76118	910.88077	911.00025	911.11961
	(0.17093)	(0.17098)	(0.17103)	(0.17108)	(0.17114)	(0.17119)
170.	900.63285	900.75901	900.88504	901.01093	901.13669	901.26232
	(0.16012)	(0.16017)	(0.16022)	(0.16027)	(0.16032)	(0.16037)
180.	890.28426	890.41762	890.55082	890.68387	890.81677	890.94952
	(0.15061)	(0.15066)	(0.15072)	(0.15077)	(0.15082)	(0.15087)
190.	879.44920	879.59079	879.73220	879.87343	880.01449	880.15537
	(0.14221)	(0.14226)	(0.14231)	(0.14236)	(0.14241)	(0.14246)
200.	868.09363	868.24469	868.39555	868.54619	868.69663	868.84686
	(0.13472)	(0.13477)	(0.13482)	(0.13487)	(0.13492)	(0.13497)
210.	856.17458	856.33664	856.49845	856.66001	856.82133	856.98240
	(0.12800)	(0.12805)	(0.12810)	(0.12815)	(0.12820)	(0.12825)
220.	843.63805	843.81298	843.98761	844.16195	844.33600	844.50976
	(0.12191)	(0.12196)	(0.12201)	(0.12206)	(0.12212)	(0.12217)
230.	830.41586	830.60603	830.79584	830.98530	831.17440	831.36315
	(0.11635)	(0.11640)	(0.11645)	(0.11651)	(0.11656)	(0.11661)
240.	816.42111	816.62957	816.83759	817.04517	817.25231	817.45901
	(0.11121)	(0.11127)	(0.11132)	(0.11137)	(0.11143)	(0.11148)
250.	801.54146	801.77220	802.00239	802.23202	802.46110	802.68963
	(0.10641)	(0.10647)	(0.10653)	(0.10658)	(0.10664)	(0.10670)
260.	785.62863	785.88704	786.14473	786.40171	786.65797	786.91353
	(0.10188)	(0.10194)	(0.10199)	(0.10205)	(0.10211)	(0.10217)
270.	768.48170	768.77529	769.06792	769.35959	769.65032	769.94011
	(0.09752)	(0.09758)	(0.09765)	(0.09771)	(0.09777)	(0.09784)

			Absolute pressure, kPa			
	6000.	6200.	6400.	6600.	6800.	7000.
Temp., °C						
325.	25.41102 (0.02098)	26.43721 (0.02098)	27.48009 (0.02097)	28.54039 (0.02097)	29.61892 (0.02096)	30.71652 (0.02096)
350.	23.68449 (0.02214)	24.60150 (0.02213)	25.52944 (0.02213)	26.46865 (0.02213)	27.41949 (0.02213)	28.38234 (0.02213)
375.	22.28465 (0.02327)	23.12261 (0.02327)	23.96829 (0.02327)	24.82187 (0.02327)	25.68354 (0.02327)	26.55349 (0.02327)
400.	21.10651 (0.02438)	21.88341 (0.02438)	22.66601 (0.02439)	23.45442 (0.02439)	24.24876 (0.02439)	25.04913 (0.02440)
425.	20.08878 (0.02548)	20.81637 (0.02549)	21.54832 (0.02549)	22.28468 (0.02549)	23.02553 (0.02550)	23.77094 (0.02550)
450.	19.19279 (0.02657)	19.87925 (0.02657)	20.56911 (0.02658)	21.26242 (0.02659)	21.95923 (0.02659)	22.65957 (0.02660)
475.	18.39271 (0.02764)	19.04399 (0.02765)	19.69799 (0.02766)	20.35474 (0.02766)	21.01427 (0.02767)	21.67661 (0.02768)
500.	17.67050 (0.02871)	18.29112 (0.02871)	18.91396 (0.02872)	19.53902 (0.02873)	20.16633 (0.02874)	20.79591 (0.02875)
525.	17.01305 (0.02976)	17.60656 (0.02977)	18.20188 (0.02978)	18.79904 (0.02979)	19.39804 (0.02979)	19.99890 (0.02980)
550.	16.41051 (0.03080)	16.97975 (0.03081)	17.55051 (0.03082)	18.12279 (0.03083)	18.69659 (0.03084)	19.27193 (0.03085)
575.	15.85524 (0.03183)	16.40257 (0.03184)	16.95117 (0.03185)	17.50104 (0.03186)	18.05218 (0.03187)	18.60461 (0.03188)
600.	15.34117 (0.03286)	15.86856 (0.03287)	16.39701 (0.03288)	16.92654 (0.03289)	17.45714 (0.03290)	17.98882 (0.03291)
625.	14.86334 (0.03387)	15.37247 (0.03388)	15.88250 (0.03389)	16.39343 (0.03390)	16.90528 (0.03391)	17.41804 (0.03392)
650.	14.41766 (0.03487)	14.90997 (0.03488)	15.40306 (0.03489)	15.89691 (0.03490)	16.39154 (0.03491)	16.88695 (0.03493)
675.	14.00069 (0.03586)	14.47745 (0.03587)	14.95488 (0.03589)	15.43296 (0.03590)	15.91170 (0.03591)	16.39110 (0.03592)
700.	13.60949 (0.03685)	14.07182 (0.03686)	14.53471 (0.03687)	14.99817 (0.03688)	15.46219 (0.03689)	15.92678 (0.03691)
725.	13.24156 (0.03782)	13.69043 (0.03783)	14.13979 (0.03784)	14.58964 (0.03786)	15.03997 (0.03787)	15.49079 (0.03788)
750.	12.89471 (0.03879)	13.33101 (0.03880)	13.76773 (0.03881)	14.20486 (0.03882)	14.64242 (0.03883)	15.08039 (0.03885)
775.	12.56704 (0.03974)	12.99155 (0.03975)	13.41643 (0.03977)	13.84166 (0.03978)	14.26725 (0.03979)	14.69320 (0.03980)

†Absolute viscosity μ_{cP} is shown in parentheses.

SOURCE: Derived from ASME, 1977.

			Absolute pressure, kPa			
	7200.	**7400.**	**7600.**	**7800.**	**8000.**	**8200.**
Temp., °C	287.70	289.57	291.41	293.20	294.97	296.70
Sat. vap.	37.70407	38.88745	40.08226	41.28904	42.50758	43.73844
	(0.01914)	(0.01923)	(0.01931)	(0.01940)	(0.01949)	(0.01957)
Sat. liq.	736.44662	732.90423	729.38603	725.89226	722.41663	718.95969
	(0.09045)	(0.08974)	(0.08905)	(0.08837)	(0.08770)	(0.08705)
Temp., °C						
100.	961.51089	961.60515	961.69935	961.79349	961.88757	961.98158
	(0.28406)	(0.28411)	(0.28417)	(0.28422)	(0.28428)	(0.28433)
110.	954.16716	954.26438	954.36152	954.45859	954.55560	954.65253
	(0.25682)	(0.25687)	(0.25693)	(0.25698)	(0.25703)	(0.25709)
120.	946.40485	946.50544	946.60595	946.70638	946.80674	946.90702
	(0.23393)	(0.23398)	(0.23404)	(0.23409)	(0.23414)	(0.23420)
130.	938.23166	938.33608	938.44041	938.54465	938.64882	938.75290
	(0.21452)	(0.21457)	(0.21462)	(0.21468)	(0.21473)	(0.21478)
140.	929.64947	929.75821	929.86685	929.97541	930.08387	930.19223
	(0.19792)	(0.19798)	(0.19803)	(0.19808)	(0.19813)	(0.19819)
150.	920.65476	920.76838	920.88189	920.99530	921.10861	921.22181
	(0.18363)	(0.18368)	(0.18374)	(0.18379)	(0.18384)	(0.18389)
160.	911.23886	911.35798	911.47699	911.59588	911.71466	911.83332
	(0.17124)	(0.17129)	(0.17134)	(0.17139)	(0.17144)	(0.17150)
170.	901.38781	901.51317	901.63839	901.76348	901.88845	902.01327
	(0.16042)	(0.16047)	(0.16052)	(0.16057)	(0.16062)	(0.16067)
180.	891.08211	891.21455	891.34683	891.47897	891.61096	891.74279
	(0.15092)	(0.15097)	(0.15102)	(0.15107)	(0.15112)	(0.15117)
190.	880.29607	880.43660	880.57695	880.71713	880.85713	880.99696
	(0.14251)	(0.14256)	(0.14261)	(0.14266)	(0.14271)	(0.14276)
200.	868.99689	869.14671	869.29633	869.44575	869.59496	869.74397
	(0.13502)	(0.13507)	(0.13512)	(0.13517)	(0.13522)	(0.13527)
210.	857.14323	857.30382	857.46417	857.62428	857.78415	857.94379
	(0.12830)	(0.12835)	(0.12840)	(0.12845)	(0.12850)	(0.12855)
220.	844.68322	844.85640	845.02929	845.20189	845.37421	845.54625
	(0.12222)	(0.12227)	(0.12232)	(0.12237)	(0.12242)	(0.12247)
230.	831.55155	831.73960	831.92730	832.11465	832.30166	832.48833
	(0.11666)	(0.11671)	(0.11677)	(0.11682)	(0.11687)	(0.11692)
240.	817.66528	817.87112	818.07653	818.28152	818.48608	818.69021
	(0.11154)	(0.11159)	(0.11164)	(0.11170)	(0.11175)	(0.11180)
250.	802.91762	803.14506	803.37197	803.59834	803.82418	804.04948
	(0.10675)	(0.10681)	(0.10686)	(0.10692)	(0.10697)	(0.10703)
260.	787.16838	787.42254	787.67601	787.92878	788.18088	788.43229
	(0.10223)	(0.10229)	(0.10235)	(0.10241)	(0.10246)	(0.10252)
270.	770.22896	770.51690	770.80391	771.09001	771.37521	771.65952
	(0.09790)	(0.09796)	(0.09802)	(0.09808)	(0.09815)	(0.09821)

	Absolute pressure, kPa					
	7200.	**7400.**	**7600.**	**7800.**	**8000.**	**8200.**
Temp., °C						
325.	31.83413	32.97276	34.13349	35.31752	36.52613	37.76075
	(0.02096)	(0.02095)	(0.02095)	(0.02095)	(0.02095)	(0.02095)
350.	29.35760	30.34569	31.34704	32.36213	33.39145	34.43551
	(0.02213)	(0.02213)	(0.02213)	(0.02213)	(0.02213)	(0.02213)
375.	27.43194	28.31909	29.21517	30.12039	31.03500	31.95923
	(0.02327)	(0.02327)	(0.02328)	(0.02328)	(0.02328)	(0.02329)
400.	25.85565	26.66845	27.48766	28.31338	29.14577	29.98495
	(0.02440)	(0.02440)	(0.02441)	(0.02441)	(0.02442)	(0.02442)
425.	24.52098	25.27573	26.03526	26.79965	27.56897	28.34332
	(0.02551)	(0.02551)	(0.02552)	(0.02553)	(0.02553)	(0.02554)
450.	23.36350	24.07106	24.78230	25.49727	26.21602	26.93859
	(0.02660)	(0.02661)	(0.02662)	(0.02662)	(0.02663)	(0.02664)
475.	22.34178	23.00983	23.68076	24.35463	25.03146	25.71128
	(0.02769)	(0.02769)	(0.02770)	(0.02771)	(0.02772)	(0.02773)
500.	21.42777	22.06194	22.69843	23.33726	23.97846	24.62205
	(0.02875)	(0.02876)	(0.02877)	(0.02878)	(0.02879)	(0.02880)
525.	20.60162	21.20623	21.81272	22.42113	23.03145	23.64371
	(0.02981)	(0.02982)	(0.02983)	(0.02984)	(0.02985)	(0.02986)
550.	19.84881	20.42725	21.00724	21.58881	22.17195	22.75669
	(0.03086)	(0.03087)	(0.03088)	(0.03089)	(0.03090)	(0.03091)
575.	19.15833	19.71334	20.26965	20.82726	21.38618	21.94643
	(0.03189)	(0.03190)	(0.03192)	(0.03193)	(0.03194)	(0.03195)
600.	18.52158	19.05543	19.59037	20.12640	20.66353	21.20176
	(0.03292)	(0.03293)	(0.03294)	(0.03295)	(0.03296)	(0.03298)
625.	17.93171	18.44631	18.96182	19.47825	19.99561	20.51390
	(0.03393)	(0.03394)	(0.03396)	(0.03397)	(0.03398)	(0.03399)
650.	17.38313	17.88009	18.37783	18.87635	19.37566	19.87575
	(0.03494)	(0.03495)	(0.03496)	(0.03497)	(0.03498)	(0.03500)
675.	16.87117	17.35190	17.83330	18.31536	18.79809	19.28148
	(0.03593)	(0.03594)	(0.03596)	(0.03597)	(0.03598)	(0.03599)
700.	16.39194	16.85766	17.32395	17.79081	18.25824	18.72624
	(0.03692)	(0.03693)	(0.03694)	(0.03695)	(0.03697)	(0.03698)
725.	15.94209	16.39389	16.84616	17.29893	17.75218	18.20592
	(0.03789)	(0.03790)	(0.03792)	(0.03793)	(0.03794)	(0.03795)
750.	15.51878	15.95759	16.39682	16.83646	17.27652	17.71701
	(0.03886)	(0.03887)	(0.03888)	(0.03890)	(0.03891)	(0.03892)
775.	15.11952	15.54619	15.97322	16.40062	16.82837	17.25648
	(0.03981)	(0.03983)	(0.03984)	(0.03985)	(0.03986)	(0.03988)

†Absolute viscosity μ_{cP} is shown in parentheses.

SOURCE: Derived from ASME, 1977.

			Absolute pressure, kPa			
	8400.	**8600.**	**8800.**	**9000.**	**9200.**	**9400.**
Temp., °C	298.39	300.06	301.70	303.31	304.89	306.44
Sat. vap.	44.98194	46.23840	47.50817	48.79162	50.08881	51.40083
	(0.01965)	(0.01974)	(0.01982)	(0.01991)	(0.01999)	(0.02008)
Sat. liq.	715.51979	712.09538	708.68498	705.28714	701.89833	698.52193
	(0.08642)	(0.08579)	(0.08518)	(0.08458)	(0.08398)	(0.08340)
Temp., °C						
100.	962.07553	962.16941	962.26323	962.35699	962.45068	962.54431
	(0.28438)	(0.28444)	(0.28449)	(0.28455)	(0.28460)	(0.28465)
110.	954.74940	954.84619	954.94292	955.03958	955.13617	955.23269
	(0.25714)	(0.25720)	(0.25725)	(0.25730)	(0.25736)	(0.25741)
120.	947.00723	947.10736	947.20741	947.30739	947.40730	947.50713
	(0.23425)	(0.23431)	(0.23436)	(0.23441)	(0.23447)	(0.23452)
130.	938.85689	938.96080	939.06463	939.16838	939.27205	939.37563
	(0.21484)	(0.21489)	(0.21494)	(0.21500)	(0.21505)	(0.21510)
140.	930.30051	930.40869	930.51678	930.62478	930.73269	930.84051
	(0.19824)	(0.19829)	(0.19834)	(0.19840)	(0.19845)	(0.19850)
150.	921.33491	921.44791	921.56080	921.67360	921.78629	921.89887
	(0.18394)	(0.18400)	(0.18405)	(0.18410)	(0.18415)	(0.18420)
160.	911.95186	912.07028	912.18860	912.30679	912.42487	912.54284
	(0.17155)	(0.17160)	(0.17165)	(0.17170)	(0.17175)	(0.17180)
170.	902.13797	902.26254	902.38697	902.51128	902.63546	902.75950
	(0.16073)	(0.16078)	(0.16083)	(0.16088)	(0.16093)	(0.16098)
180.	891.87447	892.00601	892.13739	892.26863	892.39972	892.53066
	(0.15122)	(0.15127)	(0.15132)	(0.15137)	(0.15142)	(0.15147)
190.	881.13661	881.27610	881.41541	881.55455	881.69352	881.83232
	(0.14281)	(0.14286)	(0.14291)	(0.14296)	(0.14301)	(0.14306)
200.	869.89278	870.04139	870.18979	870.33800	870.48601	870.63382
	(0.13532)	(0.13537)	(0.13542)	(0.13547)	(0.13552)	(0.13557)
210.	858.10318	858.26234	858.42127	858.57996	858.73841	858.89664
	(0.12860)	(0.12865)	(0.12870)	(0.12875)	(0.12880)	(0.12886)
220.	845.71800	845.88947	846.06066	846.23157	846.40220	846.57256
	(0.12252)	(0.12258)	(0.12263)	(0.12268)	(0.12273)	(0.12278)
230.	832.67465	832.86063	833.04628	833.23158	833.41655	833.60119
	(0.11697)	(0.11703)	(0.11708)	(0.11713)	(0.11718)	(0.11723)
240.	818.89393	819.09723	819.30011	819.50258	819.70464	819.90629
	(0.11186)	(0.11191)	(0.11196)	(0.11202)	(0.11207)	(0.11212)
250.	804.27427	804.49852	804.72226	804.94548	805.16819	805.39038
	(0.10709)	(0.10714)	(0.10720)	(0.10725)	(0.10731)	(0.10736)
260.	788.68303	788.93309	789.18250	789.43123	789.67932	789.92674
	(0.10258)	(0.10264)	(0.10270)	(0.10276)	(0.10281)	(0.10287)
270.	771.94293	772.22546	772.50711	772.78789	773.06781	773.34687
	(0.09827)	(0.09833)	(0.09839)	(0.09846)	(0.09852)	(0.09858)

	Absolute pressure, kPa					
	8400.	**8600.**	**8800.**	**9000.**	**9200.**	**9400.**
Temp., °C						
325.	39.02290 (0.02095)	40.31429 (0.02095)	41.63676 (0.02096)	42.99235 (0.02096)	44.38332 (0.02097)	45.81214 (0.02097)
350.	35.49489 (0.02214)	36.57017 (0.02214)	37.66198 (0.02214)	38.77099 (0.02215)	39.89794 (0.02215)	41.04358 (0.02216)
375.	32.89336 (0.02329)	33.83763 (0.02330)	34.79232 (0.02330)	35.75773 (0.02331)	36.73416 (0.02332)	37.72191 (0.02332)
400.	30.83105 (0.02443)	31.68422 (0.02443)	32.54461 (0.02444)	33.41236 (0.02445)	34.28763 (0.02446)	35.17057 (0.02446)
425.	29.12275 (0.02555)	29.90737 (0.02555)	30.69726 (0.02556)	31.49249 (0.02557)	32.29317 (0.02558)	33.09937 (0.02559)
450.	27.66504 (0.02665)	28.39542 (0.02666)	29.12977 (0.02667)	29.86816 (0.02667)	30.61064 (0.02668)	31.35726 (0.02669)
475.	26.39412 (0.02774)	27.08002 (0.02775)	27.76900 (0.02775)	28.46112 (0.02776)	29.15638 (0.02777)	29.85484 (0.02778)
500.	25.26805 (0.02881)	25.91647 (0.02882)	26.56734 (0.02883)	27.22068 (0.02884)	27.87652 (0.02885)	28.53487 (0.02886)
525.	24.25791 (0.02987)	24.87408 (0.02988)	25.49222 (0.02989)	26.11235 (0.02990)	26.73449 (0.02992)	27.35864 (0.02993)
550.	23.34301 (0.03092)	23.93095 (0.03093)	24.52050 (0.03094)	25.11167 (0.03096)	25.70448 (0.03097)	26.29893 (0.03098)
575.	22.50799 (0.03196)	23.07089 (0.03197)	23.63511 (0.03198)	24.20068 (0.03199)	24.76760 (0.03201)	25.33587 (0.03202)
600.	21.74110 (0.03299)	22.28155 (0.03300)	22.82311 (0.03301)	23.36579 (0.03302)	23.90959 (0.03304)	24.45451 (0.03305)
625.	21.03312 (0.03400)	21.55327 (0.03402)	22.07436 (0.03403)	22.59638 (0.03404)	23.11935 (0.03405)	23.64325 (0.03407)
650.	20.37663 (0.03501)	20.87830 (0.03502)	21.38076 (0.03503)	21.88400 (0.03505)	22.38805 (0.03506)	22.89289 (0.03507)
675.	19.76555 (0.03601)	20.25028 (0.03602)	20.73569 (0.03603)	21.22176 (0.03604)	21.70851 (0.03606)	22.19594 (0.03607)
700.	19.19481 (0.03699)	19.66395 (0.03700)	20.13366 (0.03702)	20.60395 (0.03703)	21.07480 (0.03704)	21.54623 (0.03705)
725.	18.66015 (0.03797)	19.11486 (0.03798)	19.57007 (0.03799)	20.02576 (0.03800)	20.48194 (0.03802)	20.93861 (0.03803)
750.	18.15791 (0.03893)	18.59922 (0.03895)	19.04096 (0.03896)	19.48311 (0.03897)	19.92569 (0.03898)	20.36868 (0.03900)
775.	17.68495 (0.03989)	18.11378 (0.03990)	18.54297 (0.03991)	18.97252 (0.03993)	19.40243 (0.03994)	19.83270 (0.03995)

†Absolute viscosity μ_{cP} is shown in parentheses.

SOURCE: Derived from ASME, 1977.

Table G.1 Density (in Kilograms per Cubic Meter) and Absolute Viscosity μ_{cP} (in Centipoises) of Steam and Compressed Water† (*continued*)

			Absolute pressure, kPa			
	9600.	10100.	10600.	11100.	11600.	12100.
Temp., °C Sat. vap.	307.97 52.72774 (0.02016)	311.69 56.11366 (0.02037)	315.27 59.60311 (0.02059)	318.73 63.20571 (0.02081)	322.06 66.93371 (0.02103)	325.28 70.79768 (0.02126)
Sat. liq.	695.15409 (0.08283)	686.76604 (0.08143)	678.40203 (0.08007)	670.04613 (0.07876)	661.68777 (0.07748)	653.30027 (0.07623)
Temp., °C						
100.	962.63788 (0.28471)	962.87153 (0.28484)	963.10479 (0.28498)	963.33766 (0.28511)	963.57014 (0.28525)	963.80225 (0.28538)
110.	955.32914 (0.25747)	955.56998 (0.25760)	955.81039 (0.25774)	956.05038 (0.25787)	956.28995 (0.25801)	956.52911 (0.25814)
120.	947.60688 (0.23457)	947.85595 (0.23471)	948.10455 (0.23484)	948.35268 (0.23498)	948.60036 (0.23511)	948.84759 (0.23524)
130.	939.47913 (0.21516)	939.73752 (0.21529)	939.99540 (0.21542)	940.25277 (0.21555)	940.50964 (0.21569)	940.76601 (0.21582)
140.	930.94823 (0.19855)	931.21715 (0.19869)	931.48550 (0.19882)	931.75329 (0.19895)	932.02052 (0.19908)	932.28720 (0.19921)
150.	922.01136 (0.18425)	922.29213 (0.18438)	922.57228 (0.18451)	922.85179 (0.18464)	923.13069 (0.18477)	923.40897 (0.18490)
160.	912.66069 (0.17185)	912.95483 (0.17198)	913.24826 (0.17211)	913.54099 (0.17224)	913.83302 (0.17236)	914.12436 (0.17249)
170.	902.88342 (0.16103)	903.19265 (0.16116)	903.50108 (0.16128)	903.80872 (0.16141)	904.11557 (0.16153)	904.42164 (0.16166)
180.	892.66146 (0.15152)	892.98780 (0.15164)	893.31323 (0.15177)	893.63776 (0.15189)	893.96139 (0.15202)	894.28413 (0.15214)
190.	881.97095 (0.14311)	882.31679 (0.14324)	882.66158 (0.14336)	883.00533 (0.14348)	883.34804 (0.14361)	883.68974 (0.14373)
200.	870.78144 (0.13562)	871.14962 (0.13575)	871.51658 (0.13587)	871.88233 (0.13600)	872.24689 (0.13612)	872.61027 (0.13624)
210.	859.05463 (0.12891)	859.44860 (0.12903)	859.84114 (0.12915)	860.23227 (0.12928)	860.62199 (0.12940)	861.01032 (0.12953)
220.	846.74264 (0.12283)	847.16664 (0.12296)	847.58894 (0.12308)	848.00956 (0.12321)	848.42852 (0.12333)	848.84583 (0.12346)
230.	833.78550 (0.11729)	834.24481 (0.11741)	834.70208 (0.11754)	835.15733 (0.11767)	835.61057 (0.11780)	836.06183 (0.11793)
240.	820.10753 (0.11218)	820.60886 (0.11231)	821.10768 (0.11244)	821.60403 (0.11257)	822.09794 (0.11270)	822.58944 (0.11283)
250.	805.61207 (0.10742)	806.16407 (0.10755)	806.71295 (0.10769)	807.25876 (0.10783)	807.80153 (0.10796)	808.34131 (0.10810)
260.	790.17352 (0.10293)	790.78765 (0.10307)	791.39781 (0.10321)	792.00406 (0.10336)	792.60646 (0.10350)	793.20508 (0.10364)
270.	773.62508 (0.09864)	774.31692 (0.09879)	775.00357 (0.09894)	775.68512 (0.09909)	776.36167 (0.09924)	777.03329 (0.09939)

	Absolute pressure, kPa					
	9600.	**10100.**	**10600.**	**11100.**	**11600.**	**12100.**
Temp., °C						
325.	47.28157	51.15372	55.35834	59.97566	65.12400	654.21575
	(0.02098)	(0.02101)	(0.02104)	(0.02109)	(0.02116)	(0.07636)
350.	42.20874	45.21306	48.36093	51.67143	55.16781	58.87864
	(0.02217)	(0.02219)	(0.02222)	(0.02225)	(0.02229)	(0.02234)
375.	38.72131	41.27314	43.90589	46.62623	49.44175	52.36115
	(0.02333)	(0.02335)	(0.02338)	(0.02341)	(0.02344)	(0.02348)
400.	36.06135	38.32368	40.63882	43.00975	45.43974	47.93237
	(0.02447)	(0.02450)	(0.02452)	(0.02455)	(0.02458)	(0.02462)
425.	33.91120	35.96597	38.05798	40.18883	42.36022	44.57393
	(0.02560)	(0.02562)	(0.02565)	(0.02568)	(0.02571)	(0.02574)
450.	32.10808	34.00386	35.92713	37.87885	39.85998	41.87156
	(0.02670)	(0.02673)	(0.02676)	(0.02679)	(0.02682)	(0.02685)
475.	30.55653	32.32510	34.11462	35.92569	37.75889	39.61486
	(0.02780)	(0.02782)	(0.02785)	(0.02788)	(0.02792)	(0.02795)
500.	29.19575	30.85922	32.53906	34.23563	35.94931	37.68050
	(0.02887)	(0.02890)	(0.02893)	(0.02896)	(0.02900)	(0.02903)
525.	27.98482	29.55927	31.14675	32.74750	34.36177	35.98980
	(0.02994)	(0.02997)	(0.03000)	(0.03003)	(0.03007)	(0.03010)
550.	26.89503	28.39258	29.90067	31.41946	32.94911	34.48978
	(0.03099)	(0.03102)	(0.03105)	(0.03109)	(0.03112)	(0.03116)
575.	25.90550	27.33557	28.77428	30.22173	31.67803	33.14328
	(0.03203)	(0.03206)	(0.03210)	(0.03213)	(0.03216)	(0.03220)
600.	25.00057	26.37068	27.74794	29.13243	30.52420	31.92333
	(0.03306)	(0.03309)	(0.03313)	(0.03316)	(0.03319)	(0.03323)
625.	24.16811	25.48440	26.80667	28.13495	29.46929	30.80973
	(0.03408)	(0.03411)	(0.03414)	(0.03418)	(0.03421)	(0.03425)
650.	23.39852	24.66611	25.93872	27.21637	28.49910	29.78693
	(0.03509)	(0.03512)	(0.03515)	(0.03519)	(0.03522)	(0.03526)
675.	22.68403	23.90724	25.13468	26.36638	27.60236	28.84263
	(0.03608)	(0.03611)	(0.03615)	(0.03618)	(0.03622)	(0.03625)
700.	22.01824	23.20076	24.38688	25.57660	26.76995	27.96692
	(0.03707)	(0.03710)	(0.03713)	(0.03717)	(0.03720)	(0.03724)
725.	21.39577	22.54081	23.68891	24.84009	25.99433	27.15165
	(0.03804)	(0.03808)	(0.03811)	(0.03815)	(0.03818)	(0.03822)
750.	20.81209	21.92245	23.03542	24.15101	25.26922	26.39005
	(0.03901)	(0.03904)	(0.03908)	(0.03911)	(0.03915)	(0.03918)
775.	20.26332	21.34146	22.42183	23.50444	24.58929	25.67637
	(0.03997)	(0.04000)	(0.04003)	(0.04007)	(0.04010)	(0.04014)

†Absolute viscosity μ_{cP} is shown in parentheses.

SOURCE: Derived from ASME, 1977.

Table G.1 Density (in Kilograms per Cubic Meter) and Absolute Viscosity μ_{cP} (in Centipoises) of Steam and Compressed Water†

(*continued*)

	Absolute pressure, kPa					
	12600.	**13100.**	**13600.**	**14100.**	**14600.**	**15100.**
Temp., °C	328.40	331.42	334.36	337.21	339.97	342.66
Sat. vap.	74.81209	78.99452	83.35878	87.92349	92.70953	97.73513
	(0.02149)	(0.02173)	(0.02199)	(0.02225)	(0.02252)	(0.02281)
Sat. liq.	644.86855	636.38201	627.80844	619.12881	610.32436	601.36128
	(0.07500)	(0.07379)	(0.07260)	(0.07141)	(0.07024)	(0.06907)
Temp., °C						
100.	964.03397	964.26531	964.49627	964.72686	964.95707	965.18692
	(0.28552)	(0.28565)	(0.28579)	(0.28592)	(0.28605)	(0.28619)
110.	956.76785	957.00617	957.24409	957.48160	957.71870	957.95540
	(0.25828)	(0.25841)	(0.25855)	(0.25868)	(0.25881)	(0.25895)
120.	949.09436	949.34068	949.58655	949.83198	950.07696	950.32150
	(0.23538)	(0.23551)	(0.23565)	(0.23578)	(0.23591)	(0.23605)
130.	941.02188	941.27725	941.53214	941.78653	942.04044	942.29386
	(0.21595)	(0.21608)	(0.21622)	(0.21635)	(0.21648)	(0.21661)
140.	932.55333	932.81891	933.08394	933.34843	933.61239	933.87581
	(0.19934)	(0.19947)	(0.19960)	(0.19973)	(0.19986)	(0.19999)
150.	923.68664	923.96369	924.24014	924.51599	924.79124	925.06590
	(0.18503)	(0.18516)	(0.18528)	(0.18541)	(0.18554)	(0.18567)
160.	914.41501	914.70498	914.99427	915.28288	915.57083	915.85811
	(0.17262)	(0.17274)	(0.17287)	(0.17300)	(0.17312)	(0.17325)
170.	904.72693	905.03145	905.33521	905.63820	905.94044	906.24193
	(0.16178)	(0.16191)	(0.16203)	(0.16216)	(0.16228)	(0.16241)
180.	894.60599	894.92696	895.24707	895.56632	895.88470	896.20223
	(0.15227)	(0.15239)	(0.15252)	(0.15264)	(0.15276)	(0.15289)
190.	884.03042	884.37009	884.70877	885.04645	885.38315	885.71888
	(0.14385)	(0.14398)	(0.14410)	(0.14422)	(0.14435)	(0.14447)
200.	872.97246	873.33349	873.69336	874.05209	874.40968	874.76614
	(0.13637)	(0.13649)	(0.13661)	(0.13673)	(0.13686)	(0.13698)
210.	861.39728	861.78287	862.16711	862.55000	862.93157	863.31183
	(0.12965)	(0.12978)	(0.12990)	(0.13002)	(0.13014)	(0.13027)
220.	849.26150	849.67557	850.08803	850.49891	850.90821	851.31597
	(0.12358)	(0.12371)	(0.12383)	(0.12396)	(0.12408)	(0.12420)
230.	836.51114	836.95851	837.40396	837.84752	838.28920	838.72903
	(0.11805)	(0.11818)	(0.11831)	(0.11843)	(0.11856)	(0.11868)
240.	823.07855	823.56530	824.04973	824.53186	825.01171	825.48931
	(0.11296)	(0.11309)	(0.11322)	(0.11335)	(0.11348)	(0.11361)
250.	808.87813	809.41204	809.94308	810.47127	810.99666	811.51929
	(0.10823)	(0.10836)	(0.10850)	(0.10863)	(0.10876)	(0.10890)
260.	793.79995	794.39115	794.97873	795.56274	796.14323	796.72025
	(0.10378)	(0.10392)	(0.10406)	(0.10420)	(0.10433)	(0.10447)
270.	777.70009	778.36213	779.01951	779.67229	780.32055	780.96437
	(0.09954)	(0.09969)	(0.09983)	(0.09998)	(0.10012)	(0.10027)

	Absolute pressure, kPa					
	12600.	**13100.**	**13600.**	**14100.**	**14600.**	**15100.**
Temp., °C						
325.	656.11556	657.96541	659.76842	661.52738	663.24483	664.92308
	(0.07666)	(0.07695)	(0.07723)	(0.07750)	(0.07778)	(0.07804)
350.	62.83945	67.09502	71.70285	76.73867	82.30612	88.55575
	(0.02240)	(0.02247)	(0.02256)	(0.02266)	(0.02279)	(0.02295)
375.	55.39450	58.55343	61.85153	65.30470	68.93171	72.75476
	(0.02353)	(0.02358)	(0.02363)	(0.02370)	(0.02377)	(0.02386)
400.	50.49155	53.12160	55.82728	58.61390	61.48735	64.45422
	(0.02466)	(0.02470)	(0.02475)	(0.02480)	(0.02486)	(0.02492)
425.	46.83190	49.13613	51.48881	53.89226	56.34897	58.86161
	(0.02578)	(0.02582)	(0.02587)	(0.02591)	(0.02596)	(0.02602)
450.	43.91465	45.99036	48.09985	50.24433	52.42507	54.64339
	(0.02689)	(0.02693)	(0.02697)	(0.02702)	(0.02706)	(0.02711)
475.	41.49423	43.39765	45.32582	47.27943	49.25920	51.26588
	(0.02799)	(0.02803)	(0.02807)	(0.02811)	(0.02815)	(0.02820)
500.	39.42959	41.19699	42.98313	44.78842	46.61331	48.45824
	(0.02907)	(0.02911)	(0.02915)	(0.02919)	(0.02923)	(0.02928)
525.	37.63185	39.28817	40.95904	42.64472	44.34549	46.06161
	(0.03014)	(0.03017)	(0.03021)	(0.03026)	(0.03030)	(0.03034)
550.	36.04163	37.60483	39.17955	40.76596	42.36424	43.97456
	(0.03119)	(0.03123)	(0.03127)	(0.03131)	(0.03135)	(0.03139)
575.	34.61758	36.10105	37.59380	39.09593	40.60757	42.12881
	(0.03224)	(0.03227)	(0.03231)	(0.03235)	(0.03239)	(0.03244)
600.	33.32988	34.74393	36.16555	37.59481	39.03178	40.47653
	(0.03327)	(0.03330)	(0.03334)	(0.03338)	(0.03342)	(0.03346)
625.	32.15632	33.50911	34.86814	36.23345	37.60509	38.98311
	(0.03429)	(0.03432)	(0.03436)	(0.03440)	(0.03444)	(0.03448)
650.	31.07989	32.37801	33.68132	34.98984	36.30361	37.62265
	(0.03529)	(0.03533)	(0.03537)	(0.03541)	(0.03545)	(0.03549)
675.	30.08720	31.33610	32.58935	33.84695	35.10893	36.37530
	(0.03629)	(0.03633)	(0.03637)	(0.03640)	(0.03644)	(0.03648)
700.	29.16752	30.37178	31.57968	32.79125	34.00649	35.22542
	(0.03728)	(0.03731)	(0.03735)	(0.03739)	(0.03743)	(0.03747)
725.	28.31206	29.47555	30.64212	31.81179	32.98456	34.16043
	(0.03825)	(0.03829)	(0.03833)	(0.03836)	(0.03840)	(0.03844)
750.	27.51350	28.63956	29.76825	30.89956	32.03348	33.17003
	(0.03922)	(0.03925)	(0.03929)	(0.03933)	(0.03936)	(0.03940)
775.	26.76569	27.85723	28.95101	30.04701	31.14524	32.24569
	(0.04017)	(0.04021)	(0.04024)	(0.04028)	(0.04032)	(0.04036)

†Absolute viscosity μ_{cP} is shown in parentheses.

SOURCE: Derived from ASME, 1977.

Table G.1 Density (in Kilograms per Cubic Meter) and Absolute
Viscosity μ_{cP} (in Centipoises) of Steam and Compressed Water†
(*continued*)

	Absolute pressure, kPa					
	15600.	16100.	16600.	17100.	17600.	18100.
Temp., °C	345.28	347.83	350.32	352.74	355.11	357.41
Sat. vap.	103.01669	108.57316	114.43574	120.76237	127.55896	134.88289
	(0.02311)	(0.02343)	(0.02378)	(0.02415)	(0.02455)	(0.02499)
Sat. liq.	592.19528	582.78128	572.78068	562.97935	552.80448	542.08579
	(0.06789)	(0.06671)	(0.06547)	(0.06428)	(0.06307)	(0.06181)
Temp., °C						
100.	965.41639	965.64550	965.87423	966.10261	966.33062	966.55828
	(0.28632)	(0.28646)	(0.28659)	(0.28673)	(0.28686)	(0.28700)
110.	958.19170	958.42760	958.66310	958.89821	959.13292	959.36724
	(0.25908)	(0.25922)	(0.25935)	(0.25949)	(0.25962)	(0.25976)
120.	950.56561	950.80928	951.05251	951.29531	951.53769	951.77963
	(0.23618)	(0.23631)	(0.23645)	(0.23658)	(0.23671)	(0.23685)
130.	942.54680	942.79926	943.05125	943.30276	943.55380	943.80438
	(0.21674)	(0.21688)	(0.21701)	(0.21714)	(0.21727)	(0.21740)
140.	934.13870	934.40106	934.66290	934.92421	935.18500	935.44528
	(0.20012)	(0.20025)	(0.20038)	(0.20051)	(0.20064)	(0.20077)
150.	925.33996	925.61344	925.88633	926.15865	926.43039	926.70155
	(0.18580)	(0.18593)	(0.18605)	(0.18618)	(0.18631)	(0.18644)
160.	916.14473	916.43069	916.71600	917.00066	917.28468	917.56805
	(0.17338)	(0.17350)	(0.17363)	(0.17375)	(0.17388)	(0.17400)
170.	906.54268	906.84268	907.14195	907.44049	907.73831	908.03540
	(0.16253)	(0.16266)	(0.16278)	(0.16291)	(0.16303)	(0.16315)
180.	896.51892	896.83477	897.14978	897.46397	897.77733	898.08988
	(0.15301)	(0.15313)	(0.15326)	(0.15338)	(0.15350)	(0.15362)
190.	886.05363	886.38743	886.72027	887.05217	887.38312	887.71315
	(0.14459)	(0.14471)	(0.14484)	(0.14496)	(0.14508)	(0.14520)
200.	875.12148	875.47572	875.82885	876.18090	876.53186	876.88175
	(0.13710)	(0.13722)	(0.13735)	(0.13747)	(0.13759)	(0.13771)
210.	863.69078	864.06843	864.44481	864.81991	865.19376	865.56636
	(0.13039)	(0.13051)	(0.13063)	(0.13075)	(0.13088)	(0.13100)
220.	851.72219	852.12688	852.53007	852.93176	853.33197	853.73072
	(0.12433)	(0.12445)	(0.12457)	(0.12470)	(0.12482)	(0.12494)
230.	839.16702	839.60320	840.03757	840.47017	840.90101	841.33011
	(0.11881)	(0.11893)	(0.11906)	(0.11918)	(0.11930)	(0.11943)
240.	825.96470	826.43788	826.90890	827.37777	827.84452	828.30917
	(0.11374)	(0.11386)	(0.11399)	(0.11412)	(0.11424)	(0.11437)
250.	812.03919	812.55639	813.07093	813.58284	814.09216	814.59892
	(0.10903)	(0.10916)	(0.10929)	(0.10942)	(0.10955)	(0.10968)
260.	797.29386	797.86410	798.43102	798.99466	799.55508	800.11232
	(0.10461)	(0.10474)	(0.10488)	(0.10502)	(0.10515)	(0.10528)
270.	781.60383	782.23898	782.86991	783.49667	784.11934	784.73797
	(0.10041)	(0.10055)	(0.10070)	(0.10084)	(0.10098)	(0.10112)

			Absolute pressure, kPa			
	15600.	**16100.**	**16600.**	**17100.**	**17600.**	**18100.**
Temp., °C						
325.	666.56424	668.17024	669.74285	671.28370	672.79431	674.27606
	(0.07830)	(0.07856)	(0.07881)	(0.07906)	(0.07930)	(0.07954)
350.	95.72835	104.27565	574.64389	579.13742	583.33249	587.26913
	(0.02315)	(0.02341)	(0.06570)	(0.06629)	(0.06685)	(0.06738)
375.	76.80037	81.10043	85.69362	90.62733	95.96042	101.76706
	(0.02395)	(0.02406)	(0.02419)	(0.02433)	(0.02449)	(0.02468)
400.	67.52185	70.69853	73.99354	77.41737	80.98188	84.70048
	(0.02499)	(0.02507)	(0.02515)	(0.02524)	(0.02534)	(0.02545)
425.	61.43307	64.06645	66.76511	69.53269	72.37313	75.29070
	(0.02608)	(0.02614)	(0.02621)	(0.02628)	(0.02635)	(0.02643)
450.	56.90070	59.19846	61.53820	63.92154	66.35019	68.82595
	(0.02717)	(0.02722)	(0.02728)	(0.02734)	(0.02741)	(0.02748)
475.	53.30023	55.36304	57.45514	59.57736	61.73058	63.91571
	(0.02825)	(0.02830)	(0.02835)	(0.02841)	(0.02847)	(0.02853)
500.	50.32366	52.21005	54.11786	56.04759	57.99973	59.97478
	(0.02932)	(0.02937)	(0.02942)	(0.02947)	(0.02953)	(0.02959)
525.	47.79338	49.54107	51.30498	53.08539	54.88260	56.69691
	(0.03039)	(0.03043)	(0.03048)	(0.03053)	(0.03058)	(0.03064)
550.	45.59710	47.23203	48.87954	50.53981	52.21302	53.89935
	(0.03144)	(0.03148)	(0.03153)	(0.03158)	(0.03163)	(0.03168)
575.	43.65977	45.20058	46.75133	48.31215	49.88315	51.46444
	(0.03248)	(0.03252)	(0.03257)	(0.03262)	(0.03266)	(0.03271)
600.	41.92914	43.38968	44.85821	46.33482	47.81957	49.31254
	(0.03351)	(0.03355)	(0.03360)	(0.03364)	(0.03369)	(0.03374)
625.	40.36755	41.75846	43.15588	44.55987	45.97046	47.38770
	(0.03452)	(0.03457)	(0.03461)	(0.03466)	(0.03470)	(0.03475)
650.	38.94700	40.27667	41.61171	42.95213	44.29797	45.64925
	(0.03553)	(0.03557)	(0.03561)	(0.03566)	(0.03570)	(0.03575)
675.	37.64609	38.92130	40.20096	41.48507	42.77367	44.06676
	(0.03652)	(0.03657)	(0.03661)	(0.03665)	(0.03669)	(0.03674)
700.	36.44804	37.67436	38.90439	40.13813	41.37561	42.61682
	(0.03751)	(0.03755)	(0.03759)	(0.03763)	(0.03767)	(0.03772)
725.	35.33940	36.52148	37.70668	38.89499	40.08642	41.28097
	(0.03848)	(0.03852)	(0.03856)	(0.03860)	(0.03864)	(0.03869)
750.	34.30921	35.45100	36.59541	37.74245	38.89211	40.04439
	(0.03944)	(0.03948)	(0.03952)	(0.03956)	(0.03960)	(0.03964)
775.	33.34836	34.45324	35.56035	36.66966	37.78119	38.89493
	(0.04039)	(0.04043)	(0.04047)	(0.04051)	(0.04055)	(0.04059)

†Absolute viscosity μ_{cP} is shown in parentheses.

SOURCE: Derived from ASME, 1977.

Table G.1 Density (in Kilograms per Cubic Meter) and Absolute Viscosity μ_{cP} (in Centipoises) of Steam and Compressed Water† (*continued*)

	Absolute pressure, kPa					
	18600.	19100.	19600.	20100.	20600.	21100.
Temp., °C	359.67	361.87	364.02	366.12	368.17	370.18
Sat. vap.	142.83686	151.57602	161.34391	172.54227	185.88183	202.78981
	(0.02548)	(0.02602)	(0.02665)	(0.02737)	(0.02827)	(0.02943)
Sat. liq.	530.65543	518.30726	504.76689	489.61685	472.10814	450.66361
	(0.06049)	(0.05908)	(0.05756)	(0.05589)	(0.05399)	(0.05171)
Temp., °C						
100.	966.78557	967.01251	967.23909	967.46532	967.69120	967.91673
	(0.28713)	(0.28726)	(0.28740)	*0.28753)	(0.28767)	(0.28780)
110.	959.60117	959.83472	960.06788	960.30066	960.53305	960.76507
	(0.25989)	(0.26002)	(0.26016)	(0.26043)	(0.26043)	(0.26056)
120.	952.02116	952.26226	952.50294	952.74320	952.98305	953.22249
	(0.23698)	(0.23711)	(0.23724)	(0.23738)	(0.23751)	(0.23764)
130.	944.05448	944.30413	944.55332	944.80205	945.05032	945.29814
	(0.21753)	(0.21767)	(0.21780)	(0.21793)	(0.21806)	(0.21819)
140.	935.70504	935.96430	936.22305	936.48129	936.73903	936.99627
	(0.20090)	(0.20103)	(0.20116)	(0.20129)	(0.20142)	(0.20154)
150.	926.97215	927.24218	927.51164	927.78055	928.04890	928.31670
	(0.18656)	(0.18669)	(0.18682)	(0.18695)	(0.18707)	(0.18720)
160.	917.85079	918.13290	918.41438	918.69524	918.97548	919.25510
	(0.17413)	(0.17426)	(0.17438)	(0.17451))	(0.17463)	(0.17476)
170.	908.33178	908.62745	908.92242	909.21668	909.51025	909.80313
	(0.16328)	(0.16340)	(0.16352)	(0.16365)	(0.16377)	(0.16389)
180.	898.40162	898.71256	899.02269	899.33204	899.64059	899.94837
	(0.15375)	(0.15387)	(0.15399)	(0.15411)	(0.15423)	(0.15436)
190.	888.04225	888.37043	888.69771	889.02407	889.34955	889.67413
	(0.14532)	(0.14544)	(0.14556)	(0.14569)	(0.14581)	(0.14593)
200.	877.23057	877.57834	877.92506	878.27074	878.61539	878.95902
	(0.13783)	(0.13795)	(0.13807)	(0.13819)	(0.13831)	(0.13843)
210.	865.93772	866.30785	866.67677	867.04449	867.41100	867.77634
	(0.13112)	(0.13124)	(0.13136)	(0.131.48)	(0.13160)	(0.13172)
220.	854.12801	854.52387	854.91830	855.31131	855.70293	856.09316
	(0.12506)	(0.12518)	(0.12530)	(0.12543)	(0.12555)	(0.12567)
230.	841.75748	842.18314	942.60711	843.02941	843.45006	843.86906
	(0.11955)	(0.11967)	(0.11980)	(0.11992)	(0.12004)	(0.12016)
240.	828.77175	829.23228	829.69078	830.14727	830.60178	831.05432
	(0.11449)	(0.11462)	(0.11474)	(0.11487)	(0.11499)	(0.11512)
250.	815.10315	815.60487	816.10413	816.60094	817.09535	817.58738
	(0.10981)	(0.10994)	(0.11006)	(0.11019)	(0.11032)	(0.11045)
260.	800.66641	801.21741	801.76535	802.21028	802.85224	803.39126
	(0.10542)	(0.10555)	(0.10568)	(0.10582)	(0.10595)	(0.10608)
270.	785.35263	785.96339	786.57029	787.17340	787.77277	788.36847
	(0.10126)	(0.10140)	(0.10154)	(0.10168)	(0.10181)	(0.10195)

			Absolute pressure, kPa			
	18600.	**19100.**	**19600.**	**20100.**	**20600.**	**21100.**
Temp., °C						
325.	675.73026 (0.07978)	677.15809 (0.08001)	678.56070 (0.08024)	679.93911 (0.08046)	681.29431 (0.08069)	682.62720 (0.08091)
350.	590.97941 (0.06787)	594.49008 (0.06835)	597.82338 (0.06880)	600.99785 (0.06924)	604.02921 (0.06966)	606.92106 (0.07006)
375.	108.14259 (0.02490)	115.21243 (0.02517)	123.14679 (0.02548)	132.46505 (0.02588)	143.65667 (0.02639)	157.81666 (0.02710)
400.	88.58842 (0.02557)	92.66307 (0.020570)	96.94424 (0.02584)	101.45464 (0.02600)	106.22033 (0.02617)	111.27137 (0.02636)
425.	78.29007 (0.02652)	81.37627 (0.02661)	84.55481 (0.02671)	87.83166 (0.02682)	91.21336 (0.02693)	94.70698 (0.02706)
450.	71.35071 (0.02755)	73.92649 (0.02762)	76.55540 (0.02770)	79.23969 (0.02779)	81.98173 (0.02788)	84.78403 (0.02797)
475.	66.13366 (0.02859)	68.38541 (0.02866)	70.67195 (0.02873)	72.99432 (0.02880)	75.35358 (0.02888)	77.75085 (0.02896)
500.	61.97326 (0.02965)	63.99568 (0.02971)	66.04258 (0.02977)	68.11450 (0.02983)	70.21199 (0.02990)	72.33563 (0.02997)
525.	58.52862 (0.03069)	60.37803 (0.03075)	62.24545 (0.03081)	64.13119 (0.03087)	66.03557 (0.03093)	67.95889 (0.03100)
550.	55.59899 (0.03173)	57.31212 (0.03179)	59.03893 (0.03184)	60.77959 (0.03190)	62.53431 (0.03196)	64.30325 (0.03202)
575.	53.05614 (0.03276)	54.65837 (0.03281)	56.27124 (0.03287)	57.89485 (0.03292)	59.52933 (0.03298)	61.17479 (0.03303)
600.	50.81381 (0.03378)	52.32343 (0.03383)	53.84148 (0.03388)	55.36804 (0.03394)	56.90317 (0.03399)	58.44694 (0.03404)
625.	48.81164 (0.03479)	50.24232 (0.03484)	51.67978 (0.03489)	53.12408 (0.03494)	54.57524 (0.03499)	56.03332 (0.03504)
650.	47.00600 (0.03579)	48.36825 (0.03584)	49.73603 (0.03589)	51.10936 (0.03594)	52.48826)0.03599)	53.87276 (0.03604)
675.	46.36436 (0.03678)	46.66649 (0.03683)	47.97315 (0.03687)	49.28438 (0.03692)	50.60017 (0.03697)	51.92055 (0.03702)
700.	43.86177 (0.03776)	45.11048 (0.03780)	46.36295 (0.03785)	47.61918 (0.03790)	48.87919 (0.03794)	50.14298 (0.03799)
725.	42.47865 (0.03873)	43.67946 (0.03877)	44.88340 (0.03882)	46.09048 (0.03886)	47.30069 (0.03891)	48.51403 (0.03895)
750.	41.19929 (0.03969)	42.35682 (0.03973)	43.51696 (0.03977)	44.67972 (0.03981)	45.84510 (0.03986)	47.01310 (0.03990)
775.	40.01087 (0.04063)	41.12901 (0.04067)	42.24935 (0.04072)	43.37189 (0.04076)	44.49663 (0.04080)	45.62355 (0.04084)

†Absolute viscosity μ_{cP} is shown in parentheses.

SOURCE: Derived from ASME, 1977.

	Absolute pressure, kPa					
	21600.	**22100.**	**22600.**	**23100.**	**23600.**	**24100.**
Temp., °C	372.15	374.08	−0.00	−0.00	−0.00	−0.00
Sat. vap.	226.04553	257.76042	257.76042	257.76042	257.76042	257.76042
	(0.03112)	(0.03358)	(0.03358)	(0.03358)	(0.03358)	(0.03358)
Sat. liq.	420.80175	381.66109	381.66109	381.66109	381.66109	381.66109
	(0.04860)	(0.4468)	(0.04468)	(0.4468)	(0.04468)	(0.04468)
Temp., °C						
100.	968.14191	968.36674	968.59123	968.81538	969.03919	969.26265
	(0.28794)	(0.28807)	(0.28820)	(0.28834)	(0.28847)	(0.28861)
110.	960.99671	961.22798	961.45887	961.68939	961.91954	962.14933
	(0.26069)	(0.26083)	(0.26096)	(0.26110)	(0.26123)	(0.25136)
120.	953.46151	953.70013	953.93834	954.17615	954.41355	954.65055
	(0.23778)	(0.23791)	(0.23804)	(0.23817)	(0.23831)	(0.23844)
130.	945.54551	945.79244	946.03892	946.28495	946.53055	946.77571
	(0.21832)	(0.21845)	(0.21858)	(0.21871)	(0.21884)	(0.21897)
140.	937.25302	937.50928	937.76504	938.02032	938.27511	938.52943
	(0.20167)	(0.20180)	(0.20193)	(0.20206)	(0.20219)	(0.20232)
150.	928.58395	928.85066	929.11682	929.38244	929.64753	929.91209
	(0.18733)	(0.18745)	(0.18758)	(0.18771)	(0.18783)	(0.18796)
160.	919.53411	919.81251	920.09031	920.36751	920.64411	920.92011
	(0.17488)	(0.17500)	(0.17513)	(0.17525)	(0.17538)	(0.17550)
170.	910.09532	910.38682	910.67765	910.96781	911.25730	911.54612
	(0.16402)	(0.16414)	(0.16426)	(0.16439)	(0.16451)	(0.16463)
180.	900.25537	900.56160	900.86706	901.17176	901.47571	901.77891
	(0.15448)	(0.15460)	(0.15472)	(0.15484)	(0.15496)	(0.15508)
190.	889.99783	890.32065	890.64260	890.96368	891.28391	891.60329
	(0.14605)	(0.14617)	(0.14629)	(0.14641)	(0.14653)	(0.14665)
200.	879.30163	879.64324	879.98385	880.32347	880.66210	880.99976
	(0.13855)	(0.13867)	(0.13879)	(0.13891)	(0.13903)	(0.13915)
210.	868.14050	868.50349	868.86533	869.22602	869.58558	869.94401
	(0.13184)	(0.13196)	(0.13208)	(0.13220)	(0.13232)	(0.13243)
220.	856.48202	856.86951	857.25566	857.64047	858.02396	858.40613
	(0.12579)	(0.12591)	(0.12603)	(0.12615)	(0.12627)	(0.12638)
230.	844.28645	844.70222	845.11640	845.52901	845.94005	846.34955
	(0.12028)	(0.12041)	(0.12053)	(0.12065)	(0.12077)	(0.12089)
240	831.50492	831.95361	832.40039	832.84529	833.28833	833.72953
	(0.11524)	(0.11536)	(0.11549)	(0.11561)	(0.11573)	(0.11585)
250.	818.07705	818.56439	819.04944	819.53222	820.01275	820.49106
	(0.11057)	(0.11070)	(0.11082)	(0.11095)	(0.11108)	(0.11120)
260.	803.92738	804.46065	804.99110	805.51876	806.04368	806.56587
	(0.10621)	(0.10634)	(0.10647)	(0.10660)	(0.10673)	(0.10686)
270.	788.96053	789.54903	790.13400	790.71551	791.29359	791.86830
	(0.10209)	(0.10222)	(0.10236)	(0.10249)	(0.10263)	(0.10276)

	Absolute pressure, kPa					
	21600.	**22100.**	**22600.**	**23100.**	**23600.**	**24100.**
Temp., °C						
325.	683.93866 (0.08112)	685.22949 (0.08134)	686.50044 (0.08155)	687.75222 (0.08176)	688.98552 (0.08196)	690.20097 (0.08217)
350.	609.71495 (0.07044)	612.39100 (0.07082)	614.96789 (0.07118)	617.45377 (0.07187)	619.85540 (0.07187)	622.17849 (0.07220)
375.	177.74785 (0.02819)	215.65664 (0.03057)	413.28820 (0.04796)	452.84809 (0.05218)	472.46552 (0.05435)	486.28395 (0.05591)
400.	116.64253 (0.02658)	122.37426 (0.02682)	128.51378 (0.02708)	135.11656 (0.02739)	142.24814 (0.02773)	149.98651 (0.02811)
425.	98.32025 (0.02719)	102.06159 (0.02732)	105.94015 (0.02747)	109.96589 (0.02763)	114.14962 (0.02781)	118.50311 (0.02799)
450.	87.64925 (0.02807)	90.58019 (0.02817)	93.57983 (0.02828)	96.65130 (0.02840)	99.79792 (0.02852)	103.02319 (0.02865)
475.	80.18727 (0.02904)	82.66403 (0.02913)	85.18237 (0.02922)	87.74356 (0.02931)	90.34892 (0.02941)	92.99982 (0.02951)
500.	74.48598 (0.03005)	76.66363 (0.03012)	78.86918 (0.03020)	81.10323 (0.03028)	83.36641 (0.03036)	85.65936 (0.03045)
525.	69.90147 (0.03106)	71.86363 (0.03113)	73.84570 (0.03120)	75.84799 (0.03127)	77.87082 (0.03135)	79.91453 (0.03142)
550.	66.08661 (0.03208)	67.88457 (0.03214)	69.69731 (0.03221)	71.52501 (0.03227)	73.36786 (0.03234)	75.22603 (0.03241)
575.	62.83133 (0.03309)	64.49907 (0.03315)	66.17811 (0.03321)	67.86856 (0.03327)	69.57053 (0.03334)	71.28413 (0.03340)
600	59.99942 (0.03410)	61.56067 (0.03415)	63.13077 (0.03421)	64.70976 (0.03427)	66.29773 (0.03433)	67.89473 (0.03439)
625.	57.49835 (0.03510)	58.97038 (0.03515)	60.44944 (0.03521)	61.93558 (0.03526)	63.42882 (0.03532)	64.92921 (0.03538)
650.	55.26290 (0.03609)	56.65868 (0.03614)	58.06013 (0.03619)	59.46728 (0.03624)	60.88015 (0.03630)	62.29876 (0.03635)
675.	53.24552 (0.03707)	54.57511 (0.03712)	55.90932 (0.03717)	57.24817 (0.03722)	58.59167 (0.03727)	59.93982 (0.03732)
700.	51.41056 (0.03804)	52.68194 (0.03809)	53.95711 (0.03813)	55.23609 (0.03818)	56.51888 (0.03823)	57.80549 (0.03828)
725.	49.73052 (0.03900)	50.95015 (0.03904)	52.17293 (0.03909)	53.39884 (0.03914)	54.62790 (0.03919)	55.86011 (0.03924)
750.	48.18371 (0.03995)	49.35693 (0.03999)	50.53277 (0.04004)	51.71122 (0.04009)	52.89228 (0.04013)	54.07594 (0.04018)
775.	46.75266 (0.04089)	47.88395 (0.04093)	49.01743 (0.04098)	50.15307 (0.04102)	51.29089 (0.04107)	52.43088 (0.04111)

†Absolute viscosity μ_{cP} is shown in parentheses.

SOURCE: Derived from ASME, 1977.

Table G.1 Density (in Kilograms per Cubic Meter) and Absolute Viscosity μ_{cP} (in Centipoises) of Steam and Compressed Water† (*continued*)

	Absolute pressure, kPa					
	24600.	**25100.**	**25600.**	**26100.**	**26600.**	**27100.**
Temp., °C	−0.00	−0.00	−0.00	−0.00	−0.00	−0.00
Sat. vap.	257.76042	257.76042	257.76042	257.76042	257.76042	257.76042
	(0.03358)	(0.03358)	(0.03358)	(0.03358)	(0.03358)	(0.03358)
Sat. liq.	381.66109	381.66109	381.66109	381.66109	381.66109	381.66109
	(0.04468)	(0.04468)	(0.04468)	(0.04468)	(0.04468)	(0.04468)
Temp., °C						
100.	969.48578	969.70858	969.93104	970.15317	970.37497	970.59643
	(0.28874)	(0.28888)	(0.28901)	(0.28914)	(0.28928)	(0.28941)
110.	962.37875	962.60780	962.83650	963.06483	963.29281	963.52043
	(0.26150)	(0.26163)	(0.26176)	(0.26190)	(0.26203)	(0.26216)
120.	954.88716	955.12337	955.35919	955.59461	955.82965	956.06430
	(0.23857)	(0.23870)	(0.23884)	(0.23897)	(0.23910)	(0.23923)
130.	947.02043	947.26472	947.50858	947.75202	947.99502	948.23760
	(0.21910)	(0.21924)	(0.21937)	(0.21950)	(0.21963)	(0.21976)
140.	938.78326	939.03662	939.28950	939.54191	939.79386	940.04534
	(0.20244)	(0.20257)	(0.20270)	(0.20283)	(0.20296)	(0.20309)
150.	930.17611	930.43961	930.70258	930.96504	931.22698	931.48840
	(0.18808)	(0.18821)	(0.18834)	(0.18846)	(0.18859)	(0.18871)
160.	921.19553	921.47036	921.74461	922.01828	922.29137	922.56390
	(0.17563)	(0.17575)	(0.17587)	(0.17600)	(0.17612)	(0.17625)
170.	911.83428	912.12179	912.40864	912.69485	912.98042	913.26534
	(0.16475)	(0.16487)	(0.16500)	(0.16512)	(0.16524)	(0.16536)
180.	902.08136	902.38307	902.68405	902.98431	903.28383	903.58264
	(0.15520)	(0.15532)	(0.15544)	(0.15556)	(0.15568)	(0.15580)
190.	891.92182	892.23951	892.55636	892.87239	893.18760	893.50200
	(0.14677)	(0.14689)	(0.14700)	(0.14712)	(0.14724)	(0.14736)
200.	881.33645	881.67218	882.00696	882.34080	882.67369	883.00565
	(0.13927)	(0.13938)	(0.13950)	(0.13962)	(0.13974)	(0.13986)
210.	870.30133	870.65753	871.01264	871.36666	871.71959	872.07146
	(0.13255)	(0.13267)	(0.13279)	(0.13291)	(0.13303)	(0.13314)
220.	858.78700	859.16658	859.54489	859.92192	860.29770	860.67224
	(0.12650)	(0.12662)	(0.12674)	(0.12686)	(0.12698)	(0.12709)
230.	846.75752	847.16397	847.56892	847.97238	848.37437	848.77489
	(0.12101)	(0.12113)	(0.12125)	(0.12137)	(0.12148)	(0.12160)
240.	834.16891	834.60649	835.04228	835.47631	835.90858	836.33913
	(0.11598)	(0.11610)	(0.11622)	(0.11634)	(0.11646)	(0.11658)
250.	820.96718	821.44112	821.91293	822.38261	822.85020	823.31571
	(0.11132)	(0.11145)	(0.11157)	(0.11170)	(0.11182)	(0.11194)
260.	807.08539	807.60227	808.11653	808.62820	809.13733	809.64394
	(0.10698)	(0.10711)	(0.10724)	(0.10737)	(0.10749)	(0.10762)
270.	792.43968	793.00778	793.57265	794.13433	794.69287	795.24830
	(0.10289)	(0.10302)	(0.10316)	(0.10329)	(0.10342)	(0.10355)

			Absolute pressure, kPa			
	24600.	**25100.**	**25600.**	**26100.**	**26600.**	**27100.**
Temp., °C						
325.	691.39917	692.58070	693.74608	694.89584	696.03047	697.15041
	(0.08237)	(0.08257)	(0.08277)	(0.08296)	(0.08316)	(0.08335)
350.	624.42907	626.61161	628.73042	630.79001	632.79352	634.74434
	(0.07252)	(0.07284)	(0.07314)	(0.07344)	(0.07373)	(0.07402)
375.	497.13607	506.15902	513.92955	520.78442	526.93695	532.53223
	(0.05715)	(0.05820)	(0.05911)	(0.05992)	(0.06065)	(0.06133)
400.	158.58779	168.32403	179.36094	192.03631	206.76390	223.97781
	(0.02856)	(0.02910)	(0.02973)	(0.03051)	(0.03146)	(0.03263)
425.	123.03912	127.77147	132.71514	137.88632	143.30250	148.98253
	(0.02819)	(0.02840)	(0.02864)	(0.02889)	(0.02915)	(0.02945)
450.	106.33079	109.72457	113.20860	116.78713	120.46460	124.24562
	(0.02878)	(0.02892)	(0.02907)	(0.02923)	(0.02939)	(0.02957)
475.	95.69767	98.44395	101.24014	104.08782	106.98858	109.94406
	(0.02961)	(0.02972)	(0.02984)	(0.02996)	(0.03008)	(0.03021)
500.	87.98270	90.33710	92.72323	95.14175	97.59336	100.07876
	(0.03054)	(0.03063)	(0.03072)	(0.03082)	(0.03092)	(0.03103)
525.	81.97944	84.06587	86.17417	88.30464	90.45763	92.63347
	(0.03150)	(0.03158)	(0.03166)	(0.03175)	(0.03184)	(0.03193)
550.	77.09970	78.98906	80.89427	82.81551	84.75294	86.70675
	(0.03248)	(0.03255)	(0.03263)	(0.03271)	(0.03278)	(0.03286)
575.	73.00944	74.74658	76.49564	78.25672	80.02990	81.81529
	(0.03347)	(0.03353)	(0.03360)	(0.03367)	(0.03375)	(0.03382)
600.	69.50081	71.11604	72.74047	74.37416	76.01715	77.66951
	(0.03445)	(0.03452)	(0.03458)	(0.03465)	(0.03471)	(0.03478)
625.	66.43678	67.95157	69.47360	71.00291	72.53953	74.08348
	(0.03543)	(0.03549)	(0.03555)	(0.03562)	(0.03568)	(0.03574)
650.	63.72312	65.15326	66.58920	68.03094	69.47851	70.93193
	(0.03641)	(0.03647)	(0.03652)	(0.03658)	(0.03664)	(0.03670)
675.	61.29265	62.65015	64.01234	65.37922	66.75081	68.12710
	(0.03738)	(0.03743)	(0.03749)	(0.03754)	(0.03760)	(0.03766)
700.	59.09591	60.39016	61.68823	62.99013	64.29586	65.60542
	(0.03834)	(0.03839)	(0.03844)	(0.03849)	(0.03855)	(0.03860)
725.	57.09546	58.33396	59.57559	60.82037	62.06830	63.31935
	(0.03929)	(0.03934)	(0.03939)	(0.03944)	(0.03949)	(0.03954)
750.	55.26220	56.45106	57.64252	58.83657	60.03322	61.23244
	(0.04023)	(0.04028)	(0.04032)	(0.04037)	(0.04042)	(0.04047)
775.	53.57302	54.71733	55.86378	57.01239	58.16313	59.31601
	(0.04116)	(0.04121)	(0.04125)	(0.04130)	(0.04135)	(0.04140)

†Absolute viscosity μ_{cP} is shown in parentheses.

SOURCE: Derived from ASME, 1977.

			Absolute pressure, kPa			
	27600.	28100.	28600.	29100.	29600.	30100.
Temp., °C	−0.00	−0.00	−0.00	−0.00	−0.00	−0.00
Sat. vap.	257.76042	257.76042	257.76042	257.76042	257.76042	257.76042
	(0.03358)	(0.03358)	(0.03358)	(0.03358)	(0.03358)	(0.03358)
Sat. liq.	381.66109	381.66109	381.66109	381.66109	381.66109	381.66109
	(0.04468)	(0.04468)	(0.04468)	(0.04468)	(0.04468)	(0.04468)
Temp., °C						
100.	970.81758	971.03839	971.25888	971.47905	971.69890	971.91842
	(0.28955)	(0.28968)	(0.28981)	(0.28995)	(0.29008)	(0.29021)
110.	963.74769	963.97460	964.20116	964.42738	964.65324	964.87876
	(0.26230)	(0.26243)	(0.26256)	(0.26270)	(0.26283)	(0.26296)
120.	956.29856	956.53244	956.76594	956.99905	957.23179	957.46416
	(0.23936)	(0.23950)	(0.23963)	(0.23976)	(0.23989)	(0.24002)
130.	948.47977	948.72151	948.96283	949.20375	949.44425	949.68433
	(0.21989)	(0.22002)	(0.22015)	(0.22028)	(0.22041)	(0.22054)
140.	940.29636	940.54691	940.79701	941.04666	941.29585	941.54459
	(0.20321)	(0.20334)	(0.20347)	(0.20360)	(0.20372)	(0.20385)
150.	931.74931	932.00971	932.26961	932.52901	932.78790	933.04630
	(0.18884)	(0.18897)	(0.18909)	(0.18922)	(0.18934)	(0.18947)
160.	922.83585	923.10724	923.37807	923.64835	923.91806	924.18723
	(0.17637)	(0.17649)	(0.17662)	(0.17674)	(0.17686)	(0.17699)
170.	913.54963	913.83329	914.11632	914.39873	914.68052	914.96169
	(0.16548)	(0.16560)	(0.16573)	(0.16585)	(0.16597)	(0.16609)
180.	903.88073	904.17811	904.47479	904.77077	905.06605	905.36064
	(0.15592)	(0.15604)	(0.15616)	(0.15628)	(0.15640)	(0.15652)
190.	893.81558	894.12836	894.44034	894.75153	895.06193	895.37155
	(0.14748)	(0.14760)	(0.14772)	(0.14783)	(0.14795)	(0.14807)
200.	883.33669	883.66681	883.99602	884.32433	884.65174	884.97826
	(0.13997)	(0.14009)	(0.14021)	(0.14033)	(0.14044)	(0.14056)
210.	872.42226	872.77201	873.12070	873.46837	873.81500	874.16061
	(0.13326)	(0.13338)	(0.13349)	(0.13361)	(0.13373)	(0.13384)
220.	861.04554	861.41762	861.78848	862.15814	862.52661	862.89389
	(0.12721)	(0.12733)	(0.12745)	(0.12756)	(0.12768)	(0.12780)
230.	849.17397	849.57162	849.96785	850.36267	850.75610	851.14815
	(0.12172)	(0.12184)	(0.12196)	(0.12208)	(0.12219)	(0.12231)
240.	836.76797	837.19511	837.62057	838.04437	838.46653	838.88706
	(0.11670)	(0.11682)	(0.11694)	(0.11706)	(0.11718)	(0.11729)
250.	823.77917	824.24060	824.70003	825.15747	825.61295	826.06648
	(0.11206)	(0.11219)	(0.11231)	(0.11243)	(0.11255)	(0.11267)
260.	810.14807	810.64973	811.14897	811.64580	812.14027	812.63239
	(0.10774)	(0.10787)	(0.10799)	(0.10812)	(0.10824)	(0.10837)
270.	795.80067	796.35001	796.89638	797.43980	797.98033	798.51798
	(0.10368)	(0.10381)	(0.10394)	(0.10407)	(0.10419)	(0.10432)

			Absolute pressure, kPa			
	27600.	28100.	28600.	29100.	29600.	30100.
Temp., °C						
325.	698.25613	699.34803	700.42652	701.49198	702.54478	703.58526
	(0.08354)	(0.08373)	(0.08391)	(0.08410)	(0.08428)	(0.08446)
350.	636.64546	638.49968	640.30945	642.07707	643.80467	645.49424
	(0.07430)	(0.07457)	(0.07484)	(0.07510)	(0.07536)	(0.07561)
375.	537.67365	542.43718	546.88137	551.05086	554.98170	558.70307
	(0.06195)	(0.06253)	(0.06308)	(0.06359)	(0.06408)	(0.06455)
400.	243.91811	266.27549	290.00937	313.75254	336.40291	357.30764
	(0.03408)	(0.03580)	(0.03774)	(0.03978)	(0.04181)	(0.04376)
425.	154.94669	161.21680	167.81620	174.76993	182.10465	189.89861
	(0.02976)	(0.03011)	(0.03048)	(0.03088)	(0.03132)	(0.03181)
450.	128.13498	132.13763	136.25869	140.50338	144.87704	149.38508
	(0.02975)	(0.02994)	(0.03015)	(0.03036)	(0.03059)	(0.03083)
475.	112.95594	116.02593	119.15580	122.34731	125.60225	128.92243
	(0.03034)	(0.03048)	(0.03062)	(0.03077)	(0.03093)	(0.03109)
500.	102.59865	105.15374	107.74477	110.37245	113.03752	115.74072
	(0.03114)	(0.03125)	(0.03136)	(0.03148)	(0.03160)	(0.03173)
525.	94.83249	97.05501	99.30137	101.57190	103.86692	106.18675
	(0.03202)	(0.03211)	(0.03221)	(0.03231)	(0.03241)	(0.03252)
550.	88.67709	90.66413	92.66802	94.68894	96.72702	98.78243
	(0.03295)	(0.03303)	(0.03312)	(0.03320)	(0.03329)	(0.03338)
575.	83.61296	85.42301	87.24552	89.08055	90.92820	92.78854
	(0.03389)	(0.03397)	(0.03405)	(0.03413)	(0.03421)	(0.03429)
600.	79.33127	81.00248	82.68319	84.37343	86.07324	87.78267
	(0.03485)	(0.03492)	(0.03499)	(0.03506)	(0.03514)	(0.03521)
625.	75.63479	77.19348	78.75958	80.33310	81.91407	83.50249
	(0.03581)	(0.03587)	(0.03594)	(0.03601)	(0.03608)	(0.03615)
650.	72.39119	73.85632	75.32732	76.80421	78.28699	79.77567
	(0.03676)	(0.03682)	(0.03689)	(0.03695)	(0.03702)	(0.03708)
675.	69.50811	70.89383	72.28428	73.67945	75.07935	76.48397
	(0.03771)	(0.03777)	(0.03783)	(0.03789)	(0.03795)	(0.03801)
700.	66.91881	68.23603	69.55708	70.88196	72.21067	73.54320
	(0.03866)	(0.03871)	(0.03877)	(0.03883)	(0.03889)	(0.03894)
725.	64.57355	65.83088	67.09133	68.35492	69.62162	70.89144
	(0.03960)	(0.03965)	(0.03970)	(0.03976)	(0.03981)	(0.03987)
750.	62.43425	63.63863	64.84558	66.05509	67.26717	68.48179
	(0.04053)	(0.04058)	(0.04063)	(0.04068)	(0.04073)	(0.04079)
775.	60.47103	61.62816	62.78742	63.94879	65.11226	66.27783
	(0.04145)	(0.04150)	(0.04155)	(0.04160)	(0.04165)	(0.04170)

†Absolute viscosity μ_{cP} is shown in parentheses.

SOURCE: Derived from ASME, 1977.

Table G.2 Compressibility Factors for Air

| Temperature | | | Pressure, psia (bars) | | | | | | | | | |
°F	°R	K	14.5 (1)	72.5 (5)	145 (10)	290 (20)	580 (40)	870 (60)	1160 (80)	1450 (100)	2175 (150)	2900 (200)
−325	135	75								0.5099	0.7581	1.0025
−316	144	80								0.4887	0.7258	0.9588
−298	162	90	0.9764							0.4581	0.6779	0.8929
−280	180	100	0.9797	0.8872					0.3498	0.4337	0.6386	0.8377
−244	216	120	0.9880	0.9373	0.8660	0.6730			0.3371	0.4132	0.5964	0.7720
−208	252	140	0.9927	0.9614	0.9205	0.8297	0.5856	0.3313	0.3737	0.4340	0.5909	0.7699
−172	288	160	0.9951	0.9748	0.9489	0.8954	0.7803	0.6603	0.5696	0.5489	0.6340	0.7564
−136	324	180	0.9967	0.9832	0.9660	0.9314	0.8625	0.7977	0.7432	0.7084	0.7180	0.7986
−100	360	200	0.9978	0.9886	0.9767	0.9539	0.9100	0.8701	0.8374	0.8142	0.8061	0.8549
−10	450	250	0.9992	0.9957	0.9911	0.9822	0.9671	0.9549	0.9463	0.9411	0.9450	0.9713
80	540	300	0.9999	0.9987	0.9974	0.9950	0.9917	0.9901	0.9903	0.9930	1.0074	1.0326
170	630	350	1.0000	1.0002	1.0004	1.0014	1.0038	1.0075	1.0121	1.0183	1.0377	1.0635
260	720	400	1.0002	1.0012	1.0025	1.0046	1.0100	1.0159	1.0229	1.0312	1.0533	1.0795
350	810	450	1.0003	1.0016	1.0034	1.0063	1.0133	1.0210	1.0287	1.0374	1.0614	1.0913
440	900	500	1.0003	1.0020	1.0034	1.0074	1.0151	1.0234	1.0323	1.0410	1.0650	1.0913

SOURCE: Perry and Chilton (1973).

Table G.3 **Compressibility Factors for Argon**

Temperature			Pressure, psia (atm)						
°F	°R	K	14.7 (1)	58.8 (4)	103 (7)	147 (10)	588 (40)	1617 (70)	1470 (100)
−280	180	100	0.9782	0.9079					
−190	270	150	0.9930	0.9716	0.950	0.927			
−100	360	200	0.9971	0.9882	0.9792	0.9702	0.8978	0.7838	0.6917
−10	450	250	0.9986	0.9945	0.9905	0.9864	0.9476	0.9141	0.8878
26	486	270	0.9990	0.9960	0.9930	0.9900	0.9622	0.9388	0.9208
44	504	280	0.9991	0.9966	0.9940	0.9915	0.9679	0.9486	0.9340
62	522	290	0.9993	0.9971	0.9949	0.9927	0.9729	0.9570	0.9454
80.3	540	300	0.9994	0.9975	0.9957	0.9938	0.9773	0.9643	0.9553
98	558	310	0.9995	0.9979	0.9963	0.9948	0.9810	0.9706	0.9637
116	576	320	0.9996	0.9982	0.9969	0.9956	0.9843	0.9761	0.9710
170	630	350	0.9998	0.9990	0.9983	0.9977	0.9921	0.9888	0.9879
260	720	400	1.0000	0.9999	0.9999	0.9998	1.0002	1.0022	1.0057
350	810	450	1.0001	1.0004	1.0007	1.0011	1.0050	1.0101	1.0162
440	900	500	1.0002	1.0007	1.0013	1.0018	1.0079	1.0147	1.0224
530	990	550	1.0002	1.0009	1.0016	1.0023	1.0095	1.0174	1.0259

SOURCE: Perry and Chilton (1973).

Table G.4 Compressibility Factors for Carbon Dioxide

Temperature			Pressure, psia (bars)								
°F	°R	°C	14.5 (1)	72.5 (5)	145 (10)	290 (20)	580 (40)	870 (60)	1160 (80)	1450 (100)	2900 (200)
32	492	0	0.9933	0.9658	0.9294	0.8496					
122	582	50	0.9964	0.9805	0.9607	0.9195	0.8300	0.7264	0.5981	0.4239	
212	672	100	0.9977	0.9883	0.9764	0.9524	0.9034	0.8533	0.8022	0.7514	0.5891
302	762	150	0.9985	0.9927	0.9853	0.9705	0.9416	0.9131	0.8854	0.8590	0.7651
392	852	200	0.9991	0.9953	0.9908	0.9818	0.9640	0.9473	0.9313	0.9170	0.8649
482	942	250	0.9994	0.9971	0.9943	0.9886	0.9783	0.9684	0.9593	0.9511	0.9253
572	1032	300	0.9996	0.9982	0.9967	0.9936	0.9875	0.9822	0.9773	0.9733	0.9640
662	1122	350	0.9998	0.9991	0.9983	0.9964	0.9938	0.9914	0.9896	0.9882	0.9895
752	1212	400	0.9999	0.9997	0.9994	0.9989	0.9982	0.9979	0.9979	0.9984	1.0073
842	1302	450	1.0000	1.0000	1.0003	1.0005	1.0013	1.0023	1.0038	1.0056	1.0070

SOURCE: Perry and Chilton (1973).

Table G.5 Compressibility Factors for Carbon Monoxide

Temperature			Pressure, psia (atm)						
°F	°R	K	14.7 (1)	58.8 (4)	103 (7)	147 (10)	588 (40)	1617 (70)	1470 (100)
−100	360	200	0.9973	0.9893	0.9813	0.9734			
−10	450	250	0.9989	0.9957	0.9926	0.9896	0.9632		
80	540	300	0.9997	0.9987	0.9977	0.9968	0.9907	0.9896	0.9935
170	630	350	1.0000	1.0002	1.0003	1.0005	1.0042	1.0112	1.0216
260	720	400	1.0002	1.0010	1.0017	1.0025	1.0042	1.0112	1.0216
350	810	450	1.0003	1.0014	1.0025	1.0035	1.0152	1.0285	1.0433
440	900	500	1.0004	1.0016	1.0029	1.0041	1.0172	1.0314	1.0469
620	1080	600	1.0005	1.0018	1.0032	1.0045	1.0186	1.0332	1.0485

SOURCE: Perry and Chilton (1973).

Table G.6 Compressibility Factors for Hydrogen

Temperature			Pressure, psia (atm)						
°F	°R	K	14.7 (1)	58.8 (4)	103 (7)	147 (10)	588 (40)	1617 (70)	1470 (100)
−370	90	50	0.9919	0.9675	0.9431	0.9186			
−280	180	100	0.9998	0.9992	0.9987	0.9983	1.0029	1.0222	1.0560
−190	270	150	1.0006	1.0024	1.0041	1.0058	1.0260	1.0507	1.0796
−100	360	200	1.0007	1.0028	1.0048	1.0069	1.0283	1.0513	1.0760
−10	450	250	1.0006	1.0025	1.0044	1.0065	1.0264	1.0469	1.0682
80	540	300	1.0006	1.0024	1.0042	1.0059	1.0238	1.0420	1.0607
170	630	350	1.0005	1.0020	1.0036	1.0053	1.0213	1.0376	1.0541
260	720	400	1.0005	1.0020	1.0034	1.0048	1.0193	1.0339	1.0486
350	810	450	1.0004	1.0016	1.0030	1.0044	1.0176	1.0307	1.0439
440	900	500	1.0004	1.0016	1.0028	1.0040	1.0160	1.0280	1.0400

SOURCE: Perry and Chilton (1973).

Table G.7 Compressibility Factors for Methane

Temperature			Pressure, psia (bars)								
°F	°R	K	14.5 (1)	72.5 (5)	145 (10)	290 (20)	580 (40)	870 (60)	1160 (80)	1450 (100)	2900 (200)
−280	180	100								0.4313	0.8498
−190	270	150	0.9856	0.9243	0.8333					0.3405	0.6573
−100	360	200	0.9937	0.9682	0.9350	0.8629	0.6858	0.3755	0.3218	0.3657	0.6148
−10	450	250	0.9972	0.9841	0.9678	0.9356	0.8694	0.8035	0.7403	0.6889	0.6953
80	540	300	0.9982	0.9915	0.9828	0.9663	0.9342	0.9042	0.8773	0.8548	0.8280
170	630	350	0.9988	0.9954	0.9905	0.9821	0.9657	0.9513	0.9390	0.9293	0.9226
260	720	400	0.9995	0.9976	0.9957	0.9908	0.9833	0.9771	0.9721	0.9691	0.9783
350	810	450	0.9999	0.9996	0.9991	0.9965	0.9941	0.9923	0.9917	0.9922	1.0128
440	900	500	1.0000	1.0000	1.0000	1.0003	1.0009	1.0021	1.0043	1.0068	1.0335
620	1080	600	1.0002	1.0010	1.0021	1.0040	1.0083	1.0128	1.0175	1.0227	1.0555

SOURCE: Perry and Chilton (1973).

Table G.8 Compressibility Factors for Nitrogen

Temperature			Pressure, psia (bars)								
°F	°R	K	14.5 (1)	72.5 (5)	145 (10)	290 (20)	580 (40)	870 (60)	1160 (80)	1450 (100)	2900 (200)
−334	126	70						0.3400	0.4516	0.5623	1.1044
−316	144	80	0.9593					0.3122	0.4140	0.5148	1.0061
−298	162	90	0.9722					0.2938	0.3888	0.4826	0.9362
−280	180	100	0.9798	0.8910				0.2823	0.3720	0.4605	0.8840
−244	216	120	0.9883	0.9397	0.8732	0.7059		0.2822	0.3641	0.4438	0.8188
−208	252	140	0.9927	0.9635	0.9253	0.8433	0.6376	0.4251	0.4278	0.4799	0.7942
−172	288	160	0.9952	0.9766	0.9529	0.9042	0.8031	0.7017	0.6304	0.6134	0.8107
−136	324	180	0.9967	0.9846	0.9690	0.9381	0.8782	0.8125	0.7784	0.7530	0.8550
−100	360	200	0.9978	0.9897	0.9791	0.9592	0.9212	0.8882	0.8621	0.8455	0.9067
−10	450	250	0.9992	0.9960	0.9924	0.9857	0.9741	0.9655	0.9604	0.9589	1.0048
80	540	300	0.9998	0.9990	0.9983	0.9971	0.9964	0.9973	1.0000	1.0052	1.0559
170	630	350	1.0001	1.0007	1.0011	1.0029	1.0069	1.0125	1.0189	1.0271	1.0810
260	720	400	1.0002	1.0011	1.0024	1.0057	1.0125	1.0199	1.0283	1.0377	1.0926
350	810	450	1.0003	1.0018	1.0033	1.0073	1.0153	1.0238	1.0332	1.0430	1.0973
440	900	500	1.0004	1.0020	1.0040	1.0081	1.0167	1.0257	1.0350	1.0451	1.0984

SOURCE: Perry and Chilton (1973).

Table G.9 Compressibility Factors for Oxygen

| Temperature | | | Pressure, psia (bars) | | | | | | | | |
°F	°R	K	14.5 (1)	72.5 (5)	145 (10)	290 (20)	580 (40)	870 (60)	1160 (80)	1450 (100)	2900 (200)
−325	135	75								0.4200	0.8301
−316	144	80								0.4007	0.7912
−298	162	90								0.3696	0.7281
−280	180	100	0.9757							0.3464	0.6798
−244	216	120	0.9855	0.9246	0.8367					0.3173	0.6148
−208	252	140	0.9911	0.9535	0.9034	0.7852	0.1334	0.1940	0.2527	0.3099	0.5815
−172	288	160	0.9939	0.9697	0.9379	0.8689	0.6991	0.3725	0.2969	0.3378	0.5766
−136	324	180	0.9960	0.9793	0.9579	0.9134	0.8167	0.7696	0.5954	0.5106	0.6043
−100	360	200	0.9970	0.9853	0.9705	0.9399	0.8768	0.8140	0.7534	0.6997	0.6720
−10	450	250	0.9987	0.9938	0.9870	0.9736	0.9477	0.9237	0.9030	0.8858	0.8563
80	540	300	0.9994	0.9968	0.9941	0.9884	0.9771	0.9676	0.9597	0.9542	0.9560
170	630	350	0.9998	0.9990	0.9979	0.9961	0.9919	0.9890	0.9870	0.9870	1.0049
260	720	400	1.0000	1.0000	1.0000	1.0000	1.0003	1.0011	1.0022	1.0045	1.0305
350	810	450	1.0002	1.0007	1.0015	1.0024	1.0048	1.0074	1.0106	1.0152	1.0445
440	900	500	1.0002	1.0011	1.0022	1.0038	1.0075	1.0115	1.0161	1.0207	1.0523

SOURCE: Perry and Chilton (1973).

Table G.10 Compressibility Factors for Steam

Pressure, psia	Temperature, °F																		
	400	600	800	1000	1200	1400	1600	1800	2000	2200	2400	2600	2800	3000	3200	3400	3600	3800	4000
10	0.9965	0.9989	0.9992	0.9995	0.9999	0.9999	0.9999	1.0000	1.0000	1.0000	1.0001	1.0006	1.0012	1.0024	1.0053	1.0084	1.0145	1.0211	1.0332
15	0.9943	0.9972	0.9986	0.9993	0.9997	0.9998	0.9999	0.9999	1.0000	1.0000	1.0001	1.0004	1.0012	1.0022	1.0042	1.0072	1.0124	1.0188	1.0295
20	0.9930	0.9970	0.9981	0.9991	0.9995	0.9996	0.9998	0.9999	1.0000	1.0000	1.0001	1.0003	1.0011	1.0020	1.0036	1.0065	1.0112	1.0173	1.0269
40	0.9861	0.9940	0.9967	0.9981	0.9990	0.9994	0.9996	0.9998	0.9999	0.9999	1.0001	1.0003	1.0010	1.0018	1.0028	1.0054	1.0090	1.0139	1.0214
60	0.9788	0.9910	0.9951	0.9973	0.9984	0.9991	0.9924	0.9997	0.9999	0.9999	1.0001	1.0003	1.0009	1.0018	1.0024	1.0048	1.0080	1.0120	1.0186
80	0.9714	0.9878	0.9935	0.9963	0.9979	0.9987	0.9992	0.9996	0.9998	0.9999	1.0001	1.0003	1.0008	1.0016	1.0023	1.0044	1.0073	1.0108	1.0170
100	0.9469	0.9848	0.9919	0.9954	0.9974	0.9985	0.9990	0.9995	0.9998	0.9999	1.0001	1.0004	1.0007	1.0015	1.0022	1.0042	1.0067	1.0099	1.0157
150	0.9435	0.9770	0.9879	0.9931	0.9960	0.9976	0.9985	0.9993	0.9997	0.9998	1.0001	1.0004	1.0006	1.0014	1.0021	1.0039	1.0059	1.0087	1.0137
200	0.9216	0.9690	0.9839	0.9908	0.9947	0.9968	0.9980	0.9991	0.9996	0.9998	1.0001	1.0005	1.0007	1.0015	1.0021	1.0037	1.0055	1.0080	1.0126
400		0.9356	0.9675	0.9817	0.9893	0.9935	0.9960	0.9982	0.9982	0.9998	1.0002	1.0005	1.0011	1.0017	1.0023	1.0033	1.0049	1.0070	1.0105
600		0.8989	0.9509	0.9725	0.9839	0.9904	0.9942	0.9973	0.9988	0.9997	1.0002	1.0008	1.0014	1.0019	1.0026	1.0034	1.0048	1.0066	1.0097
800		0.8586	0.9336	0.9633	0.9790	0.9872	0.9925	0.9964	0.9985	0.9996	1.0003	1.0010	1.0016	1.0022	1.0029	1.0036	1.0049	1.0065	1.0094
1,000		0.8138	0.9162	0.9540	0.9733	0.9841	0.9905	0.9955	0.9981	0.9994	1.0004	1.0012	1.0019	1.0025	1.0032	1.0039	1.0052	1.0066	1.0092
1,500		0.6702	0.8695	0.9305	0.9600	0.9764	0.9859	0.9932	0.9971	0.9992	1.0007	1.0017	1.0026	1.0033	1.0040	1.0048	1.0059	1.0072	1.0096
2,000			0.8188	0.9067	0.9468	0.9687	0.9813	0.9900	0.9958	0.9990	1.0010	1.0023	1.0034	1.0042	1.0049	1.0058	1.0068	1.0082	1.0104
4,000			0.5608	0.8060	0.8942	0.9392	0.9647	0.9836	0.9930	0.9989	1.0024	1.0050	1.0069	1.0082	1.0093	1.0106	1.0118	1.0132	1.0149
6,000				0.7042	0.8442	0.9121	0.9497	0.9771	0.9907	0.9991	1.0048	1.0081	1.0110	1.0128	1.0138	1.0152	1.0165	1.0179	1.0195
8,000				0.6185	0.8003	0.8883	0.9371	0.9714	0.9895	1.0004	1.0075	1.0118	1.0152	1.0172	1.0188	1.0204	1.0216	1.0229	1.0242
10,000				0.5699	0.7657	0.8693	0.9274	0.9668	0.9890	1.0025	1.0105	1.0158	1.0196	1.0220	1.0240	1.0258	1.0271	1.0284	1.0298

SOURCE: Perry and Chilton (1973).

Table G.11 **Supercompressibility Factors for Natural Gas of Specific Gravity 0.555 to 0.600†**

$$F_{pv} = \left(1 + \frac{2.48 p_G \times 10^{5.0 + 2.02G}}{T_f^{3.825}} \right)^{1/2} \qquad G \leq 0.600$$

Gauge pressure‡ p_G	Specific gravity G									
	0.555	0.560	0.565	0.570	0.575	0.580	0.585	0.590	0.595	0.600
10	1.001	1.001	1.001	1.001	1.001	1.001	1.001	1.001	1.001	1.001
20	1.001	1.001	1.001	1.001	1.001	1.002	1.002	1.002	1.002	1.002
30	1.002	1.002	1.002	1.002	1.002	1.002	1.002	1.002	1.002	1.002
40	1.003	1.003	1.003	1.003	1.003	1.003	1.003	1.003	1.003	1.003
50	1.003	1.003	1.004	1.004	1.004	1.004	1.004	1.004	1.004	1.004
60	1.004	1.004	1.004	1.004	1.004	1.005	1.005	1.005	1.005	1.005
70	1.005	1.005	1.005	1.005	1.005	1.005	1.005	1.006	1.006	1.006
80	1.005	1.005	1.006	1.006	1.006	1.006	1.006	1.006	1.006	1.007
90	1.006	1.006	1.006	1.006	1.007	1.007	1.007	1.007	1.007	1.007
100	1.007	1.007	1.007	1.007	1.007	1.007	1.008	1.008	1.008	1.008
110	1.007	1.008	1.008	1.008	1.008	1.008	1.008	1.009	1.009	1.009
120	1.008	1.008	1.008	1.009	1.009	1.009	1.009	1.009	1.010	1.010
130	1.009	1.009	1.009	1.009	1.010	1.010	1.010	1.010	1.010	1.011
140	1.009	1.010	1.010	1.010	1.010	1.010	1.011	1.011	1.011	1.011
150	1.010	1.010	1.010	1.011	1.011	1.011	1.011	1.012	1.012	1.012
160	1.011	1.011	1.011	1.011	1.012	1.012	1.012	1.013	1.013	1.013
170	1.011	1.012	1.012	1.012	1.012	1.013	1.013	1.013	1.014	1.014
180	1.012	1.012	1.013	1.013	1.013	1.013	1.014	1.014	1.014	1.015
190	1.013	1.013	1.013	1.014	1.014	1.014	1.015	1.015	1.015	1.016
200	1.103	1.014	1.014	1.014	1.015	1.015	1.015	1.016	1.016	1.016
210	1.014	1.014	1.015	1.015	1.015	1.016	1.016	1.016	1.017	1.017
220	1.015	1.015	1.015	1.016	1.016	1.016	1.017	1.017	1.018	1.018
230	1.015	1.016	1.016	1.016	1.017	1.017	1.018	1.018	1.018	1.019
240	1.016	1.016	1.017	1.017	1.017	1.018	1.018	1.019	1.019	1.020
250	1.017	1.017	1.017	1.018	1.018	1.019	1.019	1.020	1.020	1.020
260	1.017	1.018	1.018	1.018	1.019	1.019	1.020	1.020	1.021	1.021
270	1.018	1.018	1.019	1.019	1.020	1.020	1.021	1.021	1.022	1.022
280	1.019	1.019	1.019	1.020	1.020	1.021	1.021	1.022	1.022	1.023
290	1.019	1.020	1.020	1.021	1.021	1.022	1.022	1.023	1.023	1.024
300	1.020	1.020	2.021	1.021	1.022	1.022	1.023	1.023	1.024	1.024
310	1.021	1.021	1.022	1.022	1.023	1.023	1.024	1.024	1.025	1.025
320	1.021	1.022	1.022	1.023	1.023	1.024	1.024	1.025	1.025	1.026
330	1.022	1.022	1.023	1.023	1.024	1.025	1.025	1.026	1.026	1.027
340	1.023	1.023	1.024	1.024	1.025	1.025	1.026	1.026	1.027	1.028
350	1.023	1.024	1.024	1.025	1.025	1.026	1.027	1.027	1.028	1.028
360	1.024	1.024	1.025	1.026	1.026	1.027	1.027	1.028	1.029	1.029
370	1.024	1.025	1.026	1.026	1.027	1.027	1.028	1.029	1.029	1.030
380	1.025	1.026	1.026	1.027	1.028	1.028	1.029	1.030	1.030	1.031

Table G.11 **Supercompressibility Factors for Natural Gas of Specific Gravity 0.555 to 0.600† (*Continued*)**

Gauge pressure‡ p_G	Specific gravity G									
	0.555	0.560	0.565	0.570	0.575	0.580	0.585	0.590	0.595	0.600
390	1.026	1.026	1.027	1.028	1.028	1.029	1.030	1.030	1.031	1.032
400	1.026	1.027	1.028	1.028	1.029	1.030	1.030	1.031	1.032	1.032
420	1.028	1.028	1.029	1.030	1.030	1.031	1.032	1.033	1.033	1.034
440	1.029	1.030	1.030	1.031	1.032	1.033	1.033	1.034	1.035	1.036
460	1.030	1.031	1.032	1.032	1.033	1.034	1.035	1.036	1.036	1.037
480	1.032	1.032	1.033	1.034	1.035	1.035	1.036	1.037	1.038	1.039
500	1.033	1.034	1.034	1.035	1.036	1.037	1.038	1.039	1.040	1.040
520	1.034	1.035	1.036	1.037	1.038	1.038	1.039	1.040	1.041	1.042
540	1.036	1.036	1.037	1.038	1.039	1.040	1.041	1.042	1.043	1.044
560	1.037	1.038	1.039	1.039	1.040	1.041	1.042	1.043	1.044	1.045
580	1.038	1.039	1.040	1.041	1.042	1.043	1.044	1.045	1.046	1.047
600	1.039	1.040	1.041	1.042	1.043	1.044	1.045	1.046	1.047	1.048

†Tabulated values are for a temperature $T_f = 520°R$ (60°F).

‡Gauge pressure is measured at the downstream tap for orifice flowmeters.

SOURCE: Derived from PEA (1977).

Table G.12 **Supercompressibility Factors for Natural Gas of Specific Gravity 0.605 to 0.650†**

$$F_{pv} = \left(1 + \frac{3.32 p_G \times 10^{5.0+1.81G}}{T_f^{3.825}} \right)^{1/2} \qquad 0.601 \le G \le 0.650$$

Gauge pressure‡ p_G	Specific gravity G									
	0.605	0.610	0.615	0.620	0.625	0.630	0.635	0.640	0.645	0.650
10	1.001	1.001	1.001	1.001	1.001	1.001	1.001	1.001	1.001	1.001
20	1.002	1.002	1.002	1.002	1.002	1.002	1.002	1.002	1.002	1.002
30	1.003	1.003	1.003	1.003	1.003	1.003	1.003	1.003	1.003	1.003
40	1.003	1.003	1.004	1.004	1.004	1.004	1.004	1.004	1.004	1.004
50	1.004	1.004	1.004	1.004	1.005	1.005	1.005	1.005	1.005	1.005
60	1.005	1.005	1.005	1.005	1.005	1.006	1.006	1.006	1.006	1.006
70	1.006	1.006	1.006	1.006	1.006	1.007	1.007	1.007	1.007	1.007
80	1.007	1.007	1.007	1.007	1.007	1.007	1.008	1.008	1.008	1.008
90	1.008	1.008	1.008	1.008	1.008	1.008	1.009	1.009	1.009	1.009
100	1.008	1.009	1.009	1.009	1.009	1.009	1.010	1.010	1.010	1.010
110	1.009	1.009	1.010	1.010	1.010	1.010	1.010	1.011	1.011	1.011
120	1.010	1.010	1.011	1.011	1.011	1.011	1.011	1.012	1.012	1.012
130	1.011	1.011	1.011	1.012	1.012	1.012	1.012	1.013	1.013	1.013

†Tabulated values are for a temperature $T_f = 520°R$ (60°F).

‡Gauge pressure is measured at the downstream tap for orifice flowmeters.

Gauge pressure‡ P_G	Specific gravity G									
	0.605	0.610	0.615	0.620	0.625	0.630	0.635	0.640	0.645	0.650
140	1.012	1.012	1.012	1.013	1.013	1.013	1.013	1.014	1.014	1.014
150	1.013	1.013	1.013	1.013	1.014	1.014	1.014	1.015	1.015	1.015
160	1.013	1.014	1.014	1.014	1.015	1.015	1.015	1.016	1.016	1.016
170	1.014	1.015	1.015	1.015	1.015	1.016	1.016	1.016	1.017	1.017
180	1.015	1.015	1.016	1.016	1.016	1.017	1.017	1.017	1.018	1.018
190	1.016	1.016	1.017	1.017	1.017	1.018	1.018	1.018	1.019	1.019
200	1.017	1.017	1.017	1.018	1.018	1.019	1.019	1.019	1.020	1.020
210	1.018	1.018	1.018	1.019	1.019	1.019	1.020	1.020	1.021	1.021
220	1.018	1.019	1.019	1.020	1.020	1.020	1.021	1.021	1.022	1.022
230	1.019	1.020	1.020	1.020	1.021	1.021	1.022	1.022	1.023	1.023
240	1.020	1.020	1.021	1.021	1.022	1.022	1.023	1.023	1.024	1.024
250	1.021	1.021	1.022	1.022	1.023	1.023	1.024	1.024	1.025	1.025
260	1.022	1.022	1.023	1.023	1.024	1.024	1.025	1.025	1.026	1.026
270	1.023	1.023	1.023	1.025	1.024	1.025	1.026	1.026	1.027	1.027
280	1.023	1.024	1.024	1.025	1.025	1.026	1.026	1.027	1.028	1.028
290	1.024	1.025	1.025	1.026	1.026	1.027	1.027	1.028	1.029	1.029
300	1.025	1.026	1.026	1.027	1.027	1.028	1.028	1.029	1.029	1.030
310	1.026	1.026	1.027	1.027	1.028	1.029	1.029	1.030	1.030	1.031
320	1.027	1.027	1.028	1.028	1.029	1.030	1.030	1.031	1.031	1.032
330	1.027	1.028	1.029	1.029	1.030	1.030	1.031	1.032	1.032	1.033
340	1.028	1.029	1.029	1.030	1.031	1.031	1.032	1.033	1.033	1.034
350	1.029	1.030	1.030	1.031	1.032	1.032	1.033	1.034	1.034	1.035
360	1.030	1.031	1.031	1.032	1.033	1.033	1.034	1.035	1.035	1.036
370	1.031	1.031	1.032	1.033	1.033	1.034	1.035	1.036	1.036	1.037
380	1.032	1.032	1.033	1.034	1.034	1.035	1.036	1.036	1.037	1.038
390	1.032	1.033	1.034	1.034	1.035	1.036	1.037	1.037	1.038	1.039
400	1.033	1.034	1.035	1.035	1.036	1.037	1.038	1.038	1.039	1.040
420	1.035	1.036	1.036	1.037	1.038	1.039	1.039	1.040	1.041	1.042
440	1.036	1.037	1.038	1.039	1.040	1.040	1.041	1.042	1.043	1.044
460	1.038	1.039	1.040	1.041	1.041	1.042	1.043	1.044	1.045	1.046
480	1.040	1.041	1.041	1.042	1.043	1.044	1.045	1.046	1.047	1.048
500	1.041	1.042	1.043	1.044	1.045	1.046	1.047	1.048	1.049	1.050
520	1.043	1.044	1.045	1.046	1.047	1.048	1.049	1.050	1.051	1.052
540	1.045	1.046	1.046	1.047	1.048	1.049	1.050	1.051	1.052	1.054
560	1.046	1.047	1.048	1.049	1.050	1.051	1.052	1.053	1.054	1.055
580	1.048	1.049	1.050	1.051	1.052	1.053	1.054	1.055	1.056	1.057
600	1.049	1.050	1.051	1.053	1.054	1.055	1.056	1.057	1.058	1.059

†Tabulated values are for a temperature T_f = 520°R (60°F).

‡Gauge pressure is measured at the downstream tap for orifice flowmeters.

SOURCE: Derived from PEA (1977).

Table G.13 **Supercompressibility Factors for Natural Gas of Specific Gravity 0.660 to 0.750†**

$$F_{pv} = \left(1 + \frac{4.66 p_G \times 10^{5.0+1.6G}}{T_f^{3.825}} \right)^{1/2} \qquad 0.651 \leq G \leq 0.750$$

Gauge pressure‡ P_G	Specific gravity G									
	0.660	0.670	0.680	0.690	0.700	0.710	0.720	0.730	0.740	0.750
10	1.001	1.001	1.001	1.001	1.001	1.001	1.001	1.001	1.001	1.002
20	1.002	1.002	1.002	1.002	1.003	1.003	1.003	1.003	1.003	1.003
30	1.003	1.003	1.003	1.004	1.004	1.004	1.004	1.004	1.004	1.005
40	1.004	1.004	1.005	1.005	1.005	1.005	1.005	1.006	1.006	1.006
50	1.005	1.006	1.006	1.006	1.006	1.006	1.007	1.007	1.007	1.008
60	1.006	1.007	1.007	1.007	1.008	1.008	1.008	1.008	1.009	1.009
70	1.008	1.008	1.008	1.008	1.009	1.009	1.009	1.010	1.010	1.011
80	1.009	1.009	1.009	1.010	1.010	1.010	1.011	1.011	1.012	1.012
90	1.010	1.010	1.010	1.011	1.011	1.012	1.012	1.013	1.013	1.013
100	1.011	1.011	1.012	1.012	1.012	1.013	1.013	1.014	1.014	1.015
110	1.012	1.012	1.013	1.013	1.014	1.014	1.015	1.015	1.016	1.016
120	1.013	1.013	1.014	1.015	1.015	1.016	1.016	1.017	1.017	1.018
130	1.014	1.015	1.015	1.016	1.016	1.017	1.017	1.018	1.019	1.019
140	1.015	1.016	1.016	1.017	1.017	1.018	1.019	1.019	1.020	1.021
150	1.016	1.017	1.017	1.018	1.019	1.019	1.020	1.021	1.022	1.022
160	1.017	1.018	1.018	1.019	1.020	1.021	1.021	1.022	1.023	1.024
170	1.018	1.019	1.020	1.020	1.021	1.022	1.023	1.024	1.024	1.025
180	1.019	1.020	1.021	1.022	1.022	1.023	1.024	1.025	1.026	1.027
190	1.020	1.021	1.022	1.023	1.024	1.024	1.025	1.026	1.027	1.028
200	1.021	1.022	1.023	1.024	1.025	1.026	1.027	1.028	1.029	1.030
210	1.022	1.023	1.024	1.025	1.026	1.027	1.028	1.029	1.030	1.031
220	1.024	1.024	1.025	1.026	1.027	1.028	1.029	1.030	1.031	1.033
230	1.025	1.026	1.026	1.027	1.028	1.030	1.031	1.032	1.033	1.034
240	1.026	1.027	1.028	1.029	1.030	1.031	1.032	1.033	1.034	1.036
250	1.027	1.028	1.029	1.030	1.031	1.032	1.033	1.034	1.036	1.037
260	1.028	1.029	1.030	1.031	1.032	1.033	1.035	1.036	1.037	1.038
270	1.029	1.030	1.031	1.032	1.033	1.035	1.036	1.037	1.039	1.040
280	1.030	1.031	1.032	1.033	1.035	1.036	1.037	1.039	1.040	1.041
290	1.031	1.032	1.033	1.034	1.036	1.037	1.038	1.040	1.041	1.043
300	1.032	1.033	1.034	1.036	1.037	1.038	1.040	1.041	1.043	1.044
310	1.033	1.034	1.036	1.037	1.038	1.040	1.041	1.043	1.044	1.046
320	1.034	1.035	1.037	1.038	1.039	1.041	1.042	1.044	1.046	1.047
330	1.035	1.036	1.038	1.039	1.041	1.042	1.044	1.045	1.047	1.049
340	1.036	1.038	1.039	1.040	1.042	1.043	1.045	1.047	1.048	1.050
350	1.037	1.039	1.040	1.041	1.043	1.045	1.046	1.048	1.050	1.051
360	1.038	1.040	1.041	1.043	1.044	1.046	1.048	1.049	1.051	1.053
370	1.039	1.041	1.042	1.044	1.045	1.047	1.049	1.051	1.052	1.054
380	1.040	1.042	1.043	1.045	1.047	1.048	1.050	1.052	1.054	1.056

Table G.13 **Supercompressibility Factors for Natural Gas of Specific Gravity 0.660 to 0.750† (Continued)**

Gauge pressure‡ P_G	Specific gravity G									
	0.660	0.670	0.680	0.690	0.700	0.710	0.720	0.730	0.740	0.750
390	1.041	1.043	1.044	1.046	1.048	1.050	1.051	1.053	1.055	1.057
400	1.042	1.044	1.046	1.047	1.049	1.051	1.053	1.055	1.057	1.059
420	1.044	1.046	1.048	1.050	1.051	1.053	1.055	1.057	1.059	1.061
440	1.047	1.048	1.050	1.052	1.054	1.056	1.058	1.060	1.062	1.064
460	1.049	1.050	1.052	1.054	1.056	1.058	1.060	1.063	1.065	1.067
480	1.051	1.053	1.054	1.056	1.059	1.061	1.063	1.065	1.068	1.070
500	1.053	1.055	1.057	1.059	1.061	1.063	1.065	1.068	1.070	1.073
520	1.055	1.057	1.059	1.061	1.063	1.066	1.068	1.070	1.073	1.076
540	1.057	1.059	1.061	1.063	1.066	1.068	1.070	1.073	1.076	1.078
560	1.059	1.061	1.063	1.066	1.068	1.070	1.073	1.076	1.078	1.081
580	1.061	1.063	1.065	1.068	1.070	1.073	1.076	1.078	1.081	1.084
600	1.063	1.065	1.068	1.070	1.073	1.075	1.078	1.081	1.084	1.087

†Tabular values are for a temperature $T_f = 520°R$ (60°F).

‡Gauge pressure is measured at the downstream tap for orifice flowmeters.

SOURCE: Derived from PEA (1977).

Table G.14 **Supercompressibility Factors for Natural Gas of Specific Gravity 0.765 to 0.900†**

$$F_{pv} = \left(1 + \frac{7.91 p_G \times 10^{5.0+1.26G}}{T_f^{3.825}} \right)^{1/2} \quad 0.751 \leq G \leq 0.900$$

Gauge pressure‡ P_G	Specific gravity G									
	0.765	0.780	0.795	0.810	0.825	0.840	0.855	0.870	0.885	0.900
10	1.001	1.002	1.002	1.002	1.002	1.002	1.002	1.002	1.002	1.002
20	1.003	1.003	1.003	1.003	1.004	1.004	1.004	1.004	1.004	1.004
30	1.004	1.005	1.005	1.005	1.005	1.006	1.006	1.006	1.006	1.007
40	1.006	1.006	1.006	1.007	1.007	1.007	1.008	1.008	1.008	1.009
50	1.007	1.008	1.008	1.008	1.009	1.009	1.010	1.010	1.010	1.011
60	1.009	1.009	1.010	1.010	1.011	1.011	1.012	1.012	1.013	1.013
70	1.010	1.011	1.011	1.012	1.012	1.013	1.013	1.014	1.015	1.015
80	1.012	1.012	1.013	1.013	1.014	1.015	1.015	1.016	1.017	1.017
90	1.013	1.014	1.014	1.015	1.016	1.017	1.017	1.018	1.019	1.020
100	1.015	1.015	1.016	1.017	1.018	1.018	1.019	1.020	1.021	1.022
110	1.016	1.017	1.018	1.018	1.019	1.020	1.021	1.022	1.023	1.024
120	1.018	1.018	1.019	1.020	1.021	1.022	1.023	1.024	1.025	1.026

Table G.14 Supercompressibility Factors for Natural Gas of Specific Gravity 0.765 to 0.900† (*Continued*)

Gauge pressure‡ P_G	Specific gravity G									
	0.765	0.780	0.795	0.810	0.825	0.840	0.855	0.870	0.885	0.900
130	1.019	1.020	1.021	1.022	1.023	1.024	1.025	1.026	1.027	1.028
140	1.021	1.022	1.022	1.023	1.024	1.026	1.027	1.028	1.029	1.030
150	1.022	1.023	1.024	1.025	1.026	1.027	1.029	1.030	1.031	1.032
160	1.024	1.025	1.026	1.027	1.028	1.029	1.030	1.032	1.033	1.035
170	1.025	1.026	1.027	1.028	1.030	1.031	1.032	1.034	1.035	1.037
180	1.026	1.028	1.029	1.030	1.031	1.033	1.034	1.036	1.037	1.039
190	1.028	1.029	1.030	1.032	1.033	1.035	1.036	1.038	1.039	1.041
200	1.029	1.031	1.032	1.033	1.035	1.036	1.038	1.040	1.041	1.043
210	1.031	1.032	1.034	1.035	1.037	1.038	1.040	1.041	1.043	1.045
220	1.032	1.034	1.035	1.037	1.038	1.040	1.042	1.043	1.045	1.047
230	1.034	1.035	1.037	1.038	1.040	1.042	1.043	1.045	1.047	1.049
240	1.035	1.037	1.038	1.040	1.042	1.043	1.045	1.047	1.049	1.051
250	1.037	1.038	1.040	1.042	1.043	1.045	1.047	1.049	1.051	1.054
260	1.038	1.040	1.041	1.043	1.045	1.047	1.049	1.051	1.053	1.056
270	1.039	1.041	1.043	1.045	1.047	1.049	1.051	1.053	1.055	1.058
280	1.041	1.043	1.044	1.046	1.048	1.050	1.053	1.055	1,057	1.060
290	1.042	1.044	1.046	1.048	1.050	1.052	1.055	1.057	1.059	1.062
300	1.044	1.046	1.048	1.050	1.052	1.054	1.056	1.059	1.061	1.064
310	1.045	1.047	1.049	1.051	1.053	1.056	1.058	1.061	1.063	1.066
320	1.047	1.049	1.051	1.053	1.055	1.058	1.060	1.063	1.065	1.068
330	1.048	1.050	1.052	1.054	1.057	1.059	1.062	1.064	1.067	1.070
340	1.049	1.051	1.054	1.056	1.058	1.061	1.064	1.066	1.069	1.072
350	1.051	1.053	1.055	1.058	1.060	1.063	1.065	1.068	1.071	1.074
360	1.052	1.054	1.057	1.059	1.062	1.064	1.067	1.070	1.073	1.076
370	1.054	1.056	1.058	1.061	1.063	1.066	1.069	1.072	1.075	1.078
380	1.055	1.057	1.060	1.062	1.065	1.068	1.071	1.074	1.077	1.080
390	1.056	1.059	1.061	1.064	1.067	1.070	1.073	1.076	1.079	1.082
400	1.058	1.060	1.063	1.066	1.068	1.071	1.074	1.078	1.081	1.084
420	1.061	1.063	1.066	1.069	1.072	1.075	1.078	1.081	1.085	1.088
440	1.063	1.066	1.069	1.072	1.075	1.078	1.082	1.085	1.089	1.093
460	1.066	1.069	1.072	1.075	1.078	1.082	1.085	1.089	1.093	1.097
480	1.069	1.072	1.075	1.078	1.082	1.085	1.089	1.093	1.096	1.101
500	1.072	1.075	1.078	1.081	1.085	1.089	1.092	1.096	1.100	1.105
520	1.075	1.078	1.081	1.085	1.088	1.092	1.096	1.100	1.104	1.109
540	1.077	1.081	1.084	1.088	1.091	1.095	1.099	1.104	1.108	1.112
560	1.080	1.083	1.087	1.091	1.095	1.099	1.103	1.107	1.112	1.116
580	1.083	1.086	1.090	1.094	1.098	1.102	1.106	1.111	1.115	1.120
600	1.086	1.089	1.093	1.097	1.101	1.105	1.110	1.114	1.119	1.124

†Tabular values are for a temperature $T_f = 520°R$ (60°F).

‡Gauge pressure is measured at the downstream tap for orifice flowmeters.

SOURCE: Derived from PEA (1977).

Table G.15   **Supercompressibility Factors for Natural Gas of Specific Gravity 0.920 to 1.100†**

$$F_{pv} = \left(1 + \frac{11.63 p_G \times 10^{5.0+1.07G}}{T_f^{3.825}} \right)^{1/2} \qquad 0.901 \leq G \leq 1.100$$

Gauge pressure‡ P_G	Specific gravity G									
	0.920	0.940	0.960	0.980	1.000	1.020	1.040	1.060	1.080	1.100
10	1.002	1.002	1.003	1.003	1.003	1.003	1.003	1.003	1.003	1.004
20	1.005	1.005	1.005	1.005	1.006	1.006	1.006	1.006	1.007	1.007
30	1.007	1.007	1.008	1.008	1.008	1.009	1.009	1.010	1.010	1.011
40	1.009	1.010	1.010	1.011	1.011	1.012	1.012	1.013	1.014	1.014
50	1.011	1.012	1.013	1.013	1.014	1.015	1.015	1.016	1.017	1.018
60	1.014	1.014	1.015	1.016	1.017	1.017	1.018	1.019	1.020	1.021
70	1.016	1.017	1.018	1.018	1.019	1.020	1.021	1.022	1.024	1.025
80	1.018	1.019	1.020	1.021	1.022	1.023	1.024	1.026	1.027	1.028
90	1.020	1.021	1.023	1.024	1.025	1.026	1.027	1.029	1.030	1.032
100	1.023	1.024	1.025	1.026	1.028	1.029	1.030	1.032	1.033	1.035
110	1.025	1.026	1.027	1.029	1.030	1.032	1.033	1.035	1.037	1.039
120	1.027	1.028	1.030	1.031	1.033	1.035	1.036	1.038	1.040	1.042
130	1.029	1.031	1.032	1.034	1.036	1.037	1.039	1.041	1.043	1.045
140	1.032	1.033	1.035	1.037	1.038	1.040	1.042	1.044	1.047	1.049
150	1.034	1.035	1.037	1.039	1.041	1.043	1.045	1.047	1.050	1.052
160	1.036	1.038	1.040	1.042	1.044	1.046	1.048	1.051	1.053	1.056
170	1.038	1.040	1.042	1.044	1.046	1.049	1.051	1.054	1.056	1.059
180	1.040	1.042	1.045	1.047	1.049	1.051	1.054	1.057	1.059	1.062
190	1.043	1.045	1.047	1.049	1.052	1.054	1.057	1.060	1.063	1.066
200	1.045	1.047	1.049	1.052	1.054	1.057	1.060	1.063	1.066	1.069
210	1.047	1.049	1.052	1.054	1.057	1.060	1.063	1.066	1.069	1.072
220	1.049	1.052	1.054	1.057	1.060	1.063	1.066	1.069	1.072	1.076
230	1.051	1.054	1.057	1.059	1.062	1.065	1.069	1.072	1.075	1.079
240	1.054	1.056	1.059	1.062	1.065	1.068	1.071	1.075	1.079	1.082
250	1.056	1.058	1.061	1.064	1.068	1.071	1.074	1.078	1.082	1.086
260	1.058	1.061	1.064	1.067	1.070	1.074	1.077	1.081	1.085	1.089
270	1.060	1.063	1.066	1.069	1.073	1.076	1.080	1.084	1.088	1.092
280	1.062	1.065	1.068	1.072	1.075	1.079	1.083	1.087	1.091	1.095
290	1.064	1.068	1.071	1.074	1.078	1.082	1.086	1.090	1.094	1.099
300	1.067	1.070	1.073	1.077	1.081	1.084	1.089	1.093	1.097	1.102
310	1.069	1.072	1.076	1.079	1.083	1.087	1.091	1.096	1.100	1.105
320	1.071	1.074	1.078	1.082	1.086	1.090	1.094	1.099	1.103	1.108
330	1.073	1.077	1.080	1.084	1.088	1.092	1.097	1.102	1.107	1.112
340	1.075	1.079	1.083	1.087	1.091	1.095	1.100	1.105	1.110	1.115
350	1.077	1.081	1.085	1.089	1.093	1.098	1.103	1.107	1.113	1.118
360	1.079	1.083	1.087	1.091	1.096	1.101	1.105	1.110	1.116	1.121
370	1.081	1.085	1.090	1.094	1.098	1.103	1.108	1.113	1.119	1.124
380	1.084	1.088	1.092	1.096	1.101	1.106	1.111	1.116	1.122	1.128

Table G.15   **Supercompressibility Factors for Natural Gas of Specific Gravity 0.920 to 1.100† (*Continued*)**

Gauge pressure‡ P_G	Specific gravity G									
	0.920	0.940	0.960	0.980	1.000	1.020	1.040	1.060	1.080	1.100
390	1.086	1.090	1.094	1.099	1.104	1.108	1.114	1.119	1.125	1.131
400	1.088	1.092	1.097	1.101	1.106	1.111	1.116	1.122	1.128	1.134
420	1.092	1.096	1.101	1.106	1.111	1.116	1.122	1.128	1.134	1.140
440	1.096	1.101	1.106	1.111	1.116	1.122	1.127	1.133	1.140	1.146
460	1.100	1.105	1.110	1.116	1.121	1.127	1.133	1.139	1.146	1.153
480	1.105	1.110	1.115	1.120	1.126	1.132	1.138	1.145	1.152	1.159
500	1.109	1.114	1.119	1.125	1.131	1.137	1.144	1.150	1.158	1.165
520	1.113	1.118	1.124	1.130	1.136	1.142	1.149	1.156	1.163	1.171
540	1.117	1.123	1.128	1.134	1.141	1.147	1.154	1.162	1.169	1.177
560	1.121	1.127	1.133	1.139	1.146	1.153	1.160	1.167	1.175	1.183
580	1.125	1.131	1.137	1.144	1.151	1.158	1.165	1.173	1.181	1.189
600	1.129	1.135	1.142	1.148	1.155	1.163	1.170	1.178	1.187	1.195

†Tabular values are for a temperature $T_f = 520°R$ (60°F).

‡Gauge pressure is measured at the downstream tap for orifice flowmeters.

SOURCE: Derived from PEA (1977).

Table G.16   **Supercompressibility Factors for Natural Gas of Specific Gravity 1.140 to 1.500†**

$$F_{pv} = \left(1 + \frac{17.48 p_G \times 10^{5.0+0.9G}}{T_f^{3.825}} \right)^{1/2} \qquad 1.101 < G < 1.500$$

Gauge pressure‡ P_G	Specific gravity G									
	1.140	1.180	1.220	1.260	1.300	1.340	1.380	1.420	1.460	1.500
10	1.004	1.004	1.004	1.005	1.005	1.006	1.006	1.007	1.007	1.008
20	1.008	1.008	1.009	1.010	1.011	1.011	1.012	1.013	1.015	1.016
30	1.011	1.012	1.013	1.014	1.016	1.017	1.019	1.020	1.022	1.024
40	1.015	1.016	1.018	1.019	1.021	1.023	1.025	1.027	1.029	1.031
50	1.019	1.020	1.022	1.024	1.026	1.028	1.031	1.033	1.036	1.039
60	1.022	1.024	1.026	1.029	1.031	1.034	1.037	1.040	1.043	1.047
70	1.026	1.028	1.031	1.033	1.036	1.039	1.043	1.046	1.050	1.054
80	1.030	1.032	1.035	1.038	1.041	1.045	1.049	1.053	1.057	1.062
90	1.034	1.036	1.039	1.043	1.046	1.050	1.055	1.059	1.064	1.070
100	1.037	1.040	1.044	1.047	1.051	1.056	1.061	1.066	1.071	1.077
110	1.041	1.044	1.048	1.052	1.057	1.061	1.066	1.072	1.078	1.084
120	1.045	1.048	1.052	1.057	1.061	1.067	1.072	1.078	1.085	1.092
130	1.048	1.052	1.057	1.061	1.066	1.072	1.078	1.084	1.091	1.099

Table G.16 Supercompressibility Factors for Natural Gas of Specific Gravity 1.140 to 1.500† (*Continued*)

Gauge pressure‡ P_G	Specific gravity G									
	1.140	1.180	1.220	1.260	1.300	1.340	1.380	1.420	1.460	1.500
140	1.052	1.056	1.061	1.066	1.071	1.077	1.084	1.091	1.098	1.106
150	1.055	1.060	1.065	1.070	1.076	1.083	1.090	1.097	1.105	1.113
160	1.059	1.064	1.069	1.075	1.081	1.088	1.095	1.103	1.112	1.121
170	1.063	1.068	1.073	1.079	1.086	1.093	1.101	1.109	1.118	1.128
180	1.066	1.072	1.078	1.084	1.091	1.098	1.107	1.115	1.125	1.135
190	1.070	1.075	1.082	1.088	1.096	1.104	1.112	1.121	1.131	1.142
200	1.073	1.079	1.086	1.093	1.101	1.109	1.118	1.127	1.138	1.149
210	1.077	1.083	1.090	1.097	1.105	1.114	1.123	1.133	1.144	1.156
220	1.080	1.087	1.094	1.102	1.110	1.119	1.129	1.139	1.151	1.163
230	1.084	1.091	1.098	1.106	1.115	1.124	1.134	1.145	1.157	1.170
240	1.087	1.094	1.102	1.111	1.120	1.129	1.140	1.151	1.163	1.176
250	1.091	1.098	1.106	1.115	1.124	1.134	1.145	1.157	1.170	1.183
260	1.094	1.102	1.110	1.119	1.129	1.139	1.151	1.163	1.176	1.190
270	1.098	1.106	1.114	1.124	1.134	1.144	1.156	1.169	1.182	1.197
280	1.101	1.109	1.118	1.128	1.138	1.150	1.162	1.174	1.188	1.203
290	1.104	1.113	1.122	1.132	1.143	1.154	1.167	1.180	1.194	1.210
300	1.108	1.117	1.126	1.137	1.148	1.159	1.172	1.186	1.201	1.216
310	1.111	1.120	1.130	1.141	1.152	1.164	1.178	1.192	1.207	1.223
320	1.115	1.124	1.134	1.145	1.157	1.169	1.183	1.197	1.213	1.229
330	1.118	1.128	1.138	1.149	1.161	1.174	1.188	1.203	1.219	1.236
340	1.122	1.131	1.142	1.154	1.166	1.179	1.193	1.209	1.225	1.242
350	1.125	1.135	1.146	1.158	1.170	1.184	1.199	1.214	1.231	1.249
360	1.128	1.139	1.150	1.162	1.175	1.189	1.204	1.220	1.237	1.255
370	1.132	1.142	1.154	1.166	1.179	1.194	1.209	1.225	1.243	1.262
380	1.135	1.146	1.158	1.170	1.184	1.198	1.214	1.231	1.249	1.268
390	1.138	1.149	1.161	1.174	1.188	1.203	1.219	1.236	1.255	1.274
400	1.142	1.153	1.165	1.179	1.193	1.208	1.224	1.242	1.260	1.280
420	1.148	1.160	1.173	1.187	1.202	1.217	1.234	1.253	1.272	1.293
440	1.155	1.167	1.181	1.195	1.210	1.227	1.244	1.263	1.284	1.305
460	1.161	1.174	1.188	1.203	1.219	1.236	1.254	1.274	1.295	1.317
480	1.168	1.181	1.196	1.211	1.228	1.245	1.264	1.285	1.306	1.329
500	1.174	1.188	1.203	1.219	1.236	1.255	1.274	1.295	1.318	1.341
520	1.181	1.195	1.211	1.227	1.245	1.264	1.284	1.306	1.329	1.353
540	1.187	1.202	1.218	1.235	1.253	1.273	1.294	1.316	1.340	1.365
560	1.194	1.209	1.225	1.243	1.262	1.282	1.303	1.326	1.351	1.377
580	1.200	1.216	1.233	1.251	1.270	1.291	1.313	1.336	1.361	1.388
600	1.206	1.222	1.240	1.258	1.278	1.299	1.322	1.346	1.372	1.400

†Tabular values are for a temperature $T_f = 520°R$ (60°F).

‡Gauge pressure is measured at the downstream tap for orifice flowmeters.

SOURCE: Derived from PEA (1977).

Temperature T_F, °F

F_{pv}	40	50	60	70	80	90	100	110	120	130	140	150	160	170	180
1.001	1.001	1.001	1.001	1.001	1.001	1.001	1.001	1.001	1.001	1.001	1.001	1.001	1.001	1.000	1.000
1.002	1.002	1.002	1.002	1.002	1.002	1.002	1.002	1.001	1.001	1.001	1.001	1.001	1.001	1.001	1.001
1.003	1.003	1.003	1.003	1.003	1.003	1.002	1.002	1.002	1.002	1.002	1.002	1.002	1.002	1.001	1.001
1.004	1.005	1.004	1.004	1.004	1.003	1.003	1.003	1.003	1.003	1.002	1.002	1.002	1.002	1.002	1.002
1.005	1.006	1.005	1.005	1.005	1.004	1.004	1.004	1.004	1.003	1.003	1.003	1.003	1.003	1.002	1.002
1.006	1.007	1.006	1.006	1.006	1.005	1.005	1.005	1.004	1.004	1.004	1.003	1.003	1.003	1.003	1.003
1.007	1.008	1.008	1.007	1.007	1.006	1.006	1.005	1.005	1.005	1.004	1.004	1.004	1.004	1.003	1.003
1.008	1.009	1.009	1.008	1.007	1.007	1.006	1.006	1.006	1.005	1.005	1.005	1.004	1.004	1.004	1.004
1.009	1.010	1.010	1.009	1.008	1.008	1.007	1.007	1.006	1.006	1.006	1.005	1.005	1.005	1.004	1.004
1.010	1.012	1.011	1.010	1.009	1.009	1.008	1.008	1.007	1.007	1.006	1.006	1.005	1.005	1.005	1.005
1.011	1.013	1.012	1.011	1.010	1.010	1.009	1.008	1.008	1.007	1.007	1.006	1.006	1.006	1.005	1.005
1.012	1.014	1.013	1.012	1.011	1.010	1.010	1.009	1.008	1.008	1.007	1.007	1.007	1.006	1.006	1.005
1.013	1.015	1.014	1.013	1.012	1.011	1.011	1.010	1.009	1.009	1.008	1.008	1.007	1.007	1.006	1.006
1.014	1.016	1.015	1.014	1.013	1.012	1.011	1.011	1.010	1.009	1.009	1.008	1.008	1.007	1.007	1.006
1.015	1.017	1.016	1.015	1.014	1.013	1.012	1.011	1.011	1.010	1.009	1.009	1.008	1.008	1.007	1.007
1.016	1.019	1.017	1.016	1.015	1.014	1.013	1.012	1.011	1.011	1.010	1.009	1.009	1.008	1.008	1.007
1.017	1.020	1.018	1.017	1.016	1.015	1.014	1.013	1.012	1.011	1.011	1.010	1.009	1.009	1.008	1.008
1.018	1.021	1.019	1.018	1.017	1.016	1.015	1.014	1.013	1.012	1.011	1.010	1.010	1.009	1.009	1.008
1.019	1.022	1.020	1.019	1.018	1.016	1.015	1.014	1.013	1.013	1.012	1.011	1.010	1.010	1.009	1.009
1.020	1.023	1.022	1.020	1.019	1.017	1.016	1.015	1.014	1.013	1.012	1.012	1.011	1.010	1.010	1.009
1.021	1.024	1.023	1.021	1.020	1.018	1.017	1.016	1.015	1.014	1.013	1.012	1.011	1.011	1.010	1.010
1.022	1.026	1.024	1.022	1.020	1.019	1.018	1.017	1.016	1.015	1.014	1.013	1.012	1.011	1.011	1.010
1.023	1.027	1.025	1.023	1.021	1.020	1.019	1.017	1.016	1.015	1.014	1.013	1.013	1.012	1.011	1.010
1.024	1.028	1.026	1.024	1.022	1.021	1.019	1.018	1.017	1.016	1.015	1.014	1.013	1.012	1.012	1.011
1.025	1.029	1.027	1.025	1.023	1.022	1.020	1.019	1.018	1.017	1.015	1.015	1.014	1.013	1.012	1.011

Table G.17 F_{pv} Correction for Temperature (continued)

	Temperature T_F, °F														
F_{pv}	40	50	60	70	80	90	100	110	120	130	140	150	160	170	180
1.026	1.030	1.028	1.026	1.024	1.023	1.021	1.020	1.018	1.017	1.016	1.015	1.014	1.013	1.013	1.012
1.027	1.031	1.029	1.027	1.025	1.023	1.022	1.020	1.019	1.018	1.017	1.016	1.015	1.014	1.013	1.012
1.028	1.032	1.030	1.028	1.026	1.024	1.023	1.021	1.020	1.019	1.017	1.016	1.015	1.014	1.014	1.013
1.029	1.034	1.031	1.029	1.027	1.025	1.023	1.022	1.020	1.019	1.018	1.017	1.016	1.015	1.014	1.013
1.030	1.035	1.032	1.030	1.028	1.026	1.024	1.023	1.021	1.020	1.018	1.017	1.016	1.015	1.015	1.014
1.031	1.036	1.033	1.031	1.029	1.027	1.025	1.023	1.022	1.021	1.019	1.018	1.017	1.016	1.015	1.014
1.032	1.037	1.034	1.032	1.030	1.028	1.026	1.024	1.023	1.021	1.020	1.019	1.018	1.016	1.015	1.015
1.033	1.038	1.036	1.033	1.031	1.029	1.027	1.025	1.023	1.022	1.020	1.019	1.018	1.017	1.016	1.015
1.034	1.039	1.037	1.034	1.032	1.029	1.028	1.026	1.024	1.023	1.021	1.020	1.019	1.017	1.016	1.016
1.035	1.041	1.038	1.035	1.033	1.030	1.028	1.026	1.025	1.023	1.022	1.020	1.019	1.018	1.017	1.016
1.036	1.042	1.039	1.036	1.034	1.031	1.029	1.027	1.025	1.024	1.022	1.021	1.020	1.019	1.017	1.016
1.037	1.043	1.040	1.037	1.034	1.032	1.030	1.028	1.026	1.025	1.023	1.022	1.020	1.019	1.018	1.017
1.038	1.044	1.041	1.038	1.035	1.033	1.031	1.029	1.027	1.025	1.024	1.022	1.021	1.020	1.018	1.017
1.039	1.045	1.042	1.039	1.036	1.034	1.032	1.030	1.028	1.026	1.024	1.023	1.021	1.020	1.019	1.018
1.040	1.046	1.043	1.040	1.037	1.035	1.032	1.030	1.028	1.027	1.025	1.023	1.022	1.021	1.019	1.018
1.041	1.047	1.044	1.041	1.038	1.036	1.033	1.031	1.029	1.027	1.025	1.024	1.022	1.021	1.020	1.019
1.042	1.049	1.045	1.042	1.039	1.036	1.034	1.032	1.030	1.028	1.026	1.025	1.023	1.022	1.020	1.019
1.043	1.050	1.046	1.043	1.040	1.037	1.035	1.033	1.030	1.029	1.027	1.025	1.024	1.022	1.021	1.020
1.044	1.051	1.047	1.044	1.041	1.038	1.036	1.033	1.031	1.029	1.027	1.026	1.024	1.023	1.021	1.020
1.045	1.052	1.048	1.045	1.042	1.039	1.036	1.034	1.032	1.030	1.028	1.026	1.025	1.025	1.022	1.021
1.046	1.053	1.049	1.046	1.043	1.040	1.037	1.035	1.033	1.031	1.029	1.027	1.025	1.024	1.022	1.021
1.047	1.054	1.051	1.047	1.044	1.041	1.038	1.036	1.033	1.031	1.029	1.027	1.026	1.024	1.023	1.022
1.048	1.056	1.052	1.048	1.045	1.042	1.039	1.036	1.034	1.032	1.030	1.028	1.026	1.025	1.023	1.022
1.049	1.057	1.053	1.049	1.046	1.043	1.040	1.037	1.035	1.033	1.031	1.029	1.027	1.025	1.024	1.022
1.050	1.058	1.054	1.050	1.047	1.043	1.041	1.038	1.035	1.033	1.031	1.029	1.027	1.026	1.024	1.023

1.023	1.024	1.026	1.027	1.029	1.031	1.033	1.035	1.038	1.041	1.043	1.047	1.050	1.054	1.058	1.050
1.023	1.025	1.026	1.028	1.030	1.032	1.034	1.036	1.039	1.041	1.044	1.047	1.051	1.055	1.059	1.051
1.024	1.025	1.027	1.029	1.030	1.032	1.035	1.037	1.039	1.042	1.045	1.048	1.052	1.056	1.060	1.052
1.024	1.026	1.027	1.029	1.031	1.033	1.035	1.038	1.040	1.043	1.046	1.049	1.053	1.057	1.061	1.053
1.025	1.026	1.028	1.030	1.032	1.034	1.036	1.038	1.041	1.044	1.047	1.050	1.054	1.058	1.062	1.054
1.025	1.027	1.028	1.030	1.032	1.034	1.037	1.039	1.042	1.045	1.048	1.051	1.055	1.059	1.064	1.055
1.026	1.027	1.029	1.031	1.033	1.035	1.037	1.040	1.042	1.045	1.049	1.052	1.056	1.060	1.065	1.056
1.026	1.028	1.029	1.031	1.033	1.036	1.038	1.040	1.043	1.046	1.050	1.053	1.057	1.061	1.066	1.057
1.027	1.028	1.030	1.032	1.034	1.036	1.039	1.041	1.044	1.047	1.050	1.054	1.058	1.062	1.067	1.058
1.027	1.029	1.031	1.032	1.035	1.037	1.039	1.042	1.045	1.048	1.051	1.055	1.059	1.063	1.068	1.059
1.028	1.029	1.031	1.033	1.035	1.037	1.040	1.043	1.046	1.049	1.052	1.056	1.060	1.064	1.069	1.060
1.028	1.030	1.032	1.034	1.036	1.038	1.041	1.043	1.046	1.049	1.053	1.057	1.061	1.066	1.071	1.061
1.028	1.030	1.032	1.034	1.036	1.039	1.041	1.044	1.047	1.050	1.054	1.058	1.062	1.067	1.072	1.062
1.029	1.031	1.033	1.035	1.037	1.039	1.042	1.045	1.048	1.051	1.055	1.059	1.063	1.068	1.073	1.063
1.029	1.031	1.033	1.035	1.038	1.040	1.043	1.045	1.049	1.052	1.056	1.060	1.064	1.069	1.074	1.064
1.030	1.032	1.034	1.036	1.038	1.041	1.043	1.046	1.049	1.053	1.056	1.061	1.065	1.070	1.075	1.065
1.030	1.032	1.034	1.036	1.039	1.041	1.044	1.047	1.050	1.054	1.057	1.061	1.066	1.071	1.076	1.066
1.031	1.033	1.035	1.037	1.039	1.042	1.045	1.048	1.051	1.054	1.058	1.062	1.067	1.072	1.077	1.067
1.031	1.033	1.035	1.037	1.040	1.042	1.045	1.048	1.052	1.055	1.059	1.063	1.068	1.073	1.079	1.068
1.032	1.034	1.036	1.038	1.040	1.043	1.046	1.049	1.052	1.056	1.060	1.064	1.069	1.074	1.080	1.069
1.032	1.034	1.036	1.039	1.041	1.044	1.047	1.050	1.053	1.057	1.061	1.065	1.070	1.075	1.081	1.070
1.033	1.035	1.037	1.039	1.042	1.044	1.047	1.050	1.054	1.058	1.062	1.066	1.071	1.076	1.082	1.071
1.033	1.035	1.037	1.040	1.042	1.045	1.048	1.051	1.055	1.058	1.063	1.067	1.072	1.077	1.083	1.072
1.034	1.036	1.038	1.040	1.043	1.046	1.049	1.052	1.055	1.059	1.063	1.068	1.073	1.078	1.084	1.073
1.034	1.036	1.038	1.041	1.043	1.046	1.049	1.053	1.056	1.060	1.064	1.069	1.074	1.079	1.086	1.074
1.035	1.037	1.039	1.041	1.044	1.047	1.050	1.053	1.057	1.061	1.065	1.070	1.075	1.081	1.087	1.075
1.035	1.037	1.039	1.042	1.045	1.048	1.051	1.054	1.058	1.062	1.066	1.071	1.076	1.082	1.088	1.076
1.036	1.038	1.040	1.043	1.045	1.048	1.051	1.055	1.059	1.063	1.067	1.072	1.077	1.003	1.089	1.077
1.036	1.038	1.041	1.043	1.046	1.049	1.052	1.056	1.059	1.063	1.068	1.073	1.078	1.084	1.090	1.078
1.036	1.039	1.041	1.044	1.046	1.049	1.053	1.056	1.060	1.064	1.069	1.074	1.079	1.085	1.091	1.079

Table G.17 F_{pv} Correction for Temperature (*continued*)

F_{pv}	40	50	60	70	80	90	100	110	120	130	140	150	160	170	180
							Temperature T_F, °F								
1.080	1.092	1.086	1.080	1.075	1.070	1.065	1.061	1.057	1.053	1.050	1.047	1.044	1.042	1.039	1.037
1.081	1.094	1.087	1.081	1.076	1.070	1.066	1.062	1.058	1.054	1.051	1.048	1.045	1.042	1.040	1.037
1.082	1.095	1.088	1.082	1.076	1.071	1.067	1.062	1.058	1.055	1.051	1.048	1.045	1.043	1.040	1.038
1.083	1.096	1.089	1.083	1.077	1.072	1.067	1.063	1.059	1.055	1.052	1.049	1.046	1.043	1.041	1.038
1.084	1.097	1.090	1.084	1.078	1.073	1.068	1.064	1.060	1.056	1.053	1.049	1.046	1.044	1.041	1.039
1.085	1.098	1.091	1.085	1.079	1.074	1.069	1.065	1.061	1.057	1.053	1.050	1.047	1.044	1.042	1.039
1.086	1.099	1.092	1.086	1.080	1.075	1.070	1.065	1.061	1.057	1.054	1.051	1.048	1.045	1.042	1.040
1.087	1.100	1.093	1.087	1.081	1.076	1.071	1.066	1.062	1.058	1.055	1.051	1.048	1.045	1.043	1.040
1.088	1.102	1.094	1.088	1.082	1.077	1.072	1.067	1.063	1.059	1.055	1.052	1.049	1.046	1.043	1.041
1.089	1.103	1.096	1.089	1.083	1.077	1.072	1.068	1.063	1.059	1.056	1.052	1.049	1.046	1.044	1.041
1.090	1.104	1.097	1.090	1.084	1.078	1.073	1.068	1.064	1.060	1.056	1.053	1.050	1.047	1.044	1.042
1.091	1.105	1.098	1.091	1.085	1.079	1.074	1.069	1.065	1.061	1.057	1.054	1.050	1.047	1.045	1.042
1.092	1.106	1.099	1.092	1.086	1.080	1.075	1.070	1.066	1.061	1.058	1.054	1.051	1.048	1.045	1.043
1.093	1.107	1.100	1.093	1.087	1.081	1.076	1.071	1.066	1.062	1.058	1.055	1.052	1.048	1.046	1.043
1.094	1.108	1.101	1.094	1.088	1.082	1.076	1.072	1.067	1.062	1.059	1.055	1.052	1.049	1.046	1.044
1.095	1.110	1.102	1.095	1.089	1.083	1.077	1.072	1.068	1.064	1.060	1.056	1.053	1.050	1.047	1.044
1.096	1.111	1.103	1.096	1.090	1.084	1.078	1.073	1.068	1.064	1.060	1.057	1.053	1.050	1.047	1.044
1.097	1.112	1.104	1.097	1.090	1.084	1.079	1.074	1.069	1.065	1.061	1.057	1.054	1.051	1.048	1.045
1.098	1.113	1.105	1.098	1.091	1.085	1.080	1.074	1.070	1.066	1.062	1.058	1.054	1.051	1.048	1.045
1.099	1.114	1.106	1.099	1.092	1.086	1.081	1.075	1.071	1.066	1.062	1.058	1.055	1.052	1.049	1.046
1.100	1.115	1.107	1.100	1.093	1.087	1.081	1.076	1.071	1.067	1.063	1.059	1.055	1.052	1.049	1.046

Table G.18 Corrections to Supercompressibility Factor F_{pv} for Air

Specific gravity	Percent air	35		135		235		335		435	
		30–89°	90–149°	30–89°	90–149°	30–89°	90–149°	30–89°	90–149°	30–89°	90–149°
0.50–0.649	1	....	....	....	....	....	....	0.001	0.000	0.001	0.001
	2	....	....	0.001	....	0.001	0.001	0.002	0.001	0.002	0.001
	3	0.001	....	0.001	0.001	0.002	0.001	0.002	0.002	0.003	0.002
	4	0.001	0.001	0.002	0.001	0.002	0.001	0.003	0.002	0.005	0.003
	5	0.001	0.001	0.002	0.001	0.003	0.002	0.004	0.003	0.006	0.004
	6	0.001	0.001	0.002	0.002	0.004	0.002	0.005	0.004	0.007	0.004
	7	0.001	0.001	0.003	0.002	0.004	0.003	0.006	0.004	0.008	0.005
	8	0.001	0.001	0.003	0.002	0.005	0.003	0.007	0.005	0.009	0.006
0.650–0.749	1	....	....	....	....	0.001	....	0.001	0.001	0.001	0.001
	2	....	....	0.001	....	0.001	0.001	0.002	0.001	0.003	0.002
	3	0.001	....	0.001	0.001	0.002	0.001	0.002	0.002	0.004	0.002
	4	0.001	0.001	0.002	0.001	0.002	0.002	0.003	0.003	0.005	0.003
	5	0.001	0.001	0.002	0.002	0.003	0.002	0.004	0.003	0.006	0.004
	6	0.001	0.001	0.003	0.002	0.004	0.003	0.005	0.004	0.008	0.005
	7	0.001	0.001	0.003	0.002	0.005	0.003	0.006	0.005	0.009	0.006
	8	0.001	0.001	0.004	0.003	0.005	0.004	0.007	0.005	0.010	0.007
0.750–0.849	1	....	....	....	....	0.001	0.000	0.001	0.001	0.001	0.001
	2	....	....	0.001	0.001	0.001	0.001	0.002	0.001	0.003	0.002
	3	0.001	....	0.001	0.001	0.002	0.001	0.003	0.002	0.004	0.003
	4	0.001	0.001	0.002	0.001	0.003	0.002	0.004	0.003	0.006	0.004
	5	0.001	0.001	0.002	0.002	0.004	0.003	0.005	0.004	0.007	0.005
	6	0.001	0.001	0.003	0.002	0.004	0.003	0.006	0.005	0.009	0.006
	7	0.001	0.001	0.003	0.002	0.005	0.004	0.007	0.005	0.010	0.007
	8	0.001	0.001	0.004	0.003	0.006	0.005	0.009	0.006	0.012	0.008

Pressure, psig

Table G.18 Corrections to Supercompressibility Factor F_{pv} for Air (continued)

Specific gravity	Percent air	Pressure, psig 35 30-89°	35 90-149°	135 30-89°	135 90-149°	235 30-89°	235 90-149°	335 30-89°	335 90-149°	435 30-89°	435 90-149°
0.850–0.949	1	....	....	....	....	0.001					
	2	....	....	0.001	0.001	0.001	0.001				
	3	0.001	....	0.002	0.001	0.002	0.002				
	4	0.001	0.001	0.002	0.002	0.003	0.002				
	5	0.001	0.001	0.003	0.002	0.004	0.003				
	6	0.001	0.001	0.003	0.002	0.005	0.004				
	7	0.001	0.001	0.004	0.003	0.005	0.004				
	8	0.002	0.001	0.004	0.003	0.006	0.005				
0.950–1.04	1	....	....	0.001	....	0.001	0.001				
	2	....	....	0.001	0.001	0.002	0.001				
	3	0.001	0.001	0.002	0.001	0.002	0.002				
	4	0.001	0.001	0.002	0.002	0.003	0.003				
	5	0.001	0.001	0.003	0.002	0.004	0.004				
	6	0.001	0.001	0.004	0.002	0.005	0.004				
	7	0.002	0.001	0.004	0.003	0.006	0.005				
	8	0.002	0.001	0.005	0.003	0.006	0.006				
1.05–1.14	1	....	....	0.001	0.001						
	2	0.001	....	0.001	0.001						
	3	0.001	0.001	0.002	0.002						
	4	0.001	0.001	0.003	0.002						
	5	0.001	0.001	0.003	0.003						
	6	0.002	0.001	0.004	0.003						
	7	0.002	0.001	0.005	0.004						
	8	0.002	0.001	0.005							

1.15–1.24

1	. . . .	. . . .	0.001	0.001
2	0.001	. . . .	0.002	0.002
3	0.001	0.001	0.002	0.002
4	0.001	0.001	0.003	0.003
5	0.002	0.001	0.004	0.003
6	0.002	0.001	0.004	0.003
7	0.002	0.001	0.005	0.004
8	0.002	0.001	0.006	0.004

1.25–1.34

1	. . . .	. . . .
2	0.001	. . . .
3	0.001	0.001
4	0.001	0.001
5	0.002	0.001
6	0.002	0.001
7	0.002	0.001
8	0.002	0.001

1.35–1.44

1	0.001	. . . .
2	0.001	. . . .
3	0.001	0.001
4	0.002	0.001
5	0.002	0.001
6	0.002	0.001
7	0.002	0.001
8	0.002	0.002

NOTE: The above corrections for given amounts of air are deductible from the corrected supercompressibility factors in Tables G.11 through G.17. Each heading is inclusive to next higher heading. Do not interpolate.

Table G.19 Corrections to Supercompressibility Factor F_{pv} for Carbon Dioxide

Specific gravity	Percent CO_2	35		135		235		335		435	
		30–89°	90–149°	30–89°	90–149°	30–89°	90–149°	30–89°	90–149°	30–89°	90–149°
0.550–0.649	1	....	....	....	....	....	....	0.001	....	0.001	0.001
	2	....	0.001	0.001	0.001	0.001	0.001	0.001	0.001	0.002	0.001
	3	0.001	0.001	0.001	0.001	0.001	0.001	0.002	0.001	0.002	0.002
	4	0.001	0.001	0.002	0.001	0.002	0.002	0.002	0.002	0.003	0.002
	5	0.001	0.001	0.002	0.002	0.002	0.002	0.003	0.002	0.004	0.003
	6	0.001	0.001	0.002	0.002	0.003	0.002	0.003	0.002	0.004	0.003
	7	0.001	0.001	0.003	0.002	0.004	0.002	0.004	0.003	0.005	0.003
	8	0.002	0.001	0.003	0.002	0.004	0.002	0.005	0.003	0.006	0.004
0.650–0.749	1	....	....	....	....	0.001	0.001	0.001	0.001	0.002	0.001
	2	....	....	0.001	....	0.001	0.001	0.002	0.002	0.003	0.002
	3	0.001	0.001	0.001	0.001	0.002	0.002	0.003	0.002	0.004	0.003
	4	0.001	0.001	0.002	0.001	0.003	0.002	0.004	0.003	0.006	0.004
	5	0.001	0.001	0.002	0.002	0.003	0.003	0.005	0.004	0.007	0.004
	6	0.001	0.001	0.003	0.002	0.004	0.003	0.006	0.004	0.008	0.005
	7	0.002	0.001	0.003	0.002	0.005	0.004	0.006	0.005	0.009	0.006
	8	0.002	0.001	0.004	0.003	0.005	0.004	0.007	0.005	0.010	0.007
0.750–0.849	1	....	....	....	....	0.001	0.001	0.001	0.001	0.002	0.001
	2	....	....	0.001	0.001	0.002	0.001	0.003	0.002	0.004	0.003
	3	0.001	0.001	0.001	0.001	0.003	0.001	0.004	0.003	0.005	0.004
	4	0.001	0.001	0.002	0.002	0.003	0.003	0.005	0.004	0.006	0.005
	5	0.001	0.001	0.002	0.002	0.004	0.003	0.007	0.005	0.008	0.006
	6	0.002	0.001	0.003	0.003	0.005	0.004	0.008	0.006	0.010	0.007
	7	0.002	0.001	0.004	0.003	0.006	0.005	0.009	0.007	0.011	0.008
	8	0.002	0.002	0.004	0.003	0.007	0.005	0.010	0.007	0.013	0.009

0.850–0.949	1	....	....	0.001	0.001	0.001	0.001
	2	0.001	0.001	0.001	0.001	0.002	0.002
	3	0.001	0.001	0.002	0.001	0.003	0.002
	4	0.001	0.001	0.003	0.002	0.004	0.003
	5	0.002	0.001	0.004	0.002	0.005	0.004
	6	0.002	0.002	0.004	0.003	0.006	0.005
	7	0.002	0.002	0.005	0.003	0.007	0.005
	8	0.002	0.002	0.006	0.004	0.008	0.006
0.950–1.04	1	....	....	0.001	0.001	0.001	0.001
	2	0.001	0.001	0.001	0.001	0.002	0.002
	3	0.001	0.001	0.002	0.002	0.004	0.003
	4	0.002	0.001	0.003	0.002	0.005	0.003
	5	0.002	0.002	0.004	0.003	0.006	0.004
	6	0.002	0.002	0.004	0.003	0.008	0.005
	7	0.003	0.002	0.005	0.004	0.009	0.006
	8	0.003	0.002	0.006	0.004	0.011	0.007
1.05–1.14	1	....	....	0.001	0.001		
	2	0.001	0.001	0.002	0.001		
	3	0.001	0.001	0.003	0.002		
	4	0.001	0.001	0.003	0.003		
	5	0.002	0.002	0.004	0.003		
	6	0.002	0.002	0.005	0.004		
	7	0.003	0.002	0.006	0.004		
	8	0.003	0.002	0.007	0.005		
1.15–1.24	1	....	....	0.001	0.001		
	2	0.001	0.001	0.002	0.001		
	3	0.001	0.001	0.003	0.002		
	4	0.002	0.001	0.004	0.003		
	5	0.002	0.002	0.005	0.004		

Table G.19 Corrections to Supercompressibility Factor F_{pv} for Carbon Dioxide (continued)

Specific gravity	Percent CO_2	Pressure, psig 35		135		235		335		435	
		30–89°	90–149°	30–89°	90–149°	30–89°	90–149°	30–89°	90–149°	30–89°	90–149°
	6	0.002	0.002	0.007	0.004						
	7	0.003	0.002	0.008	0.005						
	8	0.003	0.003	0.009	0.006						
	1	....	0.001								
	2	0.001	0.001								
	3	0.001	0.001								
1.25–1.34	4	0.002	0.002								
	5	0.002	0.002								
	6	0.003	0.002								
	7	0.003	0.003								
	8	0.004	0.003								
	1	0.001									
	2	0.001	0.001								
	3	0.001	0.001								
1.35–1.44	4	0.002	0.001								
	5	0.002	0.002								
	6	0.003	0.002								
	7	0.003	0.002								
	8	0.004	0.003								

NOTE: The above corrections for given amounts of carbon dioxide are deductible from the corrected supercompressibility factors in Tables G.11 through G.17. Each heading is inclusive to next higher heading. Do not interpolate.

Table G.20 Supercompressibility Factors for Natural Gas of 0.6 Specific Gravity:† U.S. Units

							Flowing temperature T_F, °F							
P_G, Psig	−40	−20	0	20	40	60	80	100	120	140	160	180	200	220
0	1.0000	1.0000	1.0000	1.0000	1.0000	1.0000	1.0000	1.0000	1.0000	1.0000	1.0000	1.0000	1.0000	1.0000
100	1.0157	1.0137	1.0119	1.0104	1.0091	1.0079	1.0070	1.0061	1.0054	1.0047	1.0041	1.0036	1.0032	1.0028
200	1.0329	1.0284	1.0245	1.0212	1.0184	1.0160	1.0141	1.0123	1.0108	1.0094	1.0082	1.0072	1.0063	1.0055
300	1.0517	1.0441	1.0378	1.0325	1.0281	1.0243	1.0212	1.0185	1.0161	1.0141	1.0123	1.0108	1.0094	1.0082
400	1.0725	1.0610	1.0518	1.0442	1.0380	1.0327	1.0284	1.0247	1.0215	1.0187	1.0163	1.0142	1.0124	1.0108
500	1.0954	1.0792	1.0665	1.0564	1.0481	1.0413	1.0357	1.0309	1.0268	1.0233	1.0202	1.0176	1.0153	1.0132
600	1.1211	1.0988	1.0821	1.0690	1.0585	1.0499	1.0430	1.0370	1.0320	1.0277	1.0240	1.0208	1.0180	1.0156
700	1.1500	1.1202	1.0986	1.0820	1.0691	1.0587	1.0502	1.0432	1.0372	1.0321	1.0277	1.0240	1.0207	1.0178
800	1.1826	1.1431	1.1158	1.0955	1.0798	1.0674	1.0575	1.0492	1.0423	1.0364	1.0313	1.0270	1.0232	1.0199
900	1.2193	1.1678	1.1337	1.1092	1.0906	1.0761	1.0646	1.0551	1.0472	1.0405	1.0348	1.0299	1.0256	1.0219
1000	1.2597	1.1939	1.1522	1.1231	1.1014	1.0847	1.0716	1.0609	1.0520	1.0445	1.0381	1.0326	1.0279	1.0237
1100	1.3021	1.2211	1.1711	1.1370	1.1120	1.0931	1.0784	1.0665	1.0566	1.0483	1.0413	1.0352	1.0301	1.0255
1200	1.3421	1.2481	1.1898	1.1506	1.1224	1.1013	1.0850	1.0719	1.0610	1.0519	1.0443	1.0377	1.0321	1.0272
1300	1.3754	1.2735	1.2079	1.1639	1.1325	1.1091	1.0912	1.0770	1.0652	1.0554	1.0471	1.0400	1.0339	1.0287
1400	1.4007	1.2959	1.2246	1.1764	1.1421	1.1166	1.0971	1.0818	1.0691	1.0586	1.0497	1.0421	1.0356	1.0300
1500	1.4138	1.3132	1.2393	1.1879	1.1511	1.1235	1.1026	1.0863	1.0728	1.0615	1.0521	1.0440	1.0371	1.0312
1600	1.4175	1.3248	1.2512	1.1978	1.1589	1.1298	1.1076	1.0904	1.0762	1.0643	1.0543	1.0458	1.0385	1.0323
1700	1.4145	1.3311	1.2603	1.2062	1.1658	1.1353	1.1121	1.0941	1.0792	1.0667	1.0563	1.0474	1.0398	1.0333
1800	1.4068	1.3330	1.2663	1.2126	1.1715	1.1400	1.1160	1.0974	1.0819	1.0689	1.0580	1.0487	1.0409	1.0341
1900	1.3959	1.3312	1.2695	1.2173	1.1760	1.1439	1.1194	1.1002	1.0842	1.0708	1.0595	1.0499	1.0417	1.0347
2000	1.3827	1.3264	1.2704	1.2202	1.1793	1.1470	1.1221	1.1026	1.0861	1.0724	1.0607	1.0508	1.0424	1.0352

Table G.20 Supercompressibility Factors for Natural Gas of 0.6 Specific Gravity:† U.S. Units (Continued)

	Flowing temperature T_F, °F													
P_G, Psig	−40	−20	0	20	40	60	80	100	120	140	160	180	200	220
2100	1.3680	1.3186	1.2677	1.2207	1.1812	1.1493	1.1243	1.1044	1.0876	1.0734	1.0614	1.0513	1.0429	1.0355
2200	1.3527	1.3093	1.2635	1.2199	1.1820	1.1508	1.1258	1.1058	1.0887	1.0741	1.0619	1.0516	1.0431	1.0358
2300	1.3365	1.2988	1.2579	1.2175	1.1816	1.1513	1.1266	1.1066	1.0893	1.0745	1.0621	1.0517	1.0432	1.0359
2400	1.3205	1.2879	1.2511	1.2139	1.1800	1.1508	1.1266	1.1067	1.0894	1.0745	1.0691	1.0515	1.0431	1.0357
2500	1.3059	1.2763	1.2436	1.2095	1.1777	1.1496	1.1261	1.1065	1.0892	1.0741	1.0615	1.0511	1.0427	1.0355
2600	1.2906	1.2646	1.2352	1.2040	1.1742	1.1475	1.1248	1.1056	1.0885	1.0736	1.0610	1.0506	1.0422	1.0350
2700	1.2756	1.2527	1.2265	1.1980	1.1703	1.1450	1.1230	1.1043	1.0875	1.0728	1.0603	1.0499	1.0415	1.0343
2800	1.2608	1.2408	1.2174	1.1915	1.1657	1.1418	1.1208	1.1026	1.0862	1.0717	1.0593	1.0489	1.0407	1.0335
2900	1.2464	1.2289	1.2081	1.1845	1.1606	1.1381	1.1180	1.1005	1.0845	1.0703	1.0580	1.0478	1.0396	1.0324
3000	1.2324	1.2172	1.1987	1.1773	1.1552	1.1341	1.1150	1.0981	1.0825	1.0687	1.0566	1.0464	1.0383	1.0313

NOTE: For expanded tables of supercompressibility factors for natural gas, see A.G.A. Report No. 3 or A.G.A. Supercompressibility Tables (NX-19).

†For specific gravities other than 0.6, adjust pressure and temperature before entering table:

$$(p_G)_{table} = \frac{156.47 p_G}{160.8 - 7.22G + 100 x_{CO_2} - 39.2 x_{N_2}}$$

$$(T_F)_{table} = \frac{226.29}{99.15 + 211.9G - 100 x_{CO_2} - 168.1 x_{N_2}} (T_F + 460) - 460$$

SOURCE: Dieterich Standard Corp.

Table G.21   Supercompressibility Factors for Natural Gas of 0.6 Specific Gravity:† SI Units

Gauge press, kPa	Flowing temperature T_C, °C													
	−40	−29	−18	−7	4	16	27	38	49	60	71	82	93	104
0	1.0000	1.0000	1.0000	1.0000	1.0000	1.0000	1.0000	1.0000	1.0000	1.0000	1.0000	1.0000	1.0000	1.0000
689	1.0157	1.0137	1.0119	1.0104	1.0091	1.0079	1.0070	1.0061	1.0054	1.0047	1.0041	1.0036	1.0032	1.0028
1,379	1.0329	1.0284	1.0245	1.0212	1.0184	1.0160	1.0141	1.0123	1.0108	1.0094	1.0082	1.0072	1.0063	1.0055
2,068	1.0517	1.0441	1.0378	1.0325	1.0281	1.0243	1.0212	1.0185	1.0161	1.0141	1.0123	1.0108	1.0094	1.0082
2,758	1.0725	1.0610	1.0518	1.0442	1.0380	1.0327	1.0284	1.0247	1.0215	1.0187	1.0163	1.0142	1.0124	1.0108
3,447	1.0954	1.0792	1.0665	1.0564	1.0481	1.0413	1.0357	1.0309	1.0268	1.0233	1.0202	1.0176	1.0153	1.0132
4,137	1.1211	1.0988	1.0821	1.0690	1.0585	1.0499	1.0430	1.0370	1.0320	1.0277	1.0240	1.0208	1.0180	1.0156
4,826	1.1500	1.1202	1.0986	1.0820	1.0691	1.0587	1.0502	1.0432	1.0372	1.0321	1.0277	1.0240	1.0207	1.0178
5,516	1.1826	1.1431	1.1158	1.0955	1.0798	1.0674	1.0575	1.0492	1.0423	1.0364	1.0313	1.0270	1.0232	1.0199
6,205	1.2193	1.1678	1.1337	1.1092	1.0906	1.0761	1.0646	1.0551	1.0472	1.0405	1.0348	1.0299	1.0256	1.0219
6,895	1.2597	1.1939	1.1522	1.1231	1.1014	1.0847	1.0716	1.0609	1.0520	1.0445	1.0381	1.0326	1.0279	1.0237
7,584	1.3021	1.2211	1.1711	1.1370	1.1120	1.0931	1.0784	1.0665	1.0566	1.0483	1.0413	1.0352	1.0301	1.0255
8,274	1.3421	1.2481	1.1898	1.1506	1.1224	1.1013	1.0850	1.0719	1.0610	1.0519	1.0443	1.0377	1.0321	1.0272
8,963	1.3754	1.2735	1.2079	1.1639	1.1325	1.1091	1.0912	1.0770	1.0652	1.0554	1.0471	1.0400	1.0339	1.0287
9,653	1.4007	1.2959	1.2246	1.1764	1.1421	1.1166	1.0971	1.0818	1.0691	1.0586	1.0497	1.0421	1.0356	1.0300
10,340	1.4138	1.3132	1.2393	1.1879	1.1511	1.1235	1.1026	1.0863	1.0728	1.0615	1.0521	1.0440	1.0371	1.0312
11,030	1.4175	1.3248	1.2512	1.1978	1.1589	1.1298	1.1076	1.0904	1.0762	1.0643	1.0543	1.0458	1.0385	1.0323
11,720	1.4145	1.3311	1.2603	1.2062	1.1658	1.1353	1.1121	1.0941	1.0792	1.0667	1.0563	1.0474	1.0398	1.0333
12,410	1.4068	1.3330	1.2663	1.2126	1.1715	1.1400	1.1160	1.0974	1.0819	1.0689	1.0580	1.0487	1.0409	1.0341
13,100	1.3959	1.3312	1.2695	1.2173	1.1760	1.1439	1.1194	1.1002	1.0842	1.0708	1.0595	1.0499	1.0417	1.0347
13,790	1.3827	1.3264	1.2704	1.2202	1.1793	1.1470	1.1221	1.1026	1.0861	1.0724	1.0607	1.0508	1.0424	1.0352

Table G.21 Supercompressibility Factors for Natural Gas of 0.6 Specific Gravity:† SI Units (*Continued*)

Gauge press, kPa	Flowing temperature $T°C$, °C													
	−40	−29	−18	−7	4	16	27	38	49	60	71	82	93	104
14,480	1.3680	1.3186	1.2677	1.2207	1.1812	1.1493	1.1243	1.1044	1.0876	1.0734	1.0614	1.0513	1.0429	1.0355
15,170	1.3527	1.3093	1.2635	1.2199	1.1820	1.1508	1.1258	1.1058	1.0887	1.0741	1.0619	1.0516	1.0431	1.0358
15,860	1.3365	1.2988	1.2579	1.2175	1.1816	1.1513	1.1266	1.1066	1.0893	1.0745	1.0621	1.0517	1.0432	1.0359
16,550	1.3205	1.2879	1.2511	1.2139	1.1800	1.1508	1.1266	1.1067	1.0894	1.0745	1.0619	1.0515	1.0431	1.0357
17,240	1.3059	1.2763	1.2436	1.2095	1.1777	1.1496	1.1261	1.1065	1.0892	1.0741	1.0615	1.0511	1.0427	1.0355
17,930	1.2906	1.2646	1.2352	1.2040	1.1742	1.1475	1.1248	1.1056	1.0885	1.0736	1.0610	1.0506	1.0422	1.0350
18,620	1.2756	1.2527	1.2265	1.1980	1.1703	1.1450	1.1230	1.1043	1.0875	1.0728	1.0603	1.0499	1.0415	1.0343
19,300	1.2608	1.2408	1.2174	1.1915	1.1657	1.1418	1.1208	1.1026	1.0862	1.0717	1.0593	1.0489	1.0407	1.0335
19,990	1.2464	1.2289	1.2081	1.1845	1.1606	1.1381	1.1180	1.1005	1.0845	1.0703	1.0580	1.0478	1.0396	1.0324
20,680	1.2324	1.2172	1.1987	1.1773	1.1552	1.1341	1.1150	1.0981	1.0825	1.0687	1.0566	1.0464	1.0383	1.0313

NOTE: For expanded tables of supercompressibility factors for natural gas, see A.G.A. Report No. 3 or A.G.A. Supercompressibility Tables (NX-19).

†For specific gravities other than 0.6, adjust pressure and temperature before entering table:

$$(p\mathbf{\theta})_{\text{table}} = \frac{156.47 p\mathbf{\theta}}{160.8 - 7.22G + 100x_{CO_2} - 39.2x_{N_2}}$$

$$(T°C)_{\text{table}} = \frac{226.29}{99.15 + 211.9G - 100x_{CO_2} - 168.1x_{N_2}}(T°C + 273) - 273$$

SOURCE: Dieterich Standard Corp.

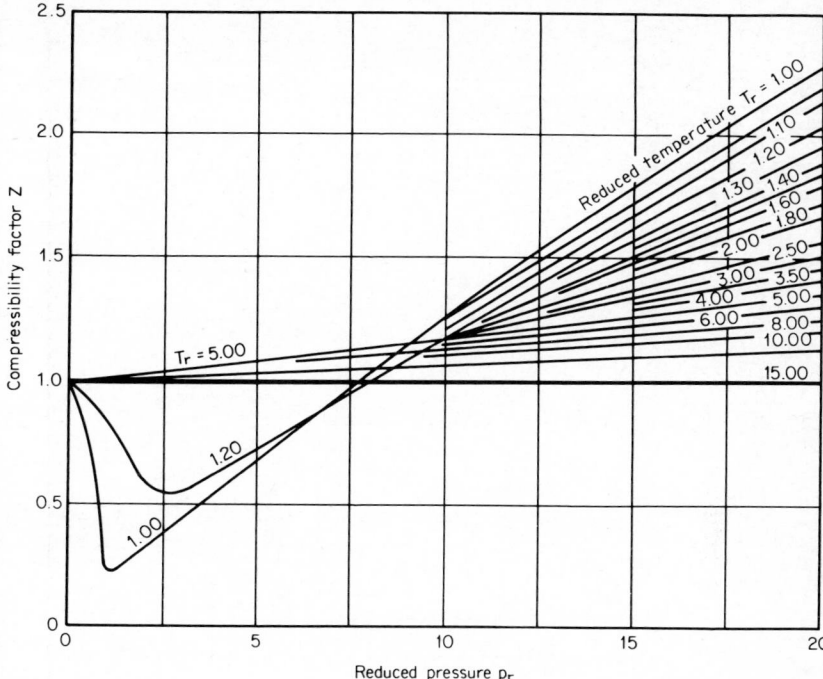

Figure G.1   Nelson-Obert compressibility chart (p_r = 0 to 20)   *(from Spink, 1967).*

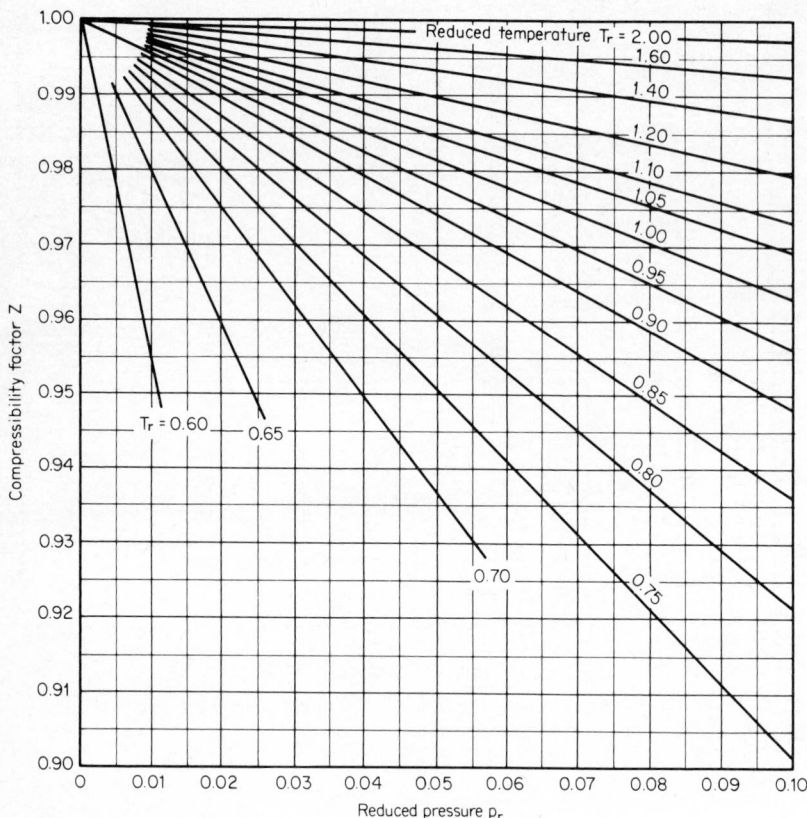

Figure G.2 Nelson-Obert compressibility diagram (p_r = 0 to 0.10) *(from Spink, 1967).*

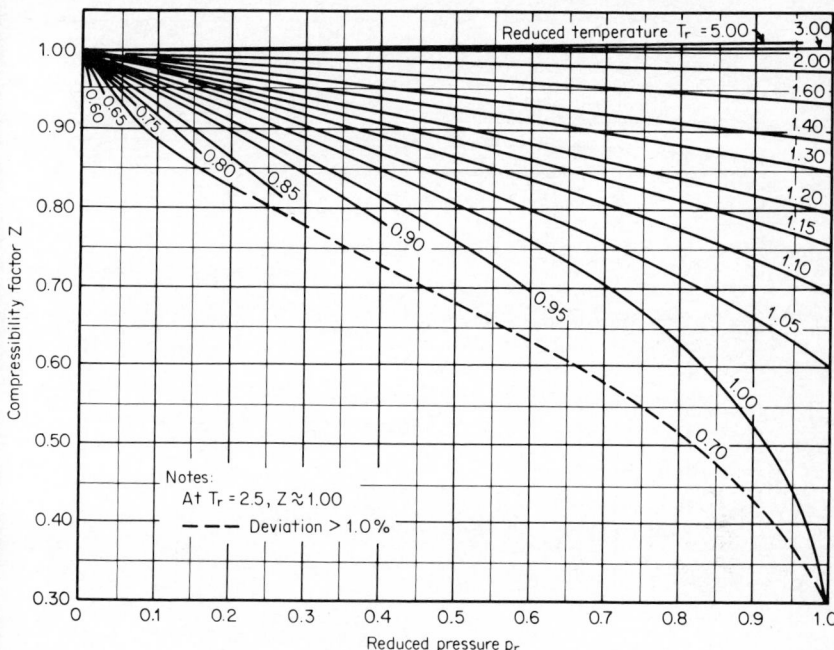

Figure G.3 Nelson-Obert compressibility chart (p_r = 0 to 1.0) *(from Spink, 1967).*

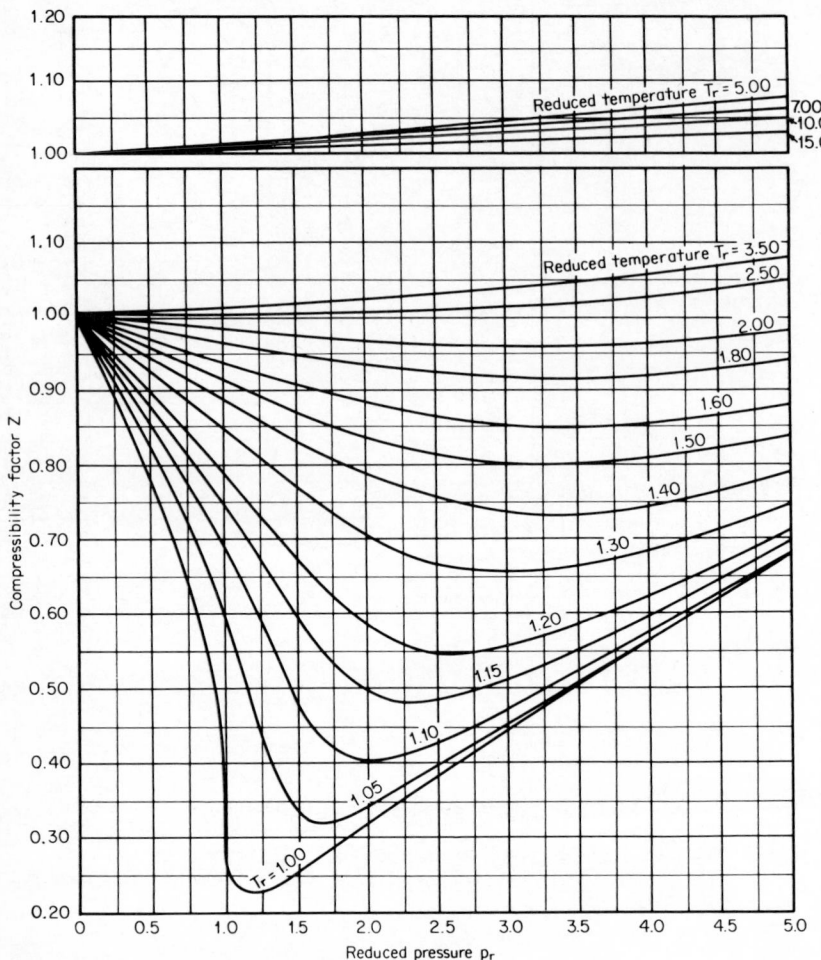

Figure G.4 Nelson-Obert compressibility chart (p_r = 0 to 5.00). *Note:* Do not use for ammonia. Do not use for methyl fluoride below T_r = 1.3 or for hydrogen or helium below T_r = 2.5. For hydrogen or helium above T_r = 2.5, use adjusted constants: T_c + 14.4°F and p_c + 117.6 lb$_f$/in². Maximum deviation, when used as above, is less than 2½ percent except near T_r = 1.0 and p_r = 1.0 *(from Spink, 1967).*

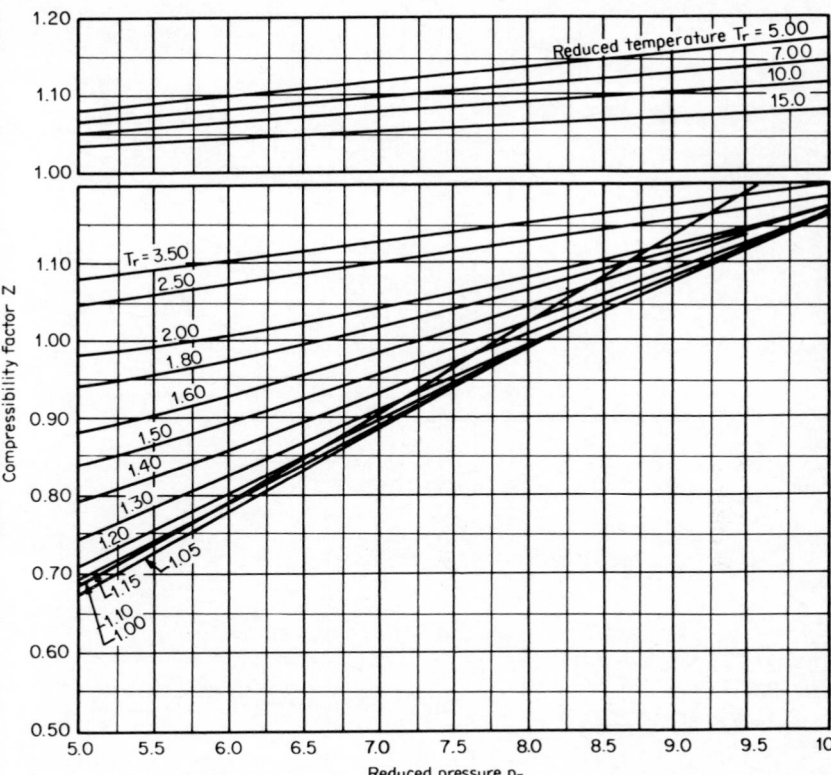

Figure G.5 Nelson-Obert compressibility chart (p_r = 5 to 10). *Note:* Do not use for ammonia. Do not use for methyl fluoride below T_r = 1.3 or for hydrogen or helium below T_r = 2.5. For hydrogen or helium above T_r = 2.5, use adjusted constants: T_c + 14.4°F and p_c + 117.6 lb_f/in^2. Maximum deviation, when used as above, is less than 2½ percent except near T_r = 1.0 and p_r = 1.0 *(from Spink, 1967).*

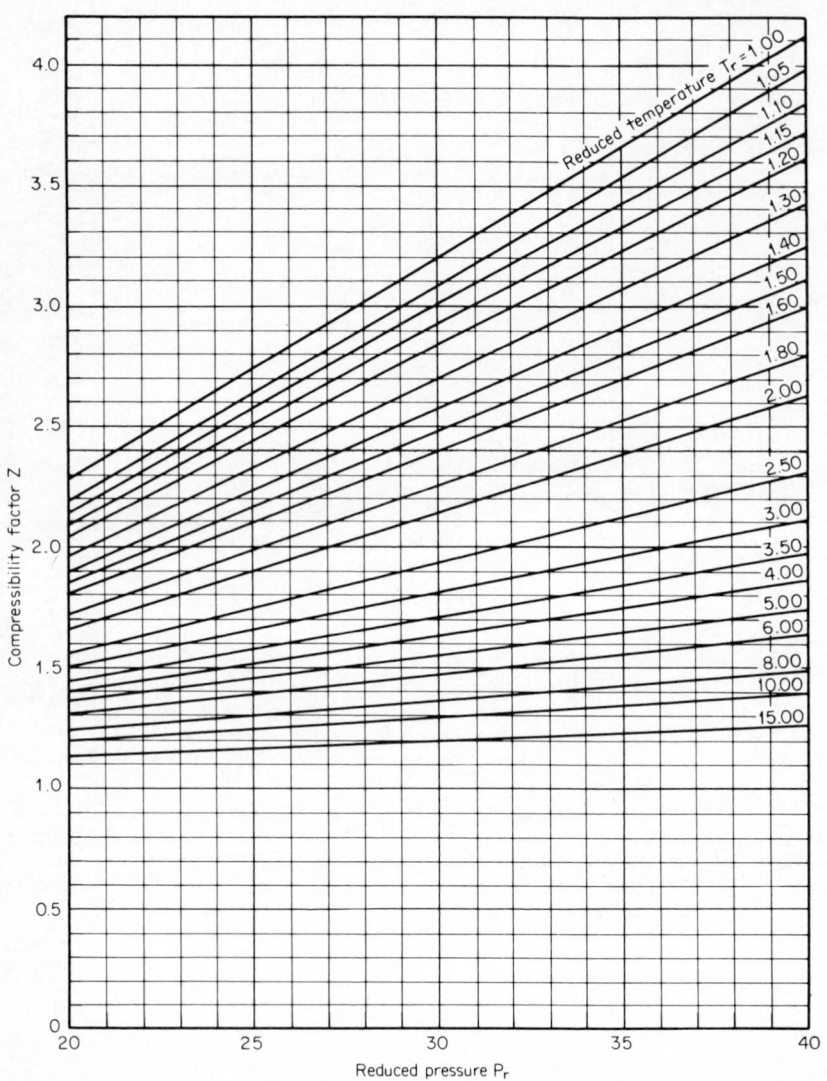

Figure G.6 Nelson-Obert compressibility chart (p_r = 20 to 40). *Note:* Maximum deviation for oxygen, argon, air, nitrogen, carbon monoxide, ethane, methane, ethylene, or propane is less than 5 percent. Also, for hydrogen or helium, if adjusted critical constants are used, then T_c + 14.4°F and p_c + 117.6 lb$_f$/in^2 *(from Spink, 1967).*

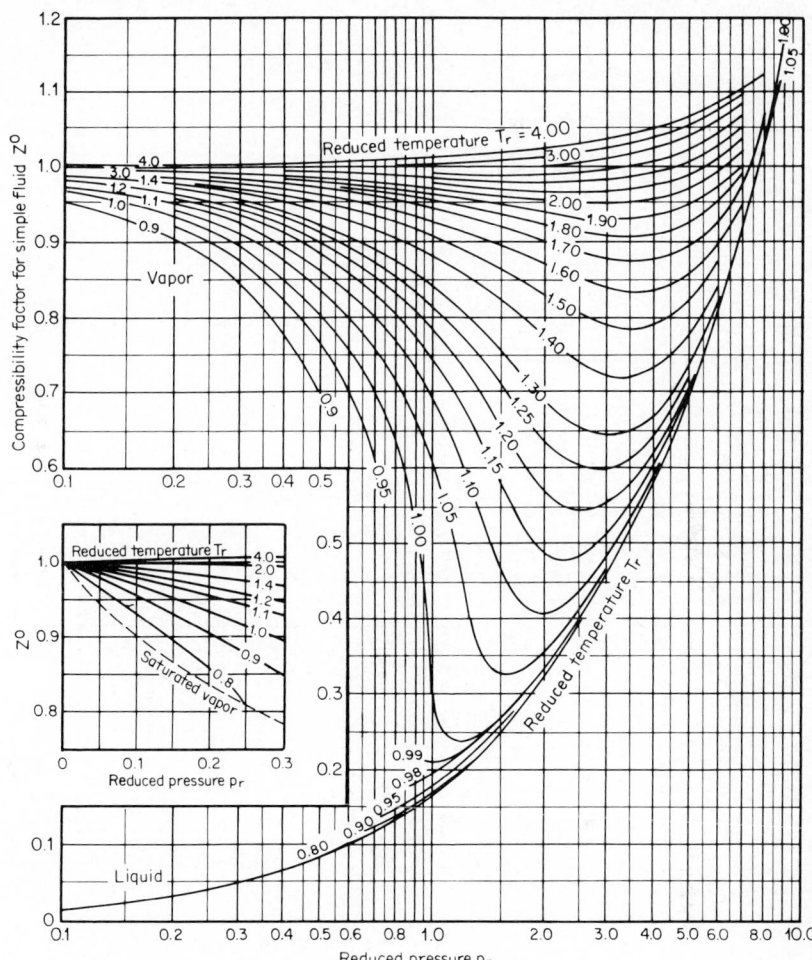

Figure G.7 **Generalized compressibility factor for the simple fluids argon, krypton, and xenon** *(from Edmister, 1974; used with permission).*

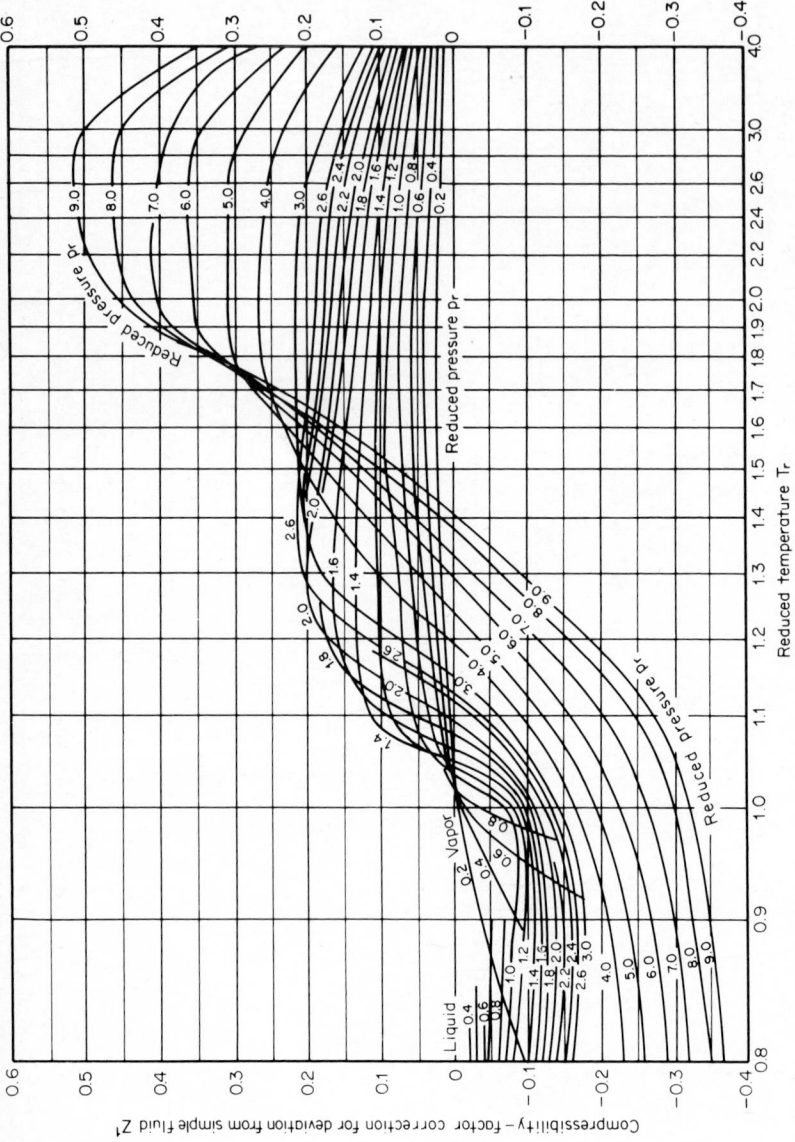

Figure G.8 Generalized compressibility-factor correction for deviation from simple fluid *(from Edmister, 1974; used with permission).*

G-148

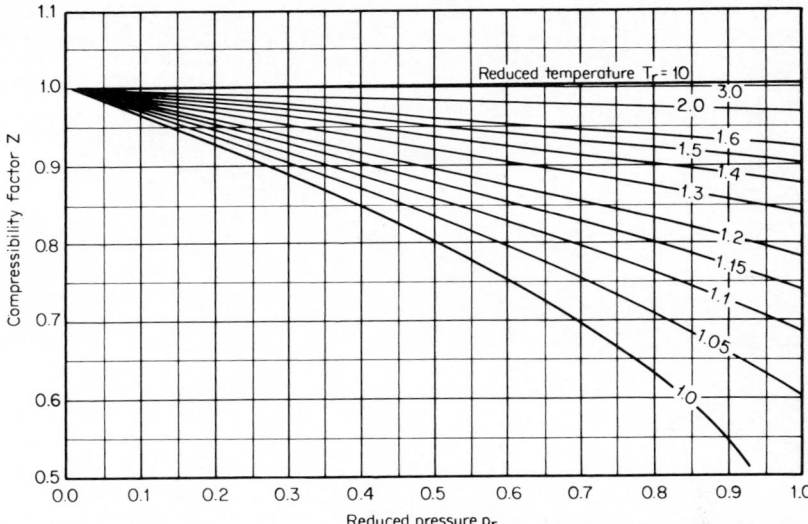

Figure G.9 Redlich-Kwong compressibility factor, based on equation of state (p_r = 0 to 1.0).

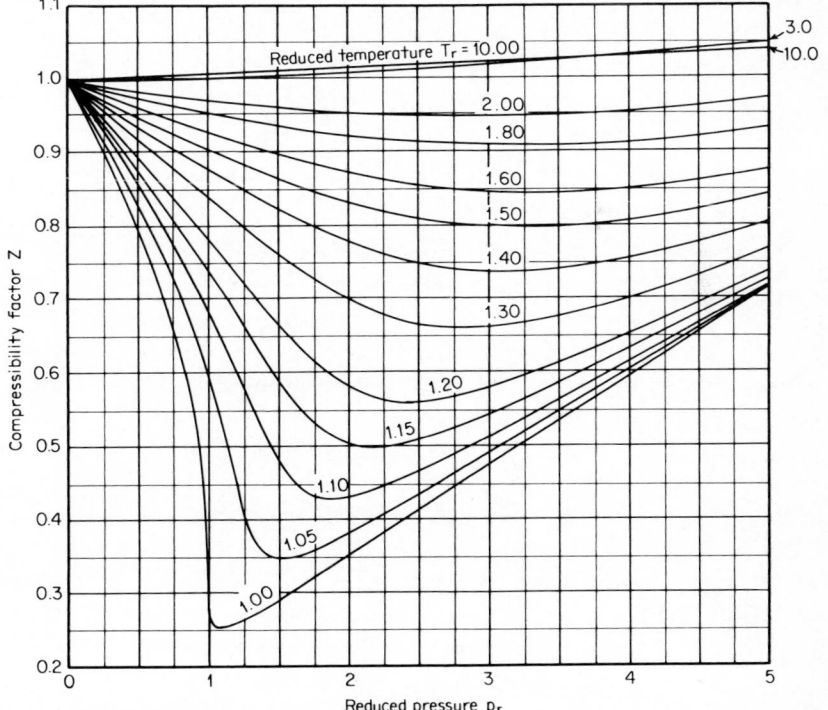

Figure G.10 Redlich-Kwong compressibility factor, based on equation of state (p_r = 0 to 5).

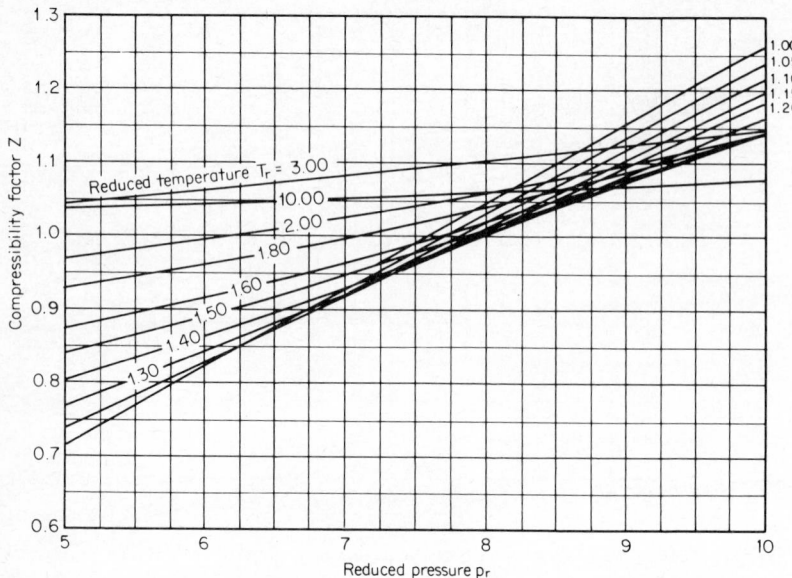

Figure G.11   Redlich-Kwong compressibility factor, based on equation of state ($p_r = 5$ to 10).

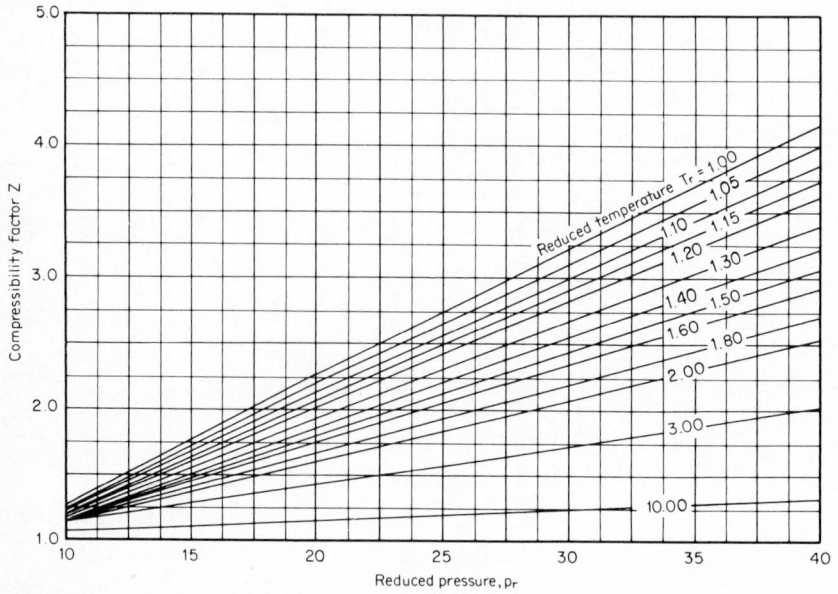

Figure G.12   Redlich-Kwong compressibility factor, based on equation of state ($p_r = 10$ to 40).

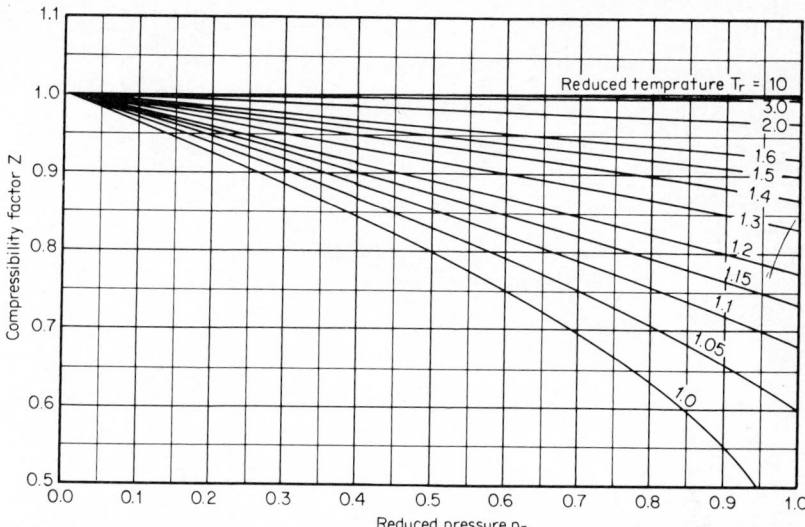

Figure G.13 Hall-Yarborough compressibility diagrams (p_r = 0 to 1.0).

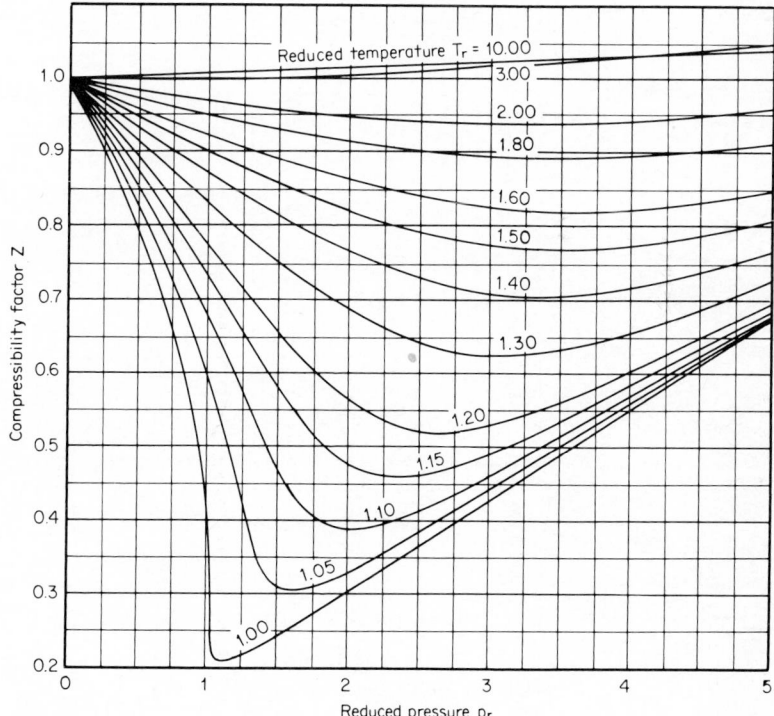

Figure G.14 Hall-Yarborough compressibility diagrams (p_r = 0 to 5).

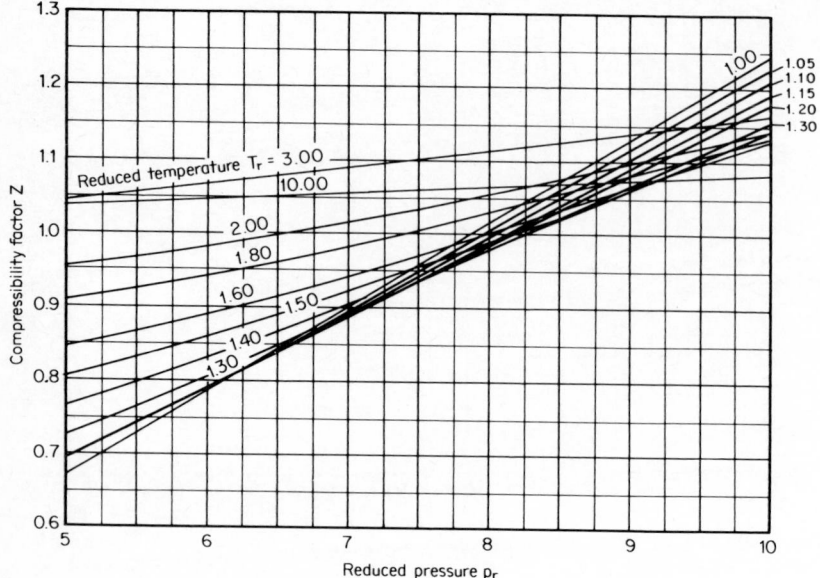

Figure G.15 Hall-Yarborough compressibility diagrams (p_r = 5 to 10).

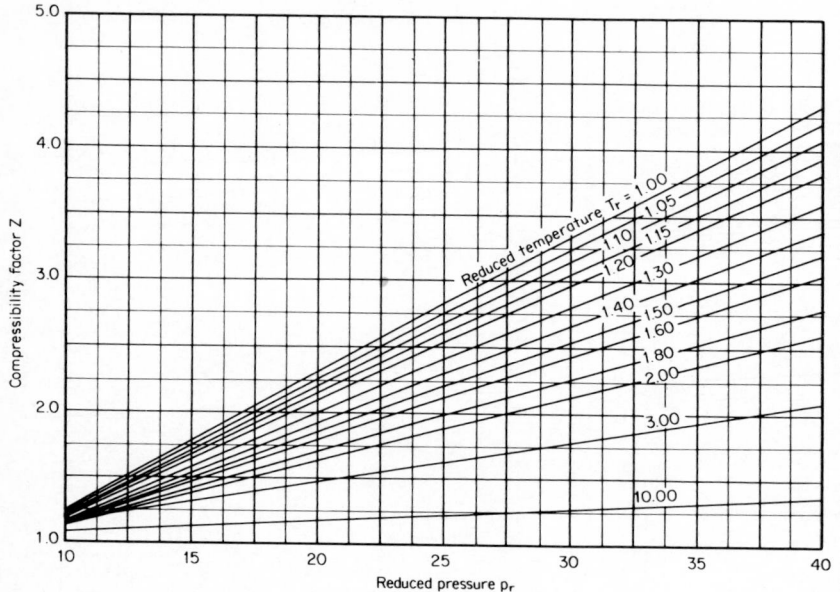

Figure G.16 Hall-Yarborough compressibility diagrams (p_r = 10 to 40).

APPENDIX H

VISCOSITIES OF GASES

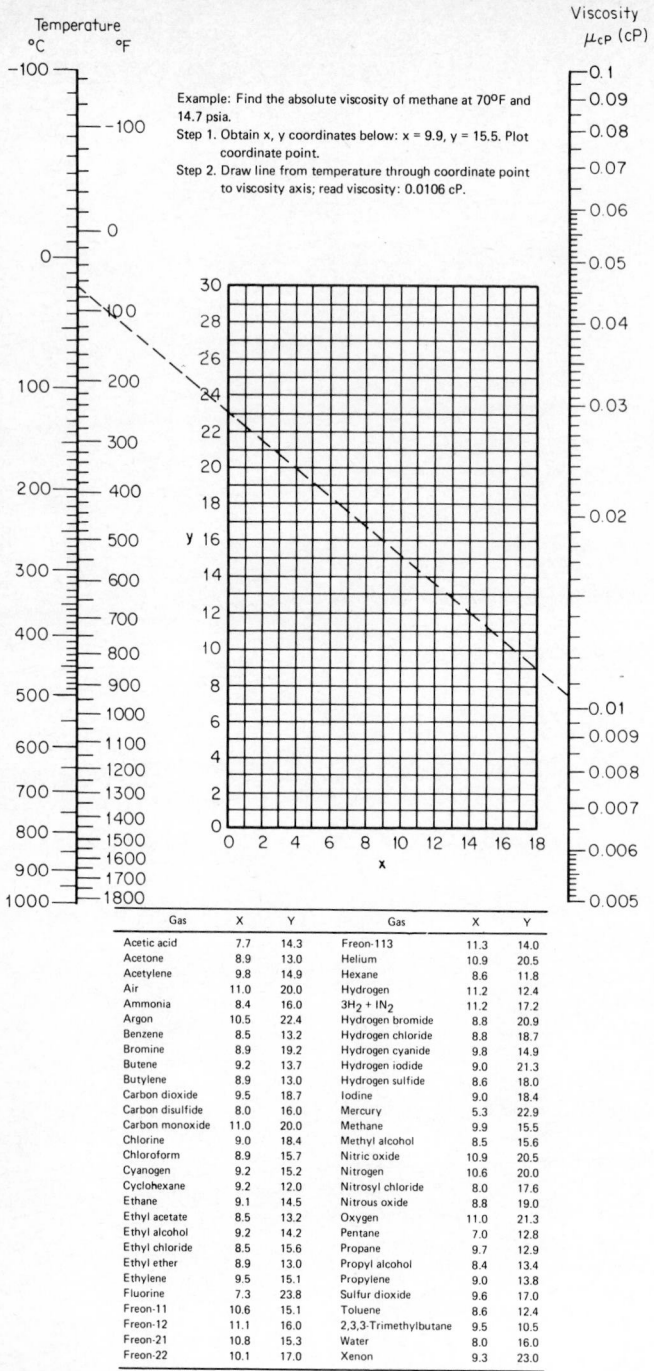

Example: Find the absolute viscosity of methane at 70°F and 14.7 psia.
Step 1. Obtain x, y coordinates below: x = 9.9, y = 15.5. Plot coordinate point.
Step 2. Draw line from temperature through coordinate point to viscosity axis; read viscosity: 0.0106 cP.

Gas	X	Y	Gas	X	Y
Acetic acid	7.7	14.3	Freon-113	11.3	14.0
Acetone	8.9	13.0	Helium	10.9	20.5
Acetylene	9.8	14.9	Hexane	8.6	11.8
Air	11.0	20.0	Hydrogen	11.2	12.4
Ammonia	8.4	16.0	$3H_2 + IN_2$	11.2	17.2
Argon	10.5	22.4	Hydrogen bromide	8.8	20.9
Benzene	8.5	13.2	Hydrogen chloride	8.8	18.7
Bromine	8.9	19.2	Hydrogen cyanide	9.8	14.9
Butene	9.2	13.7	Hydrogen iodide	9.0	21.3
Butylene	8.9	13.0	Hydrogen sulfide	8.6	18.0
Carbon dioxide	9.5	18.7	Iodine	9.0	18.4
Carbon disulfide	8.0	16.0	Mercury	5.3	22.9
Carbon monoxide	11.0	20.0	Methane	9.9	15.5
Chlorine	9.0	18.4	Methyl alcohol	8.5	15.6
Chloroform	8.9	15.7	Nitric oxide	10.9	20.5
Cyanogen	9.2	15.2	Nitrogen	10.6	20.0
Cyclohexane	9.2	12.0	Nitrosyl chloride	8.0	17.6
Ethane	9.1	14.5	Nitrous oxide	8.8	19.0
Ethyl acetate	8.5	13.2	Oxygen	11.0	21.3
Ethyl alcohol	9.2	14.2	Pentane	7.0	12.8
Ethyl chloride	8.5	15.6	Propane	9.7	12.9
Ethyl ether	8.9	13.0	Propyl alcohol	8.4	13.4
Ethylene	9.5	15.1	Propylene	9.0	13.8
Fluorine	7.3	23.8	Sulfur dioxide	9.6	17.0
Freon-11	10.6	15.1	Toluene	8.6	12.4
Freon-12	11.1	16.0	2,3,3-Trimethylbutane	9.5	10.5
Freon-21	10.8	15.3	Water	8.0	16.0
Freon-22	10.1	17.0	Xenon	9.3	23.0

Figure H.1 **Gas (vapor) absolute-viscosity alignment chart at atmospheric pressure** (from Perry and Chilton, 1973).

H-2

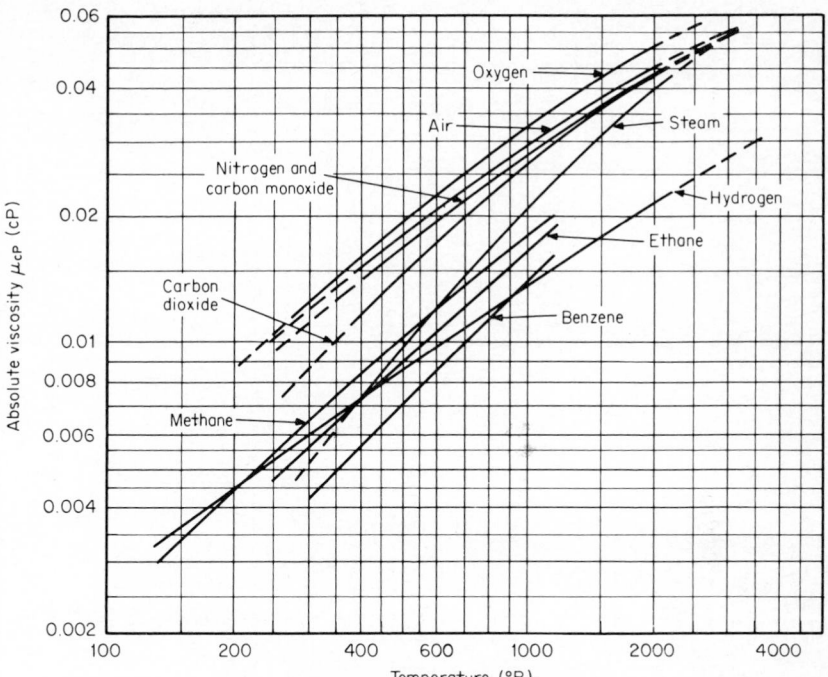

Figure H.2 Viscosities of gases at atmospheric pressure *(from Baumeister, 1978).*

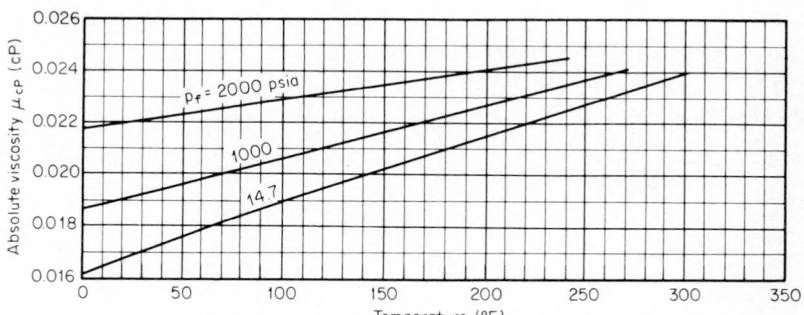

Figure H.3 Absolute viscosity of air *(from Fischer & Porter Catalog 10-A-54; used with permission).*

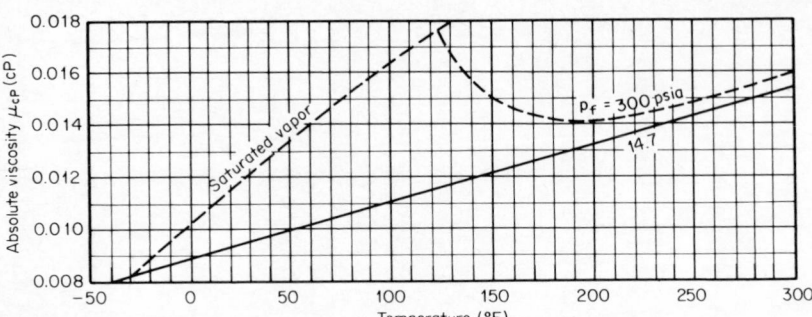

Figure H.4 **Absolute viscosity of ammonia** *(from Fischer & Porter Catalog 10-A-54; used with permission).*

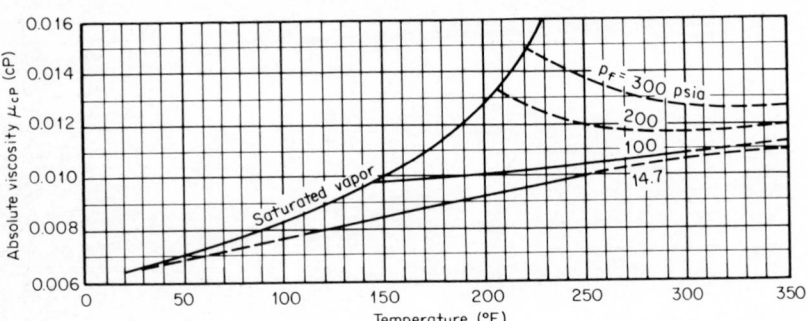

Figure H.5 **Absolute viscosity of butane** *(from Fischer & Porter Catalog 10-A-54; used with permission).*

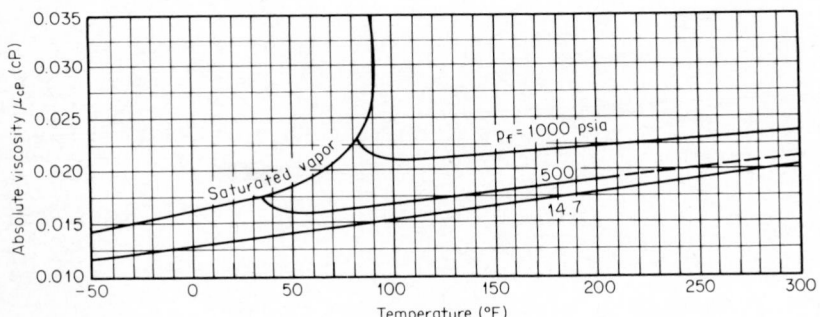

Figure H.6 **Absolute viscosity of carbon dioxide** *(from Fischer & Porter Catalog 10-A-54; used with permission).*

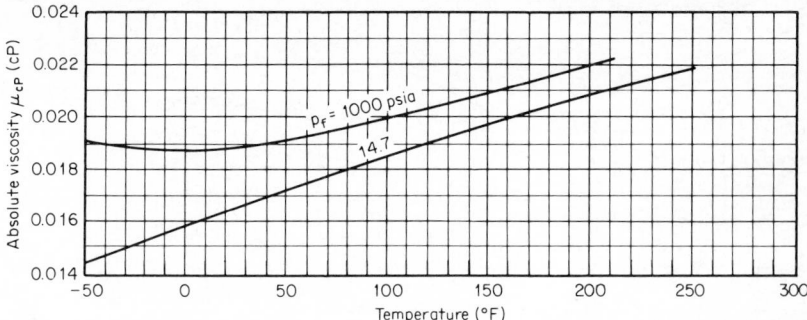

Figure H.7 **Absolute viscosity of carbon monoxide** *(from Fischer & Porter Catalog 10-A-54; used with permission).*

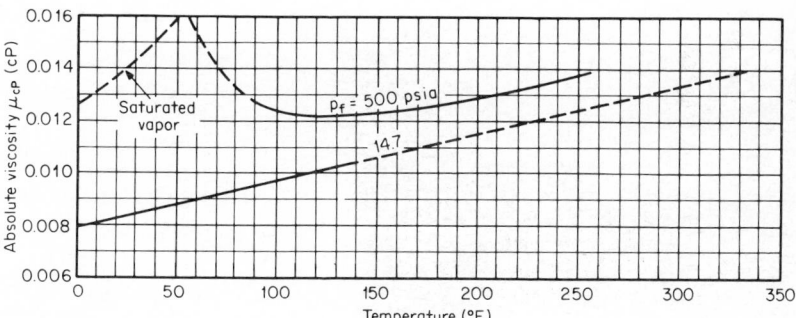

Figure H.8 **Absolute viscosity of ethane** *(from Fischer & Porter Catalog 10-A-54; used with permission).*

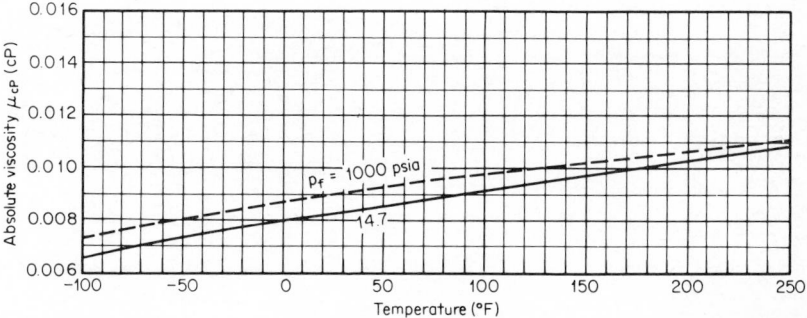

Figure H.9 **Absolute viscosity of hydrogen** *(from Fischer & Porter Catalog 10-A-54; used with permission).*

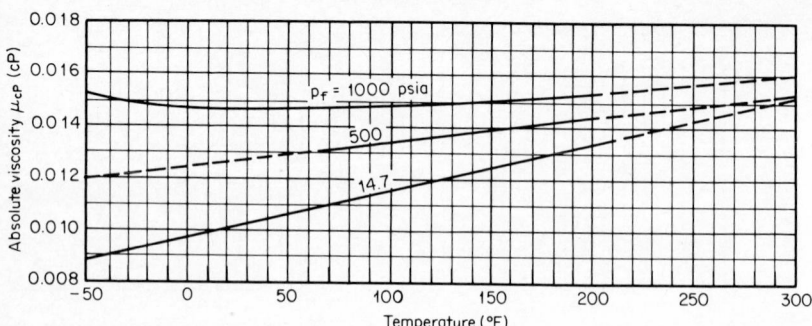

Figure H.10 **Absolute viscosity of methane** *(from Fischer & Porter Catalog 10-A-54; used with permission).*

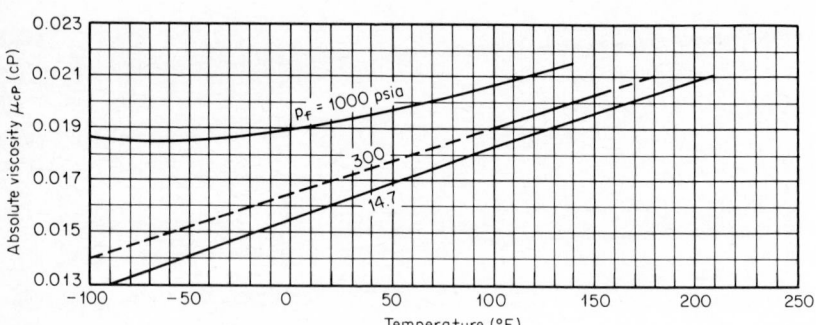

Figure H.11 **Absolute viscosity of nitrogen** *(from Fischer & Porter Catalog 10-A-54; used with permission).*

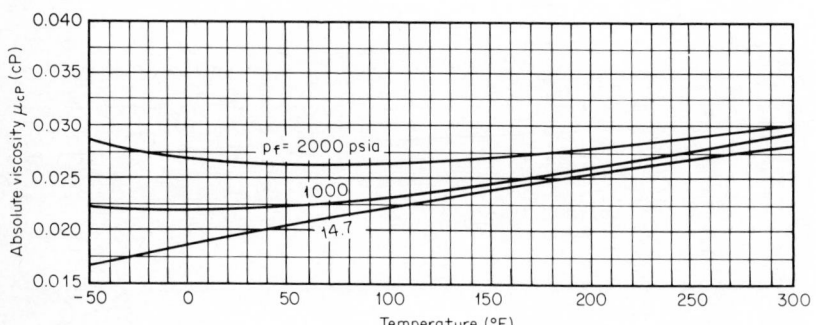

Figure H.12 **Absolute viscosity of oxygen** *(from Fischer & Porter Catalog 10-A-54; used with permission).*

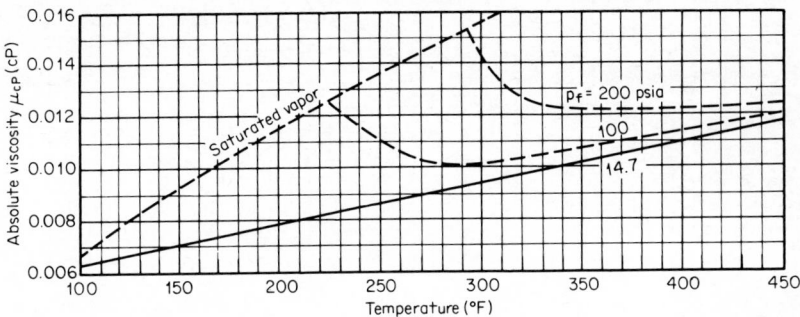

Figure H.13 **Absolute viscosity of pentane** (*from Fischer & Porter Catalog 10-A-54; used with permission*).

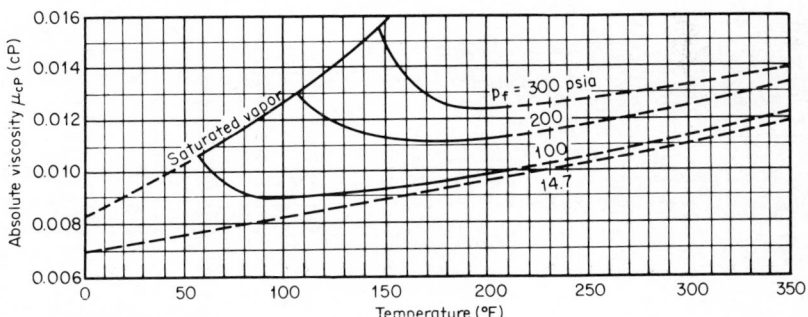

Figure H.14 **Absolute viscosity of propane** (*from Fischer & Porter Catalog 10-A-54; used with permission*).

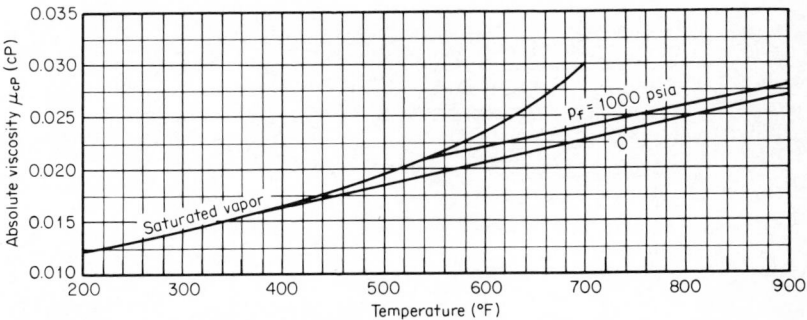

Figure H.15 **Absolute viscosity of steam** (*from Fischer & Porter Catalog 10-A-54; used with permission*).

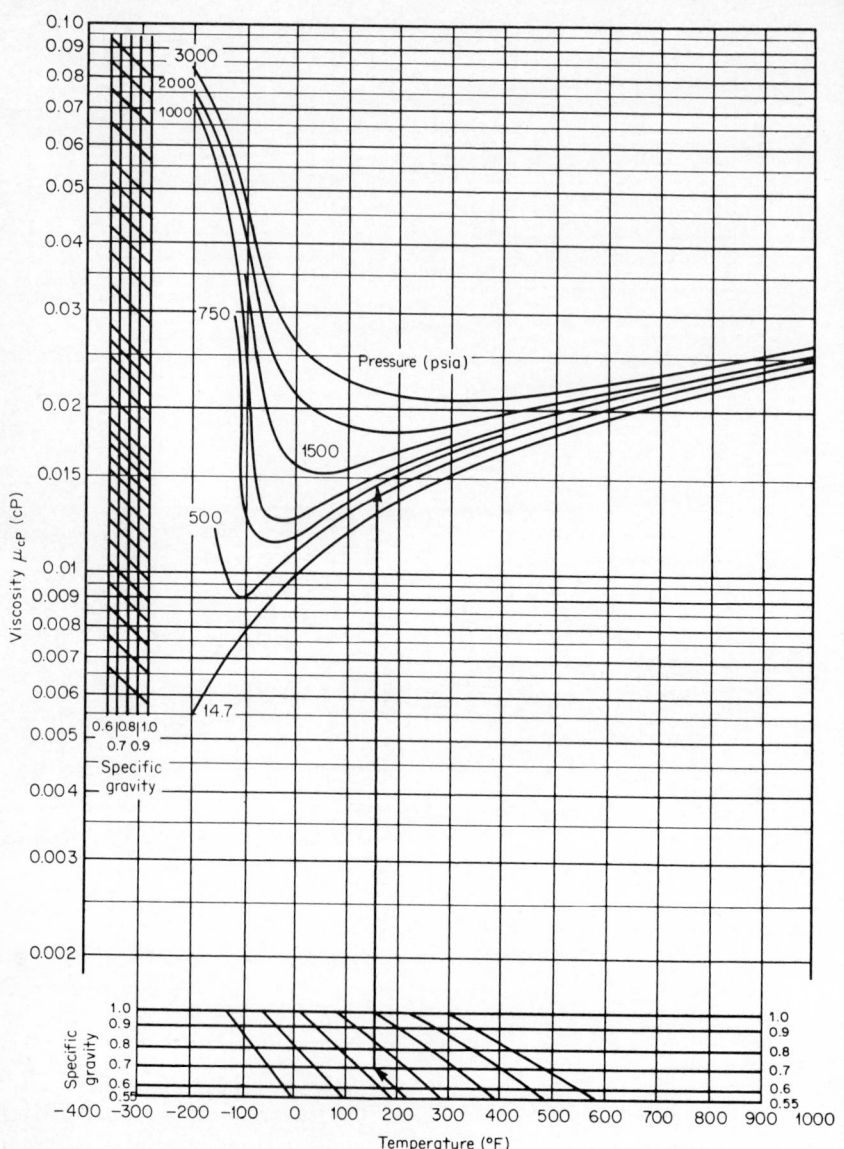

Figure H.16 **Absolute viscosity of hydrocarbon gases** *(from Gas Processors Supplier Association, 1978).*

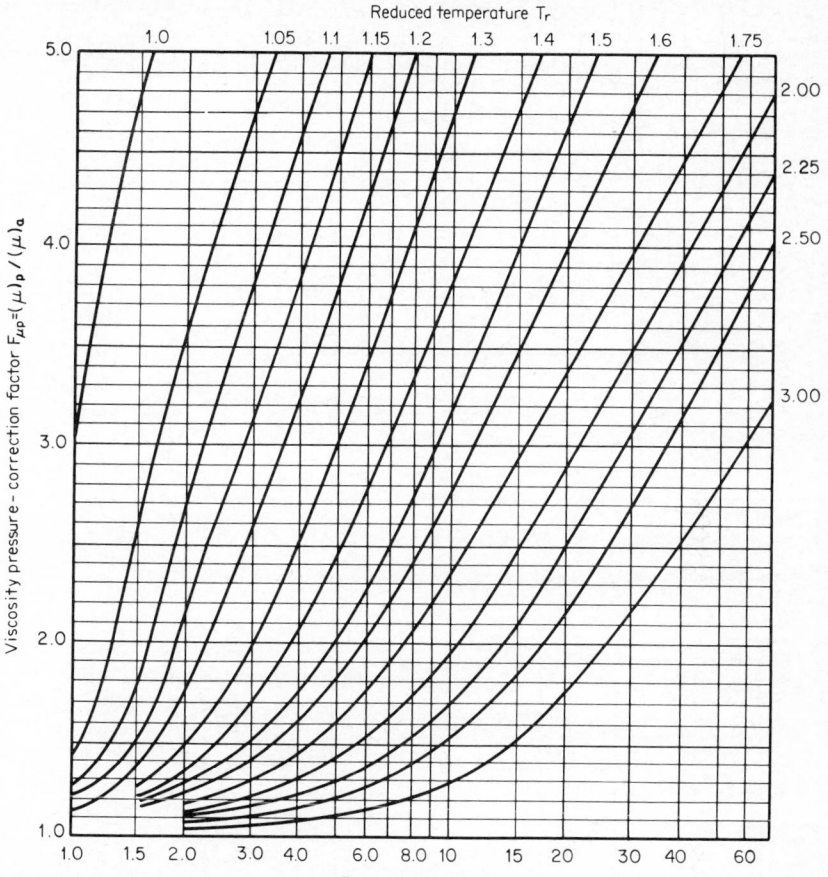

Figure H.17   Viscosity pressure correction factor: $(\mu)_a$ = absolute viscosity at atmospheric pressure; $(\mu)_p$ = absolute viscosity at operating pressure *(from Gambill, 1958).*

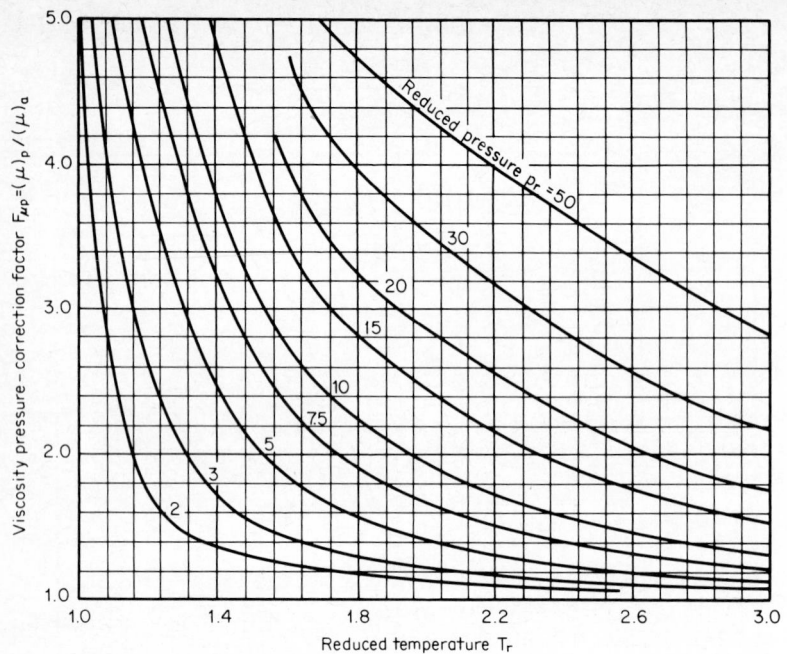

Figure H.18 Viscosity-pressure correction factor (cross-plot of Fig. H.17) *(from Gambill, 1958).*

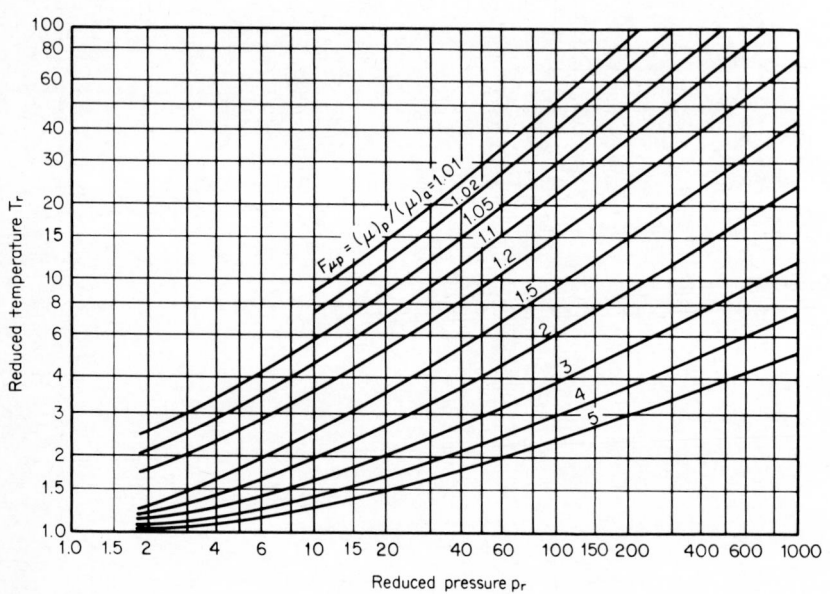

Figure H.19 Viscosity-pressure correction factor—extended pressure range *(from Gambill, 1958).*

ISENTROPIC EXPONENTS

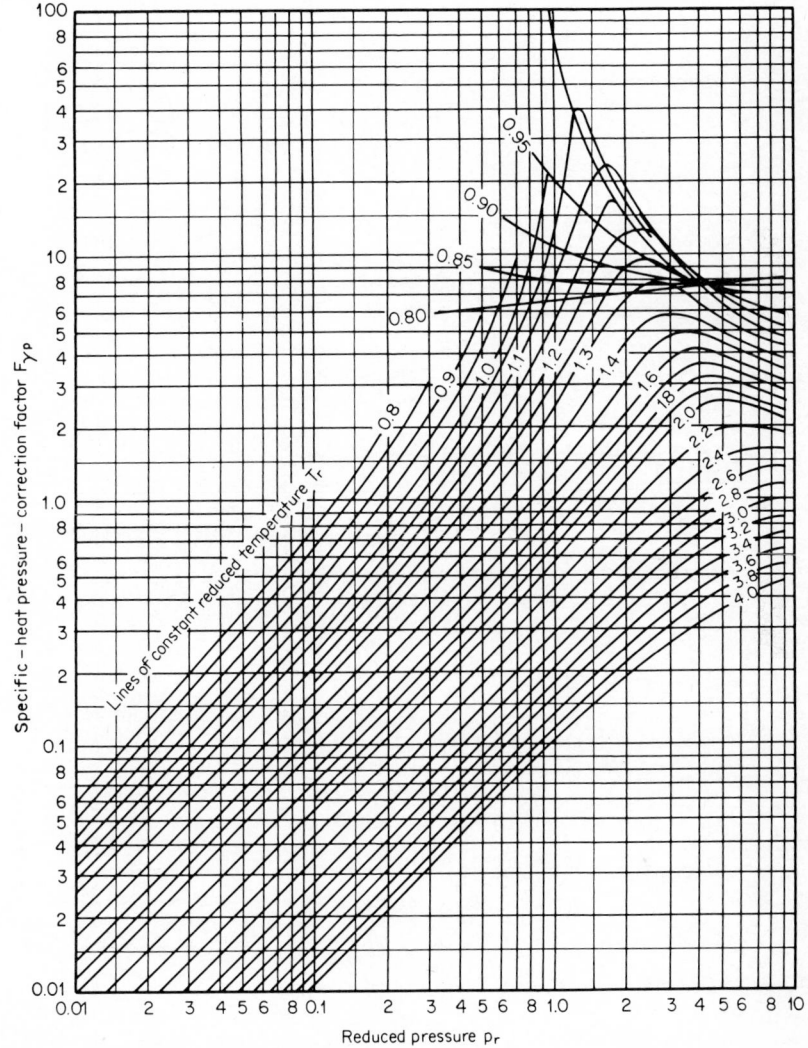

Figure I.1 Specific-heat pressure-correction factor $F_{\gamma p}$ for simple fluid, $\omega = 0$ *(from Edmister, 1974; all rights reserved; used with permission).*

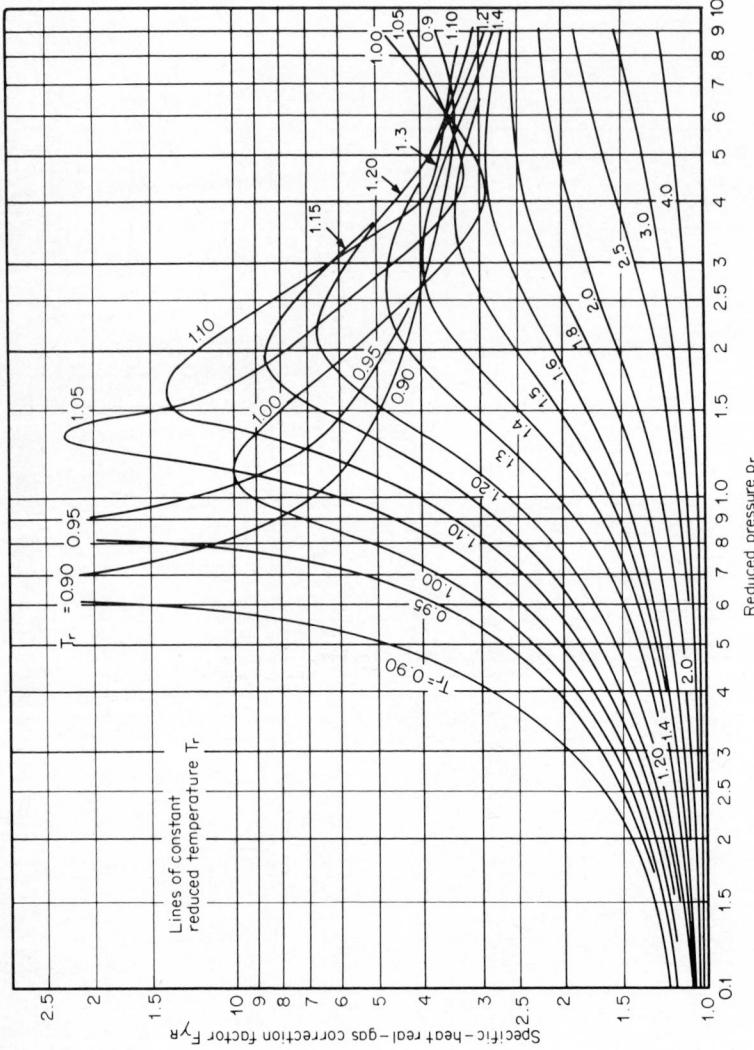

Figure I.2 Specific-heat real-gas correction factor, $F_{\gamma R}$; $\omega = 0$ *(from Edmister, 1974; all rights reserved; used with permission).*

I-2

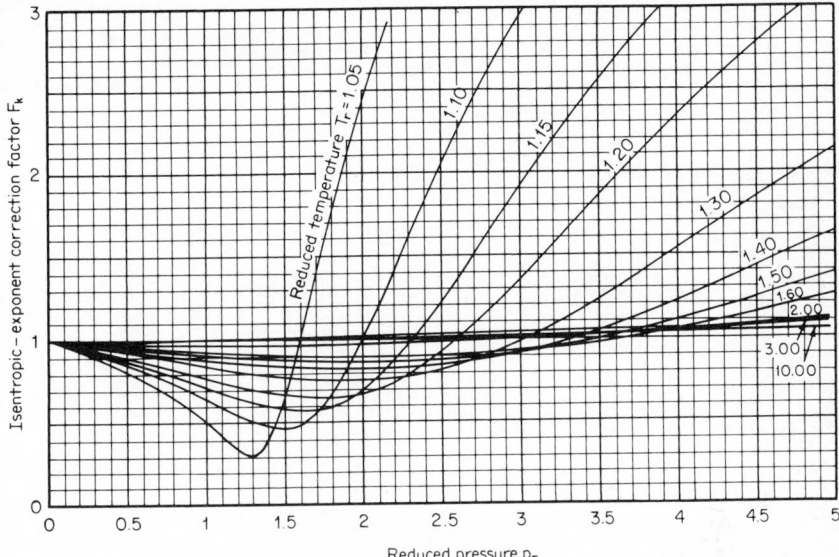

Figure I.3 Isentropic-exponent correction factor F_k.

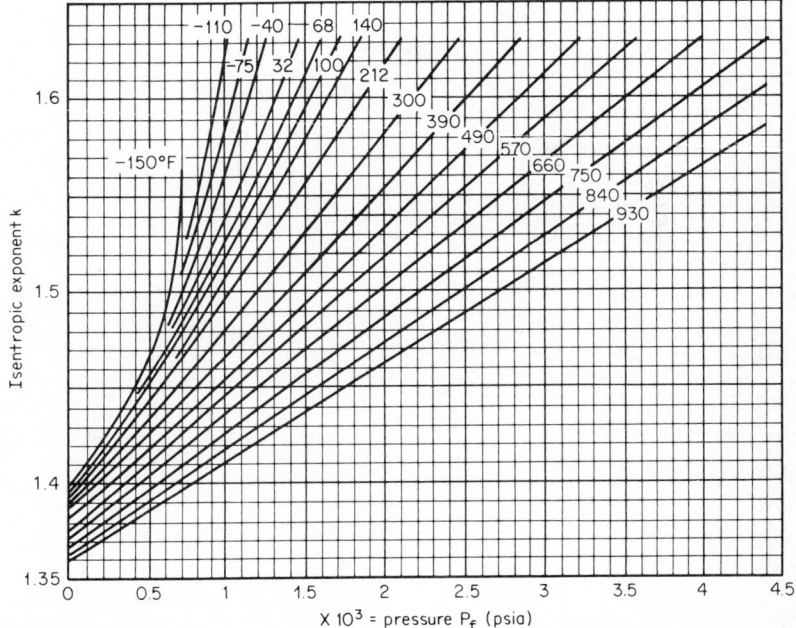

Figure I.4 Isentropic exponent k for air *(source: VDI, 1970).*

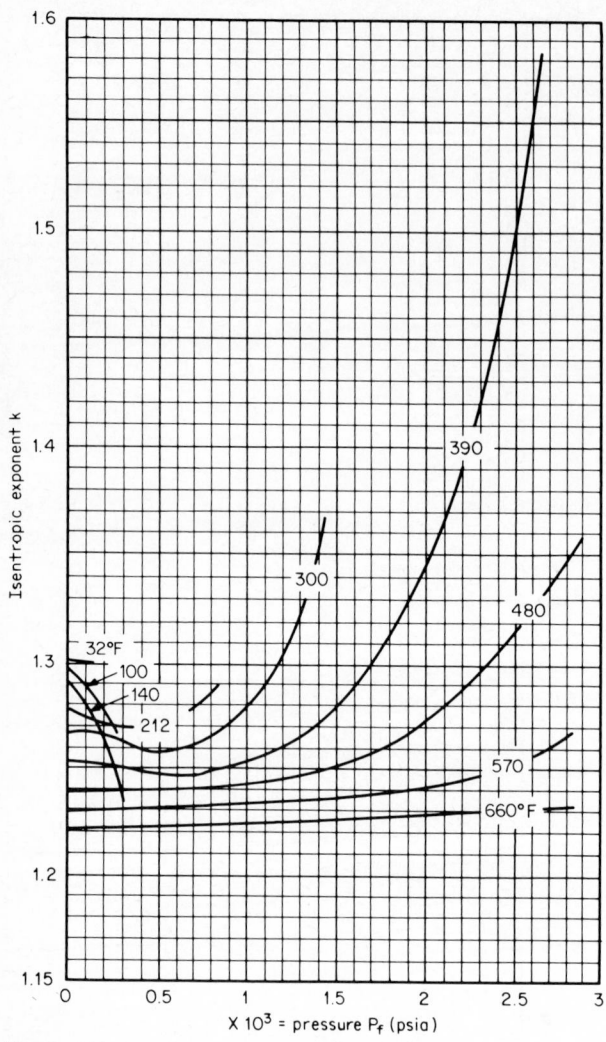

Figure I.5 Isentropic exponent *k* for ammonia *(source: VDI, 1970).*

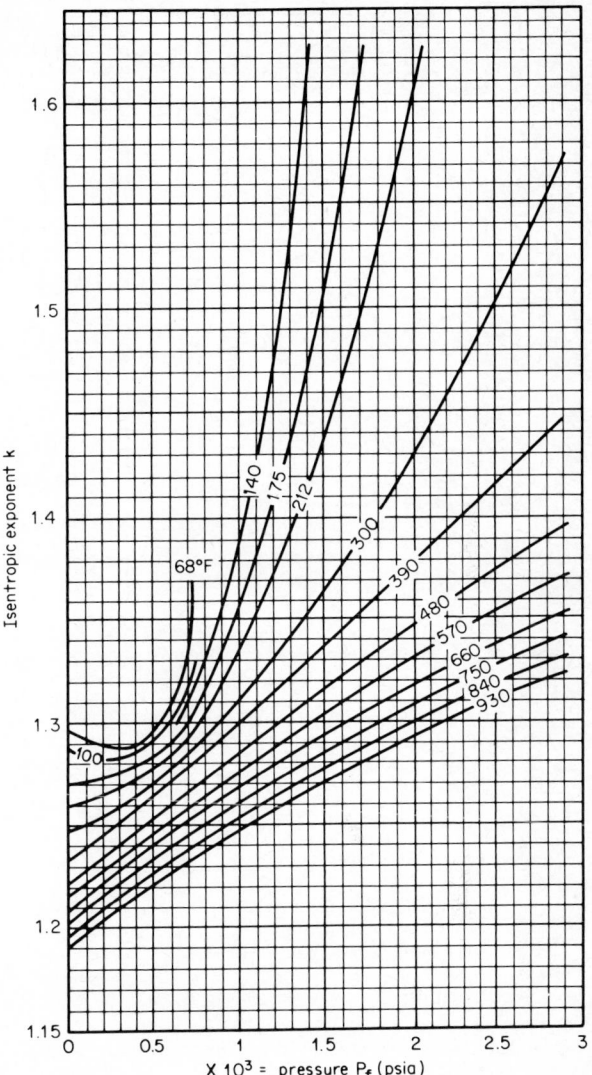

Figure I.6 Isentropic exponent *k* for carbon dioxide *(source: VDI, 1970).*

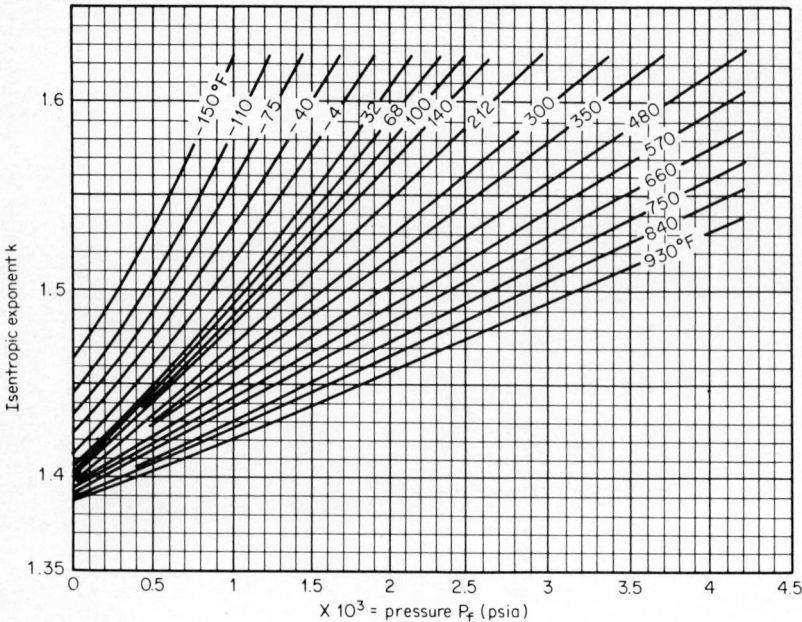

Figure I.7   **Isentropic exponent _k_ for hydrogen** *(source: VDI, 1970).*

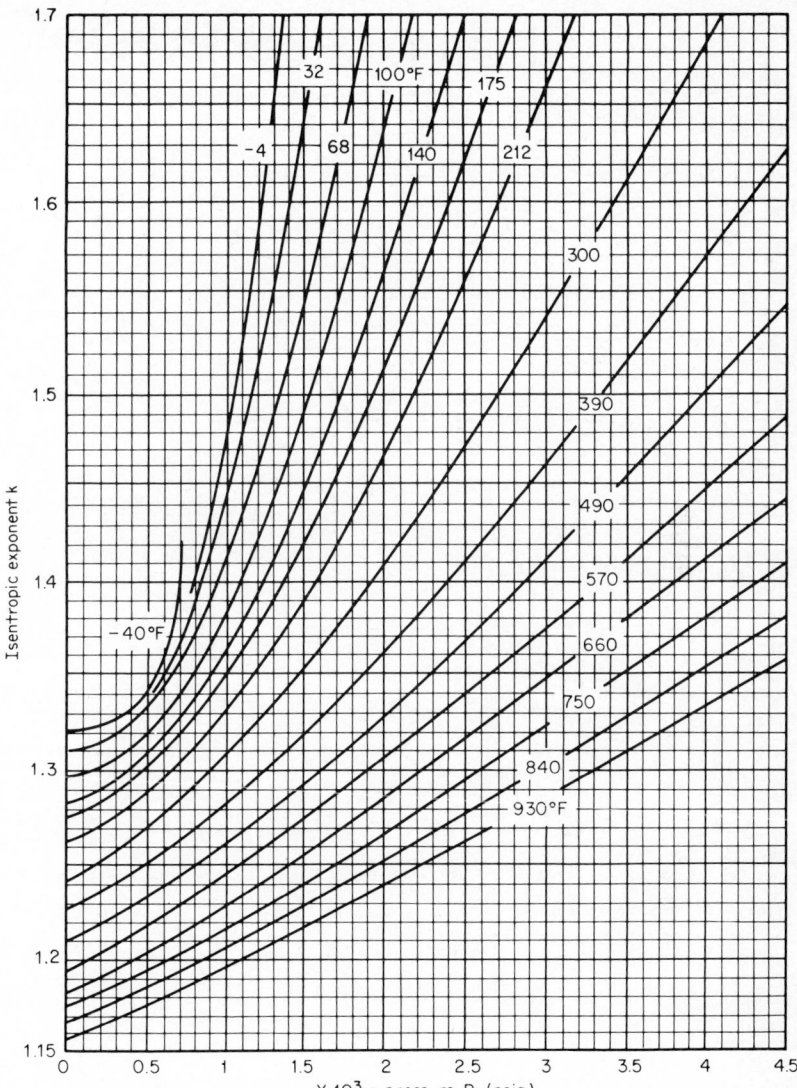

Figure I.8 Isentropic exponent *k* for methane *(source: VDI, 1970).*

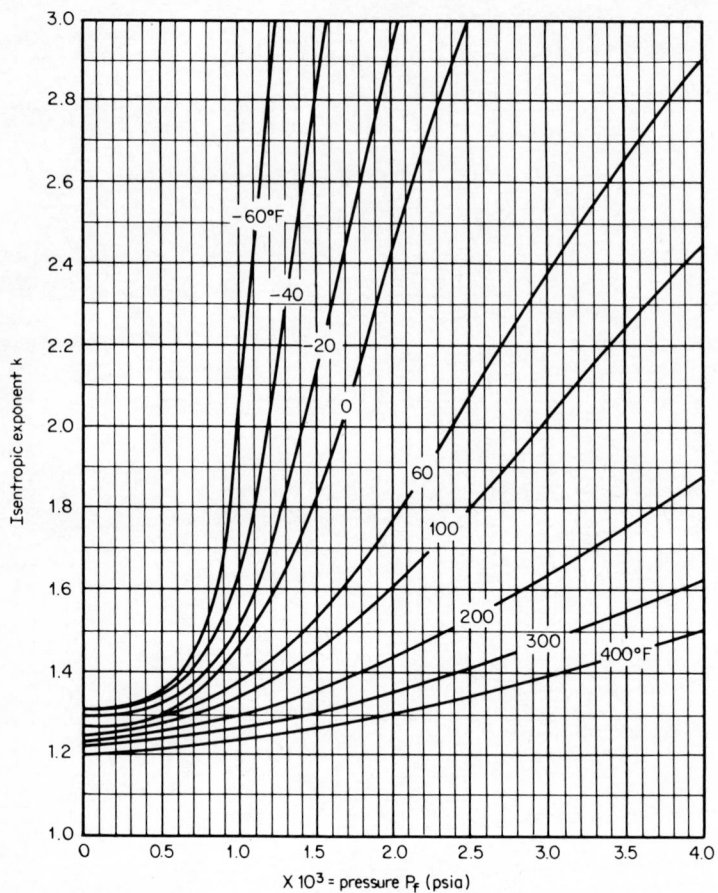

Figure I.9 Isentropic exponent for natural gas: specific gravity G = 0.6, P_{pc} = 672 psia, T_{pc} = 360°R *(from Schoenthaler, 1966; used with permission).*

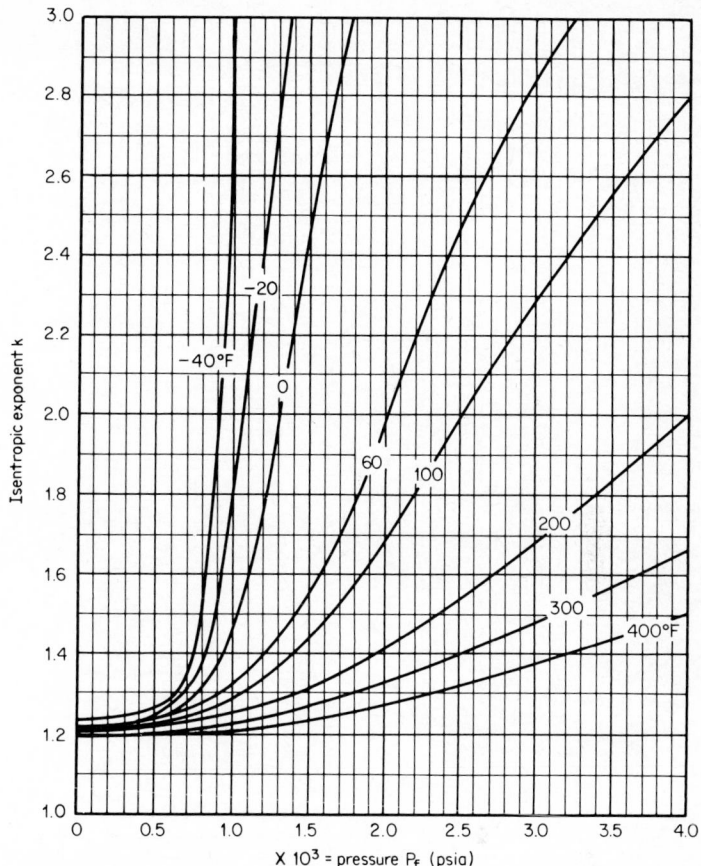

Figure I.10 **Isentropic exponent for natural gas: specific gravity** $G = 0.7$, P_{pc} = 667 psia, $T_{pc} = 382°R$ *(from Schoenthaler, 1966; used with permission).*

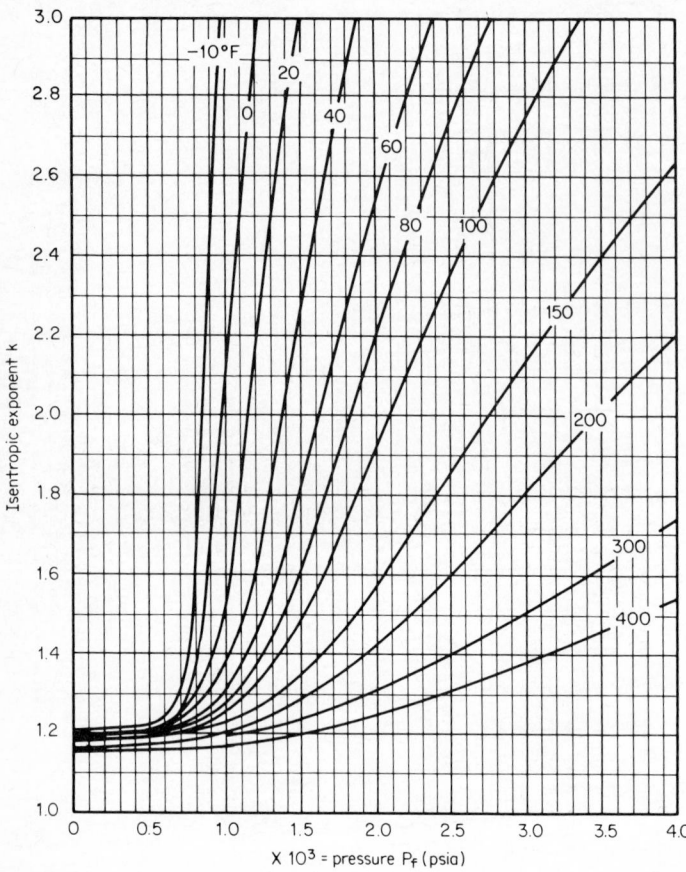

Figure I.11 Isentropic exponent for natural gas: specific gravity $G = 0.8$, P_{pc} = 662 psia, $T_{pc} = 424°R$ *(from Schoenthaler, 1966; used with permission).*

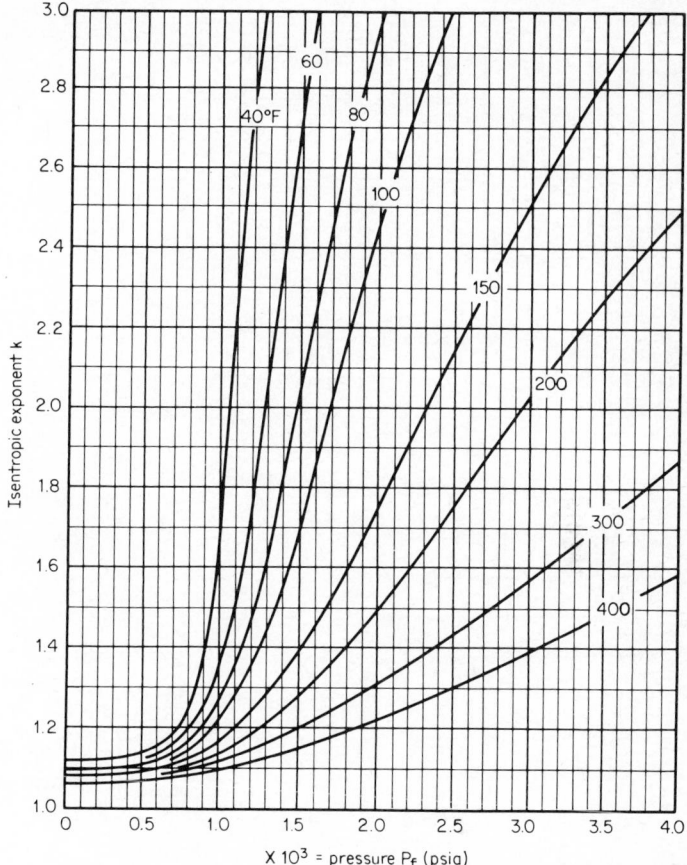

Figure I.12 **Isentropic exponent for natural gas: specific gravity** $G = 0.9$, P_{pc} = 657 **psia,** T_{pc} = 456°R   *(from Schoenthaler, 1966; used with permission).*

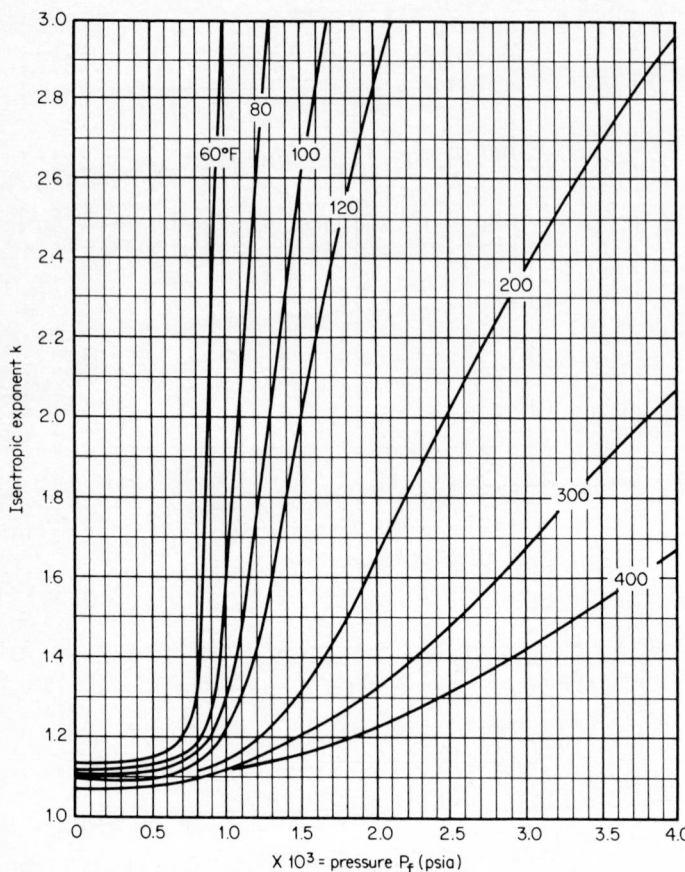

Figure I.13 Isentropic exponent for natural gas: specific gravity G = 1.0, P_{pc} = 652 psia, T_{pc} = 486°R *(from Schoenthaler, 1966; used with permission).*

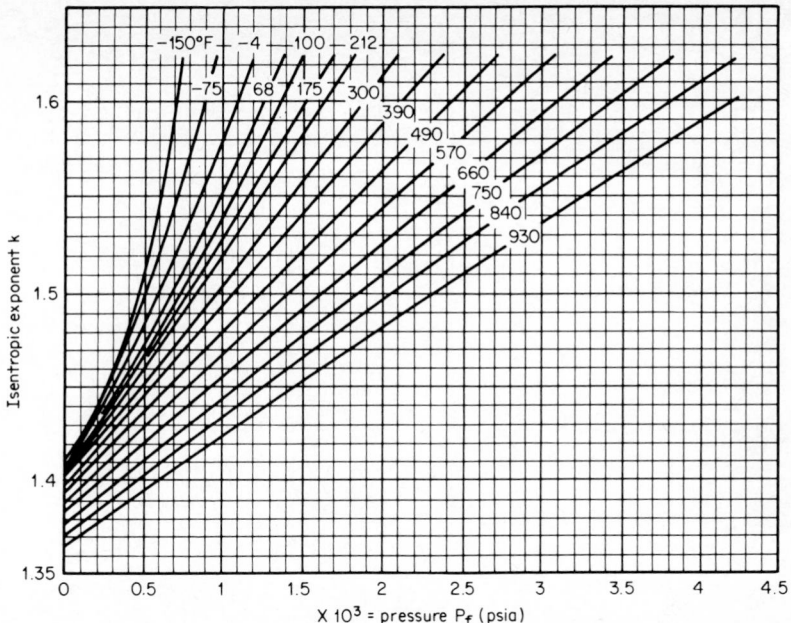

Figure I.14 Isentropic exponent *k* for nitrogen *(source: VDI, 1970)*.

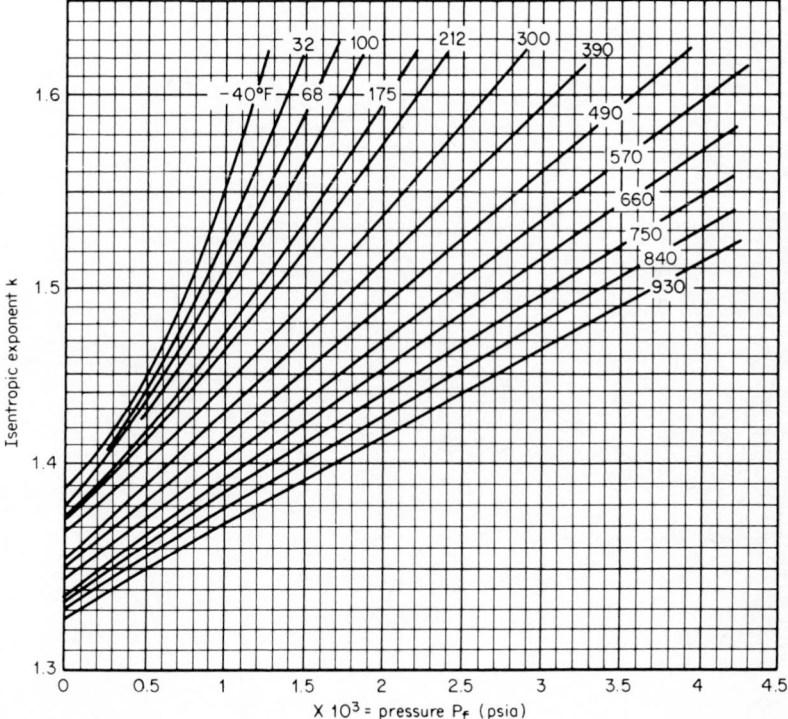

Figure I.15 Isentropic exponent *k* for oxygen *(source: VDI, 1970)*.

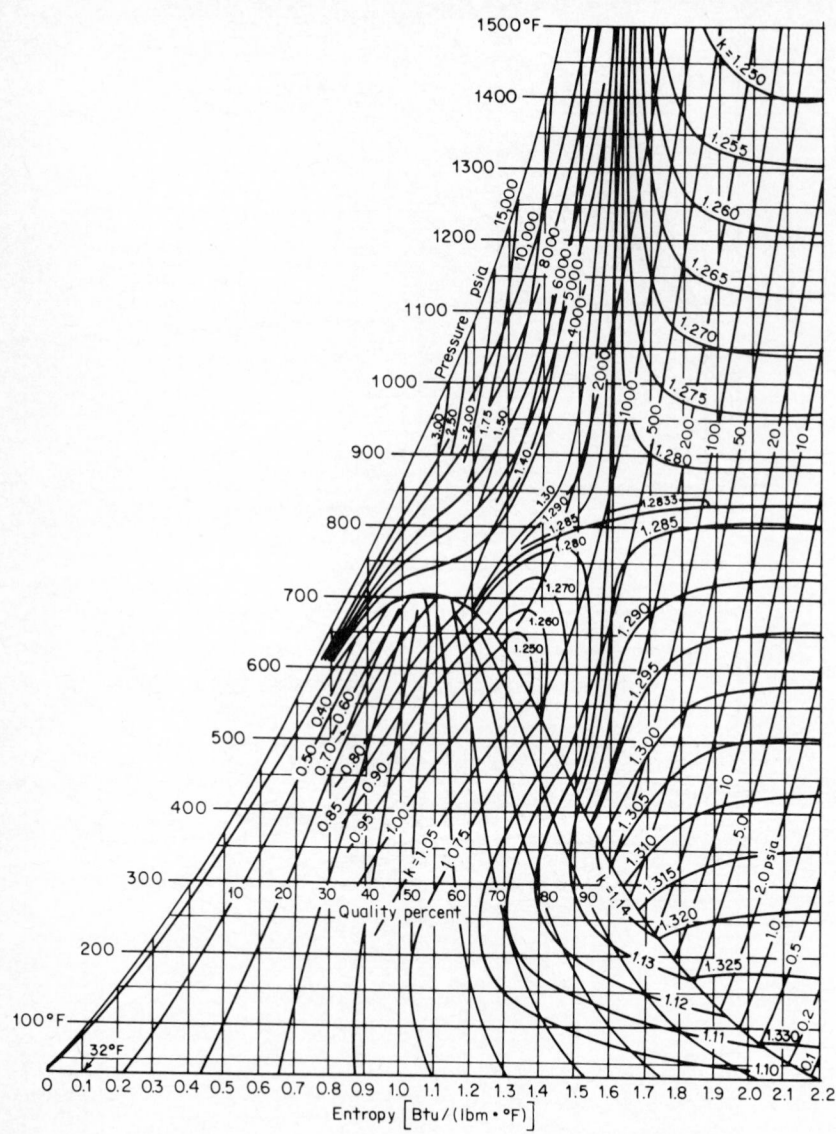

Fig. I-16 Isentropic exponent for steam *k*. (*From "ASME Steam Tables," ASME, New York, 1967.*)

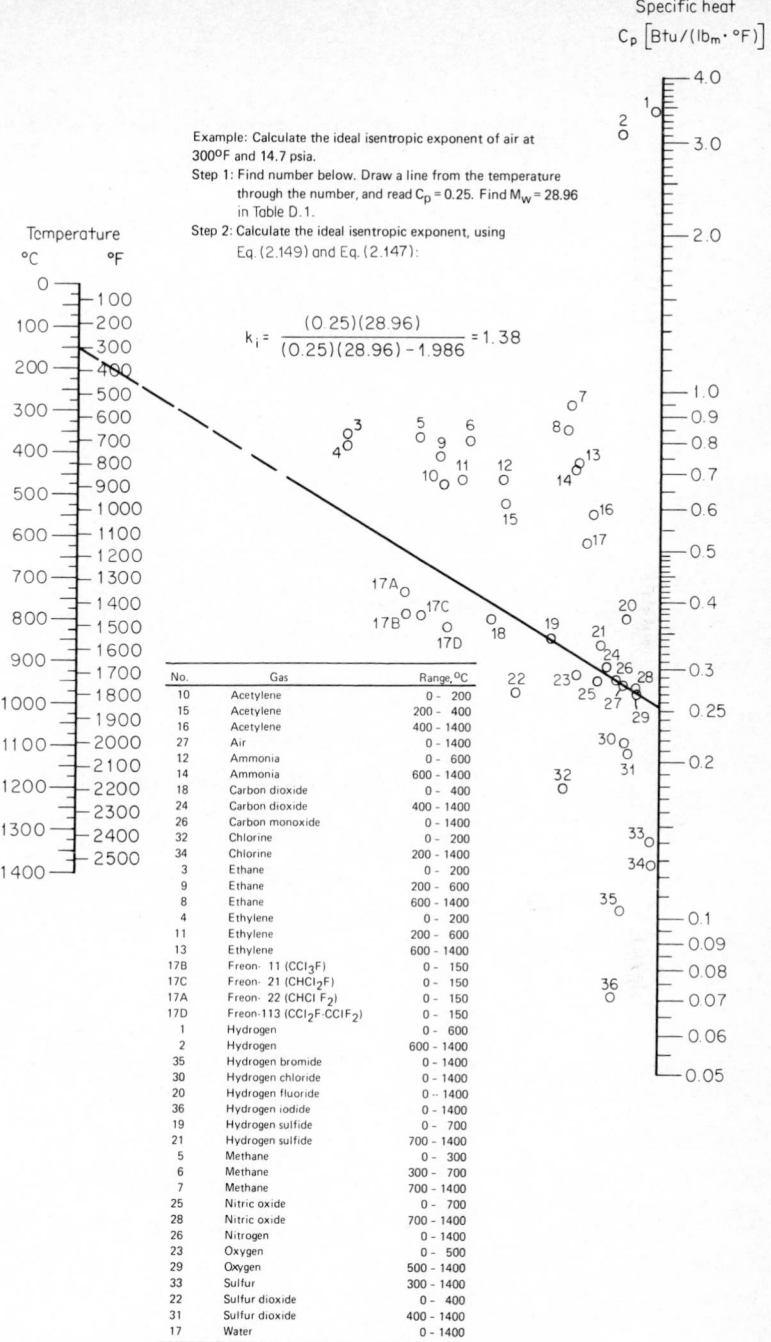

Figure I.17 Specific heats C_p of gases at 1 atm pressure *(from Perry and Chilton, 1973)*.

Example: Calculate the ideal isentropic exponent of air at 300°F and 14.7 psia.

Step 1: Find number below. Draw a line from the temperature through the number, and read $C_p = 0.25$. Find $M_w = 28.96$ in Table D.1.

Step 2: Calculate the ideal isentropic exponent, using Eq. (2.149) and Eq. (2.147):

$$k_i = \frac{(0.25)(28.96)}{(0.25)(28.96) - 1.986} = 1.38$$

Specific heat $C_p \left[\text{Btu} / (\text{lb}_m \cdot °F) \right]$

Temperature °C   °F

No.	Gas	Range, °C
10	Acetylene	0 - 200
15	Acetylene	200 - 400
16	Acetylene	400 - 1400
27	Air	0 - 1400
12	Ammonia	0 - 600
14	Ammonia	600 - 1400
18	Carbon dioxide	0 - 400
24	Carbon dioxide	400 - 1400
26	Carbon monoxide	0 - 1400
32	Chlorine	0 - 200
34	Chlorine	200 - 1400
3	Ethane	0 - 200
9	Ethane	200 - 600
8	Ethane	600 - 1400
4	Ethylene	0 - 200
11	Ethylene	200 - 600
13	Ethylene	600 - 1400
17B	Freon- 11 (CCl₃F)	0 - 150
17C	Freon- 21 (CHCl₂F)	0 - 150
17A	Freon- 22 (CHCl F₂)	0 - 150
17D	Freon-113 (CCl₂F·CClF₂)	0 - 150
1	Hydrogen	0 - 600
2	Hydrogen	600 - 1400
35	Hydrogen bromide	0 - 1400
30	Hydrogen chloride	0 - 1400
20	Hydrogen fluoride	0 -- 1400
36	Hydrogen iodide	0 - 1400
19	Hydrogen sulfide	0 - 700
21	Hydrogen sulfide	700 - 1400
5	Methane	0 - 300
6	Methane	300 - 700
7	Methane	700 - 1400
25	Nitric oxide	0 - 700
28	Nitric oxide	700 - 1400
26	Nitrogen	0 - 1400
23	Oxygen	0 - 500
29	Oxygen	500 - 1400
33	Sulfur	300 - 1400
22	Sulfur dioxide	0 - 400
31	Sulfur dioxide	400 - 1400
17	Water	0 - 1400

Table I.1 Values of Ideal-Gas State Heat Capacity (C_p), [Btu/(lb$_m$·mol·°R)] for Paraffins and Olefins

Temperature, °F	Methane, CH_4	Acetylene, C_2H_2	Ethylene, C_2H_4	Ethane, C_2H_6	Propene, C_3H_6	Propane, C_3H_8	Iso-butene, iC_4H_8	l-Butene, lC_4H_8	Iso-butane, iC_4H_{10}	n-Butane, nC_4H_{10}	Iso-pentane, iC_5H_{12}	n-Pentane, nC_5H_{12}	n-Hexane, nC_6H_{14}	n-Heptane, nC_7H_{16}	°F
−100	8.00		8.37	10.2	12.0	13.3	15.8	15.2	17.2	18.3	19.9	20.6	25.1	29.2	−100
−90	8.02		8.42	10.3	12.2	13.6	16.1	15.4	17.5	18.6	20.4	21.1	25.6	29.8	−90
−80	8.04		8.48	10.4	12.4	13.8	16.4	15.6	17.8	18.9	20.9	21.5	26.1	30.4	−80
−70	8.06		8.55	10.5	12.5	14.0	16.7	15.9	18.1	19.2	21.4	22.0	26.6	30.9	−70
−60	8.08		8.54	10.6	12.7	14.2	17.1	16.1	18.4	19.5	21.9	22.5	27.1	31.5	−60
−50	8.09		8.72	10.8	12.9	14.5	17.4	16.4	18.7	19.8	22.3	22.9	27.7	32.1	−50
−40	8.12		8.82	10.9	13.1	14.7	17.7	16.7	19.0	20.1	22.8	23.4	28.2	32.7	−40
−30	8.15		8.93	11.1	13.2	14.9	18.0	17.0	19.4	20.4	23.3	23.8	28.7	33.3	−30
−20	8.17		9.04	11.2	13.4	15.2	18.3	17.3	19.7	20.7	23.8	24.3	29.2	33.9	−20
−10	8.20		9.17	11.3	13.6	15.4	18.6	17.6	20.1	21.0	24.3	24.8	29.7	34.5	−10
0	8.23	9.68	9.32	11.4	13.7	15.6	18.9	18.0	20.4	21.3	24.8	25.2	30.2	35.1	0
10	8.26	9.79	9.46	11.6	13.8	15.9	19.2	18.3	20.7	21.5	25.2	25.7	30.7	35.7	10
20	8.29	9.90	9.60	11.7	14.1	16.1	19.5	18.6	21.1	21.8	25.7	26.2	31.3	36.3	20
30	8.33	10.0	9.74	11.9	14.3	16.4	19.9	19.0	21.5	22.1	26.2	26.6	31.8	36.9	30
40	8.37	10.1	9.88	12.0	14.5	16.6	20.2	19.3	21.8	22.4	26.6	27.0	32.3	37.5	40
50	8.41	10.2	10.0	12.2	14.7	16.9	20.5	19.6	22.2	22.6	27.1	27.5	32.8	38.1	50
60	8.45	10.3	10.2	12.2	14.9	17.1	20.8	19.9	22.5	22.9	27.6	27.9	33.3	38.7	60
70	8.49	10.4	10.3	12.5	15.1	17.4	21.1	20.2	22.9	23.2	28.0	28.4	33.8	39.3	70
80	8.54	10.5	10.4	12.6	15.3	17.6	21.4	20.5	23.2	23.4	28.5	28.8	34.4	39.8	80
90	8.59	10.6	10.6	12.8	15.5	17.9	21.6	20.9	23.6	23.7	28.9	29.3	34.9	40.4	90
100	8.65	10.7	10.7	13.0	15.8	18.2	22.0	21.2	24.0	24.1	29.4	29.7	35.4	41.1	100
110	8.70	10.8	10.9	13.1	16.0	18.4	22.2	21.5	24.3	24.4	29.8	30.1	35.9	41.7	110
120	8.76	10.9	11.0	13.3	16.2	18.7	22.5	21.8	24.7	24.8	30.3	30.6	35.4	42.2	120
130	8.82	11.0	11.1	13.4	16.4	19.0	22.8	22.1	25.0	25.1	30.8	31.0	37.0	42.8	130
140	8.88	11.1	11.3	13.6	16.6	19.2	23.1	22.4	25.4	25.5	31.2	31.5	37.5	43.4	140

150	44.0	38.0	31.9	31.7	25.8	25.8	22.7	23.4	19.5	16.8	13.8	11.4	11.1	8.95	150
160	44.6	38.5	32.3	32.1	26.1	26.1	23.0	23.7	19.8	17.0	13.9	11.5	11.2	9.01	160
170	45.2	39.0	32.7	32.6	26.5	26.5	23.3	24.0	20.1	17.2	14.1	11.7	11.3	9.08	170
180	45.8	39.5	33.2	33.0	26.8	26.9	23.6	24.3	20.4	17.4	14.3	11.8	11.4	9.14	180
190	46.4	40.0	33.6	33.4	27.2	27.2	24.0	24.6	20.6	17.6	14.5	11.9	11.5	9.21	190
200	46.9	40.5	34.0	33.9	27.5	27.6	24.2	24.9	20.9	17.8	14.6	12.1	11.5	9.28	200
210	47.5	41.0	34.5	34.3	27.9	27.9	24.5	25.2	21.2	18.1	14.8	12.2	11.6	9.35	210
220	48.1	41.5	34.9	34.7	28.2	28.3	24.8	25.4	21.4	18.3	15.0	12.4	11.7	9.42	220
230	48.7	42.0	35.3	35.2	28.5	28.7	25.1	25.7	21.7	18.5	15.1	12.5	11.8	9.49	230
240	49.2	42.5	35.7	35.6	28.9	29.0	25.4	26.0	22.0	18.7	15.3	12.6	11.8	9.56	240
250	49.5	43.0	36.1	36.0	29.2	29.4	25.7	26.3	22.2	18.9	15.5	12.8	11.9	9.64	250
260	50.4	43.4	36.5	36.4	29.6	29.7	26.0	26.5	22.5	19.1	15.7	12.9	12.0	9.71	260
270	50.9	43.9	37.0	36.9	29.9	30.1	26.3	26.8	22.8	19.3	15.8	13.0	12.0	9.78	270
280	51.5	44.4	37.4	37.3	30.2	30.4	26.6	27.1	23.0	19.5	16.0	13.1	12.1	9.85	280
290	52.0	44.9	37.7	37.7	30.6	30.8	26.9	27.4	23.3	19.7	16.2	13.3	12.2	9.92	290
300	52.6	45.4	38.2	38.1	30.9	31.1	27.1	27.6	23.6	19.9	16.3	13.4	12.2	10.0	300
310	53.2	45.8	38.6	38.5	31.2	31.5	27.4	27.9	23.8	20.1	16.5	13.5	12.3	10.1	310
320	53.7	46.3	39.0	38.9	31.6	31.8	27.7	28.1	24.1	20.3	16.7	13.7	12.3	10.2	320
330	54.2	46.8	39.4	39.4	31.9	32.1	28.0	28.4	24.3	20.5	16.8	13.8	12.4	10.2	330
340	54.8	47.3	39.8	39.8	32.2	32.5	28.3	28.7	24.6	20.7	17.0	13.9	12.4	10.3	340
350	55.3	47.7	40.2	40.2	32.5	32.8	29.5	29.0	24.8	20.9	17.2	14.0	12.5	10.4	350
360	55.9	48.2	40.5	40.5	32.8	33.1	28.8	29.2	25.1	21.1	17.3	14.2	12.6	10.5	360
370	56.4	48.7	40.9	40.9	33.2	33.4	29.1	29.4	25.3	21.3	17.5	14.3	12.6	10.6	370
380	56.9	49.1	41.3	41.3	33.5	33.7	29.4	29.7	25.6	21.5	17.7	14.4	12.7	10.6	380
390	57.5	49.6	41.7	41.7	33.8	34.0	29.6	29.9	25.8	21.7	17.8	14.5	12.7	10.7	390
400	58.0	50.0	42.1	42.1	34.1	34.4	29.9	30.2	26.0	21.8	18.0	14.7	12.8	10.8	400
410	58.5	50.5	42.4	42.5	34.4	34.7	30.1	30.4	26.3	22.0	18.2	14.8	12.8	10.9	410
420	59.0	50.9	42.8	42.9	34.7	35.0	30.4	30.7	26.5	22.2	18.3	14.9	12.9	11.0	420
430	59.5	51.3	43.2	43.3	35.0	35.3	30.6	30.9	26.8	22.4	18.5	15.0	12.9	11.0	430
440	60.0	51.8	43.6	43.7	35.3	35.6	30.9	31.2	27.0	22.6	18.6	15.2	13.0	11.1	440

Table I.1 Values of Ideal-Gas State Heat Capacity $(C_p)_i$ [Btu/(lb$_m$·mol·°R)] for Paraffins and Olefins (Continued)

Temperature, °F	Methane, CH$_4$	Acetylene, C$_2$H$_2$	Ethylene, C$_2$H$_4$	Ethane, C$_2$H$_6$	Propene, C$_3$H$_6$	Propane, C$_3$H$_8$	Iso-butane, iC$_4$H$_8$	I-Butene, iC$_4$H$_8$	Iso-butane, iC$_4$H$_{10}$	n-Butane, nC$_4$H$_{10}$	Iso-pentane, iC$_5$H$_{12}$	n-Pentane, nC$_5$H$_{12}$	n-Hexane, nC$_6$H$_{14}$	n-Heptane, nC$_7$H$_{16}$	°F
450	11.2	13.0	15.3	18.8	22.8	27.2	31.4	31.2	35.9	35.6	44.0	43.9	52.2	60.5	450
460	11.3	13.1	15.4	19.0	23.0	27.5	31.7	31.4	36.2	35.9	44.4	44.3	52.7	61.0	460
470	11.4	13.1	15.5	19.1	23.1	27.7	31.9	31.6	36.5	36.2	44.7	44.6	53.1	61.5	470
480	11.4	13.1	15.6	19.3	23.3	27.9	32.2	31.9	36.8	36.5	45.1	45.0	53.5	61.9	480
490	11.5	13.2	15.7	19.4	23.5	28.1	32.4	32.2	37.1	36.8	45.5	45.3	53.9	62.4	490
500	11.6	13.2	15.8	19.6	23.7	28.4	32.6	32.4	37.3	37.1	45.8	45.7	54.3	62.9	500
510	11.7	13.3	16.0	19.7	23.8	28.6	32.9	32.6	37.6	37.4	46.2	46.0	54.7	63.4	510
520	11.8	13.3	16.0	19.9	24.0	28.8	33.1	32.8	37.9	37.6	46.5	46.3	55.1	63.8	520
530	11.8	13.4	16.2	20.0	24.2	29.0	33.3	33.1	38.2	37.9	46.8	46.7	55.5	64.3	530
540	11.9	13.4	16.3	20.2	24.4	29.2	33.5	33.3	38.5	38.2	47.2	47.0	55.9	64.7	540
550	12.0	13.4	16.4	20.3	24.5	29.4	33.8	33.5	38.8	38.5	47.5	47.4	56.3	65.2	550
560	12.1	13.5	16.5	20.5	24.7	29.6	34.0	33.8	39.0	38.7	47.9	47.7	56.7	65.6	560
570	12.2	13.5	16.6	20.6	24.9	29.8	34.2	34.0	39.3	39.0	48.2	48.0	57.1	66.1	570
580	12.2	13.6	16.7	20.8	25.0	30.0	34.4	34.2	39.5	39.3	48.5	48.3	57.5	66.5	580
590	12.3	13.6	16.8	20.9	25.2	30.3	34.6	34.4	39.8	39.5	48.9	48.7	57.9	67.0	590
600	12.4	13.6	16.9	21.0	25.4	30.5	34.9	34.7	40.1	39.8	49.2	49.0	58.2	67.4	600
610	12.5	13.7	17.0	21.2	25.5	30.6	35.1	34.9	40.3	40.0	49.5	49.3	58.6	67.8	610
620	12.5	13.7	17.1	21.3	25.7	30.8	35.3	35.1	40.6	40.3	49.9	49.6	59.0	68.3	620
630	12.6	13.8	17.2	21.5	25.9	31.0	35.5	35.3	40.8	40.5	50.2	49.9	59.3	68.7	630
640	12.7	13.8	17.3	21.6	26.0	31.2	35.7	35.5	41.1	40.7	50.5	50.2	59.7	69.1	640
650	12.8	13.8	17.4	21.7	26.2	31.4	35.9	35.7	41.3	41.0	50.8	50.5	60.0	69.5	650
660	12.8	13.9	17.5	21.9	26.3	31.6	36.1	35.9	41.6	41.2	51.1	50.8	60.4	69.9	660
670	12.9	13.9	17.6	22.0	26.5	31.8	36.3	36.2	41.8	41.5	51.4	51.1	60.8	70.3	670
680	13.0	13.9	17.7	22.2	26.6	32.0	36.5	36.4	42.1	41.7	51.7	51.4	61.1	70.7	680
690	13.1	14.0	17.8	22.3	26.8	32.2	36.7	36.5	42.3	42.0	52.0	51.7	61.5	71.1	690

700	13.1	14.0	17.9	22.4	26.9	32.4	36.9	36.8	42.6	42.2	52.3	52.0	61.8	71.5	700
710	13.2	14.0	18.0	22.6	27.1	32.6	37.1	37.0	42.8	42.4	52.6	52.3	62.1	71.9	710
720	13.3	14.1	18.0	22.7	27.2	32.8	37.3	37.2	43.0	42.7	52.9	52.6	62.5	72.3	720
730	13.4	14.1	18.1	22.8	27.4	32.9	37.5	37.4	43.2	42.9	53.2	52.9	62.8	72.7	730
740	13.4	14.2	18.2	23.0	27.5	33.1	37.7	37.5	43.5	43.2	53.5	53.2	63.1	73.1	740
750	13.5	14.2	18.3	23.1	27.6	33.3	37.8	37.7	43.7	43.4	53.8	53.4	63.5	73.5	750
760	13.6	14.2	18.4	23.2	27.8	33.5	38.0	37.9	43.9	43.6	54.0	53.7	63.8	73.9	760
770	13.7	14.3	18.5	23.3	27.9	33.6	38.2	38.1	44.2	43.8	54.3	54.0	64.1	74.2	770
780	13.7	14.3	18.6	23.5	28.1	33.8	38.4	38.3	44.4	44.0	54.6	54.3	64.4	74.6	780
790	13.8	14.3	18.7	23.6	28.2	34.0	38.6	38.5	44.6	44.3	54.9	54.5	64.7	75.0	790
800	13.9	14.4	18.7	23.7	28.4	34.2	38.8	38.7	44.8	44.5	55.2	54.8	65.1	75.3	800
810	13.9	14.4	18.8	23.8	28.5	34.3	39.0	38.9	45.0	44.7	55.4	55.1	65.4	75.7	810
820	14.0	14.4	18.9	23.9	28.6	34.5	39.1	39.1	45.2	45.0	55.7	55.3	65.7	76.0	820
830	14.1	14.4	19.0	24.1	28.8	34.7	39.3	39.2	45.5	45.2	55.9	55.6	66.0	76.4	830
840	14.1	14.5	19.1	24.2	28.9	34.8	39.5	39.4	45.7	45.4	56.2	55.8	66.3	76.7	840
850	14.2	14.5	19.2	24.3	29.0	35.0	39.7	39.6	45.9	45.6	56.4	56.1	66.6	77.1	850
860	14.3	14.6	19.2	24.4	29.2	35.2	39.8	39.8	46.1	45.8	56.7	56.3	66.9	77.4	860
870	14.4	14.6	19.3	24.6	29.3	35.3	40.0	39.9	46.3	46.0	57.0	56.6	67.2	77.8	870
880	14.4	14.6	19.4	24.7	29.4	35.5	40.2	40.1	46.5	46.2	57.2	56.9	67.5	78.1	880
890	14.5	14.6	19.5	24.8	29.5	36.7	40.4	40.3	46.7	46.4	57.5	57.1	67.8	78.5	890

SOURCE: Edmister (1961). Copyright 1961 by Gulf Publishing Company, Houston, Texas; all rights reserved. Used with permission.

Table I.2 Values of Ideal-Gas State Heat Capacity $(C_p)_i$ [Btu/(lb$_m$·mol·°R)] for Miscellaneous Compounds

Temperature, °F	Hydrogen, H_2	Nitrogen, N_2	Oxygen, O_2	Carbon monoxide, CO	Carbon dioxide, CO_2	Water, H_2O	Hydrogen sulfide, H_2S	Sulfur dioxide, SO_2	Sulfur trioxide, SO_3	°F
−200	6.35	6.95	6.95	6.95	7.20		7.88	8.10	9.00	−200
−190	6.34	6.95	6.95	6.95	7.25		7.88	8.14	9.11	−190
−180	6.35	6.95	6.95	6.95	7.28		7.90	8.20	9.22	−180
−170	6.36	6.95	6.95	6.95	7.32		7.90	8.24	9.33	−170
−160	6.39	6.95	6.95	6.95	7.37		7.91	8.30	9.45	−160
−150	6.42	6.95	6.95	6.95	7.42		7.92	8.35	9.56	−150
−140	6.44	6.95	6.95	6.95	7.47		7.93	8.40	9.67	−140
−130	6.47	6.95	6.95	6.95	7.53		7.94	8.45	9.77	−130
−120	6.50	6.95	6.96	6.95	7.59		7.95	8.50	9.90	−120
−110	6.53	6.95	6.96	6.95	7.65		7.96	8.56	10.0	−110
−100	6.56	6.95	6.96	6.95	7.71		7.97	8.61	10.1	−100
−90	6.58	6.95	6.96	6.95	7.78		7.98	8.56	10.2	−90
−80	6.61	6.95	6.96	6.95	7.84		7.99	8.71	10.3	−80
−70	6.63	6.95	6.95	6.95	7.90		8.00	8.76	10.4	−70
−60	6.66	6.95	6.96	6.95	7.97		8.02	8.81	10.6	−60
−50	6.68	6.95	6.96	6.95	8.03		8.03	8.87	10.7	−50
−40	6.70	6.95	6.97	6.95	8.11		8.04	8.91	10.8	−40
−30	6.72	6.95	6.97	6.95	8.18		8.05	8.97	10.9	−30
−20	6.75	6.95	6.97	6.95	8.25		8.06	9.01	11.0	−20
−10	6.77	6.95	6.97	6.95	8.33		8.08	9.06	11.1	−10
0	6.78	6.95	6.98	6.95	8.38	7.98	8.09	9.12	11.2	0
10	6.80	6.95	6.98	6.95	8.45	7.99	8.10	9.18	11.3	10
20	6.82	6.95	6.98	6.95	8.51	7.99	8.12	9.23	11.4	20
30	6.83	6.95	6.99	6.96	8.58	8.00	8.13	9.28	11.6	30
40	6.85	6.95	6.99	6.96	8.65	8.00	8.15	9.33	11.7	40

50	11.8	9.38	8.16	8.00	8.70	6.96	7.00	6.95	6.86	50
60	11.9	9.43	8.17	8.01	8.76	6.96	7.00	6.95	6.87	60
70	12.0	9.48	8.19	8.02	8.83	6.96	7.01	6.95	6.88	70
80	12.1	9.53	8.20	8.02	8.89	6.96	7.02	6.96	6.89	80
90	12.2	9.59	8.22	8.03	8.95	6.96	7.02	6.96	6.90	90
100	12.3	9.64	8.24	8.03	9.01	6.96	7.03	6.96	6.90	100
110	12.4	9.69	8.25	8.04	9.06	6.96	7.03	6.96	6.91	110
120	12.6	9.74	8.27	8.05	9.12	6.97	7.04	6.96	6.92	120
130	12.7	9.79	8.28	8.06	9.19	6.97	7.05	6.96	6.92	130
140	12.8	9.84	8.30	8.06	9.24	6.97	7.06	6.96	6.93	140
150	12.9	9.89	8.32	8.07	9.30	6.97	7.07	6.96	6.93	150
160	13.0	9.94	8.34	8.08	9.35	6.97	7.08	6.96	6.94	160
170	13.1	9.98	8.36	8.09	9.40	6.98	7.09	6.96	6.94	170
180	13.2	10.0	8.38	8.10	9.46	6.98	7.10	6.96	6.94	180
190	13.3	10.1	8.39	8.11	9.51	6.98	7.11	6.96	6.95	190
200	13.4	10.1	8.41	8.12	9.57	6.98	7.12	6.97	6.95	200
210	13.5	10.2	8.43	8.13	9.62	6.99	7.13	6.97	6.96	210
220	13.6	10.2	8.45	8.14	9.66	6.99	7.14	6.97	6.96	220
230	13.7	10.2	8.47	8.15	9.72	7.00	7.15	6.98	6.96	230
240	13.8	10.3	8.49	8.16	9.77	7.00	7.16	6.98	6.97	240
250	13.9	10.4	8.51	8.17	9.82	7.00	7.18	6.98	6.97	250
260	14.0	10.4	8.53	8.18	9.86	7.01	7.19	6.99	6.97	260
270	14.1	10.4	8.55	8.19	9.91	7.01	7.20	6.99	6.97	270
280	14.2	10.5	8.57	8.20	9.96	7.02	7.21	6.99	6.97	280
290	14.3	10.5	8.59	8.22	10.0	7.02	7.22	6.99	6.98	290

Table I.2 Values of Ideal-Gas State Heat Capacity $(C_p)_i$ [Btu/(lb$_m$·mol·°R)] for Miscellaneous Compounds (*Continued*)

Temperature, °F	Hydrogen, H_2	Nitrogen, N_2	Oxygen, O_2	Carbon monoxide, CO	Carbon dioxide, CO_2	Water, H_2O	Hydrogen sulfide, H_2S	Sulfur dioxide, SO_2	Sulfur trioxide, SO_3	°F
300	6.98	7.00	7.24	7.03	10.1	8.23	8.62	10.6	14.4	300
310	6.98	7.00	7.25	7.03	10.1	8.24	8.64	10.6	14.5	310
320	6.98	7.01	7.26	7.04	10.2	8.25	8.66	10.6	14.6	320
330	6.98	7.01	7.28	7.04	10.2	8.26	8.68	10.7	14.7	330
340	6.98	7.02	7.29	7.05	10.2	8.28	8.70	10.7	14.8	340
350	6.98	7.02	7.30	7.05	10.3	8.29	8.72	10.8	14.9	350
360	6.98	7.02	7.32	7.05	10.3	8.30	8.75	10.8	15.0	360
370	6.98	7.02	7.33	7.07	10.4	8.31	8.77	10.8	15.1	370
380	6.98	7.03	7.34	7.07	10.4	8.33	8.79	10.9	15.2	380
390	6.99	7.04	7.36	7.08	10.4	8.34	8.82	10.9	15.2	390
400	6.99	7.05	7.37	7.09	10.5	8.35	8.84	11.0	15.3	400
410	6.99	7.05	7.38	7.09	10.5	8.37	8.87	11.0	15.4	410
420	6.99	7.05	7.40	7.10	10.6	8.38	8.89	11.0	15.5	420
430	6.99	7.06	7.41	7.11	10.6	8.40	8.91	11.1	15.6	430
440	6.99	7.07	7.42	7.12	10.6	8.41	8.93	11.1	15.7	440
450	6.99	7.07	7.44	7.12	10.7	8.42	8.96	11.2	15.7	450
460	6.99	7.08	7.45	7.13	10.7	8.44	8.98	11.2	15.8	460
470	6.99	7.08	7.46	7.14	10.8	8.45	9.00	11.2	15.9	470
480	6.99	7.09	7.48	7.15	10.8	8.46	9.02	11.3	16.0	480
490	6.99	7.10	7.49	7.15	10.8	8.48	9.05	11.3	16.0	490

500	6.99	7.10	7.50	7.16	10.9	8.49	9.07	11.3	16.1	500
510	7.00	7.11	7.52	7.17	10.9	8.51	9.09	11.4	16.2	510
520	7.00	7.11	7.53	7.18	11.0	8.52	9.12	11.4	16.2	520
530	7.00	7.12	7.54	7.19	11.0	8.54	9.14	11.4	16.3	530
540	7.00	7.13	7.56	7.20	11.0	8.55	9.16	11.5	16.4	540
550	7.00	7.14	7.57	7.20	11.1	8.57	9.19	11.5	16.4	550
560	7.00	7.15	7.59	7.21	11.1	8.58	9.21	11.5	16.5	560
570	7.00	7.16	7.60	7.22	11.2	8.60	9.24	11.6	16.6	570
580	7.00	7.16	7.61	7.23	11.2	8.61	9.26	11.6	16.6	580
590	7.00	7.17	7.62	7.24	11.2	8.62	9.28	11.6	16.7	590
600	7.00	7.18	7.64	7.25	11.2	8.64	9.30	11.6	16.8	600
610	7.00	7.19	7.65	7.26	11.3	8.66	9.32	11.7	16.8	610
620	7.00	7.19	7.66	7.27	11.3	8.67	9.35	11.7	16.9	620
630	7.01	7.20	7.68	7.28	11.4	8.69	9.38	11.7	17.0	630
640	7.01	7.20	7.69	7.29	11.4	8.70	9.40	11.8	17.0	640
650	7.01	7.21	7.70	7.30	11.4	8.72	9.42	11.8	17.1	650
660	7.01	7.22	7.71	7.31	11.4	8.73	9.44	11.8	17.1	660
670	7.01	7.23	7.73	7.32	11.5	8.75	9.47	11.8	17.2	670
680	7.01	7.24	7.74	7.33	11.5	8.76	9.49	11.9	17.2	680
690	7.01	7.25	7.75	7.34	11.5	8.78	9.52	11.9	17.3	690

SOURCE: Edmister (1961). Copyright 1961 by Gulf Publishing Company, Houston, Texas; all rights reserved. Used with permission.

I-23

Table I.3 **Specific-Heat Equation for Ideal Gases at 1 Atmosphere†**

Gas	Symbol	Equation for C_p, Btu/(lb$_m$·mol·°R)	Temperature range, °R
Oxygen	O_2	$11.515 - \dfrac{172}{\sqrt{T}} + \dfrac{1530}{T}$	540–5000
		$11.515 - \dfrac{172}{\sqrt{T}} + \dfrac{1530}{T} + \dfrac{0.05(T - 4000)}{1000}$	5000–9000
Nitrogen	N_2	$9.47 - \dfrac{3.47 \times 10^3}{T} + \dfrac{1.16 \times 10^6}{T^2}$	540–5000
Carbon monoxide	CO	$9.46 - \dfrac{3.29 \times 10^3}{T} + \dfrac{1.07 \times 10^6}{T^2}$	540–5000
Hydrogen	H_2	$5.76 + \dfrac{0.578T}{1000} + \dfrac{20}{\sqrt{T}}$	540–4000
		$5.76 + \dfrac{0.578T}{1000} + \dfrac{20}{\sqrt{T}} - \dfrac{0.33(T - 4000)}{1000}$	4000–9000
Water	H_2O	$19.86 - \dfrac{597}{\sqrt{T}} + \dfrac{7500}{T}$	540–5000
Carbon dioxide	CO_2	$16.2 - \dfrac{6.53 \times 10^3}{T} + \dfrac{1.41 \times 10^6}{T^2}$	540–6300
Methane	CH_4	$4.22 + 8.211 \times 10^{-3}T$	492–1800
		$27.0 - \dfrac{14.400}{T}$	1800–5940
Ethylene	C_2H_4	$6.0 + 8.33 \times 10^{-3}T$	720–1400
Ethane	C_2H_6	$6.6 + 13.33 \times 10^{-3}T$	720–1440
Ethyl alcohol	C_2H_6O	$4.5 + 21.1 \times 10^{-3}T$	680–1120
Methyl alcohol	CH_4O	$2.0 + 16.67 \times 10^{-3}T$	680–1100
Benzene	C_6H_6	$6.5 + 28.9 \times 10^{-3}T$	520–1120
Octane	C_8H_{18}	$14.4 + 53.3 \times 10^{-3}T$	720–1440
Dodecane	$C_{12}H_{26}$	$19.6 + 80.0 \times 10^{-3}T$	720–1440

† The T in this table from Baumeister is the same as T_f used throughout this handbook.

SOURCE: Baumeister (1978).

INDEX

ABOUT THE AUTHOR

R. W. Miller is a Flow Consultant with The Foxboro Company, his employer for the past twenty years. He received his BS and MS from Northeastern University, where he currently lectures on mechanical engineering. His work for The Foxboro Company has been in the field of fluid metering and fluid dynamics.

Mr. Miller has performed flowmeter research and written numerous technical articles. He is chairman of the ASME/ANSI Measurement of Fluid Flow in Closed Conduits Committee, has been the chief United States delegate at many International Standards Organization meetings on flow metering devices, and represents the United States on the ISO/TC30 steering committee. He is active as both member and chairman of several ASME and AGA flowmetering subcommittees, and the ASME Board on Standardization, ASME Board on International Standards, and serves on the U.S. OIML advisory committee.